CHILTON®

TIMING BELTS

2005 Edition
Domestic and Imported Vehicles

DELMAR
CENGAGE Learning™

Australia • Brazil • Japan • Korea • Mexico • Singapore • Spain • United Kingdom • United States

DELMAR
CENGAGE Learning™

Chilton Timing Belts: Domestic and Imported Vehicles, 2005 Edition

Vice President, Technology and Trades SBU:
Alar Elken

Executive Director, Professional Business Unit:
Gregory L. Clayton

Publisher, Professional Business Unit:
David Koontz

Marketing Director: Beth A. Lutz

Marketing Specialist: Brian McGrath

Marketing Coordinator: Marissa Maiella

Production Director: Mary Ellen Black

Production Manager: Larry Main

Production Editor: Elizabeth Hough

Publishing Coordinator: Paula Baillie

Editorial Assistant: Christine Wade

Editor: Terry Blomquist

Cover Design: Melinda Possinger

ISBN-13: 978-1-4018-9880-9

ISBN-10: 1-4018-9880-7

Delmar
Executive Woods
5 Maxwell Drive
Clifton Park, NY 12065
USA

Cengage Learning is a leading provider of customized learning solutions with office locations around the globe, including Singapore, the United Kingdom, Australia, Mexico, Brazil, and Japan. Locate your local office at **www.cengage.com/global**

Cengage Learning products are represented in Canada by Nelson Education, Ltd.

To learn more about Delmar, visit **www.cengage.com/delmar**

Purchase any of our products at your local college store or at our preferred online store **www.CengageBrain.com**

Printed in the United States of America
2 3 4 5 6 7 14 13 12 11

TABLE OF CONTENTS

TABLE OF CONTENTS

TABLE OF CONTENTS

TABLE OF CONTENTS

TABLE OF CONTENTS

TABLE OF CONTENTS

USING THIS INFORMATION

Locating Information

Timing belt service and replacement procedures are divided into individual manufacturer sections in alphabetical order. Each manufacturer section includes passenger cars, light trucks, vans and SUVs. Engines are listed in increasing numerical order based on engine liter size (i.e. 1.5L, 1.6L, etc.).

The Table of Contents at the front lists manufacturer (i.e. Ford), brand (i.e. Ford, Lincoln, Mercury), model (i.e. Aerostar, Aspire, Contour), year span (i.e. 1985-87), engine size (i.e. 2.3L) and engine VIN or code (i.e. VIN A).

To find a particular timing belt procedure, just look in the Table of Contents. Once you have found the appropriate model, year and engine size, simply turn to the indicated page to find the applicable information.

Safety Notice

Proper service and repair procedures are vital to the safe, reliable operation of all motor vehicles, as well as the personal safety of those performing the repairs. The timing belt replacement procedures include proper procedures for servicing and repairing vehicles using safe effective methods. The procedures contain many NOTES, WARNINGS and CAUTIONS which should be followed along with standard safety procedures to eliminate the possibility of personal injury or improper service which could damage the vehicle or compromise its safety.

Repair procedures, tools, parts, and technician skill and experience vary widely. It is not possible to anticipate all conceivable ways or conditions under which vehicles may be serviced, or to provide cautions for all possible hazards that may result. Standard and accepted safety precautions and equipment should be used when handling toxic or flammable fluids, and safety goggles or other protection should be used during any process that may cause sparking, material removal or projectiles.

Some procedures require the use of tools specially designed for a specific purpose. Before substituting another tool or procedure, you must be completely satisfied that neither your personal safety, nor the performance of the vehicle will be endangered.

Although information in this manual is based on industry sources and is as complete as possible at the time of publication, the possibility exists that some vehicle manufacturers made later changes which could not be included here. Information on very late models may not be available in some circumstances. While striving for total accuracy, Delmar, Cegage Learning cannot assume responsibility for any errors, changes, or omissions that may occur in the compilation of this data.

Special Tools

Special tools are recommended by the vehicle manufacturer to perform their specific job. Use has been kept to a minimum, but where absolutely necessary, they are referred to in the text by the part number of the tool manufacturer. These tools may be purchased, under the appropriate part number, from your local dealer or regional distributor, or an equivalent tool can be purchased locally from a tool supplier or parts outlet. Before substituting any tool for the one recommended, read the previous Safety Notice.

Acknowledgements

This publication contains material that is reproduced and distributed under a license from Ford Motor Company. No further reproduction or distribution of the Ford Motor Company material is allowed without the express written permission of Ford Motor Company.

Portions of materials contained herein have been reprinted with permission from General Motors Corporation, Service and Parts Operations under License Agreement #0510757.

BELT INSPECTION GUIDE

Replacement intervals for timing belts vary by manufacturer and model, and those intervals are provided herein. Inspection should be performed at least annually to spot signs of wear or defects and prevent potential engine damage.

✱✱ WARNING

Never allow antifreeze, oil or solvents to come into with a timing belt. If this occurs, immediately wash the solution from the timing belt. Also, never excessively bend or twist the timing belt; this can damage the belt so that its lifetime is severely shortened.

tccs1242

Never bend or twist a timing belt excessively, and do not allow solvents, antifreeze, gasoline, acid or oil to come into contact with the belt

Inspect both sides of the timing belt. Replace the belt with a new one if any of the following conditions exist:

- Hardening of the rubber—back side is glossy without resilience and leaves no indentation when pressed with a fingernail
- Cracks on the rubber backing
- Cracks or peeling of the canvas backing
- Cracks on rib root
- Cracks on belt sides
- Missing teeth or chunks of teeth
- Abnormal wear of belt sides—the sides are normal if they are sharp, as if cut by a knife.

If none of these conditions exist, the belt does not need replacement unless it is at the recommended interval. The belt MUST be replaced at the recommended interval.

✱✱ WARNING

On interference engines, it is very important to replace the timing belt at the recommended intervals, otherwise expensive engine damage will likely result if the belt fails.

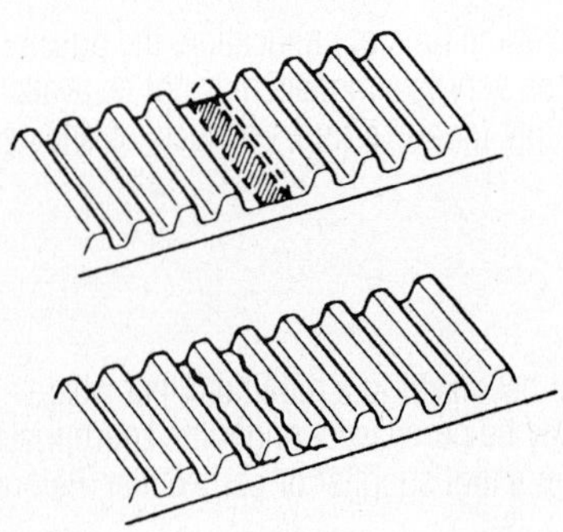

tccs1244

Broken tooth may be due to a damaged pulley

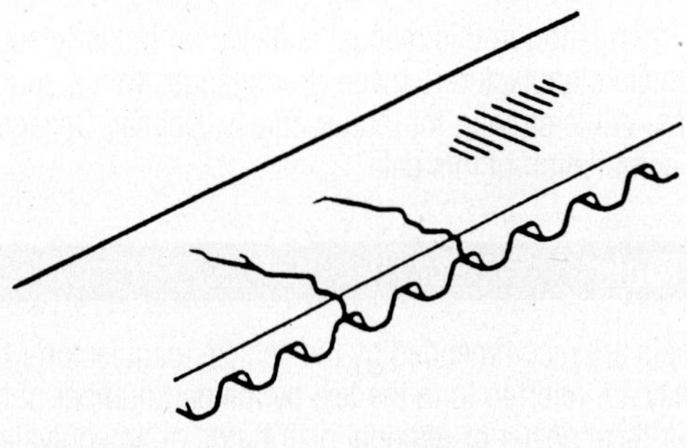

tccs1245

Back surface worn or cracked from a possible overheated engine or interference with the belt cover

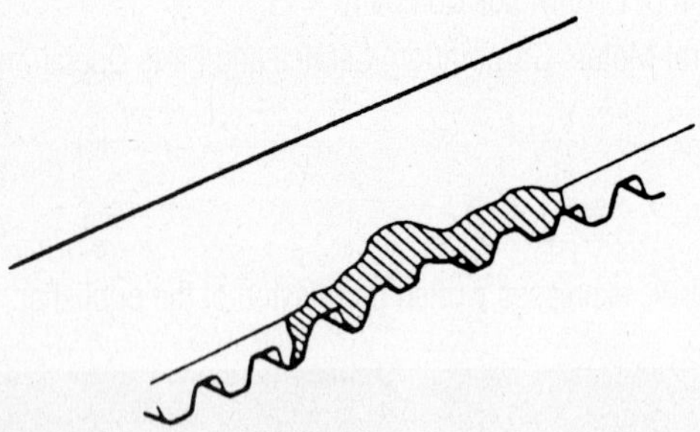

tccs1246

Side wear from improper installation

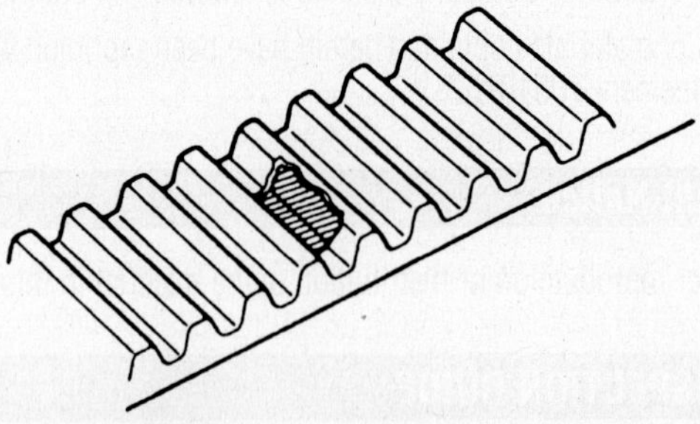

tccs1247

Worn teeth from excessive belt tension, camshaft or distributor not turning properly or fluid leaking on the belt

Acura

1.6L (D16A1) ENGINES

YEAR, MODEL & ENGINE APPLICATION	
1986-89 Acura Integra	1.6L Code D16A1

REPLACEMENT INTERVALS

Acura does not provide a replacement interval. Standard industry practice is to replace timing belts every 60,000 miles.

LABOR TIMES	
1986-89	
Replacement	3.0 hrs
Severe Service	3.4 hrs
With camshaft seal replacement add	0.5 hrs
With water pump replacement add	0.6 hrs

8807ACUg01

✻✻ CAUTION

These engines are an interference type. If the timing belt has failed, valve-to-piston damage may have occurred. A compression check should always be performed to check for engine damage.

1. Be sure to acquire the anti-theft code for the radio and the frequencies for the radio's preset buttons.
2. Disconnect the negative battery cable.
3. Remove the accessory drive belts.
4. Remove the power steering pump and position aside.
5. Turn the crankshaft pulley until cylinder No. 1 is set to Top Dead Center (TDC) on the compression stroke.
6. Remove the timing belt upper cover.
7. Raise and support the vehicle.
8. Remove the driver side wheel.
9. Remove the crankshaft pulley.
10. Remove the lower timing belt cover.
11. Loosen the timing belt tensioner bolt.
12. Remove the timing belt. If the timing belt is to be reinstalled, mark the direction of rotation on the belt with a crayon or white paint.
13. Recheck that the No. 1 piston is at TDC on its compression stroke.

To install:

■ Install the timing belt over the camshaft, crankshaft and tensioner sprockets.

Also, if installing the old belt, be sure it is turning the same direction.

14. Temporarily install the crankshaft pulley.
15. Rotate the crankshaft counterclockwise until the camshaft sprocket has moved 3 teeth from the TDC position to tension the timing belt.
16. Tighten the tensioner bolt to 33 ft. lbs. (45 Nm).
17. Remove the crankshaft pulley.

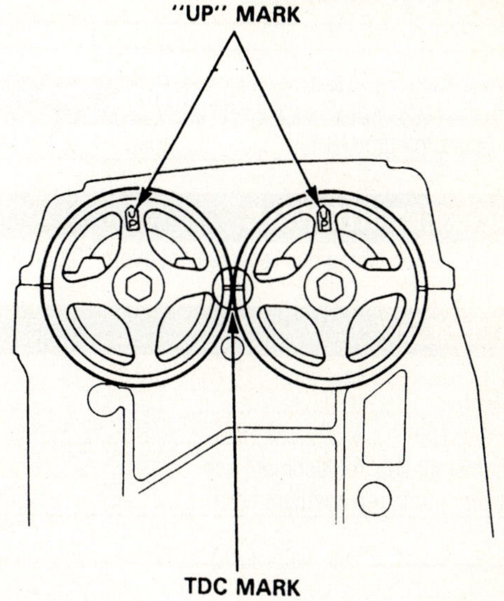

79235g17

Camshaft timing belt sprocket alignment marks for TDC—Acura 1.6L engine

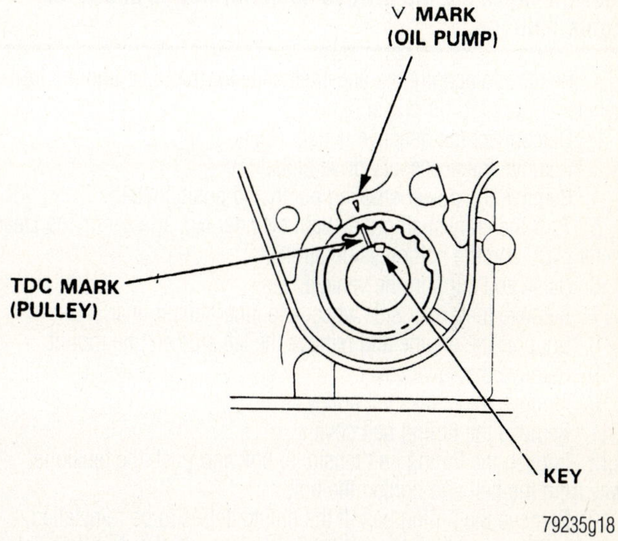

79235g18

Crankshaft timing belt sprocket alignment marks for TDC—Acura 1.6L engine

18. Install the lower belt cover. Install the crankshaft pulley, tightening the bolt to 83 ft. lbs. (115 Nm). Lubricate the threads and the flange of the bolt with engine oil before installation.
19. Install the upper timing belt cover.
20. Install the driver side wheel.
21. Lower the vehicle.
22. Install the power steering pump.
23. Install the accessory drive belts.
24. Connect the negative battery cable.
25. Enter the anti-theft code for the radio and reset the frequencies.

1.7L (B17A1) ENGINES

YEAR, MODEL & ENGINE APPLICATION

1992-93 Acura Integra GSR .... 1.7L Code B17A1

REPLACEMENT INTERVALS

Acura recommends timing belt replacement every 90,000 miles.

LABOR TIMES

1992-93	
Replacement	3.0 hrs
Severe Service	3.4 hrs
With camshaft seal replacement add	0.5 hrs
With water pump replacement add	0.6 hrs

8807ACUg02

✻✻ CAUTION

These engines are an interference type. If the timing belt has failed, valve-to-piston damage may have occurred. A compression check should always be performed to check for engine damage.

1. Be sure to acquire the anti-theft code for the radio and the frequencies for the radio's preset buttons.
2. Disconnect the negative battery cable.
3. Remove the accessory drive belts.
4. Remove the power steering pump and position aside.
5. Turn the crankshaft pulley until cylinder No. 1 is set to Top Dead Center (TDC) on the compression stroke.
6. Raise and support the vehicle.
7. Remove the driver side wheel and inner fender liner.
8. Support the engine and remove the left side engine mount.
9. Remove the valve cover.
10. Remove the crankshaft pulley.
11. Remove the timing belt covers.
12. Loosen the timing belt tensioner bolt and push the tensioner away from the belt and tighten the bolt.
13. Remove the timing belt. If the timing belt is to be reinstalled, mark the direction of rotation on the belt with a crayon or white paint.

To install:

14. Ensure the crankshaft and camshaft sprockets timing marks are aligned.
15. Temporarily install the crankshaft pulley.
16. Install the timing belt starting at the crankshaft sprocket and working clockwise, around the remaining sprockets and pulley. Also, if installing the old belt be sure it is turning the same direction.
17. Loosen the tensioner bolt and allow the tensioner to tension the belt, then tighten the bolt.
18. Rotate the crankshaft 4–6 turns counterclockwise and align the crankshaft and camshaft timing marks at the TDC position.
19. Rotate the crankshaft counterclockwise until the camshaft sprocket has moved 3 teeth from the TDC position to tension the timing belt.

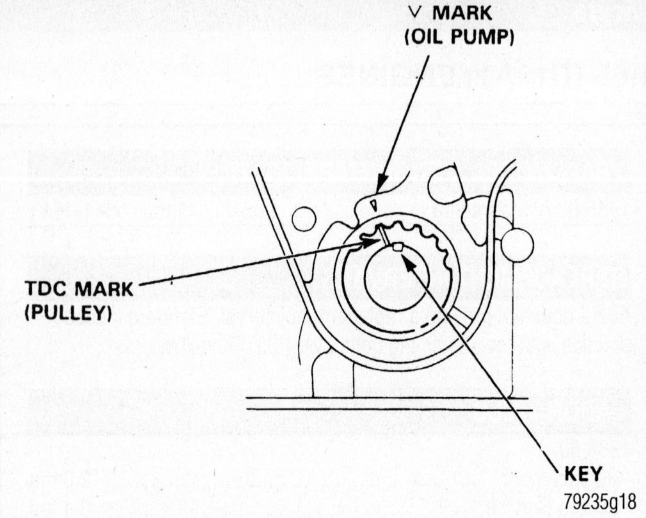

Crankshaft timing belt sprocket alignment marks for TDC—Acura 1.7L engine

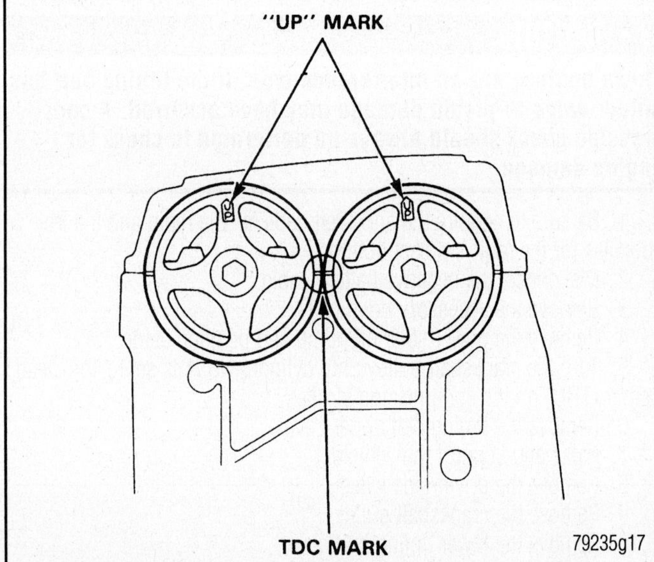

Camshaft timing belt sprocket alignment marks for TDC—Acura 1.7L engine

20. Tighten the tensioner bolt to 40 ft. lbs. (54 Nm).
21. Remove the crankshaft pulley.
22. Install timing belt covers. Install the crankshaft pulley, tightening the bolt to 130 ft. lbs. (176 Nm).
23. Install the valve cover.
24. Install the left side engine mount.
25. Install the driver side fender liner and wheel.
26. Lower the vehicle.
27. Install the power steering pump.
28. Install the accessory drive belts.
29. Connect the negative battery cable.
30. Enter the anti-theft code for the radio and reset the frequencies.

1.8L (B18A1, B18B1, B18C1 AND B18C5) ENGINES

YEAR, MODEL & ENGINE APPLICATION	
1990-93 Integra	1.8L Code B18A1
1994-01 Integra	1.8L Code B18B1
1994-01 Integra GSR	1.8L Code B18C1
1997-01 Integra Type R	1.8L Code B18C5

REPLACEMENT INTERVALS

On 1990-96 models Acura recommends timing belt replacement every 90,000 miles. On 1997-01 models Acura recommends timing belt replacement every 105,000 miles.

LABOR TIMES	
1990-01	
Replacement	3.0 hrs
Severe Service	3.4 hrs
With camshaft seal replacement add	0.5 hrs
With water pump replacement add	0.6 hrs

8807ACUg03

✻✻ CAUTION

These engines are an interference type. If the timing belt has failed, valve-to-piston damage may have occurred. A compression check should always be performed to check for engine damage.

1. Be sure to acquire the anti-theft code for the radio and the frequencies for the radio's preset buttons.
2. Disconnect the negative battery cable.
3. Turn the crankshaft pulley until cylinder No. 1 is set to Top Dead Center (TDC) on the compression stroke.
4. Remove the accessory drive belts.
5. Remove the power steering pump and position aside, if necessary.
6. Remove the cruise control actuator, if necessary.
7. Raise and support the vehicle.
8. Remove the driver side wheel and inner fender liner.
9. Support the engine and remove the left side engine mount.
10. Remove the valve cover.
11. Remove the crankshaft pulley.
12. Remove the timing belt covers.
13. Loosen the timing belt tensioner bolt and push the tensioner away from the belt and tighten the bolt.
14. Remove the timing belt. If the timing belt is to be reinstalled, mark the direction of rotation on the belt with a crayon or white paint.

To install:

15. On B18A1 engines, install the lower timing belt cover and crankshaft pulley.
16. On all engines, turn the crankshaft pulley until cylinder No. 1 is set to Top Dead Center (TDC) on the compression stroke and that the camshaft sprocket TDC marks are aligned.
17. On B18A1 and B18A2 engines, to set the camshafts to top dead center for the No. 1 cylinder, align the hole in each camshaft with the holes in the No. 1 camshaft holders, then push 5.0mm pin punches into the holes. Be sure that the **UP** arrows are pointing up and that the TDC marks on the intake and exhaust sprockets are aligned.

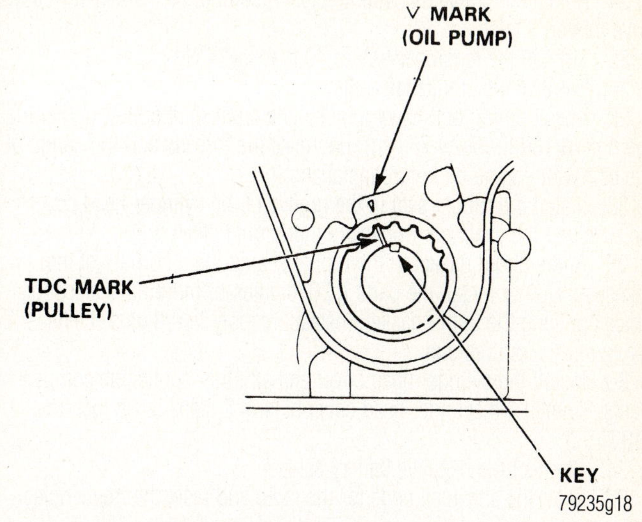

Crankshaft timing belt sprocket alignment marks for TDC—Acura 1.8L engine

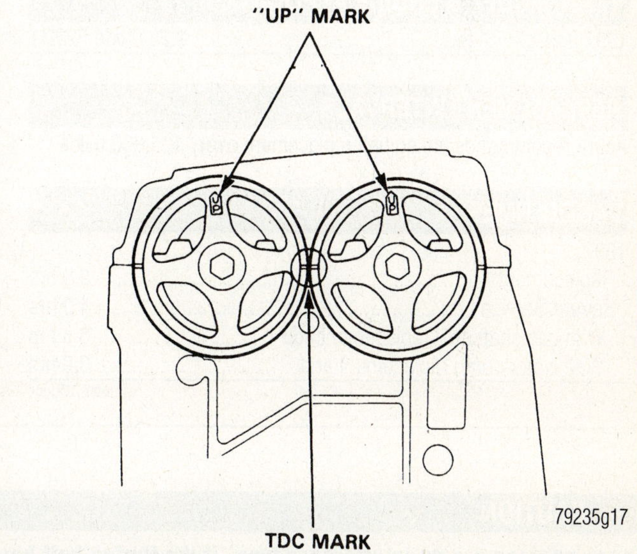

Camshaft timing belt sprocket alignment marks for TDC—Acura 1.8L engine

18. On B18A1 engines, remove the lower timing belt cover and crankshaft pulley.
19. Install the timing belt starting at the crankshaft sprocket and working clockwise, around the remaining sprockets and pulley. Also, if installing the old belt ensure it is turning the same direction.
20. Loosen the tensioner bolt and allow the tensioner to tension the belt, then tighten the bolt.
21. On B18A1 and B18A2 engines, remove the pin punches from the camshaft holders.
22. Rotate the crankshaft 4–6 turns counterclockwise and align the crankshaft and camshaft timing marks at the TDC position.
23. Set the engine to TDC compression for cylinder No. 1. Check that all timing marks for the camshaft and crankshaft are properly aligned. If any mark is out of alignment, remove the timing belt and reinstall it.

24. Rotate the crankshaft counterclockwise until the camshaft pulleys have moved 3 teeth.
25. Tighten the tensioner bolt to 40 ft. lbs. (54 Nm).
26. Remove the crankshaft pulley.
27. Install timing belt covers. Install the crankshaft pulley, tightening the bolt to 130 ft. lbs. (176 Nm). Lubricate the threads and the flange of the bolt with engine oil before installation.
28. Install the rubber seal in the groove of the cylinder head cover. Be sure that the seal and groove are thoroughly clean first.
29. Apply liquid gasket to the rubber seal at the 8 corners of the recesses. Do not install the parts if 20 minutes or more have elapsed since applying the liquid gasket. Instead, reapply liquid gasket after removing the old residue.
30. Install the cylinder head cover and all other applicable components. Tighten the cylinder head cover nuts in 2 steps to 88 inch lbs. (10 Nm).
31. Connect the negative battery cable.
32. Enter the anti-theft code for the radio and reset the frequencies.

2.2L (F22B1) ENGINES

YEAR, MODEL & ENGINE APPLICATION	
1997 2.2CL	2.2L Code F22B1

REPLACEMENT INTERVALS

Acura recommends timing belt replacement every 105,000 miles.

LABOR TIMES	
1997	
Replacement	3.5 hrs
Severe Service	4.0 hrs
With camshaft seal replacement add	0.5 hrs
With water pump replacement add	0.6 hrs

8807ACUg04

✻✻ CAUTION

These engines are an interference type. If the timing belt has failed, valve-to-piston damage may have occurred. A compression check should always be performed to check for engine damage.

1. Be sure to acquire the anti-theft code for the radio and the frequencies for the radio's preset buttons.
2. Disconnect the negative battery cable.
3. Raise and safely support the vehicle.
4. Remove the left wheel.
5. Remove the left wheel well splash shield.
6. Remove all necessary components to gain access to the cylinder head (valve) and timing belt covers.
7. Remove the valve cover and timing belt covers.
8. Turn the engine to align the timing marks and set cylinder No. 1 to Top Dead Center (TDC). The white mark on the crankshaft sprocket should align with the pointer on the timing belt cover. The words **UP** embossed on the camshaft sprocket should be aligned in the upward position. The marks on the edge of the sprocket should be aligned with the cylinder head or the back cover upper edge. Once in this position, the engine must **NOT** be turned or disturbed.
9. There are 2 belts on these engines; the belt running to the camshaft sprocket is the timing belt, the other, shorter belt drives the balance shaft and is referred to as the balancer shaft belt or timing balancer belt. Lock the timing belt adjuster in position, by installing one of the lower timing belt cover bolts in the adjuster arm.
10. Loosen the timing belt and balancer shaft tensioner adjuster nut, do not loosen the nut more than 1 revolution. Push the tensioner for the balancer belt away from the belt to relieve the tension. Hold the tensioner and tighten the adjusting nut to hold the tensioner in place.
11. Carefully remove the balancer belt. Do not crimp or bend the belt; protect it from contact with oil or coolant.
12. Remove the balancer belt sprocket from the crankshaft.

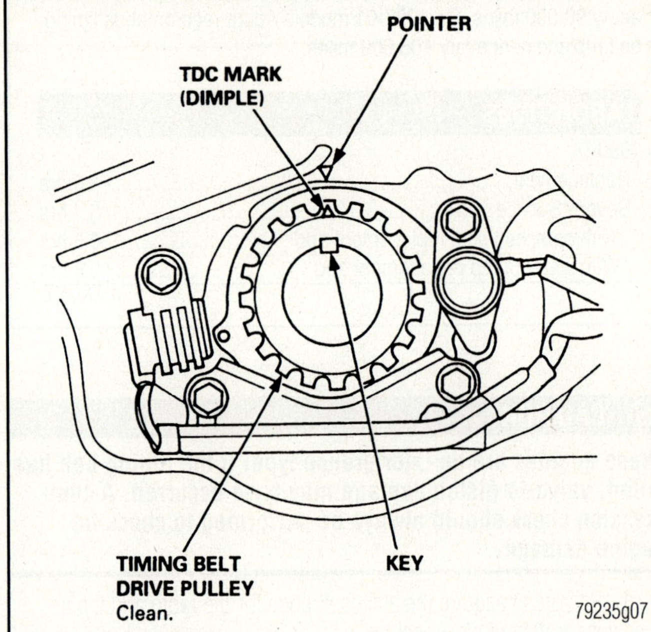

Crankshaft timing belt sprocket alignment mark locations—Acura 2.2L (F22B1) engine

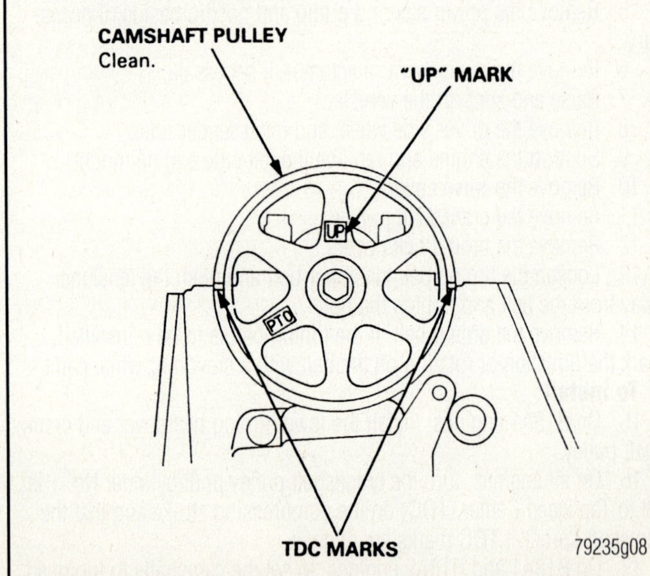

Camshaft timing belt sprocket alignment mark locations—Acura 2.2L (F22B1) engine

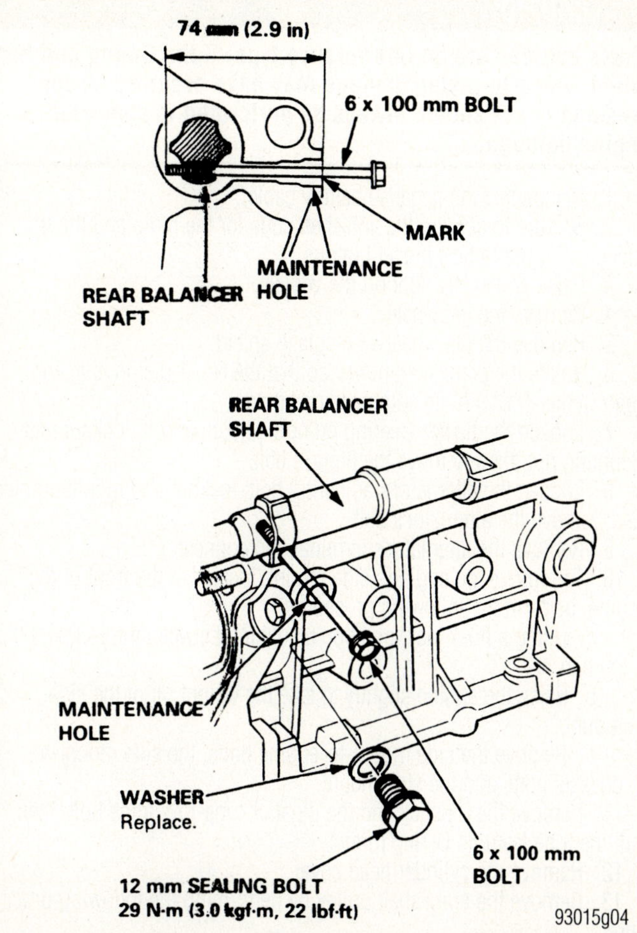

93015g04

Securing the rear balancer shaft—Acura 2.2L (F22B1) engine

13. Loosen the lockbolt, installed in the timing belt adjuster and the adjusting nut. Push on the timing belt adjuster to remove the tension on the timing belt, then tighten the adjuster nut.

14. Remove the timing belt by sliding it off the sprockets. Do not crimp or bend the belt; protect it from contact with oil or coolant.

15. If necessary, remove the belt tensioners by performing the following:

a. Remove the springs from the balancer belt and the timing belt tensioners.

b. Remove the adjusting nut from the belt tensioners.

c. Remove the bolt from the balancer belt adjuster lever, then remove the lever and the tensioner pulley.

d. Remove the lockbolt from the timing belt tensioner lever, then remove the tensioner pulley and lever from the engine.

16. This is an excellent time to check or replace the water pump. Even if the timing belt is only being replaced as part of a good maintenance schedule, consider replacing the pump at the same time.

To install:

17. If the water pump is to be replaced, install a new O-ring on the pump and make certain it is properly seated. Install the water pump and tighten the mounting bolts to 106 inch lbs. (12 Nm).

18. If the tensioners were removed, perform the following to install them:

a. Install the timing belt tensioner lever and the tensioner pulley.

b. Install the balancer belt pulley and adjuster lever.

c. Install the adjusting nut and bolt to the balancer belt adjuster lever.

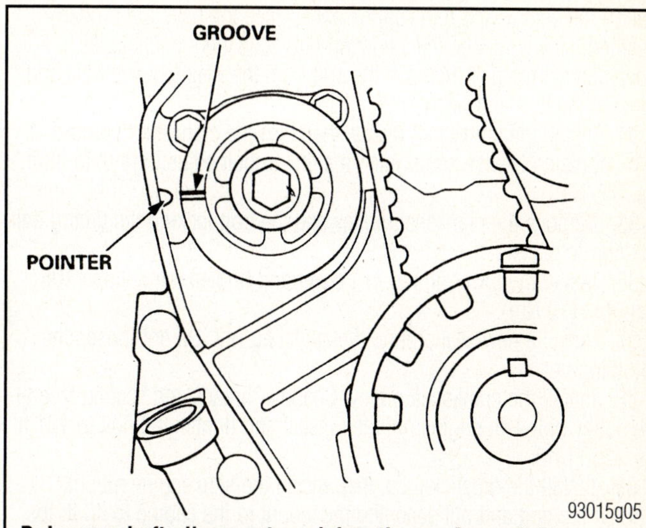

93015g05

Balancer shaft alignment mark locations—Acura 2.2L (F22B1) engine

d. Install the springs to the tensioners.

e. Install the lockbolt to the timing belt tensioner, then move the tensioner to its full deflection and tighten the lockbolt.

f. Move the balancer belt tensioner to its full deflection and tighten the adjusting nut to hold its position.

19. The pointer on the crankshaft sprocket should be aligned with the pointer on the oil pump. The camshaft sprocket must be aligned so that the word **UP** is at the top of the sprocket and the marks on the edge of the sprocket are aligned with the surfaces of the head or the back cover upper edge.

20. Install the timing belt on the crankshaft sprocket, the tensioner sprocket, water pump sprocket and the camshaft sprocket.

21. Check the timing marks to be sure that they did not move.

22. Loosen, then retighten the timing belt adjusting nut.

23. Install the timing/balancer belt drive sprocket and the lower timing belt cover.

24. Install the crankshaft pulley and bolt, tighten the bolt to 181 ft. lbs. (245 Nm). Rotate the crankshaft sprocket 5–6 revolutions to properly position the timing belt on the sprockets.

25. Set the No. 1 cylinder to TDC and loosen the timing belt adjusting nut 1 turn. Turn the crankshaft counterclockwise until the cam sprocket has moved 3 teeth; this creates the proper tension on the timing belt.

26. Tighten the timing belt adjusting nut.

27. Set the crankshaft sprocket and the camshaft sprocket to TDC. If the sprockets do not align, remove the belt to realign the marks, then install the belt.

28. Remove the crankshaft pulley and the lower cover.

29. With the timing marks aligned, lock the timing belt adjuster in place with one of the lower cover mounting bolts.

30. Loosen the adjusting nut and ensure the timing balancer belt adjuster moves freely.

31. Align the rear timing balancer sprocket using a 6 x 100mm bolt or rod. Mark the bolt or rod at a point 2.9 in. (74mm) from the end. Remove the bolt from the maintenance hole on the side of the block; insert the bolt/rod into the hole and align the 2.9 in. (74mm) mark with the face of the hole. This will hold the shaft in place during installation.

32. Align the groove on the front balancer shaft sprocket with the pointer on the oil pump.

33. Install the balancer belt. Once the belts are in place, be sure that all the engine alignment marks are still correct. If not, remove the belts,

realign the engine and reinstall the belts. Once the belts are properly installed, slowly loosen the adjusting nut, allowing the tensioner to move against the belt. Remove the bolt from the maintenance hole and reinstall the bolt and washer.

34. Install the crankshaft pulley, then turn the crankshaft sprocket 1 turn counterclockwise and tighten the timing belt adjusting nut to 33 ft. lbs. (45 Nm).
35. Remove the crankshaft pulley and the bolt locking the timing belt adjuster in place.
36. Install the lower timing belt cover and tighten the bolts to 106 inch lbs. (12 Nm).
37. Install a new seal around the adjusting nut. Do not loosen the adjusting nut.
38. Install the crankshaft pulley. Coat the threads and seating face of the pulley bolt with engine oil, then install and tighten the bolt to 181 ft. lbs. (250 Nm).
39. Install the dipstick tube, then install the side engine mount. Tighten the bolt and nut attaching the mount to the engine to 40 ft. lbs. (55 Nm). Tighten the through-bolt and nut to 47 ft. lbs. (65 Nm), then remove the jack from under the engine.
40. Install the upper timing belt cover. Tighten the bolt toward the exhaust manifold to 89 inch lbs. (10 Nm) and the bolt toward the intake manifold to 106 inch lbs. (12 Nm).
41. Install the cylinder head cover gasket in the groove of the cylinder head cover. Before installing the gasket, thoroughly clean the seal and the groove. Seat the recesses for the camshaft first, then work it into the groove around the outside edges. Be sure the gasket is seated securely in the corners of the recesses.
42. Apply liquid gasket to the 4 corners of the recesses in the cylinder head cover gasket. Do not install the parts if 5 minutes or more have elapsed since applying liquid gasket. After assembly, wait at least 20 minutes before filling the engine with oil.
43. Install the cylinder head (valve) cover and all other applicable components. Tighten the bolts attaching the cylinder head cover in 2 steps to the proper sequence of 89 inch lbs. (10 Nm).
44. Install the left fender liner and wheel.
45. Connect the negative battery cable.
46. Enter the anti-theft code for the radio and reset the frequencies.

2.3L (F23A1) ENGINES

YEAR, MODEL & ENGINE APPLICATION

1998-99 2.3CL .... 2.3L Code F23A1

REPLACEMENT INTERVALS

Acura recommends timing belt replacement every 105,000 miles.

LABOR TIMES

1998-99

Replacement	3.5 hrs
Severe Service	4.0 hrs
With camshaft seal replacement add	0.5 hrs
With water pump replacement add	0.6 hrs

8807ACUg05

✻✻ CAUTION

These engines are an interference type. If the timing belt has failed, valve-to-piston damage may have occurred. A compression check should always be performed to check for engine damage.

1. Disconnect the negative battery cable.
2. Be sure to acquire the anti-theft code for the radio and the frequencies for the radio's preset buttons.
3. Raise and safely support the vehicle.
4. Remove the left wheel.
5. Remove the left wheel-well splash shield.
6. Rotate the crankshaft pulley so that the No. 1 piston is at Top Dead Center (TDC) of its compression stroke.
7. Loosen the power steering pump's adjusting bolt, locknut and mounting nut; then, remove the pump's belt.
8. Loosen the alternator's adjusting bolt, locknut and mounting nut; then, remove the alternator's belt.
9. Remove the alternator's terminal and connector.
10. To remove the engine's side mount, located at the front of the engine, perform the following procedure:
    a. Using a floor jack, position a cushion between the jack and the oil pan.
    b. Raise the engine slightly to take the weight off of the side mount.
    c. Remove the side mount-to-engine bolts, the side mount-to-chassis bolts and the side mount.
11. Remove the dipstick and the dipstick tube-to-engine bolt; then, pull the tube from its O-ring mount.
12. Remove the cylinder head cover.
13. Remove the crankshaft pulley by performing the following procedure:

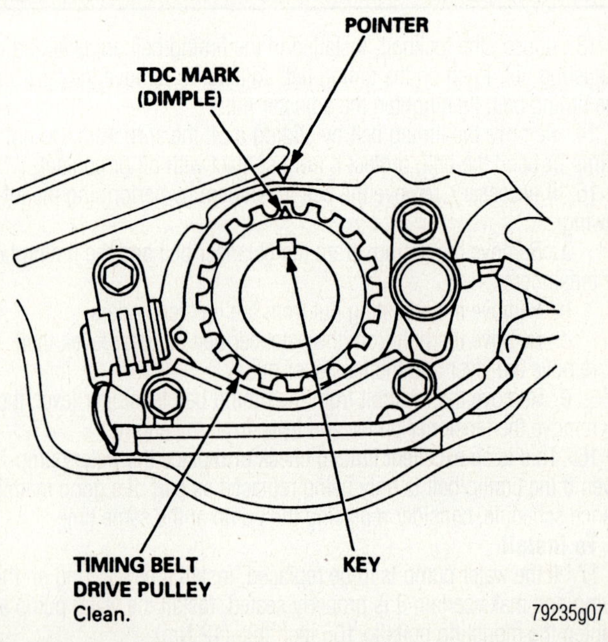

Crankshaft timing belt sprocket alignment mark locations—Acura 2.3L (F23A1) engine

a. Using the Holder Handle and the 50mm Offset Holder Attachment tool 07MAB-PY3010A or equivalent, and a 19mm socket, secure the crankshaft pulley and remove the crankshaft pulley bolt.

b. Remove the washer and pull the crankshaft pulley from the engine.

14. From the lower timing belt cover, remove the rubber seal concealing the adjusting nut. Remove the upper timing belt cover-to-engine bolts and the cover.

15. Loosen the adjusting nut 2/3–1 turn. Push the tensioner to relieve the tension from the timing belt and the balancer belt; then, retighten the adjusting nut.

16. Remove the balancer and the timing belts.

To install:

17. Clean the upper and lower timing belt covers.

18. Remove the balancer belt drive pulley from the crankshaft.

19. Position the timing belt pulley so the No. 1 piston is at TDC of its compression stroke. Align the mark on the pulley (near keyway) with the pointer mark on the oil pump.

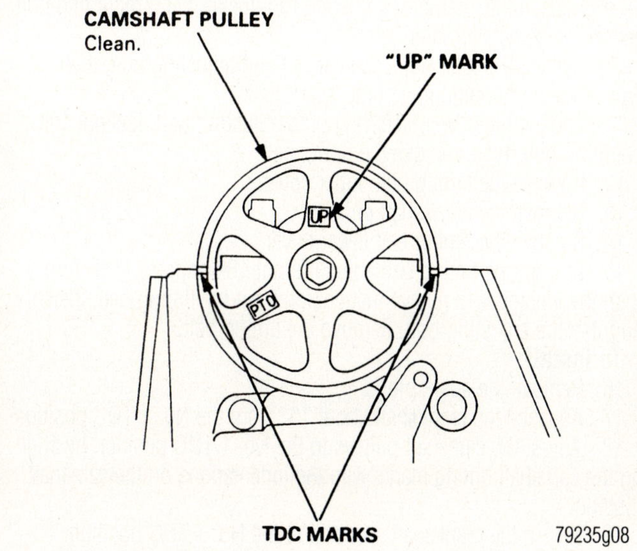

Camshaft timing belt sprocket alignment mark locations—Acura 2.3L (F23A1) engine

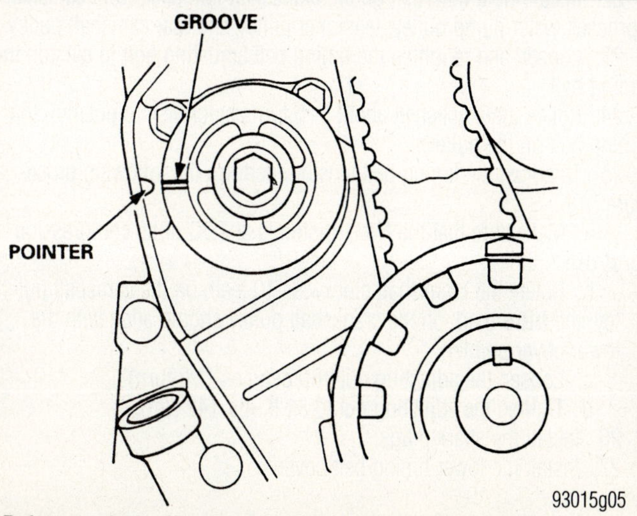

Balancer shaft alignment mark locations—Acura 2.3L (F23A1) engine

20. Adjust the camshaft pulley so that the No. 1 piston is the TDC of the compression stroke. Align the TDC marks on the pulley with the upper surface of the back cover and the **UP** mark should be facing upward.

21. Install the timing belt on the crankshaft pulley sprocket, the tensioner pulley, water pump pulley and the camshaft pulley.

✻✻ WARNING

Make sure that the camshaft and crankshaft pulleys are at TDC.

22. Loosen and retighten the adjusting nut to tension the timing belt.

23. Install the balancer belt drive pulley and the lower timing belt cover.

24. Install the crankshaft pulley and finger-tighten the bolt and washer. Using the Holder Handle and the 50mm Offset Holder Attachment tool 07MAB-PY3010A or equivalent, and a 19mm socket with a torque wrench, tighten the crankshaft pulley bolt to 181 ft. lbs. (245 Nm).

25. Rotate the crankshaft pulley about 5–6 turns counterclockwise to position the timing belt on the pulleys.

26. To adjust the timing belt tension, perform the following procedure:

a. Make sure that the No. 1 piston is at TDC of its compression stroke.

b. Loosen the adjusting nut 2/3–1 turn. Rotate the crankshaft counterclockwise 3 teeth on the camshaft pulley.

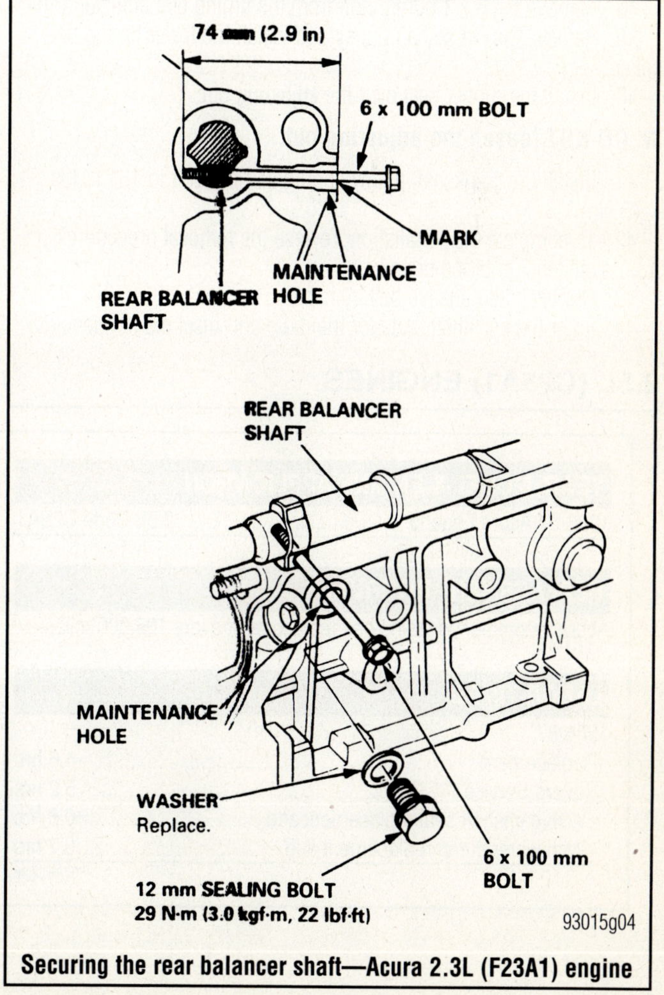

Securing the rear balancer shaft—Acura 2.3L (F23A1) engine

c. Tighten the adjusting nut to 33 ft. lbs. (44 Nm).
27. Retighten the crankshaft pulley bolt to 181 ft. lbs. (245 Nm).
28. Make sure the crankshaft and camshaft pulleys are at TDC.

■ **If the camshaft or crankshaft pulley is not at TDC, remove the timing belt and re-perform the adjustment procedure.**

29. Remove the crankshaft pulley and the timing belt lower cover.
30. Install a 6 x 1.0mm bolt to lock the timing belt adjuster arm in place.
31. Loosen the adjusting nut 2/3–1 turn and verify that the balancer belt adjuster moves freely.
32. Position the tensioner to remove tension from the balancer belt, then tighten the adjusting nut.
33. Align the rear balancer shaft pulley by performing the following procedure:
   a. Using a 6 x 100mm bolt, scribe a line 2.9 in. (74mm) from the end of the bolt.
   b. Insert the bolt into the maintenance hole to the scribed line.
   c. Align the groove on the front balancer shaft pulley with the pointer on the oil pump housing.
34. Install the balancer belt. Loosen the adjuster nut 2/3–1 turn to place tension on the balancer belt.
35. Remove the 6 x 100mm bolt from the rear balancer shaft and install the 12mm sealing bolt.
36. Install the crankshaft pulley and tighten the bolt to 181 ft. lbs. (245 Nm).
37. Rotate the crankshaft pulley 1 turn counterclockwise and tighten the adjusting nut to 33 ft. lbs. (44 Nm).
38. Remove the 6 x 1.00mm bolt from the timing belt adjuster arm.
39. Remove the crankshaft pulley and install the lower timing belt cover.
40. Install the rubber seal over the adjusting nut.

■ **DO NOT loosen the adjusting nut.**

41. Install the crankshaft pulley and tighten the bolt to 181 ft. lbs. (245 Nm).
42. To complete the installation, reverse the removal procedures. Adjust the tension of the drive belts.
43. Connect the negative battery cable.
44. Enter the anti-theft code for the radio and reset the frequencies.

2.5L (C25A1) ENGINES

YEAR, MODEL & ENGINE APPLICATION

1986-87 Acura Legend	2.5L Code C25A1

REPLACEMENT INTERVALS

Acura recommends timing belt replacement every 105,000 miles.

LABOR TIMES

1986-87	
Replacement	4.6 hrs
Severe Service	5.2 hrs
With camshaft seal replacement add	0.5 hrs
With water pump replacement add	0.7 hrs

8807ACUg06

✱✱ CAUTION

These engines are an interference type. If the timing belt has failed, valve-to-piston damage may have occurred. A compression check should always be performed to check for engine damage.

1. Be sure to acquire the anti-theft code for the radio and the frequencies for the radio's preset buttons.
2. Disconnect the negative battery cable.
3. Rotate the crankshaft pulley so that the No. 1 piston is at Top Dead Center (TDC) of its compression stroke.
4. If necessary, remove the strut brace located at the top rear of the engine.
5. Remove the pulley and harness cover from above the upper timing belt cover and remove the engine harness clamp.
6. Use a suitable device and safely support the engine.
7. Remove the engine support bolts, loosen the side mount rubber and raise the side bracket.
8. Remove the lower splash shield.
9. Loosen the alternator adjusting rod, locknut and mounting bolt; then, remove the alternator belt.
10. Loosen the idler pulley center nut and adjusting bolt; then, remove the air conditioning compressor belt.
11. Loosen the power steering pump adjusting bolt, locknut and mounting bolt; then, remove the pump belt.
12. Remove the timing belt upper cover.
13. Remove the crankshaft pulley.
14. Remove the timing belt lower cover.
15. Loosen the timing belt adjusting nut 180 degrees (1/2) turn. Push the tensioner to relieve the tension from the timing belt; then, retighten the adjusting bolt. Remove the timing belt.

To install:

16. Remove the spark plugs.
17. Advance the crankshaft about 15° from the No. 1 TDC position.
18. Adjust the camshaft pulleys to the No. 1 TDC position by aligning the camshaft timing marks with the indentations on the camshaft housing.
19. Return the crankshaft sprocket to the No. 1 TDC position.
20. Use a screwdriver and push the tensioner bracket forward to loosen the tensioner.
21. Advance the rear camshaft sprocket about (1/2 turn from TDC.
22. Install the timing belt on the crankshaft sprocket, front camshaft sprocket, water pump pulley, tensioner pulley and rear camshaft pulley.
23. Loosen and retighten the timing belt adjusting bolt to tension the timing belt.
24. Rotate the crankshaft about 5–6 turns clockwise to position the timing belt on the pulleys.
25. To adjust the timing belt tension, perform the following procedure:
   a. Make sure that the No. 1 piston is at TDC of its compression stroke.
   b. Rotate the crankshaft clockwise 10 teeth on the camshaft pulley; the **blue** mark on the crankshaft pulley should align with the lower cover pointer.
   c. Loosen the adjusting nut 180 degrees (1/2 turn).
   d. Tighten the adjusting nut to 31 ft. lbs. (42 Nm).
26. Install the spark plugs.
27. Install the lower timing belt cover.

28. Install the crankshaft pulley and tighten the bolt to 83 ft. lbs. (115 Nm).
29. Make sure the crankshaft and camshaft pulleys are at TDC.

■ **If the camshaft or crankshaft pulley is not at TDC, remove the timing belt and re-perform the adjustment procedure.**

30. To complete the installation, reverse the removal procedures. Adjust the tension of the drive belts.
31. Connect the negative battery cable.
32. Enter the anti-theft code for the radio and reset the frequencies.

2.5L (G25A1) ENGINES

YEAR, MODEL & ENGINE APPLICATION	
1992-94 Acura Vigor	2.5L Code G25A1

REPLACEMENT INTERVALS
Acura recommends timing belt replacement every 90,000 miles.

LABOR TIMES	
1992-94	
Replacement	3.5 hrs
Severe Service	4.0 hrs
With camshaft seal replacement add	0.5 hrs
With water pump replacement add	0.6 hrs

8807ACUg07

✻✻ CAUTION

These engines are an interference type. If the timing belt has failed, valve-to-piston damage may have occurred. A compression check should always be performed to check for engine damage.

1. Be sure to acquire the anti-theft code for the radio and the frequencies for the radio's preset buttons.
2. Disconnect the negative battery cable.
3. Set the No. 1 piston at Top Dead Center (TDC) on the compression stroke. The TDC mark on the crankshaft pulley must align with the pointers on the lower belt cover.
4. Disconnect the wiring harness and move aside.
5. Remove the accessory drive belts.
6. Remove the valve cover.
7. Remove the timing belt upper cover.
8. Ensure the **UP** mark and the TDC marks on the camshaft sprocket are correctly positioned.
9. Remove the crankshaft pulley.
10. Remove the timing belt lower cover.
11. Loosen the adjusting bolt 180 degrees (1/2 turn), then push the tensioner down to relieve the belt tension. Retighten the adjusting bolt to 33 ft. lbs. (44 Nm).

■ **Do not remove the adjusting bolt and tensioner pulley unless they are to be replaced. The bolt is only loosened and tightened in this procedure to tension the timing belt.**

12. Remove the timing belt.
13. Inspect the timing belt tensioner pulley and tension spring for signs of wear. Remove and replace the tensioner assembly as necessary.

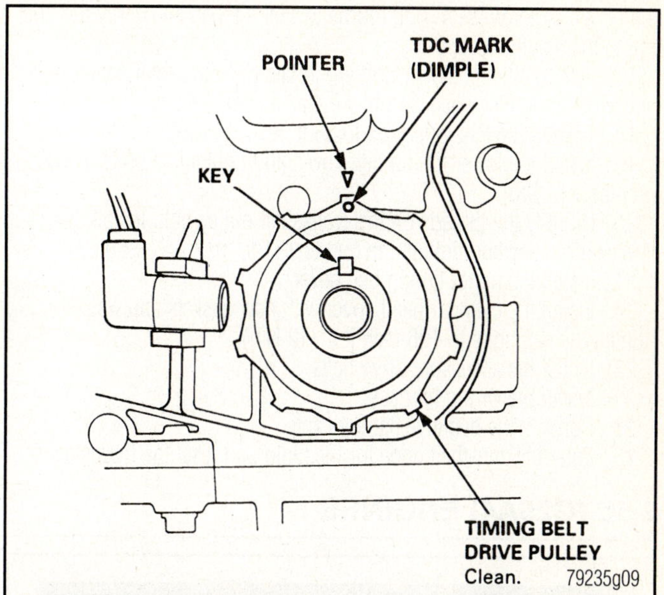

Crankshaft timing belt sprocket alignment mark locations—Acura 2.5L (G25A1) engine

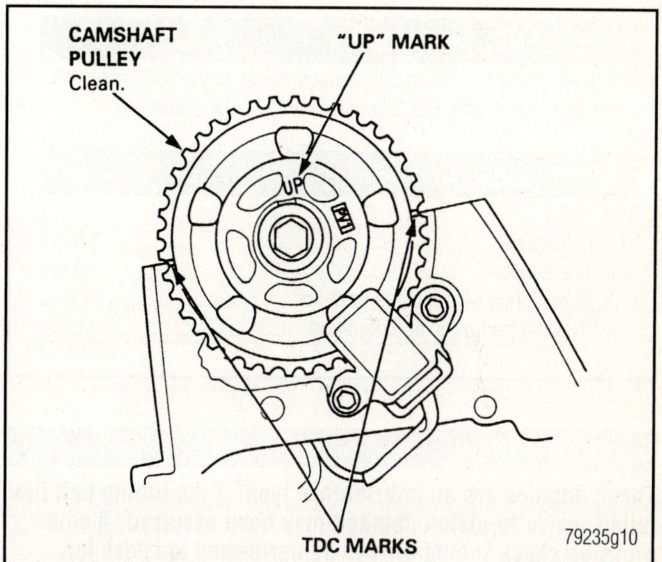

Camshaft timing belt sprocket alignment mark locations—Acura 2.5L (G25A1) engine

To install:

✻✻ WARNING

Replace the timing belt if it shows any signs of wear, damage or contamination from oil or coolant. The source of any oil or coolant contamination must be determined and corrected before the new timing belt may be installed.

14. Verify that the crankshaft and camshaft sprocket matchmarks are properly aligned.
15. Install the timing belt in the following order: first onto the crankshaft sprocket, then the tensioner pulley, the water pump sprocket and finally the camshaft sprocket.
16. Loosen the tensioner adjusting bolt to allow the spring to set the tension. Then, tighten the bolt to 33 ft. lbs. (44 Nm). Rotate the crank-

shaft counterclockwise 4–6 full turns to seat the belt and verify that the timing marks align properly.

17. Loosen the tensioner bolt and rotate the crankshaft counterclockwise until the camshaft sprocket has moved 3 teeth.

18. Tighten the tensioner bolt to 33 ft. lbs. (44 Nm).

19. Install the lower timing belt cover, and tighten the bolts to 106 inch lbs. (12 Nm).

20. Oil only the threads on the crankshaft pulley bolt. Install the crankshaft pulley and tighten the bolt to 181 ft. lbs. (245–250 Nm).

21. Install the upper timing belt cover.

22. Install the cylinder head cover with new gaskets and washers. Tighten the cap bolts to 106 inch lbs. (12 Nm).

23. Install the accessory drive belts.

24. Install the wiring harness.

25. Connect the negative battery cable.

26. Enter the anti-theft code for the radio and reset the frequencies.

2.5L (G25A4) ENGINES

YEAR, MODEL & ENGINE APPLICATION

1995-98 Acura 2.5TL .... 2.5L Code G25A4

REPLACEMENT INTERVALS

Acura recommends timing belt replacement every 90,000 miles on 1995-96 models, and 105,000 miles on 1997-98 models.

LABOR TIMES

1995-98	
Replacement	3.5 hrs
Severe Service	4.0 hrs
With camshaft seal replacement add	0.5 hrs
With water pump replacement add	0.6 hrs

8807ACUg08

⁂ CAUTION

These engines are an interference type. If the timing belt has failed, valve-to-piston damage may have occurred. A compression check should always be performed to check for engine damage.

1. Be sure to acquire the anti-theft code for the radio and the frequencies for the radio's preset buttons.

2. Disconnect the negative battery cable.

3. Set the No. 1 piston at Top Dead Center (TDC) on the compression stroke. The TDC mark on the crankshaft pulley must align with the pointers on the lower belt cover.

4. Disconnect the wiring harness and move aside.

5. Remove the accessory drive belts.

6. Remove the valve cover.

7. Remove the crankshaft pulley

8. Remove the timing belt upper and lower covers.

9. Loosen the adjusting bolt 180 degrees (1/2 turn), then push the tensioner down to relieve the belt tension. Retighten the adjusting bolt to 33 ft. lbs. (44 Nm).

■ **Do not remove the adjusting bolt and tensioner pulley unless they are to be replaced. The bolt is only loosened and tightened in this procedure to tension the timing belt.**

10. Remove the timing belt.

11. Inspect the timing belt tensioner pulley and tension spring for signs of wear. Remove and replace the tensioner assembly as necessary.

To install:

⁂ WARNING

Replace the timing belt if it shows any signs of wear, damage or contamination from oil or coolant. The source of any oil or coolant contamination must be determined and corrected before the new timing belt may be installed.

12. Verify that the crankshaft and camshaft sprocket matchmarks are properly aligned.

13. Install the timing belt in the following order: first onto the crankshaft sprocket, then the tensioner pulley, the water pump sprocket and finally the camshaft sprocket.

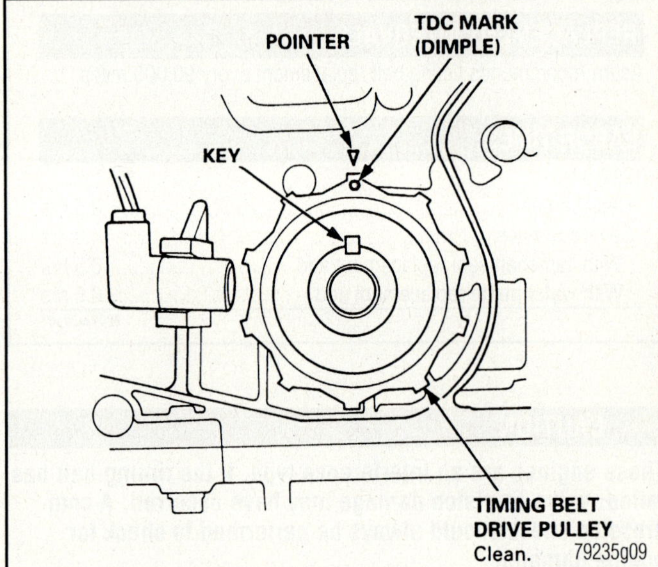

Crankshaft timing belt sprocket alignment mark locations—Acura 2.5L (G25A4) engine

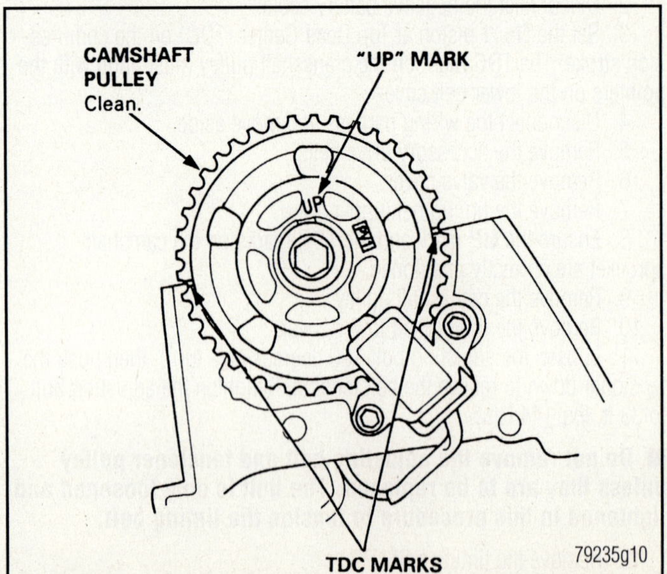

Camshaft timing belt sprocket alignment mark locations—Acura 2.5L (G25A4) engine

14. Loosen the tensioner adjusting bolt to allow the spring to set the tension. Then, tighten the bolt to 33 ft. lbs. (44 Nm). Rotate the crankshaft counterclockwise 4–6 full turns to seat the belt and verify that the timing marks align properly.

15. Loosen the tensioner bolt and rotate the crankshaft counterclockwise until the camshaft sprocket has moved 3 teeth.

16. Tighten the tensioner bolt to 33 ft. lbs. (44 Nm).

17. Install the lower timing belt cover, and tighten the bolts to 106 inch lbs. (12 Nm).

18. Oil only the threads on the crankshaft pulley bolt. Install the crankshaft pulley and tighten the bolt to 181 ft. lbs. (245–250 Nm).

19. Install the upper timing belt cover.

20. Install the cylinder head cover with new gaskets and washers. Tighten the cap bolts to 106 inch lbs. (12 Nm).

21. Install the accessory drive belts.

22. Install the wiring harness.

23. Connect the negative battery cable.

24. Enter the anti-theft code for the radio and reset the frequencies.

2.7L (C27A1) ENGINES

YEAR, MODEL & ENGINE APPLICATION

1988-90 Acura Legend .... 2.7L Code C27A1

REPLACEMENT INTERVALS

Acura does not provide a replacement interval. Standard industry practice is to replace timing belts every 60,000 miles.

LABOR TIMES

1988-90	
Replacement	4.6 hrs
Severe Service	5.2 hrs
With camshaft seal replacement add	0.5 hrs
With water pump replacement add	0.6 hrs

8807ACUg09

✻✻ CAUTION

These engines are an interference type. If the timing belt has failed, valve-to-piston damage may have occurred. A compression check should always be performed to check for engine damage.

1. Be sure to acquire the anti-theft code for the radio and the frequencies for the radio's preset buttons.

2. Disconnect the negative battery cable.

3. Rotate the crankshaft pulley so that the No. 1 piston is at Top Dead Center (TDC) of its compression stroke.

4. If necessary, remove the strut brace located at the top rear of the engine.

5. Remove the pulley and harness cover from above the upper timing belt cover and remove the engine harness clamp.

6. Use a suitable device and safely support the engine.

7. Remove the engine support bolts, loosen the side mount rubber and raise the side bracket.

8. Remove the lower splash shield.

9. Loosen the alternator adjusting rod, locknut and mounting bolt; then, remove the alternator belt.

10. Loosen the idler pulley center nut and adjusting bolt; then, remove the air conditioning compressor belt.

11. Loosen the power steering pump adjusting bolt, locknut and mounting bolt; then, remove the pump belt.

12. Remove the timing belt upper cover.

13. Remove the crankshaft pulley.

14. Remove the timing belt lower cover.

15. Loosen the timing belt adjusting nut 180 degrees (1/2) turn. Push the tensioner to relieve the tension from the timing belt; then, retighten the adjusting bolt. Remove the timing belt.

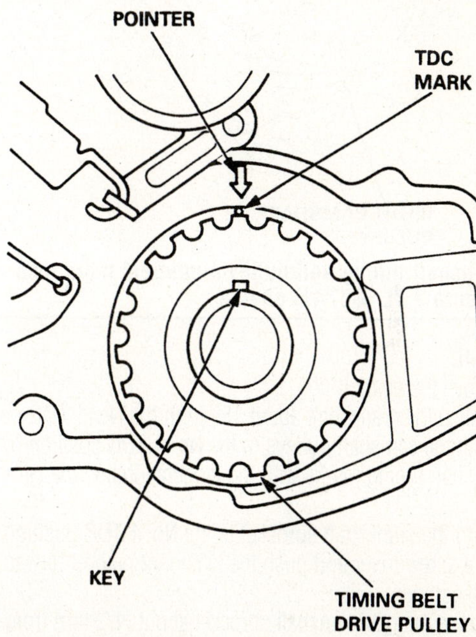

93015g10

Crankshaft timing belt pulley alignment mark locations—Acura 2.7L (C27A1) engine

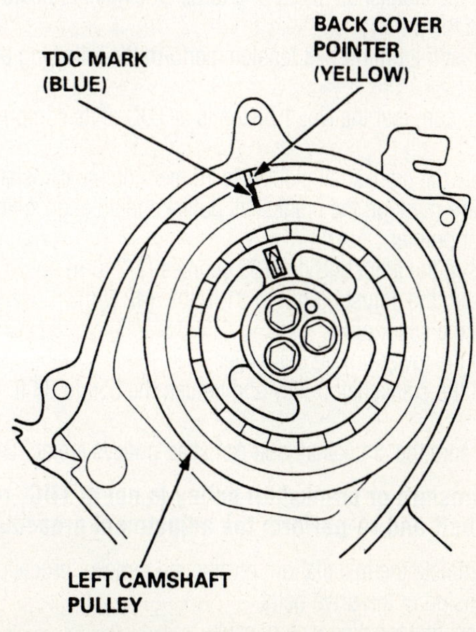

93015g11

Left camshaft timing belt pulley alignment mark locations—Acura 2.7L (C27A1) engine

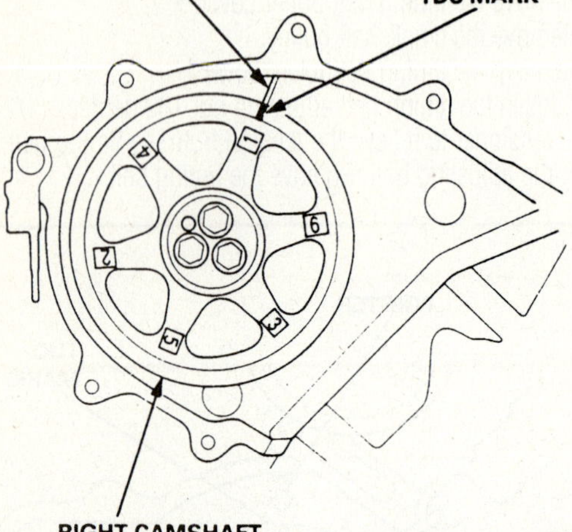

Right camshaft timing belt pulley alignment mark locations—Acura 2.7L (C27A1) engine

To install:

16. Remove the spark plugs.
17. Advance the crankshaft about 15° from the No. 1 TDC position.
18. Adjust the camshaft pulleys to the No. 1 TDC position by aligning the camshaft timing marks with the indentations on the camshaft housing.
19. Return the crankshaft sprocket to the No. 1 TDC position.
20. Use a screwdriver and push the tensioner bracket forward to loosen the tensioner.
21. Advance the rear camshaft sprocket about (1/2 turn from TDC.
22. Install the timing belt on the crankshaft sprocket, front camshaft sprocket, water pump pulley, tensioner pulley and rear camshaft pulley.
23. Loosen and retighten the timing belt adjusting bolt to tension the timing belt.
24. Rotate the crankshaft about 5–6 turns clockwise to position the timing belt on the pulleys.
25. To adjust the timing belt tension, perform the following procedure:
    a. Make sure that the No. 1 piston is at TDC of its compression stroke.
    b. Rotate the crankshaft clockwise 10 teeth on the camshaft pulley; the **blue** mark on the crankshaft pulley should align with the lower cover pointer.
    c. Loosen the adjusting nut 180 degrees (1/2 turn).
    d. Tighten the adjusting nut to 31 ft. lbs. (42 Nm).
26. Install the spark plugs.
27. Install the lower timing belt cover.
28. Install the crankshaft pulley and tighten the bolt to 83 ft. lbs. (115 Nm).
29. Make sure the crankshaft and camshaft pulleys are at TDC.

■ **If the camshaft or crankshaft pulley is not at TDC, remove the timing belt and re-perform the adjustment procedure.**

30. To complete the installation, reverse the removal procedures. Adjust the tension of the drive belts.
31. Connect the negative battery cable.
32. Enter the anti-theft code for the radio and reset the frequencies.

3.0L (C30A1) ENGINES

YEAR, MODEL & ENGINE APPLICATION	
1991-04 Acura NSX	3.0L Code C30A1

REPLACEMENT INTERVALS
Acura recommends timing belt replacement every 90,000 miles on 1991-96 models, and 105,000 miles on 1997-04 models.

LABOR TIMES	
1991-04	
Replacement	5.6 hrs
Severe Service	6.4 hrs
With camshaft seal replacement add	0.5 hrs
With water pump replacement add	0.6 hrs

8807ACUg10

✱✱ CAUTION

These engines are an interference type. If the timing belt has failed, valve-to-piston damage may have occurred. A compression check should always be performed to check for engine damage.

1. Be sure to acquire the anti-theft code for the radio and the frequencies for the radio's preset buttons.
2. Disconnect the negative battery cable.
3. Raise and safely support the vehicle. Remove the right rear wheel.
4. Remove the engine oil cooler base assembly without disconnecting the hoses.
5. Lower the vehicle to a comfortable working height and remove the rear strut brace.
6. Remove the intake manifold plate.
7. Remove the top cover.
8. Remove the injector cover and the wire harness covers.
9. Remove the ignition coil covers and the ignition coils.
10. Remove the coolant expansion tank and the air cleaner case.
11. Remove the alternator.
12. Disconnect the side engine mount and position it back into the body cavity.
13. Remove the transmission mount.
14. Remove the cylinder head covers.
15. Rotate the crankshaft so that the No. 1 cylinder is at Top Dead Center (TDC), and the timing marks are aligned and the camshaft pulley**UP** marks are at the top.
16. Using a jack, raise the engine slightly.
17. Remove the alternator bracket stiffener.
18. Remove the air conditioning adjusting pulley and belt.
19. Remove the dipstick tube mounting bolt.
20. Remove the front and rear timing belt middle covers.
21. Remove the crankshaft pulley and remove the lower timing belt cover.
22. Align the holes in the camshaft holders to the holes in the camshafts, and insert 5mm pin punches to lock the camshafts in position while the belt is removed.
23. Loosen the timing belt adjuster pulley bolt and push on the tensioner to relieve belt tension. Re-tighten the adjuster bolt.

■ **If reusing the timing belt, mark its direction of rotation on the back of the belt.**

24. Remove the timing belt.
25. This is an excellent time to check or replace the water pump. Even if the timing belt is only being replaced as part of a good maintenance schedule, consider replacing the pump at the same time.

To install:

■ **Do not rotate the crankshaft or camshafts before installing the timing belt.**

26. Noting the direction of rotation, install the timing belt. Route the belt in the following order:
   - Crankshaft pulley
   - Adjusting pulley
   - Front exhaust cam pulley
   - Front intake cam pulley
   - Water pump pulley
   - Rear intake cam pulley
   - Rear exhaust cam pulley.
27. Install the lower timing belt cover and the crankshaft pulley.
28. Remove the pin punches.
29. Adjust the timing belt tension by performing the following procedure:
   a. Loosen the tensioner bolt.
   b. While holding the crankshaft from turning, remove slack in the timing belt by rotating the camshafts slightly and in the order of belt installation.
   c. Tighten the tensioner bolt.
   d. Rotate the crankshaft clockwise 9 teeth on a camshaft pulley. The **blue** mark on the crankshaft pulley should align with the pointer on the lower timing cover.
   e. Loosen the tensioner bolt.
   f. Tighten the tensioner bolt to 31 ft. lbs. (43 Nm).
30. Rotate the crankshaft by hand 2 complete revolutions and insure the timing marks align.
31. Install the front and rear timing belt covers.
32. Install the dipstick tube mounting bolt.
33. Install the air conditioning adjustment pulley and belt.

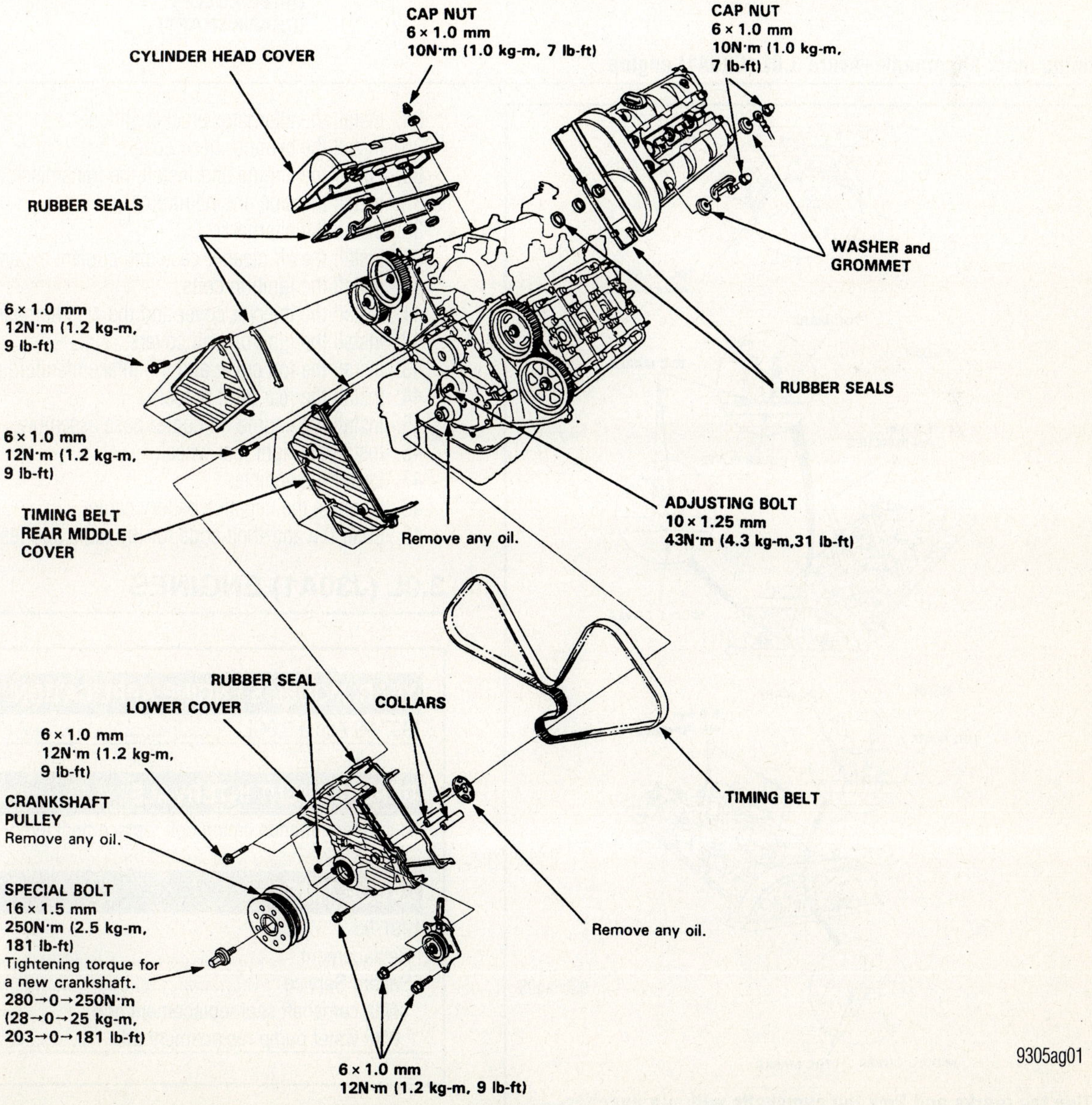

Exploded view of the timing belt assembly—Acura 3.0L (C30A1) engine

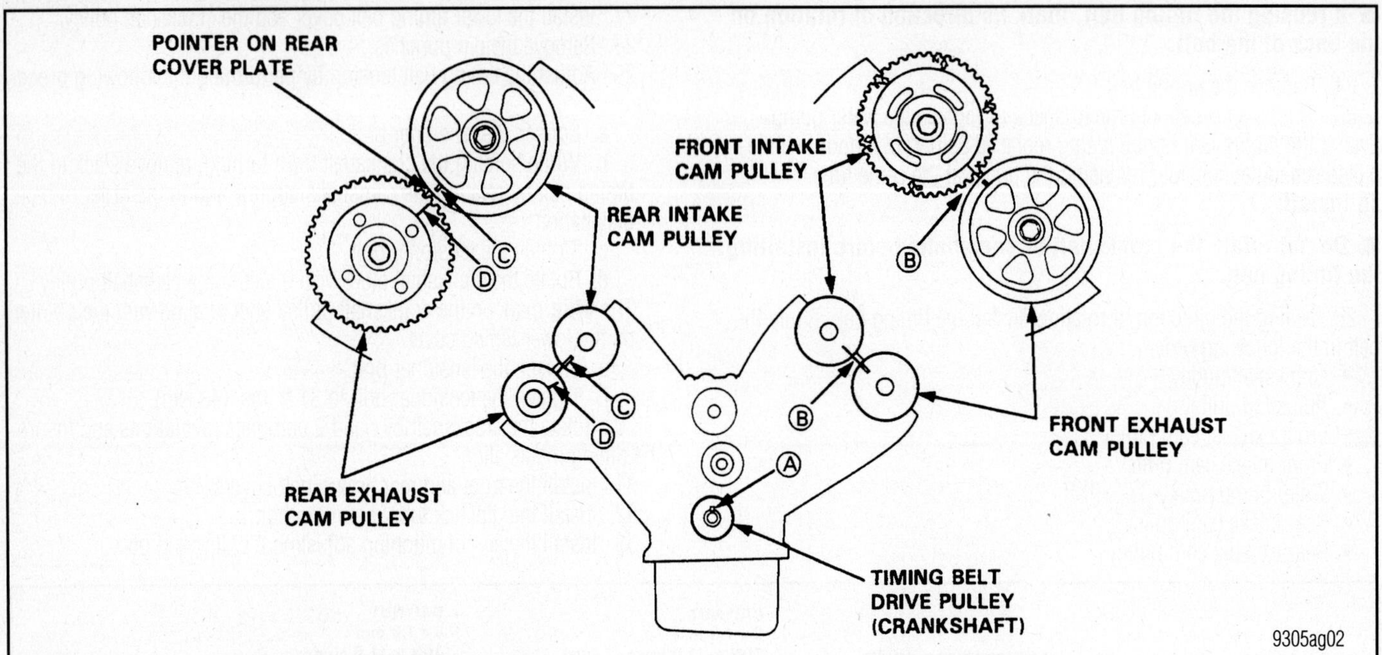

Timing mark alignment—Acura 3.0L (C30A1) engine

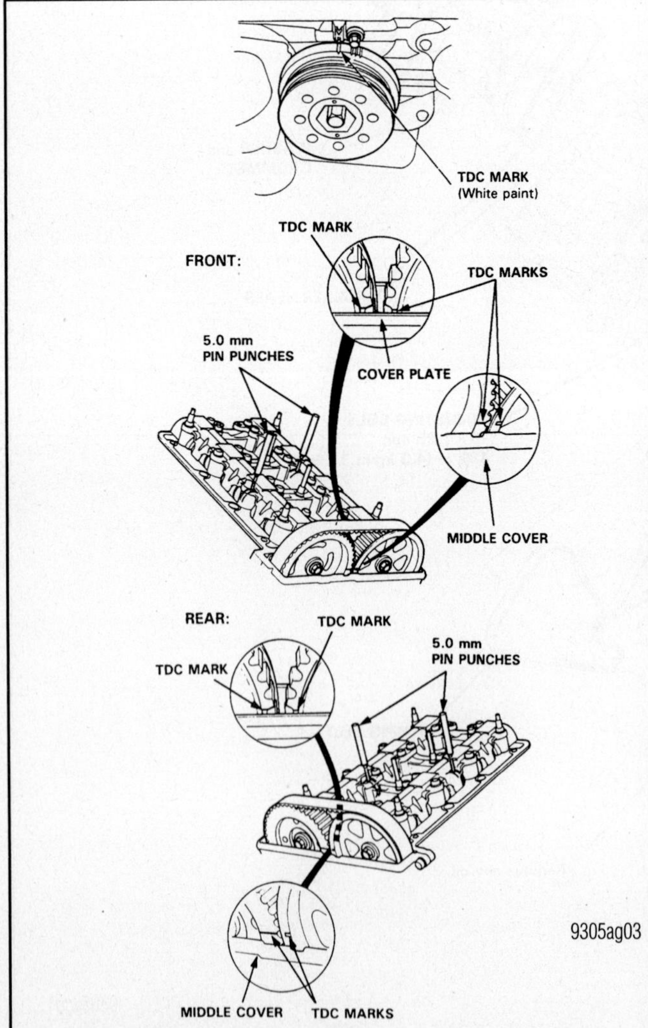

Align the marks and lock the camshafts with pin punches—Acura 3.0L (C30A1) engine

34. Install the alternator bracket stiffener.
35. Install the cylinder head covers.
36. Lower the engine and install the transmission mount.
37. Install the side engine mount.
38. Install the alternator.
39. Install the air cleaner case and coolant expansion tank.
40. Install the ignition coils.
41. Install the injector cover and the wire harness covers.
42. Install the ignition coil covers.
43. Install the top cover and the intake manifold plate.
44. Install the rear strut brace.
45. Install the engine oil cooler base assembly.
46. Install the right rear wheel.
47. Lower the vehicle.
48. Connect the negative battery cable.
49. Enter the anti-theft code for the radio and reset the frequencies.

3.0L (J30A1) ENGINES

YEAR, MODEL & ENGINE APPLICATION

1997-99 Acura 3.0CL .... 3.0L Code J30A1

REPLACEMENT INTERVALS

Acura recommends timing belt replacement every 105,000 miles.

LABOR TIMES

1997-99
Replacement .... 4.6 hrs
Severe Service .... 5.2 hrs
With camshaft seal replacement add .... 0.5 hrs
With water pump replacement add .... 0.6 hrs

8807ACUg11

✱✱ CAUTION

These engines are an interference type. If the timing belt has failed, valve-to-piston damage may have occurred. A compression check should always be performed to check for engine damage.

1. Be sure to acquire the anti-theft code for the radio and the frequencies for the radio's preset buttons.
2. Disconnect the negative battery cable.
3. Raise and safely support the vehicle.
4. Remove the engine splash shield.
5. Move the alternator's drive belt tensioner to relieve the tension; then, remove the alternator belt.
6. Loosen the power steering pump's adjusting bolt, locknut and mounting nut; then, remove the pump's belt.
7. To remove the engine-to-chassis side mount, located at the front of the engine, perform the following procedure:
   a. Using a floor jack, position a cushion between the jack and the oil pan.
   b. Raise the engine slightly to take the weight off of the side mount.
   c. Remove the side mount-to-engine bracket bolt, the side mount-to-chassis bolts and the side mount.
8. Remove the dipstick and the dipstick tube-to-engine bolt; then, pull the tube from its O-ring mount. Discard the O-ring.
9. Turn the engine to align the timing marks and set cylinder No. 1 to Top Dead Center (TDC) on the compression stroke. The white mark on the crankshaft pulley should align with the pointer on the timing belt cover. Remove the inspection caps on the upper timing belt covers to check the alignment of the timing marks. The pointers for the camshaft pulleys should align with the marks on rear upper cover mark.
10. Remove the crankshaft pulley by performing the following procedure:
    a. Using the Holder Handle and the 50mm Offset Holder Attachment tool 07MAB-PY3010A or equivalent, and a 19mm socket, secure the crankshaft pulley and remove the crankshaft pulley bolt.

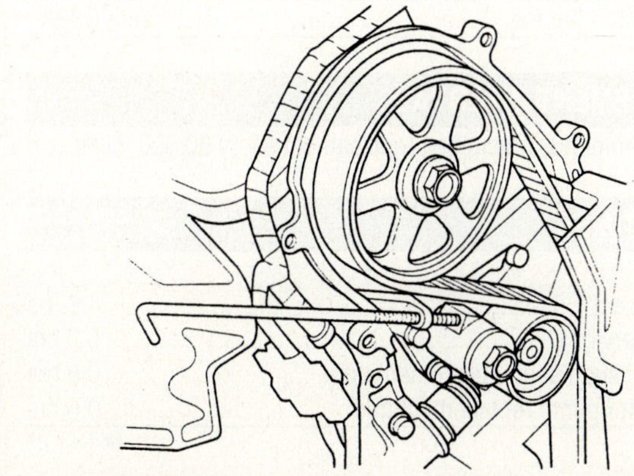

93015g06

Using battery clamp bolt to hold timing belt adjuster in position—Acura 3.0L (J30A1) engine

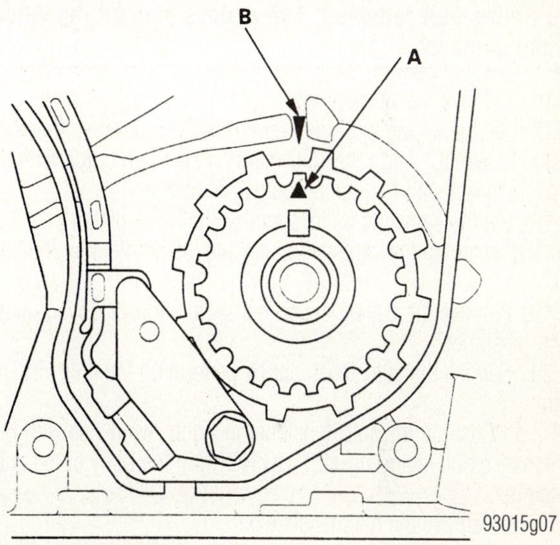

93015g07

Crankshaft sprocket alignment mark positioning for TDC—Acura 3.0L (J30A1) engine

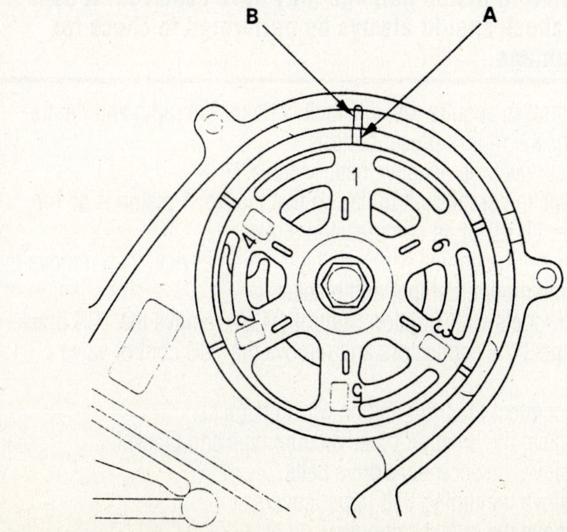

93015g08

Left camshaft sprocket alignment mark positioning for TDC—Acura 3.0L (J30A1) engine

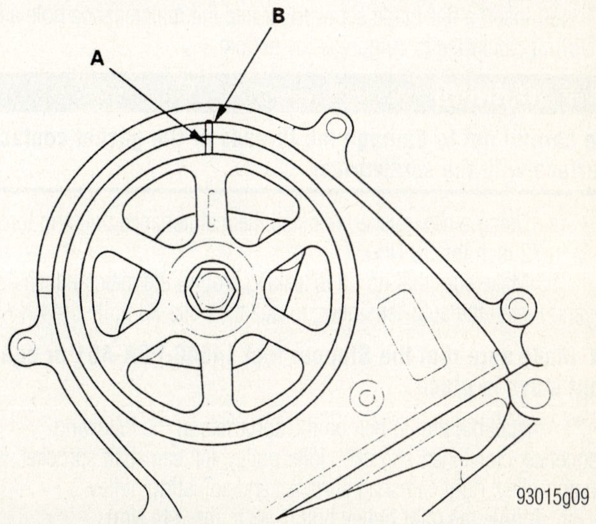

93015g09

Right camshaft sprocket alignment mark positioning for TDC—Acura 3.0L (J30A1) engine

b. Remove the washer and pull the crankshaft pulley from the engine.

11. Remove all necessary components for access to the timing belt covers.

12. Remove the upper and lower timing belt covers. Clean any dirt, oil or grease from the covers. Do not use the covers for storing removed items.

13. Remove one of the battery clamp bolts and grind a 45 degree bevel on the threaded end. Screw the battery clamp bolt into hole provided at the base of the right cylinder head to hold the timing belt adjuster in its current position; tighten the bolt by hand, DO NOT use a wrench.

14. Remove the engine mount bracket.

15. At the base of the left cylinder head, loosen the idler pulley bolt about 5–6 turns; then, remove the timing belt.

To install:

**** WARNING**

Do not rotate the crankshaft pulley or camshaft pulleys with the timing belt removed. The pistons may hit the valves and cause damage.

16. Remove the spark plugs.

17. Set the timing belt drive (crankshaft) sprocket so that the No. 1 piston is at TDC. Align the TDC mark on the tooth of the timing belt drive sprocket with the pointer on the oil pump.

18. Set the camshaft pulleys so that the No. 1 piston is at TDC. Align the TDC mark on the camshaft pulleys to the pointers on the back covers.

19. Remove the battery clamp bolt from the back cover. Remove the auto-tensioner.

20. Service the auto-tensioner by performing the following procedure:

a. Position the auto-tensioner in a soft jawed vise with the maintenance bolt facing upward. DO NOT grip the body of the auto-tensioner.

b. Remove the maintenance bolt.

c. Be careful not to spill the oil from inside the tensioner. If oil is spilled, replenish it; the total capacity is 0.22 fl. oz. (6.5 ml).

d. Using Stopper tool 14540-P8A-A01 or equivalent, position it on the auto-tensioner while turning the internal screw.

e. Insert a flat-blade screwdriver into the maintenance hole and turn it clockwise to compress the bottom.

**** WARNING**

Be careful not to damage the threads or the gasket contact surface with the screwdriver.

a. Using a new gasket, reinstall the maintenance bolt and torque it to 72 inch lbs. (8 Nm).

b. Make sure that no oil is leaking around the maintenance bolt and install the auto-tensioner; torque the bolts 33 ft. lbs. (44 Nm).

■ **Make sure that the Stopper tool 14540-P8A-A01 or equivalent stays in place.**

21. Install the timing belt on the sprockets in the following sequence: crankshaft sprocket, idler pulley, left camshaft sprocket, water pump pulley, right camshaft sprocket and adjusting pulley.

22. Torque the idler pulley bolt to 33 ft. lbs. (44 Nm).

23. Remove the Stopper tool from the auto-tensioner.

24. Install the lower and upper timing belt covers.

25. Install the engine mount-to-engine and torque the No. 10 bolts to 33 ft. lbs. (44 Nm) and the No. 6 bolt to 104 inch lbs. (12 Nm).

26. Install the crankshaft pulley and finger-tighten the bolt and washer. Using the Holder Handle and the 50mm Offset Holder Attachment tool 07MAB-PY3010A or equivalent, and a 19mm socket with a torque wrench, tighten the crankshaft pulley bolt to 181 ft. lbs. (245 Nm).

27. Rotate the crankshaft 5–6 turns clockwise so that the timing belt positions itself properly on the sprockets.

28. Set cylinder No. 1 to TDC by aligning the timing marks. If the timing marks do not align, remove the timing belt, then adjust the components and reinstall the timing belt.

29. Install all applicable components.

30. To complete the installation, reverse the removal procedures. Adjust the tension of the drive belts.

31. Disconnect the negative battery cable.

32. Enter the anti-theft code for the radio and reset the frequencies.

3.2L (C32A1) ENGINES

YEAR, MODEL & ENGINE APPLICATION	
1991-96 Acura legend	3.2L Code C32A1

REPLACEMENT INTERVALS
Acura recommends timing belt replacement every 90,000 miles.

LABOR TIMES	
1991-96	
Replacement	4.6 hrs
Severe Service	5.2 hrs
With camshaft seal replacement add	0.5 hrs
With water pump replacement add	0.6 hrs

8807ACUg12

**** CAUTION**

These engines are an interference type. If the timing belt has failed, valve-to-piston damage may have occurred. A compression check should always be performed to check for engine damage.

1. Be sure to acquire the anti-theft code for the radio and the frequencies for the radio's preset buttons.

2. Disconnect the negative battery cable.

3. Rotate the crankshaft pulley so that the No. 1 piston is at Top Dead Center (TDC) of its compression stroke.

4. Support the engine with a suitable lifting device, then remove the center engine mount, damper and bracket.

5. On models with Traction Control (TCS), remove the TCS brackets, disconnect the connectors and remove the TCS control valve assembly.

6. Disconnect the front engine wiring harness.

7. Remove the breather pipe, vacuum pipe and bracket.

8. Remove the accessory drive belts.

9. Remove the timing belt upper covers.

10. Remove the crankshaft pulley.

11. Remove the timing belt lower cover.

12. Loosen the timing belt adjusting nut 180 degrees (1/2) turn. Push the tensioner to relieve the tension from the timing belt; then, retighten the adjusting bolt. Remove the timing belt.

To install:

13. Remove the spark plugs.
14. Advance the crankshaft about 15° from the No. 1 TDC position.
15. Adjust the camshaft pulleys to the No. 1 TDC position by aligning the camshaft timing marks with the indentations on the camshaft housing.
16. Return the crankshaft sprocket to the No. 1 TDC position.
17. Install the timing belt on the crankshaft sprocket, tensioner pulley, left camshaft sprocket, water pump pulley and right camshaft pulley.

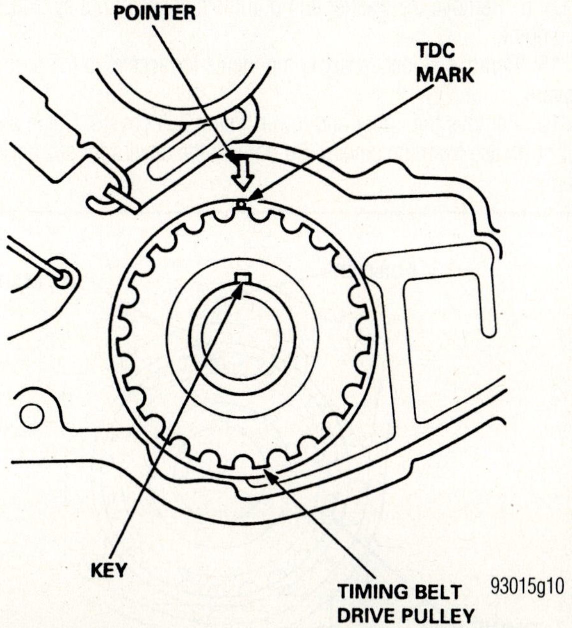

Crankshaft timing belt pulley alignment mark locations—Acura 3.2L (C32A1) engine

18. Loosen and retighten the timing belt adjusting bolt to tension the timing belt.
19. Install the lower timing belt cover.
20. Install the crankshaft pulley and tighten the bolt to 174 ft. lbs. (236 Nm).
21. Rotate the crankshaft about 5–6 turns clockwise to position the timing belt on the pulleys.
22. To adjust the timing belt tension, perform the following procedure:
    a. Make sure that the No. 1 piston is at TDC of its compression stroke.
    b. Rotate the crankshaft clockwise 10 teeth on the camshaft pulley; the **blue** mark on the crankshaft pulley should align with the lower cover pointer.
    c. Loosen the tensioner adjusting nut 180 degrees (1/2 turn).
    d. Tighten the tensioner adjusting nut to 31 ft. lbs. (42 Nm).
23. Make sure the crankshaft and camshaft pulleys are at TDC.

■ **If the camshaft or crankshaft pulley is not at TDC, remove the timing belt and re-perform the adjustment procedure.**

24. Install the spark plugs.
25. To complete the installation, reverse the removal procedures. Adjust the tension of the drive belts.
26. Connect the negative battery cable.
27. Enter the anti-theft code for the radio and reset the frequencies.

3.2L (C32A6) ENGINES

CAUTION

These engines are an interference type. If the timing belt has failed, valve-to-piston damage may have occurred. A compression check should always be performed to check for engine damage.

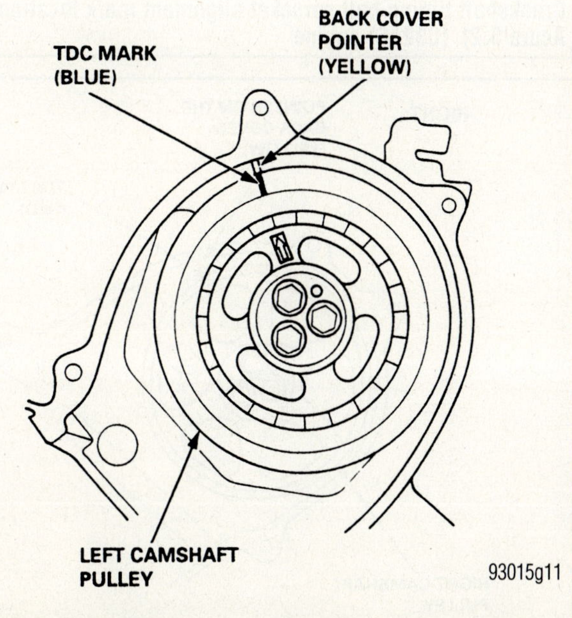

Left camshaft timing belt pulley alignment mark locations—Acura 3.2L (C32A1) engine

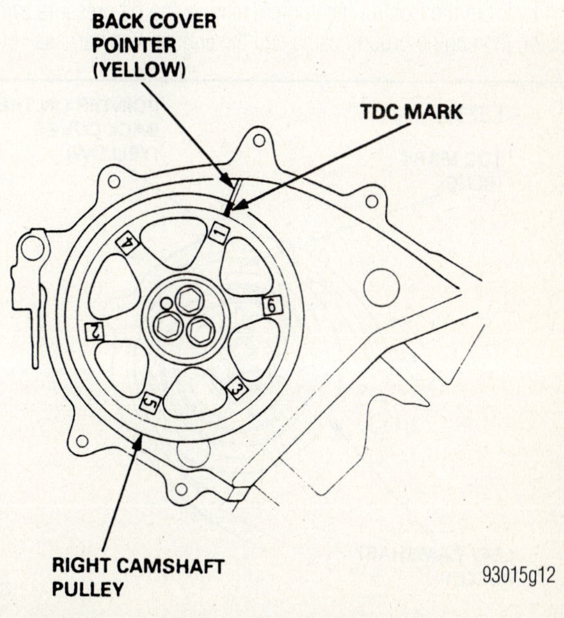

Right camshaft timing belt pulley alignment mark locations— Acura 3.2L (C32A1) engine

YEAR, MODEL & ENGINE APPLICATION

1996-98 Acura 3.2TL	3.2L Code C32A6

REPLACEMENT INTERVALS

Acura recommends timing belt replacement every 90,000 miles on 1996 models, and 105,000 miles on 1997-98 models.

LABOR TIMES

1996-98	
Replacement	4.6 hrs
Severe Service	5.2 hrs
With camshaft seal replacement add	0.5 hrs
With water pump replacement add	0.6 hrs

8807ACUg13

1. Be sure to acquire the anti-theft code for the radio and the frequencies for the radio's preset buttons.
2. Disconnect the negative battery cable.
3. Raise and support the vehicle.
4. Remove the left wheel well splash shield.
5. Loosen the power steering pump's adjusting bolt, locknut and mounting nut; then, remove the pump's belt.
6. Loosen the alternator's adjusting bolt, locknut and mounting nut; then, remove the alternator's belt.
7. Remove the alternator's terminal and connector.
8. To remove the engine-to-chassis center bracket, located at the front of the engine, perform the following procedure:
   a. Using a floor jack, position a cushion between the jack and the oil pan.
   b. Raise the engine slightly to take the weight off of the center bracket.
   c. Remove the center bracket-to-engine bolts, the center bracket-to-chassis bolts and the center bracket.
9. Remove the TCS upper and lower brackets.
10. Disconnect the TCS throttle sensor and actuator connectors and remove the TCS control valve assembly.
11. From the front of the engine, remove the oil pressure switch connector, the engine ground cable and the engine wire harness cover.
12. Remove the dipstick and the dipstick tube-to-engine bolt; then, pull the tube from its O-ring mount. Discard the O-ring.
13. Turn the engine to align the timing marks and set cylinder No. 1 to Top Dead Center (TDC) on the compression stroke. The white mark on the crankshaft pulley should align with the pointer on the timing belt cover. Remove the inspection caps on the upper timing belt covers to check the alignment of the timing marks. The pointers for the camshafts should align with the marks on the camshaft sprockets.
14. Remove the crankshaft pulley by performing the following procedure:
   a. Using the Holder Handle and the 50mm Offset Holder Attachment tool 07MAB-PY3010A or equivalent, and a 19mm socket, secure the crankshaft pulley and remove the crankshaft pulley bolt.
   b. Remove the washer and pull the crankshaft pulley from the engine.
15. Remove all necessary components for access to the timing belt covers.
16. Remove the upper and lower timing belt covers. Clean any dirt, oil or grease from the covers. Do not use the covers for storing removed items.

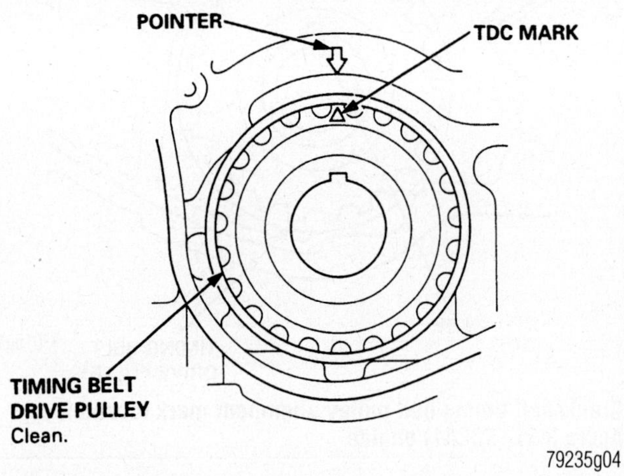

79235g04

Crankshaft timing belt sprocket alignment mark locations—Acura 3.2L (C32A6) engine

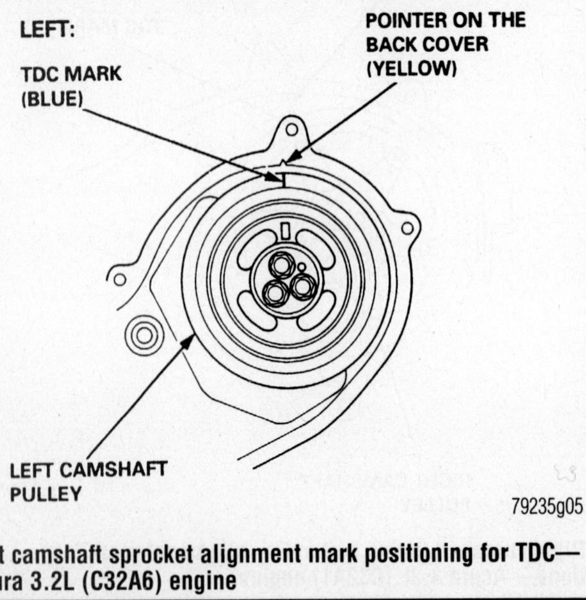

79235g05

Left camshaft sprocket alignment mark positioning for TDC—Acura 3.2L (C32A6) engine

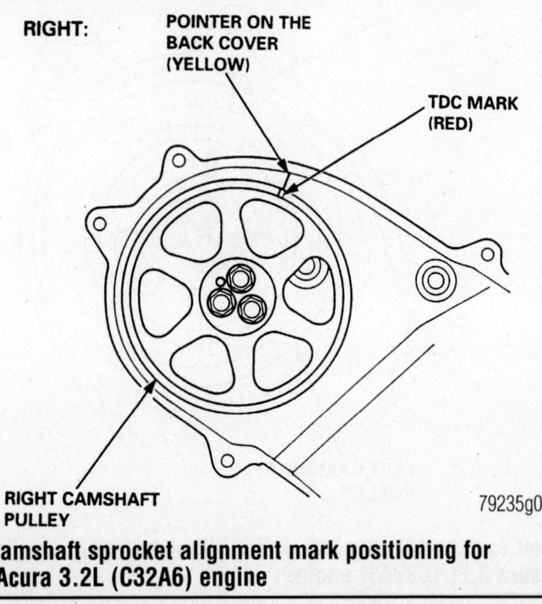

79235g06

Right camshaft sprocket alignment mark positioning for TDC—Acura 3.2L (C32A6) engine

17. Loosen the timing belt tensioner adjusting bolt 180 degrees (1/2 turn). Push on the tensioner to remove tension from the timing belt, then tighten the adjusting bolt.

18. Remove the timing belt.

19. If necessary, remove the timing belt tensioner by performing the following:

a. Remove the spring from the tensioner.

b. Remove the bolt mounting the tensioner, then remove the tensioner.

To install:

✻✻ WARNING

Do not rotate the crankshaft pulley or camshaft pulleys with the timing belt removed. The pistons may hit the valves and cause damage.

20. If necessary, install the timing belt tensioner by performing the following:

a. Install the tensioner and the attaching bolt.

b. Move the tensioner its full deflection to the left and tighten the bolt.

c. Install the spring to the tensioner.

21. Remove the spark plugs.

22. Set the timing belt drive (crankshaft) sprocket so that the No. 1 piston is at TDC. Align the TDC mark on the tooth of the timing belt drive sprocket with the pointer on the oil pump.

23. Set the camshaft pulleys so that the No. 1 piston is at TDC. Align the TDC mark on the camshaft pulleys to the pointers on the back covers.

24. Install the timing belt on the sprockets in the following sequence: crankshaft sprocket, tensioner pulley, left camshaft sprocket, water pump pulley, right camshaft sprocket.

25. Loosen, then retighten the timing belt adjuster bolt to tension the timing belt.

26. Install the lower timing belt cover.

27. Install the crankshaft pulley and finger-tighten the bolt and washer. Using the Holder Handle and the 50mm Offset Holder Attachment tool 07MAB-PY3010A or equivalent, and a 19mm socket with a torque wrench, tighten the crankshaft pulley bolt to 181 ft. lbs. (245 Nm).

28. Rotate the crankshaft 5–6 turns clockwise so that the timing belt positions itself properly on the sprockets.

29. Set cylinder No. 1 to TDC by aligning the timing marks. If the timing marks do not align, remove the timing belt, then adjust the components and reinstall the timing belt.

30. Rotate the crankshaft clockwise enough to move the camshaft pulley 9 teeth (the **blue** mark on the crankshaft pulley should align with the pointer on the lower cover).

31. Loosen the timing belt adjusting bolt 180 degrees (1/2 turn), then tighten the bolt to 31 ft. lbs. (42 Nm).

32. Install the upper timing belt covers, then install all applicable components. When installing the center bracket, tighten the bolts attaching the brackets to 40 ft. lbs. (54 Nm), then the mount through-bolt to 40 ft. lbs. (54 Nm).

33. To complete the installation, reverse the removal procedures. Adjust the tension of the drive belts.

34. Connect the negative battery cable.

35. Enter the anti-theft code for the radio and reset the frequencies.

3.2L (C32B1) ENGINES

YEAR, MODEL & ENGINE APPLICATION

1991-04 Acura NSX	3.2L Code C32B1

REPLACEMENT INTERVALS

Acura recommends timing belt replacement every 90,000 miles on 1991-96 models, and 105,000 miles on 1996-04 models.

LABOR TIMES

1991-04	
Replacement	5.6 hrs
Severe Service	6.4 hrs
With camshaft seal replacement add	0.5 hrs
With water pump replacement add	0.6 hrs

8807ACUg14

✻✻ CAUTION

These engines are an interference type. If the timing belt has failed, valve-to-piston damage may have occurred. A compression check should always be performed to check for engine damage.

1. Be sure to acquire the anti-theft code for the radio and the frequencies for the radio's preset buttons.

2. Disconnect the negative battery cable.

3. Raise and safely support the vehicle. Remove the right rear wheel.

4. Remove the engine oil cooler base assembly without disconnecting the hoses.

5. Lower the vehicle to a comfortable working height and remove the rear strut brace.

6. Remove the intake manifold plate.

7. Remove the top cover.

8. Remove the injector cover and the wire harness covers.

9. Remove the ignition coil covers and the ignition coils.

10. Remove the coolant expansion tank and the air cleaner case.

11. Remove the alternator.

12. Disconnect the side engine mount and position it back into the body cavity.

13. Remove the transmission mount.

14. Remove the cylinder head covers.

15. Rotate the crankshaft so that the No. 1 cylinder is at Top Dead Center (TDC), and the timing marks are aligned and the camshaft pulley **UP** marks are at the top.

16. Using a jack, raise the engine slightly.

17. Remove the alternator bracket stiffener.

18. Remove the air conditioning adjusting pulley and belt.

19. Remove the dipstick tube mounting bolt.

20. Remove the front and rear timing belt middle covers.

21. Remove the crankshaft pulley and remove the lower timing belt cover.

22. Align the holes in the camshaft holders to the holes in the camshafts, and insert 5mm pin punches to lock the camshafts in position while the belt is removed.

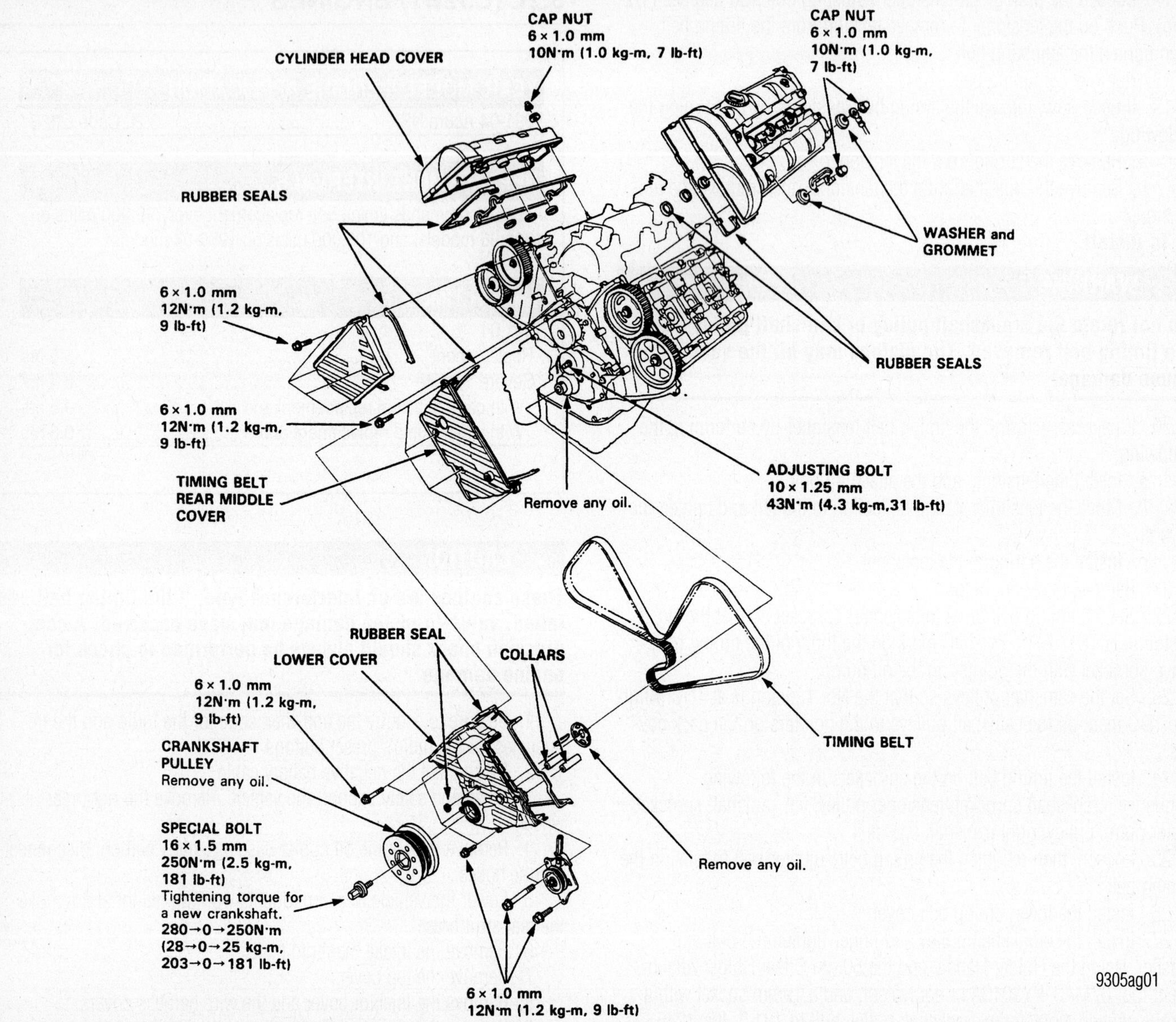

Exploded view of the timing belt assembly—Acura 3.2L (C32B1) engine

23. Loosen the timing belt adjuster pulley bolt and push on the tensioner to relieve belt tension. Re-tighten the adjuster bolt.

■ **If reusing the timing belt, mark its direction of rotation on the back of the belt.**

24. Remove the timing belt.

25. This is an excellent time to check or replace the water pump. Even if the timing belt is only being replaced as part of a good maintenance schedule, consider replacing the pump at the same time.

To install:

■ **Do not rotate the crankshaft or camshafts before installing the timing belt.**

26. Noting the direction of rotation, install the timing belt. Route the belt in the following order:

- Crankshaft pulley
- Adjusting pulley
- Front exhaust cam pulley
- Front intake cam pulley
- Water pump pulley
- Rear intake cam pulley
- Rear exhaust cam pulley.

27. Install the lower timing belt cover and the crankshaft pulley.

28. Remove the pin punches.

29. Adjust the timing belt tension by performing the following procedure:

a. Loosen the tensioner bolt.

b. While holding the crankshaft from turning, remove slack in the timing belt by rotating the camshafts slightly and in the order of belt installation.

c. Tighten the tensioner bolt.

d. Rotate the crankshaft clockwise 9 teeth on a camshaft pulley. The **blue** mark on the crankshaft pulley should align with the pointer on the lower timing cover.

e. Loosen the tensioner bolt.

f. Tighten the tensioner bolt to 31 ft. lbs. (43 Nm).

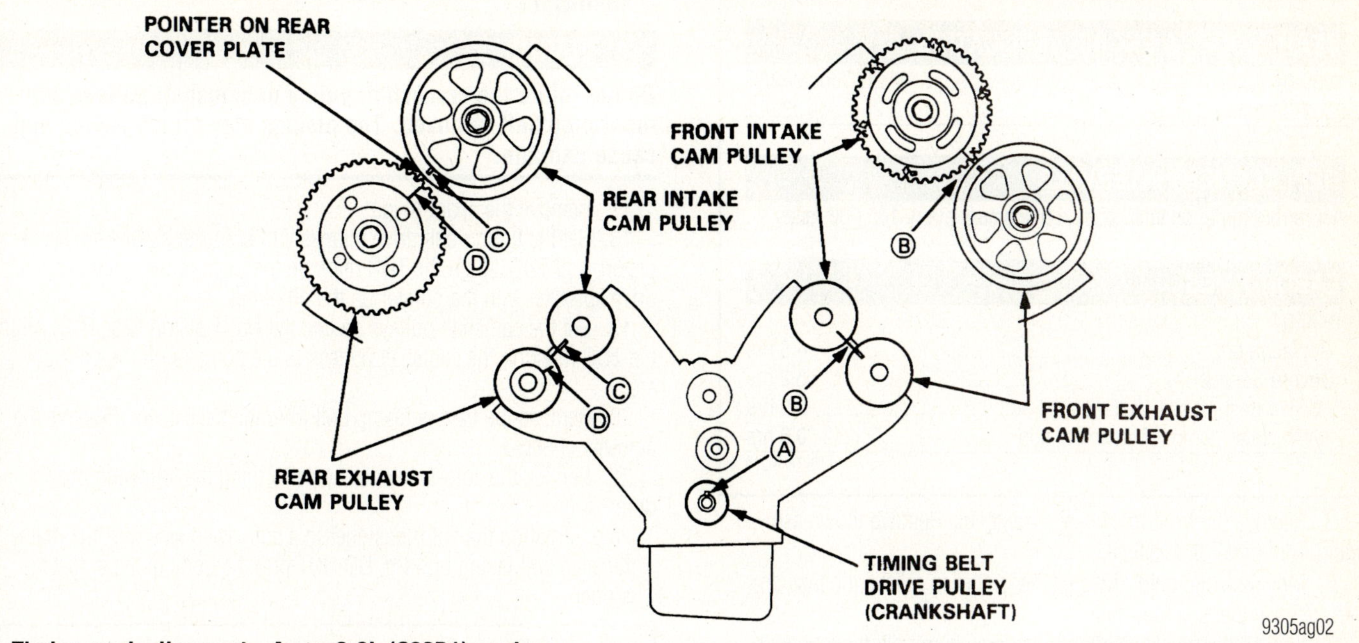

Timing mark alignment—Acura 3.2L (C32B1) engine

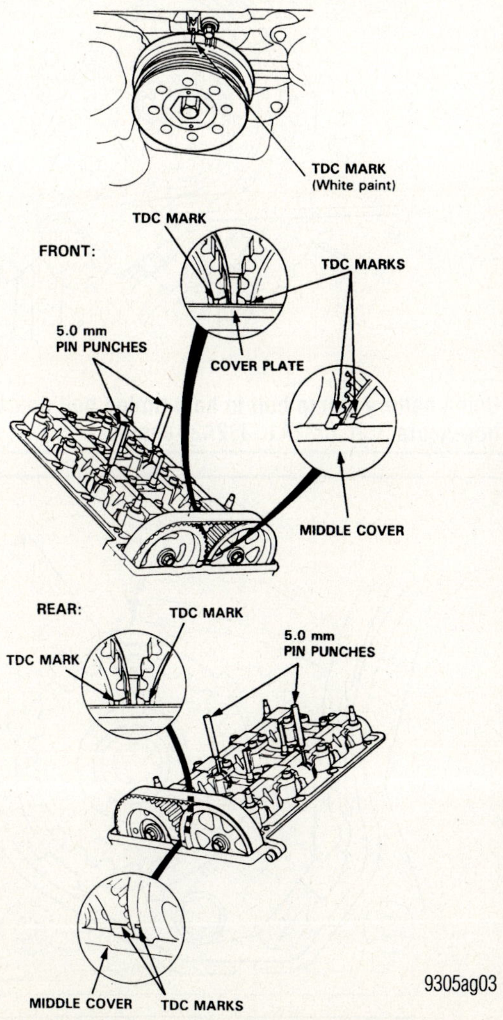

Align the marks and lock the camshafts with pin punches—Acura 3.2L (C32B1) engine

30. Rotate the crankshaft by hand 2 complete revolutions and insure the timing marks align.
31. Install the front and rear timing belt covers.
32. Install the dipstick tube mounting bolt.
33. Install the air conditioning adjustment pulley and belt.
34. Install the alternator bracket stiffener.
35. Install the cylinder head covers.
36. Lower the engine and install the transmission mount.
37. Install the side engine mount.
38. Install the alternator.
39. Install the air cleaner case and coolant expansion tank.
40. Install the ignition coils.
41. Install the injector cover and the wire harness covers.
42. Install the ignition coil covers.
43. Install the top cover and the intake manifold plate.
44. Install the rear strut brace.
45. Install the engine oil cooler base assembly.
46. Install the right rear wheel.
47. Lower the vehicle.
48. Connect the negative battery cable.
49. Enter the anti-theft code for the radio and reset the frequencies.

3.2L (J32A1 AND J32A2) ENGINES

✱✱ CAUTION

These engines are an interference type. If the timing belt has failed, valve-to-piston damage may have occurred. A compression check should always be performed to check for engine damage.

1. Be sure to acquire the anti-theft code for the radio and the frequencies for the radio's preset buttons.
2. Disconnect the negative battery cable.
3. Raise and safely support the vehicle.
4. Remove the ignition coil cover.

YEAR, MODEL & ENGINE APPLICATION

1999-04 Acura 3.2TL	3.2L Code J32A1
2001-03 Acura 3.2CL	3.2L Code J32A2

REPLACEMENT INTERVALS

Acura recommends timing belt replacement every 105,000 miles.

LABOR TIMES

1999-04	
Replacement	4.6 hrs
Severe Service	5.2 hrs
With camshaft seal replacement add	0.5 hrs
With water pump replacement add	0.6 hrs

8807ACUg15

5. Remove the front tire/wheel assemblies. Remove the splash shield from under the vehicle.

6. Move the alternator drive belt tensioner to relieve the tension; then, remove the alternator belt.

7. Loosen the power steering pump adjusting bolt, locknut and mounting nut; then, remove the pump's belt.

8. To remove the engine-to-chassis side mount, located at the front of the engine, perform the following procedure:

a. Using a floor jack, position a cushion between the jack and the oil pan.

b. Raise the engine slightly to take the weight off of the side mount.

c. Remove the side mount-to-engine bracket bolt, the side mount-to-chassis bolts and the side mount.

9. Remove the dipstick and the dipstick tube-to-engine bolt; then, pull the tube from its O-ring mount. Discard the O-ring.

10. Turn the engine to align the timing marks and set cylinder No. 1 to Top Dead Center (TDC) on the compression stroke. The white mark on the crankshaft pulley should align with the pointer on the timing belt cover. Remove the inspection caps on the upper timing belt covers to check the alignment of the timing marks. The pointers for the camshaft pulleys should align with the marks on rear upper cover mark.

11. Remove the crankshaft pulley by performing the following procedure:

a. Using the Holder Handle and the 50mm Offset Holder Attachment tool 07MAB-PY3010A or equivalent, and a 19mm socket, secure the crankshaft pulley and remove the crankshaft pulley bolt.

b. Remove the washer and pull the crankshaft pulley from the engine.

12. Remove all necessary components for access to the timing belt covers.

13. Remove the upper and lower timing belt covers. Clean any dirt, oil or grease from the covers. Do not use the covers for storing removed items.

14. Remove one of the battery clamp bolts and grind a 45 degree bevel on the threaded end. Screw the battery clamp bolt into hole provided at the base of the right cylinder head to hold the timing belt adjuster in its current position; tighten the bolt by hand, DO NOT use a wrench.

15. Remove the engine mount bracket.

16. At the base of the left cylinder head, loosen the idler pulley bolt about 5–6 turns; then, remove the timing belt.

To install:

✻✻ WARNING

Do not rotate the crankshaft pulley or camshaft pulleys with the timing belt removed. The pistons may hit the valves and cause damage.

17. Remove the spark plugs.

18. Set the timing belt drive (crankshaft) sprocket so that the No. 1 piston is at TDC. Align the TDC mark on the tooth of the timing belt drive sprocket with the pointer on the oil pump.

19. Set the camshaft pulleys so that the No. 1 piston is at TDC. Align the TDC mark on the camshaft pulleys to the pointers on the back covers.

20. Remove the battery clamp bolt from the back cover. Remove the auto-tensioner.

21. Service the auto-tensioner by performing the following procedure:

a. Position the auto-tensioner in a soft jawed vise with the maintenance bolt facing upward. DO NOT grip the body of the auto-tensioner.

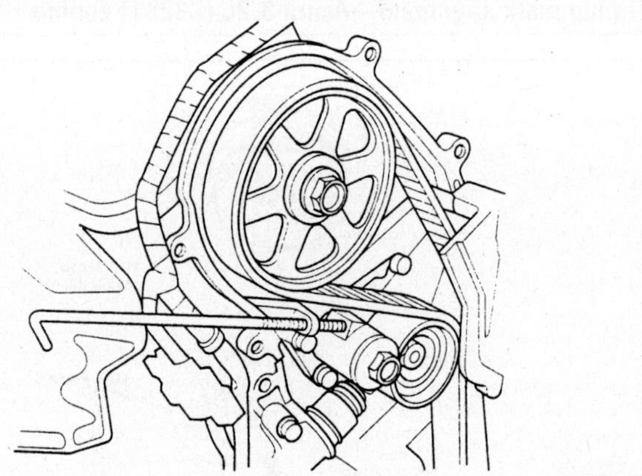

93015g06

Using battery clamp bolt to hold timing belt adjuster in position-Acura 3.2L (J32A1, J32A2) engine

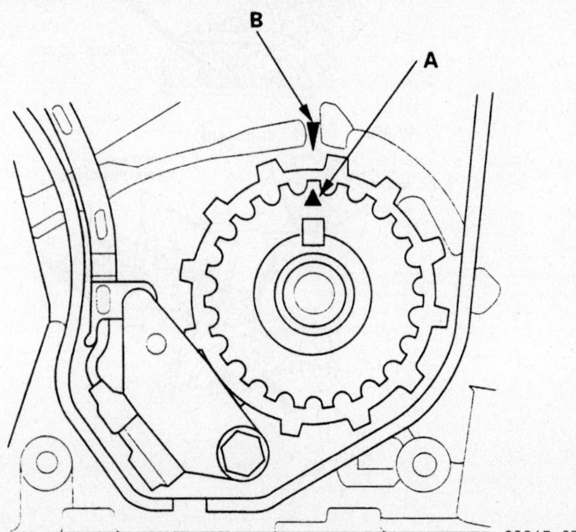

93015g07

Crankshaft sprocket alignment mark positioning for TDC-Acura 3.2L (J32A1, J32A2) engine

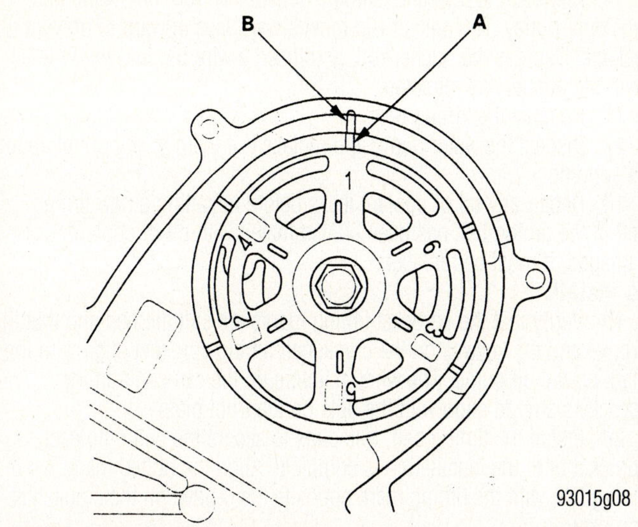

93015g08

Left camshaft sprocket alignment mark positioning for TDC-Acura 3.2L (J32A1, J32A2) engine

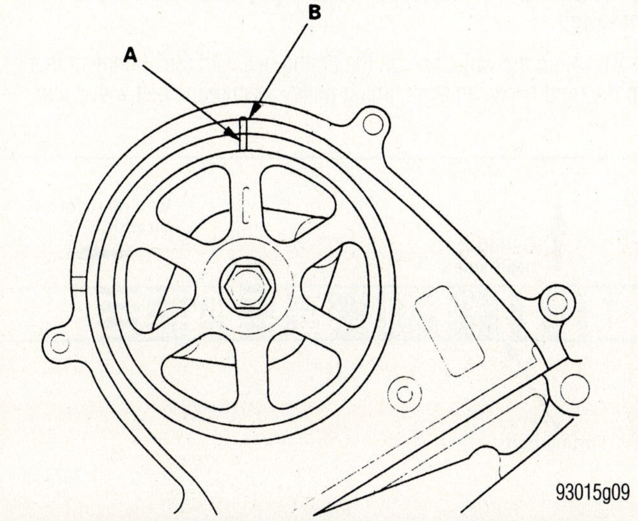

93015g09

Right camshaft sprocket alignment mark positioning for TDC-Acura 3.2L (J32A1, J32A2) engine

b. Remove the maintenance bolt.

c. Be careful not to spill the oil from inside the tensioner. If oil is spilled, replenish it; the total capacity is 0.22 fl. oz. (6.5 ml).

d. Using Stopper tool 14540-P8A-A01 or equivalent, position it on the auto-tensioner while turning the internal screw.

e. Insert a flat-blade screwdriver into the maintenance hole and turn it clockwise to compress the bottom.

⁂ WARNING

Be careful not to damage the threads or the gasket contact surface with the screwdriver.

f. Using a new gasket, reinstall the maintenance bolt and torque it to 72 inch lbs. (8 Nm).

g. Make sure that no oil is leaking around the maintenance bolt and install the auto-tensioner; torque the bolts 33 ft. lbs. (44 Nm).

■ **Make sure that the Stopper tool 14540-P8A-A01 or equivalent stays in place.**

22. Install the timing belt on the sprockets in the following sequence: crankshaft sprocket, idler pulley, left camshaft sprocket, water pump pulley, right camshaft sprocket and adjusting pulley.
23. Torque the idler pulley bolt to 33 ft. lbs. (44 Nm).
24. Remove the Stopper tool from the auto-tensioner.
25. Install the engine mount-to-engine and torque the No. 10 bolts to 33 ft. lbs. (44 Nm) and the No. 6 bolt to 104 inch lbs. (12 Nm).
26. Install the lower and upper timing belt covers.
27. Install the crankshaft pulley and finger-tighten the bolt and washer. Using the Holder Handle and the 50mm Offset Holder Attachment tool 07MAB-PY3010A or equivalent, and a 19mm socket with a torque wrench, tighten the crankshaft pulley bolt to 181 ft. lbs. (245 Nm).
28. Rotate the crankshaft 5–6 turns clockwise so that the timing belt positions itself properly on the sprockets.
29. Set cylinder No. 1 to TDC by aligning the timing marks. If the timing marks do not align, remove the timing belt, then adjust the components and reinstall the timing belt.
30. Install all applicable components.
31. To complete the installation, reverse the removal procedures. Adjust the tension of the drive belts.
32. Disconnect the negative battery cable.
33. Enter the anti-theft code for the radio and reset the frequencies.

3.2L (6VD1) AND 3.5L (6VE1) ENGINES

YEAR, MODEL & ENGINE APPLICATION

1996-97 Acura SLX	3.2L Code 6VD1
1998-99 Acura SLX	3.5L Code 6VE1

REPLACEMENT INTERVALS

Acura recommends timing belt replacement every 60,000 miles.

LABOR TIMES

1996-97	
Replacement	4.1 hrs
Severe Service	4.6 hrs
1998-99	
Replacement	3.6 hrs
Severe Service	4.1 hrs

8807ACUg16

⁂ CAUTION

These engines are an interference type. If the timing belt has failed, valve-to-piston damage may have occurred. A compression check should always be performed to check for engine damage.

1. Be sure to acquire the anti-theft code for the radio and the frequencies for the radio's preset buttons.
2. Disconnect the negative battery cable.
3. Drain the engine coolant into a sealable container.
4. Remove the air cleaner assembly and intake air duct.
5. Disconnect the upper radiator hose from the coolant inlet.
6. Remove the upper fan shroud from the radiator.
7. Remove the 4 nuts retaining the cooling fan assembly. Remove the cooling fan from the fan pulley.

8. Loosen and remove the drive belts.
9. Remove the upper timing belt covers.
10. Remove the fan pulley assembly.
11. Rotate the crankshaft to align the camshaft timing marks with the pointer dots on the back covers. Verify that the pointer on the crankshaft aligns with the mark on the lower timing cover.

■ **When the timing marks are aligned, the No. 2 piston is at Top Dead Center (TDC) of the compression.**

**** WARNING**

Align the camshaft and crankshaft sprockets with their alignment marks before removing the timing belt. Failure to align the belt and sprocket marks may result in valve damage.

12. Use tool No. J-8614-01 or a suitable pulley holding tool to remove the crankshaft pulley center bolt. Remove the crankshaft pulley.
13. If present, disconnect the 2 oil cooler hose bracket bolts on the timing cover. Move the oil cooler hoses and bracket off of the lower timing cover.
14. Remove the lower timing belt cover.

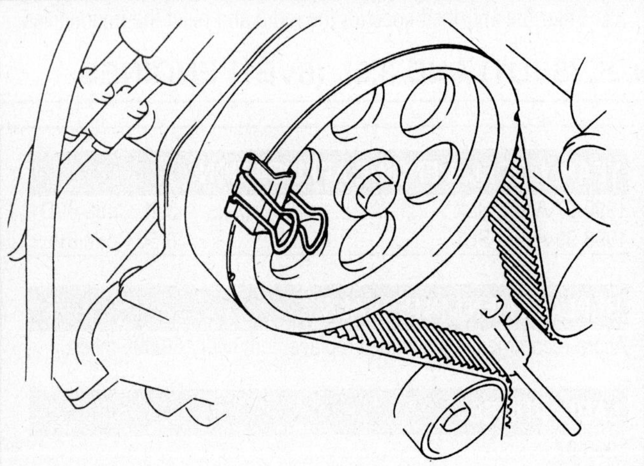

79245g08

Using a double clip to hold the belt in place—Acura 3.2L (6VD1) and 3.5L (6VE1) engine

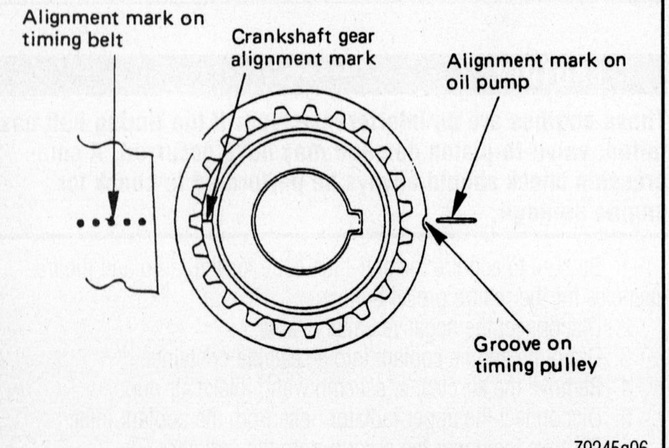

79245g06

Proper crankshaft alignment marks for timing belt installation— Acura 3.2L (6VD1) and 3.5L (6VE1) engine

15. Remove the pusher assembly (tensioner) from below the belt tensioner pulley. The pusher rod must always face upward to prevent oil leakage. Depress the pusher rod, and insert a wire pin into the hole to keep the pusher rod retracted.
16. Remove the timing belt.
17. Inspect the water pump and replace it if there is any doubt about its condition.
18. Repair any oil or coolant leaks before installing a new timing belt. If the timing belt has been contaminated with oil or coolant, or is damaged, it must be replaced.

To install:

19. Verify that the sprocket timing marks are still aligned and that the groove and the keyway on the crankshaft timing sprocket align with the mark on the oil pump. The white pointers on the camshaft timing sprockets should align with the dots on the front plate.
20. Install the timing belt. Use clips to secure the belt onto each sprocket until the installation is complete. Align the dotted marks on the timing belt with the timing mark opposite the groove on the crankshaft sprocket.

■ **The arrows on the timing belt must follow the belt's direction of rotation. The manufacturer's trademark on the belt's spine should be readable left-to-right when the belt is installed.**

21. Align the white line on the timing belt with the alignment mark on the right bank camshaft timing pulley. Secure the belt with a clip.

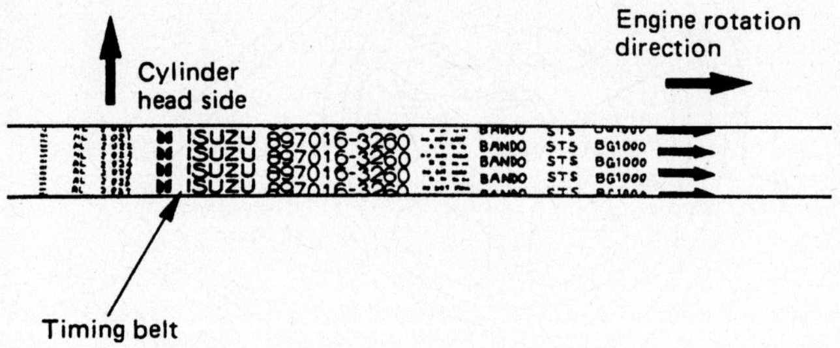

79245g07

For maximum timing belt life, install the belt as shown— Acura 3.2L (6VD1) and 3.5L (6VE1) engine

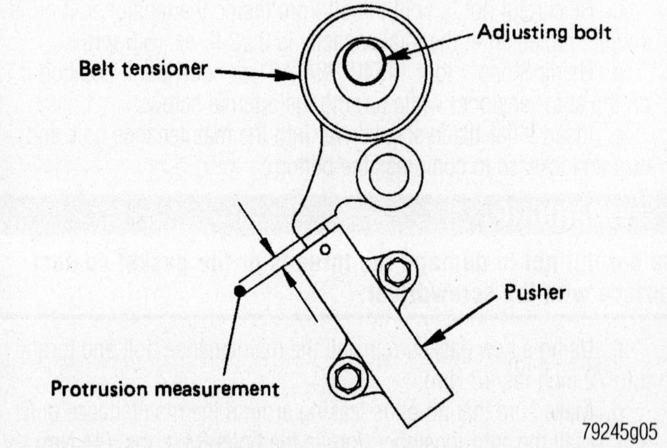

79245g05

View of timing belt tensioner and pusher— Acura 3.2L (6VD1) and 3.5L (6VE1) engine

⁂ WARNING

If any binding is felt when adjusting the timing belt tension by turning the crankshaft, STOP turning the engine, because the pistons may be hitting the valves.

22. Rotate the crankshaft counterclockwise to remove the slack between the crankshaft sprocket and the right camshaft timing belt sprocket.
23. Install the belt around the water pump pulley.
24. Install the belt on the idler pulley.
25. Align the white alignment mark on the timing belt with the alignment mark on the left bank camshaft timing belt sprocket.
26. Install the crankshaft pulley and tighten the center bolt by hand. Rotate the crankshaft pulley clockwise to give slack between the crankshaft timing belt pulley and the right bank camshaft timing belt pulley.
27. Insert a 1.4mm piece of wire through the hole in the pusher to hold the rod in. Install the pusher assembly while pushing the tension pulley toward the belt.
28. Pull the pin out from the pusher to release the rod.
29. Remove the clamps from the sprockets. Rotate the crankshaft pulley clockwise 2 turns. Measure the rod protrusion to ensure it is between 0.16–0.24 in. (4–6mm).
30. If the tensioner pulley bracket pivot bolt was removed, tighten it to 31 ft. lbs. (42 Nm).
31. Tighten the pusher bolts to 14 ft. lbs. (19 Nm).
32. Remove the crankshaft pulley. Install the lower and upper timing belt covers and tighten their bolts to 12 ft. lbs. (17 Nm).
33. Fit the oil cooler hose onto the timing cover and tighten its mounting bracket bolts to 16 ft. lbs. (22 Nm).
34. Install the crankshaft pulley and tighten the pulley bolt to 123 ft. lbs. (167 Nm).
35. Install fan pulley assembly and tighten the bolts to 16 ft. lbs. (22 Nm).
36. Install and adjust the accessory drive belts.
37. Install the cooling fan assembly and tighten the bolts to 72 inch lbs. (8 Nm).
38. Install the upper fan shroud.
39. Install the air cleaner assembly and intake air duct.
40. Connect the upper radiator hose to the coolant inlet.
41. Refill and bleed the cooling system.
42. Connect the negative battery cable.
43. Enter the anti-theft code for the radio and reset the frequencies.

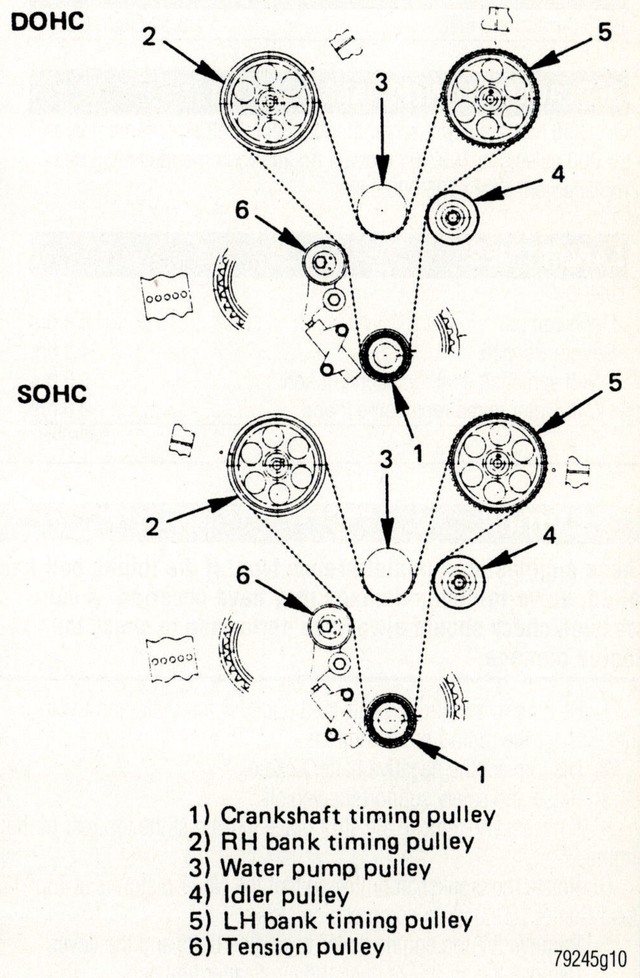

Timing belt routing— Acura 3.2L (6VD1) and 3.5L (6VE1) engine

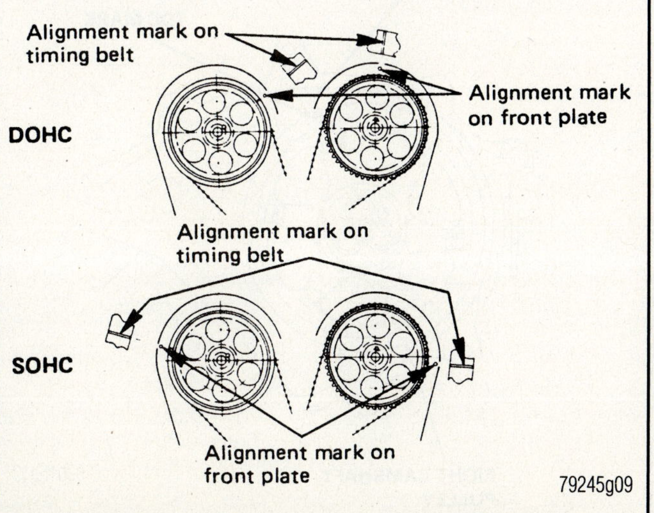

Proper camshaft alignment marks for timing belt installation— Acura 3.2L (6VD1) and 3.5L (6VE1) engine

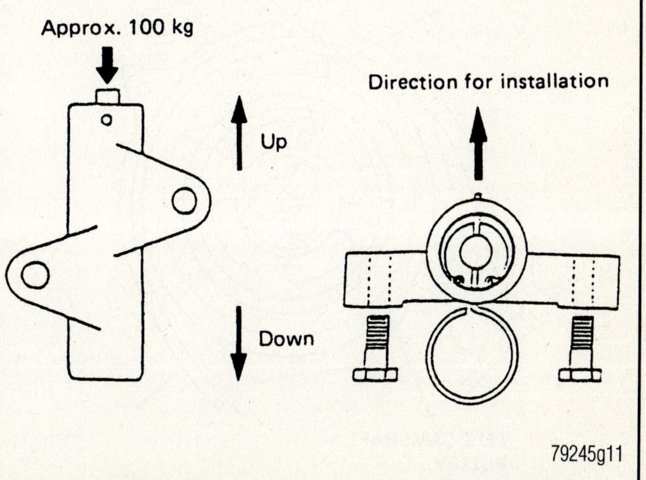

View of timing belt tensioner pusher— Acura 3.2L (6VD1) and 3.5L (6VE1) engine

3.5L (C35A1) ENGINES

YEAR, MODEL & ENGINE APPLICATION

1996-04 Acura 3.5RL .... 3.5L Code C35A1

REPLACEMENT INTERVALS

On 1996 models Acura recommends timing belt replacement every 90,000 miles. On 1997-04 models Acura recommends timing belt replacement every 105,000 miles.

LABOR TIMES

1996-04	
Replacement	3.8 hrs
Severe Service	4.3 hrs
With camshaft seal replacement add	0.5 hrs
With water pump replacement add	0.6 hrs

8807ACUg17

✳✳ CAUTION

These engines are an interference type. If the timing belt has failed, valve-to-piston damage may have occurred. A compression check should always be performed to check for engine damage.

1. Be sure to acquire the anti-theft code for the radio and the frequencies for the radio's preset buttons.
2. Disconnect the negative battery cable.
3. Raise and safely support the vehicle.
4. If necessary, remove the strut brace located at the top rear of the engine.
5. Rotate the crankshaft pulley so that the No. 1 piston is at Top Dead Center (TDC) of its compression stroke.
6. Remove the top engine cover-to-engine bolts and the cover.
7. Remove the air intake duct and air cleaner housing.

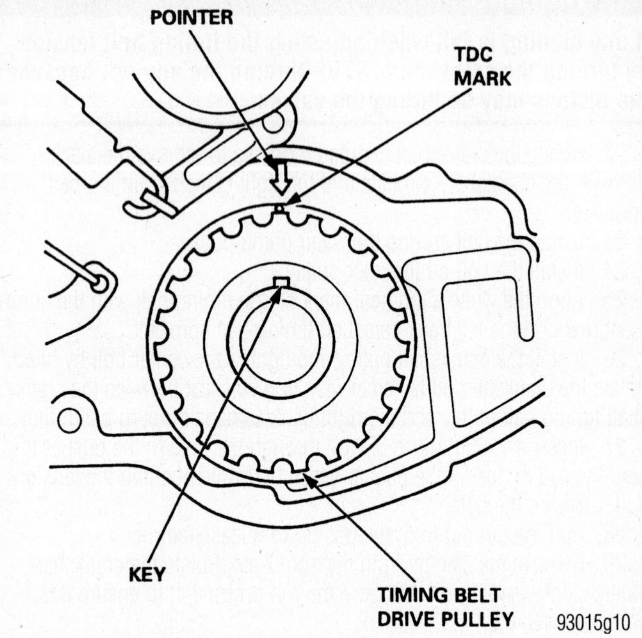

Crankshaft timing belt pulley alignment mark locations—Acura 3.5L (C35A1) engine

8. Loosen the alternator adjusting rod, locknut and mounting bolt; then, remove the alternator belt.
9. Loosen the idler pulley center nut and adjusting bolt; then, remove the air conditioning compressor belt.
10. Remove the TCS control valve upper and lower brackets.
11. Loosen the power steering pump adjusting bolt, locknut and mounting bolt; then, remove the pump belt.
12. Disconnect the TCS throttle sensor connector, the TCS throttle actuator connector and the Throttle Position (TP) sensor; then, remove the TCS control valve assembly.
13. Disconnect the Vehicle Speed Sensor (VSS) harness connector and remove the wire harness holder.

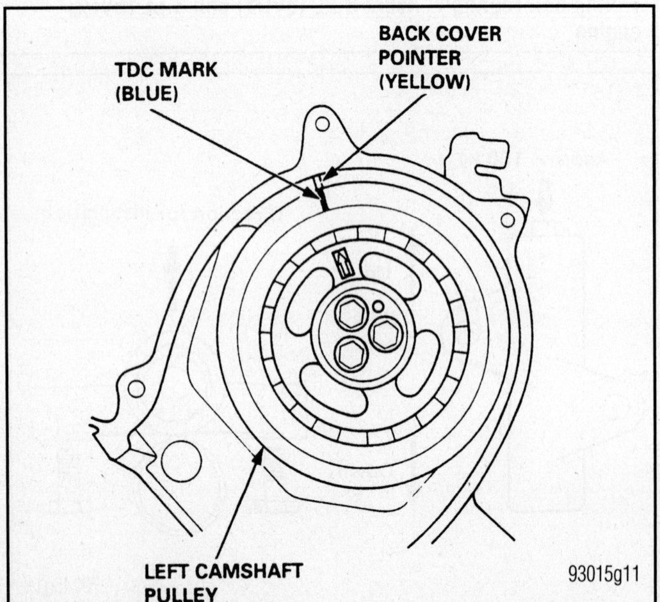

Left camshaft timing belt pulley alignment mark locations—Acura 3.5L (C35A1) engine

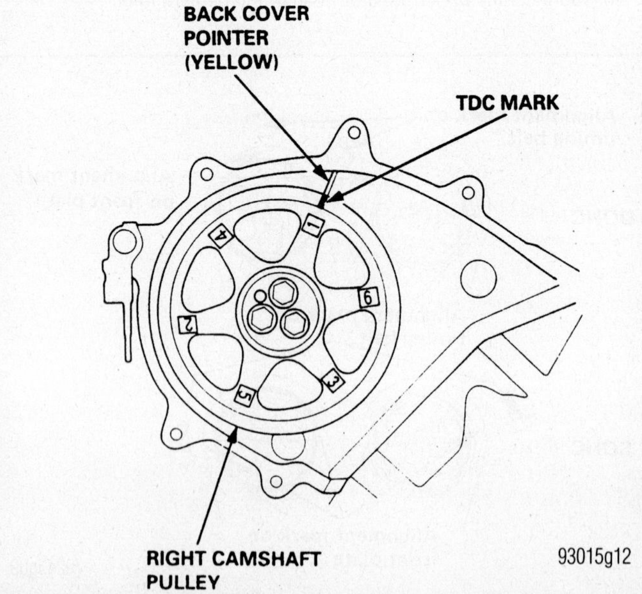

Right camshaft timing belt pulley alignment mark locations—Acura 3.5L (C35A1) engine

14. Remove the breather and vacuum hoses.
15. From the right timing belt cover, remove the Ignition Control Module (ICM) bracket.
16. From the left side of the crankshaft pulley, remove the idler pulley bracket.
17. Remove the dipstick and the dipstick tube-to-engine bolt; then, pull the tube from its O-ring mount. Discard the O-ring.

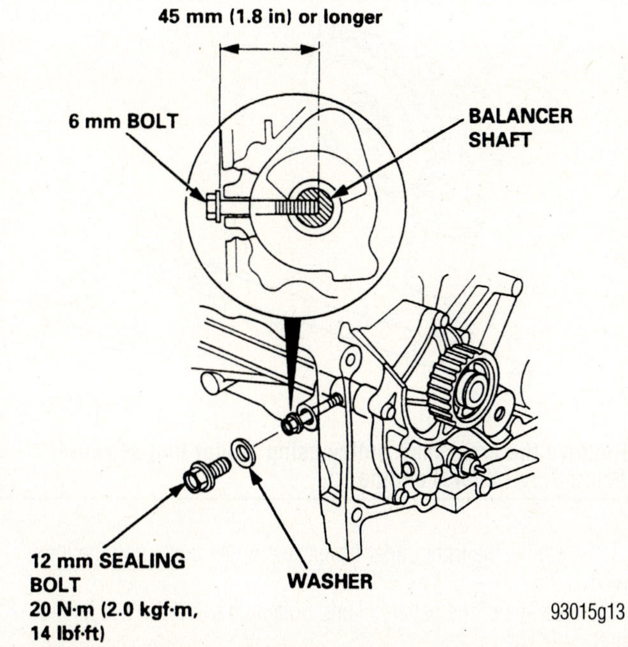

Securing the balancer shaft—Acura 3.5L (C35A1) engine

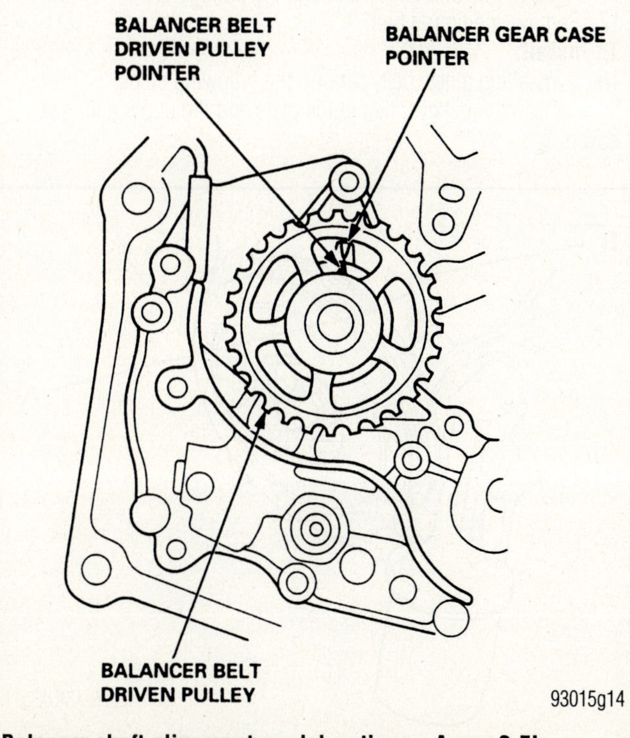

Balancer shaft alignment mark locations—Acura 3.5L (C35A1) engine

18. Remove the crankshaft pulley by performing the following procedure:
   a. Using the Holder Handle and the 50mm Offset Holder Attachment tool 07MAB-PY3010A or equivalent, and a 19mm socket, secure the crankshaft pulley and remove the crankshaft pulley bolt.
   b. Remove the washer and pull the crankshaft pulley from the engine.
19. From the upper and lower timing belt cover.
20. Loosen the balancer belt adjusting nut 180 degrees (1/2) turn. Push the tensioner to relieve the tension from the balancer belt; then, retighten the adjusting bolt. Remove the balancer belt.
21. Loosen the timing belt adjusting nut 180 degrees (1/2) turn. Push the tensioner to relieve the tension from the timing belt; then, retighten the adjusting bolt. Remove the timing belt.

To install:

22. Remove the spark plugs.
23. Remove the balancer belt drive pulley and the timing belt guide plate from the crankshaft.
24. Clean the upper and lower timing belt covers.
25. Position the timing belt pulley so the No. 1 piston is at TDC of its compression stroke. Align the mark on the pulley (near keyway) with the pointer mark on the oil pump.
26. Adjust the camshaft pulley so that the No. 1 piston is the TDC of the compression stroke. Align the TDC marks on the pulley with the upper surface of the back cover; the arrow mark on the left camshaft pulley and the **1** on the right camshaft pulley should be facing the back cover pointer.
27. Install the timing belt in the following sequence:
   a. Crankshaft timing belt pulley sprocket.
   b. Adjusting pulley.
   c. Left camshaft pulley
   d. Water pump pulley
   e. Right camshaft pulley

**** WARNING**

Make sure that the camshaft and crankshaft pulleys are at TDC.

■ **For easier installation, turn the right camshaft pulley about 1/2 tooth from TDC.**

28. Loosen and retighten the timing belt adjusting bolt to tension the timing belt.
29. Install the lower cover and the crankshaft pulley.
30. Rotate the crankshaft pulley about 5–6 turns clockwise to position the timing belt on the pulleys.
31. To adjust the timing belt tension, perform the following procedure:
   a. Make sure that the No. 1 piston is at TDC of its compression stroke.
   b. Rotate the crankshaft clockwise 10 teeth on the camshaft pulley; the **blue** mark on the crankshaft pulley should align with the lower cover pointer.
   c. Loosen the adjusting nut 180 degrees (1/2 turn).
   d. Tighten the adjusting nut to 31 ft. lbs. (42 Nm).
32. Remove the crankshaft pulley and the lower cover; then, install the timing belt guide plate and the balancer belt drive pulley.
33. Align the balancer shaft pulley by performing the following procedure:
   a. Using a 6 x 45mm bolt, insert it into the maintenance hole and the balancer shaft.

b. Align the pointer on the balancer belt pulley with the pointer on the balancer gear case.

34. Adjust the timing belt drive pulley so that the No. 1 piston is at TDC of the compression stroke.

35. Install the balancer belt drive pulley and the balancer belt.

36. Loosen and retighten the balancer adjuster bolt to place tension on the balancer belt.

37. Remove the 6mm bolt and install the sealing bolt in the maintenance hole using a new washer.

38. Install the crankshaft pulley. Rotate the crankshaft pulley about 5–6 turns clockwise to position the timing belt on the pulleys.

39. Loosen the balancer belt adjuster bolt 180 degrees (1/2 turn) and retighten the bolt to 33 ft. lbs. (44 Nm).

40. Remove the crankshaft pulley.

41. Install the upper and lower timing belt covers and the crankshaft pulley.

42. Install the crankshaft pulley and finger-tighten the bolt and washer. Using the Holder Handle and the 50mm Offset Holder Attachment tool 07MAB-PY3010A or equivalent, and a 19mm socket with a torque wrench, tighten the crankshaft pulley bolt to 181 ft. lbs. (245 Nm).

43. Make sure the crankshaft and camshaft pulleys are at TDC.

■ **If the camshaft or crankshaft pulley is not at TDC, remove the timing belt and re-perform the adjustment procedure.**

44. To complete the installation, reverse the removal procedures. Adjust the tension of the drive belts.

45. Connect the negative battery cable.

46. Enter the anti-theft code for the radio and reset the frequencies.

3.5L (J35A3) ENGINES

YEAR, MODEL & ENGINE APPLICATION

2001-04 Acura MDX .... 3.5L Code J35A3

REPLACEMENT INTERVALS

Acura recommends timing belt replacement every 105,000 miles.

LABOR TIMES

2001-04

Replacement .... 4.6 hrs

Severe Service .... 5.2 hrs

8807ACUg18

✱✱ CAUTION

These engines are an interference type. If the timing belt has failed, valve-to-piston damage may have occurred. A compression check should always be performed to check for engine damage.

1. Be sure to acquire the anti-theft code for the radio and the frequencies for the radio's preset buttons.
2. Disconnect the negative battery cable.
3. Turn the crankshaft so the white mark aligns with the pointer.
4. Make sure the number 1 piston is at Top Dead center (TDC).
5. Disconnect the negative battery cable.
6. Raise and support the vehicle.
7. Remove the front wheels and splash shield.
8. Remove the drive belt.
9. Support the engine with a block of wood and a jack under the oil pan.
10. Remove the upper side engine mount.
11. Remove the crankshaft pulley using holder tool shown in the accompanying illustration and a breaker and socket, loosen the 19mm bolt and remove the pulley.

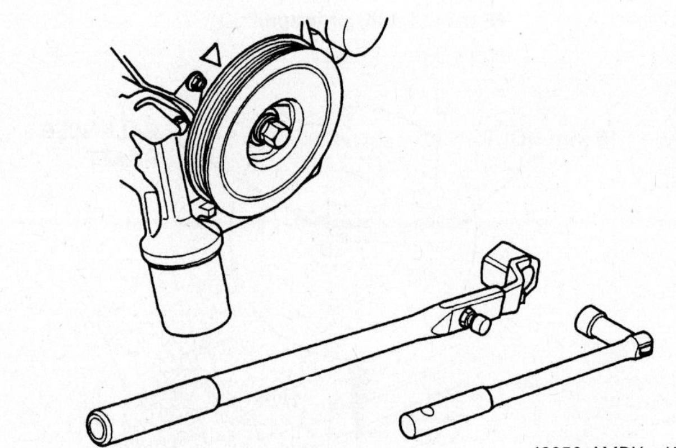

42356-AMDX-g15

Remove the crankshaft pulley using holder tool shown— Acura 3.5L (J35A3) engine

12. Remove the front upper cover, rear upper cover and the lower cover.
13. Use one of the battery clamp bolts and grind the end as illustrated.
14. Screw the battery clamp bolt as illustrated to hold the belt adjuster in position. Do not use a wrench, hand tighten only.
15. Remove the lower side engine mount.
16. Remove the idler pulley bolt and the pulley.
17. Remove the timing belt.

To install:

18. If installing a new belt, perform the following steps:

a. Clean the pulleys, belt guide plate and the upper and lower covers.

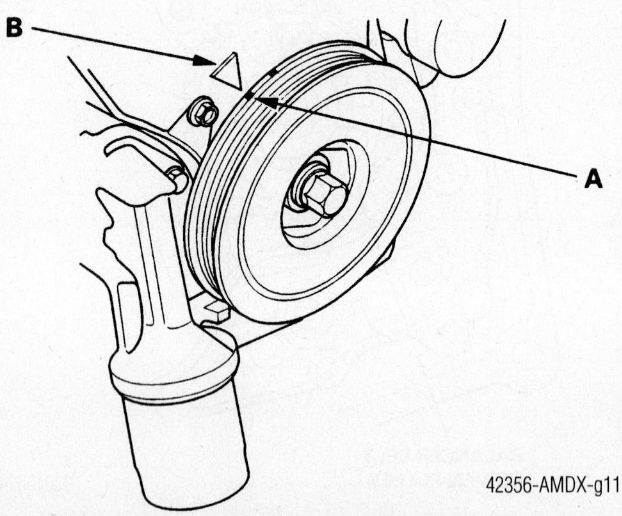

42356-AMDX-g11

Turn the crankshaft so the white mark (A) aligns with the pointer (B)— Acura 3.5L (J35A3) engine

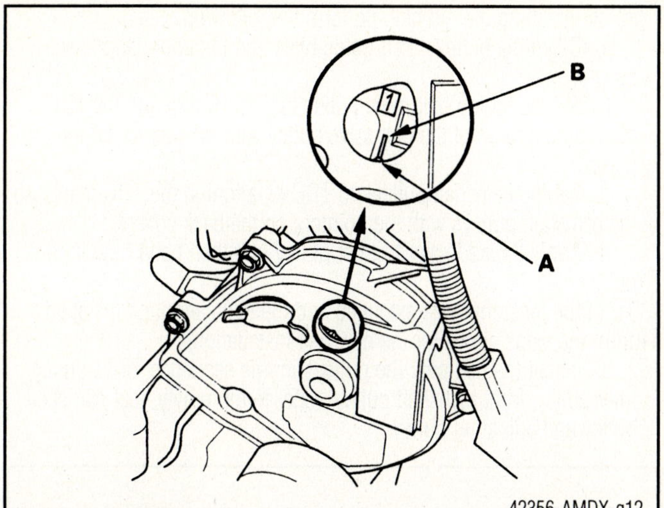

42356-AMDX-g12

Make sure the number 1 piston is at top dead center (A) on the front camshaft pulley and pointer (B)— Acura 3.5L (J35A3) engine

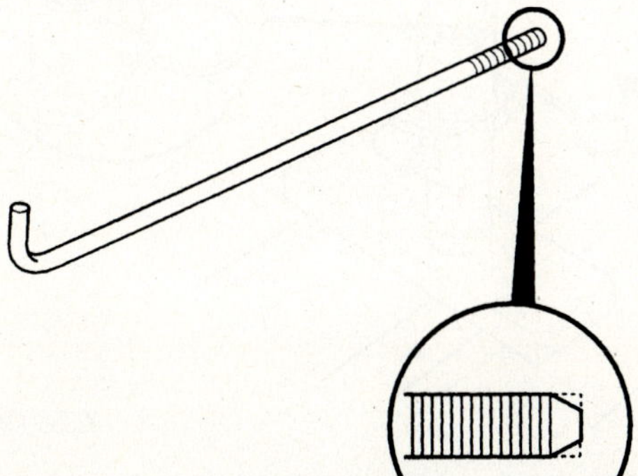

42356-AMDX-g16

Remove a battery clamp bolt and grind the end as shown— Acura 3.5L (J35A3) engine

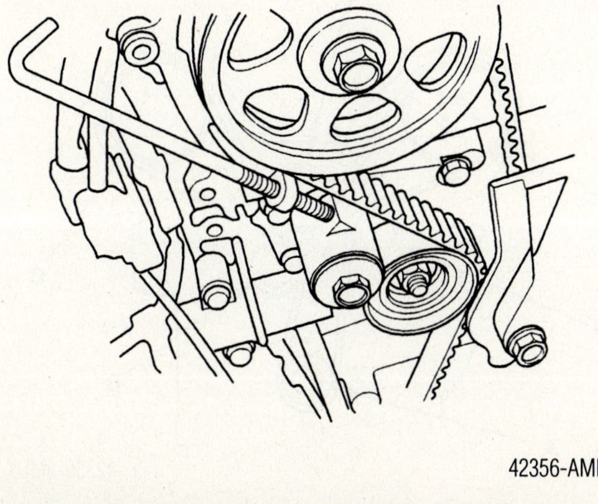

42356-AMDX-g17

Install the battery clamp bolt as shown to hold the belt adjuster in position— Acura 3.5L (J35A3) engine

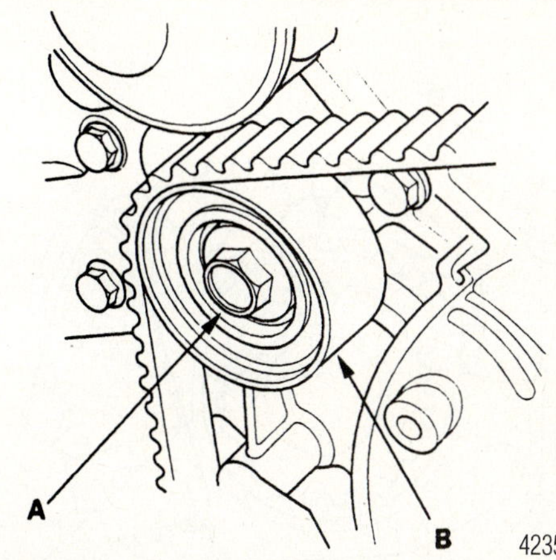

42356-AMDX-g18

Remove the idler pulley bolt (A), pulley (B) and the timing belt— Acura 3.5L (J35A3) engine

b. Set the timing belt drive pulley to TDC by aligning the TDC mark on the tooth of the belt drive pulley with the pointer on the oil pump.

c. Set the camshaft pulleys to TDC by aligning the TDC marks on the camshaft pulleys with the pointers on the back covers.

d. Remove the battery clamp bolt.

e. Remove the belt tensioner.

f. Align the holes on the rod and housing of the tensioner.

g. Using a press or other suitable device, slowly compress the tensioner and insert a 0.08 inch (2mm) pin through the housing and rod.

h. Install the tensioner making sure the pin is still installed.

i. Apply thread locker to idler pulley bolt then hand tighten the bolt.

j. Install the belt over the pulleys in this sequence; drive pulley, idler pulley, front camshaft pulley, water pump pulley, rear camshaft pulley and adjusting pulley.

k. Tighten the idler pulley bolt to 33 ft. lbs. (44 Nm).

l. Remove the pin from the tensioner.

19. Install the lower half of the side mount and tighten the 3 long bolts to 33 ft. lbs. (44 Nm) and the one short bolt to 9 ft. lbs. (12 Nm)

20. Install the timing belt guide plate as illustrated.

21. Install the lower timing cover and tighten the bolts to 9 ft. lbs. (12 Nm).

22. install the front and rear upper timing covers and tighten the bolts to 9 ft. lbs. (12 Nm).

23. Install the crankshaft pulley and tighten the bolts to 181 ft. lbs. (245 Nm), using the holding tool to prevent the unit from turning.

24. Rotate the crankshaft pulley about 5 or 6 degrees clockwise so the belt positions itself on the pulleys.

25. Turn the crankshaft pulley so the white mark aligns with the pointer.

26. Check the camshaft pulley marks are aligned. If the marks are aligned, proceed to the next step. If the marks are not aligned, remove the timing belt and reinstall using the steps outlined before this step.

27. Install the drive belt.

28. Install the upper side mount and tighten the bolts to 40 ft. lbs. (50 Nm).

29. Using a suitable scan tool, perform the Powertrain Control Module (PCM) reset and the Crankshaft position (CKP) pattern clear/learn procedures, following the scan tool manufactures instructions.

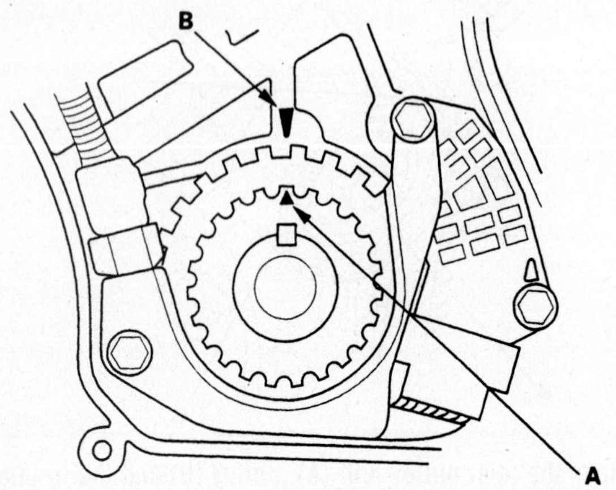

42356-AMDX-g19

Set the timing belt pulley to TDC by aligning the TDC mark (A) on the tooth of the belt pulley with the pointer (B) on the oil pump— Acura 3.5L (J35A3) engine

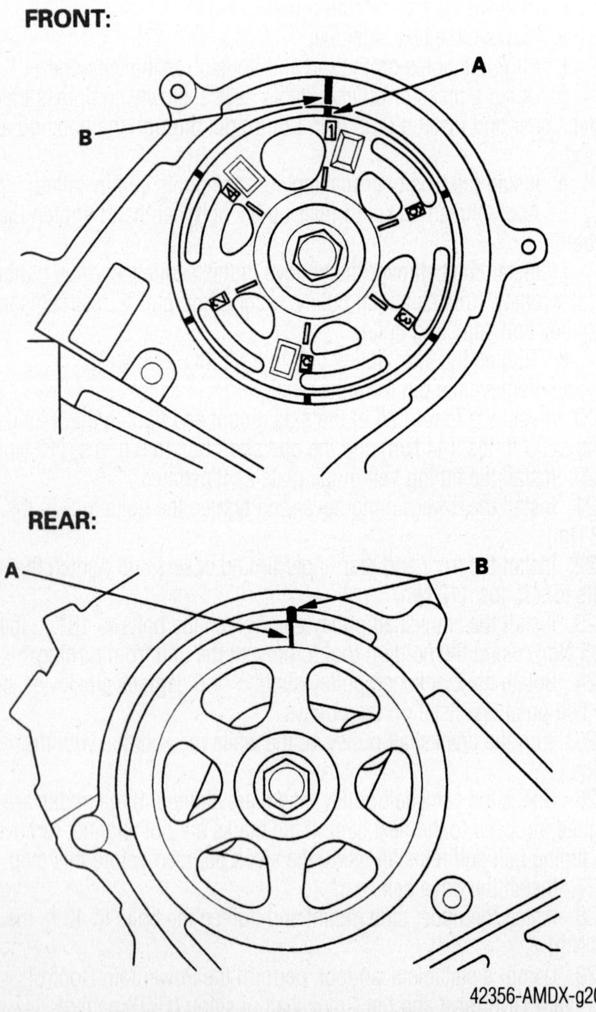

42356-AMDX-g20

Set the camshaft pulleys to TDC by aligning the TDC marks (A) on the camshaft pulleys with the pointers (B) on the back covers— Acura 3.5L (J35A3) engine

30. If installing the old belt, perform the following steps:

a. Clean the pulleys, belt guide plate and the upper and lower covers.

b. Set the timing belt drive pulley to TDC by aligning the TDC mark on the tooth of the belt drive pulley with the pointer on the oil pump.

c. Set the camshaft pulleys to TDC by aligning the TDC marks on the camshaft pulleys with the pointers on the back covers.

d. Apply thread locker to idler pulley bolt then hand tighten the bolt.

31. If the tensioner was extended and the belt cannot be installed, perform the steps above for the new belt installation.

a. Install the belt over the pulleys in this sequence; drive pulley, idler pulley, front camshaft pulley, water pump pulley, rear camshaft pulley and adjusting pulley.

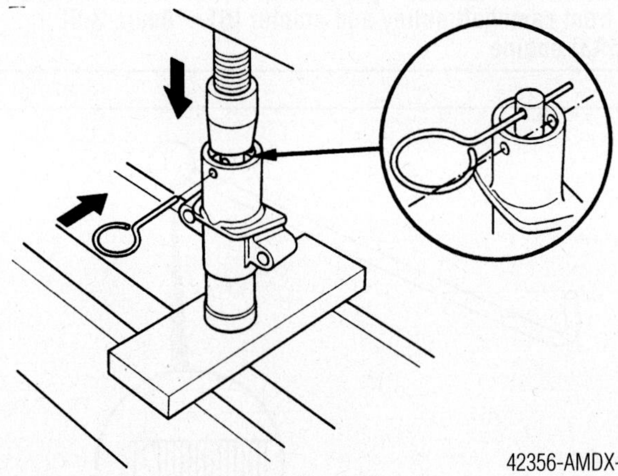

42356-AMDX-g21

Insert a 0.08 inch (2mm) pin through the tensioner housing and rod— Acura 3.5L (J35A3) engine

1 Drive pulley (A).
2 Idler pulley (B).
3 Front camshaft pulley (C).
4 Water pump pulley (D).
5 Rear camshaft pulley (E).
6 Adjusting pulley (F).

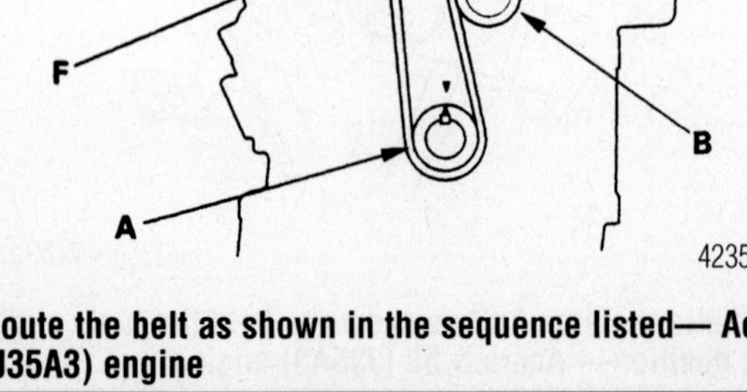

42356-AMDX-g22

Route the belt as shown in the sequence listed— Acura 3.5L (J35A3) engine

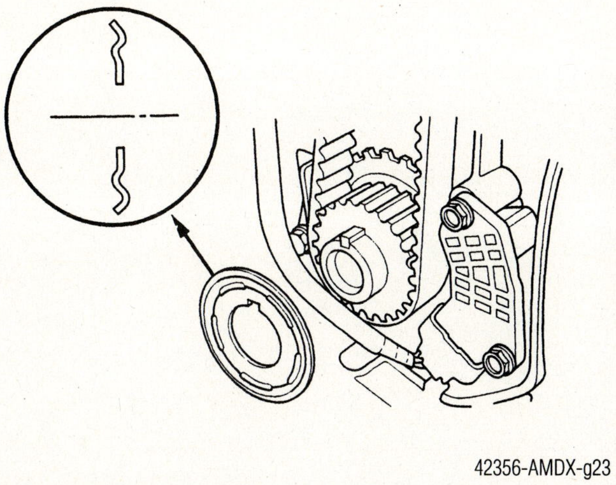

Install the timing belt guide plate as shown— Acura 3.5L (J35A3) engine

b. Tighten the idler pulley bolt to 33 ft. lbs. (44 Nm).

c. Remove the battery clamp bolt.

32. Install the lower half of the side mount and tighten the 3 long bolts to 33 ft. lbs. (44 Nm) and the one short bolt to 9 ft. lbs. (12 Nm).

33. Install the timing belt guide plate as illustrated.

34. Install the lower timing cover and tighten the bolts to 9 ft. lbs. (12 Nm).

35. Install the front and rear upper timing covers and tighten the bolts to 9 ft. lbs. (12 Nm).

36. Install the crankshaft pulley and tighten the bolts to 181 ft. lbs. (245 Nm), using the holding tool.

37. Rotate the crankshaft pulley about 5 or 6 degrees clockwise so the belt positions itself on the pulleys.

38. Turn the crankshaft pulley so the white mark aligns with the pointer.

39. Check the camshaft pulley marks are aligned. If the marks are aligned, proceed to the next step. If the marks are not aligned, remove the timing belt and reinstall using the steps outlined before this step.

40. Install the dive belt.

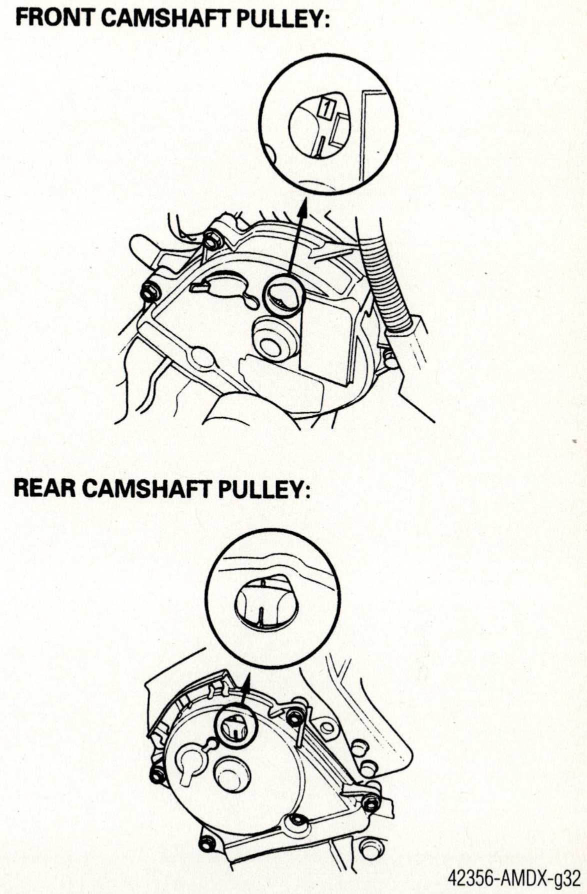

Check that the camshaft pulley marks are aligned as shown— Acura 3.5L (J35A3) engine

41. Install the upper side mount and tighten the bolts to 40 ft. lbs. (50 Nm).

42. Using a suitable scan tool, perform the Powertrain Control Module (PCM) reset and the Crankshaft position (CKP) pattern clear/learn procedures, following the scan tool manufacturers instructions.

Audi

1.8L (AEB AND AMB) ENGINES

YEAR, MODEL & ENGINE APPLICATION

1997-02 Audi A4	1.8L Code AEB
2003-04 Audi A4	1.8L Code AMB

REPLACEMENT INTERVALS

Audi recommends timing belt replacement every 105,000 miles.

LABOR TIMES

1997-04	
Replacement	2.0 hrs
Severe Service	2.3 hrs

8807AUDg01

✻✻ CAUTION

These engines are an interference type. If the timing belt has failed, valve-to-piston damage may have occurred. A compression check should always be performed to check for engine damage.

1. Disconnect the negative battery cable.
2. Raise and safely support the vehicle.
3. From under the vehicle, remove the splash shield.
4. Remove the front bumper.
5. Remove the intake air duct between the grille/front end assembly and the air cleaner housing.
6. Remove the grille/front end assembly-to-chassis bolts and the grille/front end assembly-to-vehicle fasteners.
7. If installed, remove the wiring harness retaining clamps from the left side of the radiator frame.
8. Install Support tools 3369 or equivalent bolts into the grille/front end assembly-to-chassis holes; then, pull the grille/front end assembly forward until it hits the stops.

■ **If necessary to secure the grille/front end assembly, install M6 bolts into the rear bored holes of the grille/front end assembly and the fender.**

9. Loosen the air conditioning compressor serpentine belt tensioner bolts; release the belt tension and remove the belt.
10. Place an open-end wrench on the alternator belt tensioner and rotate it clockwise toward the alternator to release the belt tension. Remove the alternator serpentine drive belt and release the tensioner.

■ **If necessary to lock the alternator tensioner in position, align the housing holes and insert an Allen wrench into the holes to secure its movement.**

11. Using a 5 x 60mm bolt, secure the viscous fan pulley. Using a hex wrench, remove the viscous fan-to-pulley bolts. Remove the viscous fan assembly.
12. Remove the upper timing belt cover.

■ **If reusing the timing belt, mark its rotational direction so it may be installed in its original position.**

13. Using the center bolt, rotate the crankshaft in the direction of engine rotation to position the No. 1 cylinder at Top Dead Center (TDC) of its compression stroke.
14. Remove the damper pulley-to-crankshaft bolts and the damper.
15. Remove the lower timing belt cover.
16. Using a Torx Wrench T45 or equivalent, loosen the timing belt tensioner, push the tensioner downward and remove the timing belt.

To Install:

17. Align the camshaft sprocket timing mark with the cylinder head cover mark.
18. Install the timing belt on the crankshaft sprocket with the arrow facing the rotational direction.
19. Install the lower timing belt cover.
20. Using a bolt, secure the damper/belt pulley on the crankshaft.
21. Align the crankshaft damper/belt pulley with the housing timing mark so that the No. 1 cylinder is at TDC of its compression stroke.
22. Install the timing belt on the camshaft sprocket and belt tensioner.
23. Using a 2-pin Spanner Matra V159 Wrench or equivalent, lift (turn clockwise) the timing belt tensioner cylinder No. 1 until it is fully extended and tensioner cylinder No. 2 is raised approx. 1mm; then, hand-tighten the mounting bolt.
24. Rotate the crankshaft 2 complete rotation in the running direction.
25. Inspect area "A" for proper alignment with the upper edge of piston No. 2 and adjust if necessary.
   - Area "A": adjustment OK
   - Area "B": wear limit

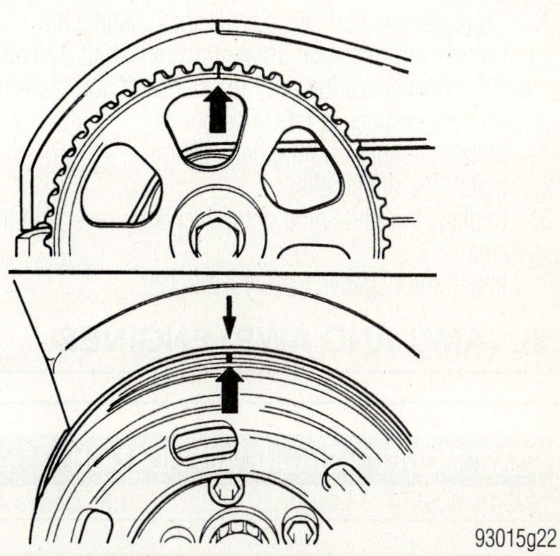

93015g22

Crankshaft pulley and camshaft sprocket alignment locations—Audi A4 1.8L (AEB & AMB) engine

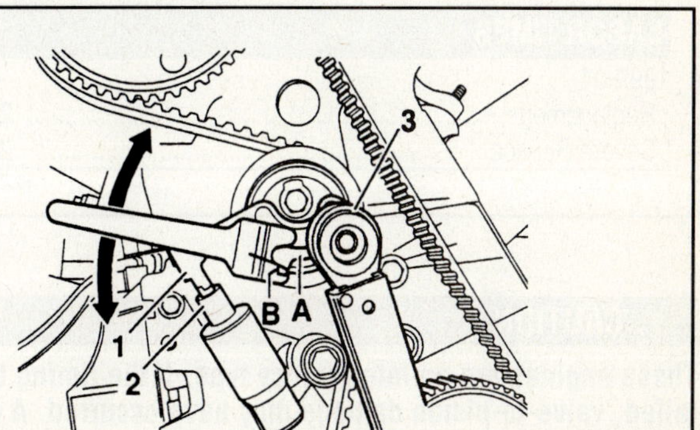

93015g23

Timing belt tension adjustment— Audi A4 1.8L (AEB & AMB) engine

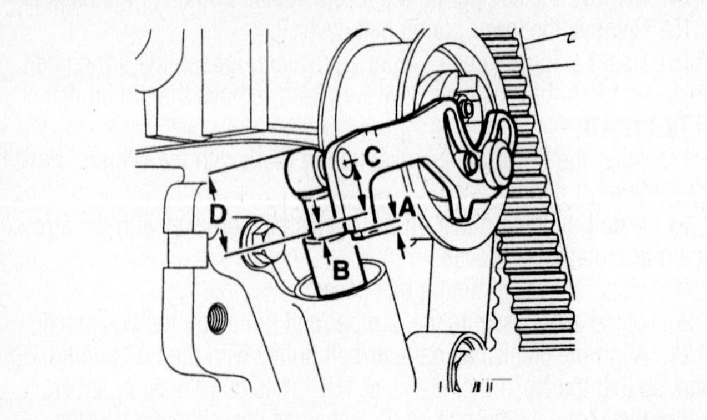

Timing belt tension wear limits— Audi A4 1.8L (AEB & AMB) engine

- Area "C": re-adjust and check belt drive including tensioner for wear.

■ If the piston edge is located in area "A", measurement "D" is 0.984–1.142 in. (25–29mm).

26. After adjustment has been verified, secure the tensioner with a 2-pin Spanner Matra V159 Wrench or equivalent, and tighten the mounting bolt.
27. Complete the damper to crankshaft installation.
28. Using the center bolt, rotate the crankshaft 2 rotations in the direction of engine rotation until the camshaft and crankshaft marks align with their respective reference points.
29. Install the upper timing belt cover.
30. Install the drive belts.
31. Replace the remaining components by reversing the removal procedures.
32. Install the negative battery cable last.

1.8L (AMU AND AWP) ENGINES

YEAR, MODEL & ENGINE APPLICATION

1999-04 Audi TT .... 1.8L Codes AMU & AWP

REPLACEMENT INTERVALS

Audi recommends timing belt replacement every 105,000 miles.

LABOR TIMES

1999-04

Replacement	2.7 hrs
Severe Service	2.9 hrs

8807AUDg02

✲✲ CAUTION

These engines are an interference type. If the timing belt has failed, valve-to-piston damage may have occurred. A compression check should always be performed to check for engine damage.

1. Disconnect the negative battery cable.
2. From under the vehicle, remove the splash shield and right side noise insulation panel.
3. Remove the intake air duct at the bottom of the right crossmember.
4. Remove the accessory drive belt and belt tensioner.
5. Remove coolant expansion tank and power steering reservoir with hoses attached and place to one side.
6. Disconnect throttle valve and charcoal canister vacuum hoses.
7. Disconnect electrical connectors from coolant tank and charcoal canister.
8. Remove the upper timing belt cover.
9. Install Engine Support bracket 10-222A and supports 10-222A1.
10. Install Retainer 3180 to right engine lifting eye and attach it to support bracket 10-222A.
11. Lift engine slightly with spindle of support bracket 1022A.

■ If reusing the timing belt, mark its rotational direction so it may be installed in its original position.

12. Using the center bolt, rotate the crankshaft in the direction of engine rotation to position the No. 1 cylinder at Top Dead Center (TDC) of its compression stroke.
13. Remove the damper pulley-to-crankshaft bolts and the damper.
14. Remove the center and lower timing belt covers.
15. Unbolt front engine support from engine console and engine console from body.
16. Disconnect connector between engine console and engine support.
17. Remove engine console.
18. Raise or lower the engine enough to remove engine support bolts. Do not remove support.
19. Using a Torx Wrench T45, loosen the timing belt tensioner, push the tensioner downward and remove timing belt.

To Install:

20. Align the camshaft sprocket timing mark with the cylinder head cover mark.
21. Install the timing belt on the crankshaft sprocket with the arrow facing the rotational direction.
22. Install the lower timing belt cover. Torque the bolts to 84 inch lbs. (10 Nm).
23. Using a bolt, secure the damper/belt pulley on the crankshaft.

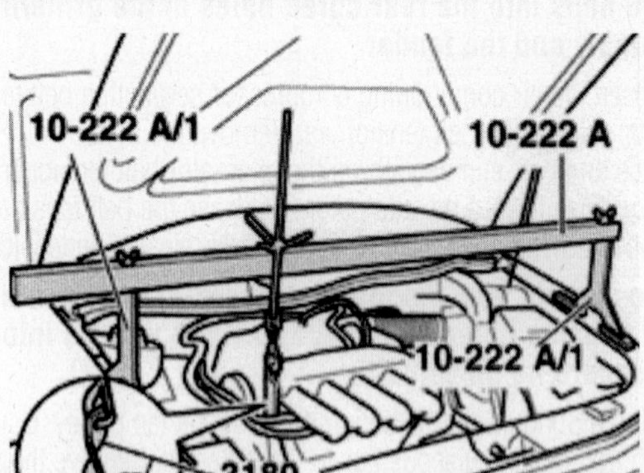

Attaching engine support and lifting tools—Audi 1.8L (AMU & AWP) engine

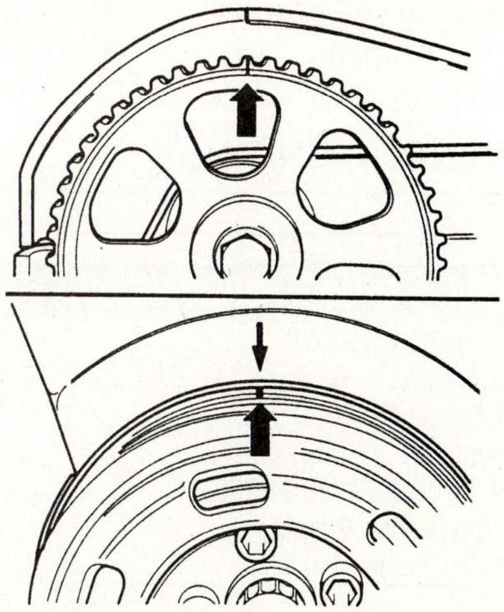
93015g22

Crankshaft pulley and camshaft sprocket alignment locations—Audi 1.8L (AMU & AWP) 4-Cylinder engine

Torque the bolt to 18 ft. lbs. (25 Nm).

24. Align the crankshaft damper/belt pulley with the housing timing mark so that the No. 1 cylinder is at TDC of its compression stroke.
25. Install the timing belt on the water pump, tensioning pulley and finally the camshaft sprocket.
26. Rotate the crankshaft 2 complete rotations in the running direction, and then set to TDC on no. 1 cylinder.
27. Check that timing belt is properly aligned.
28. Tighten engine support bolts to engine. Torque the bolts to 33 ft. lbs. (45 Nm).
29. Install engine console to body. Torque the bolts to 40 ft. lbs. (30 Nm) plus 90 degrees.
30. Install engine console to engine support. Torque the bolts to 44 ft. lbs. (60 Nm) plus 90 degrees.
31. Install the center and upper timing belt covers. Torque the bolts to 84 inch lbs. (10 Nm).
32. Replace the remaining components by reversing the removal procedures.
33. Install the negative battery cable last.

1.8L (VIN N/A) ENGINES

YEAR, MODEL & ENGINE APPLICATION

1985-87 Audi 4000S .... 1.8L SOHC Code N/A

REPLACEMENT INTERVALS

Audi does not provide a replacement interval. Standard industry practice is to replace timing belts every 60,000 miles.

LABOR TIMES

1985-87

Replacement	2.7 hrs
Severe Service	3.0 hrs

8807AUDg03

✻✻ CAUTION

These engines are an interference type. If the timing belt has failed, valve-to-piston damage may have occurred. A compression check should always be performed to check for engine damage.

1. Disconnect the negative battery cable.
2. Place the No. 1 cylinder at TDC. The "O" mark on the flywheel is aligned with the bellhousing casting and the camshaft sprocket mark is aligned with the top of the valve cover gasket.
3. Remove the accessory drive belts.
4. Remove the crankshaft pulley.
5. Remove the upper and lower timing belt covers.
6. Loosen the timing belt tensioner nut and turn the nut counterclockwise to relieve the timing belt tension.
7. Remove the timing belt.

To Install:

8. Install the timing belt on the crankshaft sprocket and the intermediate sprocket.
9. Install the crankshaft pulley.
10. Tighten the tensioner nut until the timing belt is tensioned.
11. Rotate the crankshaft two revolutions in the direction of rotation.

■ **The timing belt is correctly tensioned when it can be twisted 90 degrees along the straight run between the camshaft sprocket and intermediate sprocket.**

12. Remove the crankshaft pulley.
13. Install the lower timing belt cover.
14. Install the crankshaft pulley.
15. Install the upper timing belt cover.
16. Install the accessory drive belts.
17. Connect the negative battery cable.

2.0L (3A) ENGINES

YEAR, MODEL & ENGINE APPLICATION

1988-90 Audi 80 & 90 .... 2.0L Code 3A

REPLACEMENT INTERVALS

Audi does not provide a replacement interval. Standard industry practice is to replace timing belts every 60,000 miles.

LABOR TIMES

1988-90

Replacement	2.3 hrs
Severe Service	2.6 hrs

8807AUDg04

✻✻ CAUTION

These engines are considered freewheeling. If the timing belt fails, valve-to-piston damage is unlikely, however a precautionary compression check should be performed to verify no engine damage.

1. Disconnect the negative battery cable.
2. Using the large bolt on the crankshaft sprocket, rotate the engine until the No. 1 cylinder is at Top Dead Center (TDC) of the compression stroke. Align the TDC mark **0** with the cast mark on the bell housing.
3. Remove the accessory drive belts.
4. Remove the valve cover.
5. Remove the retaining nuts and remove the upper timing belt cover. Take care not to lose any of the washers or spacers.
6. Remove the crankshaft balancer center bolt.

■ To remove the crankshaft balancer bolt on manual transaxle vehicles, place the vehicle in 5th gear and have an assistant apply the brake. The will stop the engine from rotating while loosening the bolt. On automatic transaxle vehicles, remove the starter and hold the flywheel from turning, using a flywheel holding tool VW 10-201 or equivalent. This bolt is extremely tight.

7. Remove the crankshaft pulley.
8. Remove the water pump pulley.
9. Remove the lower timing belt cover bolts and remove the cover.
10. Slide the timing belt off the sprockets.

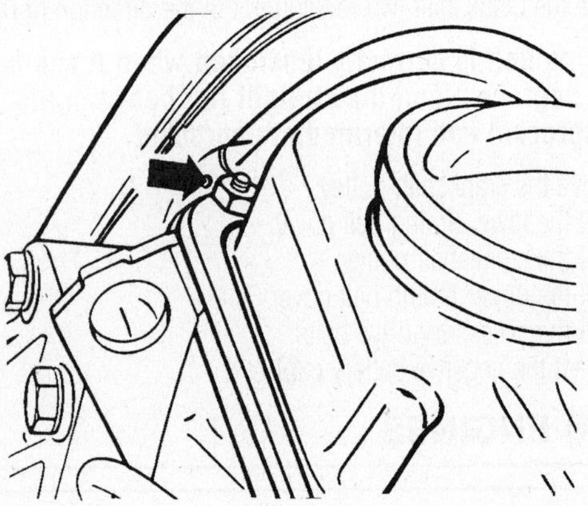

9305bg04

Camshaft sprocket alignment mark location—Audi 2.0L (3A) engine

To Install:

11. Install the timing belt on the crankshaft sprocket and intermediate sprocket.
12. Install the crankshaft pulley.
13. The timing mark on the rear face of the camshaft sprocket should be aligned with the upper left edge of the valve cover.
14. The crankshaft timing mark should be aligned with the intermediate shaft sprocket mark.
15. Install the timing belt on the camshaft sprocket and tensioner.
16. Tighten the tensioner to tension the timing belt.

■ The timing belt is correctly tensioned when it can be twisted 90 degrees along the straight run between the camshaft and intermediate sprockets.

17. Rotate the crankshaft 2 revolutions in the direct of rotation and re-check the timing belt tension.
18. Remove the crankshaft pulley.
19. Use the same procedure to install the crankshaft center bolt as when removing the bolt. Apply a locking compound on the bolt threads.
20. Install the lower timing belt cover.
21. Install the crankshaft pulley and tighten to 15 ft. lbs. (20 Nm).
22. Install the upper timing belt cover.
23. Install the accessory drive belts.
24. Connect negative battery cable.
25. Check and set the ignition timing.

2.2L 5-CYLINDER ENGINES

YEAR, MODEL & ENGINE APPLICATION	
1985-87 Audi 4000S & 4000CS	2.2L Code JT
1985 Audi 5000 Turbo	2.2L Code KH
1985-87 Audi Coupe GT	2.2L Code KX
1985-87 Audi 5000S	2.2L Code KZ
1986-88 Audi 5000CS & 1989-91 200	2.2L Code MC
1985 Audi Turbo Quattro Coupe, 1986-87 4000CS Quattro	2.2L Code WX

REPLACEMENT INTERVALS

Audi does not provide a replacement interval. Standard industry practice is to replace timing belts every 60,000 miles.

LABOR TIMES	
200 Models	
Replacement	3.1 hrs
Severe Service	3.3 hrs
All Other Models	
Replacement	2.7 hrs
Severe Service	2.9 hrs

8807AUDg05

✱✱ CAUTION

These engines are an interference type. If the timing belt has failed, valve-to-piston damage may have occurred. A compression check should always be performed to check for engine damage.

1. Disconnect the negative battery cable.
2. Using the large bolt on the crankshaft sprocket, rotate the engine until the No. 1 cylinder is at Top Dead Center (TDC) of the compression stroke. Align the TDC mark **0** with the cast mark on the bell housing.
3. The timing mark on the rear face of the camshaft sprocket should be aligned with the upper left edge of the valve cover.
4. Remove the alternator, power steering, and air conditioner compressor drive belts.
5. Position the power steering pump aside, leaving the lines attached.
6. Remove the retaining nuts and remove the upper timing belt cover. Take care not to lose any of the washers or spacers.
7. Loosen, but do not remove the crankshaft sprocket center bolt. On manual transaxle vehicles, place the vehicle in 5th gear and have an assistant apply the brake. The will stop the engine from rotating while loosening the bolt. On automatic transaxle vehicles, remove the starter and hold the flywheel from turning, using a flywheel holding tool VW 10-201 or equivalent. This bolt is extremely tight.
8. Remove the lower timing belt cover bolts and remove the cover.
9. Loosen the water pump bolts only enough to turn the pump clockwise.

■ **By loosening the water pump bolts, the coolant may drain from the engine at the water pump. If necessary, drain the cooling system, remove the water pump and reinstall it with a new O-ring.**

10. Slide the timing belt off the sprockets.
11. Remove the crankshaft damper.

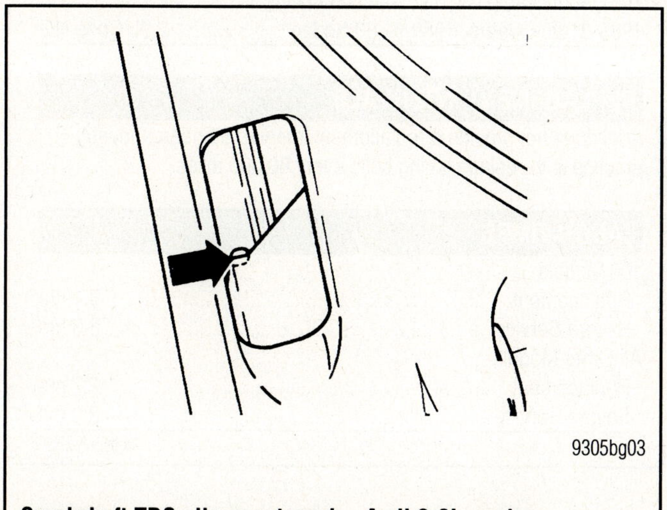

Crankshaft TDC alignment mark—Audi 2.2L engine

To Install:

12. Align crankshaft and camshaft timing marks.
13. Install the timing belt on the crankshaft sprocket and install the sprocket.
14. Use the same procedure to install the crankshaft center bolt as when removing the bolt. Apply a locking compound on the bolt threads. Torque the bolt to 258 ft. lbs. (350 Nm) in several steps.

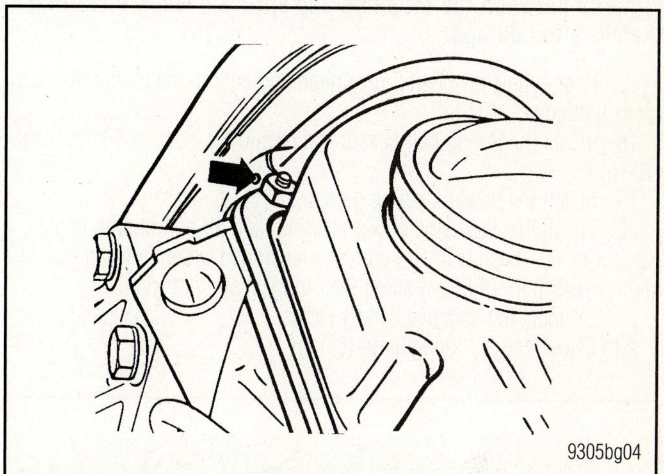

Camshaft sprocket alignment mark location—Audi 2.2L engine

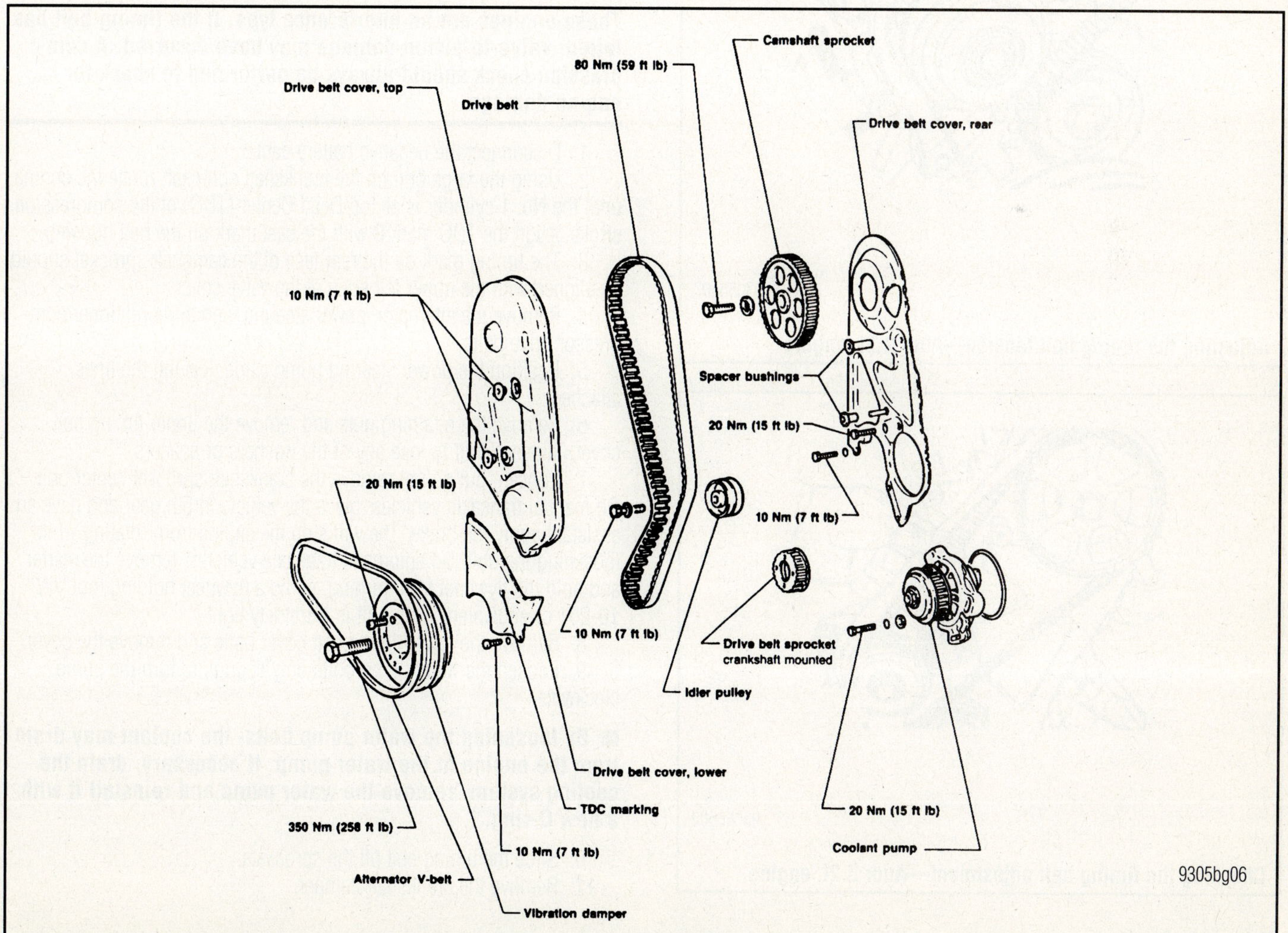

Exploded view of the timing belt and related components—Audi 2.2L engine

15. Install the timing belt on the camshaft and water pump sprockets.
16. Torque the water pump bolts to 15 ft. lbs. (20 Nm).

■ **The timing belt is correctly tensioned when it can be twisted 90 degrees along the straight run between the camshaft sprocket and water pump. The belt must not be jammed between the oil pump and sprocket when installing the vibration damper.**

17. Rotate the crankshaft 2 revolution and ensure the timing marks are still aligned.
18. Install the timing belt covers and torque the bolts to 89 inch lbs. (10 Nm).
19. Install the power steering pump.
20. Install the alternator, power steering, and air conditioning compressor belts. These belts are correctly tensioned when they can be depressed 3/8 in. (9.5mm) along their longest straight run.
21. Connect the negative battery cable.
22. Check and set the ignition timing.

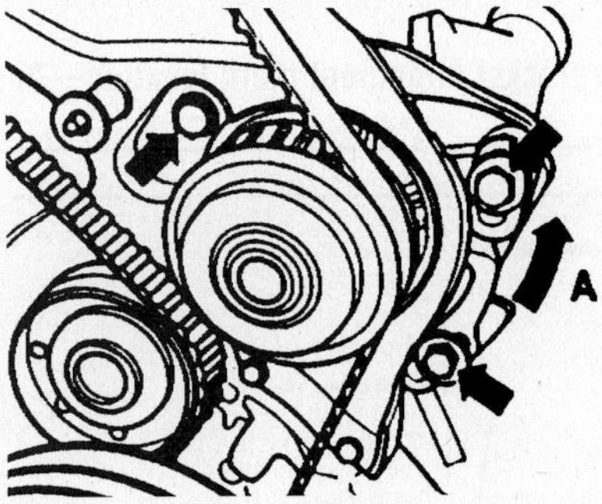

9305bg07

Adjusting the timing belt tension—Audi 2.2L engine

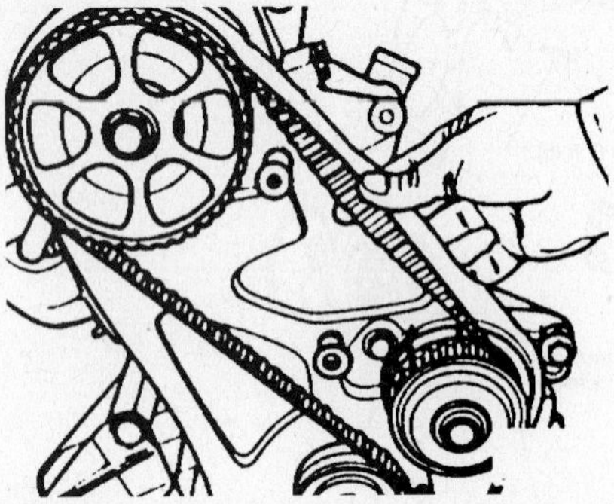

9305bg08

Checking the timing belt adjustment—Audi 2.2L engine

2.3L (NF, NG) SOHC ENGINES

YEAR, MODEL & ENGINE APPLICATION	
1989-91 Audi 100, 1989-90 Quattro, 1987-88 5000S	2.3L Code NF
1987 Audi Coupe GT, 1988-92 80 Quattro, 1988-91 90 Quattro, 1989-91 90 Sedan	2.3L Code NG

REPLACEMENT INTERVALS

Audi does not provide a replacement interval. Standard industry practice is to replace timing belts every 60,000 miles.

LABOR TIMES	
100 Models	
Replacement	3.1 hrs
Severe Service	3.3 hrs
All Other Models	
Replacement	2.7 hrs
Severe Service	3.0 hrs

8807AUDg06

✲✲ CAUTION

These engines are an interference type. If the timing belt has failed, valve-to-piston damage may have occurred. A compression check should always be performed to check for engine damage.

1. Disconnect the negative battery cable.
2. Using the large bolt on the crankshaft sprocket, rotate the engine until the No. 1 cylinder is at Top Dead Center (TDC) of the compression stroke. Align the TDC mark **0** with the cast mark on the bell housing.
3. The timing mark on the rear face of the camshaft sprocket should be aligned with the upper left edge of the valve cover.
4. Remove the alternator, power steering, and air conditioner compressor drive belts.
5. Position the power steering pump aside, leaving the lines attached.
6. Remove the retaining nuts and remove the upper timing belt cover. Take care not to lose any of the washers or spacers.
7. Loosen, but do not remove the crankshaft sprocket center bolt. On manual transaxle vehicles, place the vehicle in 5th gear and have an assistant apply the brake. The will stop the engine from rotating while loosening the bolt. On automatic transaxle vehicles, remove the starter and hold the flywheel from turning, using a flywheel holding tool VW 10-201 or equivalent. This bolt is extremely tight.
8. Remove the lower timing belt cover bolts and remove the cover.
9. Loosen the water pump bolts only enough to turn the pump clockwise.

■ **By loosening the water pump bolts, the coolant may drain from the engine at the water pump. If necessary, drain the cooling system, remove the water pump and reinstall it with a new O-ring.**

10. Slide the timing belt off the sprockets.
11. Remove the crankshaft damper.

To Install:

12. Align crankshaft and camshaft timing marks.
13. Install the timing belt on the crankshaft sprocket and install the sprocket.
14. Use the same procedure to install the crankshaft center bolt as when removing the bolt. Apply a locking compound on the bolt threads. Torque the bolt to 258 ft. lbs. (350 Nm) in several steps.
15. Install the timing belt on the camshaft and water pump sprockets.
16. Torque the water pump bolts to 15 ft. lbs. (20 Nm).

■ **The timing belt is correctly tensioned when it can be twisted 90 degrees along the straight run between the camshaft sprocket and water pump. The belt must not be jammed between the oil pump and sprocket when installing the vibration damper.**

17. Rotate the crankshaft 2 revolution and ensure the timing marks are still aligned.

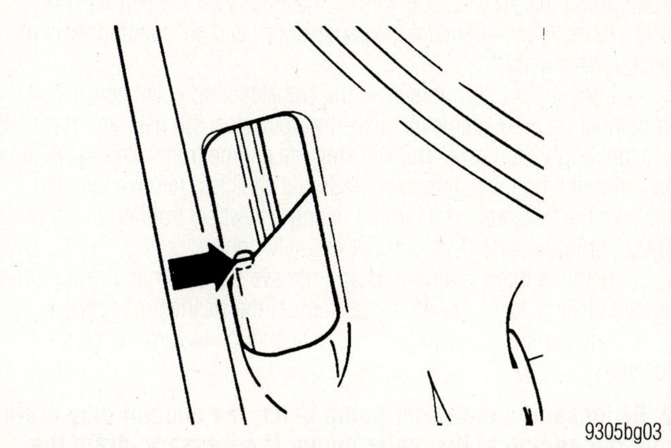

9305bg03

Crankshaft TDC alignment mark—Audi 2.3L (NF, NG) engine

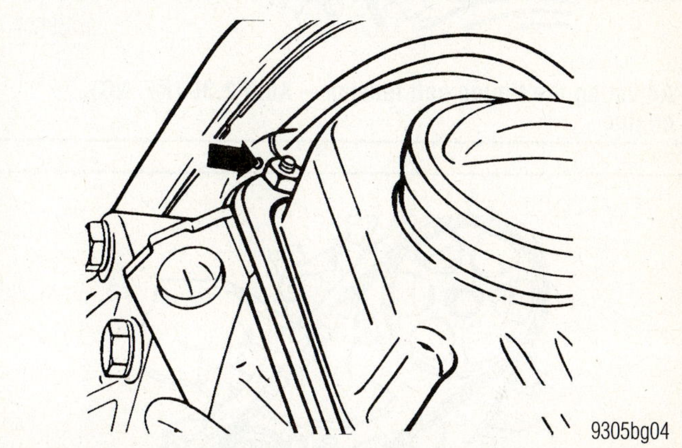

9305bg04

Camshaft sprocket alignment mark location—Audi 2.3L (NF, NG) engine

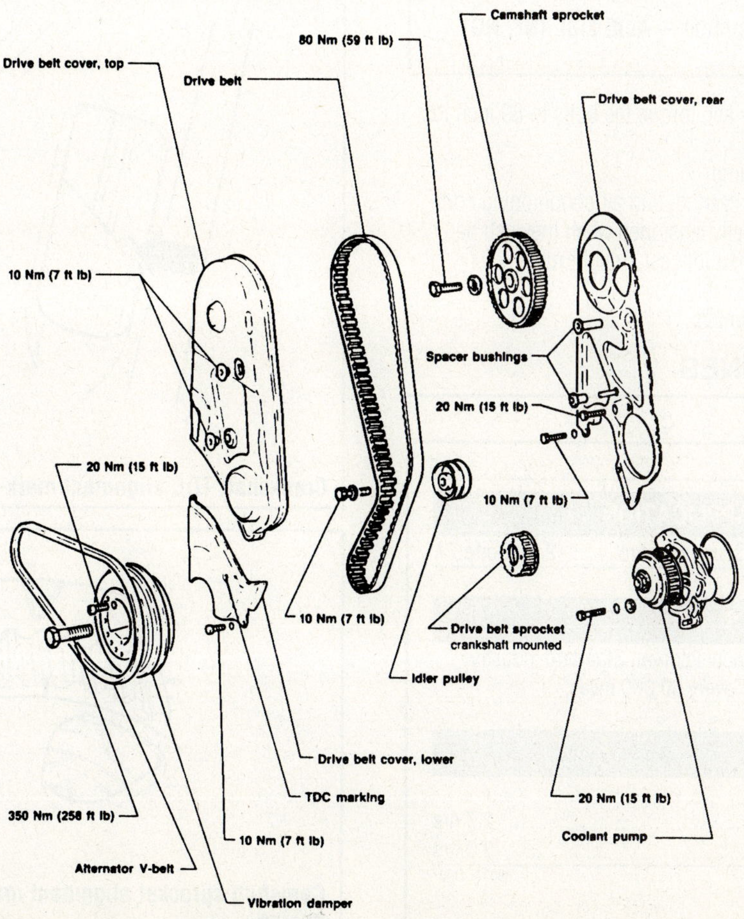

9305bg06

Exploded view of the timing belt and related components—Audi 2.3L (NF, NG) engine

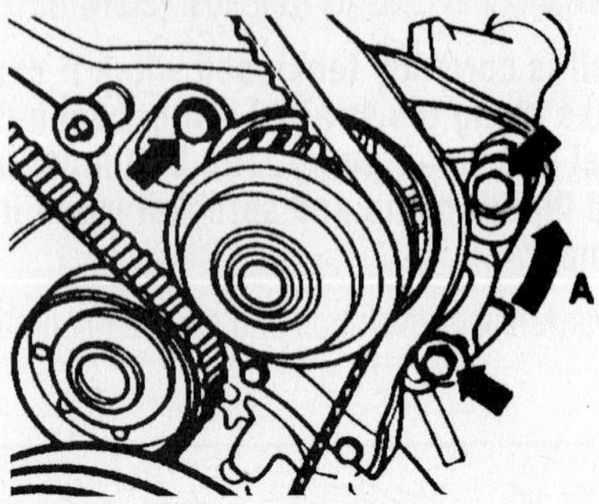

9305BG07

Adjusting the timing belt tension— Audi 2.3L (NF, NG) engine

9305BG08

Checking the timing belt adjustment— Audi 2.3L (NF, NG) engine

18. Install the timing belt covers and torque the bolts to 89 inch lbs. (10 Nm).
19. Install the power steering pump.
20. Install the alternator, power steering, and air conditioning compressor belts. These belts are correctly tensioned when they can be depressed 3/8 in. (9.5mm) along their longest straight run.
21. Connect the negative battery cable.
22. Check and set the ignition timing.

2.3L (7A) DOHC ENGINES

YEAR, MODEL & ENGINE APPLICATION

1988-90 Audi 80 & 90, 1990-91 Coupe Quattro .. 2.3L Code 7A

REPLACEMENT INTERVALS

Audi does not provide a replacement interval. Standard industry practice is to replace timing belts every 60,000 miles.

LABOR TIMES

1988-91

Replacement	2.7 hrs
Severe Service	3.0 hrs

8807AUDg07

✱✱ CAUTION

These engines are considered freewheeling. If the timing belt fails, valve-to-piston damage is unlikely, however a precautionary compression should be performed to verify no engine damage.

1. Disconnect the negative battery cable.
2. Using the large bolt on the crankshaft sprocket, rotate the engine until the No. 1 cylinder is at Top Dead Center (TDC) of the compression stroke. Align the TDC mark **0** with the cast mark on the bell housing.
3. Remove the alternator, power steering, and air conditioner compressor drive belts.
4. Loosen, but do not remove the crankshaft sprocket center bolt. On manual transaxle vehicles, place the vehicle in 5th gear and have an assistant apply the brake. The will stop the engine from rotating while loosening the bolt. On automatic transaxle vehicles, remove the starter and hold the flywheel from turning, using a flywheel holding tool VW 10-201 or equivalent. This bolt is extremely tight.
5. Remove the retaining nuts and remove the upper and lower timing belt covers. Take care not to lose any of the washers or spacers.
6. Loosen the water pump bolts only enough to turn the pump clockwise.

■ **By loosening the water pump bolts, the coolant may drain from the engine at the water pump. If necessary, drain the cooling system, remove the water pump and reinstall it with a new O-ring.**

9305bg03

Crankshaft TDC alignment mark—Audi 2.3L (7A) engine

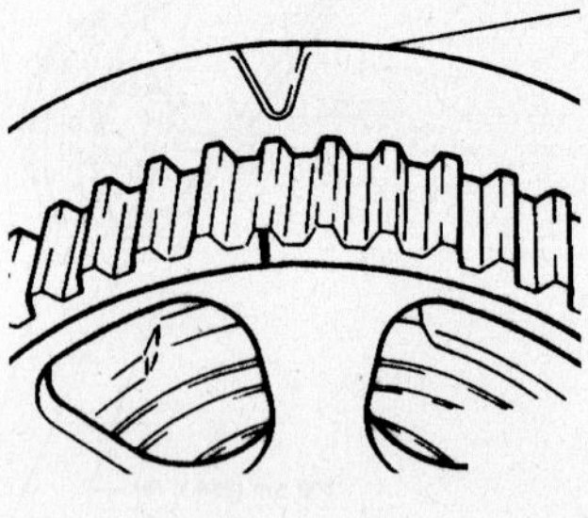

9305bg05

Camshaft sprocket alignment mark location—Audi 2.3L (7A) engine

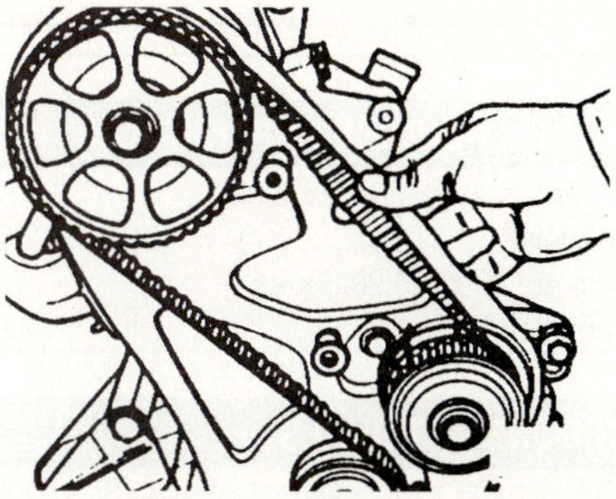

9305bg08

Checking the timing belt adjustment— Audi 2.3L (7A) engine

7. Remove the crankshaft damper and sprocket with the timing belt.

To Install:

8. Install the timing belt on the crankshaft sprocket and install the sprocket. Ensure the oil pump housing mark is aligned with the crankshaft pulley mark.

9. Align the camshaft sprocket marks, then install the timing belt on the idler pulley, water pump and camshaft sprockets.

10. Torque the water pump bolts to 15 ft. lbs. (20 Nm).

■ **The timing belt is correctly tensioned when it can be twisted 90 degrees along the straight run between the camshaft and idler sprockets.**

11. Rotate the crankshaft 2 revolution and ensure the timing marks are still aligned.

12. Install the upper and lower timing belt covers.

13. Use the same procedure to install the crankshaft center bolt as when removing the bolt. Apply a locking compound on the bolt threads. Torque the bolt to 258 ft. lbs. (350 Nm) in several steps.

14. Install the alternator, power steering, and air conditioning compressor belts. These belts are correctly tensioned when they can be depressed 3/8 in. (9.5mm) along their longest straight run.

15. Connect the negative battery cable.

16. Check and set the ignition timing.

2.7L (APB) ENGINES

YEAR, MODEL & ENGINE APPLICATION

2003-04 Audi A6 Quattro,
2003-04 Allroad Quattro, 2000-02 S4 .... 2.7L Code APB

REPLACEMENT INTERVALS

Audi recommends timing belt replacement every 105,000 miles.

LABOR TIMES

A6 & Allroad	
Replacement	3.1 hrs
Severe Service	3.3 hrs
S4	
Replacement	2.0 hrs
Severe Service	2.2 hrs

8807AUDg08

✱✱ CAUTION

These engines are an interference type. If the timing belt has failed, valve-to-piston damage may have occurred. A compression check should always be performed to check for engine damage.

1. Disconnect the negative battery cable.
2. Remove the front bumper.
3. Place the hood lock carrier in the service position.
4. Remove the serpentine drive belt by performing the following procedure:
5. Using Spanner Wrench No. 3212, secure the viscous fan pulley. Using an Open-end Spanner Wrench 3312, remove the viscous fan bearing housing by turning it clockwise.

■ **The viscous fan is mounted with a left-handed thread; turn it clockwise to loosen it.**

6. Place a 17mm box wrench on the serpentine drive belt tensioner and turn the tensioner clockwise until the 2 holes align; insert drift 3204 into the holes to secure the tensioner in place.
7. Mark the running direction of the serpentine drive belt and remove it from the pulleys.
8. Remove the top engine covers.
9. Remove the pressure hoses between charge air coolers and pressure lines.
10. Remove the pressure hoses located at the front of the engine.
11. Remove the timing belt tensioner.
12. Remove all three timing belt covers.
13. Rotate the crankshaft by hand to align the crankshaft pulley mark with the arrow on the engine housing and the large hole in each camshaft sprocket must face inward and must align; this should be Top Dead Center (TDC) of the No. 1 cylinder's compression stroke. If these conditions are not correct, rotate the crankshaft one complete revolution and realign.
14. On the left side of the cylinder block near the crankshaft, remove the sealing plug.
15. Insert Crankshaft Holder tool No. 3242 into the sealing plug hole to secure the crankshaft.
16. Remove the damper-to-crankshaft bolts and the damper.

■ **It is not necessary to remove the center bolt when removing the crankshaft damper.**

17. Remove the serpentine belt idler and the crankshaft damper guard.
18. Using a 8mm Allen® wrench, rotate the timing belt tensioner roller clockwise until the tensioner is compressed; then, insert a 2mm spring pin through the tensioner housing and tensioner plunger to secure it in place. When the plunger is secure, release the wrench tension.
19. Mark the running direction of the timing belt and remove it from the pulleys.

To Install:

20. Insert camshaft clamp 3391 in the securing plates of both camshaft sprockets.
21. Loosen both camshaft bolts and remove approximately 5 turns.
22. Remove the camshaft clamp.
23. Pull off both camshaft sprockets with special tool T40001.
24. Reinstall both camshaft sprockets with securing plates and tighten hand-tight again.

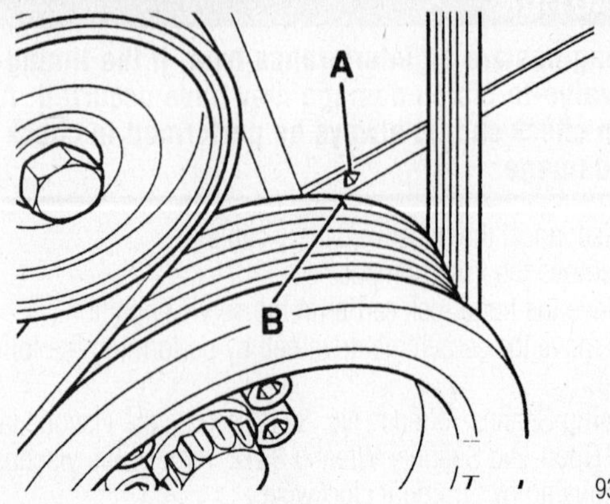

93015g15

Crankshaft pulley alignment location for TDC—Audi 2.7L (APB) engine

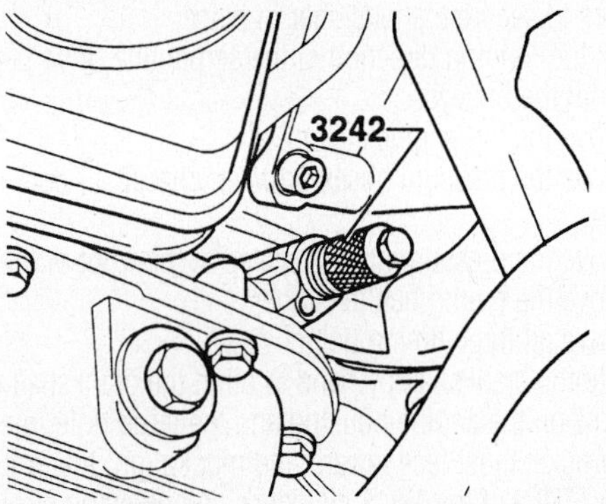

93015g17

View of crankshaft holding tool installed—Audi 2.7L (APB) engine

■ **The camshaft sprockets should be just tight enough on the camshaft tapers so that they can still be turned but do not move axially.**

25. Install the timing belt; make sure the timing belt is installed in the correct running direction from which it was removed.
26. Reinstall the camshaft clamp.
27. Using a 8mm Allen® wrench, rotate the timing belt tensioner roller clockwise until the tensioner is compressed; then, remove the 2mm spring pin from the tensioner housing. Using a torque wrench, turn tensioner roller counterclockwise to 11 ft. lbs. (15 Nm) to put pressure on the timing belt.
28. Tighten he camshaft sprocket bolts. Torque the bolts to 41 ft. lbs. (55 Nm).
29. Install the crankshaft damper guard and the serpentine belt idler pulley; torque the idler pulley bolts to 33 ft. lbs. (45 Nm).
30. Install the crankshaft damper and torque the damper-to-crankshaft bolts to 15 ft. lbs. (20 Nm). If the damper-to-crankshaft center bolt was removed, torque it to 147 ft. lbs. (200 Nm) plus 180°.
31. Remove the Crankshaft Holder tool No. 3242 and install the sealing plug.
32. Replace the remaining components by reversing the removal procedures.
33. Connect the electrical connectors. Install the negative battery cable last.
34. Test drive the vehicle.

2.8L (AAH AND AFC) ENGINES

YEAR, MODEL & ENGINE APPLICATION

1992-96 90 Series, 1992-94 100 Series, 1994-96 Cabriolet	2.8L Code AAH
1995-98 Audi A6, 1995-96 90 Series, 1995-98 Cabriolet	2.8L Code AFC

REPLACEMENT INTERVALS

Audi does not provide a replacement interval. Standard industry practice is to replace timing belts every 60,000 miles.

LABOR TIMES

1992-98	
Replacement	2.0 hrs
Severe Service	2.2 hrs

8807AUDg09

✳✳ CAUTION

These engines are an interference type. If the timing belt has failed, valve-to-piston damage may have occurred. A compression check should always be performed to check for engine damage.

1. Disconnect the negative battery cable.
2. Place a 15mm box wrench on the serpentine drive belt tensioner and turn the tensioner clockwise until the 2 holes align; insert a drift into the holes to secure the tensioner in place.
3. Mark the running direction of the serpentine drive belt and remove it from the pulleys.
4. Remove the drive belt tensioner.
5. Disconnect both timing belt upper covers.
6. Rotate the crankshaft by hand to align the crankshaft pulley mark with the arrow on the engine housing and the large hole in each camshaft sprocket must face inward and must align; this should be Top Dead Center (TDC) of the No. 1 cylinder's compression stroke. If these conditions are not correct, rotate the crankshaft one complete revolution and realign.

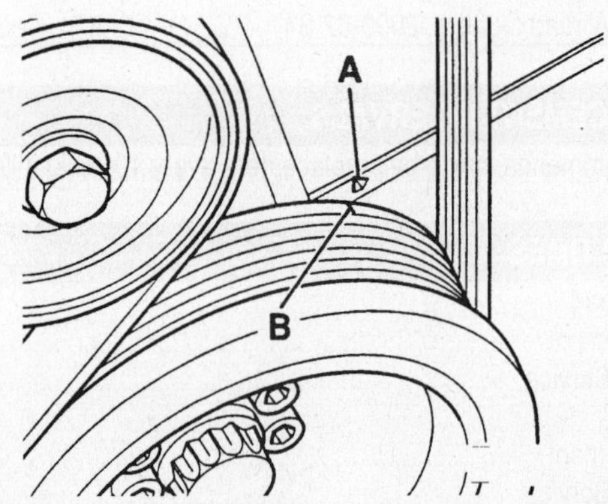

93015g15

Crankshaft pulley alignment location for TDC—Audi 2.8L (AAH, AFC) engine

7. On the left side of the cylinder block near the crankshaft, remove the sealing plug or crankshaft position sensor..

8. Insert Crankshaft Holder tool No. 3242 into the hole to secure the crankshaft.

9. Remove the crankshaft pulley bolts and the pulley.

■ **It is not necessary to remove the center bolt when removing the crankshaft pulley.**

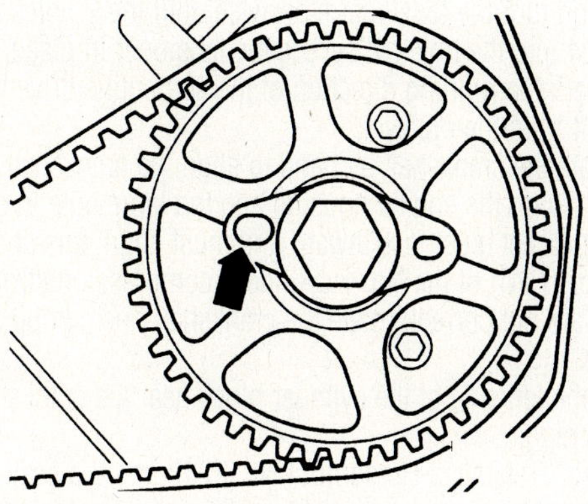

93015g16

Left camshaft sprocket alignment position for TDC; right camshaft position is similar—Audi 2.8L (AAH, AFC) engine

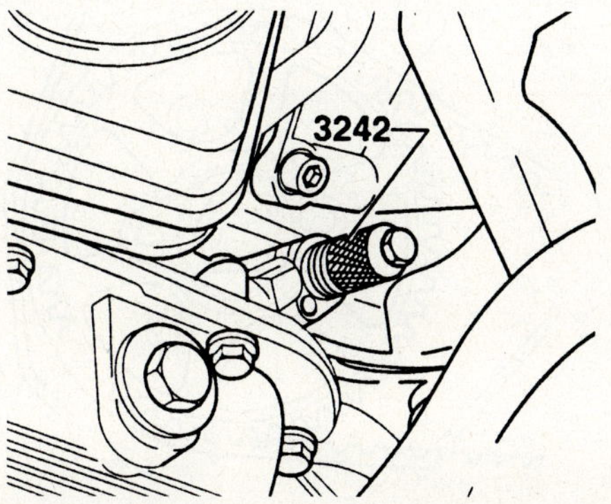

93015g17

View of crankshaft holding tool installed—Audi 2.8L (AAH, AFC) engine

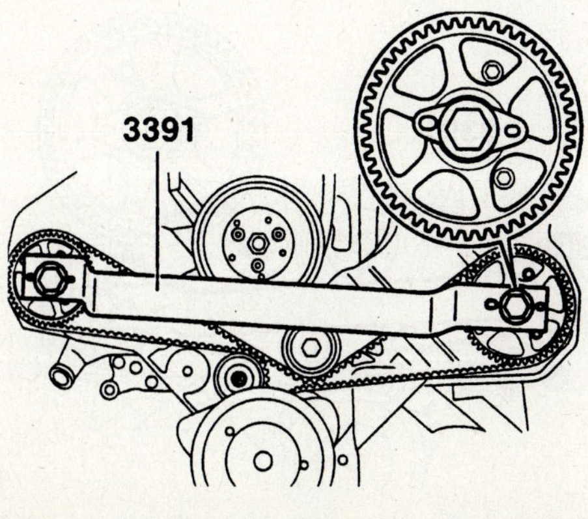

93015g18

View of camshaft locator bar installed—Audi 2.8L (AAH, AFC) engine

10. Remove the timing belt lower cover.

11. Loosen the camshaft sprocket bolts, but do not remove the sprockets.

12. Loosen the timing belt tensioner bolt and rotate the tensioner counterclockwise to relieve belt tension. Tighten the tensioner bolt.

13. Remove the timing belt from the sprockets.

To Install:

14. Ensure the camshaft sprockets can turn but not tilt.

15. Install the timing belt over the camshaft sprockets, then install camshaft sprocket holding tool 3391.

16. Install the timing belt over water pump pulley, roller pulley, crankshaft sprocket and tensioner sprocket.

17. Using a 8mm Allen® wrench, rotate the timing belt tensioner roller clockwise until the tensioner is compressed and release the tensioner's spring pressure to put pressure on the timing belt.

18. Tighten the tensioner bolt to 33 ft. lbs. (45 Nm).

19. Check the timing belt tension by twisting the belt by hand between the water pump pulley and the right camshaft sprocket. The belt should turn about 90° if tensioned properly.

20. Make sure that the camshaft pulleys and the crankshaft pulley are in alignment with TDC of the No. 1 cylinder's compression stroke.

21. Tighten the camshaft sprockets to 52 ft. lbs. (70 Nm).

22. Remove the crankshaft holding tool and the camshaft sprocket holding tool.

23. Install the crankshaft pulley and torque the bolts to 15 ft. lbs. (20 Nm).

24. Replace the remaining components by reversing the removal procedures.

25. Install the negative battery cable last.

2.8L (AHA) ENGINES

YEAR, MODEL & ENGINE APPLICATION

1997-02 Audi A4 & A6	2.8L Code AHA

REPLACEMENT INTERVALS

Audi recommends timing belt replacement every 90,000 miles on 1997 engines and 105,000 miles on 1998-02 engines.

LABOR TIMES

1997-02 A4	
Replacement	2.0 hrs
Severe Service	2.3 hrs
A6	
1997	
Replacement	2.0 hrs
Severe Service	2.2 hrs
1998-02	
Replacement	3.1 hrs
Severe Service	3.3 hrs

8807AUDg10

✱✱ CAUTION

These engines are an interference type. If the timing belt has failed, valve-to-piston damage may have occurred. A compression check should always be performed to check for engine damage.

1. Disconnect the negative battery cable.
2. Remove the upper engine cover.
3. Raise and safely support the vehicle.
4. From under the vehicle, remove the splash shield.
5. Remove the front bumper.
6. Disengage the hood lock cable
7. Remove the intake air duct between the lock carrier and the air cleaner housing at the grille/front end assembly.
8. Remove the grille/front end assembly-to-chassis bolts.
9. Disconnect the electrical connectors from the grille/front end assembly.
10. Drain the engine coolant and disconnect the coolant hoses from the radiator.
11. Detach the A/C condenser from the grille/front end assembly and suspend it on a wire at the front wheel.

**** WARNING**

DO NOT suspend the condenser by its lines. The condenser lines must not be bent or kinked.

12. Drain the automatic transmission fluid from the transmission and the transmission cooler. Disconnect the hydraulic lines from the transmission cooler.
13. If equipped, remove the charge air cooler.
14. Remove the grille/front end assembly-to-vehicle fasteners and the grille/front end assembly from the vehicle.
15. Remove the serpentine drive belt by performing the following procedure:
    a. Using Spanner Wrench No. 3212, secure the viscous fan pulley. Using an Open-end Spanner Wrench 3212, remove the viscous fan bearing housing by turning it clockwise.

■ The viscous fan is mounted with a left-handed thread; turn it clockwise to loosen it.

    b. Place a 17mm box wrench on the serpentine drive belt tensioner and turn the tensioner clockwise until the 2 holes align; insert drift 3204 into the holes to secure the tensioner in place.
    c. Mark the running direction of the serpentine drive belt and remove it from the pulleys.
16. Rotate the crankshaft by hand to align the crankshaft pulley mark with the arrow on the engine housing and the large hole in each camshaft sprocket must face inward and must align; this should be Top Dead Center (TDC) of the No. 1 cylinder's compression stroke. If these conditions are not correct, rotate the crankshaft one complete revolution and realign.
17. On the left side of the cylinder block near the crankshaft, remove the sealing plug.
18. Insert Crankshaft Holder tool No. 3242 into the sealing plug hole to secure the crankshaft.
19. Using a 8mm Allen® wrench, rotate the timing belt tensioner roller clockwise until the tensioner is compressed; then, insert a 2mm spring pin through the tensioner housing and tensioner plunger to

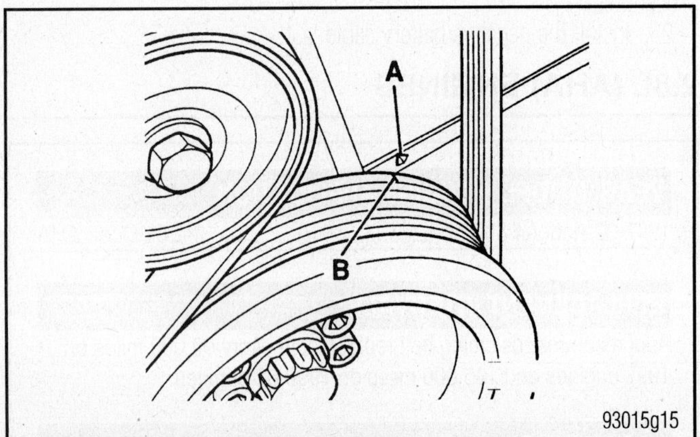

Crankshaft pulley alignment location for TDC—Audi 2.8L (AHA) engine

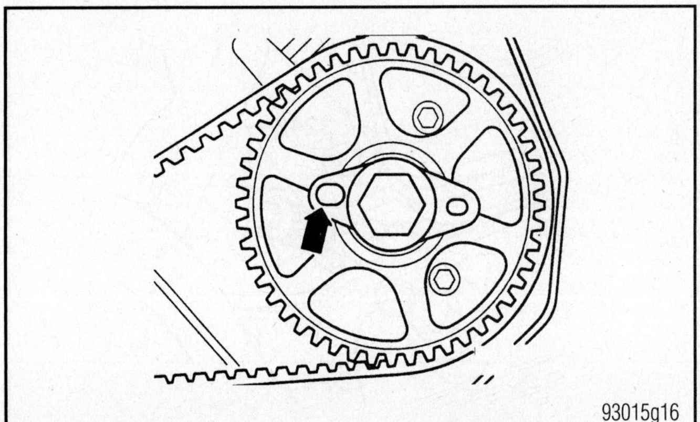

Left camshaft sprocket alignment position for TDC; right camshaft position is similar—Audi 2.8L (AHA) engine

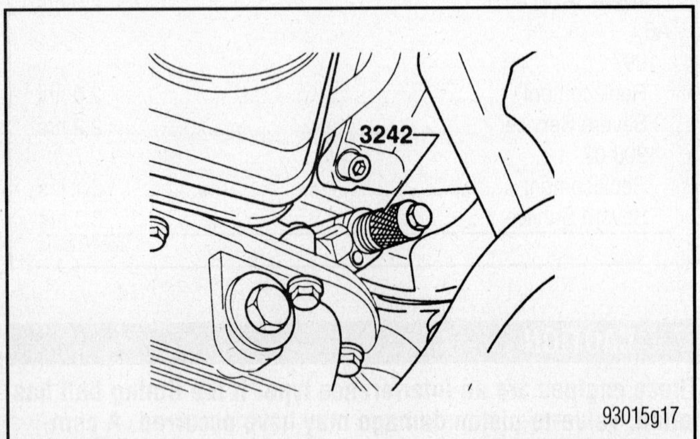

View of crankshaft holding tool installed—Audi 2.8L (AHA) engine

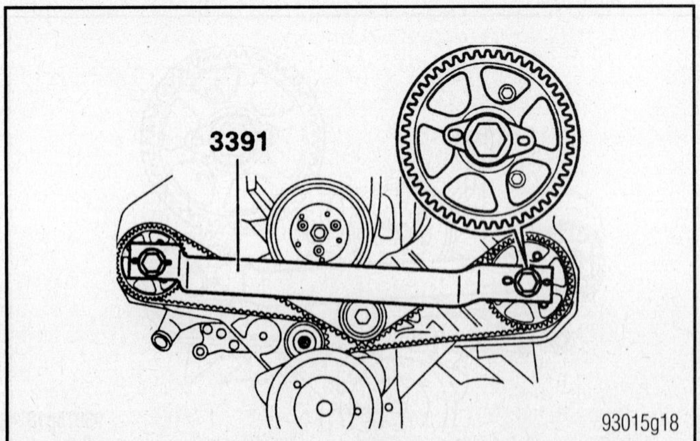

View of camshaft locator bar installed—Audi 2.8L (AHA) engine

secure it in place. When the plunger is secure, release the wrench tension.

20. Remove the damper-to-crankshaft bolts and the damper.

■ **It is not necessary to remove the center bolt when removing the crankshaft damper.**

21. Remove the serpentine belt idler and the crankshaft damper guard.

22. Mark the running direction of the timing belt and remove it from the pulleys.

To Install:

23. Make sure that the camshaft pulleys and the crankshaft pulley are in alignment with TDC of the No. 1 cylinder's compression stroke.

24. Install the timing belt; make sure the timing belt is installed in the correct running direction from which it was removed.

25. Using a 8mm Allen® wrench, rotate the timing belt tensioner roller clockwise until the tensioner is compressed; then, remove the 2mm spring pin from the tensioner housing. Slowly, release the tensioner's spring pressure to put pressure on the timing belt.

26. Install the crankshaft damper guard and the serpentine belt idler pulley; torque the idler pulley bolts to 33 ft. lbs. (45 Nm).

27. Install the crankshaft damper and torque the damper-to-crankshaft bolts to 15 ft. lbs. If the damper-to-crankshaft center bolt was removed, torque it to 147 ft. lbs. (200 Nm) plus 180° 1/2 turn).

28. Remove the Crankshaft Holder tool No. 3242 and install the sealing plug.

29. Replace the remaining components by reversing the removal procedures.

30. Refill the cooling system and the automatic transaxle. Connect the electrical connectors. Install the negative battery cable last.

31. Test drive the vehicle.

3.0L (AVK) ENGINES

YEAR, MODEL & ENGINE APPLICATION

2003-04 Audi A4 & A6	3.0L Code AVK

REPLACEMENT INTERVALS

Audi recommends timing belt replacement every 105,000 miles.

LABOR TIMES

2003-04	
Replacement	3.1 hrs
Severe Service	3.3 hrs

8807AUD11

✳✳ CAUTION

These engines are an interference type. If the timing belt has failed, valve-to-piston damage may have occurred. A compression check should always be performed to check for engine damage.

1. Disconnect negative battery cable.
2. Remove the upper engine cover.
3. Raise and safely support the vehicle.
4. From under the vehicle, remove the splash shield.
5. Remove the front bumper.
6. Disengage the hood lock cable
7. Remove the intake air duct between the lock carrier and the air cleaner housing at the grille/front end assembly.
8. Remove the grille/front end assembly-to-chassis bolts.
9. Disconnect the electrical connectors from the grille/front end assembly.
10. Drain the engine coolant and disconnect the coolant hoses from the radiator.

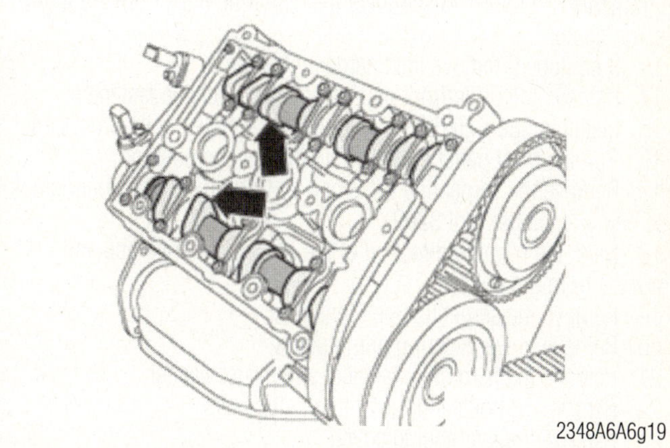

2348A6A6g19

Right side camshaft lobes positioning—Audi 3.0L (AVK) engine

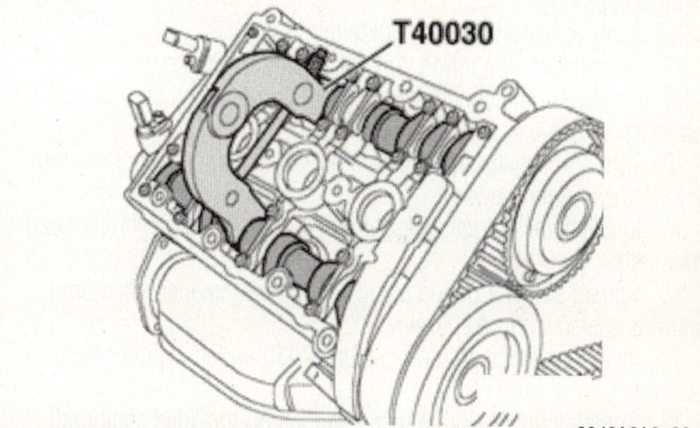

2348A6A6g20

Installing camshaft adjuster gauge on right cylinder head—Audi 3.0L (AVK) engine

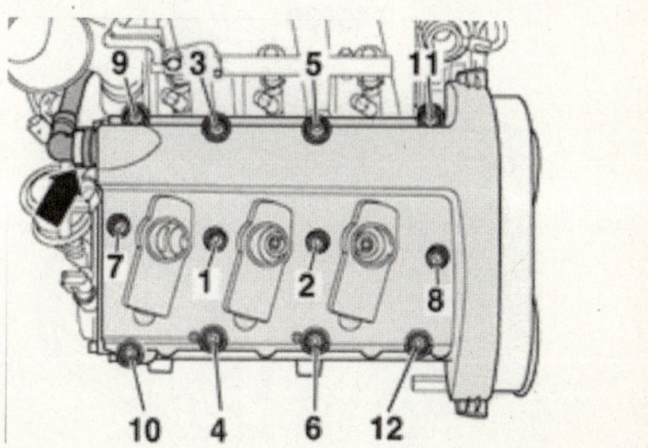

2348A6A6g17

Cylinder head cover removal sequence—Audi 3.0L (AVK) engine

11. Detach the A/C condenser from the grille/front end assembly and suspend it on a wire at the front wheel.

✱✱ WARNING

DO NOT suspend the condenser by its lines. The condenser lines must not be bent or kinked.

12. Drain the automatic transmission fluid from the transmission and the transmission cooler. Disconnect the hydraulic lines from the transmission cooler.
13. If equipped, remove the charge air cooler.
14. Remove the grille/front end assembly-to-vehicle fasteners.
15. Install support guides 3369 through the holes in the front panel.
16. Pull the front panel outward into the service position.
17. Remove the serpentine drive belt by turning belt tensioner clockwise using tensioner tool 3299.
18. Mark the running direction of the serpentine drive belt and remove it from the pulleys.
19. Remove the serpentine belt tensioner.
20. Remove power steering pump pulley.
21. Remove the harmonic balancer and thrust washer.
22. Remove 12 bolts and the timing belt covers.
23. Remove the rear engine cover.
24. Remove the coolant expansion tank and set to one side with hoses attached.
25. Remove all necessary wiring connectors and hoses, and remove air filter housing.
26. Remove the oil dipstick guide tube.
27. Disconnect all ignition coil electrical connectors.
28. Remove both cylinder head covers by removing screws in the sequence shown.
29. Turn crankshaft until camshaft lobes at cylinder no. 3 on right cylinder head point upward.
30. Install camshaft adjuster gauge T40030 on right cylinder head camshafts.
31. Spread adjuster gauge using wrench on threaded shaft until gauge is seated without end play.
32. Install camshaft adjuster gauge T4030 on left cylinder head camshafts.
33. Spread adjuster gauge using wrench on threaded shaft until gauge is seated without end play.
34. On the left side of the cylinder block near the crankshaft, remove the sealing plug.

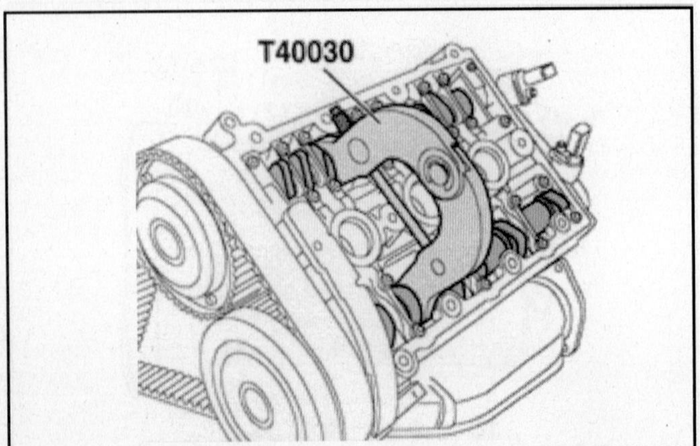

2348A6A6g21

Installing camshaft adjuster gauge on left cylinder head—Audi 3.0L (AVK) engine

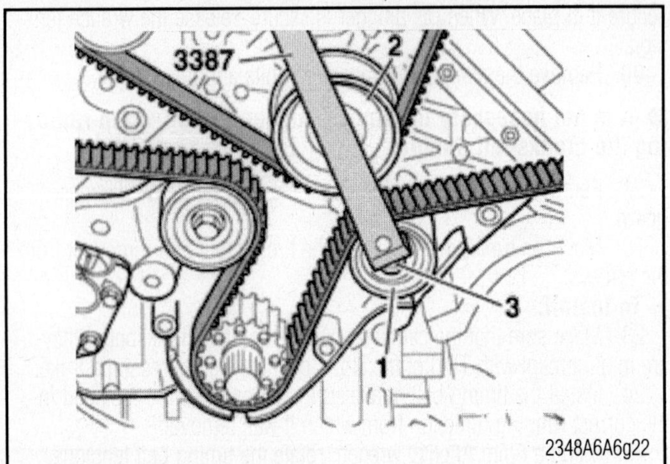

2348A6A6g22

Positioning eccentric pulley for timing belt installation—Audi 3.0L (AVK) engine

35. Insert Crankshaft Holder tool No. T40026 into the sealing plug hole to secure the crankshaft.
36. Remove securing ring from all camshaft gears.
37. Loosen, but do not remove all camshaft gear mounting bolts.
38. Using a 8mm Allen® wrench, rotate the timing belt tensioner roller clockwise until the tensioner is compressed; then, insert a 2mm spring pin through the tensioner housing and tensioner plunger to secure it in place. When the plunger is secure, release the wrench tension.
39. Mark the running direction of the timing belt, loosen eccentric pulley bolt and remove timing belt from the pulleys.

To Install:

- **Replace gaskets, O-rings and self-locking bolts.**
- **When turning camshaft, crankshaft must not be at TDC for any cylinder. Valves and/or pistons may be damaged.**

40. Bolt in the camshaft gears 1 to 4 far enough so they can still just be turned without canting.
41. Install the timing belt so it makes exact contact with front edge of all toothed belt gears.
42. Using 3387 pin wrench, turn the eccentric pulley clockwise until the handle of the pin wrench is exactly above the center axle of coolant pump gear.
43. Hold pin wrench in this position and torque eccentric bolt to 33 ft. lbs. (45 Nm).
44. Place a torque wrench horizontally on socket head of the tensioning lever and turn bolt using 33 ft. lbs. (45 Nm) of torque, to pretension the timing belt.
45. Using a 8mm Allen® wrench, rotate the timing belt tensioner roller clockwise until the tensioner is compressed; then, remove the 2mm spring pin from the tensioner housing. Using a torque wrench, turn tensioner roller counterclockwise to 11 ft. lbs. (15 Nm) to put pressure on the timing belt.
46. Place a torque wrench horizontally on socket head of the tensioning lever and turn bolt using 18 ft. lbs. (25 Nm) of torque, to tension the timing belt.
47. Place T40028 socket at camshaft adjuster of exhaust camshaft on left cylinder bank.
48. Turn rotor of camshaft adjuster clockwise to 84 inch lbs. (10 Nm).
49. Turn rotor of camshaft adjuster at right cylinder bank clockwise to 84 inch lbs. (10 Nm).

50. Torque bolts for all four camshaft gears to 74 ft. lbs. (100 Nm).
51. Install caps with new O-rings at camshaft gears and secure them with the securing ring.
52. Remove the T40030 camshaft adjuster gauge.
53. Remove the alignment bolt and thread sealing plug of TDC opening in using new O-ring.
54. Install remaining components in reverse order of removal. Torque cylinder covers to 84 inch lbs. (10 Nm) using same sequence as removal.

3.7L (AEW) ENGINES

YEAR, MODEL & ENGINE APPLICATION

1997-02 Audi A8	3.7L Code AEW

REPLACEMENT INTERVALS

Audi recommends timing belt replacement every 90,000 miles.

LABOR TIMES

1997-02	
Replacement	6.5 hrs
Severe Service	6.7 hrs

8807AUDg12

✱✱ CAUTION

These engines are an interference type. If the timing belt has failed, valve-to-piston damage may have occurred. A compression check should always be performed to check for engine damage.

1. Drain the cooling system.

■ **The battery is located in the luggage compartment, right side, under a cover.**

2. Disconnect the negative battery cable.
3. Remove the noise insulation panel.
4. Remove the intake air grille from the bumper.
5. Remove the accessory drive belt.
6. Remove the engine appearance cover.
7. Remove the coolant hose clamp on right belt guard.
8. Remove the air intake duct between air cleaner and throttle body.
9. Remove the air shroud for viscous fan and electric coolant fan on radiator (upper-left).
10. Remove the viscous fan. Counterhold with 3212 two-hole nut driver.

■ **The viscous fan has a left-hand thread; loosen by turning clockwise.**

11. Remove the viscous fan with air shroud, from above.
12. Remove the engine support mount (right front).
13. Remove the coolant hoses at engine.
14. Loosen the coolant hose on upper-right of radiator, and turn hose to right.
15. Loosen the center bolt for vibration damper approx. 1 turn. Counterhold using special tool 3197.

■ **Loosening and removing the vibration damper with the camshaft drive belt sprocket is only necessary if the belt is to be replaced, or if other engine disassembly requires it.**

■ **When the center bolt has been loosened, it must be replaced.**

16. Rotate the crankshaft to align TDC marks on vibration damper.
17. Remove Camshaft Position (CMP) sensor housing at rear of left cylinder head.
18. If camshaft position sensor is not positioned behind sensor plate window, rotate crankshaft 360°.
19. Remove the camshaft flange at rear of right cylinder head.
20. Remove the harness connector from switch for the intake manifold change-over valve.
21. Remove the belt guard.
22. Remove the cap at the guide pulley for ribbed belt at right belt guard.
23. Remove the top part of the air cleaner housing.
24. Remove the right belt guard mounting bolts and belt guard.
25. Hold camshaft sprocket with holder 3036 and loosen camshaft sprockets bolts.
26. Remove the camshaft Position (CMP) sensor plate and housing.
27. Install camshaft locking device 3341 at rear of each cylinder head, and tighten. If necessary, rotate camshaft slightly to allow locking device to engage.

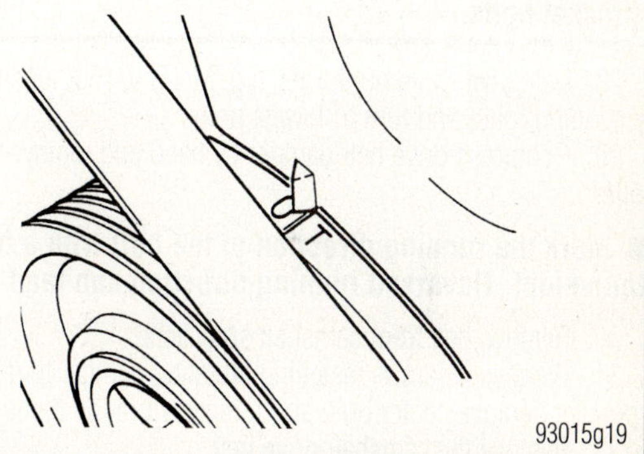

93015g19

Crankshaft pulley alignment location for TDC—Audi 3.7L (AEW) engine

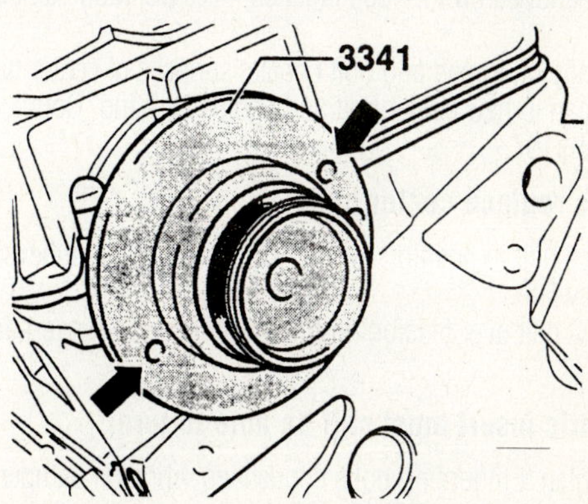

93015g20

Camshaft alignment location for TDC—Audi 3.7L (AEW) engine

93015g21

Measuring tensioning damper roller damper length—Audi 3.7L (AEW) engine

**** CAUTION**

The camshaft locking devices 3341 are not to be used as counterholds when loosening or tightening the camshaft sprocket bolts.

28. Using two-hole nut turner, e.g. Matra V/159, loosen eccentric tensioning roller and turn to lowest point.
29. Compress drive belt damper by hand and remove tensioning roller.

■ **Mark the running direction of the belt with a felt pen or equivalent. Reversed running direction can lead to damage.**

30. Remove belt from camshaft sprockets.
31. Remove 4 screws fastening vibration damper to drive belt sprocket, remove center bolt, and remove vibration damper.
32. Remove the camshaft drive belt.

To Install:

33. Place the timing belt over the crankshaft sprocket and install the vibration damper on the crankshaft.
34. By hand, screw in 4 bolts that connect vibration damper to sprocket.
35. Lightly oil thread and bolt contact surfaces of center bolt.
36. Thread in new center bolt and tighten by hand. Counterhold with special tool 3197.

■ **Always replace center bolt.**

37. Place belt in position over all sprockets, guide pulleys and water pump pulley.
38. Push belt over tensioning roller with eccentric insert and tighten nut.

■ **Eccentric insert must still be able to turn.**

39. Tighten 4 mounting bolts connecting vibration damper and crankshaft sprocket to 18 ft. lbs. (25 Nm).
40. Turn eccentric insert of tensioning roller clockwise using commercial two-hole nut turner (e.g. Matra V/159).
41. Adjust position of tensioning roller until damper length is within specified range.

- Engine cold: 5.35-5.47 inches (136-139 mm)
- Engine warm: 4.96-5.08 inches (126-129 mm)

42. Tighten tensioning roller bolt to 18 ft. lbs. (25 Nm).
43. Remove camshaft locking device from each camshaft (rear of cylinder head.
44. Install housing behind camshaft position sensor and tighten to 15 ft. lbs. (20 Nm).
45. Crank engine at least two revolutions by hand and check timing belt alignment.
46. Tighten crankshaft center bolt as follows:

- With special tool 2079: 258 ft. lbs. (350 Nm)
- Without special tool 2079: 332 ft. lbs. (450 Nm)

47. The remaining installation of the camshaft drive belt is the reverse of removal.

4.2L (ABZ) ENGINES

YEAR, MODEL & ENGINE APPLICATION

1997-02 Audi A8	4.2L Code ABZ

REPLACEMENT INTERVALS

Audi recommends timing belt replacement every 90,000 miles.

LABOR TIMES

1997-02	
Replacement	6.5 hrs
Severe Service	6.7 hrs

8807AUDg13

**** CAUTION**

These engines are an interference type. If the timing belt has failed, valve-to-piston damage may have occurred. A compression check should always be performed to check for engine damage.

1. Drain the cooling system.
2. Disconnect the negative battery cable.

■ **The battery is located in the luggage compartment, right side, under a cover.**

3. Remove the noise insulation panel.
4. Remove the intake air grille from the bumper.
5. Loosen the tensioner and remove the accessory drive belt.
6. Remove the engine appearance cover.
7. Remove the coolant hose clamp on right belt guard.
8. Remove the air intake duct between air cleaner and throttle body.
9. Remove the air shroud for the viscous fan and electric coolant fan on the radiator (upper-left).
10. Remove the viscous fan. Counterhold with 3212 two-hole nut driver

■ **The viscous fan has a left-hand thread; loosen by turning clockwise.**

11. Remove the engine support mount at the right front.
12. Remove the coolant hoses. Loosen the coolant hose on upper-right of radiator, and turn hose to right.

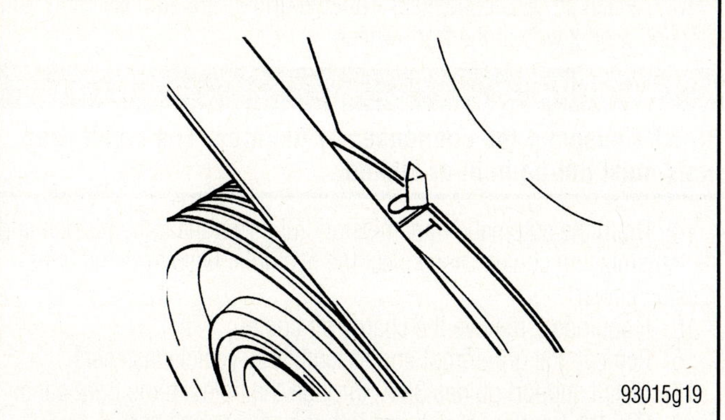

93015g19

Crankshaft pulley alignment location for TDC—Audi 4.2L (ABZ) engine

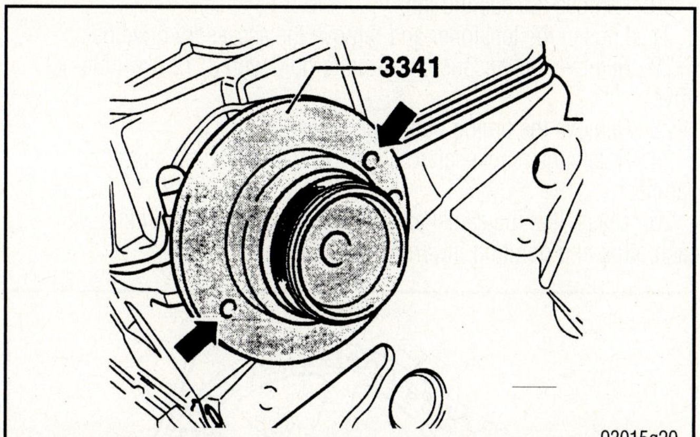

93015g20

Camshaft alignment location for TDC— Audi 4.2L (ABZ) engine

13. Loosen the center bolt for vibration damper approx. 1 turn. Counterhold using special tool 3197.

■ **Loosening and removing the vibration damper with the camshaft drive belt sprocket is only necessary if the belt is to be replaced, or if other engine disassembly requires it.**

■ **When the center bolt has been loosened, it must be replaced.**

14. Rotate crankshaft to align TDC marks on vibration damper.
15. Remove Camshaft Position (CMP) sensor housing at rear of left cylinder head.
16. If camshaft position sensor is not positioned behind sensor plate window, rotate crankshaft 360°.
17. Remove the camshaft flange at rear of right cylinder head.
18. Remove the harness connector from switch for intake manifold change-over valve.
19. Remove the belt guard.
20. Remove the cap at guide pulley for ribbed belt at right belt guard.
21. Remove the top part of air cleaner housing.
22. Remove the 6 right belt guard mounting bolts.
23. Remove the right belt guard.
24. Hold camshaft sprockets with holder 3036 and loosen camshaft sprockets bolts.
25. Remove the Camshaft Position (CMP) sensor plate and housing.

93015g21

Measuring tensioning damper roller damper length— Audi 4.2L (ABZ) engine

26. Install camshaft locking device 3341 at rear of each cylinder head, and tighten. If necessary, rotate camshaft slightly to allow locking device to engage.

**** CAUTION**

The camshaft locking devices 3341 are not to be used as counterholds when loosening or tightening the camshaft sprocket bolts.

27. Using two-hole nut turner, e.g. Matra V/159, loosen eccentric tensioning roller and turn to lowest point.
28. Compress the drive belt damper by hand and remove tensioning roller.

■ **Mark the running direction of the belt with a felt pen or equivalent. Reversed running direction can lead to damage.**

29. Remove the timing belt from the camshaft sprockets.
30. Remove 4 screws fastening vibration damper to drive belt sprocket, remove center bolt, and remove vibration damper.
31. Remove the timing belt.

To Install:

32. Place belt over crankshaft sprocket and install vibration damper on crankshaft.
33. By hand, screw in 4 bolts that connect vibration damper to sprocket.
34. Lightly oil thread and bolt contact surfaces of center bolt.
35. Thread in new center bolt and tighten by hand. Counterhold with special tool 3197.

■ **Always replace center bolt.**

36. Place belt in position over all sprockets, guide pulleys and water pump pulley.
37. Push belt over tensioning roller with eccentric insert and tighten nut.

■ **Eccentric insert must still be able to turn.**

38. Tighten 4 mounting bolts connecting vibration damper and crankshaft sprocket to 18 ft. lbs. (25 Nm).
39. Turn eccentric insert of tensioning roller clockwise using commercial two-hole nut turner (e.g. Matra V/159).

40. Adjust position of tensioning roller until damper length is within specified range. With the engine cold the range is : 5.35-5.47 inches (136-139 mm). With the engine hot, the range is: 4.96-5.08 inches (126-129 mm)
41. Tighten the tensioning roller bolt to 18 ft. lbs. (25 Nm).
42. Remove the camshaft locking device from each camshaft (rear of cylinder head.
43. Install housing behind camshaft position sensor and tighten to 15 ft. lbs. (20 Nm).
44. Crank engine at least two revolutions by hand and check timing belt alignment.
45. Tighten crankshaft center bolt as follows:
- With special tool 2079: 258 ft. lbs. (350 Nm)
- Without special tool 2079: 332 ft. lbs. (450 Nm)

46. The remaining installation of the timing belt is the reverse of removal.

4.2L (AWN, AYS) ENGINES

YEAR, MODEL & ENGINE APPLICATION	
2003-04 Audi A6 Quattro	4.2L Code AWN
2003-04 A8 Quattro	4.2L Code AYS

REPLACEMENT INTERVALS

Audi recommends timing belt replacement every 105,000 miles.

LABOR TIMES	
2003-04	
Replacement	6.5 hrs
Severe Service	6.7 hrs

8807AUDg14

✻✻ CAUTION

These engines are an interference type. If the timing belt has failed, valve-to-piston damage may have occurred. A compression check should always be performed to check for engine damage.

1. Drain the cooling system.
2. Disconnect the negative battery cable.

■ The battery is located in the luggage compartment, right side, under a cover.

3. Remove the noise insulation panel.
4. Remove the air cleaner intake air duct.
5. Remove the engine appearance cover.
6. Remove the engine splash shield.
7. On A6 models, remove the front bumper.
8. Disengage the hood lock cable
9. Remove the intake air duct between the lock carrier and the air cleaner housing at the grille/front end assembly.
10. Remove the grille/front end assembly-to-chassis bolts.
11. Disconnect the electrical connectors from the grille/front end assembly.
12. Drain the engine coolant and disconnect the coolant hoses from the radiator.
13. Detach the A/C condenser from the grille/front end assembly and suspend it on a wire at the front wheel.

✻✻ WARNING

DO NOT suspend the condenser by its lines. The condenser lines must not be bent or kinked.

14. Drain the automatic transmission fluid from the transmission and the transmission cooler. Disconnect the hydraulic lines from the transmission cooler.
15. If equipped, remove the charge air cooler.
16. Remove the grille/front end assembly-to-vehicle fasteners.
17. Install support guides 3369 through the holes in the front panel.
18. Pull the front panel outward into the service position.
19. On all models, remove the air shroud for the viscous fan and electric coolant fan on the radiator (upper-left).
20. Remove the cooling fan.
21. Loosen the tensioner and remove the accessory drive belt.
22. Remove the viscous fan. Counterhold with 3212 two-hole nut driver.
23. Remove the timing belt upper cover.
24. Rotate the crankshaft to align the TDC marks on vibration damper.
25. Check that the camshaft sprocket large holes are in line with each other and pointing inward toward the engine.

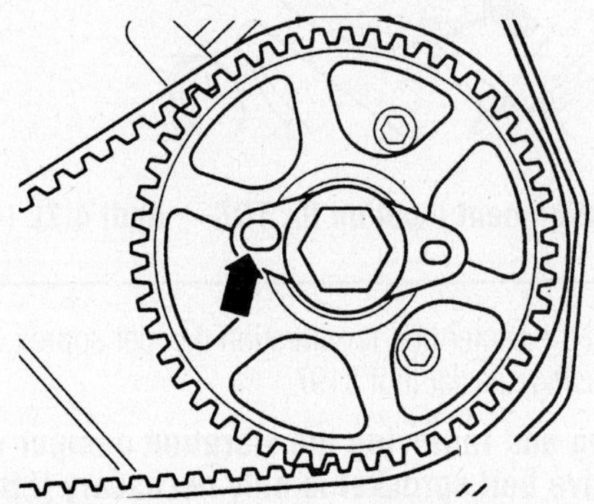

93015g16

Left camshaft sprocket alignment position for TDC; right camshaft position is similar—Audi 4.2L (AWN, AYS) engine

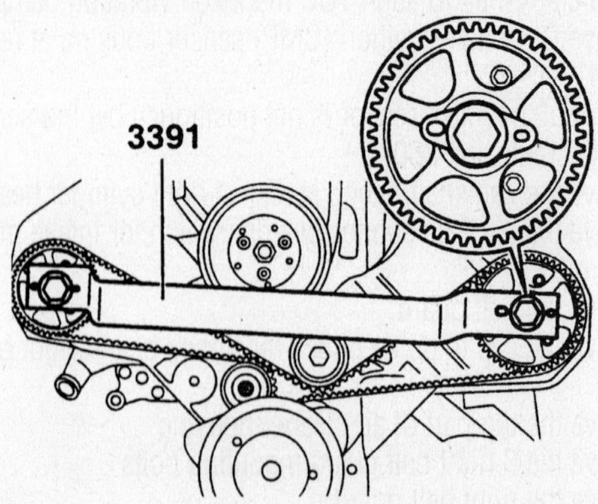

93015g18

View of camshaft locator bar installed— Audi 4.2L (AWN, AYS) engine

26. Remove the sealing plug from the lower side of the engine block.

27. With the crankshaft at TDC, insert clamping tool 3242 into the hole and tighten to lock the crankshaft.

■ **The TDC drilling in the crankshaft must be located at the sealing plug opening.**

28. Remove the drive belt tensioner.

29. Remove the crankshaft pulley outer bolts and remove the pulley.

30. Remove the fan bracket, if equipped.

31. Use a 8mm Allen wrench and turn the timing belt tensioner until the holes in the push rod and tensioner body are aligned, and insert tensioner lock T40011 into the hole.

32. Install camshaft locking device 3391, or equivalent, across the camshaft sprockets.

33. Loosen the camshaft sprocket bolts about 5 turns.

34. Remove the sprocket locking bar.

35. Press the camshaft sprockets off using puller T40001.

36. Loosen the timing belt tensioner pulley bolt, turn the pulley clockwise to release the tension and then loosely tighten the bolt.

37. Remove the right side camshaft sprocket bolt, then remove the camshaft sprocket and the timing belt as a unit.

To Install:

38. Place the timing belt over the crankshaft sprocket, tensioner pulley, left camshaft sprocket water pump and tensioner pulley

39. Install the right camshaft sprocket.

40. Install the camshaft locking device across the camshaft sprockets.

41. Install a 5mm pin between the tensioner lever and push rod.

42. Tighten the tensioner counterclockwise to 16 inch lbs. (1.8 Nm,) and then tighten the tensioner bolt to 16 ft. lbs. (22 Nm).

43. Remove the 5mm pin.

44. Use a 8mm Allen wrench and turn the timing belt tensioner clockwise until a 7mm pin can be inserted between the tensioner lever and body.

45. Tighten the camshaft sprocket bolts to 41 ft. lbs. (56 Nm).

46. Remove the camshaft sprocket locking tool and the 7mm pin.

47. Remove the crankshaft locking pin.

48. Turn the crankshaft 2 turns in the direction of rotation.

49. Ensure that the distance between the tensioner lever and body is 0.19 inch (5mm).

50. Check the clearance between the back side of the timing belt and the belt guard. Clearance should be 0.039 inch (1mm).

51. If the clearance is not as specified, turn the crankshaft until the belt guard bolts are exposed, then loosen the bolts and adjust the rear cover until the clearance is correct.

52. Install the timing belt cover.

53. Install the fan bracket.

54. Install the crankshaft pulley and tighten to 15 ft. lbs. (21 Nm).

55. Install the drive belt tensioner.

56. Install the accessory drive belt.

57. Install the cooling fan.

58. Install the air shroud for the viscous fan and electric coolant fan on the radiator (upper-left).

59. On A6 models, reinstall the grille/front end assembly-to-vehicle fasteners

60. If equipped, install the charge air cooler.

61. Connect the A/C condenser to the grille/front end assembly.

62. Install the intake air duct between the lock carrier and the air cleaner housing at the grille/front end assembly.

63. Engage the hood lock cable.

64. Install the bumper.

65. Connect the transmission oil cooler lines.

66. Install the front bumper.

67. Refill the transmission with fluid.

68. On all models, install the engine splash shield.

69. Install the air cleaner intake air duct.

70. Install the noise insulation panel.

71. Install the engine appearance cover.

72. Refill the cooling system.

73. Connect the negative battery cable.

BMW

2.4L (524 TD) DIESEL ENGINES

YEAR, MODEL & ENGINE APPLICATION	
1985-86 BMW 524td	2.4L Diesel (VIN N/A)

REPLACEMENT INTERVALS

BMW does not provide a replacement interval. Standard industry practice is to replace timing belts every 60,000 miles.

LABOR TIMES

1985-86	
Replacement	3.1 hrs
Severe Service	3.6 hrs
With tensioner replacement add	0.3 hrs

8807BMWg01

✻✻ CAUTION

These engines are an interference type. If the timing belt has failed, valve-to-piston damage may have occurred. A compression check should always be performed to check for engine damage.

1. Disconnect the negative battery cable.
2. Drain the cooling system.
3. Remove the accessory drive belts.
4. Remove the fan assembly and the water pump assembly.
5. Remove the vibration damper and pulley.
6. Disconnect the heater hose from the thermostat housing.
7. Remove the 4 bolts attaching the camshaft drive belt cover to the crankcase and remove the cover.
8. Remove the rocker arm cover.
9. Rotate the crankshaft until the No. 1 cylinder is at Top Dead Center (TDC) on its compression stroke (valves of No. 6 overlapping).
10. Install TDC Aligning Pin 13-5-340 or equivalent.

■ **Flat side of nut or cam position tool should be facing down.**

11. Using a piece of chalk or similar marker, mark the direction of engine rotation on the drive belt, unless a new belt is to be installed.
12. Loosen both belt tensioner bolts.
13. Remove the camshaft drive belt.

To install:

14. Insert a 0.098 in. (2.5mm) thick feeler gauge blade between the Cam Positioning tool No. 11-3-090 or equivalent, at the right front corner of the gasket mating surface of the cylinder head if using a new drive belt or a drive belt used less than 10,000 miles.
15. Install the Injection Pump Aligning Pin 13-5-340 or equivalent, through the injection pump sprocket.
16. Rotate the injection pump sprocket clockwise against the pin.
17. Install the camshaft drive belt. Starting at the crankshaft, route the belt around the intermediate shaft sprocket, injection pump sprocket, camshaft sprocket and tension roller, keeping slack to a minimum.
18. Hand-tighten the belt with the belt tensioner until all slack is gone.

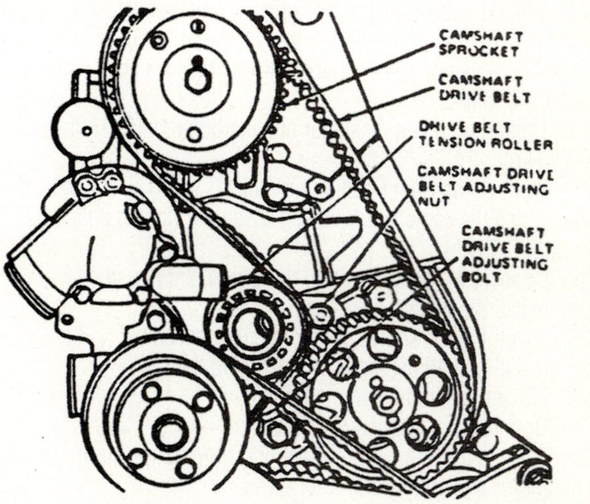

93051g23

View of the timing belt assembly— BMW 2.4L (524 TD) diesel engine

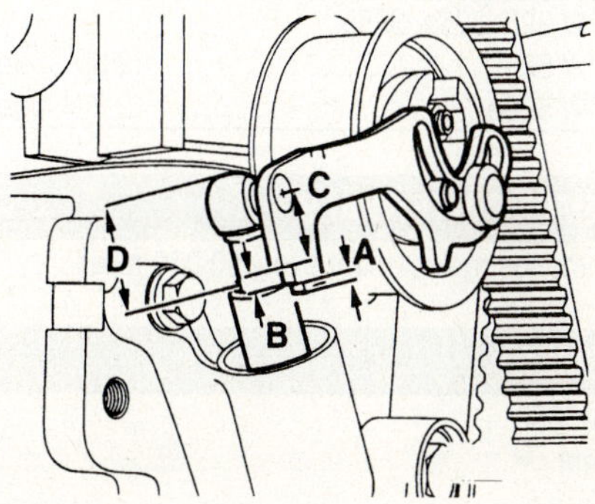

93015g24

View of the camshaft positioning tool—BMW 2.4L (524 TD) diesel engine

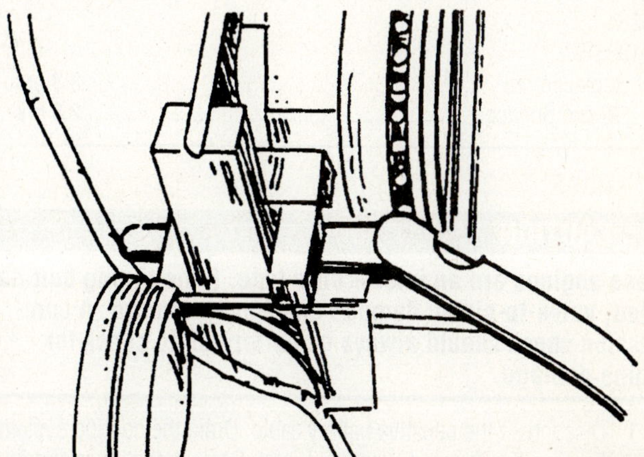

93051g25

Placement of the 2.5mm gauge under the camshaft positioning tool for installing a new belt—BMW 2.4L (524 TD) diesel engine

19. Remove the Injection Pump Aligning Pin 13-5-340 or equivalent, from the injection pump sprocket.
20. Adjust the belt tension by tightening the belt tensioner. Tighten the belt tensioner to 30.5–32.5 ft. lbs. (41.5–44 Nm)) on belts with less than 10,000 miles or 22–25 ft. lbs. (30–34 Nm) for belts with more than 10,000 miles.
21. Tighten both belt tensioner holding bolts to 15–18 ft. lbs. (20–24 Nm).
22. Remove the Cam Positioning tool 11-3-090 or equivalent.
23. Install the camshaft drive belt cover and tighten bolts to 72–84 ft. lbs. (8–9.5 Nm).
24. Connect the heater hose to the thermostat housing.
25. Install the vibration damper.
26. Install the fan and water pump pulley assembly.
27. Install and adjust the accessory drive belts.
28. Fill and bleed the cooling system.
29. Connect the negative battery cable.
30. Operate the engine to normal operating temperatures and check for oil and/or coolant leaks.
31. Check the injection pump timing.

2.5L AND 2.8L ENGINES

YEAR, MODEL & ENGINE APPLICATION

1985-92 BMW 325, 325e, 325ix,
1989-93 BMW 525i .... 2.5L Code M20-B25
1985-88 BMW 528e .... 2.8L Code M20-B27

REPLACEMENT INTERVALS

BMW recommends replacement every 50,000 miles.

LABOR TIMES

325, 325e & 525i	
Replacement	2.8 hrs
Severe Service	3.3 hrs
With tensioner replacement add	0.3 hrs
325ix	
Replacement	3.1 hrs
Severe Service	3.6 hrs
With tensioner replacement add	0.3 hrs
528e	
1985-87	
Replacement	2.3 hrs
Severe Service	2.8 hrs

✱✱ CAUTION

These engines are an interference type. If the timing belt has failed, valve-to-piston damage may have occurred. A compression check should always be performed to check for engine damage.

1. Disconnect the negative battery cable. Drain the cooling system.
2. Remove the distributor cap and rotor. Remove the inner distributor cover and seal.
3. Remove the 2 distributor guard plate attaching bolts and 1 nut. Remove the rubber guard and take out the guard plate.
4. Rotate the crankshaft to set No. 1 piston at Top Dead Center (TDC) of the compression stroke.

■ **At TDC of No. 1 piston compression stroke, the camshaft sprocket arrow should align directly with the mark on the cylinder head.**

5. Remove the radiator.
6. Remove the lower splash guard and take off the alternator, power steering and air conditioning belts.
7. Remove the crankshaft pulley and vibration damper.
8. Hold the crankshaft hub from rotating with tool 11-2-150 or equivalent. Remove the crankshaft hub bolt.
9. Install the hub bolt into the crankshaft about 3 turns and use the proper gear puller, to remove the crankshaft hub.
10. Remove the bolt from the engine end of the alternator bracket. Loosen the alternator adjusting bolt and swing the bracket aside.
11. Lift out the Top Dead Center (TDC) transmitter and set it aside.
12. Remove the remaining bolt and lift off the lower timing belt cover.
13. Loosen the tensioner pulley bolts.
14. Release the tension on the belt by pushing on the tensioner pulley bracket and tightening the upper tensioner bolt.
15. Mark the running direction of the timing belt if it is to be reused and remove the belt.

9305ag04

Timing marks alignment—BMW 2.5L and 2.8L engine

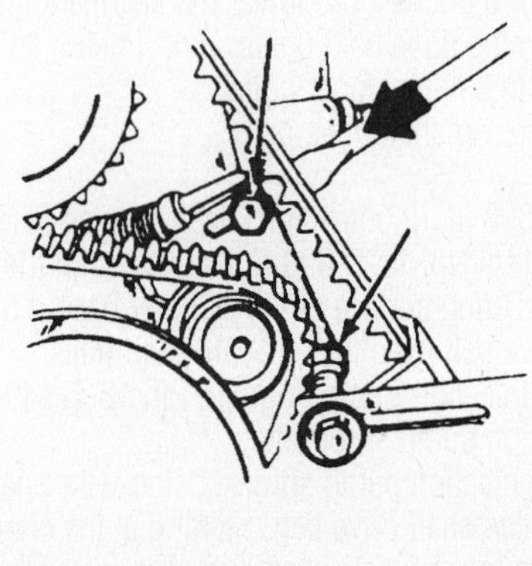

9305ag05

Release the timing belt roller tension—BMW 2.5L and 2.8L engine

To install:

16. Check the camshaft and crankshaft alignment marks and install the timing belt.
17. Loosen the upper tensioner pulley bolt.
18. Rotate the crankshaft with a wrench to tension the timing belt.
19. Tighten the tensioner pulley bolts, upper bolt first.
20. Rotate the crankshaft 2 complete revolutions and insure that the timing marks align.
21. Install the lower timing belt cover and the TDC transmitter.
22. Replace the alternator bracket.
23. Install the crankshaft hub, pulley and vibration damper.
24. Install the accessory drive belts and the lower splash guard.
25. Install the radiator.
26. Install the distributor guard plate.
27. Install the distributor inner cover and seal.
28. Install the cap and rotor.
29. Connect the negative battery cable and refill the cooling system.

DAEWOO

1.6L (VIN 6) DOHC ENGINES

YEAR, MODEL & ENGINE APPLICATION

1999-02 Daewoo Lanos .... 1.6L VIN 6

REPLACEMENT INTERVALS

Daewoo recommends timing belt replacement every 60,000 miles on 1999-01 non-California models, and 102,000 miles on California models. Replace timing belts every 60,000 miles on 2002 models.

LABOR TIMES

1999-02

Replacement	1.5 hrs
Severe Service	1.8 hrs

8807DAEg01

✻✻ CAUTION

These engines are considered freewheeling. If the timing belt fails, valve-to-piston damage is unlikely, however a precautionary compression check should be performed to verify no engine damage.

1. Relieve the fuel system pressure.
2. Remove or disconnect the following:
   - Negative battery cable
   - Intake Air Temperature (IAT) sensor electrical connector
   - Air intake tube from the throttle body
   - Breather tube from the valve cover
   - Air cleaner housing bolts and the housing
   - Right front wheel and splash shield
   - A/C compressor drive belt, (if equipped) and the alternator drive belt
   - Power steering pump pulley bolts and pulley, if equipped
   - Crankshaft pulley bolt and the pulley
   - Upper timing belt cover bolts and the cover
   - Lower timing belt cover bolts and the cover
   - Power steering pump bolts, if equipped
3. Install the crankshaft pulley bolt.
4. Using the crankshaft pulley bolt, rotate the engine clockwise until the crankshaft gear timing mark is in alignment with the notch located at the bottom of the rear timing belt cover.
5. Slightly loosen the water pump retaining bolts.
6. Using timing belt adjuster tool J 42492, rotate the water pump counterclockwise to release the tension on the timing belt, then remove the belt.

■ **Remove the timing belt behind the power steering pump.**

To install:

7. Using the crankshaft pulley bolt, rotate the engine clockwise until the crankshaft gear is in alignment with the notch located at the bottom of the rear timing belt cover.
8. Align the marks on the camshaft gears so that they are facing each other.

■ **Install the timing belt behind the power steering pump.**

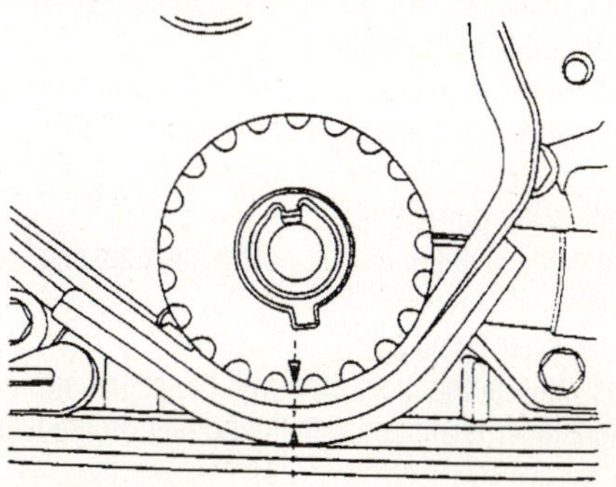

9307Zg93

Rotate the engine clockwise until the crankshaft gear timing mark is in alignment with the notch located at the bottom of the rear timing belt cover—Daewoo 1.6L (VIN 6) DOHC engine

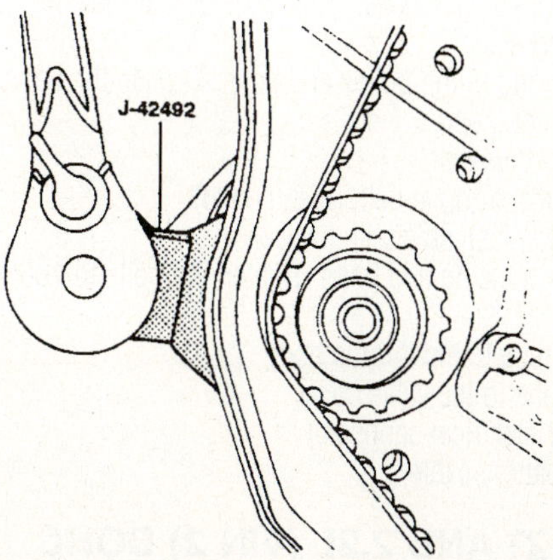

9307Zg96

Using timing belt adjuster tool J 42492, rotate the water pump counterclockwise to release the tension on the timing belt—Daewoo 1.6L (VIN 6) DOHC engine

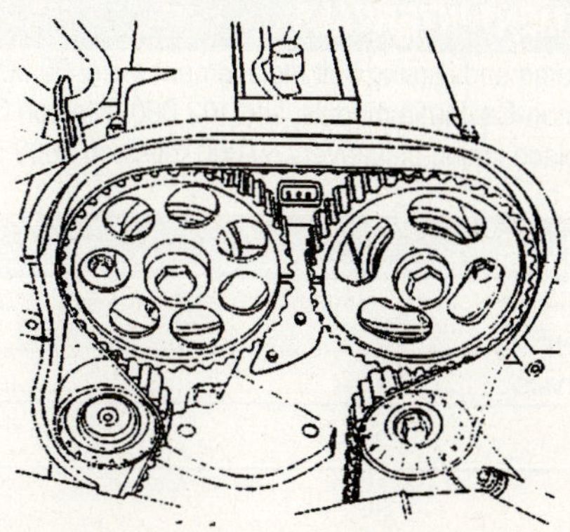

9307Zg97

Camshaft sprocket timing marks—Daewoo 1.6L (VIN 6) DOHC engine

9. Install the timing belt.
10. Using timing belt adjuster tool J 42492, rotate the water pump clockwise to tension the timing belt.
11. Rotate the water pump clockwise until the adjust arm pointer of the timing belt tensioner is aligned with the notch in the tensioner bracket.
12. Tighten the water pump bolts.
13. Using the crankshaft pulley bolt, rotate the crankshaft two full revolutions clockwise.
14. Loosen the water pump bolts.
15. Rotate the water pump clockwise until the adjust arm pointer of the timing belt tensioner is aligned with the pointer on the tensioner bracket.
16. Tighten the water pump bolts to 89 inch lbs. (10 Nm).
17. Remove the crankshaft pulley bolt.
18. Install or connect the following:

- Power steering pump bolts and tighten to 18 ft. lbs. (25 Nm), if equipped
- Upper and lower timing belt covers and bolts. Tighten the bolts to 89 inch lbs. (10 Nm).
- Crankshaft pulley and bolt. Tighten the bolt to 70 ft. lbs. (95 Nm), then using a torque angle meter, tighten the bolt 30 degrees plus an additional 15 degrees.
- Power steering pump pulley and bolts. Tighten the bolts to 18 ft. lbs. (25 Nm), if equipped
- Alternator drive belt
- A/C compressor drive belt, (if equipped)
- Right front splash shield and wheel
- Air cleaner housing and bolts. Tighten the bolts to 106 inch lbs. (12 Nm).
- Air intake tube to the throttle body
- Breather tube to the valve cover
- IAT sensor electrical connector
- Negative battery cable

2.0L (VIN Z) AND 2.2L (VIN 2) DOHC ENGINES

YEAR, MODEL & ENGINE APPLICATION

1999-2002 Daewoo Nubria	2.0L VIN Z
1999-02 Leganza	2.2L VIN 2

REPLACEMENT INTERVALS

Daewoo recommends timing belt replacement every 60,000 miles on 1999-01 non-California models, and 102,000 miles on California models. Replace timing belts every 60,000 miles on 2002 models.

LABOR TIMES

1999-02	
Replacement	1.8 hrs
Severe Service	2.1 hrs

8807DAEg02

✱✱ CAUTION

These engines are an interference type. If the timing belt has failed, valve-to-piston damage may have occurred. A compression check should always be performed to check for engine damage.

1. Relieve the fuel system pressure.
2. Remove or disconnect the following:

- Negative battery cable
- Intake Air Temperature (IAT) sensor electrical connector
- Air intake tube from the throttle body
- Breather tube from the valve cover
- Air cleaner housing bolts and the housing
- Right front wheel and splash shield
- Serpentine drive belt
- Crankshaft pulley bolt and the pulley
- Right engine mount bracket
- Power steering hose clamp bolt and move the hose aside on 2.2L engines
- Timing belt cover bolts and the cover

3. Install the crankshaft pulley bolt.
4. Using the crankshaft pulley bolt, rotate the engine clockwise until the crankshaft gear timing mark is in alignment with the notch located at the bottom of the rear timing belt cover.

✱✱ WARNING

Make sure the marks on the camshaft gears are aligned with the notches on the valve cover after rotating the crankshaft to align its timing mark or engine damage could occur.

5. Align the intake and the exhaust camshaft gears with their respective notches on the valve covers.
6. Remove the timing belt.
7. Loosen the belt tensioner bolt. Turn the hex key tab to relieve the tension.

To install:

8. Align the crankshaft gear timing mark with the notch on the bottom of the rear timing cover.
9. Align the intake and the exhaust camshaft gears with their respective notches on the valve covers.
10. Install the timing belt.
11. Turn the hex key tab clockwise to tension the belt. Turn the hex key until the pointer aligns with the notch.
12. Install the tensioner bolt and tighten to 18 ft. lbs. (25 Nm).
13. Using the crankshaft pulley bolt, rotate the crankshaft two full revolutions clockwise.
14. Recheck the tensioner position.
15. Install or connect the following:

- Timing belt cover and bolts. Tighten the bolts to 53 inch lbs. (6 Nm).
- Right engine mount bracket
- Power steering hose and clamp bolt and tighten the bolt to 71 inch lbs. (8 Nm) on 2.2L engines
- Crankshaft pulley and bolt. Tighten the bolt to 15 ft. lbs. (20 Nm).
- Serpentine drive belt

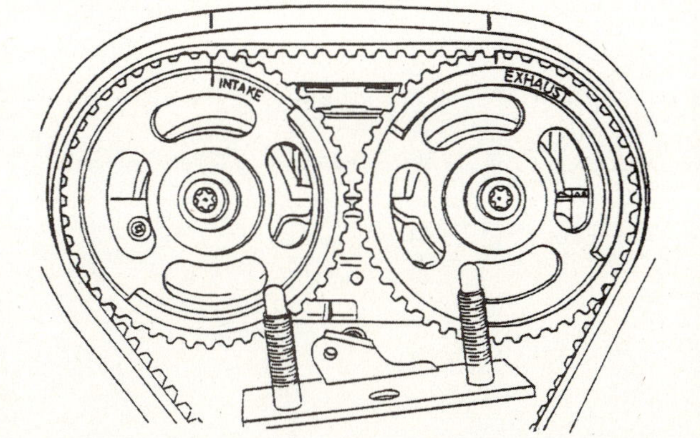

9307Zg45

Align the intake and the exhaust camshaft gears with their respective notches on the valve covers—Daewoo 2.0L (VIN Z) and 2.2L (VIN 2) DOHC engines

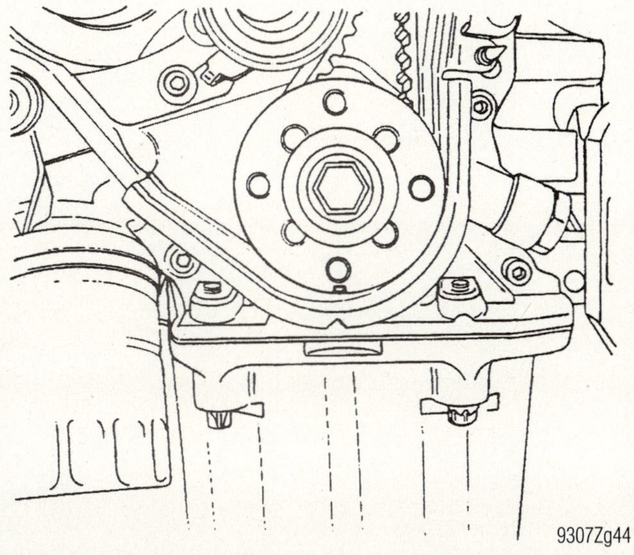

9307Zg44

Rotate the engine clockwise until the crankshaft gear timing mark is in alignment with the notch located at the bottom of the rear timing belt cover—Daewoo 2.0L (VIN Z) and 2.2L (VIN 2) DOHC engines

- Right front splash shield and wheel
- Air cleaner housing and bolts. Tighten the bolts to 71 inch lbs. (8 Nm).
- Air intake tube to the throttle body
- Breather tube to the valve cover
- IAT sensor electrical connector
- Negative battery cable

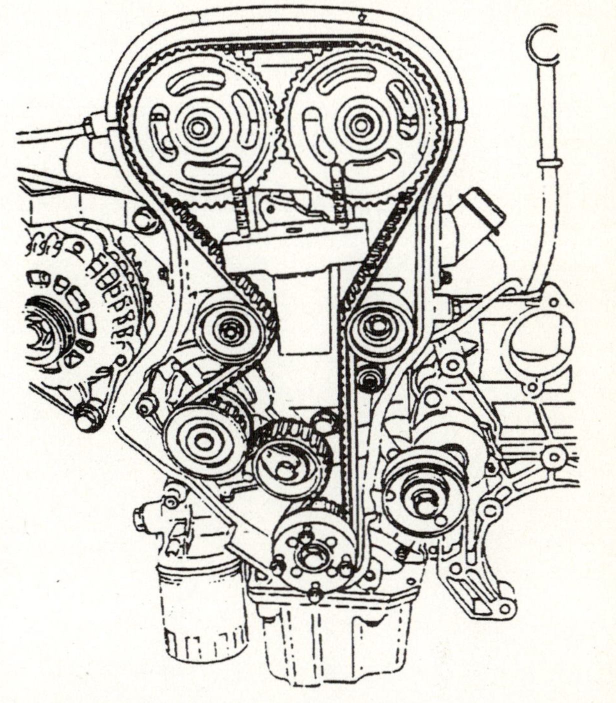

9307Zg43

View of the installed timing belt assembly—Daewoo 2.0L (VIN Z) and 2.2L (VIN 2) DOHC engines

Daihatsu

1.0L (CODE CB) ENGINES

YEAR, MODEL & ENGINE APPLICATION

1989-93 Daihatsu Charade	1.0L Code CB

REPLACEMENT INTERVALS

Daihatsu recommends timing belt replacement every 60,000 miles.

LABOR TIMES

1989-93	
Replacement	2.0 hrs
Severe Service	2.5 hrs

8807DAlg01

✱✱ CAUTION

These engines are an interference type. If the timing belt has failed, valve-to-piston damage may have occurred. A compression check should always be performed to check for engine damage.

1. Disconnect the negative battery cable.
2. Remove the air cleaner.
3. Remove the accessory drive belts.
4. Remove the right front engine undercover.
5. Support the engine with a jack and remove the right motor mount and bracket.
6. Remove the water pump pulley.
7. Remove the crankshaft pulley and remove the timing covers. Remove the crankshaft timing pulley flange.
8. Rotate the crankshaft so that the **F** on the camshaft pulley is facing up and the timing marks align.
9. Loosen the tensioner pulley lockbolt.
10. Pry the tensioner pulley away from the timing belt and tighten the lockbolt.
11. Remove the timing belt.

To install:

12. Ensure that the camshaft and crankshaft timing marks are aligned with the timing marks on the engine and install the timing belt.
13. Loosen the tensioner lockbolt and allow it to apply tension to the timing belt. Tighten the tensioner lockbolt.
14. Rotate the crankshaft 2 complete turns clockwise and ensure that the timing marks align.
15. Loosen the tensioner lockbolt and allow the tensioner to take up any remaining slack in the timing belt. Tighten the tensioner lockbolt to 25–33 ft. lbs. (33–44 Nm).
16. Install the crankshaft timing pulley flange and the timing covers.

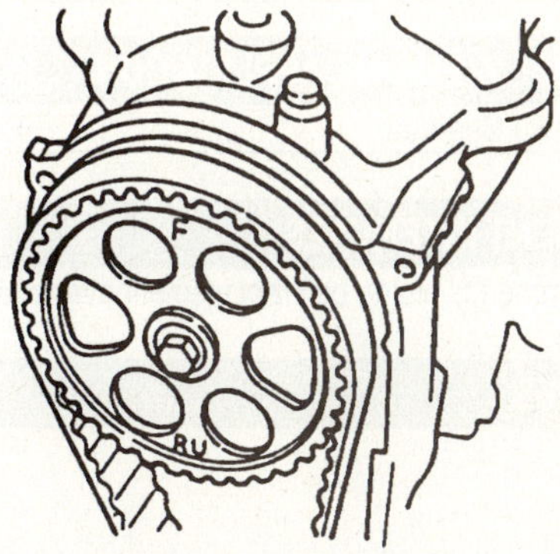

9305ag26

View of the "F" mark position on camshaft timing belt pulley—Daihatsu 1.0L (CB) engine

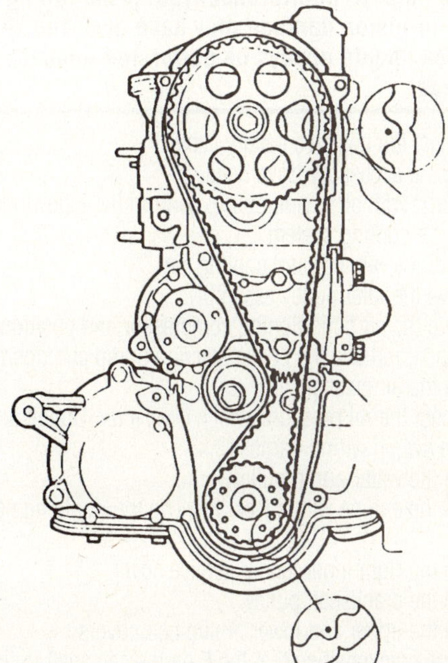

9305ag27

View of the timing mark locations—Daihatsu 1.0L (CB) engine

17. Install the crankshaft pulley and tighten the bolt to 65–72 ft. lbs. (88–98 Nm).
18. Install the right motor mount and bracket. Remove the jack.
19. Install the accessory drive belts and the engine undercover.
20. Install the air cleaner.
21. Connect the negative battery cable.

1.3L (CODE HC) ENGINES

YEAR, MODEL & ENGINE APPLICATION

1988-94 Daihatsu Charade .... 1.3L Code HC

REPLACEMENT INTERVALS

Daihatsu recommends timing belt replacement every 60,000 miles.

LABOR TIMES

1988-94
Charade
Replacement .... 2.0 hrs
Severe Service .... 2.5 hrs

8807DAIg02

✱✱ CAUTION

These engines are an interference type. If the timing belt has failed, valve-to-piston damage may have occurred. A compression check should always be performed to check for engine damage.

1. Disconnect the negative battery cable.
2. Remove the accessory drive belts.
3. If equipped with air conditioning, perform the following:
   a. Drain the cooling system.
   b. Remove the radiator and cooling fan.
   c. Remove the idler pulley assembly.
   d. Remove the air conditioning compressor and position aside.
4. If equipped, remove the power steering pump and position aside.
5. Remove the air cleaner.
6. Disconnect the oil pressure switch connector. Unbolt and remove the oil pressure switch wiring harness.
7. Remove the water pump pulley.
8. Support the engine with a jack. Remove the right engine mount and bracket.
9. Remove the right inner fender service cover.
10. Remove the crankshaft pulley.
11. Remove the upper and lower timing belt covers.
12. Rotate the crankshaft so that the **F** on the camshaft pulley is facing up and the timing marks align.
13. Loosen the tensioner pulley lockbolt.
14. Pry the tensioner pulley away from the timing belt and tighten the lockbolt.
15. Remove the timing belt.

To install:

16. Ensure that the camshaft and crankshaft timing marks are aligned with the timing marks on the engine and install the timing belt.
17. Loosen the tensioner lockbolt and allow it to apply tension to the timing belt. Tighten the tensioner lockbolt.
18. Rotate the crankshaft clockwise until the **F** mark on the camshaft pulley is 3 teeth away from aligning with the indicator mark on the engine.
19. Loosen the tensioner lockbolt.
20. Continue to rotate the crankshaft clockwise until the **F** mark aligns with the indicator mark. Tighten the tensioner lockbolt to 22–33 ft. lbs. (29–44 Nm).
21. Rotate the crankshaft 2 complete turns clockwise and ensure that the timing marks align.

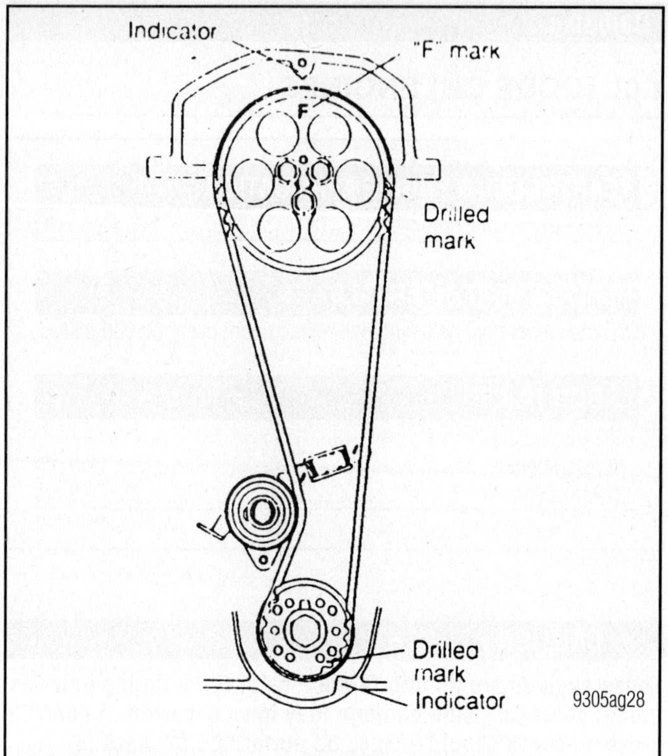

Timing mark locations—Daihatsu 1.3L (HC) engine

22. Install the upper and lower timing covers.
23. Install the crankshaft pulley. Tighten the bolts to 15–22 ft. lbs. (20–29 Nm).
24. Install the right inner fender service cover.
25. Install the right engine mount and bracket. Remove the engine support jack.
26. Install the water pump pulley.
27. Route the oil pressure switch wiring harness through the hole in the right engine mount and install the clamp bolt. Connect the oil pressure switch connector.
28. Install the power steering pump, if removed.
29. Install the air conditioning compressor and idler pulley assembly, if removed.
30. Install the radiator and cooling fan, if removed. Fill the cooling system.
31. Install the air cleaner.
32. Connect the negative battery cable.

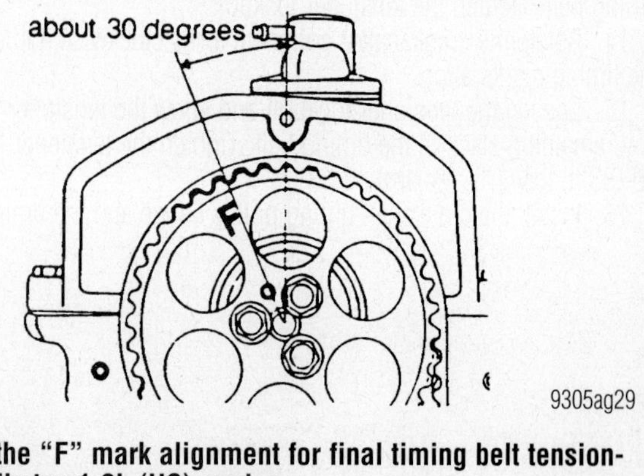

View of the "F" mark alignment for final timing belt tensioning—Daihatsu 1.3L (HC) engine

DaimlerChrysler

1.5L (VIN K) ENGINES

YEAR, MODEL & ENGINE APPLICATION

1985-87 Dodge & Plymouth Colt .... 1.5L VIN K

REPLACEMENT INTERVALS

DaimlerChrysler does not provide a replacement interval. Standard industry practice is to replace timing belts every 60,000 miles.

LABOR TIMES

1985-87	
Replacement	3.1 hrs
Severe Service	3.6 hrs
With air conditioning add	0.2 hrs

8807CHRg01

CAUTION

These engines are an interference type. If the timing belt has failed, valve-to-piston damage may have occurred. A compression check should always be performed to check for engine damage.

1. Disconnect the negative battery cable.
2. Rotate the crankshaft to position the No. 1 cylinder on Top Dead Center (TDC) of the compression stroke.
3. If necessary, remove the air cleaner assembly and intake ducting.
4. Remove the fan drive belt, the fan blades, spacer and water pump pulley.
5. If necessary, remove the valve cover.
6. Remove the timing belt covers.
7. Loosen the timing belt tensioner mounting bolt and move the tensioner away from the belt. Temporarily secure the tensioner.
8. Remove the crankshaft pulley and slide the belt off of the camshaft and crankshaft sprockets.
9. Inspect the drive sprockets for abnormal wear, cracks or damage and replace as necessary. Remove and inspect the tensioner. Check for smooth pulley rotation, excessive play or noise. Replace the tensioner, if necessary.
10. If reusing the timing belt, mark the rotational direction on the back of the belt.
11. Remove the timing belt.

To Install:

12. Make sure that the timing mark on the camshaft sprocket is aligned with the pointer on the cylinder head and that the crankshaft sprocket mark is aligned with the mark on the engine case.
13. Noting the direction of rotation, install the timing belt; make sure that there is no play between the camshaft and crankshaft sprockets.
14. Loosen the tensioner from its temporary position so that the spring pressure will allow it to contact the timing belt.
15. Rotate the crankshaft 1 complete turn in the normal rotation direction to remove any belt slack. Turn the crankshaft until the timing marks are aligned. If the timing has slipped, remove the belt and repeat the procedure.

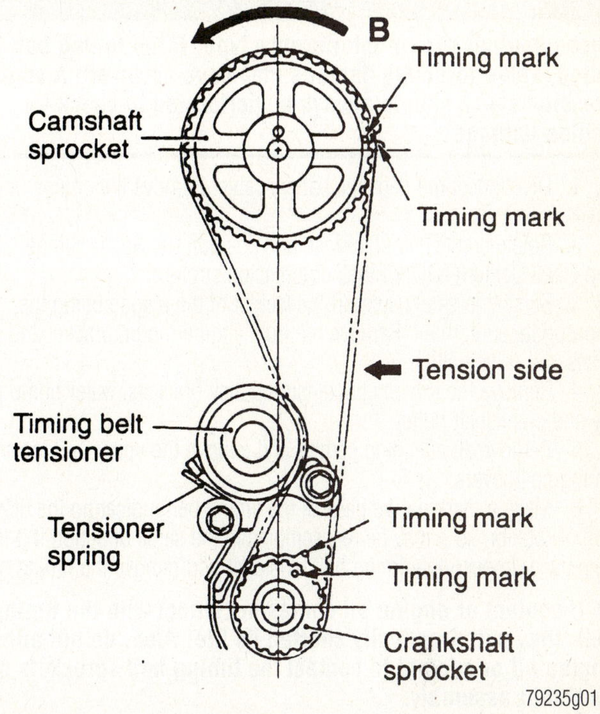

Crankshaft and camshaft timing marks—DaimlerChrysler 1.5L (VIN K) engine

■ **When the tension side of the timing belt and the tensioner are pushed in horizontally with a moderate force (about 11 lbs. [5 kg]), and the cogged side of the belt covers about a 1/4 in. of the tensioner right side mounting bolt head (across flats), the tension is correct.**

16. Install the crankshaft pulley.
17. Install the timing belt covers.
18. If removed, install the valve cover.
19. Install the water pump pulley, spacer, fan blades and drive belt.
20. If removed, install the air cleaner assembly and intake ducting.
21. Connect the negative battery cable.

1.5L (VIN A, P) ENGINES

YEAR, MODEL & ENGINE APPLICATION

1988-90 Dodge Colt Wagon,
1988-94 Dodge & Plymouth Colt,
1989-96 Eagle Summit .... 1.5L VIN A, P

REPLACEMENT INTERVALS

DaimlerChrysler does not provide a replacement interval. Standard industry practice is to replace timing belts every 60,000 miles.

LABOR TIMES

1988-96	
Replacement	3.1 hrs
Severe Service	3.6 hrs
With air conditioning add	0.2 hrs

8807CHRg02

✻✻ CAUTION

These engines are an interference type. If the timing belt has failed, valve-to-piston damage may have occurred. A compression check should always be performed to check for engine damage.

1. Disconnect the negative battery cable. Remove the engine under cover.
2. Rotate crankshaft clockwise and position the No. 1 cylinder at Top Dead Center (TDC) of the compression stroke.
3. Raise and safely support the weight of the engine using the appropriate equipment. Remove the front engine mount bracket and accessory drive belts.
4. Remove the drive belts, tension pulley brackets, water pump pulley and crankshaft pulley.
5. Remove all attaching screws and remove the upper and lower timing belt covers.
6. Make a mark on the back of the timing belt indicating the direction of rotation so it may be reassembled in the same direction if it is to be reused. Loosen the timing belt tensioner and remove the timing belt.

■ **If coolant or engine oil comes in contact with the timing belt, they will drastically shorten its life. Also, do not allow engine oil or coolant to contact the timing belt sprockets or tensioner assembly.**

7. Remove the tensioner spacer, tensioner spring and tensioner assembly.
8. Inspect the timing belt for cracks on back surface, sides, bottom and check for separation. Check the tensioner pulley for smooth rotation.

To Install:

9. Position the tensioner, tensioner spring and tensioner spacer on engine block.
10. Align the timing marks on the camshaft sprocket and crankshaft sprocket. This will position No. 1 piston on TDC on the compression stroke.

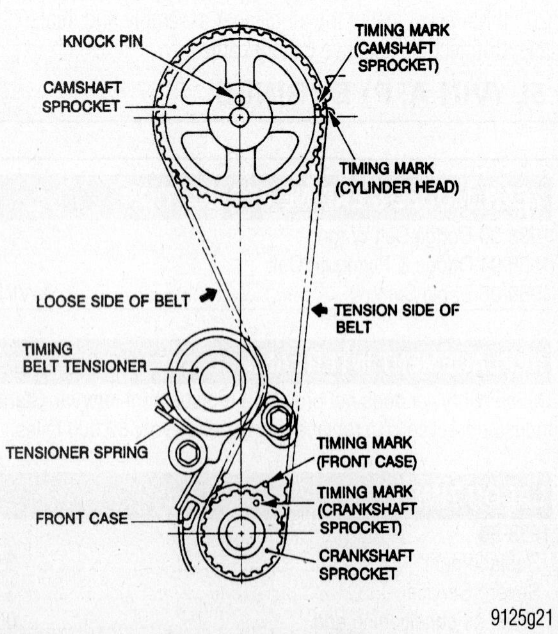

Crankshaft and camshaft sprocket timing marks—DaimlerChrysler 1.5L (VIN A, P engine)

11. Position the timing belt on the crankshaft sprocket and keeping the tension side of the belt tight, set it on the camshaft sprocket.
12. Apply counterclockwise force to the camshaft sprocket to give tension to the belt and make sure all timing marks are aligned.
13. Loosen the pivot side tensioner bolt and the slot side bolt. Allow the spring to remove the slack.
14. Tighten the slot side tensioner bolt and then the pivot side bolt. If the pivot side bolt is tightened first, the tensioner could turn with bolt, causing over tension.
15. Turn the crankshaft clockwise. Loosen the pivot side tensioner bolt and then the slot side bolt to allow the spring to take up any remaining slack. Ttighten the slot bolt and then the pivot side bolt to 14–20 ft. lbs. (20–27 Nm).
16. Install the timing belt covers and all related items.
17. Connect the negative battery cable.

1.6L (VIN F) SOHC ENGINES

YEAR, MODEL & ENGINE APPLICATION

1985-89 Dodge Colt, 1989-90 Eagle Summit .... 1.6L VIN F

REPLACEMENT INTERVALS

DaimlerChrysler recommends timing belt replacement every 60,000 miles.

LABOR TIMES

1985-90	
Replacement	2.6 hrs
Severe Service	3.1 hrs
With air conditioning add	0.2 hrs

8807CHRg03

✻✻ CAUTION

These engines are an interference type. If the timing belt has failed, valve-to-piston damage may have occurred. A compression check should always be performed to check for engine damage.

1. Disconnect the negative battery cable.
2. Disconnect the accelerator cable and breather hose. Remove the air intake duct between the turbocharger and throttle body.
3. Remove the distributor cap/spark plug wire as an assembly.
4. Remove the accessory drive belts.
5. Safely support the engine and remove the left engine mount bracket.
6. Remove the water pump and power steering pump pulleys.
7. Remove the valve cover and the upper timing belt cover.
8. Loosen the valve adjusting screws until the tip of each screw protrudes less than 0.08 inch (2mm) from the rocker arm.

■ **Loosening the valve adjusting screws is essential to provide enough free-play at the camshaft to allow for correct valve timing during the timing belt installation.**

9. Remove the damper pulley, crankshaft pulley and the lower timing belt cover.
10. Rotate the crankshaft (clockwise) until the timing marks are aligned.
11. Loosen the timing belt tensioner bolts, move the tensioner toward the water pump and retighten the bolts.

12. Mark the timing belt to indicate its direction of rotation for reinstallation purposes.

13. Remove the timing belt.

To Install:

14. Install the timing belt (in the correct direction of rotation) over the crankshaft sprocket, the oil pump sprocket and the camshaft sprocket.

■ **Be sure to keep tension on the belt as it is being installed.**

15. Loosen the timing belt tensioner bolts and move the tensioner toward the timing belt.

16. To adjust the timing belt tension, perform the following procedure:

a. Apply counterclockwise force on the camshaft sprocket to tighten the belt tension side and ensure that the timing marks remain aligned.

b. Install the crankshaft pulley and rotate the crankshaft (clockwise) until the camshaft sprocket timing mark is 2 teeth away from the alignment mark on the rear cover.

■ **DO NOT rotate the crankshaft counterclockwise or push on the belt to check its tension.**

c. Move the tensioner to apply pressure on the belt and tighten the tensioner nut; then, tighten the pivot bolt.

d. Place your thumb on the belt between the oil pump and camshaft sprockets and press the belt outward. The flex between timing belt and the cover should be 0.23 in., if not, loosen the pivot and tension the belt.

17. Rotate the crankshaft 2 revolutions and realign the timing marks to ensure that the engine is in time.

18. Adjust the valve clearances.

19. Complete the installation by reversing the removal procedure.

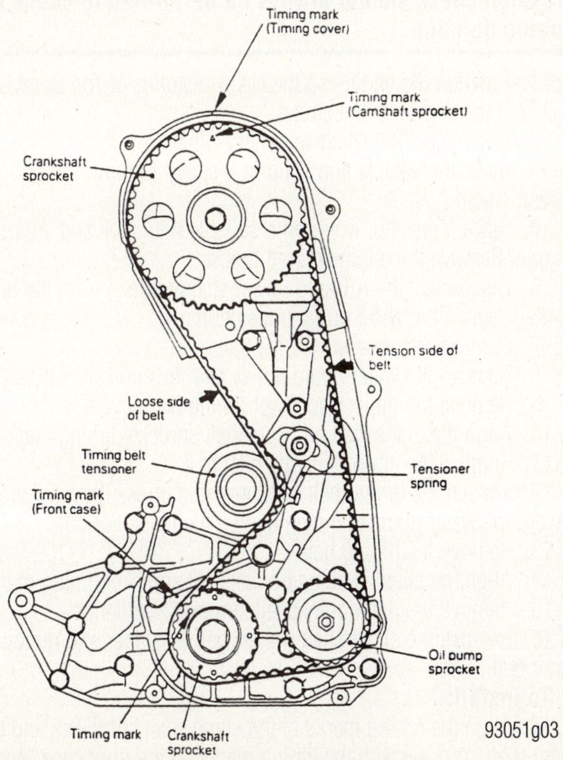

Crankshaft and camshaft sprocket timing marks—DaimlerChrysler 1.6L (VIN F) SOHC engine

1.6L (CODE 4G61) DOHC ENGINES

YEAR, MODEL & ENGINE APPLICATION	
1989-91 Dodge & Plymouth Colt	1.6L Code 4G61

REPLACEMENT INTERVALS

DaimlerChrysler recommends timing belt replacement every 60,000 miles.

LABOR TIMES	
1989-91	
Replacement	4.3 hrs
Severe Service	4.8 hrs
With air conditioning add	0.2 hrs

8807CHRg04

✳✳ CAUTION

These engines are an interference type. If the timing belt has failed, valve-to-piston damage may have occurred. A compression check should always be performed to check for engine damage.

■ **Special tools MD998752 tension pulley torque adapter and MD998738 tension pulley locker or equivalents, are required.**

1. Bring the engine to No. 1 piston at Top Dead Center (TDC) of the compression stroke. Disconnect the negative battery cable.

2. Raise the vehicle and support it safely. Remove the under engine splash shield.

3. Place a piece of wood on a suitable floor jack and support the engine. Remove the engine mount bracket.

4. Remove the alternator and power steering drive belts. Remove the air conditioner drive belt and tensioner assembly.

5. Remove the water pump pulley and the crankshaft pulley.

6. Remove the upper and lower timing belt covers.

7. Remove the engine center cover. Remove the breather hose from the rear of the rocker cover. Remove the PCV hose. Disconnect the spark plug cables from the plugs.

8. Remove the rocker cover and rear half-moon seal.

9. Confirm the engine is still at No. 1 TDC. The timing marks on the camshaft sprocket and the upper surface of the cylinder head should coincide. The dowel pin on the front of the camshafts should be in the 12 o'clock position. Remove the automatic belt tensioner. Loosen the tensioner pulley center bolt.

10. If the timing belt is to be reused, mark an arrow, on the belt, in the direction of rotation, for installation reference. Remove the timing belt.

To Install:

11. Install the automatic tensioner, after reset.

■ **To reset the tensioner: Keep the adjuster level and clamp it in a soft jawed vise. Clamp with the extended adjuster on one side and the end mounting a plug on the other side. If the plug extends out of the adjuster body, place a suitable hole sized washer over the plug so the vise jaw pushes on the washer, not the plug. Close the vise slowly, forcing the adjuster back into the body. When the hole in the adjuster**

boss aligns with the adjuster rod, insert a snug fitting pin or wire into the holes to keep the rod in the compressed position. With the locking pin or wire in place, install the tensioner.

12. Align the timing marks on the camshaft sprockets. Align the crankshaft timing marks. Align the oil pump timing marks. Place the timing belt around the intake camshaft and secure it to the sprocket with a stationary binder spring clip. Install the timing belt around the exhaust camshaft sprocket, check sprocket marks for alignment and secure the belt with a second binder clip on the exhaust sprocket.
13. Install the timing belt around the idler pulley, oil pump sprocket, crankshaft sprocket and the tensioner pulley.
14. Lift up the tensioner pulley against the belt and tighten the center bolt to hold it in position.
15. Check to see that all of the timing marks are aligned. Remove the binder clips. Rotate the crankshaft a quarter turn counter clockwise. Then turn the crankshaft clockwise until the timing marks are aligned.
16. Place special tool MD998752 or equivalent, on a torque wrench. Insert the tool into the place provided on the tension pulley. Loosen the center pulley bolt and apply 26 inch lbs. (3.0 Nm) of pressure against the timing belt with the tension pulley. While holding the required torque, tighten the center bolt. Screw in special tool MD998738 or equivalent, through the left engine support bracket until it contacts the tensioner arm bracket. Turn the tool a little more to secure the tensioner and remove the locking wire place into the automatic adjuster when it was reset.
17. Remove the special tool. Rotate the crankshaft 2 complete turns clockwise and allow it to sit for about 15 minutes. Then measure the protrusion of the automatic adjuster. It should be 0.015–0.018 in. (0.381–0.457mm). If the proper amount of protrusion is not present, repeat the tensioning process.
18. Install the upper and lower timing belt covers.
19. Install the crankshaft pulley and tighten the bolts to 18 ft. lbs. (24 Nm).

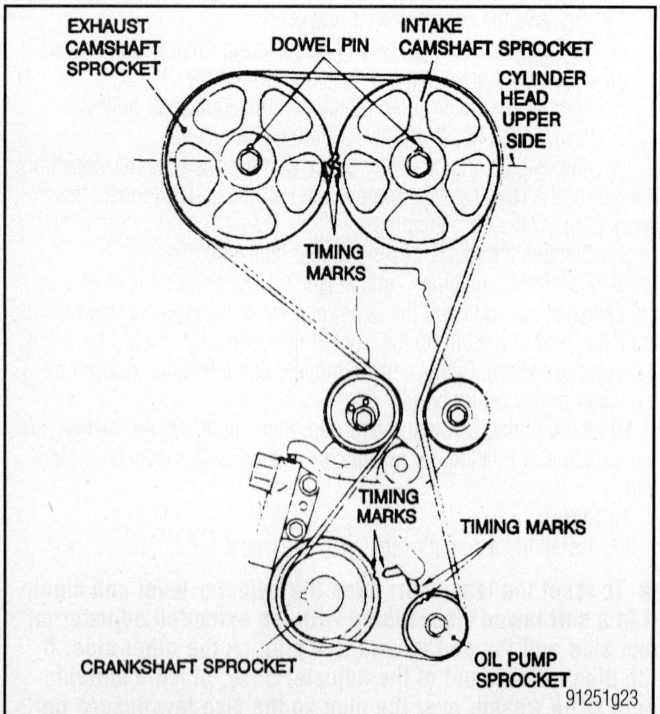

Alignment mark positioning for timing belt installation—DaimlerChrysler 1.6L DOHC engine

20. Install the water pump pulley.
21. Install the rocker cover and rear half-moon seal.
22. Install the engine center cover. Install the breather hose to the rear of the rocker cover. Connect the PCV hose. Connect the spark plug cables.
23. Install the A/C drive belt tensioner and the A/C drive belt.
24. Install the engine mount bracket.
25. Install the splash shield.
26. Connect the negative battery cable.

1.8L (VIN B) ENGINES

YEAR, MODEL & ENGINE APPLICATION	
1993-94 Eagle Talon, 1990-94 Plymouth Laser	1.8L VIN B

REPLACEMENT INTERVALS
DaimlerChrysler recommends timing belt replacement every 60,000 miles.

LABOR TIMES	
1990-94	
Replacement	2.8 hrs
Severe Service	3.2 hrs
With air conditioning add	0.3 hrs
With power steering add	0.3 hrs

8807CHRg05

✻✻ CAUTION

These engines are an interference type. If the timing belt has failed, valve-to-piston damage may have occurred. A compression check should always be performed to check for engine damage.

1. Position the engine so the No. 1 piston is at Top Dead Center (TDC) of the compression stroke.
2. Disconnect the negative battery cable.
3. Raise the vehicle and support it safely. Remove the under engine splash shield.
4. Place a piece of wood on a suitable floor jack and support the engine. Remove the engine mount bracket.
5. Disconnect the A/C and power steering lines from the body.
6. Loosen the water pump pulley bolts.
7. Remove the accessory drive belts.
8. Remove the water pump pulley and the crankshaft pulley.
9. Remove the upper and lower timing belt covers.
10. Align the crankshaft and camshaft sprocket timing marks.
11. Remove the crankshaft sprocket bolt.
12. Loosen the timing belt tensioner bolt, press the tensioner back toward the water pump and then tighten the tensioner bolt.
13. Remove the timing belt.
14. Align the silent shaft and crankshaft sprocket timing marks.
15. Remove the outer crankshaft sprocket and flange.
16. Remove the silent shaft (inner) belt tensioner and remove the inner belt.

To Install:

17. Align the timing marks of the silent shaft sprockets and the crankshaft sprocket with the timing marks on the front case. Wrap the timing belt around the sprockets so there is no slack in the upper span of the belt and the timing marks are still aligned.

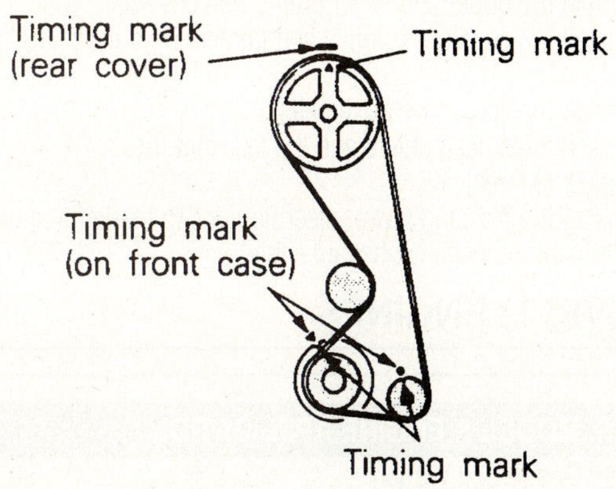

9305bg44

Crankshaft, camshaft and oil pump sprocket timing mark alignment—DaimlerChrysler 1.8L (VIN B) engine

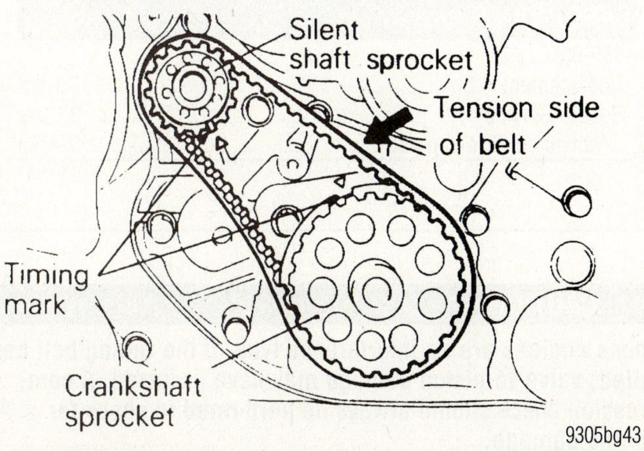

9305bg43

Silent shaft belt timing mark alignment—DaimlerChrysler 1.8L (VIN B) engine

18. Install the tensioner pulley and move the pulley by hand so the long side of the belt deflects 0.20–0.28 in. (5–7mm).
19. Hold the pulley tightly so the pulley cannot rotate when the bolt is tightened. Tighten the bolt to 15 ft. lbs. (20 Nm) and recheck the deflection amount.
20. Install the timing belt tensioner fully toward the water pump and tighten the bolts. Place the upper end of the spring against the water pump body.
21. Align the timing marks of the camshaft, crankshaft and oil pump sprockets with their corresponding marks on the front case or rear cover.

■ **There is a possibility to align all timing marks and have the oil pump sprocket out of time, causing an engine vibration during operation. If the following step is not followed exactly, there is a 50 percent chance that the oil pump shaft alignment will be 180 degrees off.**

22. Before installing the timing belt, ensure that the oil pump sprocket is in the correct position as follows:
    a. Remove the plug from the rear side of the block and insert a Phillips screwdriver with shaft diameter of 0.31 in. (8mm) into the hole.
    b. With the timing marks still aligned, the shaft of the tool must be able to go in at least 2.36 in. (60mm). If the tool can only go in 0.79–0.98 in. (20–25mm), the shaft is not in the correct orientation and will cause a vibration during engine operation. Remove the tool from the hole and turn the oil pump sprocket 1 complete revolution. Realign the timing marks and insert the tool. The shaft of the tool must go in at least 2.36 in. (60mm).
    c. Recheck and realign the timing marks.
    d. Leave the tool in place to hold the oil pump shaft while continuing.
23. Install the belt to the crankshaft sprocket, oil pump sprocket, then camshaft sprocket. While doing so, make sure there is no slack between the sprocket except where the tensioner is installed.
24. Tighten oil pump sprocket bolt to 26–29 ft. lbs. (34–40 Nm) and tighten crankshaft bolt to 80–94 ft. lbs. (110–130 Nm).
25. Recheck the timing mark alignment. If all are aligned, loosen the tensioner mounting bolt and allow the tensioner to apply tension to the belt.
26. Remove the tool that is holding the silent shaft and rotate the crankshaft a distance equal to 2 teeth on the camshaft sprocket. This will allow the tensioner to automatically apply the proper tension on the belt. Do not manually overtighten the belt or it will howl.
27. Tighten the lower mounting bolt first, then the upper spacer bolt.
28. To verify correct belt tension, check that the deflection at the longest span of the belt has 0.40 in. (12mm) clearance from the belt cover.
29. Install the upper and lower timing belt covers.
30. Install the water pump pulley and the crankshaft pulley.
31. Install the the accessory drive belts.
32. Tighten the water pump pulley bolts.
33. Connect the A/C and power steering lines to the body.
34. Install the engine mount bracket.
35. Install the under engine splash shield.
36. Connect the negative battery cable.

1.8L (VIN C) ENGINES

YEAR, MODEL & ENGINE APPLICATION

1992-94 Dodge & Plymouth Colt & Colt Vista,
1992-96 Eagle Summit & Summit Wagon .... 1.8L VIN C

REPLACEMENT INTERVALS

DaimlerChrysler recommends timing belt replacement every 60,000 miles.

LABOR TIMES

1992-96	
Replacement	3.1 hrs
Severe Service	3.6 hrs
With air conditioning add	0.2 hrs

8807CHRg06

✻✻ CAUTION

These engines are an interference type. If the timing belt has failed, valve-to-piston damage may have occurred. A compression check should always be performed to check for engine damage.

1. Disconnect the negative battery cable.
2. Disconnect the A/C and power steering lines from the left engine mount.
3. Raise and support the vehicle.
4. Remove the left side wheel and inner fender liner.
5. Remove the accessory drive belts.
6. On Summit Wagon, remove the coolant reservoir tank.
7. Remove the crankshaft pulley.
8. Remove the upper and lower timing belt covers.
9. Remove the crankshaft sprocket flange.
10. Rotate the crankshaft clockwise until the no. 1 cylinder is at TDC on the compression stroke and the crankshaft and camshaft sprockets marks are aligned.
11. Loosen the timing belt tensioner bolt, press the tensioner back and use a screwdriver to move the tensioner back and then tighten the tensioner bolt.
12. Remove the timing belt.
13. Inspect the timing belt for cracks on back surface, sides, bottom and check for separation. Check the tensioner pulley for smooth rotation.

To Install:

14. Ensure the timing marks on the camshaft sprocket and crankshaft sprocket are aligned.
15. Position the timing belt on the crankshaft sprocket, water pump sprocket, camshaft sprocket and the tensioner sprocket.

■ **Keep the tension taut on the timing belt when installing it.**

16. Loosen the tensioner bolt and allow the tensioner to tighten up the timing belt.
17. Rotate the crankshaft sprocket 2 turns clockwise until the no. 1 cylinder is at TDC on the compression stroke.
18. Ensure the timing belt is seated properly and then tighten the tensioner bolt to 18 ft. lbs. (25 Nm).
19. Check that the clearance between the timing belt and the rear cover is 1.18 inches (30mm).

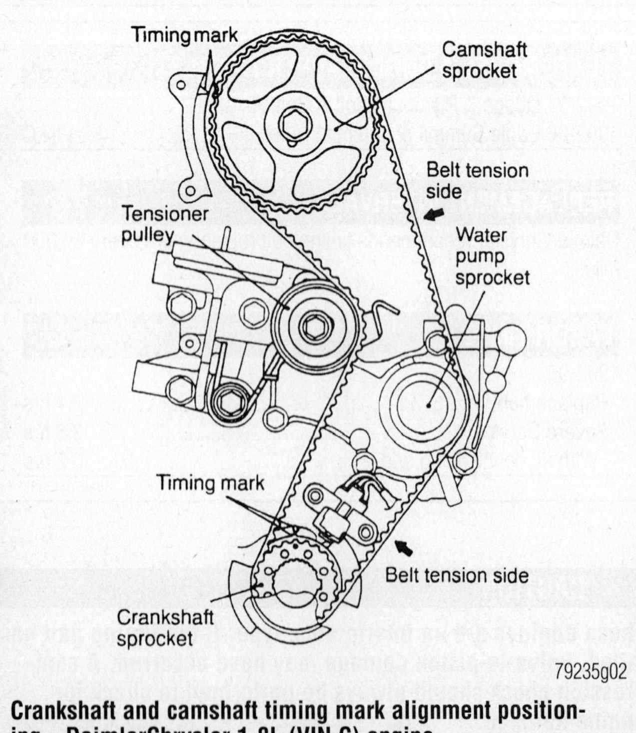

Crankshaft and camshaft timing mark alignment positioning—DaimlerChrysler 1.8L (VIN C) engine

20. Install the crankshaft flange.
21. Install the upper and lower timing belt covers.
22. Install the crankshaft pulley and tighten the bolts to 134 ft. lbs. (182 Nm).
23. Install the accessory drive belts.
24. Install the splash shield and the left front tire.
25. Lower the vehicle.
26. Install the A/C and power steering lines to the left engine mount.
27. Connect the negative battery cable.

1.8L (VIN T) ENGINES

YEAR, MODEL & ENGINE APPLICATION

1989-90 Dodge Colt Wagon	1.8L VIN T

REPLACEMENT INTERVALS

DaimlerChrysler recommends timing belt replacement every 60,000 miles.

LABOR TIMES

1989-90	
Replacement	3.1 hrs
Severe Service	3.6 hrs
With air conditioning add	0.2 hrs

8807CHRg07

✳✳ CAUTION

These engines are an interference type. If the timing belt has failed, valve-to-piston damage may have occurred. A compression check should always be performed to check for engine damage.

1. Disconnect the negative battery cable.
2. Raise and support the vehicle.
3. Remove the driver side wheel.
4. Remove the splash shield.
5. Safely support the engine with a jack and block of wood under the oil pan then remove the left engine mount bracket.
6. Remove the spark plug wires and spark plugs.
7. Remove the accessory drive belts.
8. Remove the water pump and crankshaft pulleys.
9. Remove the upper and lower timing belt covers.
10. Turn the crankshaft clockwise to place the no. 1 cylinder on TDC of the compression stroke and ensure the crankshaft and camshaft sprocket timing marks are aligned.
11. Loosen the timing belt tensioner bolts, move the tensioner toward the water pump and retighten the bolts.
12. Mark the timing belt to indicate its direction of rotation for reinstallation purposes.
13. Remove the timing belt.

To Install:

14. Install the timing belt (in the correct direction of rotation) over the crankshaft sprocket, the oil pump sprocket, camshaft sprocket and tensioner pulley.

■ **Be sure to keep tension on the belt as it is being installed.**

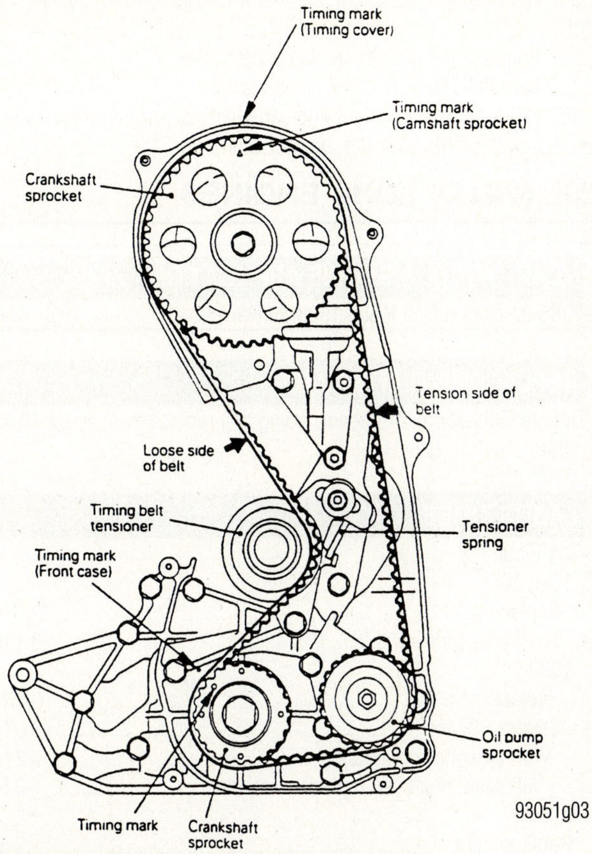

Crankshaft and camshaft sprocket timing marks—DaimlerChrysler 1.8L (VIN T) SOHC engine

15. Loosen the timing belt tensioner bolts and move the tensioner toward the timing belt.

16. To adjust the timing belt tension, perform the following procedure:

a. Apply counterclockwise force on the camshaft sprocket to tighten the belt tension side and ensure that the timing marks remain aligned.

b. Install the crankshaft pulley and rotate the crankshaft (clockwise) until the camshaft sprocket timing mark is 2 teeth away from the alignment mark on the rear cover.

■ **DO NOT rotate the crankshaft counterclockwise or push on the belt to check its tension.**

c. Move the tensioner to apply pressure on the belt and tighten the tensioner nut; then, tighten the pivot bolt.

d. Place your thumb on the belt between the oil pump and camshaft sprockets and press the belt outward. The flex between timing belt and the cover should be 0.23 in., if not, loosen the pivot and tension the belt.

17. Install the crankshaft pulley and tighten the bolts to 12 ft. lbs. (16 Nm).
18. Install the upper and lower timing belt covers.
19. Install the water pump pulley.
20. Install the accessory drive belts.
21. Install the left engine mount bracket.
22. Install the spark plug wires and spark plugs.
23. Install the splash shield.
24. Install the driver side wheel.
25. Connect the negative battery cable.

2.0L (VIN C, F) SOHC ENGINES

YEAR, MODEL & ENGINE APPLICATION

1995-99 Dodge Stratus, 1995-00 Dodge & Plymouth Neon,

1996-00 Plymouth Breeze	2.0L VIN C
2001-05 Dodge Neon	2.0L VIN F

REPLACEMENT INTERVALS

DaimlerChrysler recommends timing belt replacement every 105,000 miles.

LABOR TIMES

Breeze & Stratus	
Replacement	2.5 hrs
Severe Service	2.7 hrs
1995-99 Neon	
Replacement	2.3 hrs
Severe Service	2.6 hrs
2000-05 Neon	
Replacement	2.5 hrs
Severe Service	2.6 hrs

8807CHRg08

✳✳ CAUTION

These engines are an interference type. If the timing belt has failed, valve-to-piston damage may have occurred. A compression check should always be performed to check for engine damage.

■ **There are 2 versions of this engine. One has a mechanical timing belt tensioner and the other has a hydraulic tensioner.**

1. On Breeze, Stratus and Sebring Convertible models, disconnect the negative battery cable from the left strut tower. The ground cable is equipped with an insulator grommet which should be placed on the stud to prevent the negative battery cable from accidentally grounding.
2. On Neon models, disconnect the negative battery cable.
3. On all models, raise and support the vehicle.
4. Remove the drive belts and accessories.
5. Remove the right inner splash-shield.
6. Remove the right engine mount.
7. Remove the crankshaft damper.
8. Place a support under the engine.
9. Remove the engine mount bracket
10. Remove the timing belt cover.
11. Place the no. 1 cylinder at TDC on the compression with the crankshaft sprocket and TDC marks aligned.
12. On hydraulic tensioner models, loosen the timing belt tensioner bolts.
13. Remove the timing belt and the tensioner.

■ **When tensioner is removed from the engine it is necessary to compress the plunger into the tensioner body.**

14. Place the tensioner into a soft-jawed vise to compress the tensioner.

15. After compressing the tensioner place a pin (5/64 in. Allen wrench will work) into the plunger side hole to retain the plunger until installation.

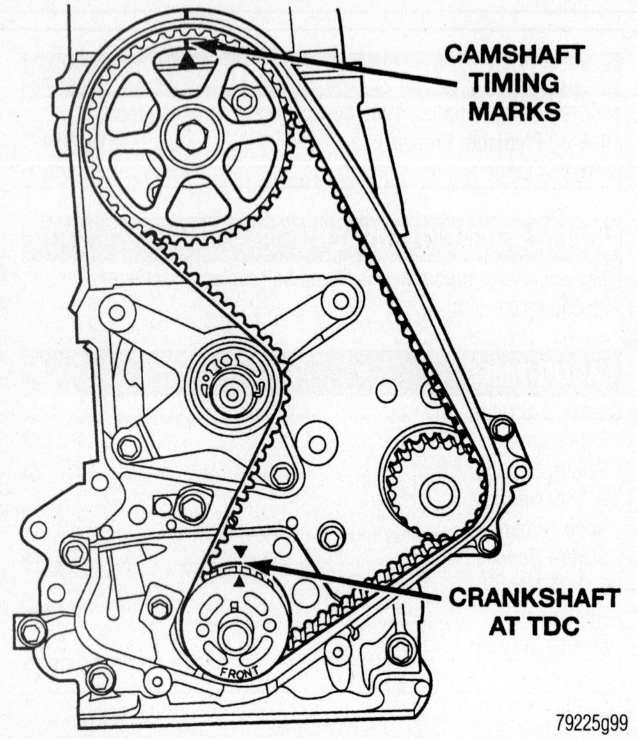

TDC alignment for timing belt installation—DaimlerChrysler 2.0L (VIN C) SOHC engine (VIN F similar)

16. On mechanical tensioner models, insert a 3 mm Allen wrench into the tensioner pulley hole and then use an 8mm Allen wrench to turn the tensioner counterclockwise until it locks in position. Remove the timing belt.

To Install:

17. Set the crankshaft sprocket to Top Dead Center (TDC) by aligning the notch on the sprocket with the arrow on the oil pump housing, then back off the sprocket 3 notches before TDC.
18. Set the camshaft to align the timing marks.
19. Move the crankshaft to 1/2 notch before TDC.
20. Install the timing belt starting at the crankshaft, around the water pump, then around the camshaft last.
21. Move the crankshaft to TDC to take up the belt slack.
22. On hydraulic tensioners, install the tensioner to the block but do not tighten it.
23. Using a torque wrench apply 21 ft. lbs. (28 Nm) of torque to the tensioner pulley.
24. With torque being applied to the tensioner pulley, move the tensioner up against the tensioner bracket and tighten the fasteners to 23 ft. lbs. (31 Nm).
25. Remove the tensioner plunger pin, the tension is correct when the plunger pin can be removed and replaced easily.
26. Rotate the crankshaft 2 revolutions and recheck the timing marks.
27. On mechanical tensioner models, remove the small Allen wrench and allow the tensioner to rotate into position.
28. Rotate the crankshaft 2 revolutions and recheck the timing marks.
29. Reinstall the timing belt cover.
30. Reinstall the engine mount bracket.
31. Reinstall the right engine mount.
32. Remove the engine support.
33. Reinstall the crankshaft damper and tighten to 105 ft. lbs. (142 Nm).
34. Reinstall the drive belts and accessories.
35. Reinstall the right inner splash-shield.
36. Perform the crankshaft and camshaft relearn alignment procedure using the DRB scan tool or equivalent.

2.0L (VIN D) SOHC ENGINES

YEAR, MODEL & ENGINE APPLICATION	
1985-91 Dodge Colt Vista, 1988-89 Ram-50	2.0L VIN D

REPLACEMENT INTERVALS

DaimlerChrysler recommends timing belt replacement every 60,000 miles.

LABOR TIMES	
Colt Vista	
1985-89	
Replacement	2.6 hrs
Severe Service	3.1 hrs
1990-91	
Replacement	3.4 hrs
Severe Service	3.9 hrs
With air conditioning add	0.2 hrs
With silent shaft belt replacement add	1.3 hrs
Ram-50	
Replacement	1.9 hrs
Severe Service	2.4 hrs
With air conditioning add	0.7 hrs
With power steering add	0.1 hrs
With silent shaft belt replacement add	2.2 hrs

8807CHRq09

✳✳ CAUTION

These engines are an interference type. If the timing belt has failed, valve-to-piston damage may have occurred. A compression check should always be performed to check for engine damage.

1. Be sure that the engine's No. 1 piston is at Top Dead Center (TDC) of the compression stroke.
2. Disconnect the negative battery cable.
3. Remove the spark plug wires from the tree on the upper cover.
4. Drain the cooling system.
5. Remove the shroud, fan and accessory drive belts.
6. Remove the radiator as required.
7. Remove the power steering pump, alternator, air conditioning compressor, tension pulley and accompanying brackets, as required.
8. Remove the upper front timing belt cover.
9. Remove the water pump pulley and the crankshaft pulley(s).
10. Remove the lower timing belt cover mounting screws and remove the cover.
11. Ensure the crankshaft sprocket, camshaft sprocket and oil pump sprocket timing marks are aligned.
12. If the belt(s) are to be reused, mark the direction of rotation on the belt.
13. Remove the timing (outer) belt tensioner and remove the belt. Unbolt the tensioner from the block and remove.

14. Remove the outer crankshaft sprocket and flange.
15. Remove the silent shaft (inner) belt tensioner and remove the inner belt. Unbolt the tensioner from the block and remove it.
16. To remove the camshaft sprockets, use SST MB990767-01 and MIT308239 or their equivalents.

To Install:

17. Install the camshaft sprockets and tighten the center bolt to 65 ft. lbs. (90 Nm).
18. Align the timing mark of the silent shaft belt sprockets on the crankshaft and silent shaft with the marks on the front case. Wrap the silent shaft belt around the sprockets so there is no slack in the upper span of the belt and the timing marks are still aligned.
19. Install the tensioner initially so the actual center of the pulley is above and to the left of the installation bolt.

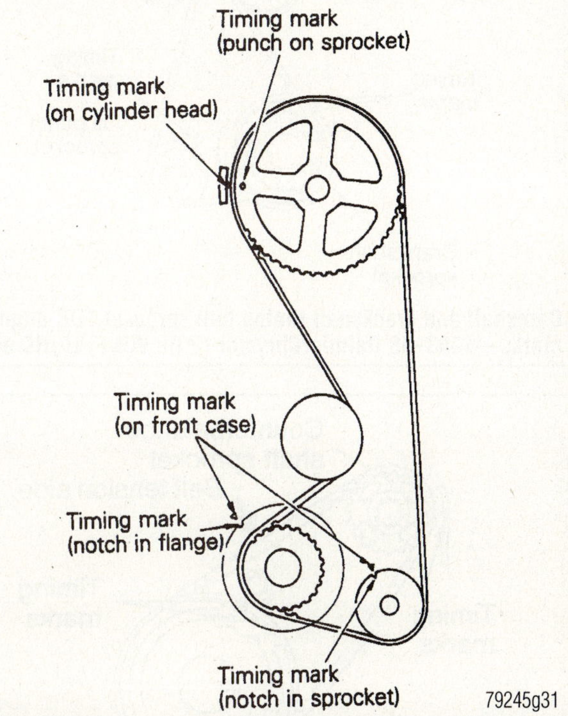

Timing belt pulley alignment marks—DaimlerChrysler 2.0L (VIN D) engine

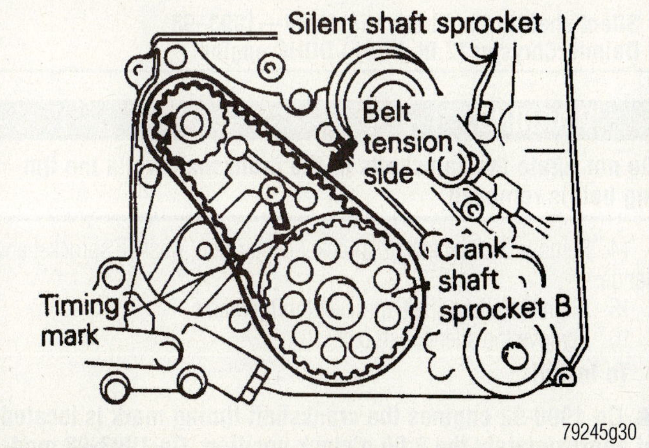

Silent shaft alignment marks. Notice the tension side of the inner (silent shaft) belt—DaimlerChrysler 2.0L (VIN D) engine

20. Move the pulley up by hand so the center span of the long side of the belt deflects about 1/4 in. (6mm).
21. Hold the pulley tightly so it does not rotate when the bolt is tightened. Tighten the bolt to 15 ft. lbs. (20 Nm). If the pulley has moved, the belt will be too tight.
22. Install the timing belt tensioner fully toward the water pump and temporarily tighten the bolts. Place the upper end of the spring against the water pump body. Align the timing marks of the camshaft, crankshaft and oil pump sprockets with the corresponding marks on the front case or head.

■ **If the following steps are not followed exactly, there is a chance that the silent shaft alignment will be 180 degrees off. This will cause a noticeable vibration in the engine and the entire procedure will have to be repeated.**

23. Before installing the timing belt, ensure that the left-side silent shaft is in the correct position.

■ **It is possible to align the timing marks on the camshaft sprocket, crankshaft sprocket and the oil pump sprocket with the left balance shaft out of alignment.**

24. With the timing mark on the oil pump pulley aligned with the mark on the front case, check the alignment of the left balance shaft to assure correct shaft timing.
25. Remove the plug located on the left-side of the block in the area of the starter.
26. Insert a tool having a shaft diameter of 0.3 in. (8mm) into the hole.
27. With the timing marks still aligned, the tool must be able to go in at least 2 1/3 in. If it can only go in about 1 in. turn the oil pump sprocket 1 complete revolution.
28. Recheck the position of the balance shaft with the timing marks realigned. Leave the tool in place to hold the silent shaft while continuing.
29. Install the belt to the crankshaft sprocket, oil pump sprocket and the camshaft sprocket, in that order. While doing so, be sure there is no slack between the sprockets except where the tensioner will take it up when released.
30. Recheck the timing mark alignments.
31. If all are aligned, loosen the tensioner mounting bolt and allow the tensioner to apply tension to the belt.
32. Remove the tool that is holding the silent shaft in place and turn the crankshaft clockwise a distance equal to 2 teeth of the camshaft sprocket. This will allow the tensioner to automatically tension the belt the proper amount.

✳✳ WARNING

Do not manually apply pressure to the tensioner. This will overtighten the belt and will cause a howling noise.

33. First tighten the lower mounting bolt and then tighten the upper spacer bolt.

✳✳ WARNING

If any binding is felt when adjusting the timing belt tension by turning the crankshaft, STOP turning the engine, because the pistons may be hitting the valves.

34. To verify that belt tension is correct, check that the deflection of the longest span (between the camshaft and oil pump sprockets) is 1/2 in. (13mm).
35. Install the lower timing belt cover. Be sure the packing is properly positioned in the inner grooves of the covers when installing.

36. Install the water pump pulley and the crankshaft pulley(s).
37. Install the upper front timing belt cover.
38. Install the power steering pump, alternator, air conditioning compressor, tension pulley and accompanying brackets, as required.
39. Install the radiator, shroud, fan and accessory drive belts.
40. Install the spark plug wires to the tree on the upper cover.
41. Refill the cooling system.
42. Connect the negative battery cable. Start the engine and check for leaks.

2.0L (VIN F) DOHC ENGINES

YEAR, MODEL & ENGINE APPLICATION	
1990-98 Eagle Talon, 1990-94 Plymouth Laser	2.0L VIN F

REPLACEMENT INTERVALS

DaimlerChrysler recommends timing belt replacement every 105,000 miles, or 84 months.

LABOR TIMES	
1990-94	
Replacement	3.2 hrs
Severe Service	3.6 hrs
With air conditioning add	0.3 hrs
With power steering add	0.3 hrs
1995-98	
Replacement	2.7 hrs
Severe Service	3.2 hrs
With air conditioning add	0.3 hrs
With power steering add	0.3 hrs

8807CHRg10

✱✱ CAUTION

These engines are an interference type. If the timing belt has failed, valve-to-piston damage may have occurred. A compression check should always be performed to check for engine damage.

1. Disconnect the negative battery cable.
2. Disconnect the A/C and power steering lines clamp from the body.
3. Raise and support the vehicle.
4. Remove the right side inner fender liner.
5. Using a suitable engine support brace, remove the stud bolt from the engine support bracket and remove the bracket and mount.
6. Remove the drive belts.
7. Remove the belt tensioner pulley.
8. Remove the water pump pulley.
9. Remove the crankshaft pulley.
10. Remove the timing belt covers.
11. Rotate the crankshaft in the direction of rotation to align the camshaft timing marks. Always turn the crankshaft in the normal direction of rotation only.
12. Remove the timing belt tensioner.

■ **If the timing belt is to be reused, mark the direction of rotation on the flat side of the belt with an arrow.**

13. Remove the timing belt.

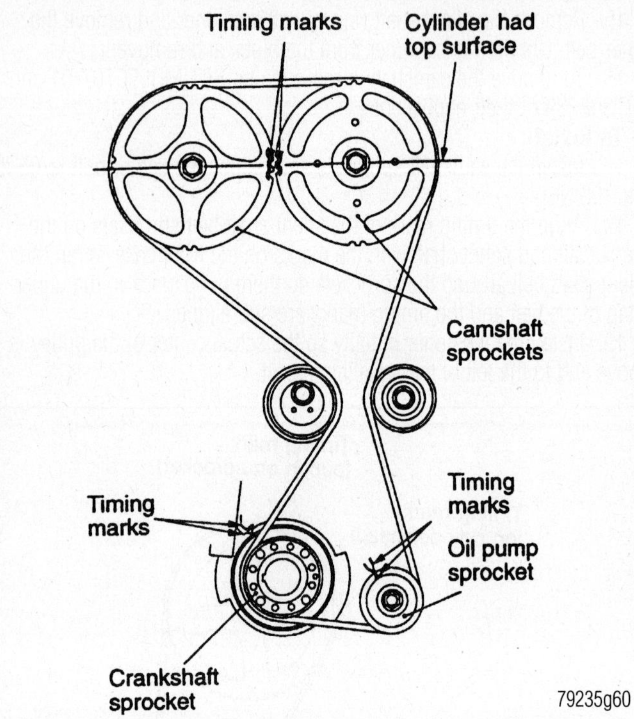

79235g60

Camshaft and crankshaft timing belt sprocket TDC alignment marks—1993–98 DaimlerChrysler (2.0L VIN F) DOHC engine

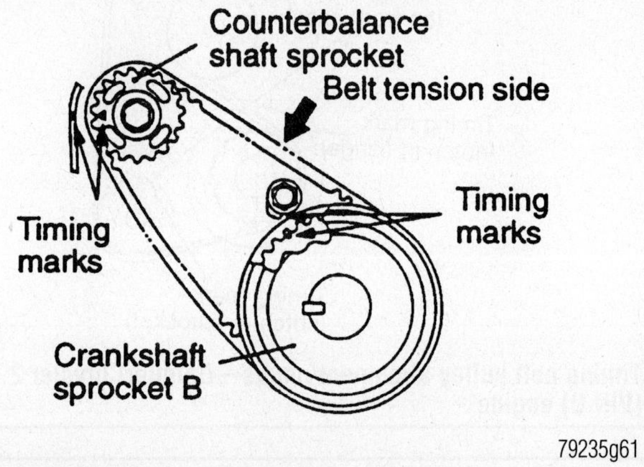

79235g61

Silent shaft timing mark alignment—1993–98 DaimlerChrysler (2.0L VIN F) DOHC engine

✱✱ WARNING

Do not rotate the camshafts or the crankshaft while the timing belt is removed.

14. Remove the crankshaft sprocket center bolt, washer, sprocket and flange.
15. Align the silent shaft sprocket timing marks.
16. Remove the silent shaft belt.

To Install:

■ **On 1990-92 engines the crankshaft timing mark is located at approximately the 2:00 o'clock position. On 1993-98 models the crankshaft timing mark is located at approximately the 10:00 o'clock position.**

17. Align the timing mark of the silent shaft belt sprockets on the crankshaft and silent shaft with the marks on the front case. Wrap the silent shaft belt around the sprockets so there is no slack in the upper span of the belt and the timing marks are still aligned.

18. Install the tensioner initially so the actual center of the pulley is above and to the left of the installation bolt.

19. Move the pulley up by hand so the center span of the long side of the belt deflects about 1/4 in. (6mm).

20. Hold the pulley tightly so it does not rotate when the bolt is tightened. Tighten the bolt to 15 ft. lbs. (20 Nm). If the pulley has moved, the belt will be too tight.

21. Install the crankshaft flange, sprocket and washer and tighten to 87 ft. lbs. (118 Nm).

22. Install the automatic tensioner, after reset.

■ **To reset the tensioner: Keep the adjuster level and clamp it in a soft jawed vise. Clamp with the extended adjuster on one side and the end mounting a plug on the other side. If the plug extends out of the adjuster body, place a suitable hole sized washer over the plug so the vise jaw pushes on the washer, not the plug. Close the vise slowly, forcing the adjuster back into the body. When the hole in the adjuster boss aligns with the adjuster rod, insert a snug fitting pin or wire into the holes to keep the rod in the compressed position. With the locking pin or wire in place, install the tensioner.**

23. Install the tensioner pulley with the hole in the shaft to the left of the center bolt, then tighten the center bolt hand tight.

24. When installing the timing belt, turn the 2 camshaft sprockets so their dowel pins are located on top. Align the timing marks facing each other with the top surface of the cylinder head. When you let go of the exhaust camshaft sprocket, it will rotate 1 tooth in the counterclockwise direction. This should be taken into account when installing the timing belts on the sprocket.

■ **Both camshaft sprockets are used for the intake and exhaust camshafts and are provided with 2 timing marks. When the sprocket is mounted on the exhaust camshaft, use the timing mark on the right with the dowel pinhole on top. For the intake camshaft sprocket, use the mark on the left with the dowel pinhole on top.**

25. Align the crankshaft sprocket and oil pump sprocket timing marks.

■ **On 1990-92 engines the crankshaft timing mark is located at approximately the 2:00 o'clock position. On 1993-98 models the crankshaft timing mark is located at approximately the 10:00 o'clock position.**

■ **If the following steps are not followed exactly, there is a chance that the silent shaft alignment will be 180 degrees off. This will cause a noticeable vibration in the engine and the entire procedure will have to be repeated.**

26. Before installing the timing belt, ensure that the left-side silent shaft is in the correct position.

■ **It is possible to align the timing marks on the camshaft sprocket, crankshaft sprocket and the oil pump sprocket with the left balance shaft out of alignment.**

27. With the timing mark on the oil pump pulley aligned with the mark on the front case, check the alignment of the left balance shaft to assure correct shaft timing.

28. Remove the plug located on the left-side of the block in the area of the starter.

29. Insert a tool having a shaft diameter of 0.3 in. (8mm) into the hole.

30. With the timing marks still aligned, the tool must be able to go in at least 2 1/3 in. If it can only go in about 1 in. turn the oil pump sprocket 1 complete revolution.

31. Recheck the position of the balance shaft with the timing marks realigned. Leave the tool in place to hold the silent shaft while continuing.

32. Ensure the camshaft sprockets timing marks are opposite each other and in the center position aligned with the top surface of the cylinder head.

33. Ensure the crankshaft and oil pump sprocket timing marks are aligned.

34. Install the timing belt on the intake camshaft. Hold the belt in place with a binder clip.

35. Install the timing belt on the exhaust camshaft. Hold the belt in place with a binder clip

36. Install the belt to the idler pulley, oil pump sprocket, crankshaft sprocket, oil pump sprocket and the tensioner.

37. Remove the binder clips.

38. Remove the screwdriver from the oil pump and install the plug.

39. Raise the tensioner pulley clockwise against the timing belt and temporarily tighten the bolt.

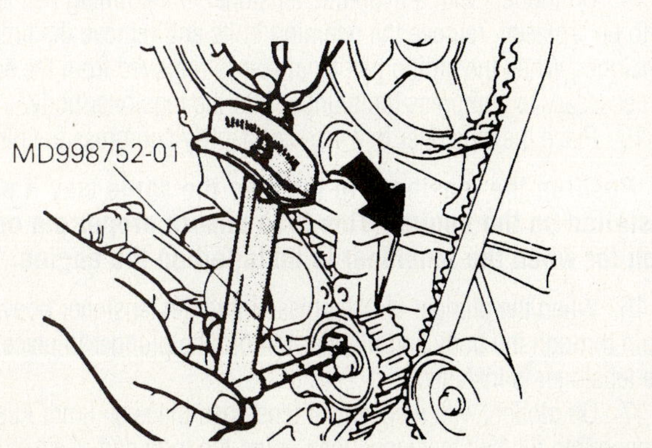

9305bg39

Applying torque to the tensioner pulley—DaimlerChrysler 1990–92 2.0L (VIN F) DOHC engine

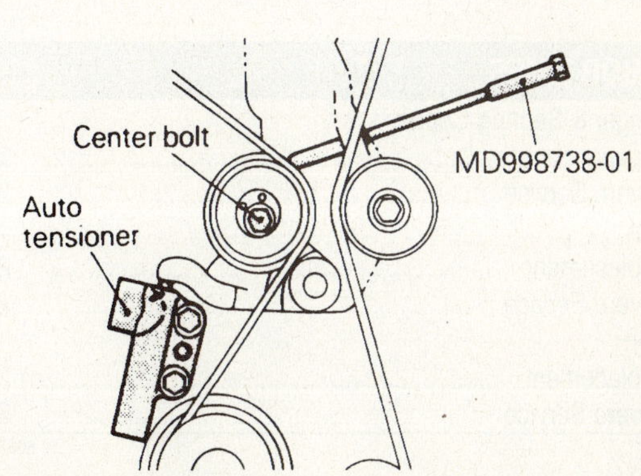

9305bg40

Inserting special tool to contact tensioner arm—DaimlerChrysler 1993–98 2.0L (VIN F) DOHC engine

40. Remove the rubber plug from the rear timing belt cover and screw special tool MD99873801 or equivalent, into the engine left support bracket until its end makes contact with the tensioner arm. At this point, screw the special tool in some more and remove the set wire attached to the auto tensioner. Then remove the special tool.
41. Tighten the tensioner pulley bolt.
42. Rotate the crankshaft 1/4 turn counterclockwise. Then, turn clockwise until the timing marks are aligned again.
43. On 1990–92 models, loosen the tensioner center bolt. Using tool MD998752 or equivalent and a torque wrench, apply a torque of 22.5–24.3 inch lbs. (2.6–2.8 Nm) to the tensioner pulley.
44. On 1993-98 models, loosen the center bolt. Install Special Tool MD998767 or equivalent, on the tensioner pulley. Turn the tensioner pulley counterclockwise with a torque of 31 inch lbs. (3.5 Nm).
45. On all models, torque the center bolt to 35 ft. lbs. (48 Nm). Do not let the tensioner pulley turn when tightening the bolt.
46. Rotate the crankshaft 2 complete turns clockwise and let it sit for approximately 15 minutes. Then, measure the auto tensioner protrusion (the distance between the tensioner arm and auto tensioner body) to ensure that it is within 0.15–0.18 in. (3.8–4.5mm).
47. Install the rubber plug to the timing belt rear cover.
48. Install the upper and lower timing belt covers and torque the bolts to 84–108 inch lbs. (9–12 Nm).
49. Install the water pump pulley and torque the bolts to 84 inch lbs. (10 Nm).
50. Install the crankshaft pulley and torque the bolts to 22 ft. lbs. (30 Nm).
51. Install the mounting brackets and the accessory drive belts.
52. Connect the A/C and power steering line clamps.
53. Install the right side inner fender liner.
54. Connect the negative battery cable.

2.0L (VIN Y) DOHC ENGINES

YEAR, MODEL & ENGINE APPLICATION

1995-99 Chrysler Sebring Coupe, 1995-98 Eagle Talon
1995-99 Dodge Avenger,
1995-99 Dodge & Plymouth Neon .... 2.0L VIN Y

REPLACEMENT INTERVALS

DaimlerChrysler recommends timing belt replacement every 105,000 miles.

LABOR TIMES

Avenger & Sebring Copue	
Replacement	2.5 hrs
Severe Service	2.7 hrs
Neon	
Replacement	2.3 hrs
Severe Service	2.6 hrs
Talon	
Replacement	2.7 hrs
Severe Service	3.1 hrs

8807CHRg11

✱✱ CAUTION

These engines are an interference type. If the timing belt has failed, valve-to-piston damage may have occurred. A compression check should always be performed to check for engine damage.

■ **There are 2 versions of this engine. One has a mechanical timing belt tensioner and the other has a hydraulic tensioner.**

Valve timing is critical to engine operation. Use care when servicing the timing belt. There are a number of timing marks that must be properly aligned or engine damage will result. If the timing belt has not broken, or jumped teeth, it is recommended that the crankshaft be turned by hand (clockwise) to Top Dead Center (TDC) of the No. 1 cylinder compression stroke (firing position) before beginning work. This should align all the timing marks and serve as a reference for later work. Some technicians will apply a small amount of white paint to all timing marks. This helps make them more visible under the low-light conditions found underhood.

1. Disconnect the negative battery cable.
2. Raise and support the vehicle.
3. Remove the right side inner splash shield or engine undercover.
4. Remove the accessory drive belts.
5. Using Crankshaft Holding tool C-3281 or equivalent, remove the crankshaft pulley center retaining bolt.
6. Using puller tools 1026 and 6827 or equivalent, remove the crankshaft pulley.
7. Remove the power steering pump from the bracket and position it aside. Do not disconnect the hoses.
8. Remove the power steering pump bracket from the engine.
9. Use a floor jack with a piece of wood on it and raise the engine to take the weight off of the engine mount.
10. Remove the engine mount and bracket.
11. Remove the front timing belt cover.
12. Align the crankshaft and camshaft sprocket timing marks.
13. Loosen the timing belt tensioner and remove the belt as follows:.

✱✱ WARNING

Do not rotate the crankshaft or camshafts after removing the timing belt or valvetrain components may be damaged. Always align the timing marks before removing the timing belt.

14. On models with a hydraulic tensioner, if the timing belt tensioner is to be replaced, remove the retaining bolts and remove the timing belt tensioner. When the timing belt tensioner is removed from the engine it is necessary to compress the plunger into the tensioner body.
15. Place the tensioner in a vise and slowly compress the plunger.

■ **Position the tensioner in the vise the same way it will be installed on the engine. This is to ensure proper pin orientation for when the tensioner is installed on the engine.**

16. When the plunger is compressed into the tensioner body, install a pin through the body and plunger to hold the plunger in place until the tensioner is installed.
17. On models with mechanical tensioner, place an 8mm Allen wrench into the belt tensioner, then using the long end of a 1/8 in. (3mm) Allen wrench, rotate the tensioner counterclockwise until it slides into the locking hole.
18. Remove the timing belt.

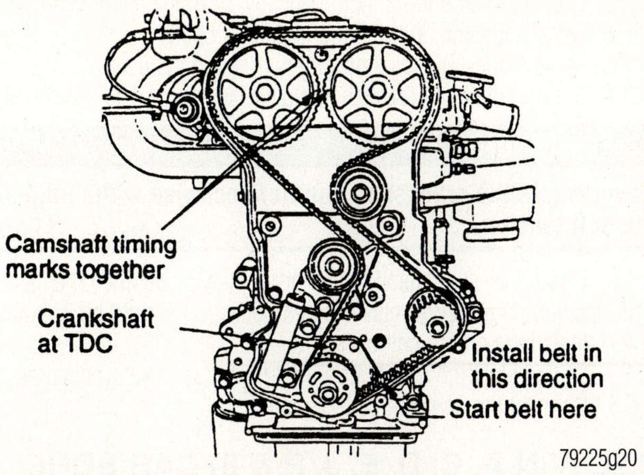

Crankshaft and camshaft timing marks—DaimlerChrysler 2.0L (VIN Y) engine with hydraulic tensioner

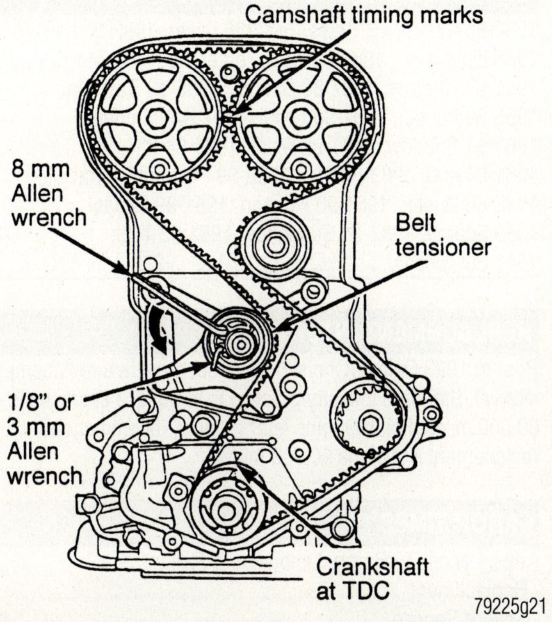

Crankshaft and camshaft timing marks —DaimlerChrysler 2.0L (VIN Y) engine with mechanical tensioner

To Install:

19. Check that all timing marks are still aligned. Bring the crankshaft sprocket to 1/2 a notch before TDC.
20. Install the timing belt. Starting at the crankshaft, route the belt around the water pump sprocket, idler pulley, camshaft sprockets, then around the tensioner pulley.
21. Move the crankshaft to TDC to take up the slack in the belt. Install the tensioner to the block but do not tighten the retaining bolts.
22. On models with a hydraulic tensioner, using a torque wrench on the tensioner pulley, apply 21 ft. lbs. (28 Nm) of torque to the pulley.
23. With the torque being applied to the tensioner pulley, move the tensioner up against the tensioner pulley bracket and tighten the retaining bolts to 23 ft. lbs. (31 Nm).
24. Remove the tensioner plunger pin. Pretension is correct when the pin can be removed and installed.
25. On moels with mechanical tensioner, remove the wrenches from the tensioner.
26. Rotate the crankshaft 2 revolutions and check the timing marks. If the timing marks are not properly aligned, remove the belt and reinstall it as described.
27. Install the front timing belt cover.
28. Lower the engine enough to install the engine mount bracket.
29. Install the bracket and remove the floor jack.
30. Install the power steering pump bracket and pump.
31. Install the crankshaft pulley using C-4685-C or an equivalent pulley installer.
32. Tighten the mounting bolt to 105 ft. lbs. (142 Nm).
33. Install the right side inner splash shield or engine undercover.
34. Install the accessory drive belts.
35. Connect the negative battery cable.
36. Start the engine. Check for leaks and proper engine operation.

2.1L (VIN B) TURBO DIESEL ENGINES

YEAR, MODEL & ENGINE APPLICATION	
1985-87 Jeep Cherokee, Comanche & Wagoneer	2.1L VIN B

REPLACEMENT INTERVALS

DaimlerChrysler does not provide a replacement interval. Standard industry practice is to replace timing belts every 60,000 miles.

LABOR TIMES	
1985-87	
Replacement	3.5 hrs
Severe Service	3.8 hrs
With tensioner replacement add	0.1

8807CHRg12

1. Disconnect the negative battery cable.
2. Rotate the crankshaft to position the No. 1 cylinder on the Top Dead Center (TDC) of the compression stroke.
3. Remove all of the necessary components in order to gain access to the timing belt cover bolts.
4. Remove the timing belt cover retaining bolts and the cover.
5. Install the Camshaft Sprocket Holding tool MOT-854 or equivalent, to retain the camshaft and injection pump sprockets.
6. Loosen the timing belt tensioner bolts, move the tensioner away from the timing belt and secure the bolts.
7. If reusing the timing belt, mark its direction of rotation.
8. Remove the timing belt.

To Install:

9. Remove the access plug on the left side of the block and install the Crankshaft Holding tool MOT-861 in the hole.
10. Slowly, rotate the crankshaft clockwise until the tool drops into the TDC locating slot into the crankshaft's counterweight.

■ **Do not use this tool as a crankshaft holding tool when tightening or loosening gear train fasteners, use a Flywheel Holding tool MOT-582 or equivalent.**

11. If removed, install the Camshaft Sprocket Holding tool MOT-854 or equivalent, to retain the camshaft and injection pump sprockets. Make sure that the timing marks are positioned correctly.
12. Noting the direction of rotation, install the timing belt. There should be a total of 19 belt teeth between the camshaft and injection pump timing marks.

13. Temporarily position the timing cover over the sprockets. The camshaft and injection pump timing marks must index with the pointer in the cover's timing slots.
14. Remove the cover.
15. Remove the Camshaft Holding tool MOT-854 or equivalent.
16. Loosen the timing belt tensioner bolts 1/2 turn, maximum.

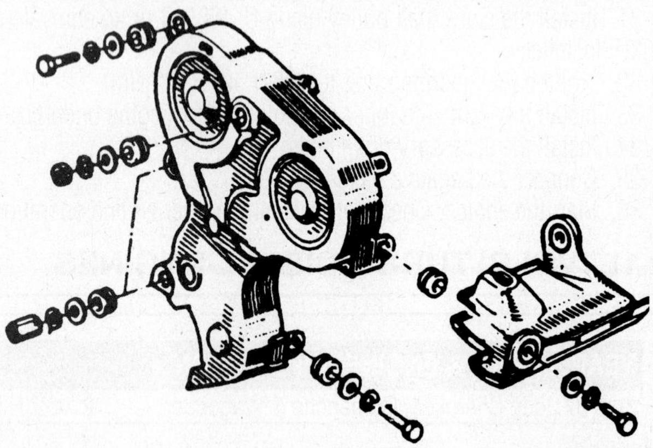

93051g15

Exploded view of the timing belt cover—DaimlerChrysler 2.1L (VIN B) turbo diesel engine

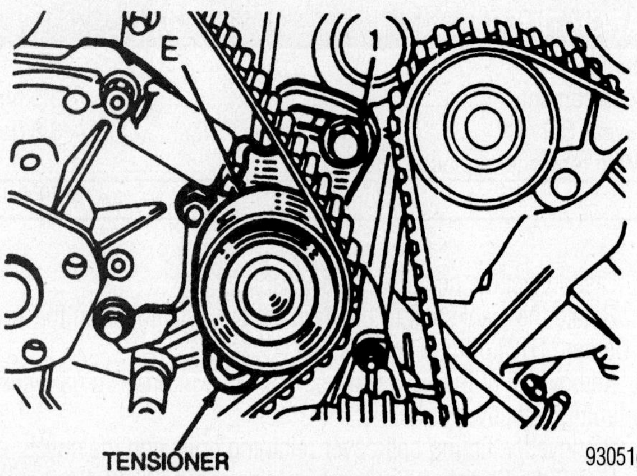

93051g16

View of the timing belt tensioner bolts— DaimlerChrysler 2.1L (VIN B) turbo diesel engine

93051g17

Camshaft and injection pump timing marks aligned—DaimlerChrysler 2.1L (VIN B) turbo diesel engine

17. The tensioner should, automatically, bear against the belt, giving the proper belt tension. Tighten the tensioner bolts.
18. Remove the TDC locating tool and install the plug.
19. Slowly, rotate the crankshaft clockwise 2 complete revolutions.

✳✳ WARNING

Never rotate the crankshaft counterclockwise while adjusting the belt tension.

20. Check the belt deflection at a point midway between the camshaft and injection pump sprockets. The belt deflection should be 3–5mm.
21. Install the timing belt cover.
22. To complete the installation, reverse the remove procedures.
23. Connect the negative battery cable.

2.2L (VIN A, C, D, E, J, P & 8) CAR SOHC ENGINES

YEAR, MODEL & ENGINE APPLICATION

1982-85 Chrysler E Class, 1982-86 Laser, 1982-90 LeBaron, 1985 Limousine, 1982-88 New Yorker, 1981-89 Dodge Aries, 1981-87 Charger, 1987-88 Dakota, 1985-93 Daytona, 1985-89 Lancer, 1981-90 Omni, 1981-84 Rampage, 1987-94 Shadow, 1991-92 Spirit, 1981-84 024, 1981-84 400, 1983-88 600, 1983-89 Plymouth Caravelle, 1985-86 Duster, 1981-90 Horizon, 1981-89 Reliant, 1983 Scamp, 1987-94 Sundance, 1981-82 TC3, 1981-86 Turismo, ... 2.2L VIN A, C, E, J, P & 8

REPLACEMENT INTERVALS

Prior to 1993 DaimlerChrysler does not provide a replacement interval. Standard industry practice is to replace timing belts every 60,000 miles. For 1993 and later models, recommended replacement interval is 90,000 miles.

LABOR TIMES

Aries, Lancer, LeBaron, Limousine & Reliant	
Replacement	2.5 hrs
Severe Service	3.0 hrs
With air conditioning add	0.3 hrs
With power steering add	0.3 hrs
Caravelle, Daytona, Daytona Shelby, Dynasty, Laser, New Yorker & 600	
Replacement	2.5 hrs
Severe Service	2.8 hrs
With air conditioning add	0.3 hrs
With power steering add	0.3 hrs
Charger, Horizon, Omni, Rampage, Scamp, TC3, Turismo & 024	
Replacement	2.6 hrs
Severe Service	2.9 hrs
With air conditioning add	0.3 hrs
With power steering add	0.3 hrs
With air injection add	0.3 hrs
Shadow & Sundance	
Replacement	2.5 hrs
Severe Service	2.9 hrs
With air conditioning add	0.3 hrs
With power steering add	0.3 hrs

8807CHRg13

✻✻ CAUTION

These engines are considered freewheeling. If the timing belt fails, valve-to-piston damage is unlikely, however a precautionary compression check should be performed to verify no engine damage.

1. Disconnect the negative battery cable.
2. Position the engine so the No. 1 piston is at Top Dead Center (TDC) of the compression stroke.
3. Remove the nuts and bolts that attach the upper cover to the valve cover, block or cylinder head.
4. Remove the bolt that attaches the upper cover to the lower cover.
5. Remove the upper cover.
6. Remove the right tire and wheel assembly. Remove the right-side inner splash shield.
7. Remove the crankshaft pulley, water pump pulley and the accessory drive belt(s).
8. Remove the lower cover attaching bolts and the cover from the engine.
9. Remove the timing belt tensioner and allow the belt to hang free.
10. Place a floor jack under the engine and separate the right motor mount.
11. Remove the air conditioning compressor belt idler pulley, if equipped, and remove the mounting stud. Remove the compressor/alternator bracket as follows:
    a. Remove the alternator pivot bolt and remove the alternator from the bracket. Turn the alternator so the wire connections are facing up and disconnect the harness connectors from the rear of the alternator.
    b. Remove the air conditioning compressor belt idler.
    c. Remove the right engine mount yoke screw securing engine mount support strut to the engine.
    d. Remove the 5 side mounting bolts retaining the bracket to the front of the engine.
    e. Remove the front mounting nut. Remove the front bolt and strut and rotate the solid mount bracket away from the engine. Slide the bracket on the stud until free of the mounting studs and remove from the engine.
    f. Remove the timing belt from the vehicle.

To Install:

12. Turn the crankshaft sprocket and intermediate shaft sprocket until the marks are aligned. Use a straight-edge from bolt to bolt to confirm alignment.
13. Turn the camshaft until the small hole in the sprocket is at the top and the arrows on the hub are aligned with the camshaft cap to cylinder head mounting lines. When looking through the hole on top of the camshaft sprocket, the uppermost center nipple of the valve cover end seal should be at the center of the hole. Use a mirror to check the alignment of the arrows so it is viewed straight on and not at an angle from above. Install the belt but let it hang free at this point.
14. Install the air conditioning compressor/alternator bracket, idler pulley and motor mount. Remove the floor jack. Raise the vehicle and support safely. Have the tensioner at an arm's reach because the timing belt will have to be held in position with one hand.
15. To properly install the timing belt, reach up and engage it with the camshaft sprocket. Turn the intermediate shaft counterclockwise slightly, then engage the belt with the intermediate shaft sprocket. Hold the belt against the intermediate shaft sprocket and turn clockwise to take up all tension; if the timing marks are out of alignment, repeat until alignment is correct.

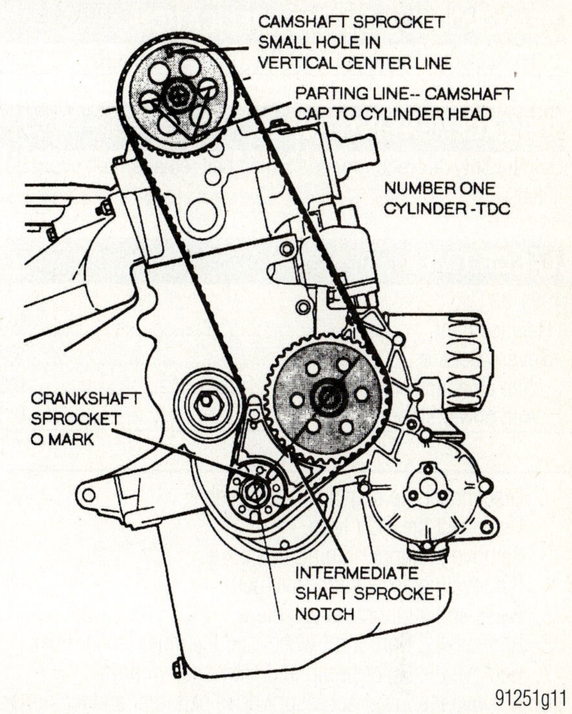

Timing belt alignment mark locations—DaimlerChrysler 2.2L SOHC engine

16. Using a wrench, turn the crankshaft sprocket counterclockwise slightly and wrap the belt around it. Turn the sprocket clockwise so there is no slack in the belt between sprockets; if the timing marks are out of alignment, repeat until alignment is correct.

■ **If the timing marks are aligned but slack exists in the belt between either the camshaft and intermediate shaft sprockets or the intermediate and crankshaft sprockets, the timing will be incorrect when the belt is tensioned. All slack must be only between the crankshaft and camshaft sprockets.**

17. Install the tensioner and install the mounting bolt loosely. Place the special tensioning tool C-4703 on the hex of the tensioner so the weight is at about the 9 o'clock position (parallel to the ground, hanging toward the rear of the vehicle) (15 degrees.
18. Hold the tool in position and tighten the bolt to 45 ft. lbs. (61 Nm). Do not pull the tool past the 9 o'clock position; this will make the belt too tight and will cause it to howl or possibly break.
19. Lower the vehicle and recheck the camshaft sprocket positioning. If it is correct install the timing belt covers and all related parts.
20. Connect the negative battery cable and road test the vehicle.

2.2L (VIN A) DOHC ENGINES

✻✻ CAUTION

These engines are an interference type. If the timing belt has failed, valve-to-piston damage may have occurred. A compression check should always be performed to check for engine damage.

YEAR, MODEL & ENGINE APPLICATION

1991-93 Dodge Daytona IROC R/T, 1991-93 Spirit R/T 2.2L VIN A

REPLACEMENT INTERVALS

DaimlerChrysler recommends timing belt replacement every 90,000 miles.

LABOR TIMES

1991-93	
Replacement	2.9 hrs
Severe Service	3.2 hrs
With air conditioning add	0.3 hrs
With power steering add	0.3 hrs

8807CHRg14

1. Disconnect the negative battery cable.
2. Disconect the PCV tube.
3. Remove the upper timing belt cover.
4. Remove the accessory drive belt.
5. Raise and support the vehicle.
6. Remove the right front wheel and the inner fender liner.
7. Remove the water pump and crankshaft pulleys.
8. Remove the lower accessory drive belt idler and tensioner pulleys.
9. Remove the lower timing belt cover.
10. Support the engine with a suitable jack and remove the right engine mount.
11. Remove the lower accessory drive belt pulley bracket.
12. Loosen the timing belt tensioner and remove the timing belt.
13. Remove the air cleaner duct.
14. Remove the ignition lead cover.
15. Remove both valve covers.
16. Loosen the rocker arm assemblies approximately 3 turns.
17. Align the small holes in the camshaft sprockets so they are opposite each other in the center of the engine and insert pin punches to hold them in position.
18. Remove the spark plugs and insert a dial indicator into the no. 1 spark plug hole.
19. Rotate the crankshaft until no. 1 cylinder is a TDC. Mark the block for reference.

To Install:

20. Install the timing belt at the lower right of the crankshaft sprocket and working counterclockwise, install the belt around the remaining sprockets and pulleys.
21. Remove the dial indicator and the pins from the camshaft sprockets.
22. Use a belt tension gauge between the camshaft pulleys and adjust the timing belt tensioner until the belt tension is 110 ft. lbs.
23. Rotate the crankshaft 2 full turns clockwise and check the alignment of the camshaft and crankshaft.
24. Recheck the belt tension and adjust, if necessary.
25. Tighten the rocker arm shafts in sequence to 105 inch lbs. (12 Nm), then again an additional 105 inch lbs. (12 Nm).
26. Install both valve covers.
27. Install the ignition lead cover.
28. Install the air cleaner duct.

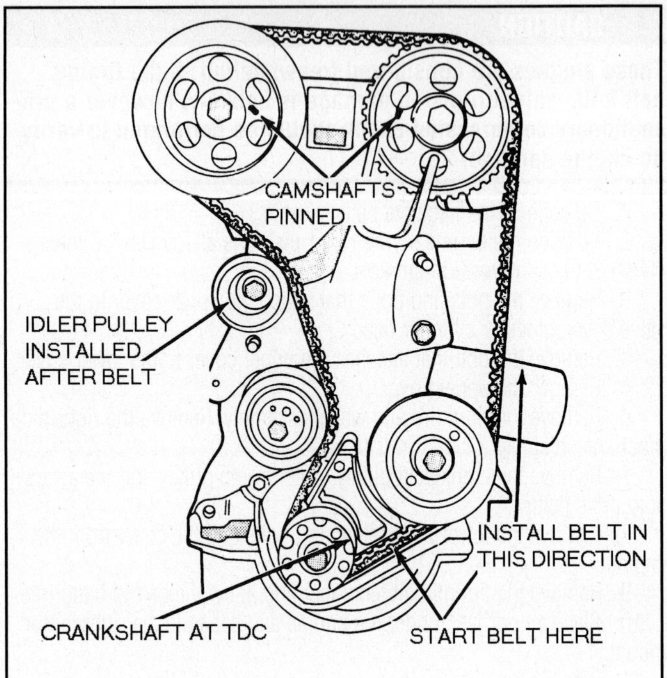

Installing timing belt and sprocket marks—DaimlerChrysler 2.2L DOHC (VIN A) engine

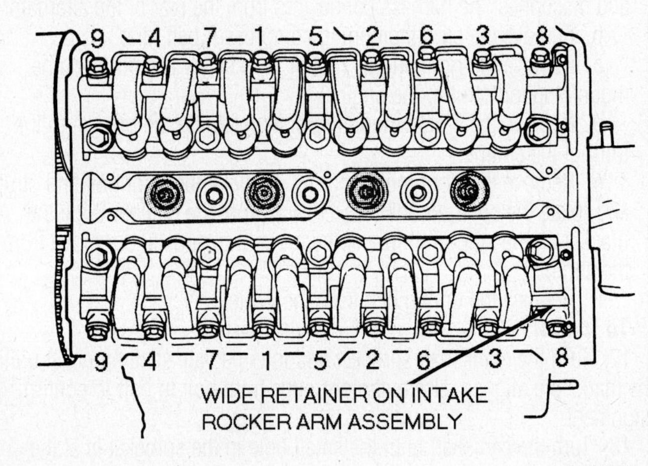

Rocker arm shaft tightening sequence—DaimlerChrysler 2.2L DOHC (VIN A) engine

29. Install the lower accessory drive belt pulley bracket.
30. Install the right engine mount.
31. Install the lower timing belt cover.
32. Install the lower accessory drive belt idler and tensioner pulleys.
33. Install the water pump and crankshaft pulleys.
34. Install the right front wheel and the inner fender liner.
35. Install the accessory drive belt.
36. Install the upper timing belt cover.
37. Connect the PCV tube.
38. Connect the negative battery cable.

2.2L (VIN D) TRUCK SOHC ENGINES

YEAR, MODEL & ENGINE APPLICATION

1985-88 Chrysler Town & Country, 1985-87 Dodge Caravan, Mini Ram Van, 1987-88 Dakota,
1985-87 Plymouth Voyager .... 2.2L VIN D

REPLACEMENT INTERVALS

DaimlerChrysler does not have a recommended replacement interval. Standard industry practice is to replace timing belts every 60,000 miles.

LABOR TIMES

1985-88 Caravan, Mini Ram Van, Town & Country & Voyager	
Replacement	2.3 hrs
Severe Service	2.6 hrs
With air conditioning add	0.2 hrs
With power steering add	0.2 hrs
1987-88 Dakota	
Replacement	1.6 hrs
Severe Service	1.9 hrs
With air conditioning add	0.5 hrs
With power steering add	0.1 hrs

8807CHRg15

✳✳ CAUTION

These engines are considered freewheeling. If the timing belt fails, valve-to-piston damage is unlikely, however a precautionary compression check should be performed to verify no engine damage.

1. Disconnect the negative battery cable.
2. Position the engine so the No. 1 piston is at Top Dead Center (TDC) of the compression stroke.
4. Remove the nuts and bolts that attach the upper cover to the valve cover, block or cylinder head.
5. Remove the bolt that attaches the upper cover to the lower cover.
6. Remove the upper cover.
7. Remove the right tire and wheel assembly. Remove the right-side inner splash shield.
8. Remove the crankshaft pulley, water pump pulley and the accessory drive belt(s).
9. Remove the lower cover attaching bolts and the cover from the engine.
10. Remove the timing belt tensioner and allow the belt to hang free.
11. Place a floor jack under the engine and separate the right motor mount.
12. Remove the air conditioning compressor belt idler pulley, if equipped, and remove the mounting stud. Remove the compressor/alternator bracket as follows:
    a. Remove the alternator pivot bolt and remove the alternator from the bracket. Turn the alternator so the wire connections are facing up and disconnect the harness connectors from the rear of the alternator.
    b. Remove the air conditioning compressor belt idler.
    c. Remove the right engine mount yoke screw securing engine mount support strut to the engine.
    d. Remove the 5 side mounting bolts retaining the bracket to the front of the engine.
    e. Remove the front mounting nut. Remove the front bolt and strut and rotate the solid mount bracket away from the engine. Slide the bracket on the stud until free of the mounting studs and remove from the engine.
    f. Remove the timing belt from the vehicle.

To Install:

13. Turn the crankshaft sprocket and intermediate shaft sprocket until the marks are aligned. Use a straight-edge from bolt to bolt to confirm alignment.
14. Turn the camshaft until the small hole in the sprocket is at the top and the arrows on the hub are aligned with the camshaft cap to cylinder head mounting lines. When looking through the hole on top of the camshaft sprocket, the uppermost center nipple of the valve cover end seal should be at the center of the hole. Use a mirror to check the alignment of the arrows so it is viewed straight on and not at an angle from above. Install the belt but let it hang free at this point.
15. Install the air conditioning compressor/alternator bracket, idler pulley and motor mount. Remove the floor jack. Raise the vehicle and support safely. Have the tensioner at an arm's reach because the timing belt will have to be held in position with one hand.
16. To properly install the timing belt, reach up and engage it with the camshaft sprocket. Turn the intermediate shaft counterclockwise slightly, then engage the belt with the intermediate shaft sprocket. Hold the belt against the intermediate shaft sprocket and turn clockwise to take up all tension; if the timing marks are out of alignment, repeat until alignment is correct.
17. Using a wrench, turn the crankshaft sprocket counterclockwise slightly and wrap the belt around it. Turn the sprocket clockwise so there is no slack in the belt between sprockets; if the timing marks are out of alignment, repeat until alignment is correct.

■ **If the timing marks are aligned but slack exists in the belt between either the camshaft and intermediate shaft sprockets or the intermediate and crankshaft sprockets, the timing will be incorrect when the belt is tensioned. All slack must be only between the crankshaft and camshaft sprockets.**

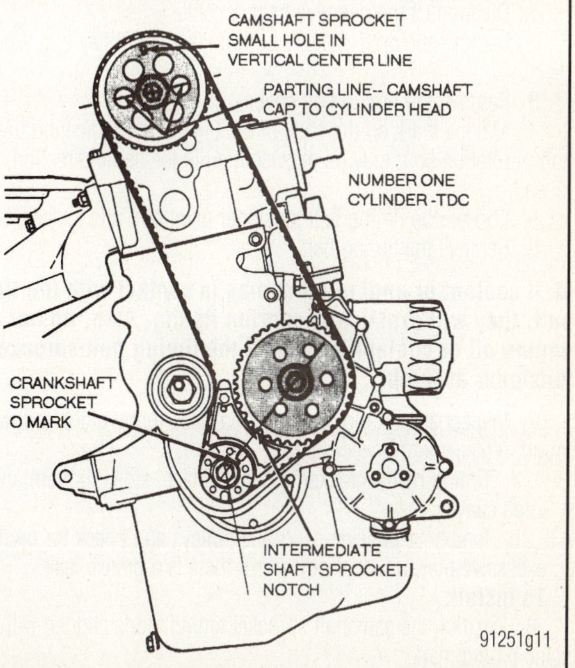

Timing belt alignment mark locations—DaimlerChrysler 2.2L (VIN D) engine

18. Install the tensioner and install the mounting bolt loosely. Place the special tensioning tool C-4703 on the hex of the tensioner so the weight is at about the 9 o'clock position (parallel to the ground, hanging toward the rear of the vehicle) (15 degrees of horizontal.)

19. Hold the tool in position and tighten the bolt to 45 ft. lbs. (61 Nm). Do not pull the tool past the 9 o'clock position; this will make the belt too tight and will cause it to howl or possibly break.

20. Lower the vehicle and recheck the camshaft sprocket positioning. If it is correct install the timing belt covers and all related parts.

21. Connect the negative battery cable and road test the vehicle.

2.2L (VIN F) ENGINES

YEAR, MODEL & ENGINE APPLICATION

1988-89 Eagle Medallion .... 2.2L VIN F

REPLACEMENT INTERVALS

DaimlerChrysler does not provide a replacement interval. Standard industry practice is to replace timing belts every 60,000 miles.

LABOR TIMES

1988-89

Replacement .... 0.8 hrs

Severe Service .... 0.9 hrs

8807CHRg16

✱✱ CAUTION

These engines are considered freewheeling. If the timing belt fails, valve-to-piston damage is unlikely, however a precautionary compression check should be performed to verify no engine damage.

1. Disconnect the negative battery cable.
2. Remove the drive belts, vibration damper, pulley and Woodruff® key.
3. Remove the timing belt cover.
4. Make a mark on the back of the timing belt indicating the direction of rotation so it may be reassembled in the same direction, if to be reused.
5. Loosen the timing belt tensioner pivot bolt and locking bolt.
6. Remove the timing belt.

■ **If coolant or engine oil comes in contact with the timing belt, they will drastically shorten its life. Also, do not allow engine oil or coolant to contact the timing belt sprockets or tensioner assembly.**

7. Inspect all parts for damage and wear. If any of the following is found, replacement is necessary:
   a. Timing belt—cracks on back surface, sides, bottom and separated canvas.
   b. Tensioner pulleys—turn the pulleys and check for binding, excessive play, unusual noise or if there is a grease leak.

To Install:

8. Position the camshaft sprocket timing index aligned with the static timing mark.
9. Position the crankshaft so that the No. 1 piston is at Top Dead Center (TDC) on the compression stroke.

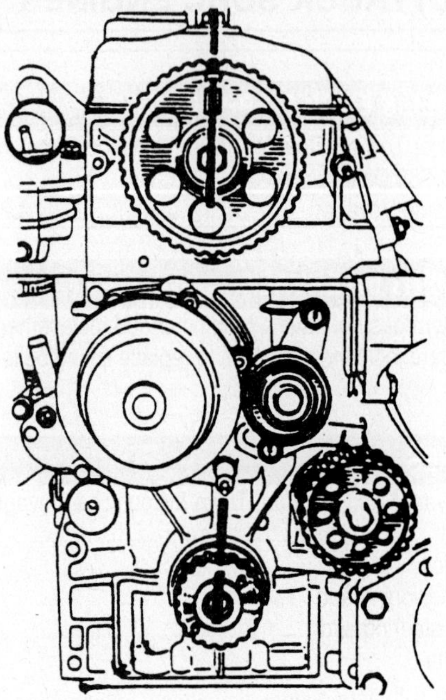

93051g04

Timing belt alignment mark locations—DaimlerChrysler 2.2L (VIN F) engine

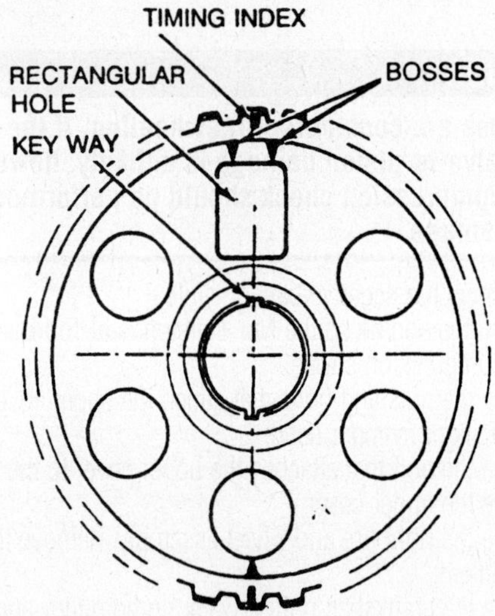

93051g05

Camshaft sprocket alignment mark locations—DaimlerChrysler 2.2L (VIN F) engine

10. Remove the access hole plug in the cylinder block and insert tool Mot. 861 (TDC Rod) or equivalent, into the TDC slot in the crankshaft counterweight.
11. Loosen the timing belt tensioner bolts. Push the tensioner pulley towards the water pump to compress the tensioner spring. Tighten the tensioner bolts. This allows for easier installation of the timing belt.
12. Install the timing belt on the sprockets. If the original timing belt is being reused, install the timing belt with the arrow previously made, pointing in the proper direction of rotation.
13. Loosen the tensioner bolts and allow the spring loaded tensioner to contact the belt. This will automatically tension the belt. Then, tighten the tensioner retaining bolts.

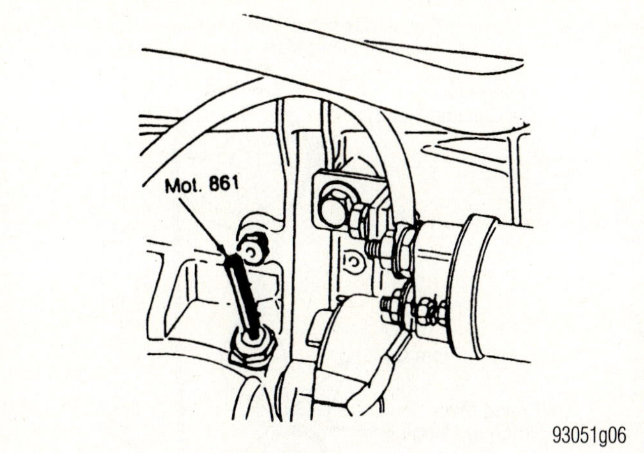

TDC access plug hole— DaimlerChrysler 2.2L (VIN F) engine

14. Position the timing belt cover over the sprockets and check the position of the camshaft sprocket timing mark with the index on the cover.

15. Remove tool Mot. 861 or equivalent, and install the cylinder block plug.

16. Check the timing belt tension adjustment by performing the following procedure:

   a. Rotate the crankshaft clockwise 2 complete revolutions.

   b. Loosen the tensioner bolts 1/4 turn.

   c. The spring loaded timing belt tensioner will automatically adjust to the correct position.

   d. Tighten the bottom tensioner bolt first. Then, tighten the upper bolt. Torque both bolts to 18 ft. lbs. (25 Nm).

   e. Check the timing belt tension with a tension gauge tool, Ele. 346, or equivalent. The deflection should be 0.216–0.276 in. (5.5–7.0mm).

17. Complete the installation by reversing the removal procedures.

2.2L (VIN R) DOHC ENGINES

YEAR, MODEL & ENGINE APPLICATION

1989-90 Chrysler TC	2.2L VIN R

REPLACEMENT INTERVALS

DaimlerChrysler recomends timing belt replacement every 105,000 miles.

LABOR TIMES

1989-90	
Replacement	1.8 hrs
Severe Service	2.2 hrs

8807CHRg17

✻✻ CAUTION

These engines are an interference type. If the timing belt has failed, valve-to-piston damage may have occurred. A compression check should always be performed to check for engine damage.

1. Disconnect the negative battery cable.
2. Position the engine so the No. 1 piston is at Top Dead Center (TDC) of the compression stroke. Disconnect the negative battery cable.
3. Remove the timing belt covers.
4. Remove the timing belt tensioner and allow the belt to hang free.
5. Place a floor jack under the engine and separate the right motor mount.
6. Remove the timing belt from the vehicle.

To Install:

7. Turn the crankshaft sprocket and intermediate shaft sprocket until the marks are aligned. Use a straight-edge from bolt to bolt to confirm alignment.
8. No. 1 and No. 6 camshaft journals have aligning pin holes to index with the blind holes in the camshaft. Turn the camshafts until the pin holes in the journals align with the aligning holes in the corresponding bearing caps. Install pin punches to secure this timing position. At this position, the sprocket timing holes on the camshaft sprockets should both be centered at the cylinder head mounting surface line.
9. Install the motor mount. Remove the floor jack. Raise the vehicle and support safely. Have the tensioner at arm's reach because the timing belt will have to be held in position with one hand.

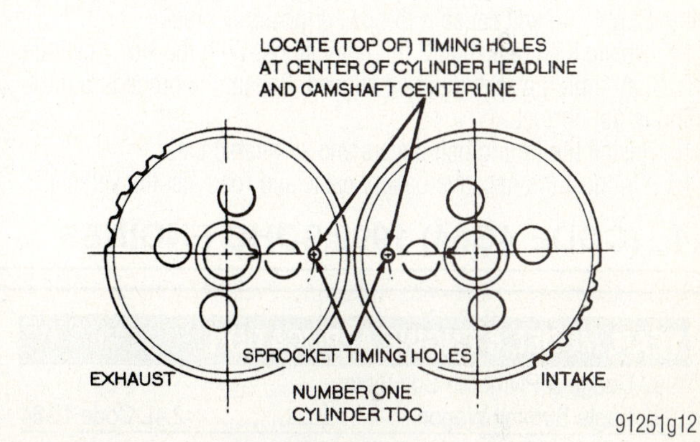

Camshaft sprocket alignment for belt service—DaimlerChrysler 2.2L (VIN R) engine

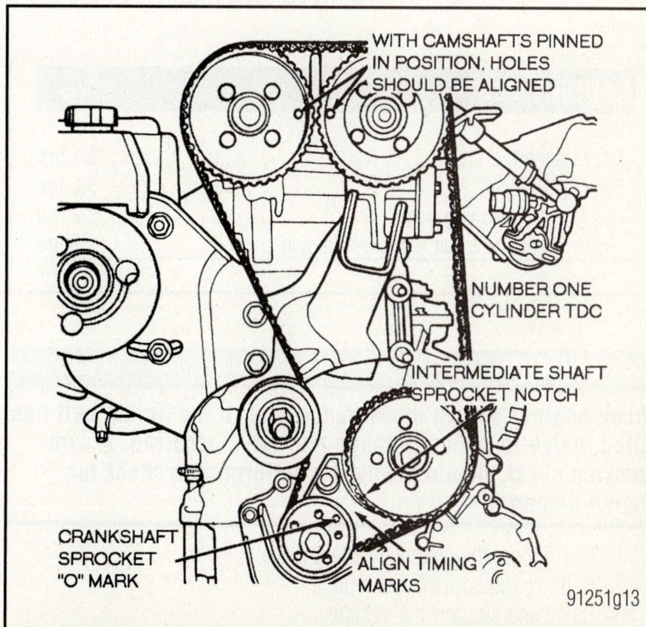

Timing belt alignment marks for proper installation—DaimlerChrysler 2.2L (VIN R) TC engine

10. To properly install the timing belt, reach up and engage it with the camshaft sprockets, leaving no tension between sprockets. Turn the intermediate shaft counterclockwise slightly, then engage the belt with the intermediate shaft sprocket. Hold the belt against the intermediate shaft sprocket and turn clockwise to take up all tension; if the timing marks are out of alignment, repeat until alignment is correct.

11. Using a wrench, turn the crankshaft sprocket counterclockwise slightly and wrap the belt around it. Turn the sprocket clockwise so there is no slack in the belt between sprockets; if the timing marks are out of alignment, repeat until alignment is correct.

■ **If the timing marks are aligned but slack exists in the belt anywhere except on the tensioner side, the timing will be incorrect when the belt is tensioned. All slack must be only between the crankshaft and exhaust camshaft sprockets.**

12. Install the tensioner and install the mounting bolt loosely. Remove the pin punches from the camshafts. Place the special tensioning tool C-4703 on the hex of the tensioner so the weight is at about the 9 o'clock position (parallel to the ground, hanging toward the rear of the vehicle) (15 degrees of horizontal.)

13. Hold the tool in position and tighten the bolt to 45 ft. lbs. (61 Nm). Do not pull the tool past the 9 o'clock position; this will make the belt too tight and will cause it to howl or possibly break.

14. Rotate the crankshaft 2 full revolutions. With the No. 1 cylinder at TDC, all timing marks must be aligned. Repeat the procedure if the timing is not correct.

15. Install the timing belt covers and all related parts.

16. Connect the negative battery cable and road test the vehicle.

2.4L (CODE 4G64) 1992 SOHC ENGINES

YEAR, MODEL & ENGINE APPLICATION

1992 Dodge & Plymouth Colt Vista, 1992 Eagle Summit Wagon	2.4L Code 4G64

REPLACEMENT INTERVALS

DaimlerChrysler recommends timing belt replacement every 60,000 miles.

LABOR TIMES

1992	
Replacement	3.1 hrs
Severe Service	3.6 hrs
With air conitioning add	0.2 hrs
With silent shaft belt replacement add	1.3 hrs

8807CHRg18

✻✻ CAUTION

These engines are an interference type. If the timing belt has failed, valve-to-piston damage may have occurred. A compression check should always be performed to check for engine damage.

1. Disconnect the negative battery cable.
2. Remove the coolant reservoir.
3. Raise and support the vehicle.
4. Remove the left wheel.
5. Remove the left side inner splash shield.

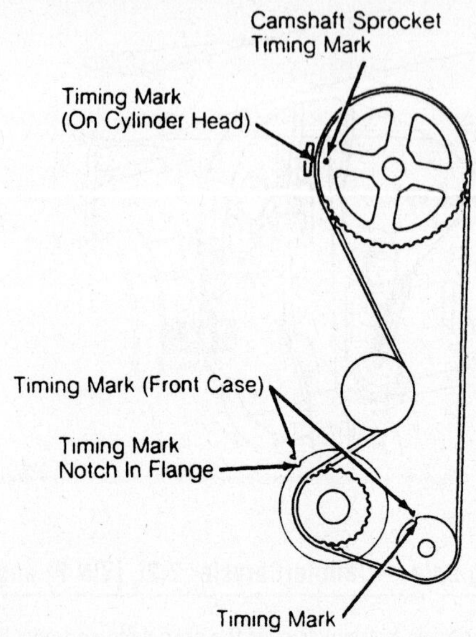

Proper timing belt mark alignment—DaimlerChrysler 1992 2.4L (Code 4G64) engine

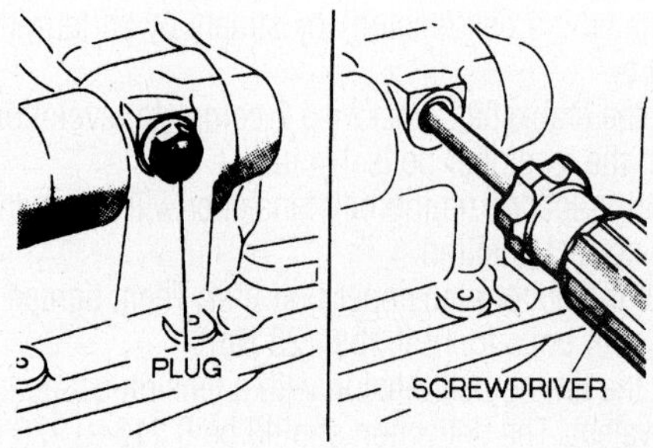

Checking the oil pump sprocket shaft for proper positioning—DaimlerChrysler 1992 2.4L (Code 4G64) engine

6. Place a floor jack under the engine and separate the right motor mount.
7. Remove the accessory drive belts.
8. Remove the drive belt tensioner pulley.
9. Remove the water pump and crankshaft pulleys
10. Remove the upper and lower timing belt covers.
11. Rotate the crankshaft sprocket clockwise and align the timing marks.
12. Loosen the tensioner bolt, move the tensioner away from the belt and tighten the bolt.
13. Remove the timing belt.
14. Align the crankshaft and silent shaft timing marks.
15. Remove the crankshaft sprocket bolt, flange and crankshaft sprocket.
16. Remove the silent shaft belt tensioner.
17. Remove the silent shaft belt.

To Install:

18. Align the timing marks of the silent shaft sprockets and the crankshaft sprocket with the timing marks on the front case. Wrap the timing belt around the sprockets so there is no slack in the upper span of the belt and the timing marks are still aligned.

19. Install the tensioner pulley and move the pulley by hand so the long side of the belt deflects 0.20–0.28 in. (5–7mm).

20. Hold the pulley tightly so the pulley cannot rotate when the bolt is tightened. Tighten the bolt to 15 ft. lbs. (20 Nm) and recheck the deflection amount.
21. Install the timing belt tensioner fully toward the water pump and tighten the bolts. Place the upper end of the spring against the water pump body.
22. Align the timing marks of the camshaft, crankshaft and oil pump sprockets with their corresponding marks on the front case or rear cover.

■ **There is a possibility to align all timing marks and have the oil pump sprocket out of time, causing an engine vibration during operation. If the following step is not followed exactly, there is a 50 percent chance that the oil pump shaft alignment will be 180 degrees off.**

23. Before installing the timing belt, ensure that the oil pump sprocket is in the correct position as follows:
    a. Remove the plug from the rear side of the block and insert a Phillips screwdriver with shaft diameter of 0.31 in. (8mm) into the hole.
    b. With the timing marks still aligned, the shaft of the tool must be able to go in at least 2.36 in. (60mm). If the tool can only go in 0.79–0.98 in. (20–25mm), the shaft is not in the correct orientation and will cause a vibration during engine operation. Remove the tool from the hole and turn the oil pump sprocket 1 complete revolution. Realign the timing marks and insert the tool. The shaft of the tool must go in at least 2.36 in. (60mm).
    c. Recheck and realign the timing marks.
    d. Leave the tool in place to hold the oil pump shaft while continuing.
24. Install the belt to the crankshaft sprocket, oil pump sprocket, then camshaft sprocket. While doing so, make sure there is no slack between the sprocket except where the tensioner is installed.
25. Tighten oil pump sprocket bolt to 26–29 ft. lbs. (34–40 Nm) and tighten crankshaft bolt to 80–94 ft. lbs. (110–130 Nm).
26. Recheck the timing mark alignment. If all are aligned, loosen the tensioner mounting bolt and allow the tensioner to apply tension to the belt.
27. Remove the tool that is holding the silent shaft and rotate the crankshaft a distance equal to 2 teeth on the camshaft sprocket. This will allow the tensioner to automatically apply the proper tension on the belt. Do not manually overtighten the belt or it will howl.
28. Tighten the lower mounting bolt first, then the upper spacer bolt.
29. To verify correct belt tension, check that the deflection at the longest span of the belt has 0.40 in. (12mm) clearance from the belt cover.
30. Install the timing belt covers and all related items.
31. Connect the negative battery cable.

2.4L (CODE 4G64) 1993–96 SOHC ENGINES

✱✱ CAUTION

These engines are an interference type. If the timing belt has failed, valve-to-piston damage may have occurred. A compression check should always be performed to check for engine damage.

YEAR, MODEL & ENGINE APPLICATION

1993-96 Eagle Summit, Plymouth Colt Vista ... 2.4L Code 4G64

REPLACEMENT INTERVALS

DaimlerChrysler does not provide a recommended replaement interval. Standard industry practice is to replace timing belts every 60,000 miles.

LABOR TIMES

1993-96	
Replacement	3.1 hrs
Severe Service	3.6 hrs
With air conditioning add	0.2 hrs

8807CHRg19

1. Disconnect the negative battery cable.
2. Remove the coolant reservoir.
3. Raise and support the vehicle.
4. Remove the left wheel.
5. Remove the left side inner splash shield.
6. Place a floor jack under the engine and separate the right motor mount.
7. Remove the accessory drive belts.
8. Remove the drive belt tensioner pulley.
9. Remove the water pump and crankshaft pulleys
10. Remove the upper and lower timing belt covers.
11. Rotate the crankshaft sprocket clockwise and align the timing marks.
12. Loosen the tensioner bolt, move the tensioner away from the belt and tighten the bolt.
13. Remove the timing belt.
14. Align the crankshaft and silent shaft timing marks.
15. Remove the crankshaft sprocket bolt, flange and crankshaft sprocket.
16. Remove the silent shaft belt tensioner.
17. Remove the silent shaft belt.

To Install:

18. Align the timing marks on the crankshaft sprocket and the silent shaft sprocket. Fit the inner timing belt over the crankshaft and silent shaft sprocket. Ensure that there is no slack in the belt.
19. While holding the inner timing belt tensioner, adjust the timing belt tension by applying a force towards the center of the belt, until the tension side of the belt is taut. Tighten the tensioner bolt.

■ **When tightening the bolt of the tensioner, ensure that the tensioner pulley shaft does not rotate with the bolt. Allowing it to rotate with the bolt can cause excessive tension on the belt.**

20. Check belt for proper tension by depressing the belt on long side and noting the belt deflection. The desired reading is 0.20–0.28 in. (5–7mm). If tension is not correct, readjust and check belt deflection.
21. Install the flange, crankshaft sprocket and washer to the crankshaft. The flange on the crankshaft sprocket must be installed towards the inner timing belt sprocket. Tighten bolt to 80–94 ft. lbs. (110–130 Nm).
22. Position the auto-tensioner into a vise with soft jaws. The plug at the rear of tensioner protrudes, be sure to use a washer as a spacer to protect the plug from contacting vise jaws.
23. Slowly push the rod into the tensioner until the set hole in rod is aligned with set hole in the auto-tensioner.

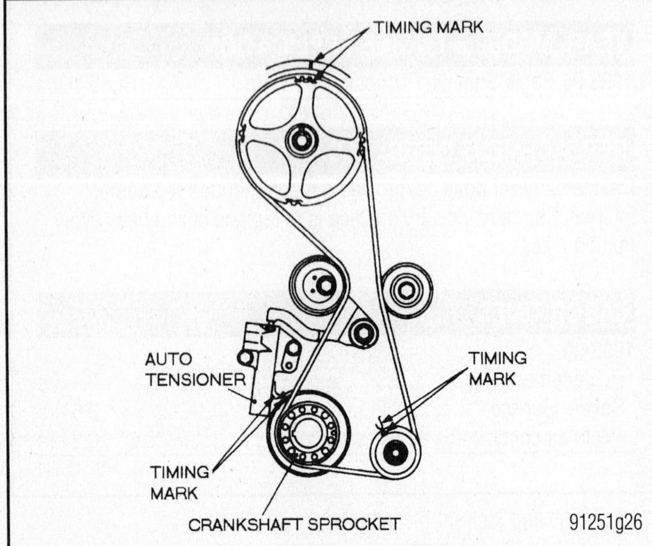

Timing belt alignment mark locations—1993–95 DaimlerChrysler 2.4L (Code 4G64) engine

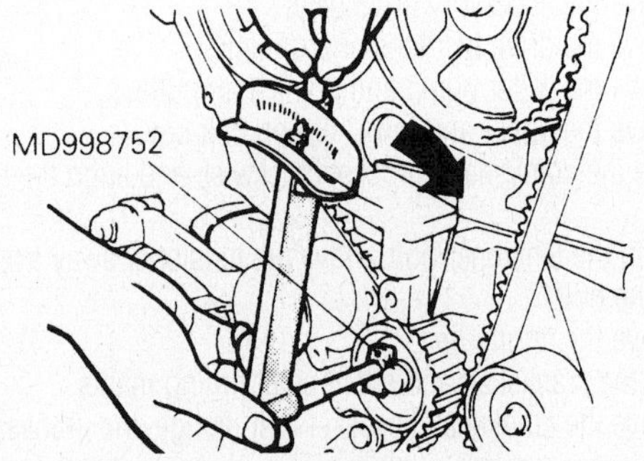

Timing belt adjusting tool No. MD998752—1993–95 DaimlerChrysler 2.4L (Code 4G64) engine

24. Insert a 0.055 in. (1.4mm) wire into the aligned set holes. Unclamp the tensioner from vise and install to vehicle. Tighten tensioner to 17 ft. lbs. (24 Nm).

■ **There is a possibility to align all timing marks and have the oil pump sprocket out of time, causing an engine vibration during operation. If the following step is not followed exactly, there is a 50 percent chance that the oil pump shaft alignment will be 180 degrees off.**

25. Before installing the timing belt, ensure that the oil pump sprocket is in the correct position as follows:
   a. Remove the plug from the rear side of the block and insert a Phillips screwdriver with shaft diameter of 0.31 in. (8mm) into the hole.
   b. With the timing marks still aligned, the shaft of the tool must be able to go in at least 2.36 in. (60mm). If the tool can only go in 0.79–0.98 in. (20–25mm), the shaft is not in the correct orientation and will cause a vibration during engine operation. Remove the tool from the hole and turn the oil pump sprocket 1 complete revolution. Realign the timing marks and insert the tool. The shaft of the tool must go in at least 2.36 in. (60mm).
   c. Recheck and realign the timing marks.

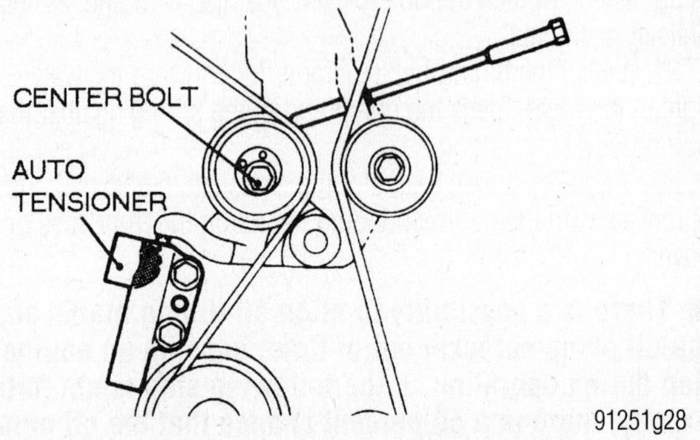

Timing belt adjusting tool No. MD998738 (auto tensioner)—1993–95 DaimlerChrysler 2.4L (Code 4G64) engine

   d. Leave the tool in place to hold the silent shaft while continuing.
26. To install the oil pump sprocket and tighten the nut to 36–43 ft. lbs. (50–60 Nm).
27. When installing timing belt, the camshaft sprocket dowel pin should be located on top. Align all timing marks.
28. Align the crankshaft sprocket, camshaft sprocket and oil pump sprocket timing marks.
29. Install the timing belt as follows:
   a. Install the timing belt around the crankshaft sprocket, oil pump sprocket, idler pulley, camshaft and the tensioner pulley.
   b. Lift upward on the tensioner pulley in a clockwise direction and tighten the center bolt. Make sure all timing marks are aligned.
   c. Rotate the crankshaft 1/4 turn counterclockwise. Then, turn in clockwise until the timing marks are aligned again.
30. Loosen the center bolt. Using tool No. MD998752 or equivalent and a torque wrench, apply a torque of 22–24 inch lbs. (2.6–2.8 Nm). Tighten the center bolt.
31. Screw the tool No. MD998738 into the engine left support bracket until its end makes contact with the tensioner arm and tighten tensioner pulley to 35 ft. lbs. (48 Nm). At this point, screw the special tool in some more and remove the set wire attached to the auto tensioner. Then remove the special tool.
32. Rotate the crankshaft 2 complete turns clockwise and let it sit for approximately 15 minutes. Then, measure the auto tensioner protrusion (the distance between the tensioner arm and auto tensioner body) to ensure that it is within 0.15–0.18 in. (3.8–4.5mm). If out of specification, repeat belt adjustment procedure until the specified value is obtained.
33. If the timing belt tension adjustment is being performed with the engine mounted in the vehicle, and clearance between the tensioner arm and the auto tensioner body cannot be measured, the following alternative method can be used:
   a. Screw in tool No. MD998738 or equivalent, until its end makes contact with the tensioner arm.
   b. After the tool makes contact with the arm, screw it in some more to retract the auto tensioner pushrod while counting the number of turns the tool makes until the tensioner arm is brought into contact with the auto tensioner body. Make sure the number of turns the tool makes conforms with the standard value of 2.5–3 turns.
   c. Install the rubber plug to the timing belt rear cover.
34. Install the timing belt covers and all related items.
35. Connect the negative battery cable.

2.4L (VIN B) DOHC ENGINES

YEAR, MODEL & ENGINE APPLICATION

2000-05 Chrysler PT Cruiser, 1995-04 Dodge Caravan, 1996-00 Plymouth Voyager	2.4L VIN B

REPLACEMENT INTERVALS

DaimlerChrysler recommends timing belt replacement every 105,000 miles.

LABOR TIMES

Caravan & Voyager	
Replacement	2.3 hrs
Severe Service	2.6 hrs
With air conditioning add	0.2 hrs
With power steering add	0.2 hrs
PT Cruiser	
With turbocharger	
Replacement	2.3 hrs
Severe Service	2.6 hrs
Without turbocharger	
Replacement	4.0 hrs
Severe Service	4.5 hrs

8807CHRg20

Caravan, Mini Ram Van and Voyager

✻✻ CAUTION

These engines are an interference type. If the timing belt has failed, valve-to-piston damage may have occurred. A compression check should always be performed to check for engine damage.

■ **You may need DRB scan tool to perform the crankshaft and camshaft relearn alignment procedure.**

■ **There are 2 versions of this engine. 1995–96 models use a hydraulic tensioner, and 1997–04 models use a mechanical timing belt tensioner.**

1. Disconnect the negative battery cable remote connector, located on the left strut tower.
2. Remove the engine undercover.
3. Remove the right inner splash shield.
4. Remove the accessory drive belts.
5. On 1995–96 engines, remove the crankshaft pulley using a suitable puller.
6. On 1997–04 engines remove the crankshaft pulley using special tools No, 1026 and 6827-a or equivalent.
7. Raise and support the vehicle.
8. Place a jack under the engine and raise it enough to relieve engine pressure.
9. Remove the right side engine mount and bracket.
10. Remove the timing belt cover.

■ **Do not rotate the crankshaft or the camshafts after the timing belt has been removed. Damage to the valve components may occur. Before removing the timing belt, always align the timing marks.**

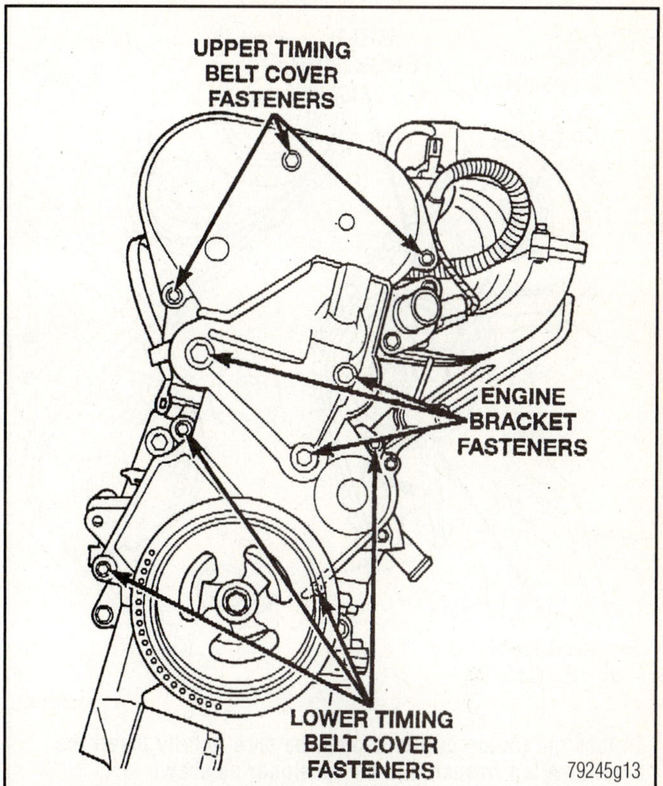

Timing cover and engine mounting bracket bolt locations—DaimlerChrysler 2.4L (VIN B) engine

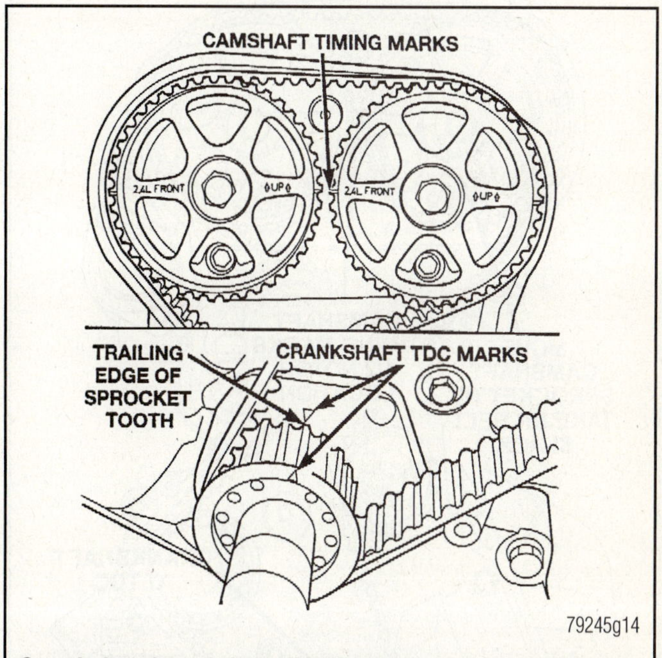

Camshaft and crankshaft alignment marks—DaimlerChrysler 2.4L (VIN B) engine

11. Align the timing marks of the timing belt sprockets to the timing marks on the rear timing belt cover and oil pump cover.
12. On hydraulic tensioner engines, loosen the timing belt tensioner bolts.
13. Remove the timing belt and the tensioner.
14. Place the tensioner into a soft-jawed vise to compress the tensioner.

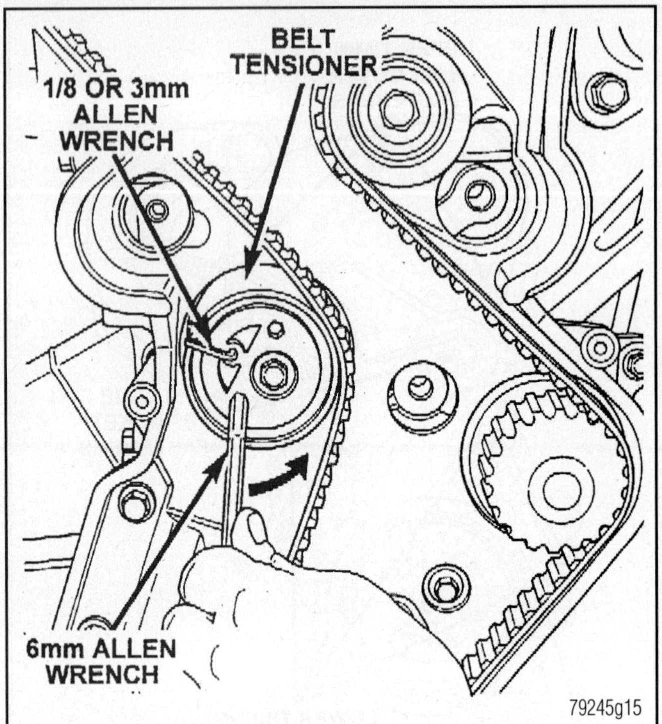

To lock the timing belt tensioner, be sure to fully insert the smaller Allen wrench into the tensioner as shown—DaimlerChrysler 2.4L (VIN B) mechanical tensioner engine

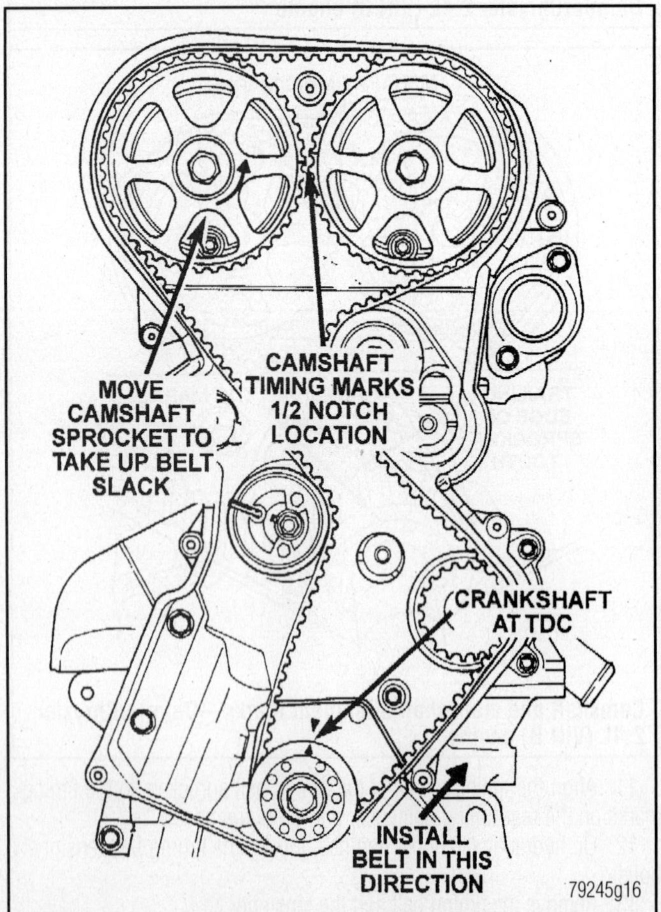

Installation of the timing belt, notice the camshaft alignment—DaimlerChrysler 2.4L (VIN B) engine

15. After compressing the tensioner, insert a pin (a 5/64 in. Allen wrench will also work) into the plunger side hole to retain the plunger until installation.
16. On mechanical tensioner engines, insert a 6 mm hex wrench in the tensioner lower hex hole.
17. Rotate the wrench tensioner counterclockwise until it locks and insert a 3mm hex wrench into the locking hole.
18. Remove the timing belt.
19. If necessary, remove the camshaft timing belt sprockets.
20. If necessary, remove the crankshaft timing belt sprocket using removal tool No. 6793, or equivalent.

To Install:

21. If necessary, use tool No. 6792, or equivalent, to install the crankshaft timing belt sprocket onto the crankshaft.
22. If necessary, install the camshaft sprockets onto the camshafts. Install and tighten the camshaft sprocket bolts to 75 ft. lbs. (101 Nm).
23. On hydraulic tensioner engines, position the tensioner into a vise with soft jaws. The plug at the rear of tensioner protrudes, be sure to use a washer as a spacer to protect the plug from contacting vise jaws.
24. Slowly push the rod into the tensioner until the set hole in rod is aligned with set hole in the auto-tensioner.
25. Insert a 0.055 in. (1.4mm) wire into the aligned set holes.
26. Set the crankshaft sprocket to Top Dead Center (TDC) by aligning the notch on the sprocket with the arrow on the oil pump housing.
27. Set the camshafts to align the timing marks on the sprockets.
28. Move the crankshaft to 1/2 notch before TDC.
29. Install the timing belt starting at the crankshaft, then around the water pump sprocket, idler pulley, camshaft sprockets and around the tensioner pulley.
30. Move the crankshaft sprocket to TDC to take up the belt slack.
31. Install the tensioner on the engine block but do not tighten.
32. Using a torque wrench on the tensioner pulley, apply 250 inch lbs. (28 Nm) of torque to the tensioner pulley.
33. With torque being applied to the tensioner pulley, move the tensioner up against the tensioner pulley bracket and tighten the fasteners to 23 ft. lbs. (31 Nm).
34. Remove the tensioner plunger pin, the tension is correct when the plunger pin can be removed and reinserted easily.

✳✳ WARNING

If any binding is felt when adjusting the timing belt tension by turning the crankshaft, STOP turning the engine, because the pistons may be hitting the valves.

35. Rotate the crankshaft 2 revolutions and recheck the timing marks. Wait several minutes and then recheck that the plunger pin can easily be removed and installed.
36. On mechanical tensioner engines, align the crankshaft sprocket timing marks.
37. Position the exhaust camshaft sprocket so it is 1/2 notch below the intake camshaft sprocket.
38. Install the timing belt starting at the crankshaft, then around the water pump sprocket, idler pulley, camshaft sprockets and around the tensioner pulley.
39. Rotate the exhaust camshaft counterclockwise to align the timing marks and remove the slack from the timing belt.
40. Remove the 3 mm wrench from the tensioner.
41. Rotate the crankshaft pulley 2 revolutions clockwise and check that the timing marks are aligned.
42. Install the front timing belt cover.
43. Install the engine mount bracket.

44. Install the right engine mount.
45. Remove the floor jack from under the vehicle.
46. Install the crankshaft damper and tighten it to 105 ft. lbs. (142 Nm).
47. Install the accessory drive belts and adjust to the proper tension.
48. Install the engine undercover.
49. Install the right inner splash shield.
50. Reconnect the negative battery cable.
51. Perform the crankshaft and camshaft relearn alignment procedure using the DRB scan tool, or equivalent.

PT Cruiser

✳✳ CAUTION

These engines are an interference type. If the timing belt has failed, valve-to-piston damage may have occurred. A compression check should always be performed to check for engine damage.

■ You may need DRB scan tool to perform the crankshaft and camshaft relearn alignment procedure.

1. Remove the A/C refrigerant using approved equipment.
2. Remove or disconnect the following to remove the upper timing cover:
   - Negative battery cable
   - Upper torque strut bolts and set the strut aside
   - A/C lines at the junction block near the upper timing belt cover, on turbocharged models
   - Upper cover bolts and the cover
3. Remove or disconnect the following to remove the lower timing cover:
   - Right front wheel and inner splash shield
   - Accessory drive belts
   - Crankshaft damper

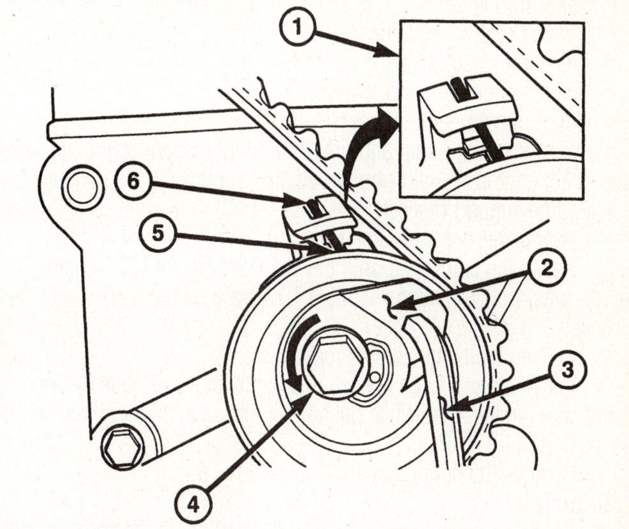

1 - ALIGN SETTING NOTCH WITH SPRING TANG
2 - TOP PLATE
3 - 6mm ALLEN WRENCH
4 - LOCK BOLT
5 - SETTING NOTCH
6 - SPRING TANG

67189-PTCR-G11

Installing the belt tensioner—DaimlerChrysler 2.4L (VIN B) engine

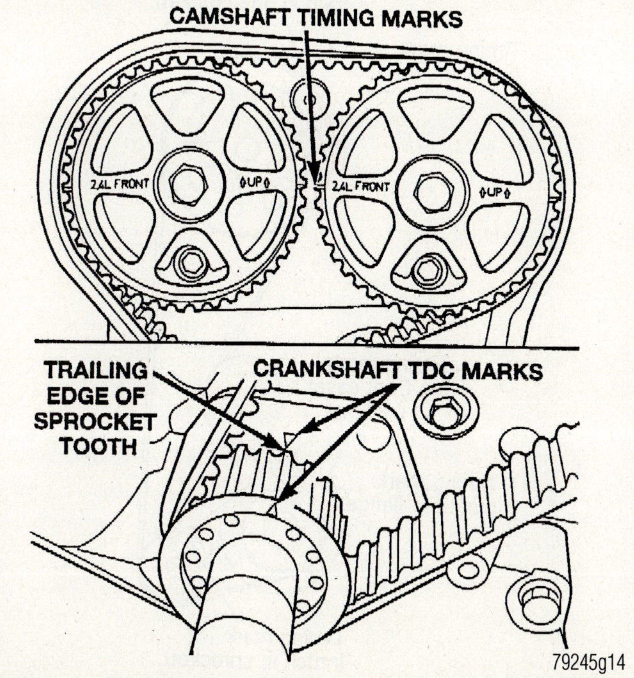

79245g14

Camshaft and crankshaft alignment marks—DaimlerChrysler 2.4L (VIN B) engine

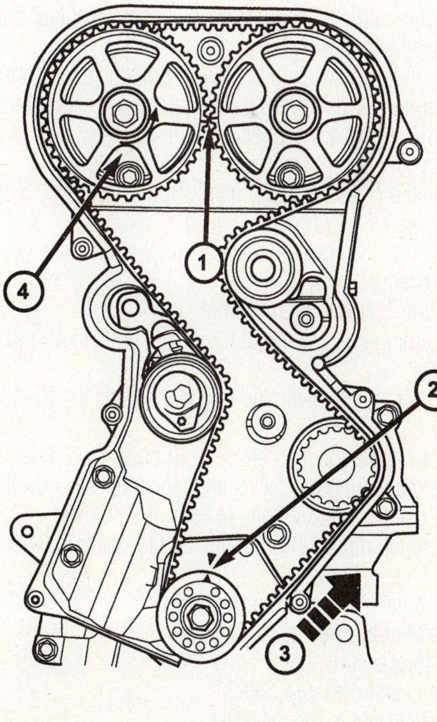

1 - CAMSHAFT TIMING MARKS 1/2 NOTCH LOCATION
2 - CRANKSHAFT AT TDC
3 - INSTALL BELT IN THIS DIRECTION
4 - ROTATE CAMSHAFT SPROCKET TO TAKE UP BELT SLACK

67189-PTCR-G12

Proper timing belt routing—DaimlerChrysler 2.4L (VIN B) engine

- Lower torque strut
- Exhaust system from the manifold
- A/C compressor switch
- Upper torque strut and bracket
- Upper radiator support crossmember
- Power steering pump and bracket without disconnecting the lines
- Right engine mount through bolt after supporting the engine
- Engine support bracket
- Lower cover bolts and the cover

4. Rotate the crankshaft until the Top Dead Center (TDC) mark on the oil pump housing aligns with the TDC mark on the crankshaft sprocket.
5. Loosen the belt tensioner lock bolt.
6. Install a 6mm Allen wrench into the tensioner. Rotate the tensioner counterclockwise while pushing on the wrench until it slides into the locking hole.
7. Remove the timing belt.

To Install:

8. Set the crankshaft sprocket at TDC by aligning the sprocket with the arrow on the oil pump housing.
9. Set the camshafts timing marks so that the exhaust camshaft sprocket is a 1/2 notch below the intake camshaft sprocket.
10. Install the timing belt by starting at the crankshaft. Go around the water pump sprocket idler pulley, camshaft sprockets and the tensioner.
11. Move the exhaust camshaft sprocket counterclockwise to align the marks and to remove any slack.
12. Insert a 6mm Allen wrench into the tensioner opening on the top plate of the tensioner pulley. Rotate the plate counterclockwise.
13. The pulley will move against the belt and the tensioner setting notch will start to move clockwise.
14. Continue to move the plate counterclockwise until the setting notch is aligned with spring tang.
15. Using the Allen wrench to prevent the tensioner from moving, tighten the tenisioner bolt to 18 ft. lbs. (25 Nm). Ensure the setting notch and spring tang are aligned after tightening.
16. Remove the Allen wrench.
17. Rotate the crankshaft two full revolutions clockwise and verify that the TDC marks are properly aligned.
18. Check that the spring tang is within the tolerance window. If not repeat the procedure.
19. Install or connect the following:
    - Lower timing belt cover and torque the bolts to 40–50 inch lbs. (5–6 Nm)
    - Upper timing belt cover and torque the bolts to 40–50 inch lbs. (5–6 Nm)
    - A/C lines at the junction block on turbocharged models.
    - Right engine support bracket and reposition the power steering pump. Torque the bracket bolts to 45 ft. lbs. (61 Nm).
    - Right engine mount through bolt and torque it to 87 ft. lbs. (118 Nm)
    - Upper radiator support crossmember
    - Torque strut bracket
    - Upper torque strut
    - A/C lines and pressure switch
    - Exhaust system to the manifold
    - Crankshaft damper
    - Accessory drive belts
    - Lower torque strut
    - Right splash shield and wheel
    - Negative battery cable
20. Recharge the A/C system.
21. Perform the camshaft/crankshaft synchronization procedure.

2.4L (VIN G) SOHC ENGINES

YEAR, MODEL & ENGINE APPLICATION

1990-93 Ram-50 .... 2.4L VIN G

REPLACEMENT INTERVALS

DaimlerChrysler recommends timing belt replacement every 60,000 miles.

LABOR TIMES

1990-93	
Replacement	2.5 hrs
Severe Service	3.0 hrs
With air conditioning add	0.5 hrs
With power steering add	0.3 hrs
With silent shaft belt replacement add	0.3 hrs

8807CHRg21

✱✱ CAUTION

These engines are an interference type. If the timing belt has failed, valve-to-piston damage may have occurred. A compression check should always be performed to check for engine damage.

1. Be sure that the engine's No. 1 piston is at Top Dead Center (TDC) of the compression stroke.
2. Disconnect the negative battery cable.
3. Remove the upper radiator shroud.
4. Remove the accessory drive belt(s).

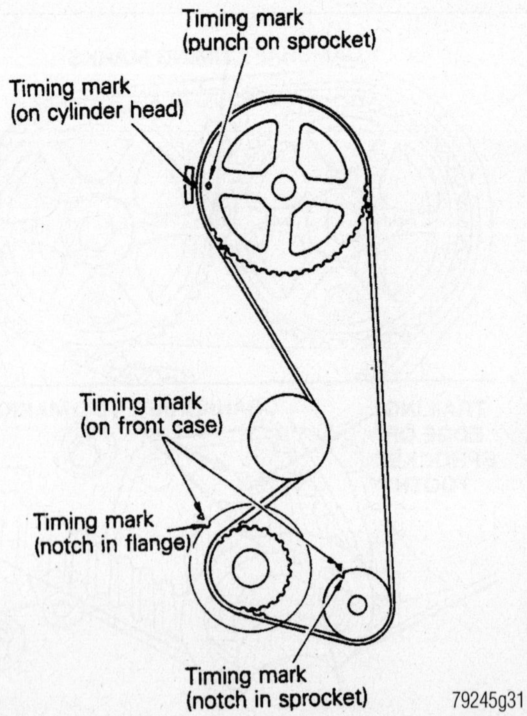

Timing belt pulley alignment marks—DaimlerChrysler 2.4L (VIN G) engine

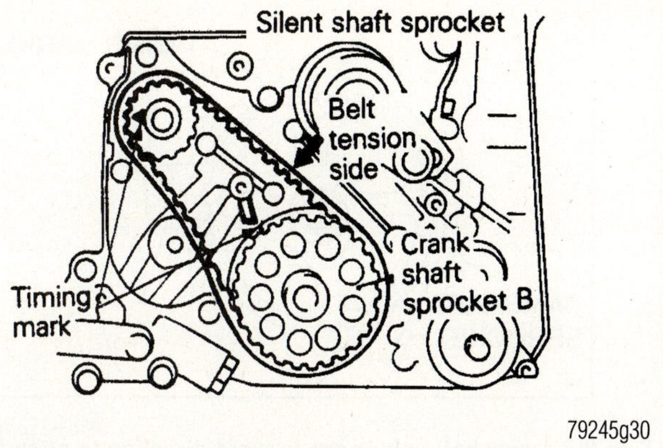

Silent shaft alignment marks. Notice the tension side of the inner (silent shaft) belt—DaimlerChrysler 2.4L (VIN G) engine

5. Remove the fan clutch and water pump pulley.
6. Remove the power steering pump pulley.
7. Remove the crankshaft pulley.
8. Remove the upper and lower timing belt covers.
9. Ensure the crankshaft sprocket, camshaft sprocket and oil pump sprocket timing marks are aligned.
10. If the belt(s) are to be reused, mark the direction of rotation on the belt.
11. Loosen the timing belt tensioner and rotate it toward the water pump, then tighten the tensioner.
12. Remove the timing belt.
13. Loosen the silent shaft belt tensioner and rotate away from the belt, then tighten the tensioner.
14. Remove the crankshaft sprocket and flange.
15. Remove the silent shaft belt.

To Install:

16. Align the timing mark of the silent shaft belt sprockets on the crankshaft and silent shaft with the marks on the front case. Wrap the silent shaft belt around the sprockets so there is no slack in the upper span of the belt and the timing marks are still aligned.
17. Install the tensioner initially so the actual center of the pulley is above and to the left of the installation bolt.
18. Move the pulley up by hand so the center span of the long side of the belt deflects about 1/4 in. (6mm).
19. Hold the pulley tightly so it does not rotate when the bolt is tightened. Tighten the bolt to 15 ft. lbs. (20 Nm). If the pulley has moved, the belt will be too tight.
20. Check the belt deflection. The belt should deflect about 0.2–03 inch (5–7 mm). If not within range, readjust the belt tension.
21. Install the crankshaft flange and sprocket. Tighten the bolt to 90–94 ft. lbs. (108–127 Nm).
22. Install the timing belt tensioner fully toward the water pump and temporarily tighten the bolts. Place the upper end of the spring against the water pump body. Align the timing marks of the camshaft, crankshaft and oil pump sprockets with the corresponding marks on the front case or head.

■ **If the following steps are not followed exactly, there is a chance that the silent shaft alignment will be 180 degrees off. This will cause a noticeable vibration in the engine and the entire procedure will have to be repeated.**

23. Before installing the timing belt, ensure that the left-side silent shaft is in the correct position.

■ **It is possible to align the timing marks on the camshaft sprocket, crankshaft sprocket and the oil pump sprocket with the left balance shaft out of alignment.**

24. With the timing mark on the oil pump pulley aligned with the mark on the front case, check the alignment of the left balance shaft to assure correct shaft timing.
25. Remove the plug located on the left-side of the block in the area of the starter.
26. Insert a tool having a shaft diameter of 0.3 in. (8mm) into the hole.
27. With the timing marks still aligned, the tool must be able to go in at least 2 1/3 in. If it can only go in about 1 in. turn the oil pump sprocket 1 complete revolution.
28. Recheck the position of the balance shaft with the timing marks realigned. Leave the tool in place to hold the silent shaft while continuing.
29. Install the belt to the crankshaft sprocket, oil pump sprocket and the camshaft sprocket, in that order. While doing so, be sure there is no slack between the sprockets except where the tensioner will take it up when released.
30. Recheck the timing mark alignments.
31. If all are aligned, loosen the tensioner mounting bolt and allow the tensioner to apply tension to the belt.
32. Remove the tool that is holding the silent shaft in place and turn the crankshaft clockwise a distance equal to 2 teeth of the camshaft sprocket. This will allow the tensioner to automatically tension the belt the proper amount.

✻✻ WARNING

Do not manually apply pressure to the tensioner. This will overtighten the belt and will cause a howling noise.

33. First tighten the lower mounting bolt and then tighten the upper spacer bolt.

✻✻ WARNING

If any binding is felt when adjusting the timing belt tension by turning the crankshaft, STOP turning the engine, because the pistons may be hitting the valves.

34. To verify that belt tension is correct, check that the deflection of the longest span (between the camshaft and oil pump sprockets) is 1/2 in. (13mm).
35. Install the lower timing belt cover. Be sure the packing is properly positioned in the inner grooves of the covers when installing.
36. Install the upper and lower timing belt covers.
37. Install the crankshaft pulley.
38. Install the power steering pump pulley.
39. Install the fan clutch and water pump pulley.
40. Install the accessory drive belt(s).
41. Install the upper radiator shroud.
42. Connect the negative battery cable. Start the engine and check for leaks.

2.4L SOHC (VIN G) ENGINES

✻✻ CAUTION

These engines are an interference type. If the timing belt has failed, valve-to-piston damage may have occurred. A compression check should always be performed to check for engine damage.

YEAR, MODEL & ENGINE APPLICATION

2001-04 Chrysler Sebring & Dodge Stratus Coupe .. 2.4L VIN G

REPLACEMENT INTERVALS

DaimlerChrysler recommends timing belt replacement every 60,000 miles.

LABOR TIMES

2001-04

Replacement	2.7 hrs
Severe Service	3.1 hrs

8807CHRg22

1. Disconnect the negative battery cable.
2. Raise and support the vehicle.
3. Place a jack under the engine/transmission assembly and raise it enough to remove the weight from the engine mount bracket insulator.
4. Remove the right front wheel.
5. Remove the right inner splash-shield.
6. Remove the right engine mount bracket and stopper
7. Remove the coolant reservoir tank.
8. Remove the accessory drive belts.
9. Remove the water pump pulley.
10. Remove the crankshaft pulley.
11. Remove the upper timing belt cover.
12. Remove the lower timing belt cover.
13. Ensure the crankshaft and camshaft timing marks are aligned.

■ **Do not rotate the crankshaft or camshafts after the timing belt has been removed. Damage to the valve components may occur. Before removing the timing belt, always align the timing marks.**

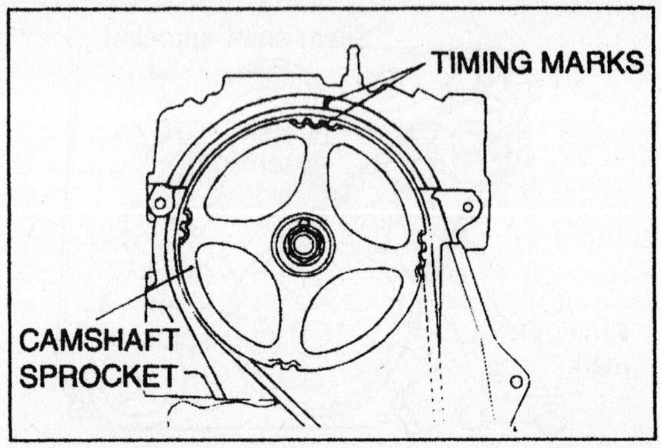

68189-SEBRR-G03

Turn the crankshaft only in the forward direction to align the camshaft sprocket timing marks—DaimlerChrysler 2.4L SOHC (VIN G) engine

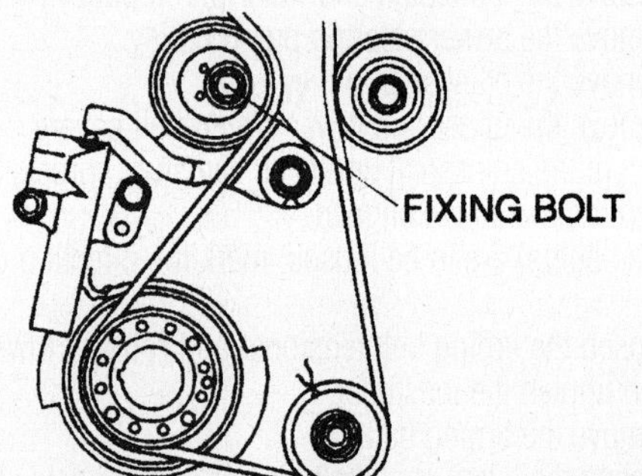

68189-SEBR-G04

Location of the tension pulley fixing bolt—DaimlerChrysler 2.4L SOHC (VIN G) engine

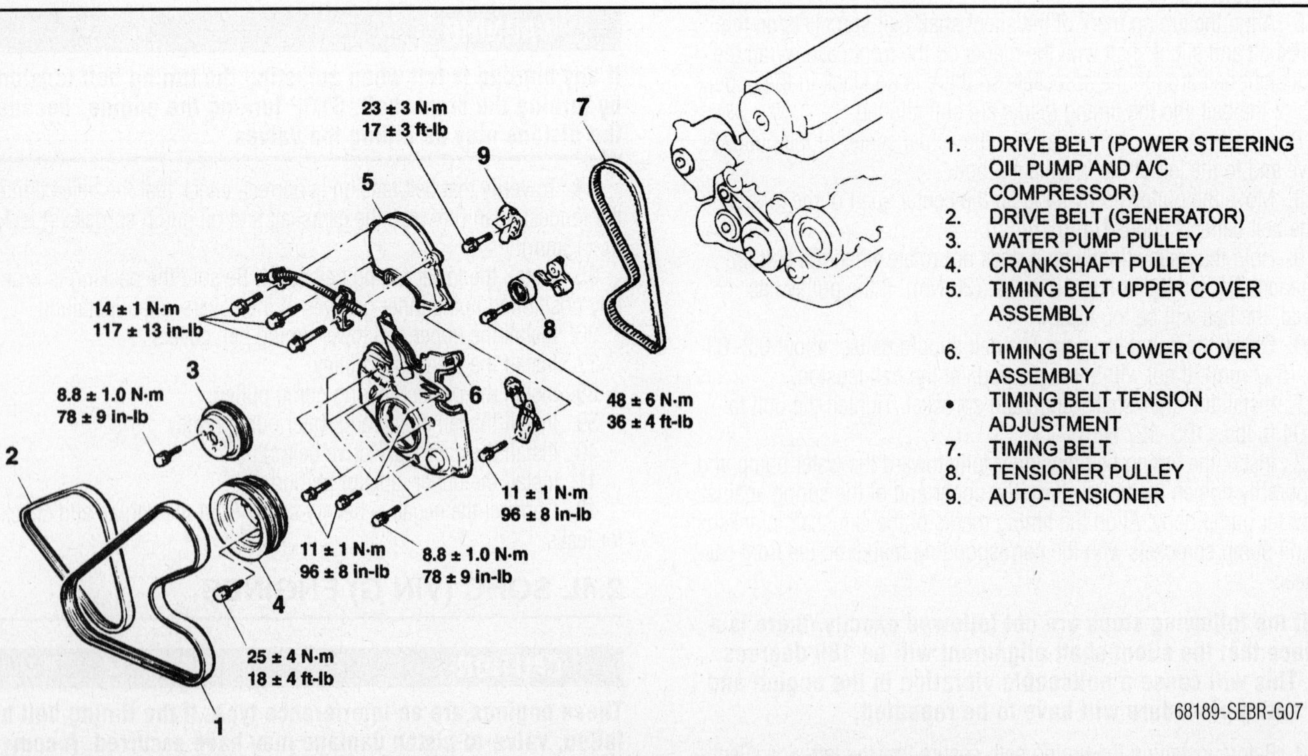

68189-SEBR-G07

Exploded view of the timing belt and related components—DaimlerChrysler 2.4L SOHC (VIN G) engine

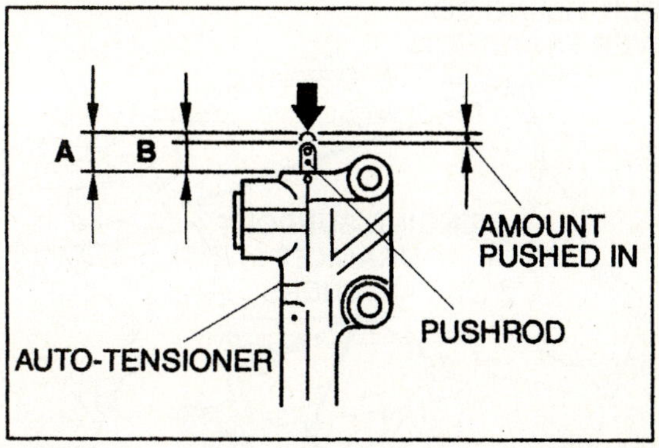

68189-SEBR-G05

Measuring the auto tensioner—DaimlerChrysler 2.4L SOHC (VIN G) engine

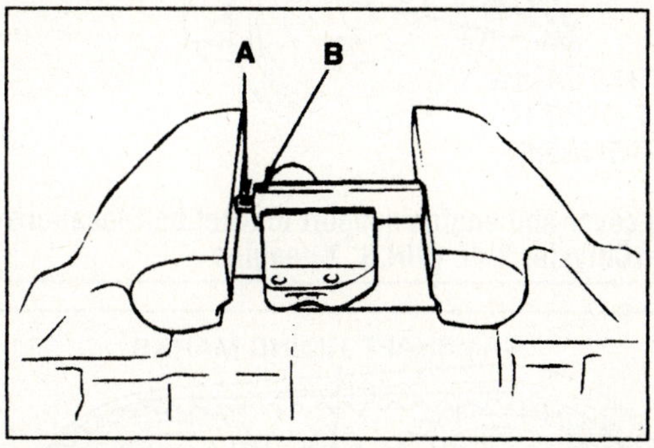

68189-SEBR-G06

Compress the pushrod until holes A and B are aligned—DaimlerChrysler 2.4L SOHC (VIN G) engine

14. Align the timing marks of the timing belt sprockets to the timing marks on the rear timing belt cover and the oil pump cover.

■ **If the timing belt is to be reused, draw an arrow indicating the direction of rotation on the back of the belt for reinstallation.**

15. Loosen the timing belt tension pulley fixing bolt, then move the tension pulley to the water pump side.
16. Remove the timing belt.
17. Remove the tensioner pulley and automatic tensioner.
18. Remove the camshaft timing belt sprockets.
19. Remove the crankshaft sprocket using special tool No. 6793.

To Install:

20. Install the crankshaft sprocket using special tool No. 6793.
21. Install the camshaft sprockets and tighten the bolts to 75 ft. lbs. (101 Nm).
22. Apply 22–44 lbs of force to the tensioner pushrod and measure the movement of the push rod. Standard value is 0.04 in. (1mm). A is the free length and B is the length when compressed. A minus B equals the movement. If the distance is outside the specification, replace the tensioner.
23. Place the tensioner into a soft-jawed vise to compress the tensioner.
24. After compressing the tensioner, place a pin into the plunger side hole to retain the plunger.

■ **When replacing the tensioner with a new part it will come with a pin.**

25. Align the timing marks on the crankshaft, oil pump and camshaft sprockets.
26. Remove the cylinder block and insert a screwdriver with a shaft diameter of 0.3 in. (8mm) into the hole.
27. With the timing marks still aligned, the tool must be able to go in at least 2.4 inches (60mm) or more. If it can only go in about 1 inch (25mm), turn the sprocket 1 complete revolution and insert the screwdriver again.
28. Install the timing belt starting at the crankshaft, oil pump sprocket and the camshaft sprocket, in that order.
29. Set the tension pulley so that the pin holes are at the bottom, press the tension pulley lightly against the timing belt and tighten the fixing bolt.
30. Adjust the belt tension by turning the crankshaft 1/4 turn counterclockwise, then turing it back clockwise until the timing marks are aligned.

**** WARNING**

When tightening the fixing bolt, make sure the tension pulley does not turn with the bolt.

31. Loosen the tension pulley fixing bolt, and use special tool MD998767 or equivalent and a torque wrench to tighten the fixing bolt to 30–40 ft. lbs. (42–54 Nm) while applying 31 inch lbs. (3.5 Nm) of tension to the timing belt.
32. Remove the set pin from the tensioner.
33. Rotate the crankshaft pulley two revolutions clockwise and allow it to settle for 15 minutes.
34. Measure the protrusion of the tensioner push rod and ensure it is 0.15–0.18 in. (3.8–4.5mm). If the measurement is not correct, repeat the tensioner adjusting procedure.
35. Install the upper and lower timing belt covers.
36. Install the engine mount stopper and bracket.
37. Install the engine mount insulator and tighten the bolt to 60 ft. lbs. (81 Nm).
38. Install the crankshaft damper and tighten to 105 ft. lbs. (142 Nm).
39. Install the water pump pulley.
40. Install the accessory drive belts.
41. Install the coolant reservoir tank.
42. Install the right engine mount bracket and stopper
43. Install the right inner splash-shield.
44. Install the right front wheel.
45. Fill the cooling system.
46. Connect the negative battery cable.
47. Start the engine and check for proper operation.

2.4L (VIN S, X) DOHC ENGINES

**** CAUTION**

These engines are an interference type. If the timing belt has failed, valve-to-piston damage may have occurred. A compression check should always be performed to check for engine damage.

■ **You may need DRB scan tool to perform the crankshaft and camshaft relearn alignment procedure.**

■ **There are 2 versions of this engine. 1995–96 models use a hydraulic tensioner, and 1997–04 models use a mechanical timing belt tensioner.**

YEAR, MODEL & ENGINE APPLICATION

1995-00 Chrysler Cirrus, 1995-04 Sebring Convertible, 2001-04 Sebring Sedan, 2003-04 Dodge Neon, 1995-04 Stratus 1996-00 Plymouth Breeze .... 2.4L VIN S, X

REPLACEMENT INTERVALS

DaimlerChrysler recommends timing belt replacement every 105,000 miles.

LABOR TIMES

Breeze, Cirrus, Stratus, Sebring	
1995-03	
Replacement	2.3 hrs
Severe Service	2.5 hrs
2004	
Replacement	5.7 hrs
Severe Service	5.7 hrs
Neon	
Replacement	2.6 hrs
Severe Service	2.6 hrs

8807CHRCARg23

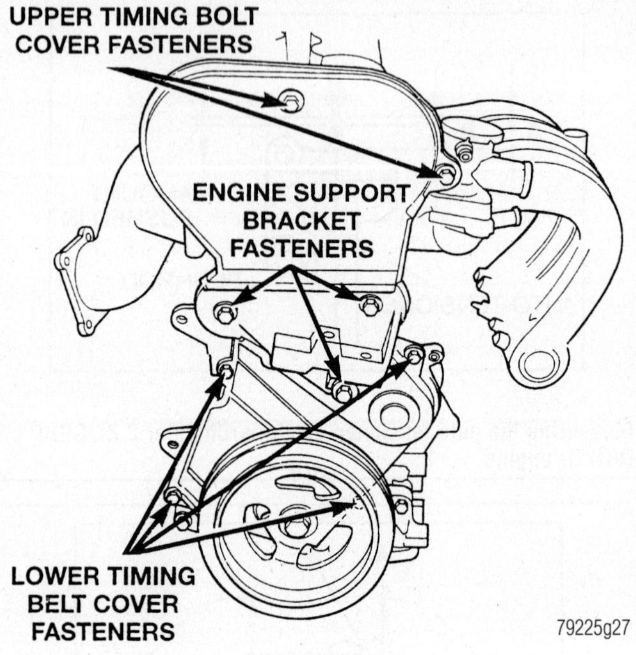

Timing cover and engine support bracket bolt locations—DaimlerChrysler 2.4L (VIN S, X) engine

1. Disconnect the negative battery cable.
2. Remove the engine undercover, if equipped.
3. Remove the right inner splash shield.
4. Remove the accessory drive belts.
5. On 1995–96 engines, remove the crankshaft pulley using a suitable puller.
6. On 1997–04 engines, remove the crankshaft pulley using special tools No, 1026 and 6827-a or equivalent.
7. Raise and support the vehicle.
8. Place a jack under the engine and raise it enough to relieve engine pressure.
9. Remove the right side engine mount and bracket.
10. Remove the timing belt cover.

■ **Do not rotate the crankshaft or the camshafts after the timing belt has been removed. Damage to the valve components may occur. Before removing the timing belt, always align the timing marks.**

11. Align the timing marks of the timing belt sprockets to the timing marks on the rear timing belt cover and oil pump cover.
12. On hydraulic tensioner engines, loosen the timing belt tensioner bolts.
13. Remove the timing belt and the tensioner.
14. Place the tensioner into a soft-jawed vise to compress the tensioner.
15. After compressing the tensioner, insert a pin (a 5/64 in. Allen wrench will also work) into the plunger side hole to retain the plunger until installation.
16. On mechanical tensioner engines, insert a 6 mm hex wrench in the tensioner lower hex hole.
17. Rotate the wrench tensioner counterclockwise until it locks and insert a 3mm hex wrench into the locking hole.
18. Remove the timing belt.
19. If necessary, remove the camshaft timing belt sprockets.
20. If necessary, remove the crankshaft timing belt sprocket using removal tool No. 6793, or equivalent.

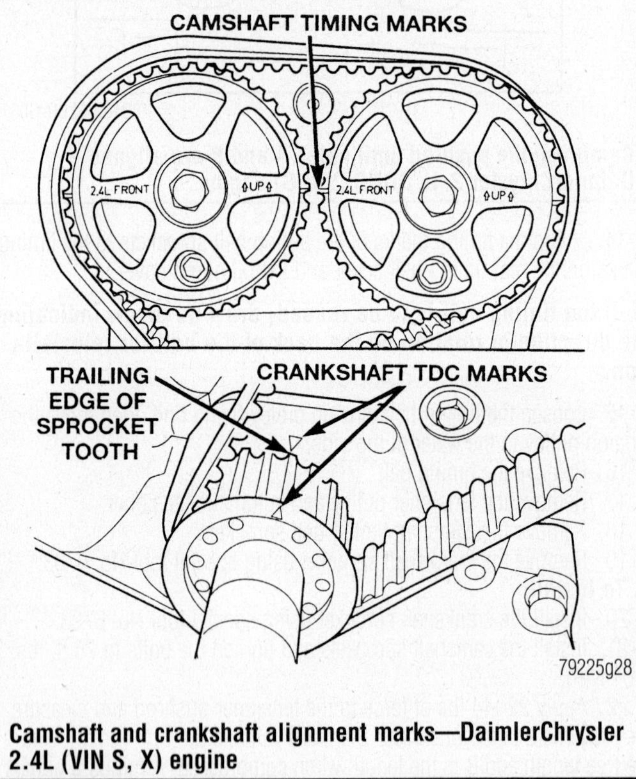

Camshaft and crankshaft alignment marks—DaimlerChrysler 2.4L (VIN S, X) engine

To Install:

21. If necessary, use tool No. 6792, or equivalent, to install the crankshaft timing belt sprocket onto the crankshaft.
22. If necessary, install the camshaft sprockets onto the camshafts. Install and tighten the camshaft sprocket bolts to 75 ft. lbs. (101 Nm).
23. On hydraulic tensioner engines, position the tensioner into a vise with soft jaws. The plug at the rear of tensioner protrudes, be sure to use a washer as a spacer to protect the plug from contacting vise jaws.

24. Slowly push the rod into the tensioner until the set hole in rod is aligned with set hole in the auto-tensioner.
25. Insert a 0.055 in. (1.4mm) wire into the aligned set holes.
26. Set the crankshaft sprocket to Top Dead Center (TDC) by aligning the notch on the sprocket with the arrow on the oil pump housing.
27. Set the camshafts to align the timing marks on the sprockets.
28. Move the crankshaft to 1/2 notch before TDC.
29. Install the timing belt starting at the crankshaft, then around the water pump sprocket, idler pulley, camshaft sprockets and around the tensioner pulley.
30. Move the crankshaft sprocket to TDC to take up the belt slack.
31. Install the tensioner on the engine block but do not tighten.
32. Using a torque wrench on the tensioner pulley, apply 250 inch lbs. (28 Nm) of torque to the tensioner pulley.
33. With torque being applied to the tensioner pulley, move the tensioner up against the tensioner pulley bracket and tighten the fasteners to 23 ft. lbs. (31 Nm).
34. Remove the tensioner plunger pin, the tension is correct when the plunger pin can be removed and reinserted easily.

✳✳ WARNING

If any binding is felt when adjusting the timing belt tension by turning the crankshaft, STOP turning the engine, because the pistons may be hitting the valves.

35. Rotate the crankshaft 2 revolutions and recheck the timing marks. Wait several minutes and then recheck that the plunger pin can easily be removed and installed.
36. On mechanical tensioner engines, align the crankshaft sprocket timing marks.
37. Position the exhaust camshaft sprocket so it is 1/2 notch below the intake camshaft sprocket.
38. Install the timing belt starting at the crankshaft, then around the water pump sprocket, idler pulley, camshaft sprockets and around the tensioner pulley.
39. Rotate the exhaust camshaft counterclockwise to align the timing marks and remove the slack from the timing belt.
40. Remove the 3 mm wrench from the tensioner.
41. Rotate the crankshaft pulley 2 revolutions clockwise and check that the timing marks are aligned.
42. Install the front timing belt cover.
43. Install the engine mount bracket.
44. Install the right engine mount.
45. Remove the floor jack from under the vehicle.
46. Install the crankshaft damper and tighten it to 105 ft. lbs. (142 Nm).
47. Install the accessory drive belts and adjust to the proper tension.
48. Install the engine undercover.
49. Install the right inner splash shield.
50. Reconnect the negative battery cable.
51. Perform the crankshaft and camshaft relearn alignment procedure using the DRB scan tool, or equivalent.

2.5L (VIN H) ENGINES

✳✳ CAUTION

These engines are an interference type. If the timing belt has failed, valve-to-piston damage may have occurred. A compression check should always be performed to check for engine damage.

YEAR, MODEL & ENGINE APPLICATION

1995-00 Chrysler Cirrus, 1996-00 Sebring Convertible, 1995-00 Dodge Stratus	2.5L VIN H

REPLACEMENT INTERVALS

DaimlerChrysler recommends timing belt replacement every 60,000 miles.

LABOR TIMES

1995-00	
Replacement	3.0 hrs
Severe Service	3.3 hrs

8807CHRg24

1. Disconnect the negative battery cable from the left strut tower. The ground cable is equipped with an insulator grommet which should be placed on the stud to prevent the negative battery cable from accidentally grounding.
2. Remove the air cleaner inlet.
3. On 1995 models, drain the cooling system, disconnect the coolant hoses and necessary connectors and remove the radiator.
4. On all models, raise and safely support the vehicle.
5. Remove the right front wheel.
6. Remove the right inner splash-shield.
7. Remove the accessory drive belts.
8. Remove the power steering pump and position it aside with the lines connected.
9. Remove the power steering pump bracket, if required.
10. Place a suitable floor jack under the vehicle to support the engine.
11. Remove the crankshaft damper.
12. Remove the right engine mount
13. Remove the right engine mount bracket
14. Remove the timing belt cover(s).
15. Loosen the timing belt tensioner bolts.

■ Before removing timing belt, be sure to align the camshaft and crankshaft sprocket timing marks.

16. If the present timing belt is going to be reused, mark the running direction of the timing belt for installation. Remove the timing belt and the tensioner.
17. Remove the camshaft timing belt sprockets from the camshaft, if necessary.
18. Remove the crankshaft timing belt sprocket and key.
19. Place the tensioner into a soft-jawed vise to compress the tensioner.
20. After compressing the tensioner place a pin into the plunger side hole to retain the plunger until installation.

To Install:

21. If removed, reinstall the camshaft sprockets onto the camshaft. Install the camshaft sprocket bolt and tighten to 65 ft. lbs. (88 Nm).
22. If removed, reinstall the crankshaft sprocket and key onto the crankshaft.
23. Set the camshafts to align the timing marks on the sprockets with the marks on the rear timing belt cover.
24. Set the crankshaft sprocket to Top Dead Center (TDC) by aligning the notch on the sprocket with the arrow on the oil pump housing, then back off the sprocket 3 notches before TDC.
25. Install the belt on the rear camshaft sprocket first.

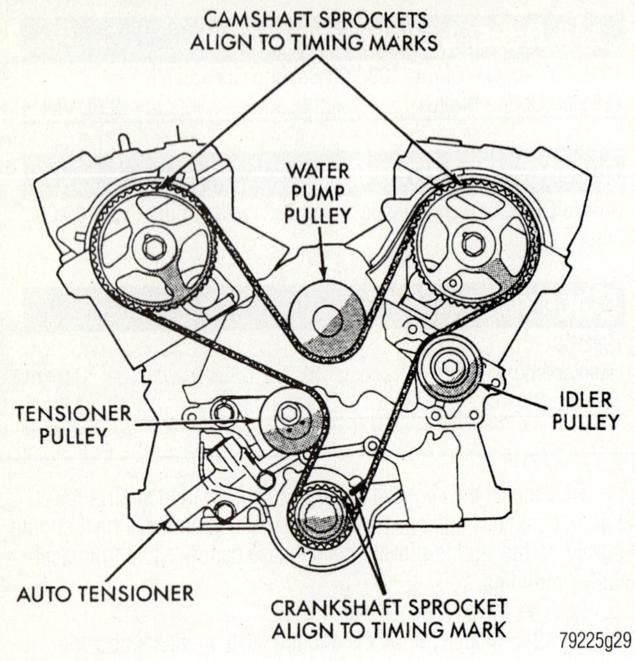

Crankshaft and camshaft sprocket timing mark alignment—DaimlerChrysler 2.5L (VIN H) engine

26. Install a binder clip on the belt to the sprocket so it won't slip out of position.
27. Keeping the belt taut, install it under the water pump pulley and around the front camshaft sprocket.
28. Install a binder clip on the front sprocket and belt.
29. Rotate the crankshaft to TDC.
30. Continue routing the belt by the idler pulley and around the crankshaft sprocket to the tensioner pulley.
31. Move the crankshaft sprocket clockwise to TDC to take up the belt slack. Check that all timing marks are in alignment.
32. Reinstall the tensioner to the block but do not tighten it at this time.
33. Using special tool No. MD998767 or equivalent, and a torque wrench on the tensioner pulley, apply 39 inch lbs. (4.4 Nm) of torque to the tensioner. Tighten the tensioner pulley bolt to 35 ft. lbs. (48 Nm).
34. With torque being applied to the tensioner pulley, move the tensioner up against the tensioner bracket and tighten the fasteners to 17 ft. lbs. (23 Nm).
35. Remove the tensioner plunger pin, the tension is correct when the plunger pin can be removed and replaced easily.
36. Rotate the crankshaft 2 revolutions clockwise and recheck the timing marks. Check to be sure the tensioner plunger pin can be easily installed and removed. If the pin does not remove and install easily, perform the procedure again.
37. Install the timing belt cover(s).
38. Install the engine mount bracket.
39. Install the right engine mount.
40. Remove the engine support.
41. Install the crankshaft damper and tighten to 134 ft. lbs. (182 Nm).
42. Install the power steering pump
43. On 1995 models, install the radiator and fill the cooling system.
44. Install the accessory drive belts and adjust them.
45. Install the right inner splash-shield.
46. Install the right front wheel.
47. Install the air cleaner inlet, if removed.
48. Install the coolant reservoir, if removed.
49. Perform the crankshaft and camshaft relearn alignment procedure using the DRB scan tool or equivalent.

2.5L (VIN K) ENGINES

YEAR, MODEL & ENGINE APPLICATION

1986 Chrysler Laser, 1986-94 LeBaron, 1986-87 New Yorker, 1986-88 Town & Country, 1984-89 Dodge Aries, 1987-95 Caravan, 1989-99 Dakota, 1988-91 Daytona Shelby, 1985-93 Daytona, 1988-93 Dynasty, 1985-89 Lancer, 1985-87 Mini Ram Van, 1988-94 Shadow, 1989-95 Spirit, 1984-88 600, 1989-95 Plymouth Acclaim, 1986-89 Caravelle, 1986-89 Reliant, 1988-94 Sundance, 1987-95 Voyager .... 2.5L VIN K

REPLACEMENT INTERVALS

Prior to 1993 DaimlerChrysler does not provide a replacement interval. Standard industry practice is to replace timing belts every 60,000 miles. For 1993 and later models, recommended replacement interval is 90,000 miles.

LABOR TIMES

Acclaim, Shadow, Spirit & Sundance	
Replacement	2.5 hrs
Severe Service	2.9 hrs
With air conditioning add	0.3 hrs
With power steering add	0.3 hrs
Aries, Lancer, LeBaron, Limousine & Reliant	
Replacement	2.5 hrs
Severe Service	3.0 hrs
With air conditioning add	0.3 hrs
With power steering add	0.3 hrs
Caravelle, Daytona, Daytona Shelby, Dynasty, Laser, New Yorker & 600	
Replacement	2.5 hrs
Severe Service	2.8 hrs
With air conditioning add	0.3 hrs
With power steering add	0.3 hrs
Caravan, Town & Country & Voyager	
Replacement	2.3 hrs
Severe Service	2.6 hrs
With air conditioning add	0.2 hrs
With power steering add	0.2 hrs
Dakota	
Replacement	1.6 hrs
Severe Service	1.9 hrs
With air conditioning add	0.5 hrs
With power steering add	0.1 hrs

8807CHRg25

** CAUTION

These engines are considered freewheeling. If the timing belt fails, valve-to-piston damage is unlikely, however a precautionary compression check should be performed to verify no engine damage.

1. Disconnect the negative battery cable.
2. Position the engine so the No. 1 piston is at Top Dead Center (TDC) of the compression stroke.
3. Remove the nuts and bolts that attach the upper cover to the valve cover, block or cylinder head.
4. Remove the bolt that attaches the upper cover to the lower cover.
5. Remove the upper cover.
6. Remove the right tire and wheel assembly. Remove the right-side inner splash shield.
7. Remove the crankshaft pulley, water pump pulley and the accessory drive belt(s).
8. Remove the lower cover attaching bolts and the cover from the engine.
9. Remove the timing belt tensioner and allow the belt to hang free.
10. Place a floor jack under the engine and separate the right motor mount.
11. Remove the air conditioning compressor belt idler pulley, if equipped, and remove the mounting stud. Remove the compressor/alternator bracket as follows:
    a. Remove the alternator pivot bolt and remove the alternator from the bracket. Turn the alternator so the wire connections are facing up and disconnect the harness connectors from the rear of the alternator.
    b. Remove the air conditioning compressor belt idler.
    c. Remove the right engine mount yoke screw securing engine mount support strut to the engine.
    d. Remove the 5 side mounting bolts retaining the bracket to the front of the engine.
    e. Remove the front mounting nut. Remove the front bolt and strut and rotate the solid mount bracket away from the engine. Slide the bracket on the stud until free of the mounting studs and remove from the engine.
    f. Remove the timing belt from the vehicle.

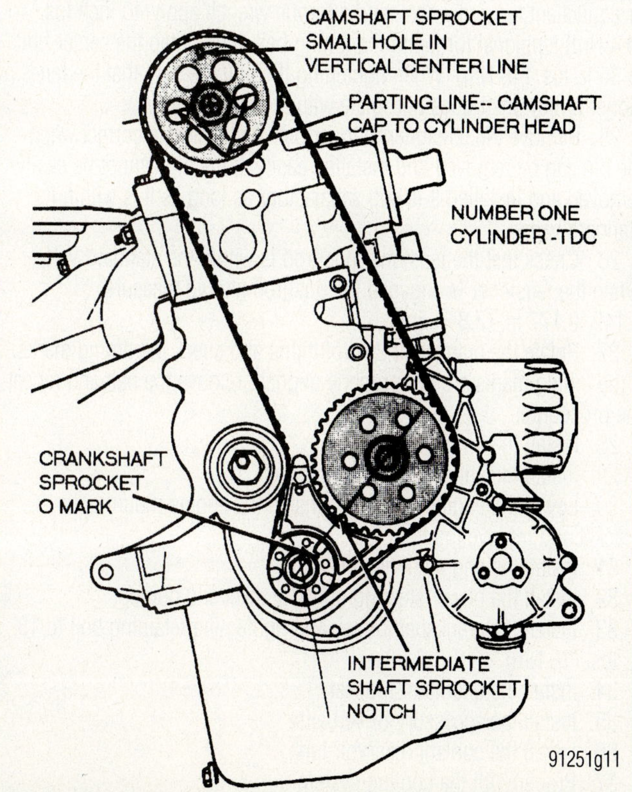

Timing belt alignment mark locations—DaimlerChrysler 2.5L (VIN K) engine

To Install:

12. Turn the crankshaft sprocket and intermediate shaft sprocket until the marks are aligned. Use a straight-edge from bolt to bolt to confirm alignment.
13. Turn the camshaft until the small hole in the sprocket is at the top and the arrows on the hub are aligned with the camshaft cap to cylinder head mounting lines. When looking through the hole on top of the camshaft sprocket, the uppermost center nipple of the valve cover end seal should be at the center of the hole. Use a mirror to check the alignment of the arrows so it is viewed straight on and not at an angle from above. Install the belt but let it hang free at this point.
14. Install the air conditioning compressor/alternator bracket, idler pulley and motor mount. Remove the floor jack. Raise the vehicle and support safely. Have the tensioner at an arm's reach because the timing belt will have to be held in position with one hand.
15. To properly install the timing belt, reach up and engage it with the camshaft sprocket. Turn the intermediate shaft counterclockwise slightly, then engage the belt with the intermediate shaft sprocket. Hold the belt against the intermediate shaft sprocket and turn clockwise to take up all tension; if the timing marks are out of alignment, repeat until alignment is correct.
16. Using a wrench, turn the crankshaft sprocket counterclockwise slightly and wrap the belt around it. Turn the sprocket clockwise so there is no slack in the belt between sprockets; if the timing marks are out of alignment, repeat until alignment is correct.

■ **If the timing marks are aligned but slack exists in the belt between either the camshaft and intermediate shaft sprockets or the intermediate and crankshaft sprockets, the timing will be incorrect when the belt is tensioned. All slack must be only between the crankshaft and camshaft sprockets.**

17. Install the tensioner and install the mounting bolt loosely. Place the special tensioning tool C-4703 on the hex of the tensioner so the weight is at about the 9 o'clock position (parallel to the ground, hanging toward the rear of the vehicle) (15 degrees of horizontal.)
18. Hold the tool in position and tighten the bolt to 45 ft. lbs. (61 Nm). Do not pull the tool past the 9 o'clock position; this will make the belt too tight and will cause it to howl or possibly break.
19. Lower the vehicle and recheck the camshaft sprocket positioning. If it is correct install the timing belt covers and all related parts.
20. Connect the negative battery cable and road test the vehicle.

2.5L (VIN N) ENGINES

YEAR, MODEL & ENGINE APPLICATION

Application	Engine
1995-00 Chrysler Sebring, Dodge Avenger	2.5L VIN N

REPLACEMENT INTERVALS

DaimlerChrysler recommends timing belt replacement every 60,000 miles.

LABOR TIMES

1995-00	
Replacement	3.1 hrs
Severe Service	3.4 hrs
With air conditioning add	0.2 hrs

8807CHRg26

✻✻ CAUTION

These engines are an interference type. If the timing belt has failed, valve-to-piston damage may have occurred. A compression check should always be performed to check for engine damage.

1. Disconnect the negative battery cable.
2. Remove the coolant reservoir tank.
3. Remove the accessory drive belts.
4. Remove the alternator bracket.
5. Remove the engine undercover.
6. Place a floor jack under the engine oil pan, with a block of wood in between, and raise the engine so that the weight of the engine is no longer being applied to the engine support bracket.
7. Using Crankshaft Holding tools MB990767 and MB998754, or their equivalents, remove the crankshaft bolt and remove the pulley.
8. Remove the heated oxygen sensor connection.
9. Remove the upper engine mount. Spraying lubricant, slowly remove the reamer (alignment) bolt and remaining bolts and remove the engine support bracket.

■ **The reamer bolt is sometimes heat-seized on the engine support bracket**

10. Remove the power steering pump with the hose attached and position it aside.
11. Remove the power steering pump bracket.
12. Remove the front timing belt covers.
13. If the timing belt is to be reused, draw an arrow indicating the direction of rotation on the back of the belt for reinstallation.
14. Align the timing marks by turning the crankshaft with MD998769 Crankshaft Turning tool or its equivalent. Loosen the center bolt on the timing belt tensioner pulley and remove the belt.

✻✻ WARNING

Do not rotate the crankshaft or camshaft after removing the timing belt, or valvetrain components may be damaged. Always align the timing marks before removing the timing belt.

15. Check the belt tensioner for leaks and check the pushrod for cracks.
16. If the timing belt tensioner is to be replaced, remove the retaining bolts and remove the timing belt tensioner. When the timing belt tensioner is removed from the engine it is necessary to compress the plunger into the tensioner body.
17. Place the tensioner in a vise and slowly compress the plunger. Take care not to damage the pushrod.

■ **Position the tensioner in the vise the same way it will be installed on the engine. This is to ensure proper pin orientation for when the tensioner is installed on the engine.**

18. When the plunger is compressed into the tensioner body, install a pin through the body and plunger to hold the plunger in place until the tensioner is installed.

To Install:

19. Check that all timing marks are still aligned.
20. Install the timing belt tensioner and tighten the retaining bolts to 17 ft. lbs. (24 Nm), but do not remove the pin at this time.

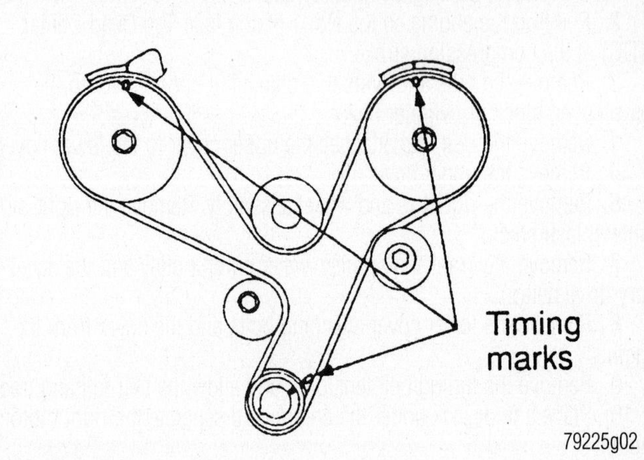

Camshaft and crankshaft alignment marks—DaimlerChrysler 2.5L (VIN N) engine

21. Use binder clips or other suitable tool to secure the timing belt and to prevent it from slacking. Install the timing belt. Starting at the crankshaft, go around the idler pulley, then the front camshaft sprocket, the water pump pulley, the rear camshaft sprocket, then around the tensioner pulley.
22. Be sure the belt is tight between the crankshaft and front camshaft sprocket, between the camshaft sprockets and the water pump. Gently raise the tensioner pulley, so that the belt does not sag, and temporarily tighten the center bolt.
23. Move the crankshaft 1/4 turn counterclockwise, then turn it clockwise to the position where the timing marks are aligned.
24. Loosen the center bolt of the tensioner pulley. Using MD998767 or equivalent tensioner tool, and a torque wrench apply 40 inch lbs. (4.4 Nm) tensional torque to the timing belt and tighten the center bolt to 35 ft. lbs. (48 Nm). When tightening the bolt, be sure that the tensioner pulley shaft does not rotate with the bolt.
25. Remove the tensioner plunger pin. Pretension is correct when the pin can be removed and installed easily. If the pin cannot be easily removed and installed it is still satisfactory as long as it is within its standard value.
26. Check that the tensioner pushrod is within the standard value. When the tensioner is engaged the pushrod should measure 0.149–0.177 in. (3.8–4.5mm).
27. Rotate the crankshaft 2 revolutions and check the timing marks. If the timing marks are not properly aligned, remove the belt and repeat the procedure.
28. Install the timing belt covers.
29. Install the engine mounting bracket.
30. Lower the engine enough to install the engine mount onto bracket and remove the floor jack.
31. Install the engine undercover.
32. Install the power steering pump bracket and pump.
33. Install the crankshaft pulley and tighten the retaining bolt to 13 ft. lbs. (18 Nm).
34. Install the alternator bracket.
35. Install the accessory drive belts.
36. Install the coolant reservoir tank.
37. Properly fill the cooling system.
38. Connect the negative battery cable.
39. Check for leaks and proper engine and cooling system operation.

3.0L (VIN 3, H AND S) SOHC ENGINES

YEAR, MODEL & ENGINE APPLICATION

1990-93 Chrysler Imperial, 1988-95 LeBaron, 1988-93 New Yorker, 1990 Town & Country, 2001-04 Sebring, 1988-00 Dodge Caravan, 1990-93 Daytona, 1988-93 Dynasty, 1990-93 Ram-50 1993-94 Shadow, 1989-95 Spirit, 1991-97 Stealth, 2001-04 Stratus, 1989-95 Plymouth Acclaim 1992-94 Sundance, 1988-00 Voyager .... 3.0L VIN 3, H, S

REPLACEMENT INTERVALS

DaimlerChrysler recommends timing belt replacement every 60,000 miles.

LABOR TIMES

Acclaim, Daytona, Spirit & LeBaron	
Replacement	2.0 hrs
Severe Service	2.3 hrs
With air conditioning add	0.3 hrs
With power steering add	0.3 hrs
Caravan, Town & Country & Voyager	
Replacement	1.9 hrs
Severe Service	2.2 hrs
With air conditioning add	0.2 hrs
With power steering add	0.2 hrs
Dynasty, Imperial & New Yorker	
Replacement	2.0 hrs
Severe Service	2.2 hrs
With air conditioning add	0.3 hrs
Ram-50	
Replacement	2.3 hrs
Severe Service	2.8 hrs
With air conditioning add	0.7 hrs
With power steering add	0.1 hrs
Sebring & Stratus	
2001-02	
Replacement	2.6 hrs
Severe Service	2.8 hrs
2003-04	
Replacement	2.9 hrs
With power steering add	3.3 hrs
Shadow & Sundance	
Replacement	1.9 hrs
Severe Service	2.1 hrs
With air conditioning add	0.3 hrs
With power steering add	0.3 hrs
Stealth	
Replacement	2.7 hrs
Severe Service	3.0 hrs
With air conditioning add	0.3 hrs

8807CHRg27

All Models Except Ram-50, Sebring, Stealth & Stratus

**** CAUTION**

These engines are an interference type. If the timing belt has failed, valve-to-piston damage may have occurred. A compression check should always be performed to check for engine damage.

1. Disconnect the negative battery cable.
2. Position the engine so the No. 1 cylinder is at Top Dead Center (TDC) of the compression stroke. Disconnect the negative battery cable.
3. Remove the accessory drive belts.
4. Remove the A/C compressor without disconnecting the lines and position aside.
5. Remove the power steering pump without disconnecting the lines and position aside.
6. Remove the A/C mounting bracket.
7. Remove the A/C and power steering drive belt tensioners.
8. Remove the crankshaft pulley and damper.
9. Using the proper equipment, slightly raise the engine to take the weight off the engine mount. Remove the engine mount bracket.
10. Remove the timing belt cover cap.
11. Remove the timing belt upper and lower covers.
12. If the same timing belt will be reused, mark the direction of the timing belt's rotation for installation in the same direction. Make sure the engine is positioned so the No. 1 cylinder is at the Top Dead Center (TDC) of its compression stroke and the timing marks are aligned with the engine's timing mark indicators.
13. Loosen the timing belt tensioner bolt and remove the belt. If the tensioner is not being removed, position it as far away from the center of the engine as possible and tighten the bolt.
14. If the tensioner is being removed, paint the outside of the spring to ensure that it is not installed backward. Unbolt the tensioner and remove it along with the spring.

**** WARNING**

Do not rotate the camshafts when the timing belt is removed from the engine. Turning the camshaft when the timing belt is removed could cause the valves to interfere with the pistons thus causing severe internal engine damage.

To Install:

15. Install the tensioner, if removed, and hook the upper end of the spring to the water pump pin and the lower end to the tensioner in exactly the same position as originally installed. If not already done, position both camshafts so the marks align with those on the rear. Rotate the crankshaft so the timing mark aligns with the mark on the front cover.
16. Install the timing belt on the crankshaft sprocket and while keeping the belt tight on the tension side, install the belt on the front camshaft sprocket.
17. Install the belt on the water pump pulley, then the rear camshaft sprocket and the tensioner.
18. Rotate the front camshaft counterclockwise to tension the belt between the front camshaft and the crankshaft. If the timing marks became misaligned, repeat the procedure.
19. Install the crankshaft sprocket flange.
20. Loosen the tensioner bolt and allow the spring to apply tension to the belt.
21. Turn the crankshaft 2 full turns in the clockwise direction until the timing marks align again. Now that the belt is properly tensioned, torque the tensioner lock bolt to 21 ft. lbs. (29 Nm). Measure the belt tension between the rear camshaft sprocket and the crankshaft with belt tension gauge. The specification is 46–68 lbs. (210–310 N).
22. Install the timing covers. Make sure all pieces of packing are positioned in the inner grooves of the covers when installing.
23. Install the engine mount bracket and remove the engine support fixture.
24. Install the crankshaft pulley. Tighten the bolt to 108–116 ft. lbs. (150–160 Nm).

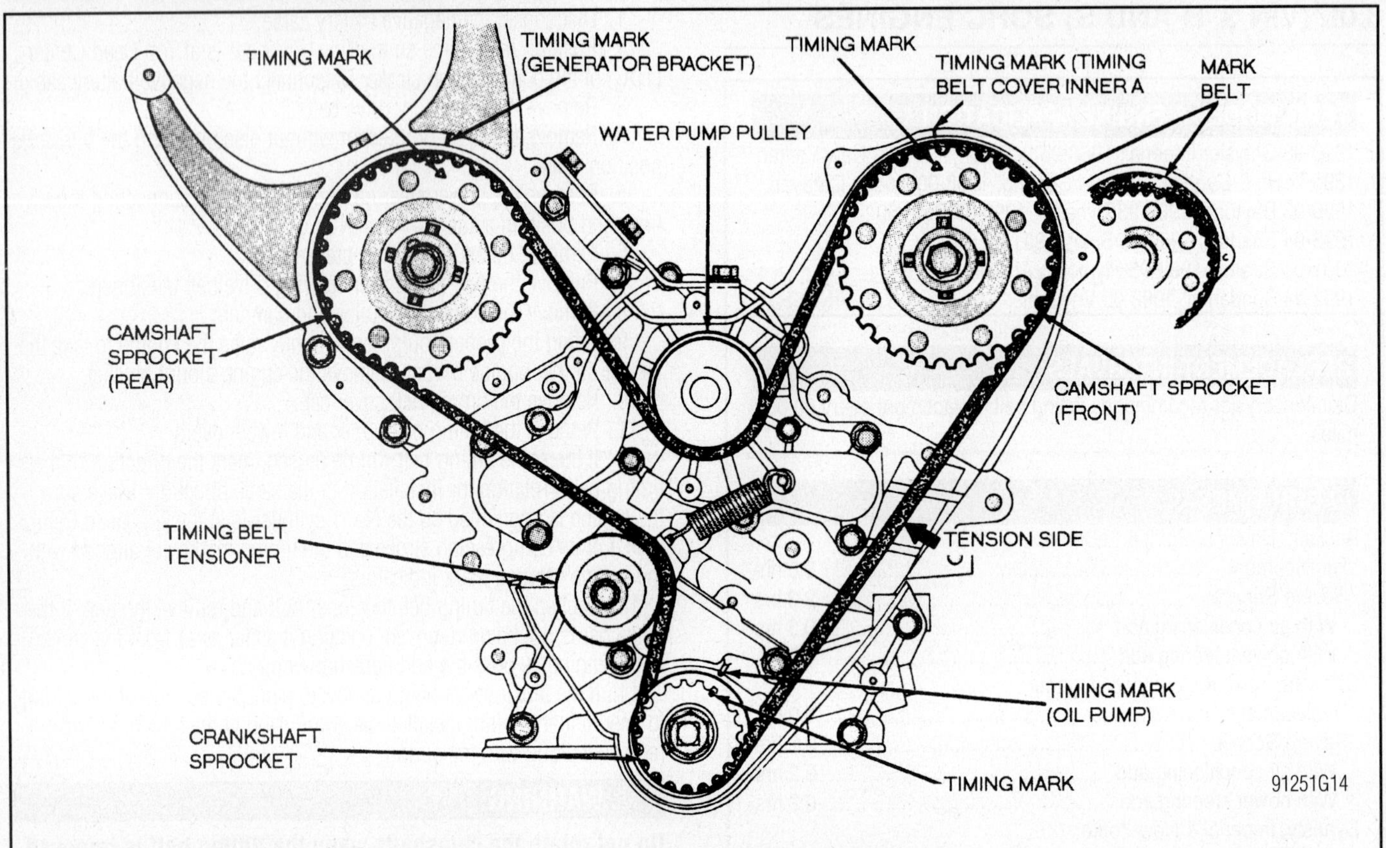

Timing belt alignment mark locations—DaimlerChrysler 3.0L (VIN 3, H and S) SOHC engine

25. Install the power steering pump and A/C compressor.
26. Install the tension pulleys and drive belts.
27. Connect the negative battery cable and road test the vehicle.

Ram-50

✻✻ CAUTION

These engines are an interference type. If the timing belt has failed, valve-to-piston damage may have occurred. A compression check should always be performed to check for engine damage.

1. Disconnect the negative battery cable.
2. Position the engine so the No. 1 cylinder is at Top Dead Center (TDC) of the compression stroke. Disconnect the negative battery cable.
3. Drain the engine coolant.
4. Remove the upper radiator hose.
5. Remove the radiator shroud.
6. Remove the cooling fan assembly.
7. Remove the accessory drive belts.
8. Remove the power steering pump without disconnecting the lines and position aside.
9. Remove the A/C compressor without disconnecting the lines and position aside.
10. Remove the oil pump brackets and tensioner bracket.
11. Remove the A/C and cooling fan mounting brackets.
12. Remove the timing belt upper and lower covers.
13. Remove the crankshaft pulley and flange.
14. If the same timing belt will be reused, mark the direction of the timing belt's rotation for installation in the same direction. Make sure the engine is positioned so the No. 1 cylinder is at the Top Dead Center (TDC) of its compression stroke and the timing marks are aligned with the engine's timing mark indicators.
15. Loosen the timing belt tensioner bolt and remove the belt.

To Install:

16. Position both camshafts so the marks align with those on the rear. Rotate the crankshaft so the timing mark aligns with the mark on the front cover.
17. Install the timing belt on the crankshaft sprocket and while keeping the belt tight on the tension side, install the belt on the right camshaft sprocket.
18. Install the belt on the water pump pulley, then the left camshaft sprocket and the tensioner.
19. Rotate the right camshaft counterclockwise to tension the belt between the camshaft and the crankshaft. If the timing marks became misaligned, repeat the procedure.
20. Install the crankshaft sprocket flange.
21. Loosen the tensioner bolt and allow the spring to apply tension to the belt.
22. Turn the crankshaft 2 full turns in the clockwise direction until the timing marks align again. Now that the belt is properly tensioned, torque the tensioner lock bolt to 21 ft. lbs. (29 Nm). Measure the belt tension between the rear camshaft sprocket and the crankshaft with belt tension gauge. The specification is 46–68 lbs. (210–310 N).
23. Install the timing belt upper and lower covers.
24. Install the A/C and cooling fan mounting brackets.
25. Install the oil pump brackets and tensioner bracket.
26. Install the A/C compressor.
27. Install the power steering pump.
28. Install the accessory drive belts.
29. Install the cooling fan assembly.

30. Install the radiator shroud.
31. Install the upper radiator hose.
32. Connect the negative battery cable and road test the vehicle.

Sebring & Stratus

✱✱ CAUTION

These engines are an interference type. If the timing belt has failed, valve-to-piston damage may have occurred. A compression check should always be performed to check for engine damage.

1. Disconnect the negative battery cable.
2. Position the engine so the No. 1 cylinder is at Top Dead Center (TDC) of the compression stroke. Disconnect the negative battery cable.
3. Remove the accessory drive belts.
4. Remove the oil dipstick and tube.
5. Remove the alternator and brace.
6. Remove the coolant reservoir.
7. Using the proper equipment, slightly raise the engine to take the weight off the engine.
8. Remove the engine mount bracket.
9. Remove the crankshaft pulley using special tools MB990767 and MB998715.
10. Remove the power steering belt tensioner.
11. Remove the front upper timing belt covers.
12. Remove the lower timing belt cover.
13. Remove the engine support bracket.
14. If the same timing belt will be reused, mark the direction of the timing belt's rotation for installation in the same direction. Make sure the engine is positioned so the No. 1 cylinder is at the Top Dead Center (TDC) of its compression stroke and the timing marks are aligned with the engine's timing mark indicators.
15. Loosen the timing belt tensioner bolt and remove the belt.

✱✱ WARNING

Do not rotate the camshafts when the timing belt is removed from the engine. Turning the camshaft when the timing belt is removed could cause the valves to interfere with the pistons thus causing severe internal engine damage.

To Install:

16. Apply 22–44 lbs of force to the tensioner pushrod and measure the movement of the push rod. Standard value is 0.04 in. (1mm). A is the free length and B is the length when compressed. A minus B equals the movement. If the distance is outside the specification, replace the tensioner.
17. Place the tensioner into a soft-jawed vise to compress the tensioner.
18. After compressing the tensioner, place a pin into the plunger side hole to retain the plunger.

■ **When replacing the tensioner with a new part it will come with a pin.**

19. Install the tensioner and tighten the bolts to 17 ft. lbs. (23 Nm).
20. Install the timing belt starting at the crankshaft, idler pulley, left camshaft sprocket, water pump pulley, right camshaft sprocket and tensioner pulley.
21. Adjust the belt tension by turning the right camshaft sprocket counterclockwise until the belt is tensioned.
22. Press the tension pulley against the timing belt and tighten the fixing bolt.
23. Rotate the crankshaft 1/4 turn counterclockwise, then turing it back clockwise until the timing marks are aligned.

✱✱ WARNING

When tightening the fixing bolt, make sure the tension pulley does not turn with the bolt.

24. Loosen the tension pulley fixing bolt, and use special tool MD998767 or equivalent and a torque wrench to tighten the fixing bolt to 30–40 ft. lbs. (42–54 Nm) while applying 31 inch lbs. (3.5 Nm) of tension to the timing belt.
25. Remove the set pin from the tensioner.
26. Rotate the crankshaft pulley two revolutions clockwise and allow it to settle for 15 minutes.
27. Measure the protrusion of the tensioner push rod and ensure it is 0.15–0.18 in. (3.8–4.5mm). If the measurement is not correct, repeat the tensioner adjusting procedure.
28. Install the right engine mount bracket.
29. Install the lower timing belt cover.
30. Install the front upper timing belt covers.
31. Install the power steering belt tensioner.
32. Install the crankshaft pulley after lubricating the bolts threads, using special tools MB990767 and MB998715. Tighten the bolt to 134 ft. lbs. (182 Nm).
33. Install the coolant reservoir.
34. Install the alternator and brace.
35. Install the oil dipstick and tube.
36. Install the accessory drive belts.
37. Connect the negative battery cable and road test the vehicle.

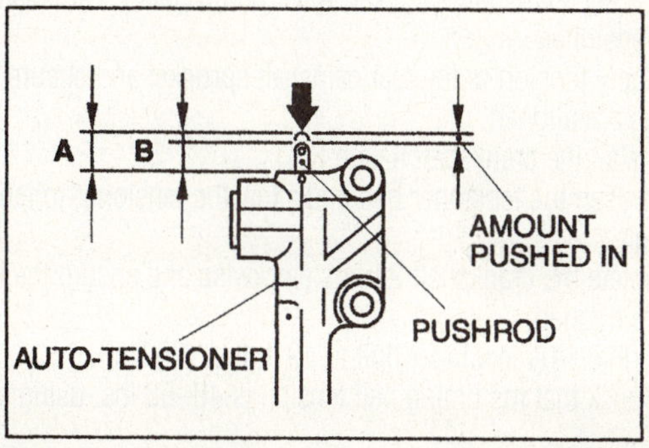

38189-SEBR-G05

Measuring the auto tensioner—DaimlerChrysler 3.0L SOHC Sebring & Stratus engine

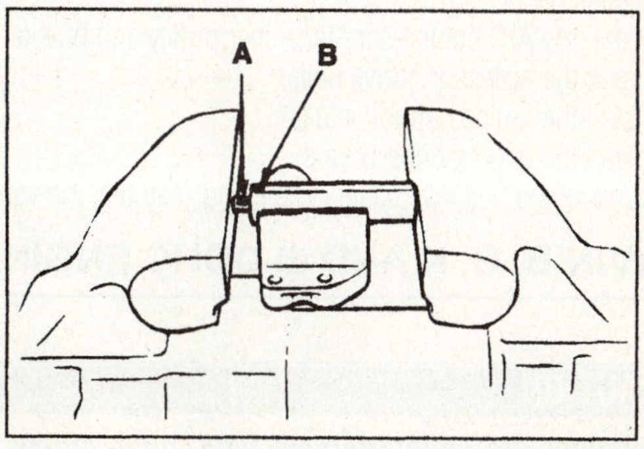

68189-SEBR-G06

Compress the pushrod until holes A and B are aligned—DaimlerChrysler 3.0L SOHC Sebring & Stratus engine

Stealth

⁂ CAUTION

These engines are an interference type. If the timing belt has failed, valve-to-piston damage may have occurred. A compression check should always be performed to check for engine damage.

1. Disconnect the negative battery cable.
2. Remove the engine undercovers.
3. Remove the cruise control actuator.
4. Place a suitable jack under the engine and raise it enough to take the weight off the engine mounts.
5. Remove the accessory drive belts.
6. Remove the A/C compressor tensioner pulley and bracket.
7. Remove the engine mount support and bracket.
8. Remove the power steering pump without disconnecting the lines and wire aside.
9. Remove the crankshaft pulley.
10. Remove the timing belt covers.
11. Remove the crankshaft flange.
12. Ensure the crankshaft and camshaft sprocket timing marks are aligned.
13. Loosen the timing belt tensioner bolt, then using a screwdriver, rotate the tensioner counterclockwise and tighten the bolt.
14. Remove the timing belt.

To Install:

15. Working counterclockwise, install the timing belt on the crankshaft sprocket, camshaft sprocket, water pump pulley, camshaft sprocket and the tensioner.
16. Apply tension to the rear camshaft sprocket and ensure the timing marks are aligned.
17. Install the crankshaft flange.
18. Loosen the tensioner bolt and allow the tensioner to tension the timing belt.
19. Rotate the crankshaft 2 turns clockwise and ensure the timing marks are aligned.
20. Tighten the tensioner bolt to 19 ft. lbs. (26 Nm).
21. Check that the timing belt tension is 46–68 lbs. using a tension gauge.
22. Install the timing belt covers.
23. Install the crankshaft pulley.
24. Install the engine mount support and bracket.
25. Install the power steering pump without disconnecting the lines and wire aside.
26. Install the A/C compressor tensioner pulley and bracket.
27. Install the accessory drive belts.
28. Install the cruise control actuator.
29. Install the engine undercovers.
30. Connect the negative battery cable and road test the vehicle.

3.0L (VIN B, C, K AND J) DOHC ENGINES

⁂ CAUTION

These engines are an interference type. If the timing belt has failed, valve-to-piston damage may have occurred. A compression check should always be performed to check for engine damage.

YEAR, MODEL & ENGINE APPLICATION	
1991-97 Dodge Stealth	3.0L VIN B, C, K, J

REPLACEMENT INTERVALS
DaimlerChrysler recommends timing belt replacement every 60,000 miles.

LABOR TIMES	
1991-97	
Replacement	3.3 hrs
Severe Service	3.6 hrs
With air conditioning add	0.3 hrs
With turbocharger add	0.3 hrs

8807CHRg28

1. Disconnect the negative battery cable.

⁂ CAUTION

Wait at least 90 seconds after the negative battery cable is disconnected to prevent possible deployment of the air bag.

2. Position the engine so the No. 1 cylinder is at Top Dead Center (TDC) of the compression stroke. Disconnect the negative battery cable.
3. Remove the left front and side engine covers.
4. Remove the cruise control actuator.
5. Remove the accessory drive belts.
6. Remove the alternator. Remove the air hose and pipe.
7. Remove the power steering belt tensioner assembly.
8. Remove the crankshaft pulley.
9. Disconnect the brake fluid level sensor.
10. Remove the timing belt upper cover.
11. Using the proper equipment, slightly raise the engine to take the weight off the side engine mount. Remove the engine mount bracket.
12. Remove the engine support bracket. The mounting bolts are different lengths; mark them for proper installation.
13. Remove the alternator and A/C idler pulley.
14. Disconnect the camshaft and crankshaft position sensor connectors.
15. Remove the timing belt lower cover. Timing bolt cover mounting bolts are different in length, note their position during removal.
16. If the same timing belt will be reused, mark the direction of the timing belt's rotation for installation in the same direction. Make sure the engine is positioned so the No. 1 cylinder is at the TDC of its compression stroke and the timing marks are aligned with the engine's timing mark indicators on the valve covers or head.
17. Loosen the center bolt of tensioner pulley and unbolt auto-tensioner assembly. The auto-tensioner assembly must be reset to correctly adjust belt tension. Remove the timing belt.

⁂ WARNING

Do not rotate the camshafts when the timing belt is removed from the engine. Turning the camshaft when the timing belt is removed could cause the valves to interfere with the pistons thus causing severe internal engine damage.

18. Remove and position the auto-tensioner into a vise with soft jaws. The plug at the rear of tensioner protrudes, be sure to use a washer as a spacer to protect the plug from contacting vise jaws.

19. Slowly push the rod into the tensioner until the set hole in rod is aligned with set hole in the auto-tensioner.

20. Insert a 0.055 in. (1.4mm) wire into the aligned set holes. Unclamp the tensioner from vise and install to vehicle. Tighten tensioner to 17 ft. lbs. (24 Nm).

21. On 1991 DOHC 3.0L engines, clean and inspect both auto tensioner mounting bolts. Coat the threads of the old bolts with Mopar thread sealer 4318034 or equivalent. If new bolts are installed, inspect the heads of the new bolts. If there is white paint on the bolt head, no sealer is required. If there is no paint on the head of the bolt, apply a coat of thread sealer to the bolt. Install both bolts and tighten to 17 ft. lbs. (24 Nm).

To Install:

**** WARNING**

Turning the camshaft sprocket when the timing belt is removed could cause the valves to interfere with the pistons.

22. Align the mark on the crankshaft sprocket with the mark on the front case. Then move the sprocket 3 teeth clockwise to lower the piston so the valves do not touch the piston if the camshafts are being moved.

23. Turn each camshaft sprocket 1 at a time to align the timing marks with the mark on the valve cover or head. If the intake and exhaust valves of the same cylinder are opened simultaneously, they could interfere with each other. Therefore, if any resistance is felt, turn the other camshaft to move the valve.

24. Using binder clips to secure the timing belt to sprockets, install the timing belt in the following order. Be sure camshafts to cylinder heads and crankshaft to front cover timing marks are aligned.
   a. Exhaust camshaft sprocket (front bank).
   b. Intake camshaft sprocket (front bank).
   c. Water pump pulley.
   d. Intake camshaft sprocket (rear bank).
   e. Exhaust camshaft sprocket (rear bank).
   f. Idler pulley.
   g. Crankshaft pulley.
   h. Tensioner pulley.

■ **Since the camshaft sprockets turn easily, secure them with box wrenches to install timing belt.**

25. Align all timing marks and raise the tensioner pulley against the belt to remove slack, and snug the tensioner bolt.

26. After ensuring all timing marks are aligned, remove the binder clips.

27. Loosen the center bolt on the tensioner pulley. Using tool MD998767 or equivalent and a torque wrench, apply a torque of 86 inch lbs. (10 Nm). Tighten the tensioner bolt to 42 ft. lbs. (58 Nm) and make sure the tensioner does not rotate with the bolt.

28. Remove the set wire attached to the auto tensioner.

29. Rotate the crankshaft 2 complete turns clockwise and let it sit for approximately 5 minutes. Then, check that the set pin can easily be inserted and removed from the hole in the auto tensioner.

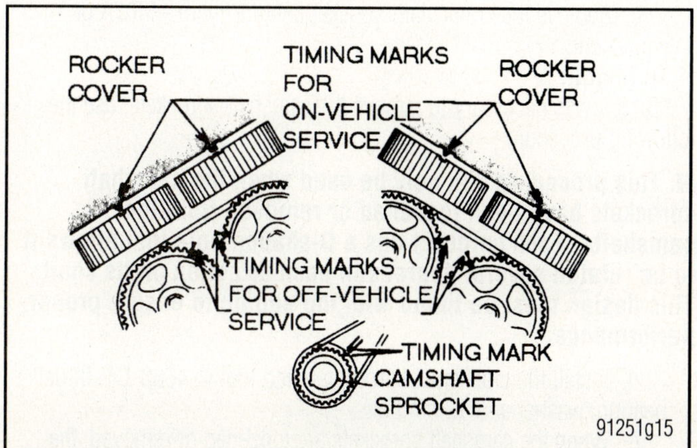

Proper timing belt alignment mark positioning for belt service—DaimlerChrysler 3.0L (VIN B, C, K and J) DOHC engine

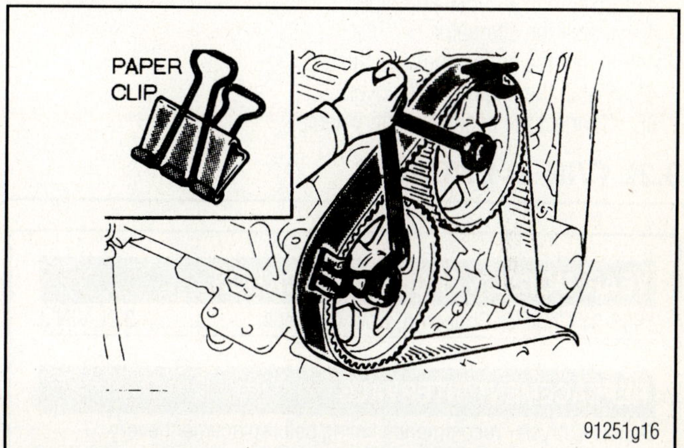

Securing camshafts for timing belt installation—DaimlerChrysler 3.0L (VIN B, C, K and J) DOHC engine

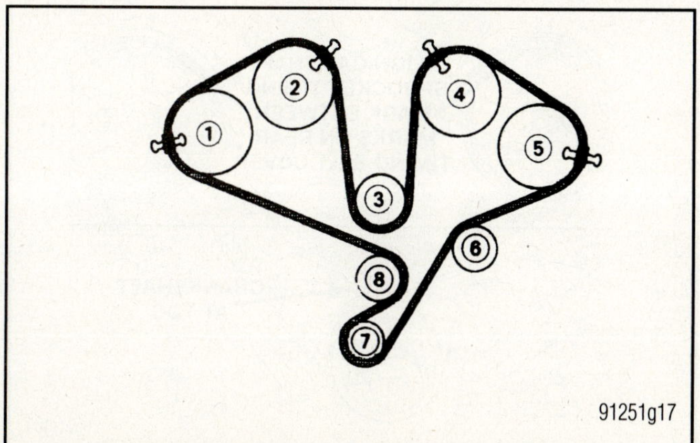

Timing belt installation sequence—DaimlerChrysler 3.0L (VIN B, C, K and J) DOHC engine

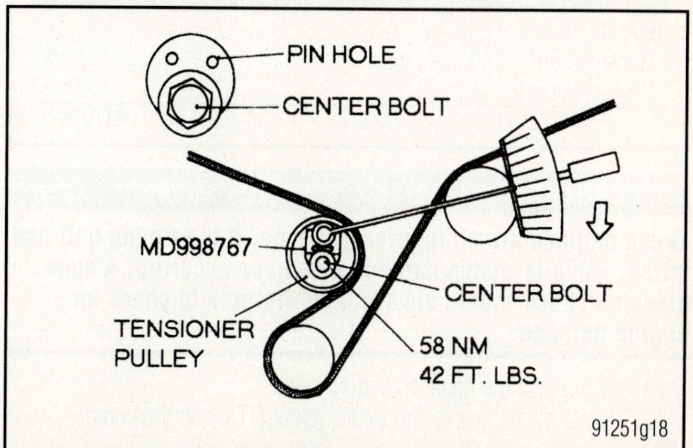

Timing belt tension adjustment—DaimlerChrysler 3.0L (VIN B, C, K and J) DOHC engine

■ **Even if the set pin cannot be easily inserted, the auto tensioner is normal if its rod protrusion is within specification.**

30. Measure the auto tensioner protrusion (the distance between the tensioner arm and auto tensioner body) to ensure that it is within 0.15–0.18 in. (3.8–4.5mm). If out of specification, repeat adjustment procedure until the specified value is obtained.
31. Check again that the timing marks on all sprockets are in proper alignment.
32. Make sure all pieces of packing are positioned in the inner grooves of the lower cover, position cover on engine and install mounting bolts in their original location.
33. Install the engine support bracket and secure using mounting bolts in their original location. Lubricate the reaming area of the reamer bolt and tighten slowly.
34. Install the idler pulley.
35. Install the engine mount bracket. Remove the engine support fixture.
36. Connect the camshaft and crankshaft position sensor connectors.
37. Install the alternator and A/C idler pulley.
38. Make sure all pieces of packing are positioned in the inner grooves of the upper cover and install.
39. Connect the brake fluid level sensor.
40. Install the crankshaft pulley. Tighten the bolt to 130–137 ft. lbs. (180–190 Nm).
41. Install the belt tensioner assembly and the power steering belt.
42. Install the air hose and pipe.
43. Install the alternator.
44. Install the cruise control actuator.
45. Install the engine undercover.
46. Connect the negative battery cable.

3.2L (VIN J) ENGINES

YEAR, MODEL & ENGINE APPLICATION

1998-01 Chrysler Concorde, Dodge Intrepid .... 3.2L VIN J

REPLACEMENT INTERVALS

DaimlerChrysler recommends timing belt replacement every 105,000 miles.

LABOR TIMES

1998-01	
Replacement	2.1 hrs
Severe Service	2.3 hrs

8807CHRg29

**** CAUTION**

These engines are an interference type. If the timing belt has failed, valve-to-piston damage may have occurred. A compression check should always be performed to check for engine damage.

1. Disconnect the negative battery cable.
2. Rotate the engine to Top Dead Center (TDC) on the compression stroke for cylinder No. 1.
3. Release the fuel system pressure using the recommended procedure.
4. Place a pan under the radiator and drain the coolant.
5. Remove the radiator and cooling fan assemblies.
6. Remove the accessory drive belts.
7. Remove the upper radiator hose.
8. Remove the crankshaft damper with a quality puller tool gripping the inside of the pulley.
9. Remove the stamped steel cover. Do not remove the sealer on the cover; it may be reusable.
10. Remove the left-side cast cover. If necessary, remove the lower belt cover, located behind the crankshaft damper.
11. If the timing belt is to be reused, mark the timing belt with the running direction for installation.
12. Align the camshaft sprockets with the marks on the rear covers.
13. Remove the timing belt and tensioner.
14. If it is necessary to service the camshaft sprockets, use the following procedure:
    a. Hold the camshaft sprocket with a 36mm box end wrench, loosen and remove the sprocket retaining bolt and washer.

■ **To remove the camshaft sprocket retainer bolt while the engine is in the vehicle, it may be necessary to raise that side of the engine due to the length of the retainer bolt. The right bolt is 8 3/8 in. (213mm) long, while the left bolt is 10 in. (254mm) long. These bolts are not interchangeable and their original location during removal should be noted.**

   b. Remove the camshaft sprocket from the camshaft. The camshaft sprockets are not interchangeable from side-to-side.
   c. Remove the crankshaft sprocket using Puller L-4407A or equivalent.

To Install:

15. If it was necessary to remove the camshaft sprockets, use the following procedure:

■ **This procedure can only be used when the camshaft sprockets have been loosened or removed from the camshafts. Each sprocket has a D-shaped hole that allows it to be rotated several degrees in each direction on its shaft. This design must be timed with the engine to ensure proper performance.**

   d. Install the crankshaft sprocket, using tool C-4685-C1, thrust bearing, washer and 12mm bolt.
   e. When the camshaft sprockets are loosened or removed, the camshafts must be timed to the engine. Install the Camshaft Alignment tools 6642-A or their exact equivalents, to the rear of the cylin-

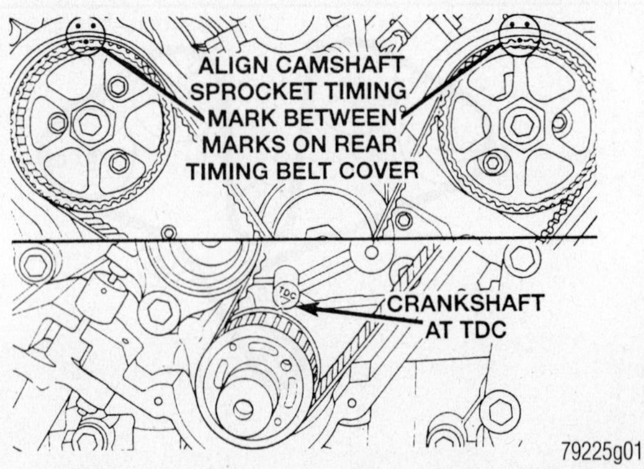

Timing belt alignment marks—DaimlerChrysler 3.2L (VIN J) engine

der heads. These tools lock the camshafts in the proper position.

16. Preload the belt tensioner as follows:

a. Place the tensioner in a vise the same way it is mounted on the engine.

b. Slowly compress the plunger into the tensioner body.

c. When the plunger is compressed into the tensioner body, install a pin through the body and plunger to retain the plunger in place until the tensioner is installed.

17. Install both camshaft sprockets to the appropriate shafts. The left camshaft sprocket has the DIS pick-up as part of the sprocket.

■ **The right bolt is 8 3/8 in. (213mm) long, while the left bolt is 10 in. (254mm) long. These bolts are not interchangeable.**

18. Apply Loctite®271 or equivalent, to the threads of the camshaft sprocket retainer bolts and install to the appropriate shafts. Do not tighten the bolts at this time.

19. Align the camshaft sprockets between the marks on the rear belt covers.

20. Align the crankshaft sprocket with the TDC mark on the oil pump cover.

21. Install the timing belt, starting at the crankshaft sprocket and going in a counterclockwise direction. After the belt is installed on the right sprocket, keep tension on the belt until it is past the tensioner pulley.

22. Holding the tensioner pulley against the belt, install the timing belt tensioner into the housing and tighten to 21 ft. lbs. (28 Nm).

23. When the tensioner is in place, pull the retainer pin out to allow tensioner to extend to the pulley bracket.

■ **Be sure that the timing marks on the cam sprockets are still between the marks on the rear cover.**

24. Remove the spark plug in the No. 1 cylinder and install a dial indicator to check for Top Dead Center (TDC) of the piston. Rotate the crankshaft until the piston is exactly at TDC.

25. Hold the camshaft sprocket hex with a 36mm wrench and tighten the right camshaft sprocket bolt to 75 ft. lbs. (102 Nm) plus an additional 90 degree turn. Tighten the left camshaft sprocket bolt to 85 ft. lbs. (115 Nm) plus an additional 90 degree turn.

26. Remove the dial indicator. Install the spark plug and tighten to 20 ft. lbs. (28 Nm).

27. Remove the camshaft alignment tools from the back of the cylinder heads and install the cam covers with new O-rings.

28. Tighten the fasteners to 20 ft. lbs. (27 Nm). Repeat this procedure on the other camshaft.

29. Rotate the crankshaft sprocket 2 revolutions and check for proper alignment of the timing marks on the camshaft and the crankshaft. If the timing marks do not align, repeat the procedure.

30. Before installing, inspect the sealer on the stamped steel cover. If some sealer is missing, use MOPAR Silicone Rubber Adhesive sealant or equivalent to replace the missing sealer.

31. Install the lower belt cover behind the crankshaft damper, if necessary.

32. Install the stamped steel cover and the left-side cast cover. Tighten the 6mm bolts to 105 inch lbs. (12 Nm), the 8mm bolts to 21 ft. lbs. (28 Nm) and the 10mm bolts to 40 ft. lbs. (54 Nm).

33. Install the crankshaft damper using special tool L-4524, a 6 in. (15cm) long bolt, thrust bearing and washer or equivalent damper installation tools. Tighten the center bolt to 85 ft. lbs. (115 Nm).

34. Install the upper radiator hose.

35. Install the accessory drive belts and adjust them to the proper tension.

36. Install the radiator and cooling fan assemblies.

37. Refill and bleed the cooling system.

38. Connect the negative battery cable.

39. With the radiator cap off so coolant can be added, run the engine. Watch for leaks and listen for unusual engine noises.

3.5L (VIN 4) ENGINES

YEAR, MODEL & ENGINE APPLICATION	
2004-05 Chrysler Pacifica	3.5L VIN 4

REPLACEMENT INTERVALS

DaimlerChrysler does not provide a replacement interval. Standard industry practice is to replace timing belts every 60,000 miles.

LABOR TIMES	
2004-05	
Replacement	2.7 hrs
Severe Service	2.9 hrs

8807CHRg30

**** CAUTION**

These engines are an interference type. If the timing belt has failed, valve-to-piston damage may have occurred. A compression check should always be performed to check for engine damage.

Use care when servicing a timing belt. Valve timing is absolutely critical to engine performance. If the valve timing marks on all drive sprockets are not properly aligned, engine damage will result. If only the belt and tensioner are being serviced, do not loosen the camshaft drive sprockets unless they are to be replaced. The sprockets have oversized openings and can be rotated several degrees in each direction on their shafts. This means the sprockets must be re-timed, requiring some special tools.

**** CAUTION**

Fuel injection systems remain under pressure, even after the engine has been turned off. The fuel system pressure must be relieved before disconnecting any fuel lines. Failure to do so may result in fire and/or personal injury.

1. Loosen the valve train rocker assemblies before servicing the timing components.
2. Release the fuel system pressure using the recommended procedure.
3. Disconnect the negative battery cable.
4. Remove the cylinder head covers.
5. Remove the accessory drive belts.
6. Remove the drive belt tensioner.
7. Remove the power steering pump bolts and position the pump aside.
8. Remove the crankshaft damper with a quality puller tool gripping the inside of the pulley
9. Remove the lower front timing belt cover retainers.
10. Support the engine with a floor jack.
11. Remove the air cleaner assembly.

12. Remove the front engine mount.
13. Remove the fuel supply line from the rail.
14. Remove the upper timing belt cover bolts and the cover.
15. If the timing belt is to be reused, mark the timing belt with the running direction for installation.

**** WARNING**

Always align the timing marks always use the crankshaft to rotate the engine. Failure to do this will result in severe engine damage.

16. Rotate the engine clockwise until the crankshaft mark aligns with the Top Dead Center (TDC) mark on the oil pump housing and the camshaft sprocket timing marks align with the marks on the rear cover.
17. Remove the timing belt tensioner and the belt.
18. Pre-load the timing belt tensioner as follows:
    a. Place tensioner in a vise the same way it is mounted on the engine.
    b. Slowly compress the plunger into the tensioner body.
    c. Once the plunger is compressed, install a pin through the body and plunger to retain it in place until the tensioner is installed.

To Install:

**** WARNING**

If the camshafts have moved from the timing marks always rotate them toward the direction nearest the to the timing marks. Never turn the camshafts a full turn or severe engine damage will occur.

19. Align the crankshaft sprocket with the TDC mark on the oil pump.
20. Align the camshaft sprockets timing marks with the mark on the rear cover.
21. Install the belt starting at the crankshaft sprocket going in a counterclockwise direction. Install the belt around the last sprocket always maintaining tension on the belt as it is positioned around the tensioner pulley.

■ **If the camshaft gears have been removed it is only necessary to have the camshaft gear retaining bolts installed to a snug torque at this time.**

22. Holding the tensioner pulley against the belt, install the tensioner into the housing and tighten to 250 inch lbs. (28 Nm) making sure all the alignment marks remain aligned.
Pull the pin from the tensioner and allow the tensioner to extend to the pulley bracket.
23. Rotate the crankshaft sprocket 2 full turns and check the timing marks on the cam and crank shafts. Make sure the marks are aligned, if not repeat the procedure.
24. if the camshafts were removed using a dial indicator, position the number 1 piston at TDC.
25. Hold the camshaft sprocket hex with a 36mm wrench and tighten the right camshaft sprocket bolt to 75 ft. lbs. (102 Nm) plus an additional 90 degree turn. Tighten the left camshaft sprocket bolt to 85 ft. lbs. (115 Nm) plus an additional 90 degree turn.
26. Remove the dial indicator. Install the spark plug and tighten to 20 ft. lbs. (28 Nm).

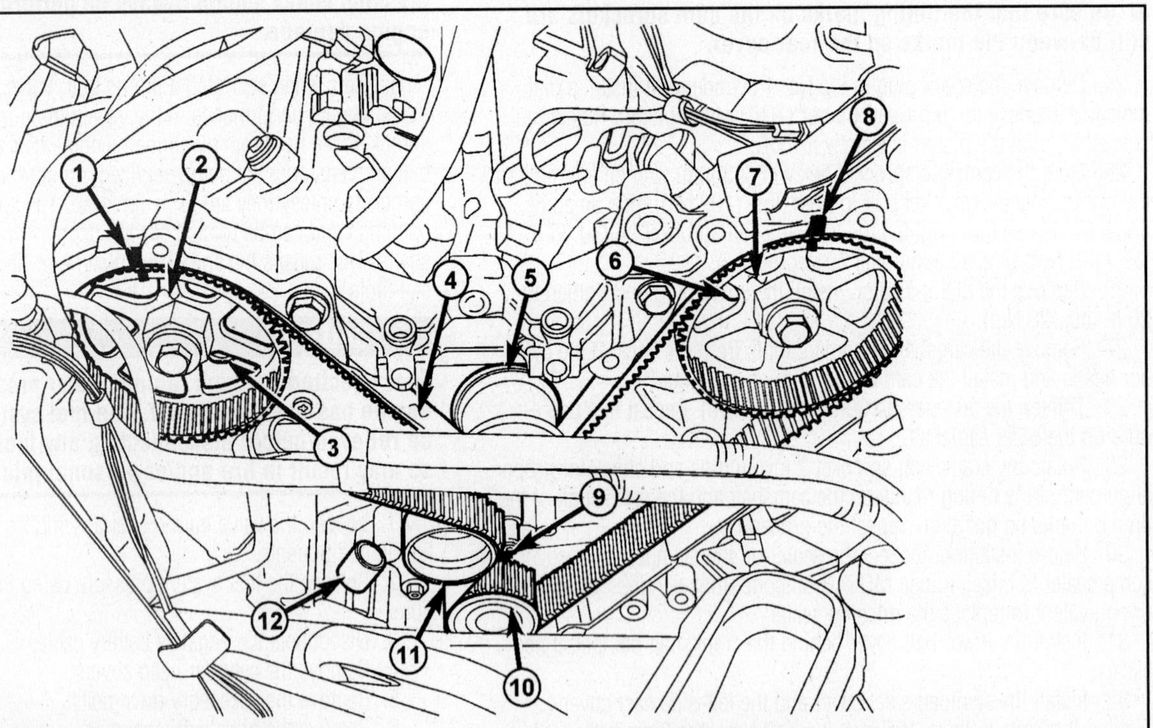

1 - RIGHT CAMSHAFT GEAR ALIGNMENT MARK
2 - RIGHT CAMSHAFT GEAR
3 - CYLINDER HEAD TO INNER TIMING BELT COVER BOLTS - RIGHT
4 - TIMING BELT
5 - WATER PUMP PULLEY
6 - CYLINDER HEAD TO INNER TIMING BELT COVER BOLTS - LEFT
7 - LEFT CAMSHAFT GEAR
8 - LEFT CAMSHAFT GEAR ALIGNMENT MARK
9 - CRANKSHAFT GEAR ALIGNMENT MARK
10 - CRANKSHAFT GEAR
11 - TIMING BELT TENSIONER PULLEY
12 - TIMING BELT TENSIONER

23990g22

View of the timing belt alignment marks—DaimlerChrysler 3.5L (VIN 4) engine

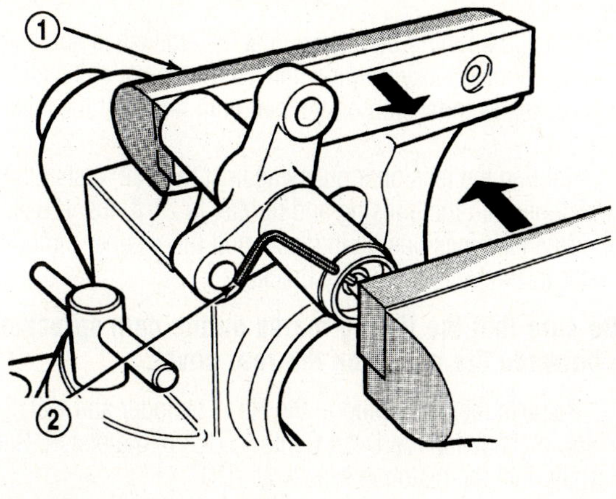

Pre-load the timing belt tensioner—DaimlerChrysler 3.5L (VIN 4) engine

27. Remove the camshaft alignment tools from the back of the cylinder heads and install the cam covers with new O-rings.
28. Install the upper timing belt cover and bolts.
29. Install the fuel supply line to the rail.
30. Install the Front engine mount
31. Install the air cleaner assembly.
32. Install the lower front timing belt cover retainers.
33. Install the crankshaft damper.
34. Install the power steering pump and bolts.
35. Install the drive belt tensioner.
36. Install the accessory drive belts.
37. Install the cylinder head covers.
38. Connect the negative battery cable.

3.5L (VIN F, G, K, V) ENGINES

YEAR, MODEL & ENGINE APPLICATION

1993-04 Chrysler Concorde, 1994-01 LHS, 1994-97 New Yorker, 1998-04 300M, 1993-04 Dodge Intrepid, 1993-97 Eagle Vision	3.5L VIN F, G, K, V

REPLACEMENT INTERVALS

DaimlerChrysler recomends timing belt replacement every 105,000 miles.

LABOR TIMES

1993-97	
Replacement	3.1 hrs
Severe Service	3.4 hrs
With air conditioning add	0.3 hrs
1998-04	
Replacement	2.1 hrs

8807CHRg31

✱✱ CAUTION

These engines are an interference type. If the timing belt has failed, valve-to-piston damage may have occurred. A compression check should always be performed to check for engine damage.

Use care when servicing a timing belt. Valve timing is absolutely critical to engine performance. If the valve timing marks on all drive sprockets are not properly aligned, engine damage will result. If only the belt and tensioner are being serviced, do not loosen the camshaft drive sprockets unless they are to be replaced. The sprockets have oversized openings and can be rotated several degrees in each direction on their shafts. This means the sprockets must be retimed, requiring some special tools.

✱✱ CAUTION

Fuel injection systems remain under pressure, even after the engine has been turned off. The fuel system pressure must be relieved before disconnecting any fuel lines. Failure to do so may result in fire and/or personal injury.

1. Disconnect the negative battery cable.
2. Rotate the engine to Top Dead Center (TDC) on the compression stroke for cylinder No. 1.
3. Release the fuel system pressure using the recommended procedure.
4. Place a pan under the radiator and drain the coolant.
5. Remove the radiator and cooling fan assemblies.
6. Remove the accessory drive belts.
7. Remove the upper radiator hose.
8. Remove the crankshaft damper with a quality puller tool gripping the inside of the pulley.
9. Remove the stamped steel cover. Do not remove the sealer on the cover; it may be reusable.
10. Remove the left-side cast cover. If necessary, remove the lower belt cover, located behind the crankshaft damper.
11. If the timing belt is to be reused, mark the timing belt with the running direction for installation.
12. Align the camshaft sprockets with the marks on the rear covers.
13. Remove the timing belt and tensioner.
14. If it is necessary to service the camshaft sprockets, use the following procedure:
    a. Hold the camshaft sprocket with a 36mm box end wrench, loosen and remove the sprocket retaining bolt and washer.

■ **To remove the camshaft sprocket retainer bolt while the engine is in the vehicle, it may be necessary to raise that side of the engine due to the length of the retainer bolt. The right bolt is 8 3/8 in. (213mm) long, while the left bolt is 10 in. (254mm) long. These bolts are not interchangeable and their original location during removal should be noted.**

    b. Remove the camshaft sprocket from the camshaft. The camshaft sprockets are not interchangeable from side-to-side.
    c. Remove the crankshaft sprocket using Puller L-4407A or equivalent.

To Install:

15. If it was necessary to remove the camshaft sprockets, use the following procedure:

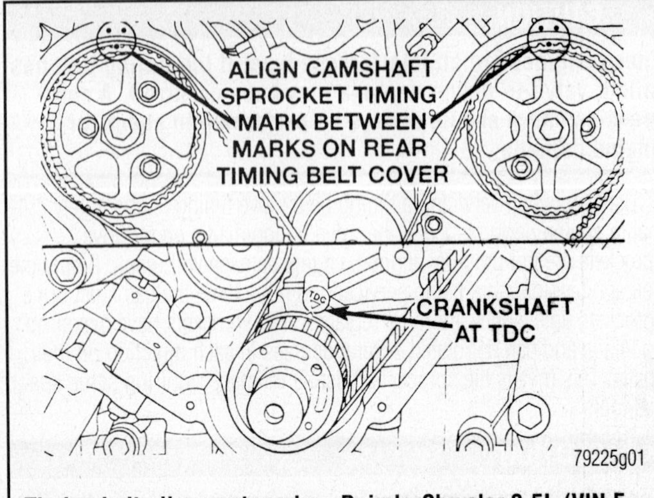

Timing belt alignment marks—DaimlerChrysler 3.5L (VIN F, G, K & V) engine

■ **This procedure can only be used when the camshaft sprockets have been loosened or removed from the camshafts. Each sprocket has a D-shaped hole that allows it to be rotated several degrees in each direction on its shaft. This design must be timed with the engine to ensure proper performance.**

a. Install the crankshaft sprocket, using tool C-4685-C1, thrust bearing, washer and 12mm bolt.

b. When the camshaft sprockets are loosened or removed, the camshafts must be timed to the engine. Install the Camshaft Alignment tools 6642-A or their exact equivalents, to the rear of the cylinder heads. These tools lock the camshafts in the proper position.

16. Preload the belt tensioner as follows:

a. Place the tensioner in a vise the same way it is mounted on the engine.

b. Slowly compress the plunger into the tensioner body.

c. When the plunger is compressed into the tensioner body, install a pin through the body and plunger to retain the plunger in place until the tensioner is installed.

17. Install both camshaft sprockets to the appropriate shafts. The left camshaft sprocket has the DIS pick-up as part of the sprocket.

■ **The right bolt is 8 3/8 in. (213mm) long, while the left bolt is 10 in. (254mm) long. These bolts are not interchangeable.**

18. Apply Loctite®271 or equivalent, to the threads of the camshaft sprocket retainer bolts and install to the appropriate shafts. Do not tighten the bolts at this time.

19. Align the camshaft sprockets between the marks on the rear belt covers.

20. Align the crankshaft sprocket with the TDC mark on the oil pump cover.

21. Install the timing belt, starting at the crankshaft sprocket and going in a counterclockwise direction. After the belt is installed on the right sprocket, keep tension on the belt until it is past the tensioner pulley.

22. Holding the tensioner pulley against the belt, install the timing belt tensioner into the housing and tighten to 21 ft. lbs. (28 Nm).

23. When the tensioner is in place, pull the retainer pin out to allow tensioner to extend to the pulley bracket.

■ **Be sure that the timing marks on the cam sprockets are still between the marks on the rear cover.**

24. Remove the spark plug in the No. 1 cylinder and install a dial indicator to check for Top Dead Center (TDC) of the piston. Rotate the crankshaft until the piston is exactly at TDC.

25. Hold the camshaft sprocket hex with a 36mm wrench and tighten the right camshaft sprocket bolt to 75 ft. lbs. (102 Nm) plus an additional 90 degree turn. Tighten the left camshaft sprocket bolt to 85 ft. lbs. (115 Nm) plus an additional 90 degree turn.

26. Remove the dial indicator. Install the spark plug and tighten to 20 ft. lbs. (28 Nm).

27. Remove the camshaft alignment tools from the back of the cylinder heads and install the cam covers with new O-rings.

28. Tighten the fasteners to 20 ft. lbs. (27 Nm). Repeat this procedure on the other camshaft.

29. Rotate the crankshaft sprocket 2 revolutions and check for proper alignment of the timing marks on the camshaft and the crankshaft. If the timing marks do not align, repeat the procedure.

30. Before installing, inspect the sealer on the stamped steel cover. If some sealer is missing, use MOPAR Silicone Rubber Adhesive sealant or equivalent to replace the missing sealer.

31. Install the lower belt cover behind the crankshaft damper, if necessary.

32. Install the stamped steel cover and the left-side cast cover. Tighten the 6mm bolts to 105 inch lbs. (12 Nm), the 8mm bolts to 21 ft. lbs. (28 Nm) and the 10mm bolts to 40 ft. lbs. (54 Nm).

33. Install the crankshaft damper using special tool L-4524, a 6 in. (15cm) long bolt, thrust bearing and washer or equivalent damper installation tools. Tighten the center bolt to 85 ft. lbs. (115 Nm).

34. Install the upper radiator hose.

35. Install the accessory drive belts and adjust them to the proper tension.

36. Install the radiator and cooling fan assemblies.

37. Refill and bleed the cooling system.

38. Connect the negative battery cable.

39. With the radiator cap off so coolant can be added, run the engine. Watch for leaks and listen for unusual engine noises.

Ford Motor Company

1.3L (VIN H, K) ENGINES

YEAR, MODEL & ENGINE APPLICATION

1994-97 Ford Aspire, 1988-94 Festiva .... 1.3L VIN H, K

REPLACEMENT INTERVALS

Ford recommends timing belt replacement every 60,000 miles.

LABOR TIMES

Aspire	
Replacement	1.9 hrs
Severe Service	2.1 hrs
Festiva	
Carbureted	
Replacement	2.4 hrs
Severe Service	2.7 hrs
Fuel Injected	
Replacement	2.1 hrs
Severe Service	2.4 hrs

8807FRDg01

1988–94

✱✱ CAUTION

These engines are considered freewheeling. If the timing belt fails, valve-to-piston damage is unlikely, however a precautionary compression check should be performed to verify no engine damage.

1. Disconnect the negative battery cable.
2. Remove the drive belts.
3. Remove the 3 water pump pulley attaching bolts and remove the water pump pulley.
4. Raise and safely support the vehicle.
5. Remove the right front wheel and tire assembly and the right inner fender panel.
6. Remove the 4 attaching bolts and the screws from the crankshaft pulley. Remove the spacer and outer pulley, if equipped. Remove the inner spacer, inner pulley and the baffle or guide plates, as required.
7. Remove the attaching bolts and the upper and lower covers.
8. Mark the direction of rotation of the timing belt, if the belt is to be reused.
9. Place No. 1 cylinder at TDC with the camshaft and crankshaft timing marks aligned.
10. Remove the timing belt tensioner pulley, spring and retaining bolt. Remove the timing belt.

To Install:

11. Align the camshaft and crankshaft timing marks with the marks located on the cylinder head and oil pump housing.
12. If reusing the original timing belt, install the timing belt with the mark made indicating the direction of rotation.
13. Install the timing belt tensioner spring and cover on the pulley. Position the tensioner and spring assembly on the engine and install the attaching bolt. Do not tighten the bolt at this time.
14. Reconnect the free end of the spring to the spring anchor. Torque the tensioner bolt to 14–19 ft. lbs. (19–26 Nm).
15. Install the upper and lower covers. Install the attaching bolts and tighten to 69–95 inch lbs. (8–11 Nm).
16. Install the crankshaft pulley baffle with the curved lip facing outward or install the large guide plate and then the small guide plate, as required.
17. Install the inner pulley with the deep recess facing outward. Install the spacer and then the outer pulley, spacer and screws. Install the pulley bolts and tighten to 9–13 ft. lbs. (12–17 Nm).
18. Install the inner fender panel and the wheel and tire assembly. Lower the vehicle.
19. Install the water pump pulley and tighten the bolts to 36–45 ft. lbs. (49–61 Nm).
20. Install the drive belts. Connect the negative battery cable.

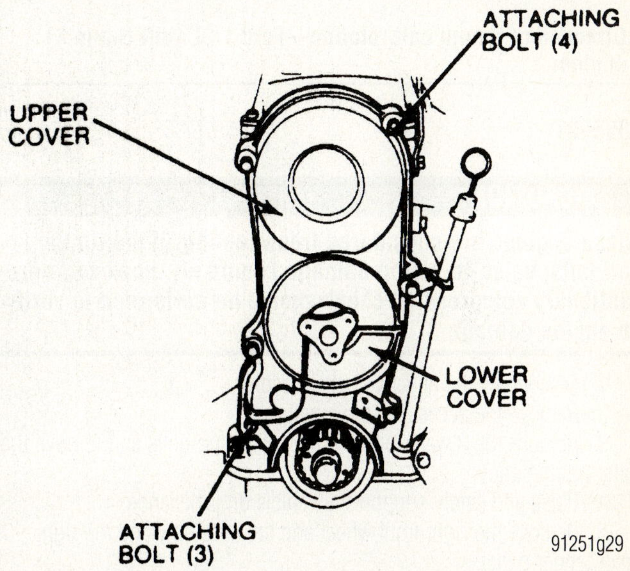

Upper and lower timing belt cover locations, showing the attaching bolts—Ford 1.3L (VIN H and K) engine

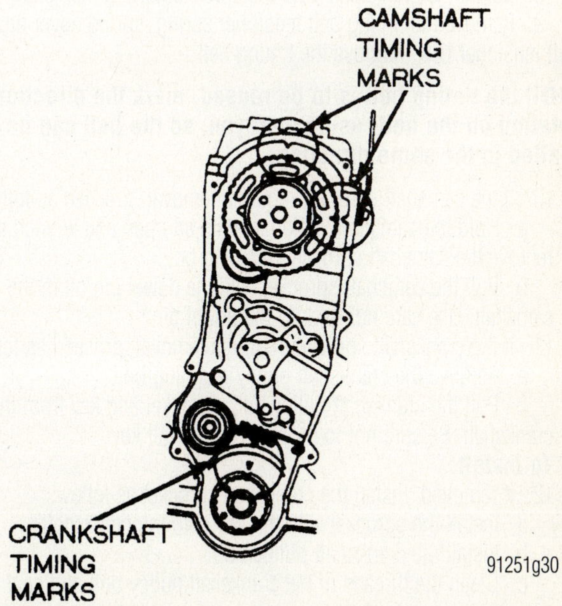

Camshaft and crankshaft timing mark locations—Ford 1.3L (VIN H and K) engine

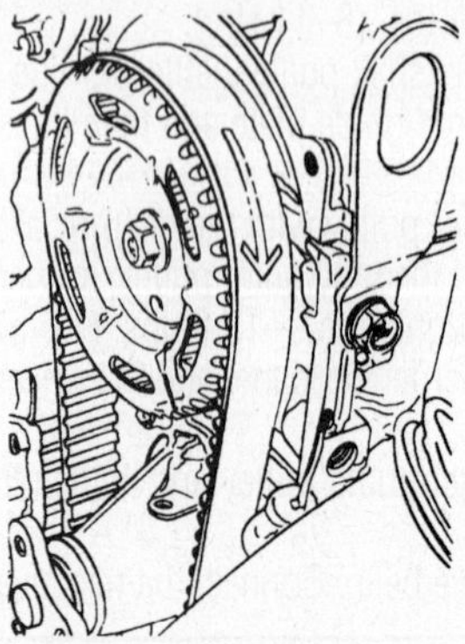
91251g31

Direction of timing belt rotation—Ford 1.3L (VIN H and K) engine

1995–97

✻✻CAUTION

These engines are considered freewheeling. If the timing belt fails, valve-to-piston damage is unlikely, however a precautionary compression check should be performed to verify no engine damage.

1. Disconnect the negative battery cable.
2. Remove the accessory drive belts.
3. Remove the 3 water pump pulley attaching bolts and remove the water pump pulley.
4. Raise and safely support the vehicle on jackstands.
5. Remove the right front wheel and tire assembly and the right inner fender panel.
6. Remove the 4 attaching bolts and the screws from the crankshaft pulley. Remove the spacer and outer pulley, if equipped. Remove the inner spacer, inner pulley and the baffle or guide plates, as required.
7. Remove the attaching bolts and the upper and lower covers.
8. Rotate the crankshaft until the sprocket timing marks are aligned.
9. Remove the timing belt tensioner spring, spring cover and timing belt tensioner bolt. Remove the timing belt.

If the timing belt is to be reused, mark the direction of rotation on the belt, using a crayon, so the belt can be reinstalled in the same direction.

10. If the camshaft sprocket requires removal, proceed as follows:
    a. Hold the camshaft stationary with an open end wrench and remove the camshaft sprocket retaining bolt.
    b. Pull the camshaft sprocket with the dowel pin off of the camshaft. Use care not to drop the dowel pin.
11. If the crankshaft sprocket requires removal, proceed as follows:
    a. Remove the crankshaft pulley retaining bolt.
    b. Pull the crankshaft pulley hub, sprocket and key from the crankshaft. Be sure not to drop the crankshaft key.

To install:

12. If removed, install the crankshaft sprocket as follows:
    a. Install the sprocket with the key onto the crankshaft.
    b. Install the crankshaft pulley hub.
    c. Clean the threads of the crankshaft pulley bolt and coat with a non-hardening sealer.
    d. Install the bolt and tighten to 80–85 ft. lbs. (108–118 Nm).
13. If removed, install the camshaft sprocket as follows:
    a. Position the sprocket and dowel pin to the camshaft and install the retaining bolt.
    b. Hold the camshaft stationary with an open end wrench and tighten the retaining bolt to 36–45 ft. lbs. (49–61 Nm).
14. Align the camshaft and crankshaft timing marks with the marks located on the cylinder head and oil pump housing.
15. If reusing the original timing belt, install the timing belt with the mark made indicating the direction of rotation.
16. Install the timing belt tensioner spring and cover on the tensioner. Position the tensioner and spring assembly on the engine and install the attaching bolt. Do not tighten the bolt at this time.
17. Rotate the crankshaft 2 turns in the direction of normal rotation and align the timing marks. Ensure all marks are still correctly aligned.
18. Reconnect the free end of the spring to the spring anchor. Tighten the tensioner bolt to 14–19 ft. lbs. (19–26 Nm).
19. Install the upper and lower covers. Install the attaching bolts and tighten to 71–97 inch lbs. (8–11 Nm).
20. Install the crankshaft pulley baffle with the curved lip facing outward, or install the large guide plate, then the small guide plate, as required.
21. Install the inner pulley with the deep recess facing outward. Install the spacer, then the outer pulley, spacer and screws. Install the pulley bolts and tighten to 9–13 inch lbs. (12–17 Nm).
22. Install the inner fender panel.
23. Install the wheel and tire assembly. Tighten the lug bolts to 65–87 ft. lbs. (88–118 Nm).
24. Lower the vehicle.
25. Install the water pump pulley and tighten the bolts to 36–45 ft. lbs. (49–61 Nm).
26. Install the accessory drive belts.
27. Connect the negative battery cable.
28. Run the engine and check for proper operation.

1.6L (VIN 5) ENGINES

YEAR, MODEL & ENGINE APPLICATION	
1985-91 Ford Escort	1.6L VIN 5

REPLACEMENT INTERVALS

Ford recommends timing belt replacement every 60,000 miles.

LABOR TIMES	
1985-91	
Replacement	3.3 hrs
Severe Service	3.8 hrs

8807FRDg02

✻✻ CAUTION

These engines are an interference type. If the timing belt has failed, valve-to-piston damage may have occurred. A compression check should always be performed to check for engine damage.

1. Disconnect the negative battery cable.
2. Remove the accessory drive belts.
3. Remove the timing belt covers.
4. Remove the timing belt tensioner spring and retaining bolt.

5. If the timing belt is to be reused, mark the rotation direction on the belt so it can be reinstalled in the same direction.
6. Remove the timing belt.

To Install:

7. Inspect the timing belt, tensioner and sprockets for signs of wear and replace, as necessary.
8. Align the marks on the camshaft and crankshaft sprockets with the cylinder head and oil pump alignment marks.
9. If reusing the timing belt, install it in the direction of the rotation mark.
10. Install the timing belt tensioner and spring. Install the spring on its anchor and hand-tighten the tensioner bolt.
11. Rotate the crankshaft 2 complete revolutions and realign the timing marks. Reaffirm that the timing marks are aligned; if not, repeat the alignment procedure.
12. Torque the tensioner bolt to 14–19 ft. lbs. (19–25 Nm) and check the timing belt deflection between the crankshaft and camshaft sprockets. The timing belt deflection should be 0.35–0.39 in. (9–10mm) at 22 lbs. (10 kg). Pressure. If the deflection is not correct, repeat Steps 10 and 11.
13. Install the remaining components in the reverse order of their removal.

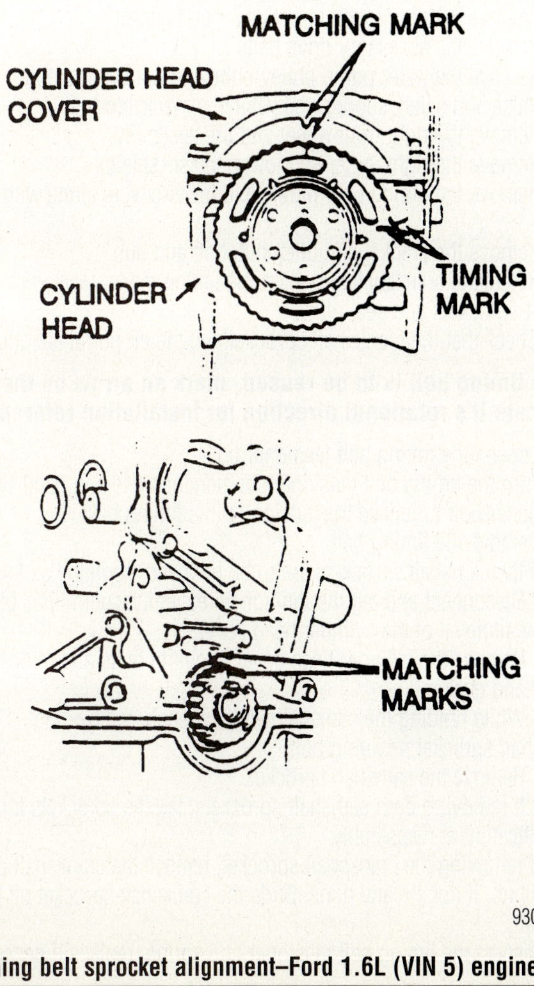

Timing belt sprocket alignment–Ford 1.6L (VIN 5) engine

1.6L (VIN Z) ENGINES

YEAR, MODEL & ENGINE APPLICATION	
1991-94 Mercury Capri	1.6L VIN Z

REPLACEMENT INTERVALS
Ford recommends timing belt replacement every 60,000 miles.

LABOR TIMES	
1991-94	
Replacement	4.1 hrs
Severe Service	4.4 hrs

8807FRDg03

✱✱ CAUTION

These engines are considered freewheeling. If the timing belt fails, valve-to-piston damage is unlikely, however a precautionary compression check should be performed to verify no engine damage.

1. Disconnect the negative battery cable. Raise and safely support the vehicle.
2. Remove the right front wheel and tire assembly and remove the right splash guard. Lower the vehicle.
3. Remove the spark plugs. Set the engine position to Top Dead Center (TDC) of the compression stroke on the No. 1 cylinder.
4. Remove the alternator and power steering belts. Remove the oil dipstick and the water pump pulley.
5. Remove the crankshaft pulley, damper and baffle plate.
6. Remove the upper, center and lower timing belt covers.
7. Remove the timing belt tension spring and loosen the timing belt tension pulley.
8. Support the engine with a floor jack and remove the right engine mount.
9. Mark the timing belt rotation direction and remove the timing belt.
10. Inspect the timing belt and timing sprockets for wear and/or damage and replace as necessary.
11. Check the free length of the timing belt tension spring. It should be 2.315 in. (58.8mm). Replace if out of specification.

To Install:

Make sure the timing marks are properly positioned on the camshafts and crankshaft. The intake camshaft should have the letter **I** aligned with the arrow on the belt cover. The exhaust camshaft should have the letter **E** aligned with the arrow on the belt covers

12. The crankshaft key should align with the arrow.
13. Tighten the tension pulley with the tension spring fully extended.
14. Install the timing belt. Keep tension on the opposite side of the tensioner as tight as possible. Make sure the rotation mark on the belt is correct.
15. Turn the crankshaft 2 full turns. Check the alignment marks. If any mark is not aligned, remove the timing belt and reset the timing.
16. Loosen the tension pulley retaining bolt to allow the tension spring to tighten the belt.
17. Tighten the tension pulley retaining bolt to 27–38 ft. lbs. (37–52 Nm). Rotate the engine 2 full turns. Make sure the timing marks are aligned.

18. Measure the timing belt tension between the camshaft pulleys. Belt deflection should be 0.33–0.45 in. (8.5–11.5mm). If incorrect, loosen the tension pulley and repeat the procedure. If proper tension cannot be achieved, replace the tension spring.
19. Install the lower, center and upper timing belt covers. Tighten the retaining bolts to 71–97 inch lbs. (8–11 Nm).
20. Install the right engine mount and lower the engine. Tighten the retaining nuts to 44–63 ft. lbs. (60–85 Nm).
21. Install the crankshaft pulley, damper and baffle. Tighten the baffle and damper retaining screws to 109–152 inch lbs. (12–17 Nm). Tighten the pulley retaining bolts to 109–152 inch lbs. (12–17 Nm).
22. Install the water pump pulley. Tighten the retaining bolts to 69–95 inch lbs. (8–11 Nm).
23. Install the alternator and power steering belts. Install the dipstick. Raise and safely support the vehicle.
24. Install the splash guard and the right front wheel and tire assembly.
25. Lower the vehicle and install the spark plugs. Start the engine and check for proper operation.

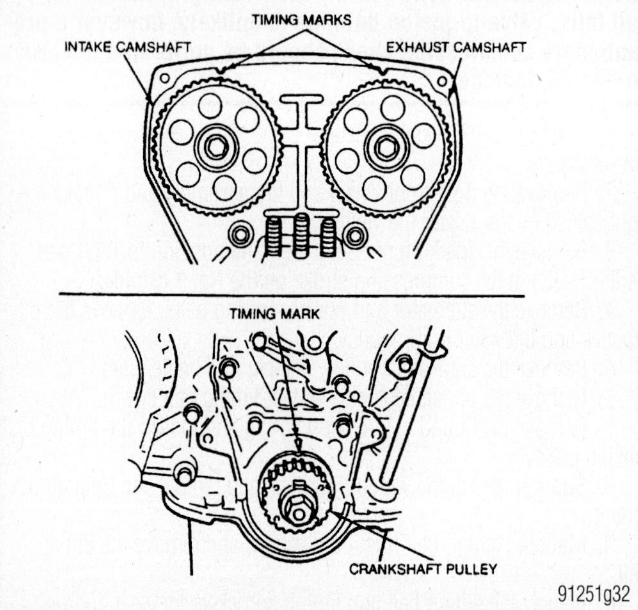

Timing belt sprocket alignment–Ford 1.6L (VIN Z) engine

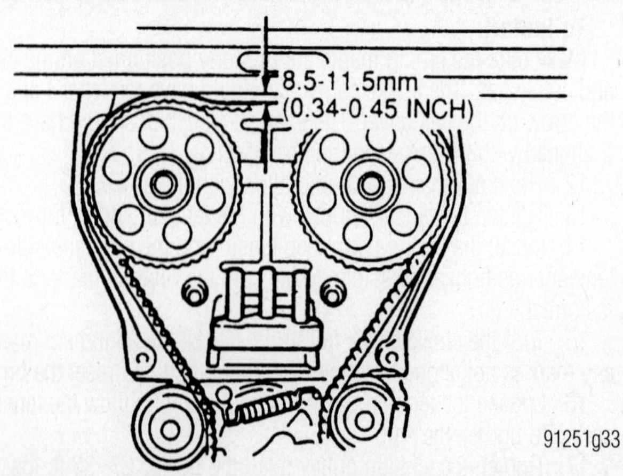

Measure the timing belt tension as indicated–Ford 1.6L (VIN Z) engine

1.8L (VIN 8) ENGINES

YEAR, MODEL & ENGINE APPLICATION

1991-96 Ford Escort GT,
1991-96 Mercury Tracer .... 1.8L VIN 8

REPLACEMENT INTERVALS

Ford recommends timing belt replacement every 60,000 miles.

LABOR TIMES

1991-96
Replacement .... 2.2 hrs
Severe Service .... 2.7 hrs

8807FRDg04

✻✻ CAUTION

These engines are considered freewheeling. If the timing belt fails, valve-to-piston damage is unlikely, however a precautionary compression check should be performed to verify no engine damage.

1. Disconnect the negative battery cable.
2. Remove the timing belt upper cover and gasket.
3. Remove the accessory drive belts.
4. Remove the water pump pulley bolts and remove the pulley.
5. Raise and safely support the vehicle on jackstands.
6. Remove the right front wheel and tire assembly.
7. Remove the right upper and lower splash-shields.
8. Remove the timing belt middle and lower covers along with the gaskets.
9. Remove the crankshaft pulley hub bolt and hub.
10. Rotate the crankshaft and align the timing marks located on the camshaft sprockets and seal plate.
11. Check that the crankshaft sprocket and the oil pump are aligned.

■If the timing belt is to be reused, mark an arrow on the belt to indicate it's rotational direction for installation reference.

12. Loosen the timing belt tensioner bolt.
13. Turn the timing belt tensioner counterclockwise and hand-tighten the tensioner bolt to relieve the tension on the timing belt.
14. Remove the timing belt.
15. If the camshaft sprockets are to be removed, continue as follows:
   a. Disconnect and tag the ignition wires and vacuum lines blocking the removal of the cylinder head cover.
   b. Remove the cylinder head cover retaining bolts and remove the cover and gasket.
   c. While holding the camshaft with a wrench, remove the camshaft sprocket retaining bolt.
   d. Remove the camshaft sprocket.
   e. If removing both camshaft sprockets, tag the sprockets for identification at reassembly.
16. If removing the crankshaft sprocket, remove the crankshaft pulley bolt and hub, if not already done. Slide the crankshaft sprocket off the crankshaft.
17. Inspect the timing belt tensioner and spring, replace if necessary.

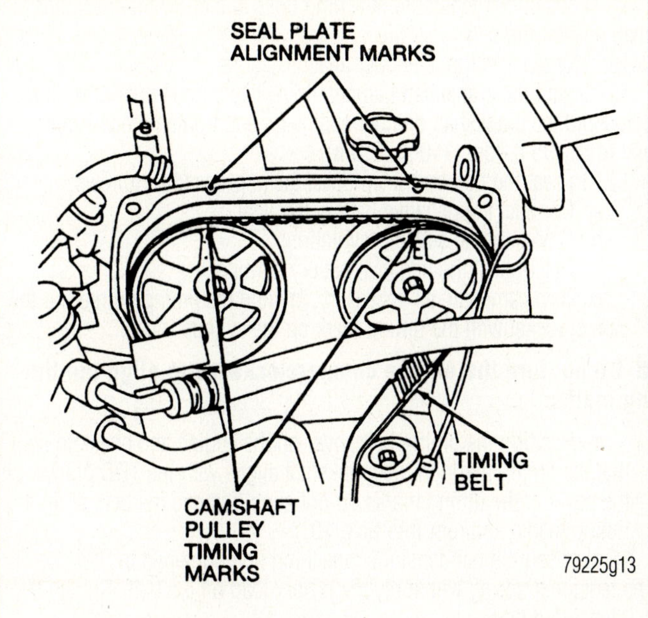

Camshaft timing alignment marks—Ford 1.8L (VIN 8) engine

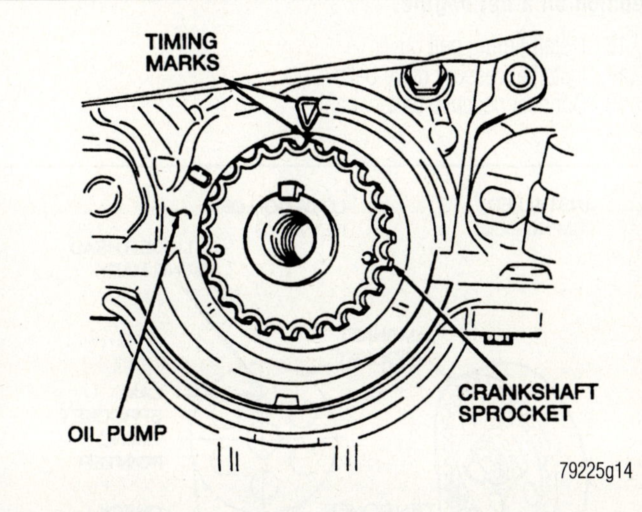

Crankshaft timing mark position—Ford 1.8L (VIN 8) engine

To Install:

18. If the crankshaft sprocket was removed, install the crankshaft key with the tapered end facing the oil pump. Install the crankshaft sprocket onto the crankshaft while making sure to match the alignment grooves.

19. If removed, install the camshaft sprockets as follows

a. Turn the camshaft until the dowel pins face straight up.

b. Install the camshaft sprocket with the **I** mark straight up for the intake camshaft or with the **E** mark straight up for the exhaust camshaft.

c. Align the camshaft sprockets with the timing marks on the seal plate.

d. While holding each camshaft with a wrench, install the camshaft sprocket retaining bolts. Tighten the bolts to 36–45 ft. lbs. (49–61 Nm).

e. Install a new cylinder head cover gasket onto the cylinder head.

f. Place the cylinder head cover into its mounting position and install the retaining bolts. Tighten the cylinder head cover bolts to 43–78 inch lbs. (4.9–8.8 Nm).

g. Install the ignition wires to the spark plugs and connect the vacuum hoses to the cylinder head cover.

20. Temporarily secure the timing belt tensioner in the far left position.

21. Verify that the timing marks on the crankshaft sprocket and the oil pump are aligned.

22. Verify that the timing marks on the camshaft sprockets and the seal plate are aligned.

23. Install the timing belt in a counterclockwise motion. Be sure there is no looseness on the idler side of the timing belt or between the camshaft sprockets.

■ **If using the old timing belt, be sure to install the belt in the same direction of travel as it was removed.**

24. Loosen the timing belt tensioner bolt. Allow the tensioner spring to apply tension to the timing belt.

25. Rotate the crankshaft 60° turns clockwise and align the timing belt pulley mark with the tension set mark which is located at approximately the 10 o'clock position.

26. Turn the crankshaft 2 turns clockwise and align the crankshaft sprocket with the tension set mark on the oil pump.

27. Verify that all timing marks are aligned. If not, remove the timing belt and repeat the installation procedures.

28. Apply tension to the timing belt tensioner and tighten the tensioner lockbolt to 27–38 ft. lbs. (37–52 Nm).

29. Rotate the crankshaft 2 1/6 (780 degrees) turns clockwise and verify that the camshaft and crankshaft timing marks are aligned.

30. Measure the timing belt deflection by applying 22 lbs. (10 kg) of pressure on the timing belt between the camshaft sprockets. The timing belt deflection should be 0.35–0.45 in. (9–11.5mm). If necessary to adjust the timing belt deflection, rotate the crankshaft 2 turns clockwise and ensure that the timing marks are still aligned. If the timing marks are not aligned, repeat the installation procedure.

31. Install the crankshaft pulley hub and tighten the retaining bolt to 80–87 ft. lbs. (108–118 Nm).

32. Install the crankshaft pulley and washer. Install the retaining bolts and tighten to 9–13 ft. lbs. (12–17 Nm).

33. Install the timing belt middle and lower covers with the gaskets. Tighten the middle and lower timing belt cover retaining bolts to 65–95 inch lbs. (7.8–11 Nm).

34. Install the power steering drive belt.

35. Install the water pump pulley and retaining bolts. Tighten the bolts to 69–95 inch lbs. (7.8–11.0 Nm).

36. Install the alternator/water pump drive belt.

37. Install the splash-shields. Tighten the bolts to 69–95 inch lbs. (7.8–11.0 Nm).

38. Install the right wheel and tire assembly. Tighten the lug nuts to 65–87 ft. lbs. (88–118 Nm).

39. Lower the vehicle.

40. Install the timing belt upper cover and gasket. Tighten the bolts to 69–95 inch lbs. (7.8–11.0 Nm).

41. Connect the negative battery cable.

42. Run the engine and check for leaks.

43. Road test the vehicle and check for proper engine operation.

1.9L (VIN J, K, 9) ENGINES

YEAR, MODEL & ENGINE APPLICATION

1985-96 Ford Escort
1985-87 Mercury Lynx, 1991-96 Tracer .... 1.9L VIN J, K, 9

REPLACEMENT INTERVALS

Ford does not have a recommended replacement interval. Standard industry practice is to replace timing belts every 60,000 miles.

LABOR TIMES

1985-90	
Replacement	3.3 hrs
Severe Service	3.8 hrs
1991-96	
Replacement	1.5 hrs
Severe Service	2.0 hrs

8807FRDg05

1985–90

**** CAUTION**

These engines are considered freewheeling. If the timing belt fails, valve-to-piston damage is unlikely, however a precautionary compression check should be performed to verify no engine damage.

1. Disconnect the negative battery cable.
2. Remove the timing belt cover.
3. Align the timing mark on the camshaft sprocket with the timing mark on the cylinder head.
4. Install the timing belt cover and confirm that the timing mark on the crankshaft damper aligns with the Top Dead Center (TDC) of the compression stroke on the front cover.
5. Remove the timing belt cover.
6. Loosen both timing belt tensioner attaching bolts.
7. Pry the belt tensioner away from the belt as far as possible and tighten 1 of the tensioner attaching bolts.
8. Remove crankshaft damper as follows:
   a. Properly support the engine and remove the right-side engine mount bolt.
   b. Lower the engine at the right-side until the crankshaft damper bolt clears the frame rail and remove the damper bolt.
   c. Raise the engine and remove the damper.
9. Remove the timing belt. If the belt is to be reused, mark the direction of rotation on the belt so it can be reinstalled in the same direction.

■ **With the timing belt removed and the No. 1 piston at TDC, Do not rotate the camshaft. If the camshaft must be rotated, align the crankshaft damper 90 degrees BTDC.**

To Install:

10. Install the timing belt over the sprockets in a counterclockwise direction starting at the crankshaft. Keep the belt span from the crankshaft to the camshaft tight as the belt is installed over the remaining sprocket.
11. Loosen belt tensioner attaching bolts and allow the tensioner to snap against the belt.
12. Tighten 1 of the tensioner attaching bolts.
13. Install the crankshaft damper, driveplate and damper attaching bolt. Hold the crankshaft damper stationary and tighten the attaching bolt to 81–96 ft. lbs. (110–130 Nm).
14. To seat the belt on the sprocket teeth, proceed as follows:
   a. Connect the negative battery terminal.
   b. Crank engine several revolutions.
   c. Disconnect the negative battery terminal.
   d. Turn camshaft, as necessary, to align the timing pointer on the cam sprocket with the timing mark on the cylinder head.

■ **Do not turn the engine counterclockwise to align the timing marks.**

   e. Position the timing belt cover on the engine and check to see that the timing mark on the crankshaft aligns with the TDC pointer on the cover. If the timing marks do not align, remove the belt, align the timing marks and return to Step 10.
15. Loosen the belt tensioner attaching bolt tightened in Step 12. The tensioner spring will apply the proper load on the belt. Tighten the belt tensioner bolt.

■ **The engine must be at room temperature. Do not set belt tension on a hot engine.**

16. Install timing belt cover.
17. Install accessory drive belts.
18. Connect negative battery cable.

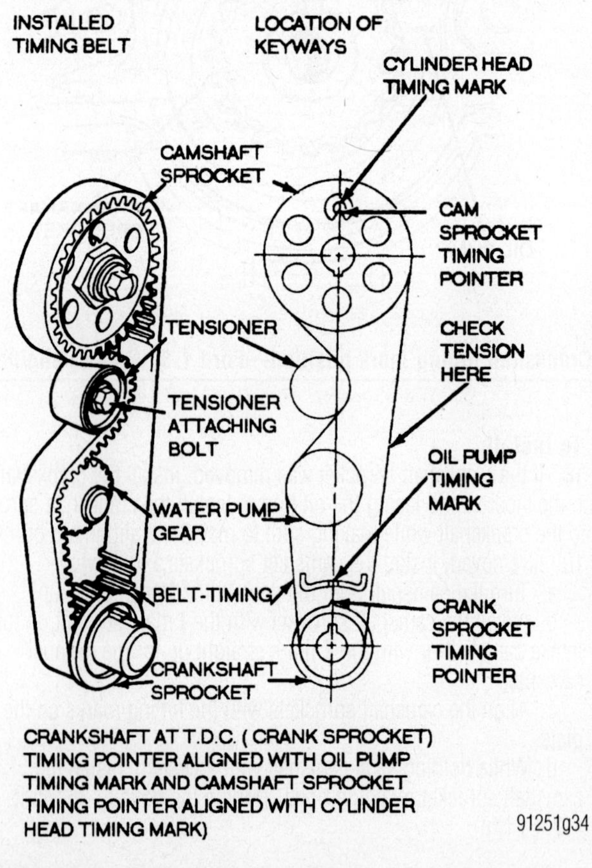

Timing belt sprocket alignment for belt service—Ford 1985–90 1.9L (VIN J, K and 9) Engine

1991–96

✻✻ CAUTION

These engines are considered freewheeling. If the timing belt fails, valve-to-piston damage is unlikely, however a precautionary compression check should be performed to verify no engine damage.

1. Disconnect the negative battery cable.
2. Remove the accessory drive belt automatic tensioner and the accessory drive belt.
3. Remove the timing belt cover.
4. Align the timing mark on the camshaft sprocket with the timing mark on the cylinder head.
5. Confirm that the timing mark on the crankshaft sprocket is aligned with the timing mark on the oil pump housing.
6. Loosen the belt tensioner attaching bolt, pry the tensioner away from the timing belt and retighten the bolt.
7. Remove the spark plugs. Remove the right engine mount.
8. Raise and safely support the vehicle on jackstands.
9. Remove the right-side splash-shield.
10. Remove the flywheel inspection shield.
11. Use a suitable tool to hold the flywheel in place.
12. Remove the crankshaft damper bolt and washer and remove the bolt.
13. Remove the timing belt.

■ **With the timing belt removed and the No. 1 piston at TDC, do not rotate the camshaft. If the camshaft must be rotated, align the crankshaft damper 90 degrees BTDC.**

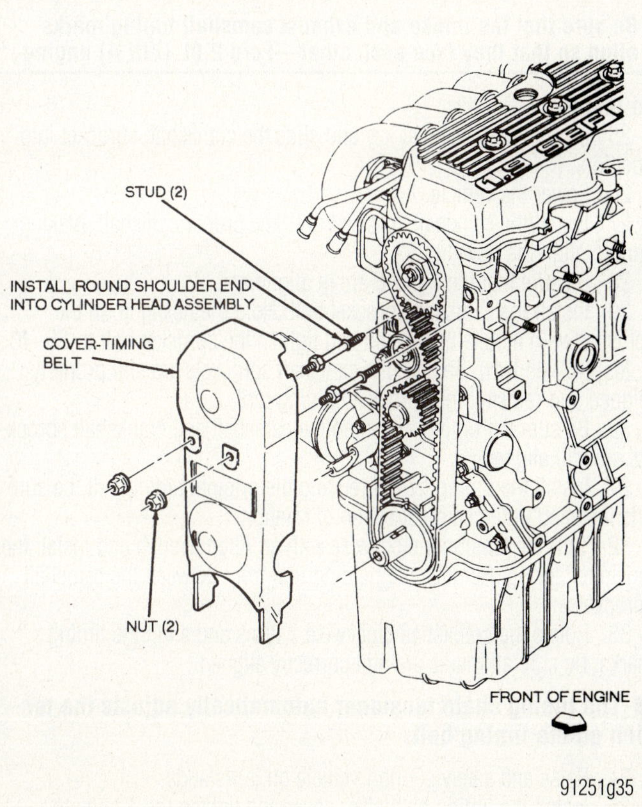

Exploded view of the timing belt cover mounting—Ford 1991–96 1.9L (VIN J, K and 9) engine

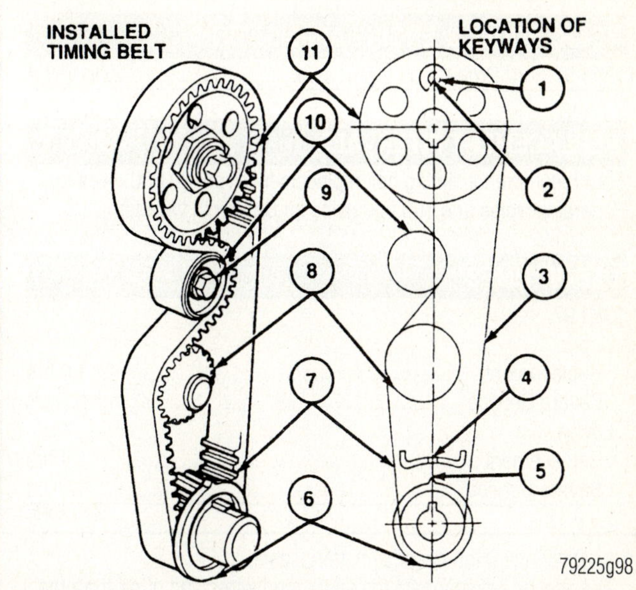

View of timing belt and alignment positions—Ford 1991–96 1.9L (VIN J, K and 9) engine

To Install:

14. Install the timing belt over the sprockets in a counterclockwise direction starting at the crankshaft. Keep the belt span from the crankshaft to the camshaft tight while installing over the remaining sprocket.
15. Loosen the belt tensioner attaching bolt, allowing the tensioner to snap against the belt.
16. Rotate the crankshaft clockwise 2 complete revolutions, stopping at TDC. This will allow the tensioner spring to load the timing belt.

■ **Do not turn the engine counterclockwise to align the timing marks. Do not rotate the crankshaft with the spark plugs installed.**

17. Recheck the camshaft and crankshaft timing marks for alignment, to be sure the timing belt has not skipped a tooth during rotation. Repeat the procedure if the timing marks are not aligned.
18. Tighten the tensioner attaching bolt to 17–22 ft. lbs. (23–30 Nm).
19. Install the crankshaft damper and the bolt and washer. Tighten the bolt to 81–96 ft. lbs. (110–130 Nm).
20. Install the flywheel inspection shield.
21. Install the splash-shield and lower the vehicle.
22. Install the right engine mount. Install the spark plugs.
23. Install the timing belt cover.
24. Install the accessory drive belt automatic tensioner and the accessory drive belt.
25. Connect the negative battery cable.

2.0L (VIN A) ENGINES

✻✻ CAUTION

These engines are an interference type. If the timing belt has failed, valve-to-piston damage may have occurred. A compression check should always be performed to check for engine damage.

YEAR, MODEL & ENGINE APPLICATION	
1993-97 Ford Probe	2.0L VIN A

REPLACEMENT INTERVALS

Ford recommends timing belt replacement every 60,000 miles for Federal vehicles and 105,000 miles for California vehicles.

LABOR TIMES	
1993-97	
Upper	
Replacement	1.0 hrs
Severe Service	1.1 hrs
Lower	
Replacement	1.4 hrs
Severe Service	1.6 hrs

8807FRDg06

1. Disconnect the negative battery cable.
2. Label and disconnect the spark plug wires and clips from the cylinder head cover. Remove the ignition distributor with wiring and set it aside.
3. Remove the power steering hose brackets from the cylinder head cover. If necessary disconnect the crankshaft position sensor.
4. Disconnect the breather tube and PCV valve from the cylinder head cover.
5. Loosen the cylinder head cover bolts in 2–3 steps. Remove the cylinder head cover.
6. Remove the power steering belt shield. Loosen the power steering adjusting bolt, lockbolt and through-bolt and remove the power steering belt.
7. Loosen the alternator adjusting bolt and upper mounting bolt. Remove the alternator belt.
8. Support the engine with Engine Support tool 014-00750 or equivalent. Raise the engine slightly with a jack and remove the right-side engine support insulator (mount).
9. Remove the oil level indicator bolt and 4 upper timing belt cover bolts and remove the upper timing belt cover.
10. Raise and safely support the vehicle.
11. Remove the right front wheel and fender liner.
12. Remove the splash-shields. Using Holder tool T92C-6316-AH, or equivalent, hold the crankshaft pulley and remove the pulley bolt. Use a suitable puller to remove the pulley, then remove the guide plate.
13. Remove the 4 lower timing belt cover bolts and remove the lower timing belt cover.
14. Temporarily install the crankshaft pulley bolt.
15. Turn the crankshaft until the timing mark on the crankshaft sprocket aligns with the timing mark on the oil pump and the camshaft sprocket timing marks, **E** and **I**, align on the camshaft sprockets.
16. Lower the vehicle.
17. Insert camshaft sprocket holding tool T92C-6256-AH or equivalent, between the camshaft sprockets.
18. Turn the timing belt tensioner with an Allen wrench and remove the tensioner spring from the tensioner spring pin.
19. If the timing belt is to be reused, mark the direction of rotation on the timing belt. Remove the timing belt.
20. If it is necessary to remove the sprockets, remove the camshaft sprocket holding tool. Hold the camshaft by placing a suitable wrench on the hexagon which is cast into the camshaft. Place another wrench onto the camshaft sprocket retaining bolt and loosen the bolt.

■ **Before removing the camshaft sprocket(s), be sure that the camshafts are still in alignment and tag each sprocket to the camshaft from which it was removed.**

21. Remove the camshaft sprocket bolt and the camshaft sprocket from the camshaft.
22. Repeat the camshaft sprocket removal procedure for the opposite camshaft if required.
23. Raise and safely support the vehicle.
24. Slide off the crankshaft sprocket and remove the crankshaft key.

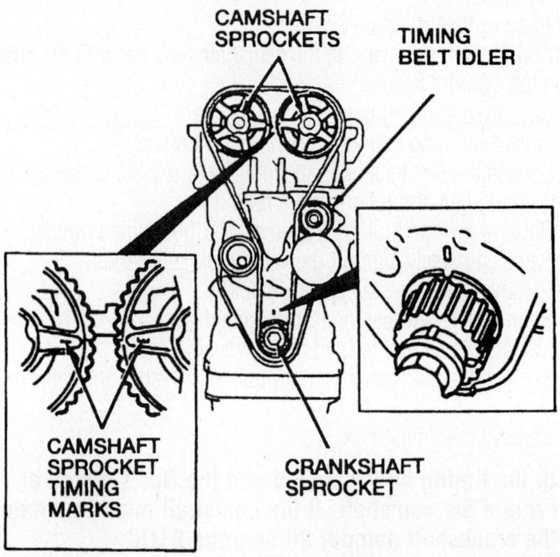

Be sure that the intake and exhaust camshaft timing marks align so that they face each other—Ford 2.0L (VIN A) engine

To Install:

25. Install the crankshaft key and slide the crankshaft sprocket into position.
26. Lower the vehicle.
27. Install the camshaft sprocket onto the proper camshaft, making sure to align the dowel pin.
28. Be sure that the **I** and **E** are in alignment.
29. Install the camshaft sprocket bolt. Hold the hexagon on the camshaft with a suitable wrench and tighten the sprocket bolt to 35–48 ft. lbs. (47–65 Nm). Be sure the camshaft sprockets are still properly aligned and reinstall the sprocket holding tool.
30. Be sure the timing marks on the camshaft and crankshaft sprockets are still aligned.
31. Install the timing belt. If reusing the original timing belt, be sure it is installed in the same direction of rotation.
32. Turn the tensioner clockwise with an Allen wrench and install the tensioner spring. Remove the holding tool from between the camshaft sprockets.
33. Rotate the crankshaft clockwise 2 turns and align the timing marks. Be sure all marks are still correctly aligned.

■ **The timing chain tensioner automatically adjusts the tension on the timing belt.**

34. Raise and safely support vehicle on jackstands.
35. Install the timing belt lower cover and tighten the 4 bolts to 71–88 inch lbs. (8–10 Nm).

36. Install the guide plate, crankshaft pulley and pulley bolt. Secure the pulley with the holder tool and tighten the bolt to 116–123 ft. lbs. (157–167 Nm).

37. Install the tire, fender liner splash-shields and lower the vehicle.

38. Raise the engine slightly with the jack and install the right-side engine mount. Tighten the mount through-bolt to 63–86 ft. lbs. (86–116 Nm) and the mount attaching nuts to 54–75 ft. lbs. (74–103 Nm). Remove the engine support tool.

39. Install the upper timing belt cover and tighten the bolts to 71–88 inch lbs. (8–10 Nm).

40. Clean the cylinder head and valve cover mating surfaces thoroughly.

41. Apply silicone sealant to the cylinder head surface in the area adjacent to the front camshaft bearing caps. Apply sealant to a new gasket and install it on the cylinder head cover.

42. Install the cylinder head cover and tighten the bolts.

43. Install the power steering hose brackets and tighten the bolts to 71–88 inch lbs. (8–10 Nm). Connect the spark plug wires and wire clips. Connect the breather tube and PCV valve. If necessary, connect the crankshaft position sensor.

44. Install the alternator belt and adjust the tension. Tighten the upper mounting bolt to 14–18 ft. lbs. (19–25 Nm) and the lower through-bolt to 27–38 ft. lbs. (37–52 Nm).

45. Install the power steering belt and adjust the tension. Tighten the through-bolt to 32–45 ft. lbs. (43–61 Nm) and the lockbolt to 23–34 ft. lbs. (31–46 Nm). Install the power steering belt shield and tighten the bolts to 61–86 inch lbs. (7–9 Nm).

46. Connect the negative battery cable.

47. Run the engine and check for leaks and proper engine operation.

2.0L (VIN B) ENGINES

YEAR, MODEL & ENGINE APPLICATION

2000-04 Ford Escape	2.0L VIN B

REPLACEMENT INTERVALS

Ford recommends timing belt replacement every 120,000 miles.

LABOR TIMES

2000-04	
Replacement	3.6 hrs
Severe Service	4.1 hrs

8807FRDg07

✱✱ CAUTION

These engines are an interference type. If the timing belt has failed, valve-to-piston damage may have occurred. A compression check should always be performed to check for engine damage.

1. Disconnect the negative battery cable.
2. Remove the air cleaner outlet pipe.
3. Disconnect cruise control and accelerator cables.
4. Remove the valve cover.
5. Remove the spark plugs.
6. Remove the catalytic converter.
7. Remove the bolt, nut, and position the coolant tube aside.
8. Remove the right wheel and tire assembly.
9. Remove the right lower splash shield.
10. Rotate the crankshaft to just before top dead center (TDC) on No. 1 cylinder.
11. Remove the stud.
12. Install the special tool 303-574 into the stud hole.

■ **Make sure the correct (second) notch in the pulley is indexed to the lower cylinder block.**

13. Rotate the crankshaft clockwise against the peg to bring it to TDC on No. 1 cylinder.
14. Loosen the water pump pulley bolts.
15. Remove the accessory drive belt.
16. Remove the crankshaft pulley.
17. Remove the bolts and the lower timing belt cover.
18. Lower the vehicle.
19. Install engine lifting bracket 303-F072.
20. Remove the ground strap.
21. Remove the engine mount upper bracket.
22. Remove the studs.
23. Remove the knock sensor connector.
24. Remove the bolts and the upper timing belt cover.
25. Remove the water pump pulley.
26. Remove the accessory drive belt idler pulley.

■ **Installation of the alignment tool into the exhaust camshaft may require the camshafts to be rotated clockwise slightly.**

27. Install special tool 303-465 and align the camshafts.
28. Raise and support the vehicle.
29. Remove the bolts and the engine mount lower bracket.
30. Loosen the timing belt tensioner pulley and allow to slide down to the bottom of its travel.

✱✱ CAUTION

If the camshaft timing belt is to be reused, mark the direction of the camshaft timing belt to rotation of camshaft prior to removal or premature wear or failure may occur.

31. Slide the timing belt off of the camshaft and crankshaft sprockets. Check the timing belt for wear. Install a new belt if necessary.

To Install:

■ **Make sure the correct (second) notch in the pulley is indexed to the lower cylinder block.**

32. Slide the crankshaft pulley onto the crankshaft and confirm the crankshaft position is at TDC (No. 1 cylinder) by rotating it clockwise against the alignment peg.
33. Remove the crankshaft pulley.
34. Lower the vehicle.
35. Confirm that the timing belt tensioner is installed correctly with the tab positioned in the slot in the inner timing cover.
36. Install the timing belt onto the timing belt sprockets.
37. Adjust the timing belt tensioner.
38. Using a 6 mm Allen wrench, rotate the adjuster counterclockwise and align the marks.
39. Tighten the tensioner pulley bolt.
40. Raise the vehicle.
41. Install the front engine mount lower bracket and tighten to 37 ft. lbs. (50Nm).
42. Install the accessory drive belt idler pulley.
43. Install the water pump pulley.

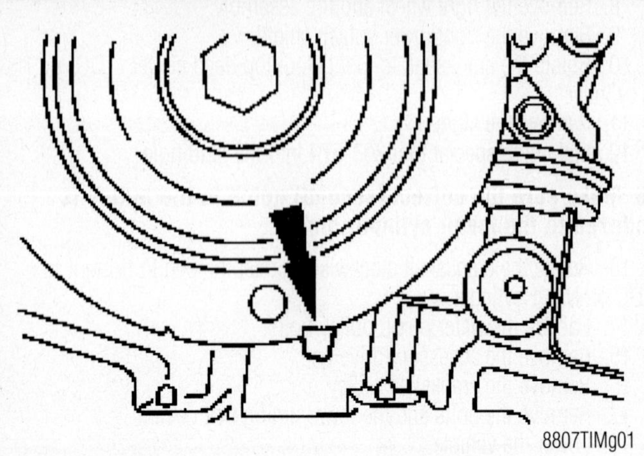

Positioning the crankshaft pulley against the alignment peg—Ford 2.0L (VIN B) engine

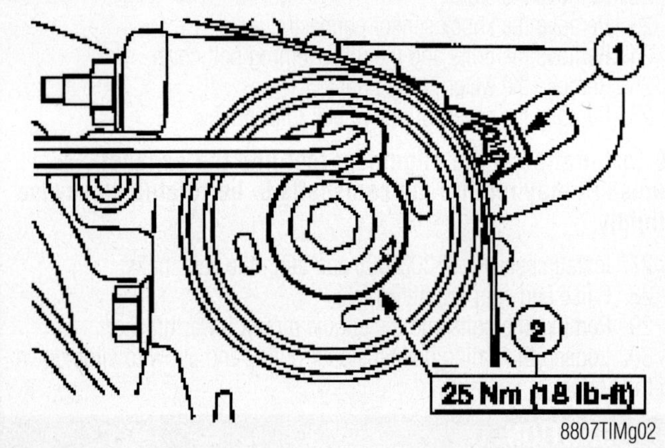

Positioning timing belt tensioner in timing cover slot—Ford 2.0L (VIN B) engine

44. Hand tighten the bolts.
45. Lower the vehicle.
46. Install the timing belt covers. Torque the cover bolts to 62 inch lbs. (7Nm).
47. Tighten the water pump pulley bolts.
48. Remove the special tool.
49. Install the stud. Torque to 25 ft. lbs. (34 Nm).
50. Install the engine mount upper bracket and tighten to 72 ft. lbs. (98Nm).
51. Install the ground strap.
52. Install the crankshaft pulley, but do not tighten the bolt.
53. Install the accessory drive belt and tighten the tensioner.
54. Tighten the crankshaft pulley bolt to 85 ft. lbs. (115Nm).
55. Install the splash shield and front wheel.
56. Lower the vehicle.
57. Install the coolant tube.
58. Install the catalytic converter.
59. Remove the special tool.
60. Install the valve cover.
61. Install the spark plugs.
62. Install the cruise control and accelerator cables.
63. Install the air cleaner outlet pipe.
64. Connect the negative battery cable.

2.0L (VIN C) ENGINES

YEAR, MODEL & ENGINE APPLICATION	
1985-88 Ford Ranger	2.0L VIN C

REPLACEMENT INTERVALS

Ford does not have a recommended replacement interval. Standard industry practice is to replace timing belts every 60,000 miles.

LABOR TIMES	
1985-88	
Replacement	3.5 hrs
Severe Service	3.8 hrs

8807FRDg08

✱✱ CAUTION

These engines are an interference type. If the timing belt has failed, valve-to-piston damage may have occurred. A compression check should always be performed to check for engine damage.

1. Disconnect the negative battery cable and drain the cooling system.
2. Remove the accessory drive belts(s).
3. Remove the upper radiator hose and thermostat housing.
4. Remove the power steering pump and position aside, if equipped.
5. Remove the 4 water pump pulley bolts.
6. Remove the timing belt outer cover retaining bolt(s). Release the cover interlocking tabs, if equipped, and remove the cover.
7. Loosen the belt tensioner adjustment screw, position belt tensioner tool T74P-6254-A or equivalent, on the tension spring roll pin and release the belt tensioner. Tighten the adjustment screw to hold the tensioner in the released position.
8. Remove the crankshaft pulley bolt and pulley.
9. Remove the crankshaft pulley, hub and belt guide. Remove the timing belt. If the belt is to be reused, mark the direction of rotation so it may be reinstalled in the same direction.

To Install:

10. Position the crankshaft sprocket to align with the TDC mark and the camshaft sprocket to align with the camshaft timing pointer. Remove the distributor cap and set the rotor to the No. 1 firing position by turning the auxiliary shaft.
11. Install the timing belt over the crankshaft sprocket and then counterclockwise over the auxiliary and camshaft sprockets. Align the belt fore-and-aft on the sprockets.
12. Loosen the tensioner adjustment bolt to allow the tensioner to move against the belt. If the spring does not have enough tension to move the roller against the belt, it may be necessary to manually push the roller against the belt and tighten the bolt.
13. To make sure the belt does not jump time during rotation, remove the spark plugs.
14. Rotate the crankshaft 2 complete turns in the direction of normal rotation to remove the slack from the belt. Tighten the tensioner adjustment to 29–40 ft. lbs. (40–55 Nm) and pivot bolts to 14–22 ft. lbs. (20–30 Nm). Check the alignment of the timing marks.

15. Install the crankshaft belt guide.

16. Install the crankshaft pulley and tighten the retaining bolt to 103–133 ft. lbs. (140–180 Nm).

17. Position the timing belt front cover. Snap the interlocking tabs into place, if necessary. Install the timing belt outer cover retaining bolt(s) and tighten to 71–106 inch lbs. (8–12 Nm).

18. Install the thermostat housing and a new gasket. Install the upper radiator hose.

19. Install the crankshaft pulley and retaining bolt. Tighten to 103–133 ft. lbs. (140–180 Nm).

20. Install the water pump pulley and the automatic belt tensioner. Install the accessory drive belt(s).

21. Install the spark plugs and remaining components.

22. Connect the negative battery cable, start the engine and check the ignition timing.

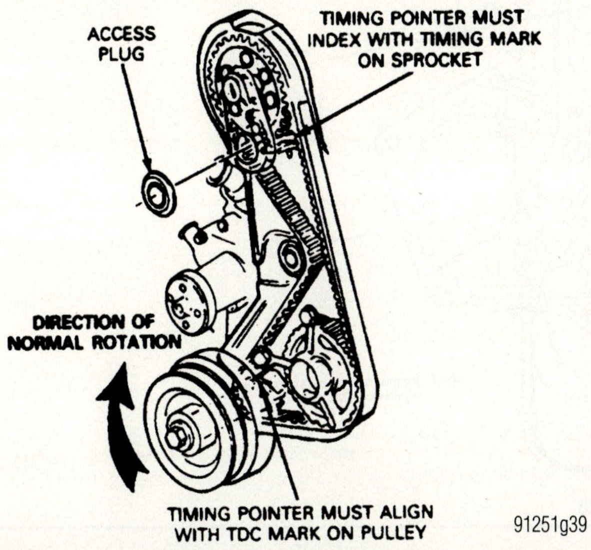

Camshaft and crankshaft sprocket positioning for timing belt installation—Ford 2.0L (VIN C) engine

2.0L (VIN H) DIESEL ENGINES

YEAR, MODEL & ENGINE APPLICATION

1985-87 Ford Escort, 1985-87 Mercury Lynx	2.0L VIN H

REPLACEMENT INTERVALS

Ford does not have a recommended replacement interval. Standard industry practice is to replace timing belts every 60,000 miles.

LABOR TIMES

1985-87	
Front Belt	
Replacement	6.3 hrs
Severe Service	6.6 hrs
Rear Belt or Tensioner	
Replacement	1.8 hrs
Severe Service	2.1 hrs

8807FRDg09

■ **The following procedures require that the engine be removed from the vehicle in order to replace the timing belt.**

FRONT BELT

✻✻ CAUTION

These engines are an interference type. If the timing belt has failed, valve-to-piston damage may have occurred. A compression check should always be performed to check for engine damage.

1. Disconnect the negative battery cable and drain the cooling system.
2. Remove the engine from the vehicle and place it on an engine stand.
3. Remove the drive belt(s) and the necessary components to allow for the front timing belt cover removal.
4. Rotate the crankshaft to position the No. 1 piston on the Top Dead Center (TDC) of its compression stroke.
5. Remove the upper timing belt cover.
6. Install a Flywheel Holding tool No. T84P6375A, or equivalent.
7. Remove the 6 crankshaft pulley-to-crankshaft sprocket bolts.
8. Install a Crankshaft Pulley Remover tool No. T58P-6316-D, or equivalent, using Adapter tool No. T74P-6700-B or equivalent, and remove the pulley.
9. Remove the front timing belt lower cover.
10. If reusing the timing belt, mark the rotational direction on it.
11. Loosen the timing belt tensioning pulley and remove the timing belt.

To Install:

12. Align the camshaft sprocket with the timing mark.

■ **Check the crankshaft sprocket to see that the timing marks are aligned.**

13. Remove the tensioner spring from the pocket in the front timing belt upper cover and install it in the slot in the tensioner lever and over the stud in the crankcase.
14. Push the tensioner lever toward the water pump as far as it will travel and tighten the lockbolt snug.
15. Noting the rotational direction, install the timing belt.
16. Install the tensioner spring in the belt tensioner lever and over the stud mounted on the front of the crankcase.
17. Adjust the timing belt tension by performing the following procedure:
    a. Loosen the tensioner pulley lockbolt.
    b. Rotate the crankshaft pulley 2 complete clockwise revolutions until the flywheel TDC timing mark aligns with the pointer on the rear cover plate.
    c. Check the front camshaft sprocket to see that it is aligned with its timing mark.
    d. Tighten the tensioner lockbolt to 23–34 ft. lbs. (31–46 Nm).
    e. Using a Belt Tension Gauge model 21-0028 or equivalent, check the timing belt tension; it should be 33–44 lbs. (15–20 kg).
18. Install the front timing belt lower cover and torque the bolts to 60–84 inch lbs. (6.8–9.5 Nm).
19. Install the crankshaft pulley and torque the bolts to 17–24 ft. lbs. (23–33 Nm).
20. Install the front timing belt upper cover and torque the bolts to 60–84 inch lbs. (6.8–9.5 Nm).
21. Install any components that were removed and install the drive belt(s).

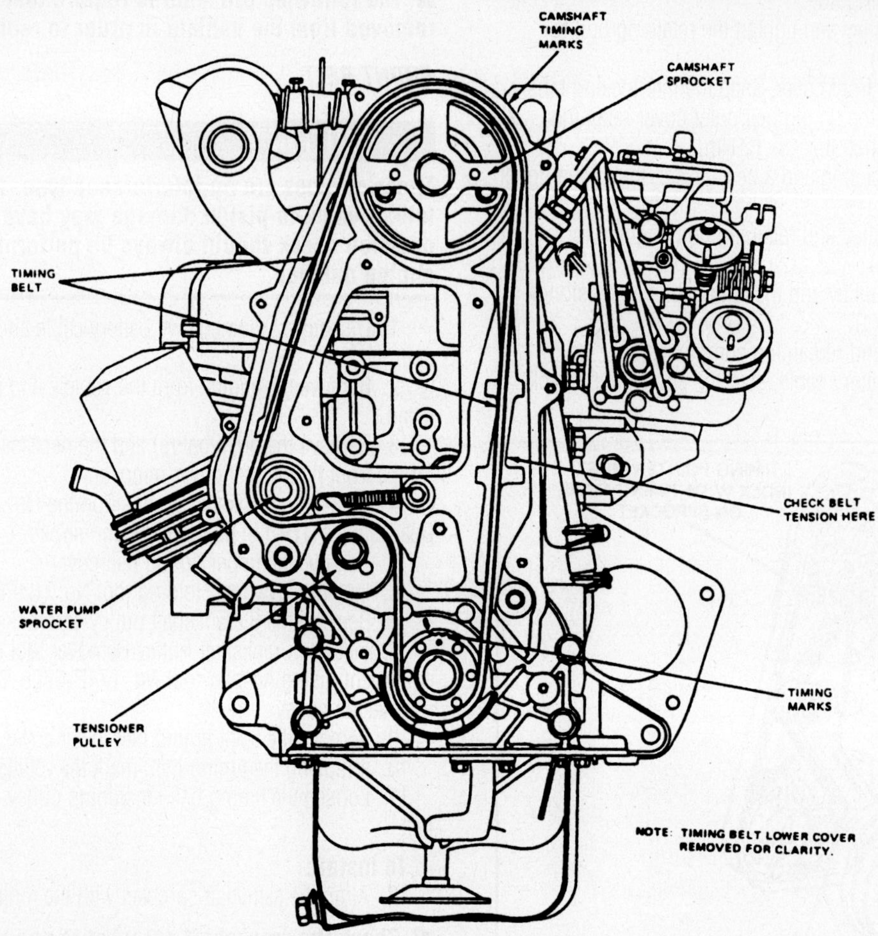

View of the front timing belt assembly—Ford 2.0L (VIN H) diesel engine

22. Install the engine into the vehicle.
23. Refill the cooling system and connect the negative battery cable.

REAR BELT

1. Disconnect the negative battery cable and drain the cooling system.
2. Remove the engine from the vehicle and place it on an engine stand.
3. Rotate the crankshaft to position the No. 1 piston on the Top Dead Center (TDC) of its compression stroke.
4. Remove the rear timing belt cover.
5. Remove the flywheel timing mark cover from the clutch housing.
6. Check that the injection pump and camshaft sprocket timing marks are aligned.
7. Loosen the tensioner locknut. With a suitable tool inserted in the slot provided, rotate the tensioner clockwise to relieve the belt tension. Tighten the locknut snug.
8. If reusing the timing belt, mark the rotational direction on it.
9. Remove the rear timing belt.

To Install:

10. Noting the rotational direction, install the rear timing belt.
11. Adjust the timing belt tension by performing the following procedure:
    a. Loosen the tensioner pulley lockbolt.
    b. Rotate the crankshaft pulley 2 complete clockwise revolutions until the flywheel TDC timing mark aligns with the pointer on the rear cover plate.

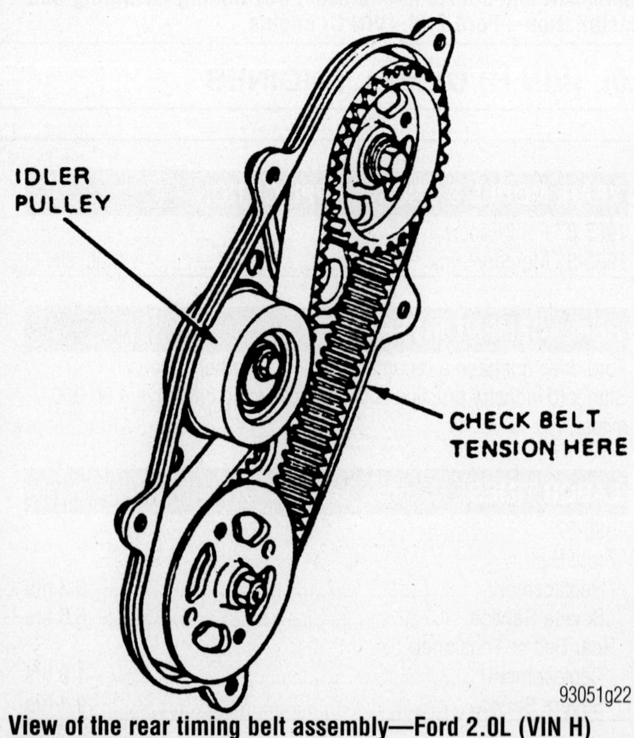

View of the rear timing belt assembly—Ford 2.0L (VIN H) diesel engine

c. Check the front camshaft sprocket and the injection pump sprocket are aligned with there timing marks.

d. Tighten the tensioner lockbolt to 15–20 ft. lbs. (20–27 Nm).

e. Using a Belt Tension Gauge model 21-0028 or equivalent, check the timing belt tension; it should be 22–33 lbs. (10–15 Nm).

12. Install the rear timing belt cover and torque the 6mm bolts to 60–84 inch lbs. (6.8–9.5 Nm) and the 8mm bolt to 12–16 ft. lbs. (16–22 Nm).

13. Install the flywheel timing mark cover.

14. Install any components that were removed and install the drive belt(s).

15. Install the engine into the vehicle.

16. Refill the cooling system and connect the negative battery cable.

2.0L (VIN P) ENGINES

YEAR, MODEL & ENGINE APPLICATION

Application	Engine
1998-00 Ford Escort, 2000-04 Ford Focus, 1997-99 Mercury Tracer	2.0L VIN P

REPLACEMENT INTERVALS

Ford does not have a recommended replacement interval. Standard industry practice is to replace timing belts every 60,000 miles.

LABOR TIMES

1997-04	
Escort & Tracer	
Replacement	1.6 hrs
Severe Service	1.9 hrs
Focus	
Replacement	2.1 hrs
Severe Service	2.3 hrs

8807FRDg10

✱✱ CAUTION

These engines are an interference type. If the timing belt has failed, valve-to-piston damage may have occurred. A compression check should always be performed to check for engine damage.

1. Disconnect the negative battery cable.
2. Remove the accessory drive belt and belt tensioner.
3. Remove the timing belt cover.
4. Raise and support the vehicle.
5. Right the right front wheel and inner splash shield.
6. Remove the crankshaft pulley.
7. Align the timing marks as shown.
8. Refer to the illustration and remove the timing belt as follows:
9. Loosen the timing belt tensioner bolt (1).
10. Use an 8mm Allen wrench, and turn the tensioner (2) counterclockwise 1/4 turn.
11. Insert a 1/8 inch drill bit in the hole (3) to lock the belt tensioner in place.
12. Remove the timing belt (4).

To Install:

13. Install the timing belt in a counterclockwise direction starting at the crankshaft.

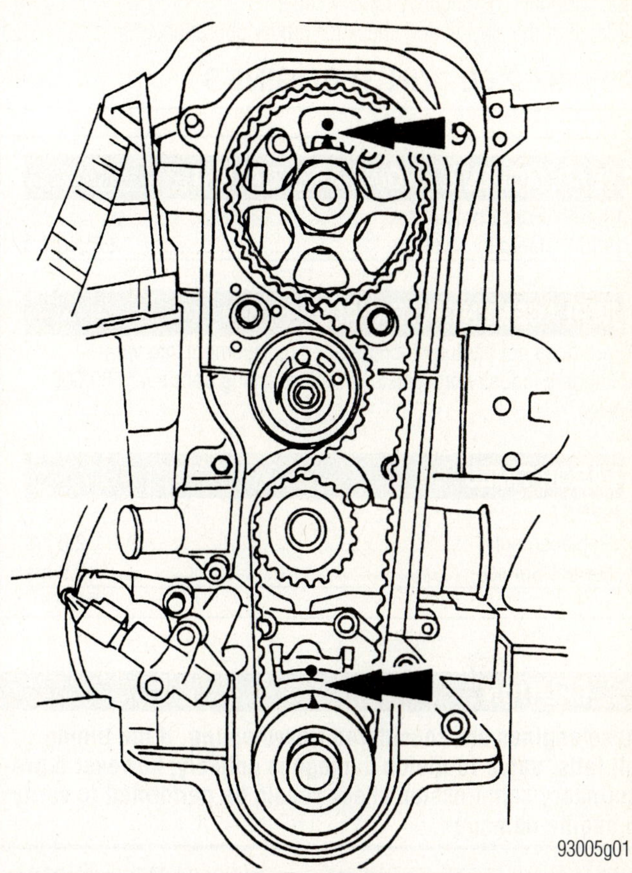

93005g01

Align the timing marks before removing or installing the timing belt—2.0L (VIN P) SOHC engine

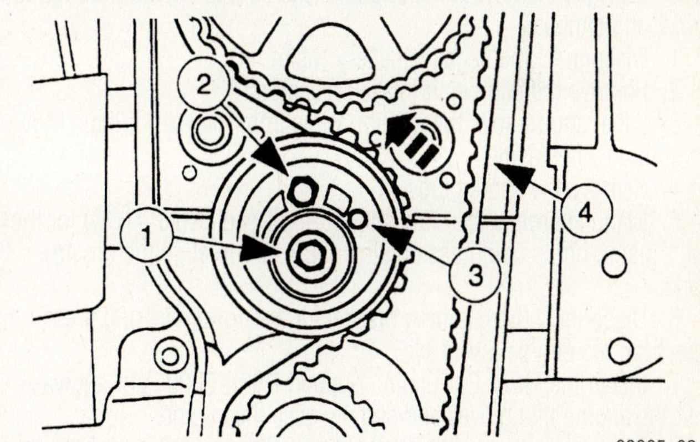

93005g02

Remove the timing belt by following these 4 numbered steps (refer to the text for an explanation)—2.0L (VIN P) SOHC engine

14. Remove the drill bit to unlock the timing belt tensioner.
15. Rotate the crankshaft two complete turns and check that the timing marks align.
16. Tighten the tensioner bolt to 15–22 ft. lbs. (20–30 Nm).
17. Install the crankshaft pulley and tighten the bolt to 81–96 ft. lbs. (110–130 Nm).
18. Install the right inner splash shield and front wheel.
19. Lower the vehicle.
20. Install the timing belt cover.
21. Install the accessory drive belt and tensioner.

22. Connect the negative battery cable.
23. Start the engine and check for proper operation.

1995–97 2.0L (VIN 3) ENGINES

YEAR, MODEL & ENGINE APPLICATION

1995-97 Ford Contour, 1995-97 Mercury Mystique	2.0L VIN 3

REPLACEMENT INTERVALS

Ford does not have a recommended replacement interval. Standard industry practice is to replace timing belts every 60,000 miles.

LABOR TIMES

1995-97	
Replacement	2.6 hrs
Severe Service	2.9 hrs

8807FRDg11

✻✻ CAUTION

These engines are considered freewheeling. If the timing belt fails, valve-to-piston damage is unlikely, however a precautionary compression check should be performed to verify no engine damage.

When installing a timing belt, tensioner spring (6L277) and retaining bolt (W700001-S309) must be purchased and properly installed on the engine. First check to see if these parts are already installed. The tensioner spring will adjust the timing belt tension and should not require further adjustments.

1. Disconnect the negative battery cable.
2. Remove the engine air intake resonators.
3. Label and remove the ignition wires from the spark plugs. Move the ignition wires aside.
4. Remove the spark plugs.
5. Manually rotate the crankshaft to Top Dead Center (TDC) for the No. 1 piston on its compression stroke. Be sure to align the timing marks.
6. Disconnect the retaining bracket for the power steering pressure hose from the engine lifting eye.
7. Install the Three Bar Engine Support D88L-6000-A or equivalent, onto the engine lifting eyes and slightly raise the engine.
8. Remove the upper camshaft timing belt cover retaining bolts and the cover from the engine.

■ **Mark the location of the upper front engine support bracket before removing it from the engine support bracket.**

9. Remove the upper front engine support bracket retainer nuts, the bracket and the upper front engine support insulator.
10. If equipped, remove the wiring harness connector from the low coolant level sensor at the radiator coolant recovery reservoir.
11. Remove the radiator coolant recovery reservoir retainers and move the reservoir aside.
12. Remove the upper front engine support insulator.
13. Set the coolant recovery reservoir back into position temporarily.
14. Loosen the water pump pulley retaining bolts. Do not remove the bolts completely.
15. Remove the accessory drive belt.
16. Remove the drive belt idler pulley retaining bolt and pulley from the alternator mounting bracket.
17. Finish removing the water pump retaining bolts and remove the water pump pulley.
18. Remove the center camshaft timing belt cover retaining bolts and the cover from the engine.
19. Raise and safely support the vehicle on jack stands.
20. Remove the right side wheel and tire and fender liner.
21. Remove the crankshaft pulley.
22. Remove the lower camshaft timing belt cover bolts and the cover from the engine.
23. Remove the valve cover as follows:
    a. Disconnect the crankcase ventilation tube from the valve cover.
    b. Remove the retaining bolt and nut for the power steering pressure hose retaining bracket and move the hose aside.
    c. Remove the valve cover retaining bolts in a standard removal sequence starting from the outside of the valve cover and working toward the inside of the valve cover.
    d. Remove the valve cover and gasket from the engine.
24. Place Camshaft Alignment Timing tool T94P-6256-CH or equivalent, into the slots of both camshafts at the rear of the cylinder head to lock the camshafts into position.
25. Loosen the camshaft timing belt tensioner pulley retaining bolt and move the tensioner pulley to relieve the tension on the timing belt.
26. Temporarily tighten the tensioner in this position.

■ **If the timing belt is to be reused, mark the belt for the direction of rotation before removing to prevent premature wear or failure.**

27. Remove the timing belt.
28. If required, remove the sprockets as follows:
    a. Hold the camshaft with the Camshaft Sprocket Holding tool T74P-6256-B or equivalent.
    b. Loosen and remove the camshaft sprocket retaining bolt.
    c. Remove the sprocket from the camshaft.
    d. Repeat the procedure for the 2nd camshaft sprocket.
    e. Remove the crankshaft sprocket.

To Install:

29. Slide the crankshaft sprocket onto the crankshaft aligning the keyway.
30. Align the camshafts using the Camshaft Alignment Timing tool T94P-6256-CH.

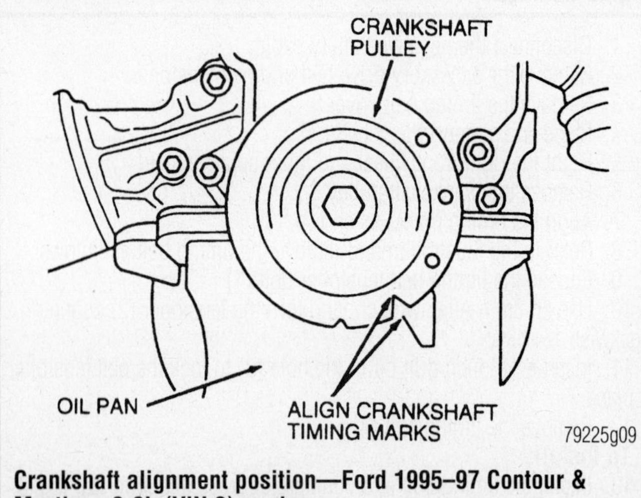

Crankshaft alignment position—Ford 1995–97 Contour & Mystique 2.0L (VIN 3) engine

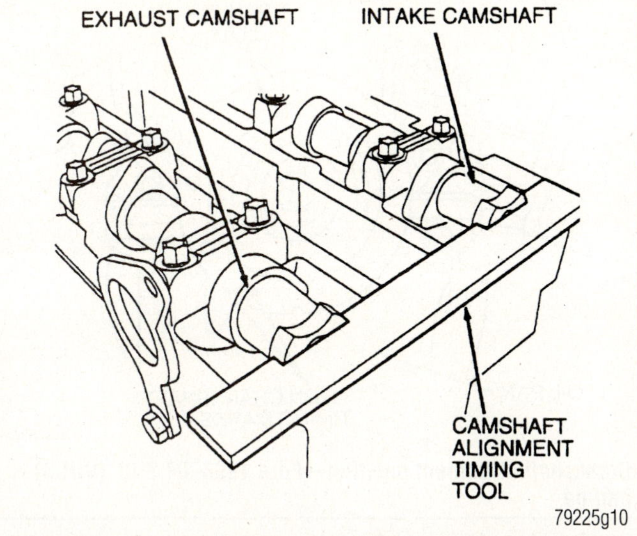

Placement of Camshaft Alignment Timing tool T94P-6256-CH or equivalent— Ford 1995–97 Contour & Mystique 2.0L (VIN 3) engine

31. Reinstall the sprockets onto the camshafts and loosely install the camshaft retaining bolts.
32. Tighten the camshaft sprocket retaining bolts to 47–53 ft. lbs. (64–72 Nm).
33. Loosely install the crankshaft pulley to verify that the engine is at TDC. Realign the marks if they have moved.
34. Verify that the camshafts are aligned.

■ **It is recommended to purchase a tensioner spring and retaining bolt through the dealer parts to apply the proper tension for used or new belt installations. The spring is bolted to the tensioner assembly and becomes a part of the engine. Ignore this notice if the tensioner spring is already installed.**

35. Reinstall the retaining bolt (W700001-S309) into the hole provided in the cylinder block and place the tensioner spring (6L277) between the bolt and the camshaft timing belt tensioner pulley.
36. Tighten the retainer bolt to 71–97 inch lbs. (8–11 Nm).
37. Remove the crankshaft pulley and install the timing belt onto the crankshaft sprocket, then onto the camshaft sprockets working in a counterclockwise direction.
38. Tighten the camshaft sprocket retaining bolts to 47–53 ft. lbs. (64–72 Nm).
39. Be sure that the span of the camshaft timing belt between the crankshaft sprocket and the exhaust camshaft sprocket is not loose.
40. Be sure that the camshaft timing belt is securely aligned on all sprockets.
41. Reinstall the lower timing belt cover and tighten the retaining bolts to 53–71 inch lbs. (6–8 Nm).
42. Apply silicone sealer to the keyway of the crankshaft pulley and install. Tighten the retaining bolt to 81–89 ft. lbs. (110–120 Nm).
43. Inspect the timing mark on the crankshaft pulley to verify that the engine is still at TDC.
44. Loosen the camshaft timing belt tensioner pulley retaining bolt and allow the tensioner spring attached to the pulley to draw the tensioner pulley against the camshaft timing belt.
45. Remove the camshaft alignment timing tool from the camshafts at the rear of the engine.
46. Turn the crankshaft 2 revolutions in a clockwise direction.
47. Tighten the camshaft timing belt tensioner pulley retaining bolt to 26–30 ft. lbs. (35–40 Nm).
48. Recheck that the crankshaft timing mark is at TDC for the No. 1 piston, and that both camshafts are in alignment using the camshaft alignment timing tool.

■ **A slight adjustment of the camshafts to allow the insertion of the camshaft alignment timing tool is permissible as long as the crankshaft stays at the TDC location.**

49. Camshaft Sprocket Holding tool T74P-6256-b or equivalent, can be used to move the camshaft sprocket(s) if a slight adjustment is required.
50. If a camshaft is not properly aligned, perform the following procedure:
    a. Loosen the retaining bolt securing the sprocket to the camshaft while holding the camshaft sprocket from turning with the sprocket holding tool.
    b. Turn the camshaft until the camshaft alignment timing tool can be installed.
    c. Verify that the crankshaft timing mark is at TDC for the No. 1 cylinder.
    d. While holding the camshaft sprocket with the camshaft sprocket holding tool, tighten the retaining bolt to 47–53 ft. lbs. (64–72 Nm).
    e. Remove the tool and rotate the crankshaft 2 revolutions (clockwise).
    f. Verify that the camshafts are aligned and that the crankshaft is at TDC for the No. 1 cylinder.
51. Reinstall the valve cover as follows:
    a. Clean the gasket sealing surfaces.
    b. Inspect the valve cover gasket and O-rings; replace as required.
    c. Reinstall the valve cover retaining bolts and tighten them in a standard sequence starting from the center and working towards the outside of the valve cover to 53–71 inch lbs. (6–8 Nm).
    d. Reinstall the power steering hose retaining bracket and the power steering hose.
    e. Reinstall the crankcase ventilation tube to the valve cover.
52. Position the center camshaft timing belt cover.
53. Reinstall the center camshaft timing belt cover retaining bolts and tighten them to 53–71 inch lbs. (6–8 Nm).
54. Reinstall the water pump pulley and the retaining bolts. Reinstall the bolts, finger-tight.
55. Reinstall the drive belt idler pulley.
56. Reinstall the drive belt idler pulley retaining bolt and tighten it to 35 ft. lbs. (48 Nm).
57. Reinstall the accessory drive belt.
58. Tighten the water pump pulley retaining bolts to 89–124 inch lbs. (10–14 Nm).
59. Move the radiator coolant recovery reservoir aside.
60. Reinstall the upper front engine support insulator.
61. Position the radiator coolant recovery reservoir and install the retainers.
62. If equipped, install the wiring harness to the low coolant level sensor on the coolant recovery reservoir.
63. Reinstall the upper front engine support bracket to the engine and the upper front engine support insulator using the mark made during the removal procedure for reference.
64. Install the upper camshaft timing belt cover.
65. Reinstall the upper camshaft timing belt cover retaining bolts and tighten to 27–44 inch lbs. (3–5 Nm).
66. Remove the engine support.

67. Reinstall the retaining bracket for the power steering pressure hose to the engine lifting eye.
68. Reinstall the spark plugs and the ignition wires.
69. Reinstall the engine air intake resonators.
70. Replace the engine oil.
71. Reconnect the negative battery cable.
72. Run the engine and check for leaks and proper operation.

1998–04 2.0L (VIN 3) ENGINES

YEAR, MODEL & ENGINE APPLICATION

1998-99 Ford Contour, 1998-04 Escort ZX2, 2000-04 Focus, 1999-01 Mercury Cougar, 1998-00 Mystique .... 2.0L VIN 3

REPLACEMENT INTERVALS

Ford does not have a recommended replacement interval. Standard industry practice is to replace timing belts every 60,000 miles.

LABOR TIMES

Contour & Mystique	
Replacement	3.2 hrs
Severe Service	3.5 hrs
Cougar & Escort ZX2	
Replacement	2.6 hrs
Severe Service	3.0 hrs
Focus	
Replacement	3.4 hrs
Severe Service	3.8 hrs

8807FRDg12

✳✳ CAUTION

These engines are an interference type. If the timing belt has failed, valve-to-piston damage may have occurred. A compression check should always be performed to check for engine damage.

1. Disconnect the negative battery cable.
2. Remove the spark plugs.
3. On ZX2, remove the catalytic converter.
4. On all models, remove the right front wheel and fender liner.
5. Remove the accessory drive belt and idler pulley.
6. Rotate the crankshaft to Top Dead Center (TDC) and install Crankshaft TDC Timing Peg T97P-6000-A.
7. Rotate the crankshaft clockwise against the peg.
8. Remove the water pump pulley.
9. Remove the valve cover.
10. Remove the crankshaft pulley.
11. Remove the timing belt covers
12. Install Camshaft Alignment Timing Tool T94P-6256-CH into the slots of both camshafts at the rear of the cylinder head to lock the camshafts into position.
13. Loosen the timing belt tensioner bolt and relieve the tension on the timing belt by disconnecting the tensioner tab from the timing cover back plate.
14. Remove the timing belt.

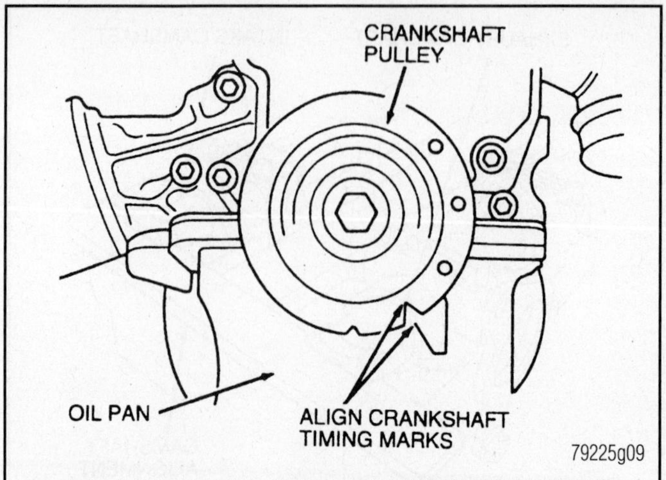

Crankshaft alignment position—Ford 1998–04 2.0L (VIN 3) engine

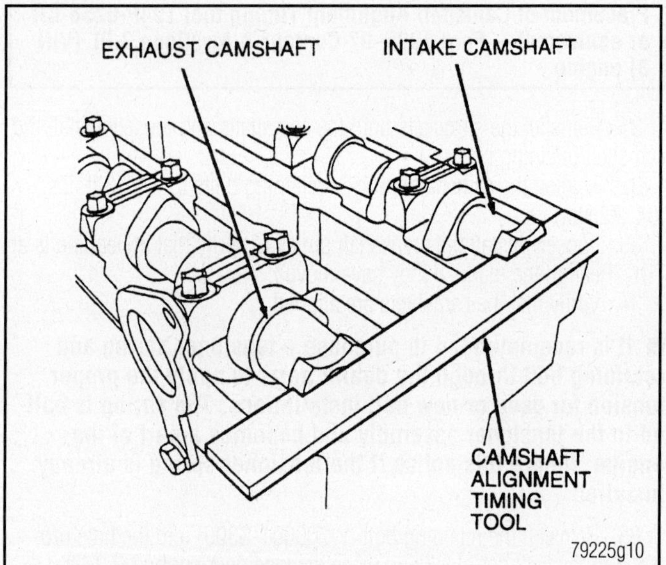

Placement of Camshaft Alignment Timing Tool T94P-6256-CH— Ford 1998–04 2.0L (VIN 3) engine

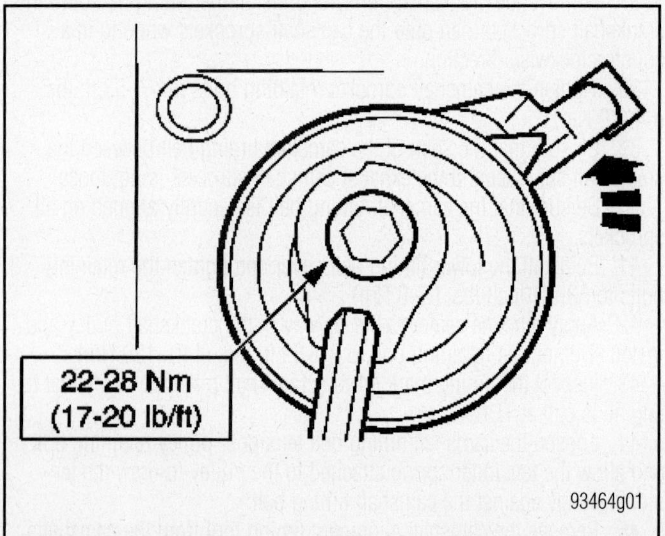

Timing belt tensioner index marks— Ford 1998–04 2.0L (VIN 3) engine

To Install:

■ **Always replace the timing belt after loosening or removing it.**

15. Install the timing belt.
16. Engage the timing belt tensioner tab into the timing cover back plate.
17. Adjust the timing belt tensioner until the index marks are aligned. Tighten the tensioner bolt to 17–20 ft. lbs. (22–28 Nm).
18. Remove the camshaft alignment tool and the crankshaft TDC peg.
19. Install the timing belt covers.
20. Install the crankshaft pulley. Tighten the bolt to 81–89 ft. lbs. (110–120 Nm).
21. Install the valve cover.
22. Install the water pump pulley.
23. Install the accessory drive belt and idler pulley.
24. Install the right inner splash shield and right front wheel.
25. On ZX2, install the catalytic converter.
26. Install the spark plugs.
27. Connect the negative battery cable.
28. Start the engine and check for proper operation.

2.0L (VIN 5) ENGINES

YEAR, MODEL & ENGINE APPLICATION	
2002-04 Ford Focus SVT	2.0L VIN 5

REPLACEMENT INTERVALS
Ford recommends timing belt replacement every 120,000 miles.

LABOR TIMES	
2002-04	
Replacement	3.4 hrs
Severe Service	3.8 hrs

<gnam "88807FRDCARg11">

✲✲ CAUTION

These engines are considered freewheeling. If the timing belt fails, valve-to-piston damage is unlikely, however a precautionary compression check should be performed to verify no engine damage.

1. Disconnect the negative battery cable.
2. Raise and support the vehicle.
3. Remove the drive belt cover.
4. Loosen the coolant pump pulley retaining bolts.
5. Remove the accessory drive belt.
6. Remove the crankshaft pulley.
7. Remove the accessory drive belt idler pulley.
8. Remove the coolant pump pulley.
9. Remove the timing belt lower cover.
10. Lower the vehicle.
11. Disconnect the coolant expansion tank and position it to one side.
12. Disconnect the power steering fluid reservoir and position it to one side.
13. Position the trolley jack with the wooden block under the oil pan and raise the engine so that the engine front mount is free from load.

■ **Mark the position of the engine front mount.**

14. Remove the engine front mount.
15. Disconnect the timing belt upper cover. Leave the timing belt cover in its installed position.
16. Remove the engine front mount bracket.
17. Remove the timing belt upper cover.
18. Remove the electronic ignition (EI) coil cover.
19. Remove the engine upper cover.
20. Disconnect the variable cam timing (VCT) valve electrical connector.

■ **Loosening sequence: from the outside to the inside, working diagonally.**

21. Disconnect the spark plug connectors.
22. Disconnect the positive crankcase ventilation (PCV) hose.
23. Remove the valve cover.
24. Remove the spark plugs.
25. Turn the engine to TDC on cylinder number 1.
26. Loosen the timing belt tensioner bolt four turns.
27. Position the tensioner so that the locating tab is at approximately the 4 o'clock position.
28. Line up the hexagonal key slot in the tensioner adjusting washer with the pointer that is located behind the pulley.
29. Remove the timing belt.

■ **Use an open-ended wrench to prevent the camshaft from rotating.**

30. Remove the intake camshaft pulley blanking plug.
31. Loosen the camshaft pulley retaining bolts.

To Install:

■ **The camshaft timing pulleys must be able to turn freely on the camshafts.**

32. Turn the camshafts to the ignition position on cylinder No. 1 and insert the special tool Camshaft Alignment Plate 303-574 into the camshafts.
33. Rotate the crankshaft to TDC on cylinder No. 1.
34. Remove the blanking plug and install the Crankshaft TDC Timing Peg 303-574.
35. Starting from the crankshaft timing belt pulley and working counterclockwise install the timing belt, keeping it under tension.

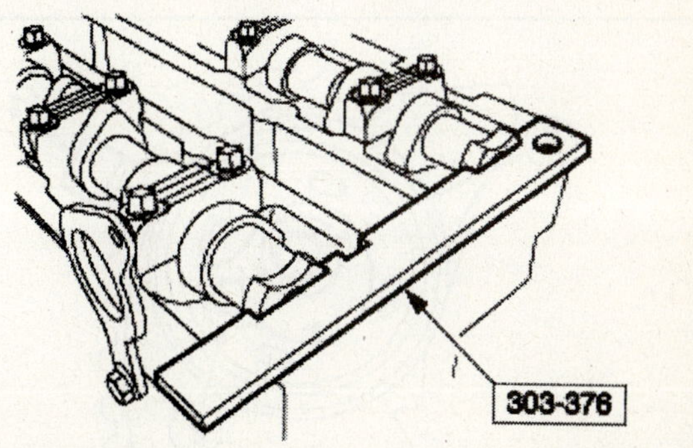

67197-focu-g25

Camshaft Alignment Tool 303-574 installation—Ford 2.0L (VIN 5) engine

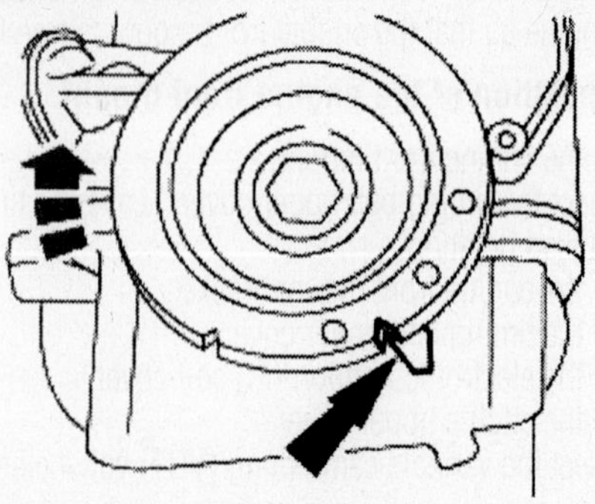

67197-focu-g26

Set the crankshaft to Top Dead Center (TDC)—Ford 2.0L (VIN 5) engine

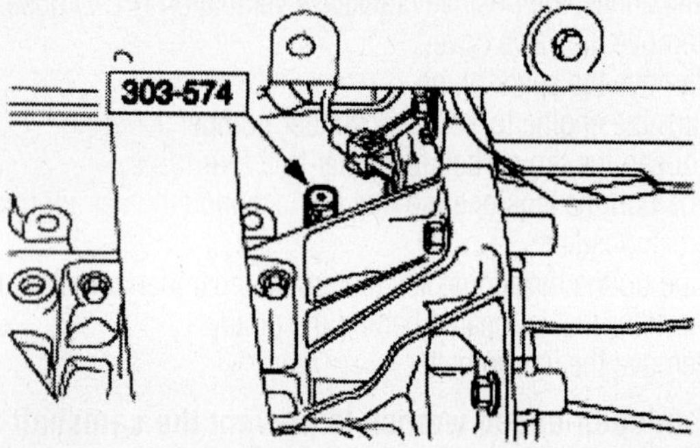

67197-focu-g27

Installation of the Crankshaft TDC Timing Peg—Ford 2.0L (VIN 5) engine

✻✻ CAUTION

Incorrect timing belt tension will result in incorrect valve timing.

36. Rotate the tensioner locating tab counterclockwise and insert the locating tab into the slot in the rear timing cover.
37. Position the hexagonal key slot in the tensioner adjusting washer to the 4 o'clock position.
38. Tighten the attaching bolt enough to seat the tensioner firmly against the rear timing cover, but still allow the tensioner adjusting washer to be rotated using a 6 mm hexagonal key.

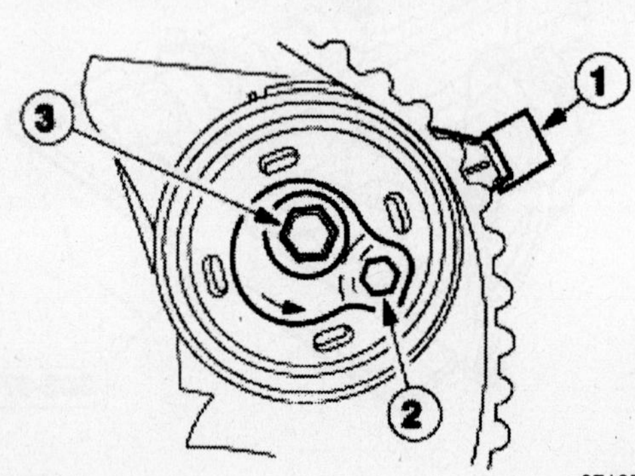

67197-focu-28

Tensioner locating tab (1), hexagonal key slot (2) and attaching bolt (3)—Ford 2.0L (VIN 5) engine

✻✻ CAUTION

Tension the timing belt working counterclockwise.

39. Using the hexagonal key, rotate the adjusting washer counterclockwise until the notch in the pointer is centered over the index line on the locating tab (the pointer will move clockwise during adjustment).
40. Tighten the bolt to 18 ft. lbs. (25 Nm), while holding the adjusting washer in position.
41. Use an open-ended wrench to prevent the camshaft from rotating and tighten the camshaft pulley retaining bolts. Tighten the intake pulley bolt to 88 ft. lbs. (120 Nm) and the exhaust pulley bolt to 50 ft. lbs. (68 Nm).
42. Install the intake camshaft pulley blanking plug and tighten it to 27 ft. lbs. (37 Nm).
43. Remove the Crankshaft TDC Timing Peg.
44. Remove the Camshaft Alignment Plate from the camshafts.
45. Turn the crankshaft two revolutions in the normal direction of rotation.
46. Check the valve timing by reinstalling the Camshaft Alignment Plate and the Crankshaft TDC Timing Peg. If necessary loosen the timing pulleys and correct the camshaft alignment.
47. Remove the Camshaft Alignment Plate and the Crankshaft TDC Timing Peg.
48. Install the blanking plug.
49. Install the valve cover and tighten the bolts as follows:
    a. Step 1: 18 inch lbs. (2 Nm)
    b. Step 2: 62 inch lbs. (7 Nm)
50. Attach the PCV hose to the valve cover.
51. Connect the VCT actuator electrical connector.
52. Coat the spark plug thread with anti-seize grease, screw in the spark plugs and push in the spark plug connector until it engages.
53. Connect the VCT valve electrical connector.
54. Install the engine upper cover.
55. Install the spark plug cover.
56. Position the upper timing belt cover and the center timing belt cover.
57. Install the engine front mount bracket. Tighten the bolts to 37 ft. lbs. (50 Nm).
58. Attach the timing belt upper cover. Tighten the bolts to 88 inch lbs. (10 Nm)
59. Install the engine front mount.
60. Attach the power steering pump fluid reservoir.
61. Attach the coolant expansion tank.
62. Remove the hydraulic jack and wooden block.

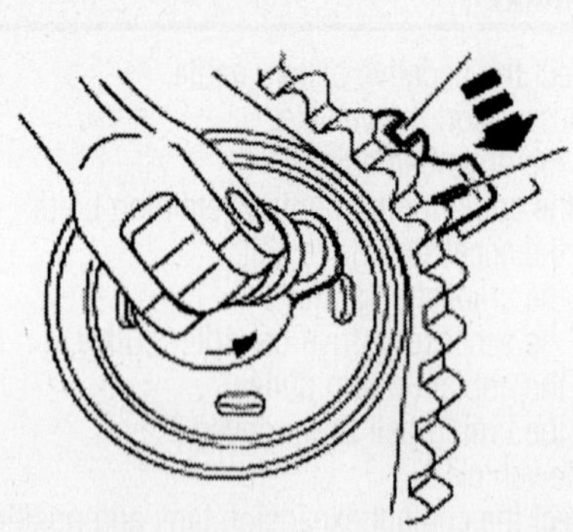

67197-focu-g29

Adjusting the timing belt tensioner—Ford 2.0L (VIN 5) engine

63. Raise and support the vehicle.
64. Attach the lower timing belt cover. Tighten the bolts to 62 inch lbs. (7 Nm).
65. Install the crankshaft pulley. Tighten the bolt to 85 ft. lbs. (115 Nm).
66. Attach the multi-groove belt idler pulley. Tighten the bolt to 29 ft. lbs. (40 Nm).
67. Attach the coolant pump pulley. Do not tighten the bolts at this stage.
68. Install the accessory drive belt.
69. Tighten the bolts on the coolant pump pulley to 17 ft. lbs. (24 Nm).
70. Attach the drive belt cover.
71. Lower the vehicle.

■ **When the battery has been disconnected and reconnected, some abnormal drive symptoms may occur while the vehicle relearns its adaptive strategy. The vehicle may need to be driven 16 km (10 miles) or more to relearn the strategy.**

72. Connect the battery negative cable.
73. Install the battery cover.
74. Check fluid levels and correct if necessary.
75. Check the routing of vacuum hoses and cables and secure them with cable ties.

2.2L (VIN C, L) ENGINES

YEAR, MODEL & ENGINE APPLICATION	
1989-92 Ford Probe	2.2L VIN C, L

REPLACEMENT INTERVALS

Ford does not have a recommended replacement interval. Standard industry practice is to replace timing belts every 60,000 miles.

LABOR TIMES	
1989-92	
Upper	
Replacement	0.3 hrs
Severe Service	0.4 hrs
Lower	
Replacement	0.8 hrs
Severe Service	1.0 hrs

8807FRDg14

✱✱ CAUTION

These engines are considered freewheeling. If the timing belt fails, valve-to-piston damage is unlikely, however a precautionary compression check should be performed to verify no engine damage.

1. Bring the No. 1 cylinder piston to Top Dead Center (TDC) on the compression stroke. The notch on the crankshaft damper should align with the TDC mark on the front cover.
2. Disconnect the negative battery cable.
3. Loosen the air conditioning compressor and alternator adjusting and pivot bolts, rotate the compressor and alternator toward the engine and remove the drive belts.
4. Raise and safely support the vehicle.
5. Remove the right front wheel and tire assembly and the right inner fender panel. Remove the 6 bolts, the crankshaft pulley and baffle plate.
6. Lower the vehicle.
7. Support the engine with a floor jack. Remove the 2 nuts and dowels from the right engine mount and remove the mount.
8. Remove the 7 bolts that retain the timing belt covers and remove the covers.
9. Remove the timing belt tensioner spring and retaining bolt. Remove the idler pulley retaining bolt.
10. If the timing belt is to be reused, mark the direction of rotation so it can be reinstalled in the same direction.
11. Remove the timing belt.

To Install:

12. Align the camshaft and crankshaft sprockets with the marks on the cylinder head front housing and the oil pump housing.
13. Install the timing belt. If reusing the old belt, observe the direction of rotation mark made during the removal procedure.
14. Place the timing belt tensioner and spring in position. Temporarily secure the tensioner with the spring fully extended. Make sure the timing belt is installed so there is no looseness at the water pump pulley at the idler side.
15. Loosen the idler bolt. Turn the crankshaft twice in the direction of rotation; align the timing marks.

■ **Always turn the crankshaft in the correct direction of rotation only. If the crankshaft is turned in the opposite direction, the timing belt may lose tension and correct belt timing may be lost.**

16. Check to see that the timing marks are correctly aligned. If they are not aligned, remove the timing belt and align the timing marks, then repeat Steps 8–11.

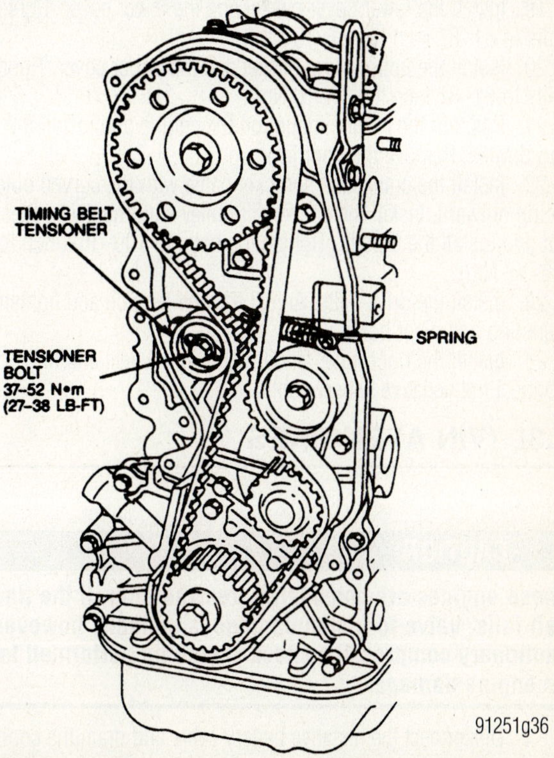

Timing belt tensioning component identification—Ford 2.2L (VIN C and L) engine

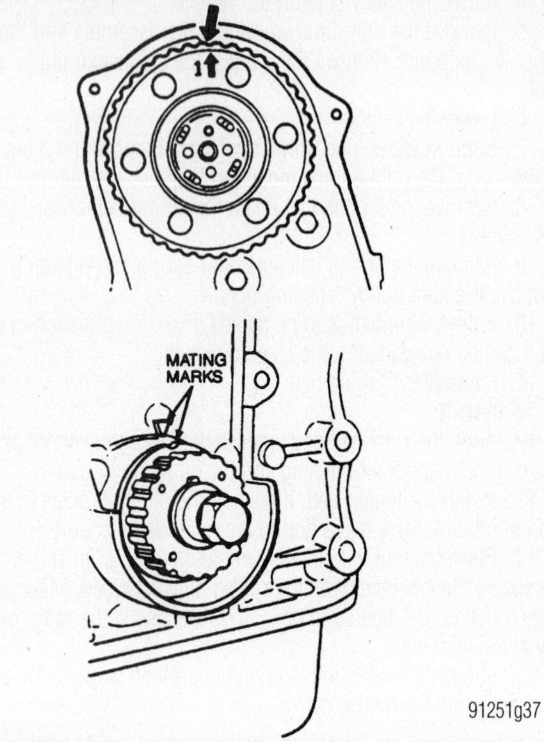

Proper timing mark locations for belt service—Ford 2.2L (VIN C and L) engine

17. Tighten the tensioner bolt to 27–38 ft. lbs. (37–52 Nm).
18. Measure the belt deflection between the crankshaft and camshaft pulleys. The correct deflection should be 0.30–0.33 (7.5–8.5mm) at 22 ft. lbs. (98 Nm) of pressure. If the deflection is not correct, loosen the tensioner bolt and repeat Steps 10 and 11.
19. Install the lower cover gasket and the lower cover. Tighten the bolts to 61–87 inch lbs. (7–10 Nm).
20. Install the upper cover gasket and the upper cover. Tighten the bolts to 61–87 inch lbs. (7–10 Nm).
21. Position the engine mount on the engine and install the 2 nuts and dowels. Remove the floor jack.
22. Install the crankshaft sprocket baffle with the curved outer lip facing outward. Install the crankshaft pulley with the deep recess facing out and install the 6 bolts. Tighten the bolts to 109–152 inch lbs. (12–17 Nm).
23. Install the drive belts. Adjust the belt tension and tighten the adjusting and pivot bolts.
24. Install the right inner fender panel and wheel and tire assembly. Connect the negative battery cable.

2.3L (VIN A) ENGINES

**** CAUTION**

These engines are considered freewheeling. If the timing belt fails, valve-to-piston damage is unlikely, however a precautionary compression check should be performed to verify no engine damage.

1. Disconnect the negative battery cable and drain the cooling system.
2. Remove the radiator shroud and cooling fan.

YEAR, MODEL & ENGINE APPLICATION

1985-87 Ford Aerostar, 1985-87 Courier, 1985-97 Ranger	2.3L VIN A

REPLACEMENT INTERVALS

Ford does not have a recommended replacement interval. Standard industry practice is to replace timing belts every 60,000 miles.

LABOR TIMES

1985-87 Aerostar	
Replacement	2.9 hrs
Severe Service	3.2 hrs
With air conditioning add	0.3 hrs
With power steering add	0.3 hrs
1985-90 Ranger	
Replacement	1.4 hrs
Severe Service	1.7 hrs
With air conditioning add	0.3 hrs
With power steering add	0.2 hrs
1991-94 Ranger	
2WD	
Replacement	1.6 hrs
Severe Service	1.9 hrs
With air conditioning add	0.3 hrs
With power steering add	0.2 hrs
4WD	
Replacement	2.2 hrs
Severe Service	2.5 hrs
With air conditioning add	0.3 hrs
With power steering add	0.2 hrs

8807FRDg15

3. Remove the 4 water pump pulley bolts and pulley.
4. Remove the automatic belt tensioner and accessory drive belt. Remove the upper radiator hose.
5. Remove the crankshaft pulley bolt and pulley.
6. Remove the thermostat housing and gasket.
7. Without disconnecting the lines, remove the A/C compressor and power steering pump reservoir and position aside, if equipped.
8. Remove the timing belt outer cover retaining bolt(s). Release the cover interlocking tabs, if equipped, and remove the cover.
9. Loosen the belt tensioner adjustment screw, position belt tensioner tool T74P-6254-A or equivalent, on the tension spring roll pin and release the belt tensioner. Tighten the adjustment screw to hold the tensioner in the released position.
10. On 1991–94 vehicles, remove the bolts holding the timing sensor in place and pull the sensor assembly free of the dowel pin.
11. Remove the crankshaft pulley, hub and belt guide. Remove the timing belt. If the belt is to be reused, mark the direction of rotation so it may be reinstalled in the same direction.

To Install:

12. Position the crankshaft sprocket to align with the TDC mark and the camshaft sprocket to align with the camshaft timing pointer. On 1985–90 vehicles, remove the distributor cap and set the rotor to the No. 1 firing position by turning the auxiliary shaft.
13. Install the timing belt over the crankshaft sprocket and then counterclockwise over the auxiliary and camshaft sprockets. Align the belt fore-and-aft on the sprockets.

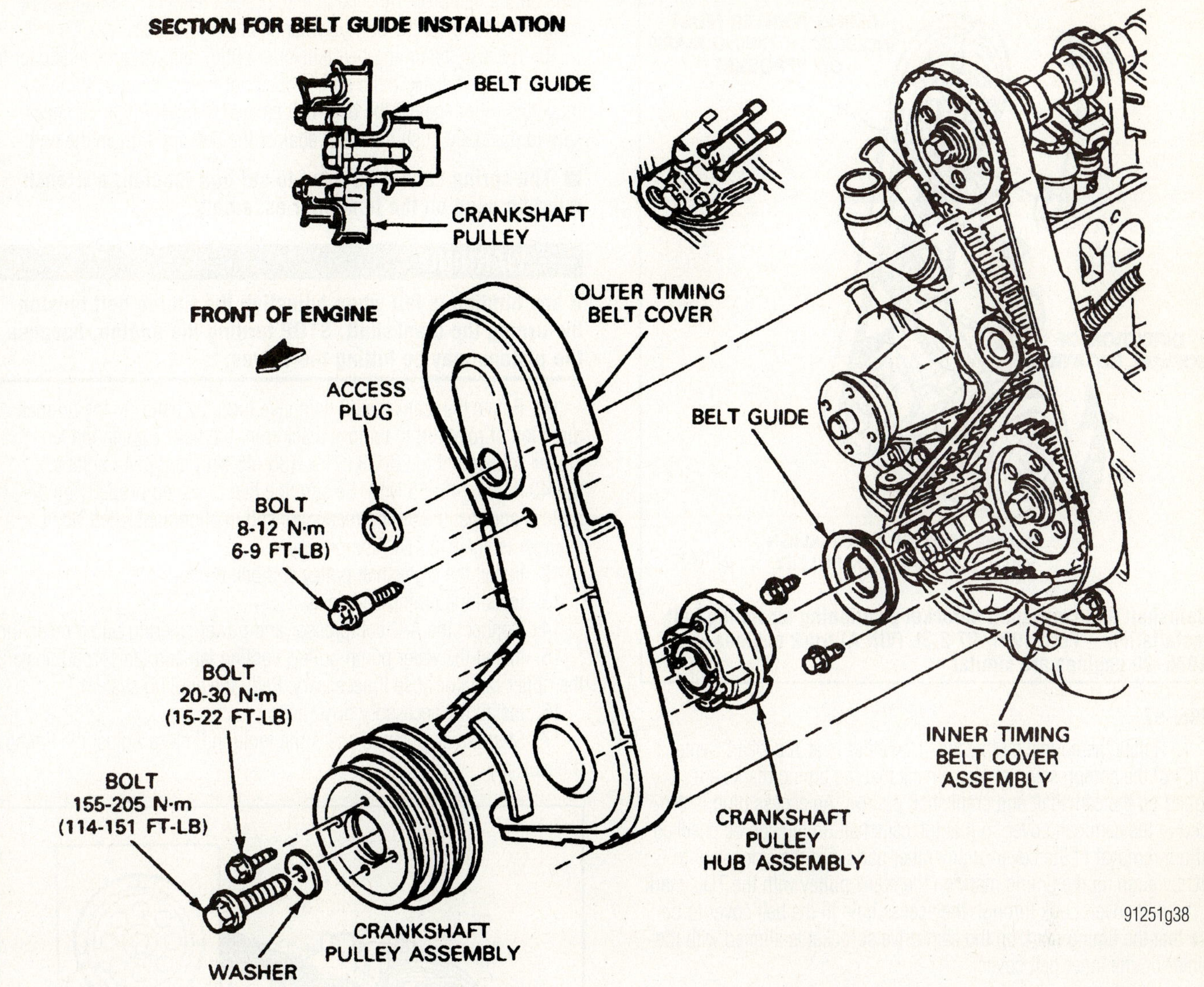

Exploded view of the timing belt front cover mounting—Ford 1991–97 2.3L (VIN A) truck engines—1985–90 engines are similar

14. Loosen the tensioner adjustment bolt to allow the tensioner to move against the belt. If the spring does not have enough tension to move the roller against the belt, it may be necessary to manually push the roller against the belt and tighten the bolt.

15. To make sure the belt does not jump time during rotation, remove the spark plugs.

16. Rotate the crankshaft 2 complete turns in the direction of normal rotation to remove the slack from the belt. Tighten the tensioner adjustment to 29–40 ft. lbs. (40–55 Nm) and pivot bolts to 14–22 ft. lbs. (20–30 Nm). Check the alignment of the timing marks.

17. Install the crankshaft belt guide.

18. On 1985–90 vehicles, install the crankshaft pulley and tighten the retaining bolt to 103–133 ft. lbs. (140–180 Nm). On 1991–94 vehicles, proceed as follows:

a. Install the timing sensor onto the dowel pin and tighten the 2 longer bolts to 14–22 ft. lbs. (20–30 Nm).

b. Rotate the crankshaft 45 degrees counterclockwise and install the crankshaft pulley and hub assembly. Tighten the bolt to 114–151 ft. lbs. (155–205 Nm).

c. Rotate the crankshaft 90 degrees clockwise so the vane of the crankshaft pulley engages with timing sensor positioner tool T89P-6316-A or equivalent. Tighten the 2 shorter sensor bolts to 14–22 ft. lbs. (20–30 Nm).

d. Rotate the crankshaft 90 degrees counterclockwise and remove the sensor positioner tool.

e. Rotate the crankshaft 90 degrees clockwise and measure the outer vane to sensor air gap. The air gap must be 0.018–0.039 in. (0.458–0.996mm).

19. Position the timing belt front cover. Snap the interlocking tabs into place, if necessary. Install the timing belt outer cover retaining bolt(s) and tighten to 71–106 inch lbs. (8–12 Nm).

20. Install the thermostat housing and a new gasket. Install the upper radiator hose.

21. Install the crankshaft pulley and retaining bolt. Tighten to 103–133 ft. lbs. (140–180 Nm) on 1985–90 vehicles or 114–151 ft. lbs. (155–205 Nm) on 1991–94 vehicles.

22. Install the water pump pulley and the automatic belt tensioner. Install the accessory drive belt.

23. Install the spark plugs and remaining components.

24. Connect the negative battery cable, start the engine and check the ignition timing.

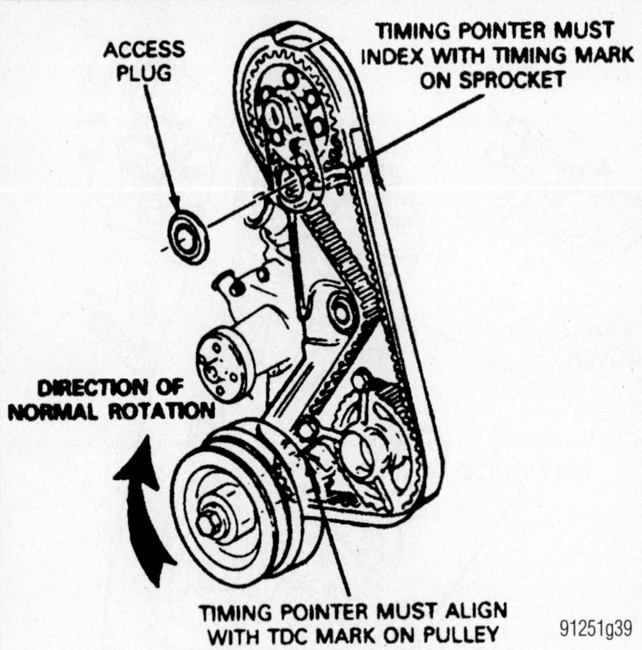

Camshaft and crankshaft sprocket positioning for timing belt installation— Ford 1991–97 2.3L (VIN A) truck engines— 1985–90 engines are similar

1995–97

1. Rotate the engine so that No. 1 cylinder is at Top Dead Center (TDC) of the compression stroke. Check that the timing marks are aligned on the camshaft and crankshaft pulleys. An access plug is provided in the cam belt cover so that the camshaft timing can be checked without removal of the cover or any other parts. Set the crankshaft to TDC by aligning the timing mark on the crank pulley with the TDC mark on the belt cover. Look through the access hole in the belt cover to be sure that the timing mark on the cam drive sprocket is aligned with the pointer on the inner belt cover.

■ **Always turn the engine in the normal direction of rotation. Backward rotation may cause the timing belt to jump time, due to the arrangement of the belt tensioner.**

2. Drain cooling system. Remove the upper radiator hose as necessary. Remove fan shroud, cooling fan and and water pump pulley bolts.
3. Loosen the alternator retaining bolts and remove the drive belt from the pulleys. Remove the water pump pulley.
4. Without disconnecting the lines, remove the A/C compressor and power steering pump reservoir and position aside, if equipped.
5. Remove the 4 timing belt outer cover retaining bolts and remove the cover. Remove the crankshaft pulley and belt guide.
6. Loosen the belt tensioner pulley assembly, then position a Camshaft Belt Adjuster tool T74P-6254-A or equivalent, on the tension spring roll pin and retract the belt tensioner away from the timing belt. Tighten the adjustment bolt to lock the tensioner in the retracted position.
7. If the belt is to be reused, mark the direction of rotation on the belt for installation reference.
8. Remove the timing belt.

To Install:

9. Install the new belt over the crankshaft sprocket and then counterclockwise over the auxiliary and camshaft sprockets, making sure the lugs on the belt properly engage the sprocket teeth on the pulleys. Be careful not to rotate the pulleys when installing the belt.
10. Release the timing belt tensioner pulley, allowing the tensioner to take up the belt slack. If the spring does not have enough tension to move the roller against the belt (belt hangs loose), it might be necessary to manually push the roller against the belt and tighten the bolt.

■ **The spring cannot be used to set belt tension; a wrench must be used on the tensioner assembly.**

**** WARNING**

If any binding is felt when adjusting the timing belt tension by turning the crankshaft, STOP turning the engine, because the pistons may be hitting the valves.

11. Rotate the crankshaft 2 complete turns by hand (in the normal direction of rotation) to remove slack from the belt. Tighten the tensioner adjustment to 26–33 ft. lbs. (35–45 Nm) and pivot bolts to 30–40 ft. lbs. (40–55 Nm). Be sure the belt is seated properly on the pulleys and that the timing marks are still in alignment when No. 1 cylinder is again at TDC/compression.
12. Install the crankshaft pulley and belt guide.
13. Install the timing belt cover.
14. Connect the A/C compressor and power steering pump reservoir.
15. Install the water pump pulley, cooling fan and fan shroud. Install the upper radiator hose if necessary. Refill the cooling system.
16. Install the accessory drive belts.
17. Start the engine and check the ignition timing. Adjust the timing, if necessary.

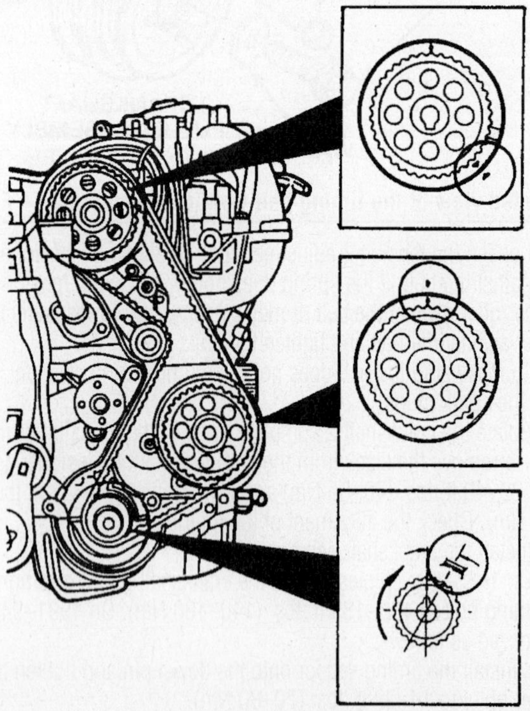

Camshaft, auxiliary shaft and crankshaft timing belt sprocket alignment mark locations—Ford 1995–97 2.3L (VIN A) truck engine

2.3L (VIN A, T, W) ENGINES

YEAR, MODEL & ENGINE APPLICATION

1985-86 Ford LTD, 1985-93 Mustang, 1985-88 Thunderbird	2.3L VIN A, T, W

REPLACEMENT INTERVALS

Ford does not have a recommended replacement interval. Standard industry practice is to replace timing belts every 60,000 miles.

LABOR TIMES

1985-93	
LTD & Thunderbird	
Replacement	1.6 hrs
Severe Service	2.0 hrs
Mustang	
Replacement	2.3 hrs
Severe Service	2.8 hrs

8807FRDg16

**** CAUTION**

These engines are considered freewheeling. If the timing belt fails, valve-to-piston damage is unlikely, however a precautionary compression check should be performed to verify no engine damage.

1. Disconnect the negative battery cable and drain the cooling system. Remove the 4 water pump pulley bolts.
2. Remove the automatic belt tensioner and accessory drive belt. Remove the upper radiator hose.
3. Remove the crankshaft pulley bolt and pulley. Remove the thermostat housing and gasket.
4. Remove the timing belt outer cover retaining bolt(s). Release the cover interlocking tabs, if equipped, and remove the cover.
5. Loosen the belt tensioner adjustment screw, position belt tensioner tool T74P-6254-A or equivalent, on the tension spring roll pin and release the belt tensioner. Tighten the adjustment screw to hold the tensioner in the released position.
6. On 1991–93 vehicles, remove the bolts holding the timing sensor in place and pull the sensor assembly free of the dowel pin.
7. Remove the crankshaft pulley, hub and belt guide. Remove the timing belt. If the belt is to be reused, mark the direction of rotation so it may be reinstalled in the same direction.

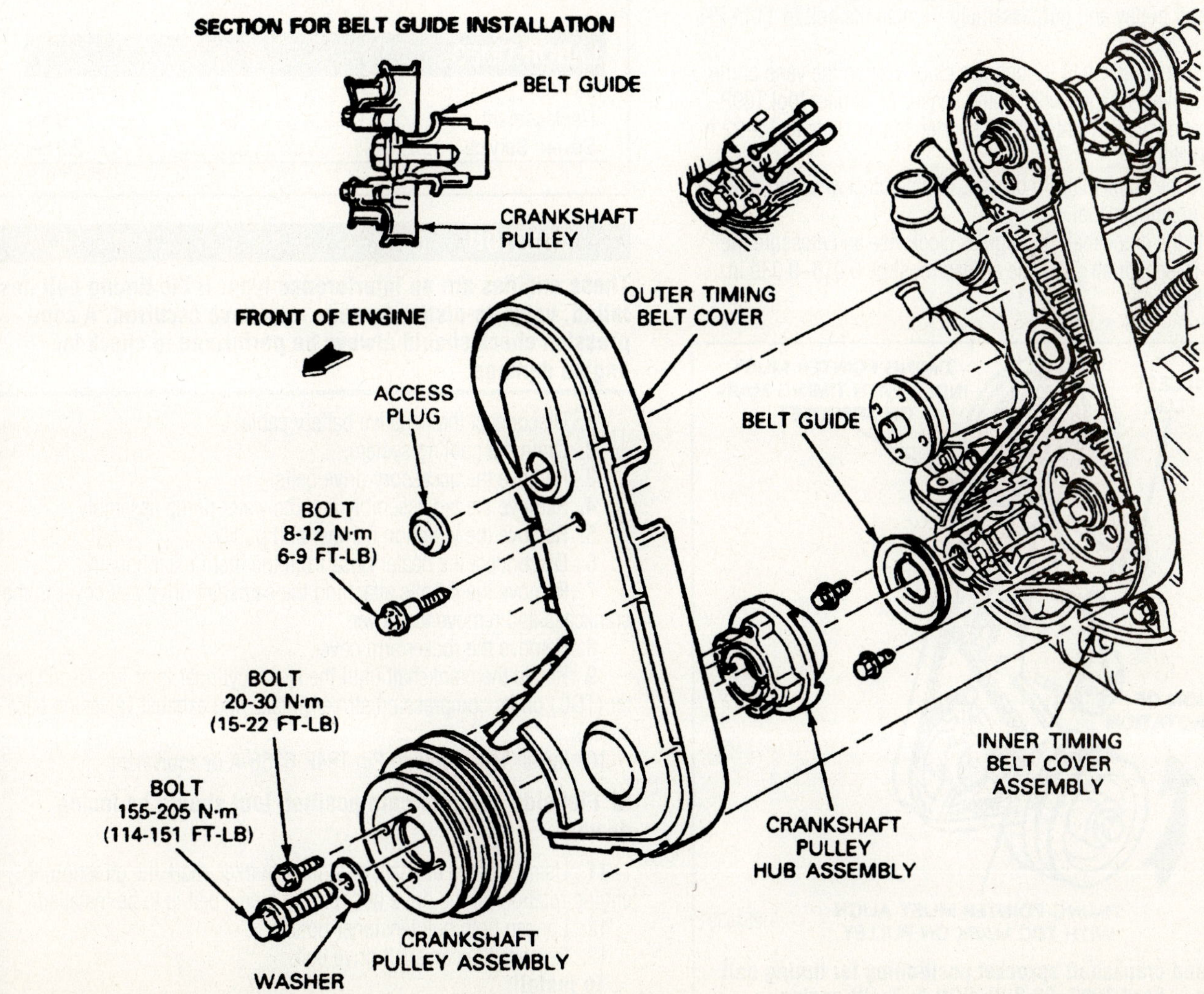

Exploded view of the timing belt front cover mounting—Ford 1991–93 2.3L (VIN A, T, W) engine—1985–90 similar

To Install:

8. Position the crankshaft sprocket to align with the TDC mark and the camshaft sprocket to align with the camshaft timing pointer. On 1985–90 vehicles, remove the distributor cap and set the rotor to the No. 1 firing position by turning the auxiliary shaft.

9. Install the timing belt over the crankshaft sprocket and then counterclockwise over the auxiliary and camshaft sprockets. Align the belt fore-and-aft on the sprockets.

10. Loosen the tensioner adjustment bolt to allow the tensioner to move against the belt. If the spring does not have enough tension to move the roller against the belt, it may be necessary to manually push the roller against the belt and tighten the bolt.

11. To make sure the belt does not jump time during rotation in Step 10, remove a spark plug from each cylinder.

12. Rotate the crankshaft 2 complete turns in the direction of normal rotation to remove the slack from the belt. Tighten the tensioner adjustment to 29–40 ft. lbs. (40–55 Nm) and pivot bolts to 14–22 ft. lbs. (20–30 Nm). Check the alignment of the timing marks.

13. Install the crankshaft belt guide.

14. On 1985–90 vehicles, install the crankshaft pulley and tighten the retaining bolt to 103–133 ft. lbs. (140–180 Nm). On 1991–93 vehicles, proceed as follows:

   a. Install the timing sensor onto the dowel pin and tighten the 2 longer bolts to 14–22 ft. lbs. (20–30 Nm).

   b. Rotate the crankshaft 45 degrees counterclockwise and install the crankshaft pulley and hub assembly. Tighten the bolt to 114–151 ft. lbs. (155–205 Nm).

   c. Rotate the crankshaft 90 degrees clockwise so the vane of the crankshaft pulley engages with timing sensor positioner tool T89P-6316-A or equivalent. Tighten the 2 shorter sensor bolts to 14–22 ft. lbs. (20–30 Nm).

   d. Rotate the crankshaft 90 degrees counterclockwise and remove the sensor positioner tool.

   e. Rotate the crankshaft 90 degrees clockwise and measure the outer vane to sensor air gap. The air gap must be 0.018–0.039 in. (0.458–0.996mm).

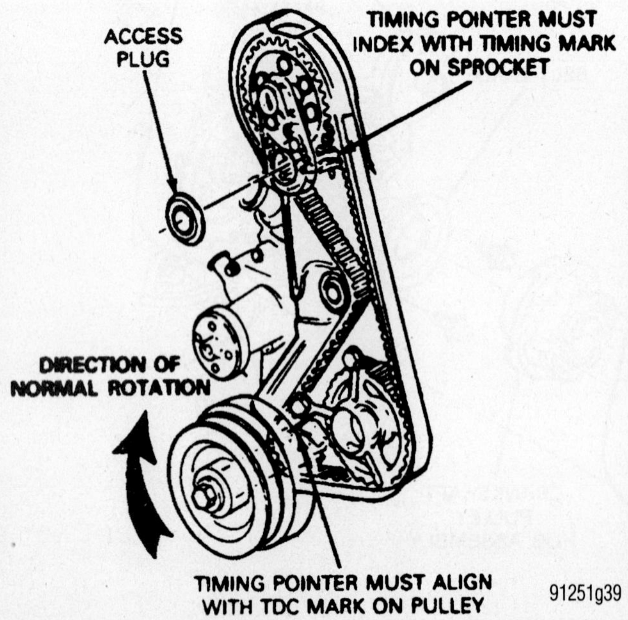

Camshaft and crankshaft sprocket positioning for timing belt installation— Ford 1991–93 2.3L (VIN A, T, W) engine—1985–90 similar

15. Position the timing belt front cover. Snap the interlocking tabs into place, if necessary. Install the timing belt outer cover retaining bolt(s) and tighten to 71–106 inch lbs. (8–12 Nm).

16. Install the thermostat housing and a new gasket. Install the upper radiator hose.

17. Install the crankshaft pulley and retaining bolt. Tighten to 103–133 ft. lbs. (140–180 Nm) on 1985–90 vehicles or 114–151 ft. lbs. (155–205 Nm) on 1991–93 vehicles.

18. Install the water pump pulley and the automatic belt tensioner. Install the accessory drive belt.

19. Install the spark plugs and remaining components.

20. Connect the negative battery cable, start the engine and check the ignition timing.

2.4L (VIN L) DIESEL ENGINES

YEAR, MODEL & ENGINE APPLICATION	
1984-85 Lincoln Continental, Mark VII	2.4L VIN L

REPLACEMENT INTERVALS

Ford does not provide a recommended replacement interval. Standard industry practice is to replace timing belts every 60,000 miles.

LABOR TIMES	
1984-85	
Replacement	1.9 hrs
Severe Service	2.2 hrs

8807FRDg17a

CAUTION

These engines are an interference type. If the timing belt has failed, valve-to-piston damage may have occurred. A compression check should always be performed to check for engine damage.

1. Disconnect the negative battery cable.
2. Drain the cooling system.
3. Remove the accessory drive belts.
4. Remove the fan assembly and the water pump assembly.
5. Remove the vibration damper and pulley.
6. Disconnect the heater hose from the thermostat housing.
7. Remove the 4 bolts attaching the camshaft drive belt cover to the crankcase and remove the cover.
8. Remove the rocker arm cover.
9. Rotate the crankshaft until the No. 1 cylinder is at Top Dead Center (TDC) on its compression stroke (intake and exhaust valves on base circle).
10. Install TDC Aligning Pin T84P-6256-A or equivalent.

■ **Flat side of nut or cam position tool should be facing down.**

11. Using a piece of chalk or similar marker, mark the direction of engine rotation on the drive belt, unless a new belt is to be installed.
12. Loosen both belt tensioner bolts.
13. Remove the camshaft drive belt.

To install:

14. Insert a 0.098 in. (2.5mm) thick feeler gauge blade between the Cam Positioning tool No. T84P-6256-A or equivalent, at the right front

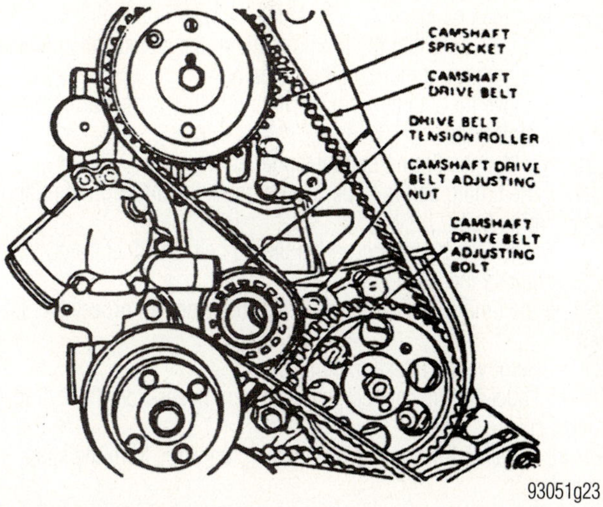

93051g23

View of the timing belt assembly—Ford 2.4L (VIN L) diesel engine

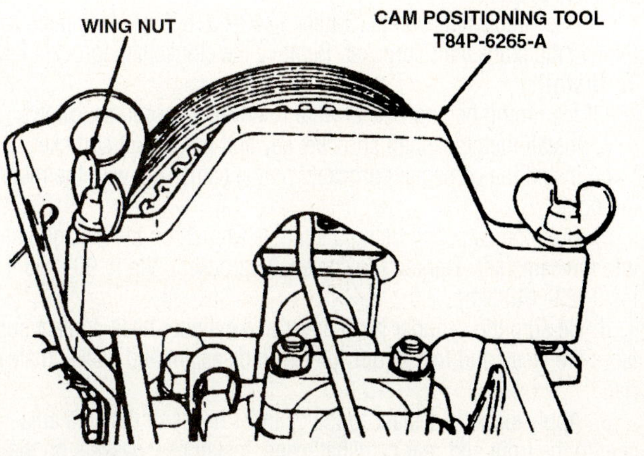

93051g24

View of the camshaft positioning tool—Ford 2.4L (VIN L) diesel engine

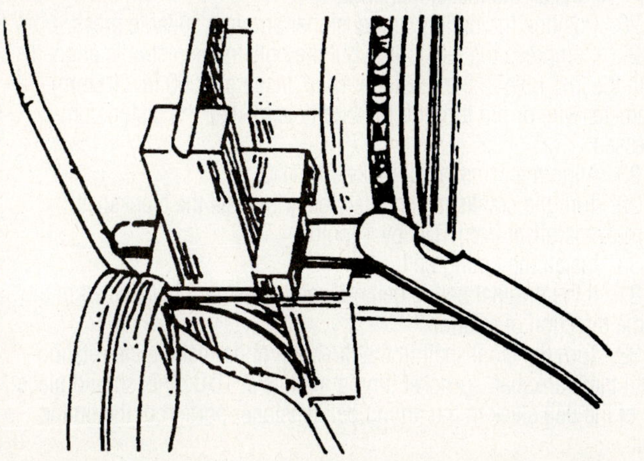

93051g25

Placement of the 2.5mm gauge under the camshaft positioning tool for installing a new belt—Ford 2.4L (VIN L) diesel engine

corner of the gasket mating surface of the cylinder head if using a new drive belt or a drive belt used less than 10,000 miles.

15. Install the Injection Pump Aligning Pin T84P-9000-A or equivalent, through the injection pump sprocket.
16. Rotate the injection pump sprocket clockwise against the pin.
17. Install the camshaft drive belt. Starting at the crankshaft, route the belt around the intermediate shaft sprocket, injection pump sprocket, camshaft sprocket and tension roller, keeping slack to a minimum.
18. Hand tighten the belt with the belt tensioner until all slack is gone.
19. Remove the Injection Pump Aligning Pin T84P-9000-A or equivalent, from the injection pump sprocket.
20. Adjust the belt tension by tightening the belt tensioner. Tighten the belt tensioner to 34–36 ft. lbs. (15.5–16.3 Nm) on belts with less than 10,000 miles or 23–25 ft. lbs. (31–34 Nm) for belts with more than 10,000 miles.
21. Tighten both belt tensioner holding bolts to 15–18 ft. lbs. (20–24.5 Nm).
22. Remove the Cam Positioning tool T84P-6265-A or equivalent.
23. Install the camshaft drive belt cover and tighten bolts to 72–84 inch lbs. (8–9.5 Nm).
24. Connect the heater hose to the thermostat housing.
25. Install the vibration damper.
26. Install the fan and water pump pulley assembly.
27. Install and adjust the accessory drive belts.
28. Fill and bleed the cooling system.
29. Connect the negative battery cable.
30. Operate the engine to normal operating temperatures and check for oil and/or coolant leaks.
31. Check the injection pump timing.

2.5L (VIN B) ENGINES

YEAR, MODEL & ENGINE APPLICATION	
1993-97 Ford Probe	2.5L VIN B

REPLACEMENT INTERVALS

Ford recommends timing belt replacement every 60,000 miles for Federal vehicles and 105,000 miles for California vehicles.

LABOR TIMES	
1993-97	
Replacement	2.1 hrs
Severe Service	2.2 hrs

8807FRDg17

✻✻ CAUTION

These engines are considered freewheeling. If the timing belt fails, valve-to-piston damage is unlikely, however a precautionary compression check should be performed to verify no engine damage.

1. Disconnect the negative battery cable.
2. Label and disengage the electrical connectors from the coolant elbow. Label and remove the electrical connectors from the knock sensor (if required) and crankshaft position sensor.
3. Loosen the drive belt tensioner locknuts and adjusting bolts. Remove the accessory drive belts.

4. Raise and safely support the vehicle on jackstands.
5. Remove the lower bolt from the air conditioning and alternator tensioner bracket.
6. Remove the right wheel and splash-shields.
7. Hold the crankshaft pulley (damper) with Holder tool T92C-6316-AH or equivalent, and remove the crankshaft pulley bolt. Remove the crankshaft pulley, using a puller if needed.
8. Remove the 5 front timing belt cover bolts.
9. Hold the water pump pulley with Holder tool T92C-6312-AH or equivalent, remove the 4 bolts and the water pump pulley.
10. Hold the power steering pump pulley with Strap Wrench D85L-6000-A or equivalent and remove the power steering pump pulley nut and pulley.
11. Lower the vehicle.
12. Remove the upper bolt from the belt idler bracket and remove the bracket.
13. Remove the engine oil dipstick tube retaining bolt and the tube.
14. Remove the 8 rear timing belt cover retaining bolts and remove the timing belt covers.
15. Temporarily reinstall the crankshaft pulley bolt.
16. Remove the 3 nuts and through-bolt from the right-hand engine support insulator and remove the support insulator. Remove the support insulator bracket.

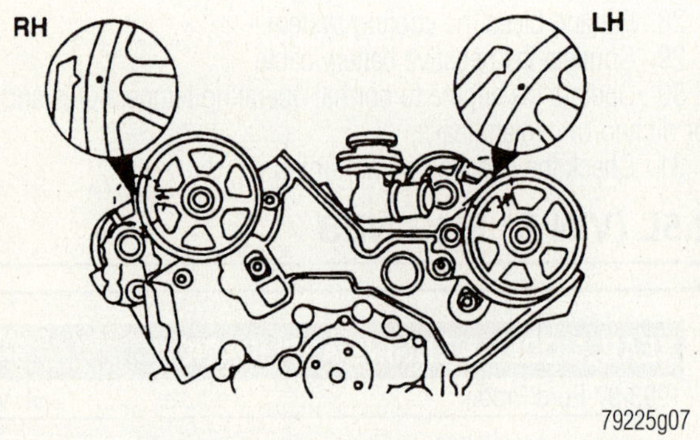

Left and right camshaft timing position—Ford 2.5L (VIN B) engine

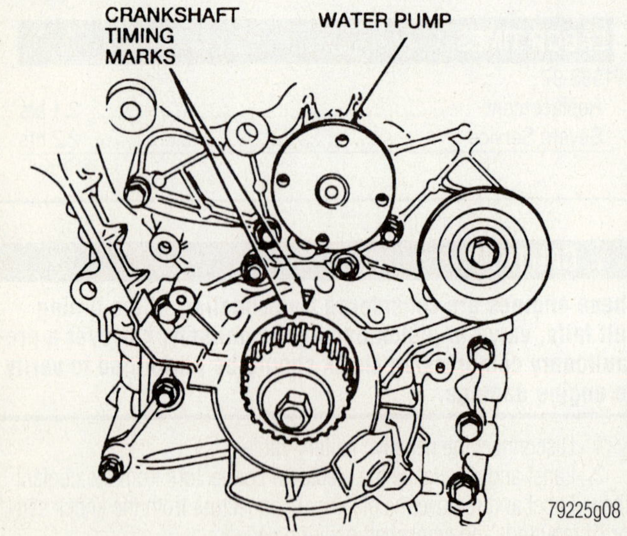

TDC alignment for the crankshaft—Ford 2.5L (VIN B) engine

17. Raise and safely support the vehicle on jack stand.
18. Turn the crankshaft to TDC No. 1 cylinder in the direction of normal rotation. Be sure that the timing mark on the crankshaft sprocket aligns with the timing mark on the oil pump.
19. Remove the 2 bolts from the timing belt tensioner arm, removing the lower bolt first.
20. Remove the timing belt tensioner arm.
21. If the timing belt is to be reused, mark the direction of rotation on the timing belt.
22. Loosen the Allen bolt on the timing belt tensioner.
23. Remove the timing belt.
24. If the timing belt sprockets are to be removed, proceed as follows:
a. Remove the intake manifold.
b. Label and disconnect the necessary hoses from the cylinder head covers.
c. Label and disconnect the spark plug wires from the spark plugs.
d. Remove the cylinder head cover retaining bolts and remove the cylinder head covers.
e. Hold the camshaft using a suitable wrench on the hexagon cast into the camshaft. Remove the camshaft sprocket bolts and the camshaft sprockets.
f. Use Crankshaft Damper Puller T74P-6316-A or equivalent, to remove the crankshaft sprocket. Remove the crankshaft sprocket key.

To Install:

25. If the timing belt sprockets were removed, proceed as follows:
a. Install the crankshaft sprocket key and crankshaft sprocket.
b. Install the camshaft sprockets on the camshafts with the retaining bolts.
c. Hold the camshaft using a suitable wrench on the hexagon cast into the camshaft. Tighten the camshaft sprocket bolts to 90–103 ft. lbs. (123–140 Nm).
d. Be sure the cylinder head cover and cylinder head contact surfaces are clean and free of dirt, oil and old sealant and gasket material.
e. Apply silicone sealant to the cylinder heads in the area adjacent to the front and rear camshaft caps. Install new gaskets on the cylinder heads.
f. Install the cylinder head covers and tighten the retaining bolts.
g. Connect the spark plug wires to the spark plugs and connect the hoses to the cylinder head covers.
h. Install the intake manifold.
26. Position the timing belt tensioner arm in a suitable press.
27. Compress the tensioner until the hole in the piston is aligned with the 2nd hole in the tensioner case. Insert a 0.060 in. (1.6mm) diameter wire or pin through the 2nd hole to keep the piston compressed.
28. Align the camshaft sprockets to TDC.
29. Turn the crankshaft counterclockwise until the crankshaft sprocket is offset from TDC by 1 tooth.
30. Install the timing belt.
31. If the original belt is being reused, be sure it is installed in the same direction of rotation.
32. Turn the crankshaft in the direction of normal engine rotation until the crankshaft sprocket timing mark is at TDC. This should place all of the belt slack in the timing belt tensioner portion of the timing belt.
33. Install the timing belt tensioner arm and 2 bolts. Tighten the bolts to 14–18 ft. lbs. (19–25 Nm).
34. Remove the wire or pin from the tensioner.

■ **When properly timed, the crankshaft timing marks will align and the crankshaft sprocket timing mark will no longer be 1 tooth off.**

35. Rotate the crankshaft 2 complete turns in the direction of normal rotation and align the timing marks. Be sure all marks are still correctly aligned. This will also set the timing belt tension.

■ **The timing belt tensioner will automatically adjust the timing belt tension.**

36. Tighten the timing belt tensioner Allen bolt to 28–32 ft. lbs. (35–51 Nm).
37. Install the right-hand engine support insulator. Tighten the 3 nuts to 54–76 ft. lbs. (74–103 Nm) and the through-bolt to 50–68 ft. lbs. (67–93 Nm).
38. Remove the crankshaft damper bolt.
39. Install the timing belt covers with the rear 8 bolts. Tighten to 71–88 inch lbs. (8–10 Nm).
40. Install the engine oil dipstick tube and retaining nut.
41. Install the belt idler bracket and the upper retaining bolt.
42. Raise and safely support the vehicle on jackstands.
43. Install the power steering pump pulley and nut. Tighten the nut to 36–43 ft. lbs. (49–59 Nm) while holding the pulley with a strap wrench.
44. Install the water pump pulley and 4 bolts. Secure the pulley with the holder tool and tighten the bolts to 71–88 inch lbs. (8–10 Nm).
45. Install the 5 front timing belt cover bolts and tighten to 71–88 inch lbs. (8–10 Nm).
46. Install the crankshaft pulley (damper) with the bolt. Hold the crankshaft pulley with the holding tool and tighten to 116–122 ft. lbs. (157–166 Nm).
47. Install the splash-shields.
48. Install the wheel and tighten the lug nuts to 65–87 ft. lbs. (88–118 Nm).
49. Install the lower bolt into the air conditioning and alternator tensioner bracket.
50. Lower the vehicle.
51. Install the accessory drive belts and adjust the tension.
52. Engage the electrical connectors to the sensors at the coolant elbow and the knock sensor and crankshaft position sensor.
53. Connect the negative battery cable.
54. Run the engine and check for leaks and proper engine operation.

2.5L (VIN C) ENGINES

YEAR, MODEL & ENGINE APPLICATION

1998-01 Ford Ranger	2.5L VIN C

REPLACEMENT INTERVALS

Ford does not have a recommended replacement interval. Standard industry practice is to replace timing belts every 60,000 miles.

LABOR TIMES

1998-01	
Replacement	1.6 hrs
Severe Service	1.9 hrs
With air conditioning add	0.8 hrs

8807FRD018

✻✻ CAUTION

These engines are considered freewheeling. If the timing belt fails, valve-to-piston damage is unlikely, however a precautionary compression check should be performed to verify no engine damage.

1. Rotate the engine so that No. 1 cylinder is at Top Dead Center (TDC) of the compression stroke. Check that the timing marks are aligned on the camshaft and crankshaft pulleys. An access plug is provided in the cam belt cover so that the camshaft timing can be checked without removal of the cover or any other parts. Set the crankshaft to TDC by aligning the timing mark on the crank pulley with the TDC mark on the belt cover. Look through the access hole in the belt cover to be sure that the timing mark on the cam drive sprocket is aligned with the pointer on the inner belt cover.

■ **Always turn the engine in the normal direction of rotation. Backward rotation may cause the timing belt to jump time, due to the arrangement of the belt tensioner.**

2. Drain cooling system. Remove the upper radiator hose as necessary. Remove the fan shroud, cooling fan and water pump pulley bolts.
3. Loosen the alternator retaining bolts and remove the drive belt from the pulleys. Remove the water pump pulley.
4. Without disconnecting the lines, remove the A/C compressor and power steering pump reservoir and position aside, if equipped.
5. Remove the 4 timing belt outer cover retaining bolts and remove the cover. Remove the crankshaft pulley and belt guide.
6. Loosen the belt tensioner pulley assembly, then position a Camshaft Belt Adjuster tool T74P-6254-A or equivalent, on the tension spring rollpin and retract the belt tensioner away from the timing belt. Tighten the adjustment bolt to lock the tensioner in the retracted posi-

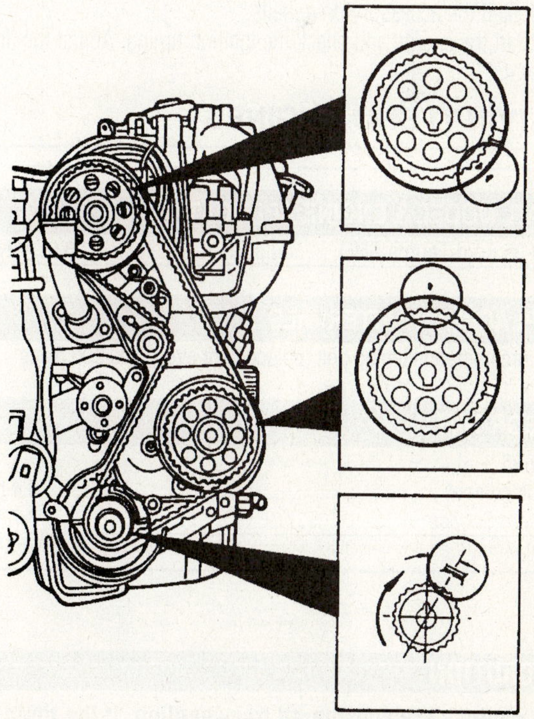

79245g20

Camshaft, auxiliary shaft and crankshaft timing belt sprocket alignment mark locations—Ford 2.5L (VIN C) engine

tion.

7. If the belt is to be reused, mark the direction of rotation on the belt for installation reference.

8. Remove the timing belt.

To Install:

9. Install the new belt over the crankshaft sprocket and then counterclockwise over the auxiliary and camshaft sprockets, making sure the lugs on the belt properly engage the sprocket teeth on the pulleys. Be careful not to rotate the pulleys when installing the belt.

10. Release the timing belt tensioner pulley, allowing the tensioner to take up the belt slack. If the spring does not have enough tension to move the roller against the belt (belt hangs loose), it might be necessary to manually push the roller against the belt and tighten the bolt.

■ **The spring cannot be used to set belt tension; a wrench must be used on the tensioner assembly.**

**** WARNING**

If any binding is felt when adjusting the timing belt tension by turning the crankshaft, STOP turning the engine, because the pistons may be hitting the valves.

11. Rotate the crankshaft 2 complete turns by hand (in the normal direction of rotation) to remove slack from the belt. Tighten the tensioner adjustment to 26–33 ft. lbs. (35–45 Nm) and pivot bolts to 30–40 ft. lbs. (40–55 Nm). Be sure the belt is seated properly on the pulleys and that the timing marks are still in alignment when No. 1 cylinder is again at TDC/compression.

12. Install the crankshaft pulley and belt guide.

13. Install the timing belt cover.

14. Connect the A/C compressor and power steering pump reservoir.

15. Install the water pump pulley,, cooling fan and fan shroud. Install the upper radiator hose if necessary. Refill the cooling system.

16. Install the accessory drive belts.

17. Start the engine and check the ignition timing. Adjust the timing, if necessary.

3.0L (VIN Y) SHO ENGINES

YEAR, MODEL & ENGINE APPLICATION

1989-95 Ford Taurus SHO	3.0L VIN Y

REPLACEMENT INTERVALS

Ford recommends timing belt replacement every 100,000 miles.

LABOR TIMES

1989-95	
Replacement	3.4 hrs
Severe Service	3.7 hrs

8807FRDg19

**** CAUTION**

These engines are considered freewheeling. If the timing belt fails, valve-to-piston damage is unlikely, however a precautionary compression check should be performed to verify no engine damage.

1. Disconnect both battery cables, negative cable first. Remove the battery.

2. Remove the right engine roll damper. Disconnect the wiring to the ignition module.

3. Remove the intake manifold crossover tube bolts. Loosen the intake manifold tube hose clamps. Remove the intake manifold crossover tube.

4. Loosen the alternator/air conditioning belt tensioner pulley and remove the drive belt.

5. Loosen the water pump/power steering belt tensioner pulley and remove the drive belt.

6. Remove the alternator/air conditioning belt tensioner pulley and bracket assembly.

7. Remove the water pump/power steering belt tensioner pulley only. Remove the upper timing belt cover.

8. Unplug the Crankshaft Position (CKP) sensor electrical connector. Place the transaxle gear selector in **N** (Neutral).

9. Rotate the crankshaft until the piston for the No. 1 cylinder is at Top Dead Center (TDC) of the compression stroke. Be sure the white mark on the crankshaft damper aligns with the **0** degree index mark on the lower timing belt cover and the marks on the intake camshaft sprockets align with the index marks on the metal timing belt cover.

10. Raise and safely support the vehicle on jackstands. Remove the right front wheel and tire assembly. Loosen the fender splash-shield and place it aside.

11. Remove the crankshaft damper and pulley retaining bolt. Using Puller T67L-3600-A or equivalent, remove the crankshaft damper and pulley.

12. Remove the lower timing belt cover.

13. Remove the center timing belt cover and disconnect the CKP sensor wire and grommet from the slot in the cover and the stud on the water pump.

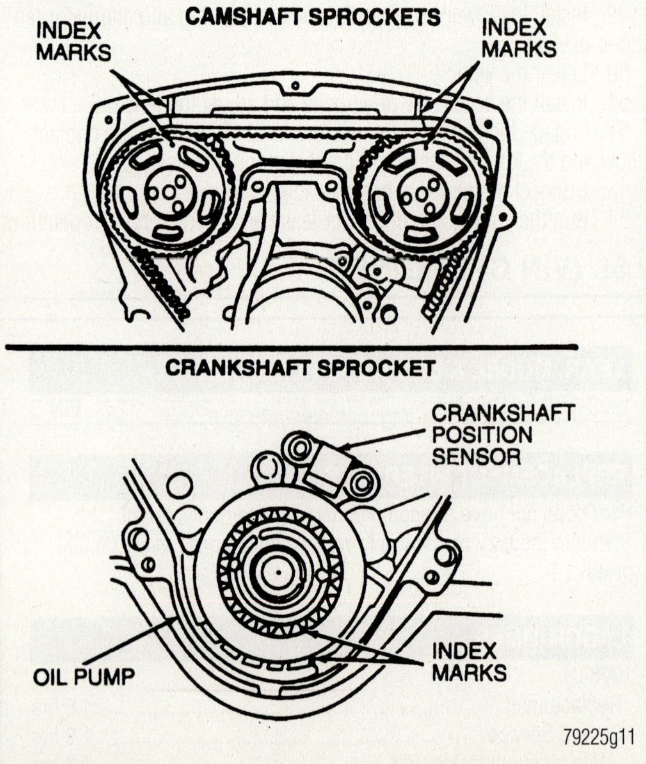

Timing marks for the camshaft and crankshaft sprockets—Ford 3.0L (VIN Y) SHO engine

14. Loosen the timing belt tensioner idler pulley. Rotate the idler pulley 180 degrees clockwise and tighten the tensioner nut to hold the pulley in an unloaded position.

15. Lower the vehicle.

16. Remove the timing belt. If the belt is to be reused, use crayon to mark an arrow on the belt to indicate the direction of rotation, for installation reference.

17. If removing one or more camshaft sprockets, remove 2 retaining bolts securing each camshaft sprocket and remove the sprocket, noting the location of the dowel pin.

18. If removing the crankshaft sprocket, install Puller T67L-3600-A or equivalent and pull the crankshaft sprocket off the crankshaft using care not to damage the pulse wheel.

To Install:

19. If the crankshaft sprocket was removed, install the crankshaft sprocket by aligning the keyway and pushing the sprocket on by hand.

20. If one or more camshaft sprockets were removed, install each camshaft sprocket by aligning the timing marks on the sprockets with the camshaft using the dowel pin as a guide. Install 2 retaining bolts and tighten to 10–13 ft. lbs. (14–18 Nm).

■ **Before installing the timing belt, inspect it for cracks, wear or other damage and replace, if necessary. Do not allow the timing belt to come into contact with gasoline, oil or coolant. Do not twist or turn the belt inside out.**

21. Be sure the engine is at TDC for the No. 1 cylinder. Check that the camshaft sprocket marks align with the index marks on the upper steel belt cover and that the crankshaft sprocket aligns with the index mark on the oil pump housing.

■ **The timing belt has 3 yellow lines. Each line aligns with the index marks.**

22. Install the timing belt over the crankshaft and camshaft sprockets. The lettering on the belt **KOA** should be readable from the rear of the engine (top of the lettering to the front of the engine). Be sure the yellow lines are aligned with the index marks on the sprockets.

23. Release the timing belt tensioner idler pulley locknut. Leave the nut loose. Raise and safely support the vehicle on jackstands.

24. Install the center timing belt cover. Be sure the CKP sensor wiring and grommet are installed and routed properly. Tighten the mounting bolts to 60–90 inch lbs. (7–11 Nm).

25. Install the lower timing belt cover. Tighten the bolts to 60–90 inch lbs. (7–11 Nm).

26. Install the crankshaft damper and pulley using Installer T88T-6701-A or equivalent. Install the retaining bolt and tighten to 113–126 ft. lbs. (152–172 Nm).

27. Rotate the crankshaft 2 revolutions in the clockwise direction until the yellow mark on the damper aligns with the **0** degree mark on the lower timing belt cover.

28. Remove the plastic door in the lower timing belt cover. Tighten the tensioner locknut to 25–37 ft. lbs. (33–51 Nm) and install the plastic door.

29. Rotate the crankshaft 60 degrees more in the clockwise direction until the white mark on the damper aligns with the **0** degree mark on the lower timing belt cover.

30. Lower the vehicle. Be sure the index marks on the camshaft sprockets align with the marks on the rear metal timing belt cover.

31. Route the CKP sensor wiring and connect it to the engine wiring harness.

32. Install the upper timing belt cover. Tighten the bolts to 60–90 inch lbs. (7–11 Nm).

33. Install the water pump/power steering tensioner pulley. Tighten the nut to 11–17 ft. lbs. (15–23 Nm).

34. Install the alternator/air conditioning tensioner pulley and bracket assembly. Tighten the bolts to 11–17 ft. lbs. (15–23 Nm).

35. Install the water pump/power steering and alternator/air conditioning drive belts, and set the tension. Tighten the idler pulley nut to 25–36 ft. lbs. (34–50 Nm).

36. Install the intake manifold crossover tube. Tighten the bolts to 11–17 ft. lbs. (15–23 Nm).

37. Install the engine roll damper. Install the battery. Connect the wiring to the ignition module.

38. Connect both battery cables, negative cable last. Raise and safely support the vehicle on jack stands.

39. Install the splash-shield. Install the right front wheel and tire assembly.

40. Lower the vehicle. Run the engine and check for proper operation.

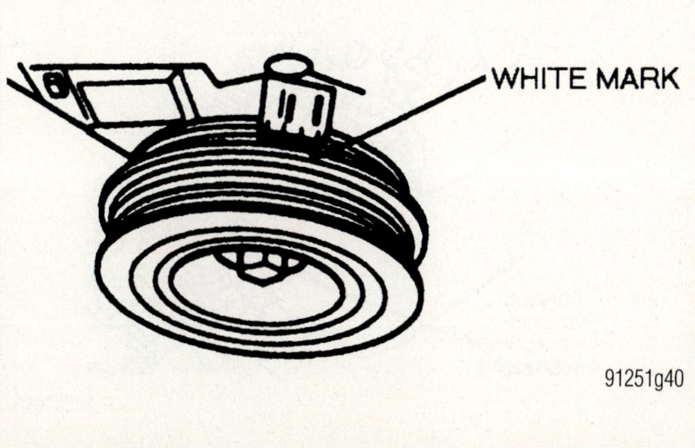

Crankshaft damper-to-lower timing cover index mark alignment—Ford 3.0L (VIN Y) SHO engine

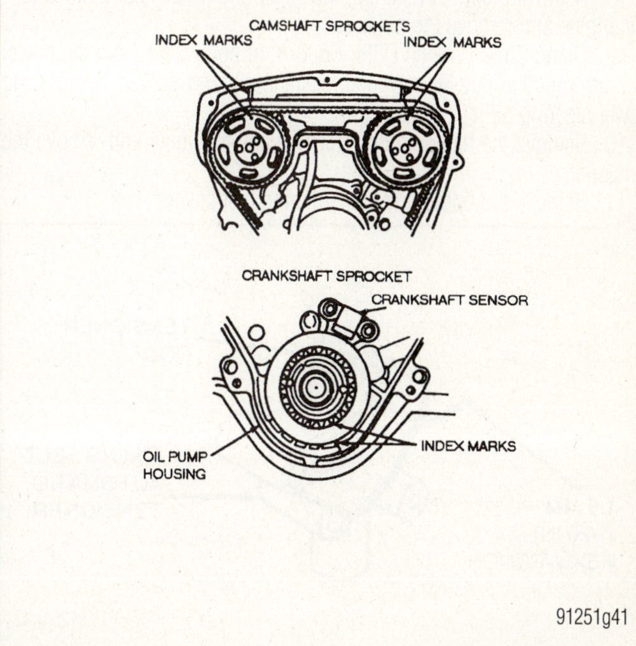

Timing belt index mark positioning—Ford 3.0L (VIN Y) SHO engine

3.2L (VIN P) SHO ENGINES

YEAR, MODEL & ENGINE APPLICATION	
1992-95 Ford Taurus SHO	3.2L VIN P

REPLACEMENT INTERVALS

Ford recommends timing belt replacement every 100,000 miles.

LABOR TIMES	
1992-95	
Replacement	3.4 hrs
Severe Service	3.7 hrs

8807FRDg20

✳✳ CAUTION

These engines are considered freewheeling. If the timing belt fails, valve-to-piston damage is unlikely, however a precautionary compression check should be performed to verify no engine damage.

1. Disconnect the battery cables, negative cable first.
2. Remove the battery.
3. Remove the right engine roll damper.
4. Disconnect the wiring to the ignition module.
5. Remove the intake manifold crossover tube bolts. Loosen the intake manifold tube hose clamps. Remove the intake manifold crossover tube.
6. Rotate the accessory drive belt tensioner clockwise to relieve belt tension. Remove the accessory drive belt.
7. Disconnect the surge tank fitting.
8. Remove the bolts retaining the upper and lower idler pulleys to the engine and remove the pulleys.
9. Using Strap Wrench D85L-6000-A, or equivalent, hold the power steering pump pulley. Remove the retaining nut and washer. Remove the power steering pulley.
10. Remove the retaining bolt from the belt tensioner and remove the tensioner.
11. Remove the upper and center timing belt covers.

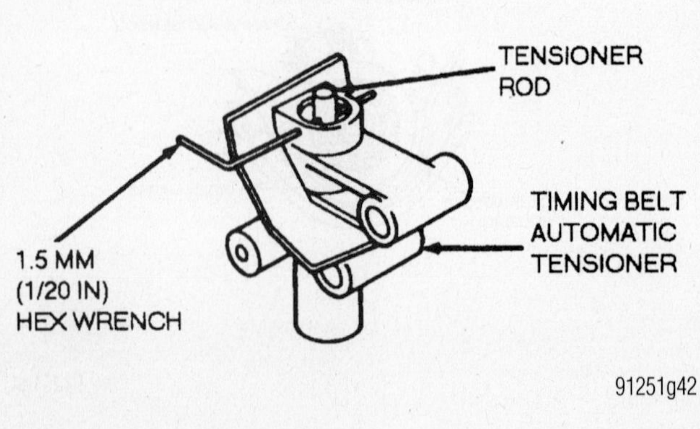

Timing belt automatic tensioner—Ford 3.2L (VIN P) SHO engine

12. Disengage the Crankshaft Position (CKP) sensor electrical connector.
13. Place the transaxle selector in **N** (Neutral).
14. Rotate the crankshaft until the piston for No. 1 cylinder is at Top Dead Center (TDC) of the compression stroke. Be sure the white mark on the crankshaft damper aligns with the **0** degree index mark on the lower timing belt cover and the marks on the intake camshaft sprockets align with the index marks on the metal timing belt cover.
15. Raise and safely support the vehicle on jackstands.
16. Remove the right front wheel and tire assembly.
17. Loosen the fender splash-shield and place it aside.
18. Remove the crankshaft pulley and damper using Puller T67L-3600-A with the appropriate adapters or equivalent.
19. Remove the lower timing belt cover and belt guide.
20. Remove the upper timing belt tensioner bolt.
21. Slowly loosen the lower timing belt tension bolt and remove the tensioner.
22. Lower the vehicle.
23. Remove the timing belt. If the belt is to be reused, use crayon to mark an arrow on the belt to indicate the direction of rotation, for installation reference.
24. If removing one or more camshaft sprockets, remove 2 retaining bolts securing each camshaft sprocket and remove the sprocket noting the location of the dowel pin.
25. If removing the crankshaft sprocket, install Puller T67L-3600-A or equivalent and pull the crankshaft sprocket off the crankshaft using

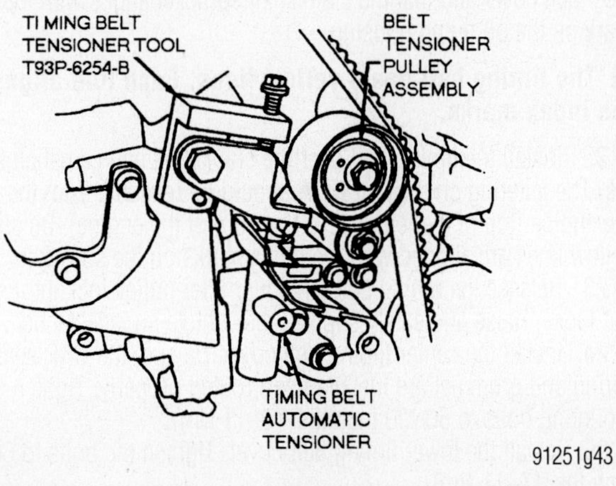

Timing belt tensioner pulley adjustment—Ford 3.2L (VIN P) SHO engine

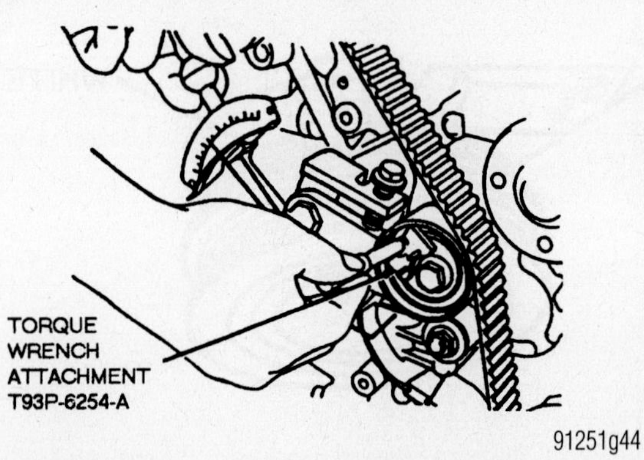

Timing belt tensioner pulley torque adjustment—Ford 3.2L (VIN P) SHO engine

care not to damage the pulse wheel.

To Install:

26. If the crankshaft sprocket was removed, install the crankshaft sprocket by aligning the keyway and pushing the sprocket on by hand.
27. If one or more camshaft sprockets were removed, install each camshaft sprocket by aligning the timing marks on the camshaft sprocket with the camshaft using the dowel pin as a guide. Install 2 retaining bolts and tighten to 10–13 ft. lbs. (14–18 Nm).

■ **Before installing the timing belt, inspect it for cracks, wear or other damage and replace as necessary. Do not allow the timing belt to come into contact with gasoline, oil or coolant. Do not twist or turn the belt inside out.**

28. Slowly compress the timing belt tensioner in a soft-jawed vise until the hole in the tensioner housing aligns with the hole in the tensioner rod.

**** CAUTION**

Use care when compressing the timing belt tensioner in the vise to insure that the tensioner does not slip from the vise.

29. Insert a 1/20 in. (1.5mm) hex wrench through the holes in the tensioner.
30. Release the tension from the vise.
31. If a new timing belt is being installed, loosen the timing belt idler bolt.
32. Ensure that the No. 1 cylinder is at TDC on its compression stroke. Check that the camshaft sprocket marks align with the index marks on the upper steel belt cover and that the crankshaft sprocket aligns with the index mark on the oil pump housing.

■ **The timing belt has 3 yellow lines. Each line aligns with the index marks.**

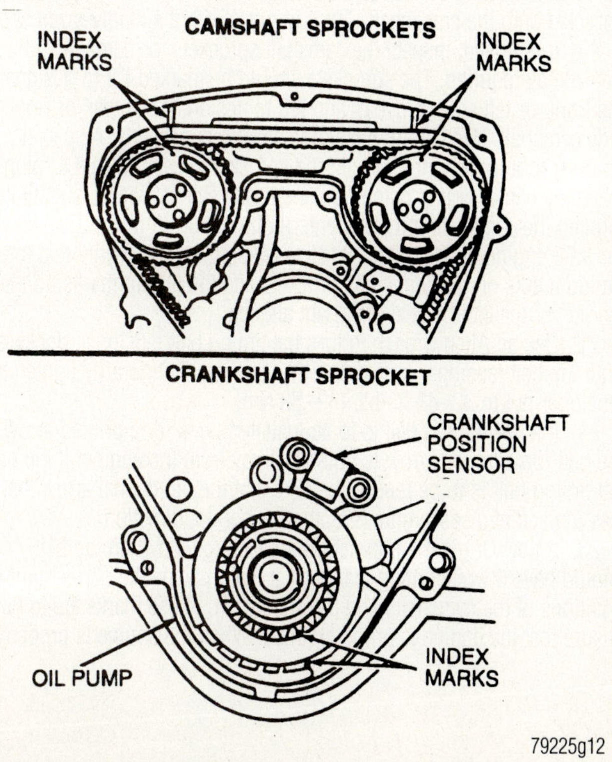

Set the engine to TDC before removing the front pulley—Ford 3.2L (VIN P) SHO engine

33. Install the timing belt over the crankshaft and camshaft sprockets. The lettering on the belt **KOB** should be readable from the rear of the engine (top of the lettering to the front of the engine). Be sure the yellow lines are aligned with the index marks on the sprockets.

**** WARNING**

Do not install the timing belt tensioner with the tensioner rod extended.

34. Install the timing belt tensioner on the cylinder block while pushing the timing belt idler toward the timing belt. Install and tighten the tensioner bolts to 12–17 ft. lbs. (16–23 Nm).
35. Install the grommets between the timing belt tensioner and the oil pump.
36. Remove the hex wrench from the timing belt tensioner.
37. If a new timing belt is being installed, perform the following steps:
    a. Remove the hex wrench from the timing belt tensioner, if installed.
    b. Mount Timing Belt Tensioner tool T93P-6254-B or equivalent, using the holes in the power steering pump support.
    c. Hand-tighten the timing belt idler pulley bolt.
    d. Using an inch pound torque wrench with attachment T93P-6254-A or equivalent, rotate the timing belt tensioner clockwise to 4.3 inch lbs. (0.5 Nm).
    e. Tighten the timing belt idler pulley bolt to 27–37 ft. lbs. (36–50 Nm). Remove both timing belt tensioning tools.
38. Raise and safely support the vehicle securely on jackstands.
39. Install the belt guide and lower timing belt cover. Tighten the retaining bolts to 12–17 ft. lbs. (16–23 Nm).
40. Using a suitable tool, install the crankshaft damper. Tighten the damper attaching bolt to 113–126 ft. lbs. (152–172 Nm).
41. Rotate the crankshaft 2 revolutions clockwise until the yellow mark on the damper aligns with the **0** degree mark on the lower timing belt cover.
42. Lower the vehicle.
43. Ensure that the index marks on the camshaft sprockets align with the marks on the rear metal timing belt cover.
44. Route the CKP sensor wiring and connect with the engine wiring harness.
45. Install the center and upper timing belt covers. Tighten the bolts to 12–17 ft. lbs. (16–23 Nm).
46. Install the steering pump pulley. Tighten the nut to 12–17 ft. lbs. (16–23 Nm).
47. Install the accessory drive belt while rotating the accessory drive belt tensioner clockwise.
48. Install the surge tank fitting.
49. Install the intake manifold crossover tube. Tighten the bolts to 11–17 ft. lbs. (15–23 Nm).
50. Install the engine roll damper
51. Install the battery.
52. Connect the wiring to the ignition module.
53. Connect both battery cables, negative cable last.
54. Raise and safely support the vehicle.
55. Install the splash-shield.
56. Install the right front wheel and tire assembly.
57. Lower the vehicle.
58. Run the engine and check for leaks and proper engine operation.

3.0L (VIN W) AND 3.3L (VIN T) ENGINES

YEAR, MODEL & ENGINE APPLICATION	
1993-98 Mercury Villager	3.0L VIN W
1999-02 Mercury Villager	3.3L VIN T

REPLACEMENT INTERVALS

Ford recommends timing belt replacement every 105,000 miles.

LABOR TIMES	
Villager	
3.0L	
Replacement	3.9 hrs
Severe Service	4.4 hrs
3.3L	
Replacement	3.2 hrs
Severe Service	3.7 hrs

8807FRDg21

**** CAUTION**

These engines are an interference type. If the timing belt has failed, valve-to-piston damage may have occurred. A compression check should always be performed to check for engine damage.

On this vehicle, right side refers to the "rear" components (near the firewall) and left side refers to the "front" components (near the radiator).

1. If the timing belt is to be removed, it is good practice to turn the crankshaft until the engine is at Top Dead Center (TDC) of the No. 1 cylinder, compression stroke (firing position), before beginning work. This should align all timing marks and serve as a reference for all work that follows. After verifying that the engine is at TDC for the No. 1 cylinder, do not crank the engine or allow the crankshaft or camshaft sprockets to be turned otherwise engine timing will be lost.
2. Drain the cooling system.
3. Disconnect the negative battery cable.
4. Remove the alternator drive belt, water pump and power steering pump belt and the air conditioning compressor belt, if equipped.
5. If equipped with air conditioning, remove the 3 air conditioning compressor drive belt idler pulley bolts and remove the idler pulley.

4. Remove the upper radiator hose bracket bolt. Remove the upper hose with the bracket from the vehicle.

5. Remove the water bypass hose from between the thermostat housing and the lower water hose connection.

6. Remove the main wiring harness from the upper engine front cover.

7. Remove the 8 upper engine front cover bolts and remove the upper cover.

8. Raise and safely support the vehicle.

9. Remove the right side front wheel and tire assembly.

10. Remove the 4 right side engine and transmission splash shield bolts and 2 screws, and remove the right side outer engine and transaxle splash shield.

11. Use a strap wrench to hold the water pump pulley. Remove the 4 pulley bolts, and the water pump pulley.

12. Use a strap wrench to hold the crankshaft pulley. Remove the center pulley bolt, and the crankshaft pulley using a harmonic balancer (damper) puller to draw the pulley from the front of the crankshaft.

13. Remove the 5 lower engine front cover bolts and remove the lower engine front cover.

14. Be sure that the timing marks between the crankshaft sprocket and the oil pump housing align.

15. If the timing belt is to be reused, mark an arrow on the belt indicating the direction of rotation. The directional arrow is necessary to ensure that the timing belt, if it to be reused, can reinstalled in the same direction.

16. Loosen the timing belt tensioner nut and slip the timing belt off of the sprockets.

17. If necessary, the camshaft sprockets can be removed. A special spanner tool is designed to hold the sprocket to keep it from turning while the center bolt is being loosened. Use care if using substitutes.

■ **The sprockets are not interchangeable.**

18. If necessary, the crankshaft sprocket can be removed. The outer timing belt guide (looks like a large washer) and the crankshaft sprocket simply pull off the front of the crankshaft.

■ **Be careful, there are 2 crankshaft keys. Use care not to lose them.**

To Install:

19. Clean all parts well. If removed, inspect the crankshaft sprocket for warping or abnormal wear. Check the sprocket teeth for wear, deformation, chipping or other damage. Replace as necessary. Clean the sprocket mounting surface to ease installation. Install the key. Slip the sprocket onto the crankshaft. Tap it in place with a suitably-sized socket.

201.If removed, inspect the camshaft sprockets for damage and wear. Replace as required. The sprockets should be marked **L3** to designate the front, or left side camshaft and **R3** to designate the rear, or right side camshaft. Use care to install the sprockets properly. A special spanner tool is designed to hold the sprocket to keep it from turning while the center bolt is being tightened. Use care if using a substitute. Tighten the camshaft sprocket center bolts to 58–65 ft. lbs. (78–88 Nm) for 3.0L engine or 61 ft. lbs. (83 Nm) for 3.3L engine. Verify that the timing marks on the camshaft sprockets and the timing marks on the rear cover (called the seal plate) are aligned.

21. Use an Allen wrench to turn the timing belt tensioner clockwise until the belt tensioner spring is fully extended. Temporarily tighten the tensioner nut to 32–43 ft. lbs. (43–58 Nm).

22. If a new timing belt is to be installed, look for a printed arrow on the belt. Be sure the arrow is pointing away from the engine. If the original timing belt is to be reused, be sure that the directional arrow that was marked at disassembly is facing the correct direction.

23. A new Original Equipment Manufacture (OEM) timing belt should have 3 white timing marks on it that indicate the correct timing positions of the camshafts and the crankshaft. These marks are to help ensure that the engine is properly timed. When the engine is properly

timed, each white timing mark on the timing belt will be aligned with the corresponding camshaft and crankshaft timing mark on the sprocket. Because the white timing marks are not evenly spaced, the technician needs to use care in installing the belt. There should be 40 timing belt teeth between the timing marks on the front and rear camshaft sprockets and 43 teeth between the timing mark on the front camshaft sprocket and the timing mark on the crankshaft sprocket.

24. Verify that the camshaft timing marks are aligned with the timing marks on the rear cover (seal plate) and that the crankshaft sprocket timing mark is aligned with the timing mark on the oil pump housing.

25. Install the timing belt starting at the crankshaft sprocket and moving around the camshaft sprockets following a counterclockwise path. Do not allow any slack in the timing belt between the sprockets. After all of the timing marks are aligned with the timing belt installed, slip the timing belt onto the belt tensioner.

26. While holding the timing belt tensioner with an Allen wrench, loosen the tensioner nut. Allow the tensioner to put pressure on the timing belt. Use an Allen wrench to turn the timing belt tensioner 70–80 degrees clockwise and tighten the timing belt tensioner nut to 32–43 ft. lbs. (43–58 Nm).

✻✻ WARNING

If any binding is felt when adjusting the timing belt tension by turning the crankshaft, STOP turning the engine, because the pistons may be hitting the valves.

27. Rotate the crankshaft clockwise twice and align the No. 1 piston to TDC on the compression stroke (firing position).

28. Apply 22 lbs. (10kg) of force on the timing belt between the rear camshaft sprocket and the timing belt tensioner. An assistant may be needed. While holding the timing belt tensioner steady with an Allen wrench, loosen the timing belt tensioner nut. Remove the Allen wrench and adjust the timing belt tensioner using the following procedure:

a. Install a 0.0138 in. (0.35mm) thick and 0.500 in. (12.7mm) wide feeler gauge where the timing belt just starts to go around the tensioner (approximately the 4 o'clock position, looking at the tensioner).

b. Turn the crankshaft sprocket clockwise, which should force the feeler gauge between the timing belt and the tensioner, up to a position on the tensioner of about 1 o'clock.

c. Tighten the timing belt tensioner nut to 32–43 ft. lbs. (43–58 Nm) for the 3.0L engine or 61 ft. lbs. (83 Nm) for the 3.3L engine.

d. Turn the crankshaft clockwise to rotate the feeler gauge out from between the timing belt tensioner and the timing belt.

29. Rotate the crankshaft clockwise twice, and once again align the No. 1 piston to TDC on the compression stroke (firing position).

30. Apply 22 lbs. (10kg) of force on the timing belt between the front and rear camshaft sprockets. Measure the amount of belt deflection. Belt deflection should be between 0.51–0.59 in. (13–15mm). If belt deflection is out of specification, repeat the procedure. If the timing belt deflection cannot be adjusted into specification, the timing belt will have to be replaced.

31. Position the lower engine front cover and install the 5 lower cover bolts. Do not over tighten. Tighten to 27–44 inch lbs. (3–5 Nm).

32. Install the outer timing belt guide next to the crankshaft sprocket with the dished side facing away from the cylinder block. Install the crankshaft pulley. Use a strap wrench to keep the crankshaft pulley from turning and tighten the center bolt to 90–98 ft. lbs. (123–132 Nm) for the 3.0L engine or 148 ft. lbs. (201 Nm) for the 3.3L engine.

33. Position the water pump pulley on the pump. Install the 4 bolts. Use a strap wrench to keep the water pump pulley from turning and tighten the 4 water pump pulley bolts to 12–15 ft. lbs. (16–21 Nm) for

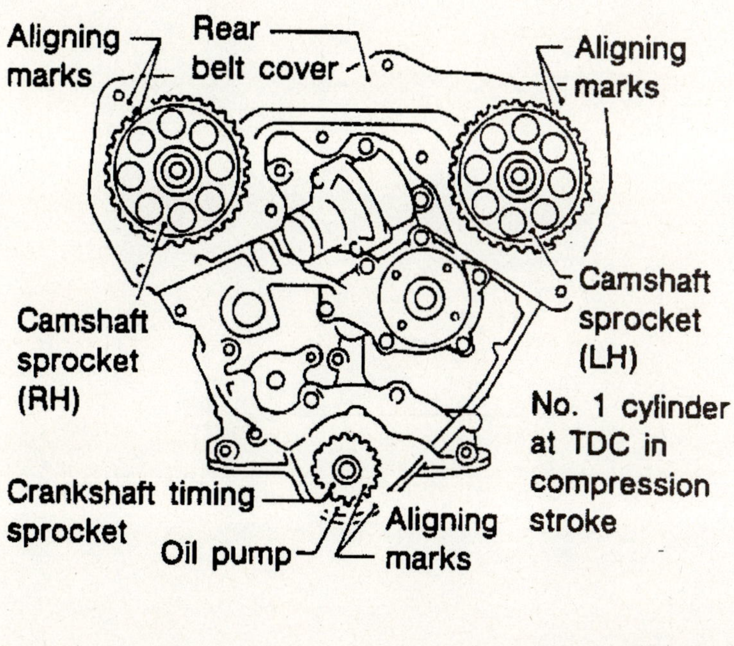

Use a shop rag to clean the alignment marks for the timing belt— Ford 3.0L (VIN W) and 3.3L (VIN T) engine

the 3.0L engine or 89 inch lbs. (10 Nm) for the 3.3L engine.

34. Position the right side outer engine and transaxle splash shield, and secure with the 4 bolts and 2 screws.

35. Install the right side front wheel. Tighten the lug nuts to 72–87 ft. lbs. (98–118 Nm).

36. Lower the vehicle.

37. Position the upper engine timing belt front cover, and tighten the 8 bolts to 27–44 inch lbs. (3–5 Nm).

38. Install the main wiring harness on the upper engine front cover.

39. Position the water bypass hose between the thermostat housing and water connection. Install the upper radiator hose between the radiator and the water hose connection. Secure the hoses with clamps. Install the upper radiator hose bracket. Tighten the bracket bolt to 34–58 ft. lbs. (46–65 Nm).

40. If equipped, position the air conditioning compressor drive belt idler pulley and install the 3 bolts. Tighten to 15 ft. lbs. (21 Nm).

41. Install and adjust the alternator drive belt, the water pump and power steering pump drive belt and the air conditioning compressor drive belt, if equipped.

42. Connect the battery cable.

43. Fill the cooling system.

44. Start the engine and allow it to warm to operating temperature. Check and adjust the ignition timing. Road test to verify correct engine operation.

General Motors/GEO /Saturn

1.0L (VIN 2, 5, 6) ENGINES

YEAR, MODEL & ENGINE APPLICATION

1985-88 Chevrolet Sprint, 1989-01 Chevrolet/GEO Metro	1.0L VIN 2, 5, 6

REPLACEMENT INTERVALS

General Motors recommends timing belt replacement every 60,000 miles.

LABOR TIMES

Metro	
1989-93	
Replacement	1.4 hrs
Severe Service	1.6 hrs
With air conditioning add	0.3 hrs
1994-01	
Replacement	1.6 hrs
Severe Service	2.0 hrs
With air conditioning add	0.3 hrs
Sprint	
Replacement	3.1 hrs
Severe Service	3.5 hrs

8807GMg01

✱✱ CAUTION

These engines are an interference type. If the timing belt has failed, valve-to-piston damage may have occurred. A compression check should always be performed to check for engine damage.

1. Disconnect the negative battery cable.
2. Remove the air cleaner and valve cover, if necessary.
3. Raise and safely support the vehicle.
4. Remove the right front wheel.
5. Remove the clips and right-side splash-shield.
6. Remove the lower alternator cover plate.
7. Remove the alternator drive belt, if equipped, the air conditioning drive belt.
8. Remove the water pump pulley.

■ **It is not necessary to remove the crankshaft timing sprocket bolt (center bolt) to remove the crankshaft pulley.**

9. Remove the crankshaft pulley bolts and crankshaft pulley.
10. Remove the retaining bolts and nut from the timing belt outside cover.
11. Remove the timing belt outside cover.
12. Turn the crankshaft to align the timing marks. The mark on the crankshaft sprocket should align with the arrow mark on the oil pump housing. The mark on the camshaft sprocket should align with the **V** notch on the timing belt inner cover or cylinder head cover.
13. If the timing belt is to be reused, mark the direction of rotation on the belt.
14. Remove the timing belt tensioner, tensioner plate, tensioner spring, spring damper and timing belt.

■ **Never turn the camshaft or crankshaft independently after the timing belt has been removed. Interference may occur between the pistons and valves, and parts may be damaged.**

15. Inspect the timing belt for wear or cracks, and replace as necessary. Check the tensioner for smooth rotation.

To install:

16. Install the tensioner plate to the tensioner.
17. Insert the lug of the tensioner plate into the hole of the tensioner.
18. Install the tensioner, tensioner plate and spring. Do not fully tighten the tensioner bolt and stud at this time.
19. Move the tensioner plate in a counterclockwise direction. This should cause the tensioner to move in the same direction. If it does not, remove the tensioner and tensioner plate, and reinsert the tensioner plate lug in the timing plate tensioner hole.
20. Check that the camshaft timing marks are aligned. If not, align the 2 marks by turning the camshaft.
21. Check that the punch mark on the crankshaft timing belt sprocket is aligned with the arrow mark on the oil pump case. If not, align the 2 marks by turning the crankshaft.
22. With the timing marks aligned, install the timing belt on the 2 sprockets. If the old belt is being reused, be sure to install it running in the same direction of original rotation.
23. Install the tensioner spring and spring damper. Turn the timing belt 2 rotations clockwise after installing the tensioner spring and damper to remove any belt slack. Tighten the tensioner stud to 96 inch lbs. (11 Nm), then the tensioner bolt to 20 ft. lbs. (27 Nm).

■ **Confirm that both sets of timing marks are aligned properly.**

24. Using a new seal, install the timing belt cover and tighten the bolts and nut to 97 inch lbs. (11 Nm).
25. Install the crankshaft pulley. Fit the keyway on the pulley to the crankshaft timing belt sprocket and tighten the bolts to 8–12 ft. lbs. (11–16 Nm).
26. Install the water pump pulley.

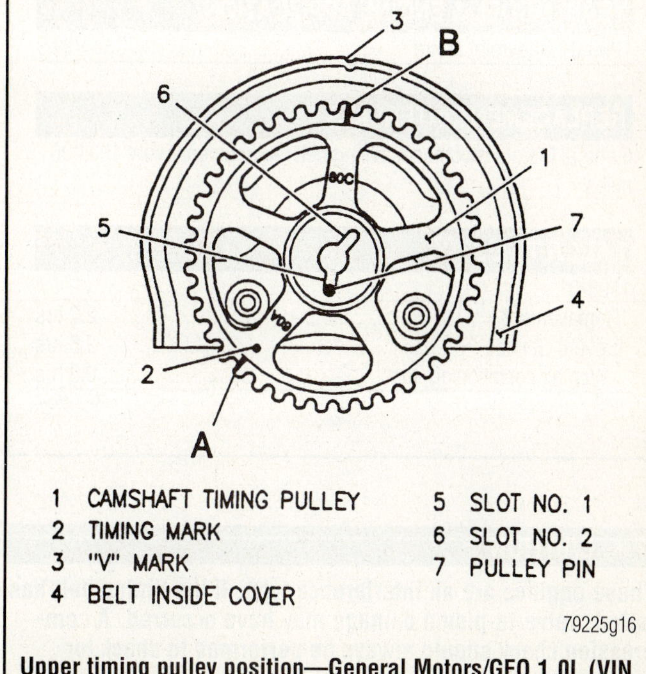

Upper timing pulley position—General Motors/GEO 1.0L (VIN 2, 5, 6) engine

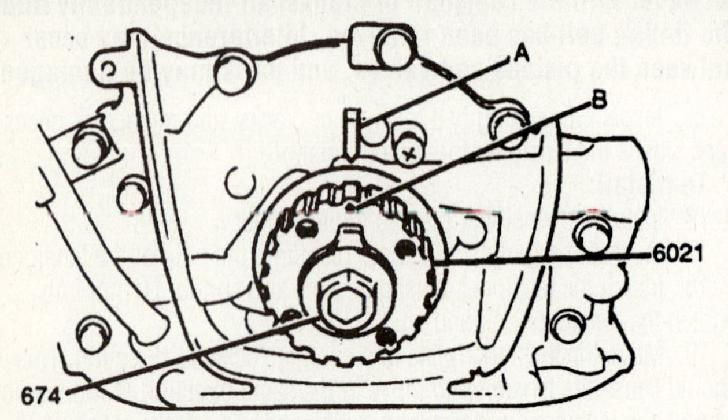

A ARROW MARK ON OIL PUMP CASE
B PUNCH MARK ON CRANKSHAFT TIMING GEAR
674 CRANKSHAFT PULLEY TIMING GEAR BOLT
6021 CRANKSHAFT TIMING GEAR

79225g17

Crankshaft timing mark—General Motors/GEO 1.0L (VIN 2, 5, 6) engine

27. Install the alternator drive belt, if equipped, the air conditioning drive belt.
28. Install the lower alternator cover plate. Tighten the bolts to 89 inch lbs. (10 Nm).
29. Install the right-side splash-shield and clips.
30. Install the right front wheel.
31. Install the air cleaner and valve cover, if removed.
32. Lower the vehicle and connect the negative battery cable.
33. Run the engine. Check for leaks.

1.3L (VIN 9) ENGINES

YEAR, MODEL & ENGINE APPLICATION

1998-01 Chevrolet Metro	1.3L VIN 9

REPLACEMENT INTERVALS

General Motors recomends timing belt replacement every 100,000 miles.

LABOR TIMES

1998-01	
Replacement	2.2 hrs
Severe Service	2.5 hrs
With air conditioning add	0.3 hrs

8807GMg02

✻✻ CAUTION

These engines are an interference type. If the timing belt has failed, valve-to-piston damage may have occurred. A compression check should always be performed to check for engine damage.

1. Disconnect the negative battery cable.
2. Remove the air cleaner.
3. Remove the A/C pipe bracket, if equipped.
4. Raise and safely support the vehicle.
5. Remove the right front wheel.
6. Remove the clips and right-side splash-shield.
7. Remove the lower alternator cover plate.
8. Remove the alternator drive belt, if equipped, the air conditioning drive belt.
9. Remove the flywheel cover and lock the flywheel using a large screwdriver.
10. Remove the crankshaft pulley bolts and crankshaft pulley.
11. Remove the retaining bolts and nut from the timing belt outside cover.
12. Remove the timing belt outside cover.
13. Turn the crankshaft to align the timing marks. The mark on the crankshaft sprocket should align with the arrow mark on the oil pump housing. The mark on the camshaft sprocket should align with the **V** notch on the timing belt inner cover or cylinder head cover.
14. If the timing belt is to be reused, mark the direction of rotation on the belt.
15. Remove the timing belt tensioner, tensioner plate, tensioner spring, spring damper and timing belt.

■ **Never turn the camshaft or crankshaft independently after the timing belt has been removed. Interference may occur between the pistons and valves, and parts may be damaged.**

16. Inspect the timing belt for wear or cracks, and replace as necessary. Check the tensioner for smooth rotation.

To install:

17. Install the tensioner plate to the tensioner.
18. Insert the lug of the tensioner plate into the hole of the tensioner.
19. Install the tensioner, tensioner plate and spring. Do not fully tighten the tensioner bolt and stud at this time.
20. Move the tensioner plate in a counterclockwise direction. This should cause the tensioner to move in the same direction. If it does not, remove the tensioner and tensioner plate, and reinsert the tensioner plate lug in the timing plate tensioner hole.
21. Check that the camshaft timing marks are aligned. If not, align the 2 marks by turning the camshaft.
22. Check that the punch mark on the crankshaft timing belt sprocket is aligned with the arrow mark on the oil pump case. If not, align the 2 marks by turning the crankshaft.
23. With the timing marks aligned, install the timing belt on the 2 sprockets. If the old belt is being reused, be sure to install it running in the same direction of original rotation.
24. Install the tensioner spring and spring damper. Turn the timing belt 2 rotations clockwise after installing the tensioner spring and damper to remove any belt slack. Tighten the tensioner stud to 96 inch lbs. (11 Nm), then the tensioner bolt to 18 ft. lbs. (25 Nm).

■ **Confirm that both sets of timing marks are aligned properly.**

25. Using a new seal, install the timing belt cover and tighten the bolts and nut to 97 inch lbs. (11 Nm).
26. Install the crankshaft pulley. Fit the keyway on the pulley to the crankshaft timing belt sprocket and tighten the bolts to 8–12 ft. lbs. (11–16 Nm).
27. Install the water pump pulley.
28. Install the alternator drive belt, if equipped, the air conditioning drive belt.

29. Install the lower alternator cover plate. Tighten the bolts to 89 inch lbs. (10 Nm).
30. Install the right-side splash-shield and clips.
31. Install the right front wheel.
32. Install the air cleaner.
33. Connect the A/C pipe bracket.
34. Lower the vehicle and connect the negative battery cable.
35. Run the engine. Check for leaks.

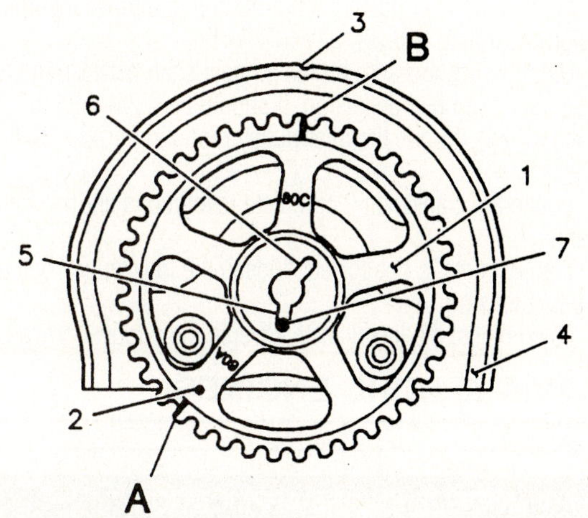

1 CAMSHAFT TIMING PULLEY
2 TIMING MARK
3 "V" MARK
4 BELT INSIDE COVER
5 SLOT NO. 1
6 SLOT NO. 2
7 PULLEY PIN

79225g16

Upper timing pulley position—General Motors/GEO 1.3L (VIN 9) engine

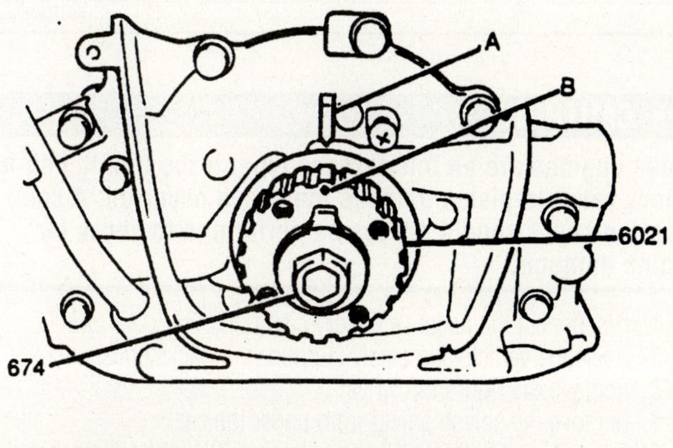

A ARROW MARK ON OIL PUMP CASE
B PUNCH MARK ON CRANKSHAFT TIMING GEAR
674 CRANKSHAFT PULLEY TIMING GEAR BOLT
6021 CRANKSHAFT TIMING GEAR

79225g17

Crankshaft timing mark—General Motors/GEO 1.3L (VIN 9) engine

1.5L (VIN 7, 9) ENGINES

YEAR, MODEL & ENGINE APPLICATION

1985-89 Chevrolet Spectrum	1.5L VIN 7, 9

REPLACEMENT INTERVALS

General Motors does not provide a replacement interval. Standard industry practice is to replace timing belts every 60,000 miles.

LABOR TIMES

1985-89	
With turbocharger	
Replacement	9.8 hrs
Severe Service	10.6 hrs
Without turbocharger	
Replacement	7.0 hrs
Severe Service	11.7 hrs

8807GMg03

✻✻ CAUTION

These engines are an interference type. If the timing belt has failed, valve-to-piston damage may have occurred. A compression check should always be performed to check for engine damage.

1. Disconnect the negative battery cable and drain the cooling system.
2. Remove the engine and mount the engine on an engine stand.
3. Remove the accessory drive belts.
4. Remove the engine mount bracket from the timing cover.
5. Rotate the crankshaft until the notch on the crankshaft pulley aligns with the **0** degree mark on the timing cover and the No. 4 cylinder is on Top Dead Center (TDC) of the compression stroke.
6. Remove the starter and install the Flywheel Holding tool No. J-35271 or equivalent.
7. Remove the crankshaft bolt and pulley.
8. Remove the timing cover bolts and the timing cover.
9. Loosen the tension pulley bolt.
10. Insert an Allen wrench into the tension pulley hexagonal hole and loosen the timing belt by turning the tension pulley clockwise.
11. If reusing the timing belt, mark the direction of rotation on the back of the belt.
12. Remove the timing belt.

To install:

13. Align the groove on the crankshaft timing gear with the mark on the oil pump.
14. Align the camshaft timing gear mark with the upper surface of the cylinder head and the dowel pin in its uppermost position.
15. Place the timing belt arrow in the direction of the engine rotation and install the timing belt. Tighten the tension pulley bolt.
16. Turn the crankshaft 2 complete revolutions and realign the crankshaft timing gear groove with the mark on the oil pump.
17. Loosen the tension pulley bolt and apply tension to the belt with an Allen wrench. Torque the pulley bolt to 37 ft. lbs. (50 Nm) while holding the pulley stationary.
18. To complete the installation, reverse the removal procedures. Torque the crankshaft pulley-to-crankshaft bolt to 109 ft. lbs. (148 Nm).

1.6L (VIN 4) SOHC ENGINES

YEAR, MODEL & ENGINE APPLICATION

1985-88 Chevrolet Nova	1.6L VIN 4

REPLACEMENT INTERVALS

General Motors recommends timing belt replacement every 60,000 miles.

LABOR TIMES

1985-88	
Replacement	3.1 hrs
Severe Service	3.5 hrs
With air conditioning add	0.2 hrs
With power steering add	0.1 hrs

8807GMg04

**** CAUTION**

These engines are considered freewheeling. If the timing belt fails, valve-to-piston damage is unlikely, however a precautionary compression check should be performed to verify no engine damage.

1. Disconnect the negative battery cable.
2. Drain the cooling system.
3. Remove the air cleaner.
4. Loosen the water pump pulley bolts and remove the alternator/water pump drive belt. If equipped with power steering, remove the power steering pump.
5. Remove the water pump pulley bolts and pulley.
6. If equipped with air conditioning, loosen the idler pulley mounting bolt. Loosen the adjusting nut, then remove the air conditioning drive belt and the idler pulley with the adjusting bolt.
7. Raise and support the vehicle.
8. Raise and safely support the vehicle.
9. Remove the right-side undercover, the flywheel cover, the crankshaft pulley-to-crankshaft bolt and the crankshaft pulley.
10. Disconnect the engine mount center bracket.
11. Disconnect the upper radiator hose from the water pump outlet. Label and disconnect all vacuum hoses that may be in the way.
12. Remove the upper timing belt cover front cover-to-engine bolts.
13. Remove the upper timing belt front cover and gasket.
14. Remove the alternator bolts and move the alternator aside.
15. Remove the middle timing belt front cover-to-engine bolts and the cover.
16. Remove the lower timing belt front cover-to-engine bolts and the cover.
17. Place a suitable jack under the engine.
18. Remove the No. 1 spark plug. Using a socket wrench on the crankshaft pulley bolt, rotate the engine (clockwise) to position the No. 1 cylinder on Top Dead Center (TDC) on its compression stroke.

■ **The TDC of the No. 1 cylinder is located when air is expelled from the cylinder.**

19. Disconnect the right side engine mount.
20. If reusing the timing belt, mark an arrow showing the direction of rotation and matchmark the belt to both pulleys.
21. Loosen the idler pulley mounting bolt and push the idler pulley to relieve the belt tension, then retighten the mounting bolt.
22. Remove the timing belt.

To install:

23. Ensure the TDC marks on the oil pump body and the crankshaft pulley are aligned.
24. Ensure the timing marks dots on the camshaft pulley and rear housing are aligned.
25. Noting the direction of rotation, install the timing belt.
26. Loosen the idler pulley mounting bolt and move the idler pulley to tension the timing belt.
27. Using finger pressure on the longest span between the pulleys, measure the timing belt deflection; it should be 0.24–0.28 in. (6–7mm) at 4.4 lbs. (2 kg). Torque the idler pulley mounting bolt to 27 ft. lbs. (37 Nm).
28. Rotate the crankshaft 2 complete revolutions and realign the timing marks.
29. If adjustment is not correct, loosen the idler pulley bolt and correct the belt tension.
30. To complete the installation, reverse the removal procedures.

1.6L (VIN 5) DOHC ENGINES

YEAR, MODEL & ENGINE APPLICATION

1985-88 Chevrolet Nova, 1990-92 GEO Prism	1.6L VIN 5

REPLACEMENT INTERVALS

General Motors recommends timing belt replacement every 60,000 miles.

LABOR TIMES

1985-88	
Replacement	3.5 hrs
Severe Service	4.0 hrs
1989-92	
Replacement	2.8 hrs
Severe Service	3.2 hrs

8807GMg05

**** CAUTION**

These engines are an interference type. If the timing belt has failed, valve-to-piston damage may have occurred. A compression check should always be performed to check for engine damage.

1. Disconnect the negative battery cable.
2. Raise the vehicle and safely support it on jackstands.
3. Remove the right front wheel.
4. Remove the splash shield from under the car.
5. Drain the coolant into clean containers. Close the draincocks when the system is empty.
6. Lower the car to the ground. Disconnect the accelerator cable and, if equipped, the cruise control cable.
7. Remove the cruise control actuator, if equipped.
8. Carefully remove the ignition coil.
9. Disconnect the radiator hose at the water outlet.
10. Remove the power steering drive belt and the alternator drive belt.

11. Remove the spark plugs.
12. Rotate the crankshaft clockwise and set the engine to Top Dead Center (TDC) on the compression stroke on No. 1 cylinder. Align the crankshaft marks at zero; look through the oil filler hole and make sure the small hole in the end of the camshaft can be seen.
13. Raise and safely support the vehicle. Disconnect the center engine mount.
14. Lower the vehicle to the ground.
15. Support the engine either from above or below. Disconnect the right engine mount from the engine.
16. Raise the engine and remove the mount.
17. Remove the water pump pulley.
18. Remove the crankshaft pulley.
19. Remove the 10 bolts and remove the timing belt covers with their gaskets.

■ **The bolts are different lengths; they must be returned to their correct location at reassembly. Label or diagram the bolts during removal.**

20. Remove the timing belt guide from the crankshaft pulley.
21. Loosen the timing belt idler pulley, move it to the left (to take tension off the belt) and tighten its bolt.
22. Make matchmarks on the belt and all pulleys showing the exact placement of the belt. Mark an arrow on the belt showing its direction of rotation.
23. Carefully slip the timing belt off the pulleys.

■ **Do not disturb the position of the camshafts or the crankshaft during removal.**

24. Remove the idler pulley bolt, pulley and return spring.
25. Remove the PCV hose and the valve cover.
26. Use an adjustable wrench to counterhold the camshaft. Be careful not to damage the cylinder head. Loosen the center bolt in each camshaft pulley and remove the pulley. Label the pulleys and keep them clean.
27. Check the timing belt carefully for any signs of cracking or deterioration. Pay particular attention to the area where each tooth or cog attaches to the backing of the belt. If the belt shows signs of damage, check the contact faces of the pulleys for possible burrs or scratches.
28. Check the idler pulley by holding it in your hand and spinning it. It should rotate freely and quietly. Any sign of grinding or abnormal noise indicates the pulley should be replaced.

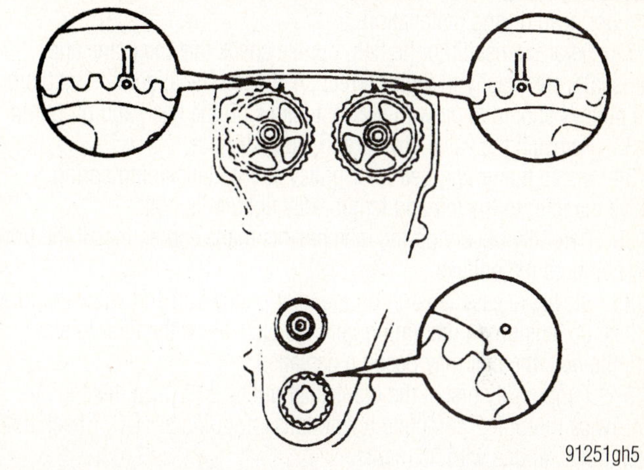

Camshaft and crankshaft timing belt sprocket alignment for proper belt replacement—General Motors 1.6L (VIN 5) Nova and Prism DOHC engine

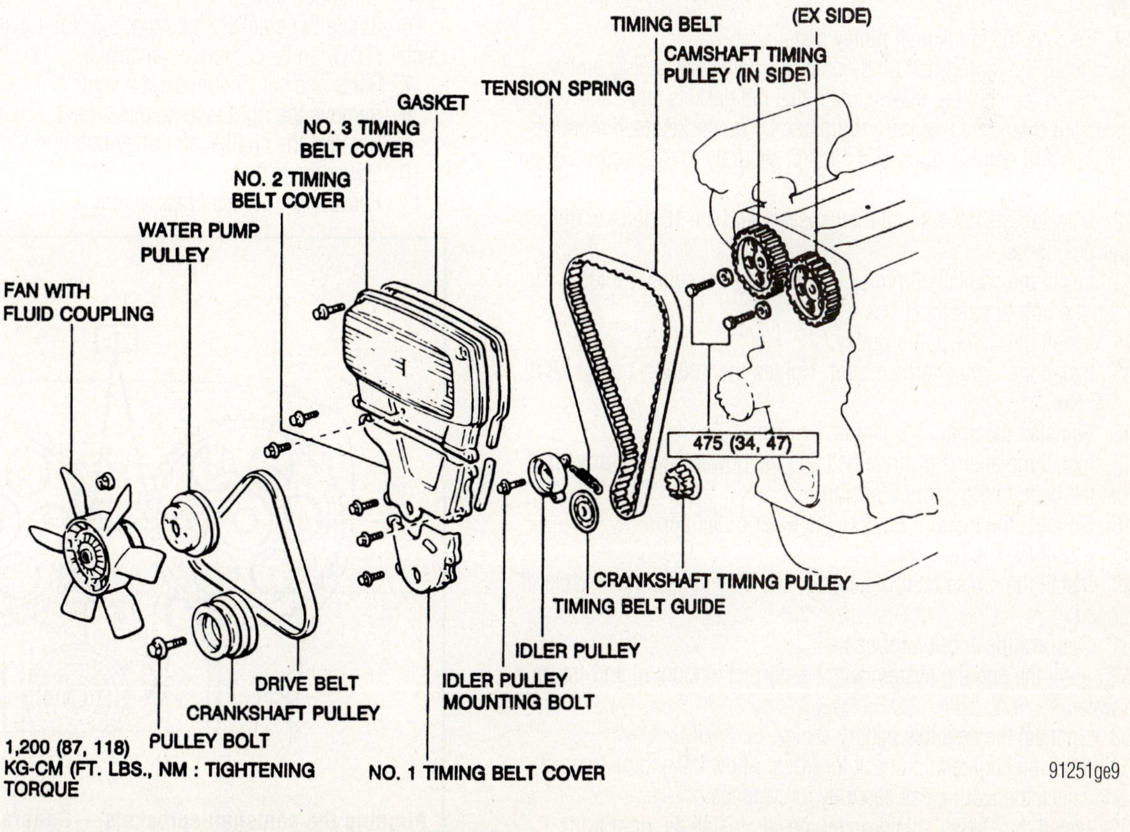

Exploded view of the timing belt and related components—General Motors 1.6L (VIN 5) Nova and Prism DOHC engine

29. Check the free length of the tension spring. Correct length is 43.5mm measured at the inside faces of the hooks. A spring which has stretched during use will not apply the correct tension to the pulley; If in doubt, replace the spring.

30. If you can test the tension of the spring, look for 22 lbs. (10 kg) of tension at 2 in. (50mm) of length. If in doubt, replace the spring.

To install:

31. Align the camshaft knock pin and the pulley. Reinstall the camshaft timing belt pulleys, making sure the pulley fits properly on the shaft and that the timing marks align correctly. Tighten the center bolt on each pulley to 43 ft. lbs. (58.5 Nm). Be careful not to damage the cylinder head during installation.

32. Before reinstalling the belt, double check that the crank and camshafts are exactly in their correct positions. The alignment marks on the pulleys should align with the cast marks on the head and oil pump.

33. Reinstall the valve covers and the PCV hose.

34. Install the timing belt idler pulley and its tensioning spring. Move the idler to the left and temporarily tighten its bolt.

35. Carefully observing the matchmarks made earlier, install the timing belt onto the pulleys.

36. Slowly release tension on the idler pulley bolt and allow the idler to take up tension on the timing belt. DO NOT allow the idler to slam into the belt; the belt may become damaged.

37. Temporarily install the crankshaft pulley bolt. Turn the engine clockwise through 2 complete revolutions, stopping at TDC. Check that each pulley aligns with its marks.

38. Check the tension of the timing belt at a top point halfway between the 2 camshaft sprockets. The correct deflection is 0.16 in. (4mm) at 4.4 lbs. (2 kg) pressure. If the belt tension is incorrect, readjust it. If the tension is correct, tighten the idler pulley bolt to 27 ft. lbs. (37 Nm).

39. Remove the crankshaft pulley bolt.

40. Install the timing belt guide onto the crankshaft timing pulley.

41. When reinstalling, make certain that the gaskets and their mating surfaces are clean and free from dirt and oil. The gasket itself must be free of cuts and deformations and must fit securely in the grooves of the covers.

42. Reinstall the covers and their gaskets and the 10 bolts in their proper positions.

43. Install the crankshaft pulley, again using the counterholding tool. Tighten the bolt to specifications.

44. Install the water pump pulley.

45. Install the right engine mount. Tighten the through-bolt to 58 ft. lbs. (79 Nm).

46. Reinstall the spark plugs and their wires.

47. Install the alternator drive belt and the power steering drive belt. Adjust the belts to the correct tension.

48. Connect the radiator hose to the water outlet port.

49. Install the ignition coil.

50. Install the cruise control actuator and the cruise control cable, if equipped.

51. Connect the accelerator cable.

52. Refill the cooling system with the correct amount of anti-freeze and water.

53. Connect the negative battery cable.

54. Start the engine and check for leaks. Allow the engine to warm up and check the work areas carefully for seepage.

55. Install the splash shield under the car. Install the right front wheel.

1.6L (VIN 5) DOHC ENGINES

YEAR, MODEL & ENGINE APPLICATION

1990-92 GEO Storm .... 1.6L VIN 5

REPLACEMENT INTERVALS

General motors recommends timing belt replacement every 60,000 miles.

LABOR TIMES

1990-92

Replacement .... 2.0 hrs

Severe Service .... 2.2 hrs

8807GMg06

✳✳ CAUTION

These engines are an interference type. If the timing belt has failed, valve-to-piston damage may have occurred. A compression check should always be performed to check for engine damage.

1. Disconnect the negative battery cable.
2. Remove the alternator belt, then the power steering belt.
3. Using the Engine Support Fixture tool J-28467-A or equivalent, support the engine.
4. Remove the right engine mount.
5. Remove the upper timing belt cover wiring harness and the upper timing belt cover.
6. Rotate the crankshaft so that the No. 1 cylinder is at Top Dead Center (TDC) on its compression stroke.
7. Raise and safely support the vehicle.
8. Remove the right side splash shield.
9. Remove the crankshaft pulley bolt and the crankshaft pulley.
10. Lower the vehicle.
11. Remove the exhaust heat shield.

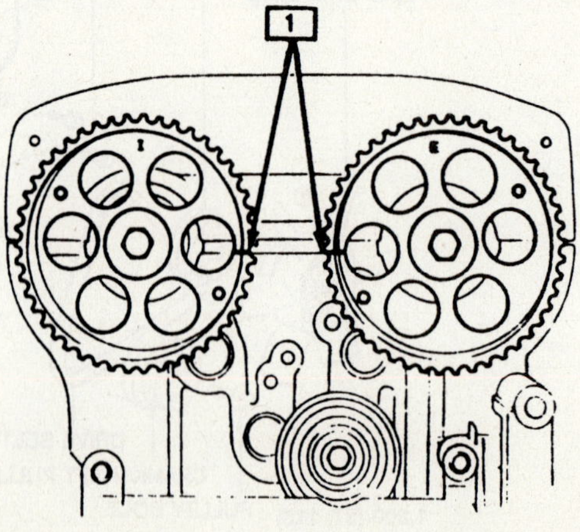

93051g29

Aligning the camshaft sprockets— General Motors 1.6L (VIN 5) Storm DOHC engine

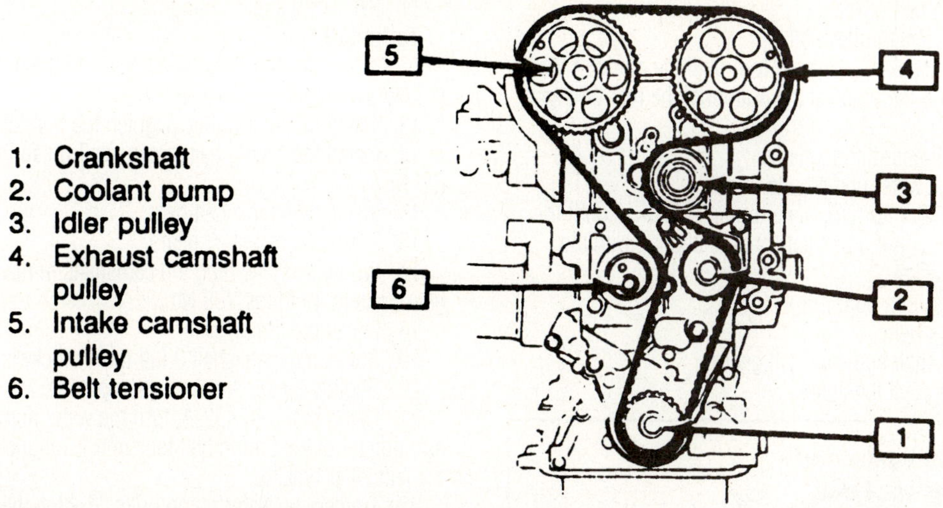

View of the timing belt assembly— General Motors 1.6L (VIN 5) Storm DOHC engine

12. Remove the power steering pump and position aside, without disconnecting the lines.
13. Remove the accessory drive belt pulley and bracket.
14. Remove the right engine mount brace.
15. Remove the lower timing belt cover.
16. Loosen the timing belt tensioner pulley bolt, then move the tensioner away from the belt and secure it with the bolt.
17. If reusing the timing belt, mark the rotational direction on the back of the belt.
18. Remove the timing belt.

To install:

19. Noting the timing belt's direction of rotation, install the belt on the crankshaft sprocket, water pump sprocket, the idler pulley, the exhaust camshaft sprocket, the intake camshaft sprocket and tensioner pulley,.
20. Adjust the timing belt tension by performing the following procedure:
    a. Loosen the belt tensioner bolt and move it toward the timing belt.
21. Using an Allen wrench, insert it into the hexagonal hole of the tension pulley. Hold the pulley stationary and temporarily tighten the tension pulley-to-engine bolt.
    a. Rotate the crankshaft 2 complete revolutions and align the crankshaft timing pulley and the camshaft sprocket marks.
    b. Loosen the tension pulley-to-engine bolt.
    c. Using the Allen wrench and a timing belt tension gauge, apply 31 ft. lbs. (42 N) of tension on the timing belt.
    d. Hold the pulley stationary and torque the tension pulley-to-engine bolt to 31 ft. lbs. (42 Nm).
22. Check that the crankshaft and camshaft sprockets are still aligned.
23. Install the lower timing belt cover.
24. Install the right engine mount brace.
25. Install the accessory drive belt pulley and bracket.
26. Install the power steering pump and position aside, without disconnecting the lines.
27. Install the exhaust heat shield.
28. Raise and safely support the vehicle.
29. Install the crankshaft pulley bolt and torque the bolt to 87 ft. lbs. (118 Nm).
30. Lower the vehicle. Install the upper timing belt cover.
31. Install the right engine mount.
32. Install the power steering pump drive belt, alternator drive belt and any other drive belts that may have been removed.
33. Remove the Engine Support Fixture tool J-28467-A or equivalent from the engine.
34. Connect the negative battery cable.

1.6L (VIN 6) ENGINES

YEAR, MODEL & ENGINE APPLICATION

2004-05 Chevrolet Aveo .... 1.6L VIN 6

REPLACEMENT INTERVALS

General Motors does not provide a replacement interval, but recommends inspecting the timing belt every 30,000 miles.

LABOR TIMES

2004-05
General Motors does not provide a replacement time.

8807GMg07

**** CAUTION**

These engines are an interference type. If the timing belt has failed, valve-to-piston damage may have occurred. A compression check should always be performed to check for engine damage.

1. Remove or disconnect the following:
- Negative battery cable
- Coolant
- Engine oil
- Radiator hoses
- Air cleaner assembly and air intake duct
- Valve cover breather tube
- Right front wheel and splash shield
- Accessory drive belt
- Crankshaft pulley

- Upper front timing belt cover
- Lower front timing belt cover
- Reinstall the crankshaft pulley bolt

2. Rotate the engine clockwise one full turn until the mark on the crankshaft gear aligns with the notch at the bottom of the rear timing cover.
3. Align the camshaft gear timing marks.
4. Slightly loosen the water pump bolts.
5. Using tool J-42492-A, turn the water pump clockwise until the adjuster arm pointer of the timing belt tensioner is aligned with the notch in the tensioner bracket.
6. Tighten the water pump bolts.
7. Remove the timing belt.
8. Remove the crankshaft and camshaft gears.
9. Remove the timing belt tensioner.

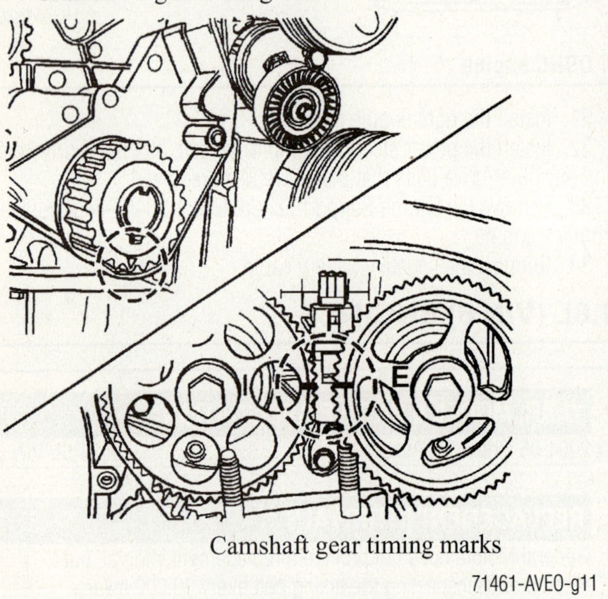

Aligning the crankshaft and camshaft gear timing marks— General Motors 1.6L (VIN 6) Aveo engine

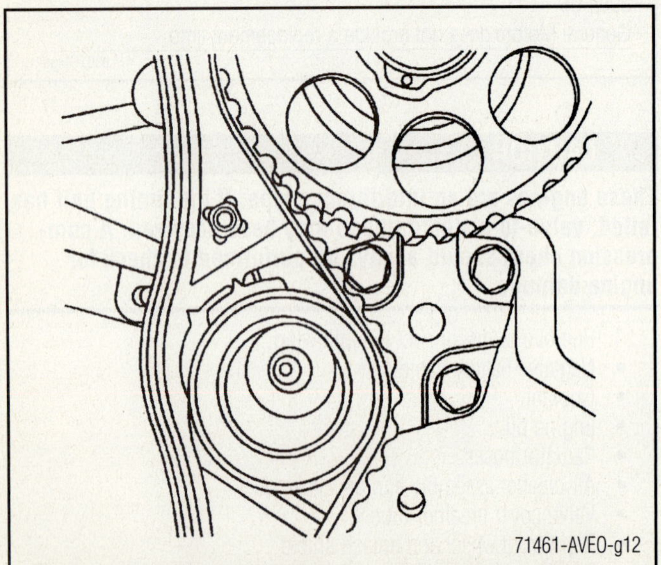

Aligning the timing belt tensioner pointer with the notch in the bracket— General Motors 1.6L (VIN 6) Aveo engine

10. Remove the idler pulley.
11. Remove the rear timing belt cover.

To install:

12. Install the rear timing belt cover. Tighten the bolts to 89 inch lbs. (10 Nm).
13. Install the idler pulley. Tighten the bolt to 30 ft. lbs. (40 Nm).
14. Install the timing belt tensioner. Tighten the bolts to 18 ft. lbs. (25 Nm).
15. Install the crankshaft gear.
16. Install the camshaft gears.
17. While firmly holding the camshafts in place, tighten the camshaft gear bolts to 49 ft. lbs. (68 Nm).
18. Install the timing belt.
19. Rotate the crankshaft 2 full turns clockwise using the pulley bolt.
20. Slightly loosen the water pump bolts.
21. Using tool J-42492-A, turn the water pump until the adjuster arm pointer of the timing belt tensioner is aligned with the pointer on the tensioner bracket.
22. Tighten the water pump bolts. Tighten the bolts to 89 inch lbs. (10 Nm).
23. Remove the crankshaft pulley bolt.
24. Install the upper and lower timing belt covers. Tighten the bolts to 89 inch lbs. (10 Nm).
25. Install the crankshaft pulley bolt. Tighten the bolt to 70 ft. lbs. (95 Nm) plus an additional 15°.
26. Install or connect the following:

- Accessory drive belt
- Right front wheel and splash shield
- Valve cover breather tube
- Air cleaner assembly and air intake duct
- Radiator hoses
- Engine oil
- Coolant
- Negative battery cable

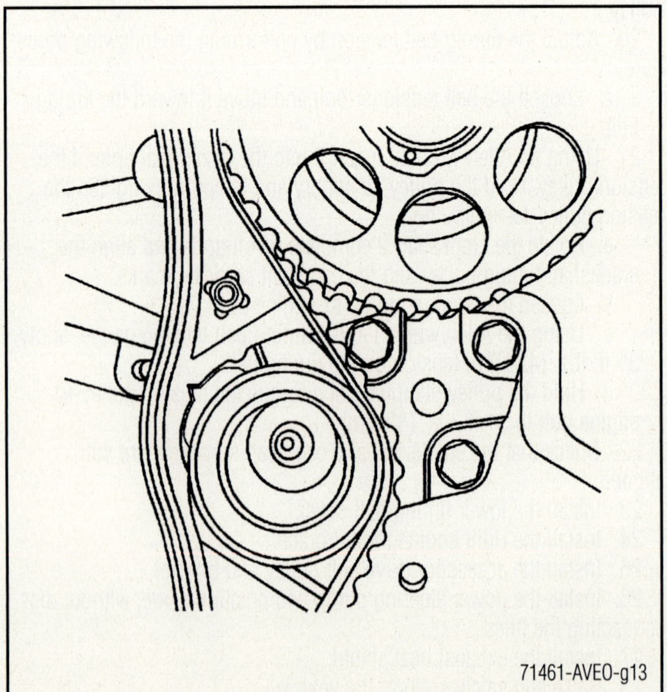

Aligning the timing belt tensioner pointer with the pointer on the bracket— General Motors 1.6L (VIN 6) Aveo engine

1.6L (VIN 6) ENGINES

YEAR, MODEL & ENGINE APPLICATION

1988-93 Pontiac LeMans	1.6L VIN 6

REPLACEMENT INTERVALS

General Motors recomends timing belt replacement every 60,000 miles.

LABOR TIMES

1988-93	
Replacement	1.6 hrs
Severe Service	1.9 hrs
With air conditioning add	0.2 hrs

8807GMg08

1988–90

✱✱ CAUTION

These engines are an interference type. If the timing belt has failed, valve-to-piston damage may have occurred. A compression check should always be performed to check for engine damage.

1. Disconnect the negative battery cable.
2. Rotate the crankshaft to position the No. 1 cylinder on the Top Dead Center (TDC) of its compression stroke.
3. Remove the accessory drive belts.
4. Unsnap the front cover, the upper half first and remove it from the engine.
5. Remove the air cleaner.
6. Remove the valve cover.
7. Remove the power steering pump lines and mounting bolts, remove the pump.
8. Mark the timing belt's direction of rotation for reinstallation purposes.
9. Loosen the water pump-to-engine bolts, rotate the water pump to release the tension on the timing belt.
10. Remove the timing belt.

To install:

11. Align the crankshaft and camshaft sprocket timing marks.
12. Noting the direction of rotation, install the timing belt onto the crankshaft, camshaft and water pump sprockets.
13. Rotate the water pump to place tension on the timing belt and tighten the water pump bolts.
14. Install the front cover to the engine.
15. Install the power steering pump, alternator and the air conditioning compressor (if equipped).
16. Install the valve cover.
17. Install the air cleaner.
18. Install the front cover.
19. Install the accessory drive belts.
20. Connect the negative battery cable.

1991–93

✱✱ CAUTION

These engines are an interference type. If the timing belt has failed, valve-to-piston damage may have occurred. A compression check should always be performed to check for engine damage.

1. Disconnect the negative battery cable.
2. Remove the accessory drive belts.
3. Remove the power steering pump pulley.
4. Remove the power steering pump lines and mounting bolts, and position the pump aside.
5. Remove the upper timing belt cover.
6. Raise and support the vehicle.
7. Remove the right side wheel and the inner splash shield.
8. Rotate the crankshaft and position the crankshaft pulley timing mark with the timing belt cover mark.
9. Remove the crankshaft pulley.
10. Remove the timing belt cover.
11. Mark the timing belt's direction of rotation for reinstallation purposes.
12. Loosen the timing belt tension and remove the timing belt.

To install:

13. Align the camshaft sprocket timing mark to the 12:00 position.
14. Install the timing belt on the camshaft sprocket.
15. Move the tensioner back and install the belt on the crankshaft sprocket.
16. Install the crankshaft pulley and tighten the bolt to 40 ft. lbs. (54 Nm).
17. Install the right side inner splash shield and the wheel.
18. Lower the vehicle.
19. Install the upper timing belt cover.
20. Install the power steering pump.
21. Install the power steering pump pulley.
22. Install the accessory drive belts.
23. Connect the negative battery cable.

1.6L (VIN 6) ENGINES

YEAR, MODEL & ENGINE APPLICATION

1990-97 GEO Prism	1.6L VIN 6

REPLACEMENT INTERVALS

General Motors recommends timing belt replacement every 60,000 miles.

LABOR TIMES

1990-97	
Replacement	2.7 hrs
Severe Service	3.1 hrs
With air conditioning add	0.2 hrs
With cruise control add	0.2 hrs
With power steering add	0.1 hrs

8807GMg09

1990–94 Engines

✻✻ CAUTION

These engines are an interference type. If the timing belt has failed, valve-to-piston damage may have occurred. A compression check should always be performed to check for engine damage.

1. Disconnect the negative battery cable.
2. Remove the accessory drive belts.
3. Remove the A/C compressor and position aside, without disconnecting the lines.
4. Remove the power steering pump and position aside, without disconnecting the lines.
5. Remove the spark plugs and valve cover.
6. Rotate the crankshaft pulley so the **0** mark is aligned with the groove in the front cover.
7. Check that the lifters on the No. 1 cylinder are loose. If not, rotate the crankshaft 360 degrees and recheck.
8. Raise and support the vehicle.
9. Remove the right side tire and splash shield.
10. Support the engine with a suitable jack and remove the right side engine mounting insulator.
11. Remove the water pump and crankshaft pulleys. The crankshaft pulley will require a 2-jaw puller.
12. Loosen the 9 bolts and remove the 3 timing belt covers. Remove the timing belt guide.
13. Loosen the bolt on the idler pulley, push it left as far as it will go then tighten it.
14. If reusing the timing belt, mark the rotational direction on the back of the belt.
15. Remove the timing belt.
16. Remove the idler pulley bolt, pulley and spring.
17. Remove the crankshaft pulley.

To install:

18. Check the free length of the tensioner spring. It should be 1.5 in.
19. Align the mark on the No. 1 bearing cap with the center of the small hole in the pulley.
20. Install the crankshaft pulley and align the marks on the pulley with the oil pump marks.
21. Install the idler pulley and tension spring. Rotate the pulley left and tighten the bolt temporarily.
22. Install the timing belt.
23. Loosen the idler pulley bolt and allow the pulley to tension the belt.
24. Rotate the crankshaft 2 revolutions clockwise.
25. Ensure the crankshaft and camshaft timing marks are aligned.
26. Tighten the idler pulley bolt to 27 ft. lbs. (37 Nm). Measure the timing belt deflection which should be 0.20–0.24 in. If tension is not correct, readjust the idler pulley.
27. Install the timing belt guide with the cupped side outward.
28. Install the 9 bolts and the 3 timing belt covers.
29. Install the water pump and crankshaft pulleys.
30. Install the right side engine mounting insulator.
31. Install the right side tire and splash shield.
32. Lower the vehicle.
33. Install the spark plugs and valve cover.
34. Connect the power steering pump.
35. Connect the A/C compressor.
36. Install the accessory drive belts.
37. Connect the negative battery cable.

1995–97 Engines

✻✻ CAUTION

These engines are an interference type. If the timing belt has failed, valve-to-piston damage may have occurred. A compression check should always be performed to check for engine damage.

1. Disconnect the negative battery cable.
2. Remove the air cleaner assembly and disconnect the throttle cable.
3. Remove the windshield washer reservoir.
4. Remove the accessory drive belts.
5. Remove the cruise control actuator, if equipped.
6. Remove the alternator, alternator bracket and right engine mounting bracket.
7. Raise and support the vehicle.
8. Remove the right side tire and splash shield.
9. Support the engine with a suitable jack and remove the right side engine mounting insulator.
10. Remove the valve cover and spark plugs.
11. Remove the engine front covers.
12. Rotate the crankshaft pulley so the **0** mark is aligned with the groove in the front cover.
13. Check that the lifters on the No. 1 cylinder are loose. If not, rotate the crankshaft 360 degrees and recheck.
14. Remove the water pump and crankshaft pulleys.
15. Loosen the 9 bolts and remove the timing belt covers. Remove the timing belt guide.
16. Ensure the timing marks are aligned.
17. Remove the timing belt tensioner bolt, tensioner and tension spring.
18. If reusing the timing belt, mark the rotational direction on the back of the belt.
19. Remove the timing belt.
20. Check the free length of the tensioner spring. It should be 1.45 in.

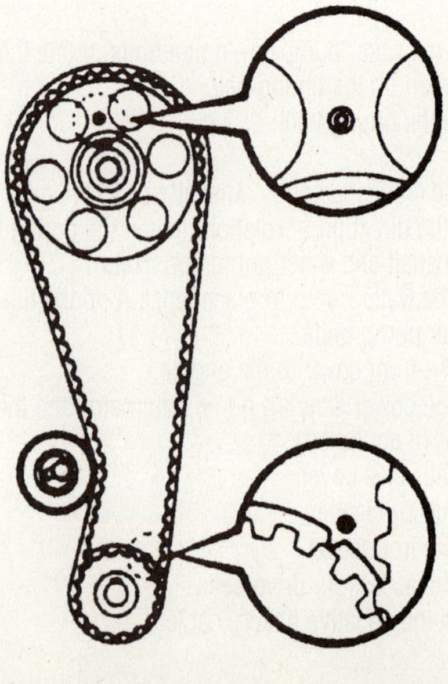

Crankshaft and camshaft sprocket timing mark alignment—General Motors 1.6L (VIN 6) 1990–94 Prism engine

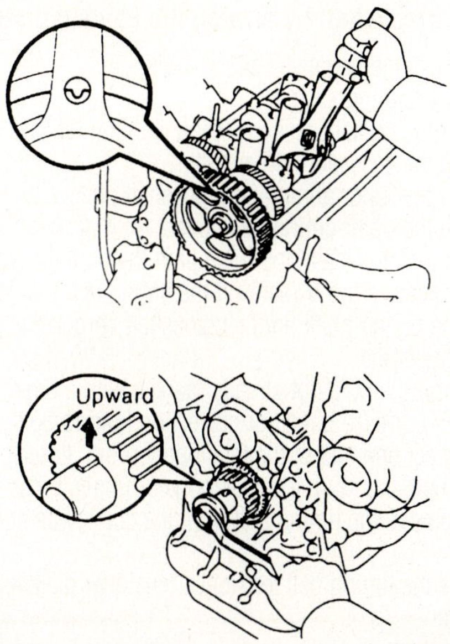

79235g71

Crankshaft and camshaft sprocket timing mark alignment—General Motors 1.6L (VIN 6) 1995–97 Prism engine

21. Align the notch on the No. 1 bearing cap with the center of the small hole in the pulley.
22. Install the tensioner and tension spring. Rotate the tensioner left and tighten the bolt temporarily.
23. Install the timing belt.
24. Loosen the tensioner bolt and allow the tensioner to tension the belt.
25. Rotate the crankshaft 2 revolutions clockwise.
26. Ensure the crankshaft and camshaft timing marks are aligned.
27. Tighten the idler pulley bolt to 27 ft. lbs. (37 Nm). Measure the timing belt deflection which should be 0.20–0.24 in. If tension is not correct, readjust the idler pulley.
28. Install the timing belt guide with the cupped side outward.
29. Install the 9 bolts and the timing belt covers. Tighten the bolts to 62 inch lbs. (7 Nm).
30. Install all remaining components. Tighten the crankshaft pulley to 87 ft. lbs. (118 Nm), the mounting bracket-to-engine mount bolt to 47 ft. lbs. (64 Nm), the mounting bracket-to-engine mount nuts to 38 ft. lbs. (52 Nm) and the engine mount bolts to 19 ft. lbs. (25 Nm).
31. Install the accessory drive belts.
32. Connect the negative battery cable.

1.6L (VIN 6) ENGINES

YEAR, MODEL & ENGINE APPLICATION	
1990-93 GEO Storm	1.6L VIN 6

REPLACEMENT INTERVALS

General Motors recommends timing belt replacement every 60,000 miles.

LABOR TIMES	
1990-93	
Replacement	2.0 hrs
Severe Service	2.2 hrs

8807GMg10

✻✻ CAUTION

These engines are an interference type. If the timing belt has failed, valve-to-piston damage may have occurred. A compression check should always be performed to check for engine damage.

1. Disconnect the negative battery cable.
2. Remove the alternator belt, then the power steering belt.
3. Raise and safely support the vehicle.
4. Using the Engine Support Fixture tool J-28467-A or equivalent, support the engine.
5. Remove the right engine mount.
6. Remove the engine mounting bracket from the timing belt cover.
7. Remove the crankshaft pulley.
8. Remove the timing belt cover and the timing belt cover.
9. Rotate the crankshaft so that the No. 4 cylinder is at Top Dead Center (TDC) on its compression stroke by aligning the camshaft pulley timing mark to the 9 o'clock position.
10. Remove the crankshaft pulley center hub.
11. Lower the vehicle. Loosen the timing belt tensioner pulley bolt, then move the tensioner away from the belt and secure it with the bolt.
12. If reusing the timing belt, mark the rotational direction on the back of the belt.
13. Remove the timing belt.

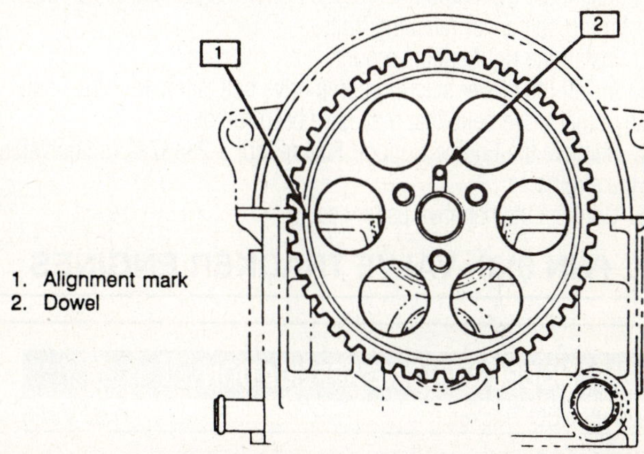

93051g26

View of the camshaft alignment—General Motors 1.6L (VIN 6) SOHC Storm engine

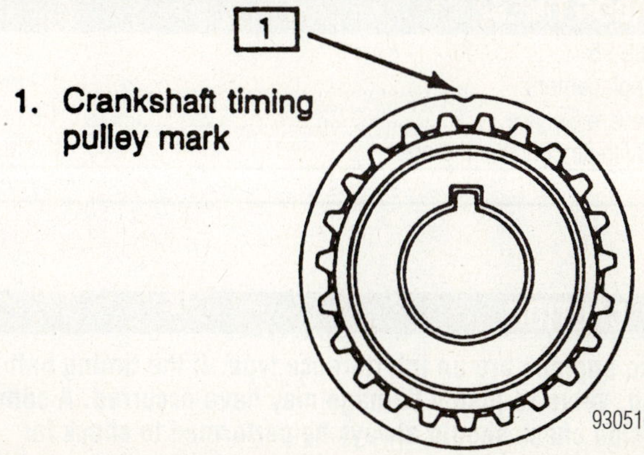

93051g27

View of the crankshaft alignment—General Motors 1.6L (VIN 6) SOHC Storm engine

To install:

14. Noting the timing belt's direction of rotation, install the belt on the crankshaft sprocket, tensioner pulley, water pump sprocket and the camshaft sprocket.
15. Adjust the timing belt tension by performing the following procedure:
    a. Loosen the belt tensioner bolt and move it toward the timing belt.
16. Using an Allen wrench, insert it into the hexagonal hole of the tension pulley. Hold the pulley stationary and temporarily tighten the tension pulley-to-engine bolt.
    a. Rotate the crankshaft 2 complete revolutions and align the crankshaft timing pulley groove with the mark on the oil pump.
    b. Loosen the tension pulley-to-engine bolt.
    c. Using the Allen wrench and a timing belt tension gauge, apply 31 ft. (42 N) of tension on the timing belt.
    d. Hold the pulley stationary and torque the tension pulley-to-engine bolt to 37 ft. lbs. (50 Nm).
17. Check that the camshaft is still aligned to the 9 o'clock position.
18. Install the timing belt cover, secure it with the 2 lower mounting bolts and torque the bolts to 89 inch lbs. (10 Nm).
19. Install the crankshaft pulley to the crankshaft damper and secure the center bolt and the 4 side bolts. Torque the center bolt to 87 ft. lbs. (118 Nm) and the 4 side bolts to 17 ft. lbs. (23 Nm).
20. Install the engine mounting bracket from the timing belt cover.
21. Install the right side undercover. Lower the vehicle. Install the 4 upper timing belt cover retaining bolts.
22. Install the right engine mount.
23. Install the power steering pump drive belt, alternator drive belt and any other drive belts that may have been removed.
24. Remove the Engine Support Fixture tool J-28467-A or equivalent from the engine.
25. Connect the negative battery cable.

1.6L (VIN 6) 8–VALVE TRACKER ENGINES

YEAR, MODEL & ENGINE APPLICATION	
1989-95 GEO Tracker	1.6L VIN 6
REPLACEMENT INTERVALS	
General Motors recommends timing belt replacement every 60,000 miles.	
LABOR TIMES	
1989-95	
Replacement	1.7 hrs
Severe Service	1.8 hrs
With air conditioning add	0.7 hrs

8807GMg11

✻✻ CAUTION

These engines are an interference type. If the timing belt has failed, valve-to-piston damage may have occurred. A compression check should always be performed to check for engine damage.

■ **Do not rotate the crankshaft counterclockwise or attempt to rotate the crankshaft by turning the camshaft sprocket.**

1. Disconnect the negative battery cable.
2. Drain the cooling system.
3. Remove the coolant reservoir tank, cooling fan and shroud.
4. Remove the accessory drive belts.
5. Remove the water pump and crankshaft pulleys.
6. Remove the outer timing belt cover.
7. Rotate the crankshaft clockwise until the timing mark on the camshaft sprocket and the **V** mark on the timing belt inside cover are aligned, and the punch mark on the crankshaft sprocket is aligned with the mark on the engine.
8. If the timing belt is not already marked with a directional arrow, use white paint, a grease pencil or correction fluid to do so.
9. Disconnect one end of the tensioner spring. Loosen the timing belt tensioner bolt and stud, then, using your finger, press the tensioner plate up and remove the timing belt from the crankshaft and camshaft sprockets.
10. Remove the timing belt tensioner, tensioner plate and spring from the engine.

To install:

11. Assemble the timing belt tensioner plate and the tensioner, making sure that the lug of the tensioner plate engages the tension.
12. Install the timing belt tensioner, tensioner plate and spring on the engine. Tighten the mounting bolt and stud only finger-tight at this time. Ensure that when the tensioner is moved in a counterclockwise direction, the tensioner moves in the same direction. If the tensioner does not move, remove it and the tensioner plate to reassemble them properly.
13. Remove the air cleaner and valve cover.
14. Loosen all rocker arm valve lash locknuts and adjusting screws. This will permit movement of the camshaft without any rocker arm associated drag, which is essential for proper timing belt tensioning. If the camshaft does not rotate freely (free of rocker arm drag), the belt will not be properly tensioned.
15. Rotate the camshaft sprocket clockwise until the timing mark on the sprocket and the V mark on the timing belt inside cover are aligned.

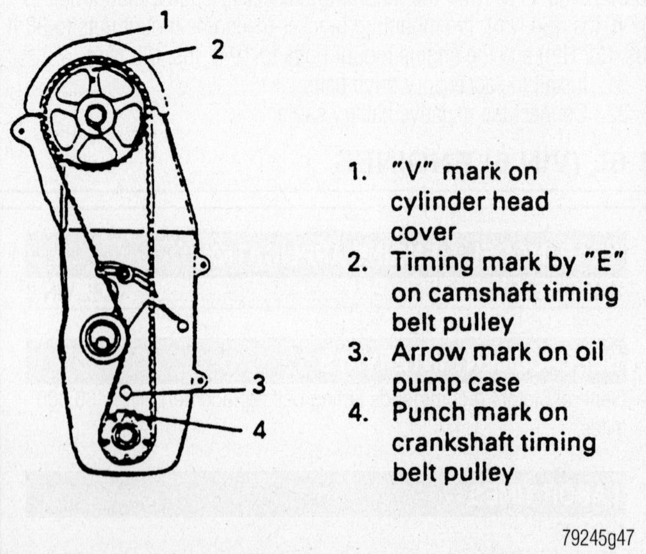

Crankshaft and camshaft and timing mark alignment—General Motors 1.6L (VIN 6) 8–valve Tracker engine

16. With the camshaft and crankshaft marks properly aligned, push the tensioner up with your finger and install the timing belt on the 2 sprockets, ensuring that the drive side of the belt is free of all slack. Release the tensioner. Be sure to install the timing belt so that the directional arrow is pointing in the appropriate direction.

■ **In this position, the No. 4 cylinder is at Top Dead Center (TDC) on the compression stroke.**

17. Rotate the crankshaft clockwise 2 full revolutions, then tighten the tensioner stud to 97 inch lbs. (11 Nm). Then, tighten the tensioner bolt to 18 ft. lbs. (24 Nm).
18. Ensure that all 4 timing marks are still aligned as before; if they are not, remove the timing belt, and install and tension it again.
19. Install the timing belt cover.
20. Install the crankshaft pulley and tighten to 97 inch lbs. (11 Nm).
21. Install the water pump and crankshaft pulleys.
22. Install the accessory drive belts.
23. Install the coolant reservoir tank, cooling fan and shroud.
24. Fill the cooling system.
25. Adjust the valve clearances. Exhaust valve clearance should be 0.0063–0.0075 inch. Intake valve clearance should be 0.0051–0.0067 inch.
26. Install the valve cover and air cleaner.
27. Connect the negative battery cable.

1.6L (VIN 6) 16–VALVE TRACKER ENGINES

YEAR, MODEL & ENGINE APPLICATION	
1994-97 GEO Tracker	1.6L VIN 6

REPLACEMENT INTERVALS

General Motors recommends timing belt replacement every 60,000 miles.

LABOR TIMES	
1994-97	
Replacement	1.6 hrs
Severe Service	1.7 hrs
With air conditioning add	0.7 hrs

8807GMg12

✻✻ CAUTION

These engines are an interference type. If the timing belt has failed, valve-to-piston damage may have occurred. A compression check should always be performed to check for engine damage.

■ **Do not rotate the crankshaft counterclockwise or attempt to rotate the crankshaft by turning the camshaft sprocket.**

1. Disconnect the negative battery cable.
2. Drain the cooling system.
3. Loosen the A/C compressor bolts.
4. Remove the accessory drive belts.
5. Remove the cooling fan and water pump pulley.
6. Raise and support the vehicle.
7. Remove the engine undercover and skid plate.
8. Remove the lower radiator shroud bolts.
9. Lower the vehicle.
10. Disconnect the A/C refrigerant lines from the radiator support.
11. Remove the upper radiator shroud bolts and the shroud.
12. Ensure the No. 1 cylinder is at TDC, then remove the crankshaft pulley, without removing the center bolt.
13. Remove the timing belt cover.
14. Ensure the crankshaft and camshaft sprocket timing marks are aligned.
15. If the timing belt is not already marked with a directional arrow, use white paint, a grease pencil or correction fluid to do so.
16. Loosen the timing belt tensioner bolt and stud, then, using your finger, press the tensioner up and remove the timing belt from the crankshaft and camshaft sprockets.

To install:

17. Install the timing belt on the crankshaft and camshaft sprockets and the tensioner.
18. Ensure the belt is taut on the non-tensioned side.
19. Connect the timing tensioner spring.
20. Rotate the crankshaft clockwise 2 full revolutions, and ensure the timing marks are still aligned.
21. Tighten the tensioner stud to 80–106 inch lbs. (9–12 Nm). Then, tighten the tensioner bolt to 18–21 ft. lbs. (24–30 Nm).

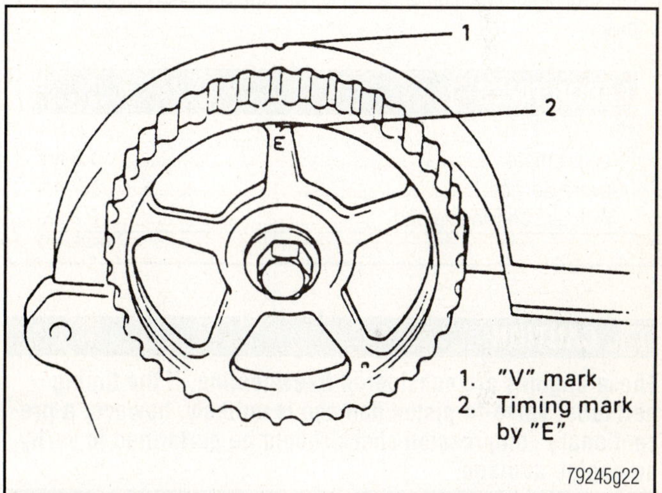

Camshaft sprocket timing marks—General Motors 1.6L Tracker 16–valve engine

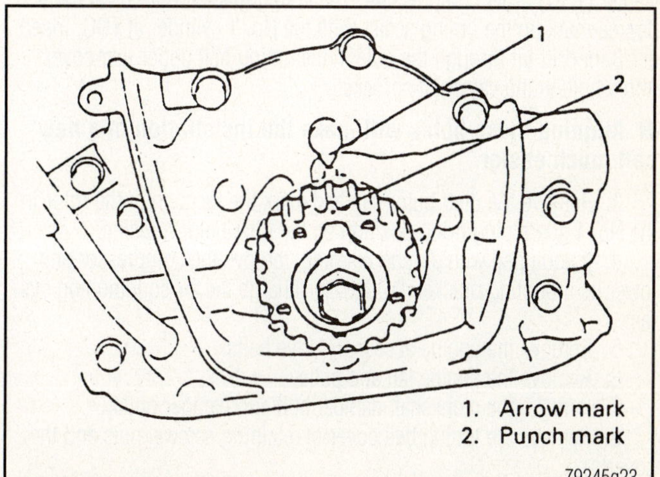

Align the crankshaft punch mark with the arrow for proper timing belt installation— General Motors 1.6L Tracker 16–valve engine

22. Install the timing belt cover.
23. Install the crankshaft pulley bolts and tighten to 12 ft. lbs.
24. Install the upper radiator shroud bolts and the shroud.
25. Connect the A/C refrigerant lines to the radiator support.
26. Raise and support the vehicle.
27. Install the lower radiator shroud bolts.
28. Install the engine undercover and skid plate.
29. Lower the vehicle.
30. Install the accessory drive belts.
31. Tighten the A/C compressor bolts.
32. Install the cooling fan and water pump pulley.
33. Connect the negative battery cable.
34. Fill the cooling system.

1.6L (VIN C) ENGINES

YEAR, MODEL & ENGINE APPLICATION

1985-87 Chevrolet Chevette, Pontiac T1000 .... 1.6L VIN C

REPLACEMENT INTERVALS

General Motors recommends timing belt replacment every 60,000 miles.

LABOR TIMES

1985-87	
Replacement	1.6 hrs
Severe Service	1.7 hrs
With air conditioning add	0.2 hrs

8807GMg13

✻✻ CAUTION

These engines are considered freewheeling. If the timing belt fails, valve-to-piston damage is unlikely, however a precautionary compression check should be performed to verify no engine damage.

1. Disconnect the negative battery cable.
2. Rotate the engine to position the No. 1 cylinder at Top Dead Center (TDC) or its compression. The timing mark should be at the **0** degree mark on the timing scale. With the No. 1 cylinder at TDC, insert a 1/8 in. drill bit through the hole in the timing belt upper rear cover into a hole in the camshaft sprocket.

■ **Aligning these holes will make the installation of a new belt much easier.**

3. Remove the distributor cap and mark the location of the rotor in the No. 1 spark plug firing position on the distributor housing.
4. If equipped with air conditioning, remove the compressor and lower its mounting bracket; DO NOT discharge the air conditioning system.
5. Remove the engine accessory drive belts.
6. Remove the engine fan and pulley.
7. Remove the crankshaft damper bolt and damper pulley.
8. Remove the timing belt cover(s) retaining screws, nuts and the cover(s).
9. Mark the direction of rotation on the timing belt, if to be reused.
10. Loosen the idler pulley adjustment bolt and loosen the tension of the timing belt.
11. Remove the timing belt from the sprockets.

To install:

12. Align the crankshaft pulley timing mark with the **0** mark on the timing scale and the distributor rotor with the scribed mark on the distributor housing.

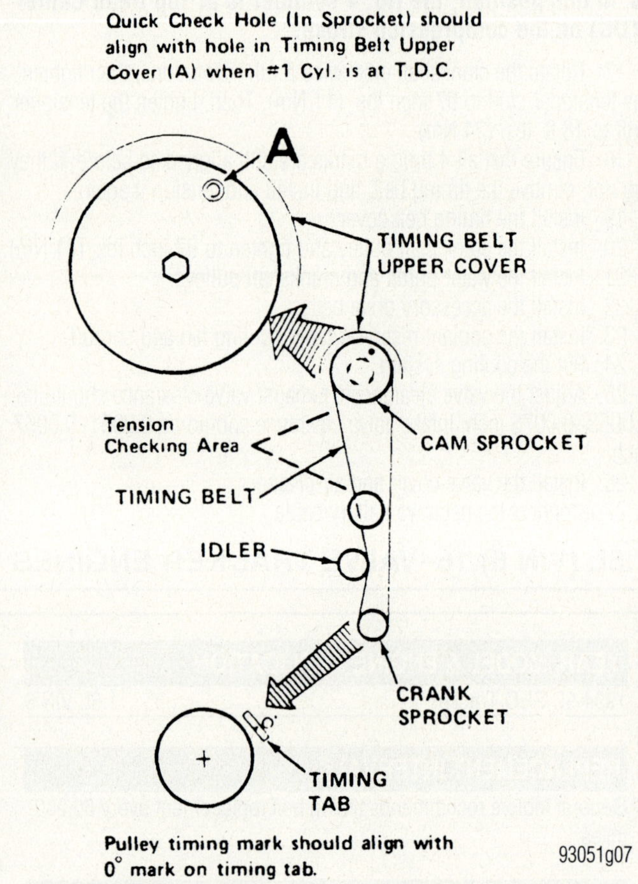

93051g07

View of the timing sprockets and belt positions—General Motors 1.6L (VIN C) engine

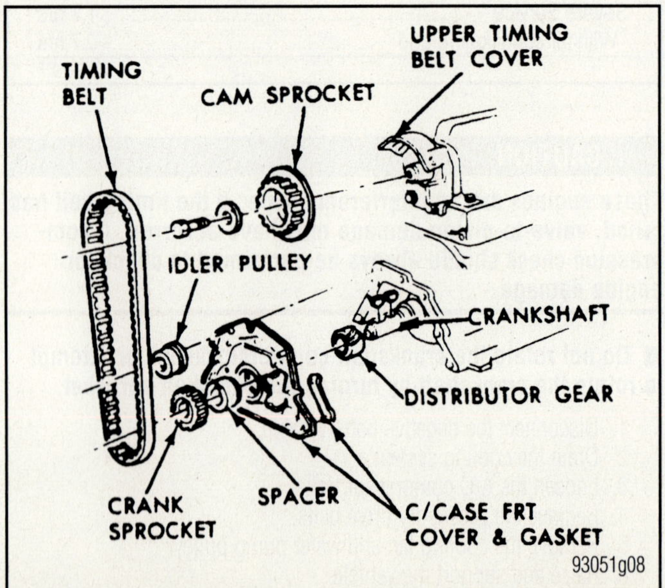

93051g08

Exploded view of the timing belt assembly—General Motors 1.6L (VIN C) engine

13. Align the hole in the camshaft sprocket with the hole in the upper rear timing belt cover. Then, insert a 1/8 in. drill bit to hold the sprocket in alignment.

14. Noting the timing belt's direction of rotation, install it onto the crankshaft and camshaft sprockets.

15. Install the crankshaft damper pulley and bolt.

16. Adjust the timing belt tension by performing the following procedure:

a. Using a Timing Belt Tension Gauge, place it on the belt on the same side as the idler pulley midway between the camshaft sprocket and the idler pulley.

■ **Be sure that the center finger of the gauge extension fits in a notch between the teeth on the belt.**

b. Rotate the idler pulley to place 70 lbs. (32 kg) of tension on the belt; then, tighten the idler pulley bolt to 15 ft. lbs. (20 Nm).

17. Install the distributor cap.

18. If equipped with air conditioning, install the compressor bracket and compressor.

19. Install the timing belt covers, the engine fan and pulley.

20. Install and adjust the accessory drive belts.

21. Connect the negative battery cable.

1.8L (VIN 8) DOHC ENGINES

YEAR, MODEL & ENGINE APPLICATION

1993-97 GEO Prism	1.8L VIN 8

REPLACEMENT INTERVALS

General Motors recommends timing belt replacement every 60,000 miles.

LABOR TIMES

1993-97	
Replacement	2.7 hrs
Severe Service	3.1 hrs
With air conditioning add	0.2 hrs
With cruise control add	0.2 hrs
With power steering add	0.1 hrs

8807GMg14

✻✻ CAUTION

These engines are an interference type. If the timing belt has failed, valve-to-piston damage may have occurred. A compression check should always be performed to check for engine damage.

1. Disconnect the negative battery cable.
2. Remove the air cleaner assembly and disconnect the throttle cable.
3. Remove the windshield washer reservoir.
4. Remove the accessory drive belts.
5. Remove the cruise control actuator, if equipped.
6. Remove the alternator, alternator bracket and right engine mounting bracket.
7. Raise and support the vehicle.
8. Remove the right side tire and splash shield.
9. Support the engine with a suitable jack and remove the right side engine mounting insulator.
10. Remove the valve cover and spark plugs.
11. Remove the engine front covers.
12. Rotate the crankshaft pulley so the 0 mark is aligned with the groove in the front cover.
13. Check that the lifters on the No. 1 cylinder are loose. If not, rotate the crankshaft 360 degrees and recheck.
14. Remove the water pump and crankshaft pulleys.
15. Loosen the 9 bolts and remove the timing belt covers. Remove the timing belt guide.
16. Ensure the timing marks are aligned.
17. Remove the timing belt tensioner bolt, tensioner and tension spring.
18. If reusing the timing belt, mark the rotational direction on the back of the belt.
19. Remove the timing belt.
20. Check the free length of the tensioner spring. It should be 1.45 in.
21. Align the notch on the No. 1 bearing cap with the center of the small hole in the pulley.
22. Install the tensioner and tension spring. Rotate the tensioner left and tighten the bolt temporarily.
23. Install the timing belt.
24. Loosen the tensioner bolt and allow the tensioner to tension the belt.
25. Rotate the crankshaft 2 revolutions clockwise.
26. Ensure the crankshaft and camshaft timing marks are aligned.
27. Tighten the idler pulley bolt to 27 ft. lbs. (37 Nm). Measure the timing belt deflection which should be 0.20–0.24 in. If tension is not correct, readjust the idler pulley.
28. Install the timing belt guide with the cupped side outward.
29. Install the 9 bolts and the timing belt covers. Tighten the bolts to 62 inch lbs. (7 Nm).
30. Install all remaining components. Tighten the crankshaft pulley to 87 ft. lbs. (118 Nm), the mounting bracket-to-engine mount bolt to

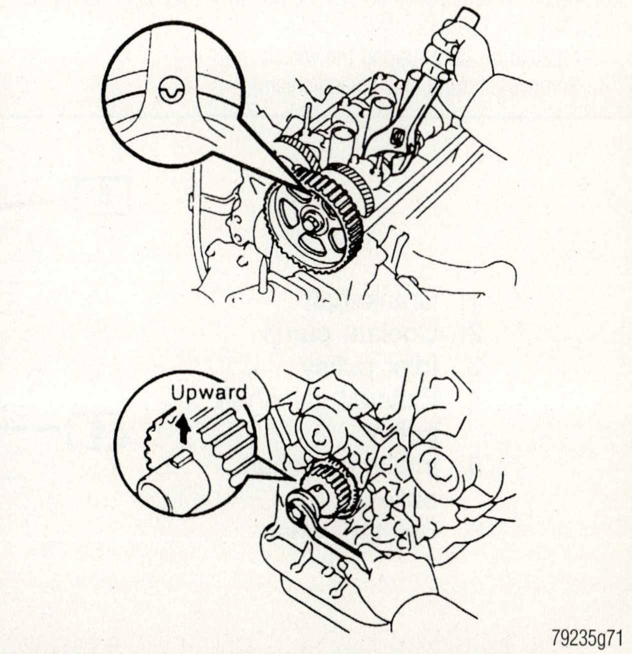

79235g71

Crankshaft and camshaft sprocket timing mark alignment—General Motors 1.6L (VIN 6) 1995–97 Prism engine

47 ft. lbs. (64 Nm), the mounting bracket-to-engine mount nuts to 38 ft. lbs. (52 Nm) and the engine mount bolts to 19 ft. lbs. (25 Nm).

31. Install the accessory drive belts.
32. Connect the negative battery cable.

1.8L (VIN 8) DOHC ENGINES

YEAR, MODEL & ENGINE APPLICATION	
1992-93 GEO Storm	1.8L VIN 8

REPLACEMENT INTERVALS

General Motors recommends timing belt replacement every 60,000 miles.

LABOR TIMES	
1992-93	
Replacement	2.0 hrs
Severe Service	2.2 hrs

8807GMg15

✲✲ CAUTION

These engines are an interference type. If the timing belt has failed, valve-to-piston damage may have occurred. A compression check should always be performed to check for engine damage.

1. Disconnect the negative battery cable.
2. Remove the alternator belt, then the power steering belt.
3. Using the Engine Support Fixture tool J-28467-A or equivalent, support the engine.
4. Remove the right engine mount.
5. Remove the upper timing belt cover wiring harness and the upper timing belt cover.
6. Rotate the crankshaft so that the No. 1 cylinder is at Top Dead Center (TDC) on its compression stroke.
7. Raise and safely support the vehicle.
8. Remove the right side splash shield.
9. Remove the crankshaft pulley bolt and the crankshaft pulley.
10. Lower the vehicle.
11. Remove the exhaust heat shield.
12. Remove the power steering pump and position aside, without disconnecting the lines.
13. Remove the accessory drive belt pulley and bracket.
14. Remove the right engine mount brace.
15. Remove the lower timing belt cover.
16. Loosen the timing belt tensioner pulley bolt, then move the tensioner away from the belt and secure it with the bolt.
17. If reusing the timing belt, mark the rotational direction on the back of the belt.
18. Remove the timing belt.

To install:

19. Noting the timing belt's direction of rotation, install the belt on the crankshaft sprocket, water pump sprocket, the idler pulley, the exhaust camshaft sprocket, the intake camshaft sprocket and tensioner pulley,.

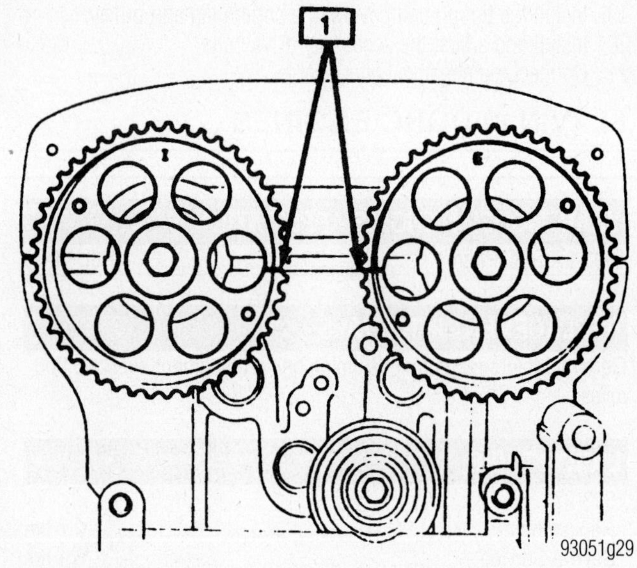

Aligning the camshaft sprockets— General Motors 1.8L (VIN 8) Storm DOHC engine

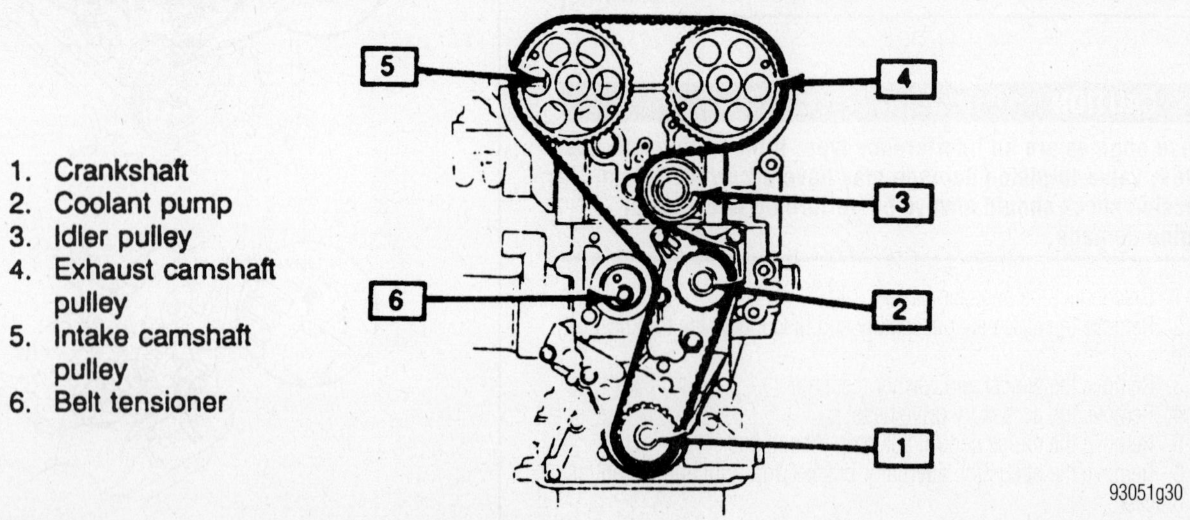

View of the timing belt assembly— General Motors 1.8L (VIN 8) Storm DOHC engine

20. Adjust the timing belt tension by performing the following procedure:
   a. Loosen the belt tensioner bolt and move it toward the timing belt.
21. Using an Allen wrench, insert it into the hexagonal hole of the tension pulley. Hold the pulley stationary and temporarily tighten the tension pulley-to-engine bolt.
   a. Rotate the crankshaft 2 complete revolutions and align the crankshaft timing pulley and the camshaft sprocket marks.
   b. Loosen the tension pulley-to-engine bolt.
   c. Using the Allen wrench and a timing belt tension gauge, apply 31 ft. (42 N) of tension on the timing belt.
   d. Hold the pulley stationary and torque the tension pulley-to-engine bolt to 31 ft. lbs. (42 Nm).
22. Check that the crankshaft and camshaft sprockets are still aligned.
23. Install the lower timing belt cover.
24. Install the right engine mount brace.
25. Install the accessory drive belt pulley and bracket.
26. Install the power steering pump and position aside, without disconnecting the lines.
27. Install the exhaust heat shield.
28. Raise and safely support the vehicle.
29. Install the crankshaft pulley bolt and torque the bolt to 87 ft. lbs. (118 Nm).
30. Lower the vehicle. Install the upper timing belt cover.
31. Install the right engine mount.
32. Install the power steering pump drive belt, alternator drive belt and any other drive belts that may have been removed.
33. Remove the Engine Support Fixture tool J-28467-A or equivalent from the engine.
34. Connect the negative battery cable.

1.8L (VIN D) DIESEL ENGINES

YEAR, MODEL & ENGINE APPLICATION	
1985-86 Chevrolet Chevette, Pontiac T1000	1.8L VIN D

REPLACEMENT INTERVALS

General Motors does not provide a replacement interval. Standard industry practice is to replace timing belts every 60,000 miles.

LABOR TIMES	
1985-86	
Replacement	2.4 hrs
Severe Service	2.7 hrs

8807GMg16

✱✱ CAUTION

These engines are an interference type. If the timing belt has failed, valve-to-piston damage may have occurred. A compression check should always be performed to check for engine damage.

1. Disconnect the negative battery cable.
2. Drain the cooling system.
3. Remove the fan shroud, cooling fan V-belt and pulley.
4. Disconnect the bypass hose and then remove the upper half of the front timing belt cover.
5. Rotate the crankshaft to position the No. 1 piston at Top Dead Center (TDC) of its compression stroke. Make sure that the notch on the injection pump gear is aligned with the index mark on the front plate. Thread a 8mm x 1.25mm lock bolt through the sprocket and into the front plate.
6. Remove the valve cover and install a Fixing Plate tool J-29761, or equivalent, in the slot at the rear of the camshaft; this will prevent the camshaft from rotating during the procedure.
7. Remove the crankshaft damper pulley and check to make sure that the No. 1 piston is still at TDC.
8. Remove the lower half of the front cover and then remove the timing belt holder from the bottom of the front plate.
9. Mark the direction of rotation on the timing belt for reinstallation purposes.
10. Remove the tension spring from behind the front plate, near to the injection pump.
11. Loosen the tension pulley and slide the timing belt off of the sprockets.

To install:

12. Noting the direction of rotation, slide the timing belt over the sprockets. The belt should be properly tensioned between the sprockets, the cogs on the belt; also, the sprockets should properly engaged and the crankshaft should not be turned.

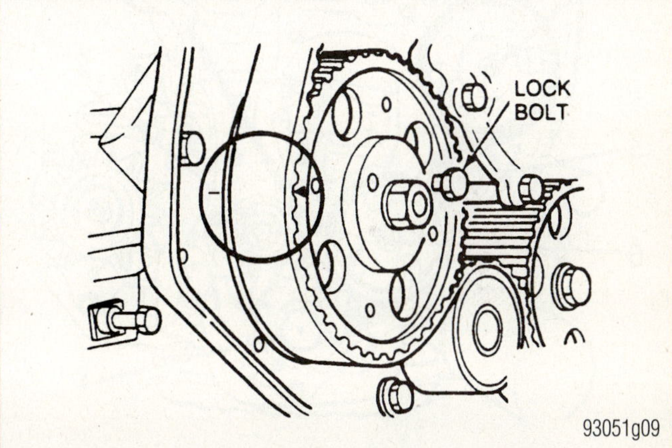

Location of the injection sprocket mark and lock bolt—General Motors 1981–86 1.8L (VIN D) Diesel engine

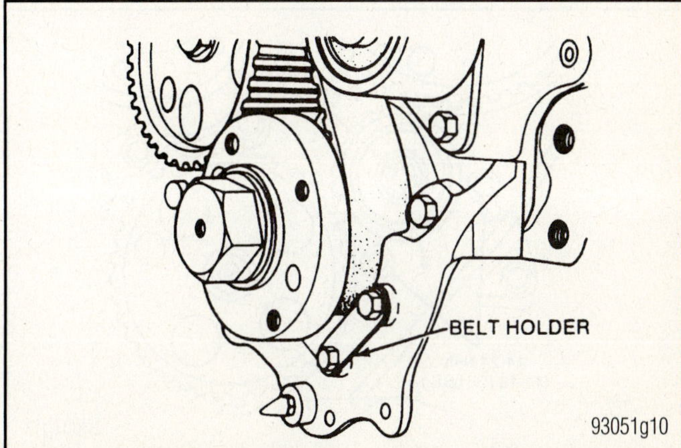

Location of the timing belt holder—General Motors 1981–86 1.8L (VIN D) Diesel engine

13. Partially, tighten the bolts in numerical order to prevent movement of the tensioner pulley.

14. Remove the injection pump sprocket lock bolt and the fixing plate on the end of the camshaft.

15. Make sure the mark on the injection pump pulley is aligned with the mark on the plate. The fixing plate should fit smoothly into the slot at the rear of the camshaft, then remove the fixing plate.

16. Install the damper pulley on the hub and check that the No. 1 cylinder is at TDC; DO NOT turn the crankshaft.

17. Loosen the tensioner pulley and plate attaching bolts. Concentrate looseness of the belt on the tensioner, then tighten the bolts in the numerical order to 11–18 ft. lbs. (14–24 Nm) for bolts No. 1 and 2; then 50–60 ft. lbs. (67–81 Nm) for No. 3.

18. Using tool J-29771 or equivalent, check the belt tension between the camshaft sprocket and the injection pump sprocket. Remove the damper pulley and install the belt holder in position away from the timing bolt.

19. Complete the installation by reversing the removal procedures.

20. Connect the negative battery cable.

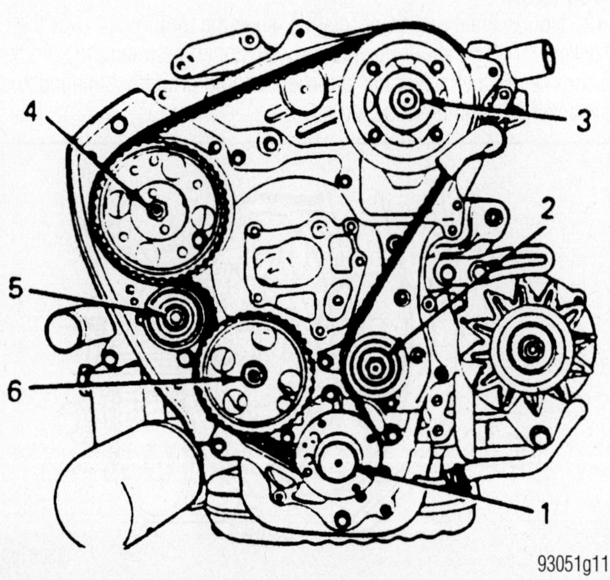

93051g11

View of the timing belt installation sequence—General Motors 1981–86 1.8L (VIN D) Diesel engine

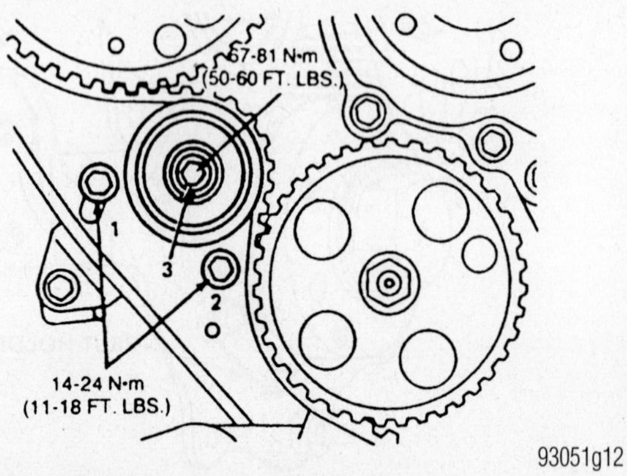

93051g12

View of the tensioner pulley torquing sequence—General Motors 1981–86 1.8L (VIN D) Diesel engine

2.0L (VIN H) ENGINES

YEAR, MODEL & ENGINE APPLICATION

1989-92 Pontiac LeMans	2.0L VIN H

REPLACEMENT INTERVALS

General Motors does not provide a replacement interval. Standard industry practice is to replace timing belts every 60,000 miles.

LABOR TIMES

1989-92	
Replacement	1.6 hrs
Severe Service	1.9 hrs
With air conditioning add	0.2 hrs

8807GMg17

✳✳ CAUTION

These engines are considered freewheeling. If the timing belt fails, valve-to-piston damage is unlikely, however a precautionary compression check should be performed to verify no engine damage.

1. Disconnect the negative battery cable and drain the engine cooling system to a level below the water pump.

2. Remove the coolant reservoir tank.

3. Remove the serpentine belt from the engine. Remove the fuel vapor pipe assembly.

4. Loosen the drive belt tensioner bolt and the tensioner will swing downward. If necessary, remove the bolt and tensioner from the engine.

5. Remove the cover attaching bolts, then remove the timing belt cover from the engine.

6. Turn the engine to align the timing marks on the gears with the marks on the timing belt rear cover.

7. Loosen the water pump bolts and release timing belt tension with Tension Adjusting tool J-33039 or equivalent. If equipped, remove the air conditioning drive belt.

8. Raise and support the vehicle safely, then remove the right splash shield.

9. Remove the pulley retaining bolt, then remove the pulley from the end of the crankshaft.

10. Lower the vehicle and remove the timing belt. If necessary, remove the retainers and the belt tensioner from the engine.

To install:

11. If removed, install the timing belt tensioner and tighten the retainers to 18 ft. lbs. (25 Nm).

12. If necessary, turn the crankshaft and/or the camshaft gears clockwise to align the timing marks on the gears with the timing marks on the rear cover.

13. Install the timing belt, making sure the portion between the camshaft gear and crankcase gear is in tension.

14. Using tool J-33039 or equivalent, turn the water pump eccentric clockwise until the tensioner contacts the high torque stop. Tighten the water pump screws slightly.

15. Turn the engine clockwise 720 degrees using the crankshaft gear bolt in order to fully seat the belt into the gear teeth.

16. Turn the water pump eccentric counterclockwise until the hole in the tensioner arm is aligned with the hole in the base. In order for the

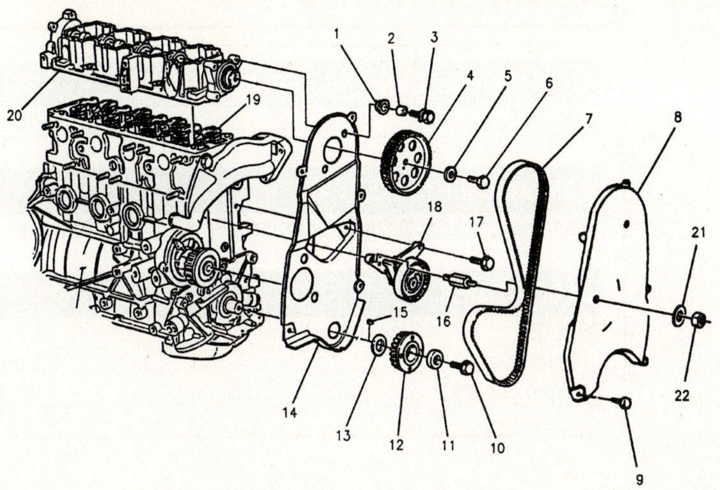

Exploded view of the timing covers, gears and belt assembly—General Motors 2.0L (VIN H) engine

holes to properly align, the engine must be at or near room temperature of 68° F (20° C).

17. Tighten the water pump screws to 18 ft. lbs. (25 Nm) while checking that the tensioner holes remain as adjusted in the prior step.
18. Raise and support the vehicle safely, then install the crankshaft pulley. Coat the pulley retainer bolt threads with Loctite® 242 or equivalent threadlock. Install the retainer and tighten to 15 ft. lbs. (21 Nm).
19. Install the right splash shield and lower the vehicle.
20. Install the timing belt cover and attaching bolts. Tighten to 62 inch lbs. (7 Nm).
21. Pivot or install the timing belt tensioner into position and tighten the retaining bolt to 35 ft. lbs. (48 Nm).
22. Install fuel vapor pipes.
23. Install the serpentine belt and, if equipped, the air conditioning drive belt. Install the coolant reservoir.
24. Connect the negative battery cable and properly fill the engine cooling system.

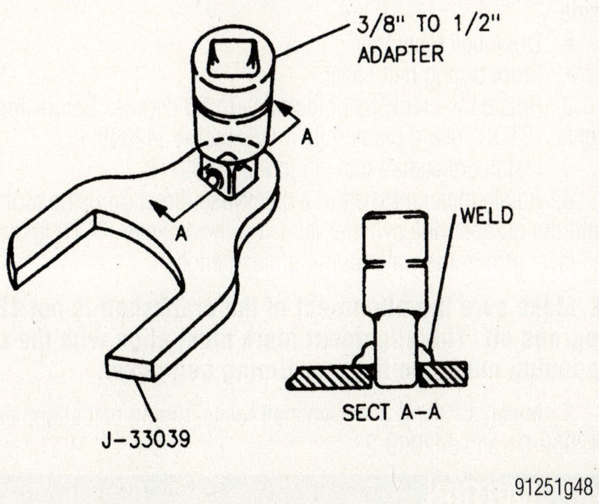

The timing belt tension adjuster may be fabricated or modified to suit the engine—General Motors 2.0L (VIN H) engine

2.0L (VIN K, M) ENGINES

YEAR, MODEL & ENGINE APPLICATION

1987-94 Buick Skyhawk, Oldsmobile Firenza, Pontiac Grand Am, J2000 & Sunbird .... 2.0L VIN K, M

REPLACEMENT INTERVALS

General Motors does not provide a replacement interval. Standard industry practice is to replace timing belts every 60,000 miles.

LABOR TIMES

1987-94	
Replacement	2.1 hrs
Severe Service	2.3 hrs
With air conditioning add	0.4 hrs

8807GMg18

⁂ CAUTION

These engines are considered freewheeling. If the timing belt fails, valve-to-piston damage is unlikely, however a precautionary compression check should be performed to verify no engine damage.

1. Disconnect the negative battery cable and drain the engine cooling system to a level below the water pump.
2. Remove the drive belt(s) from the engine.
3. Remove the fuel vapor pipes as necessary.
4. Loosen the drive belt tensioner bolt and the tensioner will swing downward. If necessary, remove the bolt and tensioner from the engine.
5. Remove the cover attaching bolts, then remove the timing belt cover from the engine.
6. Turn the engine to align the timing marks on the gears with the marks on the timing belt rear cover.
7. Remove the crankshaft pulley.
8. Remove the lower timing belt cover.
9. Loosen the water pump bolts and release timing belt tension with Tension Adjusting tool J-33039 or equivalent. If equipped, remove the air conditioning drive belt.

10. Raise and support the vehicle safely, then remove the right splash shield.
11. Remove the pulley retaining bolt, then remove the pulley from the end of the crankshaft.
12. Lower the vehicle and remove the timing belt. If necessary, remove the retainers and the belt tensioner from the engine.

To install:

13. If removed, install the timing belt tensioner and tighten the retainers to 18 ft. lbs. (25 Nm).
14. If necessary, turn the crankshaft and/or the camshaft gears clockwise to align the timing marks on the gears with the timing marks on the rear cover.
15. Install the timing belt, making sure the portion between the camshaft gear and crankcase gear is in tension.
16. Using tool J-33039 or equivalent, turn the water pump eccentric clockwise until the tensioner contacts the high torque stop. Tighten the water pump screws slightly.
17. Turn the engine clockwise 720 degrees using the crankshaft gear bolt in order to fully seat the belt into the gear teeth.
18. Turn the water pump eccentric counterclockwise until the hole in the tensioner arm is aligned with the hole in the base. In order for the holes to properly align, the engine must be at or near room temperature of 68° F (20° C).
19. Tighten the water pump screws to 18 ft. lbs. (25 Nm) while checking that the tensioner holes remain as adjusted in the prior step.
20. Rotate the crankshaft 2 revolutions clockwise to seat the belt.
21. Raise and support the vehicle safely, then install the crankshaft pulley. Coat the pulley retainer bolt threads with Loctite® 242 or equivalent threadlock. Install the retainer and tighten to 20 ft. lbs. (27 Nm).
22. Install the right splash shield and lower the vehicle.
23. Install the timing belt cover and attaching bolts. Tighten to 89 inch lbs. (10 Nm).
24. Pivot or install the timing belt tensioner into position and tighten the retaining bolt to 40 ft. lbs. (54 Nm).
25. Install the drive belt(s).
26. Install fuel vapor pipes.
27. Install the drive belt.
28. Connect the negative battery cable and properly fill the engine cooling system.

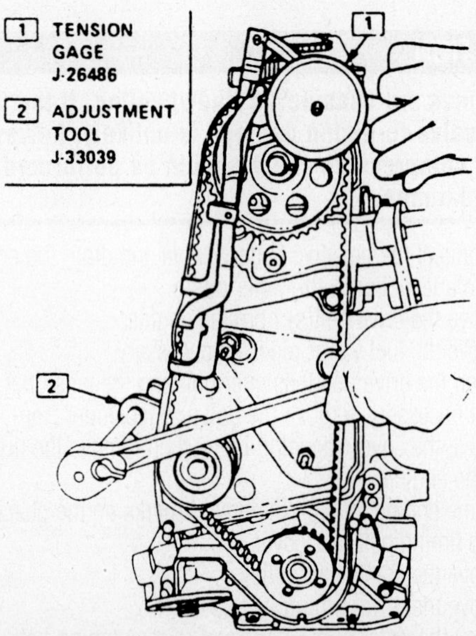

93051g14

Tensioning the timing belt assembly—General Motors 2.0L (VIN K and M) engine

3.0L (VIN B) ENGINES

YEAR, MODEL & ENGINE APPLICATION

2002-05 Saturn VUE .... 3.0L VIN B

REPLACEMENT INTERVALS

Saturn recommends timing belt replacement every 100,000 miles.

LABOR TIMES

2002-05

Replacement .... 3.5 hrs

Severe Service .... 4.0 hrs

8807GMg19

CAUTION

These engines are an interference type. If the timing belt has failed, valve-to-piston damage may have occurred. A compression check should always be performed to check for engine damage.

■ **The steps in this procedure are critical in preventing catastrophic engine damage, adhering to this sequence is imperative. There are special tools needed to perform this procedure. It is a good idea to read this procedure several times before attempting to perform this job. This is an interference engine.**

1. Remove or disconnect the following:
   - Negative battery cable
   - Intake air resonator
   - Right front wheel and splash shield
   - Water pump and idler bolts, loosen only
   - Accessory drive belt
   - Right engine mount
   - Water pump pulley
   - Idler pulley
   - Electrical harness connectors at the harness channel
   - Electrical harness channel from the front cover and position aside
   - Drive belt tensioner
   - Front timing belt cover
2. Rotate the crankshaft clockwise to 60 degrees Before Top Dead Center (BTDC) using crank hub Torx®socket J42098.
3. Install crankshaft locking tool J42069-10.
4. Rotate the crankshaft in a clockwise direction using tool J42098 until the number one cylinder is at Top dead center (TDC) and tighten the lever arm to the water pump pulley flange.

■ **Make sure the alignment of the crankshaft is not 180 degrees off. The alignment mark must align with the corresponding marks on the rear timing belt cover.**

5. Install 1–2 and 3–4 camshaft locks, timing belt alignment kit J42069-01 and J42069-2.

CAUTION

Do not rotate the crankshaft unless crankshaft is at 60 degrees BTDC or the valve could hit the crankshaft.

6. Loosen the timing belt tensioner and idler pulleys.
7. Remove the timing belt.

To install:

8. Position the crank at 60 degrees BTDC.
9. Install the lower idler pulley and tighten the bolt to 30 ft. lbs. (40 Nm).
10. Install the upper idle pulley and hand tighten the bolt.
11. Install the timing belt tensioner and hand tighten the nut.
12. With the camshaft locks installed use the green test belt from the alignment kit J42069 to make sure the distance between the camshaft sprockets.
13. Start at the number 1 and 2 sprockets, route the belt over the sprockets while aligning the belt marks to marks on the sprocket and rear cover.
14. Install the belt around the tensioner and upper idler pulley.
15. Route the belt over the 3 and 4 sprockets while aligning the belt marks to marks on the sprocket and rear cover.
16. Install the belt around the crankshaft hub sprockets while aligning marks on the belt to marks on the sprocket.
17. Install wedge tool J42069-30 to lock the belt onto the crankshaft hub sprocket.
18. Carefully rotate the crankshaft counterclockwise to get belt slack on the lower idler pulley, then route the belt around the lower idler pulley and remove the wedge tool.
19. Install the crankshaft locking tool and tighten the lever arm to the water pump pulley flange.
20. Install camshaft checking gauge to cams 3 and 4 and make sure the alignment marks are within 2mm of each other. Install 3 and 4 camshaft lock if properly aligned, then install checking gauge on cams 1 and 2.

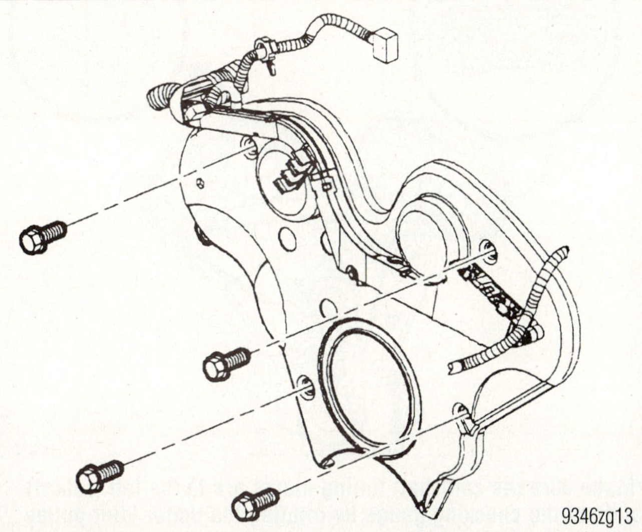

Timing belt cover bolt locations–General Motors 3.0L (VIN B) engine

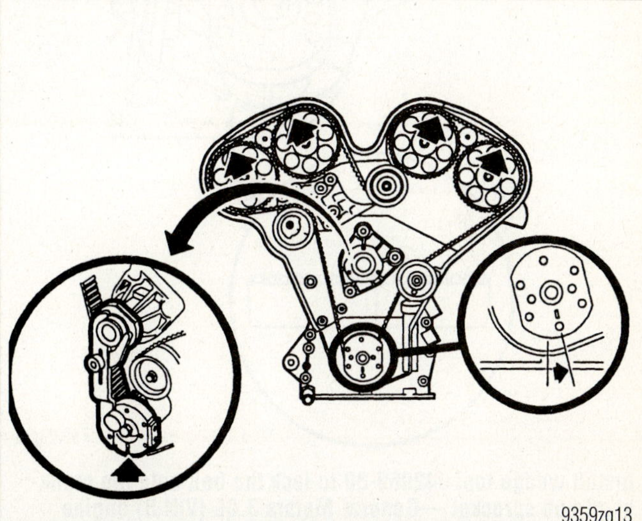

Rotate the crankshaft clockwise to 60 degrees Before Top Dead Center (BTDC) using crank hub Torx®socket J42098—General Motors 3.0L (VIN B) engine

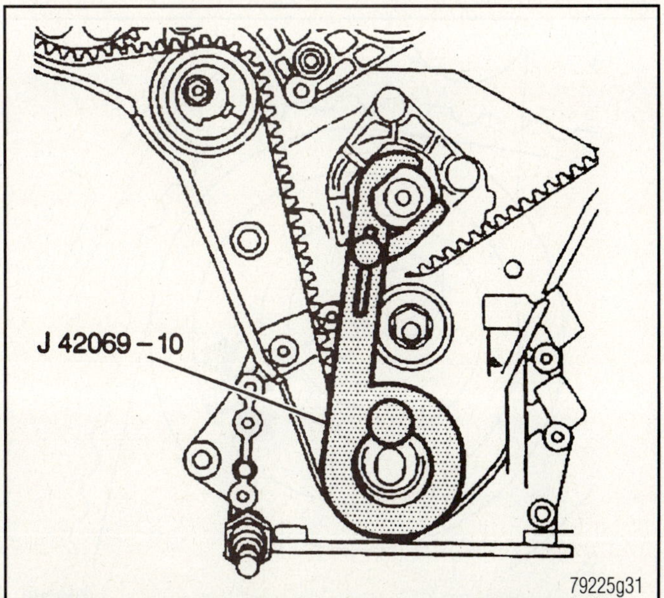

Rotate the crankshaft clockwise using tool J-42098 until No. 1 cylinder is at TDC and install locking tool J-42069-10 and tighten the lever arm to the water pump pulley flange—General Motors 3.0L (VIN B) engine

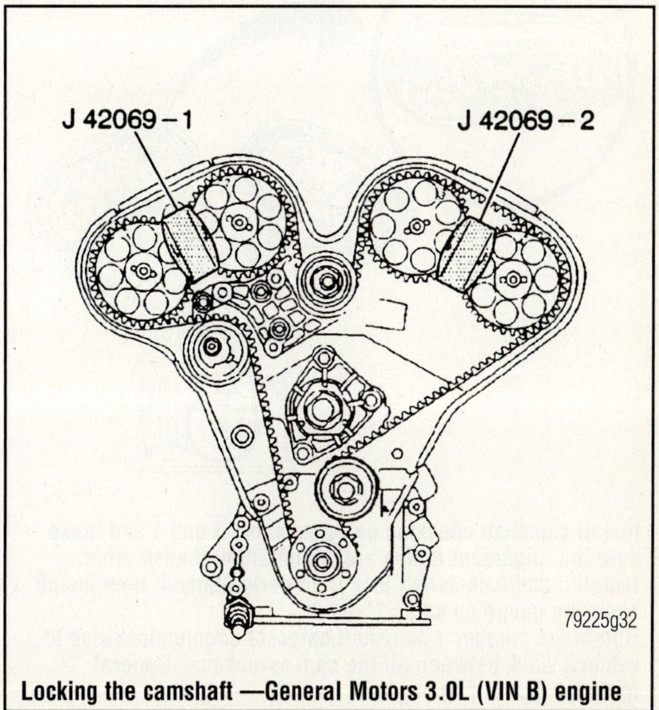

Locking the camshaft —General Motors 3.0L (VIN B) engine

Rotate the number 1 camshaft sprocket counterclockwise to remove slack between all the cam sprockets.

21. Make sure the camshaft timing marks are to the left (retard) side of the checking gauge as follows:

   a. Rotate the upper idler pulley counterclockwise until marks are properly aligned within 2mm of each other and install cam lock 1 and 2.

22. Tighten the upper idler pulley bolt to 30 ft. lbs. (40 Nm).

23. Adjust the belt tensioner so that the center alignment mark is about 1/8 inch (3mm) above the mark on the spring loader.

24. Remove 1 and 2, 3 and 4 camshaft locks, belt alignment kit J42069-1 and J42069-2 and camshaft locking tool J42069-10.

■ **If TDC is passed, do not rotate the crankshaft counterclockwise. this will not allow proper slack to be taken up between the belt tensioner and crank sprocket. To correct, rotate the crank an additional 2 turns.**

25. Rotate the crankshaft 1 3/4 turns clockwise and install crankshaft locking tool J42069-10 and stop at TDC. Tighten the lever arm to the water pump pulley flange.

26. Install gauge J42069-20 and verify the marks are within 2mm.

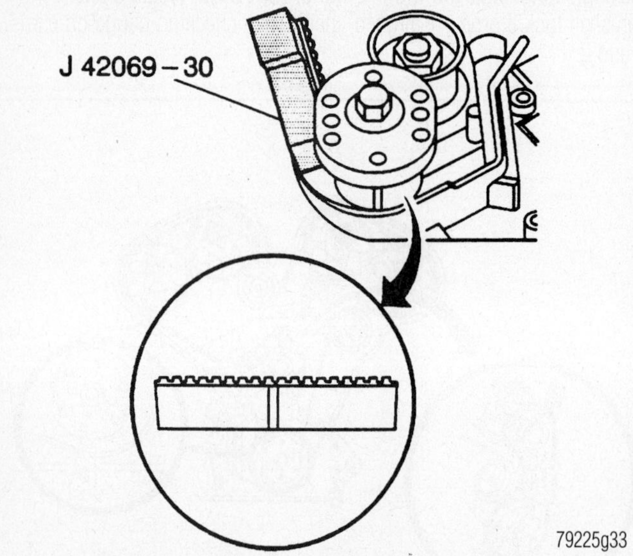

79225g33

Install wedge tool J42069-30 to lock the belt onto the crankshaft hub sprocket —General Motors 3.0L (VIN B) engine

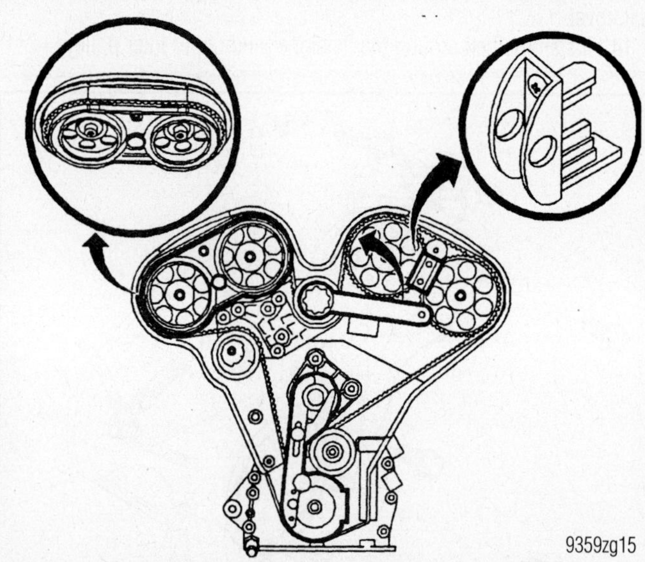

9359zg15

Make sure the camshaft timing marks are to the left (retard) side of the checking gauge by rotating the upper idler pulley counterclockwise until marks are properly aligned within 2mm of each other and install cam lock 1 and 2—General Motors 3.0L (VIN B) engine

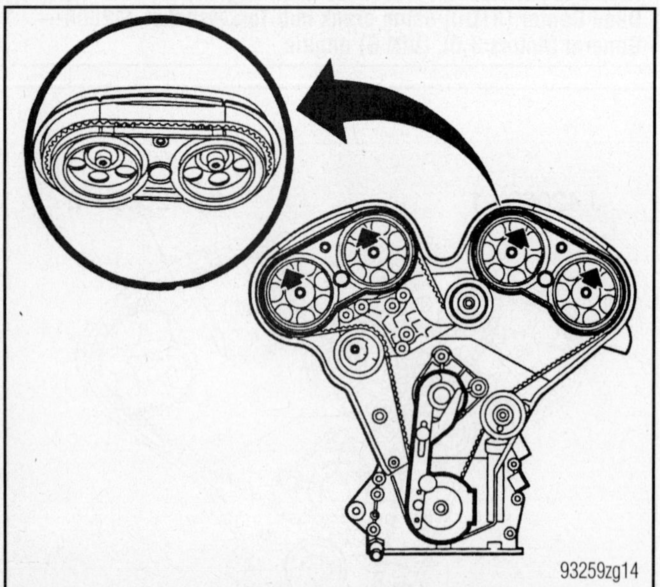

93259zg14

Install camshaft checking gauge to cams 3 and 4 and make sure the alignment marks are within 2mm of each other. Install 3 and 4 camshaft lock if properly aligned, then install checking gauge on cams 1 and 2.
Rotate the number 1 camshaft sprocket counterclockwise to remove slack between all the cam sprockets—General Motors 3.0L (VIN B) engines

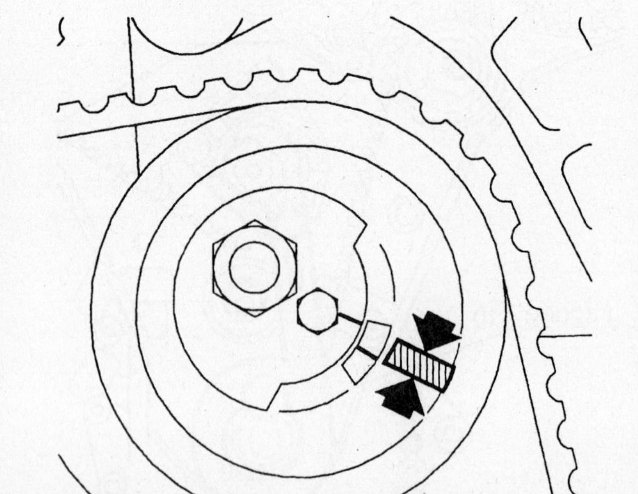

9359zg16

Adjust the belt tensioner so that the center alignment mark is about 1/8 inch (3mm) above the mark on the spring loader—General Motors 3.0L (VIN B) engine

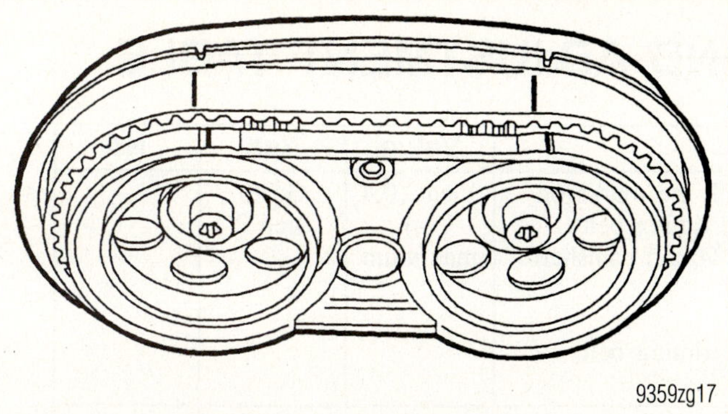

9359zg17

Install gauge J42069-20 and verify the marks are within 2mm—General Motors 3.0L (VIN B) engine

27. Adjust the belt tensioner so that the center alignment mark is about 1/8 inch (3mm) above the mark on the spring loaded idler.
28. Tighten the belt tensioner nut to 15 ft. lbs. (20 Nm).
29. Remove the crankshaft locking tool.
30. Install or connect the following:
    - Front timing belt cover. Torque the bolts to 71 inch lbs. (8 Nm).
    - Drive belt tensioner. Torque the bolts to 30 ft. lbs. (40 Nm).
    - Idler pulley and snug the bolt.
    - Electrical harness.
    - Water pump pulley
    - Accessory drive belt
31. Tighten the water pump pulley bolts to 71 inch lbs. (8 Nm) and the idler puller bolt to 15 ft. lbs. (20 Nm).
    - Engine mount bracket and tighten the bolt to 41 ft. lbs. (55 Nm)
    - Engine mount-to-frame rail and engine mount bracket, tighten the fasteners to 37 ft. lbs. (50 Nm)
32. Remove the engine support.
    - Splash shield
    - Wheel
    - Intake air resonator
    - Negative battery cable

3.0L (VIN R) ENGINES

YEAR, MODEL & ENGINE APPLICATION

1998-01 Cadillac Catera,
2000-04 Saturn L300, LW300 .... 3.0L VIN R

REPLACEMENT INTERVALS

General Motors recommends timing belt replacement every 100,000 miles.

LABOR TIMES

Catera	
Replacement	2.3 hrs
Severe Service	2.6 hrs
Saturn	
Replacement	3.6 hrs
Severe Service	4.1 hrs

8807GMg20

**** CAUTION**

These engines are an interference type. If the timing belt has failed, valve-to-piston damage may have occurred. A compression check should always be performed to check for engine damage

■ **The steps in this procedure are critical in preventing catastrophic engine damage, adhering to this sequence is imperative. There are special tools needed to perform this procedure. It is a good idea to read this procedure several times before attempting to perform this job. This is an interference engine.**

Always turn the crankshaft in the direction of rotation (clockwise), never against engine rotation. Never remove the timing belt without first setting the camshaft gears and crankshaft drive gear to TDC and locking them in place with tool J42069 or equivalent.

The "Timing Belt Installation and Adjustment Table" provides an overview of the steps needed to properly install and adjust the timing belt. Use this table as a reference, not as a substitution, for the steps in this procedure.

1. Disconnect the negative battery cable.
2. Remove the resonance chamber.
3. Remove the front timing belt cover.
4. Remove the harmonic balancer from the crankshaft.
5. Rotate the crankshaft clockwise to 60 degrees BTDC.
6. Install J42069-10, or equivalent, to the crankshaft drive gear with knurled bolt.
7. Turn the engine clockwise, with J42098 or equivalent, until the lever of J42069-10 or equivalent, firmly contacts the water pump pulley flange. Secure the lever to the water pump.

■ **Be sure the engine is not 180 degrees off. The camshaft marks must align with the rear timing cover.**

8. Lock the camshaft gears, using J42069-1, or equivalent, and J42069-2 or equivalent. It may be necessary to loosen the relevant idler pulley to lock the gears, then tighten the idler pulley bolt to 30 ft. lbs. (40 Nm).

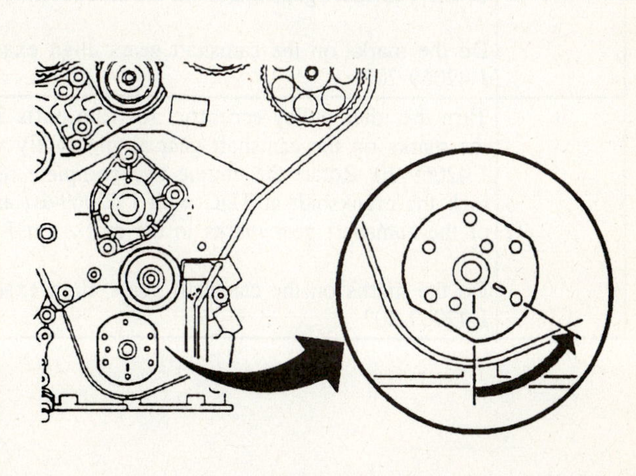

79225g30

Crankshaft alignment to 60 degrees BTDC—General Motors 3.0L (VIN R) engine

TIMING BELT INSTALLATION AND ADJUSTMENT TABLE

Step	Action	Value	Yes	No
1	Install the timing belt and align marks on the belt with the marks on the camshaft gears and the crankshaft gear. Check the timing belt deflection between the idler pulley for camshafts 3 & 4 and camshaft number 4. Is the timing belt installed, the marks aligned and the timing belt deflection adjusted?	1 cm (0.4 in) maximum	Go to Step 2	—
2	Set the initial timing belt tension at the timing belt tensioner. Is the initial timing belt tension set?	—	Go to Step 3	—
3	Rotate the engine two complete revolutions and secure the crankshaft at Top Dead Center (TDC) with the J 42069-10. Has the engine been rotated and the crankshaft secured to TDC?	—	Go to Step 4	—
4	Starting with camshafts 3 and 4, check the alignment of the marks on the camshaft gears with the marks on the J 42069-20 checking gauge. Do the marks on the camshaft gears align exactly with the marks on J 42069-20?	—	Go to Step 5	Go to Step 6
5	Check the alignment of the marks on camshafts gears 1 and 2 with the marks on the J 42069-20 checking gauge. Do the marks on the camshaft gears align exactly with the marks on J 42069-20?	—	Go to Step 14	Go to Step 10
6	Do the camshaft gear marks line up to the left (BTDC) of the marks on the J 42069-20 checking gauge?	—	Go to Step 8	Go to Step 7
7	Do the camshaft gear marks line up to the right (ATDC) of the marks on the J 42069-20 checking gauge?	—	Go to Step 9	—
8	Turn the idler pulley eccentric, for camshafts 3 and 4, counterclockwise until the marks on the camshaft gear align exactly with the marks on J 42069-20. Rotate the engine two complete revolutions, lock the crankshaft at TDC with J 42069-10 and recheck the alignment of the camshaft gear marks to the marks on J 42069-20. Do the marks on the camshaft gears align exactly with the marks on J 42069-20?	—	Go to Step 5	Go to Step 6
9	Turn the idler pulley eccentric, for camshafts 3 and 4, clockwise until the marks on the camshaft gear align exactly with the marks on J 42069-20. Rotate the engine two complete revolutions, lock the crankshaft at TDC with J 42069-10 and recheck the alignment of the camshaft gear marks to the marks on J 42069-20. Do the marks on the camshaft gears align exactly with the marks on J 42069-20?	—	Go to Step 5	Go to Step 6

79225g35

Timing belt installation and adjustment table—General Motors 3.0L (VIN R) engine

Step	Action	Value	Yes	No
10	Do the camshaft gear marks line up to the left (BTDC) of the marks on the J 42069-20 checking gauge?	—	Go to Step 12	Go to Step 11
11	Do the camshaft gear marks line up to the right (ATDC) of the marks on the J 42069-20 checking gauge?	—	Go to Step 13	—
12	Turn the idler pulley eccentric, for camshafts 1 and 2, counterclockwise until the marks on the camshaft gear align exactly with the marks on J 42069-20. Rotate the engine two complete revolutions, lock the crankshaft at TDC with J 42069-10 and recheck the alignment of the camshaft gear marks to the marks on J 42069-20. Do the marks on the camshaft gears align exactly with the marks on J 42069-20?	—	Go to Step 14	Go to Step 10
13	Turn the idler pulley eccentric, for camshafts 1 and 2, clockwise until the marks on the camshaft gear align exactly with the marks on J 42069-20. Rotate the engine two complete revolutions, lock the crankshaft at TDC with J 42069-10 and recheck the alignment of the camshaft gear marks to the marks on J 42069-20. Do the marks on the camshaft gears align exactly with the marks on J 42069-20?	—	Go to Step 14	Go to Step 10
14	Set the final timing belt tension at the timing belt tensioner. Is the final timing belt tension set?	—	Go to Step 15	—
15	Again, rotate the engine two complete revolutions and lock the crankshaft at TDC. Do a final inspection of the camshaft gear marks' relationship to the J 42069-20 marks. The marks must align exactly. Do the marks on the camshaft gears align exactly with the marks on the J 42069-20?	—	Go to Step 16	Go to Step 2
16	Remove all checking tools and ensure all idler pulleys and the tensioner locking nut are tightened to specifications. Continue with re-assembly of the engine.	—	—	—

Timing belt installation and adjustment table (continued)—GM 3.0L (VIN R) engine

79225g36

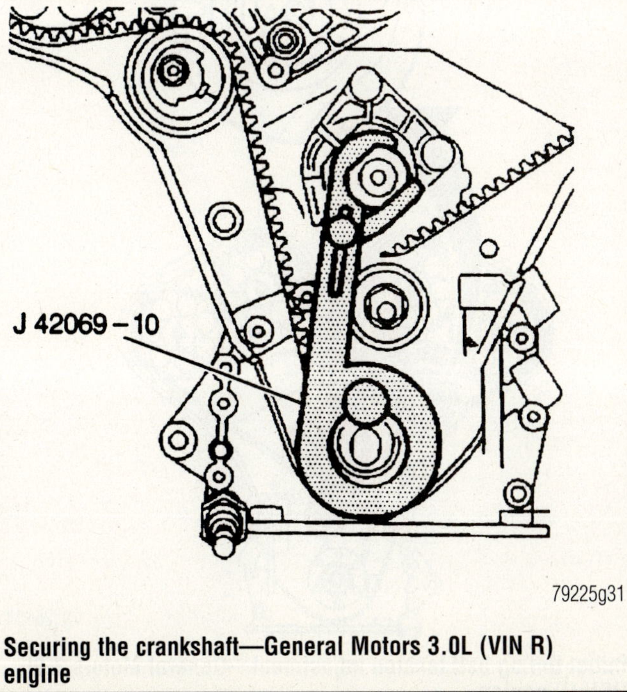

Securing the crankshaft—General Motors 3.0L (VIN R) engine

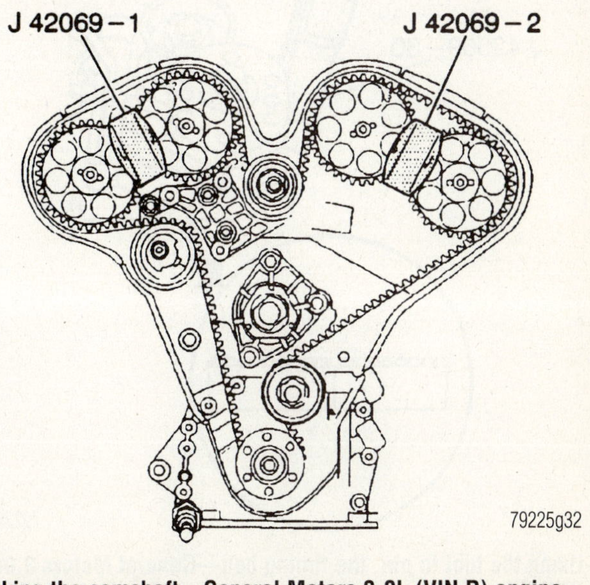

Locking the camshaft—General Motors 3.0L (VIN R) engine

9. Loosen the timing belt tensioner and remove the belt.

✻✻ WARNING

With the belt removed, do not rotate the camshaft or crankshaft, or remove the locking tools, because the pistons may contact the valves and cause internal engine damage.

To install:

10. Remove J42069-10 or equivalent.
11. Raise and safely support the vehicle securely on jackstands.
12. Install the timing belt, starting at the crankshaft gear and aligning the double dash (TDC) mark on the belt with the oil pump and belt drive gear.
13. Using J42069-30 or equivalent, secure the belt to prevent the splines from jumping.
14. Lower the vehicle.
15. Route the belt between the idler pulley for camshafts 3 and 4, then between the gears for 3 and 4.
16. If the dash marks on the belt do not align with the camshaft and rear timing cover marks, loosen the idler pulley, or move the cam gears slightly, with the camshaft gears still locked in place, until the timing belt can be properly installed. Temporarily tighten the idler pulley locking bolt; the locking bolt will be tightened to specification after final adjustments are made.

■ **The timing belt deflection must be no more than 0.4 in. (10mm) between camshaft gear 4 and the idler pulley. To adjust the deflection, rotate the timing belt idler pulley for camshafts 3 and 4 counterclockwise with tool J42069-40 or equivalent.**

17. Route the belt between the idler pulley for camshafts 1 and 2, then between the gears for 1and 2.
18. If the dash marks on the belt do not align with the camshaft and rear timing cover marks, loosen the idler pulley, or move the cam gears slightly, with the camshaft gears still locked in place, until the timing belt can be installed. Temporarily tighten the idler pulley; the locking bolt will be tightened to specification after final adjustments are made.
19. Apply tension to the belt to keep it from slipping off the gears by turning the timing belt idler pulley for camshafts 1and 2 counterclockwise with J42069-40 or equivalent. Temporarily tighten the idler pulley; the locking bolt will be tightened to specification after final adjustments are made.
20. Complete the routing of the timing belt through the belt tensioner.
21. Apply initial tension by turning the timing belt tensioner counterclockwise, with a 5mm Allen wrench, until the marks are set as shown in the initial timing belt adjustment illustration.
22. Tighten the tensioner locking nut to 15 ft. lbs. (20 Nm).
23. Ensure that the alignment marks are at their specific reference points.
24. Remove J42069-30, J42069-1 and J42069-2.
25. Rotate the engine 2 revolutions clockwise, stopping at 60 degrees BTDC.
26. Install J42069-10 to the crankshaft gear with the knurled bolt.
27. Turn the engine clockwise, with J42098 or equivalent, until the lever of J42069-10 or equivalent firmly contacts the water pump pulley flange. Secure the lever to the water pump.

■ **The alignment marks on the timing belt will no longer match the marks on the camshaft gears after 1 or more revolutions. The marks on the camshaft gears must align with the notches on the rear timing cover, and the crankshaft drive gear and the oil pump housing should match up to their mark.**

28. If timing belt adjustment is necessary, first adjust camshafts 3 and 4.
29. Check the alignment of camshafts 3 and 4, 1 and 2 with gauge J42069-20.
30. If alignment is OK, then set final belt tension as follows:
    a. Loosen the timing belt tensioner locknut.
    b. Adjust the timing belt tensioner by turning the eccentric cam, with a 5mm hex wrench, until the marks are set.

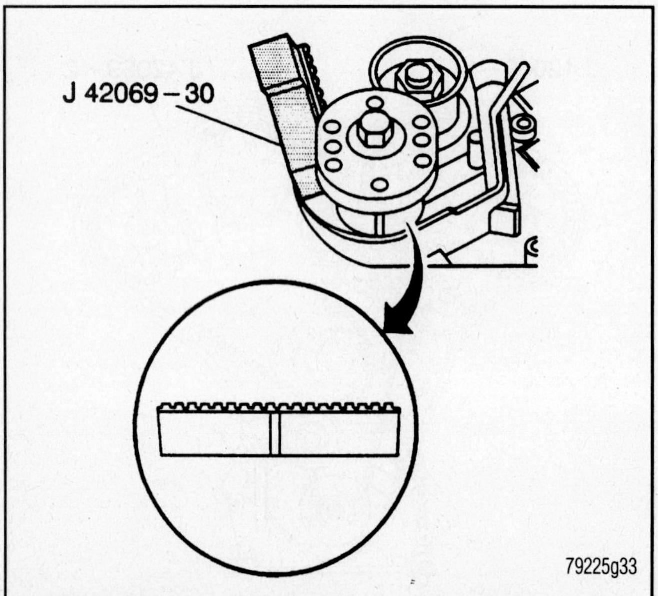

Using the tool to pin. the timing belt—General Motors 3.0L (VIN R) engine

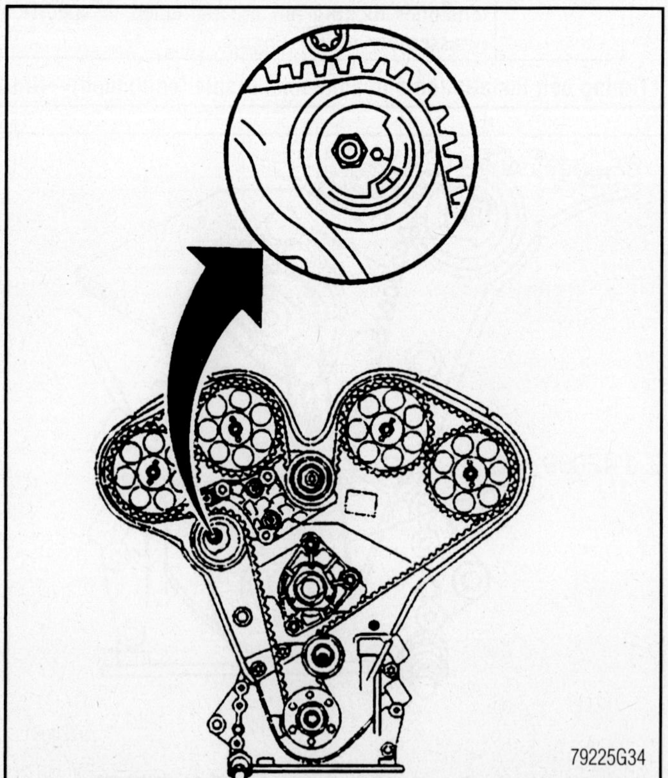

Initial timing belt tension adjustment—General Motors 3.0L (VIN R) engine

31. Tighten the tensioner locknut to 15 ft. lbs. (20 Nm).
32. Remove J42069-10 and J42069-20.
33. Rotate the engine clockwise 2 revolutions, stopping at 60 degrees BTDC.
34. If further adjustment is needed, then repeat the applicable steps.
35. Tighten the idler pulley bolts for the camshafts to 30 ft. lbs. (40 Nm).
36. Be sure all tools are removed from the engine.
37. Install the harmonic balancer and tighten the bolts to 15 ft. lbs. (20 Nm).
38. Install the timing belt covers.
39. Install the resonance chamber.
40. Connect the negative battery cable, and reprogram applicable accessories.

3.2L (VIN N) ENGINES

YEAR, MODEL & ENGINE APPLICATION

2003-04 Cadillac CTS	3.2L VIN N

REPLACEMENT INTERVALS

General Motors recommends timing belt replacement every 100,000 miles.

LABOR TIMES

2003-04

Replacement	2.8 hrs

8807GMg21

✳✳ CAUTION

These engines are an interference type. If the timing belt has failed, valve-to-piston damage may have occurred. A compression check should always be performed to check for engine damage

1. Before servicing the vehicle, refer to the precautions in the beginning of this section.
2. Remove or disconnect the following:
- Negative battery cable
- Intake air resonator
- Intake plenum

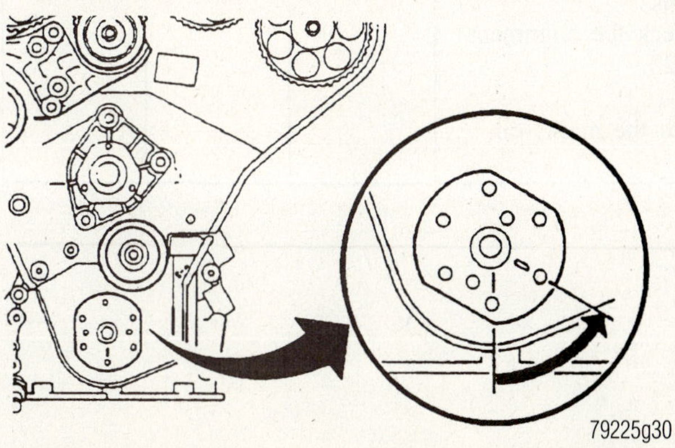

79225g30

Crankshaft alignment to 60 degrees BTDC— General Motors 3.2L (VIN N) engine

- Splash shield
- Secondary air injection (AIR) pipe
- Accessory drive belt
- Water pump and power steering pump pulleys
- Drive belt tensioner
- Front timing belt cover
- Crankshaft balancer

3. Rotate the crankshaft clockwise to 60 degrees before top dead center (BTDC).
4. Loosen the timing belt tensioner and idler pulleys.
5. Remove the timing belt.

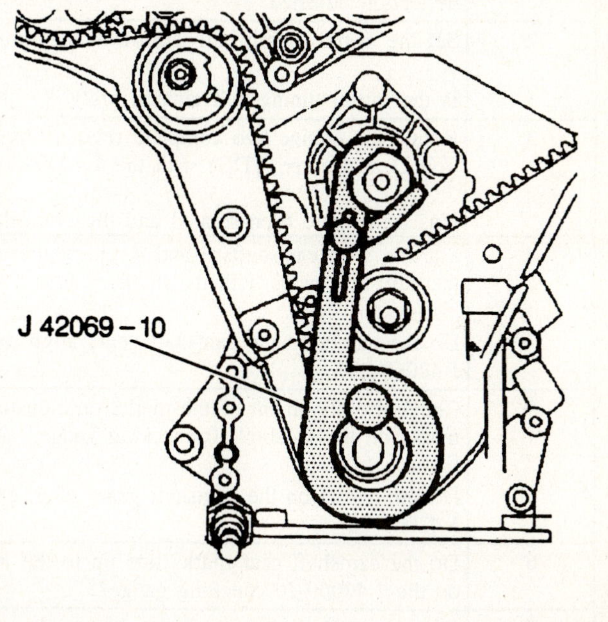

79225g31

Securing the crankshaft— General Motors 3.2L (VIN N) engine

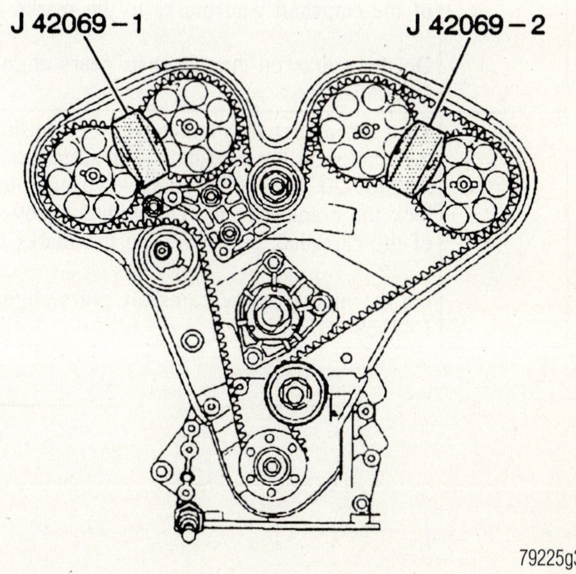

79225g32

Locking the camshaft— General Motors 3.2L (VIN N) engine

TIMING BELT INSTALLATION AND ADJUSTMENT TABLE

Step	Action	Value	Yes	No
1	Install the timing belt and align marks on the belt with the marks on the camshaft gears and the crankshaft gear. Check the timing belt deflection between the idler pulley for camshafts 3 & 4 and camshaft number 4. Is the timing belt installed, the marks aligned and the timing belt deflection adjusted?	1 cm (0.4 in) maximum	Go to Step 2	—
2	Set the initial timing belt tension at the timing belt tensioner. Is the initial timing belt tension set?	—	Go to Step 3	—
3	Rotate the engine two complete revolutions and secure the crankshaft at Top Dead Center (TDC) with the J 42069-10. Has the engine been rotated and the crankshaft secured to TDC?	—	Go to Step 4	—
4	Starting with camshafts 3 and 4, check the alignment of the marks on the camshaft gears with the marks on the J 42069-20 checking gauge. Do the marks on the camshaft gears align exactly with the marks on J 42069-20?	—	Go to Step 5	Go to Step 6
5	Check the alignment of the marks on camshafts gears 1 and 2 with the marks on the J 42069-20 checking gauge. Do the marks on the camshaft gears align exactly with the marks on J 42069-20?	—	Go to Step 14	Go to Step 10
6	Do the camshaft gear marks line up to the left (BTDC) of the marks on the J 42069-20 checking gauge?	—	Go to Step 8	Go to Step 7
7	Do the camshaft gear marks line up to the right (ATDC) of the marks on the J 42069-20 checking gauge?	—	Go to Step 9	—
8	Turn the idler pulley eccentric, for camshafts 3 and 4, counterclockwise until the marks on the camshaft gear align exactly with the marks on J 42069-20. Rotate the engine two complete revolutions, lock the crankshaft at TDC with J 42069-10 and recheck the alignment of the camshaft gear marks to the marks on J 42069-20. Do the marks on the camshaft gears align exactly with the marks on J 42069-20?	—	Go to Step 5	Go to Step 6
9	Turn the idler pulley eccentric, for camshafts 3 and 4, clockwise until the marks on the camshaft gear align exactly with the marks on J 42069-20. Rotate the engine two complete revolutions, lock the crankshaft at TDC with J 42069-10 and recheck the alignment of the camshaft gear marks to the marks on J 42069-20. Do the marks on the camshaft gears align exactly with the marks on J 42069-20?	—	Go to Step 5	Go to Step 6

79225g35

Timing belt installation and adjustment table—General Motors 3.2L (VIN N) engine

Step	Action	Value	Yes	No
10	Do the camshaft gear marks line up to the left (BTDC) of the marks on the J 42069-20 checking gauge?	—	Go to Step 12	Go to Step 11
11	Do the camshaft gear marks line up to the right (ATDC) of the marks on the J 42069-20 checking gauge?	—	Go to Step 13	—
12	Turn the idler pulley eccentric, for camshafts 1 and 2, counterclockwise until the marks on the camshaft gear align exactly with the marks on J 42069-20. Rotate the engine two complete revolutions, lock the crankshaft at TDC with J 42069-10 and recheck the alignment of the camshaft gear marks to the marks on J 42069-20. Do the marks on the camshaft gears align exactly with the marks on J 42069-20?	—	Go to Step 14	Go to Step 10
13	Turn the idler pulley eccentric, for camshafts 1 and 2, clockwise until the marks on the camshaft gear align exactly with the marks on J 42069-20. Rotate the engine two complete revolutions, lock the crankshaft at TDC with J 42069-10 and recheck the alignment of the camshaft gear marks to the marks on J 42069-20. Do the marks on the camshaft gears align exactly with the marks on J 42069-20?	—	Go to Step 14	Go to Step 10
14	Set the final timing belt tension at the timing belt tensioner. Is the final timing belt tension set?	—	Go to Step 15	—
15	Again, rotate the engine two complete revolutions and lock the crankshaft at TDC. Do a final inspection of the camshaft gear marks' relationship to the J 42069-20 marks. The marks must align exactly. Do the marks on the camshaft gears align exactly with the marks on the J 42069-20?	—	Go to Step 16	Go to Step 2
16	Remove all checking tools and ensure all idler pulleys and the tensioner locking nut are tightened to specifications. Continue with re-assembly of the engine.	—	—	—

79225g36

Timing belt installation and adjustment table (continued)— General Motors 3.2L (VIN N) engine

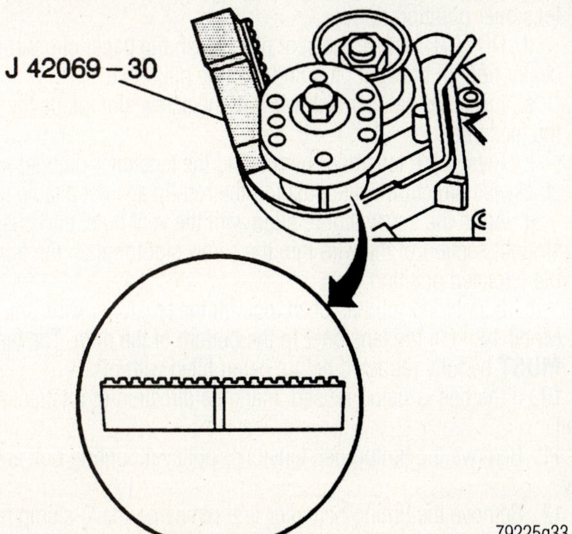

79225g33

Using the tool to pin the timing belt— General Motors 3.2L (VIN N) engine

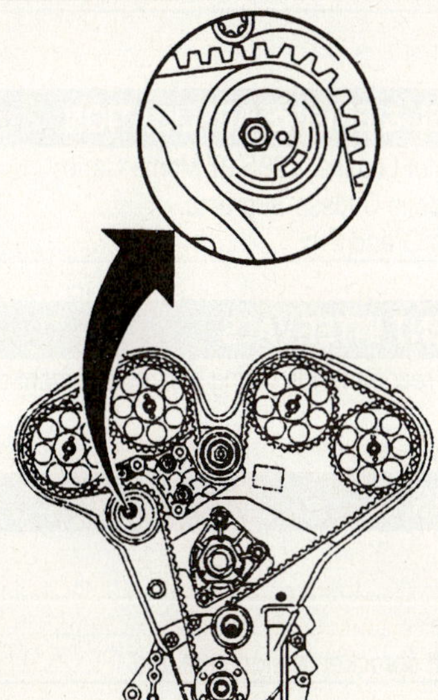

79225g34

Initial timing belt tension adjustment— General Motors 3.2L (VIN N) engine

To install:

6. Start at the crankshaft sprocket and install the timing belt with the double dash (TDC) aligned with the marks on the oil pump and on the belt drive gear.

7. Set the initial timing belt tension as follows:

a. Turn the tensioner nut COUNTERCLOCKWISE to full stop, then, turn the nut back until the reference mark is 1 mm (0.003 in) over the flange.

b. Tighten the timing belt tensioner locking nut until snug, the locking nut will be tightened to specifications after all final adjustments are made.

8. Rotate the engine in the clockwise direction two revolutions stopping at 60 degrees BTDC.

9. Inspect the alignment of the reference marks on the camshaft gears with the notches on the rear timing belt cover, as well as, the mark on the crankshaft sprocket and oil pump housing.

10. If timing belt adjustment IS NOT required, set the final timing belt tension:

a. Loosen the timing belt tensioner locking nut.

b. Turn the locking nut counterclockwise to full stop, then back until the reference mark is 2-4 mm (0.078-0.157 in) ABOVE the reference mark on the flange.

11. Tighten the timing belt tensioner locking nut 15 ft. lbs. (20 Nm).

12. Tighten the idler pulley bolts to 30 ft. lbs. (40 Nm).

13. Install or connect the following:

- Crankshaft balancer. Torque the bolts to 15 ft. lbs. (20 Nm).
- Front timing belt cover. Torque the bolts to 71 inch lbs. (8 Nm).
- Drive belt tensioner. Torque the bolts to 30 ft. lbs. (40 Nm).
- Water pump pulley. Torque the bolts to 71 inch lbs. (8 Nm).
- Power steering pump pulley. Torque the bolts to 15 ft. lbs. (20 Nm).
- Accessory drive belt
- AIR injection pipe
- Splash shield
- Intake plenum
- Intake air resonator

3.4L (VIN X) ENGINES

YEAR, MODEL & ENGINE APPLICATION

1991-97 Chevrolet Lumina, 1995-97 Monte Carlo,
1991-96 Oldsmobile Cutlass Supreme,
1991-96 Pontiac Grand Prix .... 3.4L VIN X

REPLACEMENT INTERVALS

General Motors recommends timing belt replacement every 60,000 miles.

LABOR TIMES

1991-97	
Replacement	5.8 hrs
Severe Service	6.8 hrs
With camshaft sprocket realignment add	0.3 hrs

8807GMg22

✳✳ CAUTION

These engines are an interference type. If the timing belt has failed, valve-to-piston damage may have occurred. A compression check should always be performed to check for engine damage.

■ **The 3.4L (VIN X) engine uses a timing chain and camshaft timing belts.**

1. Disconnect the negative battery cable.

2. Disconnect and remove the power steering pump from the pump mounting bracket.

3. Remove the left, right and center timing belt covers.

4. Rotate the engine clockwise to align the timing marks, Top Dead Center (TDC) on the No. 1 exhaust stroke, on the camshaft sprockets and intermediate shaft.

5. Loosely clamp the 2 camshaft sprockets on each side of the engine together using clamping pliers or the equivalent. Secure the belt to the right-side cam sprocket with a C-clamp and a wide pad on the belt.

■ **When clamping the sprockets no deflection should be noticed. If any deflection is noticed, loosen the clamping devices. DO NOT mar the camshaft sprockets with the clamping device.**

6. Remove the tensioner side plate retaining bolts from the tensioner and remove the side plate from the actuator and base.

7. Rotate the actuator assembly around the arm pivot and out of the base. Removal of the tensioner from the base allows it to extend to its maximum travel.

8. Set the actuator aside on a table in a vertical position to allow the oil to drain into the boot end. The tensioner should be allowed to sit for 5 minutes prior to refilling with oil.

9. Reset the timing belt actuator as follows:

a. Straighten out a paper clip or a piece of stiff wire 0.032 in. (0.75mm) diameter to a minimum straight length of 1.85 in. (47mm). Form a double loop in the remaining end.

b. Remove the rubber end plug from the rear of the tensioner assembly. This will aid in allowing the oil in the tensioner to escape.

c. Hold the tensioner in your hand with the rubber boot end of the tensioner pointing down.

d. DO NOT remove the vent plug. Push the paper clip through the center hole in the vent plug and into the pilot hole.

e. Insert a small screwdriver into the screw slot inside the end of the tensioner.

f. Retract the tensioner by rotating the tensioner plunger in a clockwise direction while pushing the rod tip against a table top.

g. Align the screw slot to align with the vent hole, and push the straight section of the wire into the screw slot to retain the plunger in the retracted position.

h. If tensioner oil has been lost, fill the tensioner with SAE 5W30 Mobil 1®. Fill the tensioner to the bottom of the plug. The tensioner **MUST** be fully retracted before being filled with oil.

10. If the belt is being reused, mark the direction of rotation on the belt.

11. Remove the timing belt tensioner pulley mounting bolt and pulley.

12. Remove the timing belt after first removing the C-clamp retaining the belt to the right-side camshaft sprocket.

13. Remove the Torx® head bolts securing the idler pulleys, if the idlers need to be replaced.

14. Remove the intermediate shaft sprocket using the following procedure:
a. Use a suitable tool to hold the engine from turning.
b. Remove the intermediate shaft sprocket mounting bolt and washer.
c. Using J-38616 or equivalent sprocket puller, remove the sprocket from the intermediate shaft.
15. If the camshaft sprockets need to be removed, proceed as follows:
a. Remove the camshaft carrier cover(s).
b. Remove the camshaft sprocket clamping pliers.
c. Rotate the camshaft being serviced so the flats on the camshaft are face up.
d. Install a Camshaft Hold-down tool J-38613 or equivalent, and tighten to 22 ft. lbs. (30 Nm).
e. Remove the camshaft sprocket mounting bolt and washer while holding the camshaft from turning with J-38613 and J-38614 or equivalent
f. Using J-38616, or an equivalent sprocket puller remove the sprocket from the camshaft.
g. Repeat for each camshaft sprocket as necessary.
16. Drain the cooling system.
17. Disconnect the lower radiator hose from water pump inlet pipe.
18. If equipped with a manual transaxle, disconnect the front AIR hose at the AIR pipe.
19. Disconnect the heater hose at the front cover.
20. Remove the heater pipe bracket mounting bolts at the frame.
21. Raise and safely support the vehicle on jack stands. Remove the right front tire and wheel assembly and the right inner fender splashshield.
22. Remove the crankshaft pulley mounting bolts and remove the pulley from the damper.
23. Remove the crankshaft damper as follows:
a. While holding the crankshaft from turning using a suitable tool, remove the damper mounting bolt and washer.
b. Install tool J-24420-B or equivalent, and remove the damper from the crankshaft.
24. Place an oil catch pan under the oil filter and remove the oil filter.
25. Remove the air conditioning compressor mounting bracket bolts.
26. Remove the lower front cover bolts.
27. On automatic transaxle vehicles, remove the halfshaft following the recommended procedure.
28. Remove the rear alternator bracket.
29. Disconnect and remove the starter following the recommended procedure.
30. Lower the vehicle.
31. Remove the intermediate shaft drive belt sprocket retaining bolt and remove the intermediate shaft drive belt sprocket using J38616 or equivalent puller.
32. Remove the upper alternator mounting bolts.
33. Remove the forward light relay center screws and position the relay center aside.
34. Disconnect the oil cooler hose from the front cover.
35. Remove the water pump pulley.
36. Remove the upper front cover bolts and remove the front cover.
37. Mark the intermediate shaft sprocket, chain link, crankshaft sprocket and cylinder block for assembly reference. The marks should be made with paint so they won't be lost when the components are removed.
38. Retract the timing chain tensioner shoe as follows:
a. Insert J-33875 or an equivalent, on both sides of the tensioner.
b. Pull on the through-pin in the tensioner arm to retract the spring located in the tensioner arm.
c. While compressing the spring, use a suitable tool, a cotter pin or nail, and insert the pin in the hole in the tensioner assembly to

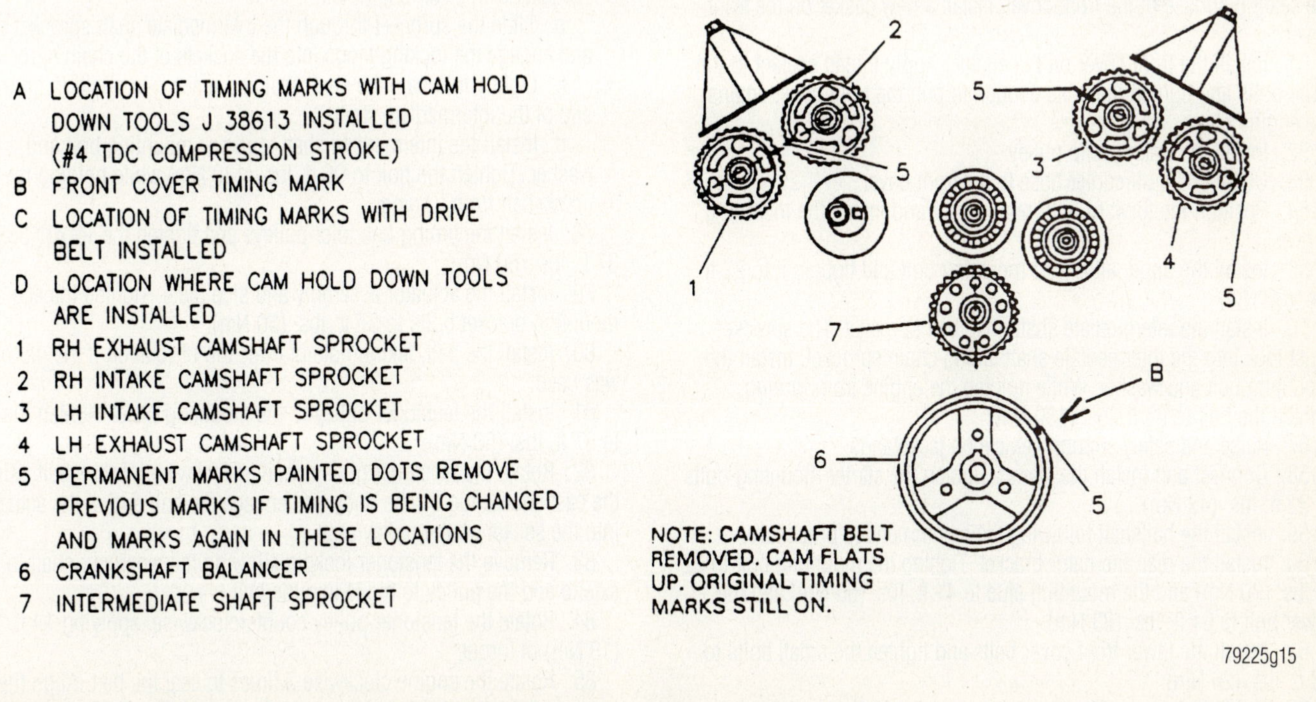

Timing marks with hold-down tool in place—General Motors 3.4L (VIN X) engine

hold the tensioner compressed. The tool used must be strong enough to hold the tensioner compressed.

■ **The timing chain, crankshaft sprocket and intermediate shaft sprocket will be removed at the same time. If, when removing the assembly, the intermediate shaft sprocket does not easily come off the intermediate shaft, rotate the crankshaft back and forth to loosen the intermediate shaft sprocket.**

39. Install a suitable puller, J-38611 and J-8433 or their equivalents.
40. Tighten the bolt on the puller and slowly pull the crankshaft sprocket off the crankshaft. Be sure the intermediate shaft sprocket is moving along with the crankshaft sprocket.
41. Remove the timing chain and sprockets.
42. Remove the tensioner mounting bolts and remove the tensioner assembly.

To install:

43. Install the tensioner assembly and tensioner assembly mounting bolts finger-tight first. Tighten the bolt in the slotted hole first to 18 ft. lbs. (25 Nm), then tighten the remainder of the bolts to 18 ft. lbs. (25 Nm).
44. Check to ensure that the crankshaft key is fully seated in the crankshaft cutout and the tensioner assembly is fully retracted.
45. Assemble the timing chain, intermediate shaft sprocket and crankshaft sprocket on a work bench. The timing marks made should be in alignment. The large chamfer and counterbore of the crankshaft sprocket are installed facing toward the engine and the intermediate shaft spline sockets are installed facing away from the engine.
46. Install the sprocket and chain assembly onto the engine. As the sprockets are installed, parallel alignment must be maintained.
47. The crankshaft sprocket will have to be pressed on the final 0.31 in. (8mm). This can be done using J-38612, or an equivalent puller.
48. Ensure timing was maintained.
49. Remove the retaining pin from tensioner. Clean all gasket surfaces completely.
50. Apply GM Sealer 1052080 or equivalent, to the lower edges of the sealing surface of the front cover. Install a new gasket on the front cover.
51. Install the front cover on the engine. Apply thread sealant to the large bolts and tighten the bolts enough to pull the front cover against the engine block.
52. Install the water pump pulley.
53. Connect the oil cooler hose to the front cover.
54. Position the forward light relay center and install the mounting screws.
55. Install the upper alternator mounting bolt and tighten it to 22 ft. lbs. (30 Nm).
56. Install the intermediate shaft drive belt sprocket. The sprocket must lock into the intermediate shaft timing chain sprocket. Install the mounting bolt and washer. While holding the engine from turning, tighten the bolt to 95 ft. lbs. (130 Nm).
57. Raise and safely support vehicle on jackstands.
58. Connect and install the starter. Tighten the starter mounting bolts to 32 ft. lbs. (43 Nm).
59. Install the halfshaft following the recommended procedure.
60. Install the rear alternator bracket. Tighten the mounting bolt to 22 ft. lbs. (30 Nm) and the mounting stud to 41 ft. lbs. (55 Nm) and the lower bolt to 61 ft. lbs. (83 Nm).
61. Install the lower front cover bolts and tighten the small bolts to 18 ft. lbs. (25 Nm).
62. Install the air conditioning compressor mounting bolts and tighten the mounting bolts to 37 ft. lbs. (50 Nm).
63. Install a new oil filter.
64. Install the crankshaft damper as follows:
    a. Coat the seal contact area on the damper with clean engine oil.
    b. Align the notch inside the damper with the crankshaft key and slide the damper on until the key is started into the notch.
    c. Using J-29113 or an equivalent puller, press the damper into position on the crankshaft.
    d. Install the crankshaft pulley and pulley mounting bolts. Tighten the pulley mounting bolts to 37 ft. lbs. (50 Nm).
    e. Install the crankshaft damper mounting bolt and washer, and tighten to 78 ft. lbs. (105 Nm).
65. Install the right-side inner fender splash-shield.
66. Install the tire and wheel assembly.
67. Lower the vehicle.
68. Tighten the upper front cover small bolts to 18 ft. lbs. (25 Nm) and the large bolts to 35 ft. lbs. (47 Nm).
69. Connect the heater hose to the front cover.
70. Install the heater pipe bracket mounting bolts.
71. If equipped with a manual transaxle, connect the front AIR hose to the AIR pipe.
72. Connect the lower radiator hose to the water pump.
73. Install the right-side cooling fan. Install the upper radiator support.
74. Position the torque strut mounting bracket and install the mounting bolts and tighten to 52 ft. lbs. (70 Nm).
75. Install the front engine lift hook and tighten the mounting bolt to 52 ft. lbs. (70 Nm).
76. To install the camshaft sprockets, proceed as follows:
    a. Wipe the camshaft noses with clean engine oil.
    b. Install the camshaft sprocket onto the nose of the camshaft.
    c. Install the lockring and shim ring.
    d. Install, but DO NOT tighten, the camshaft sprocket mounting bolts at this time.
77. Install the intermediate shaft sprocket as follows:
    a. Lubricate the seal contact area on the intermediate shaft sprocket with clean engine oil.
    b. Slide the sprocket through the intermediate shaft sprocket seal and engage the locking tangs into the sockets of the chain sprocket.
    c. Lightly lubricate the shaft seal and place it in position on the end of the intermediate shaft.
    d. Install the intermediate shaft sprocket mounting bolt and washer. Tighten the bolt to 96 ft. lbs. (130 Nm) while holding the crankshaft from turning.
78. Install the timing belt idler pulleys and tighten the Torx® bolts to 37 ft. lbs. (50 Nm).
79. Install the actuator assembly and side plate. Tighten the actuator mounting bracket bolts to 37 ft. lbs. (50 Nm).
80. Install the belt, taking note of direction of rotation if the old belt was used.
81. Install the tensioner pulley to the mounting base. Tighten the bolt to 37 ft. lbs. (50 Nm).
82. Rotate the tensioner pulley counterclockwise into the belt using the cast square lug on the body and engage the ball end of the actuator into the socket on the pulley arm.
83. Remove the tensioner lockpin allowing the tensioner shaft to extend and the pulley to move into the belt.
84. Rotate the tensioner pulley counterclockwise, applying 14 ft. lbs. (18 Nm) of torque.
85. Rotate the engine clockwise 3 times to seat the belt. Align the crankshaft reference marks during the final rotation to TDC. Do not allow the crankshaft to spring back or reverse its direction of rotation.

■ **The timing flats on the camshafts should be 180 degrees apart from the left-side to the right-side. Both camshafts on the same side should be the same.**

86. To perform the camshaft timing procedure, proceed as follows:
   a. Rotate the camshaft flats up on the right-side camshafts and install a Camshaft Hold-down tool J-38613 or equivalent, and tighten to 22 ft. lbs. (30 Nm).
   b. Seat the lockring on the right exhaust and intake camshaft sprockets by threading in the mounting bolt and washer.
   c. Hold the sprocket from turning using tool J-38614 or equivalent.

■ **Running torque of the bolts before seating should be 55 ft. lbs. (75 Nm).**

   d. If less torque is required, replace the shim ring and lockring.
   e. If more torque is required, replace the shim ring and lockring and inspect the bolts for burring.
   f. Seating of the lockring is accomplished when the edge is flush with the sprocket hub.
   g. With the lockring seated tighten the bolt to final torque of 81 ft. lbs. (110 Nm).
   h. Remove J-38613 or equivalent.
   i. Rotate the engine clockwise 1 full revolution or any number of odd revolutions. **DO NOT** rotate the engine backward.
   j. Be sure the timing mark on the damper lines up with the mark on the front cover.
   k. Repeat Substeps **a** through **i** for the left-side.
87. Install the camshaft carrier covers.
88. Install timing the belt left, right and center covers and retaining bolts.
89. Connect the negative battery cable.
90. Start the engine and verify proper operation and engine performance.

3.5L (VIN 4) ENGINES

YEAR, MODEL & ENGINE APPLICATION

2004-05 Saturn VUE	3.5L VIN 4

REPLACEMENT INTERVALS

General Motors had not provided a timing belt replacement interval.

LABOR TIMES

Labor times are not available.

8807GMg23

⁂ CAUTION

These engines are an interference type. If the timing belt has failed, valve-to-piston damage may have occurred. A compression check should always be performed to check for engine damage.

■ **The Powertrain Control Module (PCM) has to perform the crankshaft position learn process whenever the timing belt is replaced.**

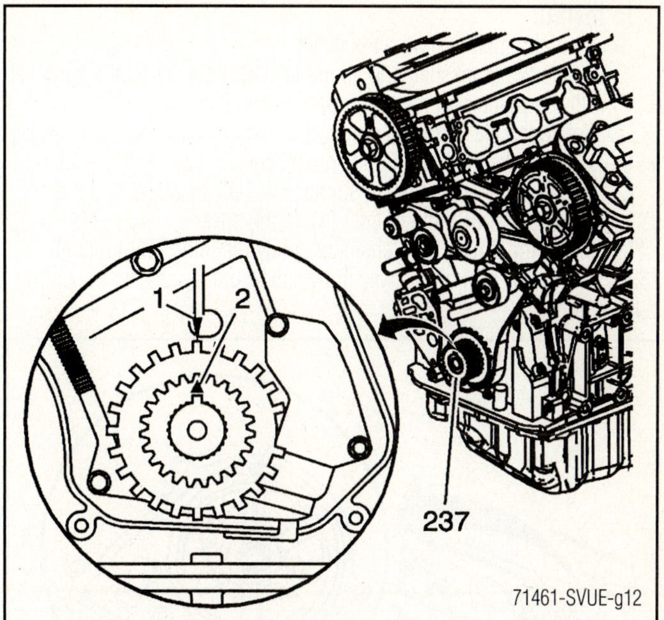

Rotate the crankshaft and align the TDC mark (2) on the sprocket (237) with the pointer on the oil pump housing (1)—General Motors 3.5L (VIN 4) engine

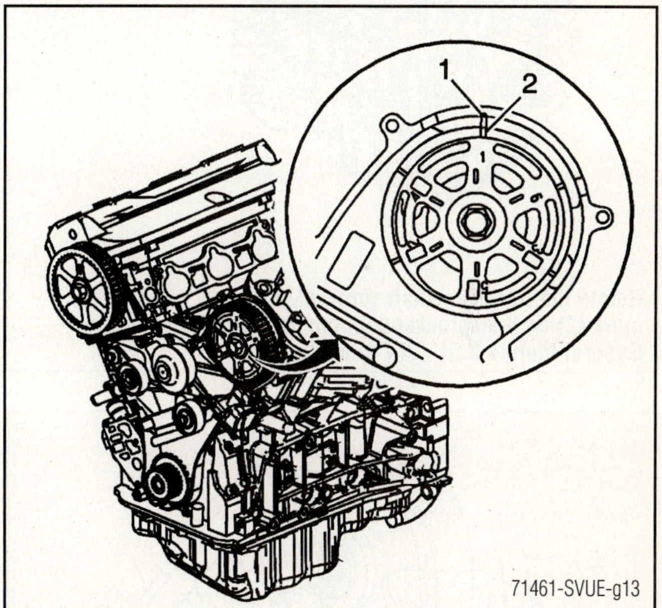

Rotate the left camshaft sprocket to TDC by aligning the mark (2) on the sprocket with the pointer (1) on the cover—General Motors 3.5L (VIN 4) engine

1. Disconnect the negative battery cable.
2. Remove the accessory drive belt.
3. Remove the right side splash shield.
4. Remove the crankshaft balancer using balancer holding tool EN 46337.
5. Remove the timing belt cover bolts and the cover.
6. Use timing belt tensioner pullet retaining bolt tool EN 36331 to retain the tensioner pulley.
7. Loosen the idler pulley about 5 or 6 turns.
8. Remove the timing belt from the pulleys.

To install:

9. Clean the belt pulleys and covers.
10. Rotate the crankshaft and align the TDC mark (2) on the sprocket (237) with the pointer on the oil pump housing. See the illustration.
11. Rotate the left camshaft sprocket to TDC by aligning the mark (2) on the sprocket with the pointer (1) on the cover.
12. Rotate the right camshaft sprocket to TDC by aligning the mark (2) on the sprocket with the pointer (1) on the cover.
13. Install the timing belt tensioner pulley retaining bolt tool EN 36331. Screw the tool all the way in by hand until it contacts the timing belt tensioner.
14. Install the timing belt on the sprockets in a clockwise direction starting with the tensioner pulley.
15. Tighten the idler pulley bolt to 33 ft. lbs. (44 Nm).
16. Remove the tools.
17. Install the guide (236), balancer (228) and bolt (227) and tighten the bolt until snug.
18. Using the balancer holding tool EN 46337, rotate the crankshaft clockwise about 5 or 6 turns to position the timing belt on the sprockets.
19. Align the TDC mark (2) on the sprocket (237) with the pointer on the oil pump housing.

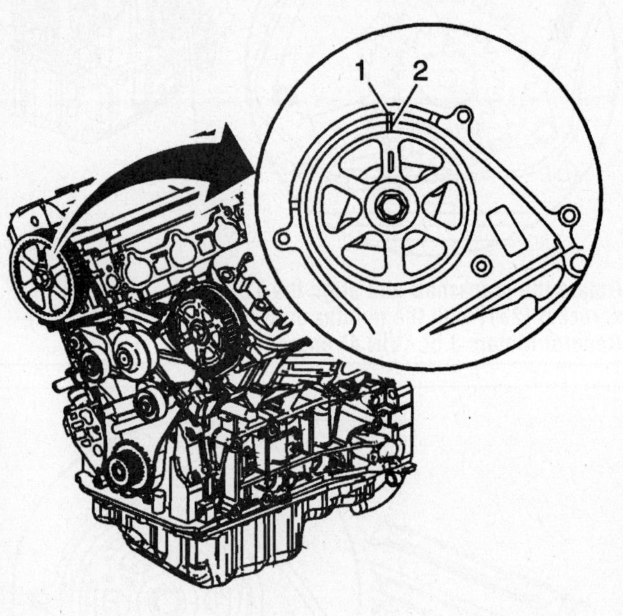

71461-SVUE-g14

Rotate the right camshaft sprocket to TDC by aligning the mark (2) on the sprocket with the pointer (1) on the cover—General Motors 3.5L (VIN 4) engine

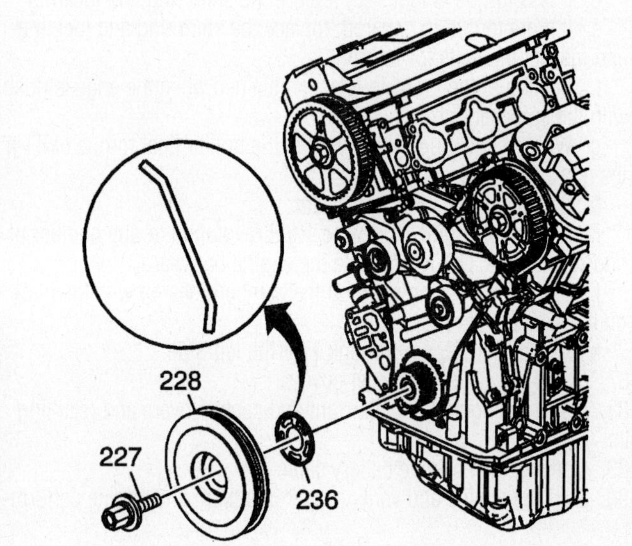

71461-SVUE-g18

Install the guide (236), balancer (228) and bolt (227)—General Motors 3.5L (VIN 4) engine

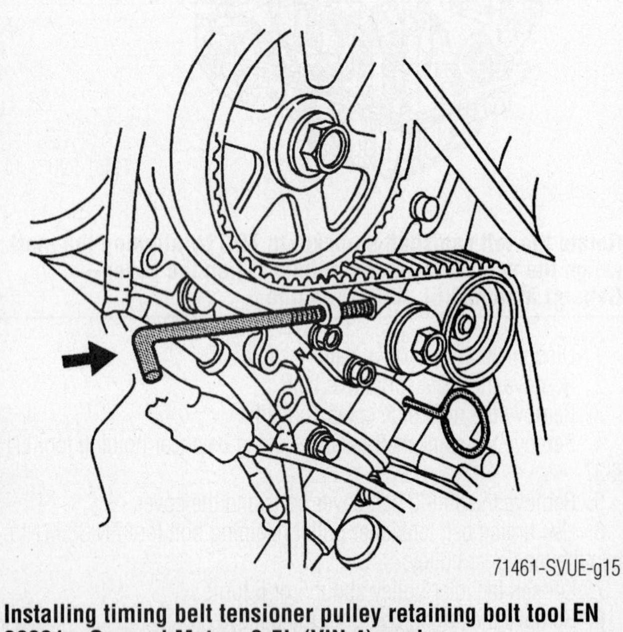

71461-SVUE-g15

Installing timing belt tensioner pulley retaining bolt tool EN 36331—General Motors 3.5L (VIN 4) engine

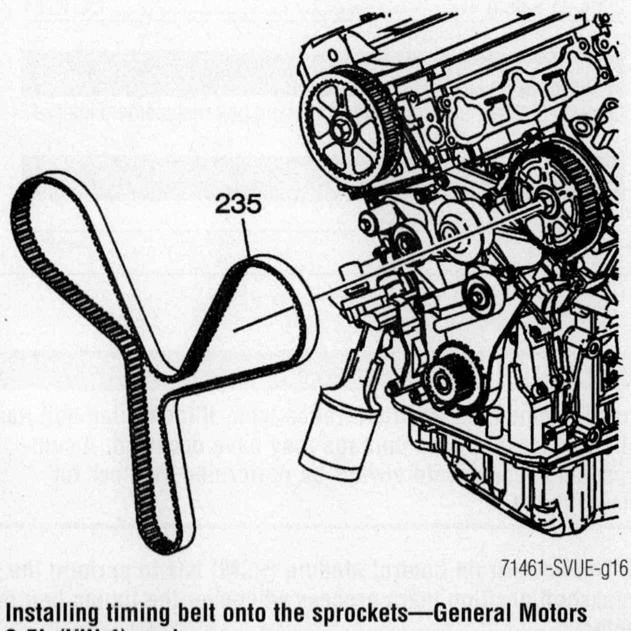

71461-SVUE-g16

Installing timing belt onto the sprockets—General Motors 3.5L (VIN 4) engine

20. Inspect both camshaft sprockets and ensure the timing marks are still aligned.
21. Remove the balancer.
22. Install the timing belt cover and tighten the bolts to 106 inch lbs. (12 Nm).
23. Install the balancer.
24. Install the splash shield.
25. Install the accessory drive belt.
26. Connect the negative battery cable.
27. Perform the crankshaft position learn process.

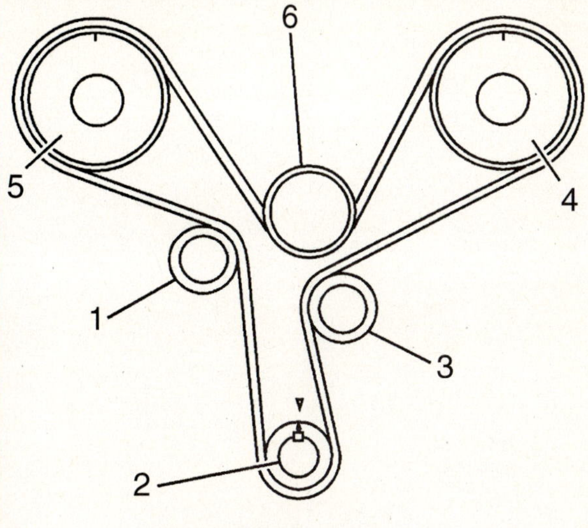

1. Tensioner pulley
2. Crankshaft sprocket
3. Idler pulley
4. Camshaft sprocket - left
5. Camshaft sprocket - right
6. Water pump

71461-SVUE-g17

Install the timing belt on the sprockets in the sequence shown—General Motors 3.5L (VIN 4) engine

Honda

1980–00 1.3L, 1.5L, 1.6L & 2.0L SOHC 4-CYLINDER ENGINES

YEAR, MODEL & ENGINE APPLICATION

1980-87 Civic & Civic Si	1.3L Codes EJ1, EV1, EW 3, EW4
1980-87 Civic 1500	1.5L Codes EM1, EW1
1988-91 Civic	1.5L Code D15B1
1988-91 Civic CRX	1.5L Code D15B2
1992-95 Civic	1.5L Codes D15B7, D15B8
1992-95 Civic VTEC	1.5L Code D15Z1
1993-95 Civic del sol VTEC	1.5L Code D15B7
1988-91 Civic, CRX & Si	1.6L Code D16A6
1992-95 Civic VTEC	1.6L Code D16Z6
1993-95 Civic del sol VTEC	1.6L Code D16Z6
1996-00 Civic	1.6L Codes D16Y5, D16Y7, Da6Y8
1987-91 Prelude, Prelude Si	2.0L Code B20A3

REPLACEMENT INTERVALS

Honda does not provide a replacement interval prior to 1990. Standard industry practice is to replace timing belts every 60,000 miles. Honda recommends timing belt replacement every 60,000 miles on 1990-96 engines, and every 105,000 miles on 1997-00 engines.

LABOR TIMES

Civic	
1980-85	
Replacement	1.7 hrs
Severe Service	2.2 hrs
1986-87	
Replacement	2.0 hrs
Severe Service	2.5 hrs
1988-00	
Replacement	2.7 hrs
Severe Service	3.1 hrs
Prelude	
Replacement	3.1 hrs
Severe Service	3.6 hrs
All Models	
With air conditioning add	1.2 hrs
With power steering add	0.2 hrs
With water pump replacement add	0.6 hrs

8807HONg01

WITHOUT BALANCE SHAFTS

✻✻ CAUTION

These engines are an interference type. If the timing belt has failed, valve-to-piston damage may have occurred. A compression check should always be performed to check for engine damage.

1. Disconnect the negative battery cable.
2. Rotate the crankshaft to set the engine at Top Dead Center (TDC) on the compression stroke for the No. 1 piston.
3. Raise and support the vehicle.
4. Remove the left front wheel and splash shield.
5. Remove the valve cover.
6. Remove the upper timing cover.
7. Remove the accessory drive belts.
8. If equipped, remove the power steering pump and bracket.
9. If equipped with air conditioning, remove the tensioner pulley and bracket. For Preludes, remove the air conditioning compressor and bracket.
10. If equipped with an accessory belt-driven water pump, remove the water pump pulley.
11. Remove the crankshaft pulley.
12. Remove the lower timing cover.
13. If equipped with an engine side-mount, support the engine and remove the mount.
14. Loosen the tensioner adjusting and pivot bolts and remove the timing belt.

To install:

15. Insure that the camshaft and crankshaft timing marks are aligned and install the timing belt.

■ **Some engines do not have an UP indication on the camshaft pulley. In these cases, align the camshaft marks with the woodruff key facing up.**

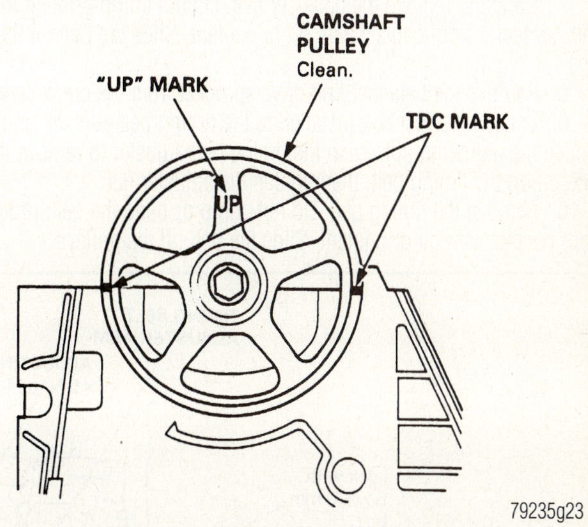

Camshaft timing belt sprocket TDC mark positioning for timing belt installation—Honda 1996–00 4-cylinder SOHC engine

16. Rotate the crankshaft counterclockwise 1/4 turn and tighten the tensioner adjustment bolt.
17. Tighten the tensioner pivot bolt.
18. Install the lower timing cover and the crankshaft pulley.
19. Install the water pump pulley, if removed.
20. If removed, install the air conditioning compressor and bracket.
21. If removed, install the engine side-mount, the power steering pump and bracket, and the air conditioning tensioner pulley and bracket.
22. Install the accessory drive belts and the upper timing cover.
23. Install the valve cover.
24. Install the splash shield and the left front wheel.
25. Lower the vehicle.
26. Connect the negative battery cable.

WITH BALANCE SHAFTS

✲✲ CAUTION

These engines are an interference type. If the timing belt has failed, valve-to-piston damage may have occurred. A compression check should always be performed to check for engine damage.

1. Disconnect the negative battery cable.
2. Turn the crankshaft to align the timing belt matchmarks and set cylinder No. 1 to Top Dead Center (TDC) on the compression stroke. Once in this position, the engine must NOT be turned or disturbed.
3. Remove all necessary components to gain access to the cylinder head and timing belt covers.
4. Remove the cylinder head and timing belt covers.
5. There are 2 belts in this system; the one running to the camshaft pulley is the timing belt. The other, shorter one drives the balance shafts and is referred to as the balancer belt or timing balancer belt. Lock the timing belt adjuster in position by installing one of the lower timing belt cover bolts to the adjuster arm.
6. Loosen the timing belt and balancer shafts tensioner adjuster nut, but do not loosen the nut more than 1 turn. Push the tensioner for the balancer belt away from the belt to relieve the tension. Hold the tensioner and tighten the adjusting nut to hold the tensioner in place.
7. Carefully remove the balancer belt. Do not crimp or bend the belt; protect it from contact with oil or coolant. Slide the belt off the pulleys.
8. Remove the balancer belt drive sprocket from the crankshaft.
9. Loosen the lockbolt installed to the timing belt adjuster and loosen the adjusting nut. Push the timing belt adjuster to remove the tension on the timing belt, then tighten the adjuster nut.
10. Remove the timing belt. Do not crimp or bend the belt; protect it from contact with oil or coolant. Slide the belt off the pulleys.
11. If defective, remove the belt tensioners by performing the following:
   a. Remove the springs from the balancer belt and the timing belt tensioners.
   b. Remove the adjusting nut.
   c. Remove the bolt from the balancer belt adjuster lever, then remove the lever and the tensioner pulley.
   d. Remove the lockbolt from the timing belt tensioner lever, then remove the tensioner pulley and lever from the engine.

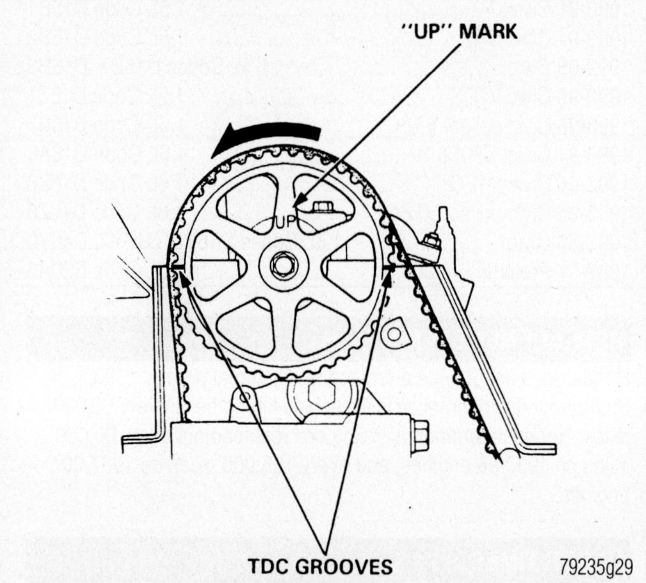

Camshaft timing belt sprocket alignment mark positioning for timing belt—Honda 1980–95 4-cylinder engine

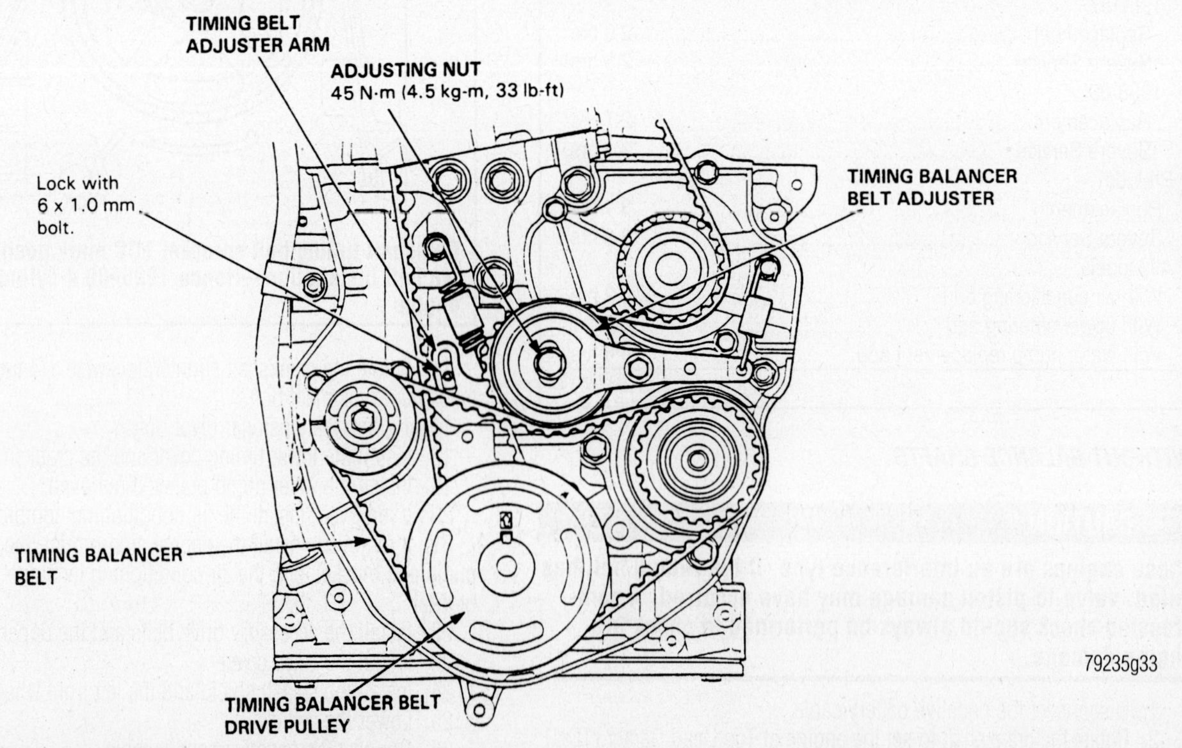

Balancer belt routing—Honda 1980–95 4-cylinder engine

12. This is an excellent time to check or replace the water pump. Even if the timing belt is only being replaced as part of a good maintenance schedule, consider replacing the pump at the same time.

To install:

13. If the water pump is to be replaced, install a new O-ring and make certain it is properly seated. Install the water pump and retaining bolts. Tighten the mounting bolts to 106 inch lbs. (12 Nm).

14. If the tensioners were removed, perform the following to install them:

a. Install the timing belt tensioner lever and tensioner pulley.

■ **The tensioner lever must be properly positioned on its pivot pin located on the oil pump. Be sure that the timing belt lever and tensioner moves freely and does not bind.**

b. Install the lockbolt to the timing belt tensioner, do not tighten the lockbolt at this time.

c. Install the balancer belt pulley and adjuster lever.

d. Install the adjusting nut and the bolt to the balancer belt adjuster lever. Do not tighten the adjuster nut or bolt at this time.

■ **Be sure that the balancer lever and tensioner moves freely and does not bind.**

e. Install the springs to the tensioners.

f. Move the timing belt tensioner its full deflection and tighten the lockbolt.

g. Move the balancer its full deflection and tighten the adjusting nut.

15. The crankshaft timing pointer must be perfectly aligned with the white mark on the flywheel or flex-plate; the camshaft pulley must be aligned so that the word **UP** is at the top of the pulley and the marks on the edge of the pulley are aligned with the surfaces of the head.

16. Install the timing belt over the pulleys and tensioners.

17. Loosen the bolt used to lock the timing belt tensioner. Loosen, then tighten the timing belt adjusting nut.

18. Turn the crankshaft counterclockwise until the cam pulley has moved 3 teeth; this creates tension on the timing belt. Loosen, then tighten the adjusting nut and tighten it to 33 ft. lbs. (45 Nm). Tighten the bolt used to lock the timing belt tensioner.

19. Realign the timing belt marks, then install the balancer belt drive sprocket on the crankshaft.

20. Align the front balancer pulley; the face of the front timing balancer pulley has a mark, which must be aligned with the notch on the oil pump body. This pulley is the one at 10 o'clock to the crank pulley when viewed from the pulley end.

21. Align the rear timing balancer pulley (2 o'clock from the crank pulley) using a 6 x 100mm bolt or rod. Mark the bolt or rod at a point 2.9 in. (74mm) from the end. Remove the bolt from the maintenance hole on the side of the block; insert the bolt or rod into the hole. Align the 2.9 in. (74mm) mark with the face of the hole. This pin will hold the shaft in place during installation.

22. Install the balancer belt. Once the belts are in place, be sure that all the engine alignment marks are still correct. If not, remove the belts, realign the engine and reinstall the belts. Once the belts are properly installed, slowly loosen the adjusting nut, allowing the tensioner to move against the belt. Remove the pin from the maintenance hole and reinstall the bolt and washer.

23. Turn the crankshaft 1 full turn, then tighten the adjuster nut to 33 ft. lbs. (45 Nm). Remove the bolt used to lock the timing belt tensioner.

24. Install the lower cover, ensuring the rubber seals are in place. Install a new seal around the adjusting nut, DO NOT loosen the adjusting nut.

25. Install the key on the crankshaft and install the crankshaft pulley. Apply oil to the bolt threads and tighten it to 181 ft. lbs. (250 Nm).

26. Install the upper timing belt cover and all applicable components. When installing the side engine mount, tighten the bolt and nut attaching the mount to the engine to 40 ft. lbs. (55 Nm) and the through-bolt and nut to 47 ft. lbs. (65 Nm).

27. Connect the negative battery cable.

1.6L (B16A3, B16A2) DOHC ENGINES

YEAR, MODEL & ENGINE APPLICATION

1999 Civic VTEC,	
1994-97 Civic del sol	1.6L Codes B16A2, B16A3

REPLACEMENT INTERVALS

Honda recommends timing belt replacement every 90,000 miles on 1994-96 models, and 105,000 miles on 1997-99 models.

LABOR TIMES

1994-99	
Replacement	2.6 hrs
Severe Service	3.1 hrs

8807HONg02

✻✻ CAUTION

These engines are an interference type. If the timing belt has failed, valve-to-piston damage may have occurred. A compression check should always be performed to check for engine damage.

1. Disconnect the negative battery cable.

2. Rotate the crankshaft to set the engine at Top Dead Center (TDC) on the compression stroke for the No. 1 piston. The white mark on the crankshaft pulley should align with the pointers on the timing cover. Once the engine is in this position, it must not be disturbed.

3. Remove the engine undercover.

4. Remove the accessory drive belts.

5. Remove the power steering pump and position aside, without disconnecting the lines.

6. Remove the power steering pump lower bracket and heat shield.

7. Remove the cruise control actuator.

8. Support the engine with a jack and a block of wood under the oil pan.

9. Remove the side engine mount.

10. Remove the spark plug wires.

11. Remove the valve cover.

12. Remove the crankshaft pulley.

13. Cover the rocker arm and shaft assemblies with a towel or sheet of plastic to keep out dust and foreign objects.

14. Remove the timing belt covers.

15. Loosen the timing belt adjusting bolt 180 degrees (1/2 turn). Push the tensioner pulley down to release the belt tension. After releasing the tension, retighten the tensioner pulley bolt until snug.

■ **Do not remove the tensioner pulley unless it is to be replaced.**

16. Remove the timing belt. Mark the direction of the belt's rotation if it is to be reinstalled.

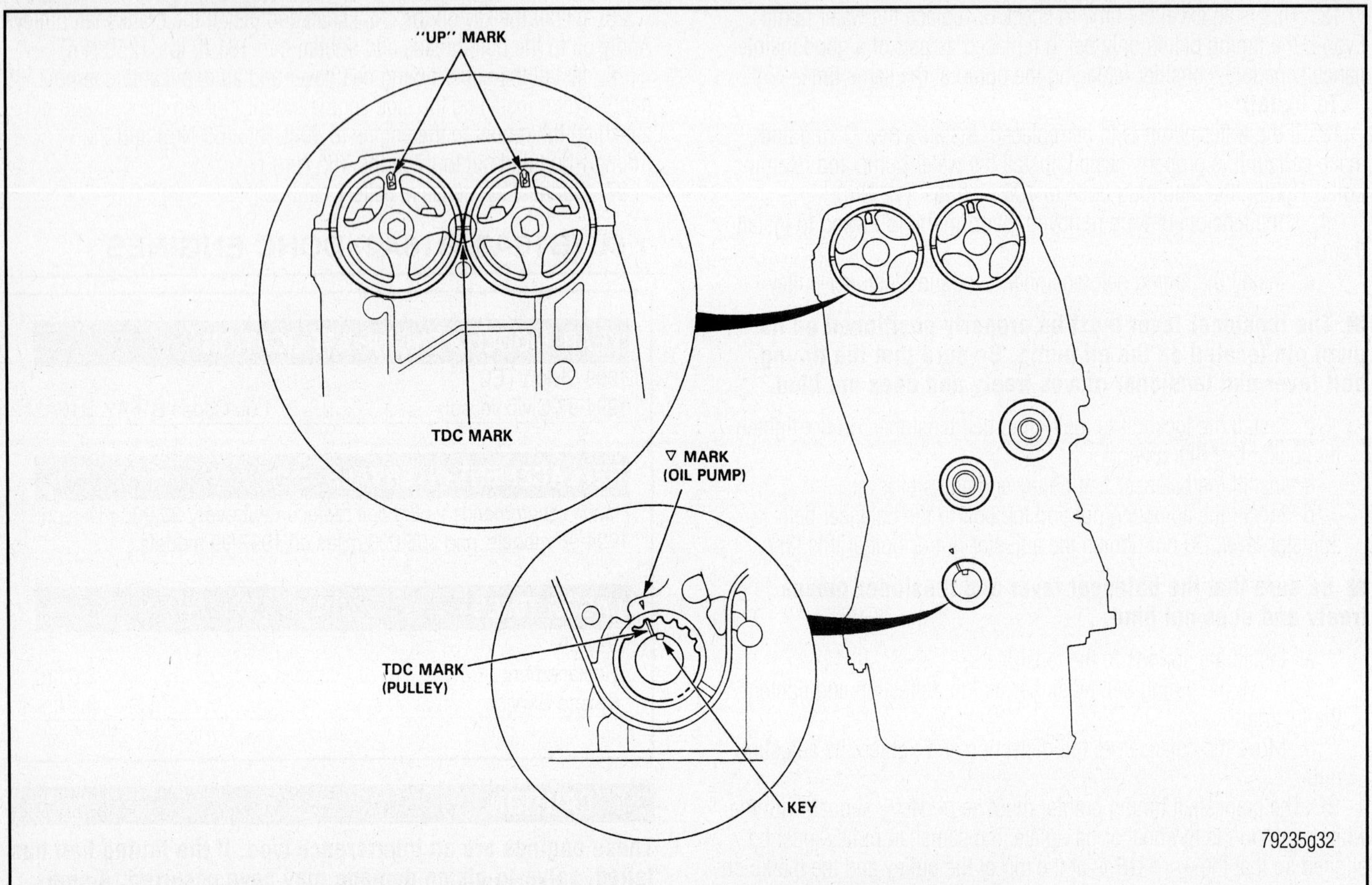

Crankshaft and camshaft timing belt alignment marks—Honda 1.6L (B16A2 & B16A3) DOHC engine

To install:

■ **Inspect the water pump when replacing the timing belt; the manufacturer recommends replacing the water pump at the timing belt's service interval. Replace the timing belt if it shows any signs of wear, or if it is contaminated with oil or coolant.**

17. Verify that the timing is set at TDC on the compression stroke for the No. 1 cylinder as follows: The groove in the crankshaft sprocket must align with the pointer on the oil pump. The TDC marks on the camshaft sprockets must align with the pointer located between the sprockets. The TDC marks will also be in line with the upper surface of the head.
18. The **UP** mark on the camshaft sprocket must point up.
19. Install the timing belt onto the crankshaft sprocket, tensioner pulley, water pump pulley, exhaust camshaft sprocket and intake camshaft sprocket.
20. Loosen the adjusting pulley bolt 180 degrees (1/2 turn). Allow the timing belt to tension and tighten the adjusting bolt to 40 ft. lbs. (55 Nm).
21. Install the lower timing belt cover and the crankshaft pulley. Apply a light coat of fresh oil to the pulley bolt threads, then tighten it to 134 ft. lbs. (181 Nm).
22. Rotate the crankshaft 5–6 turns counterclockwise to position the belt on the sprockets.
23. Adjust the timing belt tension, as follows:
    a. Set the No. 1 piston at TDC on the compression stroke for the No. 1 cylinder.
    b. Loosen the adjusting pulley bolt 180 degrees (1/2 turn).
    c. Rotate the crankshaft counterclockwise so that the camshaft sprocket moves 3 teeth from the TDC/compression position.
    d. Tighten the adjusting bolt to 33 ft. lbs. (45 Nm).
    e. Tighten the crankshaft pulley to 134 ft. lbs. (181 Nm).
24. Verify that the crankshaft and camshaft sprockets will align properly at the TDC/compression position. If the camshaft pulley is not at TDC/compression, remove the timing belt, adjust the sprocket positions and reinstall the belt.
25. Install the upper timing and cylinder head covers, and all other applicable components. When reattaching the side engine mount, tighten the support nuts to 54 ft. lbs. (75 Nm).
26. Connect the negative battery cable.

1.7L (D17A1, D17A2 & D17A6) ENGINES

✻✻ CAUTION

These engines are an interference type. If the timing belt has failed, valve-to-piston damage may have occurred. A compression check should always be performed to check for engine damage.

1. Rotate the crankshaft to set the engine at Top Dead Center (TDC) on the compression stroke for the No. 1 piston. The white mark on the crankshaft pulley should align with the pointers on the timing cover. Once the engine is in this position, it must not be disturbed.
2. Remove the necessary components for access to the cylinder head and timing belt covers. Cover the rocker arm and shaft assemblies with a towel or sheet of plastic to keep out dust and foreign objects.

YEAR, MODEL & ENGINE APPLICATION

2001-04 Honda Civic .... 1.7L Codes D17A1, D17A2, D17A6

REPLACEMENT INTERVALS

Honda recommends timing belt replacement every 110,000 miles under regular conditions, or every 60,000 miles under severe conditions.

LABOR TIMES

2001-04	
Replacement	3.0 hrs
Severe Service	3.4 hrs
With air conditioning add	1.2 hrs
With power steering add	0.2 hrs
With water pump replacement add	0.6 hrs

8807HONgo3

3. Remove the timing belt covers.

4. Loosen the timing belt adjusting bolt 180 degrees (1/2 turn). Push the tensioner pulley down to release the belt tension. After releasing the tension, retighten the tensioner pulley bolt until snug.

■ **Do not remove the tensioner pulley unless it is to be replaced.**

5. Remove the timing belt. Mark the direction of the belt's rotation if it is to be reinstalled.

To install:

■ **Inspect the water pump when replacing the timing belt; the manufacturer recommends replacing the water pump at the timing belt's service interval. Replace the timing belt if it shows any signs of wear, or if it is contaminated with oil or coolant.**

6. Verify that the timing is set at TDC on the compression stroke for the No. 1 cylinder as follows:

a. The groove in the crankshaft sprocket must align with the pointer on the oil pump.

b. The TDC marks on the camshaft sprockets must align with the pointer located between the sprockets. The TDC marks will also be in line with the upper surface of the head.

c. On other engines, the TDC mark on the camshaft sprocket must align with the pointer on the back cover.

d. The **UP** mark on the camshaft sprocket must point up.

7. Install the timing belt onto the crankshaft sprocket, then around the adjusting pulley and water pump sprocket, and finally over the camshaft sprocket.

8. Loosen the adjusting pulley bolt 180 degrees (1/2 turn). Then, tighten the adjusting bolt to 40 ft. lbs. (55 Nm).

9. Install the lower timing belt cover and the crankshaft pulley. Apply a light coat of fresh oil to the pulley bolt threads, then tighten it to 134 ft. lbs. (181 Nm).

10. Rotate the crankshaft 5–6 turns counterclockwise to position the belt on the sprockets.

11. Adjust the timing belt tension, as follows:

a. Set the No. 1 piston at TDC on the compression stroke for the No. 1 cylinder.

b. Loosen the adjusting pulley bolt 180 degrees (1/2 turn).

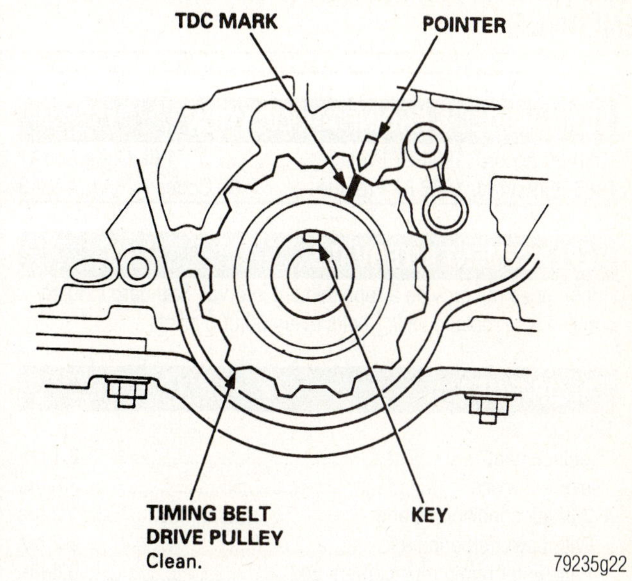

TDC alignment mark locations for the crankshaft sprocket—Honda 1.7L engine

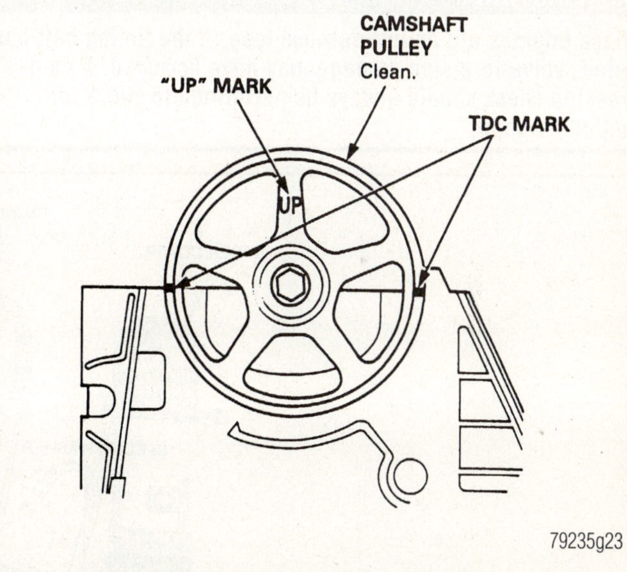

Single camshaft timing belt sprocket TDC mark positioning for timing belt installation—Honda 1.7L engine

c. Rotate the crankshaft counterclockwise so that the camshaft sprocket moves 3 teeth from the TDC/compression position.

d. Tighten the adjusting bolt to 33 ft. lbs. (45 Nm).

e. Tighten the crankshaft pulley to 134 ft. lbs. (181 Nm).

12. Verify that the crankshaft and camshaft sprockets will align properly at the TDC/compression position. If the camshaft pulley is not at TDC/compression, remove the timing belt, adjust the sprocket positions and reinstall the belt.

13. Install the upper timing and cylinder head covers, and all other applicable components. When reattaching the side engine mount, tighten the support nuts to 54 ft. lbs. (75 Nm).

1.8L (A18A1) AND 2.0L (A20A1, A20A3) ENGINES

YEAR, MODEL & ENGINE APPLICATION	
1980-85 Accord, 1984-87 Prelude	1.8L Code A18A1
1986-89 Accord, 1986-87 Prelude	2.0L Codes A20A1, A20A3

REPLACEMENT INTERVALS

Honda does not provide a replacement interval. Standard industry practice is to replace timing belts every 60,000 miles.

LABOR TIMES	
1980-89	
Replacement	3.1 hrs
Severe Service	3.6 hrs
With air conditioning add	1.2 hrs
With power steering add	0.2 hrs
With water pump replacement add	0.6 hrs

8807HONg04

**** CAUTION**

These engines are an interference type. If the timing belt has failed, valve-to-piston damage may have occurred. A compression check should always be performed to check for engine damage.

1. Disconnect the negative battery cable.
2. Support the engine with a jack and a block of wood under the oil pan.
3. Remove the engine support bolts and nuts and the side mount rubber.
4. Remove the side mount bracket.
5. Remove the engine splash shield, then the power steering pump adjuster pulley, belt, and pump.
6. Remove the alternator and belt, then the A/C compressor and belt, if equipped.
7. Remove the valve cover, then the crankshaft pulley.
8. Remove the timing belt upper or middle and lower covers, then loosen the timing belt tensioner bolts. Rotate the timing belt tensioner until tension is released from the timing belt. Temporarily tighten the timing belt tensioner in this position.
9. Ensure the No. 1 piston is at TDC on the compression stroke and the camshaft sprocket is aligned.
10. Remove the timing belt.

To install:

11. Install the new timing belt, then loosen the tensioner bolts. The tensioner will rotate, applying tension to the timing belt. Leave the bolts loose.
12. Adjust the timing belt tension as follows:
    a. Temporarily install the crankshaft pulley and bolt. Tighten the bolt to 83 ft. lbs. on 1986-87 Prelude and 1986-89 Accord or 108 ft. lbs. on 1988-90 Prelude.
    b. Rotate the crankshaft counterclockwise approximately 1/4 turn on 1986-87 Prelude and 1986-89 Accord or 3 teeth on 1988-90 Prelude to create tension on the timing belt, then torque the tensioner bolts to 31 ft. lbs.

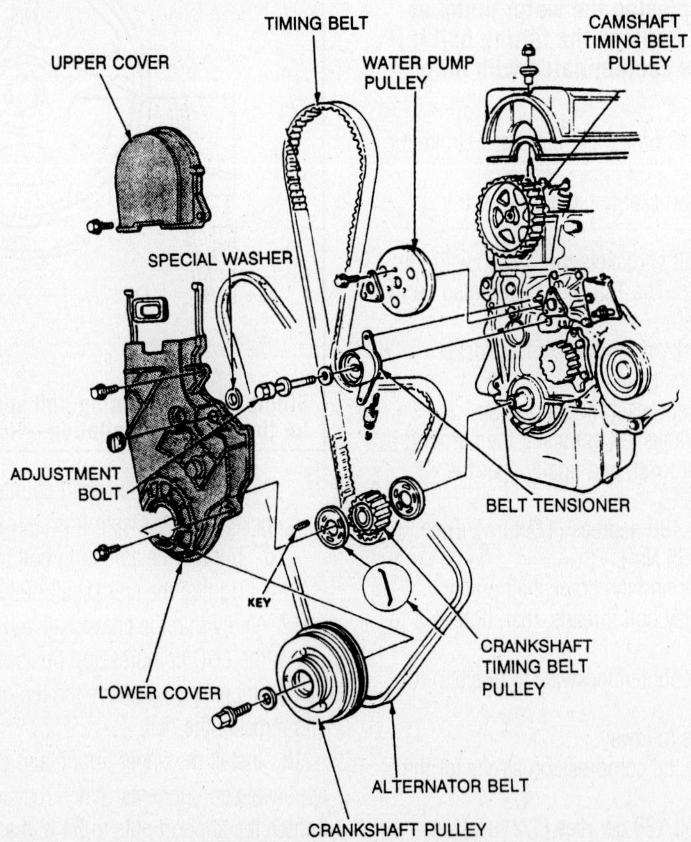

Exploded view of timing belt and cover—Honda 1.8L (A18A1) and 2.0L (A20A1, A20A3) engine

13. Remove the crankshaft pulley, then install the upper or middle and the lower timing belt covers. Tighten the cover bolts to 84 inch lbs.

14. Install the crankshaft pulley and bolt. Torque the bolt to 83 ft. lbs. on 1986-87 Prelude and 1986-89 Accord or 108 ft. lbs. on 1988-90 Prelude.

15. Install the valve cover. Tighten the nuts to 84 inch lbs.

16. Install the A/C compressor and the A/C belt and tighten the bolts to 18 ft. lbs.

17. Install the alternator and belt. Tighten the through bolt to 33 ft. lbs. and the adjusting bolt to 17 ft. lbs.

18. Install the power steering pump, belt and adjuster pulley. Tighten the through bolt to 20 ft. lbs. and the adjuster bolt to 31 ft. lbs.

19. Install the engine splash shield, then the side rubber mount. Install the side mount bracket and the cruise control actuator. Tighten the support bolts to 29 ft. lbs., support nuts to 47 ft. lbs. and the actuator bolt to 108 inch lbs.

20. Connect the battery ground cable.

2.0L (B20A5) AND 2.1L (B21A1) ENGINES

YEAR, MODEL & ENGINE APPLICATION

1988-91 Honda Prelude . 2.0L Code B20A5, 2.1L Code B21A1

REPLACEMENT INTERVALS

Honda recommends timing belt rreplacement every 60,000 miles on 1988-89 models, and every 90,000 miles on 1990-91 models.

LABOR TIMES

1988-91	
Replacement	3.1 hrs
Severe Service	3.6 hrs
With air conditioning add	1.2 hrs
With water pump replacement add	0.6 hrs

8807HONg05

✳✳ CAUTION

These engines are an interference type. If the timing belt has failed, valve-to-piston damage may have occurred. A compression check should always be performed to check for engine damage.

1. Disconnect the negative battery cable.
2. Bring the piston in No. 1 cylinder to TDC on the compression stroke.
3. Remove the valve cover and timing belt front covers.
4. Mark the direction of timing belt rotation.
5. Loosen the adjusting bolt and remove the timing belt.

To install:

6. Align the camshaft sprocket(s) and crankshaft sprocket as follows:
   a. Make sure the UP mark on the camshaft sprocket(s) is at the top most position. The sprocket timing marks should be aligned with the cylinder head upper surface.
7. Remove the timing inspection hole cover at the rear of the engine block. Make sure the TDC mark on the flywheel, indicated by a white painted mark, is aligned with the pointer in the inspection hole.

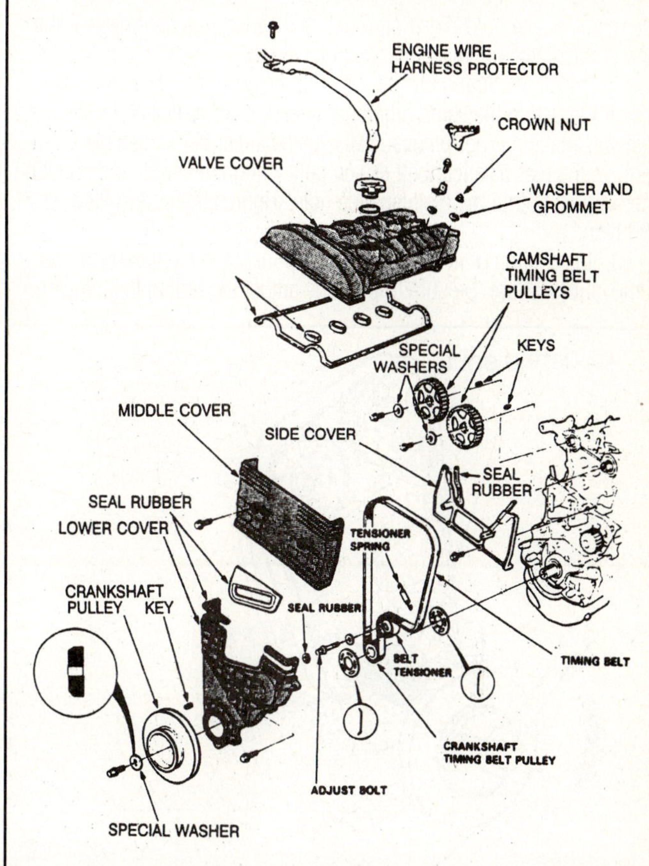

8807Timg03

Exploded view of timing belt and cover assembly–Honda 2.0L (B20A5) and 2.1L (B21A1) engines

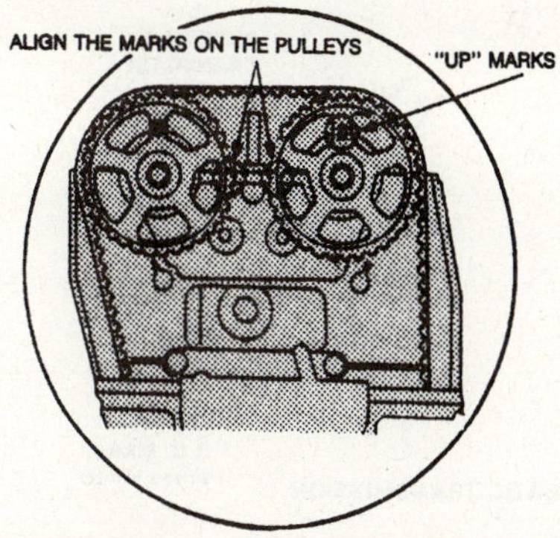

7913fo65

Aligning the camshaft sprocket timing marks–Honda 2.0L (B20A5) and 2.1L (B21A1) engines

8. Install the timing belt. If the old timing belt is reused, install the belt in the same rotational direction, as indicated by the mark that was made during removal.

9. The timing belt balancer drive pulley should already be at TDC, if the timing belt is installed correctly.

10. Align the groove on the front timing balancer belt driven pulley with the pointer on the oil pump body.

11. Remove the bolt from the maintenance hole on the cylinder block, next to the rear balancer shaft. Align the rear timing balancer belt driven pulley using a 6 X 100mm bolt, or equivalent. Mark a line at the

74mm length of the bolt. Align the pulley by inserting the bolt through the maintenance hold. The bolt must be inserted to a depth where the line is flush with the maintenance hole surface.

12. Install the timing balancer belt. If the old balancer belt is reused, install the belt in the same rotational direction, as indicated by the mark that was made during removal. After the balancer belt is installed, remove the rear balancer belt driven pulley alignment bolt and reinstall the original bolt in the maintenance hole. Tighten the bolt to 22 ft. lbs. (30 Nm).

13. Installation of the remaining components is the reverse of the removal procedure. Be sure to properly adjust the timing belt tension.

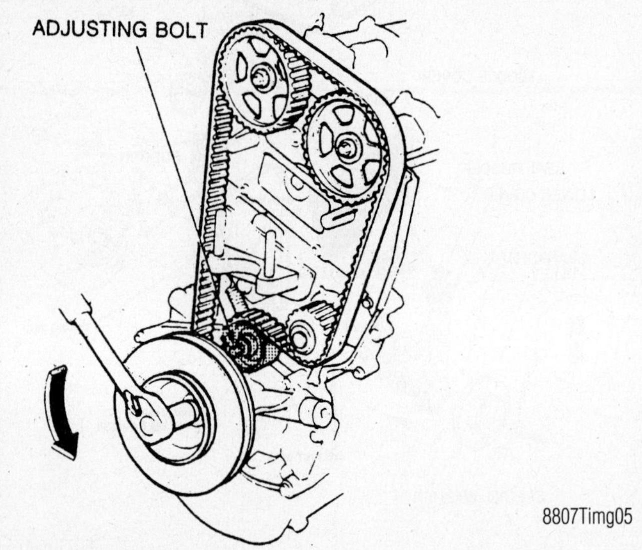

Adjusting the timing belt tensioner–Honda 2.0L (B20A5) and 2.1L (B21A1) engines

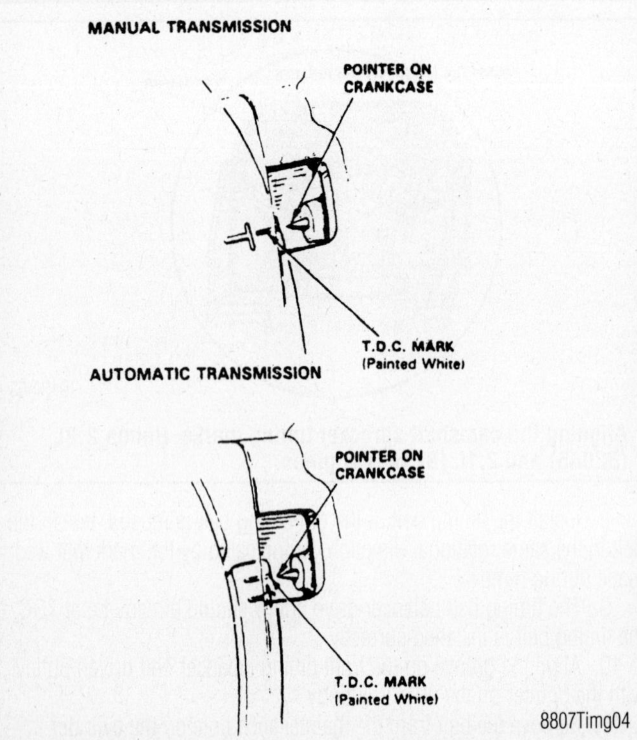

Locating flywheel TDC marks–Honda 2.0L (B20A5) and 2.1L (B21A1) engines

14. Tighten the crankshaft pulley bolt as follows: on 1988 models, tighten the bolt to 83 ft. lbs. (115 Nm), on 1989-92 models, tighten the bolt to 108 ft. lbs. (150 Nm).

2.0L (B20B4, B20Z2) ENGINES

YEAR, MODEL & ENGINE APPLICATION

1997-01 Honda CR-V .... 2.0L Codes B20B4, B20Z2

REPLACEMENT INTERVALS

Honda recommends timing belt replacement every 105,000 miles.

LABOR TIMES

1997-01	
Replacement	3.1 hrs
Severe Service	3.6 hrs

8807HONg06

✻✻ CAUTION

These engines are an interference type. If the timing belt has failed, valve-to-piston damage may have occurred. A compression check should always be performed to check for engine damage.

1. Disconnect the negative battery cable.
2. Position crankshaft so that No. 1 piston is at Top Dead Center (TDC) of the compression stroke.
3. Raise and support the vehicle.
4. Remove the left side wheel.
5. Remove the splash guard.
6. Remove the accessory drive belts.
7. If equipped, remove the cruise control actuator.
8. Place a piece of wood between the oil pan and the jack, support the engine with a jack.
9. Remove left side engine mount.
10. Remove the valve cover.
11. Remove the timing belt covers.
12. Ensure the camshaft sprocket UP marks are facing upward.
13. Remove the crankshaft pulley bolt and crankshaft pulley.
14. Remove the upper and lower timing belt covers.
15. Remove the crankshaft sprocket guide washer.
16. Loosen the adjusting bolt 180 degrees. Release the tension from the belt by pushing on the tensioner, then retighten the adjusting bolt.
17. Remove the timing belt.

To install:

18. Be sure the timing marks are properly aligned.
19. Install the timing belt on the pulleys following this sequence:
    a. Crankshaft pulley.
    b. Adjusting pulley.
    c. Water pump pulley.
    d. Exhaust camshaft pulley.
    e. Intake camshaft pulley.
20. Loosen and retighten the adjusting bolt to allow tension to be applied to the belt.
21. Install the lower and middle timing covers.
22. Install the crankshaft pulley and tighten the bolt to 130 ft. lbs. (177 Nm).

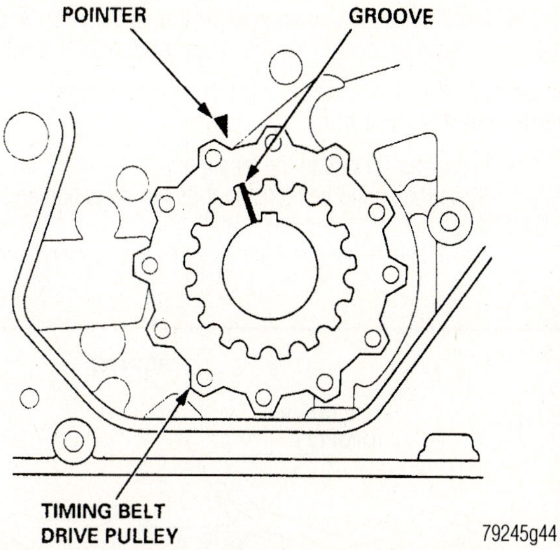

Crankshaft timing location—Honda 2.0L (B20B4, B20Z2) engine

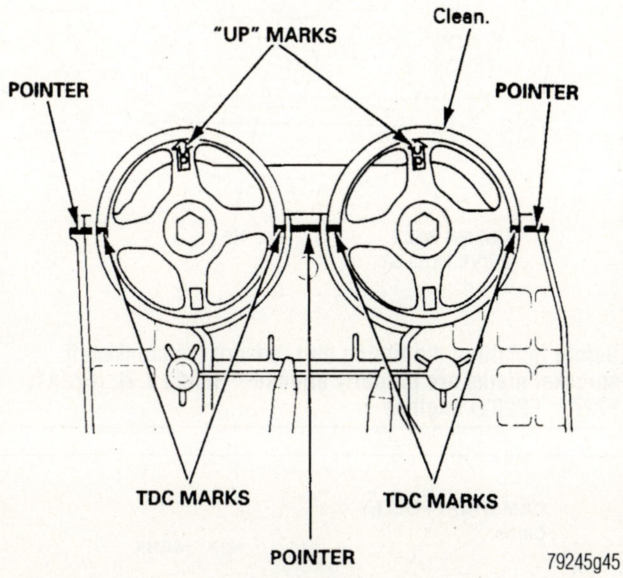

Intake and exhaust camshaft timing marks properly aligned at TDC— Honda 2.0L (B20B4, B20Z2) engine

WARNING

If any binding is felt when adjusting the timing belt tension by turning the crankshaft, STOP turning the engine, because the pistons may be hitting the valves.

23. Rotate the crankshaft about 5–6 times counterclockwise to seat the timing belt.
24. Position the No. 1 piston to TDC.
25. Loosen the adjusting bolt 1/2 turn.
26. Rotate the crankshaft counterclockwise 3 teeth on the camshaft pulley.
27. Tighten the adjusting bolt to 40 ft. lbs. (54Nm).
28. Retighten the crankshaft pulley bolt to 130 ft. lbs. (177 Nm).
29. Install the valve cover.
30. Install the engine mounting bracket, then remove the jack.
31. If removed, install the cruise control actuator.
32. Install the accessory drive belts.
33. Install the splash guard.
34. Connect the negative battery cable.
35. Check the engine operation and road test.

2.2L (F22A1, F22B1, F22B2) ENGINES

YEAR, MODEL & ENGINE APPLICATION

1990-97 Honda Accord	2.2L Codes F22A1, F22B1, F22B2
1992-96 Honda Prelude	2.2L Code F22A1

REPLACEMENT INTERVALS

Honda recommends timing belt replacement every 90,000 miles on 1990-96 models, and 105,000 miles on 1997 models.

LABOR TIMES

1990-97	
Replacement	3.5 hrs
Severe Service	4.0 hrs
With air conditioning add	1.2 hrs
With power steering add	0.2 hrs
With water pump replacement add	0.6 hrs

8807HONg07

CAUTION

These engines are an interference type. If the timing belt has failed, valve-to-piston damage may have occurred. A compression check should always be performed to check for engine damage.

1. Disconnect the negative battery cable.
2. Turn the crankshaft to align the crankshaft or flywheel timing marks and set cylinder No. 1 to Top Dead Center (TDC) on the compression stroke. Once in this position, the engine must NOT be turned or disturbed.
3. Raise and support the vehicle.
4. Remove the left side wheel.
5. Remove the lower splash shield.
6. Remove the cruise control actuator, if necessary.
7. Remove the accessory drive belts.
8. Remove the left side engine mount.
9. On 1994-97 Accord, remove the alternator and bracket.
10. Remove the power steering pump and position aside without disconnecting the lines.
11. Remove all necessary components to gain access to the cylinder head and timing belt covers.
12. Remove the cylinder head and upper timing belt cover.
13. Ensure the camshaft sprocket timing marks are aligned with the edge of the cylinder head.
14. Remove the crankshaft pulley.
15. Remove the tensioner adjusting bolt seal.
16. Remove the lower timing belt cover.
17. There are 2 belts in this system; the one running to the camshaft pulley is the timing belt. The other, shorter one drives the balance shafts and is referred to as the balancer belt or timing balancer belt. Lock the timing belt adjuster in position by installing one of the lower timing belt cover bolts to the adjuster arm.

18. Loosen the timing belt and balancer shafts tensioner adjuster nut, but do not loosen the nut more than 1 turn. Push the tensioner for the balancer belt away from the belt to relieve the tension. Hold the tensioner and tighten the adjusting nut to hold the tensioner in place.

19. Carefully remove the balancer belt. Do not crimp or bend the belt; protect it from contact with oil or coolant. Slide the belt off the pulleys.

20. Remove the balancer belt drive sprocket from the crankshaft.

21. Loosen the lockbolt installed to the timing belt adjuster and loosen the adjusting nut. Push the timing belt adjuster to remove the tension on the timing belt, then tighten the adjuster nut.

22. Remove the timing belt. Do not crimp or bend the belt; protect it from contact with oil or coolant. Slide the belt off the pulleys.

23. If defective, remove the belt tensioners by performing the following:

a. Remove the springs from the balancer belt and the timing belt tensioners.

b. Remove the adjusting nut.

c. Remove the bolt from the balancer belt adjuster lever, then remove the lever and the tensioner pulley.

d. Remove the lockbolt from the timing belt tensioner lever, then remove the tensioner pulley and lever from the engine.

24. This is an excellent time to check or replace the water pump. Even if the timing belt is only being replaced as part of a good maintenance schedule, consider replacing the pump at the same time.

To install:

25. If the water pump is to be replaced, install a new O-ring and make certain it is properly seated. Install the water pump and retaining bolts. Tighten the mounting bolts to 106 inch lbs. (12 Nm).

26. If the tensioners were removed, perform the following to install them:

a. Install the timing belt tensioner lever and tensioner pulley.

■ **The tensioner lever must be properly positioned on its pivot pin located on the oil pump. Be sure that the timing belt lever and tensioner moves freely and does not bind.**

b. Install the lockbolt to the timing belt tensioner, do not tighten the lockbolt at this time.

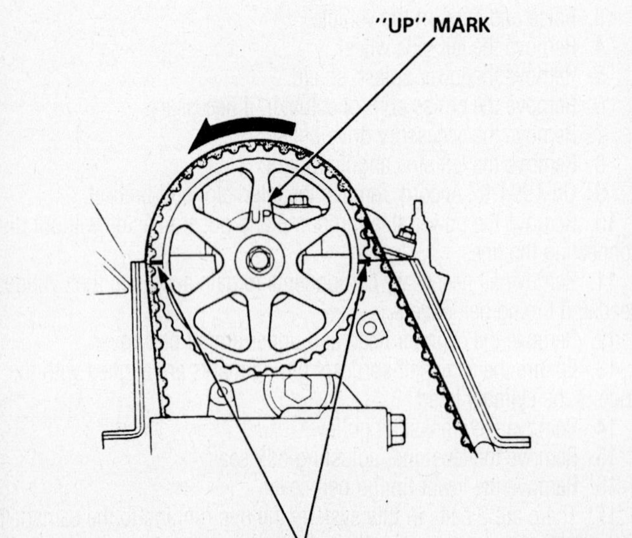

Camshaft timing belt sprocket alignment mark positioning for timing belt installation—Honda 2.2L (F22A1) engine

c. Install the balancer belt pulley and adjuster lever.

d. Install the adjusting nut and the bolt to the balancer belt adjuster lever. Do not tighten the adjuster nut or bolt at this time.

■ **Be sure that the balancer lever and tensioner moves freely and does not bind.**

e. Install the springs to the tensioners.

f. Move the timing belt tensioner its full deflection and tighten the lockbolt.

g. Move the balancer its full deflection and tighten the adjusting nut.

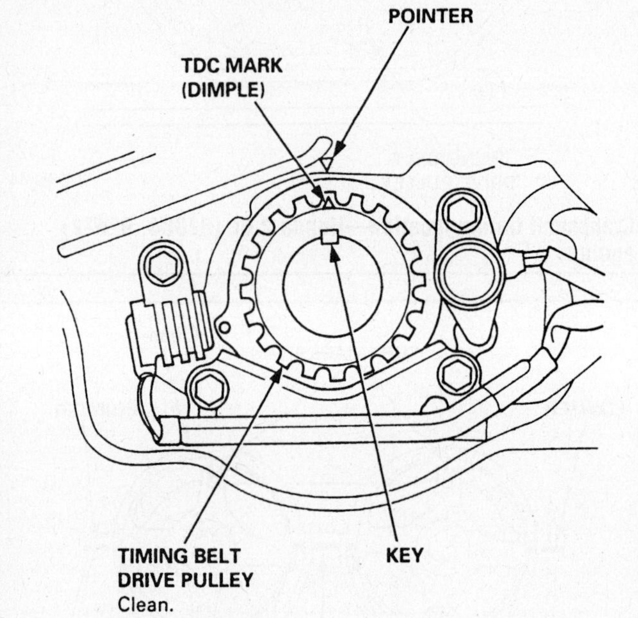

Before installing the timing belt, ensure the crankshaft sprocket marks are properly aligned—Honda 2.2L (F22A1, F22B1, F22B2) engines

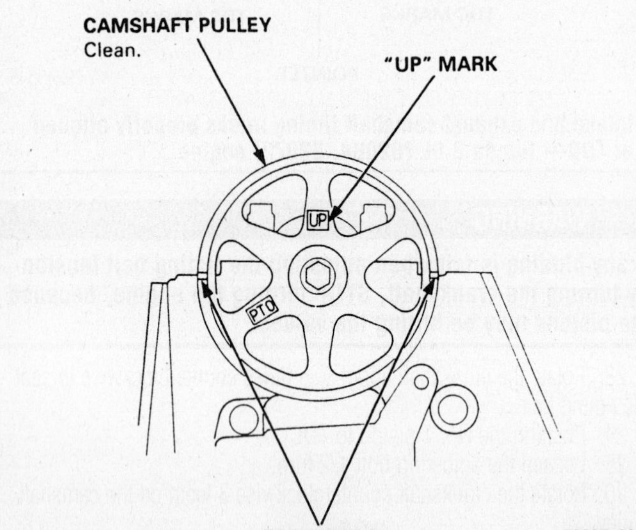

Position the camshaft sprocket as indicated for timing belt installation—Honda 2.2L (F22B1 and F22B2) engines

27. The crankshaft timing pointer must be perfectly aligned with the white mark on the flywheel, flex-plate or crankshaft sprocket; the camshaft pulley must be aligned so that the word **UP** is at the top of the pulley and the marks on the edge of the pulley are aligned with the surfaces of the head.

28. Install the timing belt over the crankshaft sprocket, camshaft sprocket, water pump and tensioner.

29. Loosen the bolt used to lock the timing belt tensioner. Loosen, then tighten the timing belt adjusting nut.

30. Turn the crankshaft counterclockwise until the cam pulley has moved 3 teeth; this creates tension on the timing belt. Loosen, then tighten the adjusting nut and tighten it to 33 ft. lbs. (45 Nm). Tighten the bolt used to lock the timing belt tensioner.

31. Realign the timing belt marks, then install the balancer belt drive sprocket on the crankshaft.

32. Align the front balancer pulley; the face of the front timing balancer pulley has a mark, which must be aligned with the notch on the oil pump body. This pulley is the one at 10 o'clock to the crank pulley when viewed from the pulley end.

33. Align the rear timing balancer pulley (2 o'clock from the crank pulley) using a 6 x 100mm bolt or rod. Mark the bolt or rod at a point 2.9 in. (74mm) from the end. Remove the bolt from the maintenance

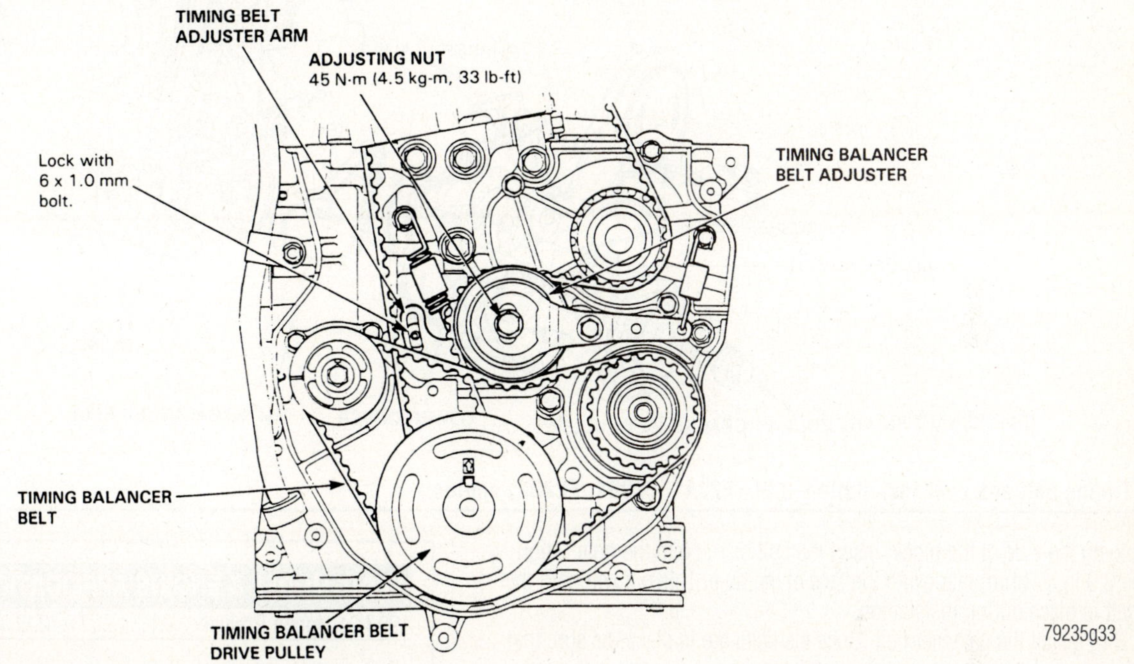

Balancer belt routing—Honda 2.2L (F22A1, F22B1, F22B2) engine

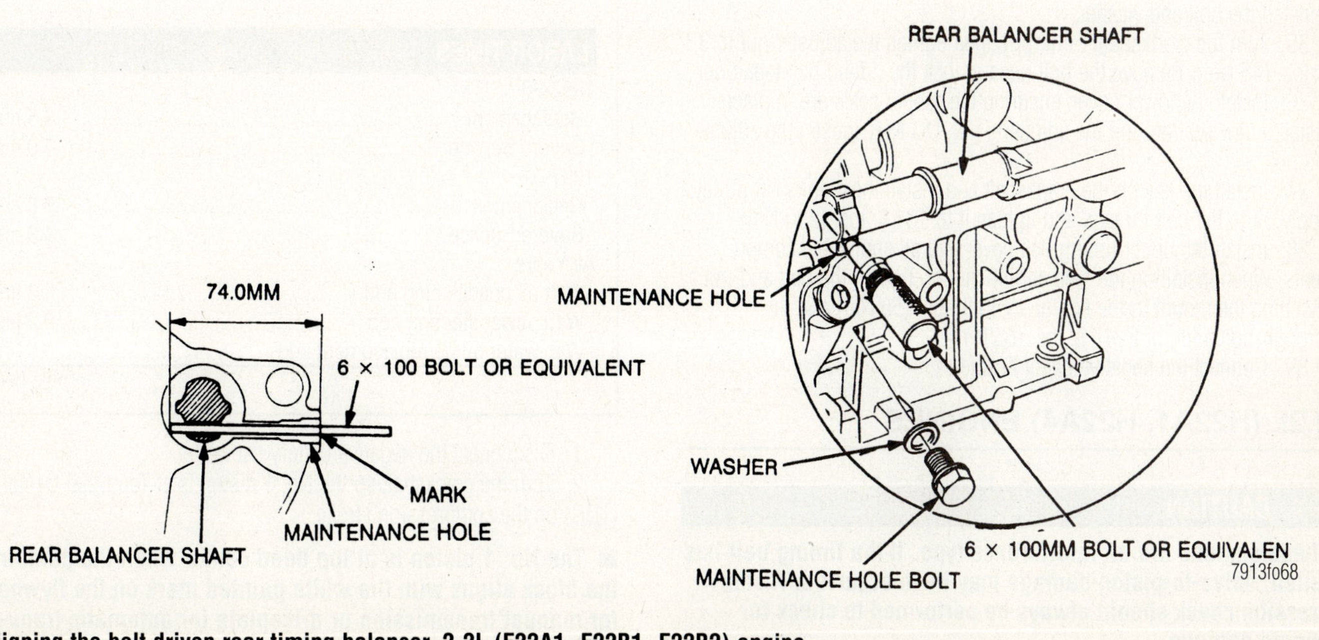

Aligning the belt driven rear timing balancer–2.2L (F22A1, F22B1, F22B2) engine

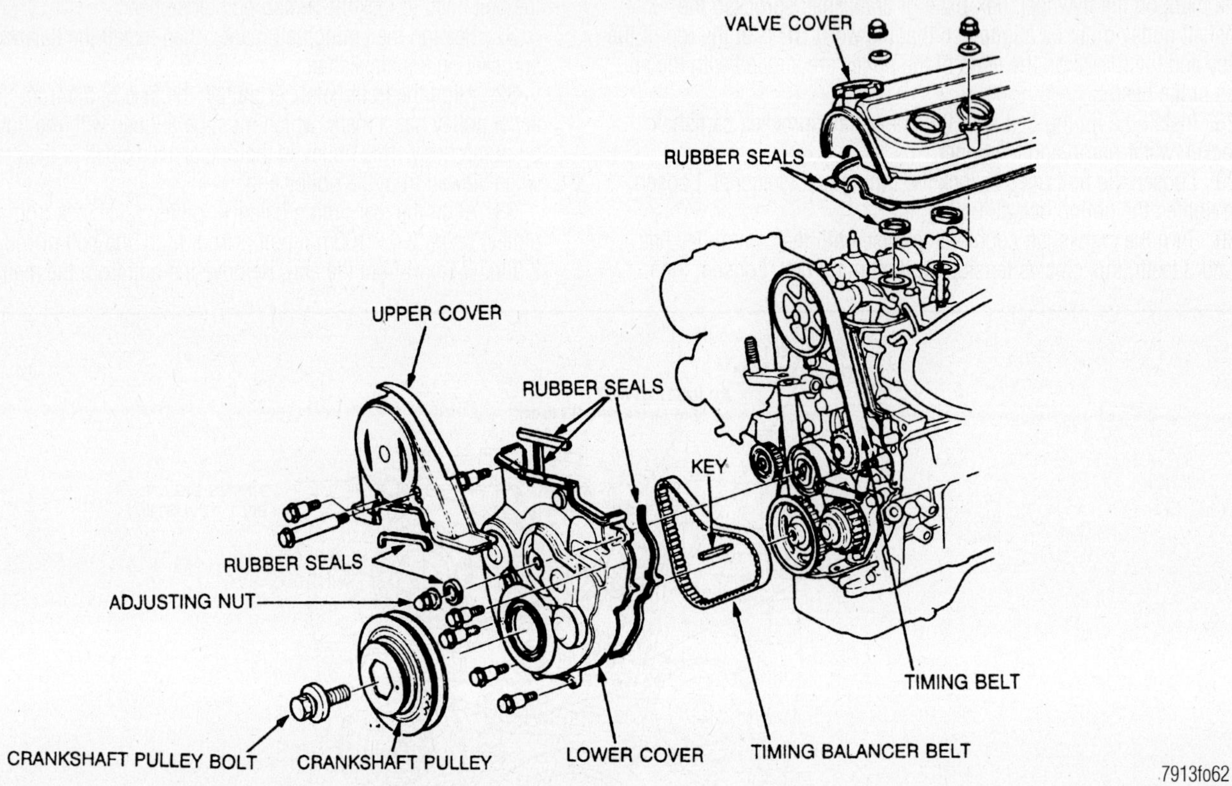

Timing belt and over installation–2.2L (F22A1, F22B1, F22B2) engine

hole on the side of the block; insert the bolt or rod into the hole. Align the 2.9 in. (74mm) mark with the face of the hole. This pin will hold the shaft in place during installation.

34. Install the balancer belt. Once the belts are in place, be sure that all the engine alignment marks are still correct. If not, remove the belts, realign the engine and reinstall the belts. Once the belts are properly installed, slowly loosen the adjusting nut, allowing the tensioner to move against the belt. Remove the pin from the maintenance hole and reinstall the bolt and washer.
35. Turn the crankshaft 1 full turn, then tighten the adjuster nut to 33 ft. lbs. (45 Nm). Remove the bolt used to lock the timing belt tensioner.
36. Install the lower cover, ensuring the rubber seals are in place. Install a new seal around the adjusting nut, DO NOT loosen the adjusting nut.
37. Install the key on the crankshaft and install the crankshaft pulley. Apply oil to the bolt threads and tighten it to 181 ft. lbs. (250 Nm).
38. Install the upper timing belt cover and all applicable components. When installing the side engine mount, tighten the bolt and nut attaching the mount to the engine to 40 ft. lbs. (55 Nm) and the through-bolt and nut to 47 ft. lbs. (65 Nm).
39. Connect the negative battery cable.

2.2L (H22A1, H22A4) ENGINES

✻✻ CAUTION

These engines are an interference type. If the timing belt has failed, valve-to-piston damage may have occurred. A compression check should always be performed to check for engine damage.

YEAR, MODEL & ENGINE APPLICATION

1993-01 Honda Prelude	2.2L Codes H22A1, H22A4

REPLACEMENT INTERVALS

Honda recommends timing belt replacement every 90,000 miles on 1993-96 models, and 105,000 miles on 1997-01 engines.

LABOR TIMES

1992-96	
Replacement	3.5 hrs
Severe Service	4.0 hrs
1997-01	
Replacement	3.8 hrs
Severe Service	4.3 hrs
All Years	
With air conditioning add	1.2 hrs
With power steering add	0.2 hrs
With water pump replacement add	0.6 hrs

8807HONg08

1. Disconnect the negative battery cable.
2. Turn the crankshaft so the No. 1 piston is at Top Dead Center (TDC) on the compression stroke.

■ **The No. 1 piston is at top dead center when the pointer on the block aligns with the white painted mark on the flywheel for manual transmission or driveplate for automatic transmission.**

3. Raise and support the vehicle.
4. Remove the left side wheel.
5. Remove the splash shield.
6. Remove all necessary components for access to the cylinder head cover, then remove the cylinder head cover.
7. Ensure the words **UP** embossed on the camshaft pulleys are aligned in the upward position.
8. Insert a 5.0mm pin punch in each of the camshaft caps, nearest to the pulleys, through the holes provided.
9. Remove the upper and middle timing belt covers.
10. Support the engine with a floor jack below the center of the center beam. Tension the jack so that it is just supporting the beam but not lifting it. Remove the 2 rear bolts from the center beam to allow the engine to drop down for clearance to remove the lower cover.
11. Remove and discard the rubber seal from the timing belt adjuster. Do not loosen the adjusting nut.
12. Remove the lower timing belt cover.
13. There are 2 belts in this system; the one running to the camshaft pulley is the timing belt. The other, shorter one drives the balance shaft and is referred to as the balancer belt or timing balancer belt. Lock the timing belt adjuster in position, by installing one of the lower timing belt cover bolts to the adjuster arm.
14. Loosen the timing belt and balancer shaft tensioner adjuster nut(s), do not loosen the nut(s) more than 1 turn. Push the tensioner for the balancer belt away from the belt to relieve the tension. Hold the tensioner and tighten the adjusting nut to hold the tensioner in place.
15. Carefully remove the balancer belt by sliding it off of the pulleys. Do not crimp or bend the belt; protect it from contact with oil or coolant.
16. Remove the balancer belt drive sprocket from the crankshaft.
17. Loosen the lockbolt installed in the timing belt adjuster and loosen the adjusting nut. Push the timing belt adjuster to remove the tension on the timing belt, then tighten the adjuster nut.
18. Remove the timing belt by sliding it off of the pulleys. Do not crimp or bend the belt; protect it from contact with oil or coolant.
19. If defective, remove the belt tensioners by performing the following:
    a. Remove the springs from the balancer belt and the timing belt tensioners.
    b. Remove the adjusting nut.
    c. Remove the bolt from the balancer belt adjuster lever, then remove the lever and the tensioner pulley.
    d. Remove the lockbolt from the timing belt tensioner lever, then remove the tensioner pulley and lever from the engine.

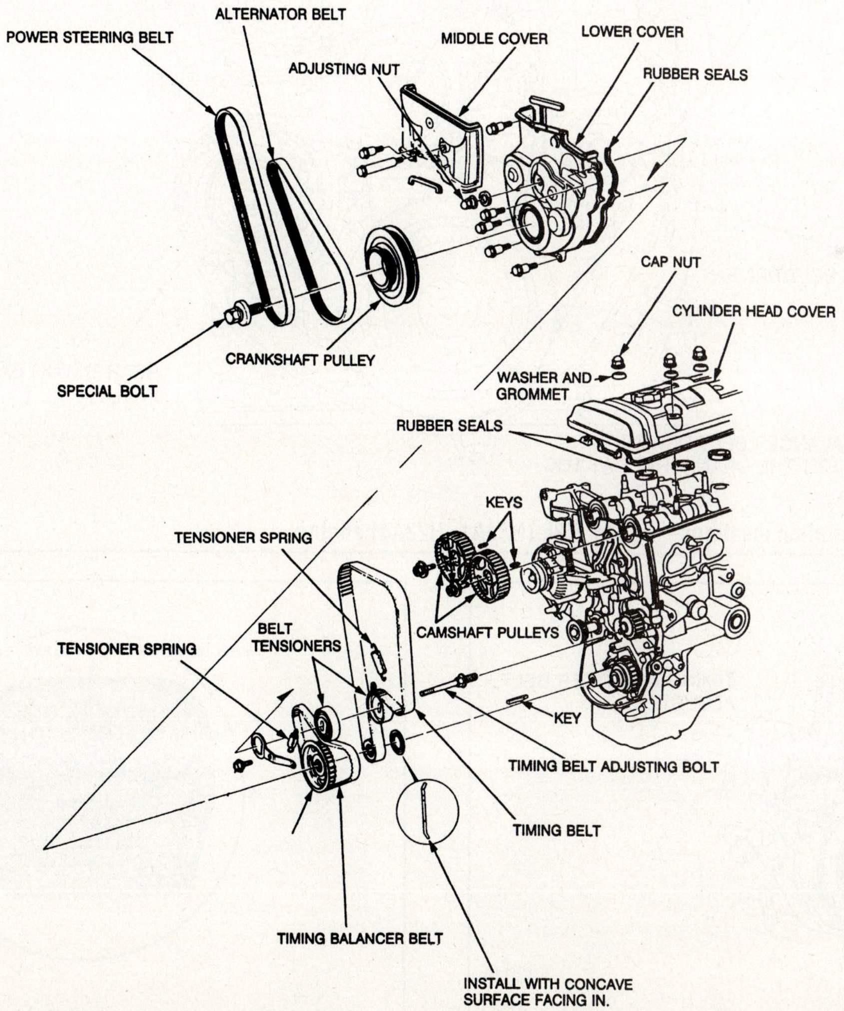

Exploded view of timing belt and timing balance assembly—Honda 2.2L (H22A1, H22A4) engine

20. This is an excellent time to check or replace the water pump. Even if the timing belt is only being replaced as part of a good maintenance schedule, consider replacing the pump at the same time.

To install:

21. If the water pump is to be replaced, install a new O-ring and make certain it is properly seated. Install the water pump and retaining bolts. Tighten the mounting bolts to 106 inch lbs. (12 Nm).

22. If the tensioners were removed perform the following to install them:

a. Install the timing belt tensioner lever and tensioner pulley.

■ **The tensioner lever must be properly positioned on its pivot pin located on the oil pump. Be sure that the timing belt lever and tensioner moves freely and does not bind.**

b. Install the lockbolt to the timing belt tensioner, do not tighten the lockbolt at this time.

c. Install the balancer belt pulley and adjuster lever.

d. Install the adjusting nut and the bolt to the balancer belt adjuster lever. Do not tighten the adjuster nut or bolt at this time.

■ **Be sure that the balancer lever and tensioner moves freely and does not bind.**

e. Install the springs to the tensioners.

f. Move the timing belt tensioner its full deflection and tighten the lockbolt.

g. Move the balancer its full deflection and tighten the adjusting nut.

23. The crankshaft timing pointer must be perfectly aligned with the white mark on the flywheel or flex-plate. The camshaft pulley must be aligned so that the word **UP** is at the top of the pulley and the marks on the edge of the pulley are aligned with the surfaces of the head.

24. Install the timing belt.

25. Install the balancer belt drive sprocket to the crankshaft.

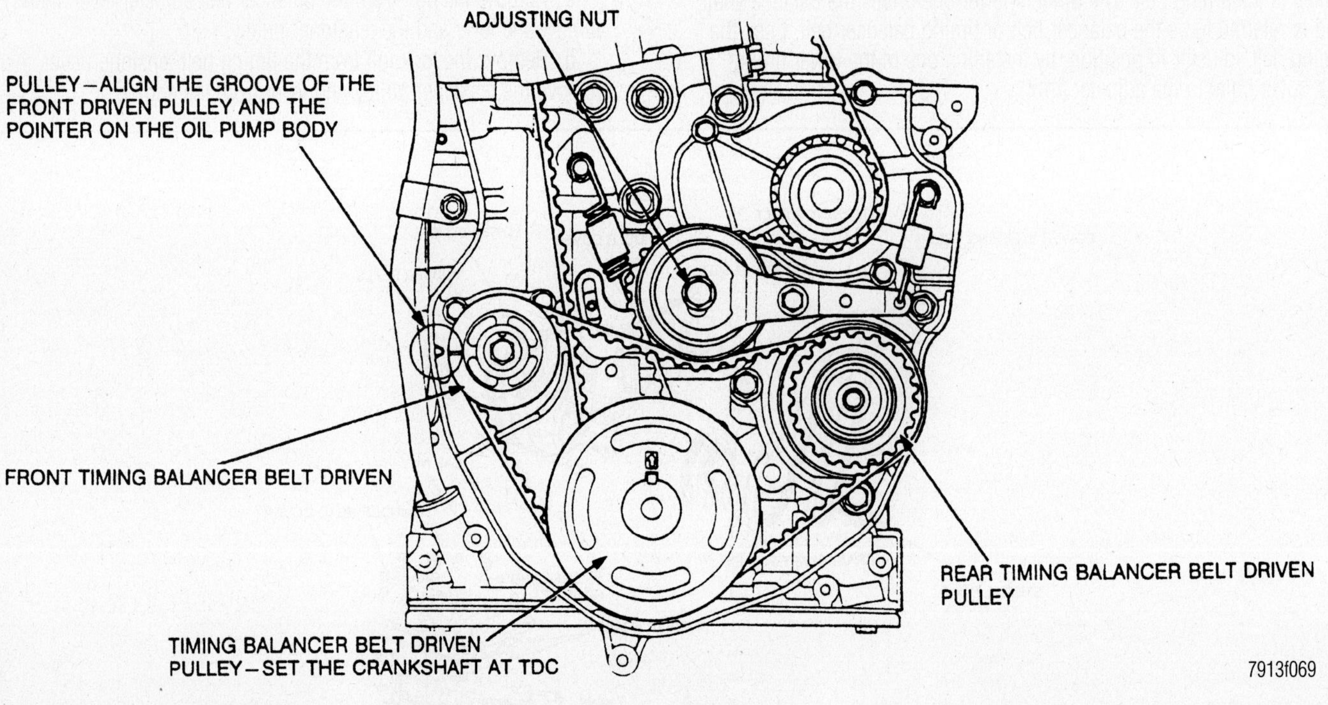

Timing balancer belt installation locations—Honda 2.2L (H22A1, H22A4) engine

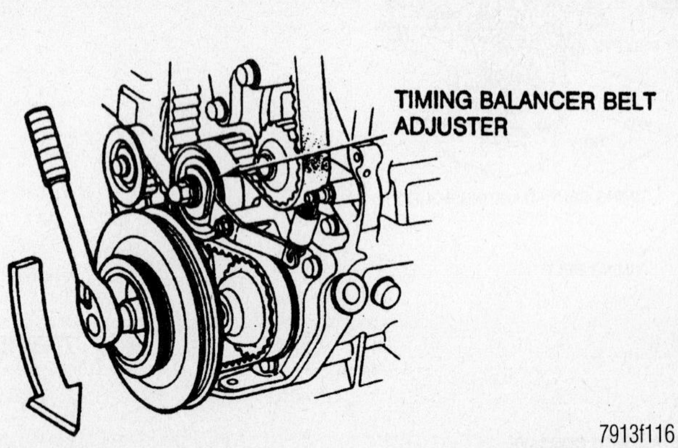

Timing balancer belt adjustment—Honda 2.2L (H22A1, H22A4) engine

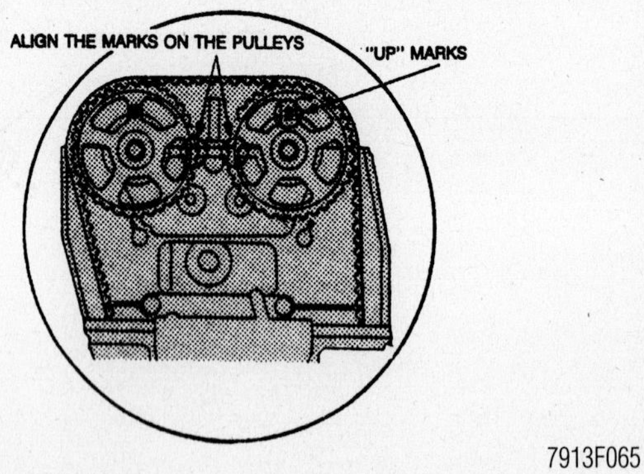

Camshaft sprocket timing mark alignment—Honda 2.2L (H22A1, H22A4) engine

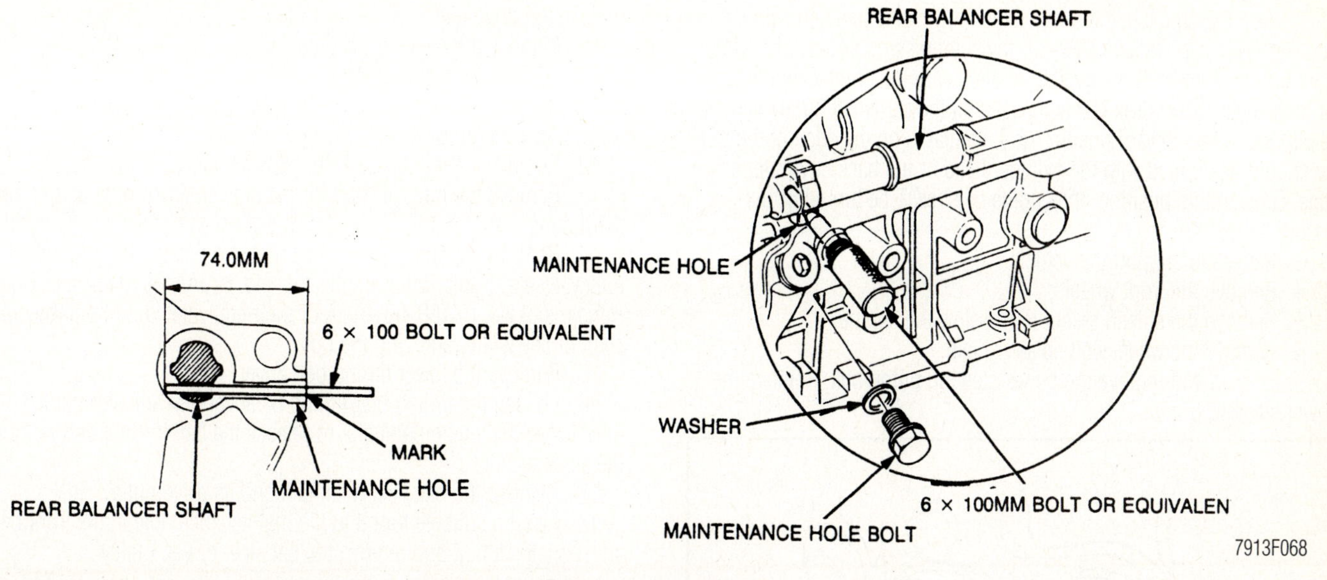

Aligning the rear timing balancer shaft—Honda 2.2L (H22A1, H22A4) engine

26. Remove the 2, 5.0mm pin punches from the camshaft bearing caps.
27. Loosen the bolt used to lock the timing belt tensioner. Loosen, then tighten the timing belt adjuster nut.
28. Turn the crankshaft counterclockwise until the cam pulley has moved 3 teeth; this creates tension on the timing belt. Loosen, then tighten the adjusting nut and tighten it to 33 ft. lbs. (45 Nm). Tighten the bolt used to lock the timing belt tensioner.
29. Realign the timing belt timing marks.
30. Align the groove on the front balancer shaft pulley with the pointer on the oil pump.
31. Align the rear timing balancer pulley using a 6 x 100mm bolt or rod. Mark the bolt or rod at a point 2.913 in. (74mm) from the end. Remove the bolt from the maintenance hole on the side of the block; insert the bolt/rod into the hole and align the 74mm mark with the face of the hole. This pin will hold the shaft in place during installation.
32. Loosen the adjusting nut and ensure the timing balancer belt adjuster moves freely.
33. Install the balancer belt. Once the belts are in place, be sure that all the engine alignment marks are still correct. If not, remove the belts, realign the engine and reinstall the belts. Once the belts are properly installed, slowly loosen the adjusting nut, allowing the tensioner to move against the belt. Remove the pin from the maintenance hole and reinstall the bolt and washer.
34. Turn the crankshaft pulley 1 full turn and tighten the adjusting nut to 33 ft. lbs. (45 Nm).

■ **Both belt adjusters are spring loaded to properly tension the belts. Do not apply extra pressure to the pulleys or tensioners while performing the adjustment.**

35. Remove the 6 x 100mm bolt from the timing belt adjuster arm.
36. Install the timing belt and cylinder head covers. Reinstall all applicable components. When installing the crankshaft pulley, coat the threads and seating face of the pulley bolt with engine oil, then install and tighten the bolt to 181 ft. lbs. (250 Nm). When installing the side engine mount, tighten the bolt and nut attaching the mount to the engine to 40 ft. lbs. (55 Nm), and the through-bolt and nut to 47 ft. lbs. (65 Nm). Remove the jack from under the center beam.
37. Connect the negative battery cable.

2.2L (F22B6) AND 2.3L (F23A1, F23A5, F23A7) ENGINES

YEAR, MODEL & ENGINE APPLICATION	
1998-02 Honda Accord	2.3L Codes F23A1, F23A4, F23A5
1995-97 Honda Odyssey	2.2L Code F22B6
1998 Honda Odyssey	2.3L Code F23A7

REPLACEMENT INTERVALS

Honda recommends timing belt replacement every 90,000 miles on 1995-97 models, and 105,000 miles on 1998-02 models.

LABOR TIMES	
Accord	
Replacement	5.0 hrs
Severe Service	5.7 hrs
Odyssey	
Replacement	3.2 hrs
Severe Service	4.0 hrs
All Models	
With air conditioning add	1.2 hrs
With water pump replacement add	0.5 hrs

8807HONg09

✱✱ CAUTION

These engines are an interference type. If the timing belt has failed, valve-to-piston damage may have occurred. A compression check should always be performed to check for engine damage.

■ **The radio may contain a coded theft protection circuit. Always make note the code number before disconnecting the battery.**

1. Disconnect the negative and positive battery cables.
2. Turn the crankshaft to align the timing marks and set cylinder No. 1 to Top Dead Center (TDC) for the compression stroke. The white mark on the crankshaft pulley should align with the pointer on the timing belt cover. The words **UP** embossed on the camshaft pulley should be aligned in the upward position and the marks on the edge of the pulley should be aligned with the cylinder head or the back cover upper edge. Once in this position, the engine must NOT be turned or disturbed.
3. Raise and support the vehicle.
4. Remove the front wheels.
5. Remove the splash shield from below the engine.
6. Remove the wheel well splash shield.
7. Loosen and remove the power steering pump belt. Remove the power steering pump.

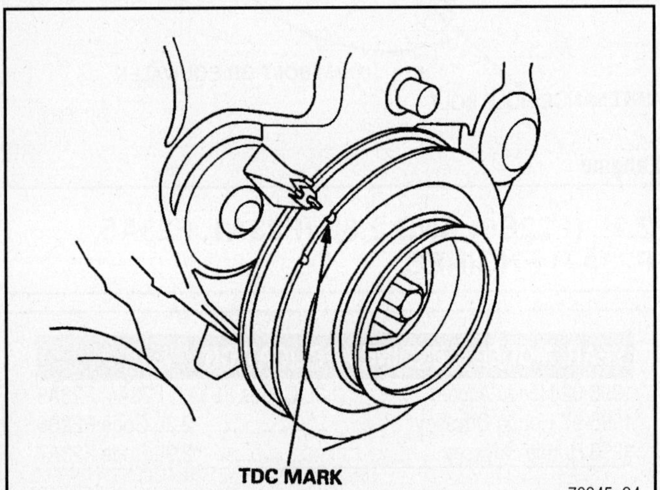

79245g24

Align the camshaft, crankshaft and engine marks before removing the timing belt and pulleys—Honda 2.2L (F22B6) and 2.3L (F23A1, F23A5, F23A7) engines

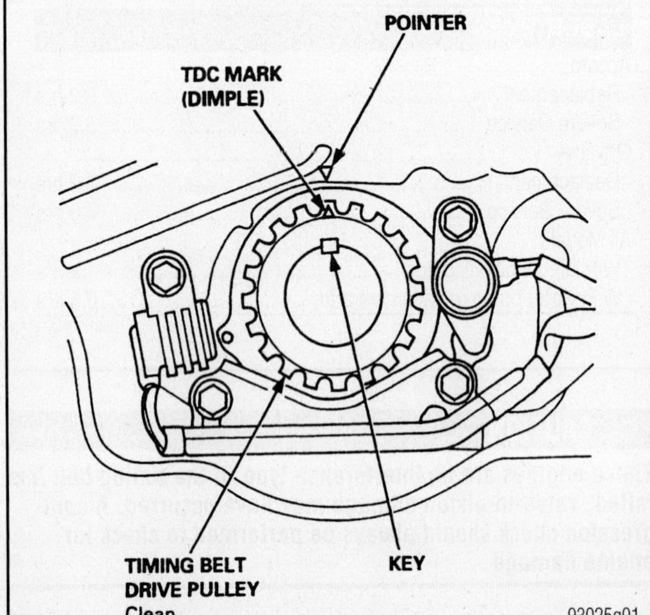

93025g01

Align the crankshaft sprocket with the oil pump pointer before installing the timing belt— Honda 2.2L (F22B6) and 2.3L (F23A1, F23A5, F23A7) engines

8. Loosen the adjusting and mounting bolts for the alternator and remove the drive belt.
9. Remove the cylinder head cover.
10. Remove the upper timing belt cover.
11. Support the engine with a floor jack cushioned with a piece of wood under the oil pan.
12. Remove the dipstick and the dipstick tube.
13. Remove the through-bolt for the side engine mount and remove the mount.
14. Remove the crankshaft pulley bolt and remove the crankshaft pulley. Use a Crank Pulley Holder tool No. 07JAA-001020A and Holder Handle tool No. 07JAB-001020A or their equivalents, to hold the crankshaft pulley while removing the bolt.
15. Remove the lower timing belt cover.
16. Loosen the timing belt/timing balancer belt adjuster nut 2/3–1 turn. Move the tension adjuster to release the belt tension and retighten the adjuster nut.
17. Remove the balancer shaft belt and its drive pulley.
18. Insert a suitable tool into the maintenance hole in the front balancer shaft. Unbolt and remove the balancer driven pulley.

■ **For servicing the balance shafts, front refers to the side of the engine facing the radiator. Rear refers to the side of the engine facing the firewall.**

19. Remove the timing belt.
20. If equipped with a TDC sensor assembly at the crankshaft sprocket, unbolt the assembly and move it to the side before removing the sprocket.
21. Remove the key and the spacers to remove the crankshaft timing sprocket.
22. Unbolt and remove the camshaft timing sprocket.

To install:

23. Install the camshaft timing sprocket so that the **UP** mark is up and the TDC marks are parallel to the cylinder head gasket surface. Install the key and tighten the bolt to 27 ft. lbs. (37 Nm).
24. Install the crankshaft sprocket so that the TDC mark aligns with the pointer on the oil pump. Install the spacers with their concave surfaces facing in. Install the key. Install the TDC sensor assembly back into position before installing the timing belt.
25. Install the timing belt on the crankshaft sprocket, tensioner pulley, water pump sprocket and camshaft sprocket.
26. Loosen the tensioner bolt and allow the tensioner to tension the timing belt, then tighten the bolt.
27. Install the balancer belt sprocket on the crankshaft.
28. Install the lower timing belt cover and the crankshaft pulley.
29. Rotate the crankshaft counterclockwise 5–6 turns to be sure the belt is properly seated.
30. Set the No. 1 piston at TDC for its compression stroke.

**** WARNING**

If any binding is felt when adjusting the timing belt tension by turning the crankshaft, STOP turning the engine, because the pistons may be hitting the valves.

31. Rotate the crankshaft counterclockwise so that the camshaft pulley moves only 3 teeth beyond its TDC mark.
32. Tighten the tensioner adjusting nut to 33 ft. lbs. (45 Nm).
33. Remove the crankshaft pulley and lower timing belt cover.
34. Align the rear balancer shaft into position by performing the following procedures:
   a. Scribe a 3 in. (74mm) line from the end of a 6 x 1.0mm bolt.

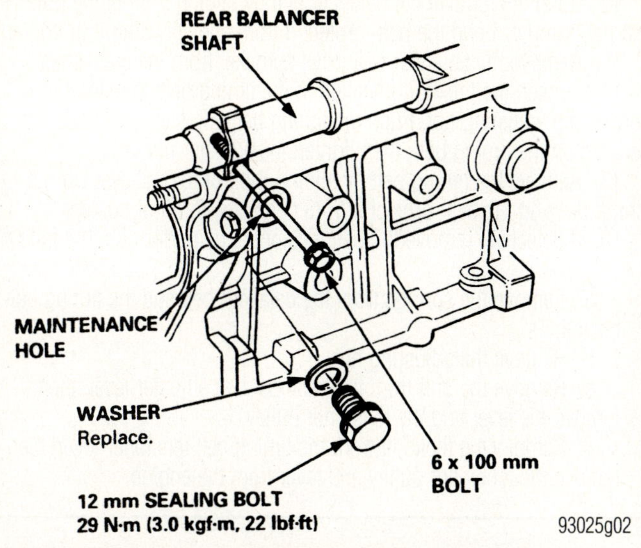

Aligning the rear timing balancer shaft— Honda 2.2L (F22B6) and 2.3L (F23A1, F23A5, F23A7) engines

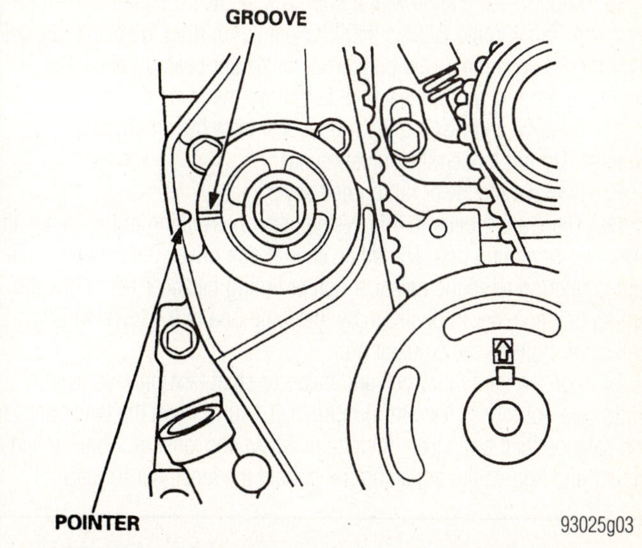

Aligning the front timing balancer shaft— Honda 2.2L (F22B6) and 2.3L (F23A1, F23A5, F23A7) engines

b. Remove the maintenance hole sealing bolt and insert the 6 x 1.0mm bolt into the maintenance hole to the scribed line.

35. Install the balancer shaft belt drive pulley.

36. Align the groove on the pulley edge with the pointer on the balancer gear case.

37. Check the alignment of the pointer on the balancer pulley to the pointer on the oil pump.

38. Install and tension the timing balancer shaft belt.

a. Remove the 6 x 1.0mm bolt from the maintenance hole and install the hole cover.

39. Be sure the timing belts have been tensioned correctly and that all TDC and alignment marks are in their proper positions.

40. Install the lower timing cover and the crankshaft pulley. Apply engine oil to the pulley bolt threads and washer surface. Install the pulley bolt and tighten it to 181 ft. lbs. (245 Nm).

41. Install the upper timing cover and the valve cover. Be sure the seals are properly seated.

42. Install the side engine mount. Tighten the through-bolt to 47 ft. lbs. (64 Nm). Tighten the mount nut and bolt to 40 ft. lbs. (55 Nm) each.

43. Remove the floor jack.

44. Install and tension the alternator belt.

45. Install the power steering pump and tension its belt.

46. Install the splash shields.

47. Reconnect the positive and negative battery cables. Enter the radio security code.

48. Check the engine operation.

2.3L (H23A1) ENGINES

✱✱ CAUTION

These engines are an interference type. If the timing belt has failed, valve-to-piston damage may have occurred. A compression check should always be performed to check for engine damage.

YEAR, MODEL & ENGINE APPLICATION

1992-96 Honda Prelude .... 2.3L Code H23A1

REPLACEMENT INTERVALS

Honda recommends timing belt replacement every 90,000 miles.

LABOR TIMES

1992-96	
Replacement	3.5 hrs
Severe Service	4.0 hrs
With air conditioning add	1.2 hrs
With power steering add	0.2 hrs
With water pump replacement add	0.6 hrs

8807HONg10

1. Disconnect the negative battery cable.

2. Turn the crankshaft so the No. 1 piston is at Top Dead Center (TDC) on the compression stroke.

■ **The No. 1 piston is at top dead center when the pointer on the block aligns with the white painted mark on the flywheel for manual transmission or driveplate for automatic transmission.**

3. Raise and support the vehicle.

4. Remove the left side wheel.

5. Remove the splash shield.

6. Remove all necessary components for access to the cylinder head cover, then remove the cylinder head cover.

7. Ensure the words **UP** embossed on the camshaft pulleys are aligned in the upward position.

8. Insert a 5.0mm pin punch in each of the camshaft caps, nearest to the pulleys, through the holes provided.

9. Remove the upper and middle timing belt covers.

10. Support the engine with a floor jack below the center of the center beam. Tension the jack so that it is just supporting the beam but not lifting it. Remove the 2 rear bolts from the center beam to allow the engine to drop down for clearance to remove the lower cover.

11. Remove and discard the rubber seal from the timing belt adjuster. Do not loosen the adjusting nut.

12. Remove the lower timing belt cover.

13. There are 2 belts in this system; the one running to the camshaft pulley is the timing belt. The other, shorter one drives the balance shaft and is referred to as the balancer belt or timing balancer belt. Lock the timing belt adjuster in position, by installing one of the lower timing belt cover bolts to the adjuster arm.

14. Loosen the timing belt and balancer shaft tensioner adjuster nut(s), do not loosen the nut(s) more than 1 turn. Push the tensioner for the balancer belt away from the belt to relieve the tension. Hold the tensioner and tighten the adjusting nut to hold the tensioner in place.

15. Carefully remove the balancer belt by sliding it off of the pulleys. Do not crimp or bend the belt; protect it from contact with oil or coolant.

16. Remove the balancer belt drive sprocket from the crankshaft.

17. Loosen the lockbolt installed in the timing belt adjuster and loosen the adjusting nut. Push the timing belt adjuster to remove the tension on the timing belt, then tighten the adjuster nut.

18. Remove the timing belt by sliding it off of the pulleys. Do not crimp or bend the belt; protect it from contact with oil or coolant.

19. If defective, remove the belt tensioners by performing the following:

a. Remove the springs from the balancer belt and the timing belt tensioners.

b. Remove the adjusting nut.

c. Remove the bolt from the balancer belt adjuster lever, then remove the lever and the tensioner pulley.

d. Remove the lockbolt from the timing belt tensioner lever, then remove the tensioner pulley and lever from the engine.

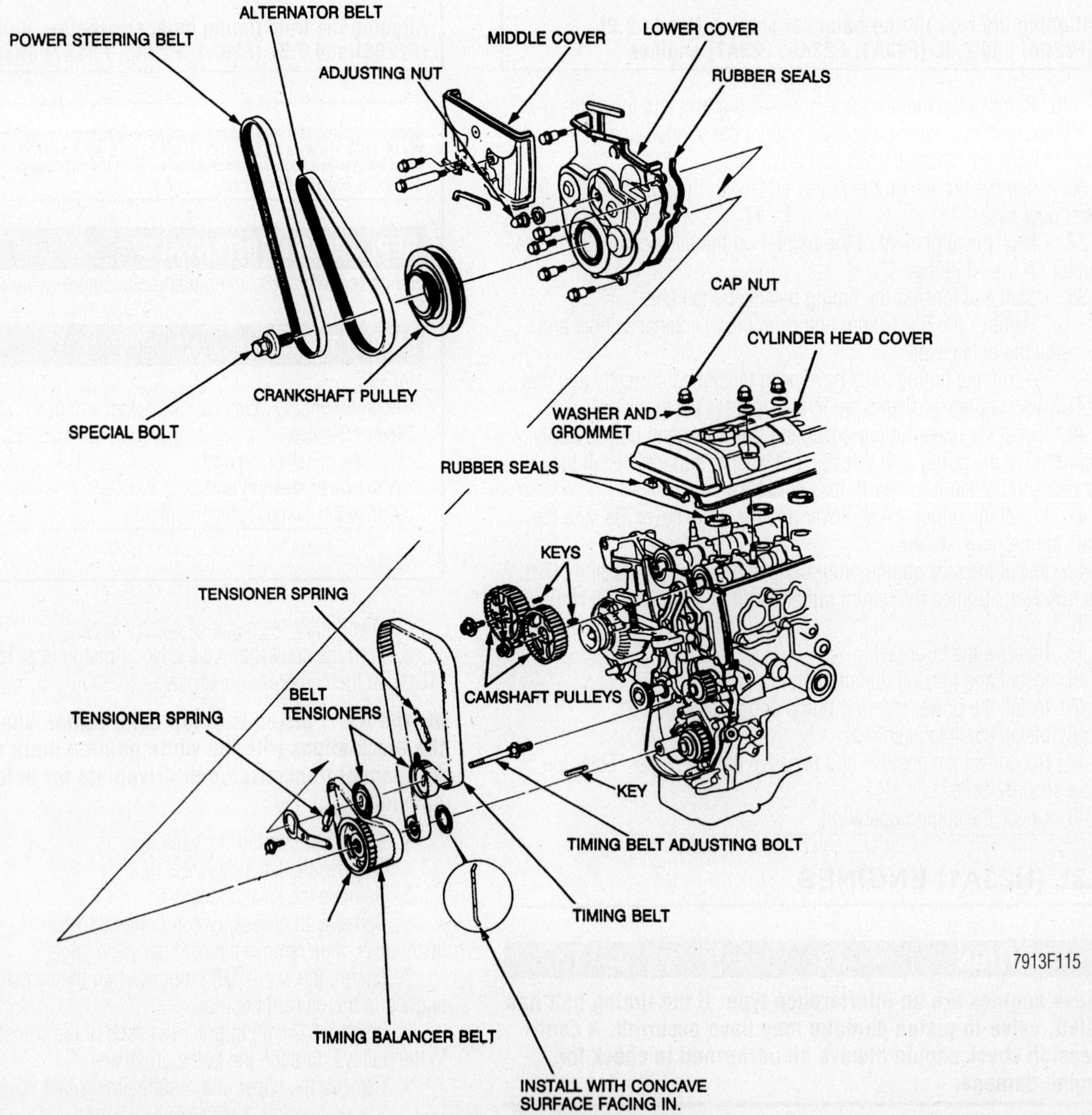

Exploded view of timing belt and timing balance assembly—Honda 2.3L (H23A1) engine

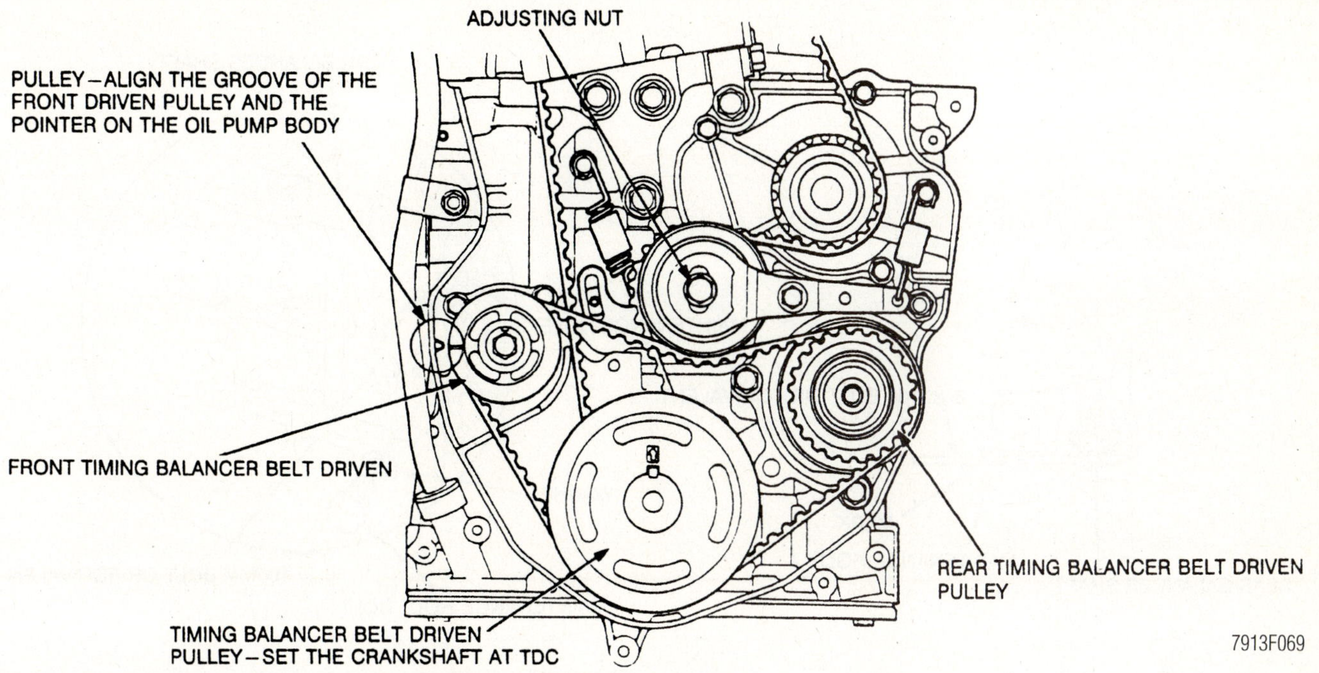

Timing balancer belt installation locations—Honda 2.3L (H23A1) engine

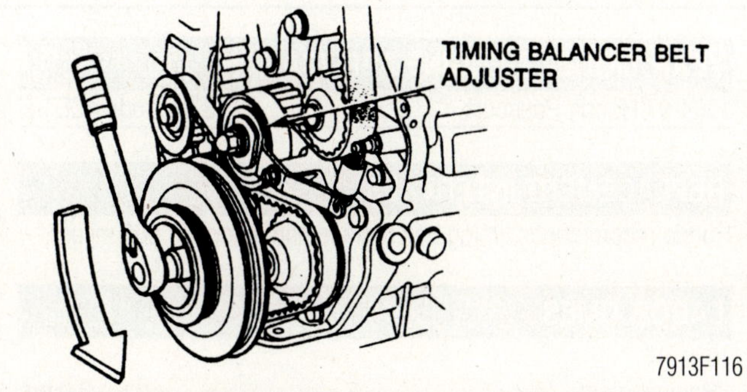

Timing balancer belt adjustment—Honda 2.3L (H23A1) engine

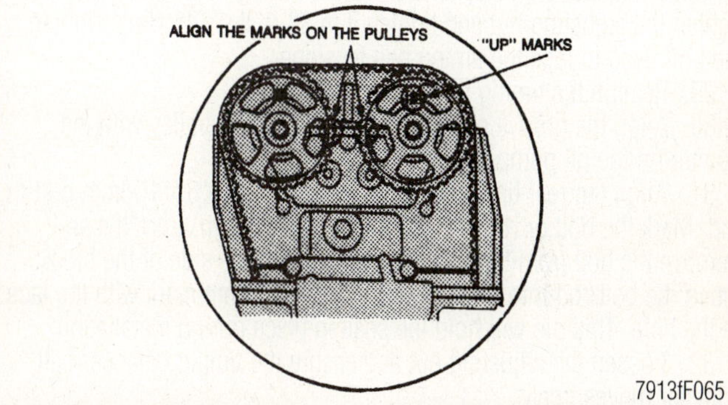

Camshaft sprocket timing mark alignment—Honda 2.3L (H23A1) engine

20. This is an excellent time to check or replace the water pump. Even if the timing belt is only being replaced as part of a good maintenance schedule, consider replacing the pump at the same time.

To install:

21. If the water pump is to be replaced, install a new O-ring and make certain it is properly seated. Install the water pump and retaining bolts. Tighten the mounting bolts to 106 inch lbs. (12 Nm).

22. If the tensioners were removed perform the following to install them:

a. Install the timing belt tensioner lever and tensioner pulley.

■ **The tensioner lever must be properly positioned on its pivot pin located on the oil pump. Be sure that the timing belt lever and tensioner moves freely and does not bind.**

b. Install the lockbolt to the timing belt tensioner, do not tighten the lockbolt at this time.

c. Install the balancer belt pulley and adjuster lever.

d. Install the adjusting nut and the bolt to the balancer belt adjuster lever. Do not tighten the adjuster nut or bolt at this time.

■ **Be sure that the balancer lever and tensioner moves freely and does not bind.**

e. Install the springs to the tensioners.

f. Move the timing belt tensioner its full deflection and tighten the lockbolt.

g. Move the balancer its full deflection and tighten the adjusting nut.

23. The crankshaft timing pointer must be perfectly aligned with the white mark on the flywheel or flex-plate. The camshaft pulley must be aligned so that the word **UP** is at the top of the pulley and the marks on the edge of the pulley are aligned with the surfaces of the head.

24. Install the timing belt.

25. Install the balancer belt drive sprocket to the crankshaft.

26. Remove the 2, 5.0mm pin punches from the camshaft bearing caps.

27. Loosen the bolt used to lock the timing belt tensioner. Loosen, then tighten the timing belt adjuster nut.

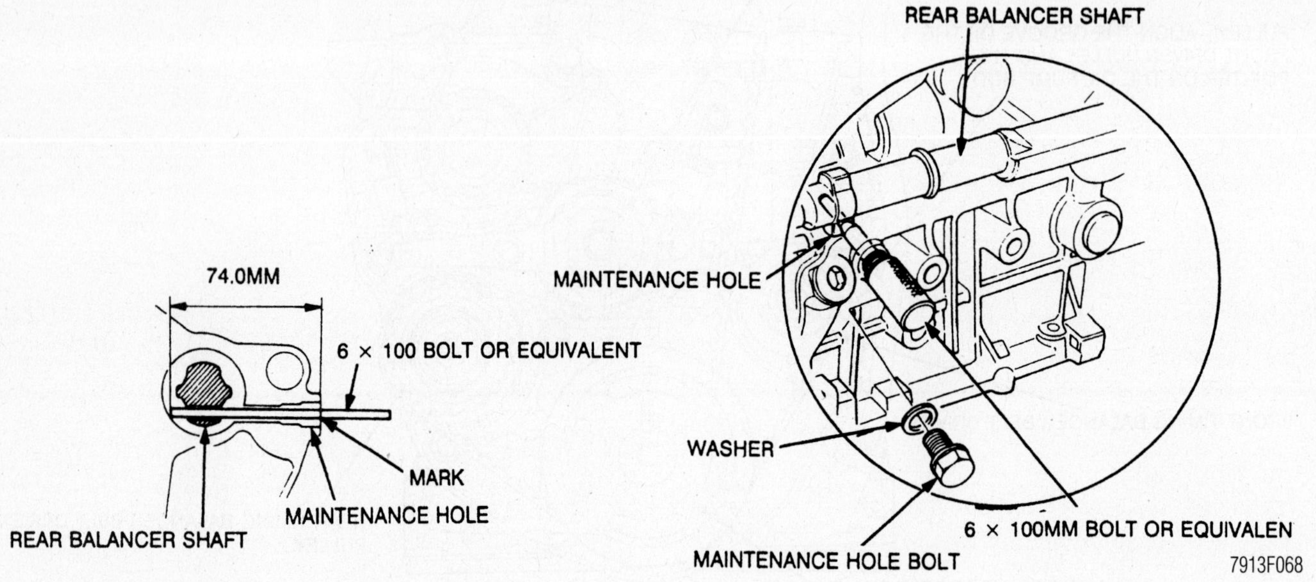

Aligning the rear timing balancer shaft—Honda 2.3L (H23A1) engine

28. Turn the crankshaft counterclockwise until the cam pulley has moved 3 teeth; this creates tension on the timing belt. Loosen, then tighten the adjusting nut and tighten it to 33 ft. lbs. (45 Nm). Tighten the bolt used to lock the timing belt tensioner.
29. Realign the timing belt timing marks.
30. Align the groove on the front balancer shaft pulley with the pointer on the oil pump.
31. Align the rear timing balancer pulley using a 6 x 100mm bolt or rod. Mark the bolt or rod at a point 2.913 in. (74mm) from the end. Remove the bolt from the maintenance hole on the side of the block; insert the bolt/rod into the hole and align the 74mm mark with the face of the hole. This pin will hold the shaft in place during installation.
32. Loosen the adjusting nut and ensure the timing balancer belt adjuster moves freely.
33. Install the balancer belt. Once the belts are in place, be sure that all the engine alignment marks are still correct. If not, remove the belts, realign the engine and reinstall the belts. Once the belts are properly installed, slowly loosen the adjusting nut, allowing the tensioner to move against the belt. Remove the pin from the maintenance hole and reinstall the bolt and washer.
34. Turn the crankshaft pulley 1 full turn and tighten the adjusting nut to 33 ft. lbs. (45 Nm).

■ **Both belt adjusters are spring loaded to properly tension the belts. Do not apply extra pressure to the pulleys or tensioners while performing the adjustment.**

35. Remove the 6 x 100mm bolt from the timing belt adjuster arm.
36. Install the timing belt and cylinder head covers. Reinstall all applicable components. When installing the crankshaft pulley, coat the threads and seating face of the pulley bolt with engine oil, then install and tighten the bolt to 181 ft. lbs. (250 Nm). When installing the side engine mount, tighten the bolt and nut attaching the mount to the engine to 40 ft. lbs. (55 Nm), and the through-bolt and nut to 47 ft. lbs. (65 Nm). Remove the jack from under the center beam.
37. Connect the negative battery cable.

2.6L (4ZE1/E) ENGINES

YEAR, MODEL & ENGINE APPLICATION	
1994-97 Honda Passport	2.6L Code 4ZE1/E
REPLACEMENT INTERVALS	
Honda recommends timing belt replacement every 60,000 miles.	
LABOR TIMES	
1994-97	
Replacement	3.0 hrs
Severe Service	3.5 hrs
With air conditioning add	0.3 hrs

88U7HUNg11

✻✻ CAUTION

These engines are an interference type. If the timing belt has failed, valve-to-piston damage may have occurred. A compression check should always be performed to check for engine damage.

1. Disconnect the negative battery cable.
2. Loosen and remove the engine accessory drive belts.
3. Remove the cooling fan assembly and the water pump pulley.
4. Drain the fluid from the power steering reservoir.
5. Unbolt and remove the power steering pump. Unbolt the hydraulic line brackets from the upper timing cover and move the pump from the work area without disconnecting the hydraulic lines.
6. Disconnect and remove the starter motor if a Flywheel Holder tool No. J-38674 or equivalent, is to be used.
7. Remove the upper timing belt cover.
8. Rotate the crankshaft to set the engine at Top Dead Center (TDC)

of the compression for the No. 1 cylinder. The arrow mark on the camshaft sprocket will be aligned with the mark on the rear timing cover.

9. Remove the crankshaft pulley.
10. Remove the lower timing belt cover.
11. Verify that the engine is set at TDC/compression for the No. 1 cylinder. The notch on the crankshaft sprocket will be aligned with the pointer on the oil seal retainer.
12. Release and remove the tensioner spring to release the timing belt tension.
13. Remove the timing belt.
14. Unbolt the tensioner pulley bracket from the engine's front cover.
15. If necessary, unbolt and remove the camshaft sprockets. If necessary, use a puller to remove the crankshaft pulley. Don't lose the crankshaft sprocket key.

To install:

16. If removed, install the camshaft and crankshaft sprockets. Align the camshaft and crankshaft timing marks and be sure to install any keys. Tighten the camshaft sprocket bolt to 43 ft. lbs. (59 Nm).
17. Install the tensioner assembly. Tighten the tensioner mounting bolt to 14 ft. lbs. (19 Nm) and the cap bolt to 108 inch lbs. (13 Nm).
18. Be sure the crankshaft and the camshaft sprockets are aligned with their timing marks. Install the timing belt onto the sprockets using the following sequence:
    - Crankshaft sprocket
    - Oil pump sprocket
    - Camshaft sprocket
    - Tensioner pulley
19. Loosen the tensioner mounting bolt. This will allow the tensioner spring to apply pressure to the timing belt.
20. After the spring has pulled the timing belt as far as possible, temporarily tighten the tensioner mounting bolt to 14 ft. lbs. (19 Nm).

■ **Remove the flywheel holder before rotating the crankshaft. Reinstall the holder to tighten the crankshaft pulley bolt.**

✱✱ WARNING

If any binding is felt when adjusting the timing belt tension by turning the crankshaft, STOP turning the engine, because the pistons may be hitting the valves.

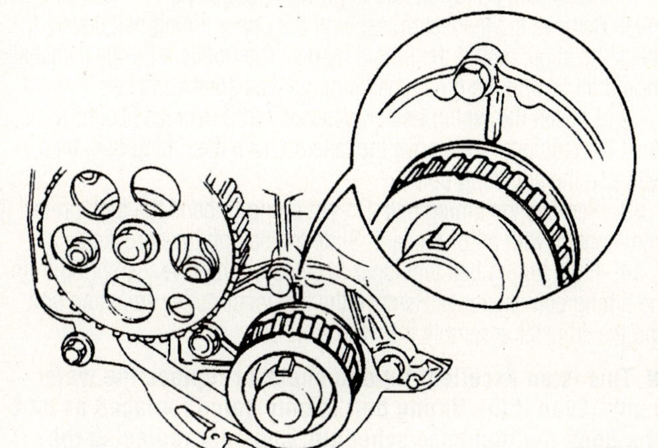
79245g01

Align the crankshaft pulley timing mark the with oil retainer setting mark—Honda 2.6L (4ZE1/E) engine

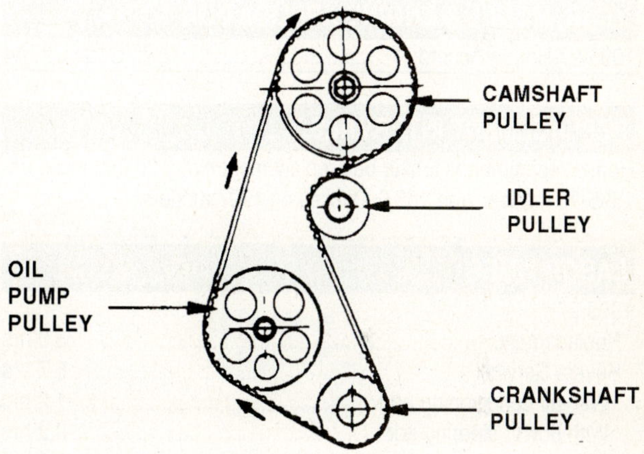

79245g02

Timing belt direction of travel—Honda 2.6L (4ZE1/E) engine

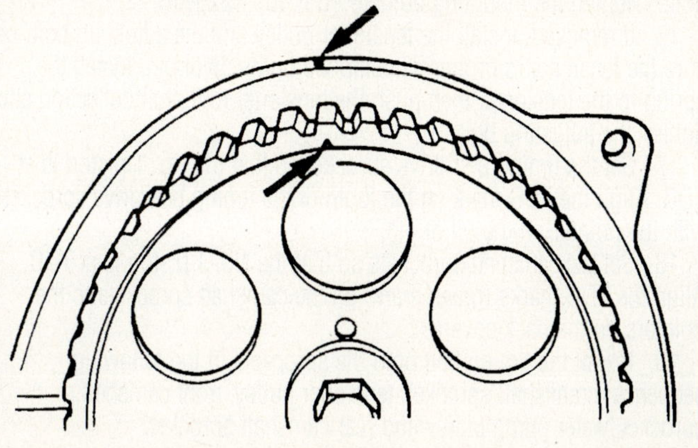
79245g04

Camshaft-to-timing cover mark—Honda 2.6L (4ZE1/E) engine

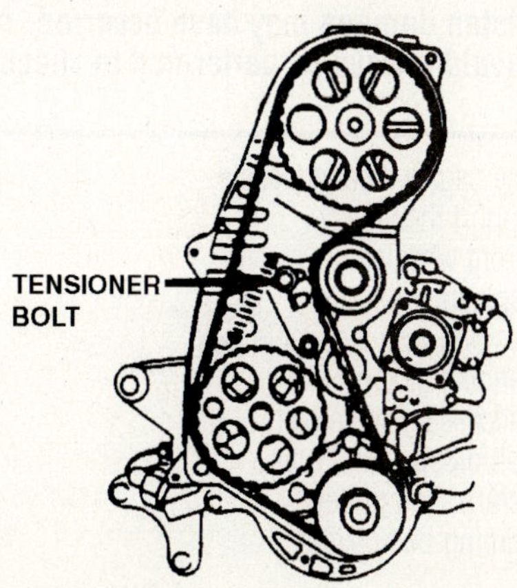

79245g03

Be sure to tighten the timing belt tensioner bolt (B)—Honda 2.6L (4ZE1/E) engine

21. Rotate the crankshaft counterclockwise 2 complete revolutions to check the rotation of the belt and the alignment of the timing marks. Listen for any rubbing noises that may mean the belt is binding.
22. Loosen the tensioner pulley bolt to allow the spring to adjust the correct tension. Then, retighten the tensioner pulley bolt to 14 ft. lbs. (19 Nm).
23. Install the lower timing belt cover and the crankshaft pulley.
24. Tighten the crankshaft pulley bolt to 87 ft. lbs. (118 Nm). Tighten the small pulley bolts to 72 inch lbs. (8 Nm).
25. Install the upper timing cover.
26. If removed, install the starter and tighten the bolts to 30 ft. lbs. (40 Nm).
27. Install the power steering pump. If the hydraulic lines were disconnected, refill and bleed the power steering system.
28. Install the water pump pulley and tighten its nut to 20 ft. lbs. (26 Nm).
29. Install the cooling fan assembly.
30. Install and adjust the accessory drive belts.
31. Connect the negative battery cable.

2.7L (C27A4) ENGINES

YEAR, MODEL & ENGINE APPLICATION	
1995-97 Honda Accord	2.7L Code C27A4

REPLACEMENT INTERVALS

Honda recommends timing belt replacement every 90,000 miles on 1995-96 models, and 105,000 miles on 1997 models.

LABOR TIMES	
1995-97	
Replacement	5.0 hrs
Severe Service	5.7 hrs
With air conditioning add	1.2 hrs
With power steering add	0.2 hrs
With water pump replacement add	0.6 hrs

8807HONg12

✻✻ CAUTION

These engines are an interference type. If the timing belt has failed, valve-to-piston damage may have occurred. A compression check should always be performed to check for engine damage.

1. Disconnect the negative battery cable.
2. Raise and support the vehicle.
3. Remove the front wheels.
4. Remove the splash shield.
5. Support the engine with a suitable jack.
6. Remove the engine side mount.
7. Remove the accessory drive belts.
8. Remove the oil dipstick and tube.
9. Remove the crankshaft pulley.
10. Remove the timing belt covers.

■ **Do not use the covers to store removed items.**

11. Turn the engine to align the timing marks and set cylinder No. 1 to Top Dead Center (TDC) on the compression stroke. The white mark on the crankshaft pulley should align with the pointer on the timing belt cover. Remove the inspection caps on the upper timing belt covers to check the alignment of the timing marks. The pointers for the camshafts should align with the green marks on the camshaft sprockets.

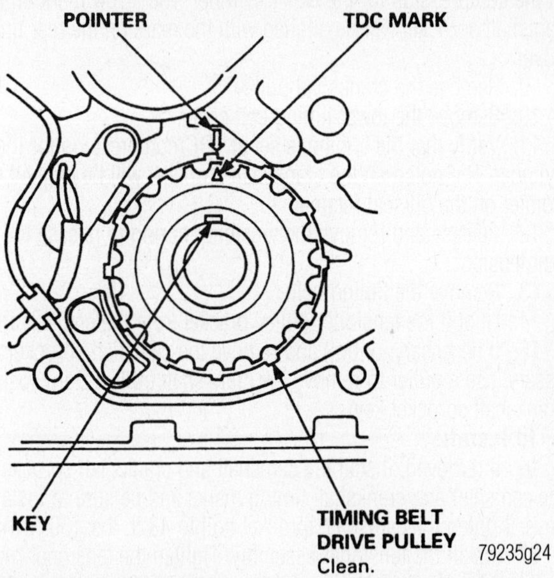

Crankshaft timing belt sprocket alignment mark locations—Honda 2.7L (C27A4) engine

12. Loosen the timing belt adjuster bolt 180 degrees (1/2 turn). Push the tensioner to remove the tension from the timing belt, then retighten the adjusting bolt.
13. Remove the timing belt. Do not crimp or bend the belt; protect it from contact with oil or coolant. Slide the belt off the sprockets.
14. If the timing belt tensioner is defective, remove the spring from the timing belt tensioner. Remove the tensioner pulley adjusting bolt and the adjuster assembly from the engine.

■ **This is an excellent time to check or replace the water pump. Even if the timing belt is only being replaced as part of a good maintenance schedule, consider replacing the pump at the same time.**

To install:

15. If the water pump is to be replaced, install a new O-ring and make certain it is properly seated. Install the water pump and retaining bolts. Tighten the mounting bolts to 16 ft. lbs. (22 Nm).
16. If removed, install the tensioner pulley and the adjusting bolt, be sure the tensioner is properly positioned on its pivot pin. Install the spring to the tensioner, then push the tensioner to its full deflection and tighten the adjusting bolt.
17. Set the timing belt drive sprocket so that the No. 1 piston is at TDC. Align the TDC mark on the tooth of the timing belt drive sprocket with the pointer on the oil pump.
18. Set the camshaft sprockets so that the No. 1 piston is at TDC. Align the TDC marks (green mark) on the camshaft sprockets to the pointers on the back covers.
19. Install the timing belt onto the sprockets in the following sequence: crankshaft sprocket, tensioner pulley, front camshaft sprocket, water pump pulley and rear camshaft sprocket.
20. Loosen, then retighten the timing belt adjuster bolt to tension the timing belt.
21. Install the lower timing belt cover.
22. Install the crankshaft sprocket and the crankshaft pulley bolt.

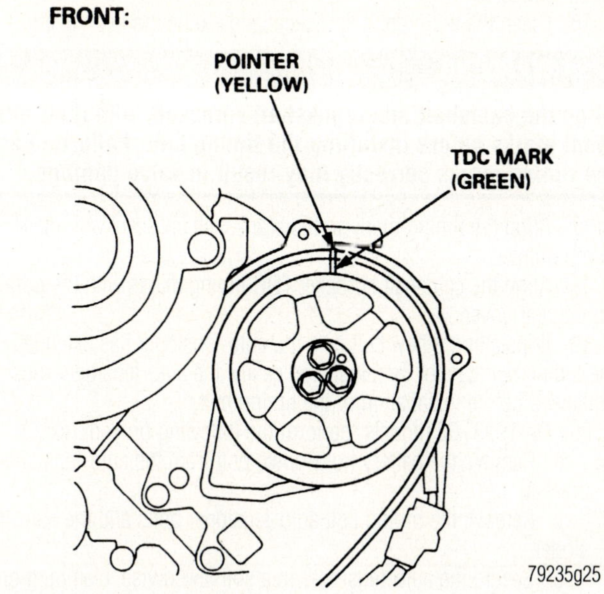

Left camshaft timing belt sprocket alignment mark location—Honda 2.7L (C27A4) engine

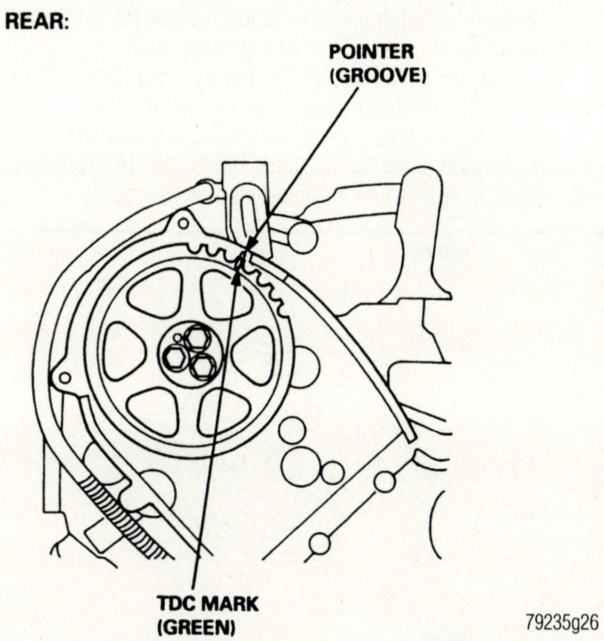

Rear camshaft timing belt sprocket alignment mark location—Honda 2.7L (C27A4) engine

Tighten the bolt to 181 ft. lbs. (245 Nm) with the aid of the crank pulley holder.

23. Rotate the crankshaft 5–6 turns clockwise so that the timing belt positions on the sprockets.
24. Set cylinder No. 1 to TDC by aligning the timing marks. If the timing marks do not align, remove the timing belt, then adjust the components and reinstall the timing belt.
25. Loosen the timing belt adjusting bolt 180 degrees (1/2 turn) and retighten the adjusting bolt. Tighten the adjusting bolt to 31 ft. lbs. (42 Nm).
26. Install the upper timing belt cover and all other applicable components. When installing the side engine mount to the engine, use 3 new attaching bolts. Tighten the new bolts to 40 ft. lbs. (54 Nm).
27. Connect the negative battery cable.

3.0L (J30A1) V6 ENGINES

YEAR, MODEL & ENGINE APPLICATION

1998-04 Honda Accord	3.0L Code J30A1

REPLACEMENT INTERVALS

Honda recommends timing belt replacement every 105,000 miles.

LABOR TIMES

1998-02	
Replacement	5.0 hrs
Severe Service	5.7 hrs
2003-04	
Replacement	4.6 hrs
Severe Service	5.2 hrs
With air conditioning add	1.2 hrs
With power steering add	0.2 hrs
With water pump replacement add	0.6 hrs

8807HONg13

✻✻ CAUTION

These engines are an interference type. If the timing belt has failed, valve-to-piston damage may have occurred. A compression check should always be performed to check for engine damage.

■ The radio may contain a coded theft protection circuit. Always make note of the code number before disconnecting the battery.

1. Disconnect the negative battery terminal.
2. Turn the crankshaft so the white mark on the crankshaft pulley aligns with the pointer on the oil pump housing cover.
3. Open the inspection plugs on the upper timing belt covers and check that the camshaft sprocket marks align with the upper cover marks.

✻✻ WARNING

Align the camshaft and crankshaft sprockets with their alignment marks before removing the timing belt. Failure to align the timing marks correctly may result in valve damage.

4. Raise and safely support the vehicle and remove both front tires/wheels.
5. Remove the front lower splash shield.
6. Remove the accessory drive belts.
7. Support the weight of the engine by placing a wood block on a floor jack and carefully lift on the oil pan.
8. Remove the bolts from the side engine mount bracket and remove the bracket.
9. Remove the dipstick, the dipstick tube and discard the O-ring.
10. Hold the crankshaft pulley with the Handle tool 07JAB-001020A and Crankshaft Holding tool 07MAB-PY3010A or equivalent. While

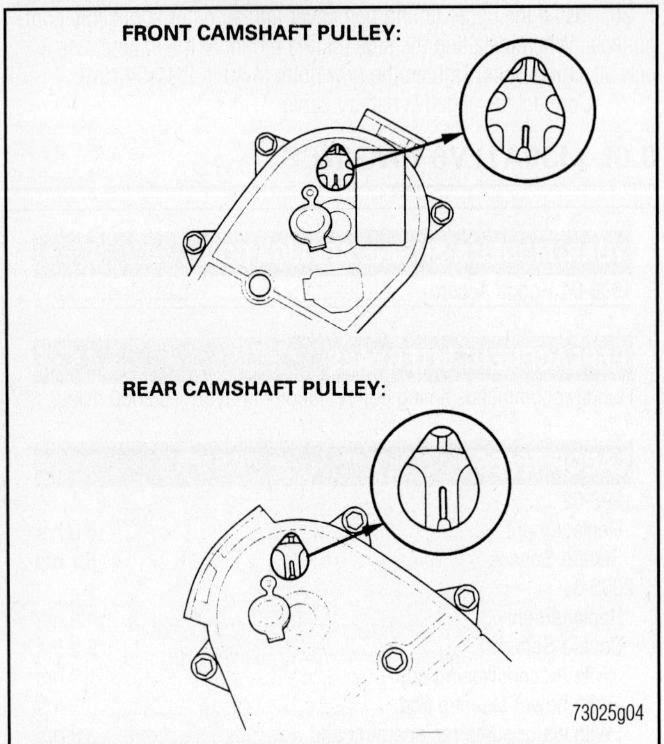

Crankshaft and camshaft timing marks at Top Dead Center (TDC)—Honda 3.0L (J30A1) engine

holding the crankshaft pulley, remove the crankshaft pulley bolt using a heavy duty 3/4 in. (19mm) socket and breaker bar.

11. Remove the crankshaft pulley, the upper timing belt covers and the lower timing belt cover.

12. Remove one of the battery clamp fasteners from the battery tray and grind a 45 degree bevel on the threaded end of the battery clamp bolt.

13. Screw in the battery hold-down bolt into the threaded bracket just above the auto-tensioner (automatic timing belt adjuster) and tighten the bolt hand-tight to hold the auto-tensioner adjuster in its current position.

14. Remove the engine mount bracket bolts and the bracket.

15. Loosen the timing belt idler pulley bolt (located on the right side across from the auto-tensioner pulley) about 5–6 revolutions and remove the timing belt.

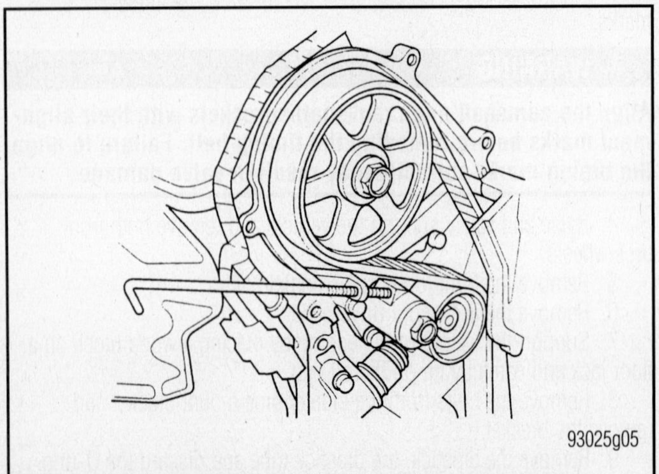

Battery hold-down bolt installed to hold auto-tensioner—Honda 3.0L (J30A1) engine

To install:

16. Clean the timing belt sprockets and the timing belt covers.

**** WARNING**

Align the camshaft and crankshaft sprockets with their alignment marks before installing the timing belt. Failure to align the timing marks correctly may result in valve damage.

17. Align the timing mark on the crankshaft sprocket with the oil pump pointer.

18. Align the camshaft sprocket TDC timing marks with the pointers on the rear cover.

19. If installing a new belt or if the auto-tensioner has extended or if the timing belt cannot be reinstalled easily, the auto-tensioner must be collapsed before installation of the timing belt.

20. On 1998–02 models, perform the following procedures:

a. Remove the battery hold-down bolt from the auto-tensioner bracket.

b. Remove the timing belt auto-tensioner bolts and the auto-tensioner.

c. Secure the auto-tensioner in a soft jawed vise, clamping onto the flat surface of one of the mounting bolt holes with the maintenance bolt facing upward.

d. Remove the maintenance bolt and use caution not to spill oil from the tensioner assembly.

e. Should oil spill from the tensioner, be sure the tensioner is filled with 0.22 ounces (6.5 ml) of fresh engine oil.

f. Using care not to damage the threads or the gasket sealing surface, insert a flat-blade screwdriver through the tensioner maintenance hole and turn the screwdriver clockwise to compress the auto-tensioner bottom while the Tensioner Holder tool 14540-P8A-A01 or equivalent, is installed on the auto-tensioner assembly.

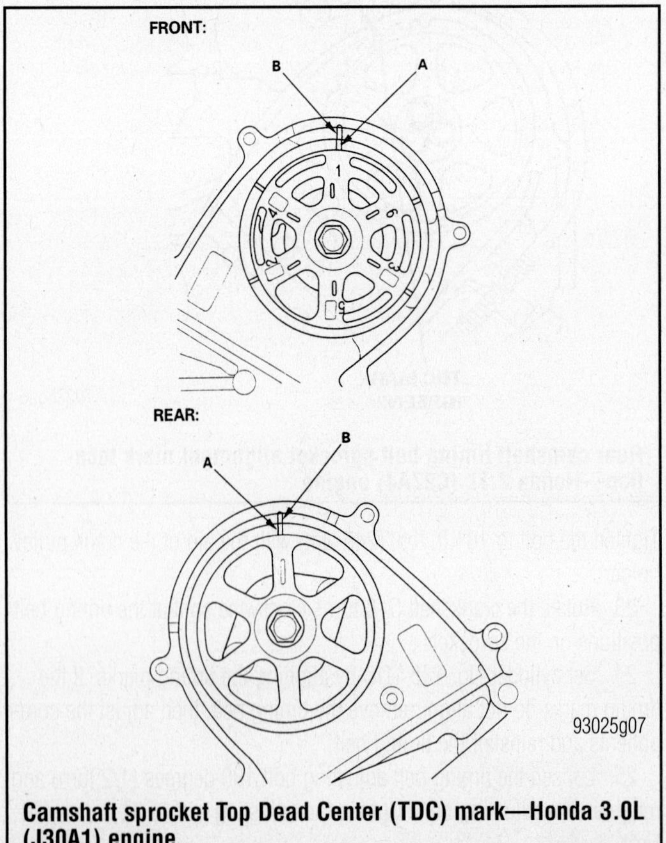

Camshaft sprocket Top Dead Center (TDC) mark—Honda 3.0L (J30A1) engine

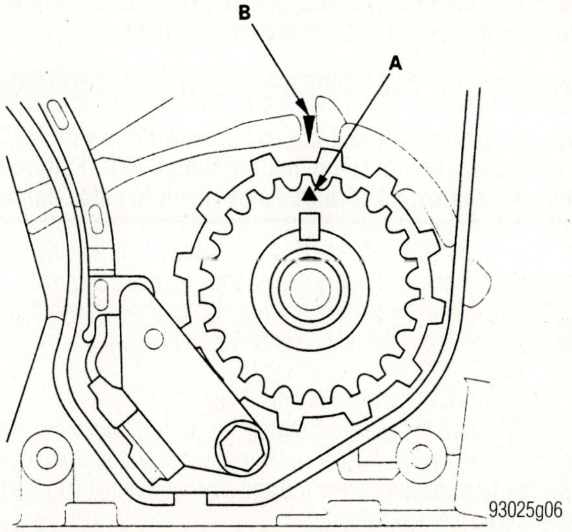

Crankshaft sprocket Top Dead Center (TDC) mark—Honda 3.0L (J30A1) engine

g. Install the auto-tensioner maintenance bolt with a new gasket and tighten to a torque 72 inch lbs. (8 Nm).

21. On 2003–04 models, align the holes of the tensioner and the rod. Compress the tensioner in a press using no more than 2200 lbs. of pressure until an 0.08 in. pin can be inserted into the tensioner holes.

22. On all models, install the auto-tensioner on the engine with the tensioner holder tool installed and torque the mounting bolts to 104 inch lbs. (12 Nm).

23. Install the timing belt in a counterclockwise pattern starting with the crankshaft drive sprocket. Install the timing belt counterclockwise in the following sequence:

- Crankshaft drive sprocket.
- Idler pulley.
- Left side camshaft sprocket.
- Water pump.
- Right side camshaft sprocket.
- Auto-tensioner adjustment pulley.

24. Torque the timing belt idler pulley bolt to 33 ft. lbs. (44 Nm).

25. Remove the auto-tensioner holding tool to allow the tensioner to extend.

26. Install the engine mount bracket to the engine and torque the bolts to 33 ft. lbs. (44 Nm).

27. Install the lower timing belt cover and both upper timing belt covers.

28. Hold the crankshaft pulley with special tools 07JAB-001020A handle and 07MAB-PY3010A crankshaft holding tool or equivalent tools. While holding the crankshaft pulley, install the crankshaft pulley bolt using a heavy duty 3/4 in. (19mm) socket and a commercially available torque wrench and torque the bolt to 181 ft. lbs. (245 Nm).

✱✱ WARNING

If any binding is felt while moving the crankshaft pulley, STOP turning the crankshaft pulley immediately because the pistons may be hitting the valves.

29. Rotate the crankshaft pulley clockwise 5–6 revolutions to allow the timing belt to be seated in the pulleys.

30. Move the crankshaft pulley to the white TDC mark and inspect the camshaft TDC marks to ensure proper timing of the camshafts.

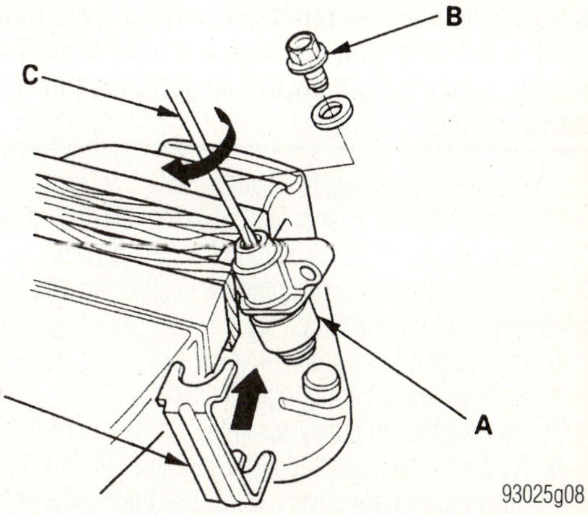

Adjusting the auto-tensioner—Honda 3.0L (J30A1) engine

✱✱ WARNING

If the timing marks do not align, the timing belt removal and installation procedure must be performed again.

31. Install the engine dipstick tube using a new O-ring.
32. Install the accessory drive belts.
33. Install both engine mount bracket bolts and torque to 33 ft. lbs. (44 Nm).
34. Install the bushing through bolt and tighten to 40 ft. lbs. (54 Nm).
35. Release and carefully remove the floor jack.
36. Install the front lower splash shield.
37. Install both front tires/wheels.
38. Carefully lower the vehicle.
39. Install the battery hold-down bolt in the battery tray.
40. Install the negative battery cable.
41. Enter the radio security code.

3.2L (6VD1) ENGINES

YEAR, MODEL & ENGINE APPLICATION

1994-02 Honda Passport	3.2L Code 6VD1

REPLACEMENT INTERVALS

Honda recommends timing belt replacement every 60,000 miles on 1994-97 models, every 75,000 miles on 1998-99 models and 100,000 miles on 2000-02 models.

LABOR TIMES

1994-02	
Replacement	3.7 hrs
Severe Service	4.2 hrs
With air conditioning add	0.3 hrs
With water pump replacement add	0.7 hrs

8807HONg14

✻✻ CAUTION

These engines are an interference type. If the timing belt has failed, valve-to-piston damage may have occurred. A compression check should always be performed to check for engine damage.

1. Disconnect the negative battery cable.
2. Remove the air cleaner assembly and intake air duct.
3. Remove the upper fan shroud from the radiator.
4. Remove the 4 nuts retaining the cooling fan assembly. Remove the cooling fan from the fan pulley.
5. Loosen and remove the drive belts.
6. Remove the upper timing belt covers.
7. Remove the fan pulley assembly.
8. Rotate the crankshaft to align the camshaft timing marks with the pointer dots on the back covers. Verify that the pointer on the crankshaft aligns with the mark on the lower timing cover.

■ **When the timing marks are aligned, the No. 2 piston is at Top Dead Center (TDC) of the compression.**

✻✻ WARNING

Align the camshaft and crankshaft sprockets with their alignment marks before removing the timing belt. Failure to align the belt and sprocket marks may result in valve damage.

9. Use tool No. J-8614-01 or a suitable pulley holding tool to remove the crankshaft pulley center bolt. Remove the crankshaft pulley.
10. If present, disconnect the 2 oil cooler hose bracket bolts on the timing cover. Move the oil cooler hoses and bracket off of the lower timing cover.
11. Remove the lower timing belt cover.
12. Remove the pusher assembly (tensioner) from below the belt tensioner pulley. The pusher rod must always face upward to prevent oil leakage. Depress the pusher rod and insert a wire pin into the hole to keep the pusher rod retracted.

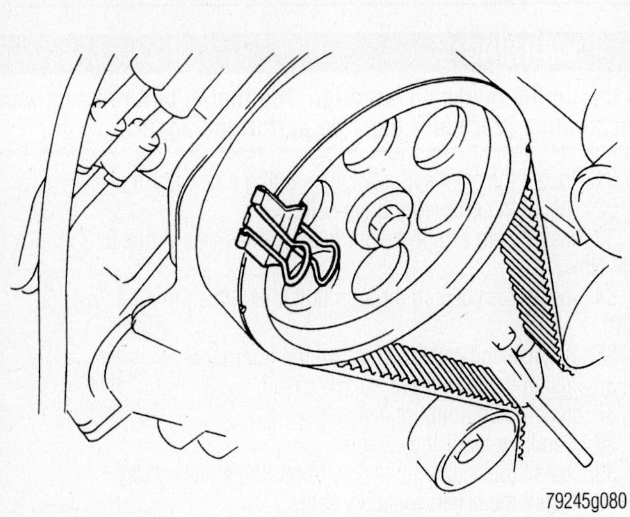

79245g080

Using a double clip to hold the belt in place—Honda 3.2L (6VD1) engine

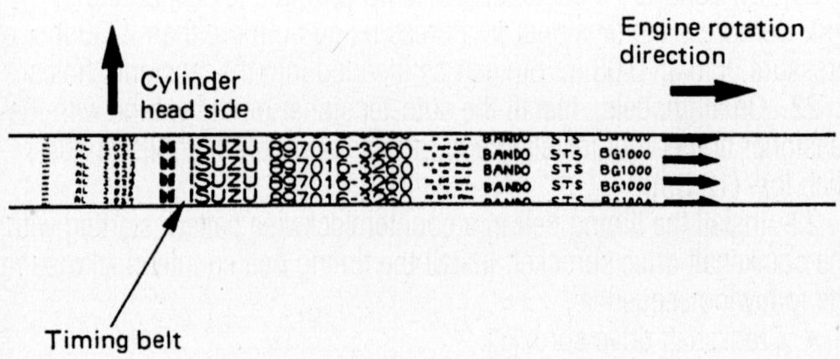

79245g07

For maximum timing belt life, install the belt as shown—Honda 3.2L (6VD1) engine

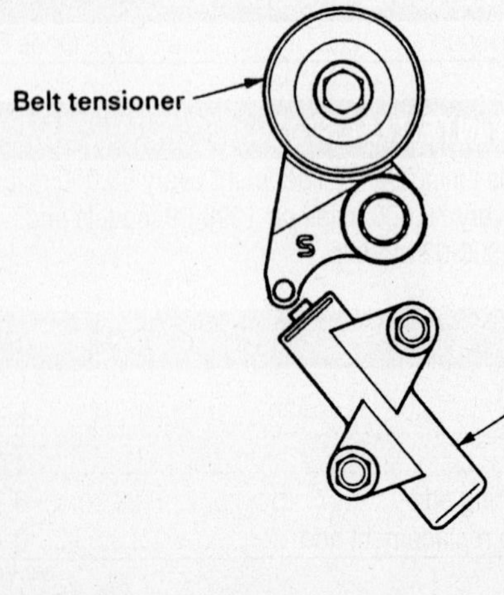

93025g10

View of timing belt tensioner and pusher—Honda 3.2L (6VD1) engine

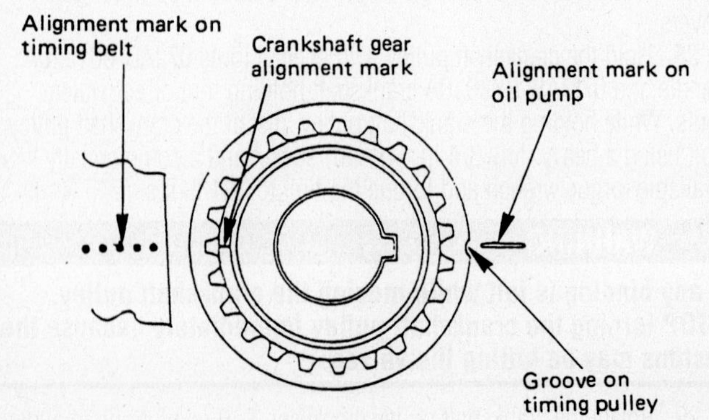

79245g06

Proper crankshaft alignment marks for timing belt installation—Honda 3.2L (6VD1) engine

13. Remove the timing belt.

14. Inspect the water pump and replace it if there is any doubt about its condition.

15. Repair any oil or coolant leaks before installing a new timing belt. If the timing belt has been contaminated with oil or coolant, or is damaged, it must be replaced.

To install:

16. Verify that the sprocket timing marks are still aligned and that the groove and the keyway on the crankshaft timing sprocket align with the mark on the oil pump. The white pointers on the camshaft timing sprockets should align with the dots on the front plate.

17. Install the timing belt. Use clips to secure the belt onto each sprocket until the installation is complete. Align the dotted marks on the timing belt with the timing mark opposite the groove on the crankshaft sprocket.

■ **The arrows on the timing belt must follow the belt's direction of rotation. The manufacturer's trademark on the belt's spine should be readable left-to-right when the belt is installed.**

18. Align the white line on the timing belt with the alignment mark on the right bank camshaft timing pulley. Secure the belt with a clip.

**** WARNING**

If any binding is felt when adjusting the timing belt tension by turning the crankshaft, STOP turning the engine, because the pistons may be hitting the valves.

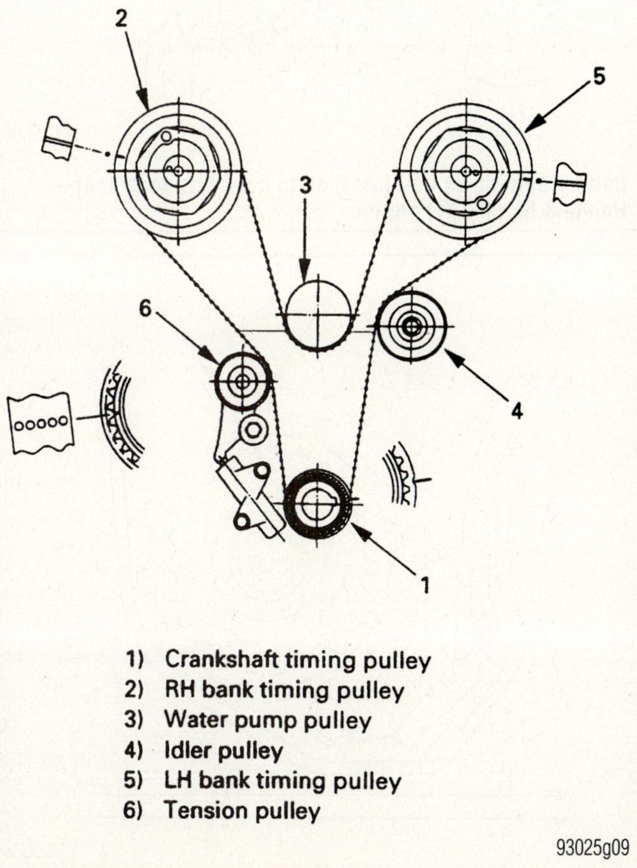

Timing mark aignment and timing belt routing —Honda 3.2L (6VD1) engine

19. Rotate the crankshaft counterclockwise to remove the slack between the crankshaft sprocket and the right camshaft timing belt sprocket.

20. Install the belt around the water pump pulley.

21. Install the belt on the idler pulley.

22. Align the white alignment mark on the timing belt with the alignment mark on the left bank camshaft timing belt sprocket.

23. Install the crankshaft pulley and tighten the center bolt by hand. Rotate the crankshaft pulley clockwise to give slack between the crankshaft timing belt pulley and the right bank camshaft timing belt pulley.

24. Insert a 1.4mm piece of wire through the hole in the pusher to hold the rod in. Install the pusher assembly while pushing the tension pulley toward the belt.

25. Pull the pin out from the pusher to release the rod.

26. Remove the clamps from the sprockets. Rotate the crankshaft pulley clockwise 2 turns. Measure the rod protrusion to ensure it is between 0.16–0.24 in. (4–6mm).

27. If the tensioner pulley bracket pivot bolt was removed, tighten it to 31 ft. lbs. (42 Nm).

28. Tighten the pusher bolts to 14 ft. lbs. (19 Nm).

29. Remove the crankshaft pulley. Install the lower and upper timing belt covers and tighten their bolts to 12 ft. lbs. (17 Nm).

30. Fit the oil cooler hose onto the timing cover and tighten its mounting bracket bolts to 16 ft. lbs. (22 Nm).

31. Install the crankshaft pulley and tighten the pulley bolt to 123 ft. lbs. (167 Nm).

32. Install fan pulley assembly and tighten the bolts to 16 ft. lbs. (22 Nm).

33. Install and adjust the accessory drive belts.

34. Install the cooling fan assembly and tighten the bolts to 72 inch lbs. (8 Nm).

35. Install the upper fan shroud.

36. Install the air cleaner assembly and intake air duct.

37. Connect the negative battery cable.

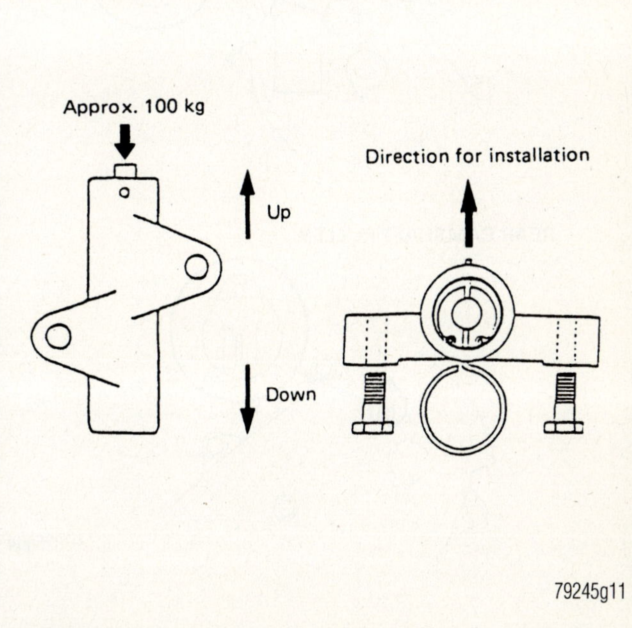

View of timing belt tensioner pusher—Honda 3.2L (6VD1) engine

3.5L (J35A1) V6 ENGINES

YEAR, MODEL & ENGINE APPLICATION

1999-04 Honda Odyssey	3.5L Code J35A1

REPLACEMENT INTERVALS

Honda recommends timing belt replacement every 105,000 miles.

LABOR TIMES

1999-04	
Replacement	4.6 hrs
Severe Service	5.2 hrs
With water pump replacement add	0.7 hrs

9907HONg15

✻✻ CAUTION

These engines are an interference type. If the timing belt has failed, valve-to-piston damage may have occurred. A compression check should always be performed to check for engine damage.

■ The radio may contain a coded theft protection circuit. Always make note of the code number before disconnecting the battery.

1. Disconnect the negative battery terminal.
2. Turn the crankshaft so the white mark on the crankshaft pulley aligns with the pointer on the oil pump housing cover.

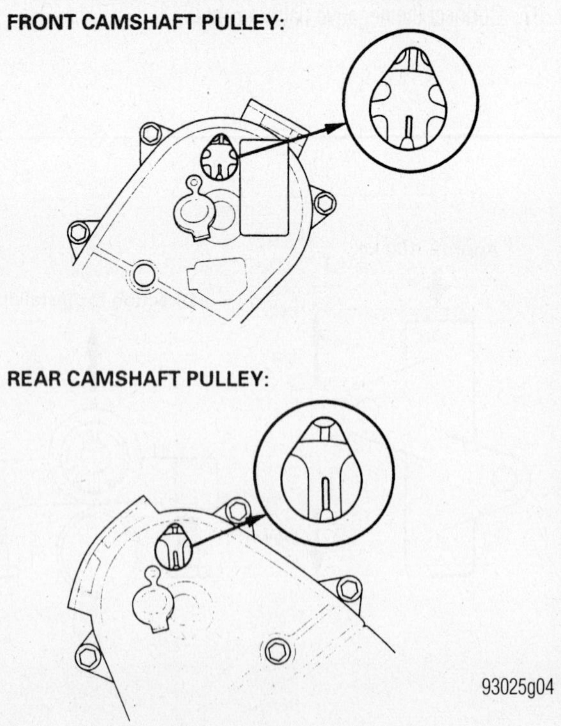

93025g04

Crankshaft and camshaft timing marks at Top Dead Center (TDC)—Honda 3.5L (J35A1) engine

3. Open the inspection plugs on the upper timing belt covers and check that the camshaft sprocket marks align with the upper cover marks.

✻✻ WARNING

Align the camshaft and crankshaft sprockets with their alignment marks before removing the timing belt. Failure to align the timing marks correctly may result in valve damage.

4. Raise and safely support the vehicle and remove both front tires/wheels.
5. Remove the front lower splash shield.
6. Move the alternator tensioner with a Belt Tensioner Release Arm tool YA9317 or equivalent, to release tension from the belt and remove the alternator drive belt.
7. Remove the alternator belt tensioner release arm.
8. Loosen the power steering pump adjustment nut, adjustment

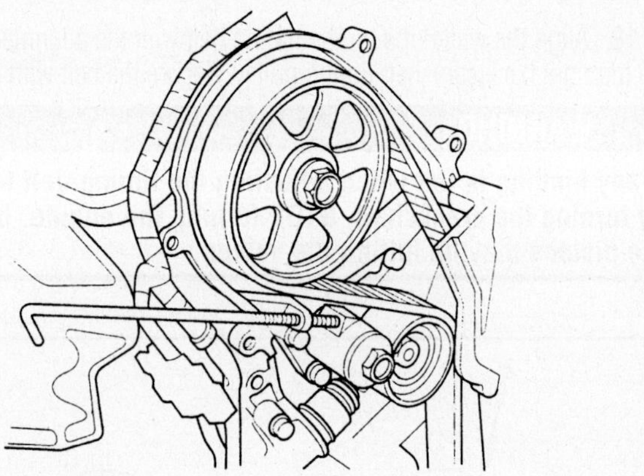

93025g05

Battery hold-down bolt installed to hold auto-tensioner—Honda 3.5L (J35A1) engine

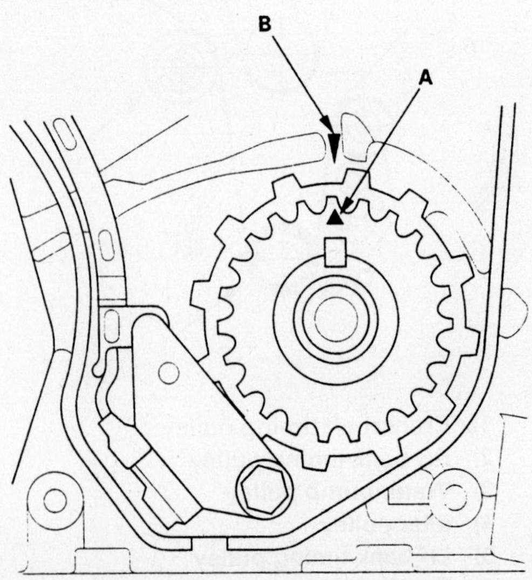

93025g06

Crankshaft sprocket Top Dead Center (TDC) mark— Honda 3.5L (J35A1) engine

locknut and mounting bolt, then remove the power steering pump with the hoses attached.

9. Support the weight of the engine by placing a wood block on a floor jack and carefully lift on the oil pan.

10. Remove the bolts from the side engine mount bracket and remove the bracket.

11. Remove the dipstick, the dipstick tube and discard the O-ring.

12. Hold the crankshaft pulley with the Handle tool 07JAB-001020A and Crankshaft Holding tool 07MAB-PY3010A or equivalent. While holding the crankshaft pulley, remove the crankshaft pulley bolt using a heavy duty 3/4 in. (19mm) socket and breaker bar.

13. Remove the crankshaft pulley, the upper timing belt covers and the lower timing belt cover.

14. Remove one of the battery clamp fasteners from the battery tray and grind a 45 degree bevel on the threaded end of the battery clamp bolt.

15. Screw in the battery hold-down bolt into the threaded bracket just above the auto-tensioner (automatic timing belt adjuster) and tighten the bolt hand-tight to hold the auto-tensioner adjuster in its current position.

16. Remove the engine mount bracket bolts and the bracket.

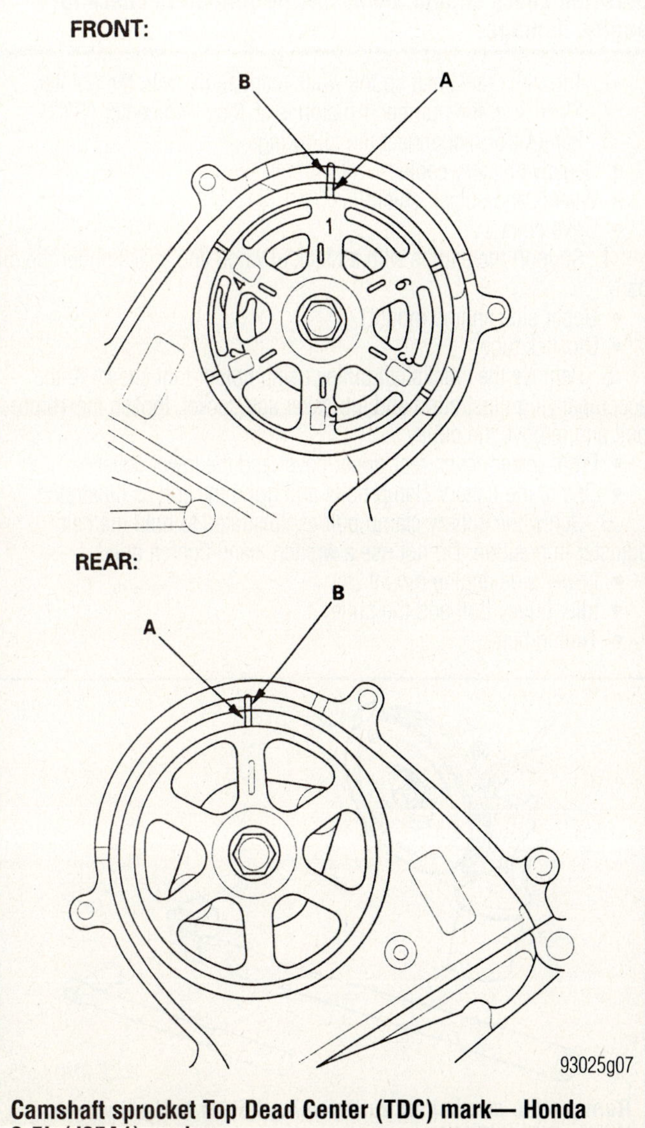

Camshaft sprocket Top Dead Center (TDC) mark— Honda 3.5L (J35A1) engine

17. Loosen the timing belt idler pulley bolt (located on the right side across from the auto-tensioner pulley) about 5–6 revolutions and remove the timing belt.

To install:

18. Clean the timing belt sprockets and the timing belt covers.

**** WARNING**

Align the camshaft and crankshaft sprockets with their alignment marks before installing the timing belt. Failure to align the timing marks correctly may result in valve damage.

19. Align the timing mark on the crankshaft sprocket with the oil pump pointer.

20. Align the camshaft sprocket TDC timing marks with the pointers on the rear cover.

21. If installing a new belt or if the auto-tensioner has extended or if the timing belt cannot be reinstalled easily, the auto-tensioner must be collapsed before installation of the timing belt.

22. On 1998–02 models, perform the following procedures:

a. Remove the battery hold-down bolt from the auto-tensioner bracket.

b. Remove the timing belt auto-tensioner bolts and the auto-tensioner.

c. Secure the auto-tensioner in a soft jawed vise, clamping onto the flat surface of one of the mounting bolt holes with the maintenance bolt facing upward.

d. Remove the maintenance bolt and use caution not to spill oil from the tensioner assembly.

e. Should oil spill from the tensioner, be sure the tensioner is filled with 0.22 ounces (6.5 ml) of fresh engine oil.

f. Using care not to damage the threads or the gasket sealing surface, insert a flat-blade screwdriver through the tensioner maintenance hole and turn the screwdriver clockwise to compress the auto-tensioner bottom while the Tensioner Holder tool 14540-P8A-A01 or equivalent, is installed on the auto-tensioner assembly.

g. Install the auto-tensioner maintenance bolt with a new gasket and tighten to a torque 72 inch lbs. (8 Nm).

23. On 2003–04 models, align the holes of the tensioner and the rod. Compress the tensioner in a press using no more than 2200 lbs. of pressure until an 0.08 in. pin can be inserted into the tensioner holes.

24. On all models, install the auto-tensioner on the engine with the tensioner holder tool installed and torque the mounting bolts to 104 inch lbs. (12 Nm).

25. Install the timing belt in a counterclockwise pattern starting with

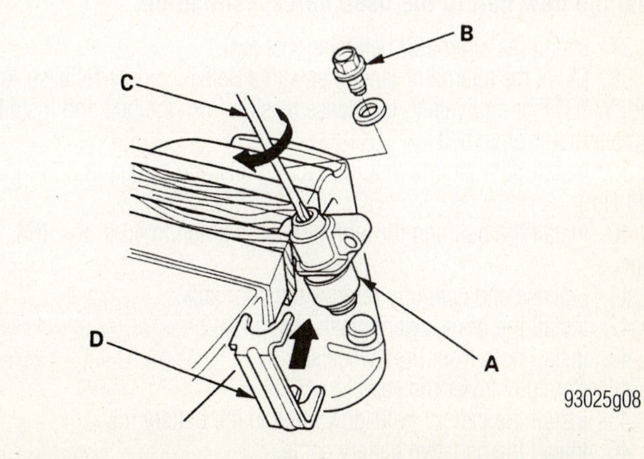

Adjusting the auto-tensioner— Honda 3.5L (J35A1) engine

the crankshaft drive sprocket. Install the timing belt counterclockwise in the following sequence:

- Crankshaft drive sprocket.
- Idler pulley.
- Left side camshaft sprocket.
- Water pump.
- Right side camshaft sprocket.
- Auto-tensioner adjustment pulley.

26. Torque the timing belt idler pulley bolt to 33 ft. lbs. (44 Nm).
27. Remove the auto-tensioner holding tool to allow the tensioner to extend.
28. Install the engine mount bracket to the engine and torque the bolts to 33 ft. lbs. (44 Nm).
29. Install the lower timing belt cover and both upper timing belt covers.
30. Hold the crankshaft pulley with special tools 07JAB-001020A handle and 07MAB-PY3010A crankshaft holding tool or equivalent tools. While holding the crankshaft pulley, install the crankshaft pulley bolt using a heavy duty 3/4 in. (19mm) socket and a commercially available torque wrench and torque the bolt to 181 ft. lbs. (245 Nm).

✳✳ WARNING

If any binding is felt while moving the crankshaft pulley, STOP turning the crankshaft pulley immediately because the pistons may be hitting the valves.

31. Rotate the crankshaft pulley clockwise 5–6 revolutions to allow the timing belt to be seated in the pulleys.
32. Move the crankshaft pulley to the white TDC mark and inspect the camshaft TDC marks to ensure proper timing of the camshafts.

✳✳ WARNING

If the timing marks do not align, the timing belt removal and installation procedure must be performed again.

33. Install the engine dipstick tube using a new O-ring.
34. Install the power steering pump, and loosely install the mounting bolt, adjustment locknut and adjustment nut.
35. Adjust the power steering belt to a tension such that a 22 lb. (98 N) pull halfway between the 2 drive pulleys will allow the belt to move 0.51–0.65 in. (13.0–16.5mm).
36. Tighten the power steering pump mounting bolt and adjustment locknut.

■ If a new belt is used, set the deflection to 0.33–0.43 in. (8.5–11.0mm) and after engine has run for 5 minutes, readjust the new belt to the used belt specification.

37. Install the alternator belt tensioner arm.
38. Move the alternator tensioner with a Belt Tensioner Release Arm tool YA9317 or equivalent, to release tension from the belt and install the alternator drive belt.
39. Install both engine mount bracket bolts and torque to 33 ft. lbs. (44 Nm).
40. Install the bushing through bolt and tighten to 40 ft. lbs. (54 Nm).
41. Release and carefully remove the floor jack.
42. Install the front lower splash shield.
43. Install both front tires/wheels.
44. Carefully lower the vehicle.
45. Install the battery hold-down bolt in the battery tray.
46. Install the negative battery cable.
47. Enter the radio security code.

3.5L (J35A4) V6 ENGINES

YEAR, MODEL & ENGINE APPLICATION

2003-05 Honda Pilot	3.5L Code J35A4

REPLACEMENT INTERVALS

Honda recommends timing belt relacement every 105,000 miles.

LABOR TIMES

2003-04	
Replacement	4.6 hrs
Severe Service	5.2 hrs
With water pump replacement add	0.7 hrs

8807HONg16

✳✳ CAUTION

These engines are an interference type. If the timing belt has failed, valve-to-piston damage may have occurred. A compression check should always be performed to check for engine damage.

1. Turn the crankshaft so the white mark aligns with the pointer.
2. Make sure the number 1 piston is at Top Dead center (TDC).
3. Remove or disconnect the following:

- Negative battery cable
- Wheels and splash shield
- Drive belts

4. Support the engine with a block of wood and a jack under the oil pan.

- Upper side engine mount
- Dipstick tube

5. Remove the crankshaft pulley using holder tool shown in the accompanying illustration and a breaker and socket, loosen the 19mm bolt and remove the pulley.

- Front upper cover, rear upper cover and the lower cover
- One of the battery clamp bolts and grind the end as illustrated

6. Screw the battery clamp bolt as illustrated to hold the belt adjuster in position. Do not use a wrench, hand tighten only.

- Lower side engine mount
- Idler pulley bolt and the pulley
- Timing belt

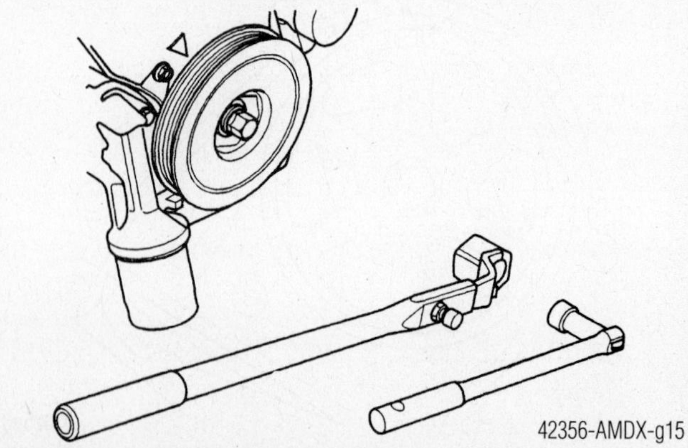

42356-AMDX-g15

Remove the crankshaft pulley using holder tool shown—Honda 3.5L (J35A4) engine

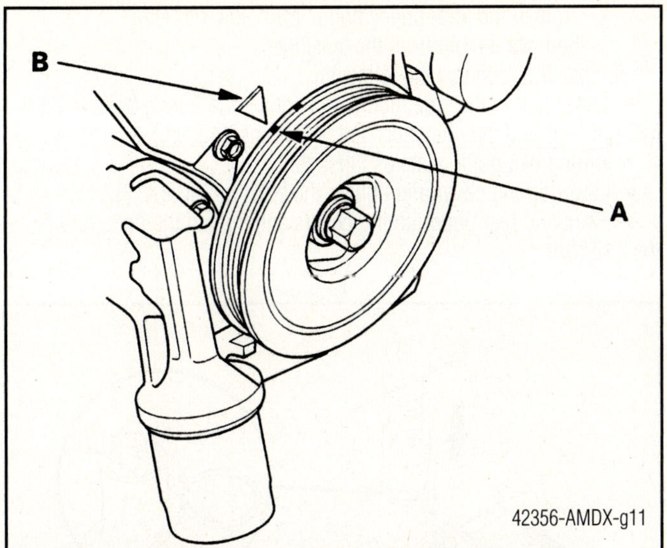

Turn the crankshaft so the white mark (A) aligns with the pointer (B) — Honda 3.5L (J35A4) engine

To install:

7. If installing a new belt, perform the following steps:
   a. Clean the pulleys, belt guide plate and the upper and lower covers.
   b. Set the timing belt drive pulley to TDC by aligning the TDC mark on the tooth of the belt drive pulley with the pointer on the oil pump.
   c. Set the camshaft pulleys to TDC by aligning the TDC marks on the camshaft pulleys with the pointers on the back covers.
   d. Remove the battery clamp bolt.
   e. Remove the belt tensioner.
   f. Align the holes on the rod and housing of the tensioner.
   g. Using a press or other suitable device, slowly compress the tensioner and insert a 0.08 inch (2mm) pin through the housing and rod.
   h. Install the tensioner making sure the pin is still installed.
   i. Apply thread locker to idler pulley bolt then hand tighten the bolt.
   j. Install the belt over the pulleys in this sequence; drive pulley, idler pulley, front camshaft pulley, water pump pulley, rear camshaft pulley and adjusting pulley.

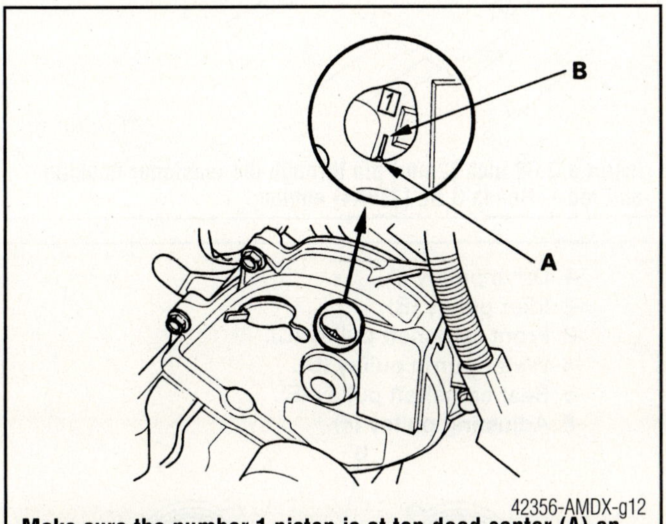

Make sure the number 1 piston is at top dead center (A) on the front camshaft pulley and pointer (B) — Honda 3.5L (J35A4) engine

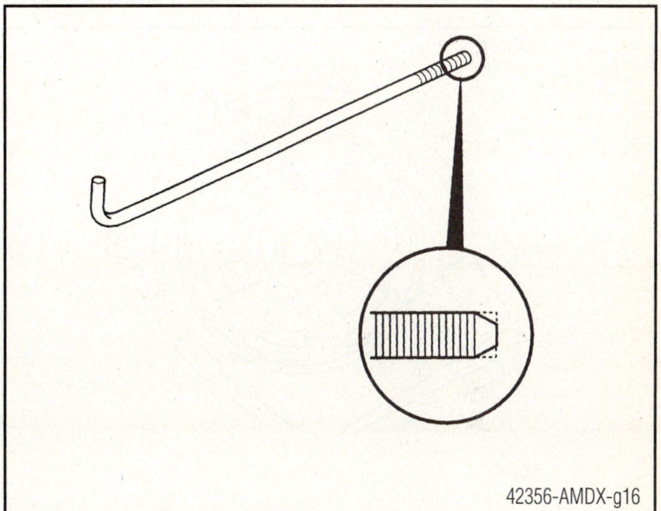

Remove a battery clamp bolt and grind the end as shown— Honda 3.5L (J35A4) engine

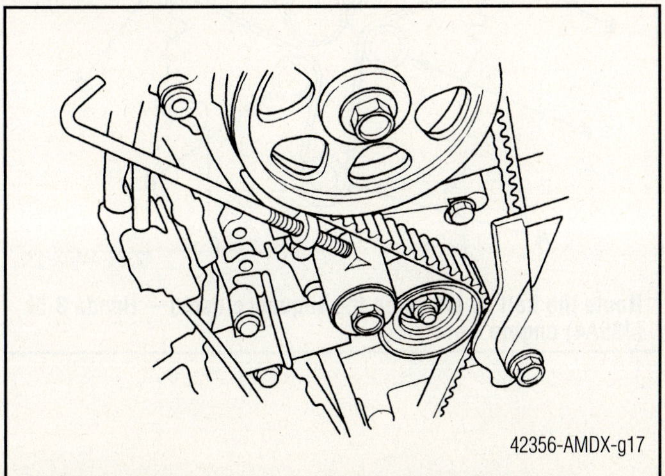

Install the battery clamp bolt as shown to hold the belt adjuster in position— Honda 3.5L (J35A4) engine

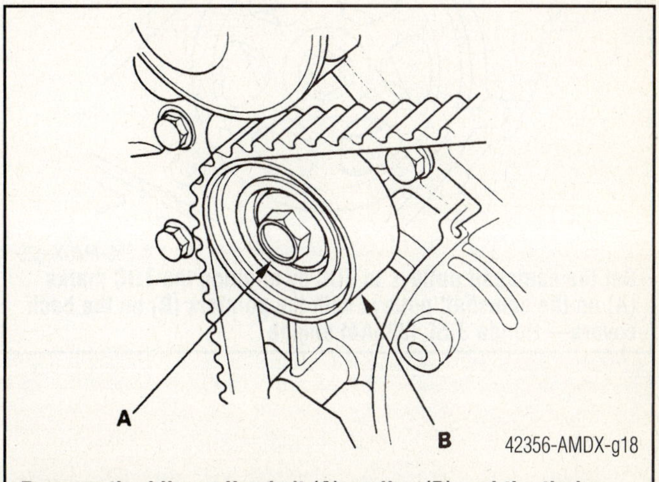

Remove the idler pulley bolt (A), pulley (B) and the timing belt— Honda 3.5L (J35A4) engine

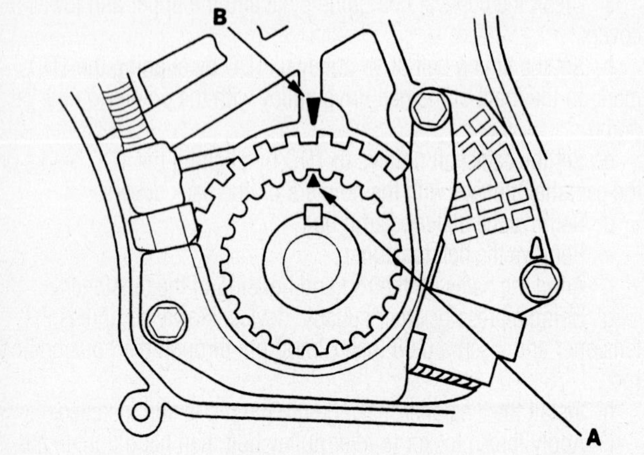

42356-AMDX-g19

Set the timing belt pulley to TDC by aligning the TDC mark (A) on the tooth of the belt pulley with the pointer (B) on the oil pump— Honda 3.5L (J35A4) engine

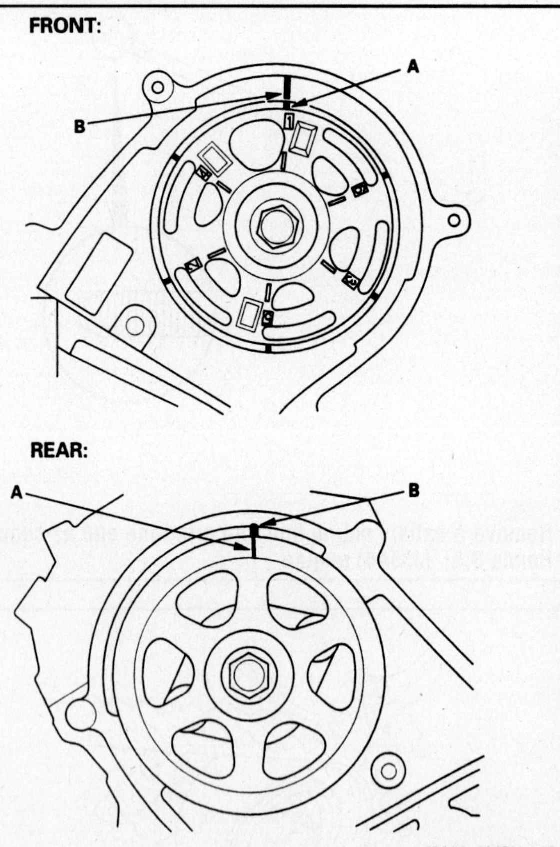

42356-AMDX-g20

Set the camshaft pulleys to TDC by aligning the TDC marks (A) on the camshaft pulleys with the pointers (B) on the back covers— Honda 3.5L (J35A4) engine

k. Tighten the idler pulley bolt to 33 ft. lbs. (44 Nm).

l. Remove the pin from the tensioner.

8. Install or connect the following:

- Lower half of the side mount and tighten the 3 long bolts to 33 ft. lbs. (44 Nm) and the one short bolt to 9 ft. lbs. (12 Nm)
- Timing belt guide plate as illustrated
- Lower timing cover and tighten the bolts to 9 ft. lbs. (12 Nm)
- Front and rear upper timing covers and tighten the bolts to 9 ft. lbs. (12 Nm)

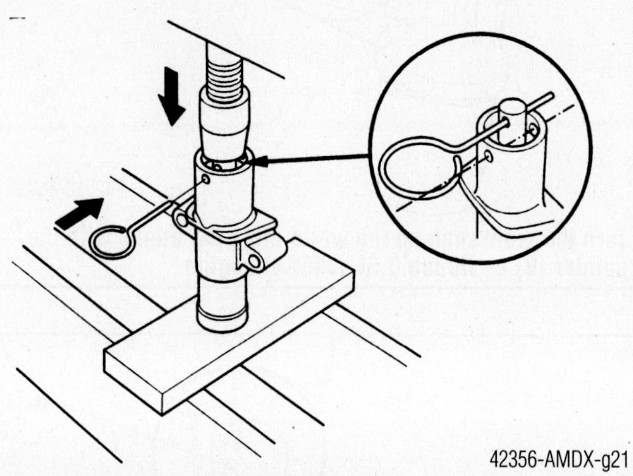

42356-AMDX-g21

Insert a 0.08 inch (2mm) pin through the tensioner housing and rod— Honda 3.5L (J35A4) engine

1 Drive pulley (A).
2 Idler pulley (B).
3 Front camshaft pulley (C).
4 Water pump pulley (D).
5 Rear camshaft pulley (E).
6 Adjusting pulley (F).

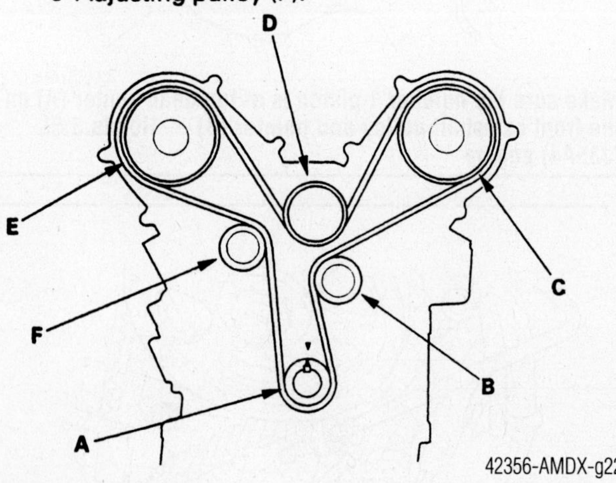

42356-AMDX-g22

Route the belt as shown in the sequence listed— Honda 3.5L (J35A4) engine

• Crankshaft pulley and tighten the bolts to 181 ft. lbs. (245 Nm), using the holding tool to prevent the unit from turning

9. Rotate the crankshaft pulley about 5 or 6 degrees clockwise so the belt positions itself on the pulleys.

10. Turn the crankshaft pulley so the white mark aligns with the pointer.

11. Check the camshaft pulley marks are aligned. If the marks are aligned, proceed to the next step. If the marks are not aligned, remove the timing belt and reinstall using the steps outlined before this step.

12. Remove or disconnect the following:

• Drive belt

• Upper side mount and tighten the bolts in the sequence illustrated to the specifications in the illustration

13. If installing the old belt, perform the following steps:

a. Clean the pulleys, belt guide plate and the upper and lower covers.

b. Set the timing belt drive pulley to TDC by aligning the TDC mark on the tooth of the belt drive pulley with the pointer on the oil pump.

c. Set the camshaft pulleys to TDC by aligning the TDC marks on the camshaft pulleys with the pointers on the back covers.

d. Apply thread locker to idler pulley bolt then hand tighten the bolt.

14. If the tensioner was extended and the belt cannot be installed, perform the steps above for the new belt installation.

a. Install the belt over the pulleys in this sequence; drive pulley, idler pulley, front camshaft pulley, water pump pulley, rear camshaft pulley and adjusting pulley.

b. Tighten the idler pulley bolt to 33 ft. lbs. (44 Nm).

c. Remove the battery clamp bolt.

15. Install or connect the following:

• Lower half of the side mount and tighten the 3 long bolts to 33 ft. lbs. (44 Nm) and the one short bolt to 9 ft. lbs. (12 Nm)

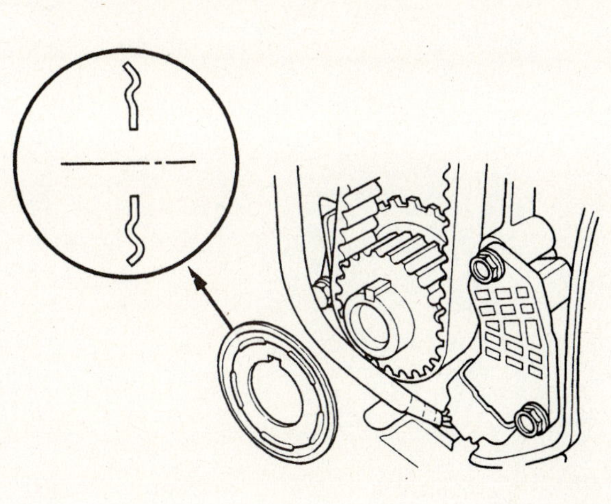

42356-AMDX-g23

Install the timing belt guide plate as shown— Honda 3.5L (J35A4) engine

• Timing belt guide plate as illustrated

• Lower timing cover and tighten the bolts to 9 ft. lbs. (12 Nm)

• Front and rear upper timing covers and tighten the bolts to 9 ft. lbs. (12 Nm)

• Crankshaft pulley and tighten the bolts to 181 ft. lbs. (245 Nm), using the holding tool to prevent the

16. Rotate the crankshaft pulley about 5 or 6 degrees clockwise so the belt positions itself on the pulleys.

17. Turn the crankshaft pulley so the white mark aligns with the pointer.

18. Check the camshaft pulley marks are aligned. If the marks are aligned, proceed to the next step. If the marks are not aligned, remove the timing belt and reinstall using the steps outlined before this step.

19. Install or connect the following:

• Drive belt

• Upper side mount and tighten the bolts in the sequence illustrated to the specifications in the illustration

• Dipstick tube

20. Connect the negative battery cable.

21. Using a suitable scan tool, perform the Powertain Control Module (PCM) reset and the Crankshaft position (CKP) pattern clear/learn procedures, following the scan tool manufacturers instructions.

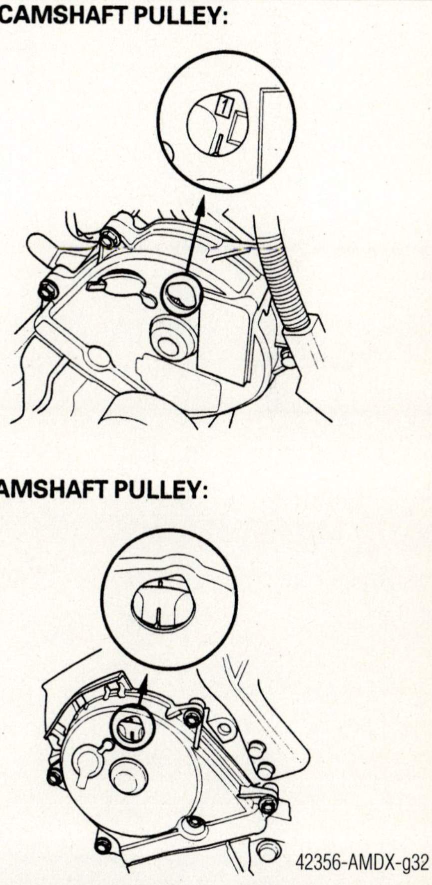

42356-AMDX-g32

Check that the camshaft pulley marks are aligned as shown— Honda 3.5L (J35A4) engine

Hyundai

1.5L (G4EB, G4EK, G4J) SOHC ENGINES

YEAR, MODEL & ENGINE APPLICATION

1995-04 Hyundai Accent	1.5L Codes G4EB, G4EK
1986-94 Hyundai Excel, S Coupe	1.5L Code G4J

REPLACEMENT INTERVALS

Hyundai recommends timing belt replacement every 60,000 miles.

LABOR TIMES

Accent	
Replacement	2.3 hrs
Severe Service	2.6 hrs
With air conditioning add	0.5 hrs
With power steering add	0.3 hrs
Excel & S Coupe	
1986-89	
Replacement	1.5 hrs
Severe Service	1.9 hrs
1990-94	
Replacement	2.6 hrs
Severe Service	2.9 hrs
All Models	
With tensioner replacement add	0.2 hrs

8807HUNg01

✳✳ CAUTION

These engines are an interference type. If the timing belt has failed, valve-to-piston damage may have occurred. A compression check should always be performed to check for engine damage.

1. Disconnect the negative battery cable.
2. Remove the engine undercover.
3. Using the proper equipment, slightly raise the engine to take the weight off the side engine mount.
4. Remove the engine mount bracket.
5. Remove the accessory drive belts.
6. Remove the water pump pulley and the alternator mounting bracket.
7. Remove the crankshaft pulley by removing the 4 retaining bolts.
8. Remove all attaching screws and remove the upper and lower timing belt covers.
9. Rotate the crankshaft clockwise and align the timing marks so the No. 1 piston will be at Top Dead Center (TDC) of the compression stroke.
10. Loosen the tensioning bolt and the pivot bolt on the timing belt tensioner. Move the tensioner as far as it will go toward the water pump. Tighten the adjusting bolt.
11. If the timing belt is to be reused, mark the belt with an arrow showing direction of rotation.
12. Remove the timing belt.
13. Inspect the belt thoroughly. The back surface must be pliable and rough. If it is hard and glossy, the belt should be replaced. Any cracks in the belt backing or teeth or missing teeth mean the belt must be replaced. The canvas cover should be intact on all the teeth. If rubber is exposed anywhere, the belt should be replaced.
14. Inspect the tensioner for grease leaking from the grease seal and any roughness in rotation. Replace a tensioner for either defect.
15. The sprockets should be inspected and replaced, if there is any sign of damaged teeth or cracking. Do not immerse sprockets in solvent, as solvent that has soaked into the metal may cause deterioration of the timing belt later. Do not clean the tensioner in solvent either, as this may wash the grease out of the bearing.

To install:

16. Align the timing marks of the camshaft sprocket and check that the crankshaft timing marks are still in alignment.
17. If removed, install the timing belt tensioner, spring and spacer with the bottom end of the spring free.

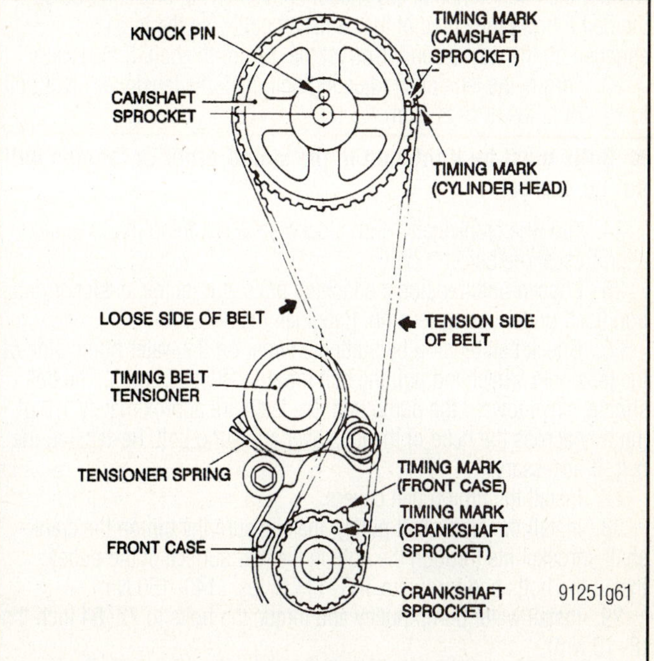

Timing belt component identification and timing mark alignment—Hyundai 1.5L (G4EB, G4EK, G4J) SOHC engine

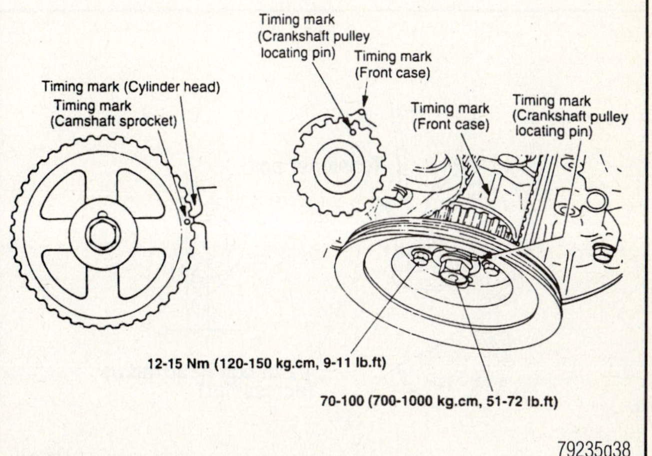

Proper pulley alignment for timing belt installation—Hyundai 1.5L (G4EB, G4EK, G4J) SOHC engine

18. Tighten the adjusting bolt slightly with the tensioner moved as far as possible away from the water pump.
19. Install the free end of the spring into the locating tang on the front case.
20. Position the timing belt over the crankshaft sprocket, then over the camshaft sprocket. Slip the back of the belt over the tensioner pulley.
21. Turn the camshaft sprocket in the opposite direction of rotation until the straight side of the belt is tight and be sure the timing marks align.

■ **If the timing marks are not properly aligned, shift the belt 1 tooth at a time in the appropriate direction until they are aligned.**

22. Loosen the tensioner mounting bolts so the tensioner works, without the interference of any friction, under spring pressure. Be sure the belt follows the curve of the camshaft pulley so the teeth are engaged all the way around. Correct the path of the belt, if necessary.
23. Torque the tensioner adjusting bolt, then the tensioner pivot bolt to 15–18 ft. lbs. (20–26 Nm).

■ **Bolts must be tightened in the stated order or tension will not be correct.**

24. Turn the crankshaft 1 turn clockwise until timing marks again align to seat the belt.
25. Loosen both tensioner attaching bolts and let the tensioner position itself under spring tension. Retighten the bolts.
26. Check belt tension by putting a finger on the water pump side of the tensioner wheel and pull the belt toward the water pump. The belt should move toward the pump until the teeth are approximately 1/2 of the way across the head of the tensioner adjusting bolt. Re-tension the belt, if necessary.
27. Install the timing belt covers.
28. Install the crankshaft pulley, making sure the pin on the crankshaft sprocket fits through the hole in the rear surface of the pulley. Install the bolts and torque to 103–111 ft. lbs. (140–150 Nm).
29. Install water pump pulley and torque the bolts to 72–84 inch lbs. (8–10 Nm).
30. Install drive belts and adjust to proper tension.
31. Install engine bracket and lower the engine.
32. Install any previously removed covers.
33. Connect the negative battery cable.
34. Check and adjust ignition timing.

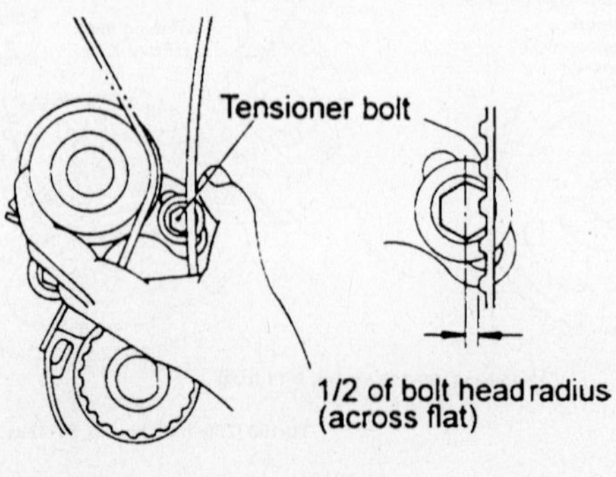

Checking timing belt tension—Hyundai 1.5L (G4EB, G4EK, GEJ) SOHC engine

1.5L (G4EK) & 1.6L (G4ED) DOHC ENGINES

YEAR, MODEL & ENGINE APPLICATION	
1996-00 Hyundai Accent	1.5L Code G4EK
2000-04 Accent	1.6L Code G4ED

REPLACEMENT INTERVALS

Hyundai recommends timing belt replacement every 60,000 miles.

LABOR TIMES	
1996-04	
Replacement	2.3 hrs
Severe Service	2.6 hrs
With air conditioning add	0.5 hrs
With power steering add	0.3 hrs
With tensioner replacement add	0.2 hrs

8807HUNg02

✱✱CAUTION

These engines are an interference type. If the timing belt has failed, valve-to-piston damage may have occurred. A compression check should always be performed to check for engine damage.

1. Disconnect the negative battery cable.
2. Remove the engine undercover.
3. Using the proper equipment, slightly raise the engine to take the weight off the side engine mount.
4. Remove the engine mount bracket.
5. Remove the accessory drive belts.
6. Remove the water pump pulley and the alternator mounting bracket.
7. Remove the crankshaft pulley by removing the 4 retaining bolts.
8. Remove all attaching screws and remove the upper and lower timing belt covers.
9. Rotate the crankshaft clockwise and align the timing marks so the No. 1 piston will be at Top Dead Center (TDC) of the compression stroke.
10. Loosen the tensioning bolt and the pivot bolt on the timing belt tensioner. Move the tensioner as far as it will go toward the water pump. Tighten the adjusting bolt.
11. If the timing belt is to be reused, mark the belt with an arrow showing direction of rotation.
12. Remove the timing belt.
13. Inspect the belt thoroughly. The back surface must be pliable and rough. If it is hard and glossy, the belt should be replaced. Any cracks in the belt backing or teeth or missing teeth mean the belt must be replaced. The canvas cover should be intact on all the teeth. If rubber is exposed anywhere, the belt should be replaced.
14. Inspect the tensioner for grease leaking from the grease seal and any roughness in rotation. Replace a tensioner for either defect.
15. The sprockets should be inspected and replaced, if there is any sign of damaged teeth or cracking. Do not immerse sprockets in solvent, as solvent that has soaked into the metal may cause deterioration of the timing belt later. Do not clean the tensioner in solvent either, as this may wash the grease out of the bearing.

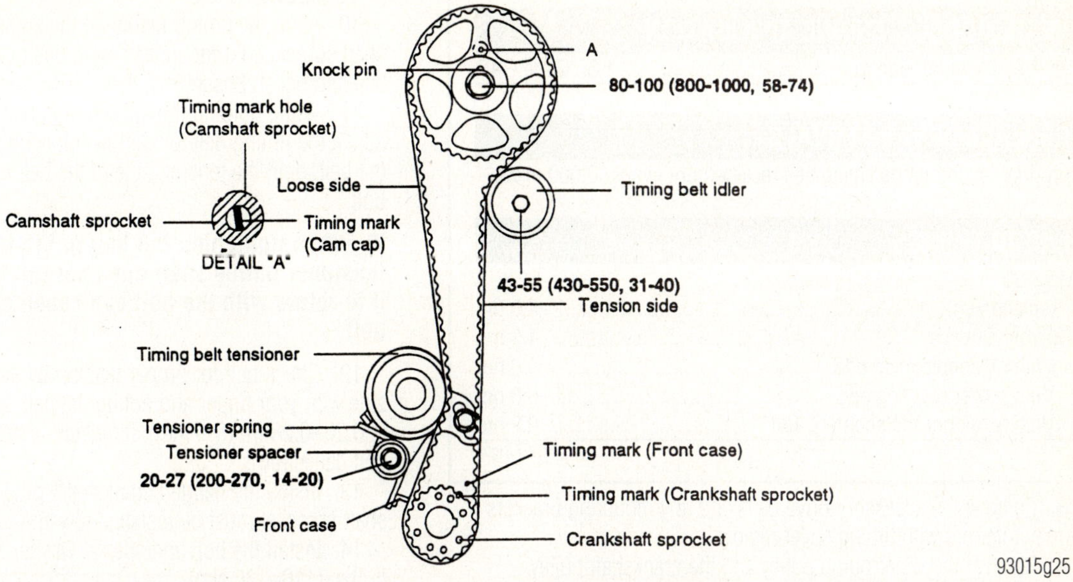

Proper pulley alignment for timing belt installation at TDC—Hyundai 1.5L & 1.6L DOHC engines

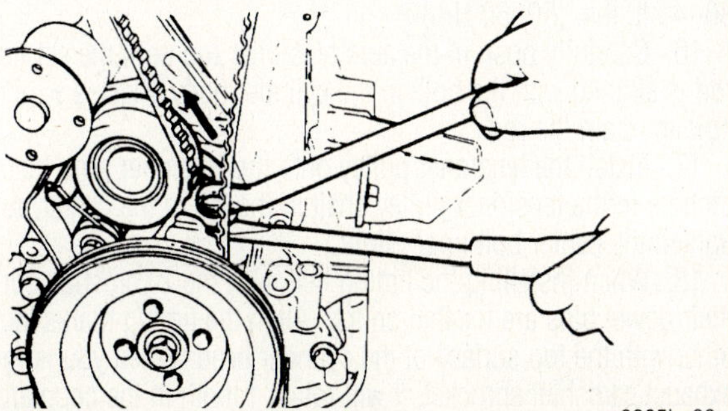

Timing belt tensioner movement—Hyundai 1.5L & 1.6L DOHC engines

To install:

16. Align the timing marks of the camshaft sprocket and check that the crankshaft timing marks are still in alignment.
17. If removed, install the timing belt tensioner, spring and spacer with the bottom end of the spring free.
18. Tighten the adjusting bolt slightly with the tensioner moved as far as possible away from the water pump.
19. Install the free end of the spring into the locating tang on the front case.
20. Position the timing belt over the crankshaft sprocket, then over the camshaft sprocket. Slip the back of the belt over the tensioner pulley.
21. Turn the camshaft sprocket in the opposite direction of rotation until the straight side of the belt is tight and be sure the timing marks align.

■ If the timing marks are not properly aligned, shift the belt 1 tooth at a time in the appropriate direction until they are aligned.

22. Loosen the tensioner mounting bolts so the tensioner works, without the interference of any friction, under spring pressure. Be sure the belt follows the curve of the camshaft pulley so the teeth are engaged all the way around. Correct the path of the belt, if necessary.
23. Torque the tensioner adjusting bolt, then the tensioner pivot bolt to 15–18 ft. lbs. (20–26 Nm).

■ Bolts must be tightened in the stated order or tension will not be correct.

24. Turn the crankshaft 1 turn clockwise until timing marks again align to seat the belt.
25. Loosen both tensioner attaching bolts and let the tensioner position itself under spring tension. Retighten the bolts.
26. Check belt tension by putting a finger on the water pump side of the tensioner wheel and pull the belt toward the water pump. The belt should move toward the pump until the teeth are approximately 1/2 of the way across the head of the tensioner adjusting bolt. Re-tension the belt, if necessary.
27. Install the timing belt covers.
28. Install the crankshaft pulley, making sure the pin on the crankshaft sprocket fits through the hole in the rear surface of the pulley. Install the bolts and torque to 103–111 ft. lbs. (140–150 Nm).
29. Install water pump pulley and torque the bolts to 72–84 inch lbs. (8–10 Nm).
30. Install drive belts and adjust to proper tension.
31. Install engine bracket and lower the engine.
32. Install any previously removed covers.
33. Connect the negative battery cable.
34. Check and adjust ignition timing.

1.6L (G4R) ENGINES

✳✳ CAUTION

These engines are an interference type. If the timing belt has failed, valve-to-piston damage may have occurred. A compression check should always be performed to check for engine damage.

1. Disconnect the negative battery cable.
2. Remove the engine undercover.

YEAR, MODEL & ENGINE APPLICATION

Application	Engine
1992-95 Hyundai Elantra	1.6L Code G4R

REPLACEMENT INTERVALS

Hyundai recommends timing belt replacement every 60,000 miles.

LABOR TIMES

1992-95	
Replacement	4.0 hrs
Severe Service	4.3 hrs
With air conditioning add	0.6 hrs
With power steering add	0.3 hrs
With tensioner replacement add	0.2 hrs

8807HUNg03

3. Remove the accessory drive belts and any mounting brackets that may interfere with timing cover removal.

4. Remove the water pump pulley and the crankshaft pulley.

5. Remove all attaching screws and remove the upper and lower timing belt covers.

6. Rotate the crankshaft clockwise and align the timing marks so No. 1 piston will be at Top Dead Center (TDC) of the compression stroke. At this time the timing marks on the camshaft sprocket and the upper surface of the cylinder head should coincide, and the dowel pin of the camshaft sprocket should be at the upper side.

■ **Always rotate the crankshaft in a clockwise direction. If the timing belt is to be reused, make a mark on the back of the timing belt indicating the direction of rotation so it may be reassembled in the same direction.**

7. Remove the auto tensioner and remove the outermost timing belt.

8. Remove the timing belt tensioner pulley, tensioner arm, idler pulley, oil pump sprocket, special washer, flange and spacer.

9. Remove the silent shaft (inner) belt tensioner and remove the inner belt.

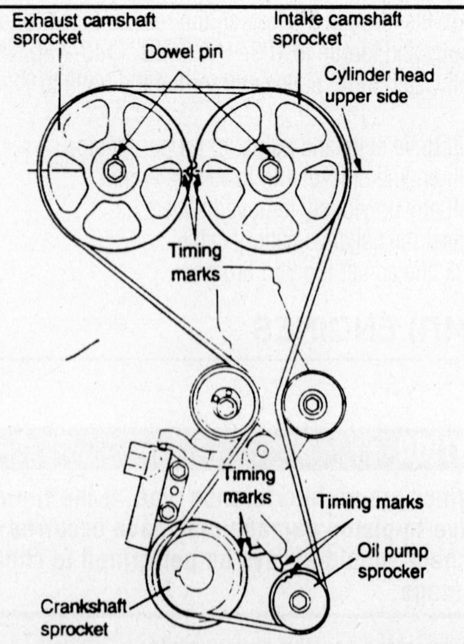

9305bg11

Alignment of engine timing marks for belt installation—Hyundai 1.6L (G4R) engine

To install:

10. Align the timing marks on the crankshaft sprocket and the silent shaft sprocket. Fit the inner timing belt over the crankshaft and silent shaft sprocket. Ensure that there is no slack in the belt.

11. While holding the inner timing belt tensioner with your fingers, adjust the timing belt tension by applying a force towards the center of the belt, until the tension side of the belt is taut. Tighten the tensioner bolt.

■ **When tightening the bolt of the tensioner, ensure that the tensioner pulley shaft does not rotate with the bolt. Allowing it to rotate with the bolt can cause excessive tension on the belt.**

12. Check belt for proper tension by depressing the belt on its long side with your finger and noting the belt deflection. The desired reading is 0.20–0.28 in. (5–7mm). If tension is not correct, readjust and check belt deflection.

13. Install the flanged crankshaft sprocket. The flange on the crankshaft sprocket must be installed toward the inner timing belt sprocket.

14. Install the bolt and special washer and torque the bolt to 80–94 ft. lbs. (110–130 Nm).

15. To install the oil pump sprocket, insert a Phillips screwdriver with a shaft 0.31 in. (8mm) in diameter into the plug hole in the left side of the cylinder block to hold the left silent shaft. Tighten the nut to 36–43 ft. lbs. (50–60 Nm).

16. Carefully push in the auto tensioner rod until the set hole in the rod is aligned with the hole in the cylinder body. Place a wire into the hole to retain the rod.

17. Install the tensioner pulley onto the tensioner arm. Locate the pinhole in the tensioner pulley shaft to the left of the center bolt. Then, tighten the center bolt finger-tight.

18. When installing the timing belt, turn the 2 camshaft sprockets so their dowel pins are located on top. Align the timing marks facing each other with the top surface of the cylinder head. When you let go of the exhaust camshaft sprocket, it will rotate 1 tooth in the counterclockwise direction. This should be taken into account when installing the timing belts on the sprocket.

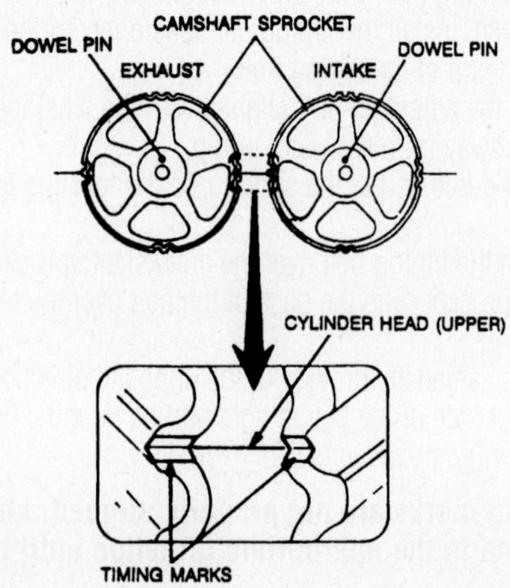

91251g63

Alignment of camshaft sprocket timing marks— Hyundai 1.6L (G4R) engine

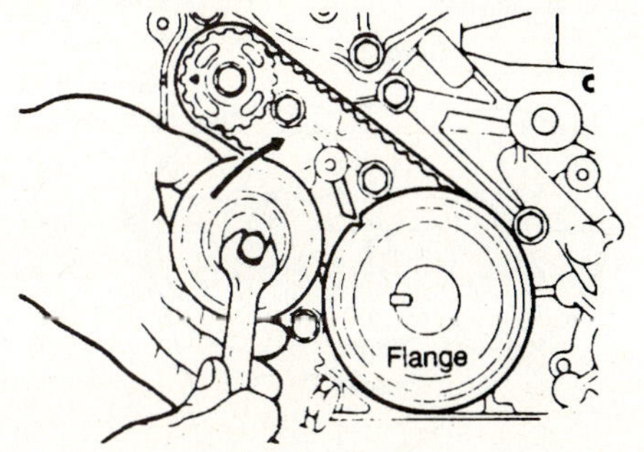

9305bg12

Adjusting the inner timing belt—Hyundai 1.6L (G4R) engine

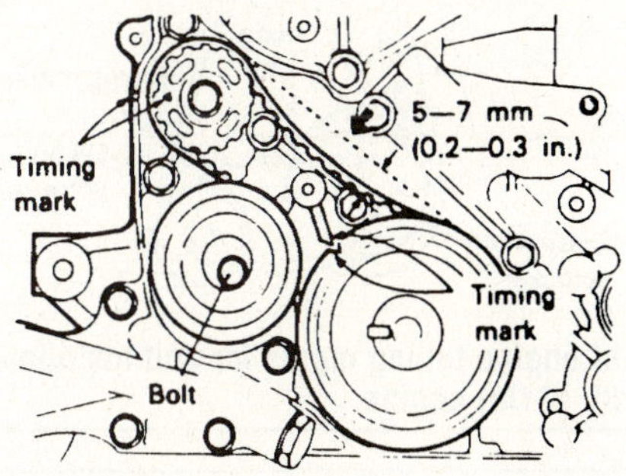

9305bg13

Checking inner belt deflection—Hyundai 1.6L (G4R) engine

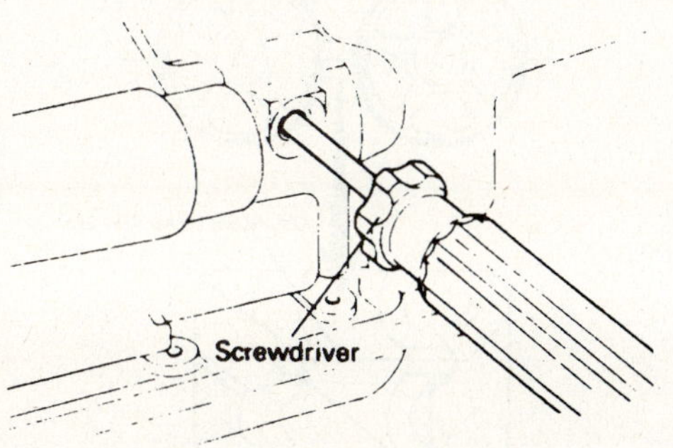

9305bg27

Holding the left side silent shaft—Hyundai 1.6L (G4R) engine

■ **Both camshaft sprockets are used for the intake and exhaust camshafts and are provided with 2 timing marks. When the sprocket is mounted on the exhaust camshaft, use the timing mark on the right with the dowel pinhole on top. For the intake camshaft sprocket, use the mark on the left with the dowel pinhole on top.**

19. Align the crankshaft sprocket and oil pump sprocket timing marks.

20. After alignment of the oil pump sprocket timing marks, remove the plug on the cylinder block and insert a Phillips screwdriver with a shaft diameter of 0.31 in. (8mm) through the hole. If the shaft can be inserted 2.4 in. (70mm) deep, the silent shaft is in the correct position. If the shaft of the tool can only be inserted 0.8—1.0 in. (20–25mm) deep, turn the oil pump sprocket 1 turn and realign the marks. Reinsert the tool making sure it is inserted 2.4 in. (70mm) deep. Keep the tool inserted in hole for the remainder of this procedure.

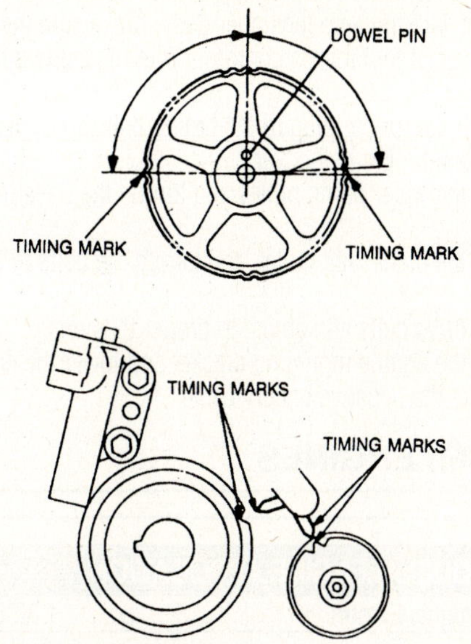

91251g64

Alignment of crankshaft and oil pump sprocket timing marks—Hyundai 1.6L (G4R) engine

■ **The above step assures that the oil pump socket is in correct orientation to the silent shafts. This step must not be skipped or a vibration may develop during engine operation.**

21. Install the timing belt as follows around the tensioner pulley, crankshaft sprocket, oil pump sprocket, idler pulley and intake camshaft sprocket.

22. Rotate the exhaust camshaft sprocket 1 tooth clockwise to align the timing mark and install the belt on the exhaust camshaft sprocket.

23. Lift upward on the tensioner pulley in a clockwise direction and tighten the center bolt. Make sure all timing marks are aligned.

24. Rotate the crankshaft 1/4 turn counterclockwise. Then, turn clockwise until the timing marks are aligned again.

25. To adjust the timing (outer) belt, turn the crankshaft 1/4 turn counterclockwise, then turn it clockwise to move No. 1 cylinder to TDC.

26. Loosen the center bolt. Using tool 09244-28100 or equivalent and a torque wrench, apply a torque of 22–24 inch lbs. (2.6–2.8 Nm). Tighten the center bolt.

27. Screw tensioner setscrew tool 09244-28000 into the engine left support bracket until its end makes contact with the tensioner arm. At this point, screw the special tool in some more and remove the set wire attached to the auto tensioner, if the wire was not previously removed. Then remove the special tool.

28. Rotate the crankshaft 2 complete turns clockwise and let it sit for approximately 15 minutes. Then, measure the auto tensioner protrusion (the distance between the tensioner arm and auto tensioner body) to ensure that it is within 0.15–0.18 in. (3.8–4.5mm). If out of specification, repeat the procedure until the specified value is obtained.

29. If the timing belt tension adjustment is being performed with the engine mounted in the vehicle, and clearance between the tensioner arm and the auto tensioner body cannot be measured, the following alternative method can be used:

a. Screw in special tool MD998738 or equivalent, until its end makes contact with the tensioner arm.

b. After the special tool makes contact with the arm, screw it in some more to retract the auto tensioner pushrod while counting the number of turns the tool makes until the tensioner arm is brought into contact with the auto tensioner body. Make sure the number of turns the special tool makes conforms with the standard value of 21/2–3 turns.

c. Install the rubber plug to the timing belt rear cover.

30. Install the timing belt covers.
31. Install the water pump pulley and torque the bolts to 84 inch lbs. (10 Nm).
32. Install the crankshaft pulley and torque the bolts to 22 ft. lbs. (30 Nm).
33. Install drive belts and adjust to proper tension.
34. Install the engine mounting bracket and lower the engine.
35. Connect the negative battery cable.

1.8L (G4M) ENGINES

YEAR, MODEL & ENGINE APPLICATION

1993-95 Hyundai Elantra	1.8L Code G4M

REPLACEMENT INTERVALS

Hyundai recommends timing belt replacement every 60,000 miles.

LABOR TIMES

1993-95	
Replacement	4.0 hrs
Severe Service	4.6 hrs
With air conditioning add	0.6 hrs
With power steering add	0.3 hrs
With tensioner replacement add	0.2 hrs

8807HUNg04

✳✳ CAUTION

These engines are an interference type. If the timing belt has failed, valve-to-piston damage may have occurred. A compression check should always be performed to check for engine damage.

1. Disconnect the negative battery cable.
2. Remove the engine undercover.
3. Remove the accessory drive belts and any mounting brackets that may interfere with timing cover removal.
4. Remove the water pump pulley and the crankshaft pulley.
5. Remove all attaching screws and remove the upper and lower timing belt covers.
6. Rotate the crankshaft clockwise and align the timing marks so No. 1 piston will be at Top Dead Center (TDC) of the compression stroke. At this time the timing marks on the camshaft sprocket and the upper surface of the cylinder head should coincide, and the dowel pin of the camshaft sprocket should be at the upper side.

■ **Always rotate the crankshaft in a clockwise direction. If the timing belt is to be reused, make a mark on the back of the timing belt indicating the direction of rotation so it may be reassembled in the same direction.**

7. Remove the auto tensioner and remove the outermost timing belt.
8. Remove the timing belt tensioner pulley, tensioner arm, idler pulley, oil pump sprocket, special washer, flange and spacer.

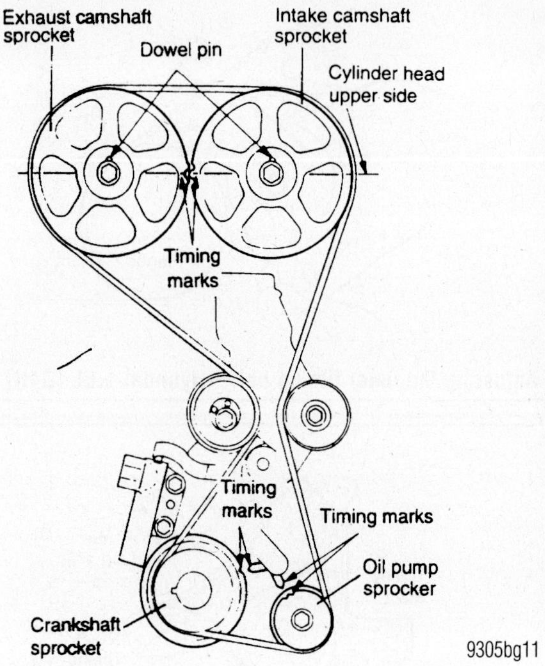

Alignment of engine timing marks for belt installation—Hyundai 1.8L (G4M) engine

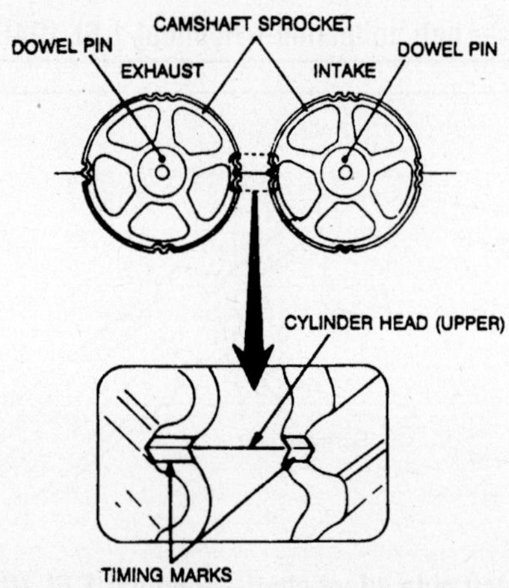

Alignment of camshaft sprocket timing marks— Hyundai 1.8L (G4M) engine

9. Remove the silent shaft (inner) belt tensioner and remove the inner belt.

To install:

10. Align the timing marks on the crankshaft sprocket and the silent shaft sprocket. Fit the inner timing belt over the crankshaft and silent shaft sprocket. Ensure that there is no slack in the belt.
11. While holding the inner timing belt tensioner with your fingers, adjust the timing belt tension by applying a force towards the center of the belt, until the tension side of the belt is taut. Tighten the tensioner bolt.

■ **When tightening the bolt of the tensioner, ensure that the tensioner pulley shaft does not rotate with the bolt. Allowing it to rotate with the bolt can cause excessive tension on the belt.**

12. Check belt for proper tension by depressing the belt on its long side with your finger and noting the belt deflection. The desired reading is 0.20–0.28 in. (5–7mm). If tension is not correct, readjust and check belt deflection.

13. Install the flanged crankshaft sprocket. The flange on the crankshaft sprocket must be installed toward the inner timing belt sprocket.

14. Install the bolt and special washer and torque the bolt to 80–94 ft. lbs. (110–130 Nm).

15. To install the oil pump sprocket, insert a Phillips screwdriver with a shaft 0.31 in. (8mm) in diameter into the plug hole in the left side of the cylinder block to hold the left silent shaft. Tighten the nut to 36–43 ft. lbs. (50–60 Nm).

16. Carefully push in the auto tensioner rod until the set hole in the rod is aligned with the hole in the cylinder body. Place a wire into the hole to retain the rod.

17. Install the tensioner pulley onto the tensioner arm. Locate the pinhole in the tensioner pulley shaft to the left of the center bolt. Then, tighten the center bolt finger-tight.

18. When installing the timing belt, turn the 2 camshaft sprockets so their dowel pins are located on top. Align the timing marks facing each other with the top surface of the cylinder head. When you let go of the exhaust camshaft sprocket, it will rotate 1 tooth in the counterclockwise direction. This should be taken into account when installing the timing belts on the sprocket.

■ **Both camshaft sprockets are used for the intake and exhaust camshafts and are provided with 2 timing marks. When the sprocket is mounted on the exhaust camshaft, use the timing mark on the right with the dowel pinhole on top. For the intake camshaft sprocket, use the mark on the left with the dowel pinhole on top.**

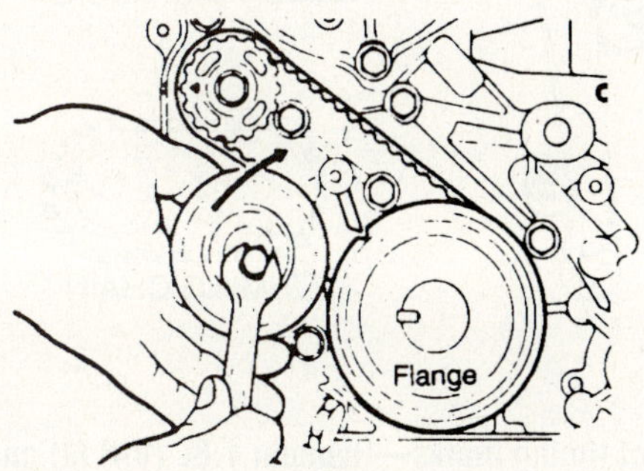

9305bg12

Adjusting the inner timing belt—Hyundai 1.8L (G4M) engine

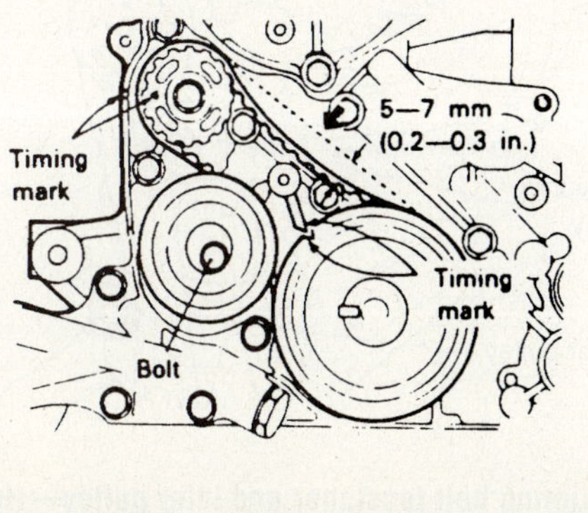

9305bg13

Checking inner belt deflection—Hyundai 1.8L (G4M) engine

19. Align the crankshaft sprocket and oil pump sprocket timing marks.

20. After alignment of the oil pump sprocket timing marks, remove the plug on the cylinder block and insert a Phillips screwdriver with a shaft diameter of 0.31 in. (8mm) through the hole. If the shaft can be inserted 2.4 in. (70mm) deep, the silent shaft is in the correct position. If the shaft of the tool can only be inserted 0.8—1.0 in. (20–25mm) deep, turn the oil pump sprocket 1 turn and realign the marks. Reinsert the tool making sure it is inserted 2.4 in. (70mm) deep. Keep the tool inserted in hole for the remainder of this procedure.

■ **The above step assures that the oil pump socket is in correct orientation to the silent shafts. This step must not be skipped or a vibration may develop during engine operation.**

21. Install the timing belt as follows around the tensioner pulley, crankshaft sprocket, oil pump sprocket, idler pulley and intake camshaft sprocket.

22. Rotate the exhaust camshaft sprocket 1 tooth clockwise to align the timing mark and install the belt on the exhaust camshaft sprocket.

23. Lift upward on the tensioner pulley in a clockwise direction and tighten the center bolt. Make sure all timing marks are aligned.

24. Rotate the crankshaft 1/4 turn counterclockwise. Then, turn clockwise until the timing marks are aligned again.

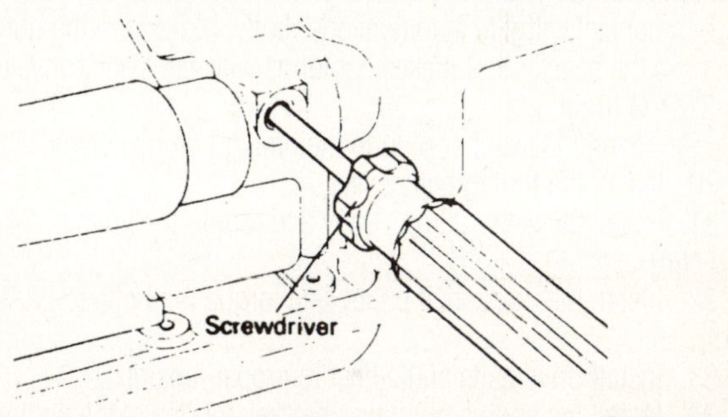

9305bg27

Holding the left side silent shaft—Hyundai 1.8L (G4M) engine

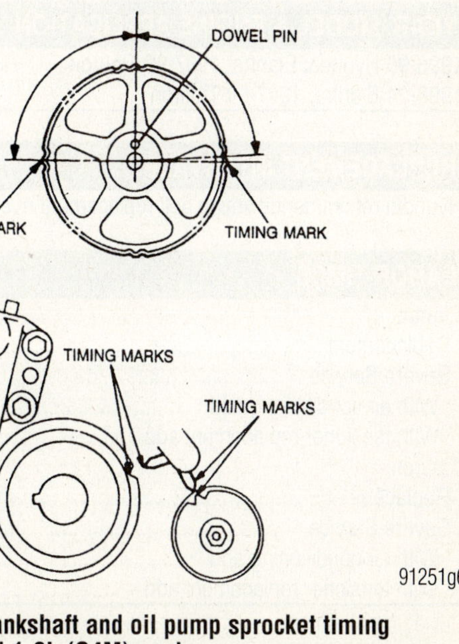

91251g64

Alignment of crankshaft and oil pump sprocket timing marks—Hyundai 1.8L (G4M) engine

25. To adjust the timing (outer) belt, turn the crankshaft 1/4 turn counterclockwise, then turn it clockwise to move No. 1 cylinder to TDC.

26. Loosen the center bolt. Using tool 09244-28100 or equivalent and a torque wrench, apply a torque of 22–24 inch lbs. (2.6–2.8 Nm). Tighten the center bolt.

27. Screw tensioner setscrew tool 09244-28000 into the engine left support bracket until its end makes contact with the tensioner arm. At this point, screw the special tool in some more and remove the set wire attached to the auto tensioner, if the wire was not previously removed. Then remove the special tool.

28. Rotate the crankshaft 2 complete turns clockwise and let it sit for approximately 15 minutes. Then, measure the auto tensioner protrusion (the distance between the tensioner arm and auto tensioner body) to ensure that it is within 0.15–0.18 in. (3.8–4.5mm). If out of specification, repeat the procedure until the specified value is obtained.

29. If the timing belt tension adjustment is being performed with the engine mounted in the vehicle, and clearance between the tensioner arm and the auto tensioner body cannot be measured, the following alternative method can be used:

a. Screw in special tool MD998738 or equivalent, until its end makes contact with the tensioner arm.

b. After the special tool makes contact with the arm, screw it in some more to retract the auto tensioner pushrod while counting the number of turns the tool makes until the tensioner arm is brought into contact with the auto tensioner body. Make sure the number of turns the special tool makes conforms with the standard value of 21/2–3 turns.

c. Install the rubber plug to the timing belt rear cover.

30. Install the timing belt covers.

31. Install the water pump pulley and torque the bolts to 84 inch lbs. (10 Nm).

32. Install the crankshaft pulley and torque the bolts to 22 ft. lbs. (30 Nm).

33. Install drive belts and adjust to proper tension.

34. Install the engine mounting bracket and lower the engine.

35. Connect the negative battery cable.

1.8L (VIN M) & 2.0L (VIN D, F) ENGINES

YEAR, MODEL & ENGINE APPLICATION

Application	Engine
1996-98 Hyundai Elantra, 1997-99 Tiburon	1.8L VIN M
1998-04 Elantra, 1997-04 Tiburon	2.0L VIN D, F

REPLACEMENT INTERVALS

Hyundai recommends timing belt replacement every 60,000 miles.

LABOR TIMES

Elantra	
Replacement	2.8 hrs
Severe Service	3.1 hrs
With air conditioning add	0.6 hrs
With tensioner replacement add	0.2 hrs
Tiburon	
Replacement	3.1 hrs
Severe Service	3.4 hrs
With air conditioning add	0.3 hrs
With tensioner replacement add	0.2 hrs

8807HUNg05

✱✱ CAUTION

These engines are an interference type. If the timing belt has failed, valve-to-piston damage may have occurred. A compression check should always be performed to check for engine damage.

1. Disconnect the negative battery cable.

2. Remove the accessory drive belts, tension pulley brackets, water pump pulley and crankshaft pulley.

3. Remove all attaching screws and remove the upper and lower timing belt covers.

■ **Always rotate the crankshaft in a clockwise direction.**

4. Rotate the crankshaft clockwise and align the timing marks so No. 1 piston will be at Top Dead Center (TDC) of the compression stroke. At this time the timing marks on the camshaft sprocket and the upper surface of the cylinder head should coincide, and the dowel pin of the camshaft sprocket should be at the upper side.

5. Remove the timing belt tensioner.

6. Remove the timing belt.

7. If timing belt is to be reused, mark the belt with an arrow to indicate the direction of rotation.

8. Inspect the belt tensioner and the idler pulley for free rotation and wear. Replace as necessary.

To install:

9. Align the timing marks on the camshaft and crankshaft sprockets.

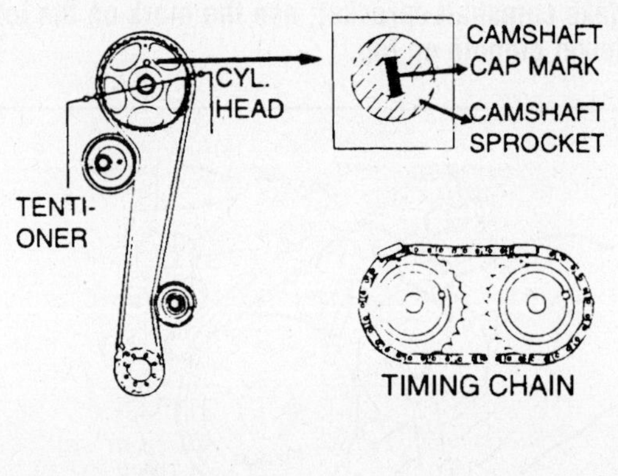

Alignment of timing marks—Hyundai 1.8L (VIN M) and 2.0L (VIN D, F) engines

View of the timing belt tensioner and idler pulley—Hyundai 1.8L (VIN M) and 2.0L (VIN D, F) engines

10. Loosely install the timing belt tensioner.
11. Install the idler pulley. Torque the bolt to 32–41 ft. lbs. (43–55 Nm).
12. Install the timing belt onto the crankshaft and then onto the crankshaft.

■ **When installing the timing belt, ensure that the tension side has no slack.**

13. Rotate the belt tensioner toward the water pump to take up any slack. Tighten the tensioner bolt.
14. Turn the crankshaft 1 full turn in a clockwise rotation and realign the crankshaft timing mark.
15. Check the belt for proper tension by depressing the belt on its long side with your finger and noting the belt deflection. The desired deflection should be 0.16–0.24 in. (4–6mm).

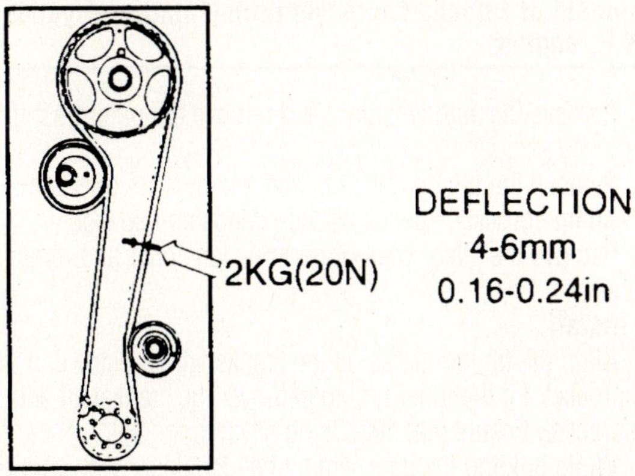

9305bg16

Checking timing belt deflection—Hyundai 1.8L (VIN M) and 2.0L (VIN D, F) engines

16. Tighten the tensioner bolt to 32–41 ft. lbs. (43–55 Nm).

■ **When tightening the bolt of the tensioner, ensure that the tensioner pulley shaft does not rotate with the bolt. Allowing it to rotate with the bolt can cause excessive tension on the belt.**

17. Ensure the timing marks are in alignment.
18. Install the timing belt covers and torque the bolts to 84 inch lbs. (10 Nm).
19. Install the crankshaft pulley and torque the bolt to 125–133 ft. lbs. (170–180 Nm).
20. Install the water pump pulley. Torque the bolts to 84 inch lbs. (10 Nm).
21. Install the drive belts and properly adjust tension.
22. Connect the negative battery cable.

2.0L (VIN P) ENGINES

✱✱ CAUTION

These engines are an interference type. If the timing belt has failed, valve-to-piston damage may have occurred. A compression check should always be performed to check for engine damage.

1. Disconnect the negative battery cable.

YEAR, MODEL & ENGINE APPLICATION

1992-98 Hyundai Sonata .... 2.0L VIN P

REPLACEMENT INTERVALS

Hyundai recommends timing belt replacement every 60,000 miles.

LABOR TIMES

1992-98	
Replacement	3.1 hrs
Severe Service	3.5 hrs
With air conditioning add	0.6 hrs
With tensioner replacement add	0.2 hrs

8807HUNg06

2. Remove the accessory drive belts and any mounting brackets that may interfere with timing cover removal.
3. Remove the water pump pulley and the crankshaft pulley.
4. Remove all attaching screws and remove the upper and lower timing belt covers.
5. Rotate the crankshaft clockwise and align the timing marks so No. 1 piston will be at Top Dead Center (TDC) of the compression stroke. At this time the timing marks on the camshaft sprocket and the upper surface of the cylinder head should coincide, and the dowel pin of the camshaft sprocket should be at the upper side.

■ **Always rotate the crankshaft in a clockwise direction. If the timing belt is to be reused, make a mark on the back of the timing belt indicating the direction of rotation so it may be reassembled in the same direction.**

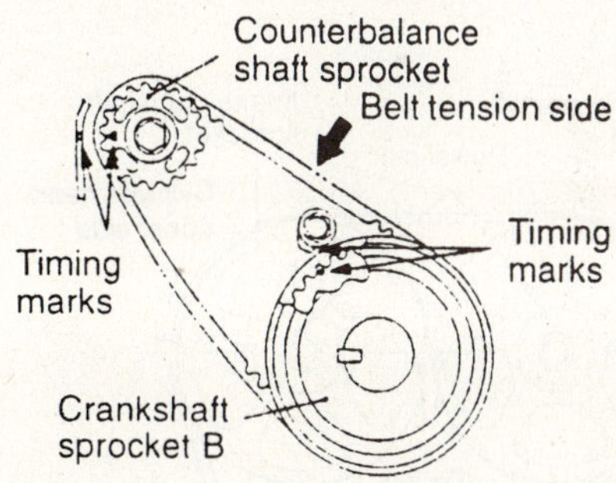

79225g22

Alignment of silent shaft timing mark locations—Hyundai 2.0L (VIN P) 1992 engine

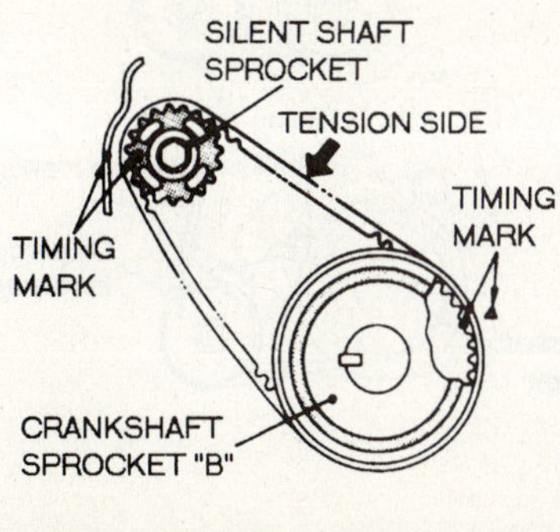

91251gg08

Alignment of silent shaft timing mark locations—Hyundai 2.0L (VIN P) 1993–98 engine

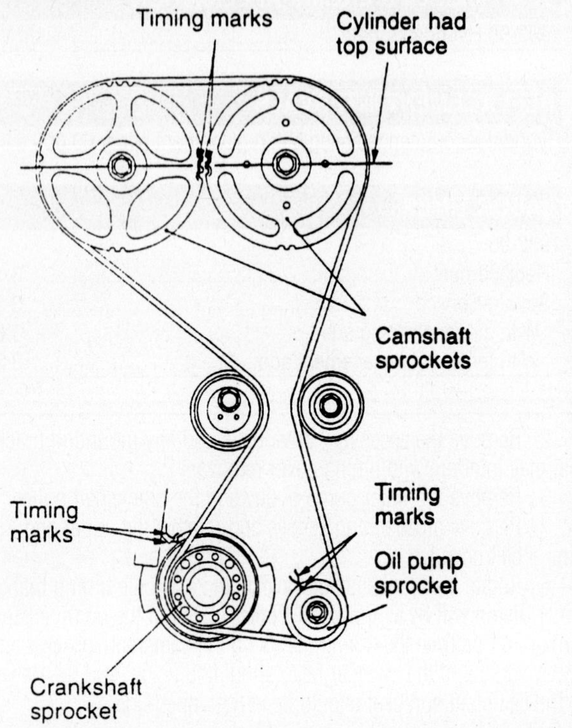

Alignment of engine timing marks for belt installation—Hyundai 2.0L (VIN P) 1992 engine

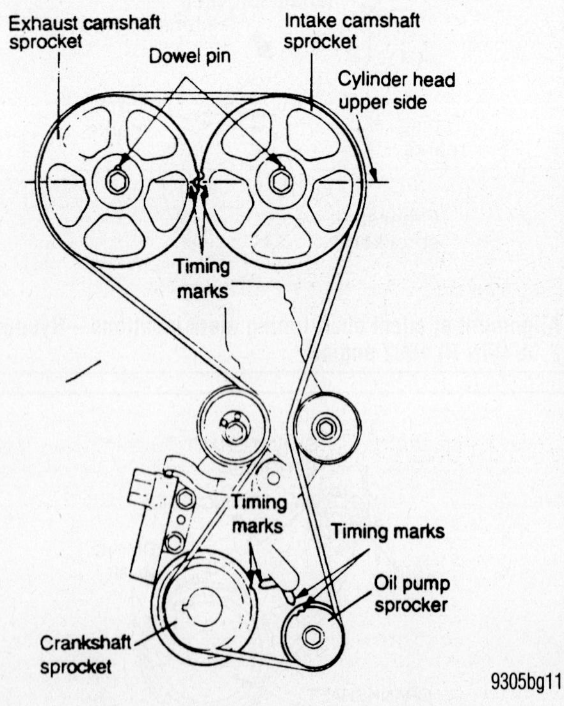

Alignment of engine timing marks for belt installation—Hyundai 2.0L (VIN P) 1993—98 engine

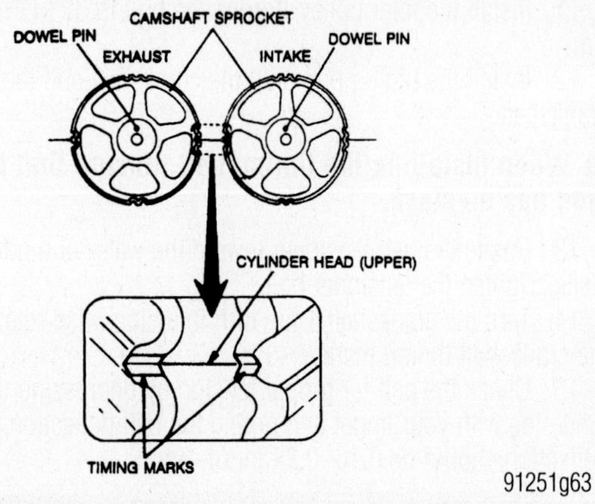

Alignment of camshaft sprocket timing marks—Hyundai 2.0L (VIN P) engine

6. Remove the auto tensioner and remove the outermost timing belt.
7. Remove the timing belt tensioner pulley, tensioner arm, idler pulley, oil pump sprocket, special washer, flange and spacer.
8. Remove the silent shaft (inner) belt tensioner and remove the inner belt.

To install:

9. Align the timing marks on the crankshaft sprocket and the silent shaft sprocket. Fit the inner timing belt over the crankshaft and silent shaft sprocket. Ensure that there is no slack in the belt.
10. While holding the inner timing belt tensioner with your fingers, adjust the timing belt tension by applying a force towards the center of the belt, until the tension side of the belt is taut. Tighten the tensioner bolt.

■ **When tightening the bolt of the tensioner, ensure that the tensioner pulley shaft does not rotate with the bolt. Allowing it to rotate with the bolt can cause excessive tension on the belt.**

11. Check belt for proper tension by depressing the belt on its long side with your finger and noting the belt deflection. The desired reading is 0.20–0.28 in. (5–7mm). If tension is not correct, readjust and check belt deflection.
12. Install the flanged crankshaft sprocket. The flange on the crankshaft sprocket must be installed towards the inner timing belt sprocket.
13. Install the bolt and special washer and torque the bolt to 80–94 ft. lbs. (110–130 Nm).
14. To install the oil pump sprocket, insert a Phillips screwdriver with a shaft 0.31 in. (8mm) in diameter into the plug hole in the left side of the cylinder block to hold the left silent shaft. Tighten the nut to 36–43 ft. lbs. (50–60 Nm).
15. Carefully push in the auto tensioner rod until the set hole in the rod is aligned with the hole in the cylinder body. Place a wire into the hole to retain the rod.
16. Install the tensioner pulley onto the tensioner arm. Locate the pinhole in the tensioner pulley shaft to the left of the center bolt. Then, tighten the center bolt finger-tight.
17. When installing the timing belt, turn the 2 camshaft sprockets so their dowel pins are located on top. Align the timing marks facing each other with the top surface of the cylinder head. When you let go of the exhaust camshaft sprocket, it will rotate 1 tooth in the counterclockwise direction. This should be taken into account when installing the timing belts on the sprocket.

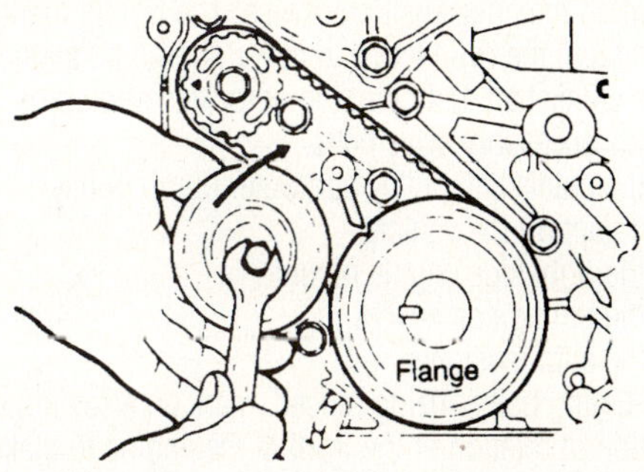

Adjusting the inner timing belt—Hyundai 2.0L (VIN P) engine

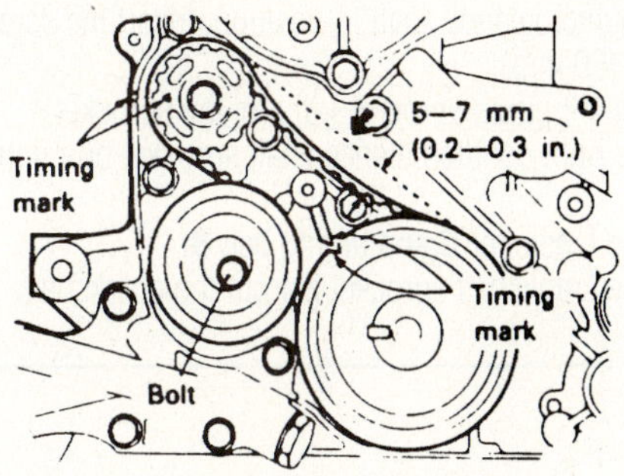

Checking inner belt deflection—Hyundai 2.0L (VIN P) engine

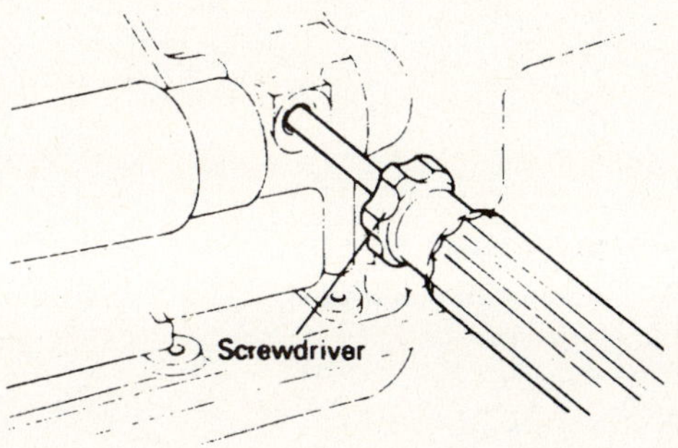

Holding the left side silent shaft—Hyundai 2.0L (VIN P) engine

■ **Both camshaft sprockets are used for the intake and exhaust camshafts and are provided with 2 timing marks. When the sprocket is mounted on the exhaust camshaft, use the timing mark on the right with the dowel pinhole on top. For the intake camshaft sprocket, use the mark on the left with the dowel pinhole on top.**

18. Align the crankshaft sprocket and oil pump sprocket timing marks.

19. After alignment of the oil pump sprocket timing marks, remove the plug on the cylinder block and insert a Phillips screwdriver with a shaft diameter of 0.31 in. (8mm) through the hole. If the shaft can be inserted 2.4 in. (70mm) deep, the silent shaft is in the correct position. If the shaft of the tool can only be inserted 0.8—1.0 in. (20–25mm) deep, turn the oil pump sprocket 1 turn and realign the marks. Reinsert the tool making sure it is inserted 2.4 in. (70mm) deep. Keep the tool inserted in hole for the remainder of this procedure.

■ **The above step assures that the oil pump socket is in correct orientation to the silent shafts. This step must not be skipped or a vibration may develop during engine operation.**

20. Install the timing belt as follows:
   a. Install the timing belt around the intake camshaft sprocket and retain it with 2 spring clips or binder clips.
   b. Install the timing belt around the exhaust sprocket, aligning the timing marks with the cylinder head top surface using 2 wrenches. Retain the belt with 2 spring clips.
   c. Install the timing belt around the idler pulley, oil pump sprocket, crankshaft sprocket and the tensioner pulley. Remove the 2 spring clips.
   d. Lift upward on the tensioner pulley in a clockwise direction and tighten the center bolt. Make sure all timing marks are aligned.
   e. Rotate the crankshaft 1/4 turn counterclockwise. Then, turn clockwise until the timing marks are aligned again.
21. To adjust the timing (outer) belt, turn the crankshaft 1/4 turn counterclockwise, then turn it clockwise to move No. 1 cylinder to TDC.
22. Loosen the center bolt. Using tool MD998738 or equivalent and a torque wrench, apply a torque of 22–24 inch lbs. (2.6–2.8 Nm). Tighten the center bolt.
23. Screw the special tool into the engine left support bracket until its end makes contact with the tensioner arm. At this point, screw the special tool in some more and remove the set wire attached to the auto tensioner, if the wire was not previously removed. Then remove the special tool.
24. Rotate the crankshaft 2 complete turns clockwise and let it sit for approximately 15 minutes. Then, measure the auto tensioner protrusion (the distance between the tensioner arm and auto tensioner body) to ensure that it is within 0.15–0.18 in. (3.8–4.5mm). If out of specification, repeat Step 1–4 until the specified value is obtained.
25. If the timing belt tension adjustment is being performed with the engine mounted in the vehicle, and clearance between the tensioner arm and the auto tensioner body cannot be measured, the following alternative method can be used:
   a. Screw in special tool MD998738 or equivalent, until its end makes contact with the tensioner arm.
   b. After the special tool makes contact with the arm, screw it in some more to retract the auto tensioner pushrod while counting the number of turns the tool makes until the tensioner arm is brought into contact with the auto tensioner body. Make sure the number of turns the special tool makes conforms with the standard value of 21/2-3 turns.
   c. Install the rubber plug to the timing belt rear cover.
26. Install the timing belt covers.
27. Install the water pump pulley and torque the bolts to 84 inch lbs. (10 Nm).
28. Install the crankshaft pulley and torque the bolts to 22 ft. lbs. (30 Nm).
29. Install drive belts and adjust to proper tension.
30. Install the engine mounting bracket and lower the engine.
31. Connect the negative battery cable.

2.4L (VIN B) ENGINES

YEAR, MODEL & ENGINE APPLICATION

2000-05 Hyundai Sante Fe .... 2.4L VIN B

REPLACEMENT INTERVALS

Hyundai recommends timing belt replacement every 60,000 miles.

LABOR TIMES

2000-05

Replacement	2.6 hrs
Severe Service	2.9 hrs
With tensioner replacement add	0.2 hrs

8807HUNg07

✳✳ CAUTION

These engines are an interference type. If the timing belt has failed, valve-to-piston damage may have occurred. A compression check should always be performed to check for engine damage.

1. Align the timing marks to set the No. 1 piston to Top Dead Center (TDC) by rotating the crankshaft clockwise. The timing marks of the camshaft sprocket and the cylinder head cover should be aligned and the marks on the crankshaft and oil pump sprockets should be aligned.
2. Remove the accessory drive belts.
3. Remove the crankshaft pulley and water pump pulley.
4. Raise and support the vehicle.
5. Remove the right side engine mount.
6. Remove the timing belt covers.
7. Remove the auto tensioner.
8. Mark the timing belt is being reused, mark an arrow on the belt noting the direction of rotation or the front of the engine to make sure the belt is reinstalled in its original position.
9. Remove the timing belt.
10. Remove the crankshaft outer sprocket and washer.
11. When removing the oil pump socket nut, first remove the plug at the side of the block and insert a 0.3 inch (8mm) diameter screwdriver to keep the left counterbalance shaft in position. Insert the screwdriver at least 2.36 inch (60 mm).
12. Remove the oil pump sprocket nut and the sprocket.
13. Loosen the right counterbalance shaft sprocket bolt until you can loosen it by hand.
14. Remove the tensioner **B** and timing belt **B**.
15. Remove the crankshaft sprocket **B** from the crankshaft.

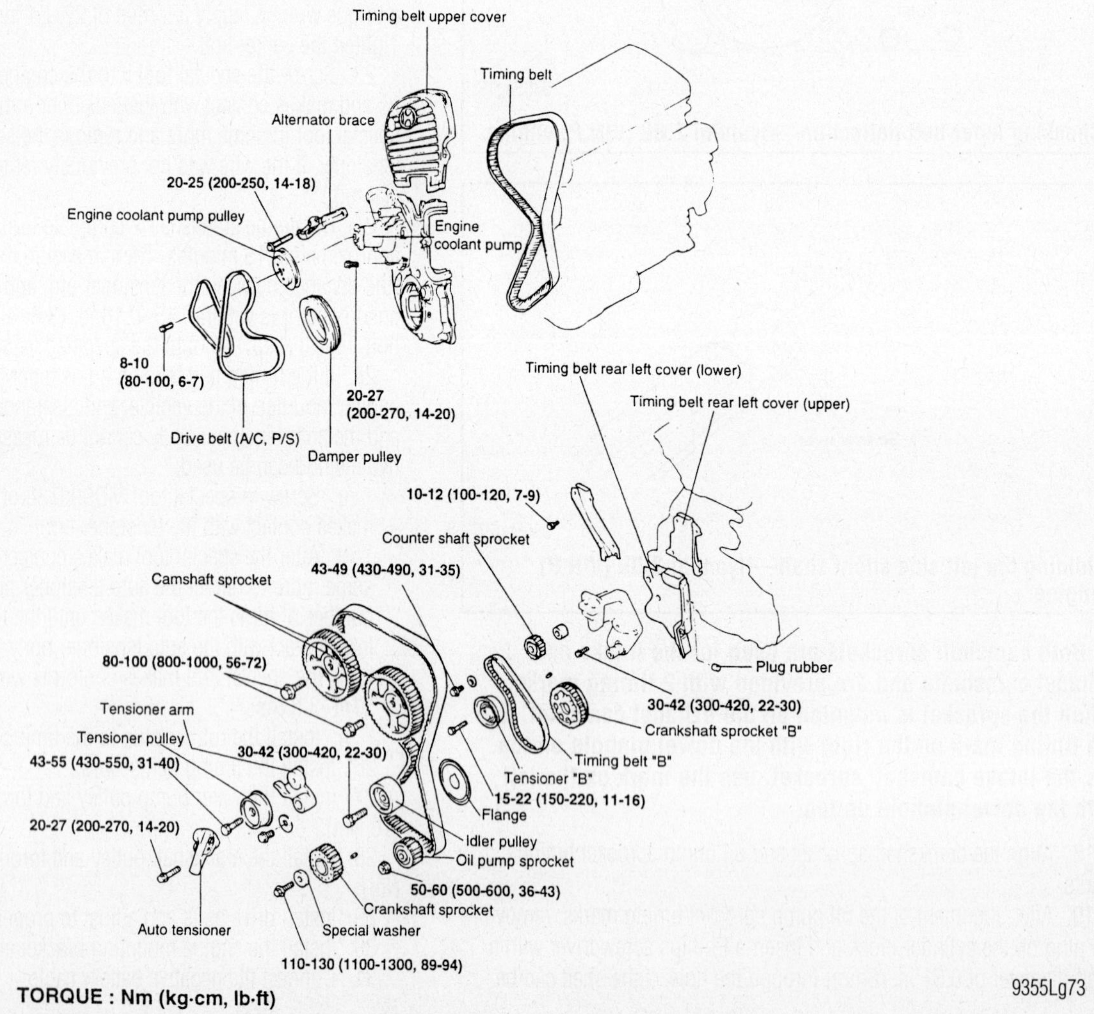

Exploded view of the timing belt assembly and related components—Hyundai 2.4L (VIN B) engine

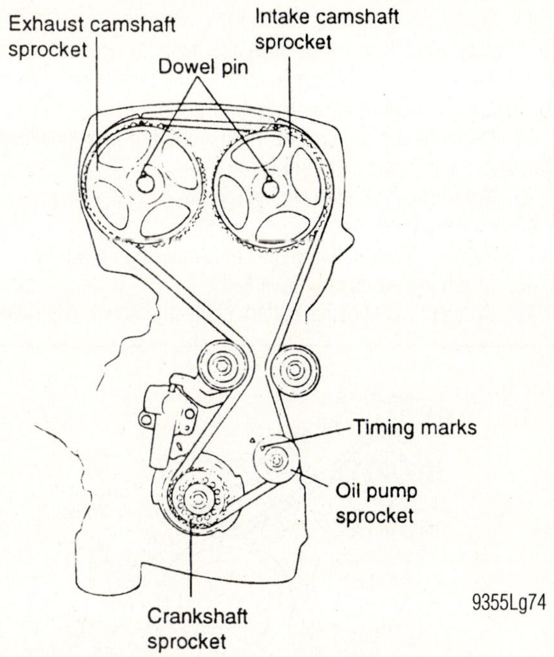

Correct sprocket alignment when the belt is installed—Hyundai 2.4L (VIN B) engine

To install:

16. Install crankshaft sprocket **B** to the crankshaft.

⁂ CAUTION

Pay attention to the direction of the flange. If it is installed in the wrong direction, the belt will break.

17. Apply engine oil to the outer surface of the spacer lightly and install the spacer to the right counterbalance shaft. Be sure to install spacer correctly.
18. Install the counterbalance shaft sprocket onto the right counterbalance shaft and then tighten the flange bolt by hand until it is tight
19. Align the timing mark on each sprocket with its corresponding timing mark on the front case.
20. Install timing belt **B** and make sure there is no slack.
21. Install the tensioner **B** so that the center of the pulley is located on the left side of the mounting bolt and the pulley flange faces the front of the engine
22. Align the timing mark on the right counterbalance shaft sprocket with the timing mark on the front case.
23. Lift the tensioner **B** to tighten tensioner **B** so that its tension side is pulled tight. Tighten the bolt on tensioner **B**. As the bolt is being tightened, make sure the shaft does not turn. If the shaft turns, the belt will be overtightened.
24. Make sure the timing marks are aligned.
25. Check the belt tension by depressing the center of the belt span with an index finger. The deflection should be 0.20–0.28 inch (5–7mm).
26. Install the flange and crankshaft sprocket making sure it is installed properly. Installing the flange incorrectly will cause the belt to break.
27. Install the crankshaft washer and bolt. Tighten the bolt to 80–94 ft. lbs. (110–130 Nm).
28. Insert a 0.3 inch (8mm) diameter screwdriver through the plug hole on the left side of the block to keep the left counterbalance shaft in position. Insert the screwdriver at least 2.36 inch (60 mm).

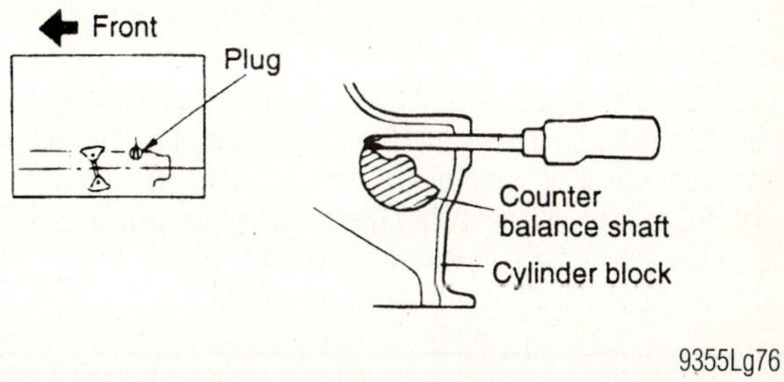

Insert a 0.3 inch (8mm) diameter screwdriver to keep the left counterbalance shaft in position—Hyundai 2.4L (VIN B) engine

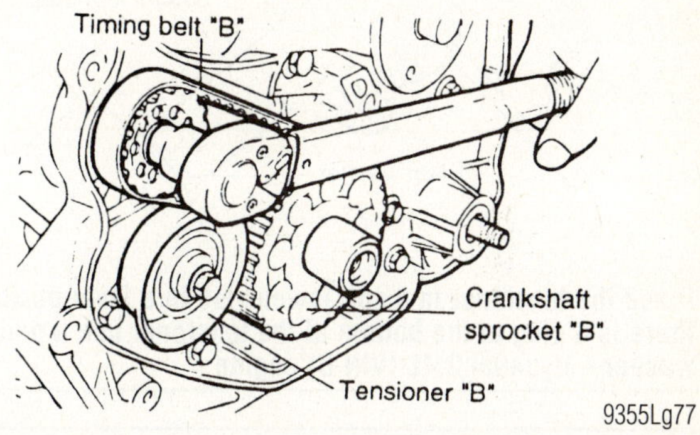

Remove tensioner B and timing belt B—Hyundai 2.4L (VIN B) engine

29. Install the oil pump sprocket and tighten the nut to 36–43 ft. lbs. (50–60 Nm).
30. Reset the auto tensioner as follows:
    a. Place the tensioner in a soft jawed vise in a level position. If there is a plug at the bottom of the tensioner use a plain washer.
    b. Compress the rod slowly using the vise until the set hole in the rod is aligned with the set hole on the cylinder.
    c. Insert a set pin through the body and rod and leave the pin installed.
31. Install the tensioner and tighten the bolts to 14–20 ft. lbs. (20–27 Nm). Leave the set pin in place.
32. Install the tensioner pulley and tighten the bolt to 31–40 ft. lbs. (43–55 Nm).
33. Rotate the camshaft sprockets so that the dowel pin is at the upper side. Set the timing mark of the sprocket correctly.

■ Before installing the belt, if the timing mark of the camshaft sprocket does not coincide with that of the rocker cover, do not rotate the cam. sprocket more than 2 teeth of the sprocket in any direction. Rotating the sprocket more that 2 teeth may make the valve and piston contact each other. If it is necessary to rotate the sprocket more that 2 teeth, rotate the crankshaft sprocket counterclockwise first based on the timing mark. After the camshaft sprocket is properly timed, return the crankshaft to TDC.

34. Align the crankshaft sprocket timing marks.
35. Align the oil pump sprocket timing marks.
36. Install the timing belt counterclockwise around the tensioner pulley and crankshaft sprocket. Hold the belt onto the tensioner pulley using your hand.

37. Pull the belt around the oil pump sprocket using your other hand.

38. Install the belt around the right-hand idler pulley, then the intake camshaft sprocket.

39. Turn the exhaust camshaft sprocket one tooth clockwise to align its timing mark with the cylinder top surface, then pull the belt around the exhaust camshaft sprocket.

40. Raise the tensioner pulley gently so that the belt does not sag and temporarily tighten the pulley center bolt.

41. Recheck that all timing marks are correct.

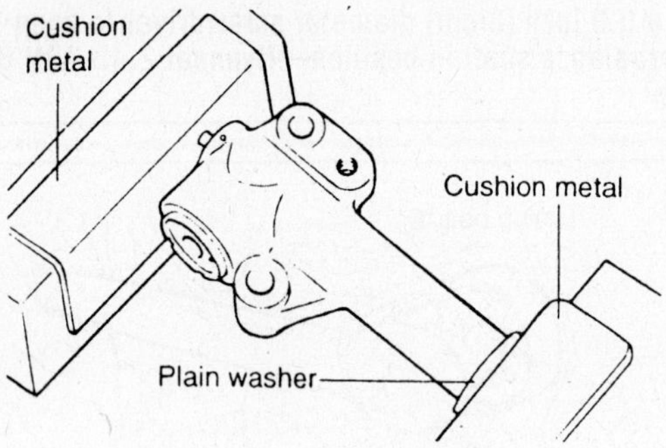

9355Lg78

Place the tensioner in a soft jawed vise in a level position. If there is a plug at the bottom of the tensioner use a plain washer—Hyundai 2.4L (VIN B) engine

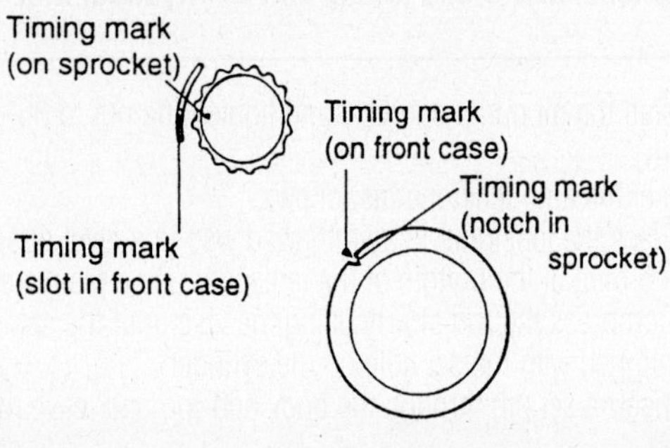

9355Lg81

Align the timing mark on each sprocket with its corresponding timing mark on the front case—Hyundai 2.4L (VIN B) engine

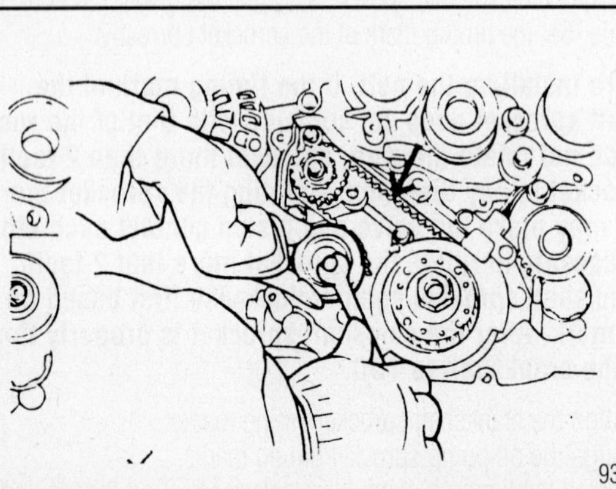

9355Lg83

Lift the tensioner B to tighten tensioner B so that its tension side is pulled tight—Hyundai 2.4L (VIN B) engine

42. Remove the set pin from the auto tensioner.

43. Rotate the crankshaft two turns clockwise and let it sit for around 15 minutes. After 15 minutes, measure the auto tensioner protrusion **A** (the distance between the tensioner arm and tensioner). The specification should be 0.24–0.35 inch (6–9mm).

44. Install the timing covers. Tighten the bolts to specification as shown in the accompanying illustration.

45. Install the right side engine mount.

46. Lower the vehicle.

47. Install the crankshaft pulley and water pump pulley.

48. Install the accessory drive belts.

49. Connect the negative battery cable and check the ignition timing.

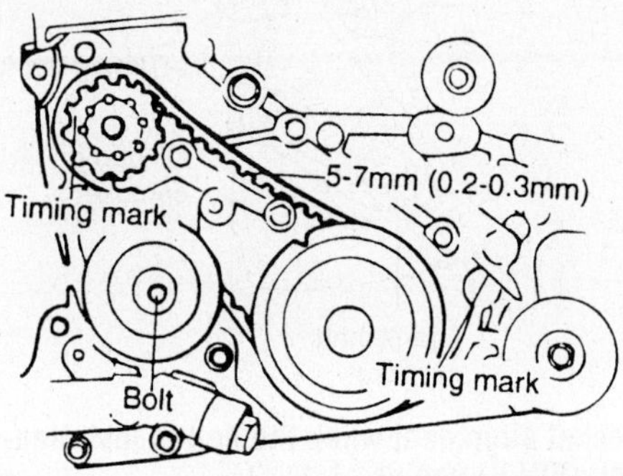

9355Lg84

Make sure the timing belt B marks are aligned and the belt tensioner is correct—Hyundai 2.4L (VIN B) engine

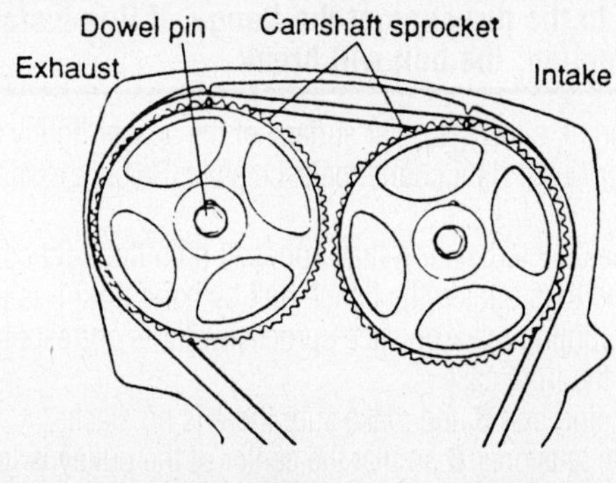

9355Lg87

Rotate the camshaft sprockets so that the dowel pin is at the upper side, set the timing mark of the sprocket correctly—Hyundai 2.4L (VIN B) engine

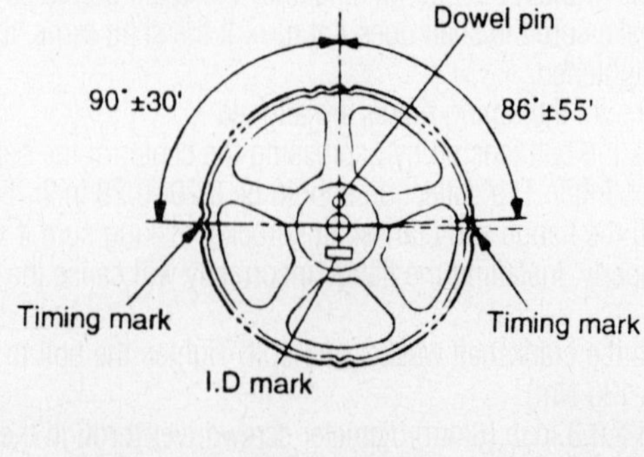

9355Lg88

Camshaft sprocket alignment—Hyundai 2.4L (VIN B) engine

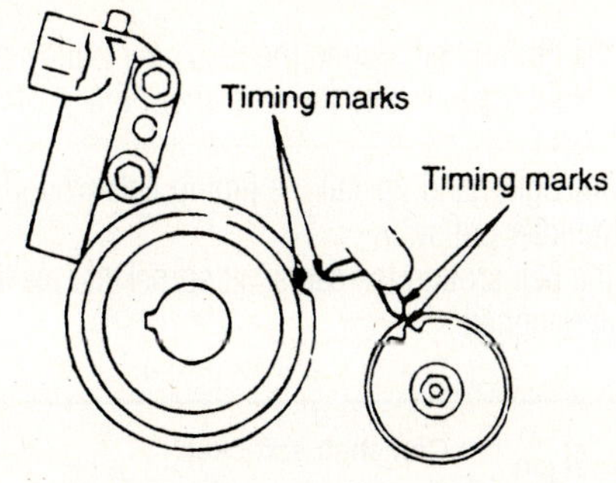

9355Lg89

Align the oil pump sprocket timing marks—Hyundai 2.4L (VIN B) engine

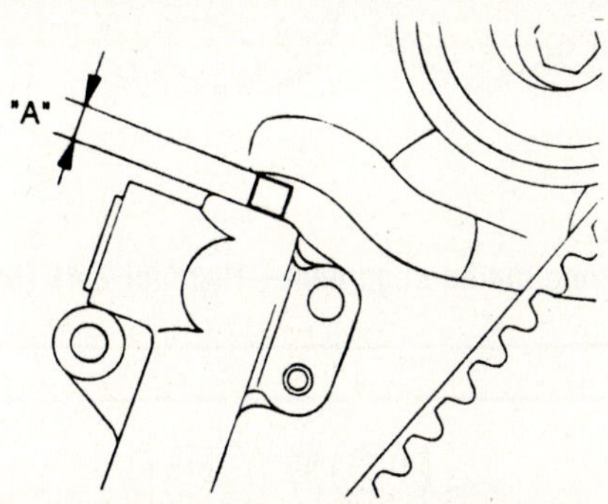

9355Lg90

Measure the auto tensioner protrusion A —Hyundai 2.4L (VIN B) engine

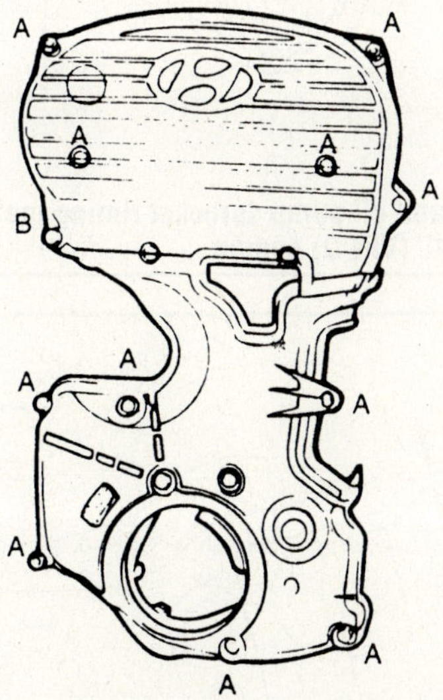

9355Lg91

Timing cover bolt location and torque specifications A —Hyundai 2.4L (VIN B) engine

2.4L (VIN D) ENGINES

YEAR, MODEL & ENGINE APPLICATION

1999-03 Hyundai Sonata	2.4L VIN D

REPLACEMENT INTERVALS

Hyundai recommends timing belt replacement every 60,000 miles.

LABOR TIMES

1999-03	
Replacement	2.8 hrs
Severe Service	3.1 hrs
With air conditioning add	0.2 hrs
With tensioner replacement add	0.2 hrs

8807HUNg08

** CAUTION

These engines are an interference type. If the timing belt has failed, valve-to-piston damage may have occurred. A compression check should always be performed to check for engine damage.

1. Disconnect the negative battery cable.
2. Remove any engine or side covers necessary to access the timing belt covers.
3. Using the proper equipment, slightly raise the engine to take the weight off the side engine mount. Remove the engine mount bracket.
4. Remove the accessory drive belts.
5. Remove the water pump pulley and crankshaft pulley.
6. Remove all attaching screws and remove the upper and lower timing belt covers.
7. Turn the crankshaft in a clockwise rotation to align the timing marks on the crankshaft and the camshafts.
8. Remove the auto tensioner.

■ **If the timing belt is to be reused, mark the belt with an arrow to indicate the direction of rotation.**

9. Remove the tensioner pulley and arm assembly.
10. Remove the timing belt.
11. Remove the balancer belt tensioner and remove the balancer belt.
12. Check camshaft and crankshaft sprockets, tensioner and idler pulleys for abnormal wear, cracks or damage. Replace as necessary.
13. Check the auto tensioner assembly for leaks and replace if necessary.

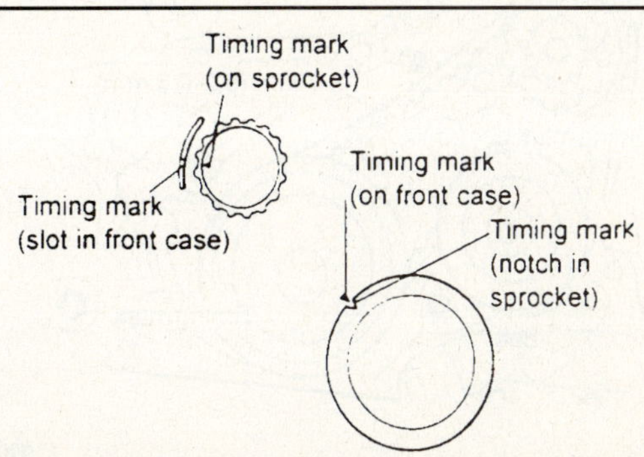

9305bg17

Inner timing belt mark alignment—Hyundai 2.4L (VIN D) engine

To install:

14. To install the balancer shaft belt, perform the following procedure:

a. Align the balancer timing belt marks with the corresponding marks on the front case.

b. Install the balancer belt onto the crankshaft and counter balance shaft sprockets.

c. Make sure the timing marks are properly aligned and that the tension side of the belt has no slack.

d. Install the balancer belt tensioner with center of pulley located on the left side of the mounting bolt and with the pulley flange toward the front of the engine.

e. Rotate the balancer belt tensioner in a clockwise direction so that the tension side of the belt will be pulled tight. Tighten the tensioner bolt.

■ **When the tensioner bolt is tightened, use care not to move the tensioner pulley. The belt will be over tightened.**

f. Check that the balancer timing belt marks are still aligned.

g. Check the balancer belt tension. Deflection should be 0.2–0.3 in. (5–7mm).

15. Using a vise with soft jaw adapters, push the rod of the auto tensioner slowly into the tensioner body until the hole in the rod is aligned with the hole in the tensioner body. Install a set pin through the holes to hold the rod in place.

16. Install the auto tensioner assembly leaving the set pin in place.

17. Install the tensioner pulley and arm assembly.

18. Align the camshaft timing marks. The marks will be at the top of each sprocket aligning with the marks on the upper surface of the rocker cover.

20. Install the timing belt around the tensioner pulley and the crankshaft sprocket. Hold the belt onto the tensioner pulley with your left hand.

21. With your right hand, install the timing belt around the oil pump sprocket and the idler pulley.

22. Install the belt around the camshaft sprockets, making sure the timing marks are aligned.

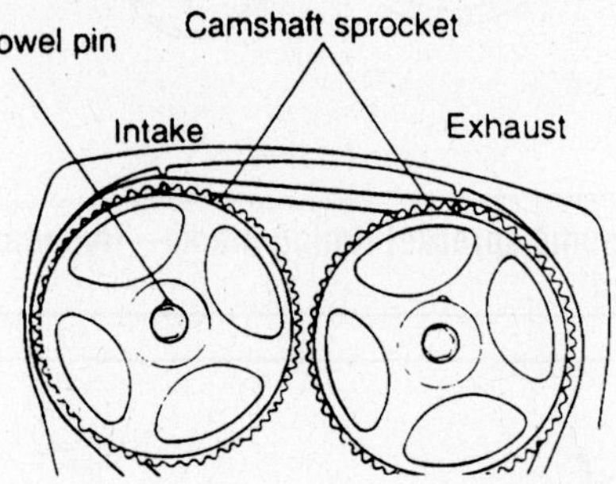

Camshaft timing marks alignment—Hyundai 2.4L (VIN D) engine

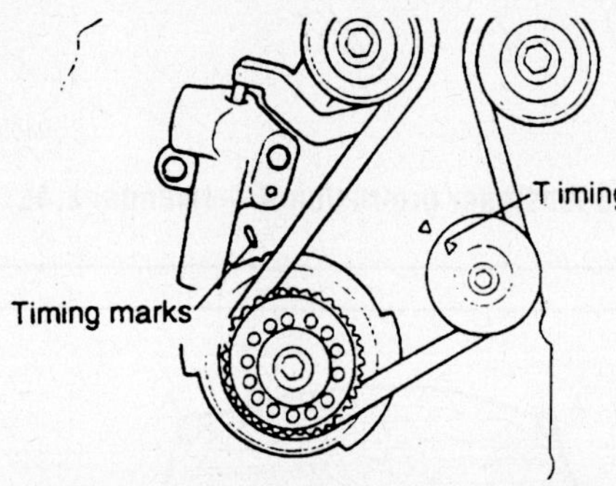

Crankshaft and oil pump sprocket timing marks alignment—Hyundai 2.4L (VIN D) engine

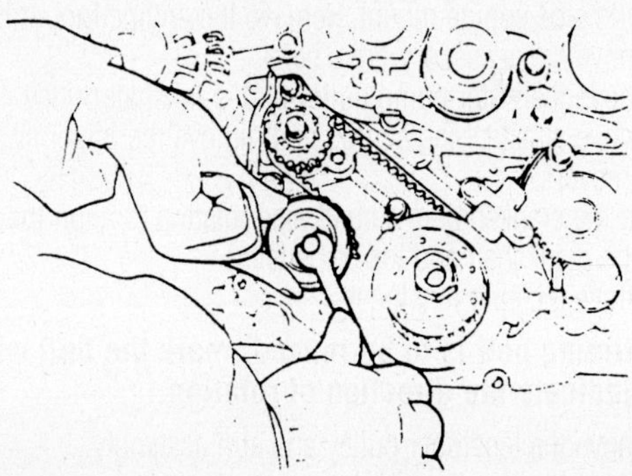

Adjusting the balancer timing belt tension—Hyundai 2.4L (VIN D) engine

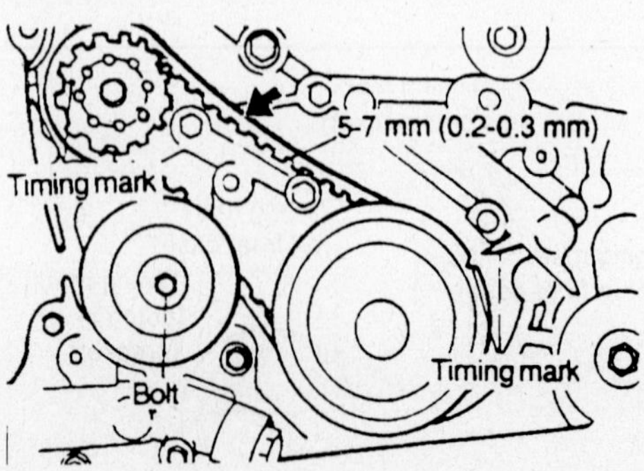

Checking the balancer timing belt tension—Hyundai 2.4L (VIN D) engine

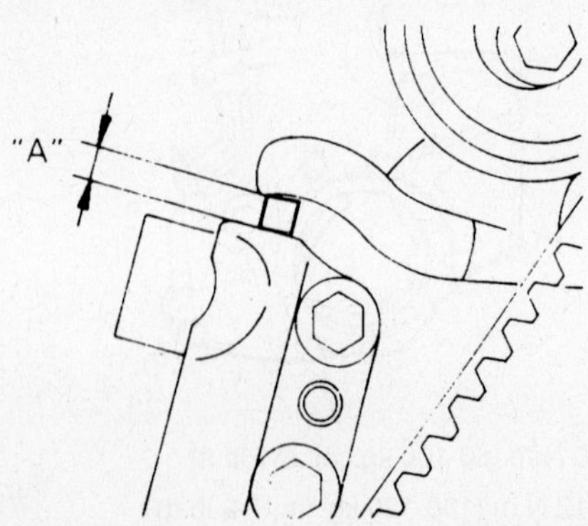

Measuring auto tensioner rod protrusion—Hyundai 2.4L (VIN D) engine

23. Raise the tensioner pulley against the timing belt so that the belt does not sag. Temporarily tighten the tensioner pulley bolt.
24. Check to make sure that all timing marks are aligned.
25. Remove the set pin from the auto tensioner body to release the rod against the tensioner pulley arm.
26. Rotate the crankshaft 2 revolutions and check that the timing marks are correctly aligned.
27. Measure the auto tensioner rod protrusion between the tensioner body and the tensioner arm. The measurement should be 0.15–0.18 in. (6–8mm).
28. Install the lower and upper timing belt covers.
29. Install the water pump pulley and crankshaft pulley.
30. Install the accessory drive belts.
31. Install the front engine bracket and lower the engine.
32. Install any covers that were previously removed for access.
33. Connect the negative battery cable.

2.4L (VIN S) ENGINES

YEAR, MODEL & ENGINE APPLICATION

1989-91 Hyundai Sonata	2.4L VIN S

REPLACEMENT INTERVALS

Hyundai recommends timing belt replacement every 60,000 miles.

LABOR TIMES

1989-91	
Replacement	3.1 hrs
Severe Service	3.5 hrs
With air conditioning add	0.6 hrs
With tensioner replacement add	0.2 hrs

8807HUNg09

✻✻ CAUTION

These engines are an interference type. If the timing belt has failed, valve-to-piston damage may have occurred. A compression check should always be performed to check for engine damage.

■ **An 8mm diameter metal bar is needed for this procedure.**

1. Disconnect the negative battery cable.
2. Remove the accessory drive belts and water pump pulley.
3. Rotate the crankshaft clockwise and align the timing marks so No. 1 piston will be at Top Dead Center (TDC) of the compression stroke. At this time the timing marks on the camshaft sprocket and the upper surface of the cylinder head should coincide, and the notch of the camshaft sprocket should be in line with the pointer.
4. Remove the crank adapter and the crankshaft pulley.
5. Remove the upper and lower timing belt covers.
6. Move the tensioner fully in the direction of the water pump and temporarily secure it there.
7. If the timing belt is to be reused, make an arrow mark on the belt to indicate the direction of rotation.
8. Remove the timing belt.
9. Remove the crankshaft sprocket bolt and pull the crankshaft sprocket and flange from the crankshaft.
10. Remove the plug on the left side of the block and insert an 8mm diameter metal bar in the opening to keep the silent shaft in position.
11. Remove the oil pump sprocket retaining nut and remove the oil pump sprocket.
12. Loosen the right silent shaft mounting bolt until it can be turned by hand.
13. Remove the belt tensioner and remove the silent shaft belt.

■ **Do not attempt to turn the silent shaft sprocket or loosen its bolt while the belt is off.**

14. Remove the silent shaft belt sprocket from the crankshaft.
15. Check the belt for wear, damage or glossing. Replace it if any cracks, damage, brittleness or excessive wear is found.
16. Check the tensioners for a smooth rate of movement.
17. Replace any tensioner that shows grease leakage through the seal.

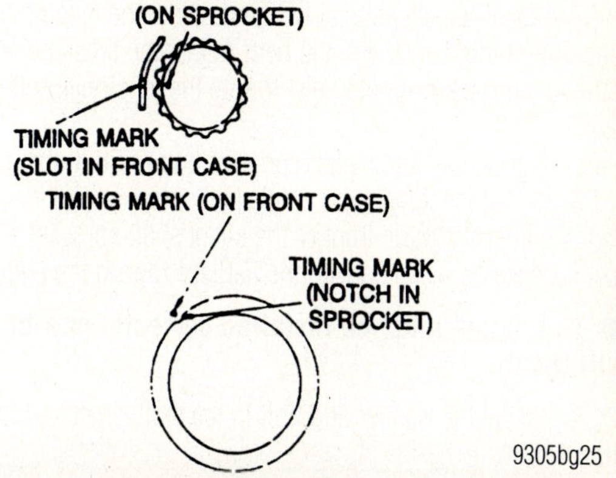

Alignment of counterbalance shaft and crankshaft sprocket timing marks—Hyundai 2.4L (VIN S) engine

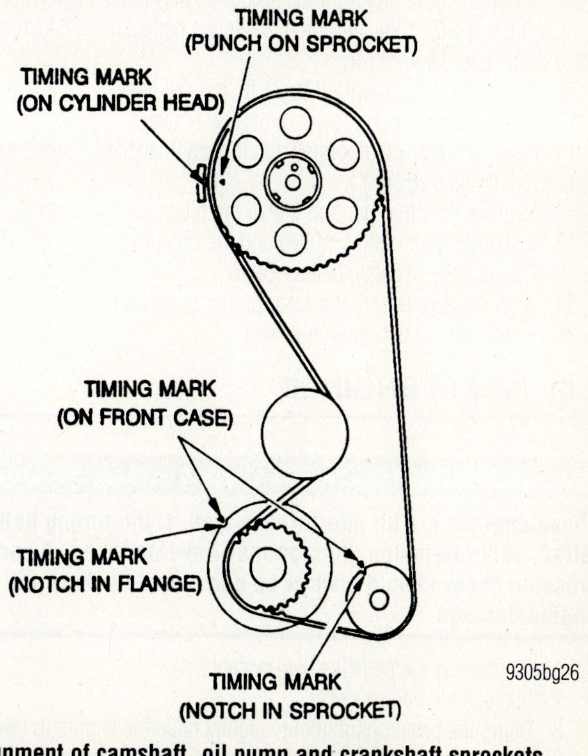

Alignment of camshaft, oil pump and crankshaft sprockets timing marks—Hyundai 2.4L (VIN S) engine

To install:

18. Install the silent shaft belt sprocket on the crankshaft, with the flat face toward the engine.
19. Apply a light coat of engine oil on the outer face of the spacer and install the spacer on the right silent shaft. The side with the rounded shoulder faces the engine.
20. Install the sprocket on the right silent shaft and install the bolt but do not tighten completely at this time.

■ **Align the silent shaft and oil pump sprockets using the timing marks. If the 8mm metal bar can not be inserted into the hole 2.36 inches (60mm) deep, the oil pump sprocket will have to be turned 1 full rotation until the bar can be to the full length. This step assures that the oil pump sprocket and the silent shafts are in correct orientation. This step must not be skipped or a vibration may develop during engine operation.**

21. Install the silent shaft belt and adjust the tension by moving the tensioner into contact with the belt. Adjust the tensioner enough to remove the slack from the belt. Torque the tensioner bolt to 21 ft. lbs. (28 Nm).
22. Torque the silent shaft sprocket bolt to 28 ft. lbs. (36 Nm).
23. Install the flange and crankshaft sprocket on the crankshaft. The flange conforms to the front of the silent shaft sprocket and the timing belt sprocket is installed with the flat face toward the engine.

■ **The flange must be installed correctly or a broken belt will result.**

24. Install the washer and bolt in the crankshaft and torque to 94 ft. lbs. (130 Nm).
25. Install the timing belt tensioner, spacer and spring.
26. Align the timing mark on each sprocket with the corresponding mark on the front case.
27. Install the timing belt on the sprockets and move the tensioner against the belt with sufficient force to allow a deflection of 5–7mm along its longest straight run.
28. Torque the tensioner bolt to 21 ft. lbs. (28 Nm).
29. Install the upper and lower timing belt covers.
30. Install the crankshaft pulley and the crank adapter. Torque the bolts to 21 ft. lbs. (28 Nm).
31. Remove the 8mm bar and install the plug.
32. Install the water pump pulley and the drive belt.
33. Connect the negative battery cable.
34. Start the engine and check for proper operation.
35. Check and adjust ignition timing.

2.5L (VIN E) ENGINES

✻✻ CAUTION

These engines are an interference type. If the timing belt has failed, valve-to-piston damage may have occurred. A compression check should always be performed to check for engine damage.

1. Disconnect the negative battery cable.
2. Remove the engine cover.
3. Using the proper equipment, slightly raise the engine to take the weight off the side engine mount. Remove the engine mount bracket.
4. Remove the accessory drive belt.

YEAR, MODEL & ENGINE APPLICATION

1999-01 Hyundai Sonata	2.5L VIN E

REPLACEMENT INTERVALS

Hyundai recommends timing belt replacement every 60,000 miles.

LABOR TIMES

1999-01	
Replacement	2.8 hrs
Severe Service	3.1 hrs
With air conditioning add	0.2 hrs
With tensioner replacement add	0.2 hrs

8807HUNg10

5. Remove the accessory belt tensioner and remove the crankshaft pulley.
6. Remove the upper and lower timing belt covers.
7. Turn the crankshaft in a clockwise direction to align the timing marks.
8. Remove the auto tensioner assembly.
9. Remove the timing belt.
10. Inspect the timing belt tensioner pulley, the idler pulley, crankshaft sprocket and the camshaft sprockets for cracks, abnormal wear or damage. Replace as necessary.

To install:

11. Using a vise with soft jaw adapters, push the rod of the auto tensioner slowly into the tensioner body until the hole in the rod is aligned with the hole in the tensioner body. Install a set pin through the holes to hold the rod in place.
12. Install the auto tensioner assembly leaving the set pin in place.
13. Install the tensioner pulley and arm assembly.
14. Make sure that all timing marks are aligned.
15. Install the timing belt in the following order:
    a. Crankshaft sprocket
    b. Idler pulley
    c. Intake cam sprocket (left side)
    d. Water pump pulley
    e. Intake cam sprocket (right side)
    f. Tensioner pulley
16. Raise the tensioner pulley against the timing belt so that the belt does not sag. Temporarily tighten the tensioner pulley bolt.
17. Check to make sure that all timing marks are aligned.
18. Remove the set pin from the auto tensioner body to release the rod against the tensioner pulley arm.
19. Rotate the crankshaft 2 revolutions and check that the timing marks are correctly aligned.
20. Measure the auto tensioner rod protrusion between the tensioner body and the tensioner arm. The measurement should be 0.23–0.31 in. (6–8mm).
21. Install the upper and lower timing covers.
22. Install the accessory belt tensioner and install the crankshaft pulley.
23. Install the accessory drive belt.
24. Install the engine mount and bracket.
25. Install the engine cover.
26. Connect the negative battery cable.

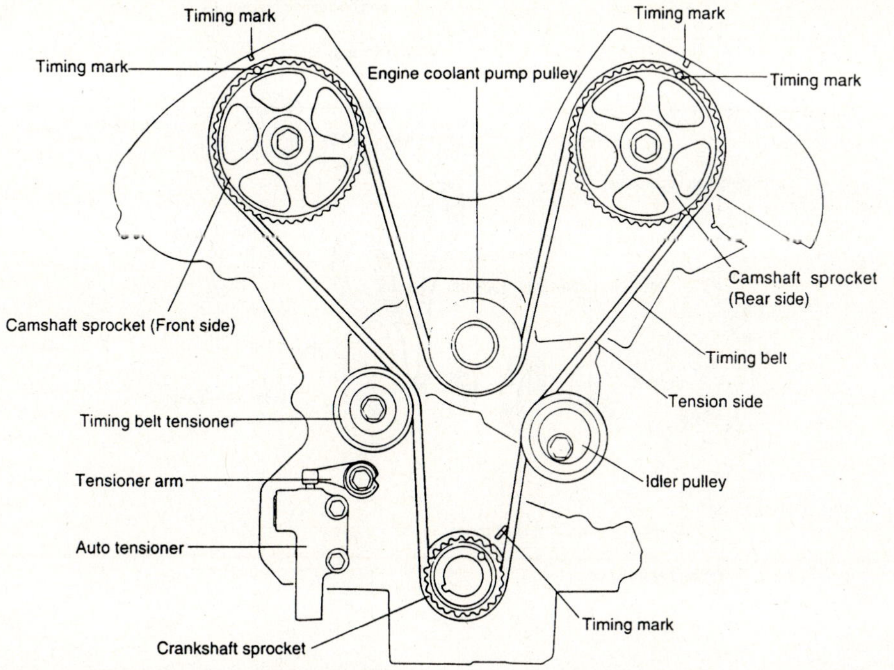

Alignment of the timing marks—Hyundai 2.5L (VIN E) engine

2.7L (VIN D, F, H) ENGINES

YEAR, MODEL & ENGINE APPLICATION	
2000-05 Hyundai Santa Fe	2.7L VIN D
2002-05 Sonata	2.7L VIN H
2003-05 Tiburon	2.7L VIN F

REPLACEMENT INTERVALS

Hyundai recommends timing belt replacement every 60,000 miles.

LABOR TIMES	
Santa Fe	
Replacement	2.9 hrs
Severe Service	3.2 hrs
With tensioner replacement add	0.2 hrs
Sonata	
Replacement	2.8 hrs
Severe Service	3.1 hrs
With air conditioning add	0.2 hrs
With tensioner replacement add	0.2 hrs
Tiburon	
Replacement	3.1 hrs
Severe Service	3.4 hrs
With air conditioning add	0.2 hrs
With tensioner replacement add	0.2 hrs

8807HUNg11

✱✱ CAUTION

These engines are an interference type. If the timing belt has failed, valve-to-piston damage may have occurred. A compression check should always be performed to check for engine damage.

1. Disconnect the negative battery cable.
2. Remove the engine appearance cover.
3. Using a 16mm wrench, rotate the tensioner arm clockwise about 14 degrees and remove the drive belt.
4. Remove the power steering pump pulley, idler pulley, tensioner pulley and crankshaft pulley.
5. Using the proper equipment, slightly raise the engine to take the weight off the side engine mount. Remove the engine mount bracket.
6. Remove the upper and lower timing covers.
7. Remove the timing belt tensioner.
8. Rotate the crankshaft clockwise and align the timing mark to set the No. 1 cylinder to Top Dead Center (TDC). Make sure the timing marks of the camshaft sprocket and cylinder head cover should align with each other.

■ **If reusing the belt, mark the direction of rotation on the belt to ensure proper belt installation.**

9. Unbolt the tensioner and remove the belt.
10. Hold the flange of the camshaft with a wrench, unfasten the sprocket bolts and remove the sprockets.

To install:

11. Instal the idler pulley on the water pump boss.
12. Install the idler pulley to the roll pin that is pressed in the water pump boss.
13. Install the tensioner arm and plain washer to the block.

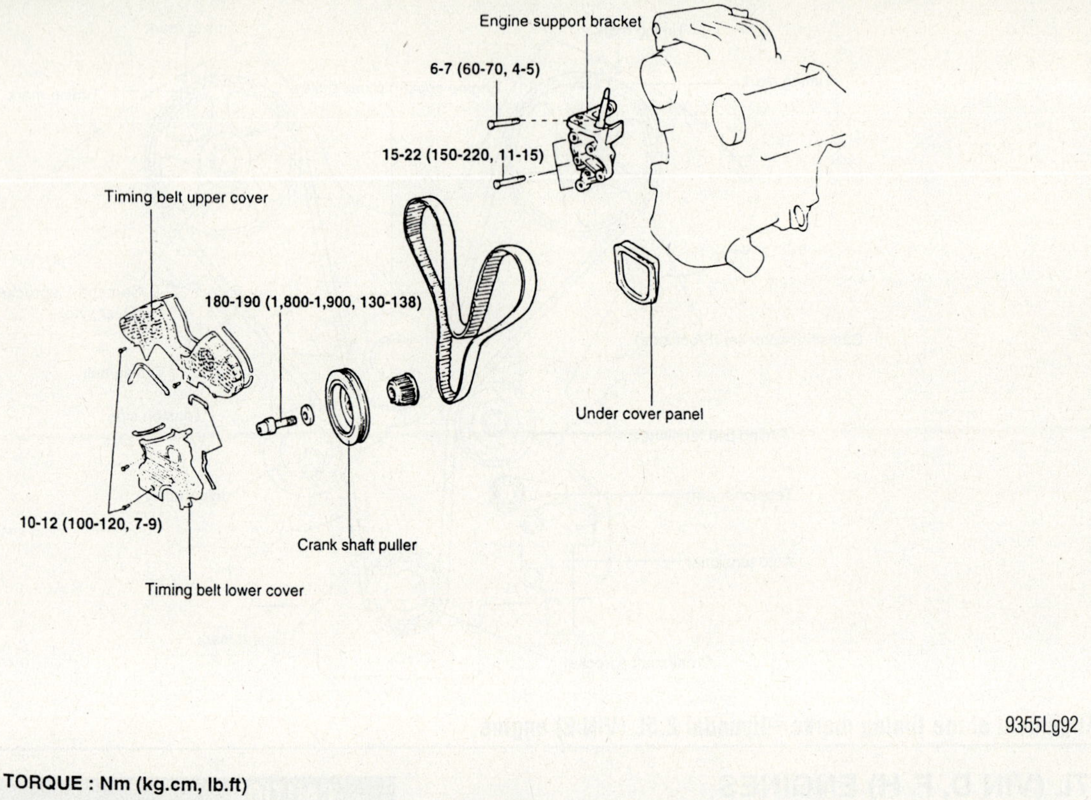

Exploded view of the timing belt assembly—Hyundai 2.7L (VIN D, F, H) engine

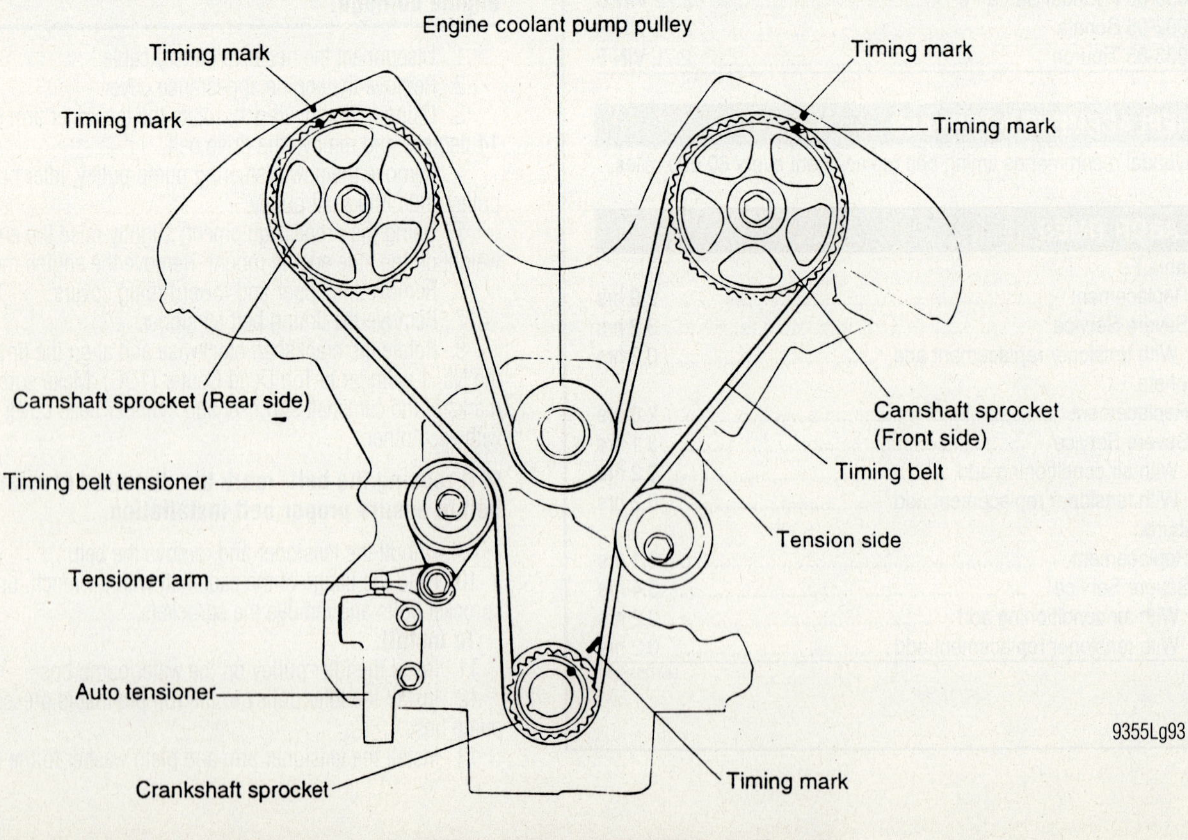

Correct timing marks alignment with the timing belt installed—Hyundai 2.7L (VIN D, F, H) engine

14. Install the tensioner pulley to the tensioner arm.
15. Compress the tensioner and install a set pin to keep the plunger in position.
16. Install the tensioner and tighten the bolt to 14–20 ft. lbs. (20–27 Nm).
17. Align the sprocket timing marks and install the belt in the following order:
   a. Crankshaft sprocket
   b. Idler pulley
   c. Camshaft sprocket on the left hand side
   e. Water pump pulley
   f. Camshaft sprocket on the right hand side
   g. Tensioner pulley
18. Remove the tensioner set pin.
19. Adjust the timing belt tension as follows:
   a. Rotate the crankshaft 2 turns clockwise and measure the projected length of the auto tensioner at TDC after being installed for 5 minutes.
   b. The projected length should be 0.23–0.31 in. (6–8mm).

Make sure the timing marks are in their specified positions or remove and reinstall the belt.

20. Install the upper and lower timing belt covers.
21. Install the power steering pump pulley, idler pulley, tensioner pulley and crankshaft pulley.
22. Install the accessory drive belt.
23. Install the engine mount and bracket.
24. Install the engine cover.
25. Connect the negative battery cable.

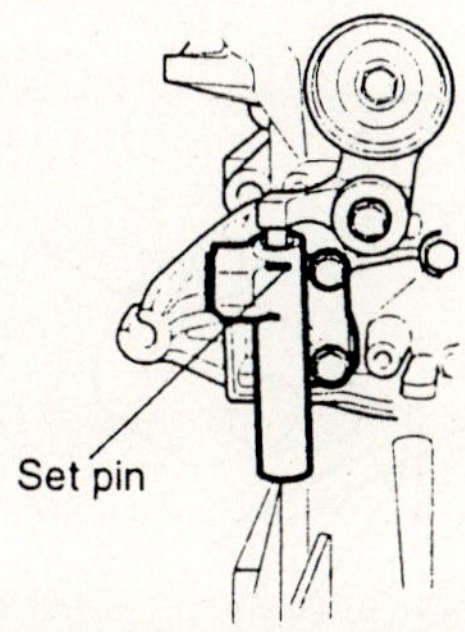

9355Lg94

Remove the set pin from the auto tensioner—Hyundai 2.7L (VIN D, F, H) engine

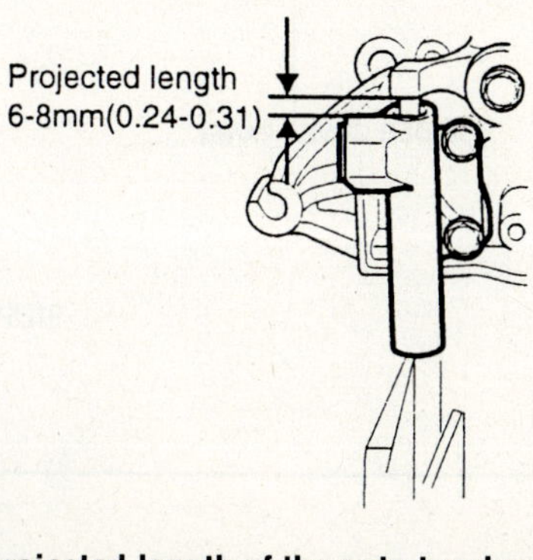

9355Lg95

The projected length of the auto tensioner at Top Dead Center (TDC) should be 6–8mm—Hyundai 2.7L (VIN D, F, H) engine

3.0L V6 (VIN T) SOHC ENGINES

YEAR, MODEL & ENGINE APPLICATION	
1990-98 Hyundai Sonata	3.0L VIN T

REPLACEMENT INTERVALS

Hyundai recommends timing belt replacement every 60,000 miles.

LABOR TIMES	
1990-98	
Replacement	3.3 hrs
Severe Service	3.7 hrs
With air conditioning add	0.6 hrs
With tensioner replacement add	0.2 hrs

8807HUNg12

1990–94

✱✱ CAUTION

These engines are an interference type. If the timing belt has failed, valve-to-piston damage may have occurred. A compression check should always be performed to check for engine damage.

1. Disconnect the negative battery cable.
2. To remove the air conditioning compressor belt, loosen the adjustment pulley locknut, turn the screw counterclockwise to reduce the drive belt tension and remove the belt.
3. To remove the serpentine drive belt, insert a 1/2 in. breaker bar in to the square hole of the tensioner pulley, rotate it counterclockwise to reduce the drive belt tension and remove the belt.
4. Remove the air conditioning compressor and the air compressor bracket, power steering pump and alternator from the mounts and support them to the side. Remove power steering pump/alternator automatic belt tensioner bolt and the tensioner.
5. Raise the vehicle and support safely. Remove the right inner fender splash shield.
6. Remove the crankshaft pulley bolt and the pulley/damper assembly from the crankshaft.
7. Lower the vehicle and place a floor jack under the engine to support it.
8. Separate the front engine mount insulator from the bracket. Raise the engine slightly and remove the mount bracket.
9. Remove the timing belt cover bolts and the upper and lower covers from the engine.
10. Turn the crankshaft until the timing marks on the camshaft sprocket and cylinder head are aligned.
11. Loosen the tensioning bolt, it runs in the slotted portion of the tensioner, and the pivot bolt on the timing belt tensioner.
12. Move the tensioner counterclockwise as far as it will go. Tighten the adjusting bolt.
13. Mark the timing belt with an arrow showing direction of rotation.
14. Remove the timing belt from the camshaft sprocket.
15. Remove the crankshaft pulley. Then, remove the timing belt. Remove the timing belt tensioner. Remove the retainer bolts from the timing sprockets and remove as required.

16. Inspect the belt thoroughly. The back surface must be pliable and rough. If it is hard and glossy, the belt should be replaced. Any cracks in the belt backing or teeth or missing teeth mean the belt must be replaced. The canvas cover should be intact on all the teeth. If rubber is exposed anywhere, the belt should be replaced.

17. Inspect the tensioner for grease leaking from the grease seal and any roughness in rotation. Replace a tensioner for either defect.

18. The sprockets should be inspected and replaced if there is any sign of damaged teeth or cracking anywhere.

19. Do not immerse sprockets in solvent, as solvent that has soaked into the metal may cause deterioration of the timing belt later.

20. Do not clean the tensioner in solvent either, as this may wash the grease out of the bearing.

To install:

21. Align the timing marks of the camshaft sprocket. Check that the crankshaft timing marks are still in alignment, the locating pin on the front of the crankshaft sprocket is aligned with a mark on the front case.

22. Mount the tensioner, spring and spacer with the bottom end of the spring free. Then, install the bolts and tighten the adjusting bolt slightly with the tensioner moved as far as possible away from the water pump. Install the free end of the spring into the locating tang on the front case. Position the belt over the crankshaft sprocket and then over the camshaft sprocket. Slip the back of the belt over the tensioner

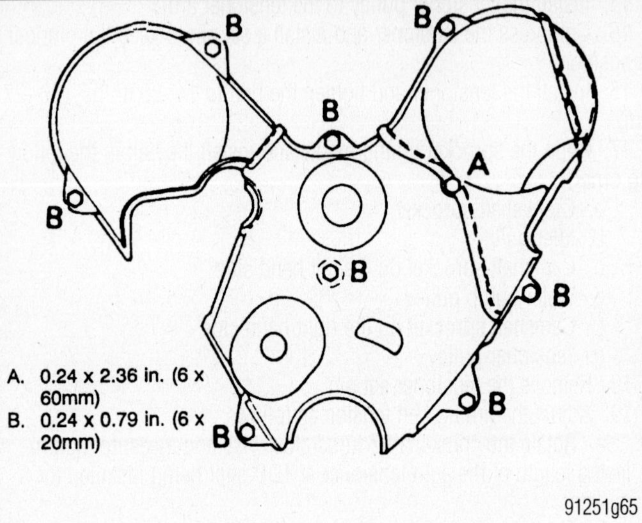

Timing belt cover bolt locations—Hyundai 1990–94 3.0L (VIN T) engine

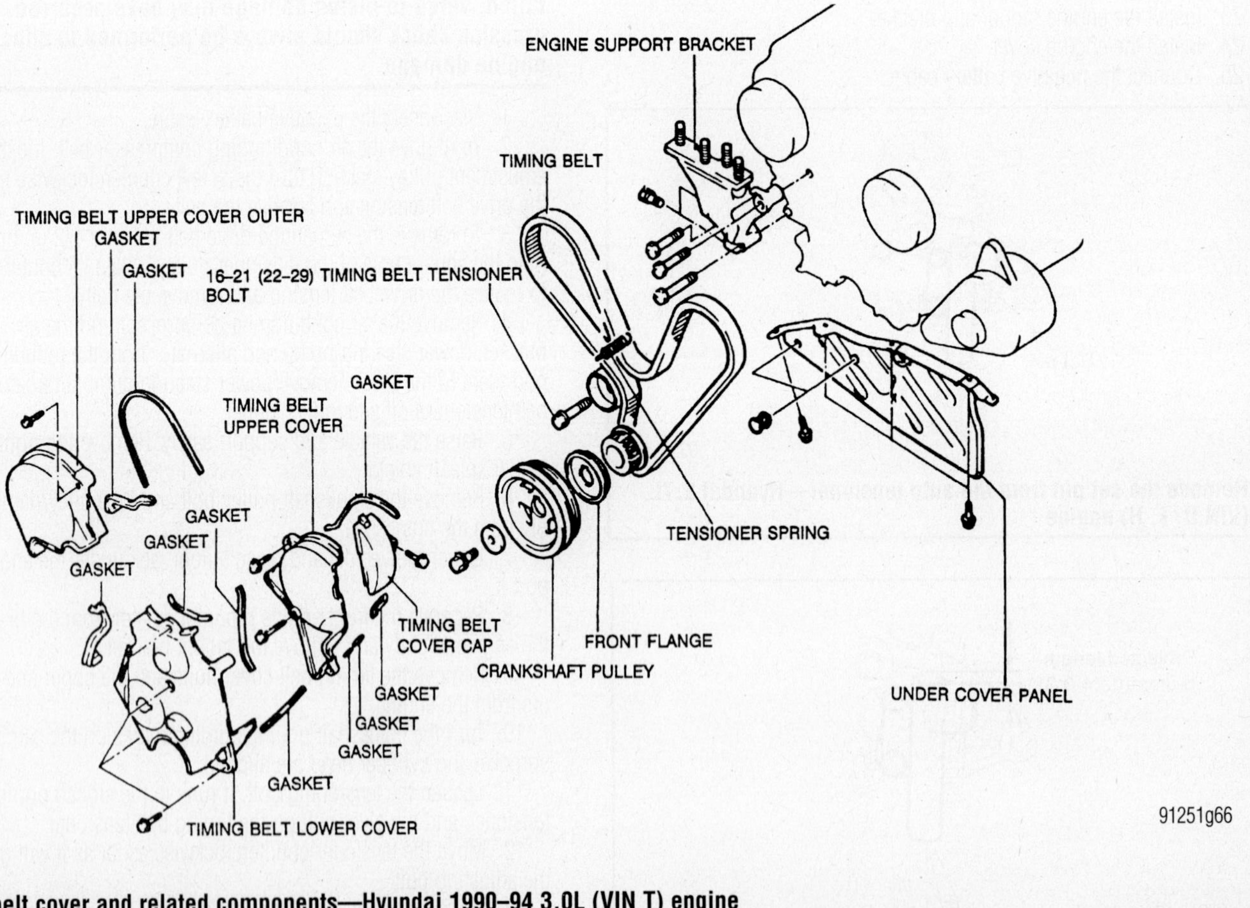

Timing belt cover and related components—Hyundai 1990–94 3.0L (VIN T) engine

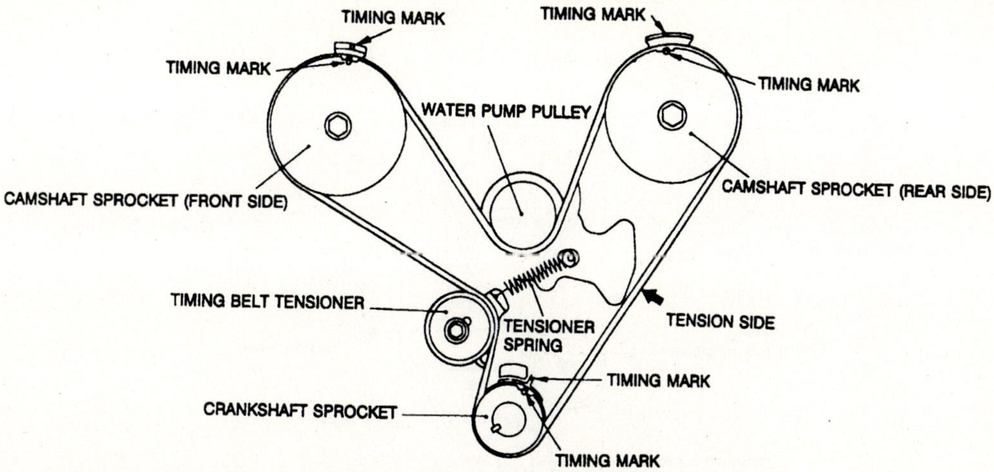

Alignment of the timing marks with the belt installed—Hyundai 1990–94 (VIN T) 3.0L engine

wheel. Turn the camshaft sprocket in the opposite of its normal direction of rotation until the straight side of the belt is tight and make sure the timing marks align. If not, shift the belt 1 tooth at a time in the appropriate direction until this occurs.

23. Loosen the tensioner mounting bolts so the tensioner works, without the interference of any friction, under spring pressure. Make sure the belt follows the curve of the camshaft pulley so the teeth are engaged all the way around.

24. Correct the path of the belt, if necessary. Torque the tensioner adjusting bolt to 16–21 ft. lbs. (22–29 Nm). Then, tighten the tensioner pivot bolt to the same figure. Bolts must be torqued in that order or tension won't be correct.

25. Turn the crankshaft 1 turn clockwise until timing marks again align to seat the belt. Then loosen both tensioner attaching bolts and let the tensioner position itself under spring tension as before. Finally, torque the bolts in the proper order exactly as before. Check belt tension by putting a finger on the water pump side of the tensioner wheel and pull the belt toward it. The belt should move toward the pump until the teeth are about 1/4 of the way across the head of the tensioner adjusting bolt. Tension the belt, if necessary.

26. Install the timing belt covers.

27. Install the crankshaft pulley, making sure the pin on the crankshaft sprocket fits through the hole in the rear surface of the pulley. Install the retaining bolt and torque to 108–116 ft. lbs. (147–157 Nm).

28. Install the engine mount bracket and secure with the mounting hardware.

29. Install the pulley damper assembly to the crankshaft. Torque the bolt to 110 ft. lbs. (149 Nm). Install the splash shield.

30. Install the power steering pump/alternator automatic belt tensioner.

31. Install the air conditioning compressor bracket, compressor, power steering pump and alternator.

32. Install the accessory drive belt.

33. Connect the negative battery cable and check all disturbed components for proper operation.

1995–98

CAUTION

These engines are an interference type. If the timing belt has failed, valve-to-piston damage may have occurred. A compression check should always be performed to check for engine damage.

1. Disconnect the negative battery cable.
2. Remove the engine undercover.
3. Remove the accessory drive belts.
4. Remove the air conditioner compressor tension pulley assembly.
5. Remove the tension pulley bracket.
6. Using the proper equipment, slightly raise the engine to take the weight off the side engine mount.
7. Disconnect the power steering pump pressure switch connector. Remove the power steering pump and wire aside.
8. Remove the engine support bracket.
9. Remove the crankshaft pulley.
10. Remove the timing belt cover cap.
11. Remove the timing belt upper and lower covers.
12. Turn the crankshaft until the timing marks on the camshaft sprocket and cylinder head are aligned.
13. Loosen the timing belt tensioner bolt and turn the tensioner counterclockwise as far as it will go. Tighten the adjusting bolt.
14. Mark the timing belt with an arrow showing direction of rotation.
15. Remove the timing belt.
16. If defective, remove the timing belt tensioner.

To install:

17. If necessary, install the timing belt tensioner.
18. Attach the top of the tensioner spring on the engine coolant pump pin. Ensure the hook on the pin is facing down and the hook on the tensioner is facing away from the engine
19. Rotate the timing belt tensioner to the extreme counterclockwise position. Temporarily lock the tensioner in place.
20. Align the timing marks of the camshaft and crankshaft sprockets.

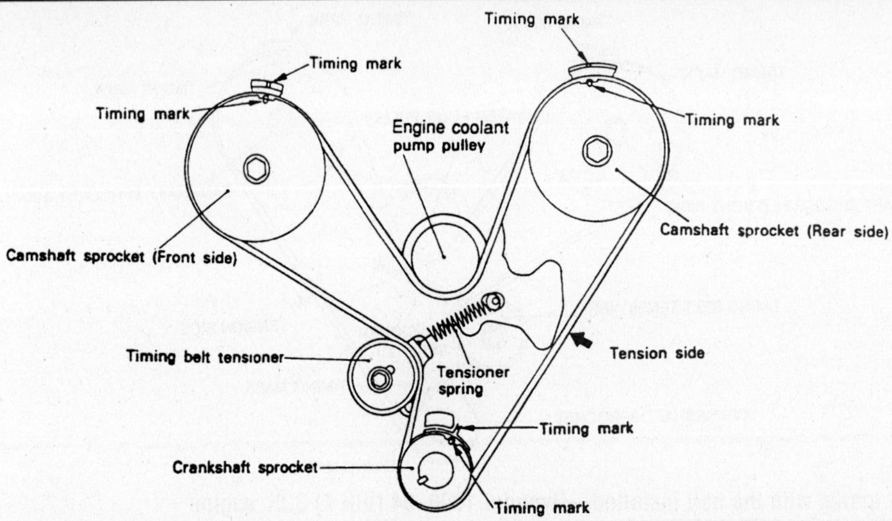

Timing belt sprocket alignment mark positioning for belt removal and installation—Hyundai 1995–98 3.0L (VIN T) engine

21. Install the timing belt on the crankshaft sprocket, then onto the rear camshaft sprocket.
22. Route the belt to the coolant pump pulley, the front camshaft sprocket and the timing belt tensioner.
23. Apply force counterclockwise to the rear camshaft sprocket with tension on the tight side of the belt and check that timing marks are aligned.
24. Loosen the tensioner bolt 1–2 turns and using a belt tension gauge, tighten the timing belt to a tension of 57–84 lbs. (260–380 N).
25. Turn the crankshaft 2 turns clockwise.
26. Readjust the sprocket timing marks and tighten the tensioner bolts.
27. Install the timing covers. Make sure all pieces of packing are positioned in the inner grooves of the covers when installing.
28. Install the crankshaft pulley. Torque the bolt to 108–116 ft. lbs. (150–160 Nm).
29. Install the engine support bracket.
30. Install the power steering pump and reconnect wire harness at the power steering pump pressure switch.
31. Install the engine mounting bracket and remove the engine support fixture.
32. Install the tension pulleys and drive belts.
33. Install the cruise control actuator.
34. Install the engine undercover.
35. Connect the negative battery cable.

3.0L (VIN D) AND 3.5L (VIN E) DOHC ENGINES

✱✱ CAUTION

These engines are an interference type. If the timing belt has failed, valve-to-piston damage may have occurred. A compression check should always be performed to check for engine damage.

YEAR, MODEL & ENGINE APPLICATION

2000-01 Hyundai XG300	3.0L VIN D
2002-04 XG350	3.5L VIN E

REPLACEMENT INTERVALS

Hyundai recommends timing belt replacement every 60,000 miles.

LABOR TIMES

2000-04	
Replacement	4.6 hrs
Severe Service	5.1 hrs
With tensioner replacement add	0.2 hrs

8807HUNg13

1. Disconnect the negative battery cable.
2. Raise and support the vehicle.
3. Remove the passenger side wheel.
4. Remove the side cover.
5. Remove the engine appearance cover.
6. Rotate the accessory drive belt tensioner clockwise and remove the drive belt using a 16mm wrench.
7. Remove the power steering pump pulley, idler pulley, tensioner pulley and crankshaft pulley.
8. Remove the upper and lower timing belt covers.
9. Turn the crankshaft until No. 1 cylinder is on TDC of the compression stroke and the timing marks on the camshaft sprocket and cylinder head are aligned.
10. Remove the auto-tensioner.
11. Remove the timing belt tensioner.
12. Mark the timing belt with an arrow showing direction of rotation.
13. Remove the timing belt.

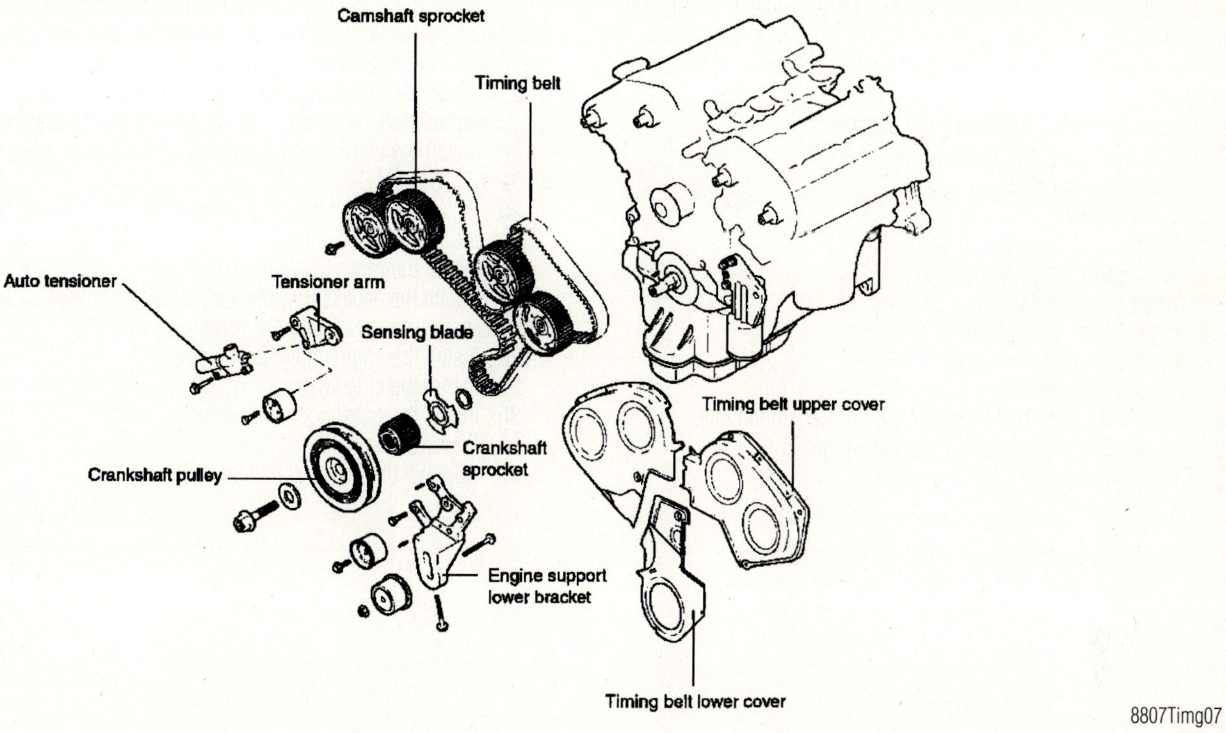

Exploded view of timing belt components—Hyundai 3.0L (VIN D) and 3.5L (VIN E) DOHC engines

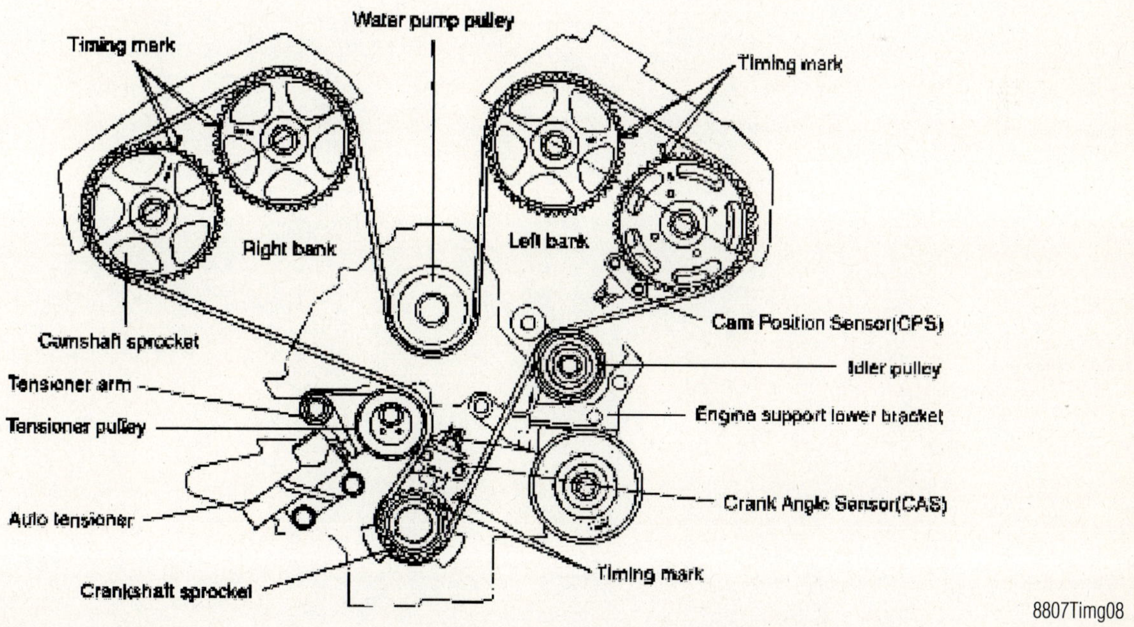

Crankshaft and camshaft sprocket timing mark locations—Hyundai 3.0L (VIN D) and 3.5L (VIN E) DOHC engines

To install:

14. Compress the tensioner and install a set pin to keep the plunger in position.
15. Install the auto-tensioner and tighten the bolts to 14–20 ft. lbs. (18–25 Nm).
16. Ensure the timing marks are still aligned.
17. Install the timing belt in the following order:
    a. Crankshaft sprocket
    b. Guide pulley
    c. Left exhaust camshaft sprocket
    e. Left intake camshaft sprocket
    f. Water pump pulley
    g. Right intake camshaft sprocket
    h. Right exhaust camshaft sprocket
    i. Tensioner pulley
18. Ensure the timing belt is taut on the non-tensioned side.
19. Remove the set pin from the auto-tensioner.
20. Turn the crankshaft 2 turns clockwise.
21. Ensure the timing marks are still aligned.
22. Allow the engine to settle for 5 minutes, then measure the protrusion of the tensioner pushrod. Projection length should be 0.15–0.18 in. If length is not as specified, remove and reinstall the timing belt.
23. Install the timing covers. Make sure all pieces of packing are positioned in the inner grooves of the covers when installing.
24. Install the crankshaft pulley. Torque the bolt to 130–138 ft. lbs. (176–187 Nm).
25. Install the power steering pump pulley, idler pulley, tensioner pulley and crankshaft pulley.
26. Install the power steering belt.
27. Rotate the accessory drive belt tensioner clockwise and install the accessory drive belt using a 16mm wrench.
28. Install the engine appearance cover.
29. Install the side cover.
30. Install the passenger side wheel.
31. Lower the vehicle.
32. Connect the negative battery cable.

Infiniti

3.0L (VG30DE) ENGINES

YEAR, MODEL & ENGINE APPLICATION	
1993-97 Infiniti J30	3.0L Code VG30DE

REPLACEMENT INTERVALS
Infiniti recommends timing belt replacement every 105,000 miles.

LABOR TIMES	
1993-97	
Replacement	4.6 hrs
Severe Service	5.1 hrs
With each timing gear replacement add	1.0 hrs

8807INFg01

1993–94

CAUTION

These engines are an interference type. If the timing belt has failed, valve-to-piston damage may have occurred. A compression check should always be performed to check for engine damage.

1. Disconnect the negative battery cable.
2. Remove the engine undercover.
3. Drain the cooling system. Remove both cylinder block drain plug to drain coolant from the block.
4. Remove the air ducts.
5. Remove the radiator, drive belts, cooling fan and coupling.
6. Remove the crankshaft pulley bolt. Remove the crankshaft pulley using a puller.
7. Remove the water inlet and outlet. Remove the starter motor.
8. Remove the front timing covers.
9. Install a suitable stopper bolt into the tensioner arm so the auto-tensioner pusher does not spread out.
10. Set the No. 1 cylinder on Top Dead Center (TDC) of the compression stroke.
11. Remove the auto-tensioner and timing belt.

To install:

12. Confirm that the No. 1 cylinder is on TDC of the compression stroke.
13. Align matchmarks on camshaft and crankshaft sprockets with aligning marks on rear belt cover and oil pump housing.
14. Remove all spark plugs.
15. Align white lines on timing belt with matchmarks on camshaft sprocket and crankshaft sprocket. Point arrow on timing belt towards front and install.
16. Adjust the tensioner arm to give 0.16 in. (4mm) clearance with pusher of auto-tensioner using a suitable vise, and then insert stopper bolt into tensioner arm so clearance does not change.
17. Install the auto-tensioner and tighten lower bolts hand tight. Push auto-tensioner toward belt until contact is made. Then push slightly more. Turn the crankshaft 10 degrees clockwise and tighten the tensioner nuts to 12–15 ft. lbs. (16–21 Nm).
18. Turn the crankshaft 120 degrees counterclockwise.
19. Loosen tensioner bolts 1/2 turn and move tensioner away from timing belt as far as it will move.
20. Turn the crankshaft clockwise and set No. 1 cylinder to TDC on its compression stroke.
21. Push the end of a pusher tool J-38387 with 13 lbs. (6 kg) of force and tighten the auto-tensioner bolts to 12–15 ft. lbs. (16–21 Nm).
22. Turn the crankshaft 120 degrees clockwise, then turn crankshaft 120 degrees counterclockwise. Set No. 1 cylinder to TDC on the compression stroke.
23. Prepare a steel plate measuring 0.12 x 0.39 in. (3 x 10mm). Set the plate on the timing belt and push it using a Pusher tool J-38387 or equivalent, with 11 lbs. (5 kg) of force at a point midway between the camshaft sprockets. Also use the tool between the camshaft sprockets and the idler/tensioner pulleys. Deflection should be 0.24–0.28 in. (6–7mm). If not, readjust the tensioner.
24. Confirm the auto-tensioner mounting nuts are tightened to 12–15 ft. lbs. (16–21 Nm). Remove the auto-tensioner stopper bolt.
25. After 5 minutes, check the clearance between the tensioner arm and the pusher stays at 0.138–0.205 in. (3.5–5.2mm).
26. Check for proper installation of the timing belt at all pulleys, then install the timing belt covers. Tighten bolts to 24–38 inch lbs. (3–5 Nm).
27. Install the water inlet and outlet.
28. Install the crankshaft pulley and bolt. Tighten the crankshaft pulley bolt to 159–174 ft. lbs. (216–235 Nm).
29. Install the cooling fan and fan coupling, drive belts and radiator.
30. Install the air ducts.
31. Fill the cooling system and install the engine undercover.

1995–97

CAUTION

These engines are an interference type. If the timing belt has failed, valve-to-piston damage may have occurred. A compression check should always be performed to check for engine damage.

1. Disconnect the negative battery cable.
2. Remove all necessary components for access to the front timing covers, then remove the covers.
3. Set the No. 1 cylinder on Top Dead Center (TDC) of the compression stroke.
4. The automatic belt tensioner is oil damped and spring operated. Install a 6mm bolt to hold the tensioner back against the spring and release tension on the belt.
5. Remove the auto-tensioner and timing belt.

WARNING

Do not rotate the crankshaft or camshaft separately because the pistons will strike the valves causing engine damage.

To install:

6. Confirm that the No. 1 cylinder is at TDC of the compression stroke.
7. Align the marks on the camshaft and crankshaft sprockets with the marks on the rear belt cover and oil pump housing.
8. With the arrows on the timing belt pointing towards the front, align the white lines on the timing belt with the marks on the sprockets and install the belt.
9. To prepare the auto-tensioner for installation, perform the following:
   a. Remove the bolt holding the tensioner in position.

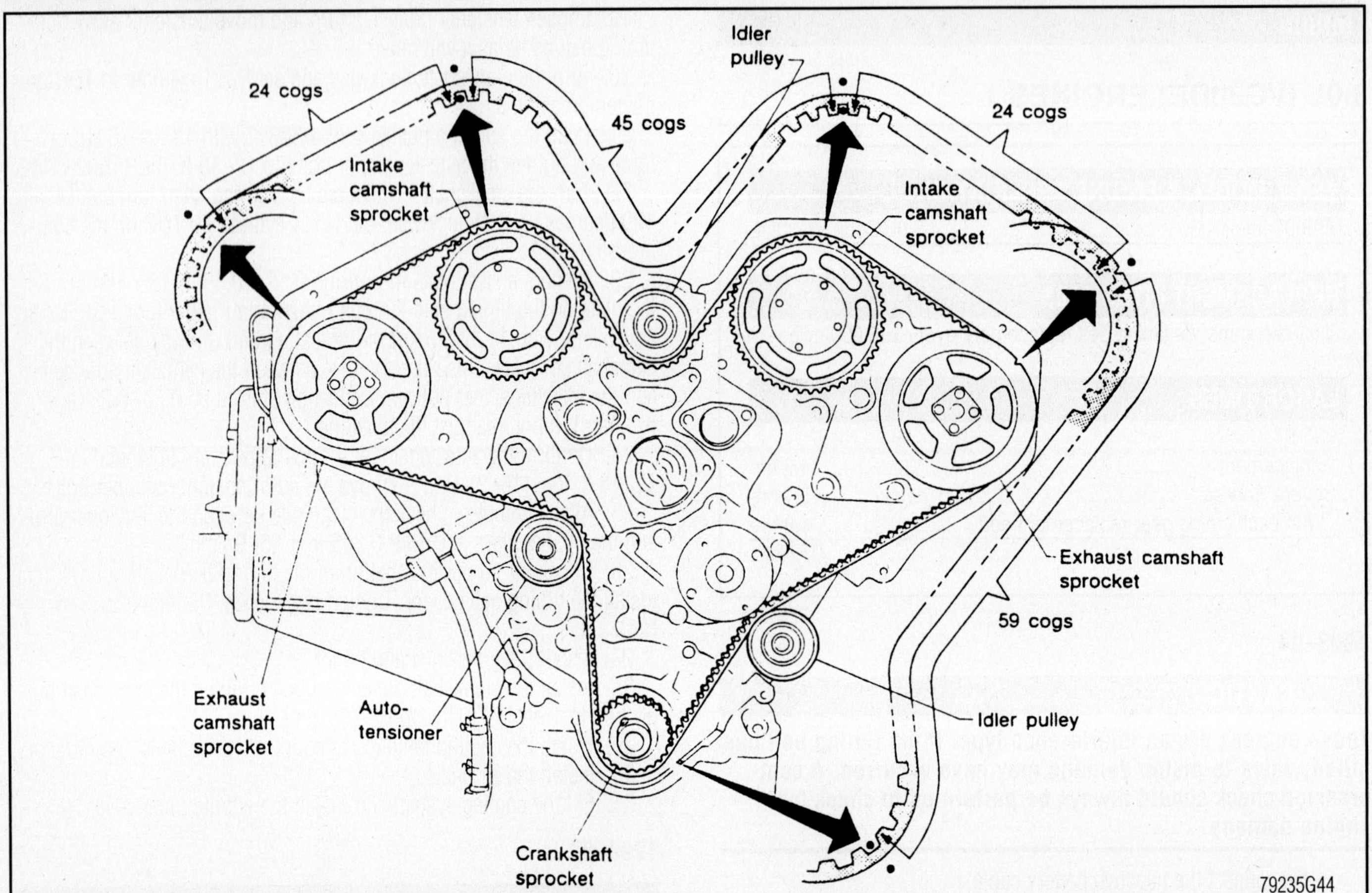

For proper timing belt positioning, ensure the number of teeth between each sprocket is as indicated—Infiniti 3.0L (VG30DE) engine

b. Use a vise to adjust the gap between the tensioner arm and pusher body to 0.160 in. (4mm).

c. Install the bolt again to hold the arm in this position. Do not try to use the bolt to adjust the gap or the threads will be damaged.

10. Install the auto-tensioner, push it toward the belt to just take up the slack, then tighten the bolts finger-tight.

11. Before adjusting the timing belt tension, the slack must be properly distributed:

a. Turn the crankshaft 10 degrees clockwise and tighten the tensioner bolts and nut to 12–15 ft. lbs. (16–21 Nm). Do not push the auto-tensioner hard or the belt will be adjusted too tight.

b. Turn the crankshaft 120 degrees (1/3 turn) counterclockwise.

c. Loosen the tensioner bolts and nut 1/2 turn and move the tensioner body away from the timing belt as far as it will move.

d. Turn the crankshaft clockwise to TDC again.

e. Push the tensioner against the belt with a force of 13 lbs. (59 N) using a spring scale or similar tool and tighten the bolts again to 12–15 ft. lbs. (16–21 Nm). The pressure specification is important and a special spring scale tool, J-38387, is available to measure the tensioner force.

12. To check the timing belt tension:

a. Turn the crankshaft 120 degrees (1/3 turn) clockwise, then turn counterclockwise and return the engine to TDC.

b. Prepare a steel plate that is approximately 3/8 in. (10mm) wide and longer than the width of the belt.

c. Set the plate on the timing belt between 2 camshaft sprockets and push against the plate with a force of 11 lbs. (49 N). Note the belt deflection.

d. Repeat the procedure between the other camshaft sprockets and between the exhaust sprockets and idler/tensioner pulleys. There will be a total of 4 measurements.

e. Add the deflection measurements and divide by 4. The average deflection must be 0.240–0.280 in. (6–7mm). If belt tension is not correct, start the entire adjustment procedure again.

13. Confirm the auto-tensioner mounting nuts are tightened to 12–15 ft. lbs. (16–21 Nm) and remove the auto-tensioner stopper bolt.

14. After 5 minutes, measure the clearance between the tensioner arm and the pusher. It should be 0.138–0.205 in. (3.5–5.2mm).

15. Be sure all the sprocket timing marks are correctly aligned. Install the timing belt covers and tighten the bolts to 24–38 inch lbs. (3–5 Nm).

16. Install all applicable components.

17. Connect the negative battery cable.

3.0L (VG30E) ENGINES

✳✳ CAUTION

These engines are considered freewheeling. If the timing belt fails, valve-to-piston damage is unlikely, however a precautionary compression check should be performed to verify no engine damage.

1. Disconnect the negative battery cable.
2. Raise and support the front of the vehicle safely.

YEAR, MODEL & ENGINE APPLICATION	
1990-92 Infiniti M30	3.0L Code VG30E

REPLACEMENT INTERVALS
Infiniti recommends timing belt replacement every 60,000 miles.

LABOR TIMES	
1990-92	
Replacement	4.9 hrs
Severe Service	5.4 hrs

8807INFg02

3. Remove the engine undercover and air duct.
4. Drain the cooling system.
5. Remove the radiator and cooling fan.
6. Remove the alternator, power steering and air conditioning compressor drive belts from the engine. When removing the power steering drive belt, loosen the idler pulley from the right-side wheel housing.
7. Remove the spark plugs, distributor cover and water pump pulley.
8. Remove the idler bracket of the compressor drive belt.
9. Remove the crankshaft pulley with a suitable puller.
10. Remove the upper and lower timing belt covers and gaskets.
11. Rotate the engine with a socket wrench on the crankshaft pulley bolt to align the punch mark on the left hand camshaft pulley with the mark on the upper rear timing belt cover. Align the punchmark on the crankshaft with the notch on the oil pump housing and temporarily install the crankshaft pulley bolt to allow for crankshaft rotation.
12. Use a hex wrench to turn the belt tensioner clockwise and tighten the tensioner locknut just enough to hold the tensioner in position. Then, remove the timing belt.

To install:

13. Before installing the timing belt, confirm that No. 1 cylinder is at Top Dead Center (TDC) of the compression stroke. Install tensioner and tensioner spring. If stud is removed, apply locking sealant to threads before installing.
14. Swing tensioner fully clockwise with hexagon wrench and temporarily tighten locknut.
15. Point the arrow on the timing belt toward the front belt cover. Align the white lines on the timing belt with the punch marks on all 3 pulleys.

■ **There are 133 total timing belt teeth. If timing belt is installed correctly there will be 40 teeth between left hand and right hand camshaft sprocket timing marks. There will be 43 teeth between left hand camshaft sprocket and crankshaft sprocket timing marks.**

16. Loosen tensioner locknut, keeping tensioner steady with a hexagon wrench.
17. Swing tensioner 70–80 degrees clockwise with hexagon wrench and temporarily tighten locknut.
18. Turn crankshaft clockwise 2–3 times, then slowly set No. 1 cylinder at TDC of the compression stroke.
19. Push middle of timing belt between right hand camshaft sprocket and tensioner pulley with a force of 22 lbs. (10 kg).
20. Loosen tensioner locknut, keeping tensioner steady with a hexagon wrench.
21. Insert a 0.138 in. (0.35mm) thick and 0.5 in. (12.7mm) wide feeler gauge between the bottom of tensioner pulley and timing belt. Turn crankshaft clockwise and position gauge completely between tensioner pulley and timing belt. The timing belt will move about 2.5 teeth.
22. Tighten tensioner locknut, keeping tensioner steady with a hexagon wrench.
23. Turn crankshaft clockwise or counterclockwise and remove the gauge.

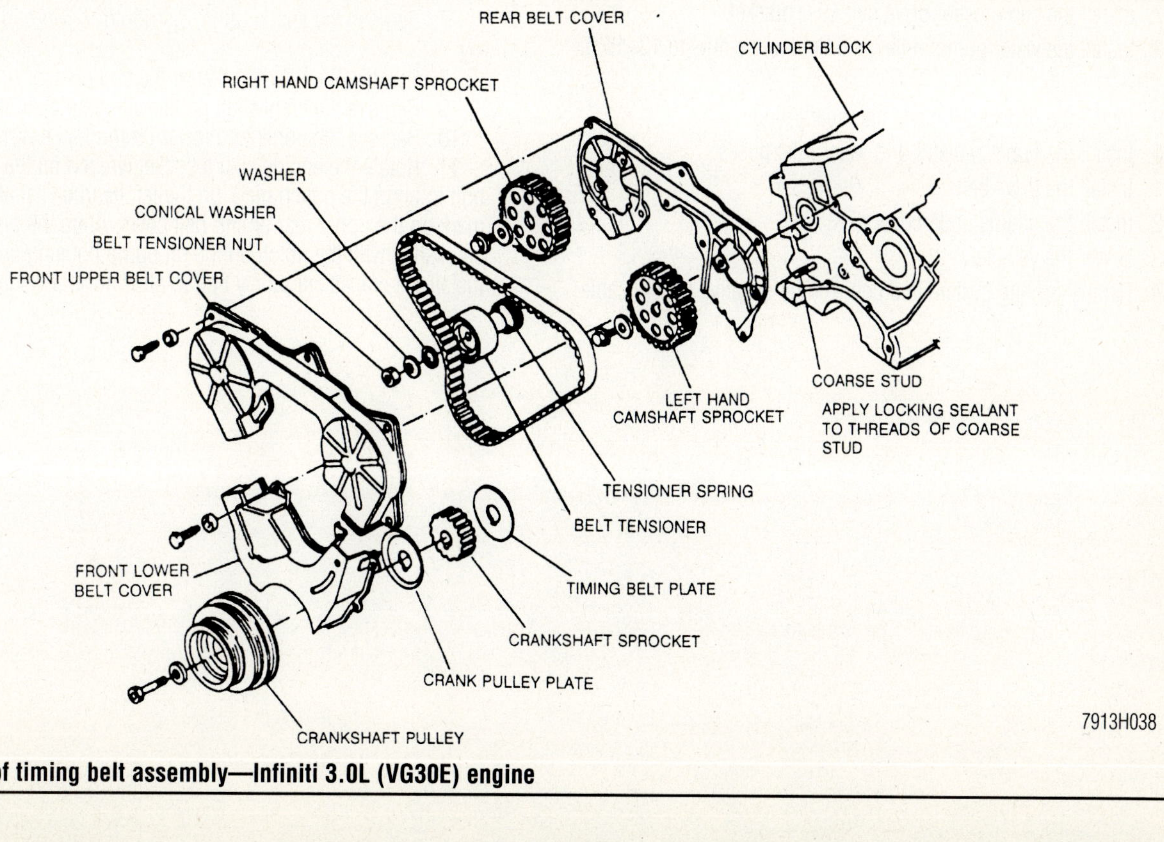

Exploded view of timing belt assembly—Infiniti 3.0L (VG30E) engine

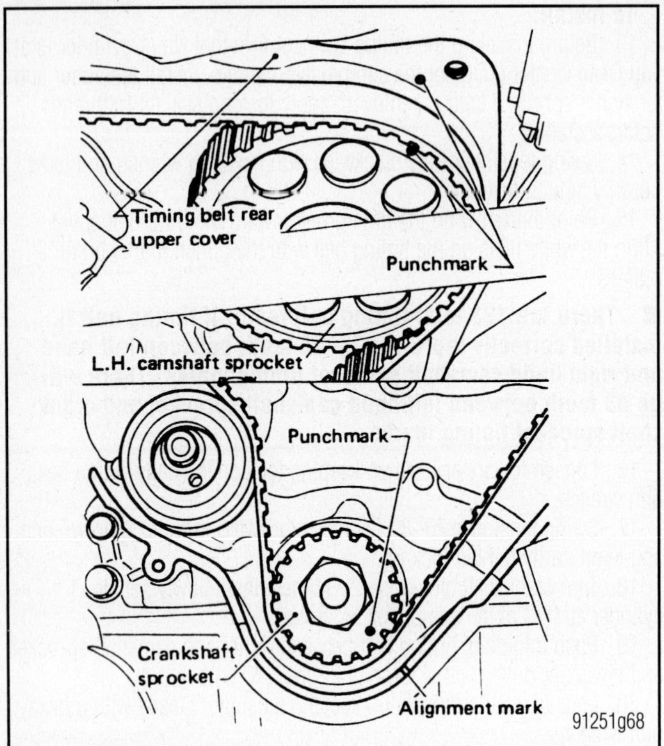

Camshaft and crankshaft sprocket timing marks—Infiniti 3.0L (VG30E) engine

24. Rotate the engine 3 times, then set No. 1 to Top Dead Center (TDC) of the compression stroke.
25. Install the upper and lower timing belt covers with new gaskets.
26. Install the crankshaft pulley. Tighten the pulley bolt to 90–98 ft. lbs. (123–132 Nm).
27. Install the compressor drive belt idler bracket.
28. Install the water pump pulley and tighten the nuts to 12–15 ft. lbs. (16–21 Nm).
29. Install the distributor cover and spark plugs.
30. Install the upper radiator and water inlet hoses.
31. Install the drive belts.
32. Install the engine undercovers.
33. Lower the vehicle.
34. Fill the cooling system and connect the negative battery cable.

3.3L (VG33E) ENGINES

YEAR, MODEL & ENGINE APPLICATION

1997-00 Infiniti QX4 .... 3.0L Code VG33E

REPLACEMENT INTERVALS

Infiniti recommends timing belt replacement every 105,000 miles.

LABOR TIMES

1997-00

Replacement	3.3 hrs
Severe Service	3.8 hrs
With air conditioning add	0.2 hrs
With power steering add	0.5 hrs
With each timing gear replacement add	0.3 hrs

8807INFg03

** CAUTION

These engines are considered freewheeling. If the timing belt fails, valve-to-piston damage is unlikely, however a precautionary compression check should be performed to verify no engine damage.

1. Disconnect the negative battery cable.
2. Raise and support the front of the vehicle safely.
3. Remove the engine undercover and air duct.
4. Drain the cooling system.
5. Remove the radiator and cooling fan.
6. Remove the alternator, power steering and air conditioning compressor drive belts from the engine. When removing the power steering drive belt, loosen the idler pulley from the right-side wheel housing.
7. Remove the spark plugs, distributor cover and water pump pulley.
8. Remove the idler bracket of the compressor drive belt.
9. Remove the crankshaft pulley with a suitable puller.
10. Remove the upper and lower timing belt covers and gaskets.
11. Rotate the engine with a socket wrench on the crankshaft pulley bolt to align the punch mark on the left hand camshaft pulley with the mark on the upper rear timing belt cover. Align the punchmark on the crankshaft with the notch on the oil pump housing and temporarily install the crankshaft pulley bolt to allow for crankshaft rotation.

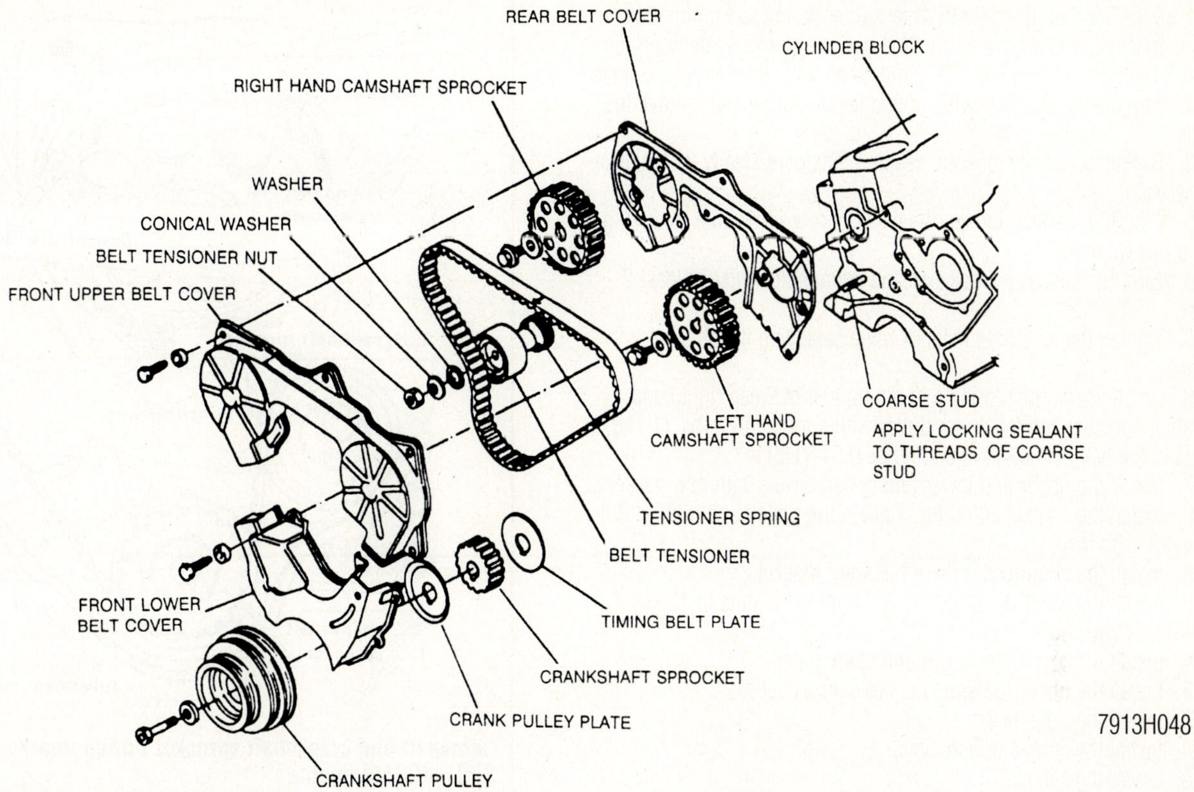

Exploded view of timing belt assembly—Infiniti 3.3L (VG33E) engine

12. Use a hex wrench to turn the belt tensioner clockwise and tighten the tensioner locknut just enough to hold the tensioner in position. Then, remove the timing belt.

To install:

13. Before installing the timing belt, confirm that No. 1 cylinder is at Top Dead Center (TDC) of the compression stroke. Install tensioner and tensioner spring. If stud is removed, apply locking sealant to threads before installing.

14. Swing tensioner fully clockwise with hexagon wrench and temporarily tighten locknut.

15. Point the arrow on the timing belt toward the front belt cover. Align the white lines on the timing belt with the punch marks on all 3 pulleys.

■ **There are 133 total timing belt teeth. If timing belt is installed correctly there will be 40 teeth between left hand and right hand camshaft sprocket timing marks. There will be 43 teeth between left hand camshaft sprocket and crankshaft sprocket timing marks.**

16. On 1997–99 models, loosen the tensioner locknut, keeping tensioner steady with a hexagon wrench.

17. Swing tensioner 70–80 degrees clockwise with hexagon wrench and temporarily tighten locknut.

18. Turn crankshaft clockwise 2–3 times, then slowly set No. 1 cylinder at TDC of the compression stroke.

19. Loosen tensioner locknut, keeping tensioner steady with a hexagon wrench.

20. Insert a 0.138 in. (0.35mm) thick and 0.5 in. (12.7mm) wide feeler gauge between the bottom of tensioner pulley and timing belt. Turn crankshaft clockwise and position gauge completely between tensioner pulley and timing belt. The timing belt will move about 2.5 teeth.
21. Turn crankshaft clockwise or counterclockwise and remove the gauge.
22. Tighten tensioner locknut, keeping tensioner steady with a hexagon wrench.
23. On 2000 models, loosen the tensioner locknut while steadying it with a hex wrench.
24. Turn the tensioner clockwise and counterclockwise at least 2 times.
25. Tighten the tensioner locknut while steadying it with a hex wrench.
26. On all years, push middle of timing belt between right hand camshaft sprocket and tensioner pulley with a force of 22 lbs. (10 kg). Measure the belt deflection. It should be 0.51–0.59 in.
27. Install the upper and lower timing belt covers with new gaskets.
28. Install the crankshaft pulley. Tighten the pulley bolt to 90–98 ft. lbs. (123–132 Nm).
29. Install the compressor drive belt idler bracket.
30. Install the water pump pulley and tighten the nuts to 12–15 ft. lbs. (16–21 Nm).
31. Install the distributor cover and spark plugs.
32. Install the upper radiator and water inlet hoses.
33. Install the drive belts.
34. Install the engine undercovers.
35. Lower the vehicle.
36. Fill the cooling system and connect the negative battery cable.

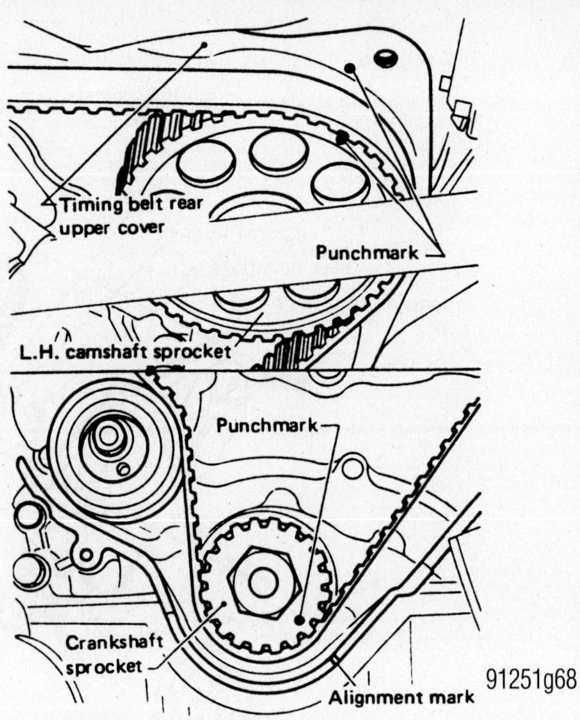

Camshaft and crankshaft sprocket timing marks—Infiniti 3.3L (VG33E) engine

Isuzu

1.5L (4XC1-U AND 4XC1-T) AND 1.6L (4XE1) SOHC ENGINES

YEAR, MODEL & ENGINE APPLICATION

1985-89 Isuzu I-Mark, I-Mark Turbo,	1.5L Codes 4XC1/T
1991-93 Stylus	1.6L Code 4XEV-1

REPLACEMENT INTERVALS

Isuzu recommends timing belt replacement every 60,000 miles.

LABOR TIMES

I-Mark	
Replacement	2.6 hrs
Severe Service	3.0 hrs
Stylus	
Replacement	1.7 hrs
Severe Service	2.1 hrs
All Models	
With air conditioning add	0.3 hrs
With automatic transmission add	0.3 hrs
With power steering add	0.3 hrs
With turbocharger add	0.5 hrs
With tensioner replacement add	0.4 hrs

8807IZUg01

✻✻ CAUTION

These engines are an interference type. If the timing belt has failed, valve-to-piston damage may have occurred. A compression check should always be performed to check for engine damage.

1. Disconnect the negative battery cable.
2. Set the crankshaft to Top Dead Center (TDC) on the compression stroke.
3. Support the engine with a jack and remove the right engine mount and brackets.
4. Remove the rear torque rod.
5. Remove the accessory drive belts.
6. Remove the crankshaft pulley.
7. Remove the timing belt cover.
8. Ensure the crankshaft and camshaft timing marks are aligned.
9. Loosen the tensioner pulley and remove the timing belt.

To install:

10. Insure that the timing marks are aligned and install the timing belt on the crankshaft sprocket, camshaft sprocket, water pump pulley and tensioner.
11. Use an Allen wrench to rotate the tensioner eccentric and torque the tensioner bolt to 37 ft. lbs. (49 Nm).
12. Rotate the crankshaft 2 complete turns and insure that the timing marks align.
13. Use a belt tension gauge to check the timing belt tension. The belt tension should be 34–42 lbs.
14. Install the timing cover and crankshaft pulley.
15. Install the accessory drive belts.
16. Install the rear torque rod and the right engine mount.
17. Remove the engine support and connect the negative battery cable.

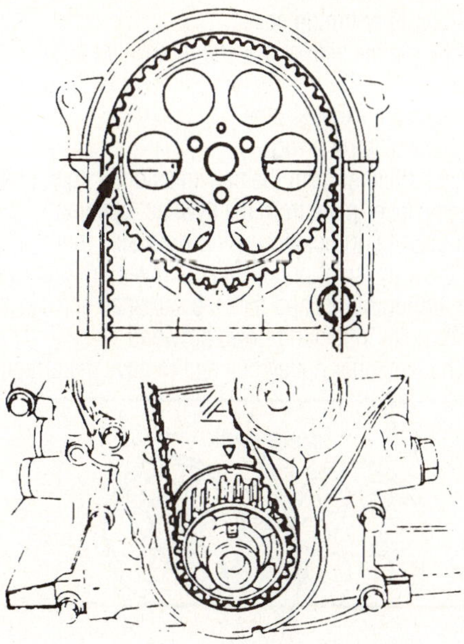

9305ag10

Camshaft and crankshaft alignment marks—Isuzu 1.5L (4XC1-U and 4XC1-T) and 1.6L (4XE1) engines

1.6L (4XE1/W/WT) DOHC ENGINES

YEAR, MODEL & ENGINE APPLICATION

1988-89 Isuzu I-Mark, 1990-92 Impulse,	1.6L Code 4XE1/W
1990-92 Impulse Turbo	1.6L Code 4XE1-WT

REPLACEMENT INTERVALS

Isuzu recommends timing belt replacement every 60,000 miles.

LABOR TIMES

I-Mark	
Replacement	2.6 hrs
Severe Service	3.0 hrs
Impulse	
Replacement	2.3 hrs
Severe Service	2.7 hrs
All Models	
With air conditioning add	0.3 hrs
With automatic transmission add	0.3 hrs
With power steering add	0.3 hrs
With turbocharger add	0.5 hrs
With tensioner replacement add	0.4 hrs

8807IZUg02

✻✻ CAUTION

These engines are an interference type. If the timing belt has failed, valve-to-piston damage may have occurred. A compression check should always be performed to check for engine damage.

1. Disconnect the negative battery cable.
2. Set the crankshaft to Top Dead Center (TDC) on the compression stroke.

3. Remove the accessory drive belts.
4. Remove the upper timing cover.
5. Support the engine with a jack and remove the right engine mount and brackets.
6. Remove the front exhaust pipe.
7. Remove the rear torque rod.
8. Remove the stiffener from the transmission, if equipped.
9. Remove the cover plate from the flywheel opening.
10. Lock the flywheel in position using a suitable tool.
11. Remove the crankshaft pulley and the lower timing cover.
12. Align the timing marks and lock the camshaft sprockets in place by inserting 6mm bolts in locking holes provided.
13. Loosen the tensioner pulley bolt and remove the timing belt.

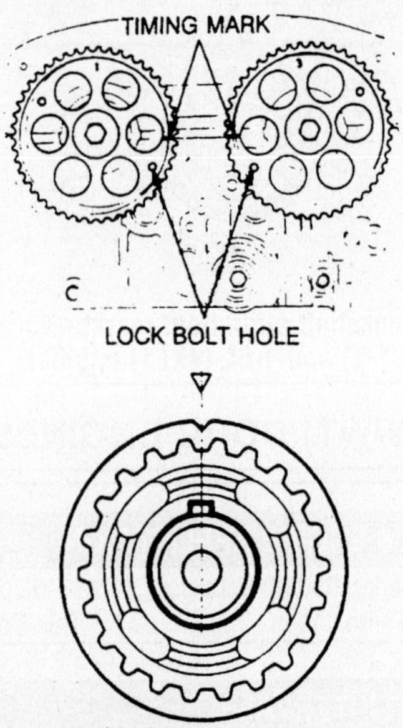

9305ag11

Camshaft and crankshaft timing marks—Isuzu 1.6L (4XE1/W/WT) DOHC engine

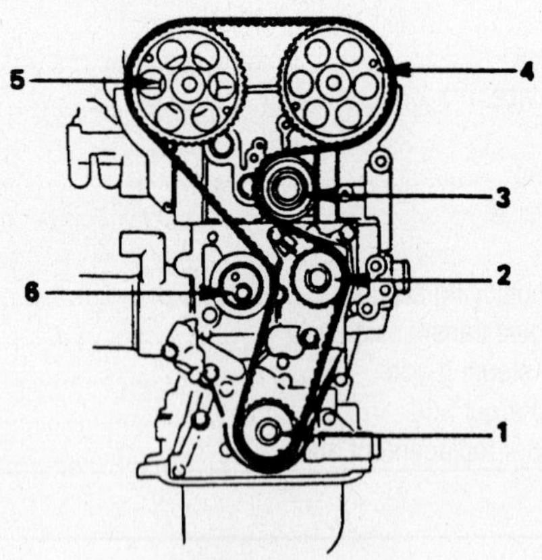

1. Crankshaft timing sprocket
2. Water pump pulley
3. Idler pulley
4. Exhaust camshaft sprocket
5. Intake camshaft sprocket
6. Tensioner pulley.

9305ag12xx

View of the timing belt assembly—Isuzu 1.6L (4XE1/W/WT) DOHC engine

To install:

14. Ensure the timing marks are still aligned.
15. Install the timing belt in the following order:
    a. Crankshaft timing pulley
    b. Water pump pulley
    c. Idler pulley
    d. Exhaust camshaft pulley
    e. Intake camshaft pulley
    f. Tensioner pulley
16. Use an Allen wrench to rotate the tensioner pulley eccentric and tighten the tensioner bolt.
17. Remove the camshaft sprocket locking bolts.
18. Rotate the crankshaft 2 complete turns and insure that the timing marks align.
19. Rotate the crankshaft 60 degrees clockwise and check the timing belt deflection. Deflection should be 0.28–0.33 in. (7.1–8.4mm). If deflection is not correct, reinstall the timing belt.
20. Install the lower timing cover and the crankshaft pulley. Tighten the crankshaft pulley bolt to 123 ft. lbs. (170 Nm) on I-Mark, or 109 ft. lbs. (148 Nm) on Impulse.
21. Remove the flywheel locking tool and install the cover plate.
22. Install the transmission stiffener, if removed.
23. Install the rear torque rod.
24. Install the front exhaust pipe.
25. Install the right engine mount and brackets. Remove the engine support.
26. Install the accessory drive belts and connect the negative battery cable.

1.8L (4FB1) DIESEL ENGINES

YEAR, MODEL & ENGINE APPLICATION	
1981-85 Isuzu I-Mark	1.8L Diesel Code 4FB1

REPLACEMENT INTERVALS

Isuzu recommends timing belt replacement every 60,000 miles.

LABOR TIMES	
I-Mark	
Replacement	3.5 hrs
Severe Service	3.9 hrs

8807IZUg03

✻✻ CAUTION

These engines are an interference type. If the timing belt has failed, valve-to-piston damage may have occurred. A compression check should always be performed to check for engine damage.

1. Disconnect the negative battery cable.
2. Remove the lower engine shrouds.
3. Remove the accessory drive belts.
4. Remove the cooling fan and shroud.
5. Remove the upper timing cover.
6. Drain the cooling system and remove the bypass hose.
7. Set the No. 1 cylinder at Top Dead Center (TDC) on the compression stroke.
8. Insure that the injection pump pulley setting mark is aligned with the front plate, then lock the pulley in place with an 8mm x 1.25 bolt.

9. Remove the valve cover and back off the valve adjustment until all rocker arms are loose.

10. Lock the camshaft in place by fitting a plate to the slot in the rear of the camshaft.

11. Remove the crankshaft pulley and the lower timing cover.

12. Remove the timing belt holder.

13. Remove the timing belt tensioner spring.

14. Loosen the tension pulley and plate bolts and remove the timing belt.

To install:

15. Loosen the camshaft pulley bolt so that the pulley turns freely.

16. Insure that the crankshaft does not turn and install the timing belt.

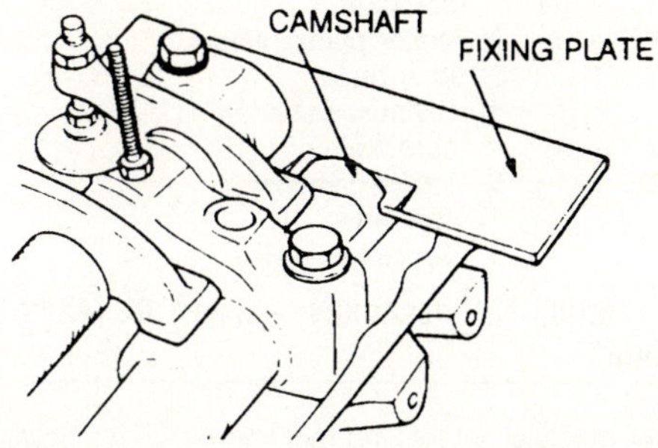

9305ag06

Positioning the camshaft fixing plate—Isuzu 1.8L (4FB1) diesel engine

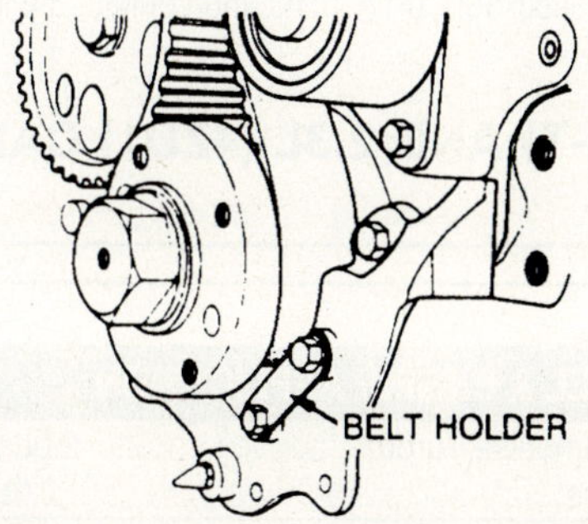

9305ag08

View of the timing belt holder—Isuzu 1.8L (4FB1) diesel engine

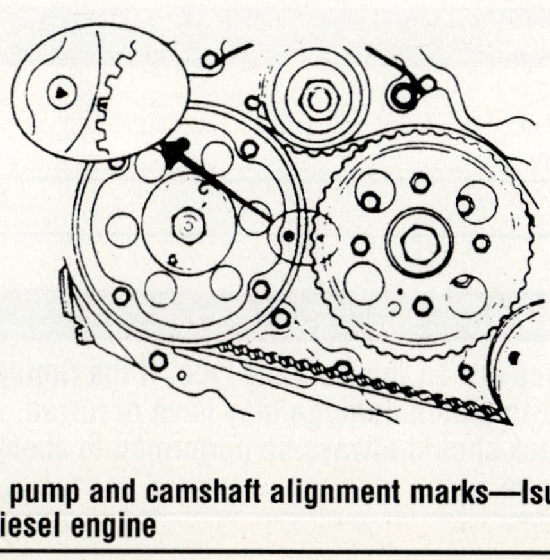

9308ag09

Injection pump and camshaft alignment marks—Isuzu 1.8L (4FB1) diesel engine

17. Gather any slack in the timing belt at the tensioner pulley and install the tensioner spring.

18. Temporarily install the crankshaft pulley. Insure that the crankshaft is at TDC.

19. Tighten the tensioner pulley and plate bolts.

20. Tighten the camshaft pulley bolt.

21. Remove the injection pump lockbolt and the camshaft fixing plate.

22. Remove the crankshaft pulley.

23. Install the timing belt holder and install the lower timing cover. Reinstall the crankshaft pulley.

24. Adjust the valves and install the valve cover.

25. Install the bypass hose and the upper timing cover.

26. Install the cooling fan and shroud.

27. Install the accessory drive belts.

28. Install the lower engine shrouds.

29. Connect the negative battery cable and refill the cooling system.

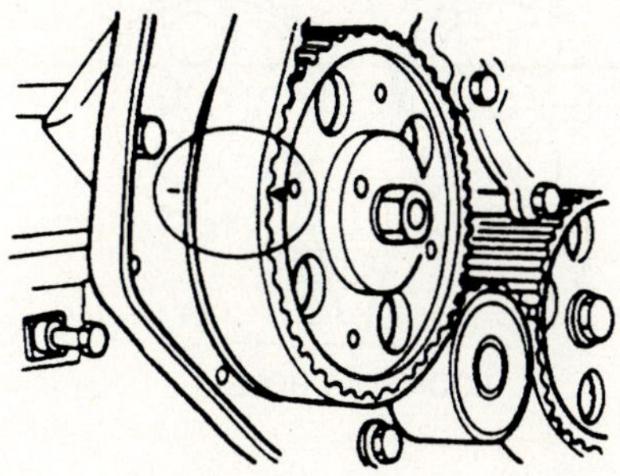

9305ag07

Injection pump alignment marks—Isuzu 1.8L (4FB1) diesel engine

1.8L (4XF1/W) DOHC ENGINES

YEAR, MODEL & ENGINE APPLICATION	
1991-92 Isuzu Impulse, 1993 Stylus	1.8L Code 4XF1/W
REPLACEMENT INTERVALS	
Isuzu recommends timing belt replacement every 60,000 miles.	
LABOR TIMES	
Impulse & Stylus	
Replacement	2.9 hrs
Severe Service	3.3 hrs
All Models	
With air conditioning add	0.3 hrs
With automatic transmission add	0.3 hrs
With power steering add	0.3 hrs
With tensioner replacement add	0.4 hrs

8807IZUg04

✱✱ CAUTION

These engines are an interference type. If the timing belt has failed, valve-to-piston damage may have occurred. A compression check should always be performed to check for engine damage.

1. Disconnect the negative battery cable.
2. Set the crankshaft to Top Dead Center (TDC) on the compression stroke.
3. Remove the accessory drive belts.
4. Remove the upper timing cover.
5. Support the engine with a jack and remove the right engine mount and brackets.
6. Remove the front exhaust pipe.
7. Remove the rear torque rod.
8. Remove the stiffener from the transmission, if equipped.
9. Remove the cover plate from the flywheel opening.
10. Lock the flywheel in position using a suitable tool.
11. Remove the crankshaft pulley and the lower timing cover.
12. Align the timing marks and lock the camshaft sprockets in place by inserting 6mm bolts in locking holes provided.
13. Loosen the tensioner pulley bolt and remove the timing belt.

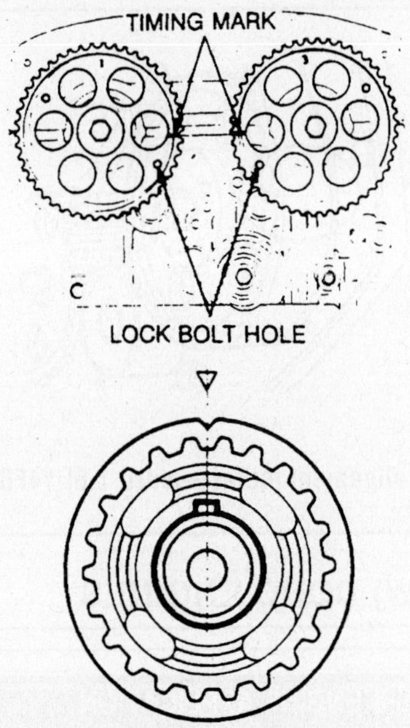

9305ag11

Camshaft and crankshaft timing marks—Isuzu 1.8L (4XF1/W) DOHC engine

To install:

14. Ensure the timing marks are still aligned.
15. Install the timing belt in the following order:
    a. Crankshaft timing pulley
    b. Water pump pulley
    c. Idler pulley
    d. Exhaust camshaft pulley
    e. Intake camshaft pulley
    f. Tensioner pulley
16. Use an Allen wrench to rotate the tensioner pulley eccentric and tighten the tensioner bolt.
17. Remove the camshaft sprocket locking bolts.
18. Rotate the crankshaft 2 complete turns and insure that the timing marks align.
19. Rotate the crankshaft 60 degrees clockwise and check the timing belt deflection. Deflection should be 0.28–0.33 in. (7.1–8.4mm). If deflection is not correct, reinstall the timing belt.
20. Install the lower timing cover and the crankshaft pulley. Tighten the crankshaft pulley bolt to 123 ft. lbs. (170 Nm) on I-Mark, or 109 ft. lbs. (148 Nm) on Impulse.

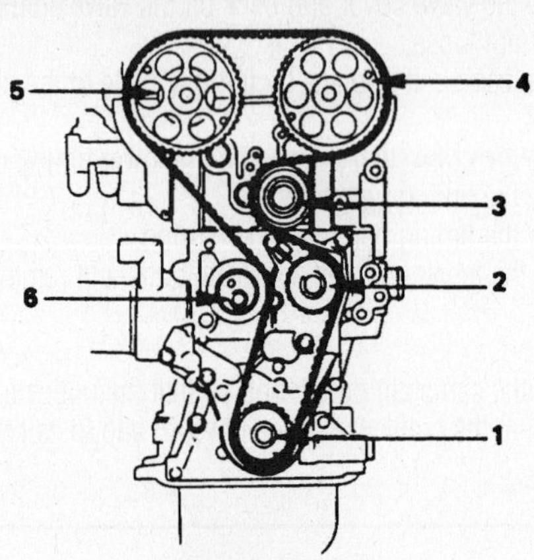

1. Crankshaft timing sprocket
2. Water pump pulley
3. Idler pulley
4. Exhaust camshaft sprocket
5. Intake camshaft sprocket
6. Tensioner pulley.

9305ag12

View of the timing belt assembly—Isuzu 1.8L (4XF1/W) DOHC engine

21. Remove the flywheel locking tool and install the cover plate.
22. Install the transmission stiffener, if removed.
23. Install the rear torque rod.
24. Install the front exhaust pipe.
25. Install the right engine mount and brackets. Remove the engine support.
26. Install the accessory drive belts and connect the negative battery cable.

2.0L (4ZC1-T) AND 2.3L (4ZD1) CAR ENGINES

YEAR, MODEL & ENGINE APPLICATION

1985-89 Isuzu Impulse Turbo	2.0L Code 4ZC1T
1988-89 Impulse	2.3L Code 4ZD1

REPLACEMENT INTERVALS

Isuzu recommends timing belt replacement every 60,000 miles.

LABOR TIMES

Impulse	
Replacement	4.0 hrs
Severe Service	4.4 hrs

8807IZUg05

✱✱ CAUTION

These engines are an interference type. If the timing belt has failed, valve-to-piston damage may have occurred. A compression check should always be performed to check for engine damage.

1. Disconnect the negative battery cable.
2. Loosen and remove the engine accessory drive belts.
3. Remove the cooling fan assembly and the water pump pulley.
4. Drain the fluid from the power steering reservoir.
5. Unbolt and remove the power steering pump. Unbolt the hydraulic line brackets from the upper timing cover and move the pump out of the work area without disconnecting the hydraulic lines.
6. Disconnect and remove the starter motor if a flywheel holder tool No. J-38674 or equivalent, is to be used.
7. Remove the upper timing belt cover.
8. Rotate the crankshaft to set the engine at Top Dead Center (TDC) of its compression for the No. 1 cylinder. The arrow mark on the camshaft sprocket will be aligned with the mark on the rear timing cover.
9. Remove the crankshaft pulley.
10. Remove the lower timing belt cover.
11. Verify that the engine is set at TDC/compression for the No. 1 cylinder. The notch on the crankshaft sprocket will be aligned with the pointer on the oil seal retainer.
12. Release and remove the tensioner spring to release the timing belt's tension.
13. Remove the timing belt.
14. Unbolt the tensioner pulley bracket from the engine's front cover.
15. If necessary, unbolt and remove the camshaft sprockets. Use a puller to remove the crankshaft pulley if necessary. Don't lose the crankshaft sprocket key.

To install:

16. If removed, install the camshaft and crankshaft sprockets. Align the camshaft and crankshaft timing marks and be sure to install any keys. Tighten the camshaft sprocket bolt to 43 ft. lbs. (59 Nm).
17. Install the tensioner assembly. Tighten the tensioner mounting bolt to 14 ft. lbs. (19 Nm) and the cap bolt to 108 inch lbs. (13 Nm).
18. Be sure the crankshaft and the camshaft sprockets are aligned with their timing marks. Install the timing belt onto the crankshaft sprocket, oil pump sprocket; camshaft sprocket and tensioner pulley.
19. Loosen the tensioner mounting bolt. This will allow the tensioner spring to apply pressure to the timing belt.
20. After the spring has pulled the timing belt as far as possible, temporarily tighten the tensioner mounting bolt to 14 ft. lbs. (19 Nm).

■ **Remove the flywheel holder before rotating the crankshaft. Reinstall the holder to tighten the crankshaft pulley bolt.**

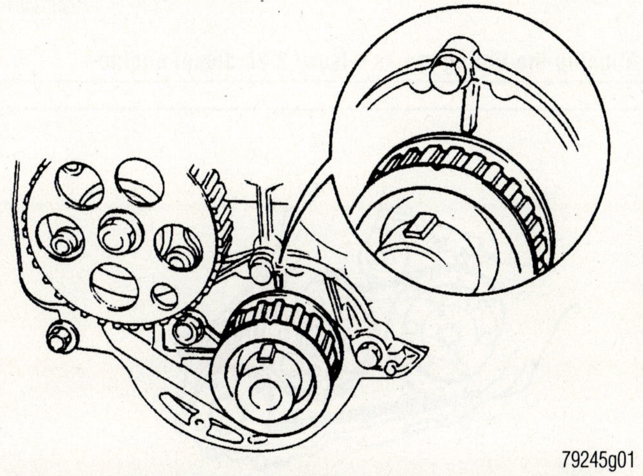

79245g01

Align the crankshaft pulley timing mark the with oil retainer setting mark—Isuzu 2.0L (4ZC1-T) and 2.3L (4ZD1) engines

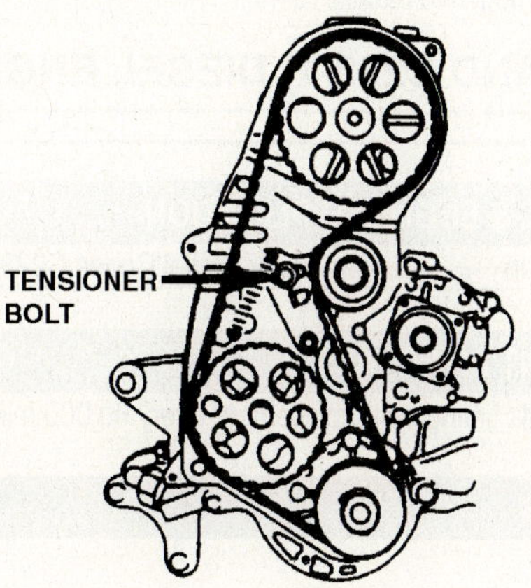

79245g03

Be sure to tighten the timing belt tensioner bolt (B)—Isuzu 2.0L (4ZC1-T) and 2.3L (4ZD1) engines

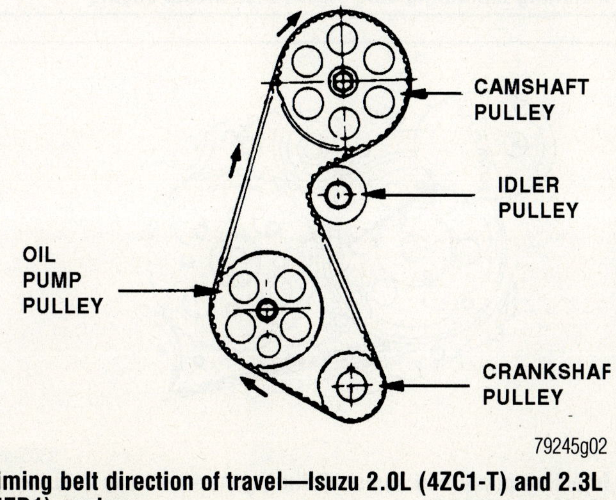

79245g02

Timing belt direction of travel—Isuzu 2.0L (4ZC1-T) and 2.3L (4ZD1) engines

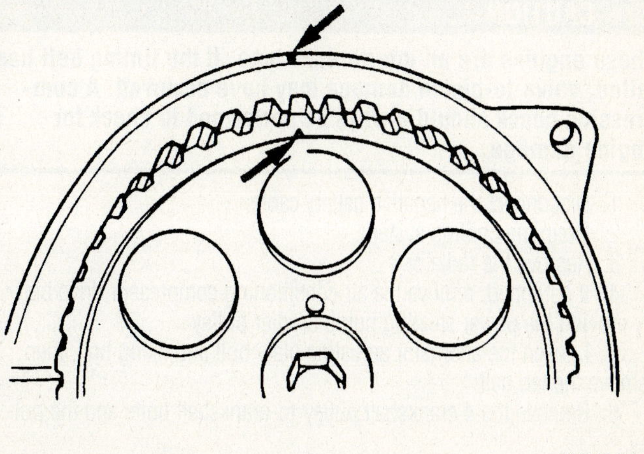

79245g04

Camshaft-to-timing cover mark—Isuzu 2.0L (4ZC1-T) and 2.3L (4ZD1) engines

✻✻ WARNING

If any binding is felt when adjusting the timing belt tension by turning the crankshaft, STOP turning the engine, because the pistons may be hitting the valves.

21. Rotate the crankshaft counterclockwise 2 complete revolutions to check the rotation of the belt and the alignment of the timing marks. Listen for any rubbing noises which may mean the belt is binding.
22. Loosen the tensioner pulley bolt to allow the spring to adjust the correct tension. Then, retighten the tensioner pulley bolt to 14 ft. lbs. (19 Nm).
23. Install the lower timing cover and the crankshaft pulley.
24. Tighten the crankshaft pulley bolt to 87 ft. lbs. (118 Nm). Tighten the small pulley bolts to 72 inch lbs. (8 Nm).
25. Install the upper timing cover.
26. If removed, install the starter and tighten the bolts to 30 ft. lbs. (40 Nm).
27. Install the power steering pump. If the hydraulic lines were disconnected, refill and bleed the power steering system.
28. Install the water pump pulley and tighten its nut to 20 ft. lbs. (26 Nm).
29. Install the cooling fan assembly.
30. Install and adjust the accessory drive belts.
31. Connect the negative battery cable.

2.2L (C223 AND C223-T) DIESEL ENGINES

YEAR, MODEL & ENGINE APPLICATION

1985-88 Isuzu P'up .... 2.2L Diesel Codes C223/C223T

REPLACEMENT INTERVALS

Isuzu recommends timing belt replacement every 60,000 miles.

LABOR TIMES

1985-88

Replacement	3.1 hrs
Severe Service	3.4 hrs

8807IZUg06

✻✻ CAUTION

These engines are an interference type. If the timing belt has failed, valve-to-piston damage may have occurred. A compression check should always be performed to check for engine damage.

1. Disconnect the negative battery cable.
2. Drain the cooling system.
3. Remove the radiator.
4. If equipped, remove the air conditioning compressor drive belt by moving the power steering pump or idler pulley.
5. Loosen the alternator adjusting plate bolt and fixing bolt, then remove the fan belt.
6. Remove the 4 crankshaft pulley-to-crankshaft bolts and the pulley.
7. Remove the timing cover bolts and the covers.
8. Remove the injection pump timing pulley flange bolts and the flange.
9. Remove the tension spring.

■When removing the tension spring, avoid using excess force or distortion of the spring will result.

10. Remove the tension pulley nut, the tension pulley and the tension center.
11. If reusing the timing belt, mark its rotational direction on the back of the belt.
12. Remove the timing belt.

93053g01

Aligning the timing marks—Isuzu 2.2L diesel engine

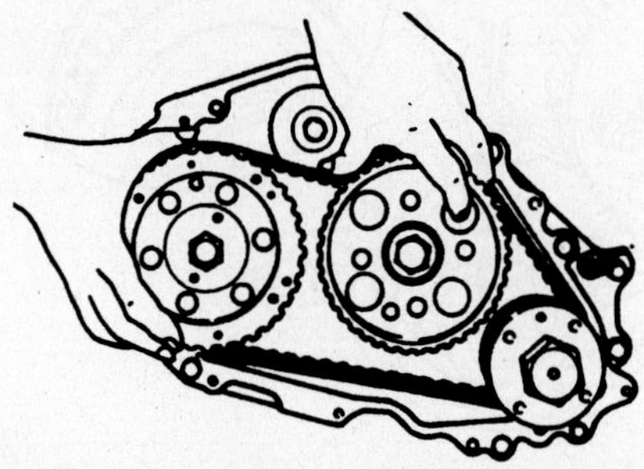

93053g02

Installing the timing belt—Isuzu 2.2L diesel engine

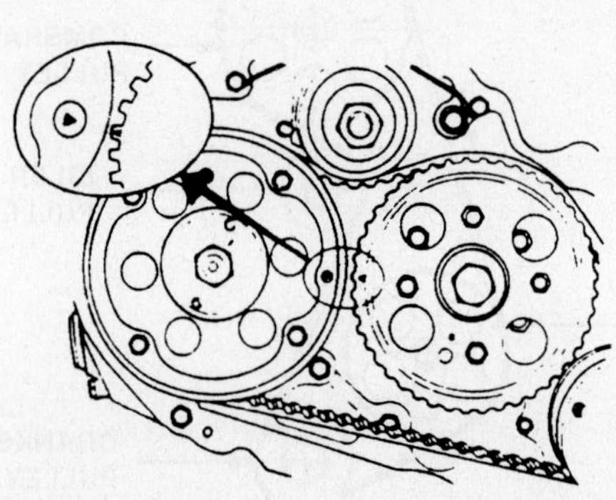

9305ag09

Injection pump and camshaft timing marks—Isuzu 2.2L (C223 and C223-T) diesel engines

To install:

✻✻ WARNING

No attempt should be made to readjust the belt tension. If the belt has been loosened through service of the timing system, it should be replaced with a new one.

13. Rotate the crankshaft sprocket, the injection pump timing sprocket and camshaft sprocket so that they are aligned with their timing marks.
14. Noting the direction of rotation, install the timing belt in the following sequence: crankshaft sprocket, camshaft sprocket and injection pump sprocket.

■ **Make an adjustment so that the belt slack is taken up by the tension pulley.**

15. Install the tension center and tension pulley by making sure that the end of the tension center is in proper contact with 2 pins on the timing pulley housing.
16. Hand-tighten the nut, so that tension pulley can slide freely.
17. Install the tension spring correctly and semi-tighten the tension pulley fixing nut.
18. Turn the crankshaft 2 turns in the normal direction of rotation to permit seating of the belt. Further rotate the crankshaft 90 degrees beyond Top Dead Center (TDC) to settle the injection pump.

✻✻ WARNING

Never attempt to turn the crankshaft in the reverse direction.

19. Loosen the tension pulley fixing nut completely, allowing the pulley to take up looseness of the belt. Then, tighten the nut to 78–95 ft. lbs. (106–129 Nm).
20. Install the flange on the injection pump pulley; the hole in the outer circumference of the flange should be aligned with the triangular mark on the injection pump sprocket.
21. Rotate the crankshaft 2 turns in normal direction of rotation to bring the No. 1 piston to TDC on the compression stroke and recheck that the triangular mark on the injection pump sprocket is aligned with the hole in the flange.
22. Using tool J-29771 or equivalent, check the timing belt tension at a point between the injection pump and crankshaft sprockets; belt tension should be 33–55 ft. lbs. (45–75 Nm).
23. Adjust the valve clearances.
24. Complete the installation by reversing the removal procedures.
25. Refill the cooling system.
26. Connect the negative battery cable.

2.2L (F22B6) AND 2.3L (F23A1, F23A7) ENGINES

✻✻ CAUTION

These engines are an interference type. If the timing belt has failed, valve-to-piston damage may have occurred. A compression check should always be performed to check for engine damage.

■ **The radio may contain a coded theft protection circuit. Always make note the code number before disconnecting the battery.**

YEAR, MODEL & ENGINE APPLICATION	
1996-97 Isuzu Oasis	2.2L Code F22B6
1998-99 Oasis	2.2L Code F23A1, F23A7

REPLACEMENT INTERVALS

Isuzu recommends timing belt replacement every 60,000 miles.

LABOR TIMES	
2.2L Engine	
Replacement	1.2 hrs
Severe Service	1.4 hrs
2.3L Engine	
Replacement	3.7 hrs
Severe Service	4.1 hrs
All Engines	
With tensioner replacement add	0.3 hrs
With water pump replacement add	0.4 hrs

8807IZUg07

1. Disconnect the negative and positive battery cables.
2. Turn the crankshaft to align the timing marks and set cylinder No. 1 to Top Dead Center (TDC) for the compression stroke. The white mark on the crankshaft pulley should align with the pointer on the timing belt cover. The words **UP** embossed on the camshaft pulley should be aligned in the upward position and the marks on the edge of the pulley should be aligned with the cylinder head or the back cover upper edge. Once in this position, the engine must NOT be turned or disturbed.
3. Raise and support the vehicle.
4. Remove the front wheels.
5. Remove the splash shield from below the engine.
6. Remove the wheel well splash shield.
7. Loosen and remove the power steering pump belt. Remove the power steering pump.
8. Loosen the adjusting and mounting bolts for the alternator and remove the drive belt.
9. Remove the cylinder head cover.
10. Remove the upper timing belt cover.
11. Support the engine with a floor jack cushioned with a piece of wood under the oil pan.
12. Remove the dipstick and the dipstick tube.
13. Remove the through-bolt for the side engine mount and remove the mount.
14. Remove the crankshaft pulley bolt and remove the crankshaft pulley. Use a Crank Pulley Holder tool No. 07JAA-001020A and Holder Handle tool No. 07JAB-001020A or their equivalents, to hold the crankshaft pulley while removing the bolt.
15. Remove the lower timing belt cover.
16. Loosen the timing belt/timing balancer belt adjuster nut 2/3–1 turn. Move the tension adjuster to release the belt tension and retighten the adjuster nut.
17. Remove the balancer shaft belt and its drive pulley.
19. Insert a suitable tool into the maintenance hole in the front balancer shaft. Unbolt and remove the balancer driven pulley.

■ **For servicing the balance shafts, front refers to the side of the engine facing the radiator. Rear refers to the side of the engine facing the firewall.**

19. Remove the timing belt.

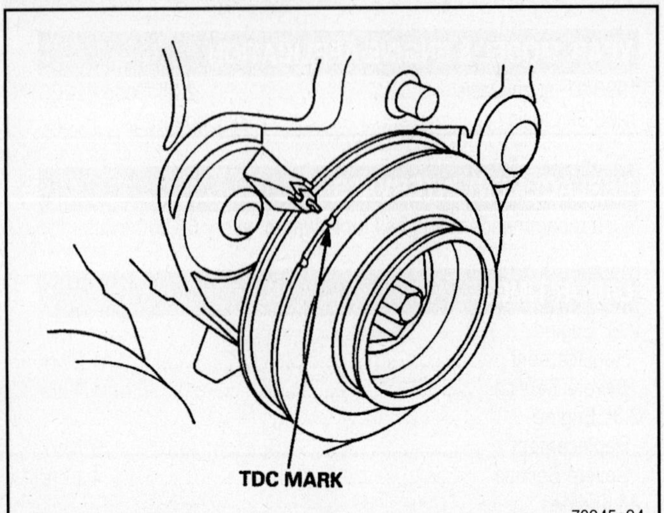

79245g24

Align the camshaft, crankshaft and engine marks before removing the timing belt and pulleys—Isuzu 2.2L (F22B6) and 2.3L (F23A1, F23A7) engines

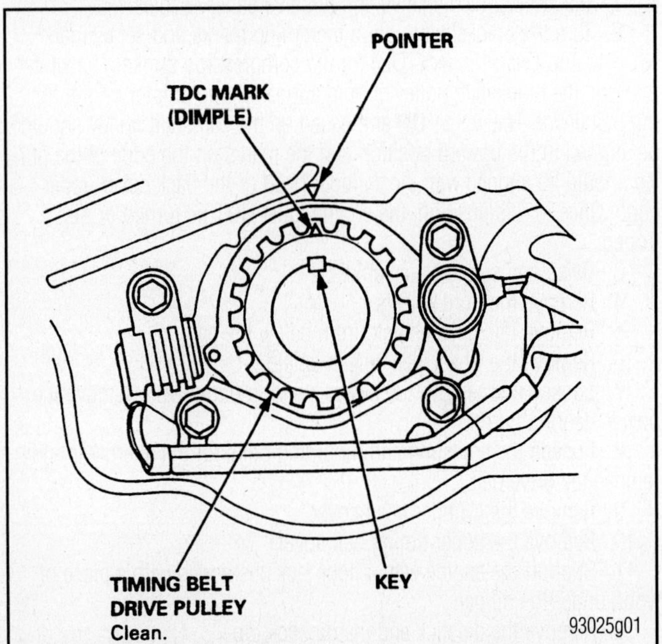

93025g01

Align the crankshaft sprocket with the oil pump pointer before installing the timing belt—Isuzu 2.2L (F22B6) and 2.3L (F23A1, F23A7) engines

20. If equipped with a TDC sensor assembly at the crankshaft sprocket, unbolt the assembly and move it to the side before removing the sprocket.
21. Remove the key and the spacers to remove the crankshaft timing sprocket.
22. Unbolt and remove the camshaft timing sprocket.

To install:

23. Install the camshaft timing sprocket so that the **UP** mark is up and the TDC marks are parallel to the cylinder head gasket surface. Install the key and tighten the bolt to 27 ft. lbs. (37 Nm).
24. Install the crankshaft sprocket so that the TDC mark aligns with the pointer on the oil pump. Install the spacers with their concave surfaces facing in. Install the key. Install the TDC sensor assembly back into position before installing the timing belt.
25. Install the timing belt on the crankshaft sprocket, tensioner pulley, water pump sprocket and camshaft sprocket.
26. Loosen the tensioner bolt and allow the tensioner to tension the timing belt, then tighten the bolt.
27. Install the balancer belt sprocket on the crankshaft.
28. Install the lower timing belt cover and the crankshaft pulley.
29. Rotate the crankshaft counterclockwise 5–6 turns to be sure the belt is properly seated.
30. Set the No. 1 piston at TDC for its compression stroke.

✳✳ WARNING

If any binding is felt when adjusting the timing belt tension by turning the crankshaft, STOP turning the engine, because the pistons may be hitting the valves.

31. Rotate the crankshaft counterclockwise so that the camshaft pulley moves only 3 teeth beyond its TDC mark.
32. Tighten the tensioner adjusting nut to 33 ft. lbs. (45 Nm).
33. Remove the crankshaft pulley and lower timing belt cover.
34. Align the rear balancer shaft into position by performing the following procedures:
    a. Scribe a 3 in. (74mm) line from the end of a 6 x 1.0mm bolt.
    b. Remove the maintenance hole sealing bolt and insert the 6 x 1.0mm bolt into the maintenance hole to the scribed line.
35. Install the balancer shaft belt drive pulley.
36. Align the groove on the pulley edge with the pointer on the balancer gear case.

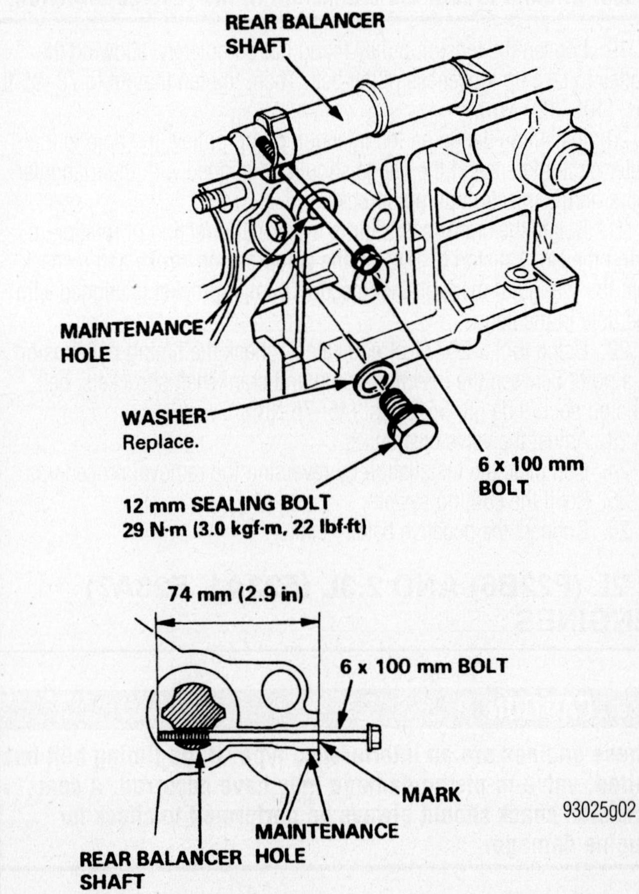

93025g02

Aligning the rear timing balancer shaft—Isuzu 2.2L (F22B6) and 2.3L (F23A1, F23A7) engines

37. Check the alignment of the pointer on the balancer pulley to the pointer on the oil pump.
38. Install and tension the timing balancer shaft belt.
   a. Remove the 6 x 1.0mm bolt from the maintenance hole and install the hole cover.
39. Be sure the timing belts have been tensioned correctly and that all TDC and alignment marks are in their proper positions.
40. Install the lower timing cover and the crankshaft pulley. Apply engine oil to the pulley bolt threads and washer surface. Install the pulley bolt and tighten it to 181 ft. lbs. (245 Nm).
41. Install the upper timing cover and the valve cover. Be sure the seals are properly seated.
42. Install the side engine mount. Tighten the through-bolt to 47 ft. lbs. (64 Nm). Tighten the mount nut and bolt to 40 ft. lbs. (55 Nm) each.
43. Remove the floor jack.
44. Install and tension the alternator belt.
45. Install the power steering pump and tension its belt.
46. Install the splash shields.
47. Reconnect the positive and negative battery cables. Enter the radio security code.
48. Check the engine operation.

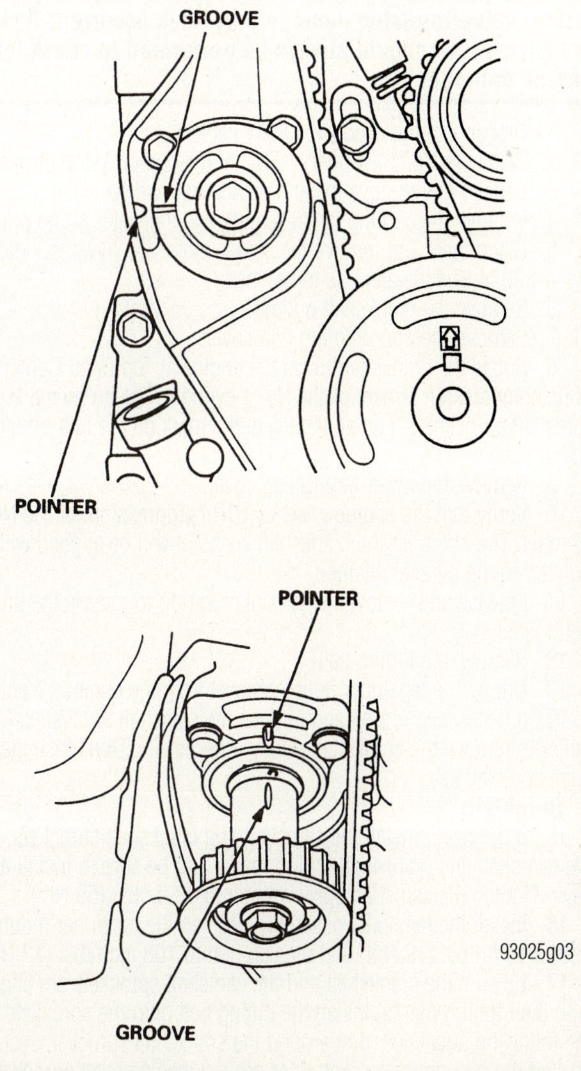

Aligning the front timing balancer shaft—Isuzu 2.2L (F22B6) and 2.3L (F23A1, F23A7) engines

2.2L (X22SE/D) ENGINES

YEAR, MODEL & ENGINE APPLICATION

1998-90 Isuzu Amigo, 1998-03 Rodeo
1998-03 Rodeo Sport .... 2.2L Code X22SED/E

REPLACEMENT INTERVALS

Isuzu recommends timing belt replacement every 100,000 miles on 2000 models, or 75,000 miles on all other years.

LABOR TIMES

2.2L Engine	
Replacement	1.2 hrs
Severe Service	1.4 hrs
With tensioner replacement add	0.3 hrs
With water pump replacement add	0.4 hrs

8807IZUg08

✳✳ CAUTION

These engines are an interference type. If the timing belt has failed, valve-to-piston damage may have occurred. A compression check should always be performed to check for engine damage.

1. Disconnect the negative battery cable.
2. Using a box-end wrench on the drive belt adjuster, turn the adjuster clockwise and remove the drive belt.
3. From the left rear of the engine compartment, disconnect the 3 electrical connectors from the chassis harness.
4. Remove the crankshaft pulley-to-crankshaft bolts and remove the pulley.
5. From the front of the engine, remove the nut and the engine harness cover.
6. Remove the timing belt cover.
7. Rotate the crankshaft to position the timing marks at Top Dead Center (TDC) of the No. 1 cylinder's compression stroke.

■ **Mark the rotational direction of the timing belt for reinstallation purposes.**

8. Remove the timing belt tensioner adjusting bolt and the tensioner from the engine.
9. Remove the timing belt.

To install:

10. Install the timing belt tensioner and finger-tighten the tensioner bolt.
11. Inspect the timing marks to be sure that the engine is positioned at TDC of the No. 1 cylinder's compression stroke.
12. Using tool J-43037 or equivalent, place it between the intake and exhaust sprockets to prevent the camshaft gear from moving during the timing belt installation.
13. Install the timing belt.
14. Position the timing belt to ensure that the tension side of the belt is taut and move the timing belt tension adjusting lever clockwise until the tensioner pointer is flowing.
15. If installing a used timing belt (used over 60 min. from new), the pointer should be positioned approximately 0.16 in. (4mm) to the left of the **V** notch when viewed from the front of the engine.

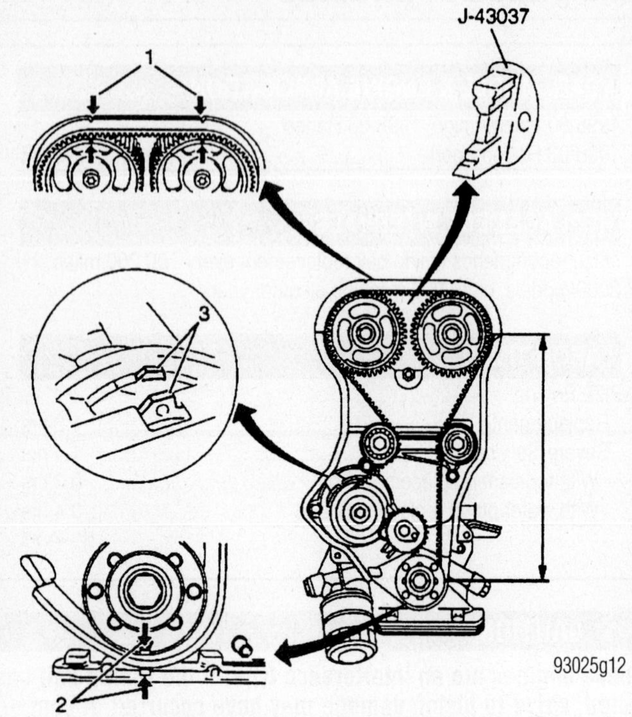

Aligning the timing marks and installing the timing belt—Isuzu 2.2L (X22SE/D) engine

16. If installing a new timing belt, the pointer should be positioned at the center of the **V** notch when viewed from the front of the engine.
17. Torque the timing belt tensioner adjusting bolt to 18 ft. lbs. (25 Nm).
18. Install the timing belt front cover and torque the bolts to 53 inch lbs. (6 Nm).
19. Install the engine harness connectors.
20. Install the crankshaft pulley and toque the pulley-to-crankshaft bolts to 14 ft. lbs. (20 Nm).
21. Move the drive belt tensioner to the loose side and install the drive belt to its normal position.
22. Connect the negative battery cable.

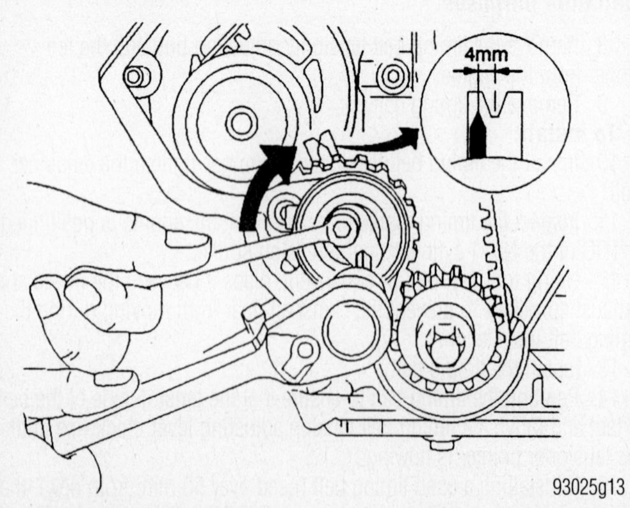

Tensioning the timing belt for a used timing belt—Isuzu 2.2L (X22SE/D) engine

2.3L (4ZD1) ENGINES

YEAR, MODEL & ENGINE APPLICATION

1989-94 Isuzu Amigo, 1986-94 Pickup,
1986-87 Trooper .... 2.3L Code 4ZD1

REPLACEMENT INTERVALS

Isuzu recommends timing belt replacement every 60,000 miles.

LABOR TIMES

1986-94	
Replacement	3.7 hrs
Severe Service	4.1 hrs
With tensioner replacement add	0.3 hrs
With water pump replacement add	0.4 hrs

8807IZUg09

✳✳ CAUTION

These engines are an interference type. If the timing belt has failed, valve-to-piston damage may have occurred. A compression check should always be performed to check for engine damage.

1. Disconnect the negative battery cable.
2. Place the No. 1 cylinder at TDC of the compression stroke.
3. Loosen and remove the engine accessory drive belts.
4. Remove the cooling fan assembly and the water pump pulley.
5. Disconnect and remove the starter motor if a flywheel holder tool No. J-38674 or equivalent, is to be used.
6. Remove the crankshaft pulley.
7. Remove the upper timing belt cover.
8. Rotate the crankshaft to set the engine at Top Dead Center (TDC) of its compression stroke for the No. 1 cylinder. The arrow mark on the camshaft sprocket will be aligned with the mark on the rear timing cover.
9. Remove the lower timing belt cover.
10. Verify that the engine is set at TDC/compression for the No. 1 cylinder. The notch on the crankshaft sprocket will be aligned with the pointer on the oil seal retainer.
11. Release and remove the tensioner spring to release the timing belt's tension.
12. Remove the timing belt.
13. Unbolt the tensioner pulley bracket from the engine's front cover.
14. If necessary, unbolt and remove the camshaft sprockets. Use a puller to remove the crankshaft pulley if necessary. Don't lose the crankshaft sprocket key.

To install:

15. If removed, install the camshaft and crankshaft sprockets. Align the camshaft and crankshaft timing marks and be sure to install any keys. Tighten the camshaft sprocket bolt to 43 ft. lbs. (59 Nm).
16. Install the tensioner assembly. Tighten the tensioner mounting bolt to 14 ft. lbs. (19 Nm) and the cap bolt to 108 inch lbs. (13 Nm).
17. Be sure the crankshaft and the camshaft sprockets are aligned with their timing marks. Install the timing belt onto the sprockets using the following sequence: first around the crankshaft sprocket; second around the oil pump sprocket; third around the camshaft sprocket.
18. Loosen the tensioner mounting bolt. This will allow the tensioner spring to apply pressure to the timing belt.

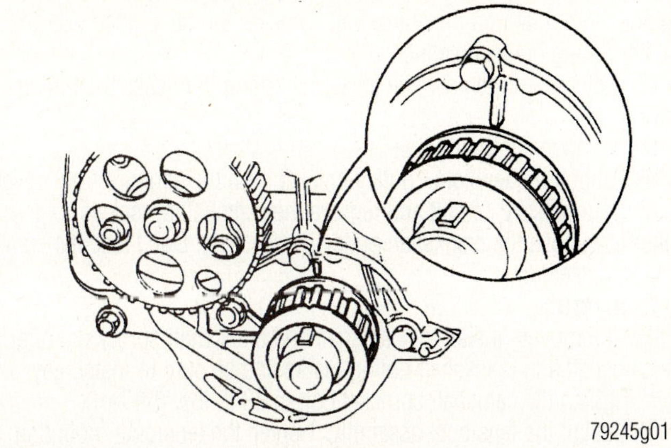

Align the crankshaft pulley timing mark the with oil retainer setting mark—Isuzu 2.3L (4ZD1) engine

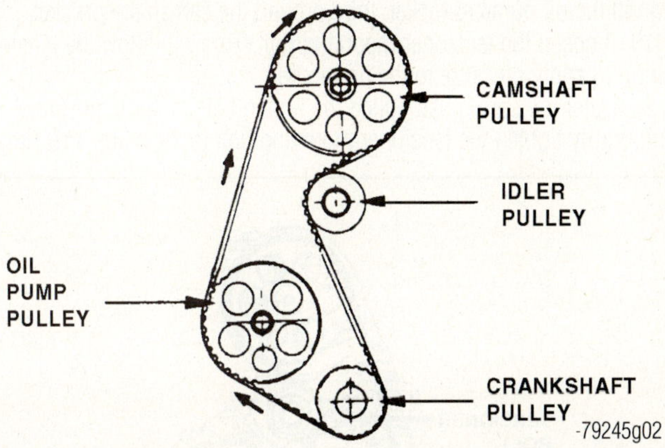

Timing belt direction of travel—Isuzu 2.3L (4ZD1) engine

19. After the spring has pulled the timing belt as far as possible, temporarily tighten the tensioner mounting bolt to 14 ft. lbs. (19 Nm).

■ **Remove the flywheel holder before rotating the crankshaft. Reinstall the holder to tighten the crankshaft pulley bolt.**

✻✻ WARNING

If any binding is felt when adjusting the timing belt tension by turning the crankshaft, STOP turning the engine, because the pistons may be hitting the valves.

20. Rotate the crankshaft counterclockwise 2 complete revolutions to check the rotation of the belt and the alignment of the timing marks.
21. Loosen the tensioner pulley bolt to allow the spring to adjust the correct tension. Then, retighten the tensioner pulley bolt to 14 ft. lbs. (19 Nm).
22. Install the lower timing cover and the crankshaft pulley.
23. Tighten the crankshaft pulley bolt to 87 ft. lbs. (118 Nm). Tighten the small pulley bolts to 72 inch lbs. (8 Nm).
24. Install the upper timing cover.
25. If removed, install the starter and tighten the bolts to 30 ft. lbs. (40 Nm).
26. Install the water pump pulley and tighten its nut to 20 ft. lbs. (26 Nm).

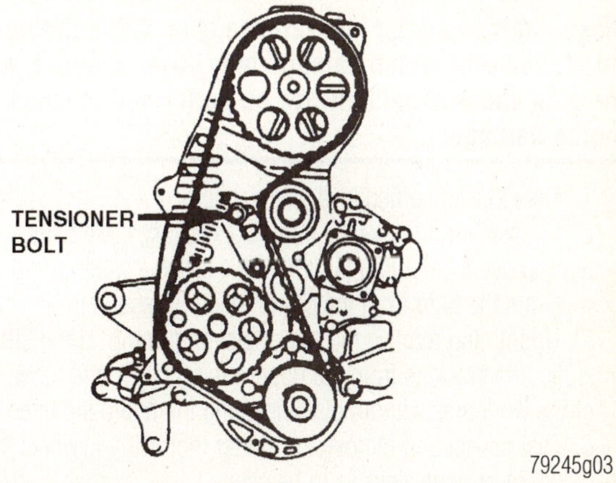

Be sure to tighten the timing belt tensioner bolt (B)—Isuzu 2.3L (4ZD1) engine

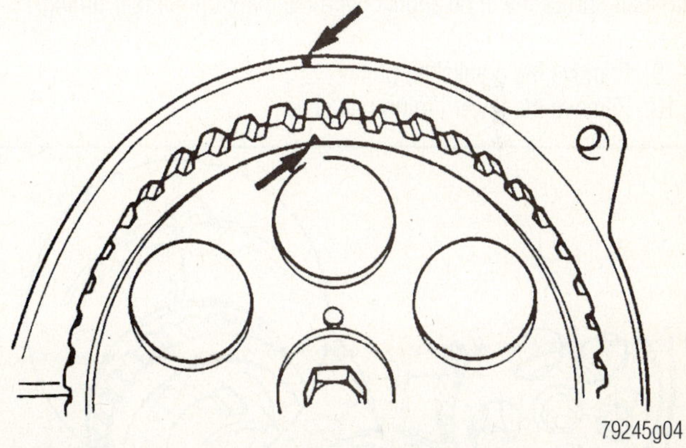

Camshaft-to-timing cover mark—Isuzu 2.3L (4ZD1) engine

27. Install the cooling fan assembly.
28. Install and adjust the accessory drive belts.
29. Connect the negative battery cable.

2.6L (4ZE1) ENGINES

YEAR, MODEL & ENGINE APPLICATION

1988-94 Amigo, 1988-95 Pickup, 1991-97 Rodeo, 1988-93 Trooper II	2.6L Code 4ZE1

REPLACEMENT INTERVALS

Isuzu recommends timing belt replacement every 60,000 miles.

LABOR TIMES

1988-97	
Replacement	2.9 hrs
Severe Service	3.2 hrs
With tensioner replacement add	0.3 hrs
With water pump replacement add	0.4 hrs

8807IZUg10

CAUTION

These engines are an interference type. If the timing belt has failed, valve-to-piston damage may have occurred. A compression check should always be performed to check for engine damage.

1. Disconnect the negative battery cable.
2. Loosen and remove the engine accessory drive belts.
3. Remove the cooling fan assembly and the water pump pulley.
4. Drain the fluid from the power steering reservoir.
5. Unbolt and remove the power steering pump. Unbolt the hydraulic line brackets from the upper timing cover and move the pump out of the work area without disconnecting the hydraulic lines.
6. Disconnect and remove the starter motor if a flywheel holder tool No. J-38674 or equivalent, is to be used.
7. Remove the upper timing belt cover.
8. Rotate the crankshaft to set the engine at Top Dead Center (TDC) of its compression stroke for the No. 1 cylinder. The arrow mark on the camshaft sprocket will be aligned with the mark on the rear timing cover.
9. Remove the crankshaft pulley.
10. Remove the lower timing belt cover.
11. Verify that the engine is set at TDC/compression for the No. 1 cylinder. The notch on the crankshaft sprocket will be aligned with the pointer on the oil seal retainer.
12. Release and remove the tensioner spring to release the timing belt's tension.
13. Remove the timing belt.
14. Unbolt the tensioner pulley bracket from the engine's front cover.
15. If necessary, unbolt and remove the camshaft sprockets. Use a puller to remove the crankshaft pulley if necessary. Don't lose the crankshaft sprocket key.

To install:

16. If removed, install the camshaft and crankshaft sprockets. Align the camshaft and crankshaft timing marks and be sure to install any keys. Tighten the camshaft sprocket bolt to 43 ft. lbs. (59 Nm).
17. Install the tensioner assembly. Tighten the tensioner mounting bolt to 14 ft. lbs. (19 Nm) and the cap bolt to 108 inch lbs. (13 Nm).
18. Be sure the crankshaft and the camshaft sprockets are aligned with their timing marks. Install the timing belt onto the sprockets using the following sequence: first around the crankshaft sprocket; second around the oil pump sprocket; third around the camshaft sprocket.
19. Loosen the tensioner mounting bolt. This will allow the tensioner spring to apply pressure to the timing belt.
20. After the spring has pulled the timing belt as far as possible, temporarily tighten the tensioner mounting bolt to 14 ft. lbs. (19 Nm).

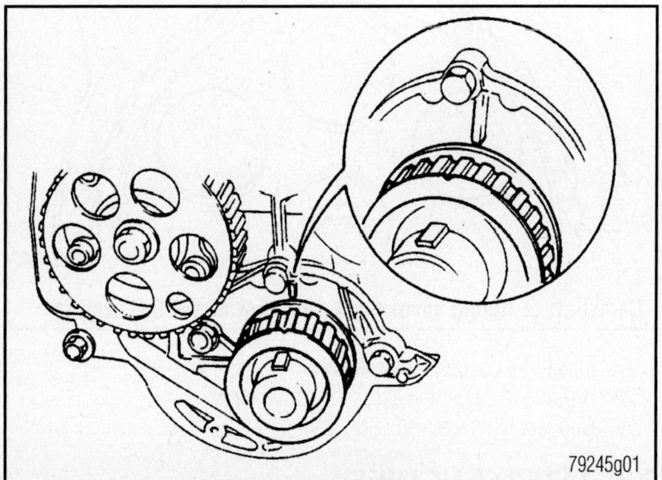

Align the crankshaft pulley timing mark the with oil retainer setting mark—Isuzu 2.6L (4ZE1) engine

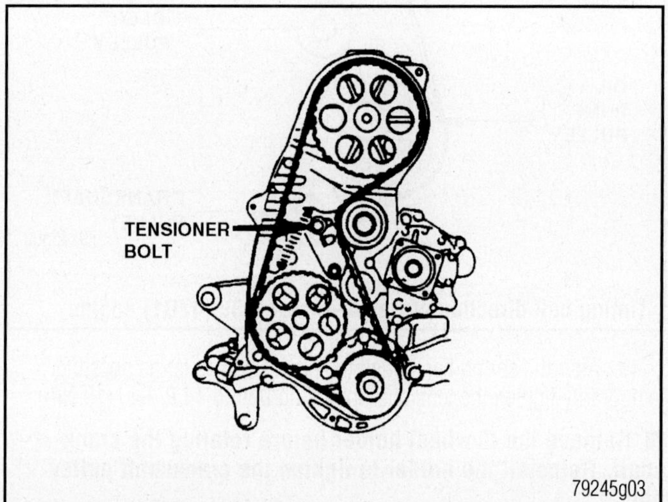

Be sure to tighten the timing belt tensioner bolt (B)—Isuzu 2.6L (4ZE1) engine

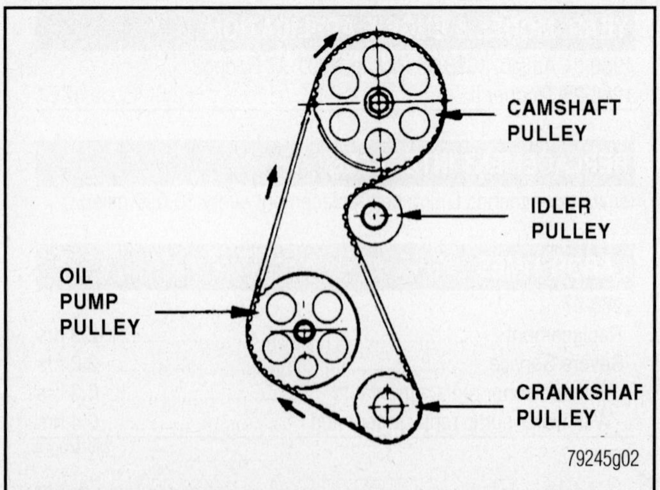

Timing belt direction of travel—Isuzu 2.6L (4ZE1) engine

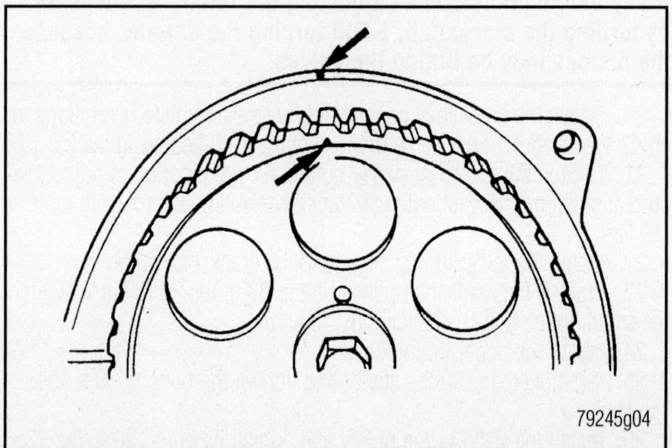

Camshaft-to-timing cover mark—Isuzu 2.6L (4ZE1) engine

■ **Remove the flywheel holder before rotating the crankshaft. Reinstall the holder to tighten the crankshaft pulley bolt.**

**** WARNING**

If any binding is felt when adjusting the timing belt tension by turning the crankshaft, STOP turning the engine, because the pistons may be hitting the valves.

21. Rotate the crankshaft counterclockwise 2 complete revolutions to check the rotation of the belt and the alignment of the timing marks.
22. Loosen the tensioner pulley bolt to allow the spring to adjust the correct tension. Then, retighten the tensioner pulley bolt to 14 ft. lbs. (19 Nm).
23. Install the lower timing cover and the crankshaft pulley.
24. Tighten the crankshaft pulley bolt to 87 ft. lbs. (118 Nm). Tighten the small pulley bolts to 72 inch lbs. (8 Nm).
25. Install the upper timing cover.
26. If removed, install the starter and tighten the bolts to 30 ft. lbs. (40 Nm).
27. Install the power steering pump. If the hydraulic lines were disconnected, refill and bleed the power steering system.
28. Install the water pump pulley and tighten its nut to 20 ft. lbs. (26 Nm).
29. Install the cooling fan assembly.
30. Install and adjust the accessory drive belts.
31. Connect the negative battery cable.

3.2L (6VD1) ENGINES

YEAR, MODEL & ENGINE APPLICATION

1998-00 Isuzu Amigo, 1998-03 Rodeo, 2001-03 Rodeo Sport	3.2L Code 6VD1

REPLACEMENT INTERVALS

Isuzu recommends timing belt replacement every 75,000 miles on 1998 models, or 100,000 miles on all other years.

LABOR TIMES

All Models	
Replacement	2.5 hrs
Severe Service	2.8 hrs
With tensioner replacement add	0.3 hrs
With water pump replacement add	0.4 hrs

8807IZUg11

**** CAUTION**

These engines are an interference type. If the timing belt has failed, valve-to-piston damage may have occurred. A compression check should always be performed to check for engine damage.

1. Disconnect the negative battery cable.
2. Drain the engine coolant.
3. Remove the air cleaner assembly and intake air duct.
4. Remove the upper fan shroud from the radiator.
5. Remove the 4 nuts retaining the cooling fan assembly. Remove the cooling fan from the fan pulley.

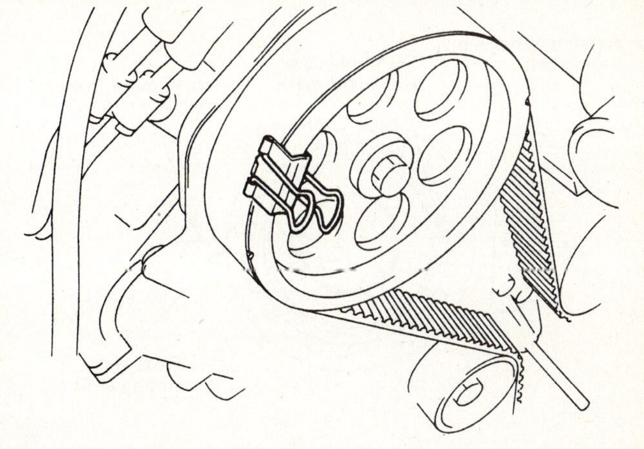

79245g08

Using a double clip to hold the belt in place—Isuzu 3.2L (6VD1) engine

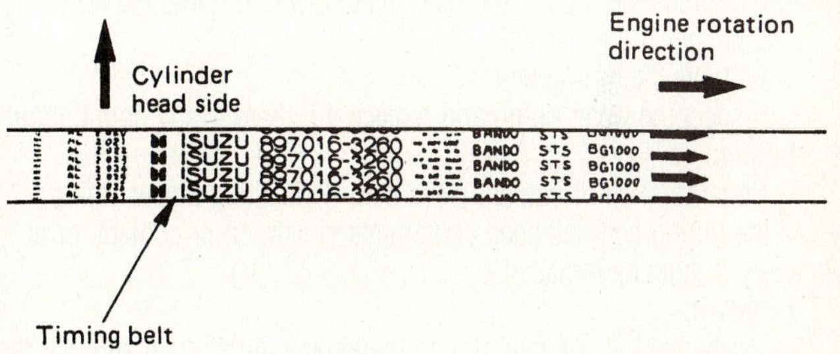

79245g07

For maximum timing belt life, install the belt correctly—Isuzu 3.2L (6VD1) engine

6. Loosen and remove the drive belts, idler pulley and belt tensioner.
7. Remove the upper radiator hose.
8. Remove the cooling fan and fan pulley assembly.
9. Drain the power steering fluid, then remove the power steering pump.
10. Remove the A/C compressor, if required.
11. Use tool No. J-8614-01, or a suitable pulley holding tool to remove the crankshaft pulley center bolt. Remove the crankshaft pulley.
12. Remove the upper timing belt cover.
13. Rotate the crankshaft to align the camshaft timing marks with the pointer dots on the back covers. Verify that the pointer on the crankshaft aligns with the mark on the lower timing cover.

■ **When the timing marks are aligned, the No. 2 piston is at Top Dead Center (TDC) of the compression stroke.**

**** WARNING**

Align the camshaft and crankshaft sprockets with their alignment marks before removing the timing belt. Failure to align the belt and sprocket marks may result in valve damage.

14. If present, disconnect the 2 oil cooler hose bracket bolts on the timing cover. Move the oil cooler hoses and bracket off of the lower timing cover.
15. Remove the lower timing belt cover.

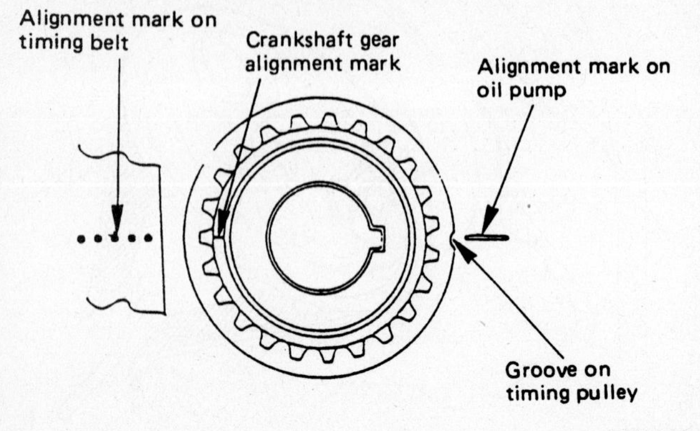

79245g06

Crankshaft alignment marks for timing belt installation— Isuzu 3.2L (6VD1) engine

16. Remove the pusher assembly (tensioner) from below the belt tensioner pulley. The pusher rod must always face upward to prevent oil leakage. Depress the pusher rod, and insert a wire pin into the hole to keep the pusher rod retracted.
17. Remove the timing belt.
18. Inspect the water pump and replace it if there is any doubt about its condition.
19. Repair any oil or coolant leaks before installing a new timing belt. If the timing belt has been contaminated with oil or coolant, or is damaged, it must be replaced.

To install:

20. Verify that the sprocket timing marks are still aligned and that the groove and the keyway on the crankshaft timing sprocket align with the mark on the oil pump. The white pointers on the camshaft timing sprockets should align with the dots on the front plate.
21. Align the right side camshaft sprocket with the mark on the cover. Rotate the right side camshaft sprocket clockwise until the timing mark springs into the 12 o'clock position with the timing marks aligned. This may require multiple turns of the sprocket.
22. Rotate the left side camshaft sprocket clockwise until the timing mark springs into the 9 o'clock position with the timing marks aligned. Rotate the sprocket an additional 90 degrees to the 12 o'clock position and align the timing marks.

■ **The arrows on the timing belt must follow the belt's direction of rotation. The manufacturer's trademark on the belt's spine should be readable left-to-right when the belt is installed.**

23. Install the timing belt. Use clips to secure the belt onto each sprocket until the installation is complete. Align the dotted marks on the timing belt with the timing mark opposite the groove on the crankshaft sprocket.

WARNING

If any binding is felt when adjusting the timing belt tension by turning the crankshaft, STOP turning the engine, because the pistons may be hitting the valves.

24. Insert a 1.4mm piece of wire through the hole in the pusher to hold the rod in. Install the pusher assembly while pushing the tension pulley toward the belt.
25. Pull the pin out from the pusher to release the rod.
26. Remove the clips from the sprockets. Install the crankshaft pulley and rotate the crankshaft pulley clockwise 2 turns. Measure the rod protrusion to ensure it is between 0.16–0.24 in. (4–6mm).
27. Remove the crankshaft pulley. Install the lower and upper timing belt covers and tighten their bolts to 12 ft. lbs. (17 Nm).
28. Install the crankshaft pulley and tighten the pulley bolt to 123 ft. lbs. (167 Nm).
29. Fit the oil cooler hose onto the timing cover and tighten its mounting bracket bolts to 16 ft. lbs. (22 Nm).
30. Install the A/C compressor, if removed.
31. Install the power steering pump.
32. Install the cooling fan assembly and tighten the bolts to 72 inch lbs. (8 Nm).
33. Install fan pulley assembly and tighten the bolts to 16 ft. lbs. (22 Nm).
34. Install the upper radiator hose.
35. Install and adjust the accessory drive belts.
36. Install the upper fan shroud.
37. Install the air cleaner assembly and intake air duct.
38. Refill the power steering fluid and engine coolant.
39. Connect the negative battery cable.

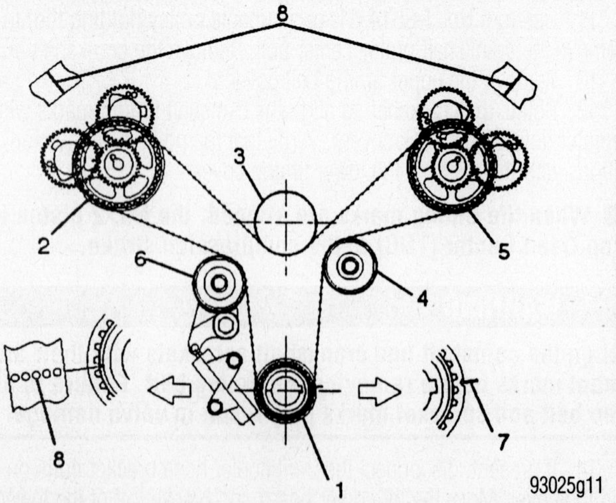

93025g11

Timing mark alignment and timing belt routing— Isuzu 3.2L (6VD1) engine

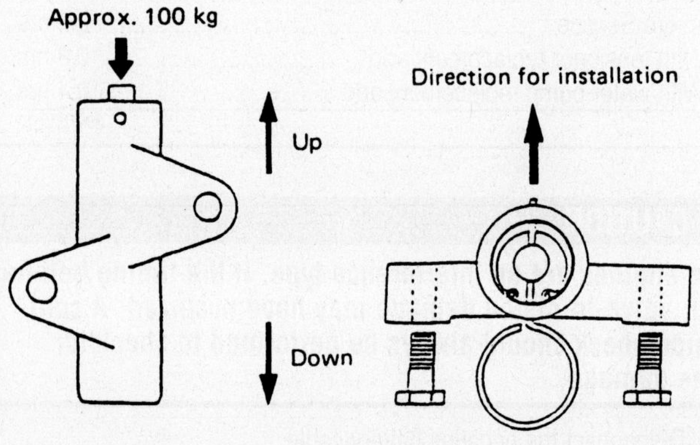

79245g11

View of timing belt tensioner pusher— Isuzu 3.2L (6VD1) engine

3.2L (VIN V) SOHC ENGINES

YEAR, MODEL & ENGINE APPLICATION

1993-97 Isuzu Rodeo, 1992-97 Trooper .... 3.2L VIN V

REPLACEMENT INTERVALS

Isuzu recommends timing belt relacement every 60,000 miles.

LABOR TIMES

Rodeo	
Replacement	3.1 hrs
Severe Service	3.4 hrs
Trooper	
Replacement	2.7 hrs
Severe Service	3.0 hrs
All Models	
With tensioner replacement add	0.3 hrs
With water pump replacement add	0.4 hrs

8807IZUg12

✻✻ CAUTION

These engines are an interference type. If the timing belt has failed, valve-to-piston damage may have occurred. A compression check should always be performed to check for engine damage.

1. Disconnect the negative battery cable.
2. Remove the air cleaner assembly and intake air duct.
3. Remove the upper fan shroud from the radiator.
4. Remove the 4 nuts retaining the cooling fan assembly. Remove the cooling fan from the fan pulley.
5. Loosen and remove the drive belts.
6. Remove the upper timing belt covers.
7. Remove the fan pulley assembly.
8. Rotate the crankshaft to align the camshaft timing marks with the pointer dots on the back covers. Verify that the pointer on the crankshaft aligns with the mark on the lower timing cover.

■ When the timing marks are aligned, the No. 2 piston is at Top Dead Center (TDC) of the compression stroke.

✻✻ WARNING

Align the camshaft and crankshaft sprockets with their alignment marks before removing the timing belt. Failure to align the belt and sprocket marks may result in valve damage.

9. Use tool No. J-8614-01, or a suitable pulley holding tool to remove the crankshaft pulley center bolt. Remove the crankshaft pulley.
10. If present, disconnect the 2 oil cooler hose bracket bolts on the timing cover. Move the oil cooler hoses and bracket off of the lower timing cover.
11. Remove the lower timing belt cover.
12. Remove the tensioner from below the belt tensioner pulley. The pusher rod must always face upward to prevent oil leakage. Depress the pusher rod, and insert a wire pin into the hole to keep the pusher rod retracted.
13. Remove the timing belt.
14. Inspect the water pump and replace it if there is any doubt about its condition.
15. Repair any oil or coolant leaks before installing a new timing belt. If the timing belt has been contaminated with oil or coolant, or is damaged, it must be replaced.

To install:

16. Verify that the sprocket timing marks are still aligned and that the groove and the keyway on the crankshaft timing sprocket align with the mark on the oil pump. The white pointers on the camshaft timing sprockets should align with the dots on the front plate.
17. Install the timing belt on the crankshaft sprocket. Use binder cip to hold it in position.
18. Install the belt on the right side camshaft sprocket. Use a binder clip to hold it in position.
19. Rotate the crankshaft counterclockwise to remove the slack between the crankshaft sprocket and the right camshaft timing belt sprocket.
20. Install the belt to the water pump and idler pulleys.
21. Install the belt on the left side camshaft sprocket. Use a binder clip to hold it in position.
22. Rotate the crankshaft clockwise to remove the slack between the crankshaft sprocket and the left camshaft timing belt sprocket.
23. Install the timing belt tensioner and tighten the bolts to 16 ft. lbs. (22 Nm).
24. Remove the pin out from the tensioner.
25. Remove the binder clips from the timing belt.

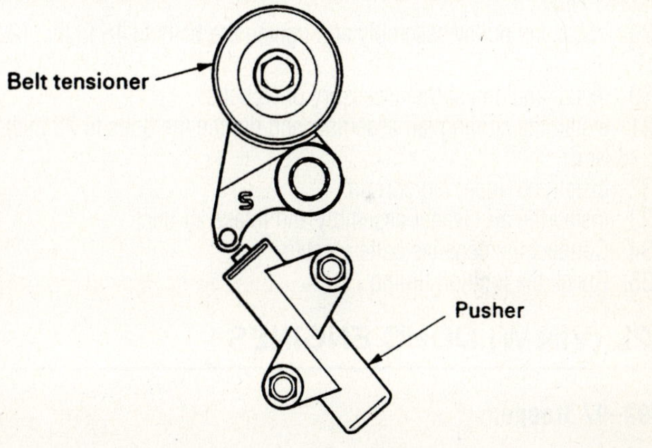

93025g10

View of timing belt tensioner and pusher—Isuzu 3.2L (VIN V) engine

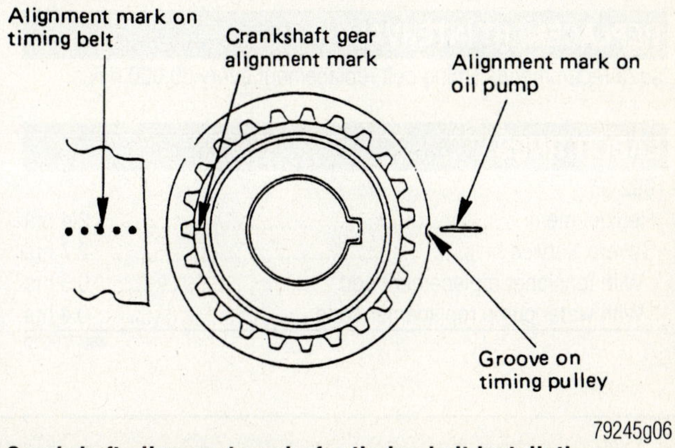

79245g06

Crankshaft alignment marks for timing belt installation—Isuzu 3.2L (VIN V) engine

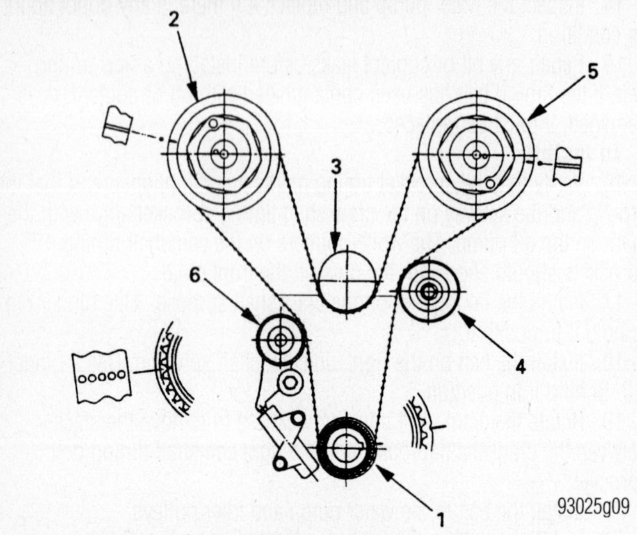

Camshaft sprocket alignment marks for timing belt installation—Isuzu 3.2L (VIN V) engine

26. Remove the crankshaft pulley. Install the lower and upper timing belt covers and tighten their bolts to 12 ft. lbs. (17 Nm).
27. Fit the oil cooler hose onto the timing cover and tighten its mounting bracket bolts to 16 ft. lbs. (22 Nm).
28. Install the crankshaft pulley and tighten the pulley bolt to 123 ft. lbs. (167 Nm).
29. Install fan pulley assembly and tighten the bolts to 16 ft. lbs. (22 Nm).
30. Install and adjust the accessory drive belts.
31. Install the cooling fan assembly and tighten the bolts to 72 inch lbs. (8 Nm).
32. Install the upper fan shroud.
33. Install the air cleaner assembly and intake air duct.
34. Connect the negative battery cable.
35. Check the ignition timing.

3.2L (VIN W) DOHC ENGINES

1992–97 Trooper

YEAR, MODEL & ENGINE APPLICATION

1992-95 Isuzu Trooper	3.2L VIN W

REPLACEMENT INTERVALS

Isuzu recommends timing belt replacement every 60,000 miles.

LABOR TIMES

1992-95	
Replacement	2.4 hrs
Severe Service	2.7 hrs
With tensioner replacement add	0.3 hrs
With water pump replacement add	0.4 hrs

8807IZUG13

✻✻ CAUTION

These engines are an interference type. If the timing belt has failed, valve-to-piston damage may have occurred. A compression check should always be performed to check for engine damage.

1. Disconnect the negative battery cable.
2. Remove the air cleaner assembly and intake air duct.
3. Remove the upper fan shroud from the radiator.
4. Remove the 4 nuts retaining the cooling fan assembly. Remove the cooling fan from the fan pulley.
5. Loosen and remove the drive belts.
6. Remove the upper timing belt covers.
7. Remove the fan pulley assembly.
8. Rotate the crankshaft to align the camshaft timing marks with the pointer dots on the back covers. Verify that the pointer on the crankshaft aligns with the mark on the lower timing cover.

■ **When the timing marks are aligned, the No. 2 piston is at Top Dead Center (TDC) of the compression stroke.**

✻✻ WARNING

Align the camshaft and crankshaft sprockets with their alignment marks before removing the timing belt. Failure to align the belt and sprocket marks may result in valve damage.

9. Use tool No. J-8614-01, or a suitable pulley holding tool to remove the crankshaft pulley center bolt. Remove the crankshaft pulley.
10. If present, disconnect the 2 oil cooler hose bracket bolts on the timing cover. Move the oil cooler hoses and bracket off of the lower timing cover.
11. Remove the lower timing belt cover.
12. Remove the tensioner from below the belt tensioner pulley. The pusher rod must always face upward to prevent oil leakage. Depress the pusher rod, and insert a wire pin into the hole to keep the pusher rod retracted.
13. Remove the timing belt.
14. Inspect the water pump and replace it if there is any doubt about its condition.
15. Repair any oil or coolant leaks before installing a new timing belt. If the timing belt has been contaminated with oil or coolant, or is damaged, it must be replaced.

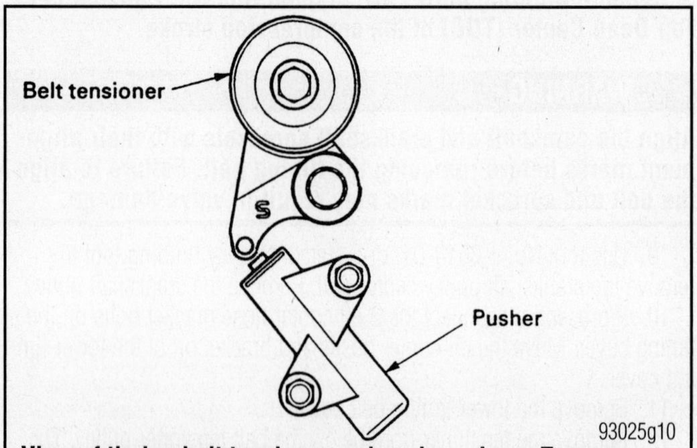

View of timing belt tensioner and pusher—Isuzu Trooper 3.2L (VIN W) engine

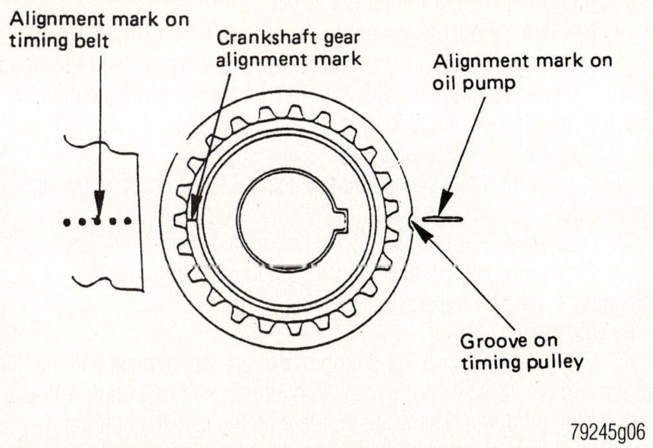

Crankshaft alignment marks for timing belt installation—Isuzu Trooper 3.2L (VIN W) engine

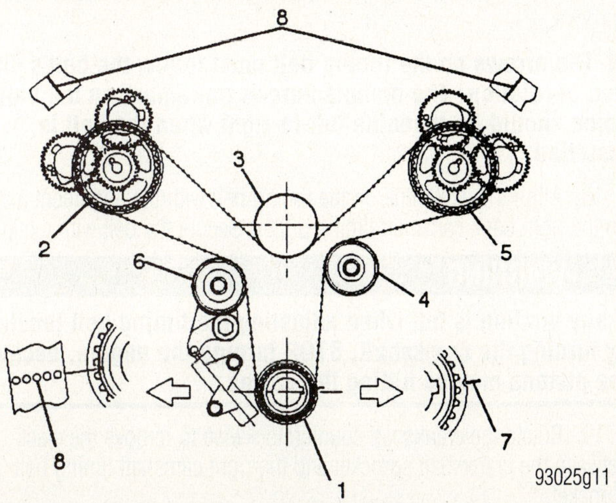

Camshaft sprocket alignment marks for timing belt installation—Isuzu Trooper 3.2L (VIN W) engine

To install:

16. Verify that the sprocket timing marks are still aligned and that the groove and the keyway on the crankshaft timing sprocket align with the mark on the oil pump. The white pointers on the camshaft timing sprockets should align with the dots on the front plate.
17. Install the timing belt on the crankshaft sprocket. Use binder cip to hold it in position.
18. Install the belt on the right side camshaft sprocket. Use a binder clip to hold it in position.
19. Rotate the crankshaft counterclockwise to remove the slack between the crankshaft sprocket and the right camshaft timing belt sprocket.
20. Install the belt to the water pump and idler pulleys.
21. Install the belt on the left side camshaft sprocket. Use a binder clip to hold it in position.
22. Rotate the crankshaft clockwise to remove the slack between the crankshaft sprocket and the left camshaft timing belt sprocket.
23. Install the timing belt tensioner and tighten the bolts to 16 ft. lbs. (22 Nm).
24. Remove the pin out from the tensioner.
25. Remove the binder clips from the timing belt.
26. Remove the crankshaft pulley. Install the lower and upper timing belt covers and tighten their bolts to 12 ft. lbs. (17 Nm).
27. Fit the oil cooler hose onto the timing cover and tighten its mounting bracket bolts to 16 ft. lbs. (22 Nm).
28. Install the crankshaft pulley and tighten the pulley bolt to 123 ft. lbs. (167 Nm).
29. Install fan pulley assembly and tighten the bolts to 16 ft. lbs. (22 Nm).
30. Install and adjust the accessory drive belts.
31. Install the cooling fan assembly and tighten the bolts to 72 inch lbs. (8 Nm).
32. Install the upper fan shroud.
33. Install the air cleaner assembly and intake air duct.
34. Connect the negative battery cable.
35. Check the ignition timing.

3.5L (6VE1) ENGINES

YEAR, MODEL & ENGINE APPLICATION

Application	Engine
2001-03 Isuzu Axiom, 1998-03 Trooper, 1999-02 VehiCROSS	3.5L Code 6VE1

REPLACEMENT INTERVALS

Isuzu recommends timing belt replacement every 75,000 miles on 1998 models, or 100,000 miles on all other years.

LABOR TIMES

All Models	
Replacement	2.4 hrs
Severe Service	2.7 hrs
With tensioner replacement add	0.3 hrs
With water pump replacement add	0.4 hrs

8807IZUg14

✻✻ CAUTION

These engines are an interference type. If the timing belt has failed, valve-to-piston damage may have occurred. A compression check should always be performed to check for engine damage.

1. Disconnect the negative battery cable.
2. Remove the air cleaner assembly and intake air duct.
3. Remove the upper fan shroud from the radiator.
4. Remove the 4 nuts retaining the cooling fan assembly. Remove the cooling fan from the fan pulley.
5. Loosen and remove the drive belts.
6. Remove the upper timing belt covers.
7. Remove the fan pulley assembly.
8. Rotate the crankshaft to align the camshaft timing marks with the pointer dots on the back covers. Verify that the pointer on the crankshaft aligns with the mark on the lower timing cover.

■ **When the timing marks are aligned, the No. 2 piston is at Top Dead Center (TDC) of the compression stroke.**

✻✻ WARNING

Align the camshaft and crankshaft sprockets with their alignment marks before removing the timing belt. Failure to align the belt and sprocket marks may result in valve damage.

9. Use tool No. J-8614-01, or a suitable pulley holding tool to remove the crankshaft pulley center bolt. Remove the crankshaft pulley.

10. If present, disconnect the 2 oil cooler hose bracket bolts on the timing cover. Move the oil cooler hoses and bracket off of the lower timing cover.

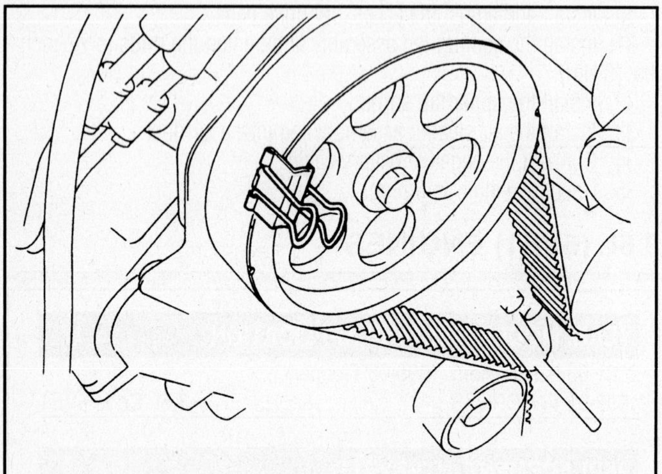

79245g08

Using a double clip to hold the belt in place—Isuzu Amigo, Rodeo and 1996–97 Trooper 3.2L and 1998–00 Trooper 3.5L (6VE1) engines

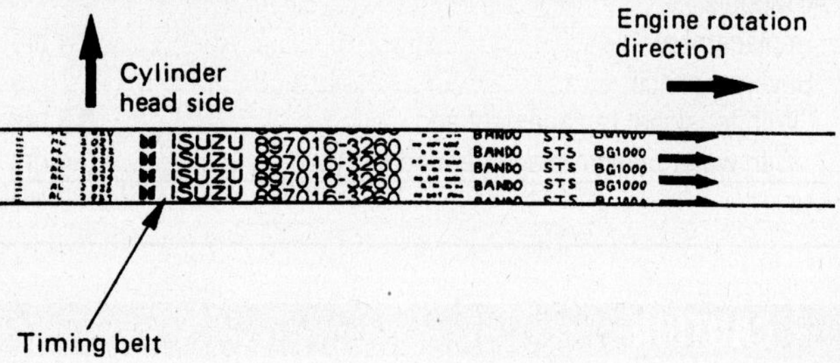

79245g07

For maximum timing belt life, install the belt—Isuzu Amigo, Rodeo and 1996–97 Trooper 3.2L (6VD1) and 1998–00 Trooper 3.5L (6VE1) engines

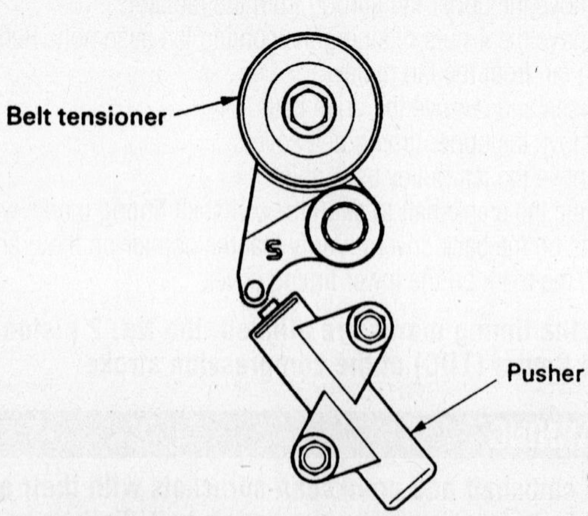

93025g10

View of timing belt tensioner and pusher—1998–00 Trooper 3.5L (6VE1) engines

11. Remove the lower timing belt cover.

12. Remove the pusher assembly (tensioner) from below the belt tensioner pulley. The pusher rod must always face upward to prevent oil leakage. Depress the pusher rod, and insert a wire pin into the hole to keep the pusher rod retracted.

13. Remove the timing belt.

14. Inspect the water pump and replace it if there is any doubt about its condition.

15. Repair any oil or coolant leaks before installing a new timing belt. If the timing belt has been contaminated with oil or coolant, or is damaged, it must be replaced.

To install:

16. Verify that the sprocket timing marks are still aligned and that the groove and the keyway on the crankshaft timing sprocket align with the mark on the oil pump. The white pointers on the camshaft timing sprockets should align with the dots on the front plate.

17. Install the timing belt. Use clips to secure the belt onto each sprocket until the installation is complete. Align the dotted marks on the timing belt with the timing mark opposite the groove on the crankshaft sprocket.

■ **The arrows on the timing belt must follow the belt's direction of rotation. The manufacturer's trademark on the belt's spine should be readable left-to-right when the belt is installed.**

18. Align the white line on the timing belt with the alignment mark on the right bank camshaft timing pulley. Secure the belt with a clip.

✱✱ WARNING

If any binding is felt when adjusting the timing belt tension by turning the crankshaft, STOP turning the engine, because the pistons may be hitting the valves.

19. Rotate the crankshaft counterclockwise to remove the slack between the crankshaft sprocket and the right camshaft timing belt sprocket.

20. Install the belt around the water pump pulley.

21. Install the belt on the idler pulley.

22. Align the white alignment mark on the timing belt with the alignment mark on the left bank camshaft timing belt sprocket.

23. Install the crankshaft pulley and tighten the center bolt by hand. Rotate the crankshaft pulley clockwise to give slack between the crankshaft timing belt pulley and the right bank camshaft timing belt pulley.

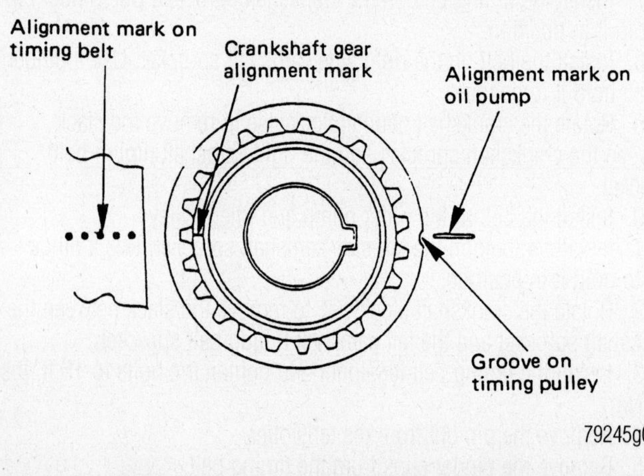

79245g06

Proper crankshaft alignment marks for timing belt installation—Isuzu Rodeo and Trooper 3.5L(6VE1) engine

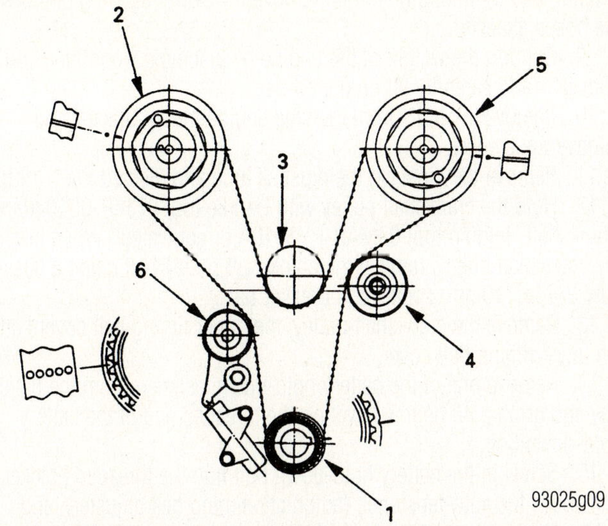

Timing mark alignment and timing belt routing—1996–97 Isuzu Rodeo and Trooper 3.5L (6VE1) engine

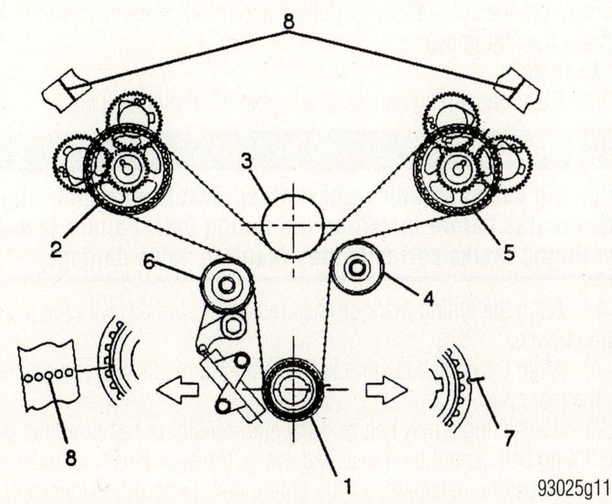

Timing mark alignment and timing belt routing—1990-00 Isuzu Rodeo and Trooper 3.5L (6VE1) engine

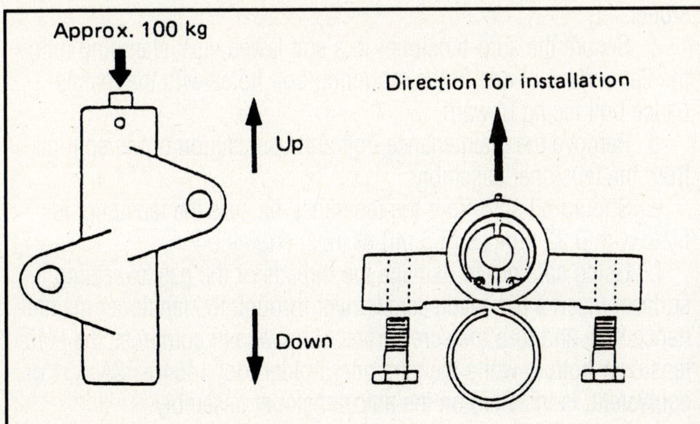

View of timing belt tensioner pusher—1998-00 Isuzu Rodeo and Trooper 3.5L (6VE1) engine

24. Insert a 1.4mm piece of wire through the hole in the pusher to hold the rod in. Install the pusher assembly while pushing the tension pulley toward the belt.
25. Pull the pin out from the pusher to release the rod.
26. Remove the clamps from the sprockets. Rotate the crankshaft pulley clockwise 2 turns. Measure the rod protrusion to ensure it is between 0.16–0.24 in. (4–6mm).
27. If the tensioner pulley bracket pivot bolt was removed, tighten it to 31 ft. lbs. (42 Nm).
28. Tighten the pusher bolts to 14 ft. lbs. (19 Nm).
29. Remove the crankshaft pulley. Install the lower and upper timing belt covers and tighten their bolts to 12 ft. lbs. (17 Nm).
30. Fit the oil cooler hose onto the timing cover and tighten its mounting bracket bolts to 16 ft. lbs. (22 Nm).
31. Install the crankshaft pulley and tighten the pulley bolt to 123 ft. lbs. (167 Nm).
32. Install fan pulley assembly and tighten the bolts to 16 ft. lbs. (22 Nm).
33. Install and adjust the accessory drive belts.
34. Install the cooling fan assembly and tighten the bolts to 72 inch lbs. (8 Nm).
35. Install the upper fan shroud.
36. Install the air cleaner assembly and intake air duct.
37. Connect the negative battery cable.

3.5L (J35A1) V6 ENGINES

YEAR, MODEL & ENGINE APPLICATION	
1999-00 Isuzu Oasis	3.5L Code J35A1

REPLACEMENT INTERVALS

Isuzu recommends timing belt replacement every 60,000 miles.

LABOR TIMES	
1999-00	
Replacement	3.1 hrs
Severe Service	3.4 hrs
With tensioner replacement add	0.3 hrs
With water pump replacement add	0.4 hrs

8807IZUg15

✻✻ CAUTION

These engines are considered freewheeling. If the timing belt fails, valve-to-piston damage is unlikely, however a precautionary compression check should be performed to verify no engine damage.

■ **The radio may contain a coded theft protection circuit. Always make note the code number before disconnecting the battery.**

1. Disconnect the negative battery terminal.
2. Turn the crankshaft so the white mark on the crankshaft pulley aligns with the pointer on the oil pump housing cover.
3. Open the inspection plug on the upper cover of the cam covers and check that the camshaft sprocket marks align with the upper cover marks.

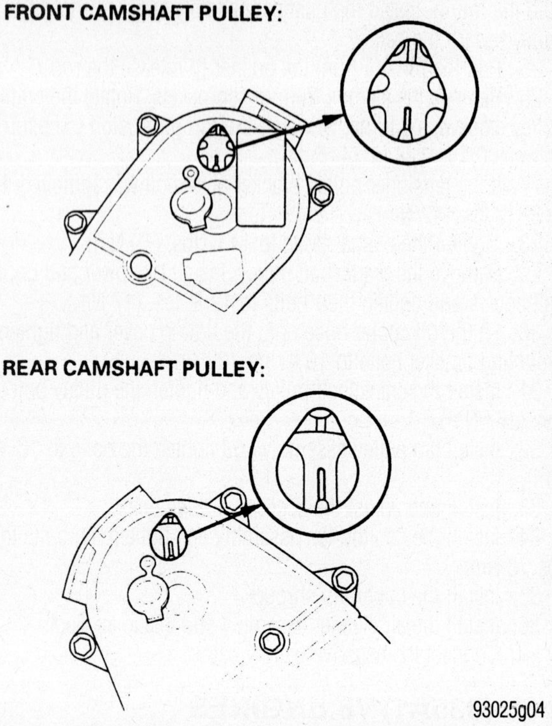

Crankshaft and camshaft timing marks at Top Dead Center (TDC)—Isuzu 1999–00 3.5L (J35A1) engine—Oasis

✻✻ WARNING

Align the camshaft and crankshaft sprockets with their alignment marks before removing the timing belt. Failure to align the timing marks correctly may result in valve damage.

4. Raise and safely support the vehicle and remove both front tires/wheels.
5. Remove the front lower splash shield.
6. Move the alternator tensioner with a Belt Tensioner Release Arm tool YA9317 or equivalent, to release tension from the belt and remove the alternator belt.
7. Remove the alternator belt tensioner release arm.
8. Loosen the power steering pump adjustment nut, adjustment locknut, and mounting bolt, then remove the power steering pump with the hoses attached.

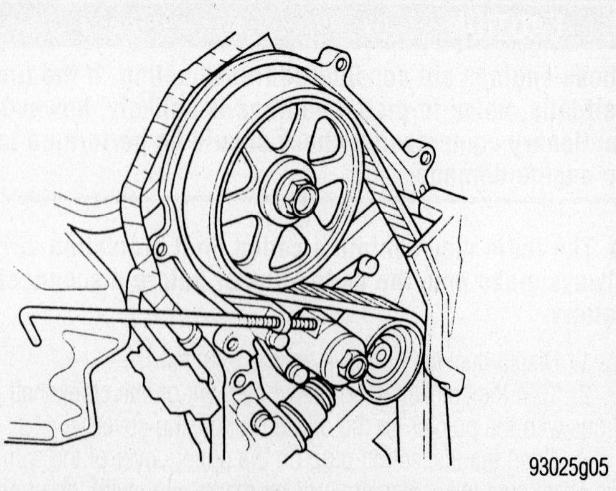

Battery hold-down bolt installed to hold auto-tensioner—Isuzu 1999–00 3.5L (J35A1) engine—Oasis

9. Support the weight of the engine by placing a wood block on a floor jack and carefully lift on the oil pan.
10. Remove the bolts from the side engine mount bracket and remove the bracket.
11. Remove the dipstick, the dipstick tube and discard the O-ring.
12. Hold the crankshaft pulley with Handle tool 07JAB-001020A and Crankshaft Holding tool 07MAB-PY3010A or equivalent. While holding the crankshaft pulley, remove the crankshaft pulley bolt using a heavy duty 3/4 in. (19mm) socket and breaker bar.
13. Remove the crankshaft pulley, the upper timing belt covers and the lower timing belt cover.
14. Remove one of the battery hold-down fasteners from the battery tray and grind a 45 degree bevel on the threaded end of the battery hold-down bolt.
15. Screw in the battery hold-down bolt into the threaded bracket just above the auto-tensioner, (automatic timing belt adjuster), and tighten the bolt hand-tight to hold the auto-tensioner adjuster in its current position.
16. Remove the engine mount bracket bolts and the bracket.
17. Loosen the timing belt idler pulley bolt (located on the right side across from the auto-tensioner pulley) about 5–6 revolutions and remove the timing belt.

To install:

18. Clean the timing belt sprockets and the timing belt covers.

✻✻ WARNING

Align the camshaft and crankshaft sprockets with their alignment marks before installing the timing belt. Failure to align the timing marks correctly may result in valve damage.

19. Align the timing mark on the crankshaft sprocket with the oil pump pointer.
20. Align the camshaft sprocket TDC timing marks with the pointers on the rear cover.
21. If installing a new belt or if the auto-tensioner has extended or if the timing belt cannot be reinstalled easily, the auto-tensioner must be collapsed before installation of the timing belt, perform the following procedures:
    a. Remove the battery hold-down bolt from the auto-tensioner bracket.
    b. Remove the timing belt auto-tensioner bolts and the auto-tensioner.
    c. Secure the auto-tensioner in a soft jawed vise, clamping onto the flat surface of one of the mounting bolt holes with the maintenance bolt facing upward.
    d. Remove the maintenance bolt and use caution not to spill oil from the tensioner assembly.
    e. Should oil spill from the tensioner, be sure the tensioner is filled with 0.22 ounces (6.5 ml) of fresh engine oil.
    f. Using care not to damage the threads or the gasket sealing surface, insert a flat-blade screwdriver through the tensioner maintenance hole and turn the screwdriver clockwise to compress the auto-tensioner bottom while the tensioner holder tool 14540-P8A-A01 or equivalent, is installed on the auto-tensioner assembly.
    g. Install the auto-tensioner maintenance bolt with a new gasket and tighten to a torque 72 inch lbs. (8 Nm).
    h. Install the auto-tensioner on the engine with the tensioner holder tool installed and torque the mounting bolts to 104 inch lbs. (12 Nm).

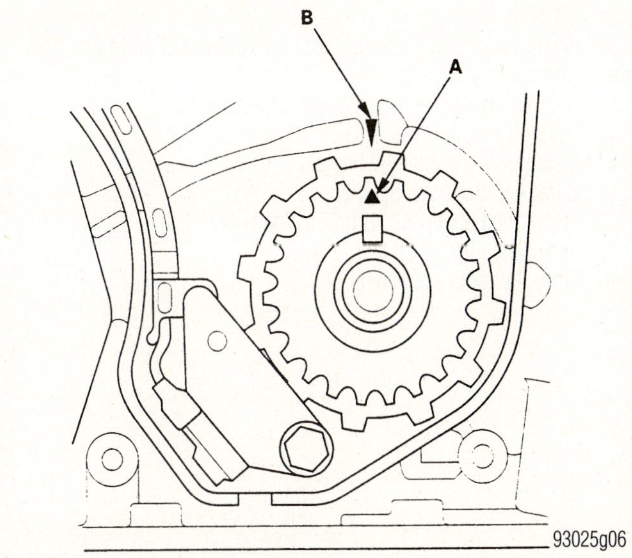

Crankshaft sprocket Top Dead Center (TDC) mark—Isuzu 1999–00 3.5L (J35A1) engine—Oasis

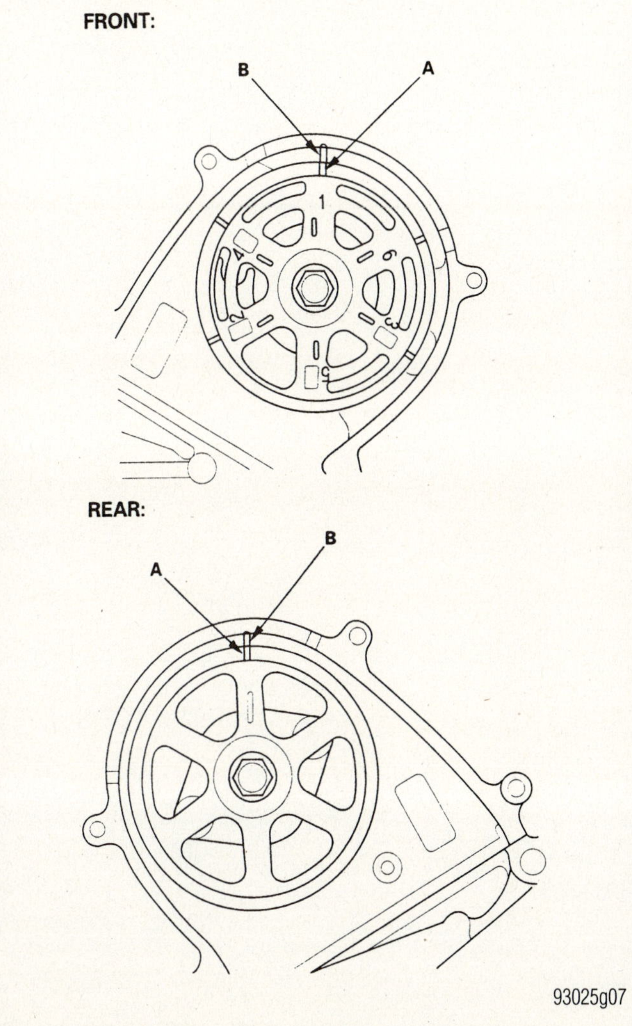

Camshaft sprocket Top Dead Center (TDC) mark—Isuzu 1999–00 3.5L (J35A1) engine—Oasis

22. Install the timing belt in a counterclockwise pattern starting with the crankshaft drive sprocket. Install the timing belt counterclockwise in the following sequence:

- Crankshaft drive sprocket.
- Idler pulley.
- Left side camshaft sprocket.
- Water pump.
- Right side camshaft sprocket.
- Auto-tensioner adjustment pulley.

23. Torque the timing belt idler pulley bolt to 33 ft. lbs. (44 Nm).

24. Remove the auto-tensioner holding tool to allow the tensioner to extend.

25. Install the engine mount bracket to the engine and torque the bolts to 33 ft. lbs. (44 Nm).

26. Install the lower timing belt cover and both upper timing belt covers.

27. Hold the crankshaft pulley with Handle tool 07JAB-001020A and Crankshaft Holding tool 07MAB-PY3010A or equivalent tools. While holding the crankshaft pulley, install the crankshaft pulley bolt using a heavy duty 3/4 in. (19mm) socket and a commercially available torque wrench and torque the bolt to 181 ft. lbs. (245 Nm).

✻✻ WARNING

If any binding is felt while moving the crankshaft pulley, STOP turning the crankshaft pulley immediately because the pistons may be hitting the valves.

28. Rotate the crankshaft pulley clockwise 5–6 revolutions to allow the timing belt to be seated in the pulleys.

29. Move the crankshaft pulley to the white TDC mark and inspect the camshaft TDC marks to ensure proper timing of the camshafts.

✻✻ WARNING

If the timing marks do not align, the timing belt removal and installation procedure must be performed again.

30. Install the engine dipstick tube using a new O-ring.

31. Install the power steering pump, and loosely install the mounting bolt, adjustment locknut and adjustment nut.

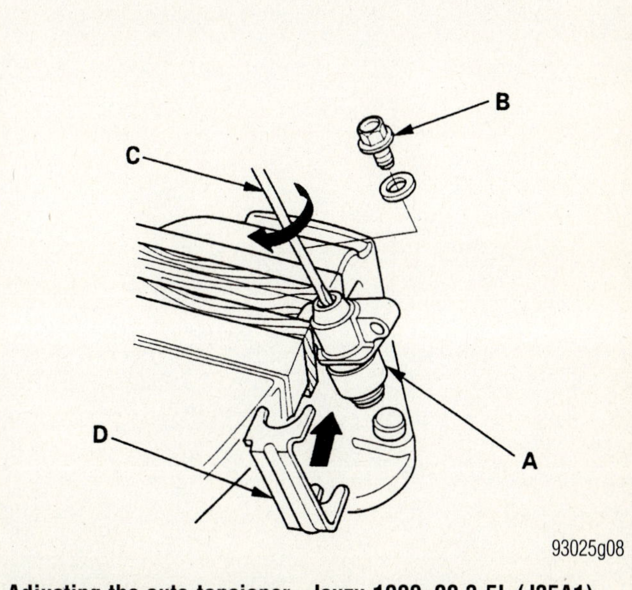

Adjusting the auto-tensioner—Isuzu 1999–00 3.5L (J35A1) engine—Oasis

32. Adjust the power steering belt to a tension such that a 22 lb. (98 N) pull halfway between the 2 drive pulleys will allow the belt to move 0.51–0.65 in. (13.0–16.5mm).

33. Tighten the power steering pump mounting bolt and adjustment locknut.

■ If a new belt is used, set the deflection to 0.33–0.43 in. (8.5–11.0mm) and after engine has run for 5 minutes, readjust the new belt to the used belt specification.

34. Install the alternator belt tensioner arm.

35. Move the alternator tensioner with a Belt Tensioner Release Arm tool YA9317 or equivalent, to release tension from the belt and install the alternator drive belt.

36. Install the 2 bolts for the engine mount bracket and torque to 33 ft. lbs. (44 Nm).

37. Install the bushing through bolt and tighten to 40 ft. lbs. (54 Nm).

38. Release and carefully remove the floor jack.

39. Install the front lower splash shield.

40. Install both front tires/wheels.

41. Carefully lower the vehicle.

42. Install the battery clamp bolt in the battery tray.

43. Install the negative battery cable.

44. Enter the radio security code.

KIA

1.5L (VIN 3) & 1.6L (VIN 5) DOHC ENGINES

YEAR, MODEL & ENGINE APPLICATION

2001-02 KIA Rio	1.5L VIN 3
2003-04 Rio	1.6L VIN 5

REPLACEMENT INTERVALS

KIA recommends timing belt replacement every 105,000 miles on California models, and 60,000 miles on all other models.

LABOR TIMES

2001-04	
Replacement	2.1 hrs
Severe Service	2.4 hrs
With air conditioning add	0.2 hrs
With power steering add	0.2 hrs

8807KIAg01

✱✱CAUTION

These engines are an interference type. If the timing belt has failed, valve-to-piston damage may have occurred. A compression check should always be performed to check for engine damage.

1. Disconnect the negative battery cable.
2. Remove the accessory drive belts.
3. Remove the water pump and crankshaft pulleys.
4. Remove the timing belt guide plate.
5. Remove the upper and lower timing belt covers.
6. Rotate the crankshaft until the camshaft sprocket and crankshaft timing marks are aligned.
7. Loosen the timing belt tensioner pulley bolt and remove the pulley.
8. Remove the timing belt.

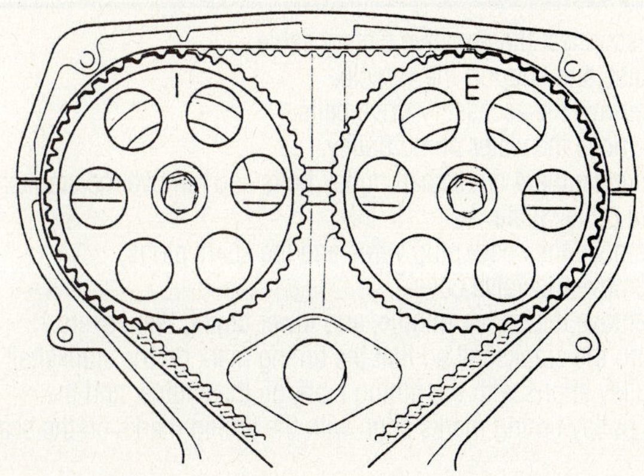

93015g03

Camshaft sprocket timing marks—KIA 1.5L (VIN 3) and 1.6L (VIN 5) engines

To install:

9. Install a new tensioner spring and install the tensioner pulley.
10. Move the tensioner away from the timing belt and tighten the bolt.
11. Ensure the timing marks are still aligned.
12. Install the timing belt over the crankshaft sprocket, guide pulley, exhaust camshaft, intake camshaft and tensioner pulley.
13. Ensure the belt is taut between sprockets on the non-tensioned side.
14. Loosen the tensioner bolt and allow the spring the tension the bolt.
15. Tighten the tensioner pulley bolt to 28–38 ft. lbs. (38–52 Nm).
16. Rotate the crankshaft 2 turns clockwise and align the timing marks.
17. Apply 22 lbs. of tension to the timing belt between the camshaft sprockets and ensure there is 0.39–0.50 in. of belt deflection. If deflection is not as specified, remove and reinstall the timing belt.
18. Install the upper and lower timing belt covers.
19. Install the timing belt guide plate.
20. Install the water pump and crankshaft pulleys.
21. Install the accessory drive belts.
22. Connect the negative battery cable.

1.6L (CODE B6) SOHC ENGINES

YEAR, MODEL & ENGINE APPLICATION

1994-97 KIA Sephia	1.6L Code B6

REPLACEMENT INTERVALS

KIA recommends timing belt replacement every 105,000 miles on California models, and 60,000 miles on all other models.

LABOR TIMES

1994-97	
Replacement	2.8 hrs
Severe Service	3.1 hrs
With air conditioning add	0.2 hrs
With power steering add	0.2 hrs

8807KIAg02

✱✱CAUTION

These engines are considered freewheeling. If the timing belt fails, valve-to-piston damage is unlikely, however a precautionary compression check should be performed to verify no engine damage.

1. Disconnect the negative battery cable.
2. Raise and support the vehicle.
3. Remove the right side wheel and inner fender panel.
4. Remove the accessory drive belts.
5. Remove the water pump pulley.
6. Remove the 4 crankshaft pulley bolts and remove the crankshaft pulley and pulley plate.
7. Remove the crankshaft pulley center bolt and remove the crankshaft pulley boss.
8. Remove the upper and lower timing belt covers.
9. Remove the spark plugs.

10. Rotate the crankshaft to align the crankshaft and camshaft timing marks.
11. Loosen the tensioner pulley lockbolt.
12. Pry the tensioner pulley back to remove the tension on the timing belt, and tighten the lockbolt.
13. Remove the timing belt.
14. Remove the tensioner lockbolt, pulley and spring.
15. Measure the spring free length. Free length should be 2.5 in. (63mm).

To install:

16. Insure that the crankshaft and camshaft timing marks are aligned.
17. Install the tensioner pulley and spring.
18. Fully extend the spring and install and tighten the lockbolt.
19. Install the timing belt in a counterclockwise direction starting at the crankshaft sprocket and ensure the belt is tensioned on the non-sprocket side.
20. Rotate the crankshaft 2 full turns clockwise and check that the timing marks on the crankshaft and camshafts align with the marks on the engine.
21.
22. Loosen the tensioner lockbolt and allow the tensioner spring to apply tension to the timing belt. Do not apply additional tension. Tighten the lockbolt to 14–19 ft. lbs. (19–25 Nm).

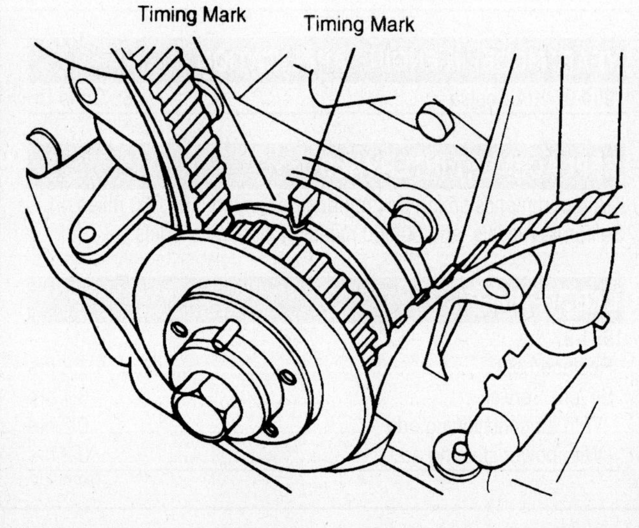

Crankshaft timing marks—KIA 1.6L (Code B6) SOHC engine

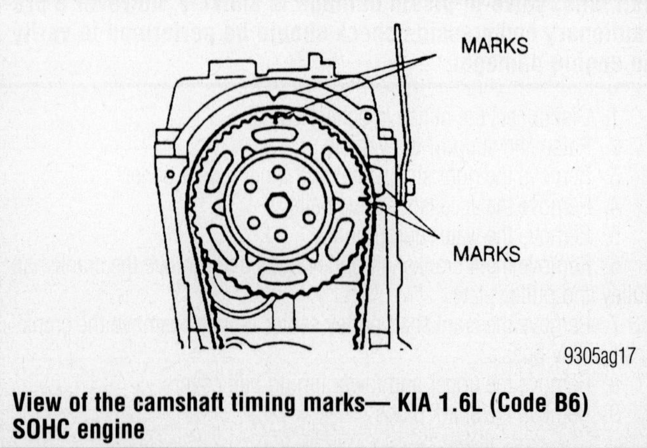

View of the camshaft timing marks— KIA 1.6L (Code B6) SOHC engine

23. Check timing belt deflection by applying 22 lbs. (10 kg) pressure to the belt midway between the crankshaft and camshaft pulleys. Belt deflection should be 0.43–0.51 inches. (11–13 Nm). If deflection is not as specified, reinstall the timing belt.
24. Install the upper and lower timing covers.
25. Install the crankshaft pulley boss and the center bolt. Hold the crankshaft from turning and tighten the center bolt to 116–122 ft. lbs. (157–166 Nm).
26. Install the water pump pulley.
27. Install the crankshaft pulley and pulley plate.
28. Install the accessory drive belts.
29. Install the spark plugs.
30. Install the right side wheel and inner fender panel.
31. Lower the vehicle.
32. Connect the negative battery cable.

1.6L (CODE B6) AND 1.8L (CODE BP) 1994–97 DOHC ENGINES

YEAR, MODEL & ENGINE APPLICATION	
1994-97 Kia Sephia	1.6L Code B6, 1.8L Code BP

REPLACEMENT INTERVALS

KIA recommends timing belt replacement every 105,000 miles on California models, and 60,000 miles on all other models.

LABOR TIMES	
1994-97	
Replacement	3.3 hrs
Severe Service	3.6 hrs
With air conditioning add	0.2 hrs
With power steering add	0.2 hrs

8807KIAg03

✻✻ CAUTION

These engines are an interference type. If the timing belt has failed, valve-to-piston damage may have occurred. A compression check should always be performed to check for engine damage.

1. Disconnect the negative battery cable.
2. Raise and support the vehicle.
3. Remove the accessory drive belts.
4. Remove the water pump pulley.
5. Remove the 4 crankshaft pulley bolts and remove the crankshaft pulley and pulley plate.
6. Remove the spark plug wires and the spark plugs.
7. Remove the valve cover.
8. Remove the upper, middle, and lower timing belt covers.
9. Turn the crankshaft so that the timing mark on the crankshaft timing pulley aligns with the timing mark on the engine, and the camshaft pulley timing marks align with the timing marks on the seal plate.
10. Loosen the tensioner pulley lockbolt.
11. Pry the tensioner pulley back to remove the tension on the timing belt, and tighten the lockbolt.
12. Remove the crankshaft pulley center bolt and remove the crankshaft pulley boss.

13. Remove the timing belt.
14. Measure the tensioner spring free length. Free length should be 2.31 in. (58.6mm) on 1.6L engines and 2.33 in. (59.1mm) on 1.8L engines.

To install:

15. Insure that the crankshaft and camshaft timing marks are aligned.
16. Install the tensioner pulley and spring.
17. Fully extend the spring and install and tighten the lockbolt.

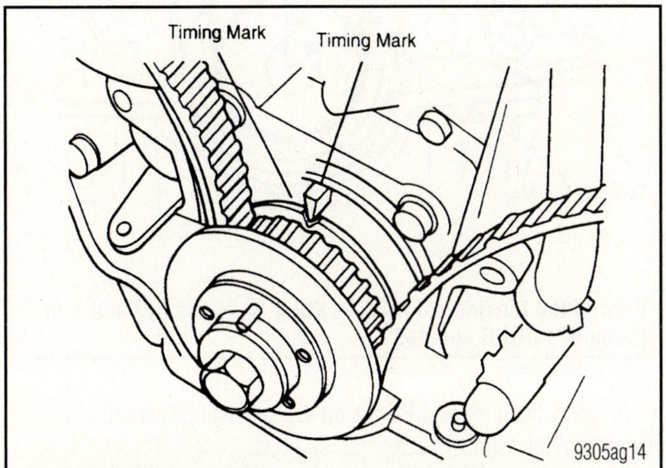

Crankshaft timing marks— KIA 1.6L (Code B6) and 1.8L (Code BP) DOHC engines

18. Install the timing belt in a counterclockwise direction starting at the crankshaft sprocket and ensure the belt is tensioned on the non-sprocket side.
19. Install the crankshaft pulley boss and the center bolt. Hold the crankshaft from turning and tighten the center bolt to 116–122 ft. lbs. (157–166 Nm).
20. Loosen the tensioner lockbolt and allow the tensioner spring to apply tension to the timing belt. Do not apply additional tension. Tighten the lockbolt to 28–38 ft. lbs. (38–51 Nm).
21. Rotate the crankshaft 2 full turns clockwise and check that the timing marks on the crankshaft and camshafts align with the marks on the engine and seal plate.

■ **Never rotate the crankshaft counterclockwise.**

22. Rotate the crankshaft 1 5/6 turns and align the crankshaft timing mark with the tension set mark.
23. Check the timing belt tension by applying 22 lbs. (10 kg) of force to the belt between the camshaft pulleys. Timing belt deflection should be 0.36–0.45 inches. (10–11 Nm).
24. Install the lower, middle, and upper timing covers.
25. Install the valve cover, spark plugs, and spark plug wires.
26. Install the water pump pulley.
27. Install the crankshaft pulley and pulley plate.
28. Install the accessory drive belts.
29. Lower the vehicle.
30. Connect the negative battery cable.

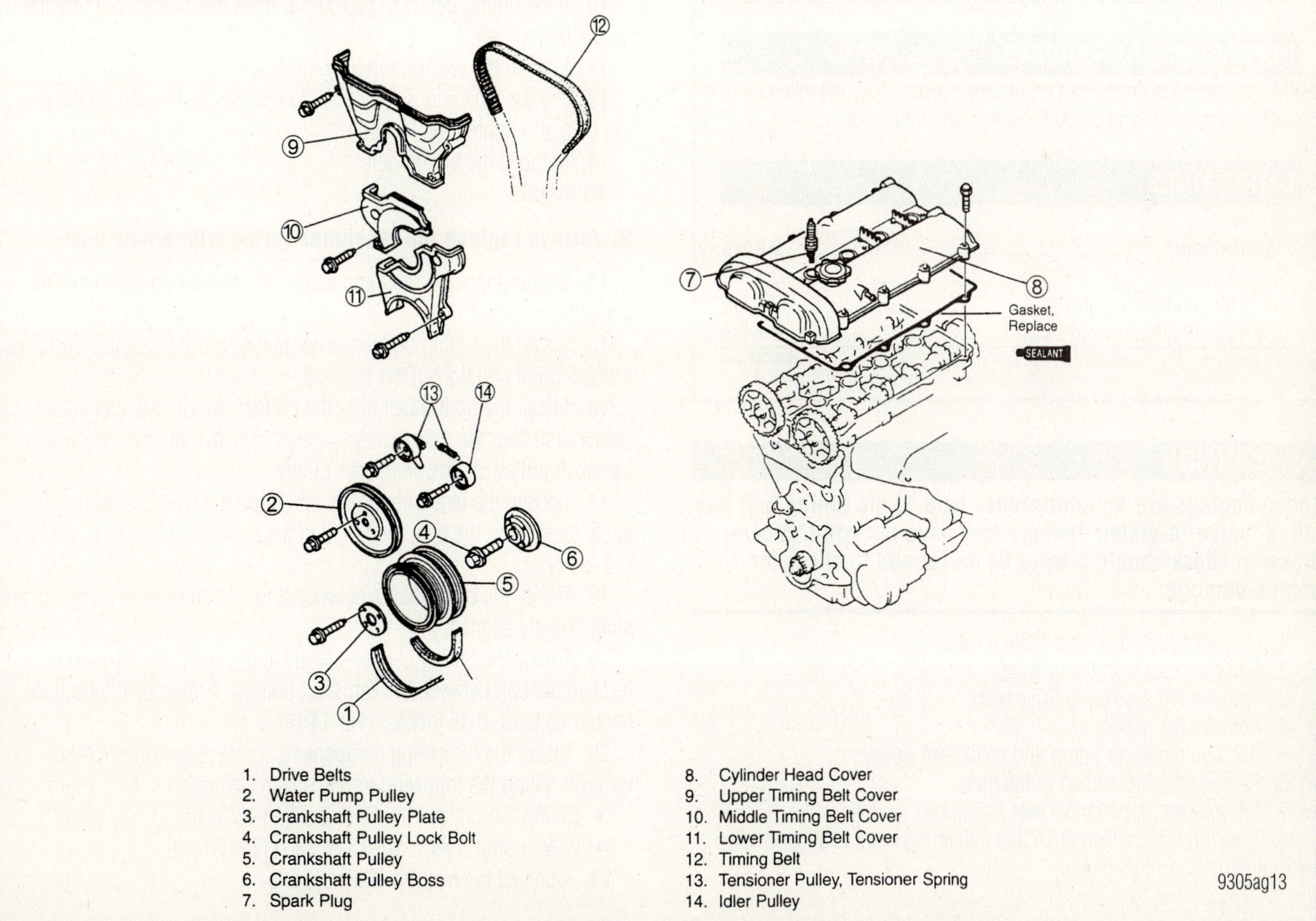

Exploded view of the timing belt assembly—KIA 1.6L (Code B6) and 1.8L (Code BP) DOHC engines

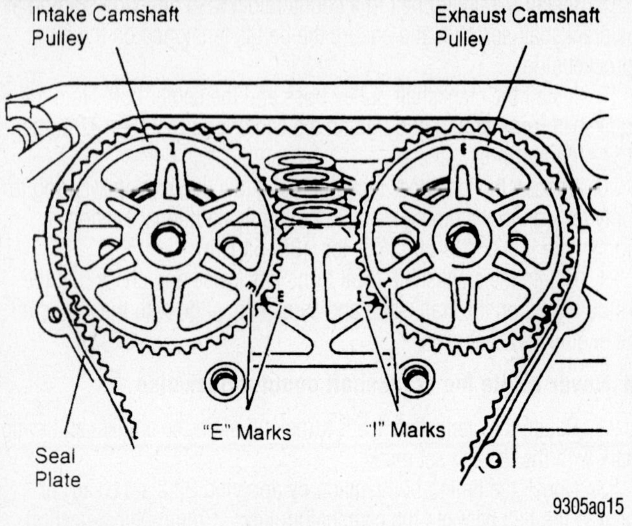

Camshaft timing marks location— KIA 1.6L (Code B6) and 1.8L (Code BP) DOHC engines

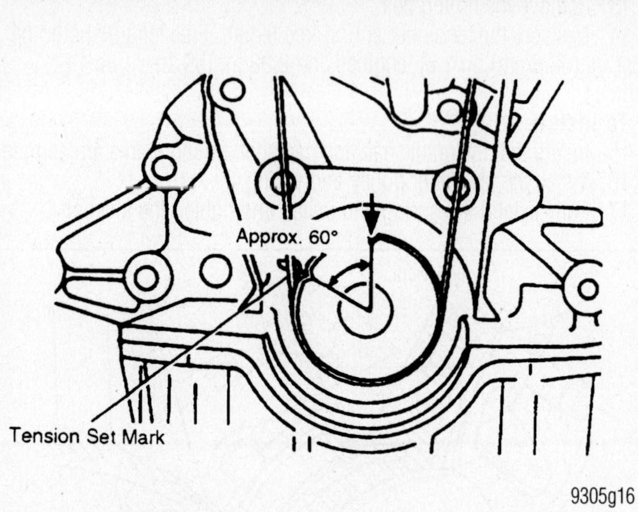

View of the tension set mark— KIA 1.6L (Code B6) and 1.8L (Code BP) DOHC engines

1.8L (VIN 1) DOHC ENGINES

YEAR, MODEL & ENGINE APPLICATION

1998-01 KIA Sephia, 2002-04 Spectra	1.8L VIN 1

REPLACEMENT INTERVALS

KIA recommends timing belt replacement every 105,000 miles on California models, and 60,000 miles on all other models.

LABOR TIMES

1994-97	
Replacement	2.8 hrs
Severe Service	3.1 hrs
With air conditioning add	0.2 hrs
With power steering add	0.2 hrs

8807KIAg04

✻✻ CAUTION

These engines are an interference type. If the timing belt has failed, valve-to-piston damage may have occurred. A compression check should always be performed to check for engine damage.

1. Disconnect the negative battery cable.
2. Raise and support the vehicle.
3. Remove the accessory drive belts.
4. Remove the generator.
5. Remove the water pump and crankshaft pulleys.
6. Remove the timing belt guide plate.
7. Remove the upper and lower timing belt covers.
8. Position the crankshaft so that the timing mark is aligned with the timing mark on the engine.
9. Verify that the **I** and **E** mark on the camshaft pulleys align with the mark on the cylinder head.

■ **Do not move the crankshaft or camshaft once the timing marks have been correctly positioned.**

10. If the timing belt is to be reused, mark the direction of rotation on the timing belt.
11. Loosen the tensioner pulley lockbolt.
12. Pry the tensioner pulley back to remove the tension on the timing belt, and tighten the lockbolt.
13. Remove the timing belt.

To install:

■ **Always replace the tensioner spring with a new one.**

14. Ensure the crankshaft and camshaft sprocket timing marks are aligned.
15. Install the timing belt tensioner pulley, move the tensioner to its furthest point and tighten the lockbolt.
16. Install the timing belt onto the pulleys, as follows, crankshaft pulley first, then the idler pulley, exhaust camshaft pulley, intake camshaft pulley and the tensioner pulley.
17. Loosen the tensioner pulley and allow the tensioner spring to apply tension on the belt, then tighten the lockbolt to 28–38 ft. lbs. (38–51 Nm).
18. Rotate the crankshaft clockwise 2 turns and be sure all marks are still correctly aligned.
19. Check the timing belt tension by applying 22 lbs. (10 kg) of force to the belt between the camshaft pulleys. Timing belt deflection should be 0.36–0.45 inches. (10–11 Nm).
20. Install the remaining components in the revere order of the removal noting the following torque specifications:
    - Crankshaft pulley: 10–13 ft. lbs. (13–17 Nm)
    - Water pump pulley: 10–13 ft. lbs. (13–17 Nm)
21. Connect the negative battery cable.

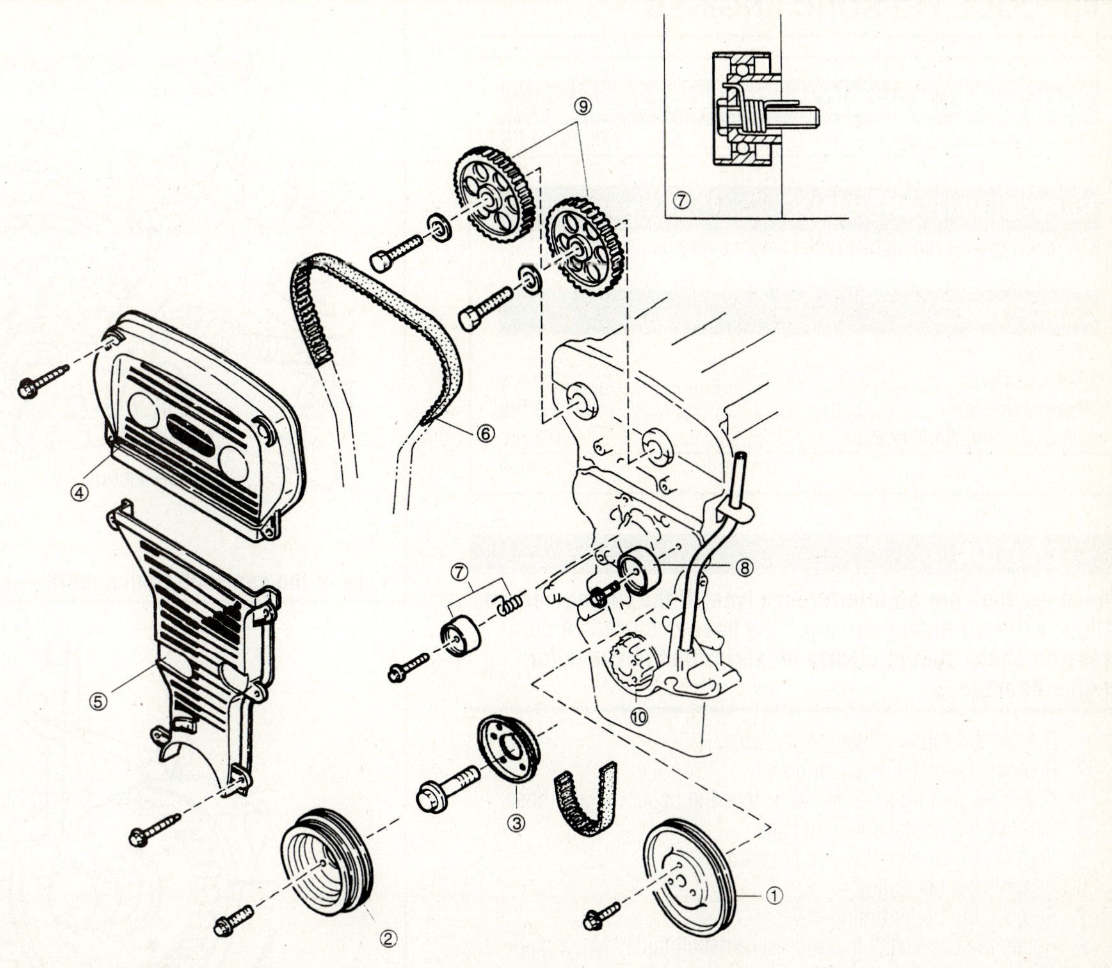

(1) Water pump pulley
(2) Crankshaft pulley
(3) Timing belt guide plate
(4) Timing belt cover (Upper)
(5) Timing belt cover (Lower)
(6) Timing belt
(7) Timing belt tensioner pulley & spring
(8) Idler pulley
(9) Camshaft pulley
(10) Timing belt pulley

93015g01

Exploded view of the timing belt cover mounting and related components—KIA 1.8L (VIN 1) 1998–04 engine

93015g02

Crankshaft sprocket timing belt alignment mark—KIA 1.8L (VIN 1) 1998–04 engine

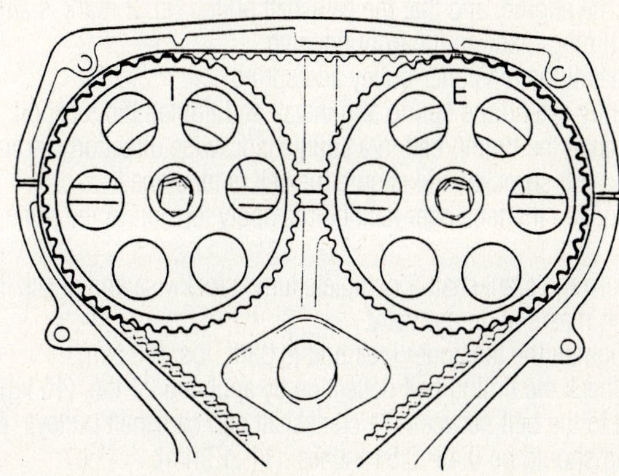

93051g03

Camshaft sprocket positioning for timing belt installation—KIA 1.8L (VIN 1) 1998–04 engine

2.0L (CODE FE) SOHC ENGINES

YEAR, MODEL & ENGINE APPLICATION	
1995-99 KIA Sportage	2.0L Code FE

REPLACEMENT INTERVALS
KIA recommends timing belt replacement every 60,000 miles.

LABOR TIMES	
1995-99	
Replacement	2.0 hrs
Severe Service	2.3 hrs
With air conditioning add	0.3 hrs

8807KIAg05

✻✻ CAUTION

These engines are an interference type. If the timing belt has failed, valve-to-piston damage may have occurred. A compression check should always be performed to check for engine damage.

1. Disconnect the negative battery cable.
2. Remove the intake fresh air duct.
3. Drain the cooling system and remove the upper radiator hose.
4. Remove the accessory drive belts.
5. Remove the cooling fan and fan shroud.
6. Remove the fan pulley.
7. Remove the upper timing cover.
8. Rotate the crankshaft to align the camshaft pulley No. **2** mark with the timing mark on the front housing.
9. Remove the crankshaft pulley and belt guide plate.
10. Remove the lower timing cover.
11. Loosen the tensioner lock bolt.
12. Pry the tensioner away from the timing belt and tighten the lock bolt.
13. Remove the timing belt.
14. Remove the tensioner spring and measure the spring free length. Free length should be 2.24 in. (56.9mm).

To install:

15. Insure that the crankshaft timing mark is aligned with the timing mark on the engine, and that the camshaft pulley No. **2** mark is aligned with the timing mark on the front housing.
16. Install the tensioner pulley and spring.
17. Fully extend the spring and install and tighten the lockbolt.
18. Install the timing belt in a counterclockwise direction starting at the crankshaft sprocket and ensure the belt is tensioned.
19. Loosen the tensioner lock bolt to apply tension to the timing belt.
20. Turn the crankshaft 2 complete turns clockwise and check that the timing marks align correctly.
21. Tighten the tensioner lock bolt to 33 ft. lbs. (45 Nm).
22. Check the timing belt deflection by applying 22 lbs. (10 kg) pressure to the belt between the crankshaft and camshaft pulleys. Belt deflection should be 0.43–0.51 inches. (11–13mm).
23. Install the upper and lower timing covers.
24. Install the crankshaft pulley and tighten the bolts to 11 ft. lbs. (15 Nm).

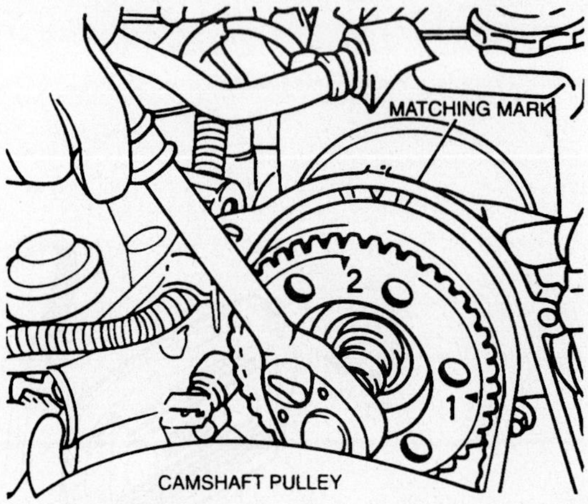

9305ag18

View of the camshaft timing marks—KIA 2.0L SOHC engine

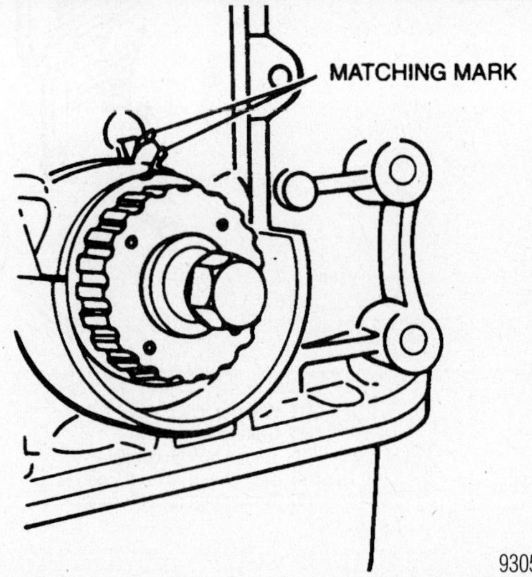

9305ag19

View of the crankshaft timing mark—KIA 2.0L SOHC engine

25. Install the cooling fan pulley. Install the cooling fan and shroud.
26. Install the accessory drive belts.
27. Install the upper radiator hose.
28. Install the intake fresh air duct.
29. Connect the negative battery cable and fill the cooling system.

2.0L (VIN 3) DOHC ENGINES

✻✻ CAUTION

These engines are an interference type. If the timing belt has failed, valve-to-piston damage may have occurred. A compression check should always be performed to check for engine damage.

1. Disconnect the negative battery cable.
2. Properly relieve the fuel system pressure.
3. Remove the alternator drive belt.

YEAR, MODEL & ENGINE APPLICATION

1995-02 KIA Sportage	2.0L VIN 3

REPLACEMENT INTERVALS

KIA recommends timing belt replacement every 105,000 miles on California models, and 60,000 miles on all other models.

LABOR TIMES

1995-02	
Replacement	2.8 hrs
Severe Service	3.1 hrs
With air conditioning add	0.2 hrs
With power steering add	0.3 hrs

8807KIAg06

4. Remove the fresh air duct from the top of the radiator.
5. Remove the upper radiator hose.
6. Remove the 4 attaching nuts to the clutch fan.
7. Remove the 5 fan shroud bolts. Remove the fan and shroud as an assembly.
8. Remove the 4 splash guard mounting bolts and the splash guard.
9. Loosen the lockbolts and loosen the air conditioning drive belt.
10. Loosen the power steering lock and mounting bolt. Remove the power steering belt.
11. Remove the 5 upper timing belt cover bolts and remove the cover.
12. Remove the 2 lower timing belt cover bolts and remove the cover.
13. Align the timing marks.

■ **When aligning the cam pulleys with the seal plate marks, align the left cam pulley I mark and the right cam pulley on the E mark.**

✻✻ WARNING

When aligning the timing marks, do not turn the timing gear counterclockwise. Damage to the engine will occur.

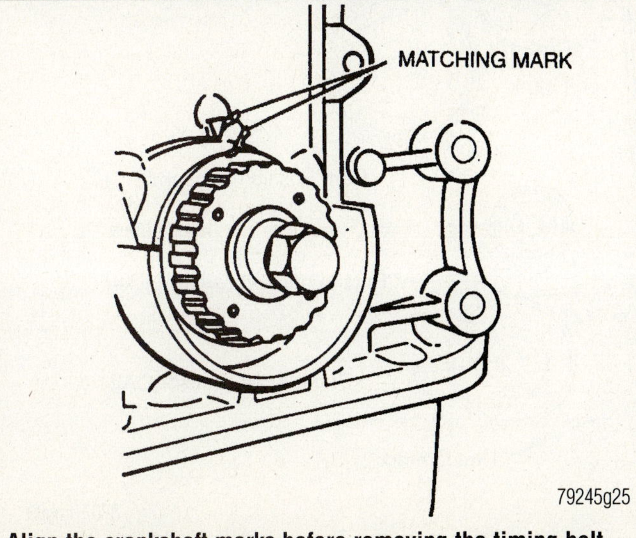

Align the crankshaft marks before removing the timing belt—KIA 2.0L DOHC engine

14. Loosen the tensioner bolt. Pry the tensioner away from the belt. Tighten the tensioner bolt to relieve the pressure against the timing belt.
15. Remove the timing belt.
16. Remove the lower timing belt pulley and locking bolt.

To install:

17. Install the lower timing belt pulley and locking bolt. Tighten the bolt to 120 ft. lbs. (162 Nm).
18. If necessary, align the timing marks.

■ **When aligning the cam pulleys with the seal plate marks, align the left cam pulley I mark and the right cam pulley on the E mark.**

✻✻ WARNING

When aligning the timing marks, do not turn the timing gear counterclockwise. Damage to the engine will occur.

19. Loosen the tensioner bolt. Pry the tensioner away from the belt. Tighten tensioner bolt to relieve the pressure against the timing belt.
20. Install the timing belt.

✻✻ WARNING

If any binding is felt when adjusting the timing belt tension by turning the crankshaft, STOP turning the engine, because the pistons may be hitting the valves.

21. Loosen the tensioner bolt and allow the tensioner to tighten the timing belt. Tighten the tensioner bolt 27–38 ft. lbs. (37–52 Nm).
22. Rotate the crankshaft sprocket 2 turns in the normal rotation direction.
23. Check the timing belt deflection. If there is more than 0.30–0.33 in. (7.5–8.5mm) replace the tensioner spring.
24. Install the 2 lower timing belt cover bolts to the cover.
25. Install the 5 upper timing belt cover bolts to the cover.
26. Install and adjust the air conditioning and power steering drive belts.
27. Install the splash guard.
28. Install and tighten the alternator belt.
29. Install the upper radiator hose.
30. Install the fan and shroud as an assembly.

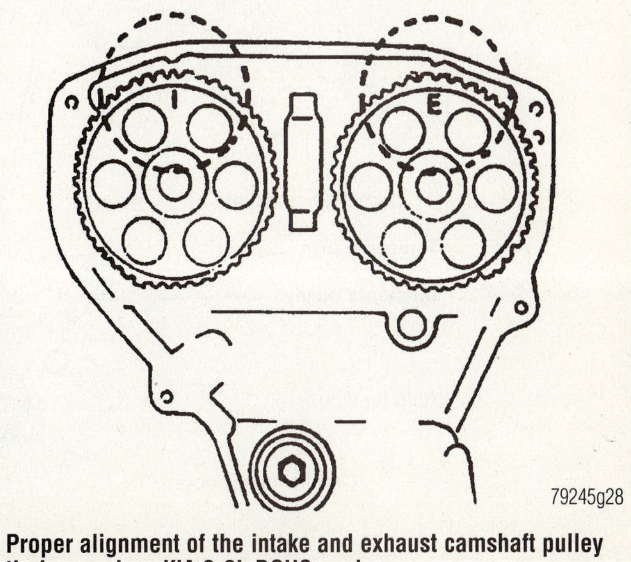

Proper alignment of the intake and exhaust camshaft pulley timing marks—KIA 2.0L DOHC engine

31. Install the 4 attaching nuts to the clutch fan.
32. Install the 5 fan shroud bolts.
33. Install the fresh air duct to the top of the radiator.
34. Properly fill the cooling system.
35. Connect the negative battery cable.
36. Start the engine and check for leaks.
37. Road test the vehicle.

2.4L (VIN S) ENGINES

YEAR, MODEL & ENGINE APPLICATION	
2001-04 KIA Optima	2.4L VIN S

REPLACEMENT INTERVALS

KIA recommends timing belt replacement every 105,000 miles on California models, and 60,000 miles on all other models.

LABOR TIMES	
2001-04	
Replacement	2.8 hrs
Severe Service	3.1 hrs
With air conditioning add	0.2 hrs
With power steering add	0.2 hrs

8807KIAg07

✱✱ CAUTION

These engines are an interference type. If the timing belt has failed, valve-to-piston damage may have occurred. A compression check should always be performed to check for engine damage.

1. Disconnect the negative battery cable.
2. Remove the accessory drive belts.
3. Remove the crankshaft pulley.
4. Remove the water pump pulley.
5. Raise and support the vehicle.
6. Remove the right side engine mount.
7. Remove the upper and lower timing belt covers.
8. Rotate the crankshaft clockwise until No. 1 cylinder is at TDC on the compression stroke. The camshaft sprockets should have the dowel pin is facing directly up.
9. Make a mark on the back of the timing belt indicating the direction of rotation so it may be reassembled in the same direction, if to be reused.
10. Remove the timing belt tensioner bolts and the tensioner.
11. Remove the timing belt.
12. Remove the crankshaft sprocket and the flange.
13. Remove the silent shaft belt tensioner pulley and silent shaft belt.

To install:

14. Measure the tensioner pushrod protrusion which should be 0.57 in. (15mm). If protrusion is not as specified, replace the tensioner.
15. Ensure the silent shaft sprocket and crankshaft sprocket marks are aligned.
16. Install the silent shaft belt tensioner pulley and hand tighten the bolt.
17. Install the silent shaft belt, keeping it taut on the tension side.
18. Rotate the tensioner pulley against the belt and tighten the bolt to 11–16 ft. lbs. (15–22 Nm).
19. Check that the belt deflection is 0.20–0.30 in. when pressed by hand.
20. Install the crankshaft sprocket and flange and tighten to 80–94 ft. lbs. (108–127 Nm).
21. Prepare the timing belt auto tensioner for installation by compressing it in a vise and installing a retaining pin.

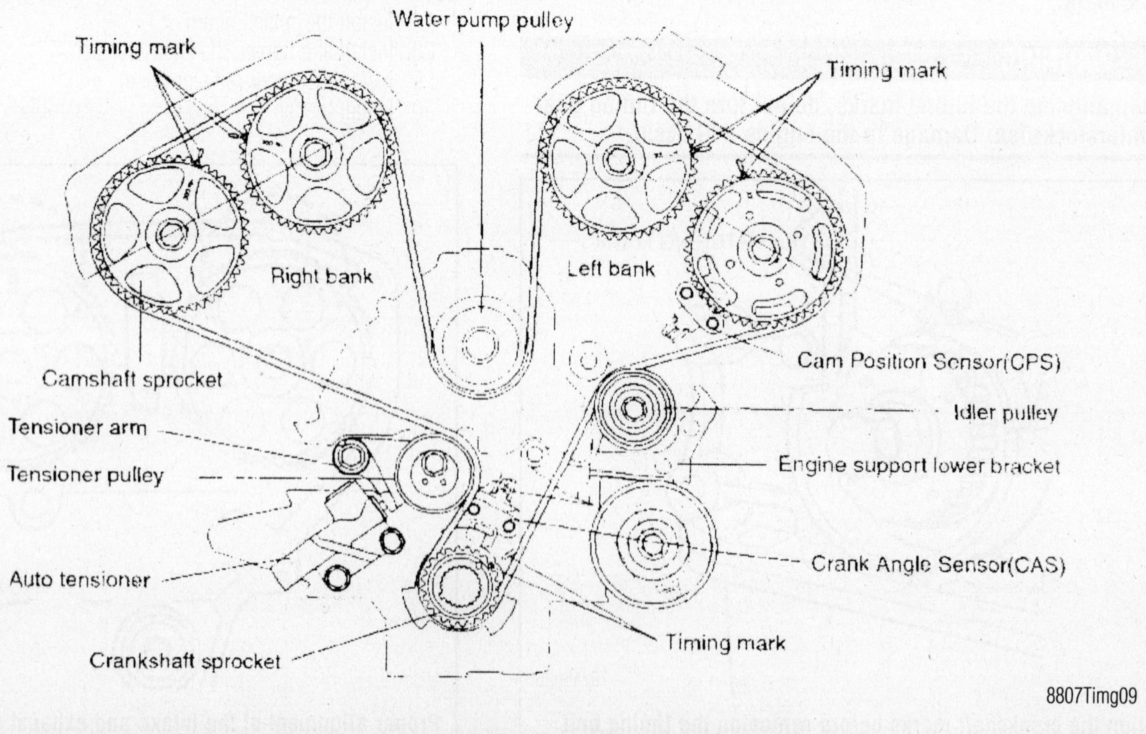

Crankshaft and camshaft sprocket timing belt alignment mark locations—KIA 2.4L (VIN S) engine

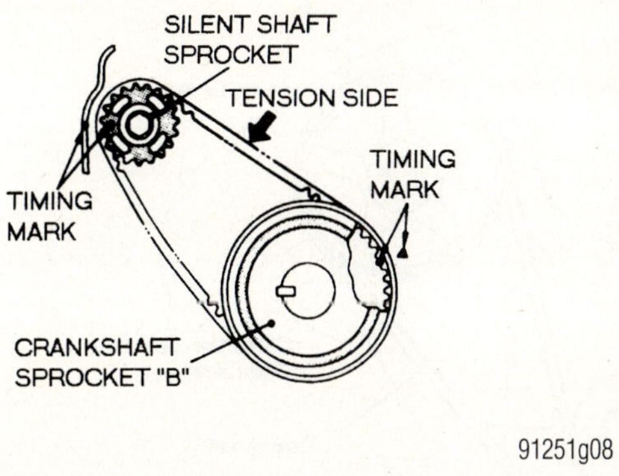

Silent shaft belt timing mark locations—KIA 2.4L (VIN S) engine

22. Install the auto tensioner. Do not remove the retaining pin at this time.
23. Check that all timing marks are still aligned.
24. Loosen the tensioner pulley bolt.
25. Remove the plug located on the front of the cylinder block.
26. Insert a tool having a shaft diameter of 0.3 in. (8mm) into the hole.
27. With the timing marks still aligned, the tool must be able to go in at least 2 1/3 in. If it can only go in about 1 in. turn the oil pump sprocket 1 complete revolution.
28. Recheck the position of the balance shaft with the timing marks realigned. Leave the tool in place to hold the silent shaft while continuing.
29. Install the timing belt on the tensioner pulley, crankshaft sprocket, oil pump sprocket, guide pulley, right camshaft sprocket and left camshaft sprocket.
30. Rotate the tensioner pulley against the belt and hand tighten the bolt.
31. Remove the pin from the tensioner.
32. Rotate the crankshaft clockwise 2 turns and check that the timing marks are aligned.
33. Wait 15 minutes for the parts to settle.
34. Check the protrusion of the auto-tensioner pushrod. If protrusion is not 0.24–0.35 in. (6–9mm), repeat the procedure.
35. Install the upper and lower timing belt covers.
36. Install the right side engine mount.
37. Lower the vehicle.
38. Install the water pump pulley.
39. Install the crankshaft pulley.
40. Install the accessory drive belts.
41. Connect the negative battery cable.

2.5L AND 2.7L (VIN 8) ENGINES

✱✱ CAUTION

These engines are an interference type. If the timing belt has failed, valve-to-piston damage may have occurred. A compression check should always be performed to check for engine damage.

1. Disconnect the negative battery cable.
2. Remove the engine appearance cover.

YEAR, MODEL & ENGINE APPLICATION

2001-04 KIA Optima	2.5L & 2.7L VIN 8

REPLACEMENT INTERVALS

KIA recommends timing belt replacement every 105,000 miles on California models, and 60,000 miles on all other models.

LABOR TIMES

2001-04	
Replacement	2.8 hrs
Severe Service	3.1 hrs
With air conditioning add	0.2 hrs
With power steering add	0.3 hrs

8807KIAg08

3. Remove the accessory drive belt.
4. Remove the power steering pump pulley, water pump pulley, idler pulley, tensioner pulley and crankshaft pulley.
5. Remove the upper and lower timing belt covers.
6. Raise and support the vehicle.
7. Remove the right side engine mount and bracket.
8. Rotate the crankshaft clockwise until No. 1 cylinder is at TDC on the compression stroke and the camshaft sprocket timing marks are aligned.
9. Remove the timing belt auto-tensioner.
10. Remove the tensioner and timing belt.

To install:

11. Ensure that the engine is set to Top Dead Center (TDC).
12. Prepare the auto tensioner for installation by compressing it in a vise and installing a retaining pin.
13. Install the auto tensioner. Do not remove the retaining pin at this time.
14. Install the timing belt in the following order:
    a. Crankshaft sprocket
    b. Idler pulley
    c. Left bank camshaft sprocket
    d. Water pump pulley
    e. Right bank camshaft sprocket
    f. Tensioner pulley
15. Ensure the timing belt is taut on the non-tension side.
16. Remove the auto tensioner retaining pin.
17. Turn the tensioner pulley clockwise so it tightens the timing belt and tighten the bolt.
18. Rotate the crankshaft 2 revolutions **Clockwise**, then wait 5 minutes for the auto tensioner to adjust.
19. Measure the auto tensioner pushrod protrusion length. If the measurement is not 0.24–0.31 in. (6–8mm), repeat the belt tensioning procedure.
20. Install the engine mount.
21. Install the upper and lower timing belt covers.
22. Install the accessory drive belt guide and tensioner pulley.
23. Install the power steering pump pulley.
24. Install the crankshaft pulley.
25. Install the water pump pulley.
26. Install the accessory drive belt.
27. Install the engine appearance cover.
28. Connect the negative battery cable.

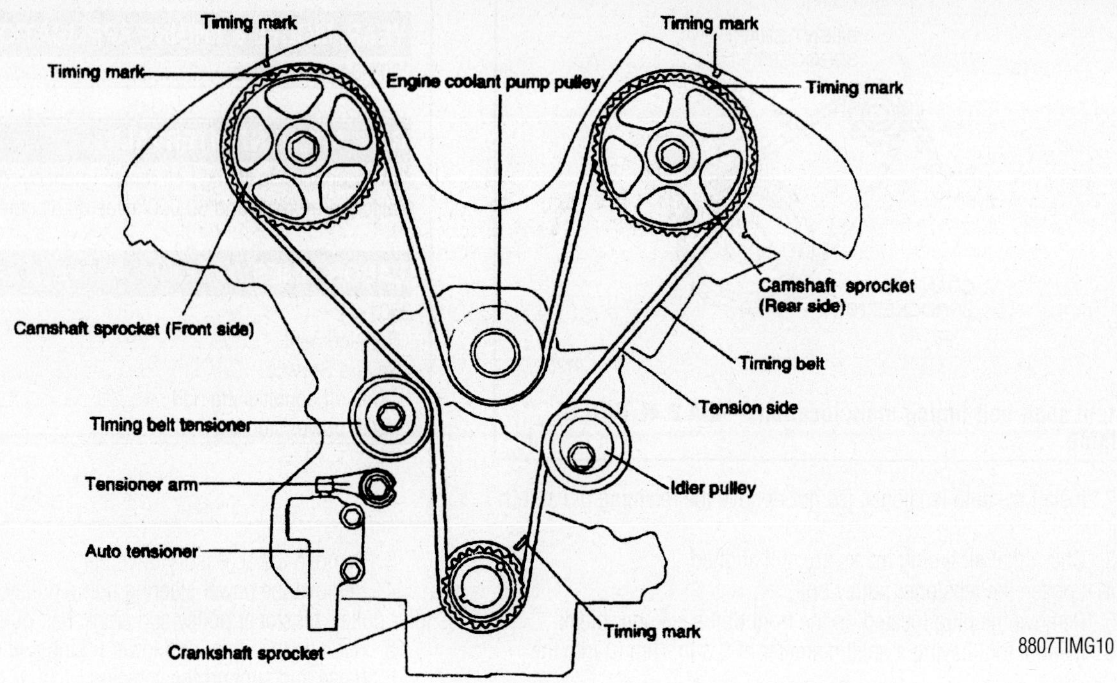

Crankshaft and camshaft sprocket timing belt alignment mark locations—KIA 2.5L and 2.7L (VIN 8) engine

3.5L (VIN 1, 3) ENGINES

YEAR, MODEL & ENGINE APPLICATION	
2002-04 KIA Sedona	3.5L VIN 1
2003-04 Sorento	3.5L VIN 3

REPLACEMENT INTERVALS

KIA recommends timing belt replacement every 105,000 miles on California models, and 60,000 miles on all other models.

LABOR TIMES	
2002-04	
Replacement	3.5 hrs
Severe Service	3.9 hrs
With air conditioning add	0.3 hrs

8807KIAg09

✳✳ CAUTION

These engines are an interference type. If the timing belt has failed, valve-to-piston damage may have occurred. A compression check should always be performed to check for engine damage.

1. Disconnect the negative battery cable.
2. Remove the engine appearance cover.
3. Remove the accessory drive belts.
4. Remove the power steering pump pulley.
5. Remove the accessory drive belt guide and tensioner pulleys.
6. Remove the crankshaft pulley.
7. Remove the upper and lower timing belt covers.
8. Support the engine with a floor jack and remove the engine mount and bracket.
9. Rotate the engine to align the camshaft sprocket timing marks with the cylinder head cover timing marks.
   a. Remove the timing belt auto tensioner.
   b. Remove the timing belt.

To install:

10. Ensure that the engine is set to Top Dead Center (TDC).
11. Prepare the auto tensioner for installation by compressing it in a vise and installing a retaining pin.
12. Install the auto tensioner. Do not remove the retaining pin at this time.
13. Install the timing belt in the following order:
   a. Crankshaft sprocket
   b. Idler pulley
   c. Left bank exhaust camshaft sprocket
   d. Left bank intake camshaft sprocket
   e. Water pump pulley
   f. Right bank intake camshaft sprocket
   g. Right bank exhaust camshaft sprocket
   h. Tensioner pulley
14. Ensure the timing belt is taut on the non-tension side.
15. Remove the auto tensioner retaining pin.
16. Turn the tensioner pulley clockwise so it tightens the timing belt and tighten the bolt.
17. Check that the crankshaft and camshaft timing marks are aligned correctly.
18. Rotate the crankshaft 1/4 turn **counterclockwise**.
19. Rotate the crankshaft 1/4 turn **clockwise** to return the engine to TDC.
20. Loosen the tensioner pulley center bolt.
21. Apply 44 inch lbs. (5 Nm) torque to the tensioner pulley as shown and tighten the center bolt to 32–41 ft. lbs. (43–55 Nm).

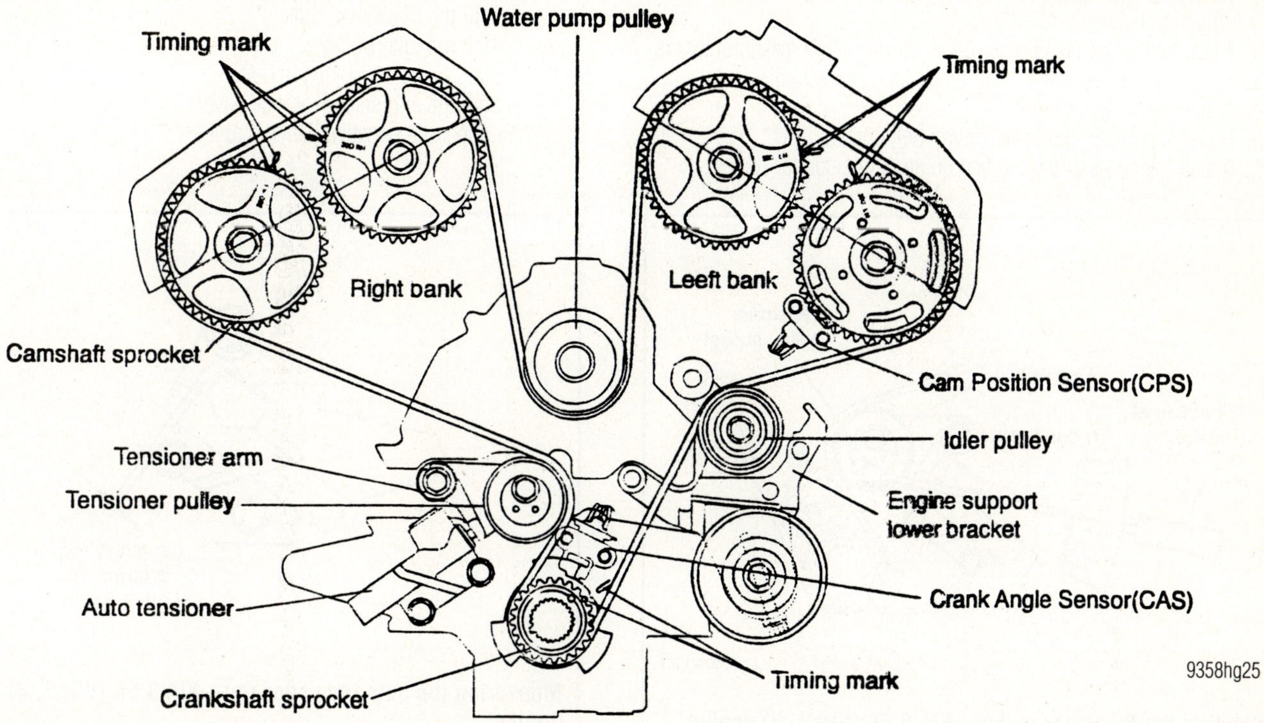

Timing belt routing and timing marks—KIA 3.5L (VIN 1, 3) engine

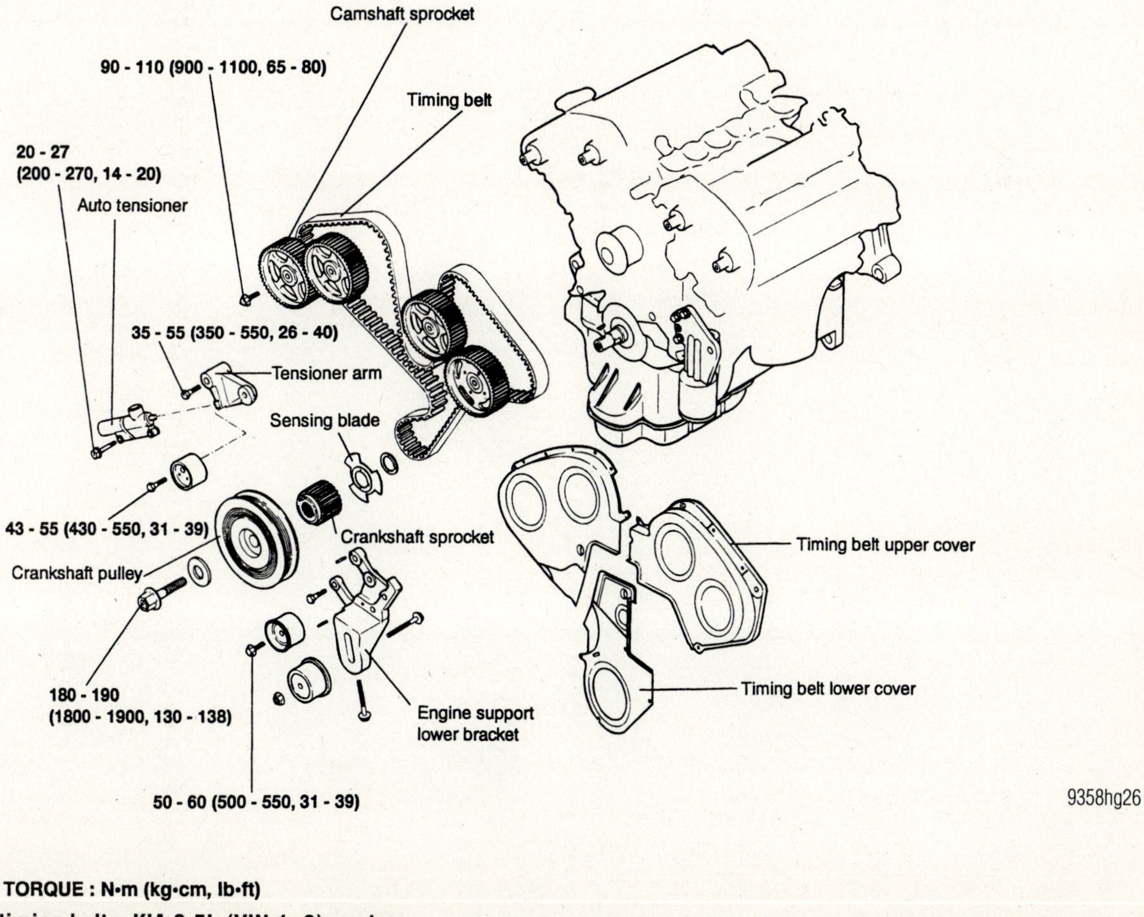

Exploded view of timing belt—KIA 3.5L (VIN 1, 3) engine

22. Rotate the crankshaft 2 revolutions **Clockwise**, then wait 5 minutes for the auto tensioner to adjust.

23. Measure the auto tensioner rod as shown. If the measurement is not 3.8-4.5 mm, then repeat the belt tensioning procedure.

24. Install the engine mount.

25. Install the upper and lower timing belt covers.

26. Install the accessory drive belt guide and tensioner pulley.

27. Install the power steering pump pulley.

28. Install the crankshaft pulley.

29. Install the idler pulley.

30. Install the accessory drive belts.

31. Install the engine appearance cover.

32. Connect the negative battery cable.

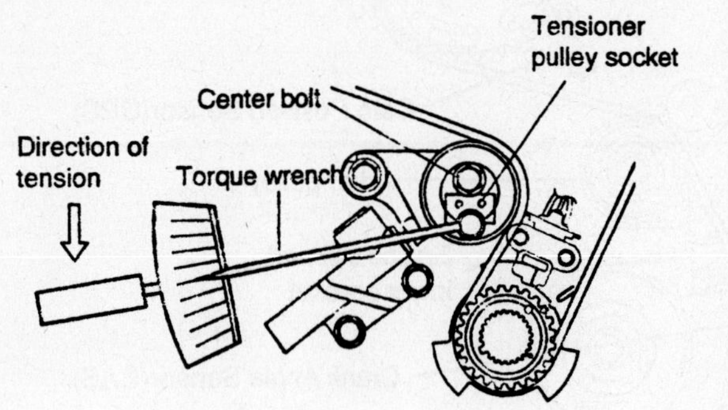

9358hg27

Adjusting the tensioner pulley—KIA 3.5L (VIN 1, 3) engine

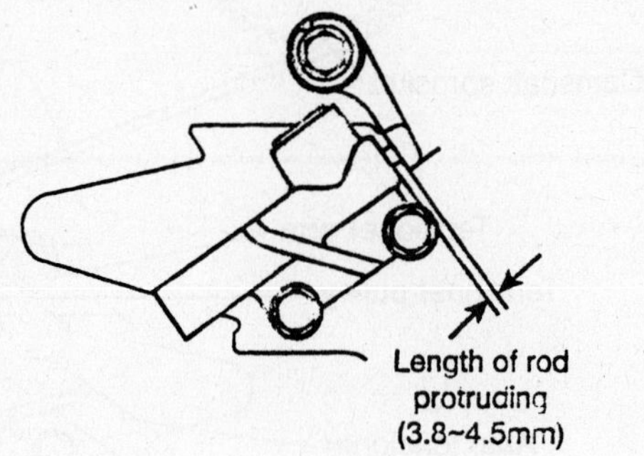

9358hg28

Measuring the auto tensioner rod—KIA 3.5L (VIN 1, 3) engine

Land Rover

2.5L (VIN 2, G) ENGINES

YEAR, MODEL & ENGINE APPLICATION	
2001-04 Land Rover Freelander	2.5L VIN 2, G

REPLACEMENT INTERVALS

Land Rover recommends timing belt replacement every 72,000 miles.

LABOR TIMES

2001-04	
Replacement	8.0 hrs
Severe Service	8.5 hrs

8807LANDg01

✱✱ CAUTION

These engines are an interference type. If the timing belt has failed, valve-to-piston damage may have occurred. A compression check should always be performed to check for engine damage.

Right Side Front Cover

1. Disconnect battery ground lead.
2. Remove auxiliary drive belt.
3. Remove 3 bolts securing camshaft timing belt right side front cover and remove cover.

To install:

4. Clean right side front timing belt cover.
5. Position right side front timing belt cover, fit and tighten bolts to 4 Nm (36 inch lbs.).
6. Fit auxiliary drive belt.
7. Connect battery ground lead.

Left Side Front Cover

8. Disconnect battery ground lead.
9. Remove auxiliary drive belt.
10. Remove 3 Torx screws securing power-assist steering (PAS) pump pulley, and remove pulley.
11. Remove Torx screw securing idler pulley to front mounting plate and remove pulley.
12. Remove bolt securing dipstick tube to cylinder block.
13. Remove 3 bolts securing camshaft timing belt left side front cover and remove cover.

To install:

14. Clean left side front timing belt cover.
15. Position left side front timing belt cover, fit and tighten bolts to 3 ft. lbs. (4 Nm).
16. Position dipstick tube bracket to cylinder block and tighten bolt to 7 ft. lbs. (9 Nm).
17. Clean idler pulley, position pulley to engine front mounting plate and tighten Torx bolt to 33 ft. lbs. (45 Nm).
18. Position PAS pump pulley, fit and tighten Torx screws to 7 ft. lbs. (9 Nm).
19. Fit auxiliary drive belt.
20. Connect battery ground lead.

RIGHT SIDE REAR COVER

21. Disconnect battery ground lead.
22. Remove intake manifold chamber.
23. Remove 2 bolts securing right side rear timing belt cover heat shield to cylinder head and remove heat shield.
24. Remove 3 bolts securing right side rear cover and remove cover.

To install:

25. Clean right side rear timing belt cover.
26. Position right side rear cover, fit bolts and tighten to 3 ft. lbs. (4 Nm).
27. Position heat shield, fit bolts and tighten M6 bolt to 7 ft. lbs. (9 Nm) and M8 bolt to 18 ft. lbs. (25 Nm).
28. Fit intake manifold chamber.
29. Connect battery ground lead.

Left Side Rear Cover

30. Disconnect battery ground lead.
31. Remove engine acoustic cover.
32. Remove 3 bolts securing camshaft timing belt left side rear cover and remove cover.

To install:

33. Clean left side rear timing belt cover.
34. Position left side rear cover, fit bolts and tighten to 3 ft. lbs. (4 Nm).
35. Fit the engine acoustic cover.
36. Connect battery ground lead.

Front Timing Belt

37. Disconnect battery ground lead.
38. Remove auxiliary drive belt.
39. Remove right side front road wheel.
40. Remove 3 bolts securing right side splash shield to body and remove shield.
41. Remove 3 bolts securing camshaft timing belt left side rear cover and remove cover.
42. Remove intake manifold chamber.
43. Remove 2 bolts from right side rear timing belt cover heat shield and remove shield.
44. Remove 3 bolts from right side rear timing belt cover and remove the cover.
45. Using a socket, rotate crankshaft pulley bolt in a clockwise direction and align the engine 'SAFE' position, notch on crankshaft pulley aligned to the 'ARROW' on front mounting plate and the timing marks aligned on the rear camshaft gears as illustrated.
46. Insert a timing pin, LRT-12-232, through hole in lower crankcase (hole immediately adjacent to side of lower crankcase). Ensure pin is located in hole in drive plate.
47. Remove bolt securing PAS pipe to engine front mounting plate.
48. Remove 3 Torx screws securing PAS pump pulley and remove pulley.
49. Remove 3 bolts securing PAS pump to front mounting plate and tie pump aside.
50. Release alternator battery lead terminal cover, loosen terminal nut and disconnect lead from terminal.
51. Disconnect multiplug from alternator.
52. Remove lower bolt and upper nut and bolt securing alternator to front mounting plate.
53. Remove alternator.

54. Remove Torx screw securing idler pulley to front mounting plate and remove pulley.

55. Loosen bolt securing lower engine steady to front sub-frame.

56. Remove bolt securing lower engine steady to sub-frame.

57. Remove bolt securing engine steady to oil pan mounting, release lower engine steady from oil pan mounting.

58. Remove 3 bolts securing camshaft timing belt right side front cover and remove cover.

59. Remove 3 bolts securing camshaft timing belt left side front cover and remove cover.

60. Assemble LRT-12-161 to LRT-12-199 and secure with clamp bolt.

61. Insert LRT-12-161 with LRT-12-199 into crankshaft pulley, loosen and remove pulley
bolt. Ensure crankshaft does not rotate during this operation.

62. Remove special tool from crankshaft pulley and remove crankshaft pulley.

63. Remove 3 bolts securing camshaft timing belt lower cover to cylinder block and remove cover.

64. Remove 2 bolts securing auxiliary drive belt tensioner and remove tensioner.

65. Note installed position of heat shield on A/C compressor and remove 3 bolts securing A/C compressor to front mounting plate and cylinder block. Release A/C compressor and heat shield and position aside.

66. Drain engine oil.

67. Remove bolt securing dipstick tube to cylinder block.

68. Depress locking collar, release and remove dipstick tube from engine oil pan.

69. Remove rubber blanking plug from around camshaft timing belt tensioner.

70. Remove 3 bolts securing engine front mounting plate and lifting bracket to cylinder block.

71. Remove bolt securing lifting bracket to right side cylinder head and remove lifting bracket.

72. Remove 2 bolts securing IRD support bracket to engine front mounting plate.

73. Remove 5 bolts and 2 pillar bolts securing engine front mounting plate to cylinder block.

74. With care, release and remove front mounting plate from engine.

75. Insert a suitable 1.5mm diameter pin through the hole in the tensioner body and into the hole in the plunger. If holes in body and plunger are not aligned, move tensioner backplate, using a suitable bread-bladed screwdriver; this will move plunger enough to enable pin to be inserted.

76. Remove and discard 2 bolts securing tensioner to cylinder block, and remove tensioner. DO NOT apply excessive force during removal; if bolts are stuck, apply anti-seize lubricant.

**** CAUTION**

DO NOT loosen Allen screw securing tensioner pulley.

77. If camshaft timing belt is to be refitted, mark direction of rotation on timing belt before removal.

78. With care, ease camshaft timing belt from gears using fingers only and remove timing belt.

**** CAUTION**

Camshaft timing belt must be replaced if cylinder head is to be removed or new drive gears, tensioner or coolant pump are to be installed. Camshaft timing belts must be stored and handled with care. Always store a camshaft timing belt on its edge with a bend radius greater than 2.0 inches (50mm). Do not use a camshaft timing belt that has been twisted or bent double as this will damage the reinforcing fibers. Do not use a camshaft timing belt if debris other than belt dust is found in timing belt covers. Do not use a camshaft timing belt if partial engine seizure has occurred. Do not use a camshaft timing belt if mileage exceeds 48,000 miles (77, 000 km). Do not use an oil or coolant contaminated timing belt, cause of contamination must be rectified.

79. Remove left side and right side exhaust camshaft cap seals.

80. Position tools LRT-12-196 to left side and right side front inlet camshaft gears and into the end of each exhaust camshaft.

**** CAUTION**

Special tools must be installed when tightening or loosening gear retaining bolts; otherwise, damage to camshafts may occur.

81. Remove and discard bolts retaining front inlet camshaft gears to camshafts.

82. Remove tools LRT-12-196 from both inlet camshaft gears and exhaust camshafts.

83. Remove camshaft drive gears and hub assemblies.

To install:

■After the front or rear timing belt installation is complete, it is possible that, after rotating the engine and positioning the crankshaft pulley to the 'SAFE' position, the timing marks on the rear timing gears may be misaligned. This misalignment is acceptable, provided that the timing belt installation procedure is carried out correctly.

84. Clean camshaft gears and hubs, crankshaft gear, tensioner and water pump pulleys.

**** CAUTION**

If the sintered metal gears have been subjected to prolonged oil contamination, thoroughly washed in clean solvent before refitting. Because of the porous construction of the sintered material, oil impregnated in the gears will emerge and contaminate the belt.

85. Install hubs to camshaft gears and install gears to camshafts. Install new bolts and tighten sufficiently to allow gears to rotate without tipping.

86. Position timing belt to gears.

■ To prevent timing belt from disengaging from crankshaft gear when installing, place a suitable wedge between the belt and oil pump belt guard.

87. Install tools LRT-12-196 to both front inlet camshaft gears and into the end of each exhaust camshaft.

88. With assistance, rotate each exhaust camshaft slightly and align timing marks on left and right rear camshaft gears.

89. Remove tool LRT-12-197 from exhaust camshaft.

90. Install tools LRT-12-175 to left and right rear camshaft gears.

91. Rotate both front intake camshaft gears fully clockwise, as viewed from front of engine.

92. Using fingers only, install timing belt to gears. Start at the crankshaft gear and work in an counterclockwise direction, keeping the belt run as taut as possible and turning the camshaft gears only a minimum amount counterclockwise to install timing belt.

✻✻ CAUTION

Gears must not be rotated counterclockwise more than 1 tooth.

93. Position an Allen key in the tensioner backplate and hold tensioner pulley against timing belt.

94. With assistance, position tensioner, install new bolts and tighten to 18 ft. lbs. (25 Nm).

95. Remove tools LRT-12-175 from rear timing belt gears.

96. Install tools LRT-12-196 to both front intake camshaft gears, and into the end of each exhaust camshaft.

✻✻ CAUTION

Special tools must be installed when tighten or loosening gear retaining bolts; otherwise, damage to camshafts may occur.

97. Tighten front intake camshaft gear bolts to the following:

- Step 1: 20 ft. lbs. (27 Nm)
- Step 2: Tighten bolts 90° further

98. Remove tools LRT-12-196 from both intake camshaft gears and exhaust camshafts.

99. Remove wedge from between drive belt and oil pump belt guard.

100. Remove 1.5mm diameter pin from tensioner.

101. Remove timing pin LRT-12-232.

102. Clean exhaust camshaft front cap seal locations and install new cap seals.

✻✻ CAUTION

The sealing edge of the cap seal and mating face must be clean and dry.

103. Position engine front mounting plate and maneuver plate into position. Install bolts, but do not tighten at this time.

104. Install bolts securing IRD support bracket to front mounting plate. Do not tighten bolts at this time.

105. Install and tighten bolts in the sequence shown, to the following specifications:

- Bolt No. 1: 18 ft. lbs. (25 Nm)
- Bolt No 5: 33 ft. lbs. (45 Nm)
- Bolts Nos. 2, 3, 4, 6 and 7: 63 ft. lbs. (85 Nm)

106. Position engine lifting bracket, install bolts securing lifting bracket and front mounting plate to cylinder block and right side cylinder head. Tighten bolts as follows:

- M10 bolts: 33 ft. lbs. (45 Nm)
- M8 bolt: 18 ft. lbs. (25 Nm)

107. Clean end of dipstick tube.

108. Position dipstick tube to oil pan and cylinder block, install bolt and tighten to 7 ft. lbs. (9 Nm).

109. Position A/C compressor to front mounting plate and cylinder block, align heat shield, install and tighten bolts to 18 ft. lbs. (25 Nm).

110. Clean lower timing belt cover.

111. Position lower timing belt cover, install and tighten bolts to 7 ft. lbs. (9 Nm).

112. Clean crankshaft pulley.

113. Install crankshaft pulley to crankshaft gear and ensure that the indent on pulley locates over the lug on crankshaft gear.

114. Install crankshaft pulley bolt and washer, position LRT-12-161 with LRT-12-199 into crankshaft pulley. Tighten pulley bolt to 118 ft. lbs. (160 Nm).

115. Remove LRT-12-161 and LRT-12-199 from crankshaft pulley.

116. Clean auxiliary drive belt tensioner.

117. Position auxiliary belt tensioner, install bolts and tighten to 18 ft. lbs. (25 Nm).

118. Clean left side and right side front timing belt covers.

119. Position left side and right side front timing belt covers, install and tighten bolts to 3 ft. lbs. (4 Nm).

120. Clean idler pulley, position pulley to engine front mounting plate and tighten Allen bolt to 33 ft. lbs. (45 Nm).

121. Position alternator to front mounting plate.

122. Install bolt and nut and bolt securing alternator to front mounting plate and tighten to 33 ft. lbs. (45 Nm).

123. Connect alternator multiplug.

124. Connect battery lead to alternator, install and tighten nut to 6 ft. lbs. (8 Nm), install terminal cover.

125. Position PAS pump to front mounting plate, install and tighten bolts to 18 ft. lbs. (25 Nm).

126. Position PAS pipe to engine front mounting plate, install and tighten bolt to 18 ft. lbs. (25 Nm).

127. Clean PAS pump pulley mating faces.

128. Position PAS pump pulley, install and tighten Torx screws to 7 ft. lbs. (9 Nm).

129. Position lower engine steady to oil pan mounting, install and tighten bolt to 74 ft. lbs. (100 Nm).

130. Tighten bolt securing lower engine steady to sub-frame to 74 ft. lbs. (100 Nm).

131. Clean left side rear timing belt cover.

132. Position left side rear timing belt cover, install and tighten bolts to 3 ft. lbs. (4 Nm).

133. Install heat shield. Install and tighten bolts as follows:

- M6 bolt: 7 ft. lbs. (9 Nm)
- M8 bolt: 18 ft. lbs. (25 Nm)

134. Install intake manifold chamber.

135. Install right side road wheel and tighten nuts to 85 ft. lbs. (115 Nm).

136. Install auxiliary drive belt.

137. Fill engine with oil.

Right Side Rear Timing Belt

138. If necessary for access, remove intake manifold chamber.

139. Remove underbelly panel.

140. Remove right side front road wheel.

141. Remove 3 bolts securing right side splash shield to body and remove shield.
142. If equipped, remove 2 bolts securing right side rear timing belt cover heat shield. Remove heat shield.
143. Remove 3 bolts securing right side rear timing belt cover. Remove cover.
144. Rotate crankshaft in a clockwise direction and align the engine 'SAFE' position, notch on crankshaft pulley aligned to the 'ARROW' on front mounting plate and the timing marks aligned on the rear camshaft gears as illustrated.
145. Remove and discard right side front exhaust camshaft cap seal from cylinder head.
146. Position LRT-12-175 to rear camshaft gears as illustrated, remove and discard bolts securing gears to camshafts.
147. Remove rear camshaft gears, timing belt and special tool as an assembly.

**** CAUTION**

Do not turn the crankshaft or the camshafts while the timing belt is removed.

148. If camshaft timing belt is to be refitted, mark direction of rotation on timing belt.
149. Remove LRT-12-175 from camshaft gears and remove timing belt from gears.

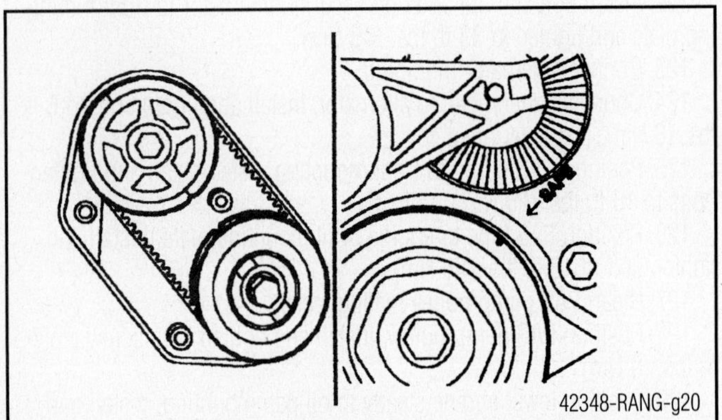

Aligning rear camshaft timing marks and crankshaft pulley marks (left side camshaft shown)—Freelander 2.5L (VIN 2, G) engine

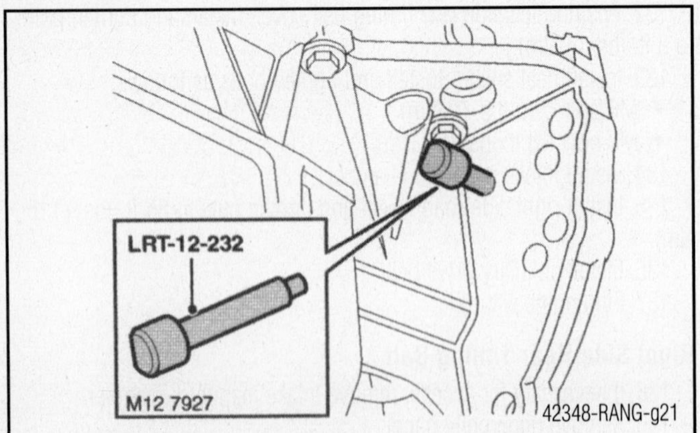

Inserting timing pin LRT-12-232 in lower crankcase—Freelander 2.5L (VIN 2, G) engine

**** CAUTION**

Camshaft timing belts must be stored and handled with care. Always store belt on its edge and with a bend radius of more than 2 inches (20mm). Do not reuse a belt that has been twisted or bent double. Do not use a belt if debris other than belt dust is found in timing belt covers, or if belt has been contaminated with coolant or oil. Do not reuse belt if partial engine seizure has occurred. Do not reused a belt if mileage is more than 45,000 miles (72,000 km).

To install:

■ **After the front or rear timing belt installation is complete, it is possible that, after rotating the engine and positioning the crankshaft pulley to the 'SAFE' position, the timing marks on the rear timing gears may be misaligned. This misalignment is acceptable, provided that the timing belt installation procedure is carried out correctly.**

**** CAUTION**

If the sintered metal gears have been subjected to prolonged oil contamination, they must be soaked in a solvent bath and then thoroughly washed in clean solvent before refitting. Because of the porous construction of the sintered material, oil impregnated in the gears will emerge and contaminate the belt.

150. Clean camshaft gears and mating faces on camshafts.
151. Place gears inverted on a flat surface, with the locating lugs on the gears positioned so they are facing each other.
152. Keeping the timing marks aligned, position timing belt onto gears.
153. Position LRT-12-195 between the gears, turn center nut sufficiently to spread drive belt and position LRT-12-175 to camshaft gears, remove LRT-12-195 from between camshaft gears.
154. Fit LRT-12-198 alignment pins into the end of each camshaft.
155. Position timing belt and gears over LRT-12-198 and locate gears onto camshafts.
156. Position LRT-12-197 into the front end of the right side exhaust camshaft.
157. With assistance, using a 30mm socket on LRT-12-197, turn the right side exhaust camshaft sufficiently to align camshaft gears to the drive slots in each camshaft.
158. Remove LRT-12-198 alignment pins and fit new camshaft gear retaining bolts.
159. Tighten camshaft gear bolts as follows:
- Step 1: 27 Nm (20 ft. lbs.)
- Step 2: Additional 90 degrees
160. Remove LRT-12-175 from camshaft gears.
161. Remove LRT-12-197 from front end of exhaust camshaft.
162. Clean right side exhaust camshaft cap seal recess and fit new cap seal.
163. Fit splash shield and secure with bolts.
164. Fit right side road wheel and tighten nuts to 85 ft. lbs. (115 Nm).
165. Fit right side rear timing belt cover.
166. Fit underbelly panel.
167. Connect battery ground lead.

Left Side Rear Timing Belt

168. Disconnect battery ground lead.
169. Remove underbelly panel.
170. Remove left side rear timing belt cover.
171. Remove right side front road wheel.
172. Remove 3 bolts securing right side splash shield to body and remove shield.
173. Rotate crankshaft in a clockwise direction and align the engine 'SAFE' position, notch on crankshaft pulley aligned to the 'ARROW' on front mounting plate and the timing marks aligned on the rear camshaft gears as illustrated.
174. Insert timing pin LRT-12-232 through hole in lower crankcase, ensuring pin is located in hole in flywheel. Hole immediately adjacent to side of lower crankcase must be used.
175. Remove and discard left side front exhaust camshaft cap seal from cylinder head.
176. Position LRT-12-175 to rear camshaft gears as illustrated, remove and discard bolts securing gears to camshafts.
177. Remove rear camshaft gears, timing belt and special tool as an assembly.

**** CAUTION**

Do not turn the crankshaft or the camshafts while the timing belt is removed.

178. If camshaft timing belt is to be refitted, mark direction of rotation on timing belt.
179. Remove LRT-12-175 from camshaft gears and remove timing belt from gears.

**** CAUTION**

Camshaft timing belts must be stored and handled with care. Always store belt on its edge and with a bend radius of more than 2 inches (20mm). Do not reuse a belt that has been twisted or bent double. Do not use a belt if debris other than belt dust is found in timing belt covers, or if belt has been contaminated with coolant or oil. Do not reuse belt if partial engine seizure has occurred. Do not reused a belt if mileage is more than 45,000 miles (72,000 km).

To install:

■ **After the front or rear timing belt installation is complete, it is possible that, after rotating the engine and positioning the crankshaft pulley to the 'SAFE' position, the timing marks on the rear timing gears may be misaligned. This misalignment is acceptable, provided that the timing belt installation procedure is carried out correctly.**

**** CAUTION**

If the sintered metal gears have been subjected to prolonged oil contamination, they must be soaked in a solvent bath and then thoroughly washed in clean solvent before refitting. Because of the porous construction of the sintered material, oil impregnated in the gears will emerge and contaminate the belt.

180. Clean camshaft gears and mating faces on camshafts.
181. Place gears inverted on a flat surface, with the locating lugs on the gears positioned facing each other.
182. Keeping the timing marks aligned, position timing belt onto gears.
183. Position LRT-12-195 between the gears, turn center nut sufficiently to spread drive belt and position LRT-12-175 to camshaft gears, remove LRT-12-195 from between camshaft gears.
184. Fit LRT-12-198 alignment pins into the end of each camshaft.
185. Position timing belt and gears over LRT-12-198 and locate gears onto camshafts.
186. Position LRT-12-197 into the front end of the left side exhaust camshaft.
187. With assistance, using a 30mm socket on LRT-12-197, turn the left side exhaust camshaft sufficiently to align camshaft gears to the drive slots in each camshaft.
188. Remove LRT-12-198 alignment pins and fit new camshaft gear retaining bolts.
189. Tighten camshaft gear bolts to 20 ft. lbs. (27 Nm) then a further 90°. Remove LRT-12-175 from camshaft gears.
190. Remove LRT-12-197 from front end of exhaust camshaft.
191. Clean left side exhaust camshaft cap seal recess and fit new cap seal.
192. Fit splash shield and secure with bolts.
193. Fit right side road wheel and tighten nuts to 85 ft. lbs. (115 Nm).
194. Fit left side rear timing belt cover.
195. Fit underbelly panel.
196. Connect battery ground lead.

Lexus

2.5L (2VZ-FE) ENGINES

YEAR, MODEL & ENGINE APPLICATION	
1990-91 Lexus ES250	2.5L Code 2VZ-FE

REPLACEMENT INTERVALS
Lexus recommends timing belt replacement every 60,000 miles.

LABOR TIMES	
1990-91	
Replacement	3.7 hrs
Severe Service	4.0 hrs

8807LEXg01

✻✻ CAUTION

These engines are considered freewheeling. If the timing belt fails, valve-to-piston damage is unlikely, however a precautionary compression check should be performed to verify no engine damage.

1. Disconnect the negative battery cable. Wait at least 90 seconds for the air bag system to deactivate.
2. Remove the strut tower brace.
3. Remove the cruise control actuator.
4. Remove the power steering reservoir without disconnecting the hoses.
5. Raise and support the vehicle.
6. Remove the right front wheel and inner splash shield.
7. Remove the accessory drive belts.
8. Support the engine with a jack and remove the right engine mount stays and the engine mount.
9. Disconnect the accelerator and throttle cables.
10. Remove the air cleaner cover, air flow meter and intake hose.
11. Remove vacuum hoses, emission lines, and electrical connectors attached to the air intake chamber, then remove the air intake chamber.
12. Remove the spark plugs.
13. Remove the No. 2 (upper) timing cover.
14. Remove the right side engine mount bracket.
15. Turn the crankshaft to align the camshaft timing marks with the timing marks on the No. 3 (rear) timing cover. Align the crankshaft pulley timing mark with the **0** mark on the No. 1 (lower) timing cover.
16. Remove the crankshaft pulley bolt and press the pulley off the crankshaft with special tool 09213-60017 or equivalent.
17. Remove the No. 1 (lower) timing cover.
18. Remove the timing belt guide.
19. Remove the timing belt tensioner and remove the timing belt.

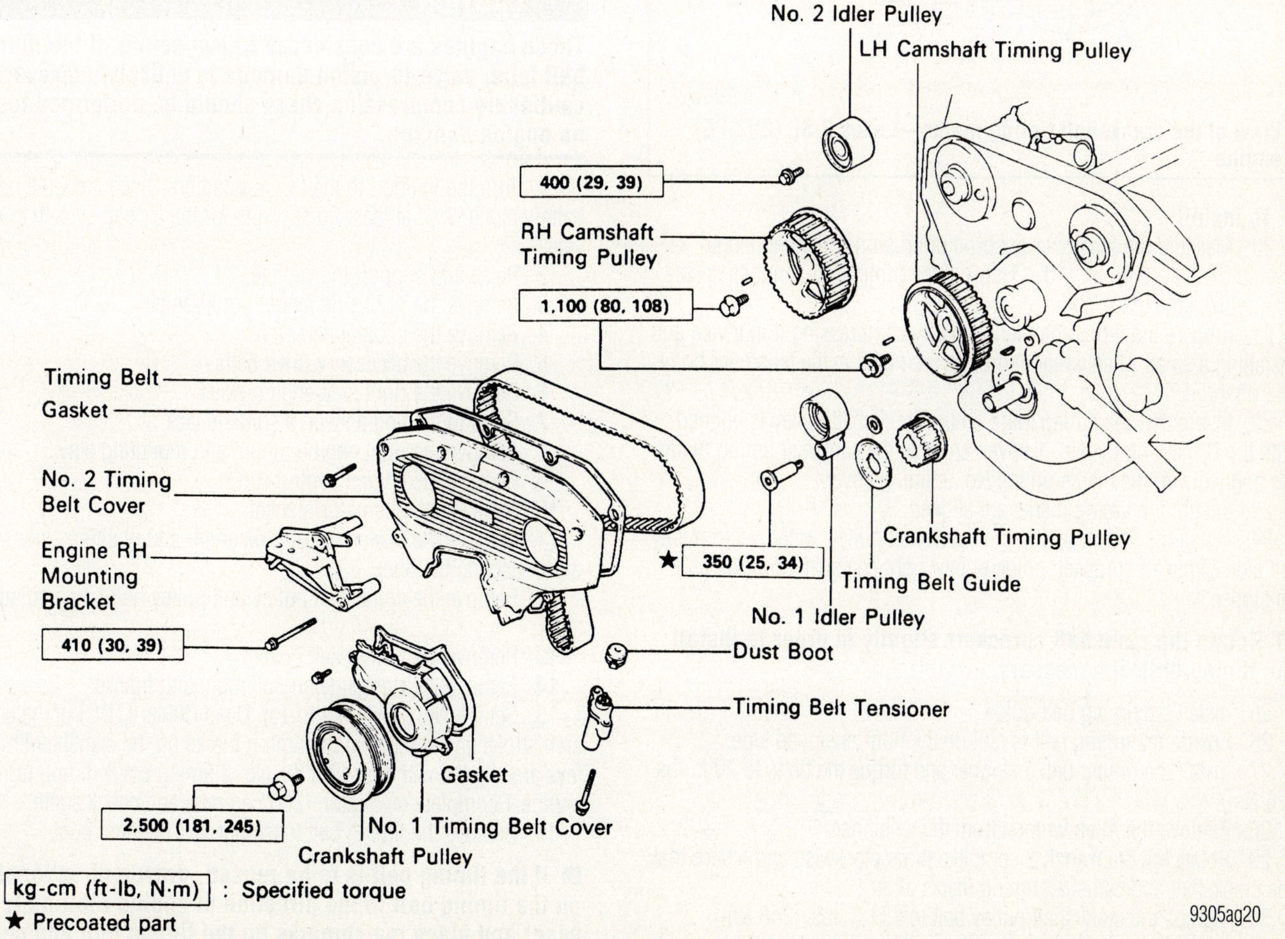

Exploded view of the timing components—Lexus 2.5L (2VZ-FE) engine

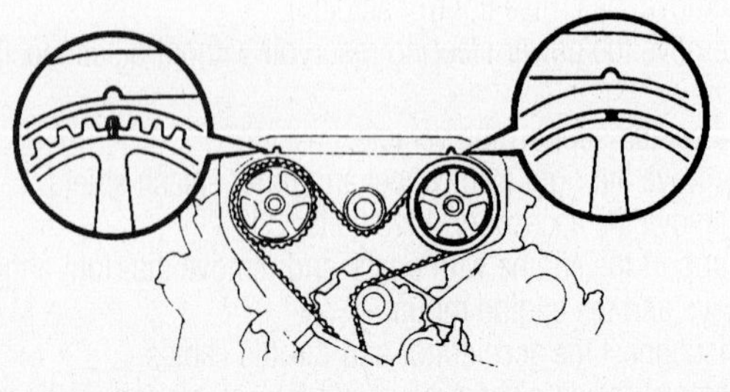

9305ag21

View of the camshaft timing marks—Lexus 2.5L (2VZ-FE) engine

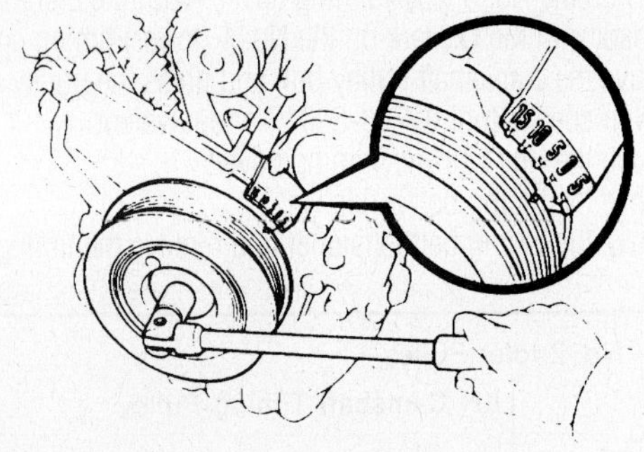

9305ag22

View of the crankshaft timing marks—Lexus 2.5L (2VZ-FE) engine

To install:

20. Measure the tensioner pushrod protrusion. Extended length should be 0.41–0.45 in. (10–11mm). If the protrusion is not as specified, replace the tensioner.
21. Prepare the timing belt tensioner by compressing it in a vise and installing a small Allen wrench through the holes in the tensioner body and pushrod.
22. Insure that the timing mark on the crankshaft pulley is aligned with the **O** mark on the No. 1 cover, and that the camshaft timing marks are aligned with the marks on the No. 3 timing cover.
23. Ensure the timing marks are aligned.
24. Install the timing belt to the crankshaft pulley, water pump pulley, left side camshaft sprocket, guide pulley, right camshaft sprocket and tensioner.

■ **Rotate the camshaft sprockets slightly in order to install the timing belt, if necessary.**

25. Install the timing belt guide.
26. Ensure the timing belt is taut on the non-tensioned side.
27. Install the timing belt tensioner and torque the bolts to 20 ft. lbs. (26 Nm).
28. Remove the Allen wrench from the tensioner.
29. Rotate the crankshaft 2 complete turns clockwise and insure that the crankshaft and camshaft timing marks align.
30. Tighten the crankshaft pulley bolt to 181 ft. lbs. (245 Nm).
31. Install the engine mount bracket and install the No. 2 timing cover.
32. Install the right engine mount and stays. Remove the engine support jack.
33. Install the accessory drive belts.
34. Install the splash shield and the right front wheel.
35. Lower the vehicle.
36. Install the power steering reservoir.
37. Install the cruise control actuator and the strut tower brace.
38. Connect the negative battery cable.

3.0L (1MZ-FE) ES 300 ENGINES

YEAR, MODEL & ENGINE APPLICATION

1994-03 Lexus ES 300 .... 3.0L Code 1MZ-FE

REPLACEMENT INTERVALS

Lexus recommends timing belt relacement every 60,000 miles on 1994-96 models, and 90,000 miles on 1997-03 models.

LABOR TIMES

1994-03	
Replacement	2.9 hrs
Severe Service	3.2 hrs

8807LEXg02

✳✳ CAUTION

These engines are considered freewheeling. If the timing belt fails, valve-to-piston damage is unlikely, however a precautionary compression check should be performed to verify no engine damage.

1. Turn the ignition to the LOCK position. Disconnect the negative battery cable. Wait at least 90 seconds for the air bag system to deactivate.
2. Raise and support the vehicle.
3. Remove the right side fender apron shield.
4. Remove the coolant reservoir.
5. Remove the accessory drive belts.
6. Remove the right side engine stays.
7. Support the engine with a suitable jack.
8. Remove the right engine mount and mounting stay.
9. Remove the engine control rod.
10. Remove the alternator bracket.
11. Remove the power steering reservoir and position aside without disconnecting the lines.
12. Remove the crankshaft pulley and pulley bolt using suitable tools.
13. Remove the timing belt covers.
14. Remove the right side engine mounting bracket.
15. Set the No. 1 cylinder to Top Dead Center (TDC) of the compression stroke and check that the timing marks on the camshaft timing pulleys are aligned with those on the No. 3 timing cover. If not, turn the engine 1 complete revolution (360 degrees) and check again.
16. Remove the timing belt tensioner and the dust boot.

■ **If the timing belt is to be reused, draw a directional arrow on the timing belt in the direction of engine rotation (clockwise) and place matchmarks on the timing belt and crankshaft gear to match the drilled mark on the pulley.**

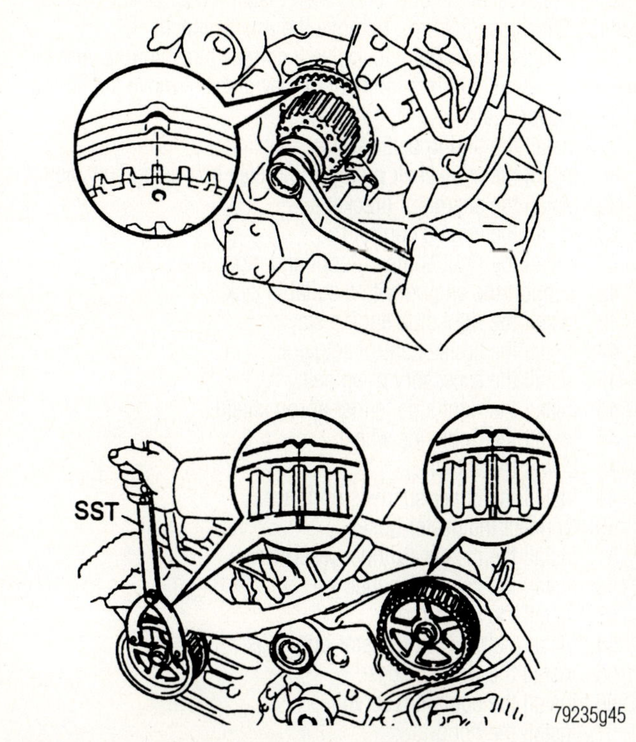

Camshaft and crankshaft pulley positioning for timing belt installation—Lexus 3.0L (1MZ-FE) engine

17. Turn the right camshaft pulley clockwise slightly to release tension, then remove the timing belt from the pulleys.
18. Remove the timing belt guide.
19. Remove the timing belt from the engine.

To install:

■ **If the old timing belt is being reinstalled, be sure the directional arrow is facing in the original direction and that the belt and crankshaft gear matchmarks are properly aligned.**

20. Measure the tensioner pushrod protrusion. Extended length should be 0.39–0.41 in. (9.9–10.4mm). If the protrusion is not as specified, replace the tensioner.
21. Ensure the timing marks are still aligned.
22. Install the timing belt in a counterclockwise starting at the crankshaft sprocket.
23. Install the timing belt guide.
24. Install the lower (No. 1) timing cover and tighten the bolts.
25. Check that the timing belt is taut between the sprockets on the non tension side.
26. Position a plate washer between the timing belt tensioner and the block, then press in the pushrod until the holes are aligned between it and the housing. Slide a 0.05 in. Allen wrench through the hole to keep the pushrod set. Install the dust boot, then install the tensioner. Tighten the bolts to 20 ft. lbs. (26 Nm). Remove the Allen wrench.
27. Turn the crankshaft clockwise 2 complete revolutions and check that all marks are still in alignment. If they aren't, remove the timing belt and start over again.
28. Install the right side engine mounting bracket.
29. Install the timing belt covers.
30. Install the crankshaft pulley and pulley bolt using suitable tools.
31. Install the power steering reservoir.
32. Install the alternator bracket.
33. Install the engine control rod.
34. Install the right engine mount and mounting stay.
35. Install the right side engine stays.
36. Install the accessory drive belts.
37. Install the coolant reservoir.
38. Install the right side fender apron shield.
39. Connect the negative battery cable.

3.0L (1MZ-FE) RX 300 ENGINES

YEAR, MODEL & ENGINE APPLICATION

1999-03 Lexus RX 300 .... 3.0L Code 1MZ-FE

REPLACEMENT INTERVALS

Lexus recommends timing belt replacement every 90, 000 miles.

LABOR TIMES

1999-03	
Replacement	3.2 hrs
Severe Service	3.6 hrs

8807LEXg03

✻✻ CAUTION

These engines are considered freewheeling. If the timing belt fails, valve-to-piston damage is unlikely, however a precautionary compression check should be performed to verify no engine damage.

1. Turn the ignition to the LOCK position. Disconnect the negative battery cable. Wait at least 90 seconds for the air bag system to deactivate.
2. Remove the hood.
3. Remove the coolant reservoir.
4. Remove the wiper blades.
5. Remove both cowl ventilator openings.
6. Remove the wiper motor.
7. Remove the cowl top panel.
8. Remove the master cylinder reservoir.
9. Disconnect the heater hoses.
10. Remove the upper suspension brace.
11. Raise and support the vehicle.
12. Remove the right side wheel.
13. Remove the right side fender apron shield.
14. Remove the accessory drive belts.
15. Remove the cruise control actuator.
16. Remove the right side engine stays.
17. Support the engine with a suitable jack.
18. Remove the right engine mount and mounting stay.
19. Remove the engine control rod.
20. Remove the alternator bracket.
21. Remove the crankshaft pulley and pulley bolt using suitable tools.
22. Remove the timing belt covers.
23. Remove the right side engine mounting bracket.
24. Set the No. 1 cylinder to Top Dead Center (TDC) of the compression stroke and check that the timing marks on the camshaft timing pul-

leys are aligned with those on the No. 3 timing cover. If not, turn the engine 1 complete revolution (360 degrees) and check again.

25. Remove the timing belt tensioner and the dust boot.

■ If the timing belt is to be reused, draw a directional arrow on the timing belt in the direction of engine rotation (clockwise) and place matchmarks on the timing belt and crankshaft gear to match the drilled mark on the pulley.

26. Turn the right camshaft pulley clockwise slightly to release tension, then remove the timing belt from the pulleys.
27. Remove the timing belt guide.
28. Remove the timing belt from the engine.

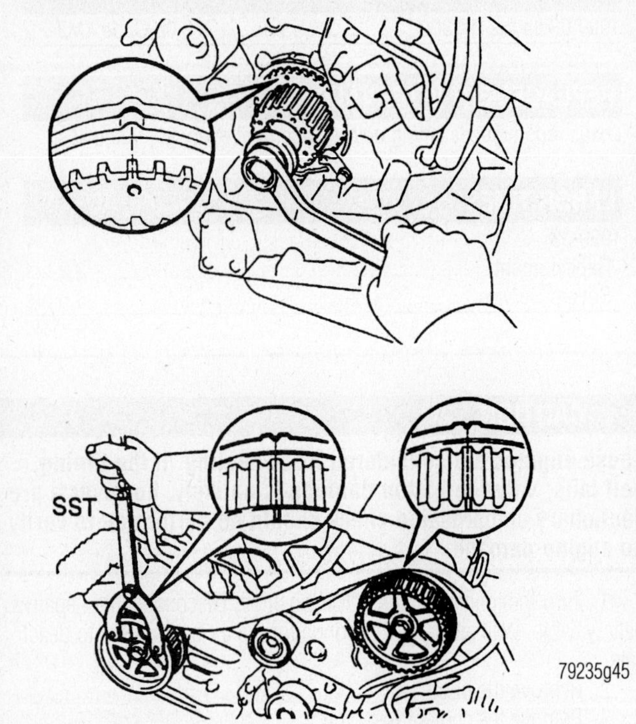

Camshaft and crankshaft pulley positioning for timing belt installation—Lexus 3.0L (1MZ-FE) engine

To install:

■ If the old timing belt is being reinstalled, be sure the directional arrow is facing in the original direction and that the belt and crankshaft gear matchmarks are properly aligned.

29. Measure the tensioner pushrod protrusion. Extended length should be 0.39–0.43 in. (9.9–11mm). If the protrusion is not as specified, replace the tensioner.
30. Ensure the timing marks are still aligned.
31. Install the timing belt to the crankshaft sprocket, water pump pulley, camshaft sprocket, guide pulley, camshaft sprocket and tensioner pulley..
32. Install the timing belt guide.
33. Install the lower (No. 1) timing cover and tighten the bolts.
34. Check that the timing belt is taut between the sprockets on the non tension side.
35. Position a plate washer between the timing belt tensioner and the block, then press in the pushrod until the holes are aligned between it and the housing. Slide an Allen wrench through the hole to keep the pushrod set. Install the dust boot, then install the tensioner. Tighten the bolts to 20 ft. lbs. (26 Nm). Remove the Allen wrench.
36. Turn the crankshaft clockwise 2 complete revolutions and check that all marks are still in alignment. If they aren't, remove the timing belt and start over again.
37. Install the timing belt covers.
38. Install the crankshaft pulley and pulley bolt using suitable tools.
39. Install the alternator bracket.
40. Install the engine control rod.
41. Install the right engine mount and mounting stay.
42. Support the engine with a suitable jack.
43. Install the right side engine stays.
44. Install the cruise control actuator.
45. Install the accessory drive belts.
46. Install the right side fender apron shield.
47. Install the right side wheel.
48. Lower the vehicle.
49. Install the upper suspension brace.
50. Connect the heater hoses.
51. Install the master cylinder reservoir.
52. Install the cowl top panel.
53. Install the wiper motor.
54. Install both cowl ventilator openings.
55. Install the wiper blades.
56. Install the coolant reservoir.
57. Install the hood.
58. Connect the negative battery cable.

3.0L (2JZ-GE) 1992–97 ENGINES

YEAR, MODEL & ENGINE APPLICATION

1993-97 Lexus GS 300, 1992-97 SC 300 ... 3.0L Code 2JZ-GE

REPLACEMENT INTERVALS

Lexus recommends timing belt relacement every 60,000 miles on 1992-96 models, and 90,000 miles on 1997 models.

LABOR TIMES

1992-97	
Replacement	3.2 hrs
Severe Service	3.7 hrs

8807LEXg04

✱✱ CAUTION

These engines are considered freewheeling. If the timing belt fails, valve-to-piston damage is unlikely, however a precautionary compression check should be performed to verify no engine damage.

1. Turn the ignition to the LOCK position. Disconnect the negative and positive battery cables. Wait at least 90 seconds for the air bag system to deactivate.
2. Drain the cooling system.
3. Raise and support the vehicle.
4. Remove the air cleaner duct.
5. Remove the engine splash shield.
6. Remove the cooling fan and clutch.
7. Remove the cooling hoses and the radiator.

8. Remove the accessory drive belts.
9. Remove the drive belt tensioner.
10. Remove the water pump pulley.
11. Remove the power steering pump pulley and bracket.
12. Remove the oil filler cap.
13. Remove the center and upper timing belt covers.
14. Rotate the crankshaft clockwise until the crankshaft and camshaft timing marks are aligned.
15. Remove the crankshaft pulley using suitable tools.
16. Remove the lower timing belt cover.
17. Ensure the timing marks are aligned.
18. Alternately loosen the 2 tensioner mounting bolts and remove them, the tensioner and the dust boot. Slide the timing belt off of the 2 camshaft sprockets. It's a good idea to matchmark the belt to the pulleys.
19. Remove the timing belt guide.
20. Remove the timing belt.

■ **If the timing belt is to be reused, draw a directional arrow on the timing belt in the direction of engine rotation (clockwise) and place matchmarks on the timing belt and crankshaft gear to match the drilled mark on the pulley.**

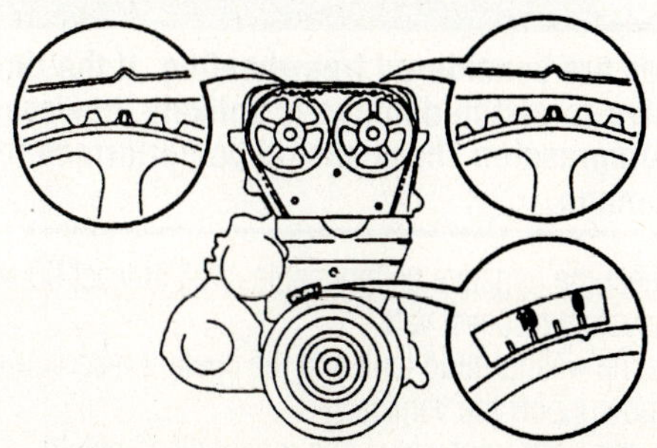

79235g46

Set the engine to TDC by aligning the marks before removing the lower timing cover—Lexus 3.0L (2JZ-GE) engine

To install:

■ **If the old timing belt is being reinstalled, be sure the directional arrow is facing in the original direction and that the belt and crankshaft gear matchmarks are properly aligned.**

21. Measure the tensioner pushrod protrusion. Extended length should be 0.32–0.35 in. (8.1–8.9mm) on GS 300 models, or 0.30–0.41 in. (7.6–10.4mm) on SC 300 models. If the protrusion is not as specified, replace the tensioner.
22. Position a plate washer between the timing belt tensioner and the block, then press in the pushrod until the holes are aligned between it and the housing. Slide a 1.5mm Allen wrench through the hole to keep the pushrod set
23. Ensure that the timing marks are aligned.
24. Install the timing belt on the crankshaft pulley and working counterclockwise around the remaining pulleys.
25. Install the dust boot, then install the tensioner. Tighten the bolts to 20 ft. lbs. (26 Nm). Remove the Allen wrench.
26. Install the lower timing belt cover.
27. Align the crankshaft pulley set key with the key groove on the pulley and slide the pulley on. Tighten the bolt to 239 ft. lbs. (324 Nm).
28. Turn the crankshaft clockwise 2 complete revolutions and check that all marks are still in alignment. If they aren't, remove the timing belt and start over again.
29. Install the center and upper timing belt covers.
30. Install the oil filler cap.
31. Install the power steering pump pulley and bracket.
32. Install the water pump pulley.
33. Install the drive belt tensioner.
34. Install the accessory drive belts.
35. Install the cooling hoses and the radiator.
36. Install the cooling fan and clutch.
37. Install the engine splash shield.
38. Install the air cleaner duct.
39. Lower the vehicle.
40. Refill the cooling system.
41. Connect the battery cables.

3.0L (2JZ-GE) 1998–04 ENGINES

YEAR, MODEL & ENGINE APPLICATION

1998-04 Lexus GS 300, 2000-04 IS 300, 1998-00 SC 300	3.0L Code 2JZ-GE

REPLACEMENT INTERVALS

Lexus recommends timing belt replacement every 90,000 miles.

LABOR TIMES

GS 300	
Replacement	3.2 hrs
Severe Service	3.6 hrs
IS 300	
Replacement	3.1 hrs
Severe Service	3.4 hrs
SC 300	
Replacement	3.3 hrs
Severe Service	3.7 hrs

8807LEXg05

⁂ CAUTION

These engines are an interference type. If the timing belt has failed, valve-to-piston damage may have occurred. A compression check should always be performed to check for engine damage.

1. Turn the ignition to the LOCK position. Disconnect the negative and positive battery cables. Wait at least 90 seconds for the air bag system to deactivate.
2. Drain the cooling system.
3. Raise and support the vehicle.
4. Remove the engine splash shield.
5. Remove the cooling hoses and the radiator.
6. Remove the accessory drive belts.
7. Remove the drive belt tensioner.
8. Remove the water pump pulley.
9. Remove the power steering pump pulley and bracket.
10. Remove the oil filler cap.
11. Remove the center and upper timing belt covers.

12. Rotate the crankshaft clockwise until the crankshaft and camshaft timing marks are aligned.
13. Remove the crankshaft pulley using suitable tools.
14. Remove the lower timing belt cover.
15. Ensure the timing marks are aligned.
16. Remove the crankshaft pulley bolt.
17. Alternately loosen the 2 tensioner mounting bolts and remove them, the tensioner and the dust boot. Slide the timing belt off of the 2 camshaft sprockets. It's a good idea to matchmark the belt to the pulleys.
18. Remove the timing belt guide.
19. Remove the timing belt.

■ **If the timing belt is to be reused, draw a directional arrow on the timing belt in the direction of engine rotation (clockwise) and place matchmarks on the timing belt and crankshaft gear to match the drilled mark on the pulley.**

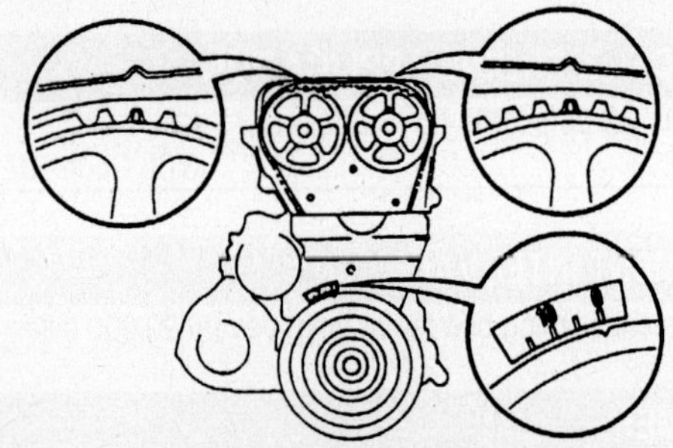

79235g46

Set the engine to TDC by aligning the marks before removing the lower timing cover—Lexus 3.0L (2JZ-GE) engine

To install:

■ **If the old timing belt is being reinstalled, be sure the directional arrow is facing in the original direction and that the belt and crankshaft gear matchmarks are properly aligned.**

20. Measure the tensioner pushrod protrusion. Extended length should be 0.32–0.35 in. (8.1–8.9mm). If the protrusion is not as specified, replace the tensioner.
21. Press in the pushrod until the holes are aligned between it and the housing. Slide a 1.5mm Allen wrench through the hole to keep the pushrod set
22. Ensure that the timing marks are aligned.
23. Install the timing belt on the crankshaft pulley and working counterclockwise around the remaining pulleys.
24. Install the dust boot, then install the tensioner. Tighten the bolts to 20 ft. lbs. (26 Nm). Remove the Allen wrench.
25. Install the lower timing belt cover.
26. Align the crankshaft pulley set key with the key groove on the pulley and slide the pulley on. Tighten the bolt to 243 ft. lbs. (329 Nm).
27. Turn the crankshaft clockwise 2 complete revolutions and check that all marks are still in alignment. If they aren't, remove the timing belt and start over again.
28. Install the center and upper timing belt covers.
29. Install the power steering pump pulley and bracket.
30. Install the water pump pulley.
31. Install the drive belt tensioner.
32. Install the accessory drive belts.
33. Install the cooling hoses and the radiator.
34. Install the engine splash shield.
35. Lower the vehicle.
36. Refill the cooling system.
37. Connect the battery cables.

3.0L (3VZ-FE) ENGINES

YEAR, MODEL & ENGINE APPLICATION

1992-93 Lexus ES 300 .... 3.0L Code 3VZ-FE

REPLACEMENT INTERVALS

Lexus recommends timing belt replacement every 60,000 miles.

LABOR TIMES

1992-93

Replacement .... 2.9 hrs

Severe Service .... 3.2 hrs

8807LEXq06

✻✻ CAUTION

These engines are considered freewheeling. If the timing belt fails, valve-to-piston damage is unlikely, however a precautionary compression check should be performed to verify no engine damage.

1. Disconnect the negative battery cable. Wait at least 90 seconds for the air bag system to deactivate.
2. Remove the washer fluid tank and the coolant reservoir tank.
3. Raise and support the vehicle.
4. Remove the right front wheel and inner splash shield.
5. Remove the accessory drive belts.
6. Remove the engine torque rod.
7. Support the engine with a jack and remove the right engine mount stays and the engine mount.
8. Remove the power steering reservoir without disconnecting the hoses.
9. Remove the No. 2 (upper) timing cover.
10. Remove the engine mount bracket.
11. Turn the crankshaft to align the camshaft timing marks with the timing marks on the No. 3 (rear) timing cover. Align the crankshaft pulley timing mark with the **0** mark on the No. 1 (lower) timing cover.
12. Remove the crankshaft pulley bolt and press the pulley off the crankshaft with special tool 09213-60017 or equivalent.
13. Remove the No. 1 (lower) timing cover.
14. Remove the timing belt guide.
15. Remove the timing belt tensioner and remove the timing belt.

To install:

16. Measure the tensioner pushrod protrusion. Extended length should be 0.39–0.41 in. (9.9–10.4mm). If the protrusion is not as specified, replace the tensioner.
17. Prepare the timing belt tensioner by compressing it in a vise and installing a small Allen wrench through the holes in the tensioner body and pushrod.
18. Install the timing belt to the crankshaft sprocket, water pump pulley and tensioner pulley.
19. Install the timing belt guide.

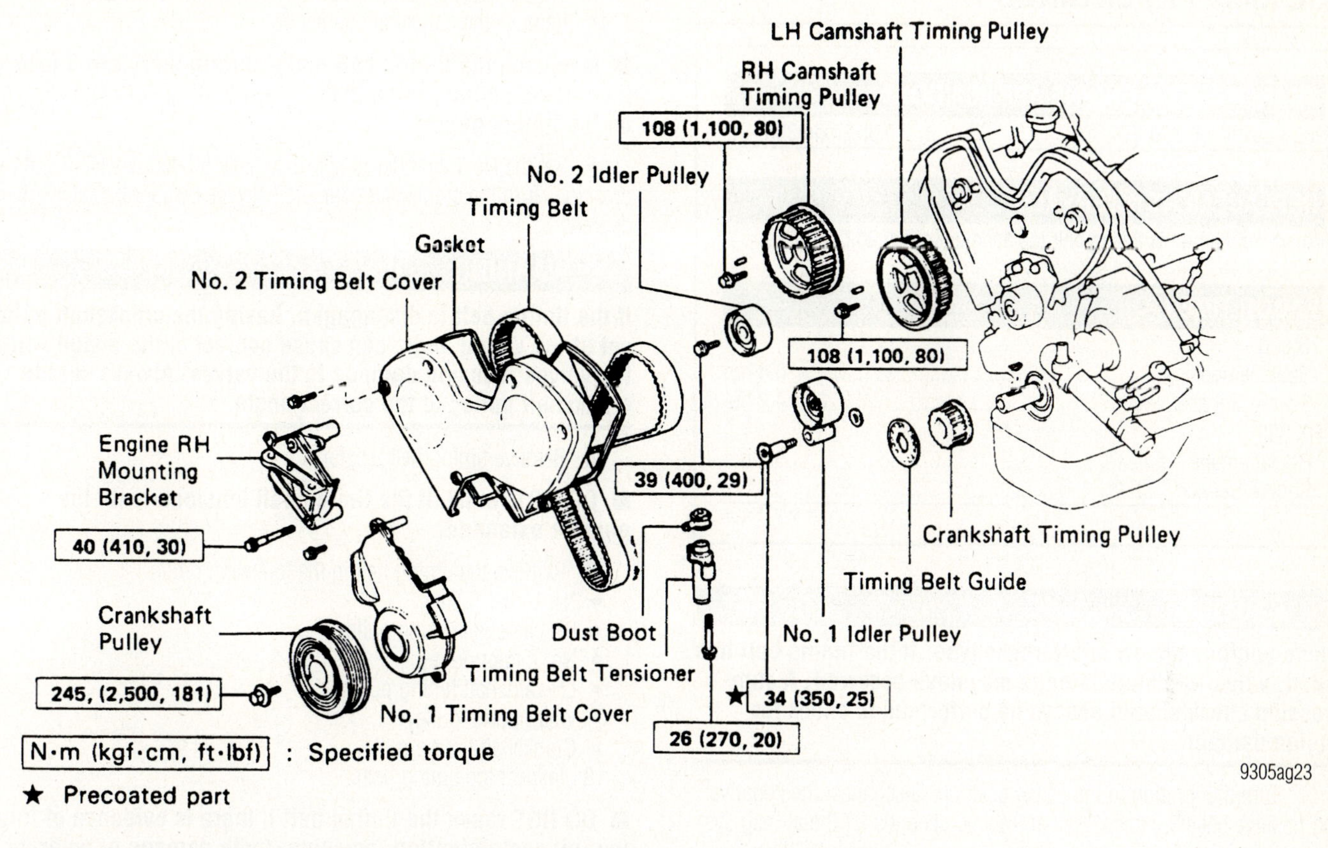

Exploded view of the timing belt and related components—Lexus 3.0L (3VZ-FE) engine

20. Install the No. 1 timing cover and install the crankshaft pulley.
21. Insure that the timing mark on the crankshaft pulley is aligned with the **O** mark on the No. 1 cover, and that the camshaft timing marks are aligned with the marks on the No. 3 timing cover.
22. Install the timing belt on the camshaft sprockets and guide pulley.
23. Ensure the belt is taut between sprockets on the non-tension side.
24. Install the timing belt tensioner and torque the bolts to 20 ft. lbs. (26 Nm).
25. Remove the Allen wrench from the tensioner.
26. Rotate the crankshaft 2 complete turns clockwise and insure that the crankshaft and camshaft timing marks align.
27. Tighten the crankshaft pulley bolt to 181 ft. lbs. (245 Nm).
28. Install the engine mount bracket and install the No. 2 timing cover.
29. Install the power steering reservoir.
30. Install the right engine mount and stays.
31. Install the engine torque rod. Remove the engine support jack.
32. Install the accessory drive belts.
33. Install the splash shield and the right front wheel.
34. Lower the vehicle.
35. Install the coolant reservoir tank and the washer fluid tank.
36. Connect the negative battery cable.

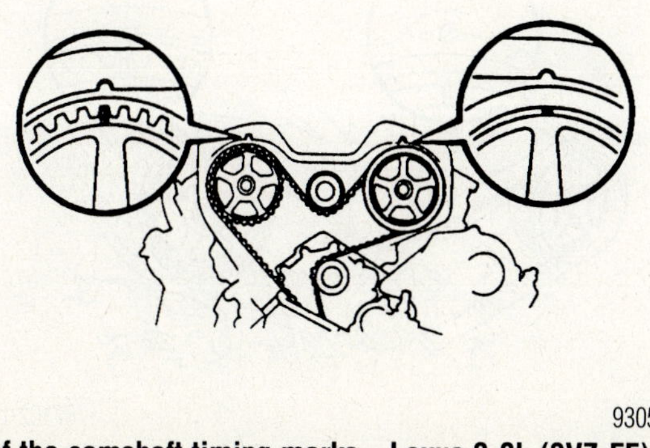

View of the camshaft timing marks—Lexus 3.0L (3VZ-FE) engine

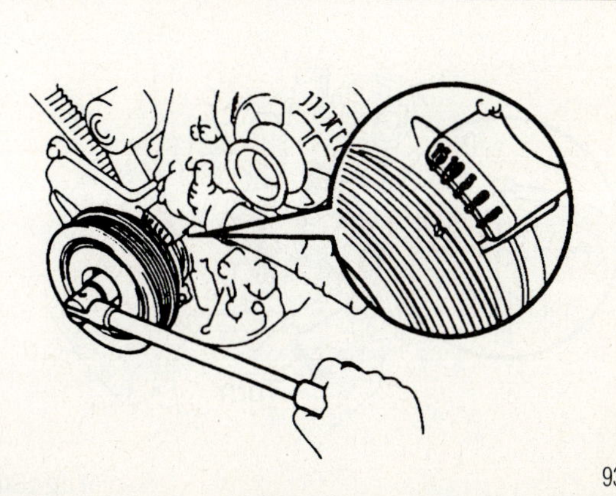

View of the crankshaft timing marks—Lexus 3.0L (3VZ-FE) engine

3.3L (3MZ-FE) ENGINES

YEAR, MODEL & ENGINE APPLICATION

2004 Lexus ES 330, RX 330 .... 3.3L Code 3MZ-FE

REPLACEMENT INTERVALS

Lexus recommends timing belt replacement every 90,000 miles.

LABOR TIMES

ES 330	
Replacement	3.8 hrs
Severe Service	4.2 hrs
RX 330	
Replacement	3.2 hrs
Severe Service	3.6 hrs

8807LEXg07

✱✱ CAUTION

These engines are an interference type. If the timing belt has failed, valve-to-piston damage may have occurred. A compression check should always be performed to check for engine damage.

1. Turn the ignition to the LOCK position. Disconnect the negative and positive battery cables. Wait at least 90 seconds for the air bag system to deactivate.
2. Remove the right side front fender apron seal.
3. Remove the accessory drive belts.
4. Remove the engine control rod and bracket.
5. Remove the engine mounting stay and bracket.
6. Remove the alternator bracket.
7. Remove the crankshaft pulley.
8. Remove the upper and lower timing belt covers.
9. Remove the right side engine mounting bracket, if necessary.
10. Remove the timing belt guide.
11. Rotate the crankshaft so No. 1 cylinder is at TDC on the compression stroke.
12. Temporarily install the crankshaft pulley and turn the crankshaft clockwise and align the crankshaft timing mark with the oil pump body.
13. Ensure that the camshaft pulley timing marks are aligned.
14. Remove the crankshaft pulley bolt.

■ **If reusing the timing belt make certain there are 3 location marks on the timing belt corresponding with the marks on the timing gears.**

15. Set the No 1 cylinder to approximately 60 degrees BTDC/compression. Turn the crankshaft counterclockwise by approximately 60 degrees.

✱✱ WARNING

If the timing belt is disengaged, having the crankshaft pulley set at the wrong angle can cause contact of the piston with the valves, causing damage to the valves. Always set the crankshaft pulley at the correct angle.

16. Remove timing belt tensioner

■ **DO NOT reinstall the timing belt tensioner with the plunger extended.**

17. Remove the timing belt in the following order.
   - No 1 idler pulley
   - RH camshaft timing pulley
   - No. 2 idler pulley
   - LH camshaft timing pulley
   - Water pump pulley
   - Crankshaft timing pulley
18. Inspect the timing belt

■ **DO NOT reuse the timing belt if there is evidence of fraying, oil contamination, cracking, tooth damage or wear, or belt distortion. If there is any doubt in the belt condition, replace the belt.**

To install:

19. After inspecting the pulleys for wear and checking for oil leaks install the timing belt using the following procedure.
   - Temporarily install the crankshaft pulley bolt
   - Make sure the crankshaft pulley is set at 60° BTDC
   - Align the timing marks on the camshaft pulley with the respective marks on the inside timing cover.
   - If reusing the old belt, align the marks previously made on the belt with the marks on the timing gears.
   - Install the belt in the reverse order used when removing the belt
20. Install the timing belt tensioner using the following procedure

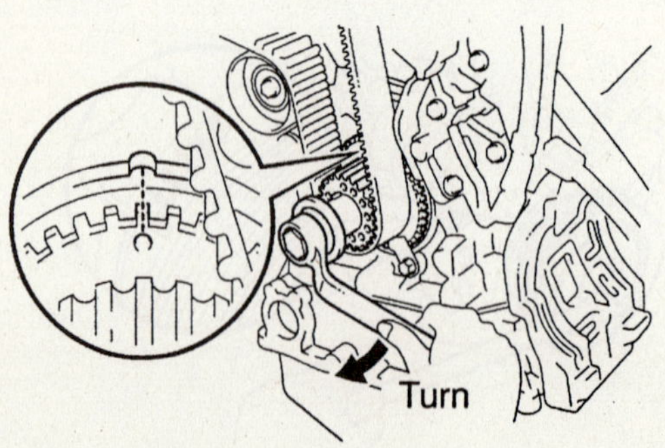

67162-LEXU-g55

Aligning crankshaft pulley—3.3L (3MZ-FE) engine

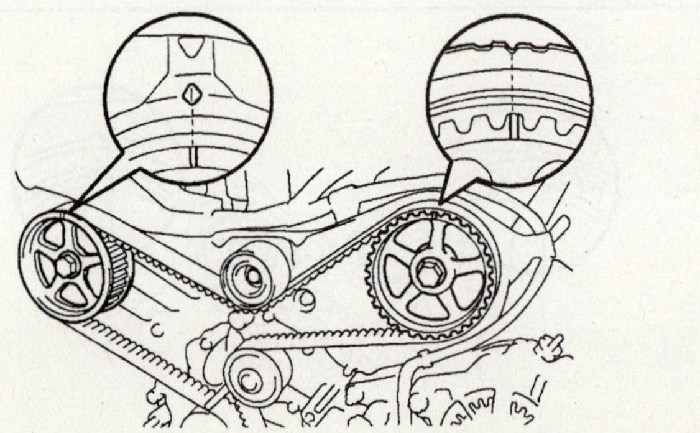

67162-LEXU-g56

Aligning camshaft pulleys—3.3L (3MZ-FE) engine

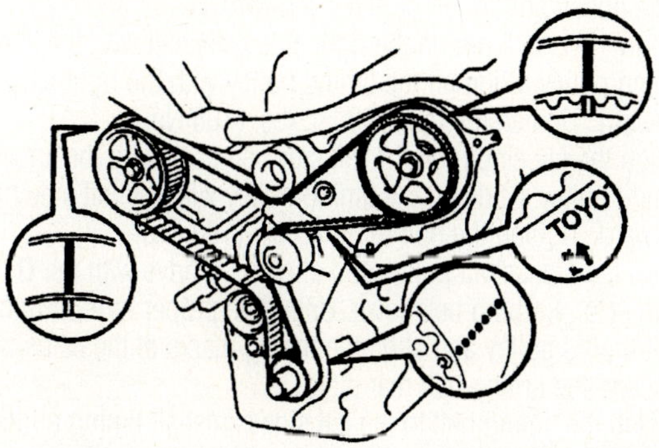

67162-LEXU-g57

Marking the timing belt for reuse—3.3L (3MZ-FE) engine

■ Keep the tensioner in an upright position

- Slowly depress the push rod and align the hole with the hole in the housing
- Insert a 1.5 mm hexagon wrench through the holes to maintain the setting

  position of the push rod.
- Install the tensioner with the 2 bolts and torque to 20 ft. lbs (27 Nm).
- Remove the hexagon wrench

21. Slowly turn the crankshaft 2 revolutions clockwise.
22. Check to see that all timing marks are in alignment.
23. Install the crankshaft bolt
24. Install the timing belt guide
25. Install the right side engine mounting bracket, if removed.
26. Install the upper and lower timing belt covers.
27. Install the crankshaft pulley.
28. Install the alternator bracket.
29. Install the engine mounting stay and bracket.
30. Install the engine control rod and bracket.
31. Install the accessory drive belts.
32. Install the right side front fender apron seal.
33. Connect the negative battery cable.

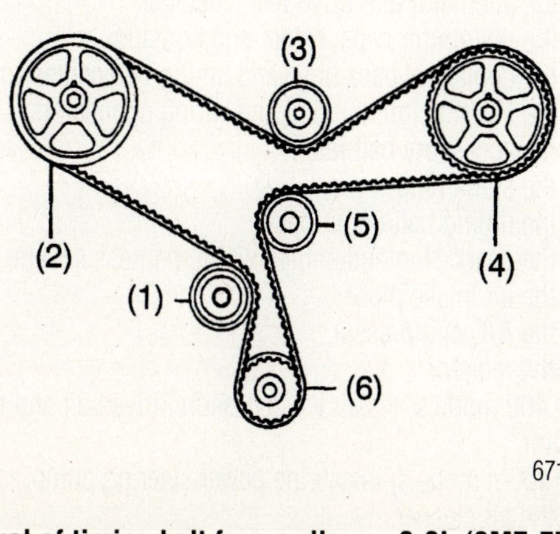

67162-LEXU-g58

Removal of timing belt from pulleys—3.3L (3MZ-FE) engine

4.0L (1UZ-FE) 1990–97 ENGINES

YEAR, MODEL & ENGINE APPLICATION

1990-97 Lexus LS 400, 1992-97 SC 400 .... 4.0L Code 1UZ-FE

REPLACEMENT INTERVALS

Lexus recommends timing belt relacement every 60,000 miles on 1990-96 models, and 90,000 miles on 1997 models.

LABOR TIMES

LS 400	
1990-94	
Replacement	4.6 hrs
Severe Service	5.1 hrs
1995-97	
Replacement	3.8 hrs
Severe Service	4.2 hrs
SC 400	
Replacement	4.8 hrs
Severe Service	5.4 hrs

8807LEXg08

✻✻ CAUTION

These engines are considered freewheeling. If the timing belt fails, valve-to-piston damage is unlikely, however a precautionary compression check should be performed to verify no engine damage.

1. Disconnect the negative battery cable. Wait at least 90 seconds for the air bag system to deactivate.
2. Drain the engine coolant.
3. Remove the engine splash shield.
4. On LS 400 models, remove the air intake duct and engine appearance covers.
5. Remove the accessory drive belt.
6. Remove the fan clutch and pulley.
7. Remove the air cleaner.
8. Remove the power steering pump without disconnecting the lines and position aside.
9. On SC 400 models, remove the accessory drive belt and the coolant reservoir.
10. On all models, remove the radiator.
11. Remove the A/C compressor without disconnecting the lines and position aside.
12. Remove the air intake pipe.
13. Remove the spark plug and engine wiring harness covers.
14. Remove the timing belt upper cover.
15. Remove the spark plug wires.
16. Remove the accessory belt idler.
17. Remove the left side ignition coil and wiring connectors.
18. Remove the coolant bypass hose and timing belt center covers.
19. Remove the distributor caps, rotors and housings.
20. Remove the alternator and drive belt tensioner.
21. Rotate the crankshaft clockwise until No. 1 cylinder is at TDC of the compression stroke and the timing marks are aligned.
22. Remove the bolt and the crankshaft pulley with the proper tool. On LS 400 remove the fan bracket. On SC 400, remove the hydraulic pump.

23. Remove the timing belt tensioner.
24. Remove the timing belt from the camshaft sprockets.

■ **It may be necessary to rotate the left camshaft sprocket slightly to ease tension on the belt.**

25. Remove the camshaft sprockets
26. Remove the timing belt lower cover.
27. Remove the timing belt guide (No. 1 crank position sensor plate).
28. Remove the timing belt.

■ **If the timing belt is to be reused, draw a directional arrow on the timing belt in the direction of engine rotation (clockwise) and place matchmarks on the timing belt and crankshaft gear to match the drilled mark on the pulley.**

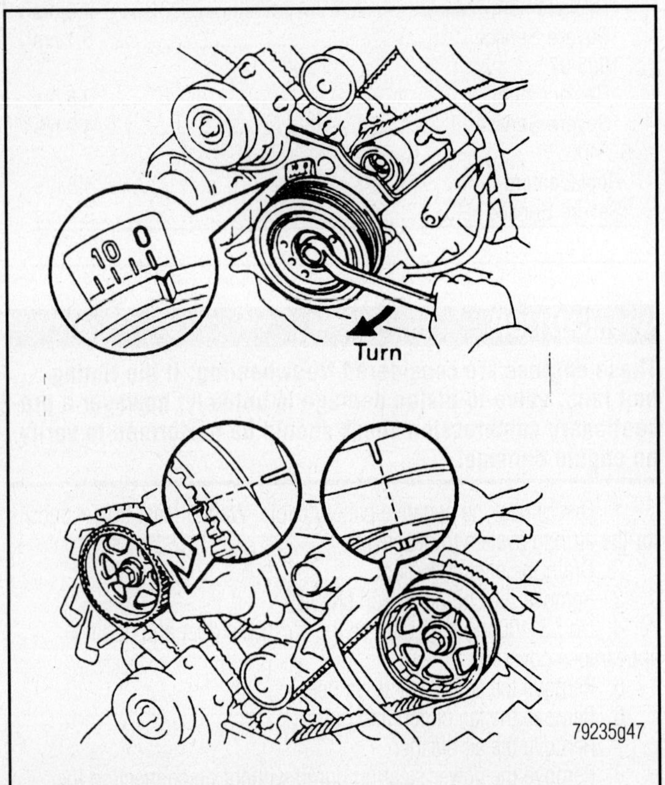

Timing belt sprocket mark alignment for belt installation—Lexus 4.0L (1UZ-FE) engine

To install:

29. Measure the tensioner pushrod protrusion. Extended length should be 0.41–0.45 in. (10.4–11.4mm). If the protrusion is not as specified, replace the tensioner.
30. Align the installation mark on the timing belt with the drilled mark of the crankshaft timing pulley. Install the timing belt on the crankshaft timing pulley, No. 1 idler pulley and the No. 2 idler pulley.

■ **If the old timing belt is being reinstalled, be sure the directional arrow is facing in the original direction and that the belt and crankshaft gear matchmarks are properly aligned.**

31. Install the timing belt guide (No. 1 crank angle sensor plate) with the cup side facing forward. Replace the timing belt cover spacer.
32. Install the No. 1 timing belt cover and tighten the mounting bolts. On the SC400, install the hydraulic pump. Install the fan bracket.
33. Align the pulley set key on the crankshaft with the key groove of the pulley. Install the pulley, using the proper tool to tap in the pulley. Tighten the pulley bolt to 181 ft. lbs. (245 Nm).
34. Align the knock pin on the right side camshaft with the knock pin of the timing pulley. Slide on the timing pulley with the right side mark facing forward. Tighten the bolt to 80 ft. lbs. (108 Nm).
35. Align the knock pin on the left side camshaft with the knock pin of the timing pulley. Slide on the timing pulley with the left side mark facing forward. Tighten the bolt to 80 ft. lbs. (108 Nm).
36. Turn the crankshaft pulley and align its groove with the **0** timing mark on the No. 1 timing belt cover. Using the proper tool, turn the crankshaft timing pulley and align the timing marks of the camshaft timing pulley and the timing belt rear plate.
37. Install the timing belt to the left side camshaft timing pulley by:
    a. Using the proper tool, slightly turn the left side timing pulley clockwise. Align the installation mark of the timing belt with the timing mark of the camshaft timing pulley and hang the timing belt on the left side camshaft pulley.
    b. Using the proper tool, align the timing marks of the left side camshaft pulley and the timing belt rear plate.
    c. Check that the timing belt has tension between crankshaft timing pulley and the left side camshaft pulley.
38. Install the timing belt to the right side camshaft timing pulley by:
    a. Using the proper tool, slightly turn the right side timing pulley clockwise. Align the installation mark of the timing belt with the timing mark of the camshaft timing pulley and hang the timing belt on the right side camshaft pulley.
    b. Using the proper tool, align the timing marks of the right side camshaft pulley and the timing belt rear plate.
    c. Check that the timing belt has tension between the crankshaft timing pulley and the right side camshaft pulley.
39. The timing belt tensioner must be set prior to installation. The tensioner can be set as follows:
    a. Place a plate washer between the tensioner and a block. Using a suitable press, press in the pushrod using 220–2205 lbs. (100–1000kg) of pressure.
    b. Align the holes of the pushrod and housing, pass the proper tool (0.05 in. Allen wrench) through the holes to keep the setting position of the pushrod.
    c. Release the press and install the dust boot on the tensioner.
40. Install the tensioner and tighten the bolts to 20 ft. lbs. (26 Nm). Remove the tool from the tensioner.
41. Turn the crankshaft pulley 2 complete revolutions from TDC-to-TDC. Always turn the crankshaft clockwise. Check that each pulley aligns with the timing marks.
42. Install the alternator and drive belt tensioner.
43. Install the distributor caps, rotors and housings.
44. Install the coolant bypass hose and timing belt center covers.
45. Install the left side ignition coil and wiring connectors.
46. Install the accessory belt idler.
47. Install the spark plug wires.
48. Install the timing belt upper cover.
49. Install the spark plug and engine wiring harness covers.
50. Install the air intake pipe.
51. Install the A/C compressor.
52. Install the radiator.
53. On SC 400 models, install the accessory drive belt and the coolant reservoir.
54. On LS 400m models, install the power steering pump.
55. Install the air cleaner.
56. Install the fan clutch and pulley.

57. Install the accessory drive belt.
58. Install the air intake duct and engine appearance covers.
59. Install the engine splash shield.
60. Refill the engine coolant.
61. Connect the negative battery cable.

4.0L (1UZ-FE) 1998–00 ENGINES

YEAR, MODEL & ENGINE APPLICATION

1998-00 Lexus GS 400, LS 400, SC 400 .... 4.0L Code 1UZ-FE

REPLACEMENT INTERVALS

Lexus recomends timing belt replacement every 90,000 miles.

LABOR TIMES

GS 400	
Replacement	4.3 hrs
Severe Service	4.7 hrs
LS 400	
Replacement	4.5 hrs
Severe Service	5.0 hrs
SC 400	
Replacement	4.8 hrs
Severe Service	5.4 hrs

8807LEXg09

✱✱ CAUTION

These engines are an interference type. If the timing belt has failed, valve-to-piston damage may have occurred. A compression check should always be performed to check for engine damage.

1. Turn the ignition to the LOCK position. Disconnect the negative and positive battery cables. Wait at least 90 seconds for the air bag system to deactivate.
2. Drain the engine coolant.
3. Remove the engine splash shield.
4. On GS 400 models, remove the air intake duct and engine appearance cover.
5. Remove the air cleaner.
6. Remove the intake air pipe.
7. Remove the accessory drive belt.
8. On LS 400 models, remove the oil pan shield.
9. Remove the battery clamp cover.
10. Remove the air intake duct and engine appearance cover.
11. Remove the air cleaner.
12. Remove the accessory drive belt.
13. On SC 400 models, remove the battery and battery tray.
14. Remove the engine appearance cover.
15. Remove the air intake connector.
16. Remove the accessory drive belt.
17. On all models, remove the radiator.
18. Remove the alternator.
19. Disconnect all electrical connectors and necessary hose clamps and remove the left and right side timing belt covers.
20. Remove the A/C compressor without disconnecting the lines and position aside.
21. Remove the fan bracket on LS 400 models.
22. Remove the hydraulic pump on SC 400 models.
23. Rotate the crankshaft clockwise until No. 1 cylinder is at TDC of the compression stroke and the timing marks are aligned.
24. Loosen the crankshaft pulley bolt with the proper tools.
25. Rotate the crankshaft pulley about 50 degrees clockwise until the pulley timing mark is aligned with the guide pulley mark.
26. Remove the crankshaft pulley bolt.
27. Remove the timing belt tensioner.
28. Remove the timing belt from the camshaft sprockets.

■ **It may be necessary to rotate the left camshaft sprocket slightly to ease tension on the belt.**

29. Remove the crankshaft pulley.
30. Remove the timing belt lower cover.
31. Remove the timing belt guide (No. 1 crank position sensor plate).
32. Remove the timing belt.

■ **If the timing belt is to be reused, draw a directional arrow on the timing belt in the direction of engine rotation (clockwise) and place matchmarks on the timing belt and crankshaft gear to match the drilled mark on the pulley.**

To install:

33. Measure the tensioner pushrod protrusion. Extended length should be 0.41–0.45 in. (10.4–11.4mm). If the protrusion is not as specified, replace the tensioner.
34. The timing belt tensioner must be set prior to installation. The tensioner can be set as follows:
    a. Place a plate washer between the tensioner and a block. Using a suitable press, press in the pushrod using 220–2205 lbs. (100–1000kg) of pressure.

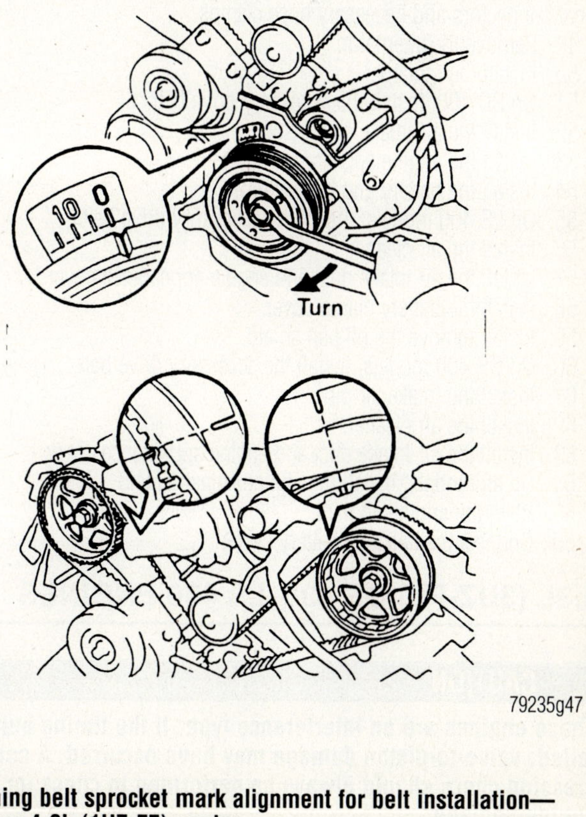

Timing belt sprocket mark alignment for belt installation—Lexus 4.0L (1UZ-FE) engine

b. Align the holes of the pushrod and housing, pass the proper tool (1.27 in. Allen wrench) through the holes to keep the setting position of the pushrod.

c. Release the press and install the dust boot on the tensioner.

■ If the old timing belt is being reinstalled, be sure the directional arrow is facing in the original direction and that the belt and crankshaft gear matchmarks are properly aligned.

35. Install the timing belt on the crankshaft timing pulley, No. 1 idler pulley and the No. 2 idler pulley.
36. Install the timing belt guide (No. 1 crank angle sensor plate) with the cup side facing forward. Replace the timing belt cover spacer.
37. Install the No. 1 timing belt cover and tighten the mounting bolts.
38. Align the pulley set key on the crankshaft with the key groove of the pulley. Install the pulley, using the proper tool to tap in the pulley.
39. Ensure that the crankshaft pulley timing mark is aligned at 50 degrees ATDC.
40. Install the timing belt to the left side camshaft timing pulley by slightly turning the left side timing pulley counterclockwise until there is tension between it and the crankshaft sprocket.
41. Place the timing belt around the right camshaft sprocket.
42. Install the tensioner and tighten the bolts to 20 ft. lbs. (26 Nm). Remove the tool from the tensioner.
43. Turn the crankshaft pulley 2 complete revolutions from TDC-to-TDC. Always turn the crankshaft clockwise. Check that each pulley aligns with the timing marks.
44. Tighten the crankshaft pulley bolt to 181 ft. lbs. (245 Nm).
45. On SC 400 models, install the hydraulic pump.
46. Install the fan bracket on LS 400 models.
47. Install the A/C compressor.
48. Install the left and right side timing covers, then connect all electrical connectors and necessary hose clamps.
49. Remove the alternator.
50. Install the radiator.
51. On SC 400, install the accessory drive belt.
52. Install the air intake connector.
53. Install the engine appearance cover.
54. Install the battery and battery tray.
55. On LS 400 models, install the accessory drive belt.
56. Install the air cleaner.
57. Install the air intake duct and engine appearance cover.
58. Install the battery clamp cover.
59. Install remove the oil pan shield.
60. On GS 400 models, install the accessory drive belt.
61. Install the intake air pipe.
62. Install the air cleaner.
63. Install the air intake duct and engine appearance cover.
64. On all models, install the engine splash shield.
65. Refill the engine coolant.
66. Connect the negative battery cable.

4.3L (3UZ-FE) GS 430–LS 430 ENGINES

✻✻ CAUTION

These engines are an interference type. If the timing belt has failed, valve-to-piston damage may have occurred. A compression check should always be performed to check for engine damage.

YEAR, MODEL & ENGINE APPLICATION

Application	Engine
2001-04 Lexus GS 430, LS 430	4.3L Code 3UZ-FE

REPLACEMENT INTERVALS

Lexus recommends timing belt replacement every 90,000 miles.

LABOR TIMES

GS 430	
Replacement	4.3 hrs
Severe Service	4.7 hrs
LS 430	
Replacement	3.3 hrs
Severe Service	3.7 hrs

8807LEXg10

1. Turn the ignition to the LOCK position. Disconnect the negative and positive battery cables. Wait at least 90 seconds for the air bag system to deactivate.
2. Drain the engine coolant.
3. Remove the air cleaner inlet.
4. Remove the engine appearance covers.
5. Remove the air cleaner.
6. Remove the intake air connector.
7. Remove the radiator hoses.
8. Disconnect the A/T cooler hoses from the radiator.
9. Remove the electrical connectors from the radiator area, then remove the radiator and cooling fan.
10. On GS 430 models, remove the No. 2 drive belt idler pulley and cover plate.
11. On all models, remove the water bypass pipe, engine wiring harness clamps, and electrical connectors from the No. 1 and 2 timing belt covers.
12. Remove the No. 2 and 3 timing belt covers.
13. Remove the A/C compressor without disconnecting the hoses and position aside.
14. Remove the No. 1 idler belt pulley.
15. Rotate the crankshaft until the pulley groove is in line with the No. 1 timing belt cover.
16. Ensure that the crankshaft and camshaft sprockets timing marks are aligned.
17. Rotate the crankshaft pulley about 50 degrees clockwise until the pulley timing mark is aligned with the guide pulley mark.
18. Remove the timing belt tensioner and dust boot.
19. Slightly turn the right side camshaft sprocket pulley clockwise to ease the timing belt tension.
20. Remove the timing belt from the camshaft sprockets.
21. Remove the accessory drive belt tensioner.
22. Remove the crankshaft pulley with a puller.
23. Remove the No. 1 timing belt cover.
24. Remove the timing belt guide and cover spacer.
25. Remove the timing belt.

■ If the timing belt is to be reused, draw a directional arrow on the timing belt in the direction of engine rotation (clockwise) and place matchmarks on the timing belt and crankshaft gear to match the drilled mark on the pulley.

To install:

26. Measure the tensioner pushrod protrusion. Extended length should be 0.38–0.41 in. (9.6–10.4mm). If the protrusion is not as specified, replace the tensioner.

27. The timing belt tensioner must be set prior to installation. The tensioner can be set as follows:

a. Place a plate washer between the tensioner and a block. Using a suitable press, press in the pushrod using 220–2205 lbs. (100–1000kg) of pressure.

b. Align the holes of the pushrod and housing, pass the proper tool (1.27 in. Allen wrench) through the holes to keep the setting position of the pushrod.

c. Release the press and install the dust boot on the tensioner.

■ **If the old timing belt is being reinstalled, be sure the directional arrow is facing in the original direction and that the belt and crankshaft gear matchmarks are properly aligned.**

28. Install the timing belt on the crankshaft timing pulley, No. 1 idler pulley and the No. 2 idler pulley.

29. Install the timing belt guide (No. 1 crank angle sensor plate) with the cup side facing forward. Replace the timing belt cover spacer.

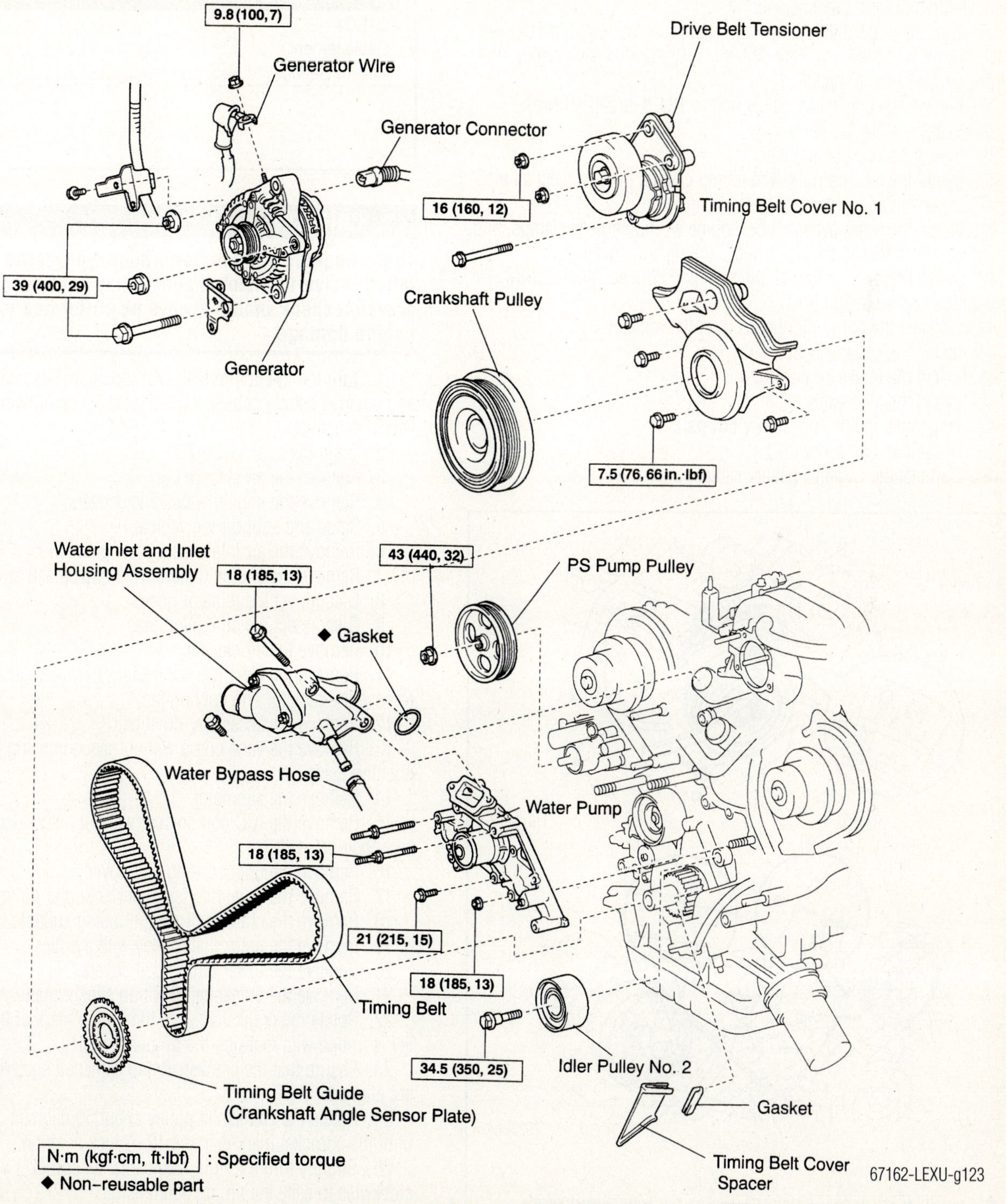

Exploded view of timing belt components—Lexus 4.3L (3UZ-FE) engine–GS 400 and LS 400

30. Install the No. 1 timing belt cover and tighten the mounting bolts.
31. Align the pulley set key on the crankshaft with the key groove of the pulley. Install the pulley, using the proper tool to tap in the pulley.
32. Ensure that the crankshaft pulley timing mark is aligned with the No.2 idler pulley bolt.
33. Install the timing belt to the left side camshaft timing pulley by slightly turning the left side timing pulley counterclockwise until there is tension between it and the crankshaft sprocket.
34. Place the timing belt around the right camshaft sprocket.
35. Install the tensioner and tighten the bolts to 20 ft. lbs. (26 Nm). Remove the tool from the tensioner.
36. Turn the crankshaft pulley 2 complete revolutions from TDC-to-TDC. Always turn the crankshaft clockwise. Check that each pulley aligns with the timing marks.
37. Tighten the crankshaft pulley bolt to 181 ft. lbs. (245 Nm).
38. Install the No. 2 idler pulley.
39. Install the A/C compressor.
40. Install the left and right side timing covers, then connect all electrical connectors and necessary hose clamps.
41. Install the water bypass pipe, engine wiring harness clamps, and electrical connectors to the No. 1 and 2 timing belt covers.
42. Install the radiator and cooling fan and connect the electrical connectors to the radiator area.
43. Connect the A/T cooler hoses from the radiator.
44. Install the radiator hoses.
45. Install the intake air connector.
46. Install the air cleaner.
47. Install the engine appearance covers.
48. Install the air cleaner inlet.
49. Connect the negative battery cable.

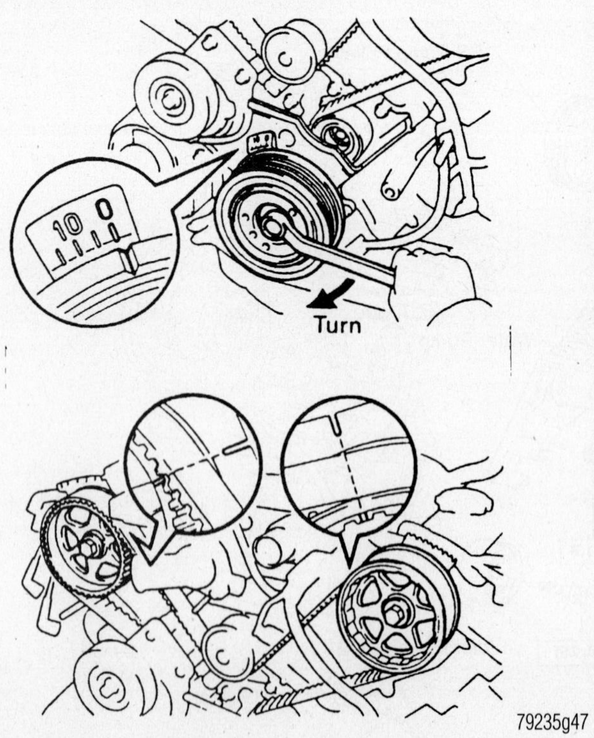

79235g47

Timing belt sprocket mark alignment for belt installation—Lexus 4.3L (3UZ-FE) engine–GS 400 and LS 400

4.3L (3UZ-FE) SC 430 ENGINES

YEAR, MODEL & ENGINE APPLICATION

2001-04 Lexus SC 430 .... 4.3L Code 3UZ-FE

REPLACEMENT INTERVALS

Lexus recomends timing belt replacement every 90,000 miles.

LABOR TIMES

2001-04

Replacement	4.1 hrs
Severe Service	4.6 hrs

8807LEXg11

✱✱ CAUTION

These engines are an interference type. If the timing belt has failed, valve-to-piston damage may have occurred. A compression check should always be performed to check for engine damage.

1. Turn the ignition to the LOCK position. Disconnect the negative and positive battery cables. Wait at least 90 seconds for the air bag system to deactivate.
2. Drain the engine coolant.
3. Remove the air cleaner inlet.
4. Remove the engine appearance covers.
5. Raise and support the vehicle.
6. Remove the air intake connector pipe.
7. Remove the engine undercovers and splash shield.
8. Disconnect the radiator hoses.
9. Disconnect the oil cooler lines.
10. Remove the air cleaner.
11. Remove the electrical connectors from the radiator area, then remove the radiator and cooling fan.
12. Remove the accessory drive belt.
13. Remove the vane pump without disconnecting the lines and position aside.
14. Remove the alternator.
15. Remove the A/C compressor without disconnecting the hoses and position aside.
16. Remove the upper timing belt cover.
17. Remove the drive belt tensioner and idler pulley assembly.
18. Remove the crankshaft damper using suitable tools.
19. Remove the crankshaft pulley with a puller.
20. Remove the lower timing belt covers.
21. Remove the crankshaft position sensor plate.
22. Rotate the crankshaft until the timing mark of the crankshaft pulley is in line with mark on the oil pump pulley.
23. Ensure that the crankshaft and camshaft sprockets timing marks are aligned.
24. Rotate the crankshaft pulley about 50 degrees counterclockwise until the sprocket mark is at the 12 o'clock position.
25. Slightly turn the right side camshaft sprocket pulley counterclockwise to ease the timing belt tension.
26. Remove the timing belt.

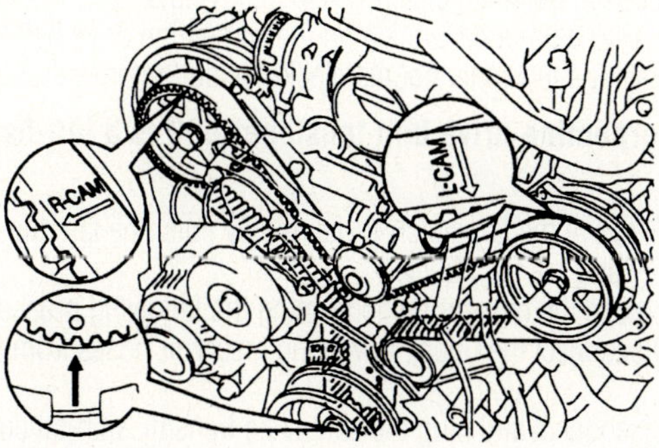

93025g28

Crankshaft and camshaft sprocket timing marks—Lexus 4.3L (3UZ-FE) engine–SC 430

93025g33

Checking TDC alignment marks after rotating crankshaft 2 revolutions—Lexus 4.3L (3UZ-FE) engine–SC 430

■ **If the timing belt is to be reused, draw a directional arrow on the timing belt in the direction of engine rotation (clockwise) and place matchmarks on the timing belt and crankshaft gear to match the drilled mark on the pulley.**

To install:

27. Measure the tensioner pushrod protrusion. Extended length should be 0.38–0.41 in. (9.6–10.4mm). If the protrusion is not as specified, replace the tensioner.

28. The timing belt tensioner must be set prior to installation. The tensioner can be set as follows:

a. Place a plate washer between the tensioner and a block. Using a suitable press, press in the pushrod using 220–2205 lbs. (100–1000kg) of pressure.

b. Align the holes of the pushrod and housing, pass the proper tool (1.27 in. Allen wrench) through the holes to keep the setting position of the pushrod.

c. Release the press and install the dust boot on the tensioner.

■ **If the old timing belt is being reinstalled, be sure the directional arrow is facing in the original direction and that the belt and crankshaft gear matchmarks are properly aligned.**

29. Set the crankshaft back to align with the mark on the oil pump housing and the camshaft sprockets back to the 12 o'clock position.

30. Install the timing belt on the crankshaft sprocket, guide pulley, left camshaft sprocket, water pump pulley, right camshaft sprocket and tensioner pulley.

■ **It may be necessary to turn the camshaft sprockets slightly to install the timing belt.**

31. Ensure the timing belt marks L-CAM and R-CAM are aligned with the camshaft sprocket marks.

32. Ensure the timing belt is taut on the non-tension side.

33. Install the tensioner and tighten the bolts to 20 ft. lbs. (26 Nm). Remove the tool from the tensioner.

34. Turn the crankshaft pulley 2 complete revolutions from TDC-to-TDC. Always turn the crankshaft clockwise. Check that each pulley aligns with the timing marks.

35. Install the crankshaft positon sensor plate.

36. Install the timing belt lower covers.

37. Install the crankshaft pulley.

38. Install the crankshaft damper and tighten the bolt to 181 ft. lbs. (245 Nm).

39. Install the drive belt tensioner and idler pulley assembly.

40. Install the upper timing belt cover.

41. Install the A/C compressor.

42. Install the alternator.

43. Install the vane pump.

44. Install the accessory drive belt.

45. Install the radiator and cooling and connect the electrical connectors to the radiator area.

46. Install the air cleaner.

47. Connect the oil cooler lines.

48. Connect the radiator hoses.

49. Install the engine undercovers and splash shield.

50. Install the air intake connector pipe.

51. Lower the vehicle.

52. Install the engine appearance covers.

53. Install the air cleaner inlet.

54. Connect the negative battery cable.

4.7L (2UZ-FE) ENGINES

YEAR, MODEL & ENGINE APPLICATION

2003-04 Lexus GX 470, 1998-04 LX 470 ... 4.7L Code 2UZ-FE

REPLACEMENT INTERVALS

Lexus recommends timing belt replacement every 90,000 miles.

LABOR TIMES

GX 470	
Replacement	3.6 hrs
Severe Service	4.1 hrs
LX 470	
Replacement	4.0 hrs
Severe Service	4.4 hrs

8807LEXg12

✻✻ CAUTION

These engines are an interference type. If the timing belt has failed, valve-to-piston damage may have occurred. A compression check should always be performed to check for engine damage.

1. Turn the ignition to the LOCK position. Disconnect the negative and positive battery cables. Wait at least 90 seconds for the air bag system to deactivate.
2. Raise and safely support the vehicle.
3. Remove the oil pan protector and the engine under cover.
4. Drain the cooling system and store the coolant for refilling purposes.
5. Lower the vehicle and remove the battery clamp cover.
6. From the top of the engine, remove the fuel return hose, the engine cover nuts/bolts and the cover.
7. Remove the air cleaner and the intake air connector assembly.
8. Remove the cooling fan pulley by performing the following procedures:
   a. Loosen the 4 fan clutch-to-fan pulley nuts.
   b. Using a box-end wrench on the serpentine drive belt tensioner bolt, rotate the tensioner counterclockwise and remove the drive belt.

■ **The serpentine drive belt tensioner bolt is a left-hand thread.**

   c. Remove the fan clutch-to-fan pulley nuts, the fan, the clutch assembly and the fan pulley.
9. Remove the radiator by performing the following procedures:
   a. Disconnect the upper, lower and reservoir hoses from the radiator.
   b. Disconnect and plug the automatic transmission oil cooler at the radiator. Disconnect the automatic transmission oil cooler hoses from the fan shroud clamp.
   c. Remove the radiator reservoir tank.
   d. Remove the fan shroud-to-radiator bolts and the shroud.

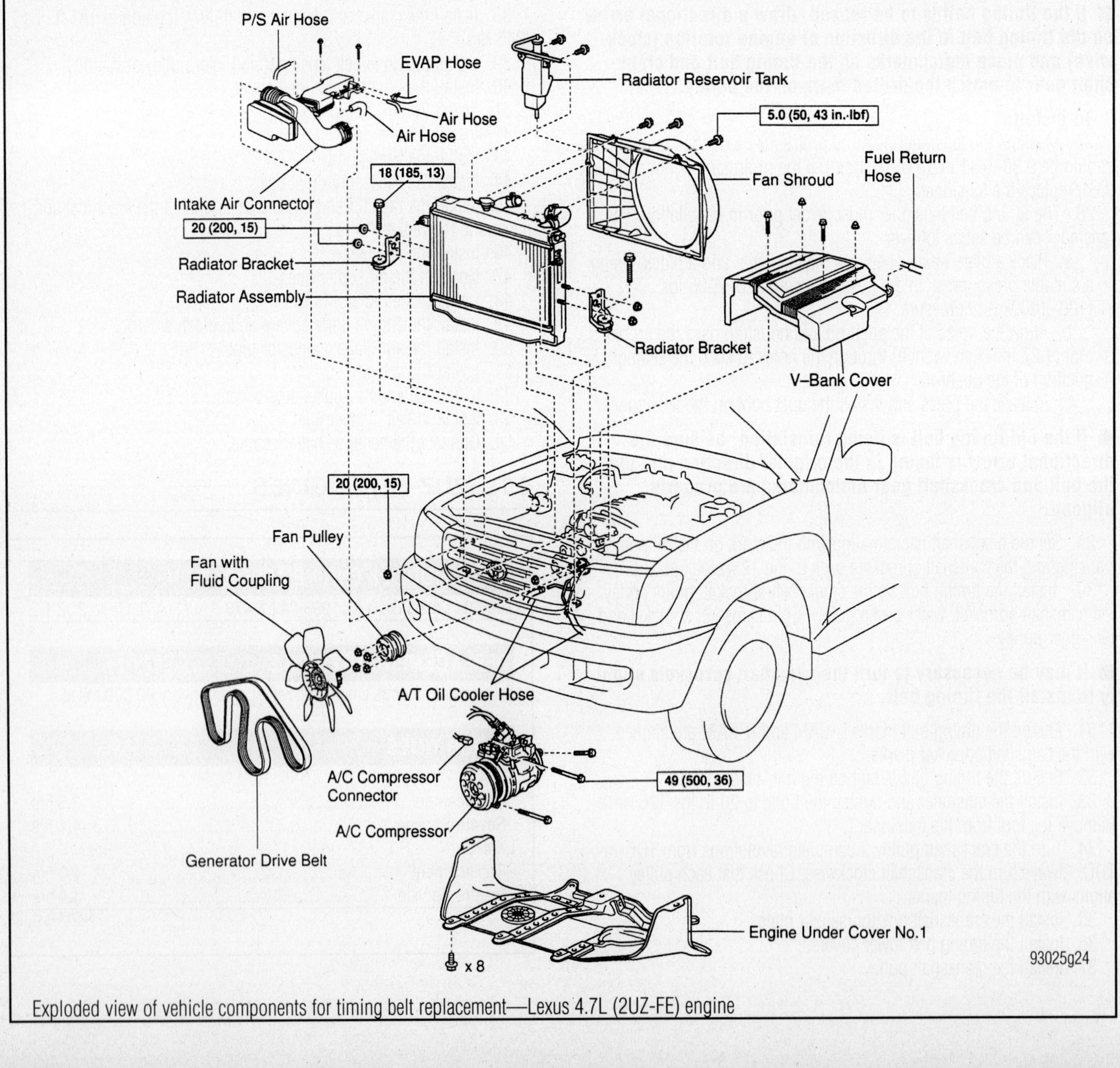

Exploded view of vehicle components for timing belt replacement—Lexus 4.7L (2UZ-FE) engine

e. Remove the 2 upper radiator-to-chassis nuts.

f. Remove the middle radiator-to-chassis nut/bolts and brackets.

g. Carefully, lift the radiator from the vehicle.

10. Remove the serpentine drive belt idler pulley bolt, cover plate and pulley.

11. Remove the right side (No. 3) timing belt cover.

12. Remove the left side (No. 3) timing belt cover by performing the following procedures:

a. Disconnect the engine wire from both wire clamps.

b. Disconnect the camshaft position sensor wire from the wire clamp on the left-side (No.3) timing belt cover.

c. Disconnect the sensor connector from the connector bracket.

d. Disconnect the sensor connector.

e. Remove the wire grommet from the left-side (No. 3) timing belt cover.

f. Remove the oil cooler tube bolts and tube.

13. Remove the middle (No. 2) timing belt cover bolts and cover.

14. Remove the cooling fan bracket nuts/bolts and bracket.

■ **If reusing the timing belt, make sure that there are 3 installation marks on the belt; if there are none, install them.**

15. Using the Crankshaft Pulley Holding tool 09213-70010, Bolt tool 90105-08076 and Companion Flange Holding tool 09330-00021, or equivalent, loosen the crankshaft pulley bolt.

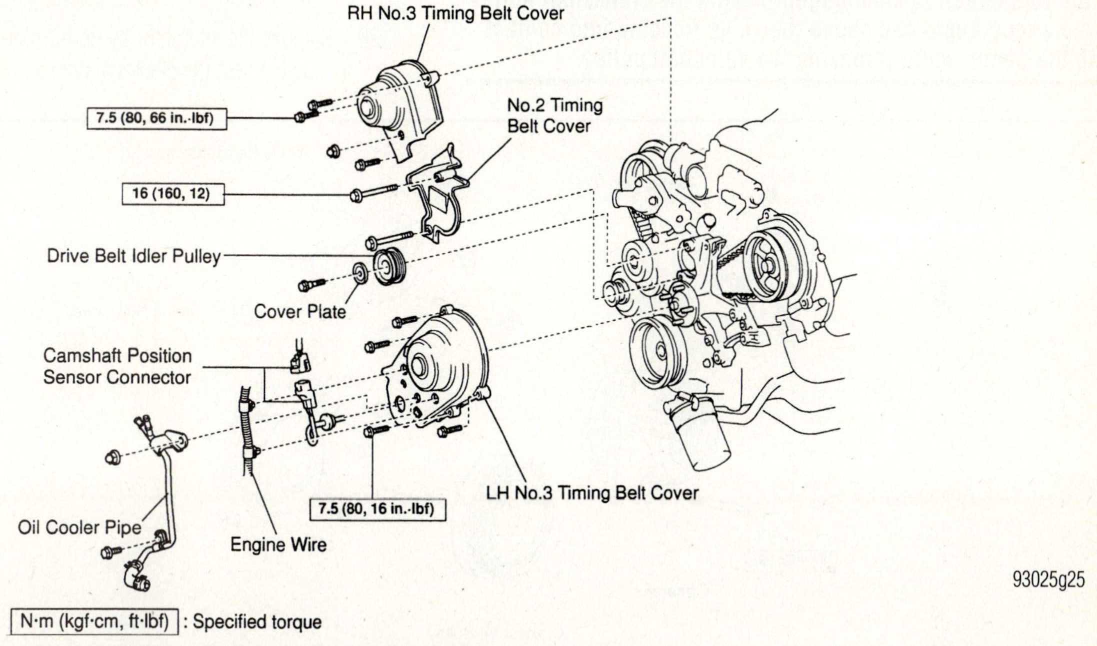

Exploded view of upper timing belt covers—Lexus 4.7L (2UZ-FE) engine

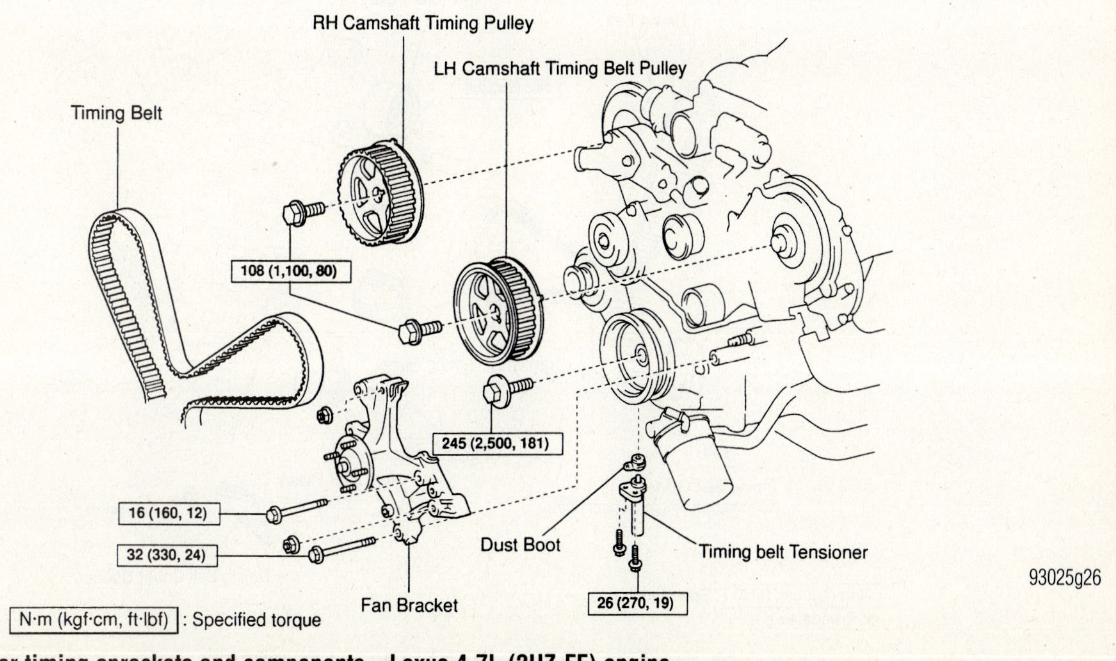

Exploded view of upper timing sprockets and components—Lexus 4.7L (2UZ-FE) engine

16. Position the No. 1 cylinder to approximately 50 degrees After Top Dead Center (ATDC) of the compression stroke by performing the following procedures:

a. Rotate the crankshaft pulley (CLOCKWISE) to align its groove with the timing mark "0" on the lower (No. 1) timing belt cover.

b. Check that the camshaft sprocket timing marks are aligned with the rear timing belt plate marks; if not, rotate the crankshaft 1 revolution (360 degrees).

c. Rotate the crankshaft pulley approximately 50 degrees (CLOCKWISE) and align the crankshaft pulley timing mark between the centers of the crankshaft pulley bolt and the idler pulley bolt.

✻✻ WARNING

If the timing belt is disengaged, having the crankshaft pulley in the wrong angle can cause the valve to come into contact with the piston when removing the camshaft pulley.

17. Remove the crankshaft pulley bolt.

■ **If reusing the timing belt and the installation marks have disappeared, place new installation marks on the timing belt to match the camshaft timing sprocket marks.**

■ **To avoid meshing the timing sprocket and the timing belt, secure one with a string; then, place matchmarks on the timing belt and the right-side camshaft timing sprocket.**

18. Remove the timing belt tensioner bolts and the tensioner.

19. Using the Camshaft Holding tool 09960-10010, or equivalent, slightly turn the left-side camshaft sprocket clockwise to loosen the tension spring. Then, disconnect the timing belt from the camshaft sprockets.

20. Remove the alternator by performing the following procedures:

a. Disconnect the electrical connector from the alternator.

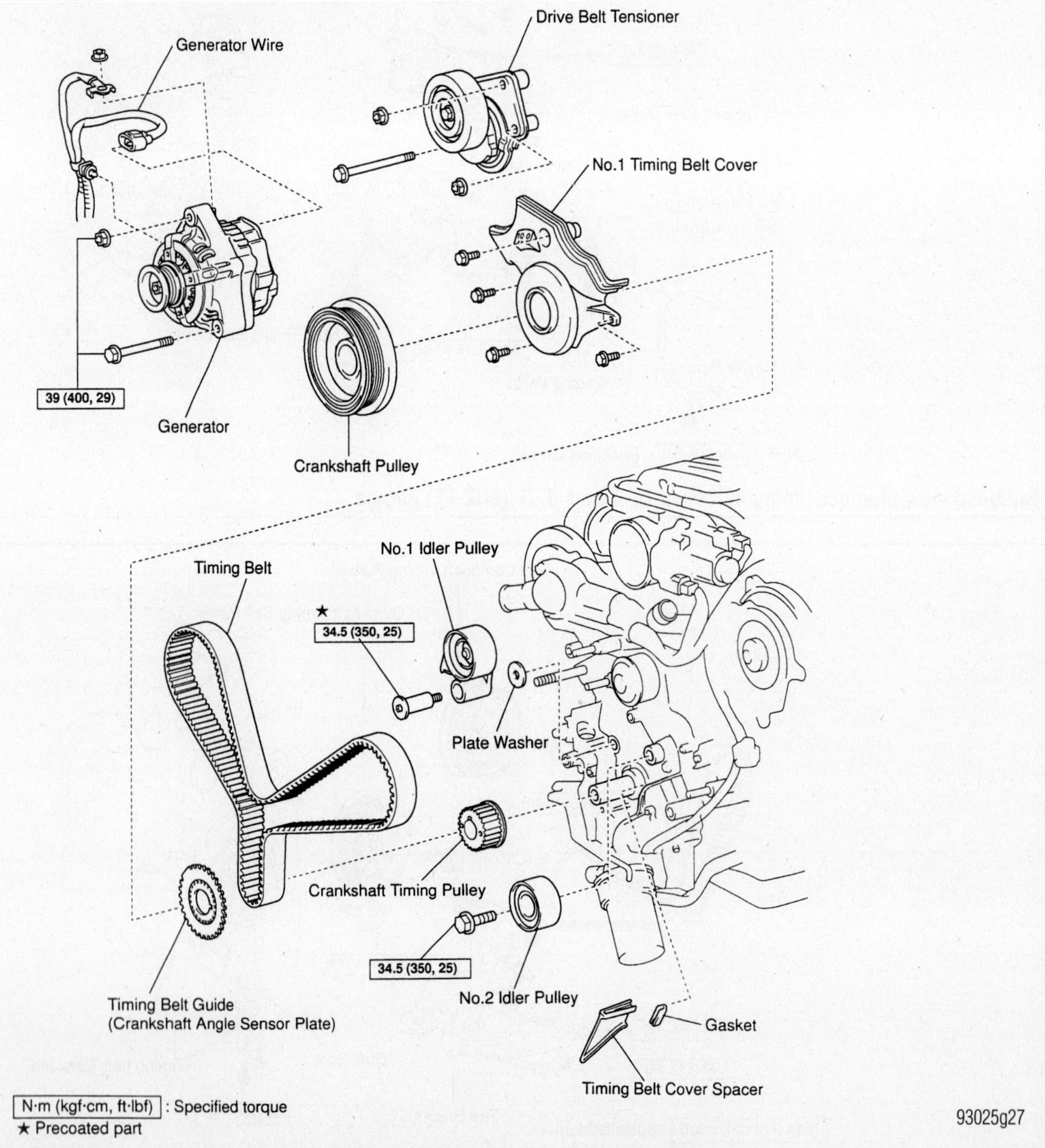

Exploded view of lower timing belt cover, sprockets and components—Lexus 4.7L (2UZ-FE) engine

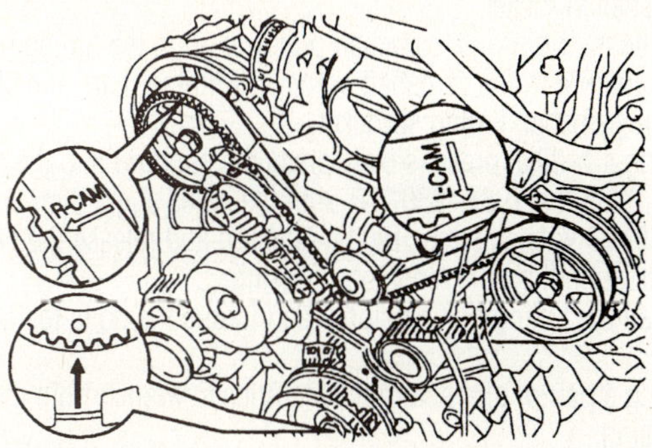

93025g28

Alignment of timing belt with the timing sprockets—Lexus 4.7L (2UZ-FE) engine

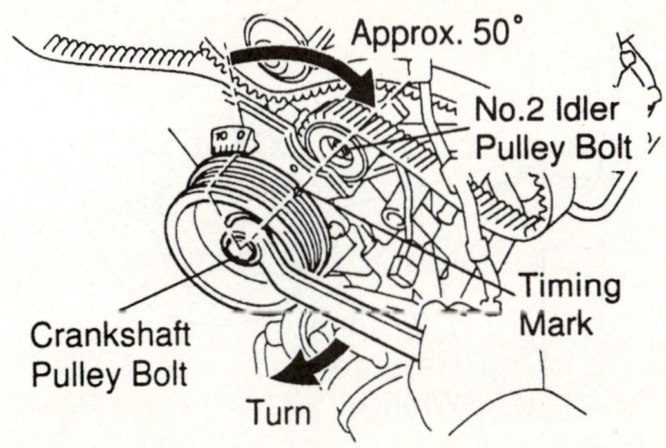

93025g29

Aligning of crankshaft pulley timing mark with the center line of the crankshaft pulley bolt and the idler pulley bolt—Lexus 4.7L (2UZ-FE) engine

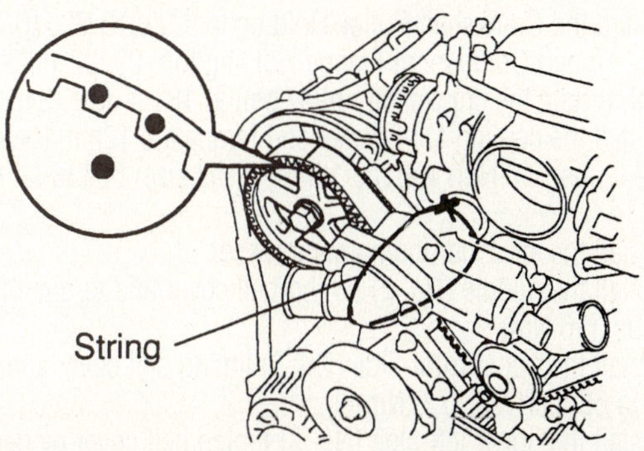

93025g30

Securing the timing belt with string and matchmarking the camshaft with the timing belt—Lexus 4.7L (2UZ-FE) engine

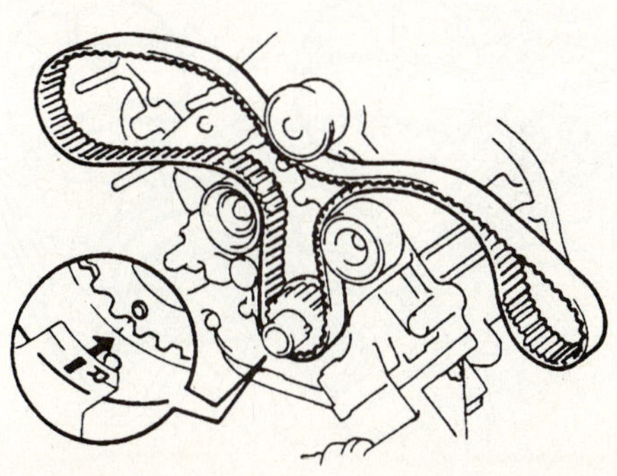

93025g31

Installing the timing belt on the crankshaft sprocket—Lexus 4.7L (2UZ-FE) engine

b. Remove the rubber cap/nut and disconnect the battery wire from the alternator.

c. Disconnect the wire clamp from the alternator cord clip.

d. Remove the alternator-to-engine nuts/bolts and the alternator.

21. Remove the serpentine drive belt tensioner nuts/bolts and the tensioner.

22. Using the Crankshaft Puller Assembly tool 09950-50012, or equivalent, press the crankshaft pulley from the crankshaft.

**** WARNING**

DO NOT rotate the crankshaft pulley.

23. Remove the lower (No. 1) timing belt cover bolts and the cover.

24. Remove the timing belt guide, spacer and the timing belt.

To install:

■ **With the timing belt removed, this is a perfect opportunity to inspect and/or replace the water pump.**

25. Inspect the timing belt tensioner by performing the following procedures:

a. Inspect the seal for leakage; if leakage is suspected, replace the tensioner.

b. Using both hands to hold the tensioner facing upward, strongly press the pushrod against a solid surface. If the pushrod moves, replace the tensioner.

**** WARNING**

Never hold the tensioner with the pushrod facing downward.

c. Measure the pushrod protrusion from the housing end, it should be 0.413–0.453 in. (10.5–11.5mm). If the protrusion is not as specified, replace the tensioner.

26. Temporarily install the timing belt by performing the following procedures:

d. Align the timing belt's installation mark with the crankshaft timing sprocket.

e. Install the timing belt on the crankshaft timing sprocket, the No. 1 idler pulley and the No. 2 idler pulley.

27. Install the gasket to the timing belt cover spacer and install the cover spacer.

28. Install the timing belt guide with the cup side facing outward.

29. Install the lower (No. 1) timing belt cover.

30. Install the crankshaft pulley by performing the following procedures:

a. Align the crankshaft pulley with the crankshaft key.

b. Using the Crankshaft Installer tool 09223-46011, or equivalent, and a hammer, tap the crankshaft pulley into position.

31. Install the serpentine drive belt tensioner and torque the tensioner-to-engine bolts to 12 ft. lbs. (16 Nm).

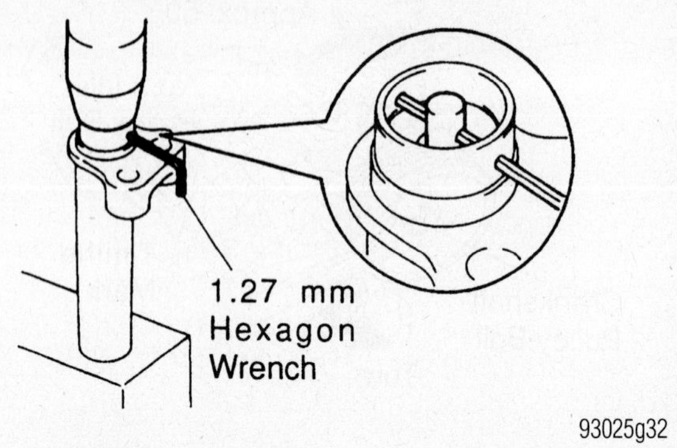

93025g32

Securing the timing belt tensioner pushrod<em dash>Lexus 4.7L (2UZ-FE) engine

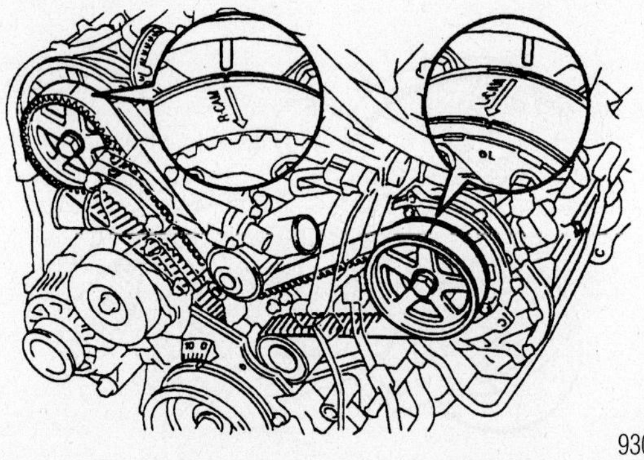

93025g33

Checking the TDC alignment marks after rotating the crankshaft 2 revolutions<em dash>Lexus 4.7L (2UZ-FE) engine

■ **To install the serpentine drive belt tensioner, use a bolt 4.18 in. (106mm) in length.**

32. Check that the crankshaft pulley's timing mark is aligned with the centers of the idler pulley and crankshaft pulley bolts.
33. Install the alternator and torque the alternator-to-engine nuts/bolts to 29 ft. lbs. (39 Nm). Connect the alternator's electrical connectors and clip.
34. Install the timing belt to the left-side camshaft by performing the following procedures:
    a. Rotate the left-side camshaft pulley to align the timing belt installation mark with the camshaft sprocket's timing mark and slide the belt onto the camshaft timing sprocket.
    b. Using the Camshaft Holding tool 09960-10010, or equivalent, slightly turn the left-side camshaft sprocket counterclockwise to place tension on the timing belt between the crankshaft sprocket and the camshaft sprocket.
35. Rotate the right-side camshaft pulley to align the timing belt installation mark with the camshaft sprocket's timing mark and slide the belt onto the camshaft timing sprocket.
36. Using a vertical press, slowly press the pushrod into the housing using 200–2205 lbs. (981–9807 N) until the holes align, then, install a 1.27mm Allen® wrench to secure the pushrod and release the press. Install the dust boot on the tensioner housing.
37. Install the timing belt tensioner and torque the bolts to 19 ft. lbs. (26 Nm).
38. Using a pair of pliers, remove the Allen® wrench from the tensioner housing.
39. Check the valve timing by performing the following procedure:
    a. Temporarily install the crankshaft pulley bolt.
    b. Slowly, rotate the crankshaft pulley 2 revolutions (CLOCKWISE) and realign the TDC marks.

■ **If the pulley/sprocket timing marks do not realign, remove the timing belt and reinstall it.**

40. Using the Crankshaft Pulley Holding tool 09213-70010, Bolt tool 90105-08076 and Companion Flange Holding tool 09330-00021, or equivalent, torque the crankshaft pulley bolt to 181 ft. lbs. (245 Nm).
41. Install the cooling fan bracket and torque the 12mm (head size) bolt to 12 ft. lbs. (16 Nm) and the 14mm (head size) bolt to 24 ft. lbs. (32 Nm).
42. Install the air conditioning compressor.
43. Install the middle (No. 2) timing belt cover and torque the bolts to 12 ft. lbs. (16 Nm).
44. Install the upper right-side (No. 3) timing belt cover and torque the bolts to 66 inch lbs. (7.5 Nm).
45. Install the upper left-side (No. 3) timing belt cover by performing the following procedures:
    a. Install the oil cooler tube and bolt.
    b. Feed the Camshaft Position Sensor (CPS) through the left-side (No. 3) timing belt cover hole.
    c. Install the left-side (No. 3) timing belt cover and torque the bolts to 66 inch lbs. (7.5 Nm).
    d. Install the wire grommet to the left-side (No. 3) timing belt cover.
    e. Install the sensor connector to the connector bracket and connect the sensor connector.
    f. Install the sensor wire and the engine wire to the clamps on the left-side (No. 3) timing belt cover.
46. Install the drive belt idler pulley and cover plate; then, torque the pulley bolt to 27 ft. lbs. (37 Nm).
47. To complete the installation, reverse the removal procedures.
48. Refill the cooling system and connect the negative battery cable.

Mazda

1.5L (Z5) DOHC ENGINES

YEAR, MODEL & ENGINE APPLICATION	
1995-98 Mazda Protege	1.5L Code Z5

REPLACEMENT INTERVALS

Mazda recommends timing belt replacement every 105,000 miles on California models, and 60,000 miles on all other models.

LABOR TIMES	
1995-98	
Replacement	2.7 hrs
Severe Service	2.9 hrs
With air conditioning add	0.2 hrs
With power steering add	0.1 hrs

8807MAZg01

✱✱ CAUTION

These engines are an interference type. If the timing belt has failed, valve-to-piston damage may have occurred. A compression check should always be performed to check for engine damage.

1. Disconnect the negative battery cable.
2. Raise and safely support the vehicle.
3. Remove the right front wheel and the right side splash shield.
4. Lower the vehicle.
5. Label and disconnect the spark plug wires. Remove the spark plugs.

■ **Spark plugs are removed to make it easier to rotate the engine.**

6. Remove the cylinder head cover.
7. Remove the accessory drive belts.
8. Remove the water pump pulley.
9. Raise and safely support the vehicle.
10. Remove the crankshaft pulley bolts and remove the crankshaft pulley and baffle plate.
11. Remove the oil dipstick and tube, if necessary.
12. Using a suitable engine holding brace, remove the front engine mount bracket.
13. Remove the upper, middle and lower timing belt covers.
14. Using a suitable tool, hold the crankshaft pulley boss and remove the pulley lockbolt. Remove the crankshaft pulley boss.
15. Temporarily reinstall the crankshaft pulley boss lockbolt.
16. Turn the crankshaft until the timing mark on the crankshaft sprocket aligns with the timing mark on the oil pump housing and the camshaft sprocket timing marks align with the surface of the cylinder head.
17. Loosen the tensioner pulley lockbolt and pull the tensioner pulley away from the center of the engine to reduce the tension on the timing belt.
18. If the timing belt is to be reused, mark the direction of rotation on the timing belt.
19. Remove the timing belt.
20. Remove the crankshaft pulley boss lockbolt.
21. Remove the tensioner and idler pulleys and check for free movement and abnormal noises. Replace as needed.

To install:

22. Check the free length of the tensioner spring. Free length should be 2.80 in (71 mm). If length is not as specified, replace the spring.
23. Be sure the timing marks on the camshaft and crankshaft sprockets are still aligned.
24. Install the idler pulleys and torque the bolts to 28–38 ft. lbs. (38–51 Nm).
25. Install the tensioner pulley and position the tensioner with the spring fully extended and temporarily torque the lockbolt.
26. Install the timing belt in the following order:
    a. Crankshaft pulley
    b. Lower idler pulley
    c. Upper idler pulley
    d. Camshaft pulley
    e. Tensioner
27. Install the crankshaft pulley boss and lockbolt.
28. Remove the holding tool from between the camshaft sprockets.
29. Rotate the crankshaft in the direction of rotation 1 5/6 turns to align the crankshaft timing mark with the tension set mark.
30. Loosen the tensioner lockbolt to apply tension to the timing belt. Torque the mounting bolt to 28–38 ft. lbs. (38–51 Nm).
31. Rotate the crankshaft in the direction of rotation 2 1/6 turns and be sure all marks are correctly aligned.
32. Torque the crankshaft pulley boss lockbolt to 116–122 ft. lbs. (157–166 Nm).
33. Install the upper, middle and lower timing belt covers.
34. Install the front engine mount bracket and remove the engine brace.
35. Install the oil dipstick and tube, if removed.
36. Install the crankshaft pulley, baffle plate and bolts. Torque the bolts to 12 ft. lbs. (17 Nm).
37. Lower the vehicle.
38. Install the water pump pulley.
39. Install the accessory drive belts.
40. Install the cylinder head cover.
41. Install the spark plugs and plug wires.
42. Raise and safely support the vehicle.
43. Install the right front wheel and the right side splash shield.
44. Connect the negative battery cable.

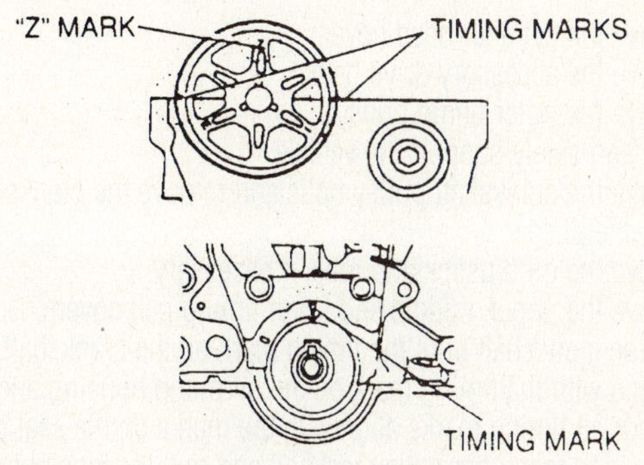

Proper timing belt sprocket mark alignment for belt installation—Mazda 1.5L (Z5) DOHC engine

1.6L (B6) DOHC ENGINES

YEAR, MODEL & ENGINE APPLICATION

1988-89 Mazda 323, 1989-93 Miata,
1994-95 MX-3 .... 1.6L Code B6

REPLACEMENT INTERVALS

Mazda recommends timing belt replacement every 60,000 miles.

LABOR TIMES

323	
Replacement	3.4 hrs
Severe Service	3.6 hrs
Miata	
Replacement	3.1 hrs
Severe Service	3.2 hrs
MX-3	
Replacement	3.1 hrs
Severe Service	3.3 hrs
All Models	
With air conditioning add	0.2 hrs
With power steering add	0.1 hrs

8807MAZa02

✳✳CAUTION

These engines are considered freewheeling. If the timing belt fails, valve-to-piston damage is unlikely, however a precautionary compression check should be performed to verify no engine damage.

1. Disconnect the negative battery cable.
2. Raise and safely support the vehicle.
3. Remove the right front wheel, side cover and splash shield.
4. Using a suitable engine holding brace, remove the front engine mount bracket.
5. Lower the vehicle.
6. Label and disconnect the spark plug wires. Remove the spark plugs.

■ Spark plugs are removed to make it easier to rotate the engine.

7. Remove the cylinder head cover.
8. Remove the accessory drive belts.
9. Remove the water pump pulley.
10. Raise and safely support the vehicle.
11. Remove the crankshaft pulley bolts and remove the crankshaft pulley and baffle plate.
12. Remove the oil dipstick and tube, if necessary.
13. Remove the upper, middle and lower timing belt covers.
14. Turn the crankshaft until the timing mark on the crankshaft sprocket aligns with the timing mark on the oil pump housing and the camshaft sprocket timing marks align with the marks on the seal plate.
15. Loosen the tensioner pulley lockbolt and pull the tensioner pulley away from the center of the engine to reduce the tension on the timing belt.

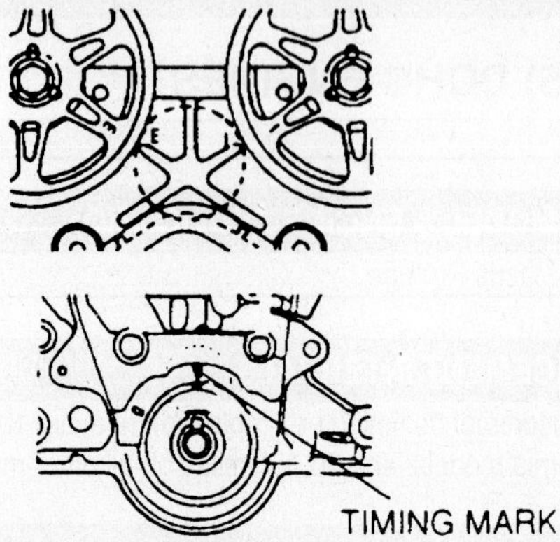

9305bg29

Proper timing belt sprocket mark alignment for belt installation—Mazda 1.6L (B6) DOHC engine

16. If the timing belt is to be reused, mark the direction of rotation on the timing belt.
17. Remove the timing belt.
18. Remove the tensioner and idler pulleys and check for free movement and abnormal noises. Replace as needed.

To install:

19. Be sure the timing marks on the camshaft and crankshaft sprockets are still aligned.
20. Install the idler pulleys and torque the bolts to 28–38 ft. lbs. (38–51 Nm).
21. Install the tensioner pulley and position the tensioner with the spring fully extended and temporarily torque the lockbolt.
22. Install the timing belt in the following order:
    a. Crankshaft pulley
    b. Idler pulley
    c. Left side camshaft pulley
    d. Right side camshaft pulley
    e. Tensioner
23. Install the crankshaft pulley.
24. Rotate the crankshaft in the direction of rotation 1 5/6 turns to align the crankshaft timing mark with the tension set mark.
25. Loosen the tensioner lockbolt to apply tension to the timing belt. Torque the mounting bolt to 28–38 ft. lbs. (38–51 Nm).

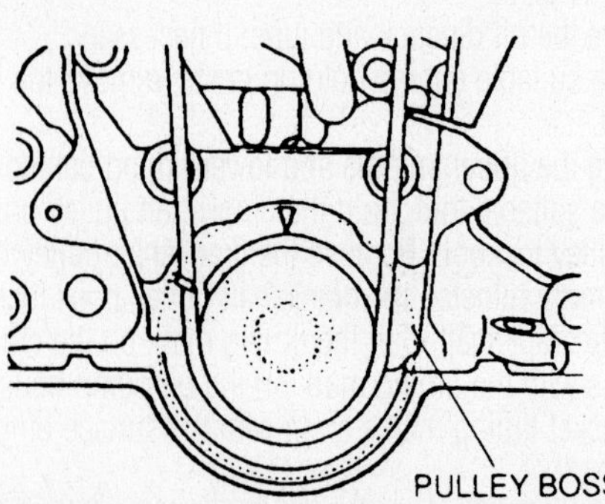

9305bg30

Alignment of crankshaft and tension set marks—Mazda 1.6L (B6) DOHC engine

26. Rotate the crankshaft in the direction of rotation 2 1/6 turns and be sure all marks are correctly aligned.
27. Torque the crankshaft pulley boss lockbolt to 116–122 ft. lbs. (157–166 Nm).
28. Install the upper, middle and lower timing belt covers.
29. Install the front engine mount bracket and remove the engine brace.
30. Install the oil dipstick and tube, if removed.
31. Install the crankshaft pulley, baffle plate and bolts. Torque the bolts to 12 ft. lbs. (17 Nm).
32. Lower the vehicle.
33. Install the water pump pulley.
34. Install the accessory drive belts.
35. Install the cylinder head cover.
36. Install the spark plugs and plug wires.
37. Raise and safely support the vehicle.
38. Install the right front wheel, side cover and the splash shield.
39. Connect the negative battery cable.

1.6L (B6E) SOHC ENGINES

YEAR, MODEL & ENGINE APPLICATION	
1986-94 Mazda 323, 1992-93 MX-3	1.6L Code B6E

REPLACEMENT INTERVALS

Mazda recommends timing belt replacement every 60,000 miles.

LABOR TIMES	
1986-94	
Replacement	2.4 hrs
Severe Service	2.6 hrs
With air conditioning add	0.2 hrs
With power steering add	0.1 hrs

8807MAZg03

✻✻ CAUTION

These engines are considered freewheeling. If the timing belt fails, valve-to-piston damage is unlikely, however a precautionary compression check should be performed to verify no engine damage.

1. Disconnect the negative battery cable.
2. Raise and safely support the vehicle.
3. Remove the right front wheel and right side splash shield.
4. Remove the accessory drive belts.
5. Remove the water pump pulley.
6. Remove the crankshaft pulley bolts and remove the crankshaft pulley and baffle plate.
7. Using a suitable tool to hold the crankshaft pulley boss, remove the pulley boss lockbolt. Remove the crankshaft pulley boss.
8. Remove the upper and lower timing belt covers.
9. Label and disconnect the spark plug wires. Remove the spark plugs.

■ **Spark plugs are removed to make it easier to rotate the engine.**

10. Temporarily reinstall the crankshaft pulley boss and lockbolt.
11. Turn the crankshaft until the camshaft sprocket and crankshaft sprocket timing marks are aligned.
12. If the timing belt is to be reused, mark the direction of rotation on the timing belt.
13. Loosen the timing belt tensioner lockbolt. Move the tensioner away from the belt to relieve tension. Torque the lockbolt.
14. Remove the timing belt.

■ **Do not rotate the engine after the timing belt has been removed.**

15. Inspect the belt for wear, peeling, cracking, hardening or signs of oil contamination. Inspect the tensioner for free and smooth rotation. Inspect the sprocket teeth for wear or damage. Replace parts, as necessary.

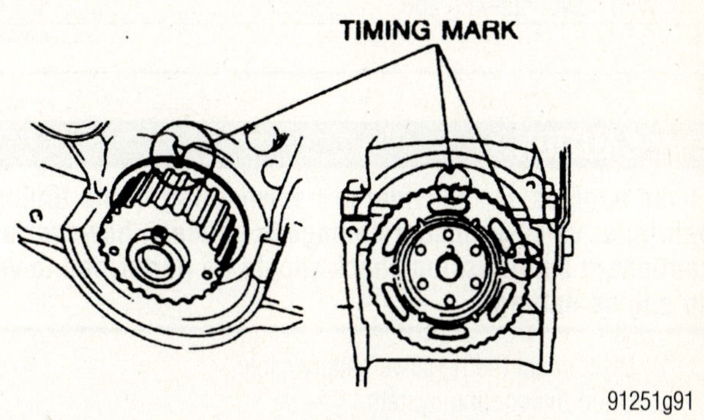

Crankshaft and camshaft sprocket timing mark alignment—Mazda 1.6L (B6E) SOHC engine

To install:

16. Make sure the timing marks on the camshaft and crankshaft sprockets are properly aligned.
17. Install the timing belt so there is no slack on the tension side. If reusing the old timing belt, make sure it is reinstalled in the same direction of rotation as marked.
18. Loosen the tensioner lockbolt to set tension against the belt, then torque the bolt to 19 ft. lbs. (25 Nm).
19. Turn the crankshaft 2 turns in the direction of rotation and check the timing mark alignment.
20. Apply approximately 22 lbs. (10 kg) pressure to the timing belt on the side opposite the tensioner, at a point midway between the sprockets. The belt should deflect 0.43–0.51 in. (11–13mm). If the tension is not as specified, replace the tensioner spring.
21. Install the upper and lower timing belt covers. Torque the bolts to 95 inch lbs. (11 Nm).
22. Using a suitable tool, install the crankshaft pulley boss and torque the lockbolt to 123 ft. lbs. (167 Nm).
23. Install the crankshaft pulley and baffle plate.
24. Install the water pump pulley.
25. Install the accessory drive belts.
26. Install the right front wheel and right side splash shield.
27. Lower the vehicle.
28. Install the spark plugs and connect the spark plug wires.
29. Connect the negative battery cable.
30. Start the engine and check for proper operation.
31. Check the ignition timing.

1.6L (B6ZE) DOHC ENGINES

YEAR, MODEL & ENGINE APPLICATION

1990-94 Mazda Miata	1.6L Code B6ZE

REPLACEMENT INTERVALS

Mazda recommends timing belt replacement every 60,000 miles.

LABOR TIMES

1990-94	
Replacement	3.1 hrs
Severe Service	3.2 hrs
With air conditioning add	0.2 hrs
With power steering add	0.1 hrs

8807MAZg04

✳✳ CAUTION

These engines are considered freewheeling. If the timing belt fails, valve-to-piston damage is unlikely, however a precautionary compression check should be performed to verify no engine damage.

1. Disconnect the negative battery cable.
2. Drain the cooling system.
3. Remove the air intake pipe.
4. Remove the upper radiator hose and disconnect the coolant hoses at the thermostat housing.
5. Remove the accessory drive belts.
6. Remove the water pump pulley.
7. Remove the crankshaft pulley bolts and the crankshaft pulley. On 1992–94 vehicles, hold the pulley boss with a suitable tool and remove the pulley lockbolt. Remove the pulley boss.
8. On 1990–91 vehicles, remove the outer and inner timing belt guide plates.
9. Label and disconnect the spark plug wires from the spark plugs. Remove the ignition coil and plug wires assembly. Remove the spark plugs.
10. Remove the cylinder head cover.
11. Remove the upper, middle and lower timing belt covers.
12. On 1992–94 vehicles, temporarily reinstall the pulley boss and lockbolt.
13. Turn the crankshaft until the crankshaft and camshaft sprocket timing marks are aligned. On 1992–94 vehicles, the pin on the pulley boss must face upward.
14. On 1992–94 vehicles, remove the pulley boss and lockbolt, being careful not to disturb the crankshaft.
15. If the timing belt is to be reused, mark the direction of rotation on the timing belt.
16. Loosen the tensioner lockbolt and pry the tensioner outward away from the belt. Temporarily torque the lockbolt with the tensioner spring fully extended.

■ **Protect the tensioner with a shop towel before prying on it. Do not rotate the crankshaft after the timing belt has been removed.**

17. Remove the timing belt.
18. Remove the tensioner and spring. If necessary, remove the idler pulley.
19. Inspect the belt for wear, peeling, cracking, hardening or signs of oil contamination. Inspect the tensioner for free and smooth rotation. Check the tensioner spring free length; it should not exceed 2.315 in. (58.8mm) from end to end. Inspect the sprocket teeth for wear or damage. Replace parts, as necessary.

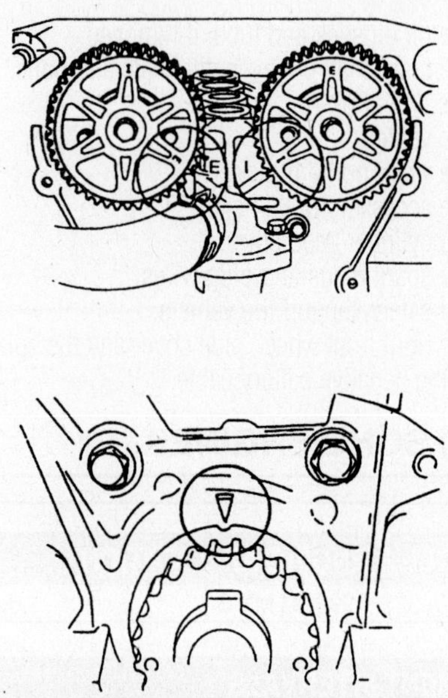

91251g94

Crankshaft and camshaft sprocket timing mark alignment—Mazda 1.6L (B6ZE) DOHC engine

To install:

20. If removed, install the idler pulley and torque the bolt to 38 ft. lbs. (52 Nm).
21. Install the tensioner and tensioner spring. Pry the tensioner outward and temporarily torque the tensioner lockbolt with the tensioner spring fully extended.
22. Make sure the crankshaft sprocket timing mark is aligned with the mark on the oil pump housing and the camshaft sprocket timing marks are aligned with the marks on the seal plate.
23. Install the timing belt so there is no looseness at the idler pulley side or between the camshaft sprockets. If reusing the old belt, make sure it is installed in the same direction of rotation as marked.
24. On 1992–94 vehicles, temporarily install the pulley boss and lockbolt.
25. Turn the crankshaft 2 turns in the direction of rotation and align the crankshaft sprocket timing mark. On 1992–94 vehicles, face the pin on the pulley boss upright. Make sure the camshaft sprocket timing marks are aligned.
26. Turn the crankshaft 1 5/6 turns clockwise and align the crankshaft sprocket timing mark with the tension set mark for proper belt tension adjustment. On 1992–94 vehicles, remove the lockbolt and pulley boss.
27. Make sure the crankshaft sprocket timing mark is aligned with the tension set mark. Loosen the tensioner lockbolt and allow the spring to apply tension to the belt. Torque the tensioner lockbolt to 38 ft. lbs. (52 Nm).
28. On 1992–94 vehicles, install the pulley boss and lockbolt.
29. Turn the crankshaft 2 1/6 turns clockwise and make sure the timing marks are correctly aligned.
30. Apply approximately 22 lbs. (10 kg) pressure to the timing belt at a point midway between the camshaft sprockets. The belt should deflect 0.35–0.45 in. (9.0–11.5mm).

31. Install the timing belt covers and torque the bolts to 95 inch lbs. (11 Nm).
32. Apply silicone sealer to the cylinder head in the area adjacent to the front and rear camshaft caps.
33. Install the cylinder head cover and torque the bolts to 78 inch lbs. (8.8 Nm).
34. Install the spark plugs. Install the ignition coil and torque the bolts to 19 ft. lbs. (25 Nm). Connect the spark plug wires.
35. On 1990–91 vehicles, install the inner timing belt guide plate with the dished side facing away from the engine. Install the outer guide plate.
36. On 1992–94 vehicles, hold the pulley boss with a suitable tool and torque the lockbolt to 123 ft. lbs. (167 Nm).
37. Install the crankshaft pulley. Torque the bolts to 13 ft. lbs. (17 Nm).
38. Install the water pump pulley.
39. Install the accessory drive belts and adjust the belt tension.
40. Connect the coolant hoses to the thermostat housing and install the upper radiator hose.
41. Install the air intake pipe.
42. Connect the negative battery cable.
43. Fill and bleed the cooling system.
44. Start the engine and bring to normal operating temperature.
45. Check for leaks and proper operation.
46. Check the ignition timing.

1.6L (ZM) ENGINES

YEAR, MODEL & ENGINE APPLICATION

1999-04 Mazda Protege/Protege5	1.6L Code ZM

REPLACEMENT INTERVALS

Mazda recommends timing belt replacement every 105,000 miles.

LABOR TIMES

1999-04	
Replacement	3.2 hrs
Severe Service	3.4 hrs
With air conditioning add	0.2 hrs
With power steering add	0.1 hrs

8807MAZg05

✻✻ CAUTION

These engines are an interference type. If the timing belt has failed, valve-to-piston damage may have occurred. A compression check should always be performed to check for engine damage.

1. Disconnect the negative battery cable.
2. Raise and safely support the vehicle.
3. Remove the right front wheel.
4. Remove the right side splash shield.
5. Lower the vehicle.
6. Remove the camshaft position sensor.
7. Remove the ignition coils.
8. Remove the accessory drive belt.
9. Remove the crankshaft pulley bolts and remove the crankshaft pulley and crankshaft position sensor plate.

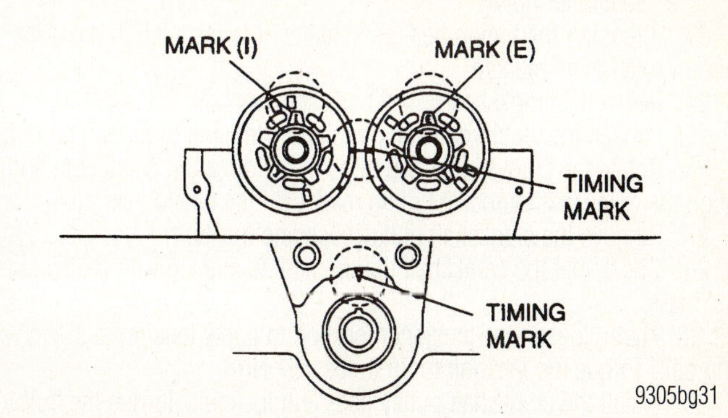

Crankshaft and camshaft sprocket timing mark alignment—Mazda 1.6L (ZM) DOHC engine

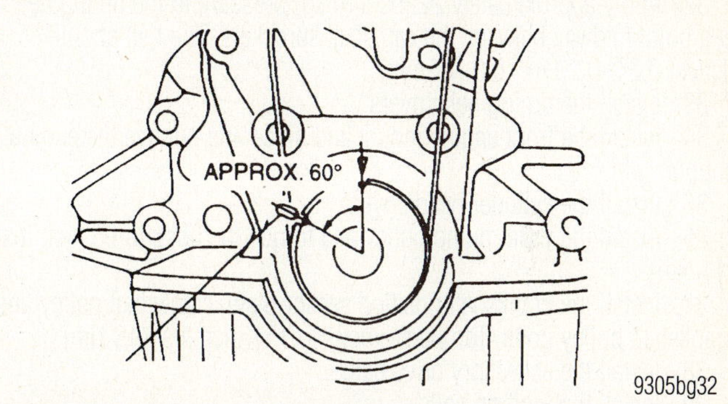

Alignment of crankshaft timing mark and tension set mark—Mazda 1.6L (ZM) DOHC engine

10. Remove the water pump pulley.
11. Remove the cylinder head cover.
12. Using special tool 49G0175A0 or equivalent, support the engine and remove the front engine mount and brace.
13. Remove the timing belt covers.
14. Rotate the engine in the direction of rotation and align the crankshaft and camshaft timing marks.
15. Using a suitable tool, hold the crankshaft pulley boss and remove the pulley bolt. Remove the pulley boss.
16. Loosen the tensioner lockbolt and move the tensioner away from the timing belt. Temporarily torque the lockbolt to hold the tensioner in place.
17. Remove the timing belt.
18. If the timing belt is to be reused, mark the direction of rotation on the belt.
19. Remove the tensioner and check the tensioner pulley for ease of movement and abnormal noises. Check the tensioner spring free length. The free length should measure 2.43 in. (61.8mm). If out of specification, replace the spring.

To install:

20. Install the tensioner and spring assembly. Move the tensioner fully to the left until the spring is in its fully extended position. Temporarily torque the lockbolt.
21. Be sure that the timing marks on the camshaft sprockets and the crankshaft sprocket are aligned.
22. Install the timing belt in the following order:
    a. Crankshaft sprocket
    b. Idler pulley
    c. Left camshaft sprocket
    d. Right camshaft sprocket

e. Tensioner pulley
23. If reusing the timing belt, be sure that it is installed in the direction of rotation as marked.
24. Install the crankshaft pulley boss and lockbolt.
25. Loosen the tensioner lockbolt to apply tension to the timing belt.
26. Rotate the crankshaft 1 5/6 turns in the direction of rotation and align the crankshaft sprocket timing mark with the tension set mark.
27. Remove the crankshaft pulley boss and lockbolt.
28. Check that the crankshaft timing mark is aligned with the tension set mark.
29. Again, loosen the tensioner lockbolt to apply tension to the timing belt. Torque the lockbolt to 38 ft. lbs. (51 Nm).
30. Install the crankshaft pulley boss and lockbolt. Torque the bolt to 122 ft. lbs. (166 Nm).
31. Turn the crankshaft 2 1/6 turns clockwise and make sure the timing marks are correctly aligned.
32. Apply approximately 22 lbs. (10 kg) pressure to the timing belt at a point midway between the camshaft sprockets. The belt should deflect 0.24–0.29 in. (6.0–7.5mm).
33. Install the timing belt covers.
34. Install the front engine mount and brace and remove the engine support tool.
35. Install the cylinder head cover.
36. Install the water pump pulley and torque the bolts to 95 inch lbs. (11 Nm).
37. Install the crankshaft position sensor plate, crankshaft pulley and crankshaft pulley bolts. Torque the bolts to 151 inch lbs. (18 Nm).
38. Install the accessory drive belt.
39. Install the ignition coils.
40. Install the camshaft position sensor.
41. Raise the vehicle.
42. Install the right side splash shield.
43. Install the right front wheel.
44. Lower the vehicle.
45. Connect the negative battery cable.

1.8L (BP) DOHC ENGINES

YEAR, MODEL & ENGINE APPLICATION

1990-94 Mazda 323, 1994-03 Miata, 1995-98 Protege .... 1.8L Code BP

REPLACEMENT INTERVALS

Mazda recommends timing belt replacement every 60,000 miles on 1990-98 models, 105,000 miles on 1999-00 models and 60,000 miles on 2001-03 models.

LABOR TIMES

323 & Protege	
Replacement	3.4 hrs
Severe Service	3.6 hrs
Miata	
Replacement	3.1 hrs
Severe Service	3.2 hrs
All Models	
With air conditioning add	0.2 hrs
With power steering add	0.1 hrs

8807MAZG06

1990–97

CAUTION

These engines are considered freewheeling. If the timing belt fails, valve-to-piston damage is unlikely, however a precautionary compression check should be performed to verify no engine damage.

1. Disconnect the negative battery cable.
2. Raise and safely support the vehicle.
3. Remove the right front wheel, engine undercover and the right side splash shield.
4. Lower the vehicle.
5. On Miata models, drain the engine coolant and remove the air intake pipe.
6. Label and disconnect the spark plug wires. Remove the spark plugs.

■ **Spark plugs are removed to make it easier to rotate the engine.**

7. Remove the cylinder head cover.
8. Remove the accessory drive belts.
9. Remove the water pump pulley.
10. Remove the crankshaft pulley bolts and remove the crankshaft pulley and baffle plate.
11. Remove the oil dipstick and tube, if necessary.

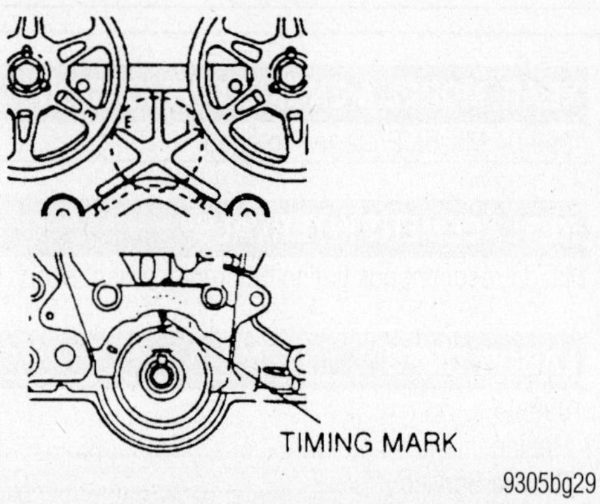

Proper timing belt sprocket mark alignment for belt installation—Mazda 1994-97 Miata and 1995-98 Protégé 1.8L (BP) DOHC engines

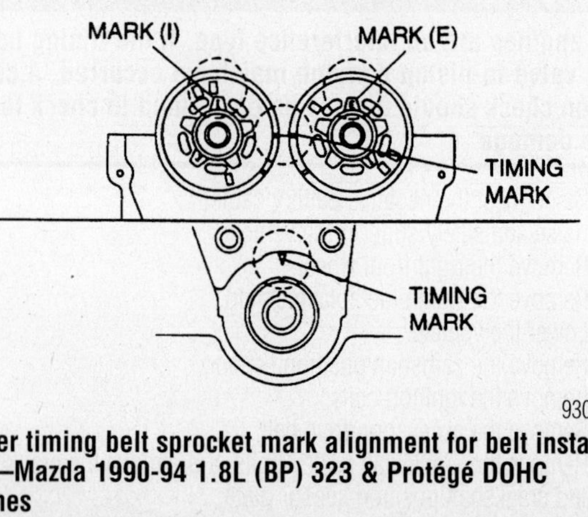

Proper timing belt sprocket mark alignment for belt installation—Mazda 1990-94 1.8L (BP) 323 & Protégé DOHC engines

12. Remove the upper, middle and lower timing belt covers.
13. Using a suitable tool, hold the crankshaft pulley boss and remove the pulley lockbolt. Remove the crankshaft pulley boss.
14. Temporarily reinstall the crankshaft pulley boss lockbolt.
15. Turn the crankshaft until the timing mark on the crankshaft sprocket aligns with the timing mark on the oil pump housing and the camshaft sprocket timing marks align with the marks on the seal plate.
16. Loosen the tensioner pulley lockbolt and pull the tensioner pulley away from the center of the engine to reduce the tension on the timing belt.
17. If the timing belt is to be reused, mark the direction of rotation on the timing belt.
18. Remove the timing belt.
19. Remove the crankshaft pulley boss lockbolt.
20. Remove the tensioner and idler pulleys and check for free movement and abnormal noises. Replace as needed.

To install:

21. Be sure the timing marks on the camshaft and crankshaft sprockets are still aligned.
22. Install the idler pulleys and torque the bolts to 28–38 ft. lbs. (38–51 Nm).
23. Install the tensioner pulley and position the tensioner with the spring fully extended and temporarily torque the lockbolt.
24. Install the timing belt in the following order:
    a. Crankshaft pulley
    b. Idler pulley
    c. Left side camshaft pulley
    d. Right side camshaft pulley
    e. Tensioner
25. Install the crankshaft pulley boss and lockbolt.
26. Remove the holding tool from between the camshaft sprockets.
27. Rotate the crankshaft in the direction of rotation 1 5/6 turns to align the crankshaft timing mark with the tension set mark.
28. Loosen the tensioner lockbolt to apply tension to the timing belt. Torque the mounting bolt to 28–38 ft. lbs. (38–51 Nm).
29. Rotate the crankshaft in the direction of rotation 2 1/6 turns and be sure all marks are correctly aligned.
30. Torque the crankshaft pulley boss lockbolt to 116–122 ft. lbs. (157–166 Nm).
31. Install the upper, middle and lower timing belt covers.
32. Install the oil dipstick and tube, if removed.
33. Install the crankshaft pulley, baffle plate and bolts. Torque the bolts to 12 ft. lbs. (17 Nm).
34. Lower the vehicle.
35. Install the water pump pulley.

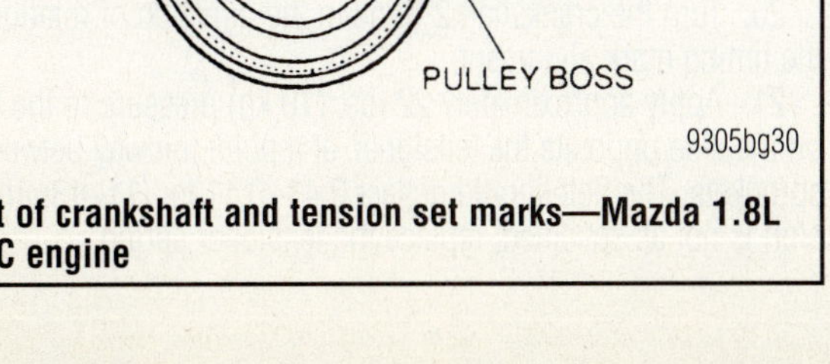

Alignment of crankshaft and tension set marks—Mazda 1.8L (BP) DOHC engine

36. Install the accessory drive belts.
37. Install the cylinder head cover.
38. Install the spark plugs and plug wires.
39. Raise and safely support the vehicle.
40. Install the right front wheel and the right side splash shield.
41. On Miata models, install the air intake pipe and refill the cooling system.
42. Connect the negative battery cable.

1998–04 Miata

✻✻ CAUTION

These engines are considered freewheeling. If the timing belt fails, valve-to-piston damage is unlikely, however a precautionary compression check should be performed to verify no engine damage.

1. Disconnect the negative battery cable.
2. Drain the engine coolant.
3. On 2001-03 models remove the front suspension tower brace.
4. On all models, remove the air intake pipe.
5. Remove the crankshaft and camshaft position sensors.
6. Disconnect the ignition coil connectors and coils.
7. Label and disconnect the spark plug wires. Remove the spark plugs.

■ **Spark plugs are removed to make it easier to rotate the engine.**

8. Remove the upper radiator hose and coolant hoses from the thermostat.
9. On 2001-03 models, remove the oil control valve.
10. On all models, remove the cylinder head cover.
11. Remove the accessory drive belts.
12. Remove the water pump pulley.
13. Remove the crankshaft pulley bolts and remove the crankshaft pulley and baffle plate.
14. Remove the crankshaft sensor wheel.
15. Using a suitable tool, hold the crankshaft pulley boss and remove the pulley lockbolt. Remove the crankshaft pulley boss.
16. Remove the timing covers.
17. Temporarily reinstall the crankshaft pulley boss lockbolt.
18. Turn the crankshaft until the timing mark on the crankshaft sprocket aligns with the timing mark on the oil pump housing and the camshaft sprocket timing marks align with the marks on the seal plate.

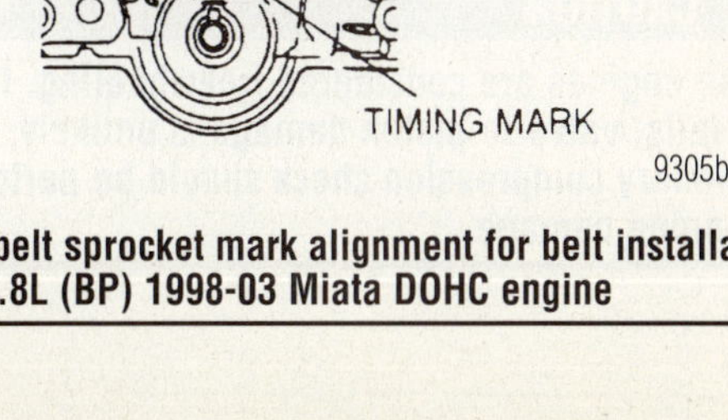

Proper timing belt sprocket mark alignment for belt installation—Mazda 1.8L (BP) 1998-03 Miata DOHC engine

19. Loosen the tensioner pulley lockbolt and pull the tensioner pulley away from the center of the engine to reduce the tension on the timing belt.
20. If the timing belt is to be reused, mark the direction of rotation on the timing belt.
21. Remove the timing belt.
22. Remove the crankshaft pulley boss lockbolt.
23. Remove the tensioner and idler pulleys and check for free movement and abnormal noises. Replace as needed.

To install:

24. Be sure the timing marks on the camshaft and crankshaft sprockets are still aligned.
25. Install the idler pulleys and torque the bolts to 28–38 ft. lbs. (38–51 Nm).
26. Install the tensioner pulley and position the tensioner with the spring fully extended and temporarily torque the lockbolt.
27. Install the timing belt in the following order:
    a. Crankshaft pulley
    b. Idler pulley
    c. Left side camshaft pulley
    d. Right side camshaft pulley
    e. Tensioner
28. Install the crankshaft pulley boss and lockbolt.
29. Remove the holding tool from between the camshaft sprockets.
30. Rotate the crankshaft in the direction of rotation 1 5/6 turns to align the crankshaft timing mark with the tension set mark.
31. Loosen the tensioner lockbolt to apply tension to the timing belt. Torque the mounting bolt to 28–38 ft. lbs. (38–51 Nm).
32. Rotate the crankshaft in the direction of rotation 2 1/6 turns and be sure all marks are correctly aligned.
33. Torque the crankshaft pulley boss lockbolt to 116–122 ft. lbs. (157–166 Nm).
34. Install the upper, middle and lower timing belt covers.
35. Using a suitable tool, hold the crankshaft pulley boss and remove the pulley lockbolt. Remove the crankshaft pulley boss.
36. Install the crankshaft sensor wheel.
37. Install the crankshaft pulley bolts, crankshaft pulley and baffle plate.
38. Install the water pump pulley.
39. Install the accessory drive belts.
40. Install remove the cylinder head cover.
41. On 2001-03 models, install the oil control valve.
42. Install the upper radiator hose and coolant hoses from the thermostat.
43. Install the spark plugs.
44. Install the ignition coil connectors and coils.
45. Install the crankshaft and camshaft position sensors.
46. Install the air intake pipe.
47. On 2001-03 models install the front suspension tower brace.
48. Refill the engine coolant.
49. Connect the negative battery cable.

1.8L (BPE) SOHC ENGINES

✱✱ CAUTION

These engines are considered freewheeling. If the timing belt fails, valve-to-piston damage is unlikely, however a precautionary compression check should be performed to verify no engine damage.

YEAR, MODEL & ENGINE APPLICATION

1990-94 Mazda 323 Protege	1.8L Code BPE

REPLACEMENT INTERVALS

Mazda recommends timing belt replacement every 60,000 miles.

LABOR TIMES

1990-94	
Replacement	3.4 hrs
Severe Service	3.6 hrs
With air conditioning add	0.2 hrs
With power steering add	0.1 hrs

8807MAZG07

1. Disconnect the negative battery cable.
2. Raise and safely support the vehicle.
3. Remove the right front wheel and right side splash shield.
4. Remove the accessory drive belts.
5. Remove the drive belt tensioner.
6. Remove the water pump pulley.
7. Remove the crankshaft pulley bolts and remove the crankshaft pulley and baffle plate.
8. Using a suitable tool to hold the crankshaft pulley boss, remove the pulley boss lockbolt. Remove the crankshaft pulley boss.
9. Remove the upper and lower timing belt covers.
10. Label and disconnect the spark plug wires. Remove the spark plugs.

■ Spark plugs are removed to make it easier to rotate the engine.

11. Temporarily reinstall the crankshaft pulley boss and lockbolt.
12. Turn the crankshaft until the camshaft sprocket and crankshaft sprocket timing marks are aligned.
13. If the timing belt is to be reused, mark the direction of rotation on the timing belt.
14. Loosen the timing belt tensioner lockbolt. Move the tensioner away from the belt to relieve tension. Torque the lockbolt.
15. Remove the timing belt.

■ Do not rotate the engine after the timing belt has been removed.

16. Inspect the belt for wear, peeling, cracking, hardening or signs of oil contamination. Inspect the tensioner for free and smooth rotation. Inspect the sprocket teeth for wear or damage. Replace parts, as necessary.

To install:

17. Make sure the timing marks on the camshaft and crankshaft sprockets are properly aligned.
18. Install the timing belt so there is no slack on the tension side. If reusing the old timing belt, make sure it is reinstalled in the same direction of rotation as marked.
19. Loosen the tensioner lockbolt to set tension against the belt, then torque the bolt to 30 ft. lbs. (40 Nm).
20. Turn the crankshaft 2 turns in the direction of rotation and check the timing mark alignment.
21. Apply approximately 22 lbs. (10 kg) pressure to the timing belt on the side opposite the tensioner, at a point midway between the sprockets. The belt should deflect 0.43–0.51 in. (11–13mm). If the tension is not as specified, replace the tensioner spring.

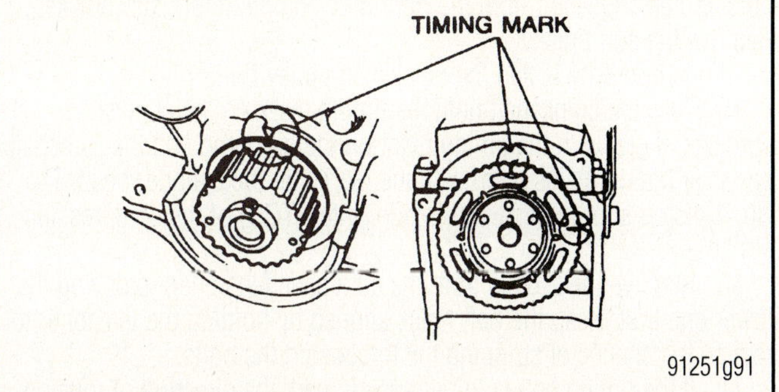

Crankshaft and camshaft sprocket timing mark alignment—Mazda 1.8L (BPE) SOHC engine

22. Install the upper and lower timing belt covers. Torque the bolts to 95 inch lbs. (11 Nm).
23. Using a suitable tool, install the crankshaft pulley boss and torque the lockbolt to 123 ft. lbs. (167 Nm).
24. Install the crankshaft pulley and baffle plate.
25. Install the water pump pulley.
26. Install the accessory drive belts.
27. Install the right front wheel and right side splash shield.
28. Lower the vehicle.
29. Install the spark plugs and connect the spark plug wires.
30. Connect the negative battery cable.
31. Start the engine and check for proper operation.
32. Check the ignition timing.

1.8L (FP) ENGINES

YEAR, MODEL & ENGINE APPLICATION

1999-01 Mazda Protege	1.8L Code FP

REPLACEMENT INTERVALS

Mazda recommends timing belt replacement every 105,000 miles.

LABOR TIMES

1999-01	
Replacement	3.2 hrs
Severe Service	3.4 hrs

8807MAZg08

✳✳ CAUTION

These engines are considered freewheeling. If the timing belt fails, valve-to-piston damage is unlikely, however a precautionary compression check should be performed to verify no engine damage.

1. Disconnect the negative battery cable.
2. Raise and safely support the vehicle.
3. Remove the right front wheel.
4. Remove the right side splash shield.
5. Lower the vehicle.
6. Remove the camshaft position sensor and the crankshaft position sensor.
7. Label and disconnect the spark plug wires.

Alignment of camshaft and crankshaft timing marks—Mazda 1.8L (FP) engine

8. Remove the spark plugs.
9. Remove the cylinder head cover.
10. Remove the accessory drive belt.
11. Remove the water pump pulley.
12. Using a suitable tool, hold the crankshaft pulley and remove the pulley bolt. Remove the crankshaft pulley and guide plate.
13. Remove the oil dipstick and tube.
14. Remove the timing belt guide plate.
15. Using special tool 49G0175AO or equivalent, support the engine and remove the front engine mount and bracket.
16. Remove the timing belt covers.
17. Reinstall the crankshaft pulley bolt. Rotate the engine in the direction of rotation to align the timing marks.
18. Using an Allen wrench, rotate the tensioner clockwise and remove the tensioner spring.
19. If timing belt is to be reused, mark the direction of rotation on the timing belt.
20. Remove the timing belt.
21. Remove the tensioner and check for ease of movement and abnormal noises. Check the tensioner spring free length. The measurement should be 1.44 in. (36.6mm). If out of specification, replace the spring.

To install:

22. Install the tensioner and lockbolt. Torque the lockbolt to 38 ft. lbs. (51 Nm).
23. Be sure that the camshaft and crankshaft timing marks are aligned.
24. Install the timing belt so that there is no slack on the idler side of the belt.
25. Turn the tensioner clockwise and install the tensioner spring.
26. Rotate the crankshaft 2 full turns in the direction of rotation. Check that the timing marks are properly aligned.
27. Install the timing belt covers.

28. Install the front engine mount and bracket and remove the engine support.
29. Install the oil dipstick and tube.
30. Install the crankshaft pulley guide plate and crankshaft pulley. Using a suitable tool, hold the crankshaft pulley, install the pulley bolt and torque to 122 ft. lbs. (166 Nm).
31. Install the water pump pulley and torque the bolts to 95 inch lbs. (11 Nm).
32. Install the timing belt guide plate.
33. Install the cylinder head cover.
34. Install the spark plugs.
35. Connect the spark plug wires.
36. Install the camshaft position sensor and the crankshaft position sensor.
37. Raise the vehicle.
38. Install the right side splash shield.
39. Install the right front wheel.
40. Lower the vehicle.
41. Install the accessory drive belt.
42. Connect the negative battery cable.

1.8L V6 (K8) ENGINES

YEAR, MODEL & ENGINE APPLICATION

1992-94 Mazda MX-3	1.8L Code K8

REPLACEMENT INTERVALS

Mazda recommends timing belt replacement every 60,000 miles.

LABOR TIMES

1992-94	
Replacement	3.9 hrs
Severe Service	4.1 hrs
With air conditioning add	0.2 hrs
With power steering add	0.1 hrs

8807MAZg09

✻✻ CAUTION

These engines are an interference type. If the timing belt has failed, valve-to-piston damage may have occurred. A compression check should always be performed to check for engine damage.

1. Disconnect the negative battery cable.
2. Raise and support the vehicle.
3. Remove the right side wheel and splash shield.
4. Remove the accessory drive belts.
5. Remove the water pump pulley.
6. Remove the idler pulley and bracket.
7. Remove the power steering pump without disconnecting the lines and position aside.
8. Remove the crankshaft pulley bolt and remove the crankshaft pulley.
9. Remove the oil dipstick and tube assembly.
10. Remove the crankshaft position sensor.
11. Remove the engine wiring harness from the timing belt cover.
12. Remove left and right side timing belt covers.
13. Support the engine using a suitable engine brace. Remove the nuts and through-bolt from the right-side engine mount sub bracket. Remove the sub bracket.
14. Temporarily install the crankshaft pulley bolt.
15. Turn the crankshaft until the timing mark on the crankshaft sprocket aligns with the timing mark on the oil pump and the camshaft sprocket timing marks align with the marks on the cylinder head. The No. 1 piston should be at Top Dead Center (TDC) of the compression stroke.
16. Remove the 2 bolts from the automatic tensioner, removing the lower one first. Keep the bolt holes aligned by holding the tensioner to reduce the chance of stripping the threads on the bolts.
17. If the timing belt is to be reused, mark the direction of rotation on the timing belt.
18. Remove the upper and lower idler pulleys and remove the timing belt.

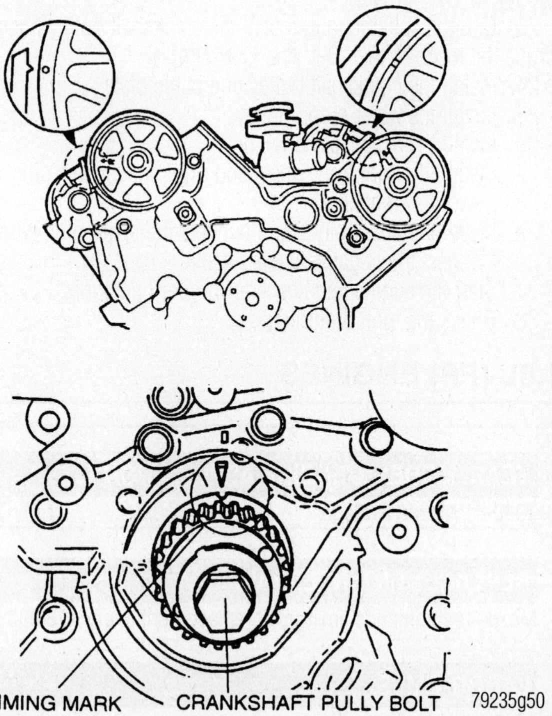

Timing belt sprocket positioning for proper timing belt installation—Mazda 1.8L V6 (K8) engine

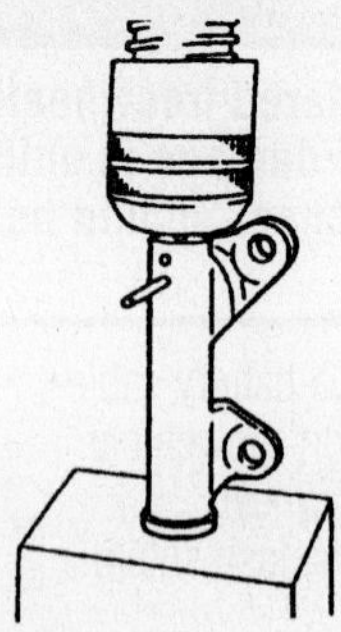

Insert a pin in the tensioner body after pressing in the tensioner rod—Mazda 1.8L V6 (K8) engine

To install:

19. Position the automatic tensioner in a suitable press. Set a flat washer under the tensioner body to prevent damage to the body plug.
20. Compress the tensioner until the hole in the piston is aligned with the 2nd hole in the tensioner case. Insert a 0.060 in. (1.6mm) diameter wire or pin through the 2nd hole to keep the piston compressed.
21. Install the lower idler pulley and torque the bolt to 28–38 ft. lbs. (38–51 Nm).
22. Install the automatic belt tensioner and snugly torque the upper bolt.
23. Make sure that all timing marks are aligned and install the timing belt in the following order:
    a. Crankshaft pulley
    b. Lower idler pulley
    c. Left side camshaft pulley
    d. Tensioner pulley
    e. Right side camshaft pulley
24. If the timing belt is being reused, be sure it is installed in the same direction of rotation as marked.
25. Make sure that there is no slack on the tension side of the belt.
26. Install the upper idler pulley while applying pressure on the timing belt. Torque the bolt to 28–38 ft. lbs. (38–51 Nm).
27. Push the tensioner body to the left and install the lower mounting bolt. Torque the upper and lower mounting bolts to 18 ft. lbs. (25 Nm).
28. Remove the wire or pin from the tensioner.
29. Rotate the crankshaft 2 turns in the normal direction of rotation and align the timing marks. Be sure all marks are aligned correctly aligned.
30. Check the timing belt deflection between the crankshaft sprocket and the tensioner pulley. Deflection should be 0.24–0.31 in. (6–8mm). If it is out of specification, replace the auto-tensioner.
31. Install the right-side engine mount sub bracket. Torque the nuts to 55–77 ft. lbs. (75–104 Nm) and the through-bolt to 63–86 ft. lbs. (86–116 Nm). Remove the engine support.
32. Install the left and right side timing belt covers.
33. Install the engine wiring harness.
34. Install the crankshaft position sensor.
35. Install the oil dipstick and tube assembly.
36. Install the water pump pulley.
37. Install the crankshaft pulley and bolt. Torque the bolt to 116–122 ft. lbs. (157–166 Nm).
38. Install the accessory drive belts and adjust tension.
39. Connect the negative battery cable.

2.0L (FE) AND 2.0L DIESEL (RF) ENGINES

**** CAUTION**

These engines are an interference type. If the timing belt has failed, valve-to-piston damage may have occurred. A compression check should always be performed to check for engine damage.

1. Disconnect the negative battery cable.
2. Remove any side or undercovers necessary for access to timing belt cover.
3. Remove the accessory drive belts.
4. Remove the crankshaft pulley and baffle plate.

YEAR, MODEL & ENGINE APPLICATION

1985 Mazda 626	2.0L Diesel Codes FE/RF

REPLACEMENT INTERVALS

Mazda does not provide a replacement interval. Standard industry practice is to replace timing belts every 60,000 miles.

LABOR TIMES

1985	
Replacement	2.0 hrs
Severe Service	2.2 hrs

8807MAZg10

5. Remove the water pump pulley.
6. Remove the upper and lower timing belt covers.
7. Turn crankshaft in the direction of rotation to align timing marks on the camshaft and crankshaft sprockets.
8. If timing belt is to be reused, mark the belt with an arrow as to indicate direction of rotation.
9. Remove the tensioner pulley lockbolt and remove the tensioner and spring assembly.
10. Remove the timing belt.
11. Inspect the timing belt for wear, damage or cracks. If the belt is coated with oil or grease, it must be replaced.
12. Check belt tensioner and idler pulley for smooth movement and abnormal noises. Replace as necessary.

To install:

13. Make sure that the timing mark on the camshaft sprocket is aligned with the mark on the upper seal plate and that the mark on the crankshaft sprocket is aligned with the mark on the oil pump housing.
14. Install the belt tensioner and spring assembly and install the lockbolt.
15. Position the tensioner so that the spring is in its fully extended position. Temporarily torque the lockbolt.
16. Install the timing belt onto the crankshaft sprocket and then onto the camshaft sprocket from the tension side.
17. If reusing the timing belt. Make sure the belt is installed in the marked direction.
18. Loosen the tensioner lockbolt to apply pressure to the belt. Make sure that there is equal pressure on both sides of the belt and that the timing marks are still aligned.
19. Turn the crankshaft 2 full revolutions in the direction of rotation. Make sure the timing marks are correctly aligned.
20. Loosen the tensioner lockbolt to allow the tensioner spring to apply pressure to the timing belt.

    Torque the lockbolt and torque to 38 ft. lbs. (52 Nm).
21. Check for proper belt tension by measuring the deflection midway between the camshaft sprocket and the crankshaft sprocket on the tension side. Deflection should measure 0.43–0.51 in. (11–13mm).
22. Install the upper and lower timing belt covers.
23. Install the water pump pulley.
24. Install the crankshaft pulley and the baffle plate.
25. Install the drive belts and adjust to proper tension.
26. Install side covers and undercovers.
27. Connect the negative battery cable.

2.0L (VIN B) ENGINES

YEAR, MODEL & ENGINE APPLICATION	
2000-04 Mazda Tribute	2.0L VIN B

REPLACEMENT INTERVALS
Mazda recommends timing belt replacement every 90,000 miles.

LABOR TIMES	
2000-04	
Replacement	4.4 hrs
Severe Service	4.9 hrs
With air conditioning add	1.2 hrs

8807MAZg11

✳✳ CAUTION

These engines are an interference type. If the timing belt has failed, valve-to-piston damage may have occurred. A compression check should always be performed to check for engine damage.

1. Disconnect the negative battery cable.
2. Remove the air cleaner outlet pipe.
3. Disconnect cruise control and accelerator cables.
4. Remove the valve cover.
5. Remove the spark plugs.
6. Remove the catalytic converter.
7. Remove the bolt, nut, and position the coolant tube aside.
8. Remove the right wheel and tire assembly.
9. Remove the right lower splash shield.
10. Rotate the crankshaft to just before top dead center (TDC) on No. 1 cylinder.
11. Remove the stud.
12. Install the special tool 303-574 into the stud hole.

■ **Make sure the correct (second) notch in the pulley is indexed to the lower cylinder block.**

13. Rotate the crankshaft clockwise against the peg to bring it to TDC on No. 1 cylinder.
14. Loosen the water pump pulley bolts.
15. Remove the accessory drive belt.
16. Remove the crankshaft pulley.
17. Remove the bolts and the lower timing belt cover.
18. Lower the vehicle.
19. Install engine lifting bracket 303-F072.
20. Remove the ground strap.
21. Remove the engine mount upper bracket.
22. Remove the studs.
23. Remove the knock sensor connector.
24. Remove the bolts and the upper timing belt cover.
25. Remove the water pump pulley.
26. Remove the accessory drive belt idler pulley.

■ **Installation of the alignment tool into the exhaust camshaft may require the camshafts to be rotated clockwise slightly.**

27. Install special tool 303-465 and align the camshafts.

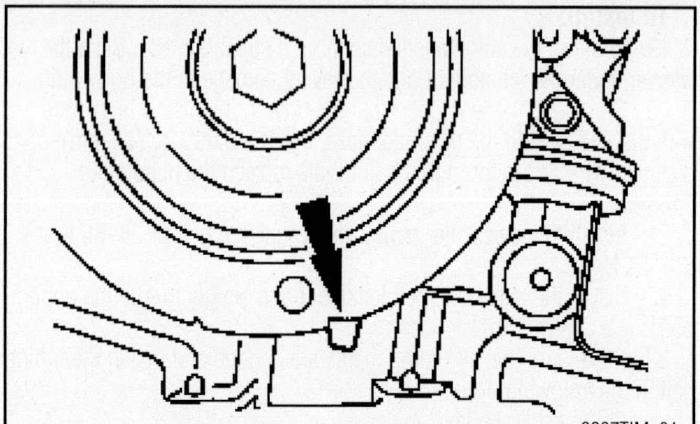

8807TIMg01

Positioning the crankshaft pulley against the alignment peg—Mazda 2.0L (VIN B) engine

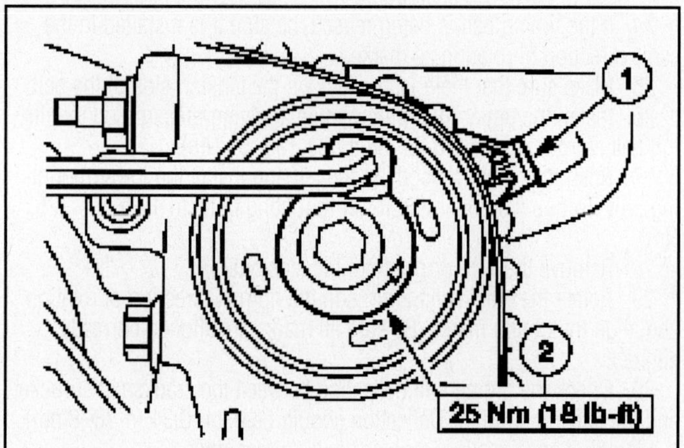

8807TIMg02

Positioning timing belt tensioner in timing cover slot—Mazda 2.0L (VIN B) engine

28. Raise and support the vehicle.
29. Remove the bolts and the engine mount lower bracket.
30. Loosen the timing belt tensioner pulley and allow to slide down to the bottom of its travel.

✳✳ CAUTION

If the camshaft timing belt is to be reused, mark the direction of the camshaft timing belt to rotation of camshaft prior to removal or premature wear or failure may occur.

31. Slide the timing belt off of the camshaft and crankshaft sprockets. Check the timing belt for wear. Install a new belt if necessary.

To install:

■ **Make sure the correct (second) notch in the pulley is indexed to the lower cylinder block.**

32. Slide the crankshaft pulley onto the crankshaft and confirm the crankshaft position is at TDC (No. 1 cylinder) by rotating it clockwise against the alignment peg.
33. Remove the crankshaft pulley.
34. Lower the vehicle.
35. Confirm that the timing belt tensioner is installed correctly with the tab positioned in the slot in the inner timing cover.
36. Install the timing belt onto the timing belt sprockets.

37. Adjust the timing belt tensioner.
38. Using a 6 mm Allen wrench, rotate the adjuster counterclockwise and align the marks.
39. Tighten the tensioner pulley bolt.
40. Raise the vehicle.
41. Install the front engine mount lower bracket and tighten to 37 ft. lbs. (50Nm).
42. Install the accessory drive belt idler pulley.
43. Install the water pump pulley.
44. Hand tighten the bolts.
45. Lower the vehicle.
46. Install the timing belt covers. Torque the cover bolts to 62 inch lbs. (7Nm).
47. Tighten the water pump pulley bolts.
48. Remove the special tool.
49. Install the stud. Torque to 25 ft. lbs. (34 Nm).
50. Install the engine mount upper bracket and tighten to 72 ft. lbs. (98Nm).
51. Install the ground strap.
52. Install the crankshaft pulley, but do not tighten the bolt.
53. Install the accessory drive belt and tighten the tensioner.
54. Tighten the crankshaft pulley bolt to 85 ft. lbs. (115Nm).
55. Install the splash shield and front wheel.
56. Lower the vehicle.
57. Install the coolant tube.
58. Install the catalytic converter.
59. Remove the special tool.
60. Install the valve cover.
61. Install the spark plugs.
62. Install the cruise control and accelerator cables.
63. Install the air cleaner outlet pipe.
64. Connect the negative battery cable.

2.0L (FS) ENGINES

YEAR, MODEL & ENGINE APPLICATION

1993-02 Mazda 626, 1993-97 MX-6, 2001-04 Protege/Protege5	2.0L Code FS

REPLACEMENT INTERVALS

Mazda recommends timing belt replacement every 60,000 miles on 1993-98 non-California models, and every 105,000 miles on all other years and models.

LABOR TIMES

626	
Replacement	3.0 hrs
Severe Service	3.2 hrs
MX-6	
Replacement	3.1 hrs
Severe Service	3.3 hrs
Protege & Protege5	
Replacement	3.1 hrs
Severe Service	3.3 hrs
All Models	
With air conditioning add	0.2 hrs
With power steering add	0.1 hrs

8807MAZg12

1993–97 Models

✻✻ CAUTION

These engines are an interference type. If the timing belt has failed, valve-to-piston damage may have occurred. A compression check should always be performed to check for engine damage.

1. Disconnect the negative battery cable.
2. Raise and safely support the vehicle.
3. Remove the right front wheel and tire assembly.
4. Remove the right side splash shield.
5. Remove the accessory drive belts and the water pump pulley.
6. Remove the power steering pump pulley shield.
7. Remove the power steering pump and position aside, leaving the hoses connected.
8. Hold the crankshaft pulley using a suitable tool and remove the pulley bolt. Remove the crankshaft pulley and the guide plate.
9. Disconnect the spark plug wires and remove the spark plugs.
10. Remove the cylinder head cover.
11. Remove the engine oil dipstick and dipstick tube.
12. Remove the upper and lower timing belt covers.
13. Support the engine using engine support tool 49 G017 5A0. Remove the right-side engine mount bracket.
14. Turn the crankshaft, in the normal direction of rotation, until the crankshaft and camshaft sprocket timing marks are aligned.
15. If the timing belt is to be reused, mark the direction of rotation on the belt.

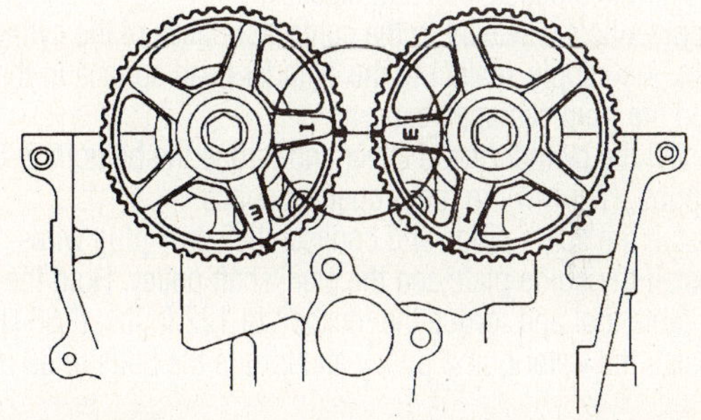

9305bg34

Camshaft sprockets timing mark alignment—Mazda 1993–97 2.0L (FS) engine

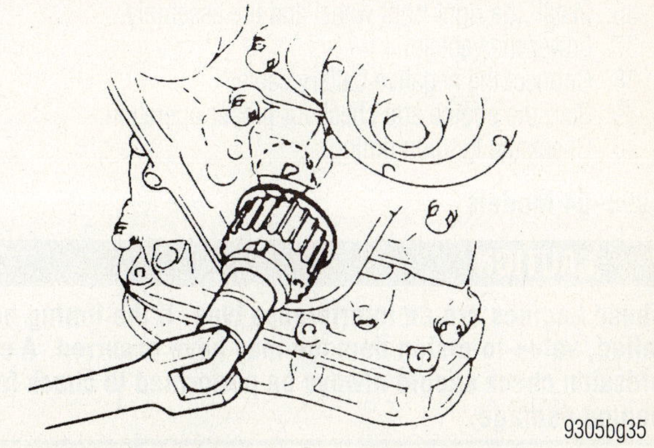

9305bg35

Crankshaft sprocket timing mark alignment— Mazda 1993–97 2.0L (FS) engine

16. Turn the belt tensioner clockwise and disconnect the tensioner spring from the hook pin. Remove the timing belt.

■ **Do not rotate the engine after the timing belt has been removed.**

17. Inspect the belt for wear, peeling, cracking, hardening or signs of oil contamination. Inspect the tensioner pulley for free and smooth rotation and for oil leaks. Check the spring bracket and grommet for looseness or damage. Measure the tensioner spring free length; it should not exceed 1.441 in. (36.6mm) from end to end. Check the sprockets for worn teeth or other damage.

To install:

18. Make sure the crankshaft and camshaft timing marks are aligned.
19. Install the timing belt so there is no looseness at the idler side or between the camshaft sprockets. If reusing the old timing belt, make sure it is installed in the same direction of rotation.
20. Turn the crankshaft 2 turns in the direction of rotation and make sure the timing marks are correctly aligned.
21. Turn the tensioner clockwise and connect the tensioner spring to the hook pin. Make sure tension is applied to the timing belt.
22. Turn the crankshaft 2 turns again in the direction of rotation and make sure the timing marks are correctly aligned.
23. Install the right-side engine mount bracket. Torque the mount-to-engine nuts to 76 ft. lbs. (102 Nm) and the mount through bolt to 86 ft. lbs. (116 Nm). Install the ground harness and torque the nut to 65 ft. lbs. (89 Nm).
24. Remove the engine support tool.
25. Install the timing belt covers and torque the bolts to 95 inch lbs. (10.7 Nm).
26. Install the dipstick tube and dipstick.
27. Apply silicone sealant to the contact surfaces of the cylinder head cover. Also apply sealant to the cylinder head surface in the area adjacent to the front camshaft caps.
28. Install the cylinder head cover and torque the bolts in 2–3 steps to 69 inch lbs. (7.8 Nm), in the proper sequence.
29. Install the spark plugs and connect the spark plug wires.
30. Install the guide plate and the crankshaft pulley. Hold the pulley with a suitable tool and torque the lockbolt to 122 ft. lbs. (166 Nm).
31. Install the water pump pulley and torque the bolts to 95 inch lbs. (10 Nm).
32. Install the power steering pump and torque the bolts to 33 ft. lbs. (46 Nm).
33. Install the power steering pump and pulley shield.
34. Install the accessory drive belts and adjust the belt tension.
35. Install the right side splash shield.
36. Install the right front wheel and tire assembly.
37. Lower the vehicle.
38. Connect the negative battery cable.
39. Start the engine and check for proper operation.
40. Check the ignition timing.

1998–04 Models

**** CAUTION**

These engines are an interference type. If the timing belt has failed, valve-to-piston damage may have occurred. A compression check should always be performed to check for engine damage.

1. Refer to the illustration of the exploded view of the timing belt assembly for component location and torque specifications.
2. Remove or disconnect the following:

- Negative battery cable
- Crankshaft Position (CKP) sensor
- Camshaft Position (CMP) sensor on Protégé models
- Spark plugs
- Power steering pump and bracket, leave the lines attached and position the pump aside
- Drive belt
- Water pump pulley

**** CAUTION**

The CKP sensor rotor is on the rear of the crankshaft pulley and can be damaged if the pulley is not removed carefully.

- Crankshaft pulley using special tools 49 G011 103, 49 E011 1A1 and 49 S120 710. Refer to the illustration for tool positioning.
- Guide plate from behind the crankshaft pulley
- Rocker arm cover
- Oil dipstick and tube
- Timing belt cover

3. Support the engine assembly with a hoist and remove the number 3 engine mount rubber.
4. Turn the crankshaft until the timing mark on the crankshaft sprocket aligns with the timing mark on the oil pump and the camshaft sprocket timing marks **E** and **I** align on the camshaft sprockets. Refer to the illustration for proper timing mark alignment.
5. Lower the vehicle. Insert a camshaft sprocket holding tool between the camshaft sprockets.

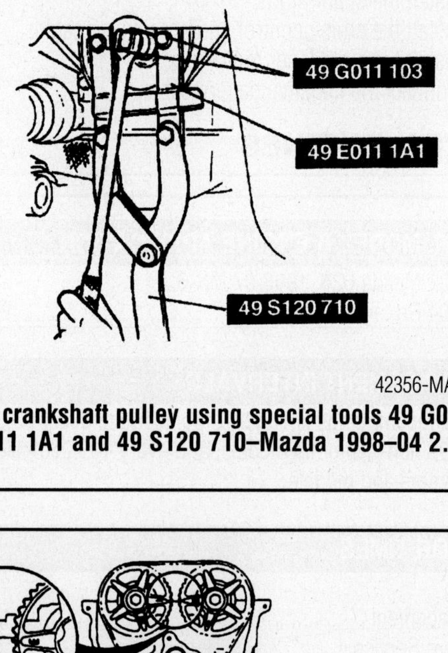

42356-MAZC-g34

Remove the crankshaft pulley using special tools 49 G011 v103, 49 E011 1A1 and 49 S120 710–Mazda 1998–04 2.0L (FS) engine

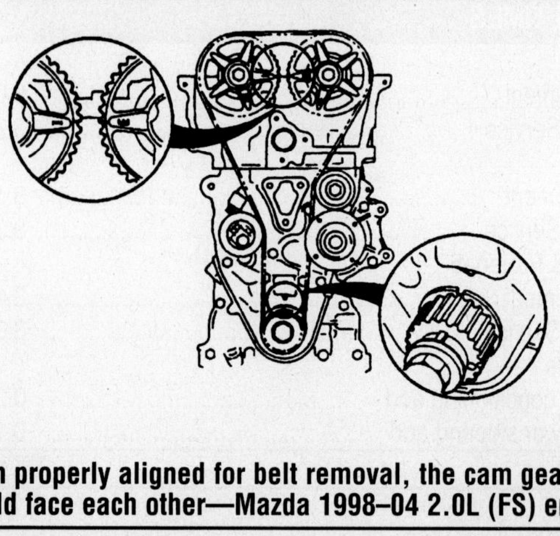

79235g52

When properly aligned for belt removal, the cam gear marks should face each other—Mazda 1998–04 2.0L (FS) engine

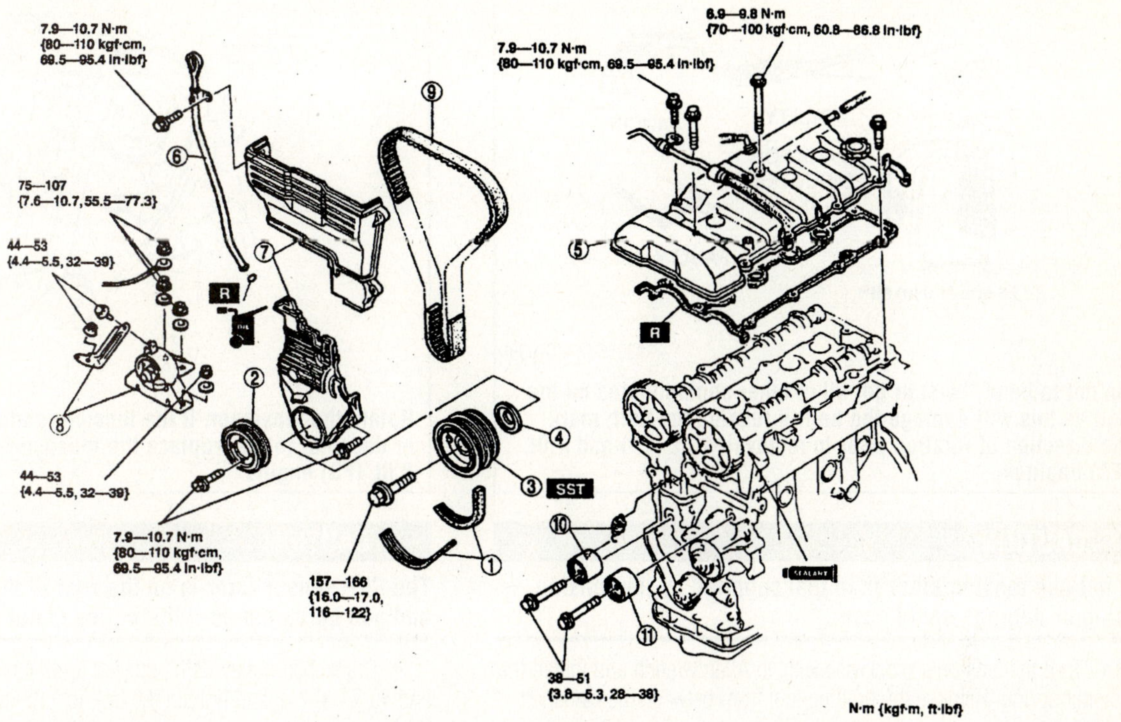

1 Drive belt (See 01-10A-2 DRIVE BELT ADJUSTMENT [FS])
2 Water pump pulley
3 Crankshaft pulley
4 Guide plate
5 Cylinder head cover
6 Dipstick and pipe
7 Timing belt cover
8 No.3 engine mount rubber
9 Timing belt
10 Tensioner, tensioner spring
11 Idler

42356-MAZC-g33

Exploded view of the timing belt assembly–Mazda 1998–04 2.0L (FS) engine

6. Turn the timing belt tensioner with an Allen wrench and remove the tensioner spring from the hook pinch

✻✻ CAUTION

Be careful not to bend, twist or get oil or other contaminates on the belt as this will damage the belt.

7. If the timing belt is to be reused, mark the direction of rotation on the timing belt. Remove the timing belt.

8. Rotate the tensioner, if the tension rotates with no resistance or does not rotate; replace the tensioner

To install:

9. Install the crankshaft sprocket bolt. Install the flywheel locking tool, if equipped with automatic transaxle, or place the shift lever in **4th** gear and apply the parking brake, if equipped with manual transaxle. Tighten the bolt to 116–122 ft. lbs. (157–166 Nm).

10. Be sure the timing marks on the camshaft and crankshaft sprockets are still aligned.

11. Install the timing belt. If reusing the original timing belt, be sure it is installed in the same direction of rotation. Make sure there is no looseness on the idler side.

12. Rotate the crankshaft 2 turns in the normal direction of rotation and align the timing marks. Be sure all marks are still correctly aligned. If the marks are not aligned, remove the belt and then reinstall it and make sure the marks are properly aligned.

13. Check the tensioner spring length, if the free length is not 1.44 inch (36.5mm), replace the spring.

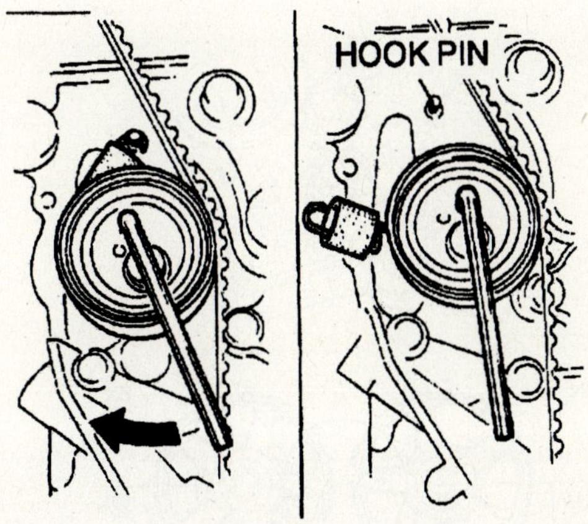

42356-MAZC-g35

Remove the tensioner spring from the hook pin–Mazda 1998–04 2.0L (FS) engine

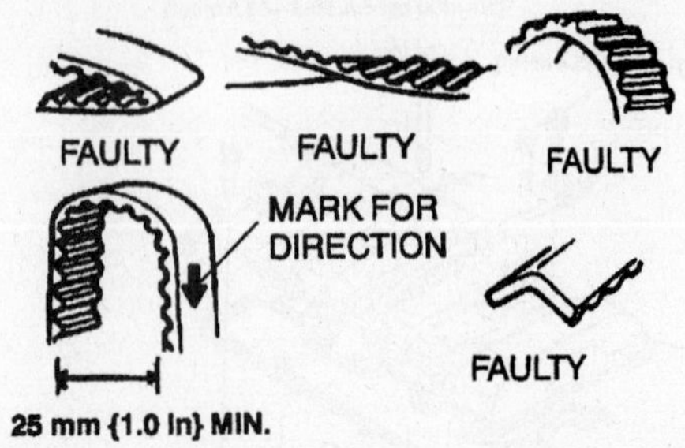

42356-MAZC-g36

Do not to bend, twist or get oil or other contaminates on the belt as this will damage the belt, if reusing the belt; mark the direction of rotation prior to removal–1.8L (FP) and 2.0L (FS) engines

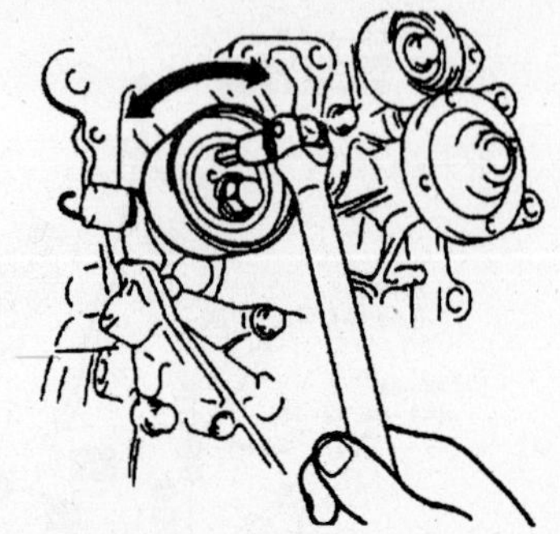

42356-MAZC-g37

Rotate the tensioner, if the tension rotates with no resistance or does not rotate; replace the tensioner–Mazda 1998–04 2.0L (FS) engine

✱✱ CAUTION

Do not use tension other than that supplied by the tension spring or damage could occur

14. Turn the tensioner clockwise with an Allen wrench and install the tensioner spring. Remove the holding tool from between the camshaft sprockets.
15. Rotate the crankshaft 2 turns in the normal direction of rotation and align the timing marks. Be sure all marks are still correctly aligned.
16. Install the number 3 engine mount rubber and tighten the fasteners to the specifications shown in the illustration. Remove the engine hoist.
17. Install or connect the following:

- Timing belt cover. Refer to the exploded view of the timing belt components for torque specifications.
- Oil dipstick and tube. Refer to the exploded view of the timing belt components for torque specifications.
- Rocker arm cover
- Guide plate

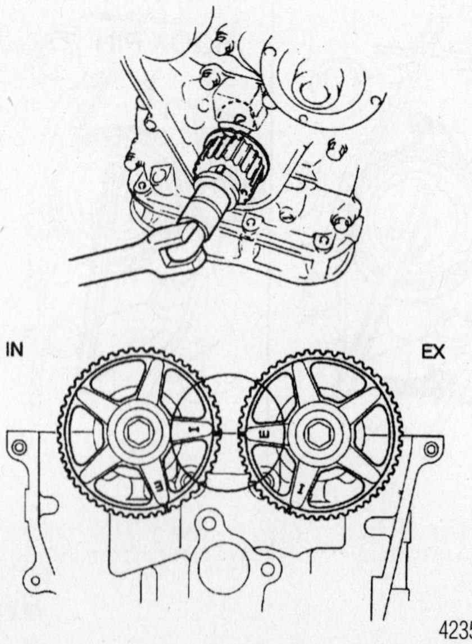

42356-MAZC-g38

Be sure the timing marks on the camshaft and crankshaft sprockets are still aligned–Mazda 1998–04 2.0L (FS) engine

✱✱ CAUTION

The CKP sensor rotor is on the rear of the crankshaft pulley and can be damaged if the pulley is not installed carefully.

- Crankshaft pulley using special tools 49 G011 v103, 49 E011 1A1 and 49 S120 710 and tighten the bolt to 116–122 ft. lbs. (157–166 Nm).
- Water pump pulley. Refer to the exploded view of the timing belt components for torque specifications.
- Drive belt
- Power steering bracket and pump, if removed

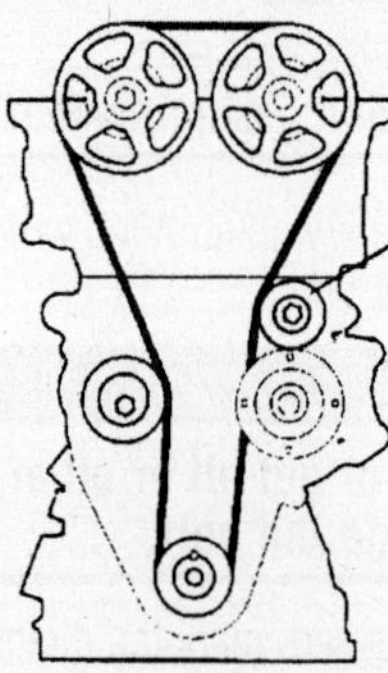

42356-MAZC-g39

Make sure there is no looseness on the idler side when the belt is installed–Mazda 1998–04 2.0L (FS) engine

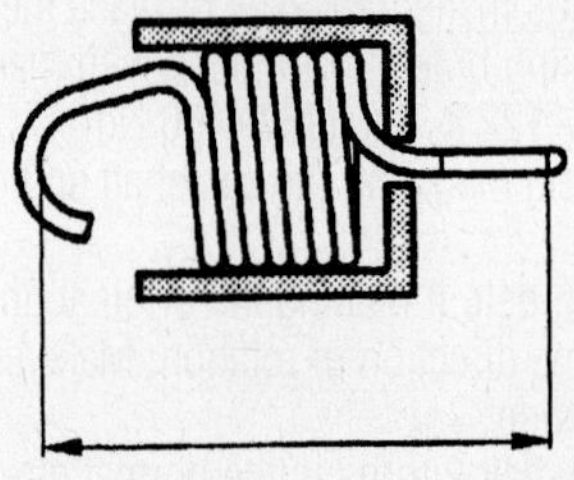

42356-MAZC-g40

Check the tensioner spring length, if the free length is not 1.44 inch (36.5mm); replace the spring–Mazda 1998–04 2.0L (FS) engine

- Spark plugs
- CKP sensor
- Negative battery cable

2.2L (F2) ENGINES

YEAR, MODEL & ENGINE APPLICATION

1986-93 Mazda B2200, 1988-92 MX-6/Turbo,
1988-93 626/Turbo .... 2.2L Code F2

REPLACEMENT INTERVALS

Mazda recommends timing belt replacement every 60,000 miles.

LABOR TIMES

B2200	
Replacement	3.1 hrs
Severe Service	3.6 hrs
626 & MX-6	
Replacement	3.5 hrs
Severe Service	3.7 hrs
With air conditioning add	0.2 hrs
With power steering add	0.1 hrs

8807MAZg13

✻✻ CAUTION

These engines are an interference type. If the timing belt has failed, valve-to-piston damage may have occurred. A compression check should always be performed to check for engine damage.

1. Disconnect the negative battery cable.
2. On B2200 models, drain the cooling system.
3. Remove the cooling fan, fan pulley and bracket.
4. Remove the distributor.
5. Remove the alternator drive belt and alternator.
6. Remove the secondary air pipe.
7. On 626 and MX-6 models, remove the right splash shield.
8. On all models, remove the accessory drive belts.
9. Remove the retaining bolts and remove the crankshaft pulley.
10. Remove the upper and lower timing belt covers and remove the baffle plate from in front of the crankshaft sprocket.
11. Turn the crankshaft clockwise until the "arrow" **1** mark on the camshaft is aligned with the mark on top of the seal plate.
12. Loosen the tensioner lockbolt. Position the tensioner with the spring in a fully extended position. Torque the lockbolt.
13. Remove the timing belt.
14. If the timing belt is to be reused, mark the direction of rotation.

■ Do not rotate the engine after the timing belt has been removed.

15. Inspect the belt for wear, peeling, cracking, hardening or signs of oil contamination. Inspect the tensioner pulley for free and smooth rotation. Measure the tensioner spring free length; it should not exceed 2.480 in. (63mm) from end to end. Check the camshaft and crankshaft sprockets for worn teeth or other damage.

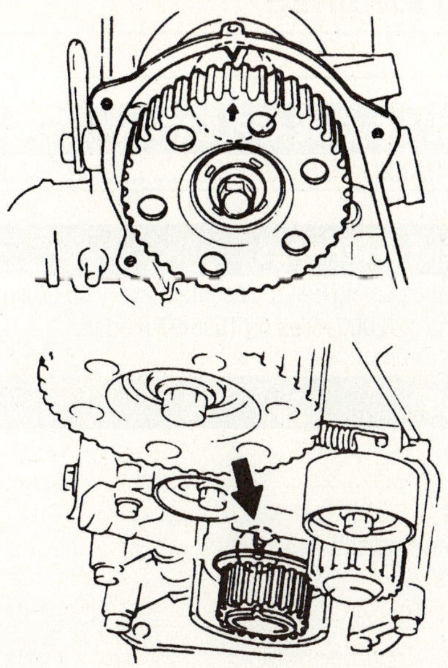

91251gj3

Camshaft and crankshaft sprocket alignment for proper timing belt installation—Mazda 2.2L (F2) engine

To install:

16. Make sure the crankshaft and camshaft timing marks are aligned.
17. Install the timing belt tensioner and spring. Move the tensioner until the spring is fully extended and temporarily torque the tensioner bolt to hold it in place.
18. Install the timing belt. Make sure there is no slack at the side of the water pump and idler pulleys. If reusing the old belt, be sure it is installed in the original direction of rotation as marked.
19. Loosen the tensioner lockbolt to apply tension to the belt. Torque the tensioner bolt to 38 ft. lbs. (52 Nm).
20. Turn the crankshaft 2 turns in the direction of rotation and make sure the timing marks are aligned.
21. Turn the crankshaft again 2 turns in the direction of rotation and make sure the timing marks are aligned.
22. Apply approximately 22 lbs. (98 N) pressure to the timing belt at a point midway between the idler pulley and camshaft sprocket. The belt should deflect 0.31–0.35 in. (8–9mm) on 626 and MX-6 and 0.43–0.51 in. (11–13mm) on B2200. If the deflection is not as specified, replace the tensioner spring.
23. Install the baffle plate with the dished side facing away from the engine.
24. Install the timing belt covers and torque the bolts to 87 inch lbs. (10 Nm).
25. Install the crankshaft pulley and torque the bolts to 13 ft. lbs. (17 Nm).
26. Install the water pump pulley.
27. Install the accessory drive belts and adjust the belt tension.]
28. Install the right side splash shield.
29. On B2200 models, install the secondary air pipe.
30. Install the alternator drive belt and alternator.
31. Install the distributor.
32. Install the cooling fan, fan pulley and bracket.
33. Refill the cooling system.
34. Connect the negative battery cable.
35. Start the engine and check for proper operation.
36. Check the ignition timing.

2.3L (VIN A) ENGINES

YEAR, MODEL & ENGINE APPLICATION

1994-97 Mazda B2300	2.3L VIN A

REPLACEMENT INTERVALS

Mazda recommends timing belt replacement every 60,000 miles on 1994 models, and 120,000 miles on 1995-97 models.

LABOR TIMES

1994 Engines	
Replacement	2.4 hrs
Severe Service	2.8 hrs
With air conditioning add	0.3 hrs
1995-97 Engines	
Replacement	2.7 hrs
Severe Service	3.2 hrs
With air conditioning add	1.2 hrs
With power steering add	0.5 hrs

8807MAZg14

**** CAUTION**

These engines are an interference type. If the timing belt has failed, valve-to-piston damage may have occurred. A compression check should always be performed to check for engine damage.

1. Disconnect the negative battery cable.
2. Rotate the engine so that No. 1 cylinder is at Top Dead Center (TDC) on the compression stroke. Check that the timing marks are aligned on the camshaft and crankshaft pulleys. An access plug is provided in the cam belt cover so that the camshaft timing can be checked without removal of the cover. Set the crankshaft to TDC by aligning the timing mark on the crank pulley with the TDC mark on the belt cover. Look through the access hole in the belt cover to be sure that the timing mark on the cam drive sprocket is aligned with the pointer on the inner belt cover.

**** WARNING**

Always turn the engine in the normal direction of rotation. Backward rotation may cause the timing belt to jump time, due to the arrangement of the belt tensioner.

3. Drain the cooling system.
4. Remove the upper radiator hose.
5. Remove the cooling fan and water pump pulley bolts.
6. Loosen the alternator retaining bolts and remove the drive belt from the pulleys.
7. Remove the water pump pulley.
8. Remove the power steering pump and set it aside.
9. Remove the A/C compressor and set it aside, if equipped.
10. Label and disconnect the spark plug wires.
11. Remove the spark plugs.
12. Remove the 4 timing belt outer cover retaining bolts and remove the cover.
13. Remove the crankshaft pulley and belt guide.
14. Loosen the belt tensioner pulley assembly, then position a Camshaft Belt Adjuster tool T74P-6254-A or equivalent, on the tension spring roll pin and retract the belt tensioner away from the timing belt. Tighten the adjustment bolt to lock the tensioner in the retracted position.
15. If the belt is to be reused, mark the direction of rotation on the timing belt for installation reference.
16. Remove the timing belt.

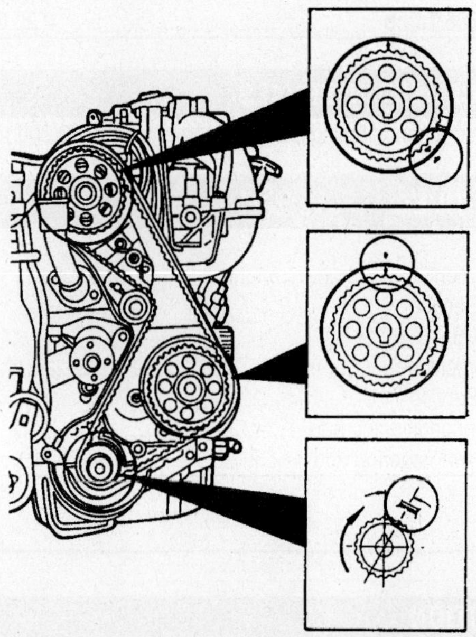

79245G20

Camshaft, auxiliary shaft and crankshaft timing belt sprocket alignment mark locations—Mazda 2.3L (VIN A) engine

To install:

17. Install the new belt over the crankshaft sprocket and then counterclockwise over the auxiliary and camshaft sprockets, making sure the lugs on the belt properly engage the sprocket teeth on the pulleys. Be careful not to rotate the pulleys when installing the belt.
18. Release the timing belt tensioner pulley, allowing the tensioner to take up the belt slack. If the spring does not have enough tension to move the roller against the belt, it may be necessary to manually push the roller against the belt and tighten the bolt.

■ **The spring cannot be used to set belt tension; a wrench must be used on the tensioner assembly.**

**** WARNING**

If any binding is felt when adjusting the timing belt tension by turning the crankshaft, STOP turning the engine, because the pistons may be hitting the valves.

19. Rotate the crankshaft 2 complete turns by hand (in the normal direction of rotation) to remove slack from the belt. Torque the tensioner adjustment to 26–33 ft. lbs. (35–45 Nm) and pivot bolts to 30–40 ft. lbs. (40–55 Nm). Be sure the belt is seated properly on the pulleys and that the timing marks are in alignment.
20. Install the crankshaft pulley and belt guide. Torque the bolt to 93–121 ft. lbs. (125–165 Nm).
21. Install the timing belt cover.
22. Install the spark plugs and plug wires.
23. Install the A/C compressor, if equipped.

24. Install the water pump pulley and the cooling fan. Torque the bolts to 13–17 ft. lbs. (17–23 Nm).
25. Install the accessory drive belts and adjust tension.
26. Install the upper radiator hose.
27. Refill and bleed the cooling system.
28. Disconnect the negative battery cable.
29. Start the engine and check for proper operation.
30. Check ignition timing and adjust if necessary.

2.3L (KJ) V6 ENGINES

YEAR, MODEL & ENGINE APPLICATION	
1995-04 Mazda Millenia	2.3L Code KJ

REPLACEMENT INTERVALS

Mazda recommends timing belt replacement every 60,000 miles on 1995-98 models, 105,000 miles on 1999-00 models and 60,000 miles on 2001-04 models.

LABOR TIMES	
1995-04	
Replacement	5.8 hrs
Severe Service	6.0 hrs
With air conditioning add	0.2 hrs
With power steering add	0.1 hrs

8807MAZg15

✱✱ CAUTION

These engines are considered freewheeling. If the timing belt fails, valve-to-piston damage is unlikely, however a precautionary compression check should be performed to verify no engine damage.

1. Refer to the illustration of the exploded view of the timing belt assembly for component location and torque specifications.
2. Remove or disconnect the following:
- Negative battery cable
- Right front wheel
- Splash shield from the right hand side
- Dust cover
- Drive belt

✱✱ CAUTION

The Crankshaft Position (CKP) sensor rotor is on the rear of the crankshaft pulley and can be damaged if the pulley is not removed carefully.

3. Turn the crankshaft pulley clockwise until the pin on the pulley is facing down.
- Crankshaft pulley using special tools 49 E120 710. Refer to the illustration for tool positioning.
- Power steering pump, leave the lines attached and position the pump aside
- Water pump pulley
- Alternator drive belt tensioner
- Camshaft Position (CMP) sensor
- Oil dipstick and tube
- Vacuum pipe
- Upper radiator hose

4. Support the engine assembly with a hoist and remove the number 3 engine mount rubber.
- Number 3 engine mount sub bracket, rubber and bracket
- Timing belt cover
- Power steering pump drive belt tensioner

5. Turn the crankshaft until the timing mark on the crankshaft sprocket aligns with the timing mark on the oil pump and the camshaft sprocket timing marks align with the marks on the cylinder head. The No. 1 piston should be at Top Dead Center (TDC) of the compression stroke.
6. Remove the two bolts from the automatic tensioner, removing the lower one first. Keep the bolt holes aligned by holding the tensioner to reduce the chance of stripping the threads on the bolts.
7. If the timing belt is to be reused, mark the direction of rotation on the timing belt.
8. Remove the timing belt.

To install:

9. Install the crankshaft sprocket bolt. Install the flywheel locking tool. Tighten the bolt to 116–122 ft. lbs. (157–166 Nm).
10. Check the tension rod projection, it should be 0.563–0.594 inch (14.3–15.1mm), if not, replace the tensioner
11. Position the automatic tensioner in a press. Set a flat washer under the tensioner body to prevent damage to the body plug.
12. Compress the tensioner until the hole in the piston is aligned with the 2nd hole in the tensioner case. Insert a 0.055 inch (1.4mm) diameter wire or pin through the 2nd hole to keep the piston compressed.

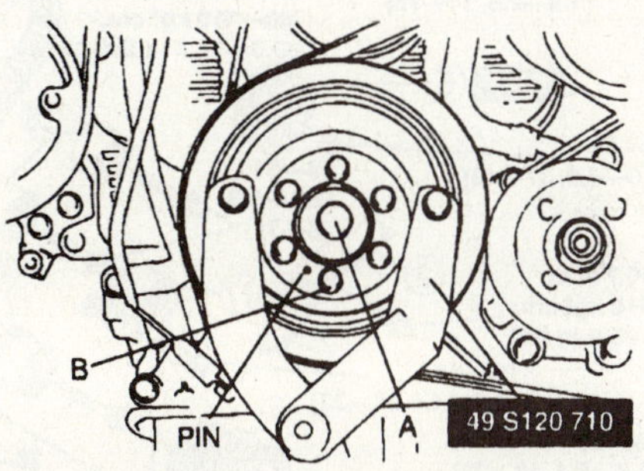

42356-MAZC-ED

Remove the crankshaft pulley using special tools 49 E120 710–Mazda 2.3L (KJ) engine

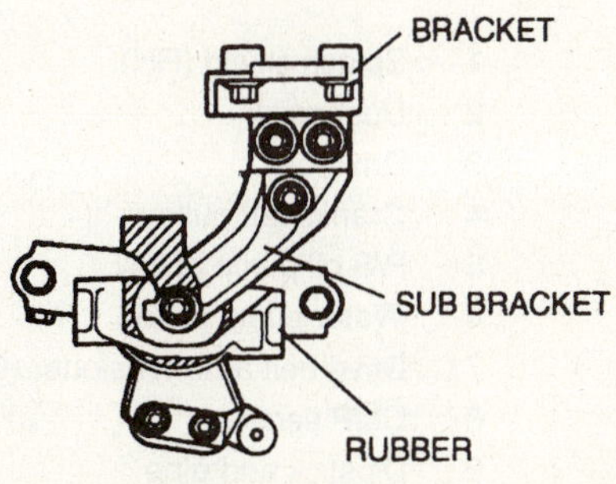

42356-MAZC-EF

Remove the number 3 engine mount sub bracket, rubber and bracket–Mazda 2.3L (KJ) engine

13. Be sure the camshaft sprocket timing marks are still aligned. Turn the crankshaft clockwise until the timing sprocket is aligned.

14. Install the timing belt. If the original belt is being reused, be sure it is installed in the same direction of rotation. The order of installation is: timing belt (crankshaft) sprocket, No. 2 idler pulley, left-hand camshaft sprocket, both No. 1 idler pulleys, right-hand camshaft sprocket and the tensioner pulley.

15. Install the automatic belt tensioner and tighten the bolts to 14–18 ft. lbs. (19–25 Nm). Remove the wire or pin from the tensioner.

16. Turn the crankshaft clockwise, until the crankshaft sprocket timing mark is again at TDC. This should place all of the belt slack in the automatic tensioner portion of the belt.

17. Rotate the crankshaft two turns in the normal direction of rotation and align the timing marks. Be sure all marks are still correctly aligned.

18. Inspect timing belt deflection, 0.24–0.31 inch (6–8mm), between the crankshaft sprocket and the tensioner pulley. If it is out of specification, replace the auto-tensioner.

19. Install or connect the following:

- Power steering pump drive belt tensioner
- Upper timing belt cover and tighten the bolts to 95 inch lbs. (11 Nm) in the sequence illustrated
- Right and left hand timing belt cover and tighten the bolts to 95 inch lbs. (11 Nm) in the sequence illustrated

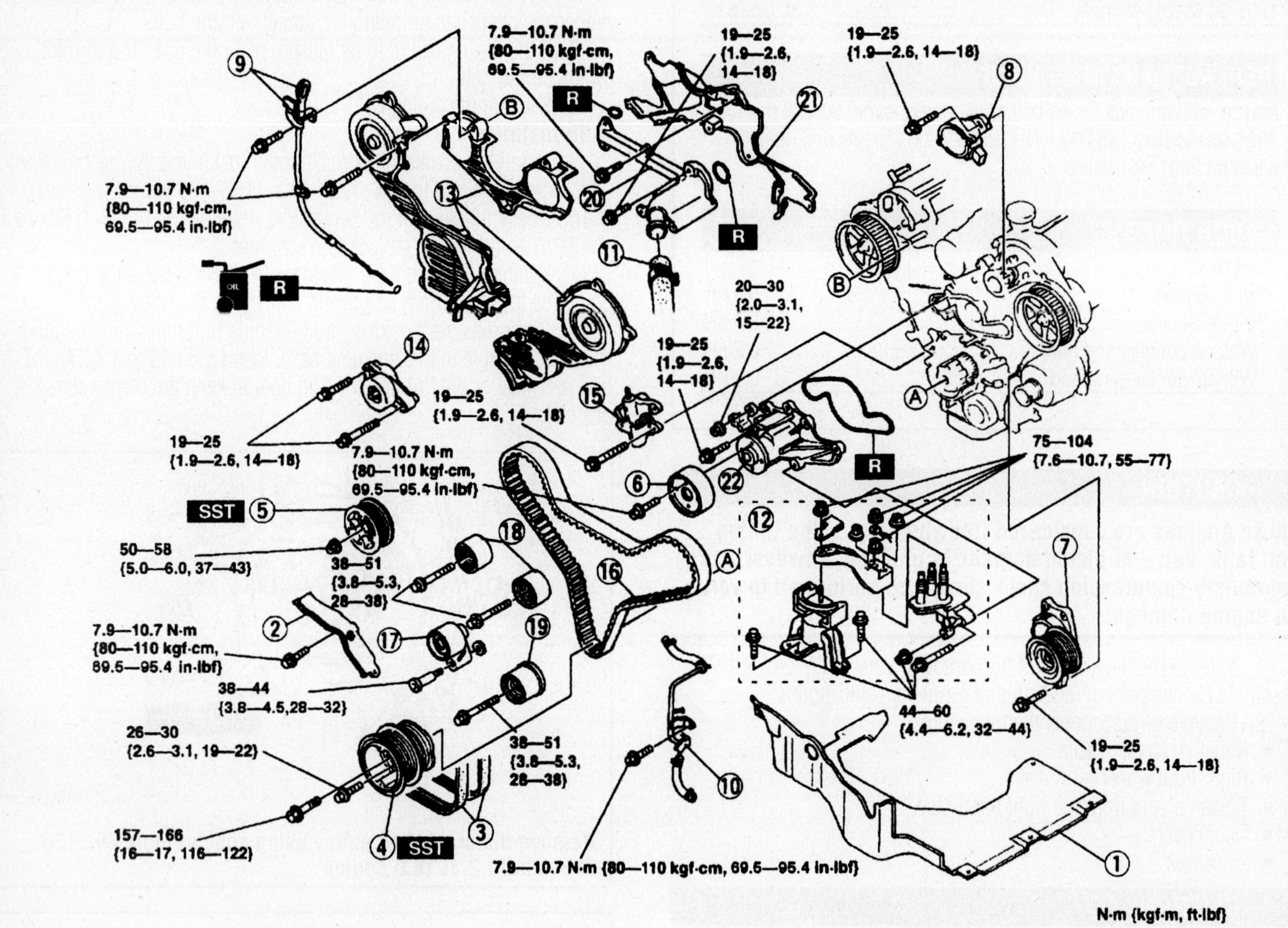

1	Splash shield (RH)	12	No.3 engine mount
2	Dust cover	13	Timing belt cover
3	Drive belt	14	Drive belt auto tensioner (P/S)
4	Crankshaft pulley	15	Timing belt auto tensioner
5	P/S oil pump pulley	16	Timing belt
6	Water pump pulley	17	Tensioner pulley
7	Drive belt auto tensioner (Generator)	18	No.1 idler pulley
8	CMP sensor	19	No.2 idler pulley
9	Dipstick and pipe	20	Water outlet pipe
10	Vacuum pipe	21	Seal plate
11	Upper radiator hose	22	Water pump

42356-MAZC-EC

Exploded view of the timing belt assembly–2.3L (KJ) engines

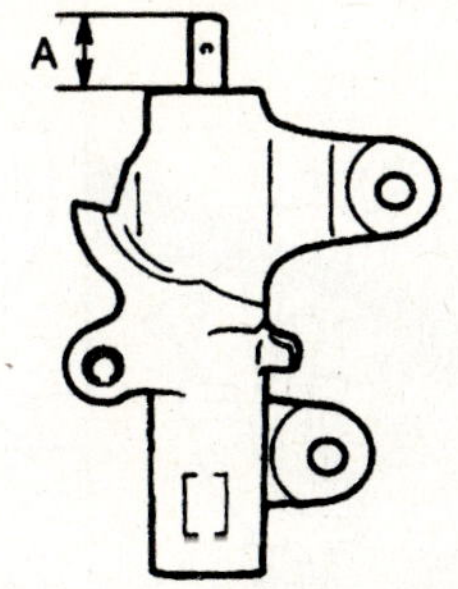

42356-MAZC-EI

Check the tension rod projection, it should be 0.563–0.594 inch (14.3–15.1mm–Mazda 2.3L (KJ) engine

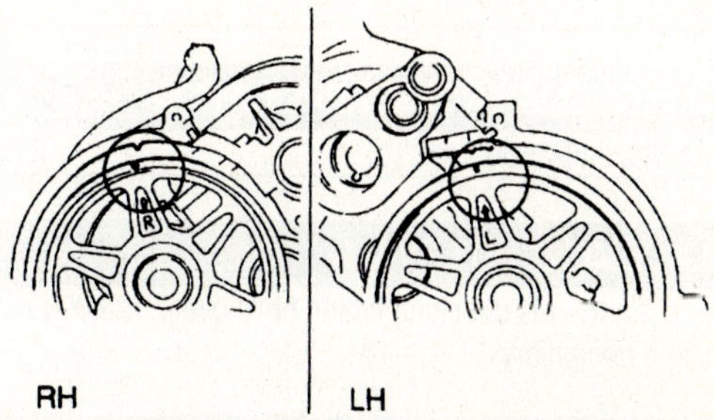

42356-MAZC-EG

Be sure the camshaft sprocket timing marks are still aligned–Mazda 2.3L (KJ) engine

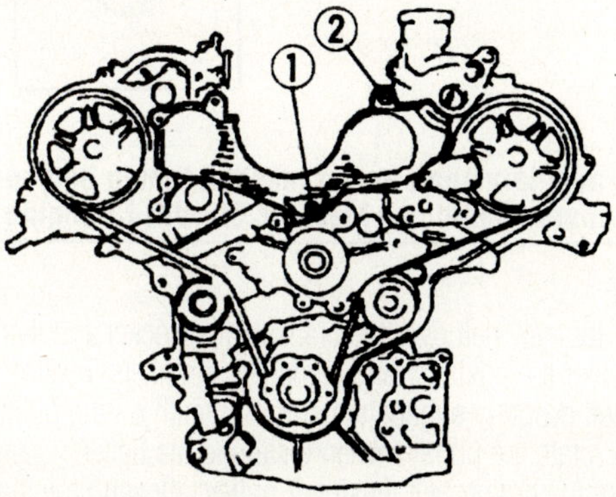

42356-MAZC-EL

Tighten the upper timing belt cover in this sequence–Mazda 2.3L (KJ) engine

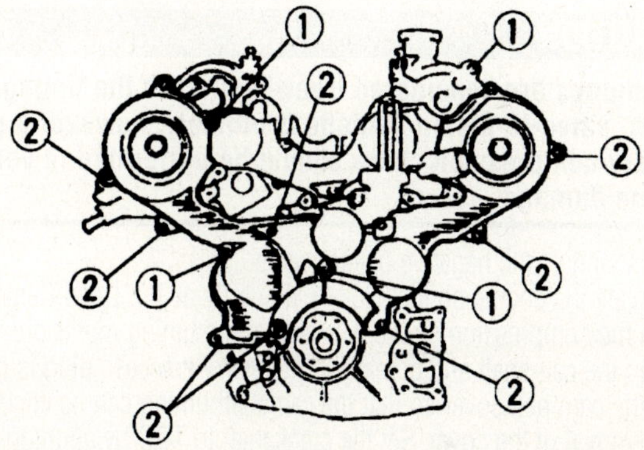

42356-MAZC-EM

Tighten the upper timing belt cover in this sequence-2.3L (KJ) engine

- Number 3 engine mount rubber, bracket and sub bracket. Tighten bolts **A** to 44 ft. lbs. (60 Nm) and bolts **B** to 77 ft. lbs. (104 Nm).
- Upper radiator hose
- Vacuum pipe
- Oil dipstick and tube
- CMP sensor
- Alternator drive belt tensioner
- Water pump pulley
- Power steering pump

20. Remove the timing belt pulley **A**, install the crankshaft pulley. Hand tighten bolts **A** and **B**. Tighten the bolt **A** to 116–122 ft. lbs. (157–166 Nm) and bolt **B** to 22 ft. lbs. (30 Nm).

- Drive belt
- Dust cover
- Splash shield from the right hand side
- Right front wheel
- Negative battery cable

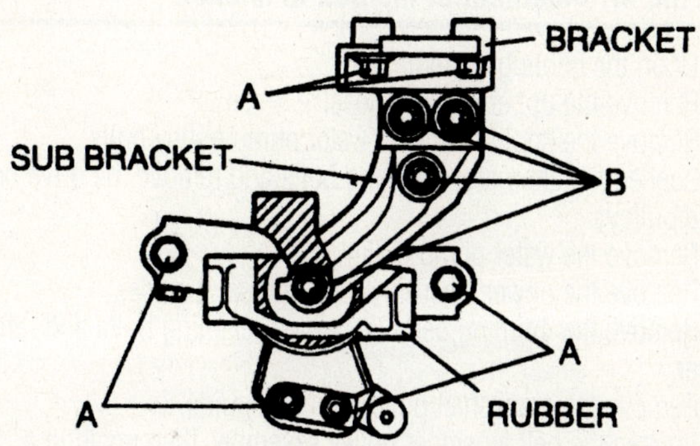

42356-MAZC-EK

Install the number 3 engine mount rubber, bracket and sub bracket and tighten to specification–Mazda 2.3L (KJ) engine

2.5L (VIN C) ENGINES

YEAR, MODEL & ENGINE APPLICATION

1998-01 Mazda B2500	2.5L VIN C

REPLACEMENT INTERVALS

Mazda recommends checking timing belts every 120,000 miles and replacing if necessary.

LABOR TIMES

1998-01	
Replacement	1.4 hrs
Severe Service	1.9 hrs
With air conditioning add	1.2 hrs
With power steering add	0.5 hrs

8807MAZg16

**** CAUTION**

These engines are considered freewheeling. If the timing belt fails, valve-to-piston damage is unlikely, however a precautionary compression check should be performed to verify no engine damage.

1. Disconnect the negative battery cable.
2. Rotate the engine so that No. 1 cylinder is at Top Dead Center (TDC) on the compression stroke. Check that the timing marks are aligned on the camshaft and crankshaft pulleys. An access plug is provided in the cam belt cover so that the camshaft timing can be checked without removal of the cover. Set the crankshaft to TDC by aligning the timing mark on the crank pulley with the TDC mark on the belt cover. Look through the access hole in the belt cover to be sure that the timing mark on the cam drive sprocket is aligned with the pointer on the inner belt cover.

**** WARNING**

Always turn the engine in the normal direction of rotation. Backward rotation may cause the timing belt to jump time, due to the arrangement of the belt tensioner.

3. Drain the cooling system.
4. Remove the upper radiator hose.
5. Remove the cooling fan and water pump pulley bolts.
6. Loosen the alternator retaining bolts and remove the drive belt from the pulleys.
7. Remove the water pump pulley.
8. Remove the power steering pump and set it aside.
9. Remove the 4 timing belt outer cover retaining bolts and remove the cover.
10. Remove the crankshaft pulley and belt guide.
11. Loosen the belt tensioner pulley assembly, then position a Camshaft Belt Adjuster tool T74P-6254-A or equivalent, on the tension spring roll pin and retract the belt tensioner away from the timing belt. Tighten the adjustment bolt to lock the tensioner in the retracted position.
12. If the belt is to be reused, mark the direction of rotation on the timing belt for installation reference.
13. Remove the timing belt.

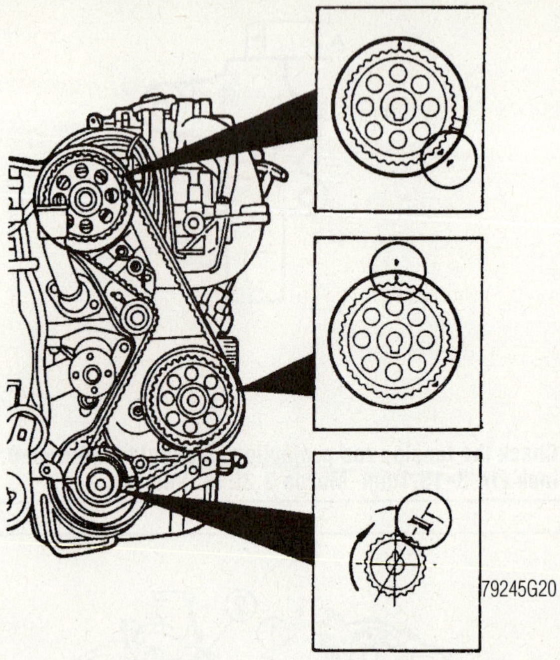

Camshaft, auxiliary shaft and crankshaft timing belt sprocket alignment mark locations—Mazda 2.5L (VIN C) engine

To install:

14. Install the new belt over the crankshaft sprocket and then counterclockwise over the auxiliary and camshaft sprockets, making sure the lugs on the belt properly engage the sprocket teeth on the pulleys. Be careful not to rotate the pulleys when installing the belt.
15. Release the timing belt tensioner pulley, allowing the tensioner to take up the belt slack. If the spring does not have enough tension to move the roller against the belt, it may be necessary to manually push the roller against the belt and tighten the bolt.

■ **The spring cannot be used to set belt tension; a wrench must be used on the tensioner assembly.**

**** WARNING**

If any binding is felt when adjusting the timing belt tension by turning the crankshaft, STOP turning the engine, because the pistons may be hitting the valves.

16. Rotate the crankshaft 2 complete turns by hand (in the normal direction of rotation) to remove slack from the belt. Torque the tensioner adjustment to 26–33 ft. lbs. (35–45 Nm) and pivot bolts to 30–40 ft. lbs. (40–55 Nm). Be sure the belt is seated properly on the pulleys and that the timing marks are in alignment.
17. Install the crankshaft pulley and belt guide. Torque the bolt to 93–121 ft. lbs. (125–165 Nm).
18. Install the timing belt cover.
19. Install the water pump pulley and the cooling fan. Torque the bolts to 13–17 ft. lbs. (17–23 Nm).
20. Install the accessory drive belts and adjust tension.
21. Install the upper radiator hose.
22. Refill and bleed the cooling system.
23. Disconnect the negative battery cable.
24. Start the engine and check for proper operation.
25. Check ignition timing and adjust if necessary.

2.5L (KL) ENGINES

YEAR, MODEL & ENGINE APPLICATION

1995-04 Mazda 626, 1995-04 Millenia,
1993-97 MX-6 .... 2.5L Code KL

REPLACEMENT INTERVALS

Mazda recommends timing belt replacement every 60,000 miles on 1995-98 non-California models, and 120,000 miles on all other models and years.

LABOR TIMES

626 & MX-6	
Replacement	4.2 hrs
Severe Service	4.4 hrs
Millenia	
Replacement	4.1 hrs
Severe Service	4.3 hrs
All Models	
With air conditioning add	0.2 hrs
With power steering add	0.1 hrs

8807MAZg17

✲✲ CAUTION

These engines are considered freewheeling. If the timing belt fails, valve-to-piston damage is unlikely, however a precautionary compression check should be performed to verify no engine damage.

1. Refer to the illustration of the exploded view of the timing belt assembly for component location and torque specifications.
2. Remove or disconnect the following:

- Negative battery cable
- Crankshaft Position (CKP) sensor
- Spark plugs
- Power steering pump, leave the lines attached and position the pump aside
- Drive belt
- Water pump pulley
- Idler pulley bracket

✲✲ CAUTION

The CKP sensor rotor is on the rear of the crankshaft pulley and can be damaged if the pulley is not removed carefully.

- Crankshaft pulley using special tools 49 E011 1A1 and 49 S120 710. Refer to the illustration for tool positioning.
- Rocker arm cover
- Oil dipstick and tube
- Timing belt cover

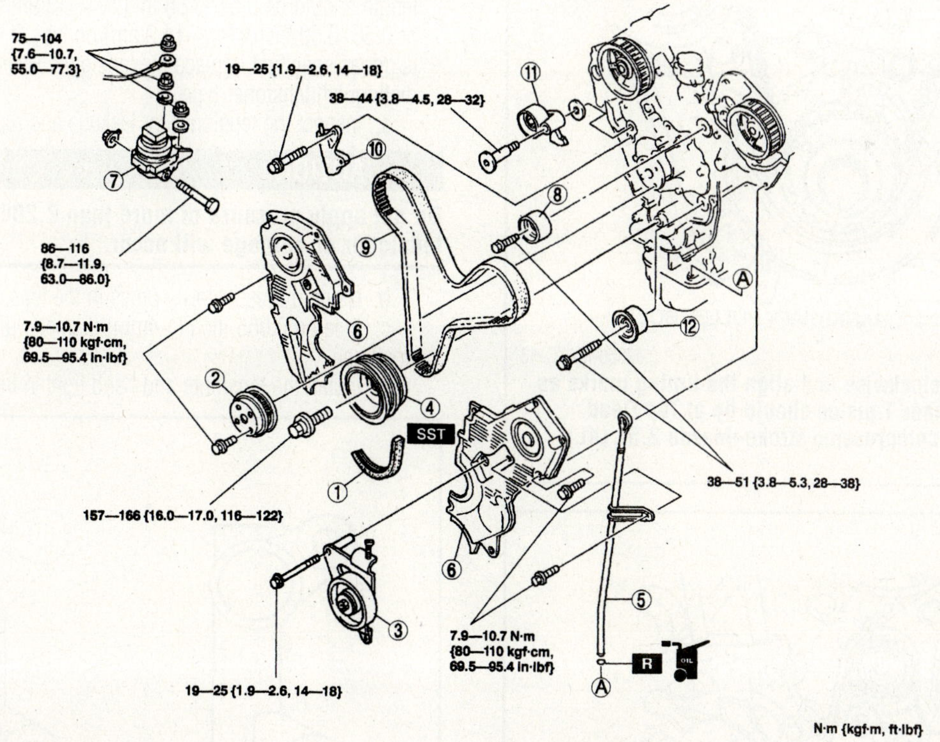

1	Drive belt	7	No.3 engine mount rubber
2	Water pump pulley	8	No.1 idler
3	Idler pulley bracket	9	Timing belt
4	Crankshaft pulley	10	Timing belt auto tensioner
5	Dipstick and pipe	11	Tensioner
6	Timing belt cover	12	No.2 idler

42356-MAZC-41

Exploded view of the timing belt assembly–Mazda 2.5L (KL) engine–626 shown; other models similar

3. Support the engine assembly with a hoist and remove the number 3 engine mount rubber.

4. Install the crankshaft pulley bolt and turn the crankshaft clockwise and align the timing marks as illustrated. The number 1 piston should be at Top Dead Center (TDC) of the compression stroke.

✲✲ CAUTION

When removing the number 1 idler pulley bolt, hold the pulley so that the threads are aligned or the threads can be damaged.

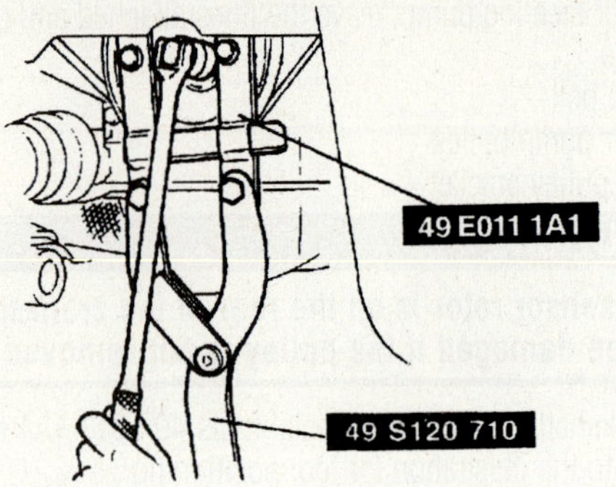

42356-MAZC-42

Remove the crankshaft pulley using special tools 49 E011 1A1 and 49 S120 710–Mazda 2.5L (KL) engine

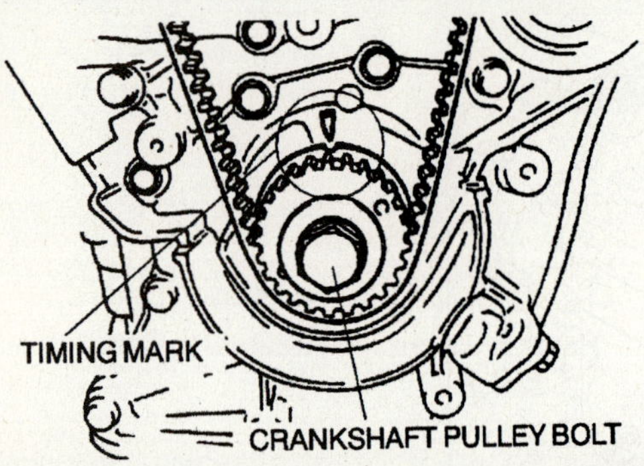

42356-MAZC-43

Turn the crankshaft clockwise and align the timing marks as illustrated. The number 1 piston should be at Top Dead Center (TDC) of the compression stroke–Mazda 2.5L (KL) engine

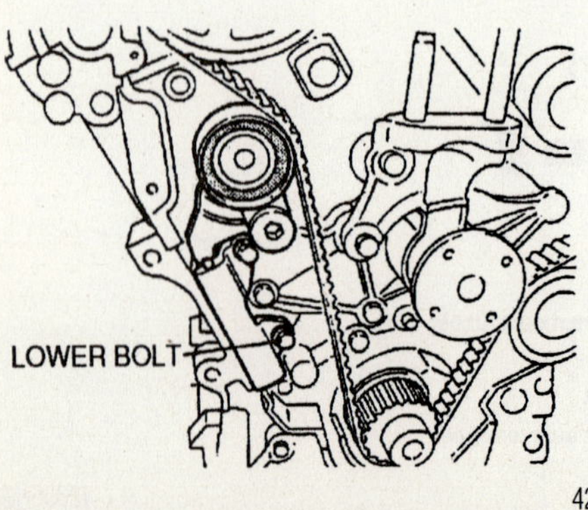

42356-MAZC-45

Loosen the auto tensioner bolts and remove the lower bolt–Mazda 2.5L (KL) engine

5. Remove the number 1 idler pulley

✲✲ CAUTION

When removing the auto tensioner bolts, hold the tensioner so that the threads are aligned or the threads can be damaged.

6. Loosen the auto tensioner bolts and remove the lower bolt.

✲✲ CAUTION

Be careful not to bend, twist or get oil or other contaminates on the belt as this will damage the belt.

7. If the timing belt is to be reused, mark the direction of rotation on the timing belt. Remove the timing belt.
8. Remove the belt.
9. If necessary, remove the auto tensioner.

To install:

✲✲ CAUTION

There are two type of auto tensioners and they are interchangeable.

10. If removed install the auto tensioner as follows:
   a. Measure the tensioner rod projection length. On 1993–96 the length should be 0.56–0.62 in. (14.2–15.mm), on 1997–00 Millenia and 1997–04 626/MX-6 the length should be 0.47–0.51 in. (11.9–12.9.mm), on type **A** tensioners, on 2001–04 Millenia the length should be 0.51–0.55 in. (12.9–13.9mm) on type **A** tensioners, or 0.56–0.59 inch (14.2–14.9mm) on type **B** tensioners. I the length is not as specified, replace the tensioner. Refer to the illustration to distinguish tensioner type.
   a. Inspect the tensioner for leakage and replace if defective.

✲✲ CAUTION

Do not apply pressure of more than 2,200 lbf. (9.8 Kn) to the tensioner or damage will occur.

   b. Using a press, slowly press in the tensioner rod.
   c. Insert a 0.055 inch (1.4mm) diameter pin to hold the rod in position.
   d. Install the tensioner and hand tighten the bolt.

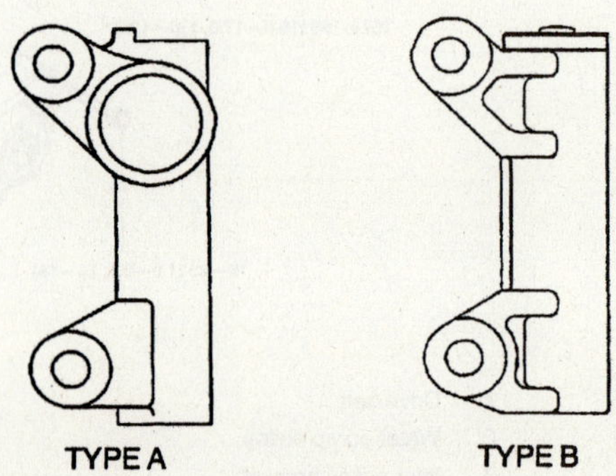

42356-MAZC-47

There are two type of auto tensioners and they are interchangeable–Mazda 2.5L (KL) engine

11. Turn the camshafts clockwise and align the timing marks as illustrated.
12. Install the crankshaft pulley bolt and turn the crankshaft clockwise and align the marks as illustrated.
13. With the number 1 idler pulley removed, install the belt on the pulleys in this order:

- Timing belt pulley
- Number 2 idler pulley
- Left hand camshaft pulley
- Tensioner pulley
- Right hand camshaft pulley

**** CAUTION**

Make sure the belt has no looseness at the tension side.

14. Make sure there is tension between pulleys (3) and (1) and pulleys (1) and (5). Refer to the illustration for location of the pulleys.
15. Install the number 1 idler pulley while applying the pressure on the timing belt. Tighten the bolt to 28–38 ft. lbs. (38–51 Nm).
16. Push the auto tensioner in the direction of arrow (refer to illustration) and hand tighten the lower bolt, then tighten the bolts. Refer to the exploded view of the timing belt components for torque specifications. Then remove the retaining pin.
17. Turn the crankshaft clockwise and align the timing marks. Make sure the timing marks are aligned, if not remove the belt and repeat the installation process.
18. Install or connect the following:

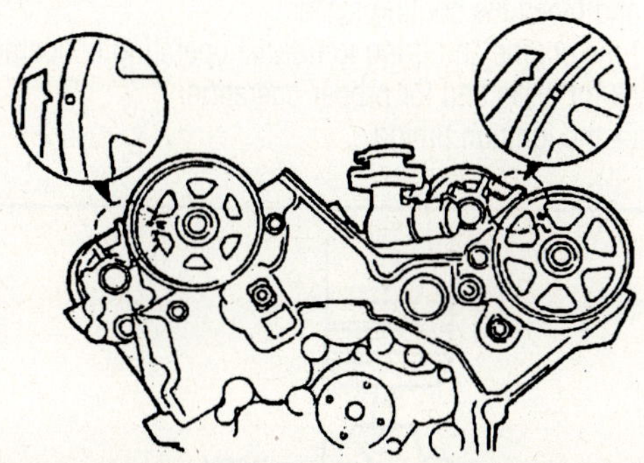

42356-MAZC-451

Turn the camshafts clockwise and align the timing marks–Mazda 2.5L (KL) engine

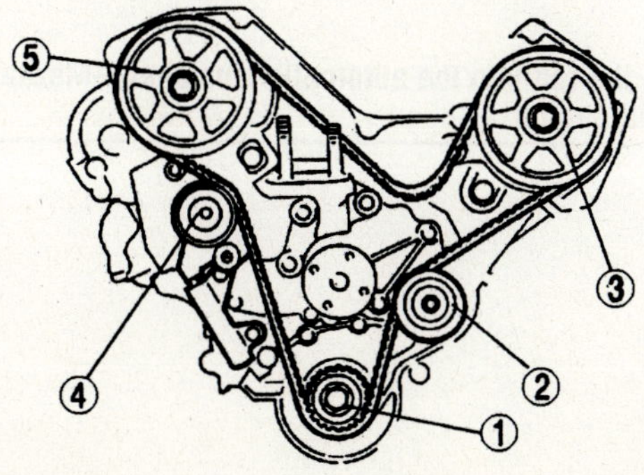

42356-MAZC-42

Make sure there is tension between pulleys (3) and (1) and pulleys (1) and (5–Mazda 2.5L (KL) engine

- Number 3 engine mount rubber. Refer to the exploded view of the timing belt components for torque specifications. Remove the engine hoist.
- Timing belt cover. Refer to the exploded view of the timing belt components for torque specifications.
- Oil dipstick and tube. Refer to the exploded view of the timing belt components for torque specifications.
- Rocker arm cover

**** CAUTION**

The CKP sensor rotor is on the rear of the crankshaft pulley and can be damaged if the pulley is not removed carefully.

- Crankshaft pulley using special tools 49 E011 1A1 and 49 S120 710. Refer to the illustration for tool positioning. Tighten the bolt to 16–122 ft. lbs. (157–166 Nm).
- Idler pulley bracket. Refer to the exploded view of the timing belt components for torque specifications.
- Water pump pulley. Refer to the exploded view of the timing belt components for torque specifications.
- Drive belt
- Power steering pump
- Spark plugs
- CKP sensor
- Negative battery cable

3.0L (JE) SOHC ENGINES

YEAR, MODEL & ENGINE APPLICATION	
1988-91 Mazda 929, 1989-99 MPV	3.0L SOHC Code JE

REPLACEMENT INTERVALS
Mazda recommends timing belt replacement every 60,000 miles.

LABOR TIMES	
929	
Replacement	3.1 hrs
Severe Service	3.3 hrs
With air conditioning add	0.2 hrs
With power steering add	0.1 hrs
MPV	
Replacement	3.4 hrs
Severe Service	3.9 hrs

8807MAZg18

**** CAUTION**

These engines are an interference type. If the timing belt has failed, valve-to-piston damage may have occurred. A compression check should always be performed to check for engine damage.

1. Disconnect the negative battery cable.
2. Drain the cooling system.
3. Label and disconnect the spark plug wires. Remove the spark plugs.
4. Remove the fresh air duct. Remove the cooling fan and the fan shroud.

5. Remove the accessory drive belts and the air conditioning compressor idler pulley.
6. Remove the crankshaft pulley and baffle plate.
7. Remove the coolant bypass hose and the upper radiator hose.
8. Remove the timing belt covers and gaskets.
9. Turn the crankshaft, in the normal direction of rotation and align the crankshaft and camshaft sprocket timing marks. Mark the direction of rotation on the timing belt.
10. Remove the upper idler pulley and remove the timing belt. Remove the automatic tensioner.

■ **Do not rotate the engine after the timing belt has been removed.**

11. Inspect the belt for wear, peeling, cracking, hardening or signs of oil contamination. Inspect the tensioner pulley for free and smooth rotation. Check the automatic tensioner for oil leakage. Check the tensioner rod projection (free length); it should be 0.47–0.55 in. (12–14mm). Inspect the sprocket teeth for wear or damage. Replace parts, as necessary.

To install:

12. Position the automatic tensioner in a suitable press. Place a flat washer under the tensioner body to prevent damage to the body plug.
13. Slowly press in the tensioner rod, but do not exceed 2200 lbs. (998 kg) force. Insert a pin into the tensioner body to hold the rod in place.
14. Install the tensioner and torque the bolts to 19 ft. lbs. (25 Nm).
15. Make sure the crankshaft and camshaft sprocket timing marks are aligned.
16. Install the timing belt in the following order:
   a. Crankshaft sprocket
   b. Lower idler pulley
   c. Left camshaft sprocket
   d. Tensioner pulley
   e. Right camshaft sprocket
17. If reusing the old timing belt, make sure it is installed in the same direction of rotation as marked.
18. Install the upper idler pulley and torque the bolt to 38 ft. lbs. (52 Nm).
19. Turn the crankshaft 2 turns, in the normal direction of rotation, and align the timing marks
20. Remove the pin from the automatic tensioner. Turn the crankshaft 2 turns, in the normal direction of rotation, and make sure the timing marks are aligned.
21. Apply approximately 22 lbs. (98 N) pressure to the timing belt at a point midway between the right camshaft sprocket and tensioner pulley. The belt should deflect 0.20–0.28 in. (5–7mm). If the deflection is not as specified, replace the automatic tensioner.
22. Install the timing belt covers with new gaskets. Torque the bolts to 95 inch lbs. (11 Nm).
23. Install the upper radiator hose and coolant bypass hose.
24. Install the baffle plate and crankshaft pulley. Torque the bolts to 130 inch lbs. (15 Nm).
25. Install the air conditioning compressor idler pulley and torque the bolts to 19 ft. lbs. (25 Nm).
26. Install the accessory drive belts and adjust the tension.
27. Install the fan shroud and cooling fan.
28. Install the fresh air duct.
29. Install the spark plugs and connect the spark plug wires.
30. Connect the negative battery cable.
31. Fill and bleed the cooling system.
32. Start the engine and bring to normal operating temperature.
33. Check for leaks and for proper operation.
34. Check the ignition timing.

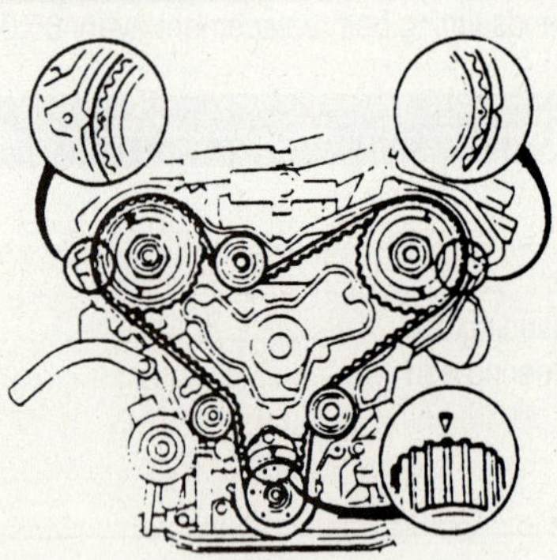

91251ga1

Crankshaft and camshaft sprocket timing mark alignment—Mazda 3.0L (JE) SOHC engine

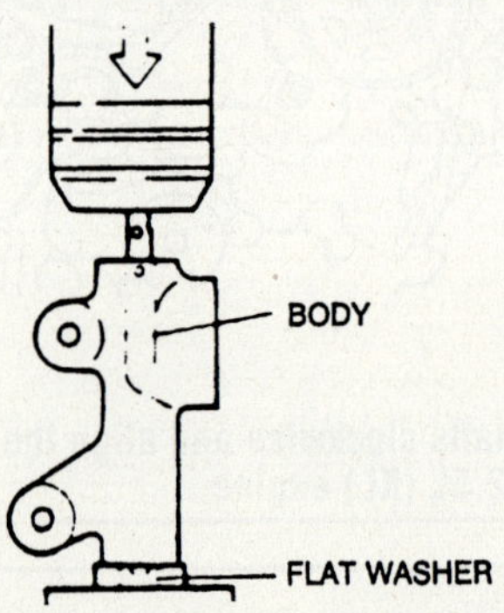

91251ga2

Pressing the rod into the automatic tensioner—Mazda 3.0L (JE) SOHC engine

3.0L (JE) DOHC ENGINES

YEAR, MODEL & ENGINE APPLICATION

1990-95 Mazda 929 .... 3.0L DOHC Code JE

REPLACEMENT INTERVALS

Mazda recommends timing belt replacement every 60,000 miles.

LABOR TIMES

1990-95	
Replacement	3.9 hrs
Severe Service	4.1 hrs
With air conditioning add	0.2 hrs
With power steering add	0.1 hrs

8807MAZg19

**** CAUTION**

These engines are an interference type. If the timing belt has failed, valve-to-piston damage may have occurred. A compression check should always be performed to check for engine damage.

1. Disconnect the negative battery cable.
2. Drain the cooling system.
3. Remove the fresh air duct.
4. Remove the cooling fan and fan shroud.
5. Remove the air intake pipe from the throttle body and air cleaner.
6. Disconnect the spark plug wires and remove the spark plugs.
7. Remove the idler pulleys and the accessory drive belts.
8. Remove the coolant bypass hose and the upper radiator hose.
9. Disconnect the electrical connector and remove the distributor.
10. Hold the crankshaft pulley with a suitable tool and remove the bolt. Remove the crankshaft pulley. Be careful not to damage the sensor rotor.
11. Remove the timing belt covers.
12. Rotate the crankshaft until the camshaft and crankshaft sprocket timing marks are aligned.
13. Remove the upper idler pulley.
14. Remove the automatic tensioner and pulley.
15. If timing belt is to be reused, mark the direction of rotation on the timing belt.
16. Remove the timing belt.

■ **Do not rotate the engine after the timing belt has been removed.**

17. Inspect the belt for wear, peeling, cracking, hardening or signs of oil contamination. Inspect the tensioner pulley for free and smooth rotation. Check the automatic tensioner for oil leakage. Check the tensioner rod projection (free length); it should be 0.47–0.55 in. (12–14mm). Inspect the sprocket teeth for wear or damage. Replace parts, as necessary.

To install:

18. Position the automatic tensioner in a suitable press. Place a flat washer under the tensioner body to prevent damage to the body plug.
19. Slowly press in the tensioner rod, but do not exceed 2200 lbs. (998 kg) force. Insert a pin into the tensioner body to hold the rod in place.
20. Install the tensioner and torque the bolts to 19 ft. lbs. (25 Nm).
21. Make sure the crankshaft and camshaft sprocket timing marks are aligned.
22. Install the timing belt in the following order:
    a. Crankshaft sprocket
    b. Lower idler pulley
    c. Left exhaust camshaft sprocket and left intake camshaft sprocket
    d. Tensioner pulley
    e. Right exhaust camshaft sprocket and right intake camshaft sprocket
23. Push the belt down and install the upper idler pulley. Torque the bolt to 38 ft. lbs. (52 Nm). Make sure the timing marks are still aligned after installing the upper idler pulley.
24. Turn the crankshaft 2 revolutions in the direction of normal rotation and realign the timing marks.
25. Remove the pin from the automatic tensioner. Again rotate the crankshaft 2 turns in the direction of rotation and make sure the timing marks are aligned.
26. Apply approximately 22 lbs. (98 N) pressure to the timing belt at a point midway between the right exhaust camshaft sprocket and the tensioner pulley. The belt should deflect 0.20–0.28 in. (5–7mm). If the deflection is not as specified, replace the automatic tensioner.
27. Install the timing belt cover and torque the bolts to 95 inch lbs. (11 Nm).
28. Install the crankshaft pulley. Hold the pulley with a suitable tool and torque the lockbolt to 123 ft. lbs. (167 Nm). Be careful not to damage the sensor rotor.
29. Install the distributor and connect the electrical connector.
30. Install the upper radiator hose and coolant bypass hose.

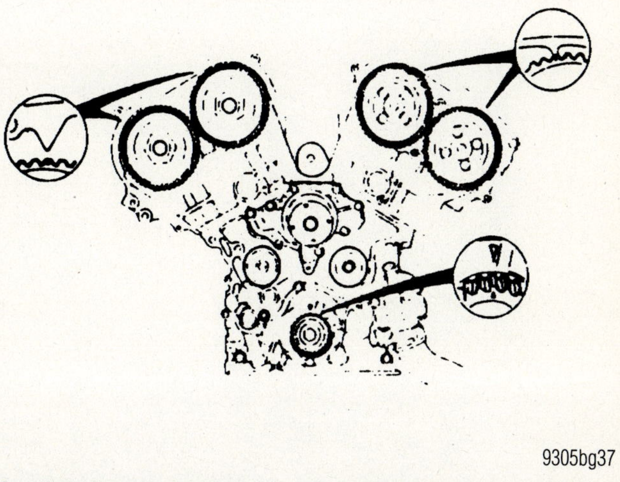

9305bg37

Crankshaft and camshaft sprocket timing mark alignment—Mazda 3.0L DOHC engine

31. Install the idler pulleys and the accessory drive belts. Adjust the belt tension.
32. Install the spark plugs and connect the spark plug wires.
33. Install the air intake pipe to the throttle body and air cleaner.
34. Install the cooling fan and radiator shroud.
35. Install the fresh air duct.
36. Connect the negative battery cable.
37. Fill and bleed the cooling system.
38. Start the engine and bring to normal operating temperature.
39. Check for leaks and for proper operation.
40. Check the ignition timing.

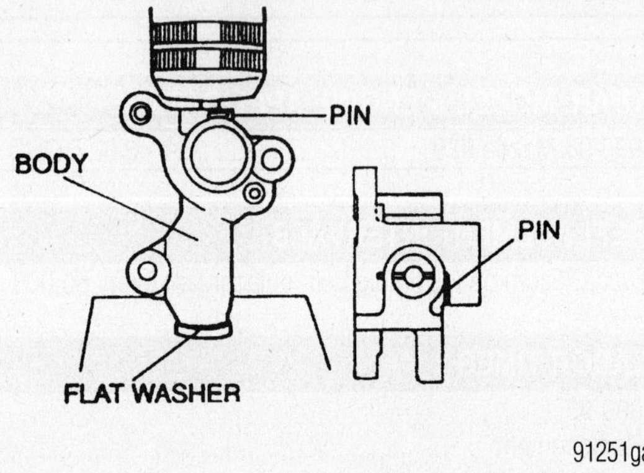

Pressing the rod into the automatic tensioner—Mazda 3.0L DOHC engine

Merkur

2.3L (VIN W) ENGINES

YEAR, MODEL & ENGINE APPLICATION

1985-89 Merkur XR4Ti	2.3L VIN W

REPLACEMENT INTERVALS

Ford recommends timing belt replacement every 120,000 miles.

LABOR TIMES

1985-89	
Replacement	2.3 hrs
Severe Service	2.5 hrs
With air conditioning add	0.5 hrs

8807MKRg01

✻✻ CAUTION

These engines are considered freewheeling. If the timing belt fails, valve-to-piston damage is unlikely, however a precautionary compression check should be performed to verify no engine damage.

1. Disconnect the negative battery cable.
2. Matchmark the distributor body to indicate the location of the No. 1 spark plug wire tower, and remove the distributor cap.
3. Rotate the crankshaft clockwise so that the timing mark is at Top Dead Center (TDC) of the compression stroke and the distributor rotor is pointing to the No. 1 spark plug wire tower mark.
4. Remove the accessory drive belts.
5. Remove the cooling fan and shroud.
6. Remove the water pump pulley.
7. Remove the front timing cover.
8. Check that the camshaft timing mark is aligned with the timing mark on the rear timing cover.
9. Hold the crankshaft from turning and remove the crankshaft pulley bolt.
10. Using crankshaft puller took T74P-6312-A, or equivalent, remove the crankshaft pulley.
11. Loosen the tensioner adjustment and pivot bolts. Using the tensioner tool T74P-6254-A, or equivalent, lever the tensioner away from the timing belt and tighten the adjustment bolt.
12. Remove the timing belt.

To install:

13. Install the timing belt.
14. Loosen the tensioner adjustment bolt, allowing it to spring back against the belt.
15. Rotate the crankshaft clockwise 2 complete turns and ensure that the timing marks align.

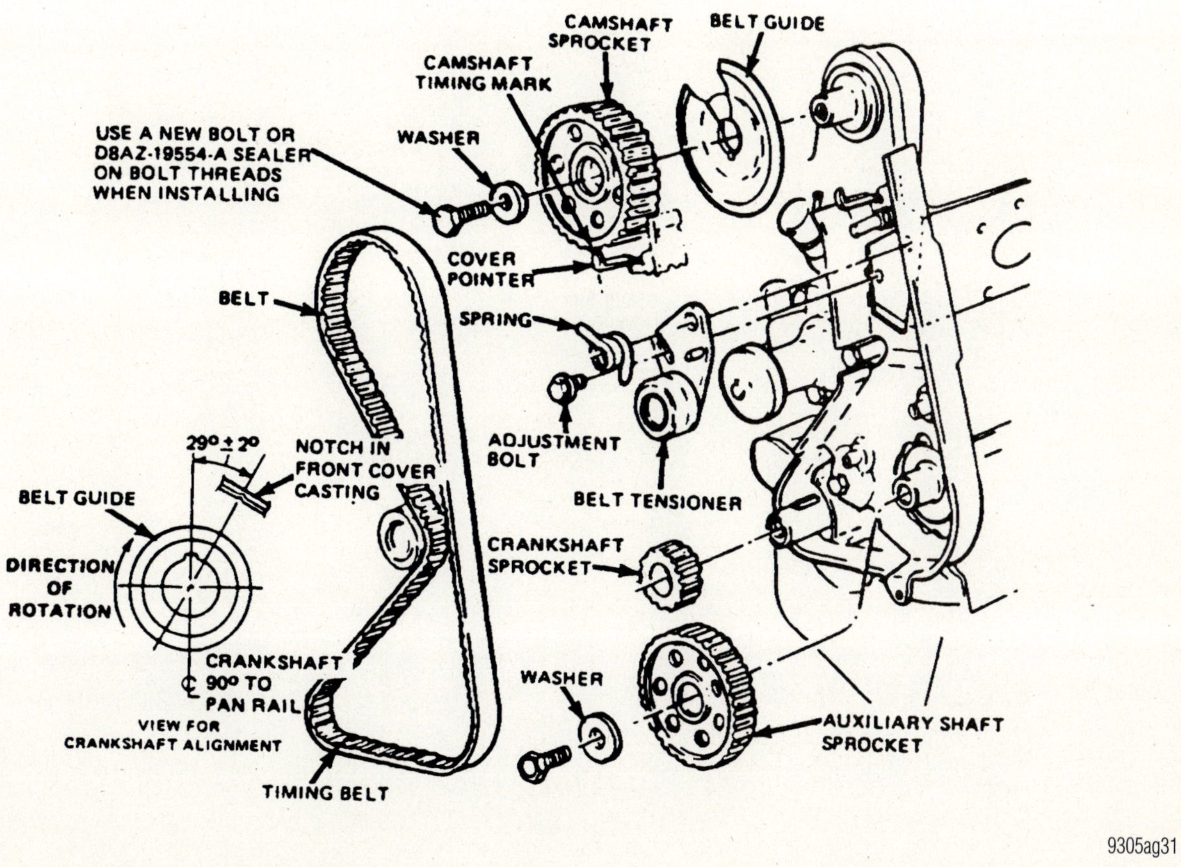

Exploded view of timing components—Merkur 2.3L (VIN W) engine

16. Tighten the tensioner adjustment bolt to 14–21 ft. lbs. (19–28 Nm).
17. Tighten the tensioner pivot bolt to 28–40 ft. lbs. (38–54 Nm).
18. Install the crankshaft pulley and tighten the center bolt to 100–120 ft. lbs. (136–163 Nm).
19. Install the front timing cover.
20. Install the water pump pulley.
21. Install the accessory drive belts.
22. Install the cooling fan and shroud.
23. Install the distributor cap.
24. Connect the negative battery cable

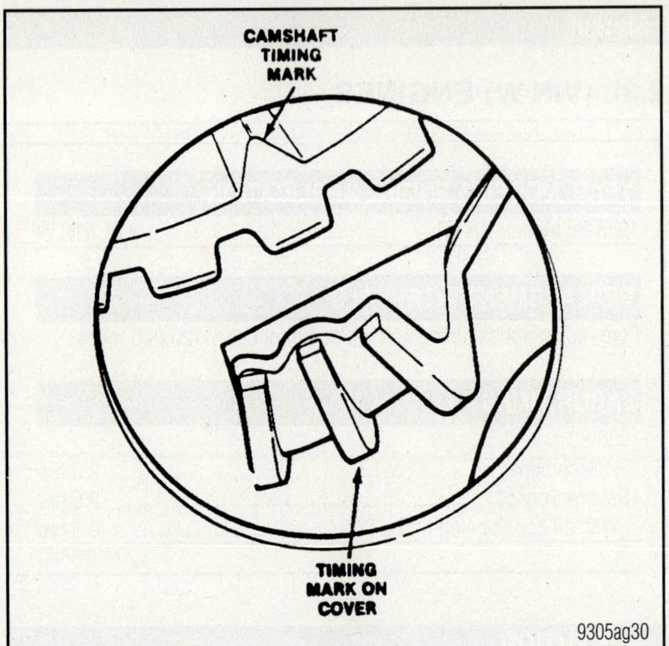

View of the camshaft timing marks—Merkur 2.3L (VIN W) engine

Mitsubishi

1.5L (G15B/4G15) AND 1.6.L (4G61) ENGINES

YEAR, MODEL & ENGINE APPLICATION

1985-87 Mitsubishi Mirage	1.5L Code G15B
1988-02 Mirage, 1987-93 Precis	1.5L Code 4G15
1991-92 Mirage	1.6L Code 4G16

REPLACEMENT INTERVALS

Mitsubishi recommends timing belt replacement every 60,000 miles.

LABOR TIMES

Mirage	
1985-96	
Replacement	2.7 hrs
Severe Service	3.2 hrs
With air conditioning add	0.5 hrs
With power steering add	0.5 hrs
1997-02	
Replacement	1.7 hrs
Severe Service	2.3 hrs
With air conditioning add	0.5 hrs
With power steering add	0.5 hrs
Precis	
Replacement	1.8 hrs
Severe Service	2.3 hrs

8807MITg01

1985–92 Models

✻✻ CAUTION

These engines are an interference type. If the timing belt has failed, valve-to-piston damage may have occurred. A compression check should always be performed to check for engine damage.

1. Disconnect the negative battery cable.
2. On Mirage, raise and safely support the vehicle.
3. Remove the left front wheel.
4. Remove the left side splash shield.
5. Lower the vehicle.
6. Using the proper equipment, slightly raise the engine to take the weight off the side engine mount.
7. Remove the engine mount bracket.
8. On Precis, remove the fan and spacer.
9. On all models, remove the accessory drive belts.
10. Remove the water pump pulley and the alternator mounting bracket.
11. Remove the crankshaft pulley by removing the 4 retaining bolts.
12. Remove all attaching screws and remove the upper and lower timing belt covers.
13. Rotate the crankshaft in the direction of rotation and align the timing marks so that the No. 1 piston is at Top Dead Center (TDC) of the compression stroke.
14. Loosen the tensioning bolt and the pivot bolt on the timing belt tensioner. Move the tensioner as far as it will go toward the water pump. Tighten the adjusting bolt.
15. If the timing belt is to be reused, mark the belt with an arrow showing direction of rotation.
16. Remove the timing belt.
17. Inspect the belt thoroughly. The back surface must be pliable and rough. If it is hard and glossy, the belt should be replaced. Any cracks or missing teeth mean the belt must be replaced.
18. Inspect the tensioner for grease leaking from the grease seal and any roughness in rotation. Replace as necessary.
19. The sprockets should be inspected for damaged teeth and cracks. Do not immerse sprockets in solvent, as solvent that has soaked into the metal may cause deterioration of the timing belt. Do not clean the tensioner in solvent, as this may wash the grease out of the bearing.

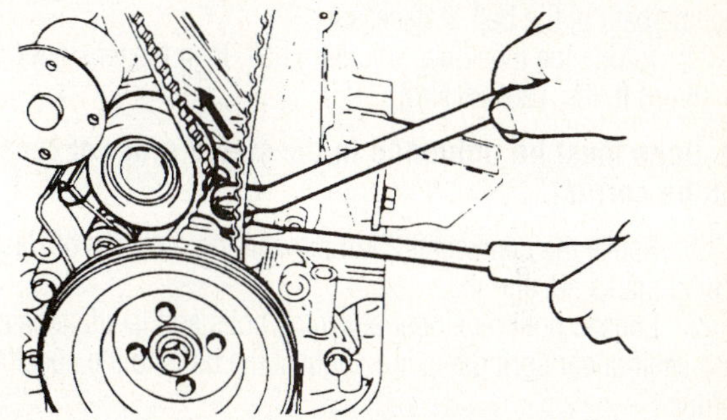

9305bg09

Timing belt tensioner movement—Mitsubishi 1.5L (G15B/4G15) and 1.6L (4G16) engines

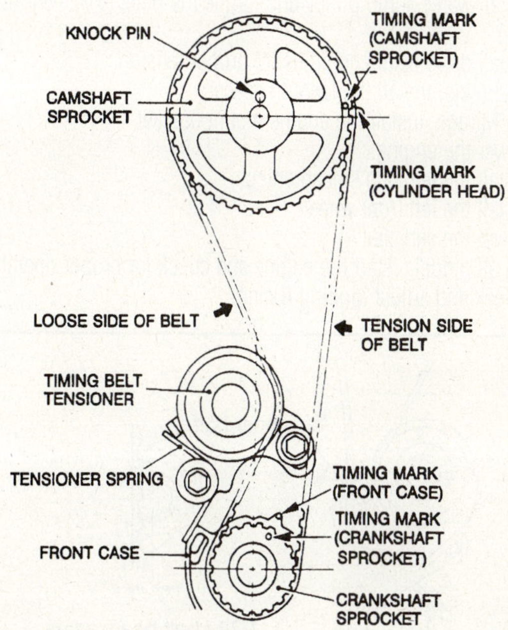

91251g61

Timing belt component identification and timing mark alignment—Mitsubishi 1.5L (G15B/4G15) and 1.6L (4G16) engines

To install:

20. Make sure the timing marks on the camshaft sprocket and crankshaft sprocket are in alignment.
21. Position the timing belt over the crankshaft sprocket, then over the camshaft sprocket. Slip the back of the belt over the tensioner pulley.
22. Rotate the camshaft sprockets in the opposite direction of rotation until the tension side of the belt is tight and that the timing marks are still aligned.

■ **If the timing marks are not properly aligned, shift the belt 1 tooth at a time in the appropriate direction until they are aligned.**

23. Loosen the tensioner mounting bolts so the tensioner applies pressure to the timing belt. Be sure the belt follows the curve of the camshaft pulley and that the teeth are engaged all the way around. Correct the path of the belt, if necessary.
24. Torque the tensioner adjusting bolt, then the tensioner pivot bolt to 15–18 ft. lbs. (20–26 Nm).

■ **Bolts must be tightened in the stated order or tension will not be correct.**

25. Rotate the crankshaft 2 full turns in the direction of rotation until timing marks are aligned.
26. Loosen both tensioner attaching bolts and let the tensioner position itself under spring tension. Tighten the bolts to the specified torque.
27. Check belt tension by putting a finger on the water pump side of the tensioner wheel and pull the belt toward the water pump. The belt should move toward the pump until the teeth are approximately 1/2 of the way across the head of the tensioner adjusting bolt.
28. Install the timing belt covers.
29. Install the crankshaft pulley, making sure the pin on the crankshaft sprocket fits through the hole in the rear surface of the pulley. Install the bolts and torque to 90–102 inch lbs. (10–11 Nm).
30. Install water pump pulley and torque the bolts to 72–84 inch lbs. (8–10 Nm).
31. Install drive belts and adjust to proper tension.
32. On Precis, install the fan and spacer.
33. On Mirage, install the engine mount bracket.
34. Lower the engine.
35. Install the left side splash shield.
36. Install the left front wheel.
37. Lower the vehicle.
38. On all models, start the engine and check for proper operation.
39. Check and adjust ignition timing.

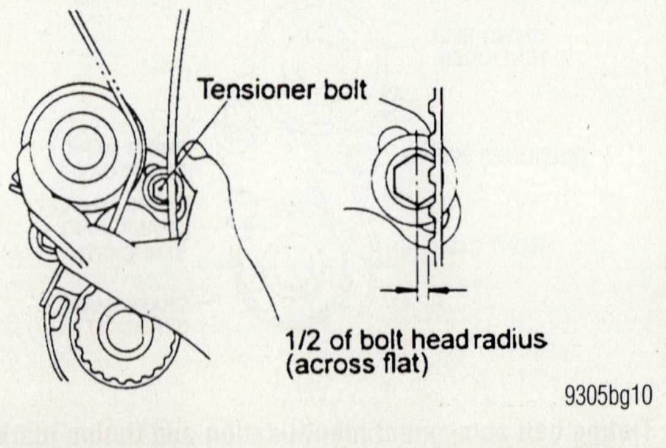

Checking timing belt tension—Mitsubishi 1.5L (G15B/4G15) and 1.6L (4G16) engines

1993–02 Models

⁂ CAUTION

These engines are an interference type. If the timing belt has failed, valve-to-piston damage may have occurred. A compression check should always be performed to check for engine damage.

1. Disconnect the negative battery cable and wait 60 seconds for the air bag system to lose power.
2. Raise and safely support the vehicle.
3. Remove the engine undercover.
4. Using the proper equipment, slightly raise the engine to take the weight off the side engine mount.
5. Remove the engine mount bracket insulator.
6. Remove all necessary components to access the right side engine mount bracket, then remove the engine mount and bracket.
7. Remove the accessory drive belts.
8. Remove the water pump pulley.
9. Remove the crankshaft pulley.
10. Remove all attaching screws and remove the upper and lower timing belt covers.
11. Rotate the crankshaft in the direction of rotation and align the timing marks so that the No. 1 piston is at Top Dead Center (TDC) of the compression stroke.
12. Loosen the tensioning bolt and the pivot bolt on the timing belt tensioner. Move the tensioner as far as it will go toward the water pump. Tighten the adjusting bolt.
13. If the timing belt is to be reused, mark the belt with an arrow showing direction of rotation.
14. Remove the timing belt.
15. Inspect the belt thoroughly. The back surface must be pliable and rough. If it is hard and glossy, the belt should be replaced. Any cracks or missing teeth mean the belt must be replaced.
16. Inspect the tensioner for grease leaking from the grease seal and any roughness in rotation. Replace as necessary.
17. The sprockets should be inspected for damaged teeth and cracks. Do not immerse sprockets in solvent, as solvent that has soaked into the metal may cause deterioration of the timing belt. Do not clean the tensioner in solvent, as this may wash the grease out of the bearing.

To install:

18. Make sure the timing marks on the camshaft sprocket and crankshaft sprocket are in alignment.
19. Position the timing belt over the crankshaft sprocket, then over the camshaft sprocket. Slip the back of the belt over the tensioner pulley.

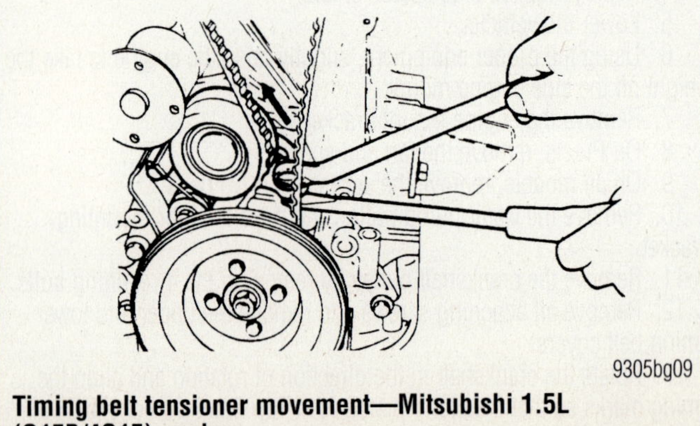

Timing belt tensioner movement—Mitsubishi 1.5L (G15B/4G15) engine

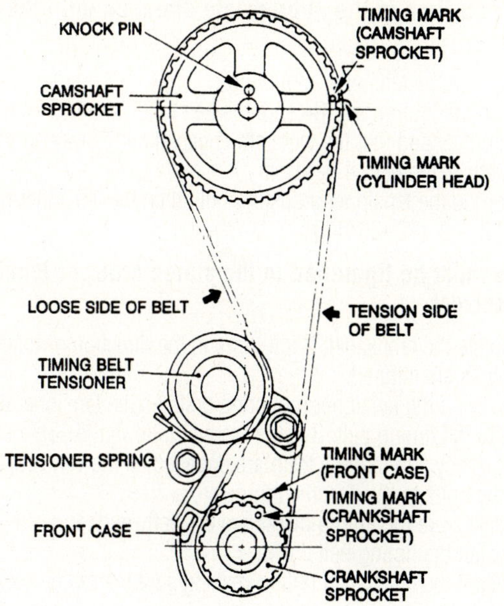

91251g61

Timing belt component identification and timing mark alignment—Mitsubishi 1.5L (G15B/4G15) engines

20. Rotate the camshaft sprockets in the opposite direction of rotation until the tension side of the belt is tight and that the timing marks are still aligned.

■ If the timing marks are not properly aligned, shift the belt 1 tooth at a time in the appropriate direction until they are aligned.

21. Loosen the tensioner mounting bolts so the tensioner applies pressure to the timing belt. Be sure the belt follows the curve of the camshaft pulley and that the teeth are engaged all the way around. Correct the path of the belt, if necessary.

22. Torque the tensioner adjusting bolt, then the tensioner pivot bolt to 15–18 ft. lbs. (20–26 Nm).

■ Bolts must be tightened in the stated order or tension will not be correct.

23. Rotate the crankshaft 2 full turns in the direction of rotation until timing marks are aligned.

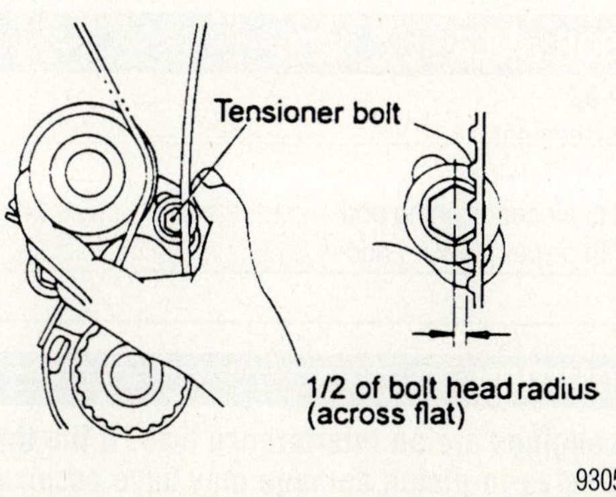

9305bg10

Checking timing belt tension—Mitsubishi 1.5L (G15B/4G15) engines

24. Loosen both tensioner attaching bolts and let the tensioner position itself under spring tension. Tighten the bolts to the specified torque.

25. Check belt tension by putting a finger on the water pump side of the tensioner wheel and pull the belt toward the water pump. The belt should move toward the pump until the teeth are approximately 1/2 of the way across the head of the tensioner adjusting bolt.

26. Install the timing belt covers.

27. Install the crankshaft pulley, making sure the pin on the crankshaft sprocket fits through the hole in the rear surface of the pulley. Install the bolts and torque to 90–102 inch lbs. (10–11 Nm).

28. Install water pump pulley and torque the bolts to 72–84 inch lbs. (8–10 Nm).

29. Install drive belts and adjust to proper tension.

30. Install engine mount and brackets.

31. Remove the engine mount bracket insulator.

32. Lower the engine.

33. Install the engine undercover.

34. Connect the negative battery cable.

35. Start the engine and check for proper operation.

36. Check and adjust ignition timing.

1.6L (G32B) TURBO ENGINES

YEAR, MODEL & ENGINE APPLICATION

1985-88 Mitsubishi Mirage Turbo	1.6L Code G32B

REPLACEMENT INTERVALS

Mitsubishi recommends timing belt replacement every 60,000 miles.

LABOR TIMES

Mirage	
Replacement	2.7 hrs
Severe Service	3.2 hrs
With air conditioning add	0.5 hrs
With power steering add	0.5 hrs

9907MITg02

✱✱ CAUTION

These engines are an interference type. If the timing belt has failed, valve-to-piston damage may have occurred. A compression check should always be performed to check for engine damage.

1. Disconnect the negative battery cable.
2. Raise and safely support the vehicle.
3. Remove the left front wheel.
4. Remove the left side splash shield.
5. Lower the vehicle.
6. Using the proper equipment, slightly raise the engine to take the weight off the side engine mount.
7. Remove the engine mount bracket.
8. Remove the accessory drive belts.
9. Remove the water pump pulley and the alternator mounting bracket.
10. Remove the crankshaft pulley by removing the 4 retaining bolts.

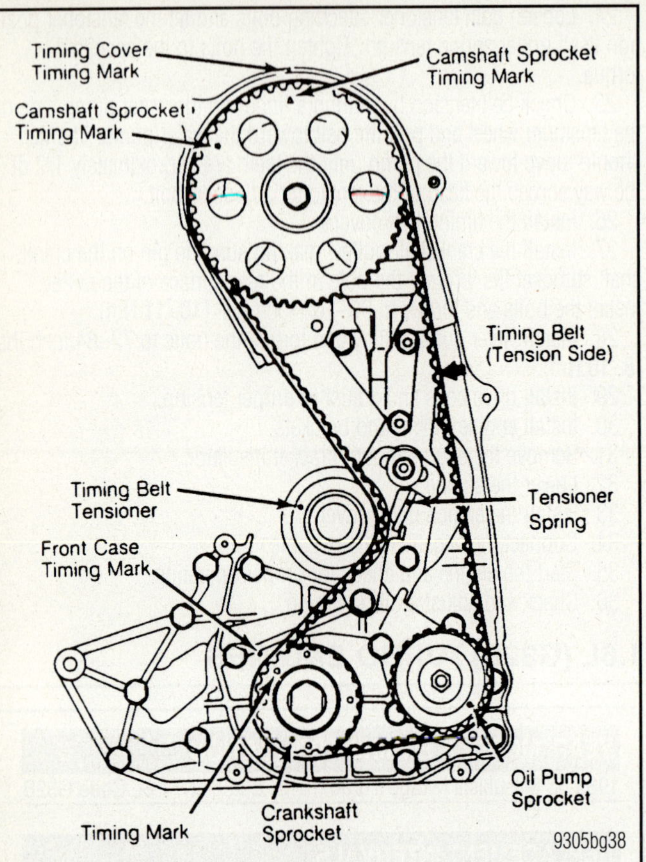

Timing mark alignment—Mitsubishi 1.6L (G32B) turbo engine

11. Remove all attaching screws and remove the upper and lower timing belt covers.
12. Rotate the crankshaft in the direction of rotation and align the timing marks so that the No. 1 piston is at Top Dead Center (TDC) of the compression stroke.
13. Loosen the tensioning bolt and the pivot bolt on the timing belt tensioner. Move the tensioner as far as it will go toward the water pump. Tighten the adjusting bolt.
14. If the timing belt is to be reused, mark the belt with an arrow showing direction of rotation.
15. Remove the timing belt.
16. Inspect the belt thoroughly. The back surface must be pliable and rough. If it is hard and glossy, the belt should be replaced. Any cracks or missing teeth mean the belt must be replaced.
17. Inspect the tensioner for grease leaking from the grease seal and any roughness in rotation. Replace as necessary.
18. The sprockets should be inspected for damaged teeth and cracks. Do not immerse sprockets in solvent, as solvent that has soaked into the metal may cause deterioration of the timing belt. Do not clean the tensioner in solvent, as this may wash the grease out of the bearing.

To install:

19. Align the timing marks of the camshaft sprocket, oil pump sprocket and crankshaft sprocket.
20. Position the timing belt over the crankshaft sprocket, oil pump sprocket, camshaft sprocket and the onto the tensioner pulley.
21. Rotate the camshaft sprockets in the opposite direction of rotation until the tension side of the belt is tight and that the timing marks are still aligned.

■ **If the timing marks are not properly aligned, shift the belt 1 tooth at a time in the appropriate direction until they are aligned.**

22. Loosen the tensioner mounting bolts so the tensioner applies pressure to the timing belt. Be sure the belt follows the curve of the camshaft pulley and that the teeth are engaged all the way around. Correct the path of the belt, if necessary.
23. Torque the tensioner adjusting bolt, then the tensioner pivot bolt to 15–18 ft. lbs. (20–26 Nm).

■ **Bolts must be tightened in the stated order or tension will not be correct.**

24. Rotate the crankshaft 2 full turns in the direction of rotation until timing marks are aligned.
25. Loosen the tensioner mounting bolts so the tensioner applies pressure to the timing belt. Be sure the belt follows the curve of the camshaft pulley and that the teeth are engaged all the way around. Tighten the bolts to the specified torque.
26. Make sure the timing marks are correctly aligned.
27. Install the timing belt covers.
28. Install the crankshaft pulley, making sure the pin on the crankshaft sprocket fits through the hole in the rear surface of the pulley. Install the bolts and torque to 90–102 inch lbs. (9–12 Nm).
29. Install water pump pulley and torque the bolts to 72–84 inch lbs. (8–10 Nm).
30. Install drive belts and adjust to proper tension.
31. Install engine mounting bracket and lower the engine.
32. Raise and safely support the vehicle.
33. Install the left side splash shield.
34. Install the left front wheel.
35. Lower the vehicle.
36. Connect the negative battery cable.
37. Start the engine and check for proper operation.
38. Check and adjust ignition timing.

1.6L (4G61) DOHC ENGINES

YEAR, MODEL & ENGINE APPLICATION

1989-92 Mitsubishi Mirage	1.6L Code 4G61

REPLACEMENT INTERVALS

Mitsubishi recommends timing belt replacement every 60,000 miles.

LABOR TIMES

1989-92	
Replacement	3.8 hrs
Severe Service	4.3 hrs
With air conditioning add	0.5 hrs
With power steering add	0.5 hrs

8807MITg03

✱✱ CAUTION

These engines are an interference type. If the timing belt has failed, valve-to-piston damage may have occurred. A compression check should always be performed to check for engine damage.

1. Disconnect the negative battery cable and wait one minute for the air bag system to drain the backup power.
2. Raise and safely support the vehicle.
3. Remove the left front wheel.
4. Remove the left front splash shield.
5. Using the proper equipment, slightly raise the engine to take the weight off the side engine mount.
6. Remove the engine mount bracket.
7. Remove the accessory drive belts and any mounting brackets that may Interfere wlth tlmlng cover removal.
8. Remove the water pump pulley.
9. Remove the crankshaft pulley bolts and remove the crankshaft pulley.
10. Remove all attaching screws and remove the upper and lower timing belt covers.
11. Label and remove the spark plug wires.

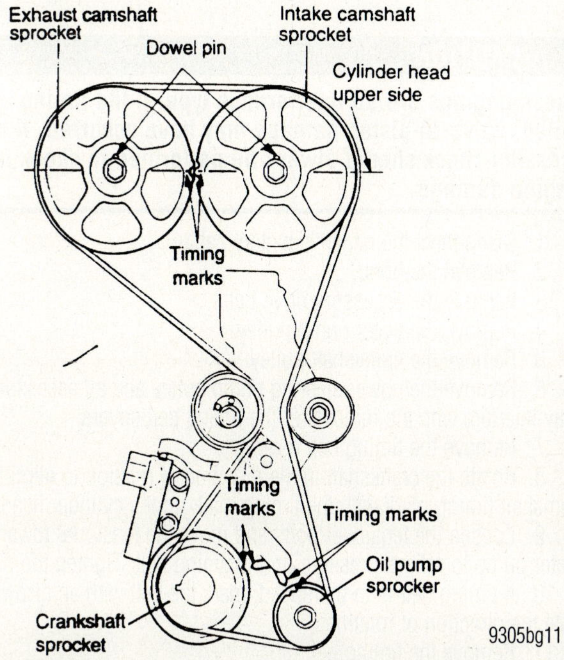

Correct alignment of engine timing marks for belt installation—Mitsubishi 1.6L (4G61) DOHC engine

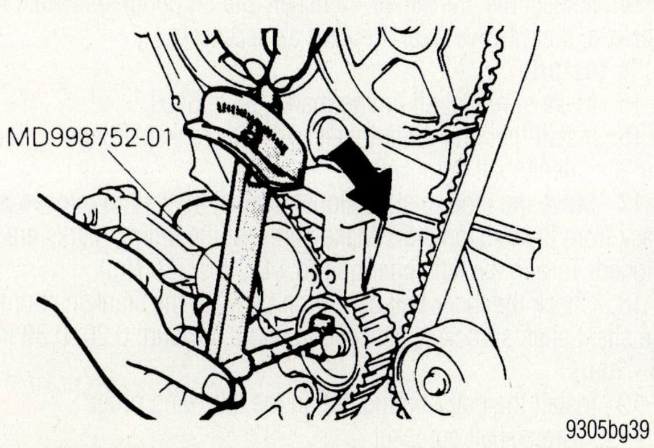

Applying torque to the tensioner pulley—Mitsubishi 1.6L (4G61) DOHC engine

12. Remove the spark plugs.

■ **Removal of the spark plugs is so that the engine can be rotated easily.**

13. Rotate the crankshaft in the direction of rotation and align the timing marks. The No. 1 piston will be at Top Dead Center (TDC) of the compression stroke. At this time the timing marks on the camshaft sprocket and the upper surface of the cylinder head should coincide, and the dowel pin of the camshaft sprocket should be at the upper side.

■ **Always rotate the crankshaft in the normal direction of rotation.**

14. While holding the auto tensioner, remove the tensioner mounting bolts. This will keep from stripping the threads on the mounting bolts. Remove the auto tensioner.
15. Remove the timing belt.
16. If the timing belt is to be reused, mark the belt with an arrow indicating the direction of rotation.
17. Check the tensioner and idler pulleys for ease of movement and abnormal noises. Replace as necessary.
18. Check crankshaft and camshaft sprockets for cracks and excessive wear. Replace as necessary.
19. Check the auto tensioner for leaks and damage. Measure the rod protrusion from the tensioner body and the tip of the rod. Measurement should be 0.47 in. (12mm). If out of specification, the tensioner must be replaced.

To install:

20. Carefully push in the auto tensioner rod until the set hole in the rod is aligned with the hole in the cylinder body. Place a wire pin into the hole to retain the rod.
21. Install the auto tensioner and torque the bolts to 14–20 ft. lbs. (20–27 Nm).
22. When installing the timing belt, turn the 2 camshaft sprockets so their dowel pins are located on top. Align the timing marks facing each other with the top surface of the cylinder head. When you let go of the exhaust camshaft sprocket, it will rotate 1 tooth in the counterclockwise direction. This should be taken into account when installing the timing belts on the sprocket.

■ **Both camshaft sprockets are used for the intake and exhaust camshafts and are provided with 2 timing marks. When the sprocket is mounted on the exhaust camshaft, use the timing mark on the right with the dowel pinhole on top. For the intake camshaft sprocket, use the mark on the left with the dowel pinhole on top.**

23. Align the crankshaft sprocket and oil pump sprocket timing marks.
24. Install the timing belt as follows:
    a. Install the timing belt around the tensioner pulley and the crankshaft sprocket while holding the timing belt onto the tensioner pulley with your left hand.
    b. Pull the belt with your right hand and install it around the oil pump sprocket.
    c. Install the belt around the idler pulley and the intake camshaft sprocket.
    d. Rotate the exhaust camshaft sprocket, 1 tooth, in the direction of rotation to align the timing mark with the intake camshaft timing mark.
    e. Install the belt around the exhaust camshaft sprocket.
    f. Raise the tensioner pulley clockwise against the timing belt and temporarily tighten the bolt.
    g. Rotate the crankshaft 1/4 turn counterclockwise. Then, turn

clockwise until the timing marks are aligned again.

25. To adjust the timing belt, turn the crankshaft 1/4 turn opposite direction of rotation, then turn it in the direction of rotation to move No. 1 cylinder to TDC.
26. Loosen the tensioner center bolt. Using tool MD998738 or equivalent and a torque wrench, apply a torque of 22.5–24.3 inch lbs. (2.6–2.8 Nm) to the tensioner pulley. Torque the center bolt to 31–40 ft. lbs. (43–55 Nm).
27. Remove the rubber plug from the rear timing belt cover and screw special tool MD99873801 or equivalent, into the engine left support bracket until its end makes contact with the tensioner arm. At this point, screw the special tool in some more and remove the set wire attached to the auto tensioner. Then remove the special tool.
28. Rotate the crankshaft 2 complete turns clockwise and let it sit for approximately 15 minutes. Then, measure the auto tensioner protrusion (the distance between the tensioner arm and auto tensioner body) to ensure that it is within 0.15–0.18 in. (3.8–4.5mm).
29. Install the rubber plug to the timing belt rear cover.
30. Install the upper and lower timing belt covers and torque the bolts to 84–108 inch lbs. (9–12 Nm).
31. Install the water pump pulley and torque the bolts to 84 inch lbs. (10 Nm).
32. Install the crankshaft pulley and torque the bolts to 22 ft. lbs. (30 Nm).
33. Install the mounting brackets and the accessory drive belts.
34. Install the engine mount bracket and engine mount. Lower the engine.
35. Install the spark plugs.
36. Install the spark plug wires.
37. Raise and safely support the vehicle.
38. Install the left front splash shield.
39. Install the left front wheel.
40. Lower the vehicle
41. Connect the negative battery cable.
42. Start the engine and check for proper operation.
43. Check the ignition timing.

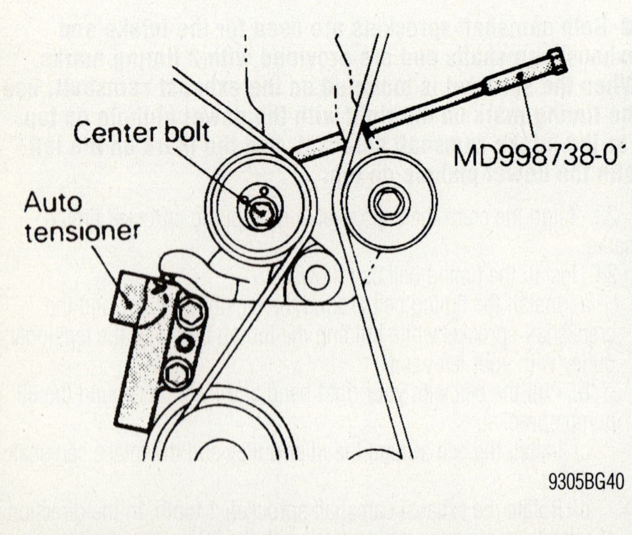

Inserting special tool to contact tensioner arm—Mitsubishi 1.6L (4G61) DOHC engine

1.8L (G62B) SOHC ENGINES

YEAR, MODEL & ENGINE APPLICATION	
1984-88 Mitsubishi Cordia, Tredia	1.8L Code G62B

REPLACEMENT INTERVALS

Mitsubishi recommends timing belt replacement every 60,000 miles.

LABOR TIMES	
1984-88	
Replacement	2.7 hrs
Severe Service	3.2 hrs
With balance shaft belt replacement add	0.3 hrs

8807MITg04

✱✱ CAUTION

These engines are an interference type. If the timing belt has failed, valve-to-piston damage may have occurred. A compression check should always be performed to check for engine damage.

1. Disconnect the negative battery cable.
2. Remove the hood.
3. Remove the accessory drive belts.
4. Remove the water pump pulley.
5. Remove the crankshaft pulley.
6. Remove the power steering pump pulley and all accessories that may interfere with the removal of the timing belt covers.
7. Remove the timing belt covers.
8. Rotate the crankshaft in the direction of rotation to align the camshaft timing mark with the timing mark on the cylinder head.
9. Loosen the tensioner bolts and move the tensioner toward the water pump to relieve pressure on the timing belt. Tighten the bolts.
10. If timing belt is to be reused, mark the belt with an arrow to indicate the direction of rotation.
11. Remove the timing belt.
12. Loosen the lockbolt for the inner timing belt. Move the tensioner away from the belt and remove the inner belt.
13. Inspect the tensioner pulley for ease of movement and abnormal noises. Replace as necessary.
14. Inspect the crankshaft, camshaft and oil pump sprockets for cracks and excessive wear. Replace as necessary.

To install:

15. Make sure that all timing marks are aligned.
16. Install the inner timing belt over the crankshaft inner sprocket and the silent shaft sprocket.
17. Move the inner belt tensioner against the belt to remove any slack from the tension side. Make sure that the timing marks are aligned. Torque the tensioner bolt to 14 ft. lbs. (19 Nm).
18. Check the inner belt deflection between the crankshaft pulley and the silent shaft sprocket. Deflection should measure 0.20–0.30 in. (5–7mm).
19. Install the outer timing belt in the following order:
    a. Crankshaft sprocket

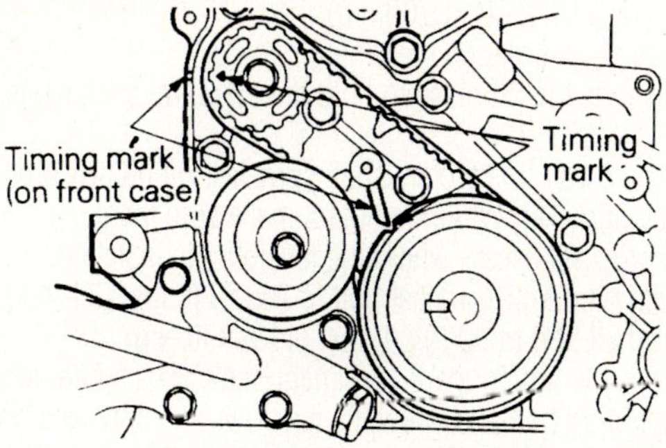

Inner timing belt mark alignment—Mitsubishi 1.8L (G62B) SOHC engine

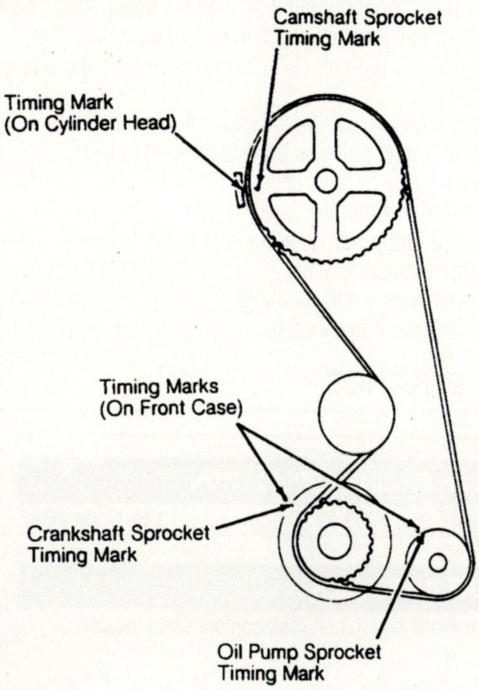

Crankshaft, camshaft and oil pump sprocket timing mark alignment—Mitsubishi 1.8L (G62B) SOHC engine

b. Oil pump sprocket
c. Camshaft sprocket
d. Tensioner pulley

20. Make sure the timing marks are aligned.
21. Loosen the tensioner bolts to apply pressure to the timing belt.
22. Rotate the crankshaft 2 full turns in the direction of rotation and align the timing marks. Make sure that there is no slack on the tension side of the belt.
23. Loosen the tensioner bolts again to apply pressure to the timing belt. Torque the bolts to 35 ft. lbs. (49 Nm).
24. Install the timing belt covers.
25. Install all accessories that were previously removed to access the timing belt covers. Install the power steering pump pulley.
26. Install the crankshaft pulley and torque the bolts to 15–21 ft. lbs. (20–29 Nm).
27. Install the water pump pulley and torque the bolts to 84 inch lbs. (9 Nm).
28. Install the accessory drive belts and adjust to proper tension.
29. Connect the negative battery cable.
30. Start the engine and check for proper operation.
31. Check ignition timing and adjust if necessary.

1.8L (4G37) ENGINES

YEAR, MODEL & ENGINE APPLICATION

1990-94 Mitsubishi Eclipse	1.8L Code 4G37

REPLACEMENT INTERVALS

Mitsubishi recommends timing belt replacement every 60,000 miles.

LABOR TIMES

1990-94	
Replacement	3.2 hrs
Severe Service	3.7 hrs
With air conditioning add	0.5 hrs
With balance shaft belt replacement add	0.3 hrs

8807MITg05

✻✻ CAUTION

These engines are an interference type. If the timing belt has failed, valve-to-piston damage may have occurred. A compression check should always be performed to check for engine damage.

1. Position the engine so the No. 1 piston is at Top Dead Center (TDC) of the compression stroke.
2. Disconnect the negative battery cable.
3. Raise the vehicle and support it safely. Remove the under engine splash shield.
4. Place a piece of wood on a suitable floor jack and support the engine. Remove the engine mount bracket.
5. Disconnect the A/C and power steering lines from the body.
6. Loosen the water pump pulley bolts.
7. Remove the accessory drive belts.
8. Remove the water pump pulley and the crankshaft pulley.
9. Remove the upper and lower timing belt covers.

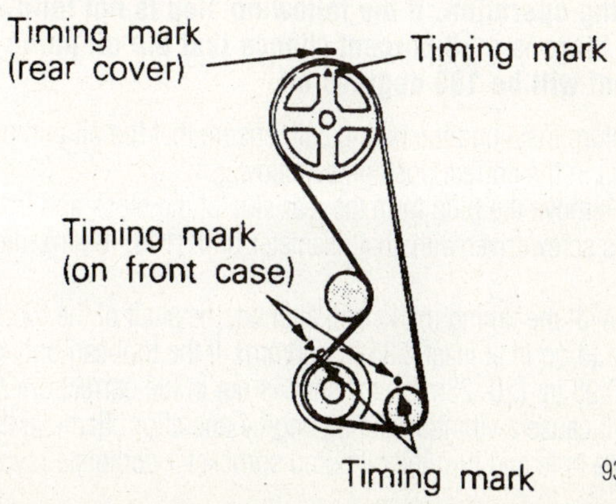

Crankshaft, camshaft and oil pump sprocket timing mark alignment—Mitsubishi 1.8L (4G37) engine

10. Align the crankshaft and camshaft sprocket timing marks.
11. Remove the crankshaft sprocket bolt.
12. Loosen the timing belt tensioner bolt, press the tensioner back toward the water pump and then tighten the tensioner bolt.
13. Remove the timing belt.
14. Align the silent shaft and crankshaft sprocket timing marks.
15. Remove the outer crankshaft sprocket and flange.
16. Remove the silent shaft (inner) belt tensioner and remove the inner belt.

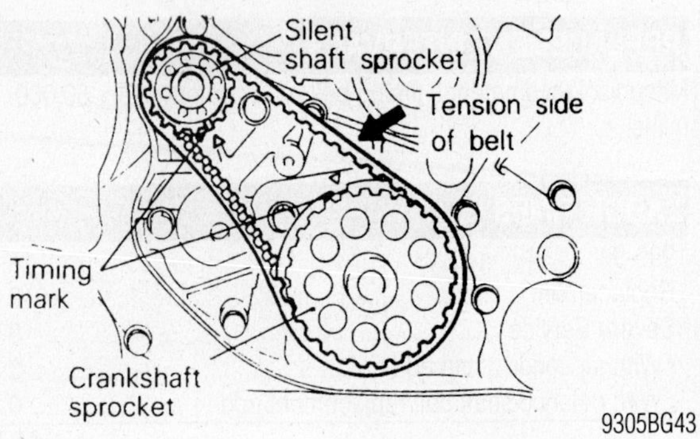

Silent shaft belt timing mark alignment—Mitsubishi 1.8L (4G37) engine

To install:

17. Align the timing marks of the silent shaft sprockets and the crankshaft sprocket with the timing marks on the front case. Wrap the timing belt around the sprockets so there is no slack in the upper span of the belt and the timing marks are still aligned.
18. Install the tensioner pulley and move the pulley by hand so the long side of the belt deflects 0.20–0.28 in. (5–7mm).
19. Hold the pulley tightly so the pulley cannot rotate when the bolt is tightened. Tighten the bolt to 15 ft. lbs. (20 Nm) and recheck the deflection amount.
20. Install the timing belt tensioner fully toward the water pump and tighten the bolts. Place the upper end of the spring against the water pump body.
21. Align the timing marks of the camshaft, crankshaft and oil pump sprockets with their corresponding marks on the front case or rear cover.

■ **There is a possibility to align all timing marks and have the oil pump sprocket out of time, causing an engine vibration during operation. If the following step is not followed exactly, there is a 50 percent chance that the oil pump shaft alignment will be 180 degrees off.**

22. Before installing the timing belt, ensure that the oil pump sprocket is in the correct position as follows:
   a. Remove the plug from the rear side of the block and insert a Phillips screwdriver with shaft diameter of 0.31 in. (8mm) into the hole.
   b. With the timing marks still aligned, the shaft of the tool must be able to go in at least 2.36 in. (60mm). If the tool can only go in 0.79–0.98 in. (20–25mm), the shaft is not in the correct orientation and will cause a vibration during engine operation. Remove the tool from the hole and turn the oil pump sprocket 1 complete revolution. Realign the timing marks and insert the tool. The shaft of the tool must go in at least 2.36 in. (60mm).
   c. Recheck and realign the timing marks.
   d. Leave the tool in place to hold the oil pump shaft while continuing.
23. Install the belt to the crankshaft sprocket, oil pump sprocket, then camshaft sprocket. While doing so, make sure there is no slack between the sprocket except where the tensioner is installed.
24. Tighten oil pump sprocket bolt to 26–29 ft. lbs. (34–40 Nm) and tighten crankshaft bolt to 80–94 ft. lbs. (110–130 Nm).
25. Recheck the timing mark alignment. If all are aligned, loosen the tensioner mounting bolt and allow the tensioner to apply tension to the belt.
26. Remove the tool that is holding the silent shaft and rotate the crankshaft a distance equal to 2 teeth on the camshaft sprocket. This will allow the tensioner to automatically apply the proper tension on the belt. Do not manually overtighten the belt or it will howl.
27. Tighten the lower mounting bolt first, then the upper spacer bolt.
28. To verify correct belt tension, check that the deflection at the longest span of the belt has 0.40 in. (12mm) clearance from the belt cover.
29. Install the upper and lower timing belt covers.
30. Install the water pump pulley and the crankshaft pulley.
31. Install the the accessory drive belts.
32. Tighten the water pump pulley bolts.
33. Connect the A/C and power steering lines to the body.
34. Install the engine mount bracket.
35. Install the under engine splash shield.
36. Connect the negative battery cable.

1.8L (4G93) ENGINES

YEAR, MODEL & ENGINE APPLICATION

1991-93 Mitsubishi Expo, 1993-02 Mirage ...... 1.8L Code 4G93

REPLACEMENT INTERVALS

Mitsubishi recommends timing belt replacement every 60,000 miles.

LABOR TIMES

Expo	
Replacement	2.7 hrs
Severe Service	3.2 hrs
With air conditioning add	0.5 hrs
Mirage	
1993-96	
Replacement	2.7 hrs
Severe Service	3.2 hrs
With air conditioning add	0.5 hrs
With power steering add	0.5 hrs
1997-02	
Replacement	1.7 hrs
Severe Service	2.3 hrs
With air conditioning add	0.5 hrs
With power steering add	0.5 hrs

8807MITg06

✻✻ CAUTION

These engines are an interference type. If the timing belt has failed, valve-to-piston damage may have occurred. A compression check should always be performed to check for engine damage.

1. Disconnect the negative battery cable and wait one minute for the air bag system to drain the backup power.
2. On Expo models, remove the coolant reservoir.
3. Raise and safely support the vehicle.
4. Remove the left front wheel.
5. Remove the left side splash shield.
6. Lower the vehicle.
7. Remove the accessory drive belts.
8. Remove the crankshaft pulley bolt and remove the crankshaft pulley.
9. Remove the upper alternator brace.
10. Remove the timing belt covers.
11. On 1997–02 models, remove the engine support bracket.
12. Reinstall the crankshaft pulley bolt and rotate the engine in the direction of rotation to align the timing marks on the camshaft sprocket and the crankshaft sprocket.
13. Loosen the tensioner lockbolt and move the tensioner counterclockwise to relieve pressure on the timing belt.
14. Temporarily tighten the tensioner lockbolt.
15. If the timing belt is to be reused, mark the belt with an arrow to indicate the direction of rotation.
16. Remove the timing belt.
17. Inspect the tensioner pulley for ease of movement and abnormal noises. Replace as necessary.
18. Inspect the crankshaft and camshaft sprockets for cracks and excessive wear. Replace as necessary.

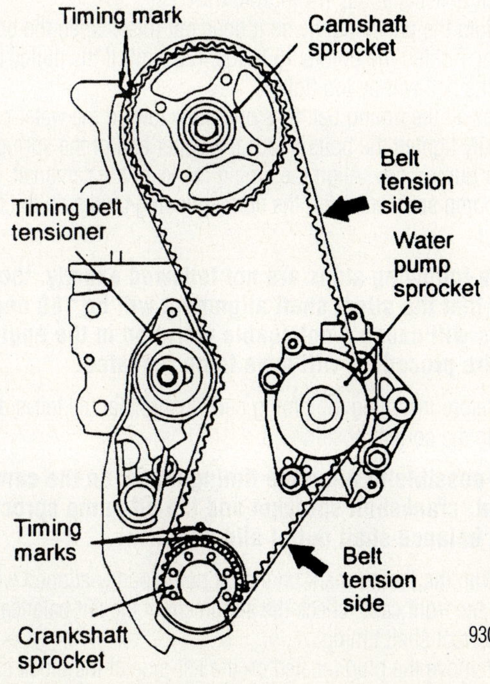

Timing mark alignment and component identification—Mitsubishi 1.8L (4G93) engine

To install:

19. Make sure that the timing marks are correctly aligned.
20. Install the timing belt in the following order:
    a. Crankshaft sprocket
    b. Water pump sprocket
    c. Camshaft sprocket
    d. Tensioner pulley
21. Make sure that here is no slack on the tension side of the belt.
22. Loosen the tensioner lockbolt to apply tension to the timing belt.
23. Temporarlly tighten the tensioner lockbolt.
24. Rotate the crankshaft 2 full turns in the direction of rotation and align the timing marks.
25. Loosen the tensioner lockbolt and make sure that there is no slack in the timing belt and that the timing marks are aligned.
26. Torque the tensioner lockbolt to 17 ft. lbs. (24 Nm).
27. Install the upper and lower timing belt covers.
28. Install the front engine mount and bracket and lower the engine.
29. Install the crankshaft pulley and torque the pulley bolt to 130–137 ft. lbs. (177–186 Nm).
30. Install the upper alternator bracket and mount the alternator.
31. Install the accessory drive belts and adjust tension.
32. Raise and safely support the vehicle.
33. Install the left side splash shield.
34. Install the left front wheel.
35. Lower the vehicle.
36. On Expo models, install the coolant reservoir.
37. Connect the negative battery cable.
38. Start the engine and check for proper operation.
39. Check and adjust the ignition timing if necessary.

2.0L (G63B/4G63) SOHC ENGINES

YEAR, MODEL & ENGINE APPLICATION

1983-88 Mitsubishi Cordia, Pickup, Tredia	2.0L Code G63B
1989-93 Galant	2.0L Code 4G63

REPLACEMENT INTERVALS

Mitsubishi recommends timing belt replacement every 60,000 miles.

LABOR TIMES

Cordia & Tredia	
Replacement	2.7 hrs
Severe Service	3.2 hrs
With balance shaft belt replacement add	0.3 hrs
Galant	
Replacement	3.3 hrs
Severe Service	3.8 hrs
With balance shaft belt replacement add	0.3 hrs
With air conditioning add	0.5 hrs
Pickup	
Replacement	2.5 hrs
Severe Service	3.0 hrs
With balance shaft belt replacement add	0.3 hrs
With air conditioning add	0.5 hrs
With power steering add	0.3 hrs

8807MITg07

Cordia, Tredia and 1989–92 Galant

✻✻ CAUTION

These engines are an interference type. If the timing belt has failed, valve-to-piston damage may have occurred. A compression check should always be performed to check for engine damage.

1. Disconnect the negative battery cable and wait one minute for the air bag system to drain the backup power.
2. Remove the hood.
3. Remove the accessory drive belts.
4. Remove the crankshaft pulley.
5. Remove the timing belt covers.
6. Rotate the crankshaft in the direction of rotation to align the camshaft timing mark with the timing mark on the cylinder head.
7. Loosen the tensioner bolts and move the tensioner toward the water pump to relieve pressure on the timing belt. Tighten the bolts.

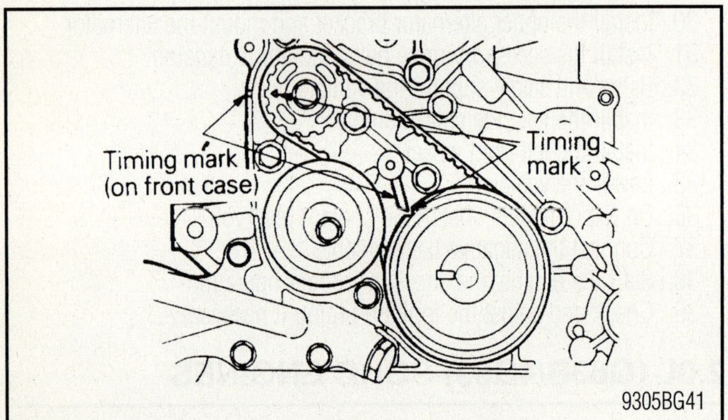

Inner timing belt mark alignment—Mitsubishi 2.0L (G63B/4G63) SOHC engine

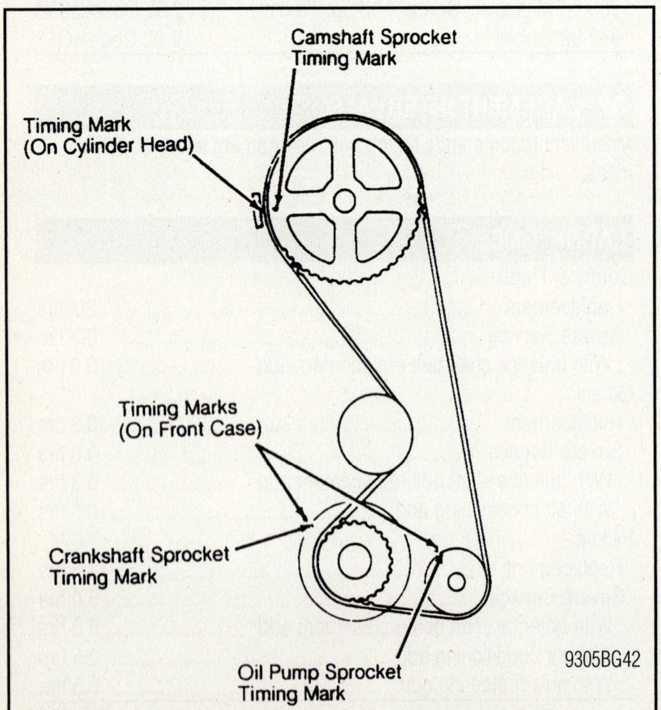

Crankshaft, camshaft and oil pump sprocket timing mark alignment—Mitsubishi 2.0L (G63B/4G63) SOHC engine

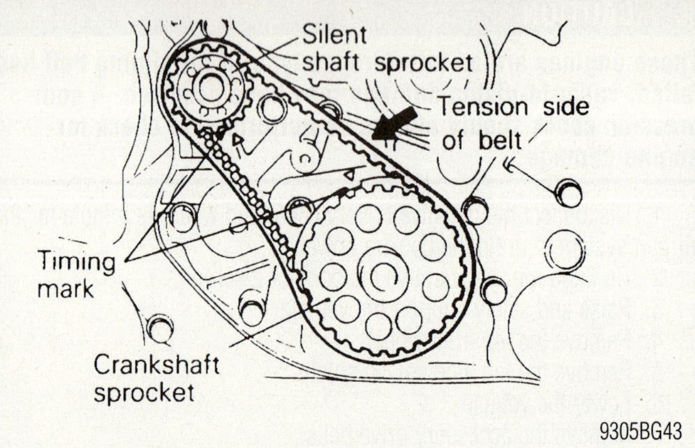

Silent shaft belt timing mark alignment—Mitsubishi 2.0L (G63B/4G63) SOHC engine

8. If timing belt is to be reused, mark the belt with an arrow to indicate the direction of rotation.
9. Remove the timing belt.
10. Loosen the lockbolt for the inner timing belt. Move the tensioner away from the belt and remove the inner belt.
11. Inspect the tensioner pulley for ease of movement and abnormal noises. Replace as necessary.
12. Inspect the crankshaft, camshaft and oil pump sprockets for cracks and excessive wear. Replace as necessary.

To install:

13. Align the timing mark of the silent shaft belt sprockets on the crankshaft and silent shaft with the marks on the front case. Wrap the silent shaft belt around the sprockets so there is no slack in the upper span of the belt and the timing marks are still aligned.
14. Install the tensioner initially so the actual center of the pulley is above and to the left of the installation bolt.
15. Move the pulley up by hand so the center span of the long side of the belt deflects about 1/4 in. (6mm).
16. Hold the pulley tightly so it does not rotate when the bolt is tightened. Tighten the bolt to 15 ft. lbs. (20 Nm). If the pulley has moved, the belt will be too tight.
17. Install the timing belt tensioner fully toward the water pump and temporarily tighten the bolts. Place the upper end of the spring against the water pump body. Align the timing marks of the camshaft, crankshaft and oil pump sprockets with the corresponding marks on the front case or head.

■ **If the following steps are not followed exactly, there is a chance that the silent shaft alignment will be 180 degrees off. This will cause a noticeable vibration in the engine and the entire procedure will have to be repeated.**

18. Before installing the timing belt, ensure that the left-side silent shaft is in the correct position.

■ **It is possible to align the timing marks on the camshaft sprocket, crankshaft sprocket and the oil pump sprocket with the left balance shaft out of alignment.**

19. With the timing mark on the oil pump pulley aligned with the mark on the front case, check the alignment of the left balance shaft to assure correct shaft timing.
20. Remove the plug located on the left-side of the block in the area of the starter.
21. Insert a tool having a shaft diameter of 0.3 in. (8mm) into the hole.

22. With the timing marks still aligned, the tool must be able to go in at least 2 1/3 in. If it can only go in about 1 in. turn the oil pump sprocket 1 complete revolution.
23. Recheck the position of the balance shaft with the timing marks realigned. Leave the tool in place to hold the silent shaft while continuing.
24. Install the belt to the crankshaft sprocket, oil pump sprocket and the camshaft sprocket, in that order. While doing so, be sure there is no slack between the sprockets except where the tensioner will take it up when released.
25. Recheck the timing mark alignments.
26. If all are aligned, loosen the tensioner mounting bolt and allow the tensioner to apply tension to the belt.
27. Remove the tool that is holding the silent shaft in place and turn the crankshaft clockwise a distance equal to 2 teeth of the camshaft sprocket. This will allow the tensioner to automatically tension the belt the proper amount.

✻✻ WARNING

Do not manually apply pressure to the tensioner. This will overtighten the belt and will cause a howling noise.

28. First tighten the lower mounting bolt and then tighten the upper spacer bolt.

✻✻ WARNING

If any binding is felt when adjusting the timing belt tension by turning the crankshaft, STOP turning the engine, because the pistons may be hitting the valves.

29. To verify that belt tension is correct, check that the deflection of the longest span (between the camshaft and oil pump sprockets) is 1/2 in. (13mm).
30. Install the lower timing belt cover. Be sure the packing is properly positioned in the inner grooves of the covers when installing.
31. Install the timing belt covers.
32. Install the crankshaft pulley and torque the bolts to 15–21 ft. lbs. (20–29 Nm).
33. Install the accessory drive belts and adjust to proper tension.
34. Connect the negative battery cable.
35. Start the engine and check for proper operation.
36. Check ignition timing and adjust if necessary.

1993 Galant

✻✻ CAUTION

These engines are an interference type. If the timing belt has failed, valve-to-piston damage may have occurred. A compression check should always be performed to check for engine damage.

1. Disconnect the negative battery cable and wait one minute for the air bag system to drain the backup power.
2. Raise and support the vehicle.
3. Remove the left side wheel and splash shield.
4. Remove the A/C and power steering line clamp from the engine mount.
5. Place a suitable jack under the engine to support it.
6. Remove the right side engine mount.
7. Remove the accessory drive belts.
8. Remove the drive belt tensioner pulley.
9. Remove the water pump pulley.
10. Remove the crankshaft pulley.
11. Remove the timing belt covers.
12. Rotate the crankshaft in the direction of rotation to align the camshaft timing mark with the timing mark on rear cover.
13. Remove the timing belt tensioner bolts and remove the tensioner.
14. If timing belt is to be reused, mark the belt with an arrow to indicate the direction of rotation.
15. Remove the timing belt.
16. Remove the crankshaft sprocket and flange.
17. Loosen the lockbolt for the inner timing belt. Move the tensioner away from the belt and remove the inner belt.
18. Inspect the tensioner pulley for ease of movement and abnormal noises. Replace as necessary.
19. Inspect the crankshaft, camshaft and oil pump sprockets for cracks and excessive wear. Replace as necessary.

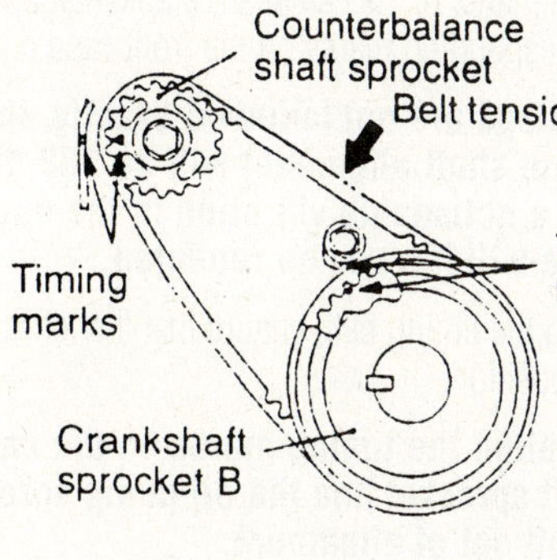

Inner timing belt mark alignment—Mitsubishi 2.0L (4G63) SOHC engine–1993 Galant

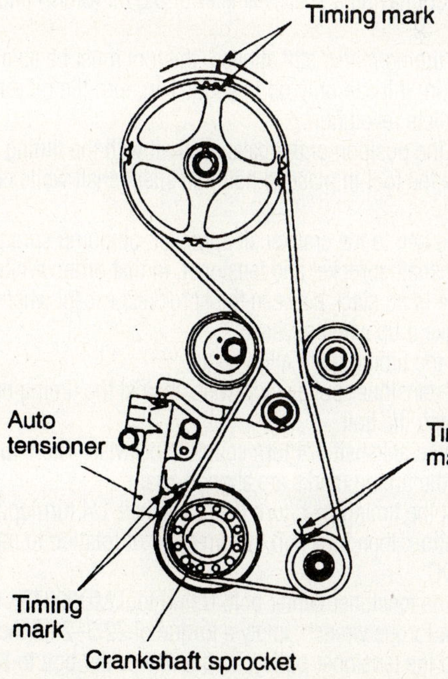

Crankshaft, camshaft and oil pump sprocket timing mark alignment—Mitsubishi 2.0L (4G63) SOHC engine–1993 Galant

To install:

20. Align the timing mark of the silent shaft belt sprockets on the crankshaft and silent shaft with the marks on the front case. Wrap the silent shaft belt around the sprockets so there is no slack in the upper span of the belt and the timing marks are still aligned.

21. Install the tensioner initially so the actual center of the pulley is above and to the left of the installation bolt.

22. Move the pulley up by hand so the center span of the long side of the belt deflects about 1/4 in. (6mm).

23. Hold the pulley tightly so it does not rotate when the bolt is tightened. Tighten the bolt to 15 ft. lbs. (20 Nm). If the pulley has moved, the belt will be too tight.

24. Install the crankshaft flange and sprocket with the raised side outward.

25. Place the tensioner in a vise and compress the plunger until a 0.55 in. pin can be inserted in the body.

26. Install the tensioner on the engine.

27. Align the timing marks of the camshaft, crankshaft and oil pump sprockets with the corresponding marks on the front case or head.

■ **If the following steps are not followed exactly, there is a chance that the silent shaft alignment will be 180 degrees off. This will cause a noticeable vibration in the engine and the entire procedure will have to be repeated.**

28. Before installing the timing belt, ensure that the left-side silent shaft is in the correct position.

■ **It is possible to align the timing marks on the camshaft sprocket, crankshaft sprocket and the oil pump sprocket with the left balance shaft out of alignment.**

29. With the timing mark on the oil pump pulley aligned with the mark on the front case, check the alignment of the left balance shaft to assure correct shaft timing.

30. Remove the plug located on the left-side of the block in the area of the starter.

31. Insert a tool having a shaft diameter of 0.3 in. (8mm) into the hole.

32. With the timing marks still aligned, the tool must be able to go in at least 2 1/3 in. If it can only go in about 1 in. turn the oil pump sprocket 1 complete revolution.

33. Recheck the position of the balance shaft with the timing marks realigned. Leave the tool in place to hold the silent shaft while continuing.

34. Install the belt to the crankshaft sprocket, oil pump sprocket, idler pulley, camshaft sprocket and tensioner, in that order. While doing so, be sure there is no slack between the sprockets except where the tensioner will take it up when released.

35. Recheck the timing mark alignments.

36. Raise the tensioner pulley clockwise against the timing belt and temporarily tighten the bolt.

37. Rotate the crankshaft 1/4 turn counterclockwise. Then, turn clockwise until the timing marks are aligned again.

38. To adjust the timing belt, turn the crankshaft 1/4 turn opposite direction of rotation, then turn it in the direction of rotation to move No. 1 cylinder to TDC.

39. Loosen the tensioner center bolt. Using tool MD998738 or equivalent and a torque wrench, apply a torque of 22.5–24.3 inch lbs. (2.6–2.8 Nm) to the tensioner pulley. Torque the center bolt to 31–40 ft. lbs. (43–55 Nm).

40. Remove the rubber plug from the rear timing belt cover and screw special tool MD99873801 or equivalent, into the engine left support bracket until its end makes contact with the tensioner arm. At this

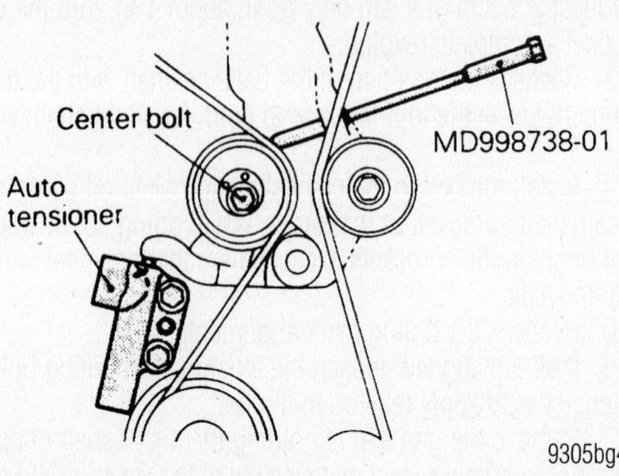

Inserting special tool to contact tensioner arm—Mitsubishi 2.0L (4G63) SOHC engine–1993 Galant

point, screw the special tool in some more and remove the set pin attached to the auto tensioner. Then remove the special tool.

41. Rotate the crankshaft 2 complete turns clockwise and let it sit for approximately 15 minutes. Then, measure the auto tensioner protrusion (the distance between the tensioner arm and auto tensioner body) to ensure that it is within 0.15–0.18 in. (3.8–4.5mm).

42. Install the rubber plug to the timing belt rear cover.

43. Install the upper and lower timing belt covers and torque the bolts to 84–108 inch lbs. (9–12 Nm).

44. Install the drive belt tensioner pulley.

45. Install the accessory drive belts.

46. Install the right side engine mount.

47. Remove the jack.

48. Install the A/C and power steering line clamp to the engine mount.

49. Install the left side wheel and splash shield.

50. Lower the vehicle

51. Connect the negative battery cable.

52. Start the engine and check for proper operation.

53. Check the ignition timing.

Pickup

**** CAUTION**

These engines are an interference type. If the timing belt has failed, valve-to-piston damage may have occurred. A compression check should always be performed to check for engine damage.

1. Disconnect and isolate the battery ground cable.
2. If equipped, remove the radiator upper shroud.
3. Remove the accessory drive belts.
4. Remove the cooling fan assembly.
5. Remove the tensioner pulley and bracket.
6. Remove the water pump pulley and crankshaft pulley bolts.
7. Remove the crankshaft power steering pump pulley.
8. Remove the crankshaft pulley
9. Remove the timing belt covers.
10. Rotate the crankshaft in the direction of rotation to align the camshaft timing mark with the timing mark on the cylinder head.
11. Loosen the tensioner bolts and move the tensioner toward the water pump to relieve pressure on the timing belt. Tighten the bolts.
12. If timing belt is to be reused, mark the belt with an arrow to indicate the direction of rotation.

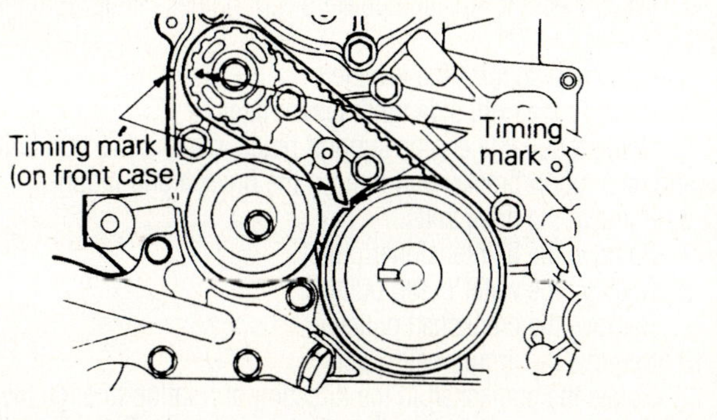

9305bg41

Inner timing belt mark alignment—Mitsubishi 2.0L (G63B/4G63) SOHC engine

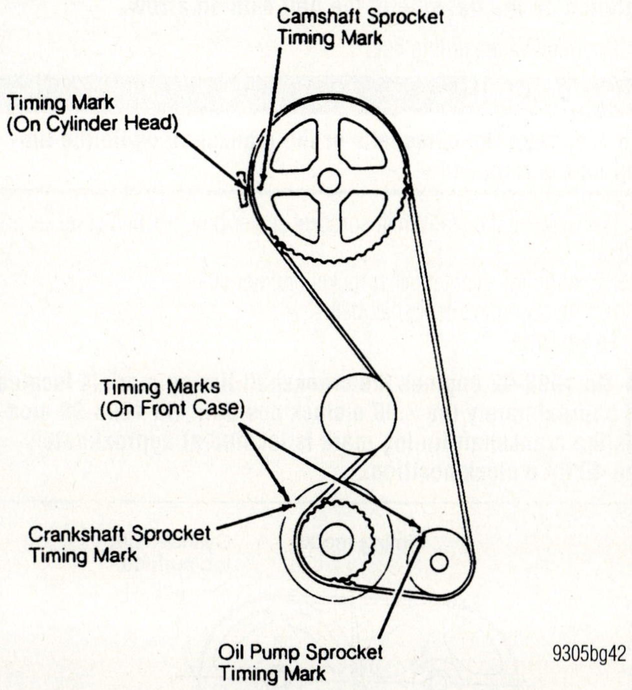

9305bg42

Crankshaft, camshaft and oil pump sprocket timing mark alignment—Mitsubishi 2.0L (G63B/4G63) SOHC engine

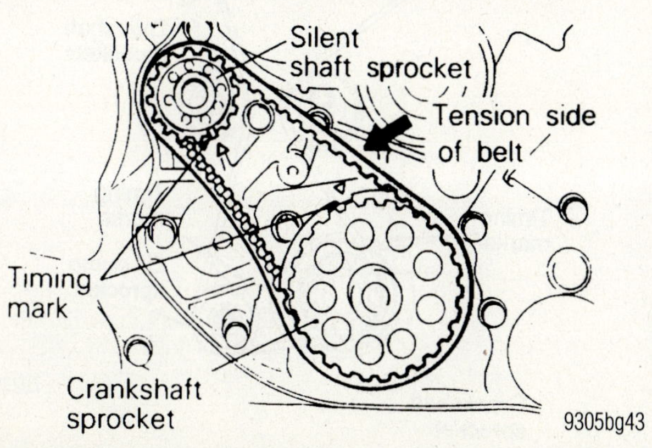

9305bg43

Silent shaft belt timing mark alignment—Mitsubishi 2.0L (G63B/4G63) SOHC engine

13. Remove the timing belt.
14. Loosen the lockbolt for the inner timing belt. Move the tensioner away from the belt and remove the inner belt.
15. Inspect the tensioner pulley for ease of movement and abnormal noises. Replace as necessary.
16. Inspect the crankshaft, camshaft and oil pump sprockets for cracks and excessive wear. Replace as necessary.

To install:

17. Align the timing mark of the silent shaft belt sprockets on the crankshaft and silent shaft with the marks on the front case. Wrap the silent shaft belt around the sprockets so there is no slack in the upper span of the belt and the timing marks are still aligned.
18. Install the tensioner initially so the actual center of the pulley is above and to the left of the installation bolt.
19. Move the pulley up by hand so the center span of the long side of the belt deflects about 1/4 in. (6mm).
20. Hold the pulley tightly so it does not rotate when the bolt is tightened. Tighten the bolt to 15 ft. lbs. (20 Nm). If the pulley has moved, the belt will be too tight.
21. Install the timing belt tensioner fully toward the water pump and temporarily tighten the bolts. Place the upper end of the spring against the water pump body. Align the timing marks of the camshaft, crankshaft and oil pump sprockets with the corresponding marks on the front case or head.

■ **If the following steps are not followed exactly, there is a chance that the silent shaft alignment will be 180 degrees off. This will cause a noticeable vibration in the engine and the entire procedure will have to be repeated.**

22. Before installing the timing belt, ensure that the left-side silent shaft is in the correct position.

■ **It is possible to align the timing marks on the camshaft sprocket, crankshaft sprocket and the oil pump sprocket with the left balance shaft out of alignment.**

23. With the timing mark on the oil pump pulley aligned with the mark on the front case, check the alignment of the left balance shaft to assure correct shaft timing.
24. Remove the plug located on the left-side of the block in the area of the starter.
25. Insert a tool having a shaft diameter of 0.3 in. (8mm) into the hole.
26. With the timing marks still aligned, the tool must be able to go in at least 2 1/3 in. If it can only go in about 1 in. turn the oil pump sprocket 1 complete revolution.
27. Recheck the position of the balance shaft with the timing marks realigned. Leave the tool in place to hold the silent shaft while continuing.
28. Install the belt to the crankshaft sprocket, oil pump sprocket and the camshaft sprocket, in that order. While doing so, be sure there is no slack between the sprockets except where the tensioner will take it up when released.
29. Recheck the timing mark alignments.
30. If all are aligned, loosen the tensioner mounting bolt and allow the tensioner to apply tension to the belt.
31. Remove the tool that is holding the silent shaft in place and turn the crankshaft clockwise a distance equal to 2 teeth of the camshaft sprocket. This will allow the tensioner to automatically tension the belt the proper amount.

✻✻ WARNING

Do not manually apply pressure to the tensioner. This will overtighten the belt and will cause a howling noise.

32. First tighten the lower mounting bolt and then tighten the upper spacer bolt.

✻✻ WARNING

If any binding is felt when adjusting the timing belt tension by turning the crankshaft, STOP turning the engine, because the pistons may be hitting the valves.

33. To verify that belt tension is correct, check that the deflection of the longest span (between the camshaft and oil pump sprockets) is 1/2 in. (13mm).
34. Install the lower timing belt cover. Be sure the packing is properly positioned in the inner grooves of the covers when installing.
35. Install the timing belt covers.
36. Install the crankshaft pulley and torque the bolts to 15–21 ft. lbs. (20–29 Nm).
37. Install the crankshaft power steering pulley, if equipped.
38. Install the accessory drive belts and adjust to proper tension.
39. Connect the negative battery cable.
40. Start the engine and check for proper operation.
41. Check ignition timing and adjust if necessary.

2.0L (4G63) DOHC NON-TURBO ENGINES

YEAR, MODEL & ENGINE APPLICATION

1990-94 Mitsubishi Eclipse, 1988-94 Galant ... 2.0L Code 4G63

REPLACEMENT INTERVALS

Mitsubishi recommends timing belt replacement every 60,000 miles.

LABOR TIMES

Eclipse	
Replacement	4.3 hrs
Severe Service	4.8 hrs
Galant	
Replacement	4.4 hrs
Severe Service	4.9 hrs
All Models	
With air conditioning add	0.3 hrs
With balance shaft belt replacement add	0.3 hrs

8807MITg08

✻✻ CAUTION

These engines are an interference type. If the timing belt has failed, valve-to-piston damage may have occurred. A compression check should always be performed to check for engine damage.

1. Disconnect the negative battery cable.
2. Disconnect the A/C and power steering lines clamp from the body.
3. Raise and support the vehicle.
4. Remove the right side inner fender liner.
5. Using a suitable engine support brace, remove the stud bolt from the engine support bracket and remove the bracket and mount.
6. Remove the drive belts.
7. Remove the belt tensioner pulley.
8. Remove the water pump pulley.
9. Remove the crankshaft pulley.
10. Remove the timing belt covers.
11. Rotate the crankshaft in the direction of rotation to align the camshaft timing marks. Always turn the crankshaft in the normal direction of rotation only.
12. Remove the timing belt tensioner.

■ **If the timing belt is to be reused, mark the direction of rotation on the flat side of the belt with an arrow.**

13. Remove the timing belt.

✻✻ WARNING

Do not rotate the camshafts or the crankshaft while the timing belt is removed.

14. Remove the crankshaft sprocket center bolt, washer, sprocket and flange.
15. Align the silent shaft sprocket timing marks.
16. Remove the silent shaft belt.

To install:

■ **On 1990-92 engines the crankshaft timing mark is located at approximately the 2:00 o'clock position. On 1993-98 models the crankshaft timing mark is located at approximately the 10:00 o'clock position.**

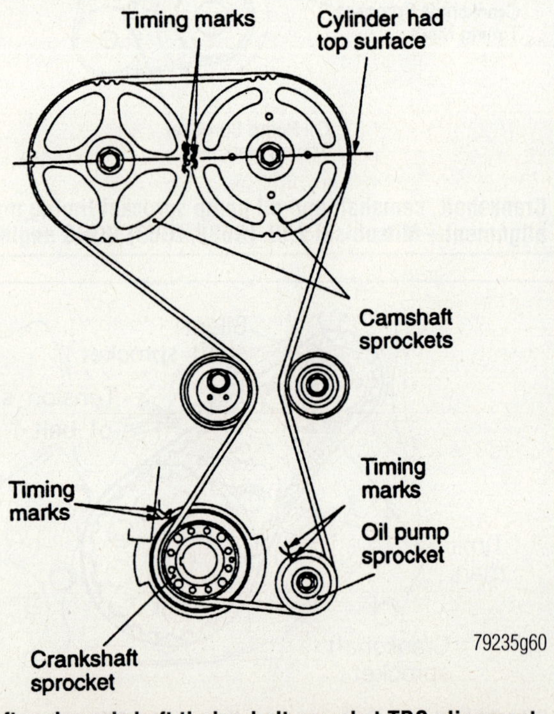

Camshaft and crankshaft timing belt sprocket TDC alignment marks—Mitsubishi 2.0L (4G63) DOHC non-turbo engine

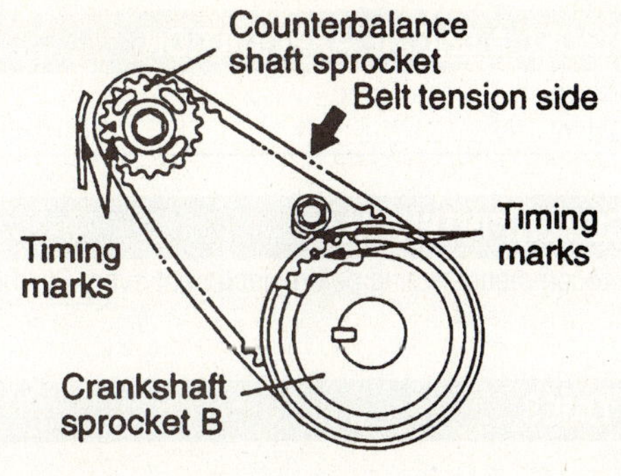

79235g61

Silent shaft timing mark alignment—Mitsubishi 2.0L (4G63) DOHC non-turbo engine

17. Align the timing mark of the silent shaft belt sprockets on the crankshaft and silent shaft with the marks on the front case. Wrap the silent shaft belt around the sprockets so there is no slack in the upper span of the belt and the timing marks are still aligned.

18. Install the tensioner initially so the actual center of the pulley is above and to the left of the installation bolt.

19. Move the pulley up by hand so the center span of the long side of the belt deflects about 1/4 in. (6mm).

20. Hold the pulley tightly so it does not rotate when the bolt is tightened. Tighten the bolt to 15 ft. lbs. (20 Nm). If the pulley has moved, the belt will be too tight.

21. Install the crankshaft flange, sprocket and washer and tighten to 87 ft. lbs. (118 Nm).

22. Install the automatic tensioner, after reset.

■ **To reset the tensioner: Keep the adjuster level and clamp it in a soft jawed vise. Clamp with the extended adjuster on one side and the end mounting a plug on the other side. If the plug extends out of the adjuster body, place a suitable hole sized washer over the plug so the vise jaw pushes on the washer, not the plug. Close the vise slowly, forcing the adjuster back into the body. When the hole in the adjuster boss aligns with the adjuster rod, insert a snug fitting pin or wire into the holes to keep the rod in the compressed position. With the locking pin or wire in place, install the tensioner.**

23. Install the tensioner pulley with the hole in the shaft to the left of the center bolt, then tighten the center bolt hand tight.

24. When installing the timing belt, turn the 2 camshaft sprockets so their dowel pins are located on top. Align the timing marks facing each other with the top surface of the cylinder head. When you let go of the exhaust camshaft sprocket, it will rotate 1 tooth in the counterclockwise direction. This should be taken into account when installing the timing belts on the sprocket.

■ **Both camshaft sprockets are used for the intake and exhaust camshafts and are provided with 2 timing marks. When the sprocket is mounted on the exhaust camshaft, use the timing mark on the right with the dowel pinhole on top. For the intake camshaft sprocket, use the mark on the left with the dowel pinhole on top.**

25. Align the crankshaft sprocket and oil pump sprocket timing marks.

■ **On 1990-92 engines the crankshaft timing mark is located at approximately the 2:00 o'clock position. On 1993-94 models the crankshaft timing mark is located at approximately the 10:00 o'clock position.**

■ **If the following steps are not followed exactly, there is a chance that the silent shaft alignment will be 180 degrees off. This will cause a noticeable vibration in the engine and the entire procedure will have to be repeated.**

26. Before installing the timing belt, ensure that the left-side silent shaft is in the correct position.

■ **It is possible to align the timing marks on the camshaft sprocket, crankshaft sprocket and the oil pump sprocket with the left balance shaft out of alignment.**

27. With the timing mark on the oil pump pulley aligned with the mark on the front case, check the alignment of the left balance shaft to assure correct shaft timing.

28. Remove the plug located on the left-side of the block in the area of the starter.

29. Insert a tool having a shaft diameter of 0.3 in. (8mm) into the hole.

30. With the timing marks still aligned, the tool must be able to go in at least 2 1/3 in. If it can only go in about 1 in. turn the oil pump sprocket 1 complete revolution.

31. Recheck the position of the balance shaft with the timing marks realigned. Leave the tool in place to hold the silent shaft while continuing.

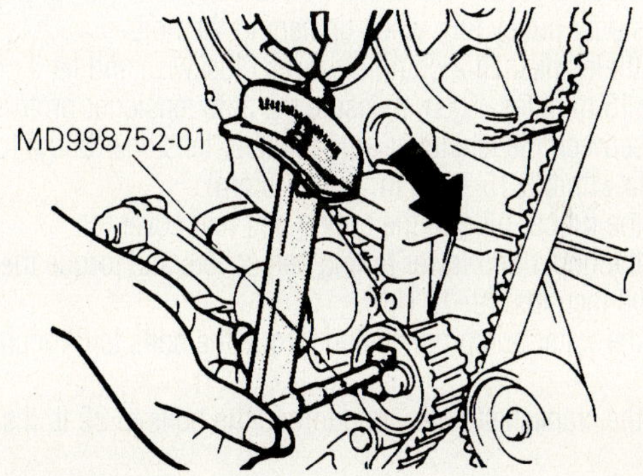

9305bg39

Applying torque to the tensioner pulley—Mitsubishi 2.0L (4G63) DOHC non-turbo engine

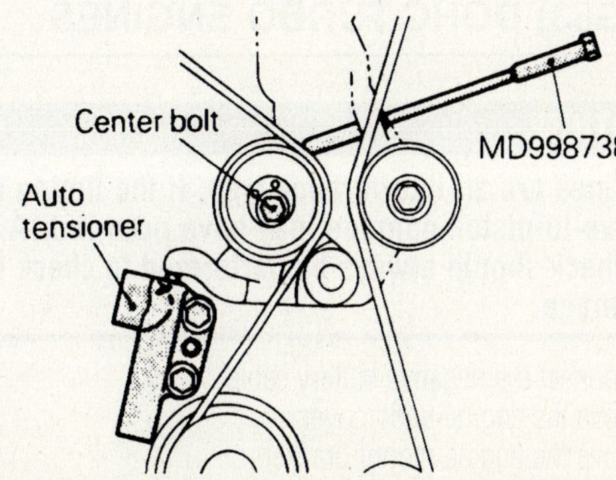

9305bg40

Inserting special tool to contact tensioner arm—Mitsubishi 2.0L (4G63) DOHC non-turbo engine

32. Ensure the camshaft sprockets timing marks are opposite each other and in the center position aligned with the top surface of the cylinder head.

33. Ensure the crankshaft and oil pump sprocket timing marks are aligned.

34. Install the timing belt on the intake camshaft. Hold the belt in place with a binder clip.

35. Install the timing belt on the exhaust camshaft. Hold the belt in place with a binder clip

36. Install the belt to the idler pulley, oil pump sprocket, crankshaft sprocket, oil pump sprocket and the tensioner.

37. Remove the binder clips.

38. Remove the screwdriver from the oil pump and install the plug.

39. Raise the tensioner pulley clockwise against the timing belt and temporarily tighten the bolt.

40. Remove the rubber plug from the rear timing belt cover and screw special tool MD99873801 or equivalent, into the engine left support bracket until its end makes contact with the tensioner arm. At this point, screw the special tool in some more and remove the set wire attached to the auto tensioner. Then remove the special tool.

41. Tighten the tensioner pulley bolt.

42. Rotate the crankshaft 1/4 turn counterclockwise. Then, turn clockwise until the timing marks are aligned again.

43. On 1990–92 models, loosen the tensioner center bolt. Using tool MD998752 or equivalent and a torque wrench, apply a torque of 22.5–24.3 inch lbs. (2.6–2.8 Nm) to the tensioner pulley.

44. On 1993-94 models, loosen the center bolt. Install Special Tool MD998767 or equivalent, on the tensioner pulley. Turn the tensioner pulley counterclockwise with a torque of 31 inch lbs. (3.5 Nm).

45. On all models, torque the center bolt to 35 ft. lbs. (48 Nm). Do not let the tensioner pulley turn when tightening the bolt.

46. Rotate the crankshaft 2 complete turns clockwise and let it sit for approximately 15 minutes. Then, measure the auto tensioner protrusion (the distance between the tensioner arm and auto tensioner body) to ensure that it is within 0.15–0.18 in. (3.8–4.5mm).

47. Install the rubber plug to the timing belt rear cover.

48. Install the upper and lower timing belt covers and torque the bolts to 84–108 inch lbs. (9–12 Nm).

49. Install the water pump pulley and torque the bolts to 84 inch lbs. (10 Nm).

50. Install the crankshaft pulley and torque the bolts to 22 ft. lbs. (30 Nm).

51. Install the mounting brackets and the accessory drive belts.

52. Connect the A/C and power steering line clamps.

53. Install the right side inner fender liner.

54. Connect the negative battery cable.

2.0L (4G63) DOHC TURBO ENGINES

✻✻ CAUTION

These engines are an interference type. If the timing belt has failed, valve-to-piston damage may have occurred. A compression check should always be performed to check for engine damage.

1. Disconnect the negative battery cable.
2. Remove the engine undercover.
3. Remove the engine mount bracket.
4. Remove the drive belts.
5. Remove the belt tensioner pulley.

YEAR, MODEL & ENGINE APPLICATION

1990-99 Mitsubishi Eclipse Turbo
1988-94 Galant Turbo .... 2.0L Code 4G63

REPLACEMENT INTERVALS

Mitsubishi recommends timing belt replacement every 60,000 miles.

LABOR TIMES

Eclipse	
Replacement	4.3 hrs
Severe Service	4.8 hrs
Galant	
Replacement	4.4 hrs
Severe Service	4.9 hrs
All Models	
With air conditioning add	0.3 hrs
With balance shaft belt replacement add	0.3 hrs

8807MITq09

6. Remove the water pump pulleys.
7. Remove the crankshaft pulley.
8. Using a suitable engine support brace, remove the stud bolt from the engine support bracket and remove the bracket and mount.
9. Remove the timing belt covers.
10. Rotate the crankshaft in the direction of rotation to align the camshaft timing marks. Always turn the crankshaft in the normal direction of rotation only.

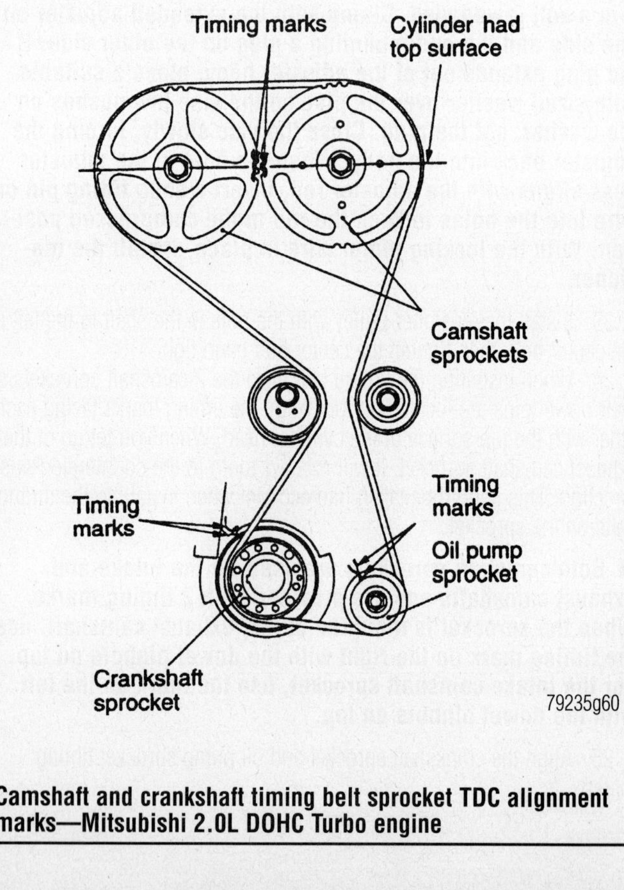

Camshaft and crankshaft timing belt sprocket TDC alignment marks—Mitsubishi 2.0L DOHC Turbo engine

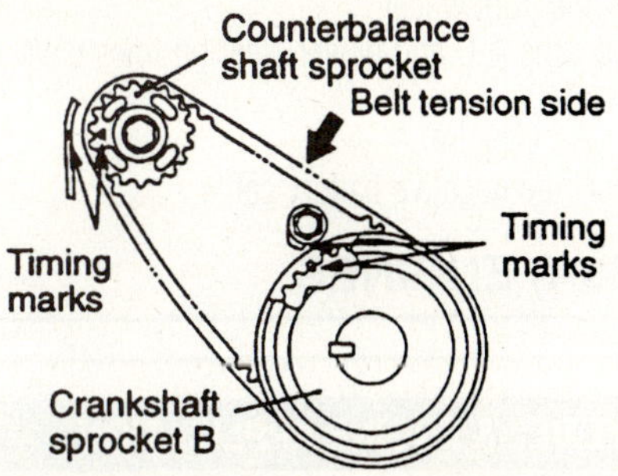

Timing belt B installation mark alignment—Mitsubishi 2.0L DOHC Turbo engine

11. Loosen the tension pulley center bolt.

■ **If the timing belt is to be reused, mark the direction of rotation on the flat side of the belt with an arrow.**

12. Move the tension pulley towards the water pump and remove the timing belt.
13. Remove the crankshaft sprocket center bolt using special tool MB990767, or equivalent, to hold the crankshaft sprocket while removing the center bolt. Then, use MB998778 or equivalent puller to remove the sprocket.
14. Mark the direction of rotation on the timing belt **B** with an arrow.
15. Loosen the center bolt on the tensioner and remove the belt.

**** WARNING**

Do not rotate the camshafts or the crankshaft while the timing belt is removed.

To install:

16. Place the crankshaft sprocket on the crankshaft. Use tool MB990767, or equivalent, to hold the crankshaft sprocket while tightening the center bolt. Torque the center bolt to 80–94 ft. lbs. (108–127 Nm).
17. Align the timing marks on the crankshaft sprocket **B** and the balance shaft.
18. Install timing belt **B** on the sprockets. Position the center of the tensioner pulley to the left and above the center of the mounting bolt.
19. Push the pulley clockwise toward the crankshaft to apply tension to the belt and torque the mounting bolt to 14 ft. lbs. (19 Nm). Do not let the pulley turn when tightening the bolt because it will cause excessive tension on the belt. The belt should deflect 0.20–0.28 in. (5–7mm) when finger pressure is applied between the pulleys.
20. Install the crankshaft sensing blade and the crankshaft sprocket. Apply engine oil to the mounting bolt and torque the bolt to 80–94 ft. lbs. (108–127 Nm).
21. Use a press or vise to compress the auto-tensioner pushrod. Insert a set pin when the holes in the rod and the tensioner body are aligned.

**** WARNING**

Do not compress the pushrod too quickly, damage to the pushrod can occur.

22. Install the auto-tensioner on the engine.
23. Align the timing marks on the camshaft sprocket, crankshaft sprocket and the oil pump sprocket.
24. After aligning the mark on the oil pump sprocket, remove the cylinder block plug and insert a prytool in the hole to check the position of the counterbalance shaft. The prytool should go in at least 2.36 in. (60mm) or more, if not, rotate the oil pump sprocket once and realign the timing mark so the prytool goes in. Do not remove the prytool until the timing belt is installed.
25. Install the timing belt on the intake camshaft and secure it with a clip.
26. Install the timing belt on the exhaust camshaft. Align the timing marks with the cylinder head top surface using 2 wrenches. Secure the belt with another clip.
27. Install the belt around the idler pulley, oil pump sprocket, crankshaft sprocket and the tensioner pulley.
28. Turn the tensioner pulley so the pinholes are at the bottom. Press the pulley lightly against the timing belt.
29. Screw the special tool into the left engine support bracket until it contacts the tensioner arm, then screw the tool in a little more and remove the pushrod pin from the auto-tensioner. Remove the special tool and torque the center bolt to 35 ft. lbs. (48 Nm).
30. Turn the crankshaft 1/4 turn counterclockwise, then clockwise until the timing marks are aligned.
31. Loosen the center bolt. Install Mitsubishi Special Tool MD998767 or equivalent, on the tensioner pulley. Turn the tensioner pulley counterclockwise with a torque of 31 inch lbs. (3.5 Nm) and torque the center bolt to 35 ft. lbs. (48 Nm). Do not let the tensioner pulley turn when tightening the bolt.
32. Turn the crankshaft clockwise 2 revolutions and align the timing marks. After 15 minutes, measure the protrusion of the pushrod on the auto-tensioner. The standard measurement is 0.150–0.177 in. (3.8–4.5mm). If the protrusion is out of specification, loosen the tensioner pulley, apply the proper torque to the belt and retighten the center bolt.
33. Install the timing belt covers and all applicable components.

2.0L (4G94) LANCER SOHC ENGINES

YEAR, MODEL & ENGINE APPLICATION	
2001-04 Mitsubishi Lancer	2.0L Code 4G94

REPLACEMENT INTERVALS
Mitsubishi recommends timing belt replacement every 60,000 miles.

LABOR TIMES	
2001-04	
Replacement	3.5 hrs
Severe Service	4.0 hrs

8807MITg10

**** CAUTION**

These engines are an interference type. If the timing belt has failed, valve-to-piston damage may have occurred. A compression check should always be performed to check for engine damage.

1. Disconnect the negative battery cable and wait one minute for the air bag system to drain the backup power.
2. Raise and support the vehicle.
3. Disconnect the A/C and power steering lines clamp from the engine mount.
4. Remove the front wheel.
5. Remove the right side inner fender liner.
6. Using a suitable engine support brace, remove the stud bolt from the engine support bracket and remove the bracket and mount.
7. Remove the drive belts.
8. Remove the crankshaft pulley.
9. Remove the crankshaft flange.
10. Remove the timing belt covers.
11. Rotate the crankshaft in the direction of rotation to align the camshaft timing marks. Always turn the crankshaft in the normal direction of rotation only.
12. Loosen the timing belt tensioner bolt and move the tensioner back and tighten the bolt.
13. Remove the timing belt.

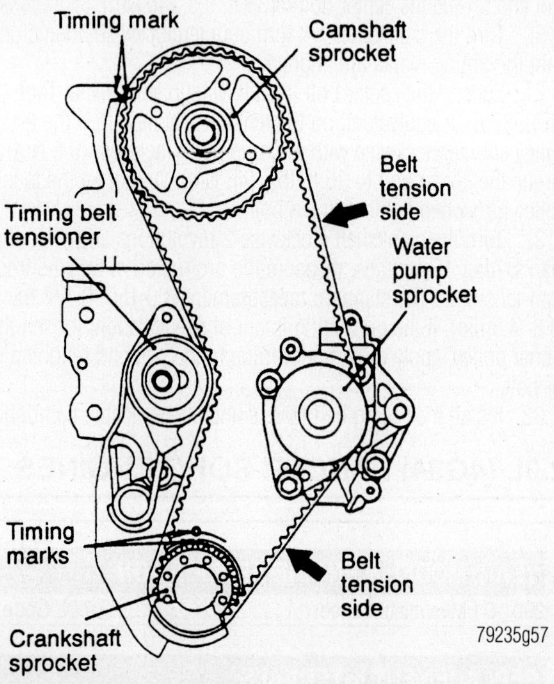

Crankshaft and camshaft sprocket timing marks—Mitsubishi 2.0L (4G94) engine

To install:

14. Install the timing belt on the crankshaft sprocket, water pump sprocket, camshaft sprocket and tensioner pulley.
15. Loosen the tensioner bolt and allow the tensioner to tension the belt.
16. Rotate the crankshaft in the normal direction of rotation 2 turns and ensure the timing marks are still aligned.
17. Tighten the tensioner bolt to 17 ft. lbs. (24 Nm).
18. Install the engine support bracket and mount.
19. Install the timing belt covers.
20. Install the crankshaft flange.
21. Install the crankshaft pulley.
22. Install the accessory drive belts.
23. Install the right side inner fender liner.
24. Install the front wheel.
25. Connect the A/C and power steering lines clamp to the engine mount.
26. Lower the vehicle.
27. Connect the negative battery cable.

2.0L (420A) ENGINES

YEAR, MODEL & ENGINE APPLICATION

1995-99 Mitsubishi Eclipse	2.0L Code 420A

REPLACEMENT INTERVALS

Mitsubishi recommends timing belt replace every 100,000 miles on California models and 60,000 miles on all other models.

LABOR TIMES

1995-99	
Replacement	3.6 hrs
Severe Service	4.1 hrs
With air conditioning add	0.5 hrs

8807MITg11

✱✱ CAUTION

These engines are an interference type. If the timing belt has failed, valve-to-piston damage may have occurred. A compression check should always be performed to check for engine damage.

■ **There are 2 versions of this engine. One has a mechanical timing belt tensioner and the other has a hydraulic tensioner.**

1. Disconnect the negative battery cable.
2. Raise and support the vehicle.
3. Remove the drive belts.
4. Remove the right side wheel.
5. Remove the right inner splash-shield.
6. Place a support under the engine.
7. Remove the engine mount bracket
8. Remove the right engine mount.
9. Remove the crankshaft pulley using a puller.
10. Remove the power steering pump without disconnect the lines and position aside.
11. Remove the timing belt cover.
12. Place the no. 1 cylinder at TDC on the compression with the crankshaft sprocket and TDC marks aligned.
13. On hydraulic tensioner models, loosen the timing belt tensioner bolts.
14. Remove the timing belt and the tensioner.

■ **When tensioner is removed from the engine it is necessary to compress the plunger into the tensioner body.**

15. Place the tensioner into a soft-jawed vise to compress the tensioner.
16. After compressing the tensioner place a pin (5/64 in. Allen wrench will work) into the plunger side hole to retain the plunger until installation.

17. On mechanical tensioner models, insert a 3 mm Allen wrench into the tensioner pulley hole and then use an 8mm Allen wrench to turn the tensioner counterclockwise until it locks in position. Remove the timing belt.

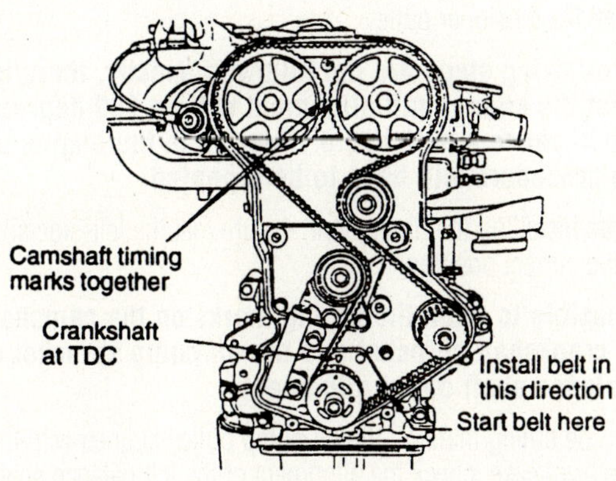

TDC alignment for timing belt installation—Mitsubishi 2.0L (420A) engine

To install:

18. Set the crankshaft sprocket to Top Dead Center (TDC) by aligning the notch on the sprocket with the arrow on the oil pump housing, then back off the sprocket 3 notches before TDC.
19. Set the camshaft to align the timing marks.
20. Move the crankshaft to 1/2 notch before TDC.
21. Install the timing belt starting at the crankshaft, around the water pump, then around the camshaft last.
22. Move the crankshaft to TDC to take up the belt slack.
23. On hydraulic tensioners, install the tensioner to the block but do not tighten it.
24. Using a torque wrench apply 21 ft. lbs. (28 Nm) of torque to the tensioner pulley.
25. With torque being applied to the tensioner pulley, move the tensioner up against the tensioner bracket and tighten the fasteners to 23 ft. lbs. (31 Nm).
26. Remove the tensioner plunger pin, the tension is correct when the plunger pin can be removed and replaced easily.
27. Rotate the crankshaft 2 revolutions and recheck the timing marks.
28. On mechanical tensioner models, remove the small Allen wrench and allow the tensioner to rotate into position.
29. Rotate the crankshaft 2 revolutions and recheck the timing marks.
30. Install the timing belt cover.
31. Install the power steering pump.
32. Install the crankshaft pulley.
33. Install the right engine mount.
34. Install the engine mount bracket
35. Install the right inner splash-shield.
36. Install the right side wheel.
37. Install the drive belts.
38. Lower the vehicle.
39. Connect the negative battery cable.

2.0L (VIN F) LANCER DOHC ENGINES

YEAR, MODEL & ENGINE APPLICATION

2003-04 Mitsubishi Lancer Evolution	2.0L VIN F

REPLACEMENT INTERVALS

Mitsubishi recommends timing belt replacement every 60,000 miles.

LABOR TIMES

2003-04	
Replacement	3.5 hrs
Severe Service	4.0 hrs

8807MITg12

✻✻ CAUTION

These engines are an interference type. If the timing belt has failed, valve-to-piston damage may have occurred. A compression check should always be performed to check for engine damage.

1. Disconnect the negative battery cable.
2. Disconnect the A/C and power steering lines clamp from the body.
3. Remove the engine under cover and side cover.
4. Remove the power steering hose clamp.

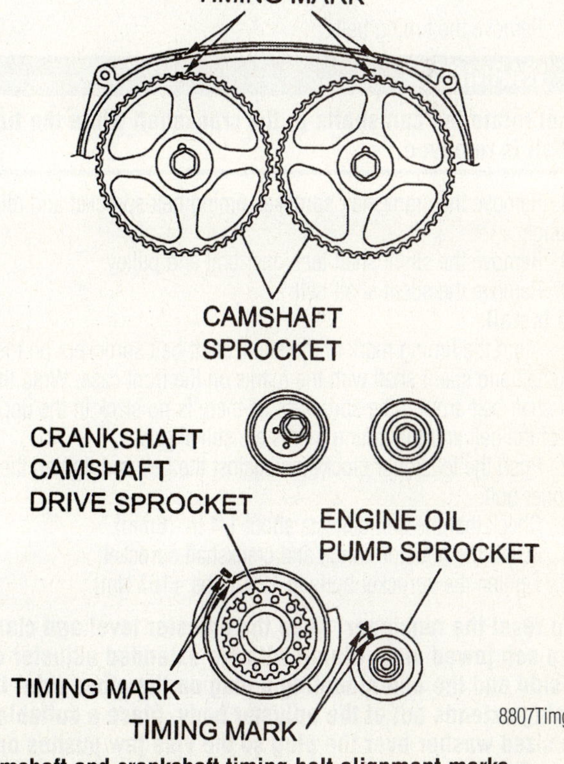

Camshaft and crankshaft timing belt alignment marks—Mitsubishi 2.0L (VIN F) engine

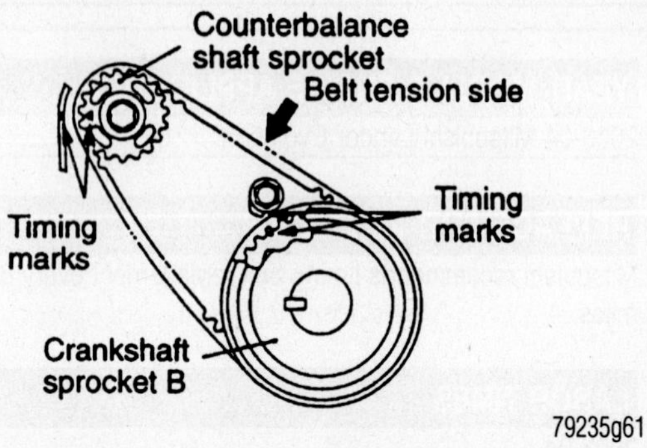

Silent shaft timing mark alignment—Mitsubishi 2.0L (VIN F) engine

5. Remove the exhaust system.
6. Remove the front crossmember.
7. Remove the accessory drive belts.
8. Remove the drive belt tensioner and bracket.
9. Remove the upper timing belt cover.
10. Remove the water pump pulley.
11. Remove the accessory drive belt idler.
12. Remove the crankshaft pulley.
13. Remove the lower timing belt cover.
14. Rotate the crankshaft in the direction of rotation to align the crankshaft and camshaft timing marks. Always turn the crankshaft in the normal direction of rotation only.
15. Remove the timing belt tensioner.
16. Remove the tensioner pulley.

■ **If the timing belt is to be reused, mark the direction of rotation on the flat side of the belt with an arrow.**

17. Remove the timing belt.

✳✳ WARNING

Do not rotate the camshafts or the crankshaft while the timing belt is removed.

18. Remove the crankshaft sprocket, timing belt sprocket and reluctor flange.
19. Remove the silent shaft tensioner bolt and pulley.
20. Remove the silent shaft belt.

To install:

21. Align the timing mark of the silent shaft belt sprockets on the crankshaft and silent shaft with the marks on the front case. Wrap the silent shaft belt around the sprockets so there is no slack in the upper span of the belt and the timing marks are still aligned.
22. Push the tensioner clockwise against the belt and tighten the tensioner bolt.
23. Check that the belt deflects about 1/4 in. (6mm).
24. Install the reluctor flange and crankshaft sprocket.
25. Tighten the sprocket bolts to 123 ft. lbs. (167 Nm).

■ **To reset the tensioner: Keep the adjuster level and clamp it in a soft jawed vise. Clamp with the extended adjuster on one side and the end mounting a plug on the other side. If the plug extends out of the adjuster body, place a suitable hole sized washer over the plug so the vise jaw pushes on the washer, not the plug. Close the vise slowly, forcing the adjuster back into the body. When the hole in the adjuster boss aligns with the adjuster rod, insert a snug fitting pin or wire into the holes to keep the rod in the compressed position. With the locking pin or wire in place, install the tensioner.**

26. Install the automatic tensioner and tighten the bolts to 18 ft. lbs. (25 Nm).
27. Install the tensioner pulley.

■ **If the following steps are not followed exactly, there is a chance that the silent shaft alignment will be 180 degrees off. This will cause a noticeable vibration in the engine and the entire procedure will have to be repeated.**

28. Before installing the timing belt, ensure that the left-side silent shaft is in the correct position.

■ **It is possible to align the timing marks on the camshaft sprocket, crankshaft sprocket and the oil pump sprocket with the left balance shaft out of alignment.**

29. With the timing mark on the oil pump pulley aligned with the mark on the front case, check the alignment of the left balance shaft to assure correct shaft timing.
30. Remove the plug located on the left-side of the block in the area of the starter.
31. Insert a tool having a shaft diameter of 0.3 in. (8mm) into the hole.
32. With the timing marks still aligned, the tool must be able to go in at least 2 1/3 in. If it can only go in about 1 in. turn the oil pump sprocket 1 complete revolution.
33. Recheck the position of the balance shaft with the timing marks realigned. Leave the tool in place to hold the silent shaft while continuing.
34. Install the timing belt on the crankshaft sprocket, oil pump sprocket, guide pulley, exhaust camshaft sprocket, intake camshaft sprocket and tensioner pulley.
35. Push the tensioner pulley against the belt and hand tighten the bolt.
36. Remove the timing belt cover rubber plug and install tool MD998738. Screw in the tool and align the tensioner adjuster rod set hole with the tensioner adjuster set hole.
37. Rotate the crankshaft 1/4 turn counterclockwise. Then, turn clockwise until the timing marks are aligned again.
38. Loosen the tensioner pulley bolt.
39. Loosen the center bolt. Install Special Tool MD998767 or equivalent, on the tensioner pulley. Turn the tensioner pulley counterclockwise with a torque of 31 inch lbs. (3.5 Nm).
40. Tighten the center bolt to 35 ft. lbs. (48 Nm). Do not let the tensioner pulley turn when tightening the bolt.
41. Remove the set pin from the tensioner.
42. Rotate the crankshaft 2 complete turns clockwise and let it sit for approximately 15 minutes. Then, measure the auto tensioner protrusion (the distance between the tensioner arm and auto tensioner body) to ensure that it is within 0.15–0.18 in. (3.8–4.5mm).
43. Install the rubber plug to the timing belt rear cover.
44. Install the upper and lower timing belt covers and torque the bolts to 84–108 inch lbs. (9–12 Nm).
45. Install the crankshaft pulley.
46. Install the accessory drive belt idler.
47. Install the water pump pulley.
48. Install the drive belt tensioner and bracket.
49. Install the accessory drive belts.
50. Install the front crossmember.
51. Install the exhaust system.
52. Install the power steering hose clamp.

53. Install the engine under cover and side cover.
54. Connect the A/C and power steering lines clamp.
55. Lower the vehicle.
56. Connect the negative battery cable.

2.3L DIESEL ENGINES

YEAR, MODEL & ENGINE APPLICATION

1983-84 Mitsubishi Pickup .... 2.3L Diesel VIN N/A

REPLACEMENT INTERVALS

Mitsubishi recommends timing belt replacement every 60,000 miles.

LABOR TIMES

1983-84	
Replacement	1.9 hrs
Severe Service	2.4 hrs
With air conditioning add	0.5 hrs

8807MITg13

⁂ CAUTION

These engines are an interference type. If the timing belt has failed, valve-to-piston damage may have occurred. A compression check should always be performed to check for engine damage.

1. Disconnect the negative battery cable.
2. Rotate the crankshaft to position the No. 1 cylinder to Top Dead Center (TDC) of its compression stroke. Make sure that the timing marks are aligned.
3. Remove the drive belt(s) and/or accessories that may interfere with the timing belt removal.

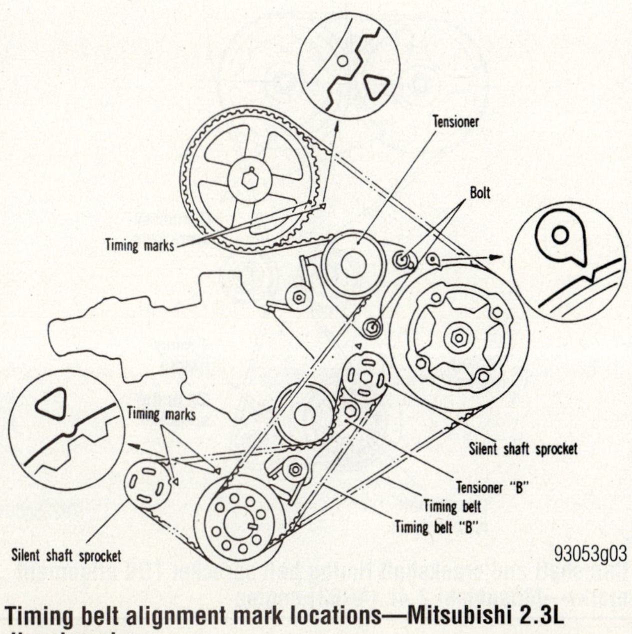

Timing belt alignment mark locations—Mitsubishi 2.3L diesel engine

4. Remove the timing belt covers and the crankshaft pulley.
5. If reusing the timing belt, mark its direction of rotation of the back of the belt.
6. Slightly loosen both belt tensioner bolts, move the tensioners toward the water pump and secure them by tightening the tensioner bolts.
7. Remove the timing belt(s) and inspect for damage, wear or deterioration. Do not use solvents or detergent to clean the timing belts, sprockets or tensioners. Replace any component that is excessively contaminated with dirt, oil or grease.

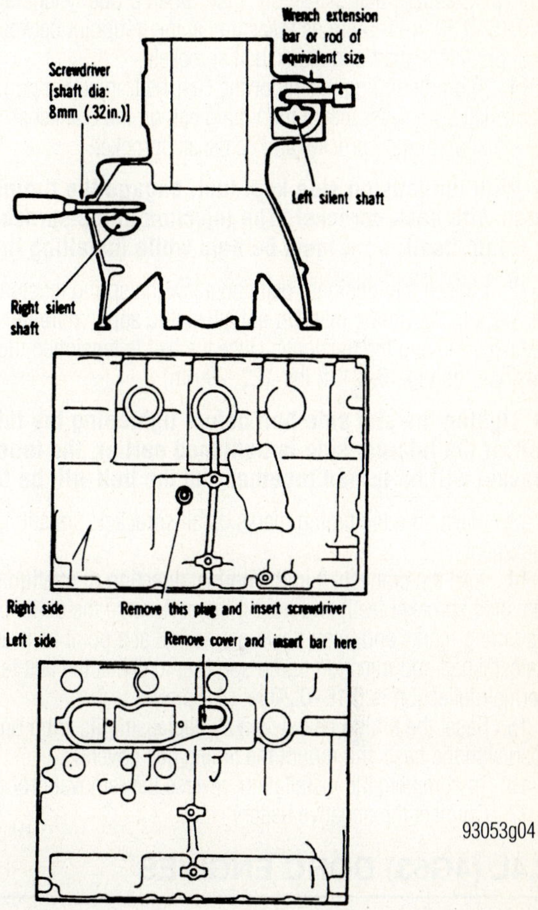

Preventing the silent shaft from rotating—Mitsubishi 2.3L diesel engine

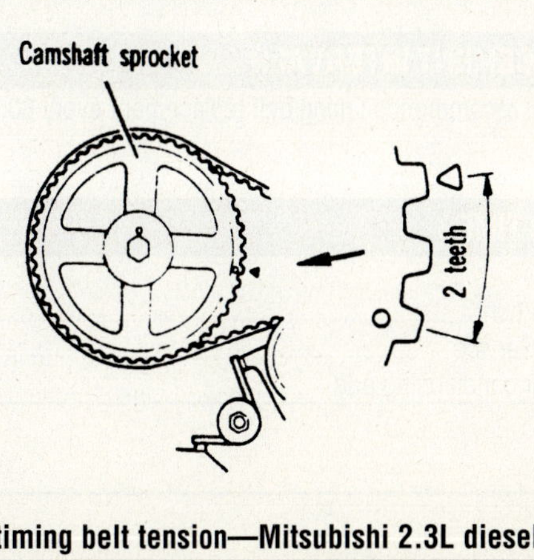

Adjusting timing belt tension—Mitsubishi 2.3L diesel engine

To install:

8. Align the timing marks of the crankshaft and silent shaft drive sprockets.
9. Noting the direction of rotation, install the silent shaft timing belt.
10. Adjust the silent shaft timing belt by performing the following procedure:
    a. Loosen the silent shaft belt tensioner and allow the spring tension to take up any belt slack. Do not attempt to move the tensioner to obtain more tension. Tighten the tensioner bolt to 25–28 ft. lbs. (34–38 Nm).
    b. Standard belt deflection "play" when properly tensioned is 0.16–0.20 in. 4–5mm as measured at the midpoint between the upper silent shaft and crankshaft sprockets.
11. Align the timing marks for the camshaft, injection pump and crankshaft sprockets. Install the timing belt on the crankshaft sprocket, the injection pump sprocket and camshaft sprocket.

■ **With the tension side kept taut, engage the timing belt teeth with each sprocket. The injection pump sprocket tends to rotate itself, so it must be held while installing the belt.**

12. Loosen the tensioner belt and allow the spring tension to take up the slack in the timing belt. Do not attempt to apply more tension by manually moving the tensioner. Once the belt is tensioned, tighten the tensioner bolt to 16–21 ft. lbs. (22–29 Nm).

■ **Tighten the slot side bolt before tightening the fulcrum bolt. If the fulcrum side is tightened earlier, the tension bracket will be turned together and the belt will be too tight.**

13. Make sure the timing marks on all sprockets are still in correct alignment.
14. Turn the crankshaft in the normal direction of rotation through 2 camshaft sprocket teeth and hold it there. Reverse the direction to align the timing marks and push down on the belt at a point halfway between the camshaft and injection pump sprocket to check the belt tension. Normal deflection is 0.16–0.20 in. (4–5mm).
15. Reset the timing belt switch by depressing its knob unit it is flush with the base and mount the timing belt covers.
16. To complete the installation, reverse the removal procedures.
17. Connect the negative battery cable.

2.4L (4G63) DOHC ENGINES

YEAR, MODEL & ENGINE APPLICATION

1994 Mitsubishi Galant .... 2.4L Code 4G63

REPLACEMENT INTERVALS

Mitsubishi recommends timing belt replacement every 60,000 miles.

LABOR TIMES

1994	
Replacement	4.4 hrs
Severe Service	4.9 hrs
With air conditioning add	0.5 hrs

8807MITg14

✻✻ CAUTION

These engines are an interference type. If the timing belt has failed, valve-to-piston damage may have occurred. A compression check should always be performed to check for engine damage.

1. Disconnect the negative battery cable.
2. Remove the engine undercover.
3. Remove the engine mount bracket.
4. Remove the drive belts.
5. Remove the belt tensioner pulley.
6. Remove the water pump pulleys.
7. Remove the crankshaft pulley.
8. Using a suitable engine support brace, remove the stud bolt from the engine support bracket and remove the bracket and mount.
9. Remove the timing belt covers.
10. Rotate the crankshaft in the direction of rotation to align the camshaft timing marks. Always turn the crankshaft in the normal direction of rotation only.
11. Loosen the tension pulley center bolt.

■ **If the timing belt is to be reused, mark the direction of rotation on the flat side of the belt with an arrow.**

12. Move the tension pulley towards the water pump and remove the timing belt.
13. Remove the crankshaft sprocket center bolt using special tool MB990767, or equivalent, to hold the crankshaft sprocket while removing the center bolt. Then, use MB998778 or equivalent puller to remove the sprocket.
14. Mark the direction of rotation on the timing belt **B** with an arrow.
15. Loosen the center bolt on the tensioner and remove the belt.

✻✻ WARNING

Do not rotate the camshafts or the crankshaft while the timing belt is removed.

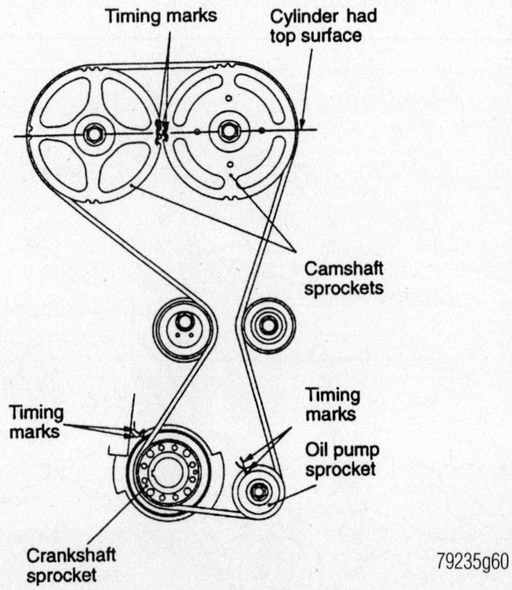

Camshaft and crankshaft timing belt sprocket TDC alignment marks—Mitsubishi 2.4L (4G63) engine

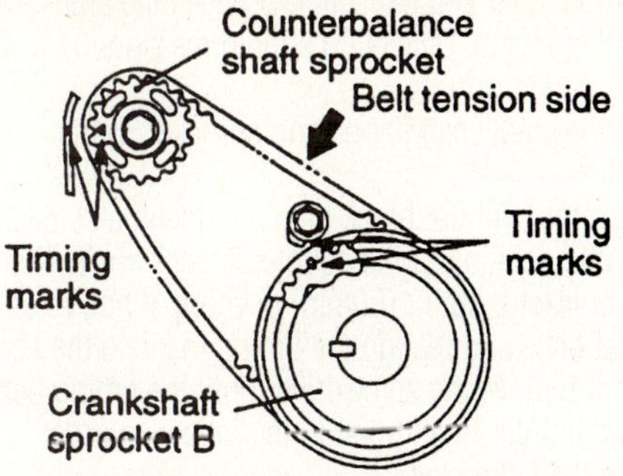

79235g61

Timing belt B installation mark alignment—Mitsubishi 2.4L (4G63) engine

To install:

16. Place the crankshaft sprocket on the crankshaft. Use tool MB990767, or equivalent, to hold the crankshaft sprocket while tightening the center bolt. Torque the center bolt to 80–94 ft. lbs. (108–127 Nm).
17. Align the timing marks on the crankshaft sprocket **B** and the balance shaft.
18. Install timing belt **B** on the sprockets. Position the center of the tensioner pulley to the left and above the center of the mounting bolt.
19. Push the pulley clockwise toward the crankshaft to apply tension to the belt and torque the mounting bolt to 14 ft. lbs. (19 Nm). Do not let the pulley turn when tightening the bolt because it will cause excessive tension on the belt. The belt should deflect 0.20–0.28 in. (5–7mm) when finger pressure is applied between the pulleys.
20. Install the crankshaft sensing blade and the crankshaft sprocket. Apply engine oil to the mounting bolt and torque the bolt to 80–94 ft. lbs. (108–127 Nm).
21. Use a press or vise to compress the auto-tensioner pushrod. Insert a set pin when the holes in the rod and the tensioner body are aligned.

**** WARNING**

Do not compress the pushrod too quickly, damage to the pushrod can occur.

22. Install the auto-tensioner on the engine.
23. Align the timing marks on the camshaft sprocket, crankshaft sprocket and the oil pump sprocket.
24. After aligning the mark on the oil pump sprocket, remove the cylinder block plug and insert a prytool in the hole to check the position of the counterbalance shaft. The prytool should go in at least 2.36 in. (60mm) or more, if not, rotate the oil pump sprocket once and realign the timing mark so the prytool goes in. Do not remove the prytool until the timing belt is installed.
25. Install the timing belt on the intake camshaft and secure it with a clip.
26. Install the timing belt on the exhaust camshaft. Align the timing marks with the cylinder head top surface using 2 wrenches. Secure the belt with another clip.
27. Install the belt around the idler pulley, oil pump sprocket, crankshaft sprocket and the tensioner pulley.
28. Turn the tensioner pulley so the pinholes are at the bottom. Press the pulley lightly against the timing belt.
29. Screw the special tool into the left engine support bracket until it contacts the tensioner arm, then screw the tool in a little more and remove the pushrod pin from the auto-tensioner. Remove the special tool and torque the center bolt to 35 ft. lbs. (48 Nm).
30. Turn the crankshaft 1/4 turn counterclockwise, then clockwise until the timing marks are aligned.
31. Loosen the center bolt. Install Mitsubishi Special Tool MD998767 or equivalent, on the tensioner pulley. Turn the tensioner pulley counterclockwise with a torque of 31 inch lbs. (3.5 Nm) and torque the center bolt to 35 ft. lbs. (48 Nm). Do not let the tensioner pulley turn when tightening the bolt.
32. Turn the crankshaft clockwise 2 revolutions and align the timing marks. After 15 minutes, measure the protrusion of the pushrod on the auto-tensioner. The standard measurement is 0.150–0.177 in. (3.8–4.5mm). If the protrusion is out of specification, loosen the tensioner pulley, apply the proper torque to the belt and retighten the center bolt.
33. Install the timing belt covers and all applicable components.

2.4L (G64B/4G64) SOHC ENGINES

YEAR, MODEL & ENGINE APPLICATION

1985-88 Mitsubishi Galant	2.4L Code G64B
1996-04 Eclipse, Eclipse Spyder, 1992-95 Expo, 1989-94 Galant, 1990-96 Mighty Max/Pickup, 1997-99 Montero Sport, 2002-04 Outlander, 1987-90 Van	2.4L Code 4G64

REPLACEMENT INTERVALS

Mitsubishi recommends timing belt replacement every 60,000 miles.

LABOR TIMES

Eclipse	
Replacement	3.3 hrs
Severe Service	3.8 hrs
Expo	
Replacement	2.7 hrs
Severe Service	3.2 hrs
Galant	
1985-98	
Replacement	4.4 hrs
Severe Service	4.9 hrs
1999-04	
Replacement	3.3 hrs
Severe Service	3.8 hrs
Mighty Max/Pickup	
Replacement	3.5 hrs
Severe Service	4.0 hrs
Montero Sport	
Replacement	3.2 hrs
Severe Service	3.7 hrs
Outlander	
Replacement	3.1 hrs
Severe Service	3.6 hrs
Van	
Replacement	3.0 hrs
Severe Service	3.5 hrs
All Models	
With balance shaft replacement add	0.3
All Models Except Montero Sport & Outlander	
With air conditioning add	0.5
With power steering add	0.3

8807MITg15

1992 Expo, 1985–87 Galant, 1990–96 Mighty Max/Pickup, 1987–Van

✱✱ CAUTION

These engines are an interference type. If the timing belt has failed, valve-to-piston damage may have occurred. A compression check should always be performed to check for engine damage.

1. Disconnect the negative battery cable.
2. Raise and support the vehicle.
3. Remove the engine splash shield.
4. Remove the engine mount bracket.
5. Remove the accessory drive belts.
6. Remove the accessory drive belt tensioner pulley and bracket.
7. Remove the water pump pulley.
8. Remove the crankshaft pulley.
9. Remove the power steering pump and A/C hose mounting bracket.
10. Remove the power steering pump pulley.
11. Remove the timing belt covers.
12. Rotate the crankshaft in the direction of rotation to align the camshaft timing mark with the timing mark on the cylinder head.
13. Loosen the tensioner bolts and move the tensioner toward the water pump to relieve pressure on the timing belt. Tighten the bolts.
14. If timing belt is to be reused, mark the belt with an arrow to indicate the direction of rotation.
15. Remove the timing belt.
16. Remove the crankshaft sprocket bolt, sprocket and flange.
17. Remove balance shaft belt tensioner bolt and pulley.
18. Remove the balance shaft belt.
19. Inspect the tensioner pulley for ease of movement and abnormal noises. Replace as necessary.
20. Inspect the crankshaft, camshaft and oil pump sprockets for cracks and excessive wear. Replace as necessary.

To install:

21. Ensure that all timing marks are aligned.
22. Install the balance shaft tensioner pulley.
23. Install the inner timing belt over the crankshaft inner sprocket and the balance shaft sprocket.
24. Move the inner belt tensioner against the belt to remove any slack from the tension side. Make sure that the timing marks are aligned. Torque the tensioner bolt to 14 ft. lbs. (19 Nm).

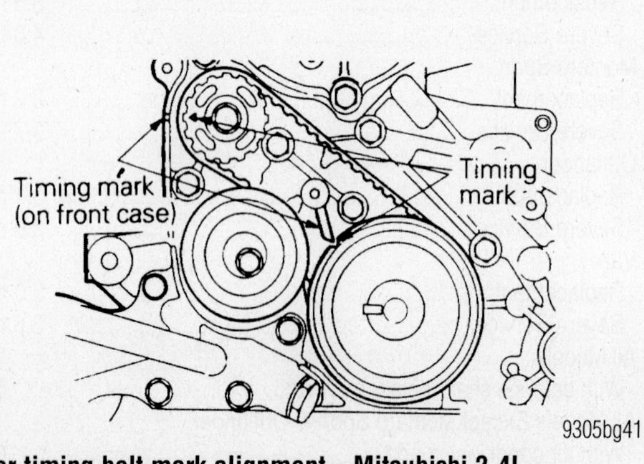

Inner timing belt mark alignment—Mitsubishi 2.4L engine–1992 Expo, 1985–87 Galant, 1990–96 Mighty Max/Pickup, 1987–Van

25. Check the inner belt deflection between the crankshaft pulley and the silent shaft sprocket. Deflection should measure 0.20–0.30 in. (5–7mm).
26. Install the crankshaft flange and sprocket. Tighten the bolts to 87 ft. lbs. (118 Nm).
27. Remove the cylinder block plug and insert a screwdriver in the hole to check the position of the counterbalance shaft. The screwdriver should go in at least 2.36 in. (60mm) or more, if not, rotate the oil pump sprocket once and realign the timing mark so the screwdriver goes in. Do not remove the screwdriver until the timing belt is installed.
28. Install the outer timing belt in the following order:
    a. Crankshaft sprocket
    b. Tensioner pulley
    c. Camshaft sprocket
    d. Oil pump sprocket
29. Make sure the timing marks are aligned.
30. Loosen the tensioner bolts to apply pressure to the timing belt.
31. Rotate the crankshaft 2 full turns in the direction of rotation and align the timing marks. Make sure that there is no slack on the tension side of the belt.
32. Loosen the tensioner bolts again to apply pressure to the timing belt. Torque the bolts to 35 ft. lbs. (49 Nm).
33. Install the timing belt covers.
34. Install the power steering pump pulley.
35. Install the power steering pump and A/C hose mounting bracket.
36. Install the crankshaft pulley.
37. Install the water pump pulley.
38. Install the accessory drive belt tensioner pulley and bracket.
39. Install the accessory drive belts.
40. Install the engine mount bracket.
41. Install the engine splash shield.
42. Lower the vehicle.
43. Connect the negative battery cable.
44. Start the engine and check for proper operation.
45. Check ignition timing and adjust if necessary.

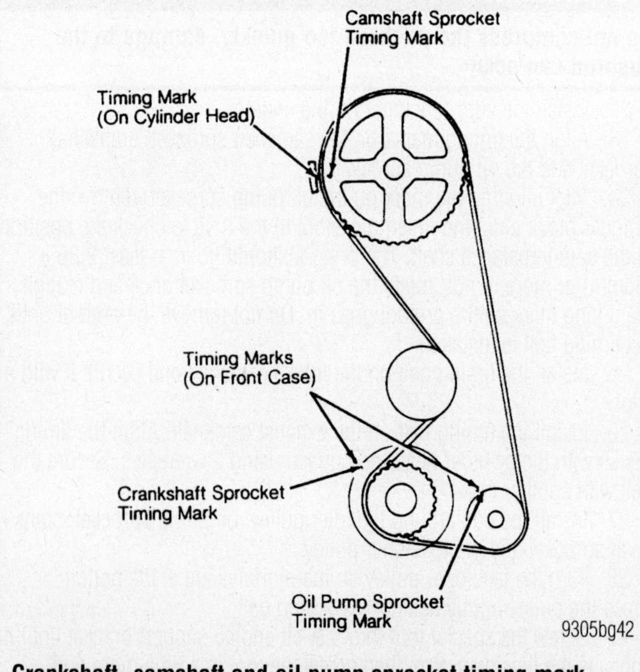

Crankshaft, camshaft and oil pump sprocket timing mark alignment—Mitsubishi 2.4L engine–1992 Expo, 1985–87 Galant, 1990–96 Mighty Max/Pickup, 1987–Van

1992–95 Expo, 1994–04 Galant, 2002–04 Outlander

✱✱ CAUTION

These engines are an interference type. If the timing belt has failed, valve-to-piston damage may have occurred. A compression check should always be performed to check for engine damage.

1. Disconnect the negative battery cable and if equipped with an air bag, wait one minute for the backup power to dlsslpate.
2. Raise and support the vehicle.
3. On 1994–97 Galant and Expo, remove the left front wheel and fender splash shield.
4. On 1998–04 Galant, remove the right front wheel and fender splash shield.
5. On Outlander, remove the power steering pressure hose and position aside.
6. On all models, remove the engine splash shield.
7. Remove the engine mount bracket.
8. Remove the accessory drive belts.
9. Remove the accessory drive belt tensioner pulley and bracket.
10. Remove the water pump pulley.
11. Remove the crankshaft pulley.
12. Remove the power steering pump pulley.
13. Remove the timing belt covers.
14. Rotate the crankshaft in the direction of rotation to align the camshaft timing mark with the timing mark on the cylinder head.
15. Loosen the tensioner bolts and move the tensioner toward the water pump to relieve pressure on the timing belt. Tighten the bolts.
16. If timing belt is to be reused, mark the belt with an arrow to indicate the direction of rotation.
17. Remove the timing belt.
18. Remove the crankshaft sprocket bolt, sprocket and flange.
19. Remove balance shaft belt tensioner bolt and pulley.
20. Remove the balance shaft belt.
21. Inspect the tensioner pulley for ease of movement and abnormal noises. Replace as necessary.
22. Inspect the crankshaft, camshaft and oil pump sprockets for cracks and excessive wear. Replace as necessary.

To install:

■ **To reset the tensioner: Keep the adjuster level and clamp it in a soft jawed vise. Clamp with the extended adjuster on one side and the end mounting a plug on the other side. If the plug extends out of the adjuster body, place a suitable hole sized washer over the plug so the vise jaw pushes on the washer, not the plug. Close the vise slowly, forcing the adjuster back into the body. When the hole in the adjuster boss aligns with the adjuster rod, insert a snug fitting pin or wire into the holes to keep the rod in the compressed position. With the locking pin or wire in place, install the tensioner.**

23. Install the automatic tensioner and tighten the bolts to 18 ft. lbs. (25 Nm).
24. Install the tensioner pulley.

■ **If the following steps are not followed exactly, there is a chance that the silent shaft alignment will be 180 degrees off. This will cause a noticeable vibration in the engine and the entire procedure will have to be repeated.**

25. Before installing the timing belt, ensure that the left-side silent shaft is in the correct position.

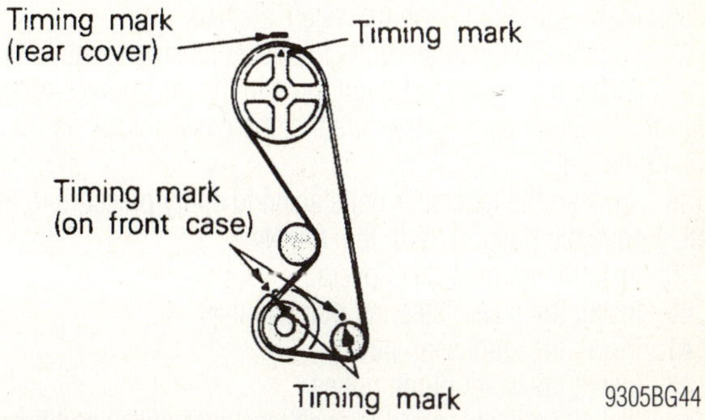

Inner timing belt mark alignment—Mitsubishi 2.4L engine–1992–95 Expo, 1994–04 Galant, 2002–04 Outlander

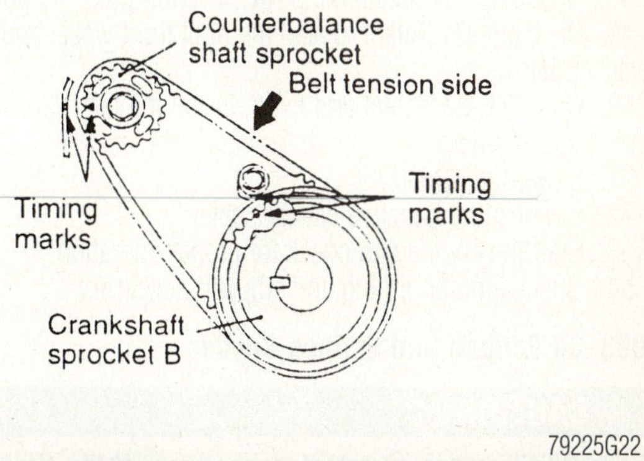

Crankshaft, camshaft and oil pump sprocket timing mark alignment—Mitsubishi 2.4L engine–1992–95 Expo, 1994–04 Galant, 2002–04 Outlander

■ **It is possible to align the timing marks on the camshaft sprocket, crankshaft sprocket and the oil pump sprocket with the left balance shaft out of alignment.**

26. With the timing mark on the oil pump pulley aligned with the mark on the front case, check the alignment of the left balance shaft to assure correct shaft timing.
27. Ensure that all timing marks are aligned.
28. Install the balance shaft tensioner pulley.
29. Install the inner timing belt over the crankshaft inner sprocket and the balance shaft sprocket.
30. Move the inner belt tensioner against the belt to remove any slack from the tension side. Make sure that the timing marks are aligned. Torque the tensioner bolt to 14 ft. lbs. (19 Nm).
31. Check the inner belt deflection between the crankshaft pulley and the silent shaft sprocket. Deflection should measure 0.20–0.30 in. (5–7mm).
32. Install the crankshaft flange and sprocket. Tighten the bolts to 87 ft. lbs. (118 Nm).
33. Remove the cylinder block plug and insert a screwdriver in the hole to check the position of the counterbalance shaft. The screwdriver should go in at least 2.36 in. (60mm) or more, if not, rotate the oil pump sprocket once and realign the timing mark so the screwdriver goes in. Do not remove the screwdriver until the timing belt is installed.
34. Install the outer timing belt in the following order:
    a. Crankshaft sprocket
    b. Tensioner pulley

c. Camshaft sprocket
d. Oil pump sprocket
35. Make sure the timing marks are aligned.
36. Loosen the tensioner bolts to apply pressure to the timing belt.
37. Rotate the crankshaft 2 full turns in the direction of rotation and align the timing marks. Make sure that there is no slack on the tension side of the belt.
38. Loosen the tensioner bolts again to apply pressure to the timing belt. Torque the bolts to 35 ft. lbs. (49 Nm).
39. Install the timing belt covers.
40. Install the power steering pump pulley.
41. Install the crankshaft pulley.
42. Install the water pump pulley.
43. Install the accessory drive belt tensioner pulley and bracket.
44. Install the accessory drive belts.
45. Install the engine mount bracket.
46. Install the engine mount bracket.
47. Install, remove the engine splash shield.
48. On Outlander, install the power steering pressure hose.
49. On 1998–04 Galant, install the right front wheel and fender splash shield.
50. On 1994–97 Galant and Expo, install the left front wheel and fender splash shield.
51. Lower the vehicle.
52. Connect the negative battery cable.
53. Start the engine and check for proper operation.
54. Check ignition timing and adjust if necessary.

1993–04 Eclipse and Eclipse Spyder

✱✱ CAUTION

These engines are an interference type. If the timing belt has failed, valve-to-piston damage may have occurred. A compression check should always be performed to check for engine damage.

1. Disconnect the negative battery cable and if equipped with an air bag wait one minute for the backup power to dissipate.
2. Raise and support the vehicle.
3. Remove the left side wheel.
4. Remove the left side engine splash shield.
5. Remove the engine mount bracket.
6. Remove the accessory drive belts.
7. Remove the accessory drive belt tensioner pulley and bracket.
8. Remove the water pump pulley.
9. Remove the crankshaft pulley.
10. Remove the timing belt covers.
11. Rotate the crankshaft in the direction of rotation to position the No. 1 cylinder at TDC of the compression stroke.
12. Ensure the crankshaft and camshaft sprocket timing marks are aligned.
13. Loosen the tensioner bolts and move the tensioner toward the water pump to relieve pressure on the timing belt. Tighten the bolts.
14. If timing belt is to be reused, mark the belt with an arrow to indicate the direction of rotation.
15. Remove the timing belt.
16. Remove the tensioner and tensioner pulley.
17. Remove the crankshaft sprocket bolt, sprocket and flange.
18. Remove balance shaft belt tensioner bolt and pulley.
19. Remove the balance shaft belt.
20. Inspect the tensioner pulley for ease of movement and abnormal noises. Replace as necessary.
21. Inspect the crankshaft, camshaft and oil pump sprockets for cracks and excessive wear. Replace as necessary.

To install:

■ **To reset the tensioner: Keep the adjuster level and clamp it in a soft jawed vise. Clamp with the extended adjuster on one side and the end mounting a plug on the other side. If the plug extends out of the adjuster body, place a suitable hole sized washer over the plug so the vise jaw pushes on the washer, not the plug. Close the vise slowly, forcing the adjuster back into the body. When the hole in the adjuster boss aligns with the adjuster rod, insert a snug fitting pin or wire into the holes to keep the rod in the compressed position. With the locking pin or wire in place, install the tensioner.**

22. Ensure that all timing marks are aligned.

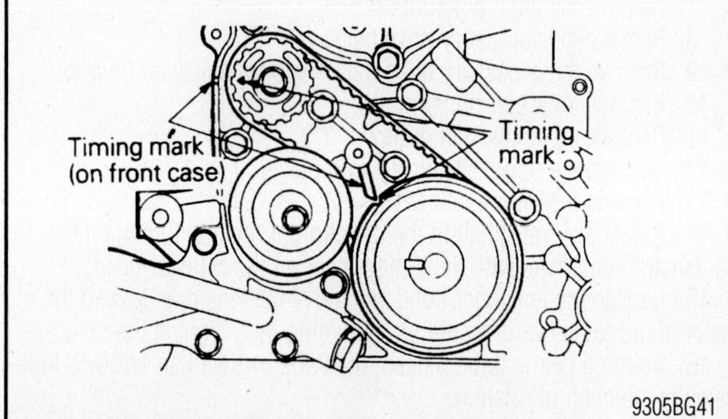

Inner timing belt mark alignment—Mitsubishi 2.4L engine–Eclipse and Eclipse Spyder

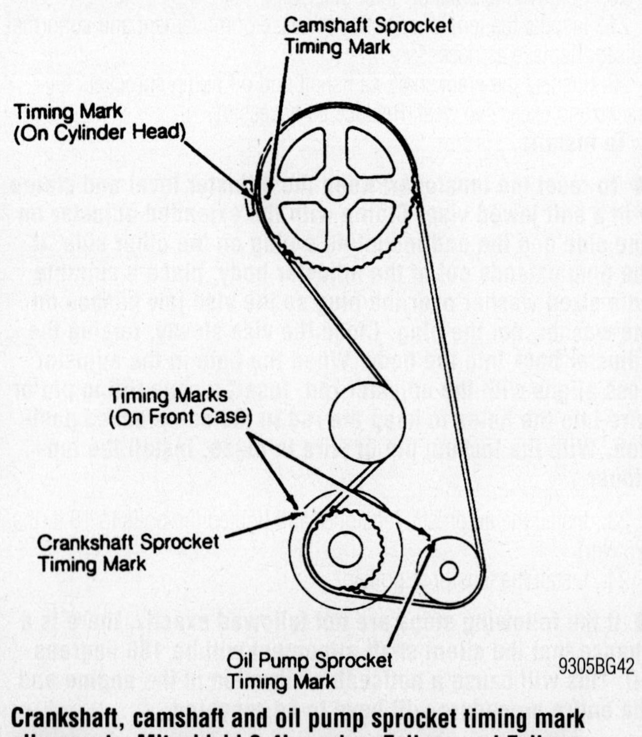

Crankshaft, camshaft and oil pump sprocket timing mark alignment—Mitsubishi 2.4L engine–Eclipse and Eclipse Spyder

23. Install the balance shaft tensioner pulley.
24. Install the inner timing belt over the crankshaft inner sprocket and the balance shaft sprocket.
25. Move the inner belt tensioner against the belt to remove any slack from the tension side. Make sure that the timing marks are aligned. Torque the tensioner bolt to 14 ft. lbs. (19 Nm).
26. Check the inner belt deflection between the crankshaft pulley and the silent shaft sprocket. Deflection should measure 0.20–0.30 in. (5–7mm).
27. Install the crankshaft flange and sprocket. Tighten the bolts to 87 ft. lbs. (118 Nm).
28. Remove the cylinder block plug and insert a screwdriver in the hole to check the position of the counterbalance shaft. The screwdriver should go in at least 2.36 in. (60mm) or more, if not, rotate the oil pump sprocket once and realign the timing mark so the screwdriver goes in. Do not remove the screwdriver until the timing belt is installed.
29. Install the outer timing belt in the following order:
   a. Crankshaft sprocket
   b. Tensioner pulley
   c. Camshaft sprocket
   d. Oil pump sprocket
30. Make sure the timing marks are aligned.
31. Loosen the tensioner bolts to apply pressure to the timing belt.
32. Rotate the crankshaft 2 full turns in the direction of rotation and align the timing marks. Make sure that there is no slack on the tension side of the belt.
33. Loosen the tensioner bolts again to apply pressure to the timing belt. Torque the bolts to 35 ft. lbs. (49 Nm).
34. Install the timing belt covers.
35. Install the power steering pump pulley.
36. Install the power steering pump and A/C hose mounting bracket.
37. Install the crankshaft pulley.
38. Install the water pump pulley.
39. Install the accessory drive belt tensioner pulley and bracket.
40. Install the accessory drive belts.
41. Install the engine mount bracket.
42. Install the engine splash shield.
43. Install the wheel.
44. Lower the vehicle.
45. Connect the negative battery cable.
46. Start the engine and check for proper operation.
47. Check ignition timing and adjust if necessary.

1997–99 Montero Sport

✳✳ CAUTION

These engines are an interference type. If the timing belt has failed, valve-to-piston damage may have occurred. A compression check should always be performed to check for engine damage.

1. Disconnect the negative battery cable and if equipped with an air bag wait one minute for the backup power to dissipate.
2. Drain the engine coolant.
3. Remove the upper radiator hose and cooling fan shroud.
4. Remove the cooling fan.
5. Remove the accessory drive belts.
6. Remove the upper timing belt cover.
7. Rotate the crankshaft in the direction of rotation to position the No. 1 cylinder at TDC of the compression stroke.
8. Remove the distributor assembly and ignition coil.
9. Remove the distributor mounting bracket.
10. Remove the crankshaft pulley.
11. Remove the lower timing belt cover.
12. Ensure the crankshaft and camshaft sprocket timing marks are aligned.
13. Remove the tensioner.
14. If timing belt is to be reused, mark the belt with an arrow to indicate the direction of rotation.
15. Remove the timing belt.
16. Remove the crankshaft sprocket bolt, sprocket and flange.
17. Remove balance shaft belt tensioner bolt and pulley.
18. Remove the balance shaft belt.
19. Inspect the tensioner pulley for ease of movement and abnormal noises. Replace as necessary.
20. Inspect the crankshaft, camshaft and oil pump sprockets for cracks and excessive wear. Replace as necessary.

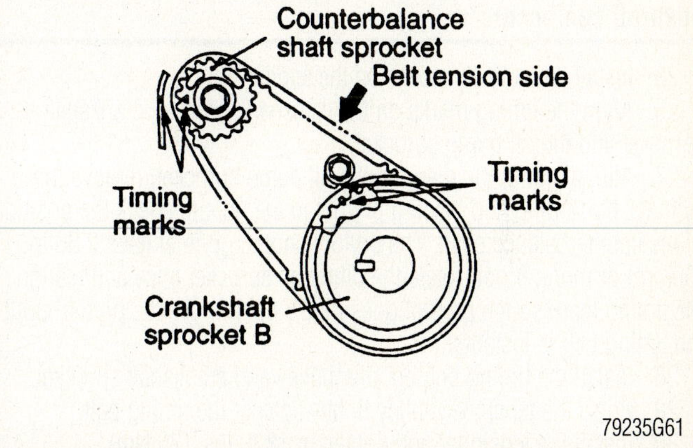

Inner timing belt mark alignment—Mitsubishi 2.4L engine–Montero Sport

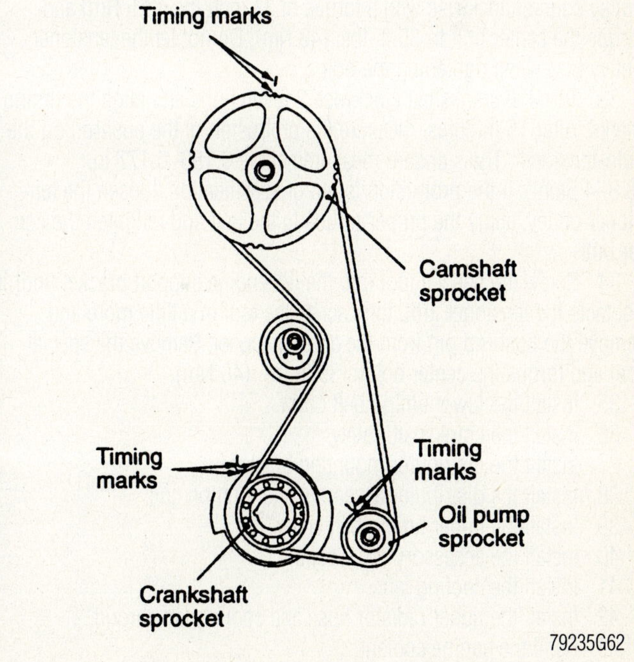

Crankshaft, camshaft and oil pump sprocket timing mark alignment—Mitsubishi 2.4L engine–Montero Sport

To install:

21. Align the timing marks on the crankshaft sprocket and the balance shaft.
22. Install the timing belt on the sprockets. Position the center of the tensioner pulley to the left and above the center of the mounting bolt.
23. Push the pulley clockwise toward the crankshaft to apply tension to the belt and torque the mounting bolt to 14 ft. lbs. (19 Nm). Do not let the pulley turn when tightening the bolt because it will cause excessive tension on the belt. The belt should deflect 0.20–0.28 in. (5–7mm) when finger pressure is applied between the pulleys.
24. Use a press or vise to compress the auto-tensioner pushrod. Insert a set pin when the holes in the rod and the tensioner body are aligned.

✱✱ WARNING

Do not compress the pushrod too quickly, damage to the pushrod can occur.

25. Install the auto-tensioner on the engine.
26. Align the timing marks on the camshaft sprocket, crankshaft sprocket and the oil pump sprocket.
27. After aligning the mark on the oil pump sprocket, remove the cylinder block plug and insert a prytool in the hole to check the position of the counterbalance shaft. The prytool should go in at least 2.36 in. (60mm) or more, if not, rotate the oil pump sprocket once and realign the timing mark so the prytool goes in. Do not remove the prytool until the timing belt is installed.
28. Install the timing belt on idler pulley and crankshaft sprocket.
29. Press the tensioner pulley lightly against the timing belt.
30. Install the tensioner and tighten to 18 ft. lbs. (25 Nm).
31. Turn the crankshaft 1/4 turn counterclockwise, then clockwise until the timing marks are aligned.
32. Loosen the center bolt. Install Mitsubishi Special Tool MD998767 or equivalent, on the tensioner pulley. Turn the tensioner pulley counterclockwise with a torque of 31 inch lbs. (3.5 Nm) and torque the center bolt to 35 ft. lbs. (48 Nm). Do not let the tensioner pulley turn when tightening the bolt.
33. Turn the crankshaft clockwise 2 revolutions and align the timing marks. After 15 minutes, measure the protrusion of the pushrod on the auto-tensioner. The standard measurement is 0.150–0.177 in. (3.8–4.5mm). If the protrusion is out of specification, loosen the tensioner pulley, apply the proper torque to the belt and retighten the center bolt.
34. Screw the special tool into the left engine support bracket until it contacts the tensioner arm, then screw the tool in a little more and remove the pushrod pin from the auto-tensioner. Remove the special tool and torque the center bolt to 35 ft. lbs. (48 Nm).
35. Install the lower timing belt cover.
36. Install the crankshaft pulley.
37. Install the distributor mounting bracket.
38. Install the distributor assembly and ignition coil.
39. Install the upper timing belt cover.
40. Install the accessory drive belts.
41. Install the cooling fan.
42. Install the upper radiator hose and cooling fan shroud.
43. Refill the engine coolant.
44. Connect the negative battery cable.
45. Start the engine and check for proper operation.
46. Check ignition timing and adjust if necessary.

3.0L (6G72) DOHC V6 ENGINES

YEAR, MODEL & ENGINE APPLICATION

Application	Engine
1993-96 Mitsubishi Diamante, 1991-99 3000GT, 3000GT Turbo	3.0L Code 6G72

REPLACEMENT INTERVALS

Mitsubishi recommends timing belt replacement every 60,000 miles.

LABOR TIMES

Diamante	
Replacement	4.3 hrs
Severe Service	4.8 hrs
3000GT	
Replacement	4.4 hrs
Severe Service	4.9 hrs
With air conditioning add	1.0 hrs
With turbocharger add	0.3 hrs

8807MITg16

✱✱ CAUTION

These engines are an interference type. If the timing belt has failed, valve-to-piston damage may have occurred. A compression check should always be performed to check for engine damage.

1. Disconnect the negative battery cable.

✱✱ CAUTION

Wait at least 90 seconds after the negative battery cable is disconnected to prevent possible deployment of the air bag.

2. Position the engine so the No. 1 cylinder is at Top Dead Center (TDC) of the compression stroke. Disconnect the negative battery cable.
3. Remove the left front and side engine covers.
4. Remove the cruise control actuator.
5. Remove the accessory drive belts.
6. Remove the alternator. Remove the air hose and pipe.
7. Remove the power steering belt tensioner assembly.
8. Remove the crankshaft pulley.
9. Disconnect the brake fluid level sensor.
10. Remove the timing belt upper cover.
11. Using the proper equipment, slightly raise the engine to take the weight off the side engine mount. Remove the engine mount bracket.
12. Remove the engine support bracket. The mounting bolts are different lengths; mark them for proper installation.
13. Remove the alternator and A/C idler pulley.
14. Disconnect the camshaft and crankshaft position sensor connectors.
15. Remove the timing belt lower cover. Timing bolt cover mounting bolts are different in length, note their position during removal.
16. If the same timing belt will be reused, mark the direction of the timing belt's rotation for installation in the same direction. Make sure the engine is positioned so the No. 1 cylinder is at the TDC of its compression stroke and the timing marks are aligned with the engine's timing mark indicators on the valve covers or head.

17. Loosen the center bolt of tensioner pulley and unbolt auto-tensioner assembly. The auto-tensioner assembly must be reset to correctly adjust belt tension. Remove the timing belt.

✱✱ WARNING

Do not rotate the camshafts when the timing belt is removed from the engine. Turning the camshaft when the timing belt is removed could cause the valves to interfere with the pistons thus causing severe internal engine damage.

18. Remove and position the auto-tensioner into a vise with soft jaws. The plug at the rear of tensioner protrudes, be sure to use a washer as a spacer to protect the plug from contacting vise jaws.

19. Slowly push the rod into the tensioner until the set hole in rod is aligned with set hole in the auto-tensioner.

20. Insert a 0.055 in. (1.4mm) wire into the aligned set holes. Unclamp the tensioner from vise and install to vehicle. Tighten tensioner to 17 ft. lbs. (24 Nm).

21. On 1991 DOHC 3.0L engines, clean and inspect both auto tensioner mounting bolts. Coat the threads of the old bolts with Mopar thread sealer 4318034 or equivalent. If new bolts are installed, inspect the heads of the new bolts. If there is white paint on the bolt head, no sealer is required. If there is no paint on the head of the bolt, apply a coat of thread sealer to the bolt. Install both bolts and tighten to 17 ft. lbs. (24 Nm).

To install:

✱✱ WARNING

Turning the camshaft sprocket when the timing belt is removed could cause the valves to interfere with the pistons.

22. Align the mark on the crankshaft sprocket with the mark on the front case. Then move the sprocket 3 teeth clockwise to lower the piston so the valves do not touch the piston if the camshafts are being moved.

23. Turn each camshaft sprocket 1 at a time to align the timing marks with the mark on the valve cover or head. If the intake and exhaust valves of the same cylinder are opened simultaneously, they could interfere with each other. Therefore, if any resistance is felt, turn the other camshaft to move the valve.

24. Using binder clips to secure the timing belt to sprockets, install the timing belt in the following order. Be sure camshafts to cylinder heads and crankshaft to front cover timing marks are aligned.

a. Exhaust camshaft sprocket (front bank).
b. Intake camshaft sprocket (front bank).
c. Water pump pulley.
d. Intake camshaft sprocket (rear bank).
e. Exhaust camshaft sprocket (rear bank).
f. Idler pulley.
g. Crankshaft pulley.
h. Tensioner pulley.

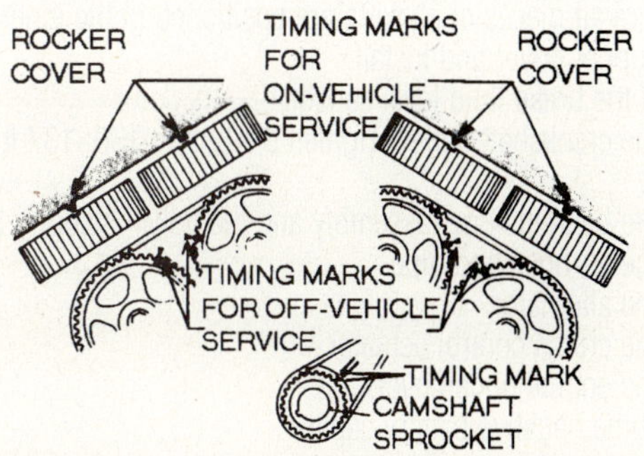

91251g15

Proper timing belt alignment mark positioning for belt service—Mitsubishi 3.0L (6G72) DOHC engine

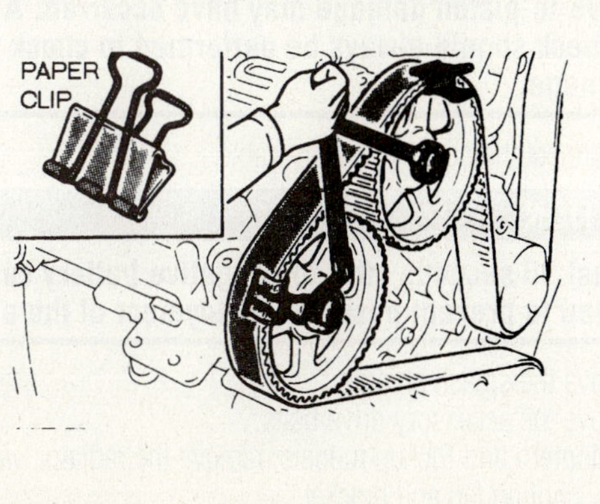

91251g16

Securing camshafts for timing belt installation—Mitsubishi 3.0L (6G72) DOHC engine

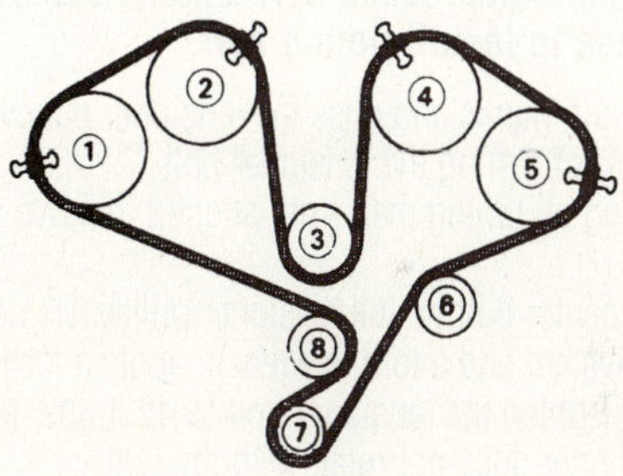

91251g17

Timing belt installation sequence—Mitsubishi 3.0L (6G72) DOHC engine

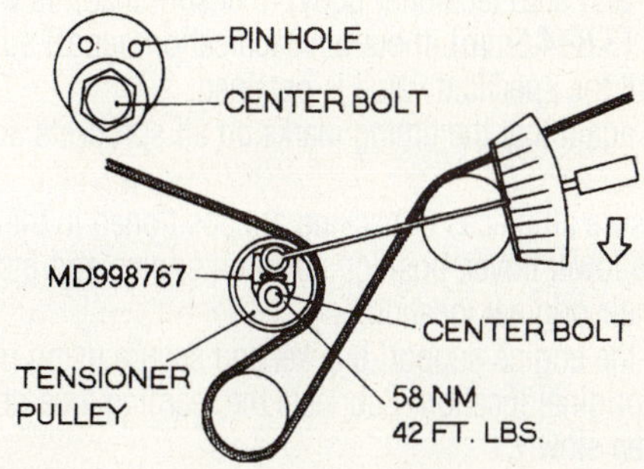

91251g18

Timing belt tension adjustment—Mitsubishi 3.0L (6G72) DOHC engine

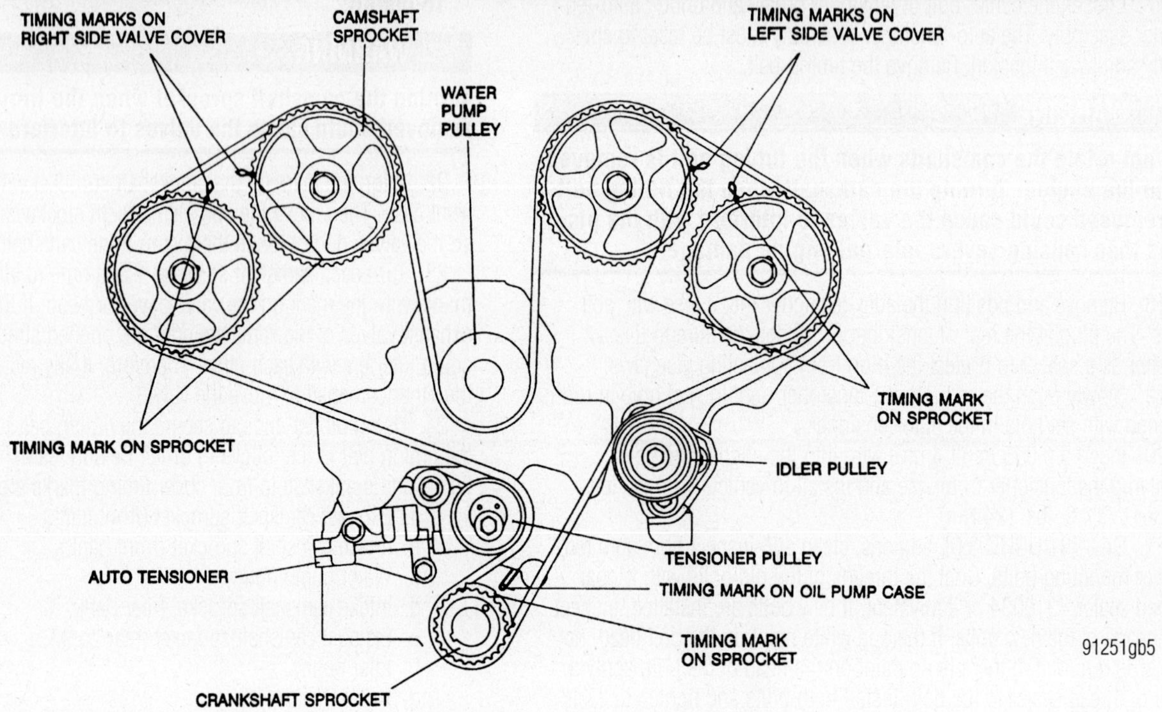

Crankshaft and camshaft sprocket timing mark alignment—Mitsubishi 3.0L (6G72) DOHC engine

■ **Since the camshaft sprockets turn easily, secure them with box wrenches to install timing belt.**

25. Align all timing marks and raise the tensioner pulley against the belt to remove slack, and snug the tensioner bolt.
26. After ensuring all timing marks are aligned, remove the binder clips.
27. Loosen the center bolt on the tensioner pulley. Using tool MD998767 or equivalent and a torque wrench, apply a torque of 86 inch lbs. (10 Nm). Tighten the tensioner bolt to 42 ft. lbs. (58 Nm) and make sure the tensioner does not rotate with the bolt.
28. Remove the set wire attached to the auto tensioner.
29. Rotate the crankshaft 2 complete turns clockwise and let it sit for approximately 5 minutes. Then, check that the set pin can easily be inserted and removed from the hole in the auto tensioner.

■ **Even if the set pin cannot be easily inserted, the auto tensioner is normal if its rod protrusion is within specification.**

30. Measure the auto tensioner protrusion (the distance between the tensioner arm and auto tensioner body) to ensure that it is within 0.15–0.18 in. (3.8–4.5mm). If out of specification, repeat adjustment procedure until the specified value is obtained.
31. Check again that the timing marks on all sprockets are in proper alignment.
32. Make sure all pieces of packing are positioned in the inner grooves of the lower cover, position cover on engine and install mounting bolts in their original location.
33. Install the engine support bracket and secure using mounting bolts in their original location. Lubricate the reaming area of the reamer bolt and tighten slowly.
34. Install the idler pulley.
35. Install the engine mount bracket. Remove the engine support fixture.
36. Connect the camshaft and crankshaft position sensor connectors.
37. Install the alternator and A/C idler pulley.
38. Make sure all pieces of packing are positioned in the inner grooves of the upper cover and install.
39. Connect the brake fluid level sensor.
40. Install the crankshaft pulley. Tighten the bolt to 130–137 ft. lbs. (180–190 Nm).
41. Install the belt tensioner assembly and the power steering belt.
42. Install the air hose and pipe.
43. Install the alternator.
44. Install the cruise control actuator.
45. Install the engine undercover.
46. Connect the negative battery cable.

3.0L (6G72) SOHC ENGINES

✱✱ CAUTION

These engines are an interference type. If the timing belt has failed, valve-to-piston damage may have occurred. A compression check should always be performed to check for engine damage.

1. Disconnect the negative battery cable.

✱✱ CAUTION

Wait at least 90 seconds after the negative battery cable is disconnected to prevent possible deployment of the air bag.

2. Remove the splash shields.
3. Remove the accessory drive belts.
4. On Montero and Pickup models, remove the radiator, water pump hoses, cooling fan and bracket.
5. On all models, remove the top engine mount and bracket.
6. Remove the alternator.

YEAR, MODEL & ENGINE APPLICATION

1992-96 Mitsubishi Diamante, 1988 Galant Sigma, 1989-91 Sigma, 1990-96 Mighty Max/Pickup, 1987-96 Montero, 1998-99 3000GT .... 3.0L Code 6G72

REPLACEMENT INTERVALS

Mitsubishi recommends timing belt replacement every 60,000 miles.

LABOR TIMES

Diamante, Galant Sigma & Sigma	
Replacement	4.3 hrs
Severe Service	4.8 hrs
Mighty Max/Pickup	
Replacement	3.5 hrs
Severe Service	4.0 hrs
With air conditioning add	0.5 hrs
Montero	
Replacement	3.8 hrs
Severe Service	4.8 hrs
With air conditioning add	0.5 hrs
3000GT	
Replacement	4.4 hrs
Severe Service	4.9 hrs

8807MITg17

7. Remove the power steering pump without disconnecting the lines, and position aside.
8. Remove the A/C compressor without disconnecting the lines, and position aside.
9. Remove the crankshaft pulley bolt, pulley and belt guide.
10. Remove the timing belt cover cap.
11. Remove the timing belt upper and lower covers.
12. Rotate the crankshaft until the No. 1 cylinder is at TDC on the compression stroke.
13. Loosen the timing belt tensioner bolt and remove the belt. If the tensioner is not being removed, position it as far away from the center of the engine as possible and tighten the bolt.
14. If the tensioner is being removed, paint the outside of the spring to ensure that it is not installed backward. Unbolt the tensioner and remove it along with the spring.
15. If the same timing belt will be reused, mark the direction of the timing belt's rotation for installation in the same direction.
16. Remove the timing belt.

To install:

17. Position both camshafts so the marks align with those on the rear. Rotate the crankshaft in the direction of rotation so the timing mark aligns with the mark on the oil pump.
18. Install the timing belt to the crankshaft sprocket, right camshaft sprocket, water pump pulley, left camshaft sprocket and tensioner pulley.
19. Ensure the timing belt is taut on the non tension side.

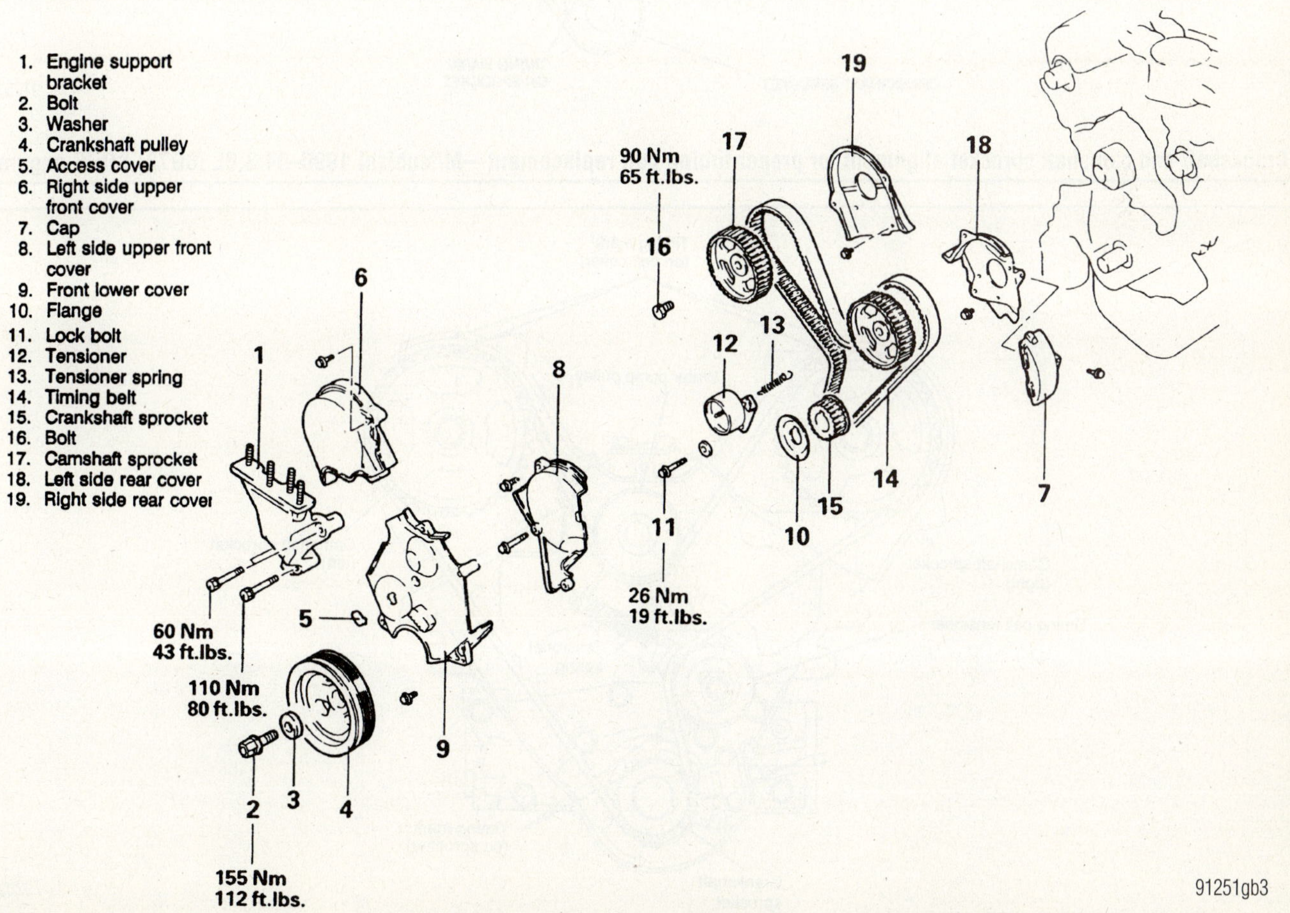

Exploded view of the timing belt covers, timing belt and related parts—Mitsubishi 3.0L (6G72) SOHC engine

20. Install the tensioner, if removed, and hook the upper end of the spring to the water pump pin and the lower end to the tensioner in exactly the same position as originally installed. If not already done,

21. Rotate the front camshaft counterclockwise to tension the belt between the front camshaft and the crankshaft. If the timing marks became misaligned, repeat the procedure.

22. Loosen the tensioner bolt and allow the spring to apply tension to the belt.

23. Turn the crankshaft 2 full turns in the clockwise direction until the timing marks align again. Now that the belt is properly tensioned, torque the tensioner lockbolt to 21 ft. lbs. (29 Nm). Measure the belt tension between the rear camshaft sprocket and the crankshaft with belt tension gauge. The specification is 46–68 lbs. (210–310 N).

24. Install the timing covers. Make sure all pieces of packing are positioned in the inner grooves of the covers when installing.

25. Install the belt guide and crankshaft pulley. Torque the bolt to 108–116 ft. lbs. (150–160 Nm).

26. Install the A/C compressor.

27. Install the power steering pump.

28. Install the alternator.

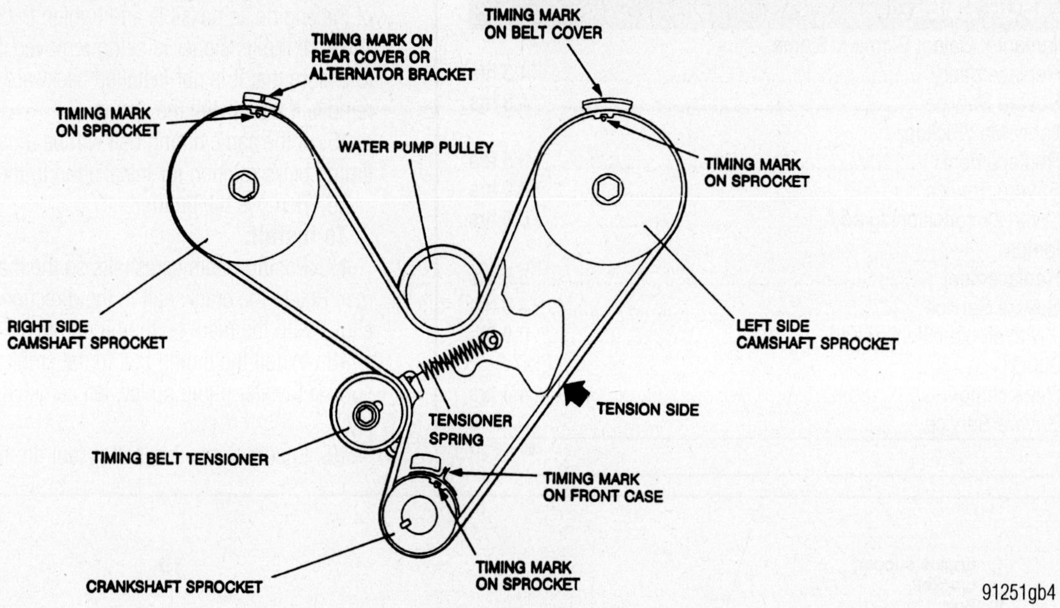

Crankshaft and camshaft sprocket alignment for proper timing belt replacement—Mitsubishi 1990–94 3.0L (6G72) SOHC engine

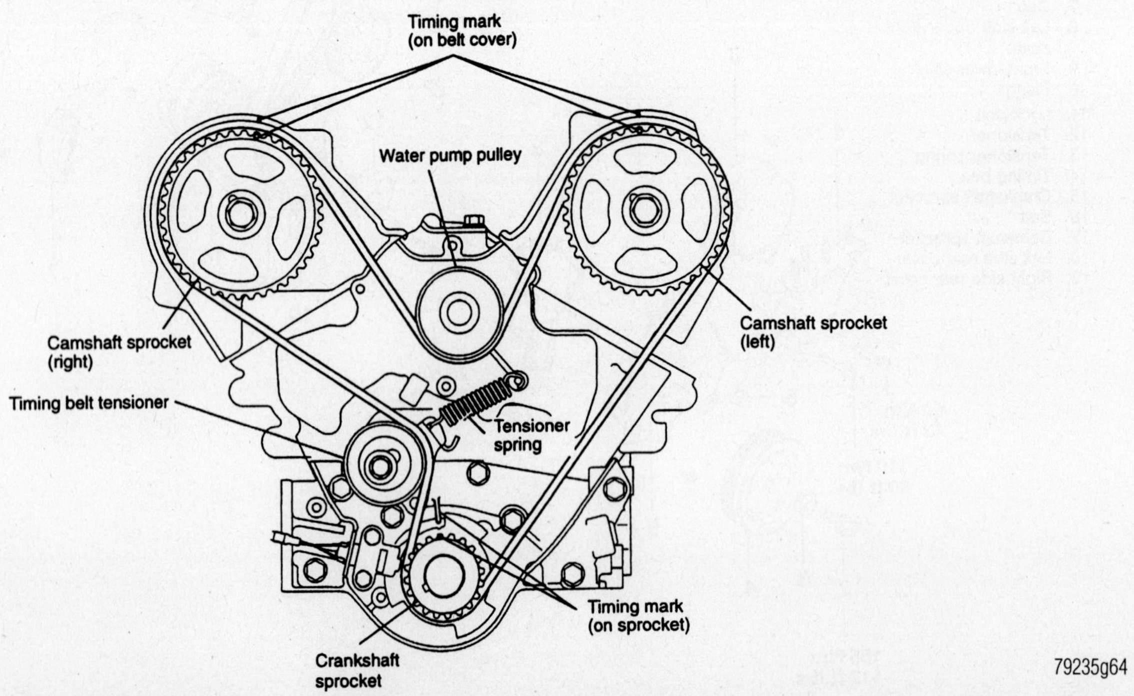

Crankshaft and camshaft sprocket alignment for proper timing belt replacement—Mitsubishi 1995–99 3.0L (6G72) SOHC engine

29. Remove the top engine mount and bracket.
30. On Montero and Pickup models, install the radiator, water pump hoses, cooling fan and bracket.
31. Install the accessory drive belts.
32. Install the splash shields.
33. Connect the negative battery cable.
34. Start the engine and check for proper operation.

3.0L (6G72) SOHC ENGINES

YEAR, MODEL & ENGINE APPLICATION

2000-04 Mitsubishi Eclipse/Eclipse Spyder, 1998-04 Galant, 1997-03 Montero Sport .... 3.0L Code 6G72

REPLACEMENT INTERVALS

Mitsubishi recommends timing belt replacement every 60,000 miles.

LABOR TIMES

Eclipse, Eclipse Spyder & Galant	
Replacement	3.3 hrs
Severe Service	3.8 hrs
Montero Sport	
Replacement	3.2 hrs
Severe Service	3.7 hrs

8807MITg18

✻✻ CAUTION

These engines are an interference type. If the timing belt has failed, valve-to-piston damage may have occurred. A compression check should always be performed to check for engine damage.

1. Disconnect the negative battery cable.

✻✻ CAUTION

Wait at least 90 seconds after the negative battery cable is disconnected to prevent possible deployment of the air bag.

2. On Montero, drain the engine coolant.
3. Remove the lower splash shield, radiator and water pump hoses.
4. Remove the viscous fan assembly.
5. On car models, raise and support the vehicle and engine and remove the right side engine mount.
6. On all models, remove the accessory drive belt tensioner and drive belts.
7. Remove the alternator.
8. Remove the power steering pump without disconnecting the lines, and position aside.
9. Remove the A/C compressor without disconnecting the lines, and position aside.
10. On Montero, remove the fan bracket.
11. On all models, remove the upper timing belt covers.
12. Remove the crankshaft pulley bolt and pulley.
13. Remove the lower timing belt cover.
14. Position the engine so the No. 1 cylinder is at Top Dead Center (TDC) of the compression stroke.

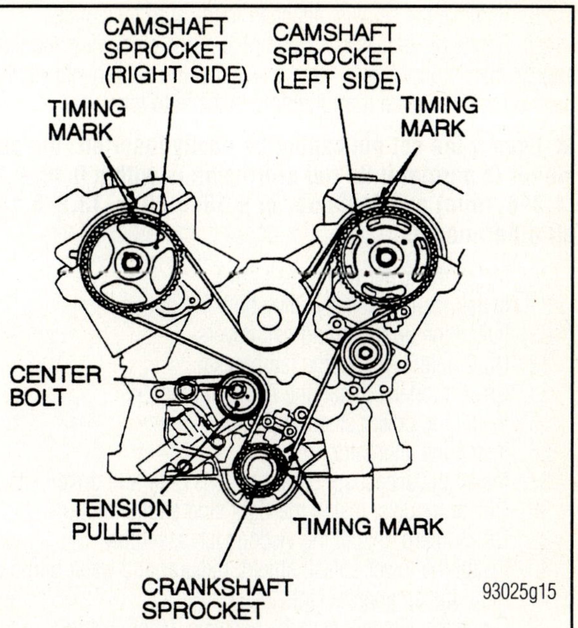

Crankshaft and camshaft sprocket timing marks—1997–04 Mitsubishi 3.0L (6G72) SOHC engine

15. Loosen the bolt of the tensioner pulley and allow the pulley to move away from the belt. Hand tighten the bolt.
16. Remove the timing belt.
17. Remove the tensioner assembly.
18. If the same timing belt will be reused, mark the direction of the timing belt's rotation for installation in the same direction.

To install:

19. Check the tensioner push rod movement. Push the tensioner against a hard surface and check that the pushrod moves no more than 0.40 in. (10mm). If movement is more than specified, replace the tensioner.
20. Position the auto-tensioner into a vise with soft jaws. The plug at the rear of tensioner protrudes, be sure to use a washer as a spacer to protect the plug from contacting vise jaws.
21. Slowly push the rod into the tensioner until the set hole in rod is aligned with set hole in the auto-tensioner.
22. Insert a 0.055 in. (1.4mm) wire into the aligned set holes. Unclamp the tensioner from vise and install to vehicle. Tighten tensioner to 17 ft. lbs. (24 Nm).
23. Ensure the timing marks are aligned.
24. Using binder clips to secure the timing belt to sprockets, install the timing belt in the following order. Be sure camshafts to cylinder heads and crankshaft to front cover timing marks are aligned.
    a. Crankshaft pulley.
    b. Tensioner pulley.
    c. Left camshaft sprocket.
    d. Water pump pulley.
    e. Right camshaft sprocket.
    f. Guide pulley.
25. Align all timing marks and raise the tensioner pulley against the belt to remove slack, and snug the tensioner bolt.
26. Rotate the crankshaft one quarter turn counterclockwise, and then back one quarter turn clockwise and align the timing marks.
27. Loosen the center bolt on the tensioner pulley. Using tool MD998767 or equivalent and a torque wrench, apply a torque of 86 inch lbs. (10 Nm). Tighten the tensioner bolt to 36 ft. lbs. (49 Nm) and make sure the tensioner does not rotate with the bolt.

28. Remove the set wire attached to the auto tensioner.
29. Rotate the crankshaft 2 complete turns clockwise and let it sit for approximately 5 minutes. Then, check that the set pin can easily be inserted and removed from the hole in the auto tensioner.

■ **Even if the set pin cannot be easily inserted, the auto tensioner is normal if its rod protrusion is within 0.19–0.24 in. (4.8–6.1mm) on Diamante, or 0.15–0.20 in. (3.8–5.1mm) on all other models.**

30. Install the lower timing belt cover.
31. Install the crankshaft pulley bolt and pulley.
32. Install the upper timing belt covers.
33. On Montero, install the fan bracket.
34. On all models, install the A/C compressor.
35. Install the power steering pump.
36. Install the alternator.
37. Install the accessory drive belt tensioner and drive belts.
38. On car models, install the right side engine mount.
39. On Montero, install the viscous fan assembly.
40. Install the lower splash shield, radiator and water pump hoses.
41. Refill the engine coolant.
42. On all models, connect the negative battery cable.
43. Start the engine and check for proper operation.

1997–04 3.5L (6G74) SOHC ENGINES

YEAR, MODEL & ENGINE APPLICATION

1997-04 Mitsuishi Diamante, 1997-00 Montero, 1998-03 Montero Sport	3.5L Code 6G74

REPLACEMENT INTERVALS

Mitsubishi recommends timing belt replacement every 60,000 miles.

LABOR TIMES

Diamante & Montero	
Replacement	3.8 hrs
Severe Service	4.3 hrs
Montero with air conditioning add	0.5 hrs
Montero Sport	
Replacement	3.2 hrs
Severe Service	3.7 hrs
With air conditioning add	0.3 hrs

8807MITg19

✱✱ CAUTION

These engines are an interference type. If the timing belt has failed, valve-to-piston damage may have occurred. A compression check should always be performed to check for engine damage.

1. Disconnect the negative battery cable.

✱✱ CAUTION

Wait at least 90 seconds after the negative battery cable is disconnected to prevent possible deployment of the air bag.

2. On Diamante, raise and support the vehicle and engine and remove the right side engine mount.
3. On truck models, drain the engine coolant and store it for reinstallation. Remove the upper radiator hose.
4. Remove the cooling fan shroud assembly.
5. Remove the cooling fan-to-clutch bolts and the fan.
6. Remove the cooling fan clutch-to-water pump nuts and the clutch assembly.
7. On all models, remove the accessory drive belt tensioner and drive belts.
8. Disconnect the electrical connectors from the alternator.
9. Remove the alternator-to-engine bolts and the alternator bracket-to-engine bolts; then, remove the alternator and bracket from the engine.
10. Remove the power steering pump cover. Remove the power steering pump-to-engine bolts and move the pump aside with the hoses and electrical connector attached.
11. Remove the air conditioning compressor-to-bracket bolts and move the compressor aside with the lines and electrical connector attached.
12. Remove the air conditioning compressor bracket-to-engine bolts and the bracket.

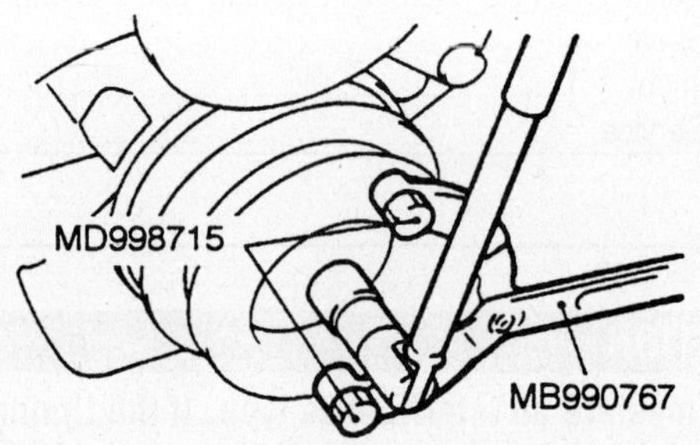

93025g14

Removing or installing the crankshaft pulley bolt—Mitsubishi 1997–04 3.5L (6G74) engine

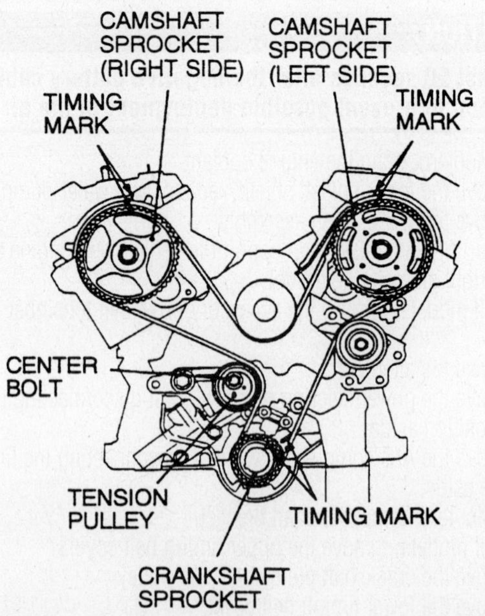

93025g15

View of the timing belt alignment marks—Mitsubishi 1997–04 3.5L (6G74) engine

13. Remove the timing indicator bracket (near crankshaft pulley) bolts and the bracket.

14. Remove the accessory mount assembly-to-engine bolts and the mount assembly.

15. Remove the upper timing belt cover assembly.

16. Using the End Yoke Holder tool MD990767 and 2 Crankshaft Pulley Holder Pin tools MD998715, or equivalent to hold the crankshaft pulley, and a socket wrench, remove the crankshaft pulley bolt and the pulley.

17. Remove the lower timing belt cover.

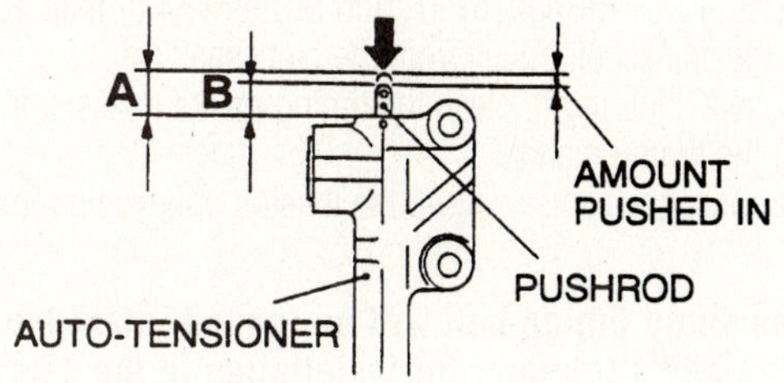

93025g16

Inspecting the auto-tensioner movement—Mitsubishi 1997–04 3.5L (6G74) engine

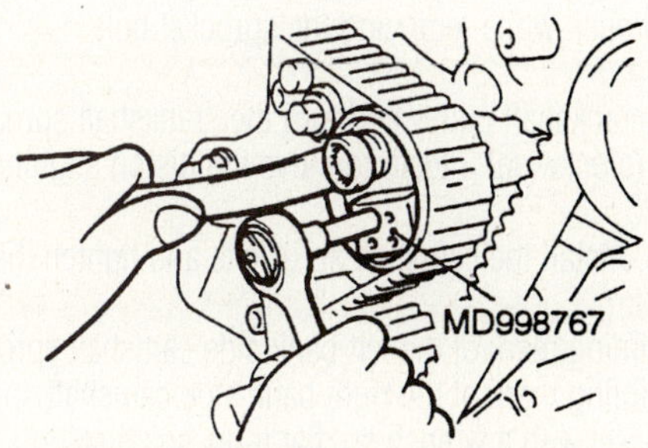

93025g17

Adjusting the timing belt tensioner pulley—Mitsubishi 1997–04 3.5L (6G74) engine

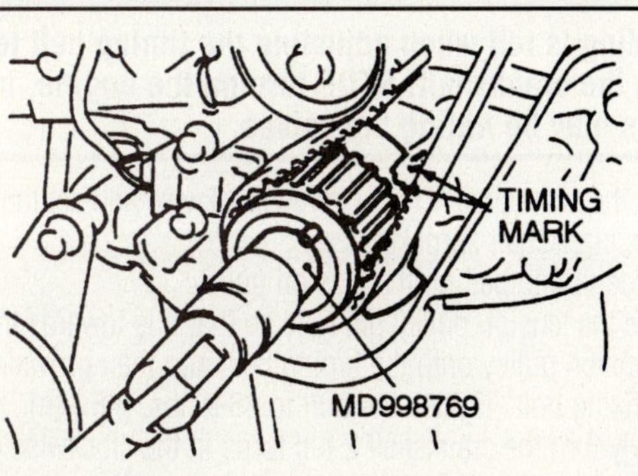

93025g18

Using crankshaft spacer tool to rotate the crankshaft—Mitsubishi 1997–04 3.5L (6G74) engine

18. Rotate the crankshaft clockwise to align the timing marks to position the No. 1 cylinder at the Top Dead Center (TDC) of its compression stroke.

19. Use chalk to mark the rotating (clockwise) direction of the timing belt for reinstallation purposes.

20. Loosen the auto-tensioner pulley center bolt and remove the timing bolt.

21. Remove the auto-tensioner pulley and the auto-tensioner arm assembly.

To install:

22. Press the end of the auto-tensioner inward with 72–145 ft. lbs. (98–196 Nm) of force and measure the distance that the pushrod is pushed in. If the standard distance is not 0.04 in. (1mm), replace the auto-tensioner.

23. Position the auto-tensioner in a soft-jawed vise and SLOWLY compress the pushrod until the pushrod and housing holes align; then, install a setting pin to secure the auto-tensioner in the retracted position.

24. Align the camshaft and crankshaft TDC timing marks.

25. Install the timing belt (noting its rotational direction) so that there is no deflection between the sprockets and pulleys in the following manner:

- Crankshaft sprocket
- Idler pulley
- Left camshaft sprocket
- Water pump pulley
- Right camshaft sprocket
- Tension pulley

26. Turn the camshaft sprocket counterclockwise until the tension side of the timing belt is firmly stretched, then, recheck the timing marks.

27. Using the Tension Pulley Socket Wrench tool MD998767, or equivalent, push the tensioner pulley into the timing belt and secure the center bolt.

28. Using the Crankshaft Pulley Spacer tool MD998769, or equivalent, rotate the crankshaft 1/4 turn counterclockwise, then, turn it again clockwise to align the timing marks.

29. Loosen the timing belt tensioner center bolt. Using the Tension Pulley Socket Wrench tool MD998767, or equivalent, and a torque wrench, apply 39 inch lbs. (4.4 Nm) pressure on the timing belt. Torque the tensioner pulley center bolt to 35 ft. lbs. (48 Nm).

30. Remove the setting pin from the auto-tensioner.

31. Rotate the crankshaft 2 complete revolutions and realign the timing marks. Then, wait for 5 minutes until the auto-tensioner's pushrod extends to its standard value. If the standard value is not 0.15–0.20 in. (3.8–5.0mm), repeat the adjustment procedure. If the standard value is still not achieved, replace the auto-tensioner.

32. Install the lower timing belt cover and crankshaft pulley.

33. Using the End Yoke Holder tool MD990767 and 2 Crankshaft Pulley Holder Pin tools MD998715, or equivalent to hold the crankshaft pulley, and a socket torque wrench, torque the crankshaft pulley bolt to 134 ft. lbs. (181 Nm).

34. Install the upper timing belt cover assembly.

35. Install the remaining items by reversing the removal procedures.

36. Refill the cooling system.

37. Connect the negative battery cable.

1994–97 3.5L (VIN M) DOHC ENGINES

YEAR, MODEL & ENGINE APPLICATION	
1994-97 Mitsubishi Montero	3.5L VIN M

REPLACEMENT INTERVALS
Mitsubishi recommends timing belt replacement every 60,000 miles.

LABOR TIMES	
1994-97	
Replacement	4.7 hrs
Severe Service	5.2 hrs
With air conditioning add	0.5 hrs

8807MITg20

**** CAUTION**

These engines are an interference type. If the timing belt has failed, valve-to-piston damage may have occurred. A compression check should always be performed to check for engine damage.

1. Disconnect the negative battery cable.
2. Drain the cooling system. Remove the drive belts.

**** CAUTION**

Never open, service or drain the radiator or cooling system when hot; serious burns can occur from the steam and hot coolant. Also, when draining engine coolant, keep in mind that cats and dogs are attracted to ethylene glycol antifreeze and could drink any that is left in an uncovered container or in puddles on the ground. This will prove fatal in sufficient quantities. Always drain coolant into a sealable container. Coolant should be reused unless it is contaminated or is several years old.

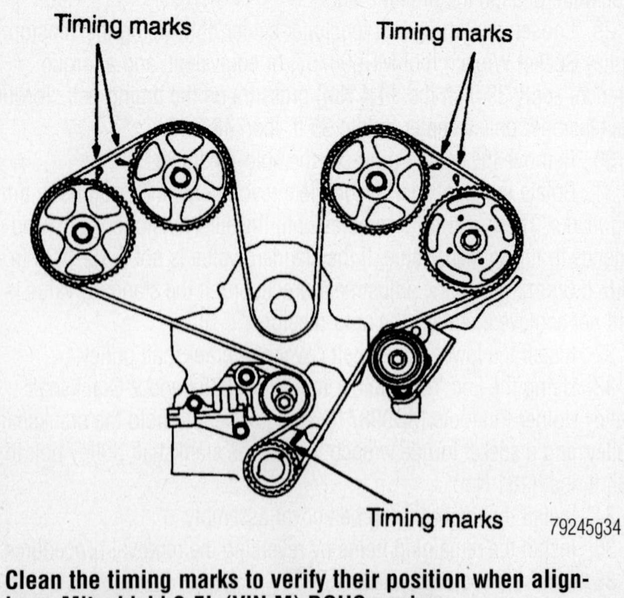

Clean the timing marks to verify their position when aligning—Mitsubishi 3.5L (VIN M) DOHC engine

3. Remove the upper radiator shroud.
4. Remove the fan and fan pulley.
5. Without disconnecting the lines, remove the power steering pump from its bracket and position it aside. Remove the pump brackets.
6. Remove the belt tensioner pulley bracket.
7. Without releasing the refrigerant, remove the air conditioning compressor from its bracket and position it aside. Remove the bracket.
8. Remove the cooling fan bracket.
9. On some vehicles, it may be necessary to remove the pulley from the crankshaft to access the lower cover bolts.
10. Remove the timing belt cover bolts and the upper and lower covers from the engine.
11. Remove the crankshaft position sensor connector.
12. Using SST MB990767-01 and MD998754 or their equivalents, remove the crankshaft pulley from the crankshaft.
13. Use a shop rag to clean the timing marks to assist in properly aligning the timing marks.
14. Loosen the center bolt on the tension pulley and remove the timing belt.

■ **If the same timing belt will be reused, mark the direction of timing belt's rotation, for installation in the same direction. Be sure engine is positioned so No. 1 cylinder is at the Top Dead Center (TDC) of its compression stroke and the sprockets timing marks are aligned with the engine's timing mark indicators.**

15. Remove the auto-tensioner, the tension pulley and the tension arm assembly.
16. Remove the sprockets by holding the hexagonal portion of the camshaft with a wrench while removing the sprocket bolt.

To install:

17. Install the crankshaft pulley and turn the crankshaft sprocket timing mark forward (clockwise) 3 teeth to move the piston slightly past No. 1 cylinder TDC.
18. If removed, install the camshaft sprockets and tighten the bolts to 64 ft. lbs. (88 Nm).
19. Align the timing mark of the left bank side camshaft sprocket.
20. Align the timing mark of the right bank side camshaft sprocket, and hold the sprocket with a wrench so that it doesn't turn.
21. Set the timing belt onto the water pump pulley.
22. Check that the camshaft sprocket timing mark of the left bank side is aligned and clamp the timing belt with double clips.
23. Set the timing belt onto the idler pulley.

**** WARNING**

If any binding is felt when adjusting the timing belt tension by turning the crankshaft, STOP turning the engine, because the pistons may be hitting the valves.

24. Turn the crankshaft 1 turn counterclockwise and set the timing belt onto the crankshaft sprocket.
25. Set the timing belt on the tension pulley.
26. Place the tension pulley pin hole so that it is towards the top. Press the tension pulley onto the timing belt, and then provisionally tighten the fixing bolt. Tighten the bolt to 35 ft. lbs. (48 Nm).
27. Slowly turn the crankshaft 2 full turns in the clockwise direction until the timing marks align. Remove the 4 double clips.
28. Install the crankshaft position sensor connector.
29. Install the upper and lower covers on the engine and secure them with the retaining screws. Be sure the packing is properly positioned in the inner grooves of the covers when installing.

30. If removed, install the crankshaft pulley and tighten the bolt to 110 ft. lbs. (150 Nm).
31. Install the air conditioning bracket and compressor on the engine. Install the belt tensioner.
32. Install the power steering pump into position. Install the fan pulley and fan.
33. Install the fan shroud on the radiator.
34. Refill the cooling system.
35. Connect the negative battery cable. Start the engine and check for fluid leaks.

2001–04 3.5L (6G74) AND 3.8L (6G75) MONTERO ENGINES

YEAR, MODEL & ENGINE APPLICATION

2001-03 Mitsubishi Montero	3.5L Code 6G74
2003-04 Montero	3.8L Code 6G75

REPLACEMENT INTERVALS

Mitsubishi recommends timing belt replacement every 60,000 miles.

LABOR TIMES

2001-04	
Replacement	4.4 hrs
Severe Service	4.9 hrs

8807MITg21

✳✳ CAUTION

These engines are an interference type. If the timing belt has failed, valve-to-piston damage may have occurred. A compression check should always be performed to check for engine damage.

1. Before servicing the vehicle, refer to the precautions in the beginning of this section.
2. Drain the engine coolant.
3. Remove or disconnect the following:
- Negative battery cable
- Skid plate and under cover
- Battery and battery tray
- Air cleaner
- Radiator shroud cover
- Drive belt
- Cooling fan and pulley
- Drive belt auto tensioner
- Accessory mount stay
- Power steering pump. Unbolt and position aside; do not disconnect the fluid lines.
- A/C compressor. Unbolt and position aside; do not disconnect the refrigerant lines.
- Compressor bracket
- Cooling fan bracket assembly
- Accessory mount assembly
- Timing belt upper cover
- Crankshaft pulley, using special tools MB991800 and MB991802
- Timing belt indicator bracket
- Timing belt lower cover
- Auto tensioner

4. Turn the crankshaft clockwise to align the timing marks and set the No. 1 cylinder at Top Dead Center (TDC). If you are reusing the timing belt, mark the flat side of the belt with an arrow showing the clockwise direction.
5. Loosen the center bolt of the tension pulley, and then remove the timing belt.
6.Remove or disconnect the following:
- Tension pulley
- Tensioner arm assembly
- Shaft
- Idler pulley

To install:

7. Install or connect the following:
- Idler pulley
- Shaft
- Tensioner arm assembly
- Tension pulley

8. Press the end of the auto-tensioner inward with 72–145 ft. lbs. (98–196 Nm) of force and measure the distance that the pushrod is pushed in. If the standard distance is not 0.04 in. (1mm), replace the auto-tensioner.
9. Position the auto-tensioner in a soft-jawed vise and SLOWLY compress the pushrod until the pushrod and housing holes align; then, install a setting pin to secure the auto-tensioner in the retracted position.
10. Align the camshaft and crankshaft TDC timing marks.
11. Install the timing belt (noting its rotational direction) so that there is no deflection between the sprockets and pulleys in the following manner:

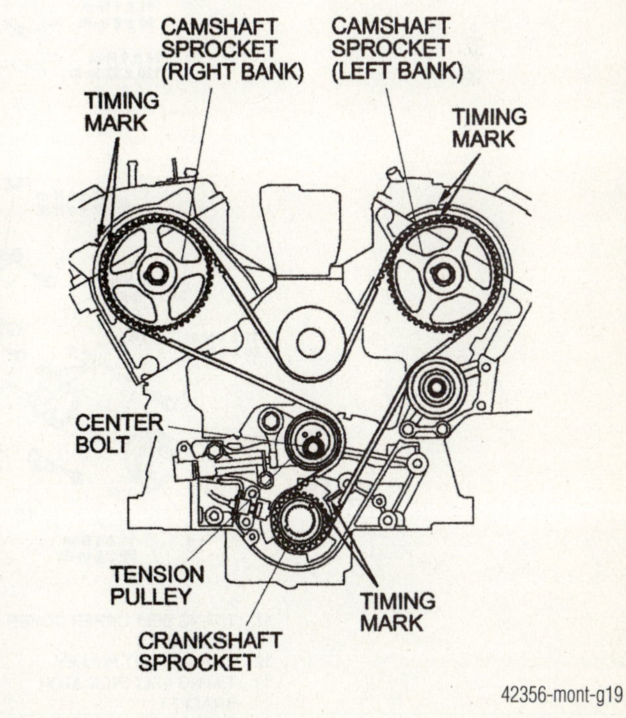

Installed view of the timing belt and alignment of the timing marks—Mitsubishi 3.5L (6G74) and 3.8L (6G75) engines

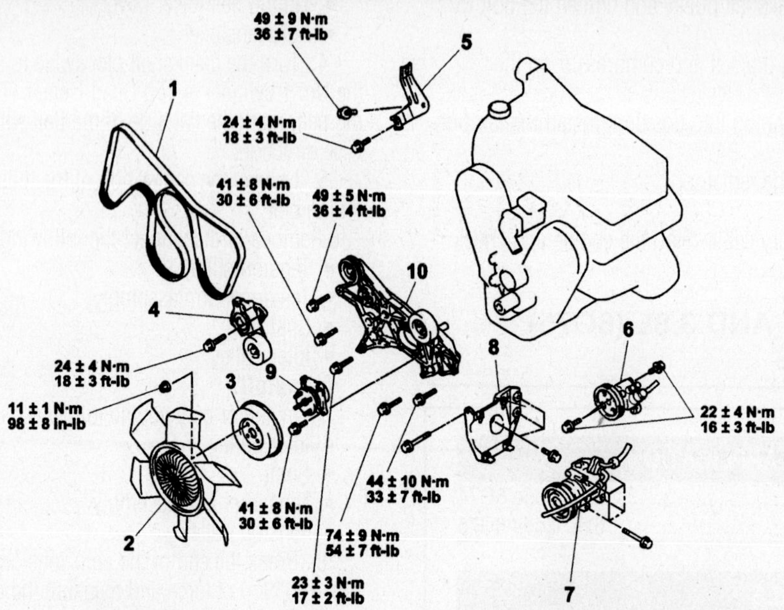

1. DRIVE BELT
2. COOLING FAN
3. COOLING FAN PULLEY
4. DRIVE BELT AUTO TENSIONER
5. ACCESSORY MOUNT STAY
6. POWER STEERING OIL PUMP ASSEMBLY
7. A/C COMPRESSOR ASSEMBLY
8. COMPRESSOR BRACKET
9. COOLING FAN BRACKET ASSEMBLY
10. ACCESSORY MOUNT ASSEMBLY

42356-mont-g17

Exploded view of the front engine components—Mitsubishi 3.5L (6G74) and 3.8L (6G75) engines

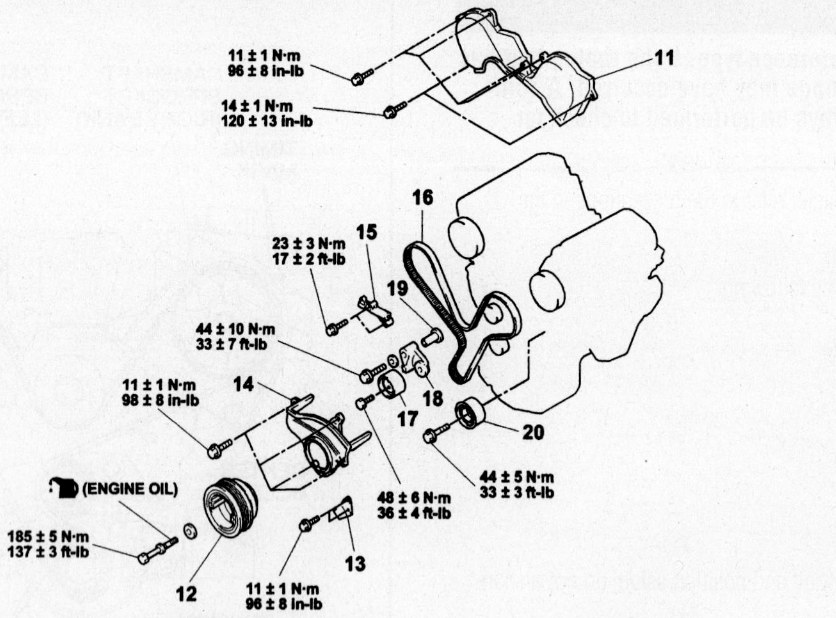

11. TIMING BELT UPPER COVER ASSEMBLY
12. CRANKSHAFT PULLEY
13. TIMING BELT INDICATOR BRACKET
14. TIMING BELT LOWER COVER ASSEMBLY
15. AUTO-TENSIONER
16. TIMING BELT
17. TENSION PULLEY
18. TENSIONER ARM ASSEMBLY
19. SHAFT
20. IDLER PULLEY

42356-mont-g18

Exploded view of the timing belt and related components—Mitsubishi 3.5L (6G74) and 3.8L (6G75) engines

- Crankshaft sprocket
- Idler pulley
- Left camshaft sprocket
- Water pump pulley
- Right camshaft sprocket
- Tension pulley

12. Turn the camshaft sprocket counterclockwise until the tension side of the timing belt is firmly stretched, then, recheck the timing marks.
13. Using the Tension Pulley Socket Wrench tool MD998767, or equivalent, push the tensioner pulley into the timing belt and secure the center bolt.
14. Using the Crankshaft Pulley Spacer tool MD998769, or equivalent, rotate the crankshaft 1/4 turn counterclockwise, then, turn it again clockwise to align the timing marks.
15. Loosen the timing belt tensioner center bolt. Using the Tension Pulley Socket Wrench tool MD998767, or equivalent, and a torque wrench, apply 39 inch lbs. (4.4 Nm) pressure on the timing belt. Torque the tensioner pulley center bolt to 35 ft. lbs. (48 Nm).
16. Remove the setting pin from the auto-tensioner.
17. Rotate the crankshaft 2 complete revolutions and realign the timing marks. Then, wait for 5 minutes until the auto-tensioner pushrod extends to its standard value. If the standard value is not 0.15–0.20 in. (3.8–5.0mm), repeat the adjustment procedure. If the standard value is still not achieved, replace the auto-tensioner.
18. Install the lower timing belt cover and crankshaft pulley.
19. Using the End Yoke Holder tool MD990767 and 2 Crankshaft Pulley Holder Pin tools MD998715, or equivalent to hold the crankshaft pulley, and a socket torque wrench, torque the crankshaft pulley bolt to 134 ft. lbs. (181 Nm).

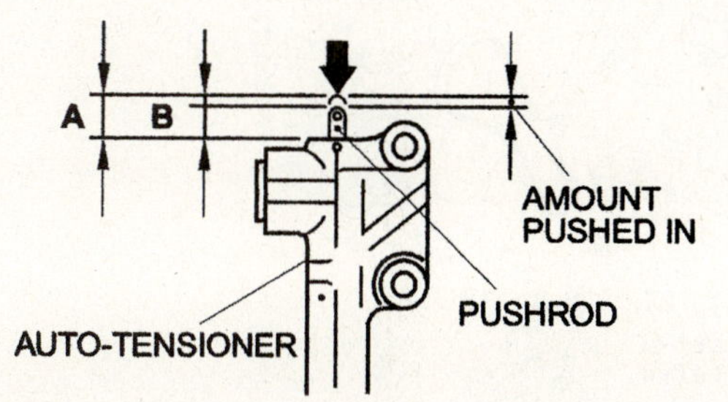

42356-mont-g20

Inspecting the auto-tensioner movement—Mitsubishi 3.5L (6G74) and 3.8L (6G75) engines

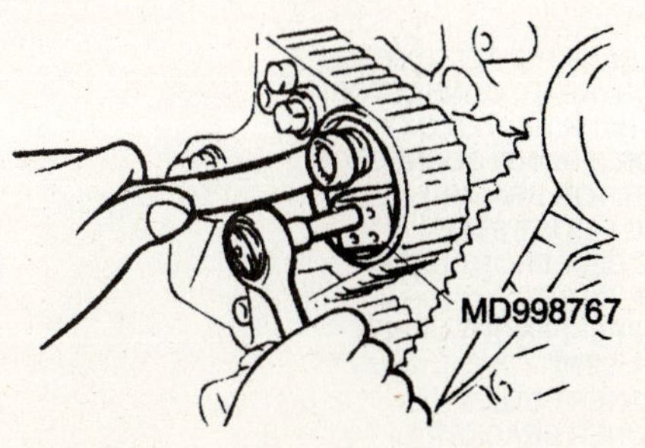

42356-mont-g21

Adjusting the timing belt tensioner pulley—Mitsubishi 3.5L (6G74) and 3.8L (6G75) engines

- Install or connect the following:
- Timing belt upper cover
- Accessory mount assembly
- Cooling fan bracket assembly
- Compressor bracket
- A/C compressor
- Power steering pump
- Accessory mount stay
- Drive belt auto tensioner
- Cooling fan pulley and fan
- Drive belt
- Radiator shroud cover
- Air cleaner
- Battery tray and battery
- Under cover and skid plate
- Negative battery cable

20. Refill the cooling system.

3.8L (6G75) ENDEAVOR ENGINES

YEAR, MODEL & ENGINE APPLICATION

2003-04 Mitsubishi Endeavor	3.8L Code 6G75

REPLACEMENT INTERVALS

Mitsubishi recommends timing belt replacement every 60,000 miles.

LABOR TIMES

2003-04	
Replacement	2.6 hrs
Severe Service	3.0 hrs

8807MITg22

✻✻ CAUTION

These engines are an interference type. If the timing belt has failed, valve-to-piston damage may have occurred. A compression check should always be performed to check for engine damage.

1. Before servicing the vehicle, refer to the precautions in the beginning of this section.
2. Drain the engine coolant.
3. Remove or disconnect the following:

- Negative battery cable
- Engine under cover and side cover
- Drive belts
- Crankshaft pulley, using special tools MB991800 and MB991802
- Manifold Differential Pressure (MDP) sensor connector
- Knock Sensor (KS) connector
- Crankshaft Position (CKP) sensor connector
- Control wiring and injector wiring harness combination connector
- Right bank oxygen sensor connectors
- Connector bracket
- Engine mount stay
- Timing belt upper covers
- Tensioner pulley
- Tensioner bracket

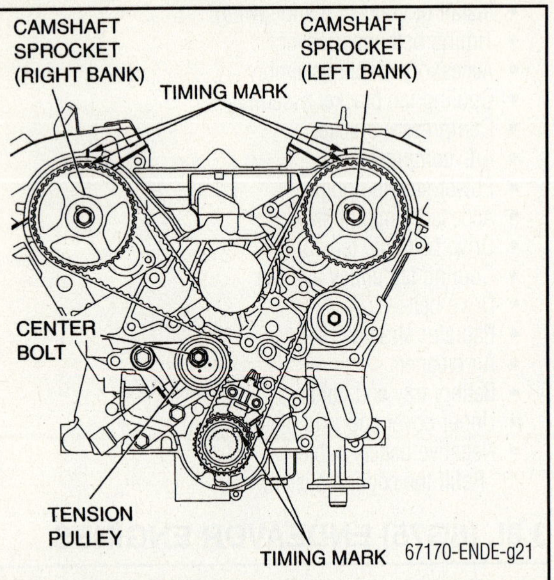

Installed view of the timing belt and alignment of the timing marks—Mitsubishi 3.8L (6G75) engine

- CKP sensor harness clap
- Lower timing belt cover
- Engine mount
- Engine support bracket

4. Turn the crankshaft clockwise to align the timing marks and set the No. 1 cylinder at Top Dead Center (TDC). If you are reusing the timing belt, mark the flat side of the belt with an arrow showing the clockwise direction.

5. Loosen the center bolt of the tension pulley, and then remove the timing belt.

6. Remove the tension pulley, tensioner arm assembly, shaft and idler pulley.

To install:

7. Install or connect the following:

- Idler pulley
- Shaft
- Tensioner arm assembly
- Tension pulley

8. Press the end of the auto-tensioner inward with 72–145 ft. lbs. (98–196 Nm) of force and measure the distance that the pushrod is pushed in. If the standard distance is not 0.04 in. (1mm), replace the auto-tensioner.

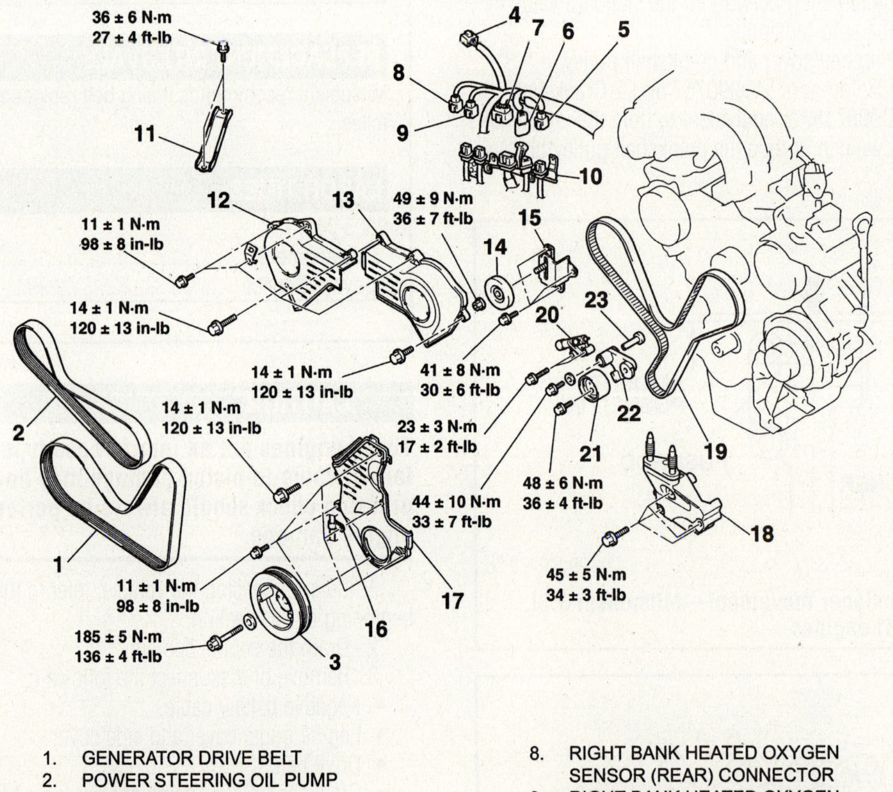

1. GENERATOR DRIVE BELT
2. POWER STEERING OIL PUMP DRIVE BELT
3. CRANKSHAFT PULLEY
4. MANIFOLD DIFFERENTIAL PRESSURE SENSOR CONNECTOR
5. KNOCK SENSOR CONNECTOR
6. CRANKSHAFT POSITION SENSOR CONNECTOR
7. CONTROL WIRING HARNESS AND INJECTOR WIRING HARNESS COMBINATION CONNECTOR
8. RIGHT BANK HEATED OXYGEN SENSOR (REAR) CONNECTOR
9. RIGHT BANK HEATED OXYGEN SENSOR (FRONT) CONNECTOR
10. CONNECTOR BRACKET
11. ENGINE MOUNT STAY
12. TIMING BELT FRONT UPPER COVER, RIGHT
13. TIMING BELT FRONT UPPER COVER, LEFT
14. TENSIONER PULLEY
15. TENSIONER BRACKET
16. CRANKSHAFT POSITION SENSOR HARNESS CLAMP
17. TIMING BELT LOWER COVER

67170-ENDE-g22

Exploded view of the timing belt and related components—Mitsubishi3.8L (6G75) engine

9. Position the auto-tensioner in a soft-jawed vise and SLOWLY compress the pushrod until the pushrod and housing holes align; then, install a setting pin to secure the auto-tensioner in the retracted position.

10. Align the camshaft and crankshaft TDC timing marks.

11. Install the auto tensioner.

12. Install the timing belt (noting its rotational direction) so that there is no deflection between the sprockets and pulleys in the following manner:

- Crankshaft sprocket
- Idler pulley
- Left camshaft sprocket
- Water pump pulley
- Right camshaft sprocket
- Tension pulley

13. Turn the camshaft sprocket counterclockwise until the tension side of the timing belt is firmly stretched, then, recheck the timing marks.

14. Using the Tension Pulley Socket Wrench tool MD998767, or equivalent, push the tensioner pulley into the timing belt and secure the center bolt.

15. Using the Crankshaft Pulley Spacer tool MD998769, or equivalent, rotate the crankshaft 1/4 turn counterclockwise, then, turn it again clockwise to align the timing marks.

16. Loosen the timing belt tensioner center bolt. Using the Tension Pulley Socket Wrench tool MD998767, or equivalent, and a torque wrench, apply 39 inch lbs. (4.4 Nm) pressure on the timing belt. Torque the tensioner pulley center bolt to 35 ft. lbs. (48 Nm).

17. Remove the setting pin from the auto-tensioner.

18. Rotate the crankshaft 2 complete revolutions and realign the timing marks. Then, wait for 5 minutes until the auto-tensioner pushrod extends to its standard value. If the standard value is not 0.19–0.24 in. (4.8–6.0mm), repeat the adjustment procedure. If the standard value is still not achieved, replace the auto-tensioner.

19. Install the lower timing belt cover and crankshaft pulley.

20. Using the End Yoke Holder tool MD990767 and 2 Crankshaft Pulley Holder Pin tools MD998715, or equivalent to hold the crankshaft pulley, and a socket torque wrench, torque the crankshaft pulley bolt to 136 ft. lbs. (185 Nm).

- Install or connect the following:
- Engine support bracket
- Engine mount
- Lower timing belt cover
- CKP sensor harness clap
- Tensioner bracket
- Tensioner pulley
- Timing belt upper covers
- Engine mount stay
- Connector bracket
- Right bank oxygen sensor connectors
- Control wiring and injector wiring harness combination connector
- Crankshaft Position (CKP) sensor connector
- Knock Sensor (KS) connector
- Manifold Differential Pressure (MDP) sensor connector
- Crankshaft pulley, using special tools MB991800 and MB991802
- Drive belts
- Engine under cover and side cover
- Negative battery cable

21. Refill the cooling system.

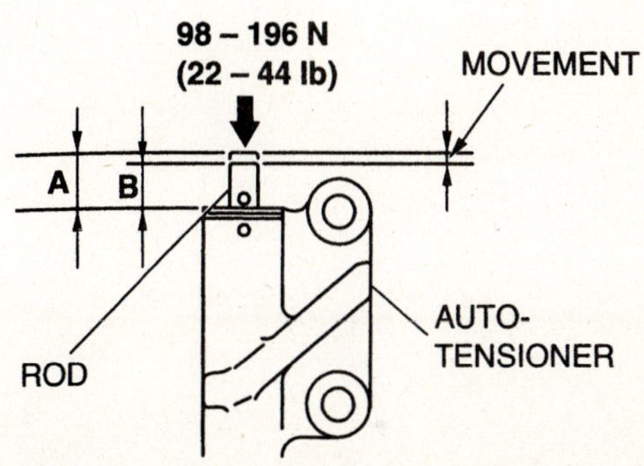

67170-ENDE-g23

Inspecting the auto-tensioner movement—Mitsubishi 3.8L (6G75) engine

Nissan/Datsun

1.5L (E15, E15ET) ENGINES

YEAR, MODEL & ENGINE APPLICATION

1982 Nissan 310, 1982-83 Sentra	1.5L Code E15
1984-85 Pulsar Turbo	1.5L Code E15ET

REPLACEMENT INTERVALS

Nissan does not provide a replacement interval. Standard industry practice is to replace timing belts every 60,000 miles.

LABOR TIMES

1982-85	
Replacement	2.2 hrs
Severe Service	2.7 hrs

8807NISc01

✱✱ CAUTION

These engines are an interference type. If the timing belt has failed, valve-to-piston damage may have occurred. A compression check should always be performed to check for engine damage.

1. Disconnect the negative battery cable.
2. Remove the accessory drive belts.
3. Support the engine with a jack and remove the right motor mount.
4. Remove the right engine splash guards.
5. Remove the water pump pulley.
6. Remove the crankshaft pulley.
7. Remove the timing covers.
8. Rotate the crankshaft so that the timing marks on the camshaft and crankshaft pulleys align with the timing marks on the rear timing cover.
9. Loosen the timing belt tensioner locknut.
10. Turn the tensioner away from the timing belt and tighten the locknut.
11. Remove the timing belt.

To install:

12. Ensure that the timing marks are aligned and install the timing belt.
13. Loosen the tensioner locknut.
14. Carefully turn the camshaft pulley clockwise 2 teeth and tighten the tensioner locknut.
15. Rotate the crankshaft 2 complete turns and check that the timing marks align.
16. Install the timing covers.
17. Install the crankshaft pulley and tighten the bolt to 83–108 ft. lbs. (113–147 Nm).
18. Install the water pump pulley.
19. Install the accessory drive belts.
20. Install the right motor mount and the splash guards. Remove the engine support jack.
21. Connect the negative battery cable.

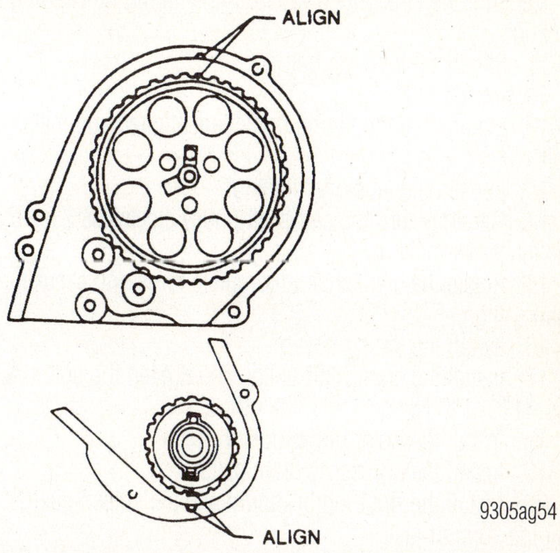

Timing mark alignment—Nissan/Datsun 1.5L (E15, E15ET) engine

1.6L (E16, E16S) ENGINES

YEAR, MODEL & ENGINE APPLICATION

1983 Nissan Pulsar, 1983-87 Sentra	1.6L Code E16
1984-85 Pulsar	1.6L Code E16S

REPLACEMENT INTERVALS

Nissan does not provide a replacement interval for 1983-85 models. Standard industry practice is to replace timing belts every 60,000 miles. Nissan recommends timing belt replacement every 60,000 miles on 1986-87 models.

LABOR TIMES

1983-87	
Replacement	2.2 hrs
Severe Service	2.7 hrs

8807NISg02

✱✱ CAUTION

These engines are an interference type. If the timing belt has failed, valve-to-piston damage may have occurred. A compression check should always be performed to check for engine damage.

1. Disconnect the negative battery cable.
2. Remove the accessory drive belts.
3. Support the engine with a jack and remove the right motor mount.
4. Remove the right engine splash guards.
5. Remove the water pump pulley.
6. Remove the crankshaft pulley.
7. Remove the timing covers.
8. Rotate the crankshaft so that the timing marks on the camshaft and crankshaft pulleys align with the timing marks on the rear timing cover.
9. Loosen the timing belt tensioner locknut.

10. Turn the tensioner away from the timing belt and tighten the locknut.
11. Remove the timing belt.

To install:

12. Ensure that the timing marks are aligned and install the timing belt.
13. Loosen the tensioner locknut.
14. Carefully turn the camshaft pulley clockwise 2 teeth and tighten the tensioner locknut.
15. Rotate the crankshaft 2 complete turns and check that the timing marks align.
16. Install the timing covers.
17. Install the crankshaft pulley and tighten the bolt to 83–108 ft. lbs. (113–147 Nm).
18. Install the water pump pulley.
19. Install the accessory drive belts.
20. Install the right motor mount and the splash guards. Remove the engine support jack.
21. Connect the negative battery cable.

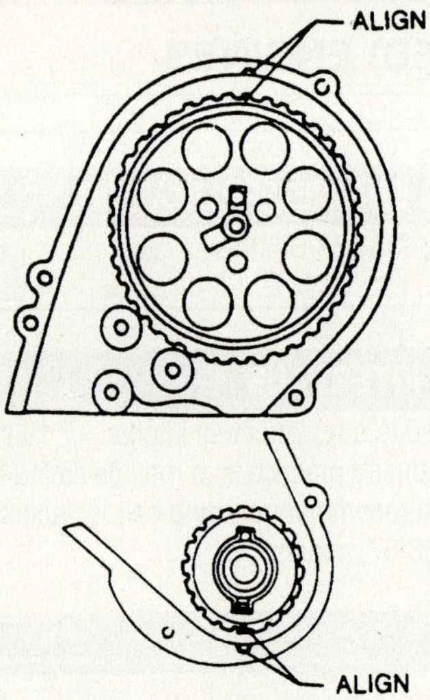

Timing mark alignment—Nissan/Datsun 1.6L (E16, E16S) engine

1.6L (CA16DE) ENGINES

YEAR, MODEL & ENGINE APPLICATION	
1987-89 Nissan Pulsar NX	1.6L Code CA16DE

REPLACEMENT INTERVALS

Nissan recommends timing belt replacement every 60,000 miles.

LABOR TIMES	
1987-89	
Replacement	4.5 hrs
Severe Service	5.0 hrs

8807NISg03

✻✻ CAUTION

These engines are an interference type. If the timing belt has failed, valve-to-piston damage may have occurred. A compression check should always be performed to check for engine damage.

1. Disconnect the negative battery cable.
2. Drain the cooling system.
3. Remove the upper radiator hose.
4. Remove the right engine undercover.
5. Remove the right front inner fender access cover.
6. Remove the accessory drive belts.
7. Remove the water pump pulley.
8. Rotate the crankshaft to set the No. 1 piston at Top Dead Center (TDC) of the compression stroke.
9. Matchmark the Camshaft Position (CMP) sensor to the upper timing cover and remove the sensor.

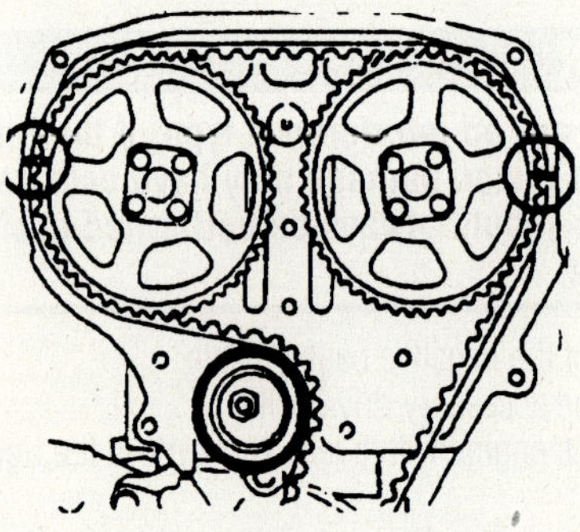

Camshaft timing marks—Nissan/Datsun 1.6L (CA16DE) engine

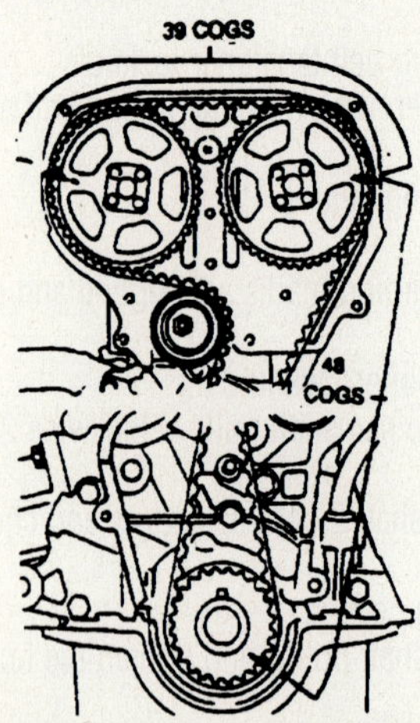

Valve timing alignment marks—Nissan/Datsun 1.6L (CA16DE) engine

10. Support the engine with a jack and remove the right engine mount and bracket.
11. Remove the upper timing cover. Check that the camshaft timing marks are aligned.
12. Remove the crankshaft pulley.
13. Remove the lower timing cover.
14. Loosen the tensioner pulley locknut. Rotate the tensioner clockwise to loosen the timing belt. Tighten the tensioner locknut
15. Remove the timing belt.

To install:

16. Ensure that the timing marks are aligned and install the timing belt. When properly installed, there will be 39 cogs between the intake and exhaust camshaft pulley timing marks, and 48 cogs between the exhaust camshaft pulley timing mark and the crankshaft timing mark.
17. Loosen the tensioner locknut.
18. Rotate the crankshaft 2 complete turns and check that the timing marks align.
19. Tighten the tensioner locknut to 16–22 ft. lbs. (22–30 Nm).
20. Install the upper and lower timing covers.
21. Install the crankshaft pulley and tighten the bolt to 105–112 ft. lbs. (143–152 Nm).
22. Install the water pump pulley.
23. Install the engine mount and bracket. Remove the engine support jack.
24. Align the matchmarks and install the CMP sensor.
25. Install the accessory drive belts.
26. Install the right engine under cover and inner fender access cover.
27. Install the upper radiator hose and refill the cooling system.
28. Connect the negative battery cable.

1.7L (CD17) DIESEL ENGINES

YEAR, MODEL & ENGINE APPLICATION

1983-87 Nissan Sentra .... 1.7L Diesel Code CD17

REPLACEMENT INTERVALS

Nissan recommends timing belt replacement every 60,000 miles.

LABOR TIMES

1983-87

Replacement	3.8 hrs
Severe Service	4.0 hrs

8807NISG04

Valve Timing Belt

✱✱ CAUTION

These engines are an interference type. If the timing belt has failed, valve-to-piston damage may have occurred. A compression check should always be performed to check for engine damage.

1. Disconnect the negative battery cable.
2. Remove the accessory drive belts.
3. Support the engine with a jack and remove the right motor mount.
4. Remove the right engine splash guards.
5. Remove the crankshaft damper pulley.
6. Remove the timing covers.
7. Rotate the crankshaft so that the timing marks on the camshaft pulley and the crankshaft timing pulley align with the marks on the inner timing cover.
8. Loosen the tensioner locknut and set the tensioner in the free position. Tighten the locknut.
9. Remove the crankshaft timing pulley and the timing belt.

To install:

10. Install the crankshaft timing pulley and timing belt.
11. Ensure that the timing marks align and loosen the tensioner locknut.
12. Carefully turn the camshaft pulley 2 teeth and tighten the tensioner locknut.
13. Rotate the crankshaft 2 complete turns and check that the timing marks align.
14. Install the timing covers.
15. Install the crankshaft damper pulley and tighten the bolt to 90–98 ft. lbs. (122–133 Nm).
16. Install the accessory drive belts.
17. Install the right splash guards and motor mount. Remove the engine support jack.
18. Connect the negative battery cable.

Injection Pump Timing Belt

1. Disconnect the negative battery cable.
2. Rotate the crankshaft to set the No. 1 cylinder at Top Dead Center (TDC) of the compression stroke. The crankshaft damper pulley mark should align with the front cover pointer.
3. Remove the rear outer timing cover.
4. Loosen the belt tensioner locknut and set the tensioner to the free position. Tighten the locknut.
5. Remove the injection pump timing belt.

To install:

6. Prepare the belt for installation by placing matchmarks 23 cogs apart. Align the matchmarks with the timing marks on the camshaft and injection pump pulleys and install the belt.
7. Rotate the crankshaft 2 complete turns and tighten the injection pump timing belt tensioner locknut.
8. Install the rear outer timing cover and connect the negative battery cable.

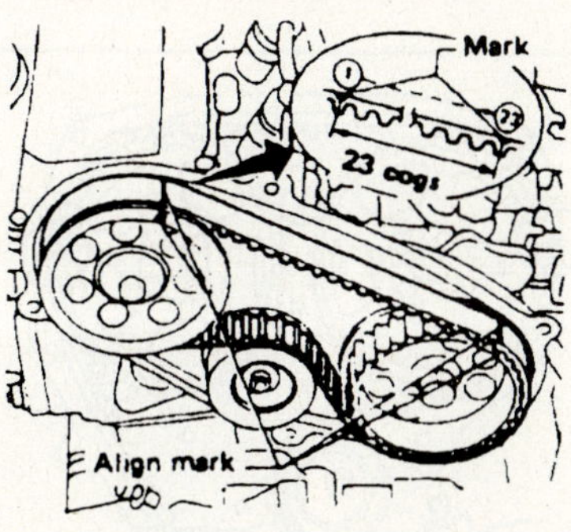

9305ag55

Injection timing marks—Nissan/Datsun 1.7L (CD17) diesel engine

1.8L (CA18DE) ENGINES

YEAR, MODEL & ENGINE APPLICATION

1988-89 Nissan Pulsar NX .... 1.8L Code CA18DE

REPLACEMENT INTERVALS

Nissan recommends timing belt replacement every 60,000 miles.

LABOR TIMES

1988-89	
Replacement	4.5 hrs
Severe Service	5.0 hrs

8807NISg05

✱✱ CAUTION

These engines are an interference type. If the timing belt has failed, valve-to-piston damage may have occurred. A compression check should always be performed to check for engine damage.

1. Disconnect the negative battery cable.
2. Drain the cooling system.
3. Remove the upper radiator hose.
4. Remove the right engine undercover.
5. Remove the right front inner fender access cover.
6. Remove the accessory drive belts.
7. Remove the water pump pulley.
8. Rotate the crankshaft to set the No. 1 piston at Top Dead Center (TDC) of the compression stroke.
9. Matchmark the Camshaft Position (CMP) sensor to the upper timing cover and remove the sensor.
10. Support the engine with a jack and remove the right engine mount and bracket.
11. Remove the upper timing cover. Check that the camshaft timing marks are aligned.
12. Remove the crankshaft pulley.
13. Remove the lower timing cover.
14. Loosen the tensioner pulley locknut. Rotate the tensioner clockwise to loosen the timing belt. Tighten the tensioner locknut
15. Remove the timing belt.

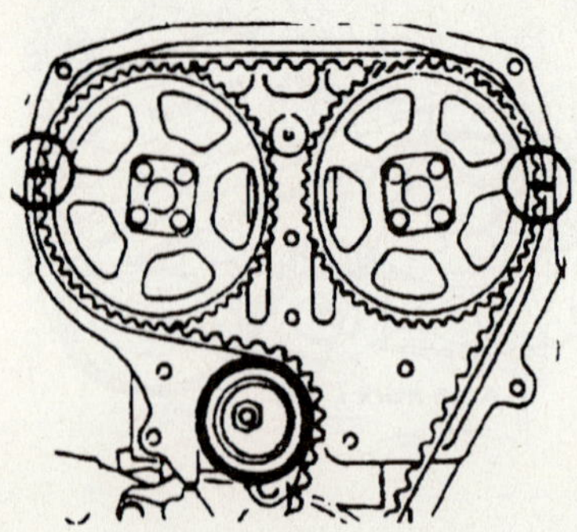

9305ag56

Camshaft timing marks—Nissan/Datsun 1.8L (CA18DE) engine

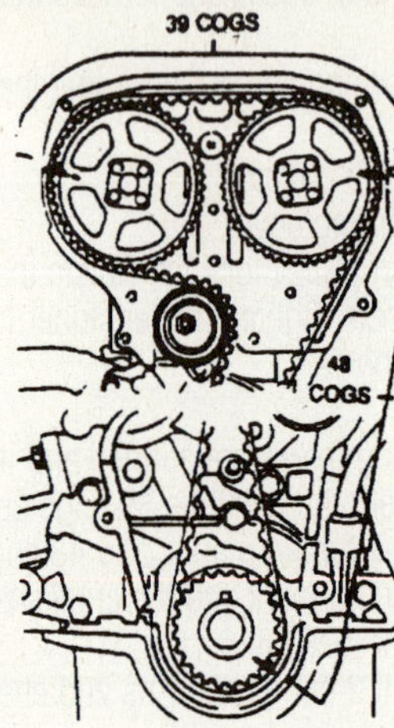

9305ag57

Valve timing alignment marks—Nissan/Datsun 1.8L (CA18DE) engine

To install:

16. Ensure that the timing marks are aligned and install the timing belt. When properly installed, there will be 39 cogs between the intake and exhaust camshaft pulley timing marks, and 48 cogs between the exhaust camshaft pulley timing mark and the crankshaft timing mark.
17. Loosen the tensioner locknut.
18. Rotate the crankshaft 2 complete turns and check that the timing marks align.
19. Tighten the tensioner locknut to 16–22 ft. lbs. (22–30 Nm).
20. Install the upper and lower timing covers.
21. Install the crankshaft pulley and tighten the bolt to 105–112 ft. lbs. (143–152 Nm).
22. Install the water pump pulley.
23. Install the engine mount and bracket. Remove the engine support jack.
24. Align the matchmarks and install the CMP sensor.
25. Install the accessory drive belts.
26. Install the right engine under cover and inner fender access cover.
27. Install the upper radiator hose and refill the cooling system.
28. Connect the negative battery cable.

1.8L (CA18ET) ENGINES

YEAR, MODEL & ENGINE APPLICATION

1984-87 200SX Turbo .... 1.8L Code CA18ET

REPLACEMENT INTERVALS

Nissan recommends timing belt replacement every 60,000 miles.

LABOR TIMES

1984-87	
Replacement	2.6 hrs
Severe Service	3.1 hrs

8807NISg06

✱✱ CAUTION

These engines are an interference type. If the timing belt has failed, valve-to-piston damage may have occurred. A compression check should always be performed to check for engine damage.

1. Disconnect the negative battery cable.
2. Remove the accessory drive belts.
3. Support the engine with a jack and remove the right motor mount.
4. Remove the right engine splash guards.
5. Remove the water pump pulley.
6. Remove the crankshaft pulley.
7. Remove the timing covers.
8. Rotate the crankshaft so that the timing marks on the camshaft and crankshaft pulleys align with the timing marks on the rear timing cover.
9. Loosen the timing belt tensioner locknut.
10. Turn the tensioner away from the timing belt and tighten the locknut.
11. Remove the timing belt.

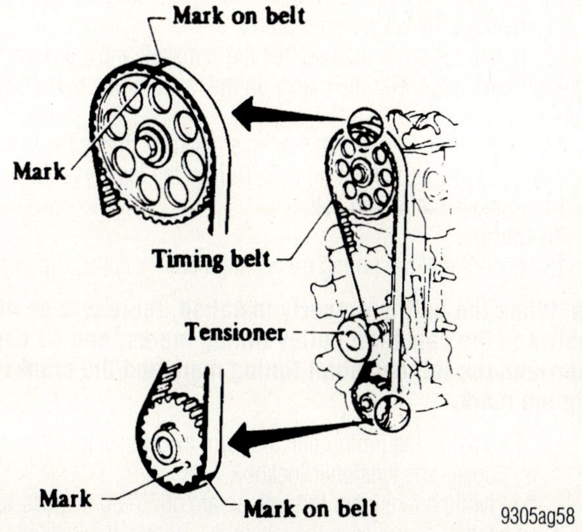

View of the timing marks—Nissan/Datsun 1.8L (CA18ET) engine

To install:

12. Ensure that the timing marks are aligned and install the timing belt.
13. Loosen the tensioner locknut.
14. Rotate the crankshaft 2 complete turns and check that the timing marks align. Tighten the tensioner locknut.
15. Install the timing covers.
16. Install the crankshaft pulley and tighten the bolt to 83–108 ft. lbs. (113–147 Nm).
17. Install the water pump pulley.
18. Install the accessory drive belts.
19. Install the right motor mount and the splash guards. Remove the engine support jack.
20. Connect the negative battery cable.

2.0L (CA20, CA20E) ENGINES

YEAR, MODEL & ENGINE APPLICATION	
1984-88 Nissan 200SX, 1982-89 Stanza	2.0L Code CA20/20E

REPLACEMENT INTERVALS
Nissan recommends timing belt replacement every 60,000 miles.

LABOR TIMES	
200SX	
Replacement	2.6 hrs
Severe Service	3.1 hrs
Stanza	
Replacement	4.9 hrs
Severe Service	5.4 hrs

8807NISg07

✱✱ CAUTION

These engines are an interference type. If the timing belt has failed, valve-to-piston damage may have occurred. A compression check should always be performed to check for engine damage.

1. Disconnect the negative battery cable.
2. Remove the accessory drive belts.
3. Support the engine with a jack and remove the right motor mount.
4. Remove the right engine splash shields.
5. Remove the water pump pulley.
6. Remove the crankshaft pulley.
7. Remove the timing covers.
8. Rotate the crankshaft so that the timing marks on the camshaft and crankshaft pulleys align with the timing marks on the rear timing cover.
9. Loosen the timing belt tensioner locknut.

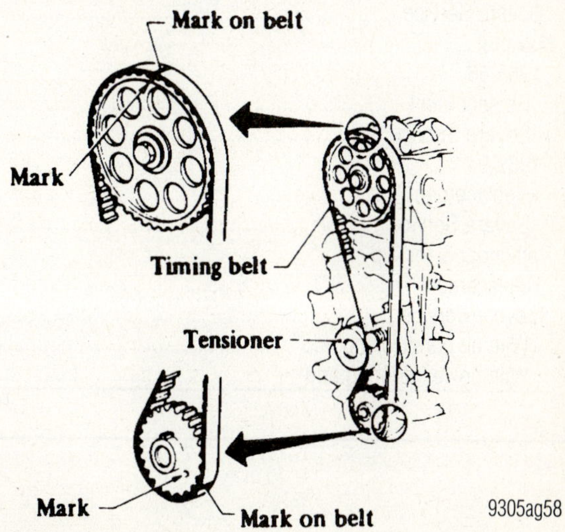

View of the timing marks—Nissan/Datsun 2.0L (CA20, CA20E) engines.

10. Turn the tensioner away from the timing belt and tighten the locknut.
11. Remove the timing belt.

To install:

12. Ensure that the timing marks are aligned and install the timing belt.
13. Loosen the tensioner locknut.
14. Rotate the crankshaft 2 complete turns and check that the timing marks align. Tighten the tensioner locknut.
15. Install the timing covers.
16. Install the crankshaft pulley and tighten the bolt to 83–108 ft. lbs. (113–147 Nm).
17. Install the water pump pulley.
18. Install the accessory drive belts.
19. Install the right motor mount and the splash shields. Remove the engine support jack.
20. Connect the negative battery cable.

3.0L (VG30E, VG30ET) ENGINES

YEAR, MODEL & ENGINE APPLICATION

1986-88 Nissan 200SX, 1984-89 300XZ, 1985-96 Maxima, 1986-95 Pathfinder, Pickup	3.0L Code VG30E
1984-89 300ZXT	3.0L Code VG30ET

REPLACEMENT INTERVALS

Nissan does not provide a replacement interval on 1984-85 models. Standard industry practice is to replace timing belts every 60,000 miles. On 1986-93 models, Nissan recommends timing belt replacement every 60,000 miles. On 1994-97 models, Nissan recommends timing belt replacement every 105,000 miles.

LABOR TIMES

200SX	
Replacement	5.8 hrs
Severe Service	6.3 hrs
300ZX/ZXT	
Replacement	7.0 hrs
Severe Service	7.5 hrs
Maxima	
1985-88	
Replacement	5.3 hrs
Severe Service	5.8 hrs
1989-94	
Replacement	3.8 hrs
Severe Service	4.3 hrs
Pathfinder & Pickup	
Replacement	3.7 hrs
Severe Service	4.2 hrs
With air conditioning add	0.3
With power steering add	0.2

8807NISg08

All models except 1986–89 Pathfinder and Pickup

✻✻ CAUTION

These engines are an interference type. If the timing belt has failed, valve-to-piston damage may have occurred. A compression check should always be performed to check for engine damage.

1. Disconnect the negative battery cable.
2. Drain the cooling system and remove the engine undercover.
3. Remove the radiator hoses and accessory drive belts.
4. On 200SX and 300ZX models, remove the radiator shroud, fan and pulleys.
5. For Maxima models, perform the following procedure:
   a. Raise and safely support the vehicle.
   b. Remove the right front wheel.
   c. Remove the right front inner fender access cover.
   d. Remove the cooling fan duct, coolant reservoir and ASCD actuator.
6. On truck models remove the radiator, fan and pulleys.
7. On all models, remove the spark plugs.
8. Place the No. 1 cylinder at TDC of the compression stroke.
9. Remove the A/C idler bracket and the crankshaft pulley.
10. Remove the upper and lower timing belt covers.
11. Remove the water pump pulley.
12. Rotate the crankshaft so that the camshaft and crankshaft pulley timing marks align with the marks on the rear timing cover and oil pump.
13. Loosen the timing belt tensioner locknut. Rotate the tensioner clockwise to loosen the timing belt. Tighten the tensioner locknut.
14. Remove the timing belt.

To install:

15. Ensure that the timing marks align and install the timing belt.

■ **When the belt is properly installed, there will be 40 cogs between the camshaft pulley timing marks, and 43 cogs between the left camshaft timing mark and the crankshaft timing mark.**

16. To tension the timing belt, perform the following:
   a. Loosen the tensioner lockbolt.
   b. Pivot the tensioner through an arc of 70–80 degrees with an Allen wrench and release the tensioner. Tighten the locknut.
   c. Rotate the crankshaft 2 complete turns and check that the timing marks align.
   d. Apply 22 lbs. (10 kg) of force to the timing belt between the camshaft pulleys and loosen the tensioner locknut.
   e. Place a 0.0138 inch x 1/2 inch thickness gauge between the tensioner pulley and the timing belt.
   f. Rotate the crankshaft slowly clockwise until the gauge is drawn into position.
   g. Hold the tensioner from turning with an Allen wrench and tighten the locknut.
   h. Rotate the crankshaft 2 complete turns and check that the timing marks align.
   i. Apply 22 lbs. (10 kg) of force to the timing belt between the camshaft pulleys and measure belt deflection. Deflection should be 0.512–0.571 inches.

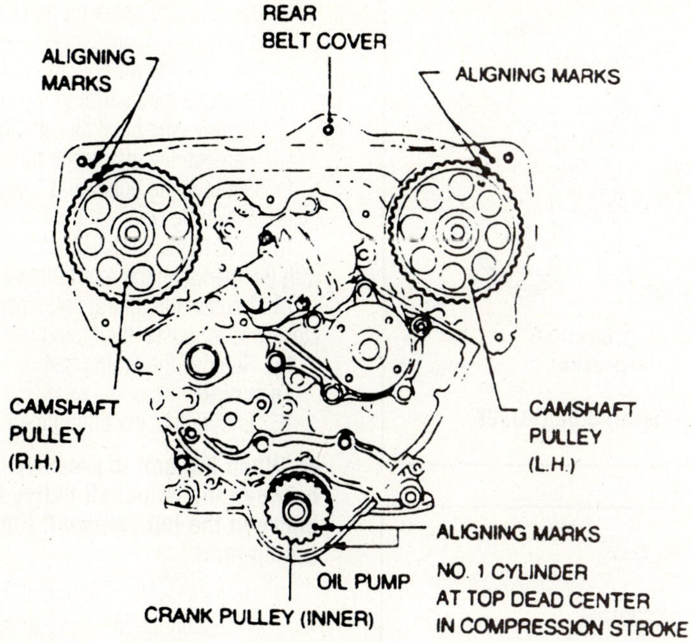

Crankshaft and camshaft marks—Nissan 3.0L (VG30E, VG30ET) engines

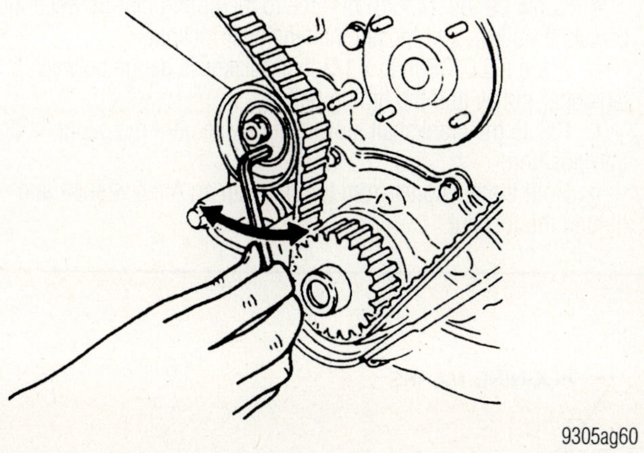

Tensioning the timing belt—Nissan 3.0L (VG30E, VG30ET) engines

17. Install the timing covers.
18. Install the crankshaft pulley and tighten the bolt to 90–98 ft. lbs. (122–136 Nm).
19. Install the water pump pulley.
20. Install the spark plugs.
21. On truck models install the radiator, fan and pulleys.
22. For Maxima models, perform the following procedure:
    a. Install the cooling fan duct, coolant reservoir and ASCD actuator.
    b. Install the right front inner fender access cover.
    c. Install the right front wheel.
    d. Lower the vehicle.
23. On 200SX and 300ZX models, install the radiator shroud, fan and pulleys.
24. On all models, install the radiator hoses and accessory drive belts.

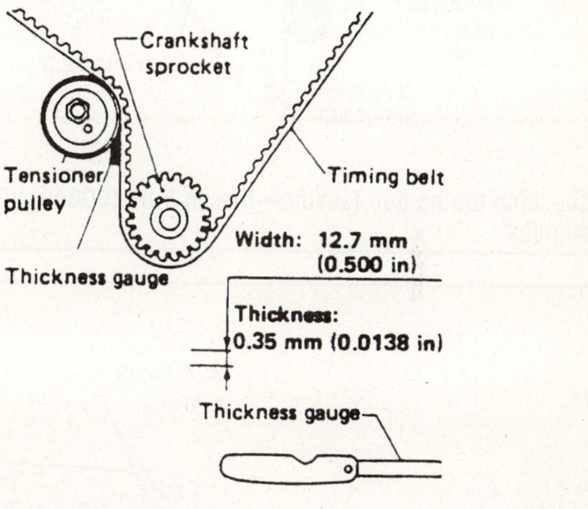

Using a thickness gauge to set timing belt tension—Nissan 3.0L (VG30E, VG30ET) engines

25. Refill the cooling system
26. Connect the negative battery cable.

1986–89 Pathfinder and Pickup

✳✳ CAUTION

These engines are an interference type. If the timing belt has failed, valve-to-piston damage may have occurred. A compression check should always be performed to check for engine damage.

1. Disconnect the negative battery cable.
2. Remove the engine splash shield.
3. Remove the radiator fan shroud and pulleys.

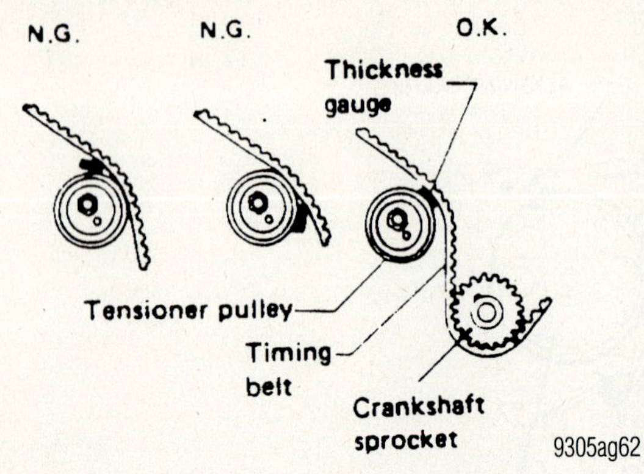

Positioning the thickness gauge—Nissan 3.0L (VG30E, VG30ET) engines

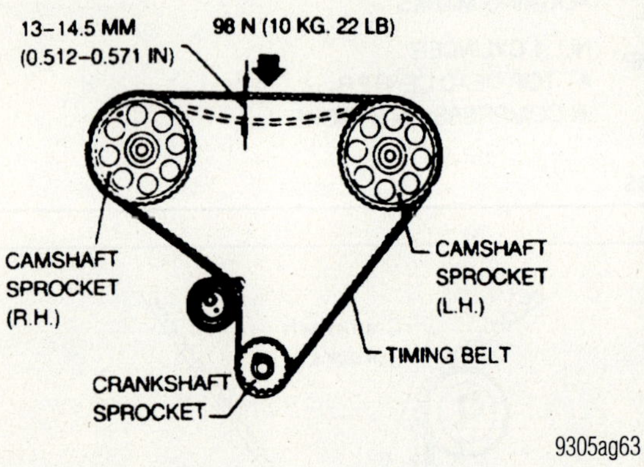

Checking timing belt tension—Nissan 3.0L (VG30E, VG30ET) engines

4. Drain the cooling system and remove the radiator and hoses.
5. Remove the spark plugs.
6. Remove the accessory drive belts.
7. Remove the water pump pulley.
8. Remove the distributor cover.
9. Remove the fresh air tube from the valve cover.
10. Remove the crankshaft pulley bolt and crankshaft pulley.
11. Remove the upper and lower timing belt covers.
12. Place the No. 1 cylinder at TDC of the compression stroke.
13. Ensure the camshaft and crankshaft pulley timing marks align with the marks on the rear timing cover and oil pump.
14. Loosen the timing belt tensioner locknut. Rotate the tensioner clockwise to loosen the timing belt. Tighten the tensioner locknut.
15. Remove the timing belt.

To install:

16. Ensure that the timing marks align and install the timing belt.

■ When the belt is properly installed, there will be 40 cogs between the camshaft pulley timing marks, and 43 cogs between the left camshaft timing mark and the crankshaft timing mark.

17. To tension the timing belt, perform the following:
    a. Loosen the tensioner lockbolt.
    b. Pivot the tensioner through an arc of 70–80 degrees with an Allen wrench and release the tensioner. Tighten the locknut.
    c. Rotate the crankshaft 2 complete turns and check that the timing marks align.
    d. Apply 22 lbs. (10 kg) of force to the timing belt between the camshaft pulleys and loosen the tensioner locknut.
    e. Place a 0.0138 inch x 1/2 inch thickness gauge between the tensioner pulley and the timing belt.
    f. Rotate the crankshaft slowly clockwise until the gauge is drawn into position.
    g. Hold the tensioner from turning with an Allen wrench and tighten the locknut.

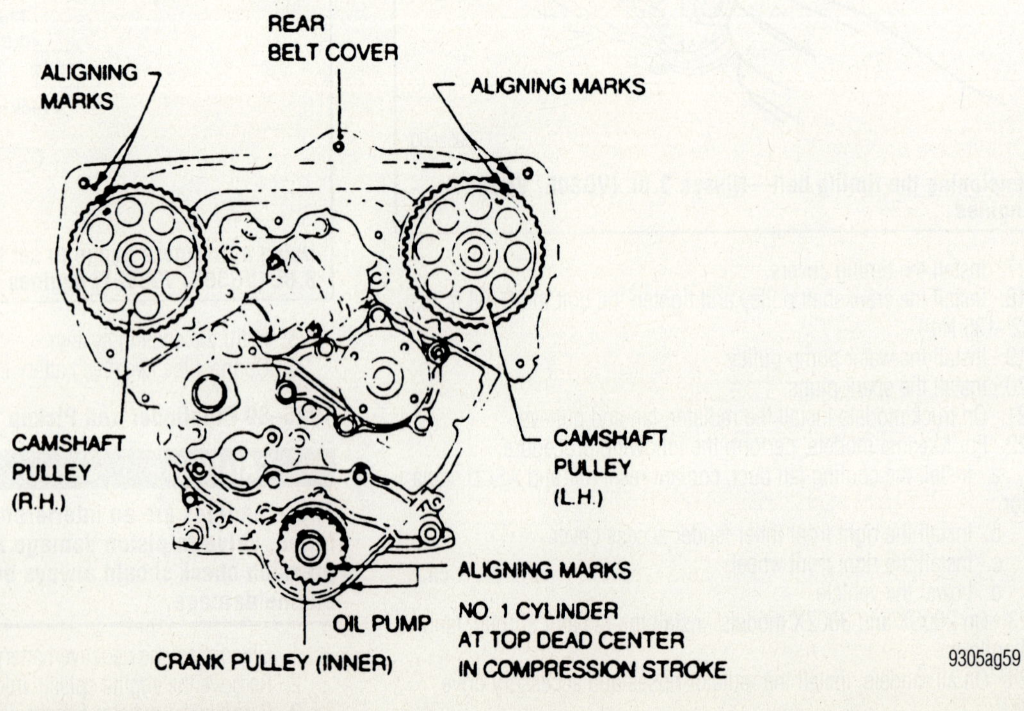

Crankshaft and camshaft marks—Nissan 3.0L (VG30E) engines

h. Rotate the crankshaft 2 complete turns and check that the timing marks align.

i. Apply 22 lbs. (10 kg) of force to the timing belt between the camshaft pulleys and measure belt deflection. Deflection should be 0.512–0.571 inches.

18. Install the timing covers.
19. Install the crankshaft pulley and tighten the bolt to 90–98 ft. lbs. (122–136 Nm).
20. Install the water pump pulley.
21. Install the spark plugs.
22. Install the fresh air tube from the valve cover.
23. Install the distributor cover.
24. Install the water pump pulley.
25. Install the accessory drive belts.
26. Install the radiator and hoses.
27. Install the radiator fan shroud and pulleys.
28. Install the engine splash shield.
29. Refill the cooling system
30. Connect the negative battery cable.

3.0L (VG30DE, VG30DETT) ENGINES

YEAR, MODEL & ENGINE APPLICATION

1990-96 Nissan 300ZX .... 3.0L Codes VG30DE, VG30DETT

REPLACEMENT INTERVALS

Nissan recommends timing belt replacement every 60,000 miles.

LABOR TIMES

1990-96	
Replacement	5.2 hrs
Severe Service	5.7 hrs

8807NISg09

✻✻ CAUTION

These engines are an interference type. If the timing belt has failed, valve-to-piston damage may have occurred. A compression check should always be performed to check for engine damage.

1. Disconnect the negative battery cable.
2. Remove the engine splash shield.
3. Drain the engine coolant.
4. Remove the radiator and water hoses.
5. Remove the cooling fan and coupling.
6. Remove the water pump pulley.
7. Remove the crankshaft pulley.
8. Remove the upper and lower timing belt covers.
9. Remove the spark plugs and position the engine so that No. 1 piston is at Top Dead Center (TDC) of the compression stroke.

✻✻ WARNING

After the timing belt has been removed, DO NOT rotate the camshafts or the crankshaft. Severe internal engine damage will result from piston and valve contact.

10. Install a suitable 6mm stopper bolt in the tensioner arm of the auto-tensioner so the length of the pusher does not change.
11. Remove the automatic tensioner and the timing belt.
12. Check the automatic tensioner for oil leaks in the pusher rod and diaphragm. If oil is evident, replace the automatic tensioner assembly.
13. Inspect the timing gear teeth for wear.

To install:

14. Verify that the No. 1 piston is at TDC of the compression stroke.
15. Align the timing marks on the camshaft and crankshaft sprockets with the timing marks on the rear timing belt cover and the oil pump housing.
16. With a feeler gauge, check the clearance between the tensioner arm and the pusher of the automatic tensioner. The clearance should be 0.16 in. (4mm) with a slight drag on the feeler gauge. If the clearance is not as specified, mount the tensioner in a vise and adjust the clearance. When the clearance is set, insert the stopper bolt into the tensioner arm to retain the adjustment.

■ **When adjusting the clearance, do not push the tensioner arm with the stopper bolt fitted, because damage to the threaded portion of the bolt will result.**

17. Mount the automatic tensioner and tighten the nuts and bolts by hand.
18. Install the timing belt starting at the crankshaft sprocket and work countercolckwise. Ensure the timing sprockets are free of oil and water. Do not bend or twist the timing belt. Align the white lines on the belt with the timing marks on the camshaft and crankshaft sprockets. Point the arrow on the belt toward the front.
19. Push the automatic tensioner slightly toward the timing belt to prevent the belt from slipping. At the same time, turn the crankshaft 10 degrees clockwise and tighten the tensioner fasteners to 12–15 ft. lbs. (16–21 Nm).

■ **Do not push the tensioner too hard because it will create excessive tension on the belt.**

20. Turn the crankshaft 120 degrees (1/3 turn) counterclockwise.
21. Back off on the automatic tensioner fasteners 1/2 turn.
22. Turn the crankshaft clockwise and set the No. 1 piston at TDC of the compression stroke.
23. Using Push-Pull gauge No. EG1486000 (J-38387) or equivalent, apply approximately 15.2–18.3 lbs. (67.7–81.4 N) of force to the tensioner.
24. Tighten the tensioner mounting bolts to 12–15 ft. lbs. (17–21 Nm).
25. Turn the crankshaft 120 degrees (1/3 turn) clockwise.
26. Turn the crankshaft 120 degrees (1/3 turn) counterclockwise and set the No. 1 piston at TDC of the compression stroke.

■ **If the timing belt deflection exceeds the specification, change the applied pushing force.**

27. Fabricate a 0.35 in. (9mm) wide x 0.10 in. (2mm) steel plate. The length of the plate should be slightly longer than the width of the belt.
28. Set the steel plate at positions mid-way between the camshaft sprockets on each head, between the left exhaust camshaft sprocket and the idler pulley, and between the right exhaust camshaft sprocket and the tensioner.
29. Using the push-pull gauge or equivalent, apply approximately 11 lbs. (49 N) of force to the tensioner.
30. Check and record the belt deflection at each position with the steel plate in place. The timing belt deflection at each position should be 0.24–0.28 in. (6–7mm). Another means of determining the belt deflection is to add all deflection readings and divide them by 4. This average deflection should be 0.24–0.28 in. (6–7mm).

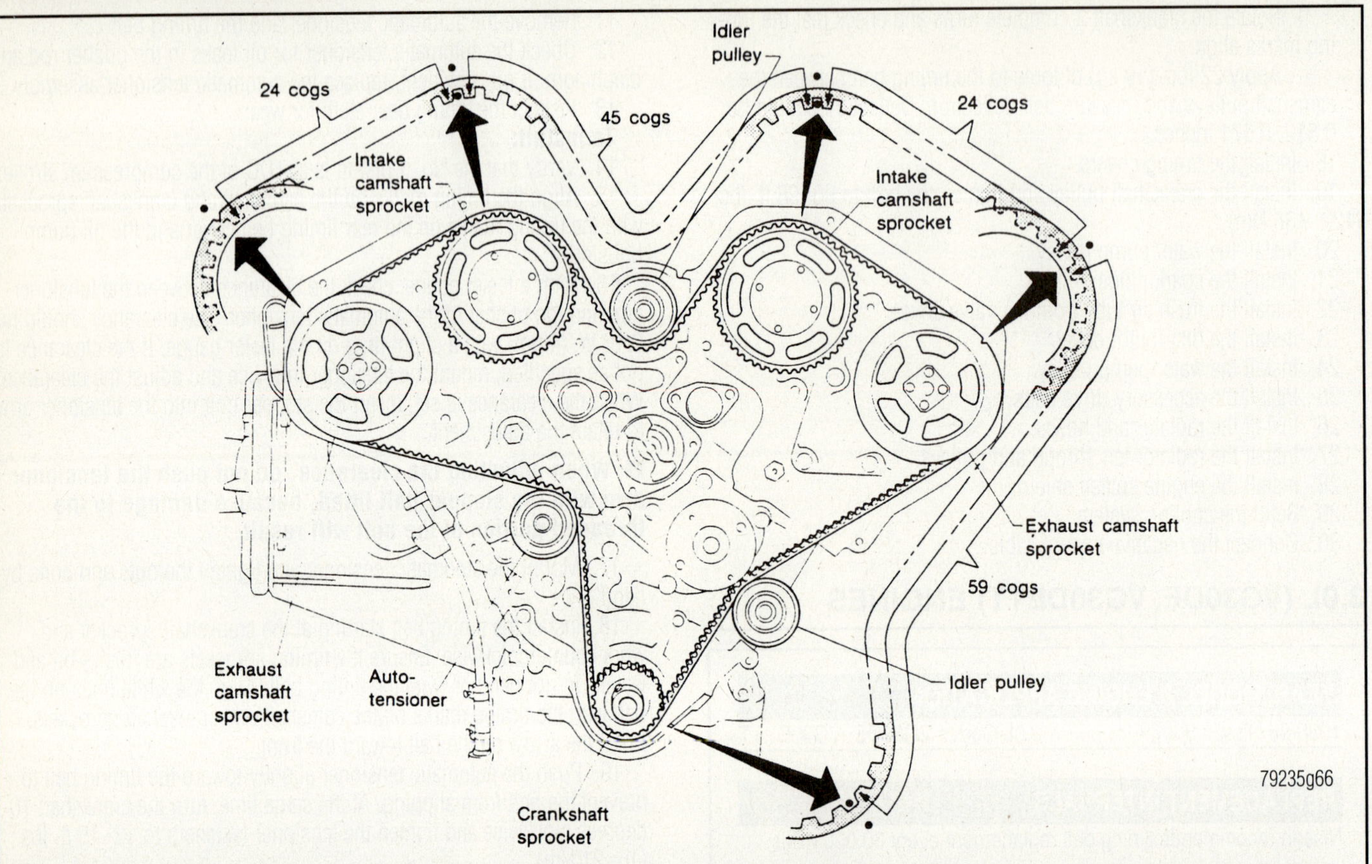

To ensure proper installation of the timing belt, be sure the proper number of belt teeth are between each sprocket, as indicated—Nissan 3.0L (VG30DE, VG30DETT) engines

31. If the belt deflection is not as specified, repeat the timing belt adjusting procedure until the belt deflection is correct.
32. Once the belt is properly tensioned, tighten the automatic tensioner fasteners to 12–15 ft. lbs. (16–21 Nm).
33. Remove the stopper bolt from the tensioner and wait 5 minutes. After 5 minutes, check the clearance between the tensioner arm and the pusher of the automatic tensioner. The clearance should remain at 0.138–0.205 in. (3.5–5.2mm).
34. Be sure the belt is installed and aligned properly on each pulley and the timing sprocket. There must be no slippage or misalignment.
35. Install the timing belt covers with new gaskets. Tighten the covers bolts to 24–48 inch lbs. (3–5 Nm).
36. Install the remaining components in the reverse order of removal.

3.0L (VG30E) AND 3.3L (VG33E) QUEST ENGINES

✻✻ CAUTION

These engines are an interference type. If the timing belt has failed, valve-to-piston damage may have occurred. A compression check should always be performed to check for engine damage.

On this vehicle, right side refers to the "rear" components (near the firewall) and left side refers to the "front" components (near the radiator).

YEAR, MODEL & ENGINE APPLICATION

1993-98 Nissan Quest	3.0L Code VG30E
1999-02 Quest	3.3L Code VG33E

REPLACEMENT INTERVALS

Nissan recommends timing belt replacement every 105,000 miles.

LABOR TIMES

1993-02	
Replacement	4.3 hrs
Severe Service	4.8 hrs

8807NISg10

1. If the timing belt is to be removed, it is good practice to turn the crankshaft until the engine is at Top Dead Center (TDC) of the No. 1 cylinder, compression stroke (firing position), before beginning work. This should align all timing marks and serve as a reference for all work that follows. After verifying that the engine is at TDC for the No. 1 cylinder, do not crank the engine or allow the crankshaft or camshaft sprockets to be turned otherwise engine timing will be lost.
2. Drain the cooling system.
3. Disconnect the negative battery cable.
4. Remove the alternator drive belt, water pump and power steering pump belt and the air conditioning compressor belt, if equipped.
5. If equipped with air conditioning, remove the 3 air conditioning compressor drive belt idler pulley bolts and remove the idler pulley.

6. Remove the upper radiator hose bracket bolt. Remove the upper hose with the bracket from the vehicle.

7. Remove the water bypass hose from between the thermostat housing and the lower water hose connection.

8. Remove the main wiring harness from the upper engine front cover.

9. Remove the 8 upper engine front cover bolts and remove the upper cover.

10. Raise and safely support the vehicle.

11. Remove the right side front wheel and tire assembly.

12. Remove the 4 right side engine and transmission splash shield bolts and 2 screws, and remove the right side outer engine and transaxle splash shield.

13. Use a strap wrench to hold the water pump pulley. Remove the 4 pulley bolts, and the water pump pulley.

14. Use a strap wrench to hold the crankshaft pulley. Remove the center pulley bolt, and the crankshaft pulley using a harmonic balancer (damper) puller to draw the pulley from the front of the crankshaft.

15. Remove the 5 lower engine front cover bolts and remove the lower engine front cover.

16. Be sure that the timing marks between the crankshaft sprocket and the oil pump housing align.

17. If the timing belt is to be reused, mark an arrow on the belt indicating the direction of rotation. The directional arrow is necessary to ensure that the timing belt, if it to be reused, can reinstalled in the same direction.

18. Loosen the timing belt tensioner nut and slip the timing belt off of the sprockets.

19. If necessary, the camshaft sprockets can be removed. A special spanner tool is designed to hold the sprocket to keep it from turning while the center bolt is being loosened. Use care if using substitutes.

■ **The sprockets are not interchangeable.**

20. If necessary, the crankshaft sprocket can be removed. The outer timing belt guide (looks like a large washer) and the crankshaft sprocket simply pull off the front of the crankshaft.

■ **Be careful, there are 2 crankshaft keys. Use care not to lose them.**

To install:

21. Clean all parts well. If removed, inspect the crankshaft sprocket for warping or abnormal wear. Check the sprocket teeth for wear, deformation, chipping or other damage. Replace as necessary. Clean the sprocket mounting surface to ease installation. Install the key. Slip the sprocket onto the crankshaft. Tap it in place with a suitably-sized socket.

22. If removed, inspect the camshaft sprockets for damage and wear. Replace as required. The sprockets should be marked **L3** to designate the front, or left side camshaft and **R3** to designate the rear, or right side camshaft. Use care to install the sprockets properly. A special spanner tool is designed to hold the sprocket to keep it from turning while the center bolt is being tightened. Use care if using a substitute. Tighten the camshaft sprocket center bolts to 58–65 ft. lbs. (78–88 Nm) for 3.0L engine or 61 ft. lbs. (83 Nm) for 3.3L engine. Verify that the timing marks on the camshaft sprockets and the timing marks on the rear cover (called the seal plate) are aligned.

23. Use an Allen wrench to turn the timing belt tensioner clockwise until the belt tensioner spring is fully extended. Temporarily tighten the tensioner nut to 32–43 ft. lbs. (43–58 Nm).

24. If a new timing belt is to be installed, look for a printed arrow on the belt. Be sure the arrow is pointing away from the engine. If the original timing belt is to be reused, be sure that the directional arrow that was marked at disassembly is facing the correct direction.

25. A new Original Equipment Manufacture (OEM) timing belt should have 3 white timing marks on it that indicate the correct timing positions of the camshafts and the crankshaft. These marks are to help ensure that the engine is properly timed. When the engine is properly timed, each white timing mark on the timing belt will be aligned with the corresponding camshaft and crankshaft timing mark on the sprocket. Because the white timing marks are not evenly spaced, the technician needs to use care in installing the belt. There should be 40 timing belt teeth between the timing marks on the front and rear camshaft sprockets and 43 teeth between the timing mark on the front camshaft sprocket and the timing mark on the crankshaft sprocket.

26. Verify that the camshaft timing marks are aligned with the timing marks on the rear cover (seal plate) and that the crankshaft sprocket timing mark is aligned with the timing mark on the oil pump housing.

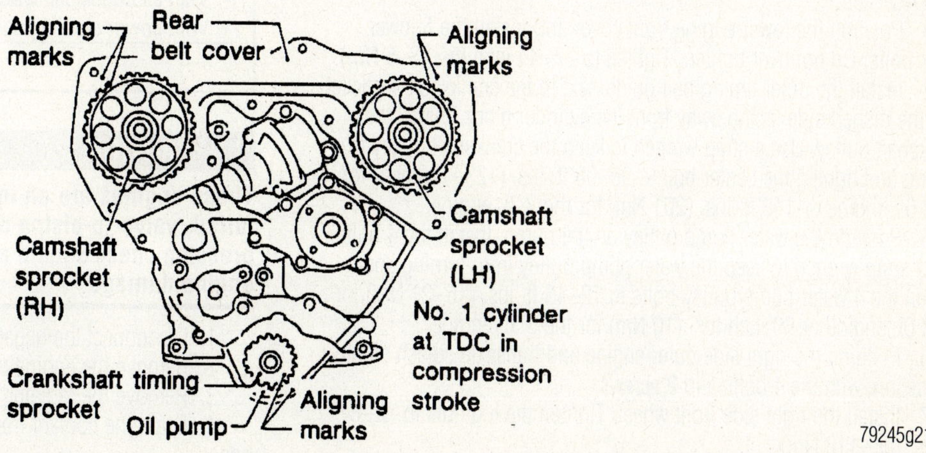

Use a shop rag to clean the alignment marks for the timing belt—Nissan 3.0L (VG30E) and 3.3L (VG33E) engines

27. Install the timing belt starting at the crankshaft sprocket and moving around the camshaft sprockets following a counterclockwise path. Do not allow any slack in the timing belt between the sprockets. After all of the timing marks are aligned with the timing belt installed, slip the timing belt onto the belt tensioner.

28. While holding the timing belt tensioner with an Allen wrench, loosen the tensioner nut. Allow the tensioner to put pressure on the timing belt. Use an Allen wrench to turn the timing belt tensioner 70–80 degrees clockwise and tighten the timing belt tensioner nut to 32–43 ft. lbs. (43–58 Nm).

**** WARNING**

If any binding is felt when adjusting the timing belt tension by turning the crankshaft, STOP turning the engine, because the pistons may be hitting the valves.

29. Rotate the crankshaft clockwise twice and align the No. 1 piston to TDC on the compression stroke (firing position).

30. Apply 22 lbs. (10kg) of force on the timing belt between the rear camshaft sprocket and the timing belt tensioner. An assistant may be needed. While holding the timing belt tensioner steady with an Allen wrench, loosen the timing belt tensioner nut. Remove the Allen wrench and adjust the timing belt tensioner using the following procedure:

a. Install a 0.0138 in. (0.35mm) thick and 0.500 in. (12.7mm) wide feeler gauge where the timing belt just starts to go around the tensioner (approximately the 4 o'clock position, looking at the tensioner).

b. Turn the crankshaft sprocket clockwise, which should force the feeler gauge between the timing belt and the tensioner, up to a position on the tensioner of about 1 o'clock.

c. Tighten the timing belt tensioner nut to 32–43 ft. lbs. (43–58 Nm) for the 3.0L engine or 61 ft. lbs. (83 Nm) for the 3.3L engine.

d. Turn the crankshaft clockwise to rotate the feeler gauge out from between the timing belt tensioner and the timing belt.

31. Rotate the crankshaft clockwise twice, and once again align the No. 1 piston to TDC on the compression stroke (firing position).

32. Apply 22 lbs. (10kg) of force on the timing belt between the front and rear camshaft sprockets. Measure the amount of belt deflection. Belt deflection should be between 0.51–0.59 in. (13–15mm). If belt deflection is out of specification, repeat the procedure. If the timing belt deflection cannot be adjusted into specification, the timing belt will have to be replaced.

33. Position the lower engine front cover and install the 5 lower cover bolts. Do not over tighten. Tighten to 27–44 inch lbs. (3–5 Nm).

34. Install the outer timing belt guide next to the crankshaft sprocket with the dished side facing away from the cylinder block. Install the crankshaft pulley. Use a strap wrench to keep the crankshaft pulley from turning and tighten the center bolt to 90–98 ft. lbs. (123–132 Nm) for the 3.0L engine or 148 ft. lbs. (201 Nm) for the 3.3L engine.

35. Position the water pump pulley on the pump. Install the 4 bolts. Use a strap wrench to keep the water pump pulley from turning and tighten the 4 water pump pulley bolts to 12–15 ft. lbs. (16–21 Nm) for the 3.0L engine or 89 inch lbs. (10 Nm) for the 3.3L engine.

36. Position the right side outer engine and transaxle splash shield, and secure with the 4 bolts and 2 screws.

37. Install the right side front wheel. Tighten the lug nuts to 72–87 ft. lbs. (98–118 Nm).

38. Lower the vehicle.

39. Position the upper engine timing belt front cover, and tighten the 8 bolts to 27–44 inch lbs. (3–5 Nm).

40. Install the main wiring harness on the upper engine front cover.

41. Position the water bypass hose between the thermostat housing and water connection. Install the upper radiator hose between the radiator and the water hose connection. Secure the hoses with clamps. Install the upper radiator hose bracket. Tighten the bracket bolt to 34–58 ft. lbs. (46–65 Nm).

42. If equipped, position the air conditioning compressor drive belt idler pulley and install the 3 bolts. Tighten to 15 ft. lbs. (21 Nm).

43. Install and adjust the alternator drive belt, the water pump and power steering pump drive belt and the air conditioning compressor drive belt, if equipped.

44. Connect the battery cable.

45. Fill the cooling system.

46. Start the engine and allow it to warm to operating temperature. Check and adjust the ignition timing. Road test to verify correct engine operation.

47. Connect the negative battery cable.

3.3L (VG33E, VG33ER) ENGINES

YEAR, MODEL & ENGINE APPLICATION

Application	Engine
1999-04 Nissan Frontier, 2000-04 Xterra, 1996-01 Pathfinder	3.3L Code VG33E
2002-04 Frontier, Xterra	3.3L Code VG33ER

REPLACEMENT INTERVALS

Nissan recommends timing belt replacement every 105,000 miles.

LABOR TIMES

Frontier & Xterra	
Replacement	3.5 hrs
Severe Service	4.0 hrs
With air conditioning add	0.3 hrs
With power steering add	0.2 hrs
Pathfinder	
Replacement	3.2 hrs
Severe Service	3.7 hrs
With air conditioning add	0.3 hrs
With power steering add	0.2 hrs

8807NISg11

**** CAUTION**

These engines are an interference type. If the timing belt has failed, valve-to-piston damage may have occurred. A compression check should always be performed to check for engine damage.

1. Disconnect the negative battery cable.
2. Remove the engine undercover.
3. Remove the radiator shroud, the fan and the pulleys.
4. Drain the coolant from the radiator and remove the water pump hose.

✻✻ CAUTION

When draining the coolant, keep in mind that cats and dogs are attracted by the ethylene glycol antifreeze, and are quite likely to drink any that is left in an uncovered container or in puddles on the ground. This will prove fatal in sufficient quantity. Always drain the coolant into a sealable container. Coolant should be reused unless it is contaminated or several years old.

5. Remove the radiator.
6. Remove the power steering, air conditioning compressor and alternator drive belts.
7. Remove the spark plugs.
8. Remove the distributor protector (dust shield).
9. Remove the air conditioning compressor drive belt idler pulley and bracket.
10. Remove the fresh air intake tube at the cylinder head cover.
11. Disconnect the radiator hose at the thermostat housing.
12. Remove the crankshaft pulley bolt, then pull off the pulley with a suitable puller.
13. Remove the bolts, then remove the front upper and lower timing belt covers.
14. Set the No. 1 piston at Top Dead Center (TDC) of its compression stroke. Align the punchmark on the left camshaft sprocket with the punchmark on the timing belt upper rear cover. Align the punchmark on the crankshaft sprocket with the notch on the oil pump housing. Temporarily install the crank pulley bolt so the crankshaft can be rotated if necessary.
15. Loosen the timing belt tensioner and return spring, then remove the timing belt.

To install:

✻✻ CAUTION

Before installing the timing belt, confirm that the No. 1 cylinder is set at the TDC of the compression stroke.

16. Remove both cylinder head covers and loosen all rocker arm shaft retaining bolts.

■ **The rocker arm shaft bolts MUST be loosened so that the correct belt tension can be obtained.**

17. Install the tensioner and the return spring. Using a hexagon wrench, turn the tensioner clockwise and temporarily tighten the locknut.
18. Be sure that the timing belt is clean and free from oil or water.
19. When installing the timing belt, align the white lines on the belt with the punchmarks on the camshaft and crankshaft sprockets. Have the arrow on the timing belt pointing toward the front belt covers.

■ **A good way (although rather tedious!) to check for proper timing belt installation is to count the number of belt teeth between the timing marks. There are 133 teeth on the belt; there should be 40 teeth between the timing marks on the left and right side camshaft sprockets, and 43 teeth between the timing marks on the left side camshaft sprocket and the crankshaft sprocket.**

20. While keeping the tensioner steady, loosen the locknut with a hex wrench.
21. Turn the tensioner approximately 70–80 degrees clockwise with the wrench, then tighten the locknut.

✻✻ WARNING

If any binding is felt when adjusting the timing belt tension by turning the crankshaft, STOP turning the engine, because the pistons may be hitting the valves.

22. Turn the crankshaft in a clockwise direction several times, then **slowly** set the No. 1 piston to TDC of the compression stroke.
23. Apply 22 lbs. (10 kg) of pressure (push it in!) to the center span of the timing belt between the right side camshaft sprocket and the tensioner pulley, then loosen the tensioner locknut.

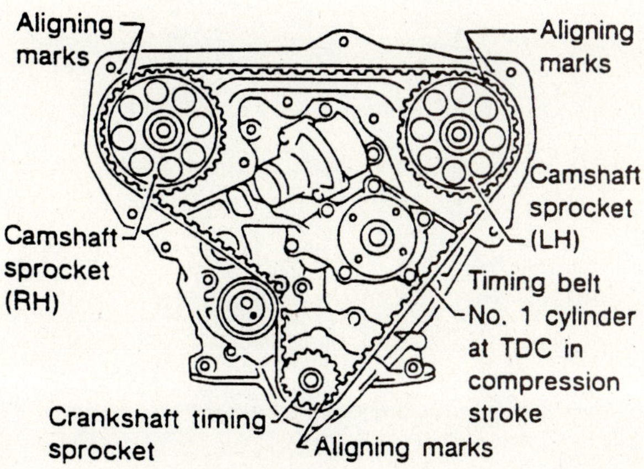

79245g35

Timing belt alignment mark locations—3.3L engines

24. Using a 0.0138 in. (0.35mm) thick feeler gauge (the actual width of the blade **must** be 1/2 in. or 13mm!), turn the crankshaft clockwise (**slowly!**). The timing belt should move approximately 21/2 teeth. Tighten the tensioner locknut, turn the crankshaft slightly and remove the feeler gauge.

25. Slowly rotate the crankshaft clockwise several more times, then set the No. 1 piston to TDC of the compression stroke.

26. Position the 2 timing covers on the block, then tighten the mounting bolts to 24 ft. lbs. (35 Nm).

27. Press the crankshaft pulley onto the shaft, then tighten the bolt to 90–98 ft. lbs. (123–132 Nm).

28. Connect the radiator hose to the thermostat housing.

29. Reconnect the fresh air intake tube at the cylinder head cover.

30. Install the air conditioning compressor drive belt idler pulley and bracket.

31. Install the distributor protector (dust shield).

32. Install the spark plugs.

33. Install the power steering, air conditioning compressor and alternator drive belts.

34. Install the radiator.

35. Reconnect the water pump hose and fill the engine with coolant. Install the fan shroud and pulleys.

36. Install the engine undercover.

37. Start the engine and check for any leaks.

Porsche

2.0L (VC AND M31) ENGINES

YEAR, MODEL & ENGINE APPLICATION	
1980-82 Porsche 924, 944	2.0L Codes M31, VC

REPLACEMENT INTERVALS

Porsche recommends timing belt replacement every 45,000 miles.

LABOR TIMES	
1980-82	
Replacement	1.2 hrs
Severe Service	1.4 hrs

8807PORg01

✱✱ CAUTION

These engines are an interference type. If the timing belt has failed, valve-to-piston damage may have occurred. A compression check should always be performed to check for engine damage.

1. Disconnect the negative battery cable.
2. Remove the accessory drive belts.
3. Remove the timing cover.
4. Rotate the crankshaft to align the Top Dead Center (TDC) mark on the flywheel with the timing mark on the clutch housing. Check that the camshaft timing mark aligns with the valve cover, and the **V** notch in the crankshaft pulley aligns with the pointer on the oil pump housing.
5. If reusing the timing belt, mark the rotational direction on the back of the belt.
6. Loosen the tensioner pulley locknut and remove the timing belt.

To install:

7. Ensure that the crankshaft and camshaft timing marks are aligned. Noting the direction of rotation, install the timing belt.
8. Rotate the tensioner pulley counterclockwise to tension the timing belt. Timing belt tension is correct when the belt can be twisted no more than 90 degrees midway between the crankshaft and camshaft pulleys.

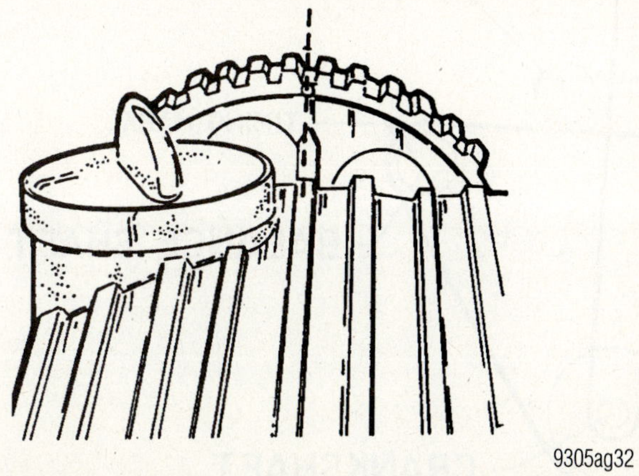

9305ag32

View of the camshaft timing mark—Porsche 2.0L (VC AND M31) engine

9305ag33

View of the crankshaft pulley timing mark—Porsche 2.0L (VC AND M31) engine

9. Tighten the tensioner locknut to 30 ft. lbs. (41 Nm).
10. Rotate the crankshaft clockwise 2 complete turns and check that the timing marks align.
11. Install the timing cover.
12. Install the accessory drive belts.
13. Connect the negative battery cable.

2.5L (M44.01–08, M44.51) ENGINES

YEAR, MODEL & ENGINE APPLICATION	
1985-87 Porsche 924	
1983-87 944	2.5L Codes M44.01-08, M44/51

REPLACEMENT INTERVALS

Porsche recommends timing belt replacement every 45,000 miles.

LABOR TIMES	
924	
Replacement	1.2 hrs
Severe Service	1.4 hrs
944	
Replacement	2.3 hrs
Severe Service	2.5 hrs

8807PORg02

✱✱ CAUTION

These engines are an interference type. If the timing belt has failed, valve-to-piston damage may have occurred. A compression check should always be performed to check for engine damage.

1. Disconnect the negative battery cable.
2. Remove the engine splash guard.
3. Remove the air cleaner and Mass Air Flow (MAF) sensor.
4. Remove the accessory drive belts.
5. Remove the distributor cap.
6. Remove the rotor arm.
7. Remove the plastic cap and remove the distributor cap mount.
8. Remove the crankshaft pulley.
9. Remove the timing covers.
10. Rotate the crankshaft to align the flywheel timing mark with the mark on the clutch housing.

11. Check that the camshaft and balance shaft timing marks are aligned with the marks on the rear timing cover.
12. Loosen the balance shaft belt guide pulley.
13. Loosen the balance shaft belt tensioner pulley and remove the balance shaft belt.
14. If reusing the timing belt, mark the rotational direction on the back of the belt.
15. Loosen the timing belt tensioner pulley and remove the timing belt.

To install:

16. Ensure that the camshaft and flywheel timing marks are aligned. Noting the direction of rotation, install the timing belt.
17. Turn the timing belt tensioner pulley counterclockwise to tension the timing belt and tighten the tensioner pulley bolt.
18. Turn the crankshaft 10 degrees counterclockwise and install the Tension Gauge No. 9201 on the timing belt between the crankshaft and camshaft pulleys.

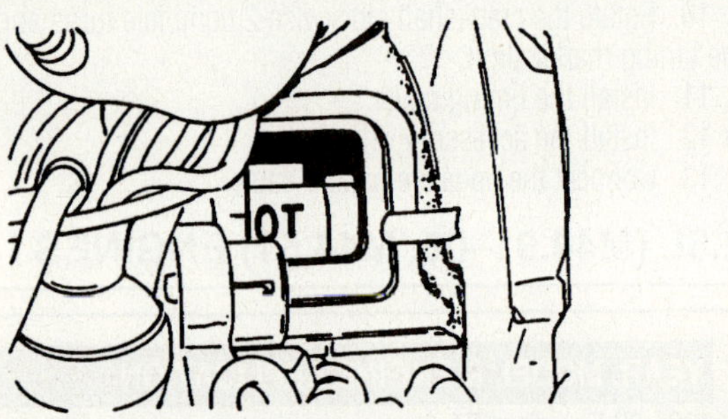

9305ag34

View of the flywheel timing marks—Porsche 2.5L (M44.01–08, M44.51) engine

19. Timing belt tension should be 3.7–4.3 gauge units for a new belt or 2.4–3.0 gauge units for a used belt.
20. If the belt tension is not correct, loosen the belt tensioner pulley lockbolt and rotate the tensioner pulley clockwise to decrease tension and counterclockwise to increase tension. Tighten the tensioner pulley lockbolt.
21. Rotate the crankshaft clockwise to Top Dead Center (TDC) and check the alignment of the flywheel and camshaft timing marks. Rotate the crankshaft counterclockwise 10 degrees and repeat the belt tension measurement.
22. When the timing belt tension is correct, tighten the tensioner pulley to 33 ft. lbs. (45 Nm). Rotate the crankshaft clockwise to TDC.

■**The balance shaft belt has teeth on both sides. Install the belt with the color coded tooth facing out.**

23. Ensure that the balance shaft timing marks are aligned and install the balance shaft belt.
24. Install the Belt Tension Gauge No. 9201 on the balance shaft belt between the tensioner pulley and the left balance shaft pulley.
25. Turn the balance shaft belt tensioner pulley clockwise until the gauge reads 2.4–3.0 units. Tighten the belt tensioner pulley lockbolt to 33 ft. lbs. (45 Nm).
26. Rotate the crankshaft 2 complete turns and align all timing marks. Recheck the balance shaft belt tension.
27. When the balance shaft belt tension is correct, adjust the guide pulley so that there is 0.5mm clearance between the guide pulley and the balance shaft belt. Tighten the guide pulley lockbolt to 33 ft. lbs. (45 Nm).
28. Install the timing covers.
29. Install the crankshaft pulley.
30. Install the distributor cap mount.
31. Install the plastic cap and the rotor arm.

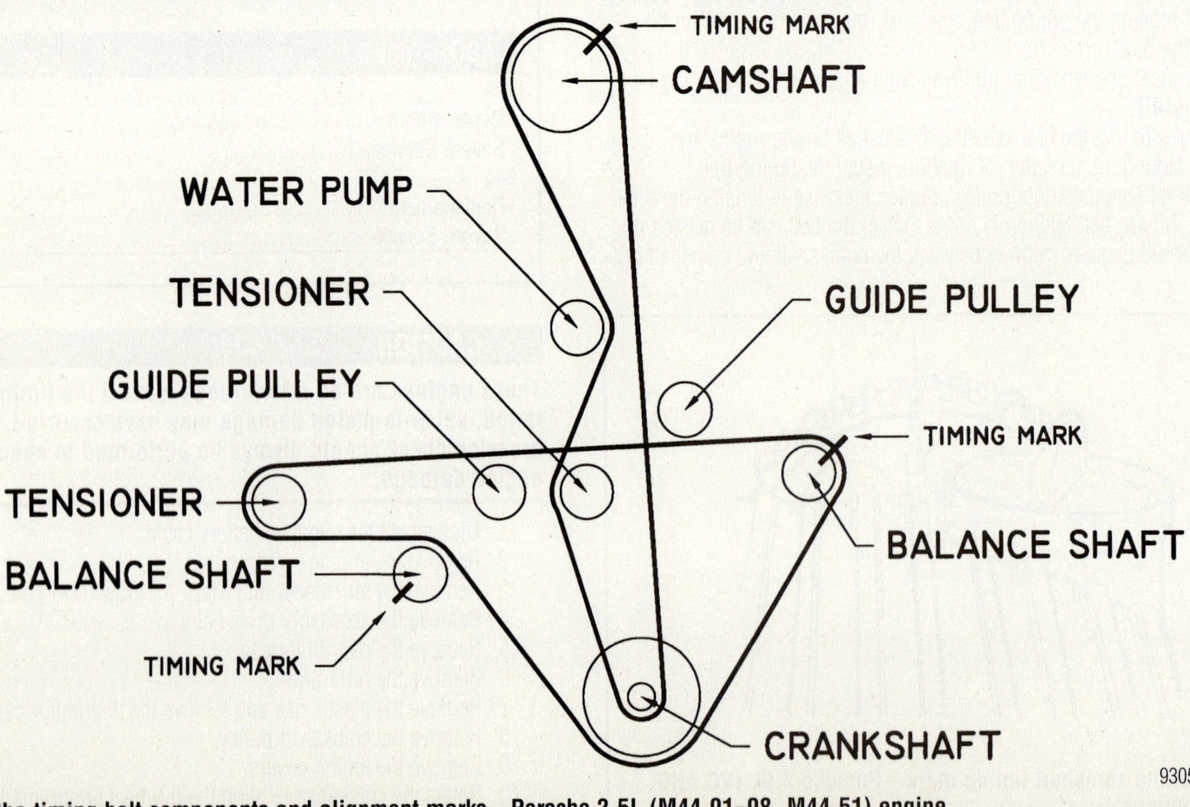

9305ag48

View of the timing belt components and alignment marks—Porsche 2.5L (M44.01–08, M44.51) engine

32. Install the distributor cap.
33. Install the accessory drive belts.
34. Install the MAF sensor and the air filter.
35. Install the engine splash guard.
36. Connect the negative battery cable.

2.5L, 2.7L, 3.0L (M44.05–12, M44.40–41, M44.52) ENGINES

YEAR, MODEL & ENGINE APPLICATION

1988-89 Porsche 944, 944 S Turbo	2.5L Codes M44.05-12
1989 944	2.7L Code M44.40-41
1989-91 944S2	3.0L Code M44.52

REPLACEMENT INTERVALS

Porsche recommends timing belt replacement every 45,000 miles.

LABOR TIMES

1988-91	
Replacement	4.0 hrs
Severe Service	4.5 hrs
With turbocharger add	0.5 hrs

8807PORg03

✱✱ CAUTION

These engines are an interference type. If the timing belt has failed, valve-to-piston damage may have occurred. A compression check should always be performed to check for engine damage.

1. Disconnect the negative battery cable.
2. Remove the engine splash guard.
3. Remove the air cleaner and Mass Air Flow (MAF) sensor.
4. Remove the accessory drive belts.
5. Remove the distributor cap.
6. Remove the rotor arm.
7. Remove the plastic cap and remove the distributor cap mount.
8. Remove the crankshaft pulley.
9. Remove the timing covers.
10. Rotate the crankshaft to align the flywheel timing mark with the mark on the clutch housing.
11. Check that the camshaft and balance shaft timing marks are aligned with the marks on the rear timing cover.
12. Loosen the balance shaft belt guide pulley.
13. Loosen the balance shaft belt tensioner pulley and remove the balance shaft belt.
14. Remove the guide plate and guide pulley.
15. If reusing the timing belt, mark the rotational direction on the back of the belt.
16. Loosen the timing belt tensioner pulley lockbolt and pivot bolt.
17. Using tool No. 9200, pivot the belt tensioner away from the timing belt and tighten the lockbolt.
18. Remove the timing belt tensioner assembly and remove the timing belt.

To install:

19. Ensure that the flywheel and camshaft timing marks are aligned. Noting the direction of rotation, install the timing belt.
20. Install the timing belt tensioner assembly.
21. Loosen the timing belt tensioner lockbolt to tension the timing belt.
22. Tighten the tensioner lockbolt.
23. Rotate the crankshaft 2 complete turns and check that the flywheel and camshaft marks align.

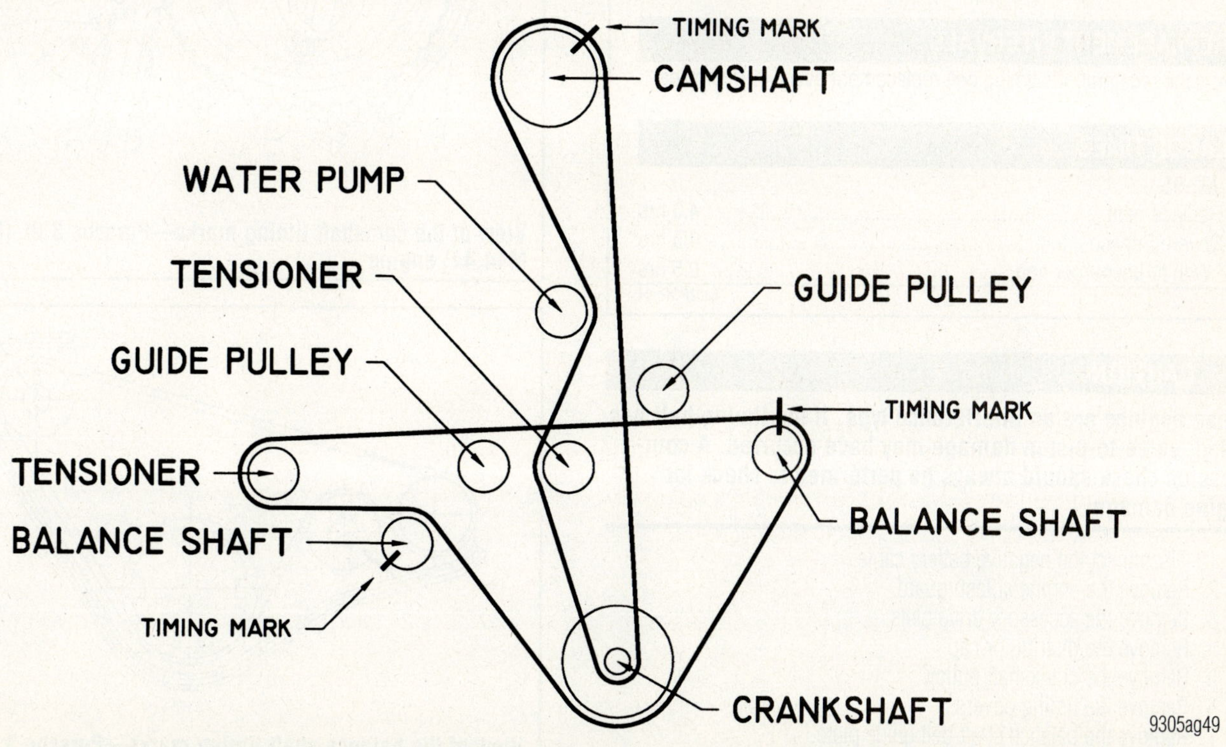

View of the timing belt components and alignment marks—Porsche 2.5L, 2.7L, 3.0L (M44.05–12, M44.40–41, M44.52) engines

24. Loosen the tensioner lockbolt and allow the tensioner to take up any remaining slack. Tighten the lockbolt and pivot bolt to 15 ft. lbs. (20 Nm).

■ **The balance shaft belt has teeth on both sides. Install the belt with the color coded tooth facing out.**

25. Ensure that the balance shaft timing marks are aligned and install the balance shaft belt.
26. Install the Belt Tension Gauge No. 9201 on the balance shaft belt between the tensioner pulley and the left balance shaft pulley.
27. Turn the balance shaft belt tensioner pulley clockwise until the gauge reads 2.4–3.0 units. Tighten the belt tensioner pulley lockbolt to 33 ft. lbs. (45 Nm).
28. Rotate the crankshaft 2 complete turns and align all timing marks. Recheck the balance shaft belt tension.
29. When the balance shaft belt tension is correct, adjust the guide pulley so that there is 0.5mm clearance between the guide pulley and the balance shaft belt. Tighten the guide pulley lockbolt to 33 ft. lbs. (45 Nm).
30. Install the timing covers.
31. Install the crankshaft pulley.
32. Install the distributor cap mount.
33. Install the plastic cap and the rotor arm.
34. Install the distributor cap.
35. Install the accessory drive belts.
36. Install the MAF sensor and the air filter.
37. Install the engine splash guard.
38. Connect the negative battery cable.

3.0L (M44.43, M44.44) ENGINES

YEAR, MODEL & ENGINE APPLICATION

1992-95 Porsche 968 .... 3.0L Codes M44.43-44

REPLACEMENT INTERVALS

Porsche recommends timing belt replacement every 45,000 miles.

LABOR TIMES

1992-95	
Replacement	4.0 hrs
Severe Service	4.5 hrs
With turbocharger add	0.5 hrs

8807PORg04

✱✱ CAUTION

These engines are an interference type. If the timing belt has failed, valve-to-piston damage may have occurred. A compression check should always be performed to check for engine damage.

1. Disconnect the negative battery cable.
2. Remove the engine splash guard.
3. Remove the accessory drive belts.
4. Remove the distributor cap.
5. Remove the crankshaft pulley.
6. Remove the timing covers.
7. Remove the balance shaft belt guide plate.
8. Rotate the crankshaft to align the flywheel timing mark with the mark on the clutch housing.
9. Check that the camshaft and balance shaft timing marks are aligned with the marks on the rear timing cover.
10. Loosen the guide pulley locknut and rotate away from the balance shaft belt.
11. Loosen the balance shaft belt tensioner locknut. Rotate the tensioner pulley counterclockwise to release belt tension.

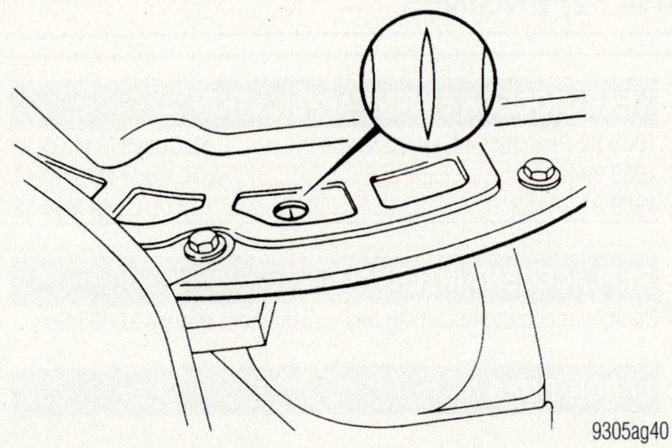

View of the flywheel timing marks—Porsche 3.0L (M44.43, M44.44) engine

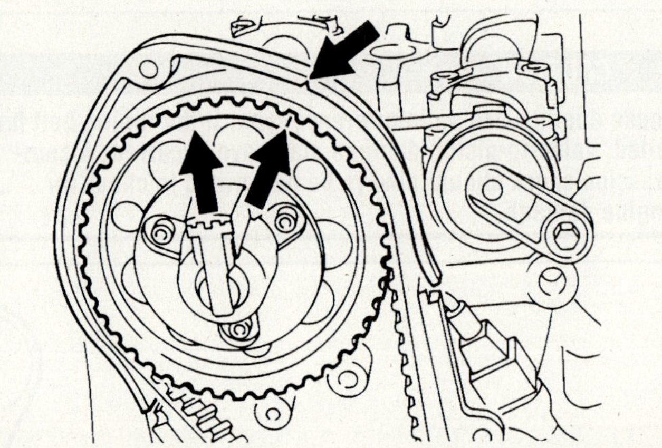

View of the camshaft timing marks—Porsche 3.0L (M44.43, M44.44) engine

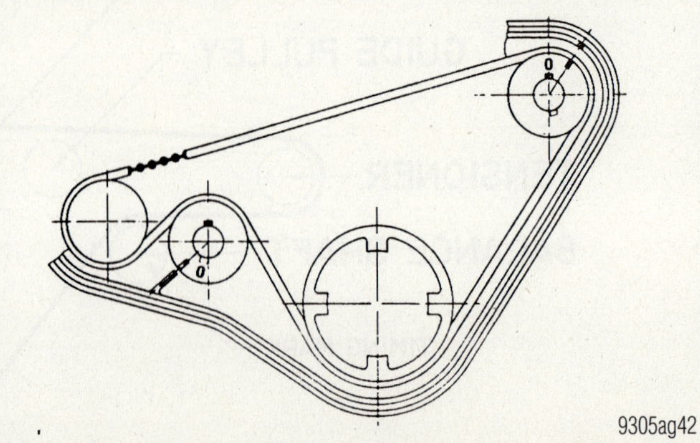

View of the balance shaft timing marks—Porsche 3.0L (M44.43, M44.44) engine

12. Remove the balance shaft belt.
13. If reusing the timing belt, mark the rotational direction on the back of the belt.
14. Remove the timing belt hydraulic tensioner.
15. Remove the tensioner pulley arm.
16. Remove the timing belt.

To install:

■ **Use a large washer to protect the bottom of the tensioner while compressing it.**

17. Compress the hydraulic tensioner slowly in a vise and install a retainer pin.
18. Ensure that the camshaft and flywheel timing marks are aligned. Noting the direction of rotation, install the timing belt.
19. Install the tensioner pulley arm.
20. Install the hydraulic tensioner and remove the retainer pin.
21. Rotate the crankshaft 2 complete turns and check that the timing marks align.

■ **The balance shaft belt has teeth on both sides. Install the belt with the color coded tooth facing out.**

22. Align the balance shaft timing marks and install the balance shaft belt.
23. Turn the balance shaft belt tensioner pulley clockwise until the gauge reads 2.4–3.0 units. Tighten the belt tensioner pulley lockbolt to 33 ft. lbs. (45 Nm).

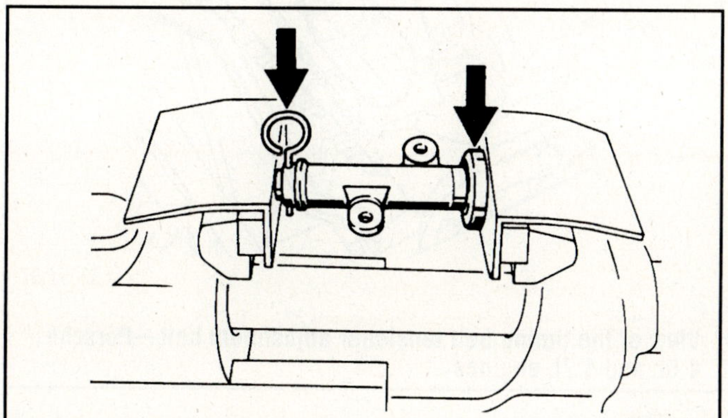

9305ag43

Preparing the hydraulic tensioner for installation—Porsche 3.0L (M44.43, M44.44) engine

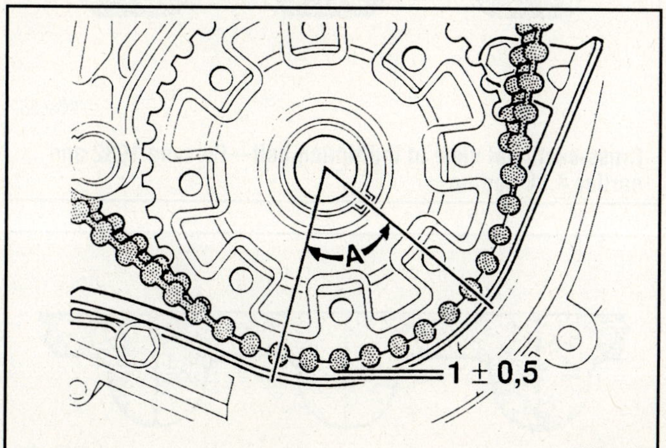

9305ag44

Balance shaft belt guide plate clearance—Porsche 3.0L (M44.43, M44.44) engine

24. Rotate the crankshaft 2 complete turns and align all timing marks. Recheck the balance shaft belt tension.
25. When the balance shaft belt tension is correct, adjust the guide pulley so that there is 0.5mm clearance between the guide pulley and the balance shaft belt. Tighten the guide pulley lockbolt to 33 ft. lbs. (45 Nm).
26. Install the balance shaft belt guide plate so that the clearance measured over 7 teeth is 0.5–1.5 mm.
27. Install the timing covers.
28. Install the crankshaft pulley and tighten the bolt to 155 ft. lbs. (210 Nm).
29. Install the distributor cap and the accessory drive belts.
30. Install the engine splash guard.
31. Connect the negative battery cable.

4.5L, 4.7L AND 5.0L (M28) ENGINES

YEAR, MODEL & ENGINE APPLICATION	
1984-87 Porsche 928S/S2, 1987-91 928S4, 1991 928GT, 1993-95 928 GTS	4.5L, 4.7L Code M28
1983-87 928	5.0L Code M28

REPLACEMENT INTERVALS

Porsche recommends timing belt replacement every 60,000 miles.

LABOR TIMES	
1983-95 8-Valve Engines	
Replacement	5.6
Severe Service	6.1
1983-95 16-Valve Engines	
Replacement	7.2
Severe Service	7.7
All Engines	
With turbocharger add	0.5

8807PORg05

✱✱ CAUTION

These engines are an interference type. If the timing belt has failed, valve-to-piston damage may have occurred. A compression check should always be performed to check for engine damage.

1. Disconnect the negative battery cable.
2. Remove the air intake hoses.
3. Remove the oil dipstick and fill tubes.
4. Disconnect the throttle cable, cruise control cable and transmission cable, if equipped with automatic transmission.
5. Disconnect the cable retainers from the fan shroud.
6. Remove the fan shroud.
7. Remove the accessory drive belts.
8. Remove the cooling fan, bracket and pulley.

✱✱ WARNING

Do not lay the fan assembly flat. The silicone oil filler will leak out and the fan will become inoperative.

9. Remove the upper timing covers.
10. Remove the power steering pump and position aside.

✳✳ WARNING

Never rotate the crankshaft counterclockwise.

11. Rotate the crankshaft clockwise to align the timing marks on the camshaft pulleys with the marks on the rear timing cover, and the Top Dead Center (TDC) mark on the crankshaft damper with the red pointer on the lower timing cover. Check that all marks align.

12. If equipped with a manual transmission, remove the clutch slave cylinder. If equipped with an automatic transmission, remove the slave cylinder hole cover.

13. Install the Crankshaft Locking tool No. 9161.

14. Remove the crankshaft pulley and the lower timing cover.

15. If reusing the timing belt, mark the rotational direction on the back of the belt.

16. Loosen the timing belt tensioner bolt and remove the timing belt.

■ **Be sure to use the correct replacement timing belt.**

To install:

17. Ensure that the camshaft timing marks are aligned. Noting the direction of rotation, install the timing belt.

18. Install the lower timing cover and the crankshaft pulley. Ensure that the crankshaft timing mark aligns with the pointer on the lower cover.

19. If all timing marks align, tighten the timing belt tensioner adjustment bolt until correct belt tension is achieved.

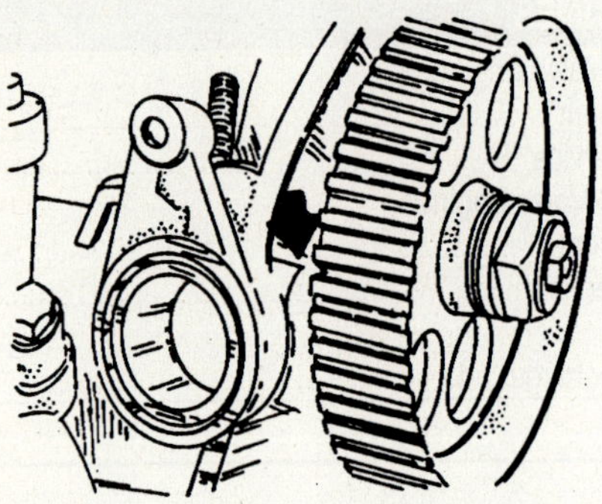

9305ag35

View of the left camshaft alignment marks—Porsche 4.5L and 4.7L engines

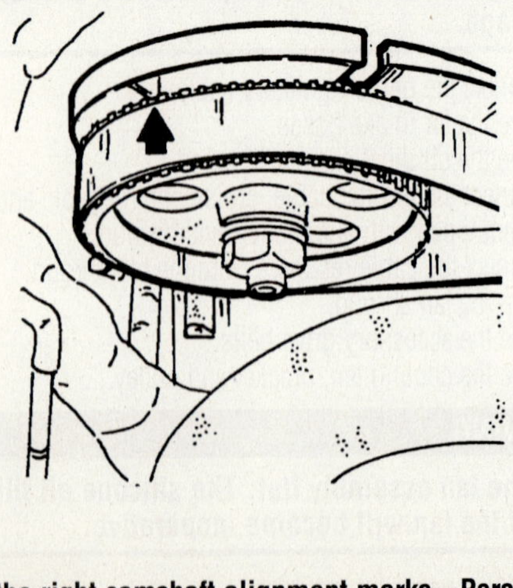

9305ag36

View of the right camshaft alignment marks—Porsche 4.5L and 4.7L engines

20. For belts with square teeth, correct tension is achieved when the belt can be twisted 90 degrees between the tensioner pulley and the right bank camshaft pulley.

21. For belts with round teeth, use Tension tool No. 9201 and adjust the tensioner until the gauge reads 4.5 units.

22. Remove the crankshaft locking tool and rotate the crankshaft 2 complete turns. Check that the timing marks align and recheck the timing belt tension. If the belt tension is correct, tighten the tensioner locknut.

23. Install the crankshaft locking tool again and tighten the crankshaft pulley bolt to 181 ft. lbs. (246 Nm). Remove the crankshaft locking tool.

24. Install the clutch slave cylinder or cover.

25. Install the upper timing covers.

26. Install the power steering pump.

27. Install the cooling fan, bracket and pulley.

28. Install the accessory drive belts.

29. Install the fan shroud.

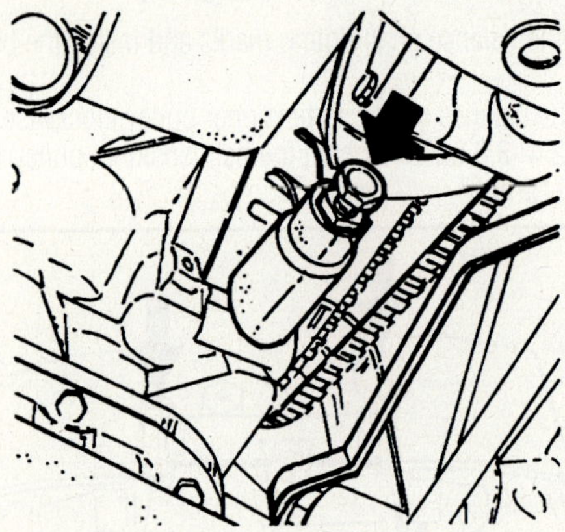

9305ag37

View of the timing belt tensioner adjustment bolt—Porsche 4.5L and 4.7L engines

9305ag38

Cross-sectional view of the timing belt—Porsche 1982 and earlier 4.5L engine

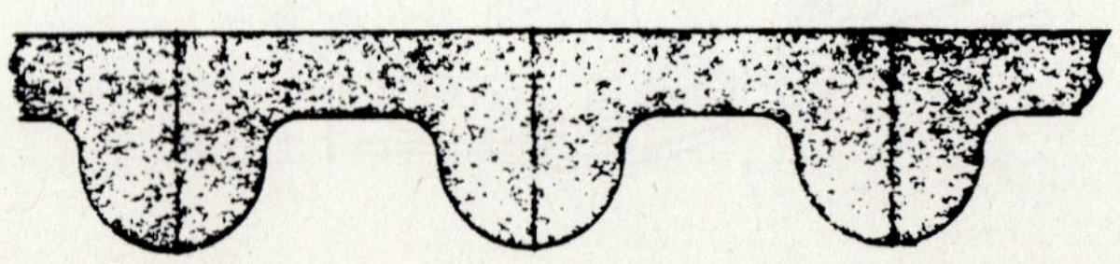

9305ag39

Cross-sectional view of the timing belt with High Torque Drive (HTD) teeth—Porsche 1983 and later 4.7L engine

30. Connect the throttle cable, the cruise control cable and the transmission cable, if equipped with automatic transmission.
31. Connect the cable retainers.
32. Install the oil dipstick and fill tubes.
33. Install the air intake hoses.
34. Connect the negative battery cable.

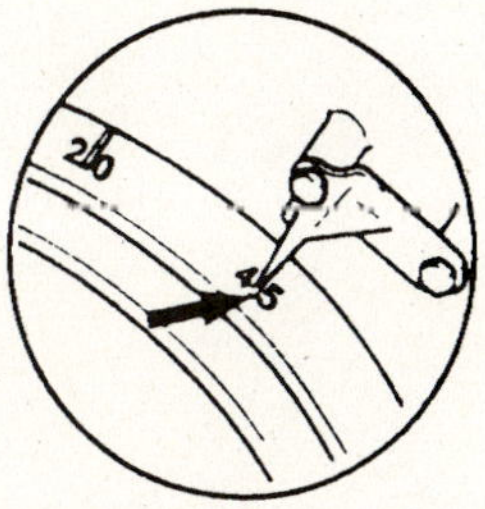

9305ag50

View of the crankshaft 45 degrees BTDC timing alignment marks—Porsche 5.0L engine

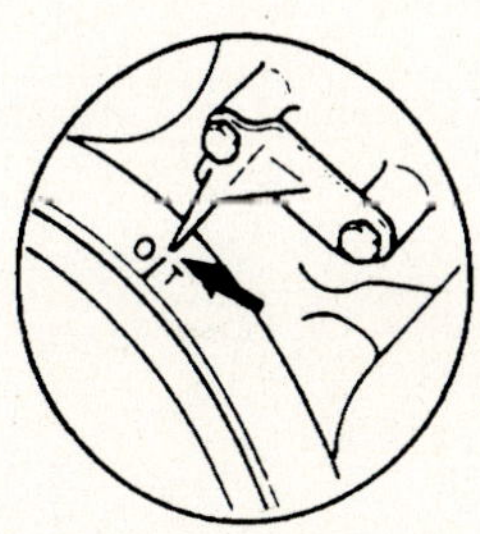

9305ag51

Top Dead Center alignment—Porsche 5.0L engine

Saab

2.5L (B258L) ENGINES

YEAR, MODEL & ENGINE APPLICATION

1994-98 Saab 900	2.5L Code B258L

REPLACEMENT INTERVALS

Saab recommends timing belt replacement every 30,000 miles.

LABOR TIMES

1994-98	
Replacement	2.5 hrs
Severe Service	3.0 hrs

8807SABg01

✳✳ CAUTION

These engines are an interference type. If the timing belt has failed, valve-to-piston damage may have occurred. A compression check should always be performed to check for engine damage.

1. Disconnect the negative battery cable.

✳✳ WARNING

To avoid damage to the valves, DO NOT rotate the camshafts once the timing belt is removed. The crankshaft may only be turned 0–60 degrees BTDC when the camshafts are locked in position with the appropriate locking tool.

2. Remove the air cleaner.
3. Remove the accessory drive belt and tensioner.
4. Remove the power steering and water pump pulleys.
5. Remove the timing belt cover.
6. Raise and support the vehicle.
7. Remove the right front wheel and the cover in the wheel well.
8. Remove the 6 bolts holding the crankshaft pulley and remove the pulley. Do not remove the center bolt.
9. Put the No. 1 cylinder in the Top Dead Center (TDC) on the compression stroke.
10. The timing marks on the crankshaft and camshafts should be in alignment with their respective marks on the engine. Insert Camshaft Locking tool 83-94-926 and install Crankshaft Locking tool 83-94-868 or their equivalents.
11. If reusing the timing belt, mark the direction of its rotation. To help with refitting, the belt can be marked at the camshaft timing marks and also at the crankshaft timing mark.
12. Remove the tensioning roller, both adjusting rollers and the timing belt.
13. Release tension from and remove the timing belt. Loosen the timing belt adjuster bolts.
14. Rotate the crankshaft back to 60 degrees BTDC, to prevent damage to the valves.
15. Remove the bracket with the upper timing belt adjuster and tensioner rollers.

To install:

16. Install the bracket with the upper timing belt adjuster and tensioner pulleys.

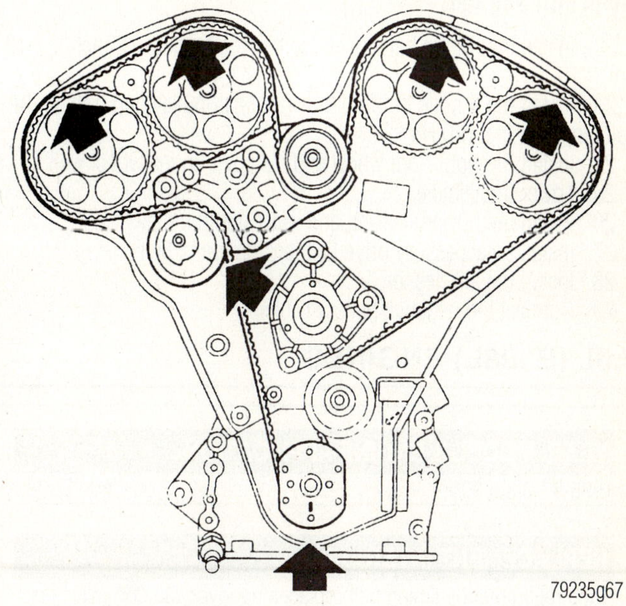

79235g67

Timing mark alignment for proper belt service—Saab 2.5L (B258L) engine

17. Install and locking tools 83-94-926 or equivalent, between the camshaft sprockets to lock the camshafts of both heads in position.
18. Rotate the crankshaft forward to just before 0 degrees TDC and install the crankshaft locking tool on the crankshaft. Carefully rotate the engine until the arm of the tool is against the water pump flange. Be sure the crankshaft is at 0 degrees TDC and all timing marks are aligned. Remove the locking tool.
19. Install the timing belt, noting the direction of rotation marked at disassembly.
    a. Adjust the tensioner lightly by hand counterclockwise to keep the timing belt from falling off. Be sure the crankshaft is at 0 degrees TDC and all timing marks are aligned.
    b. Install the camshaft locking tool.
    c. Install tool 83-93-985 or equivalent, with a cut piece from an old timing belt, to measure belt tension.
    d. Snug the center bolts of the adjusting rollers. Turn the lower adjusting roller counterclockwise, until a belt tension of 202–221 ft. lbs. (275–300 Nm) is registered on the torque wrench.
    e. Tighten the adjusting roller center bolts to 30 ft. lbs. (40 Nm).

■ **This adjustment of the timing belt is only preparatory and should, therefore, not be used as a final check.**

20. Adjust the tensioner pulley until the marks are aligned. Tighten the tensioner pulley to 15 ft. lbs. (20 Nm). Remove the camshaft locking tool on camshaft sprockets No. 1 and 2. Adjust the upper adjusting roller until sprocket No. 2 moves 0.04–0.08 in. (1–2mm) clockwise. Tighten the upper adjusting roller to 30 ft. lbs. (40 Nm) and remove the upper locking tool.
21. Rotate the engine 2 complete revolutions to just before 0 degrees TDC and install the locking tool on the crankshaft. Carefully turn the crankshaft until the arm of the locking tool is against the water pump flange and tighten the locking tool. Set the camshaft locking tool into position on the front of the camshaft sprockets. Be sure that the timing marks on the camshaft sprockets are aligned with the marks on the tool and that the edge of the timing belt is flush with the edge of the camshaft sprockets.

■ **Also check that the alignment marks on the tensioner pulley is still aligned.**

22. Install the timing belt cover. Tighten the bolts to 72 inch lbs. (8 Nm).
23. Install the 6 bolts holding the crankshaft pulley and tighten the bolts to 15 ft. lbs. (20 Nm).
24. Install the right front wheel and the cover in the wheel well.
25. Lower the vehicle.
26. Install the power steering and water pump pulleys.
27. Install he accessory drive belt tensioner and belt.
28. Install the air cleaner.
29. Connect the negative battery cable.

3.0L (B308L) ENGINES

YEAR, MODEL & ENGINE APPLICATION	
1995-97 Saab 9000	3.0L Code B308L

REPLACEMENT INTERVALS

Saab recommends timing belt replacement every 30,000 miles on 1995-96 models, and 54,000 miles on 1997 models.

LABOR TIMES	
1995-97	
Replacement	2.6 hrs
Severe Service	3.1 hrs

8807SABg02

✲✲ CAUTION

These engines are an interference type. If the timing belt has failed, valve-to-piston damage may have occurred. A compression check should always be performed to check for engine damage.

1. Disconnect the negative battery cable.

✲✲ WARNING

To avoid damage to the valves, DO NOT rotate the camshafts once the timing belt is removed from the engine. The crankshaft may only be turned between 0–60 degrees BTDC when the camshafts are locked in position with the appropriate locking tool.

2. Remove the power steering reservoir.
3. Remove the torque arm.
4. Remove the engine mounting.
5. Remove the air cleaner.
6. Remove the accessory drive belt and tensioner.
7. Remove the power steering pump and position aside.
8. Remove the timing belt cover.
9. Raise and support the vehicle.
10. Remove the right front wheel and the cover in the wheel well.
11. Remove the 6 bolts holding the crankshaft pulley and remove the pulley. Do not remove the center bolt.
12. Rotate the crankshaft to Top Dead Center (TDC) of No. 1 cylinder.
13. The timing marks on the crankshaft and camshafts should be in alignment with their respective marks on the engine. Install camshaft locking tools (such as Saab tools KM-800-1 for camshaft sprockets No. 1 and 2 and KM-800-2 for sprockets No. 3 and 4) and a flywheel locking tool (such as Saab tool 83-94-868).
14. Mark the direction of rotation of the timing belt for reassembly.
15. Release tension from and remove the timing belt. Loosen the timing belt adjuster bolts.
16. Rotate the crankshaft back to 60 degrees BTDC, to prevent damage to the valves.
17. Remove the bracket with the upper timing belt adjuster and tensioner rollers.

To install:

18. Remove the flywheel locking tool and install the flywheel inspection cover.
19. Install the bracket with the upper timing belt adjuster and tensioner pulleys.
20. Install both camshaft locking tools.
21. Rotate the crankshaft forward to just before 0 degrees TDC and install the crankshaft locking tool on the crankshaft. Carefully rotate the engine until the arm of the tool is against the water pump flange. Be sure the crankshaft is at 0 degrees TDC and all timing marks are aligned. Remove the locking tool.
22. If reusing the belt, fit the timing belt according to its marked direction of rotation and timing marks. Adjust the tensioning roller loosely by hand to prevent the belt from slipping out of the cogs. Always adjust counterclockwise.
23. Measure the belt tension with Saab tool 83-93-985 or equivalent.
24. Snug the center bolts of the adjusting rollers. Turn the lower adjusting roller counterclockwise, until a belt tension of 202–220 ft. lbs. (275–300 Nm) is reached. Tighten the adjusting roller center bolts to 30 ft. lbs. (40 Nm).

■ **This is a preliminary adjustment of the belt tension and must not be used as a check when the belt is finally adjusted.**

25. Continue to carry out the adjustment by means of the tensioning roller, mark against mark. Remove the locking tool for camshaft sprock-

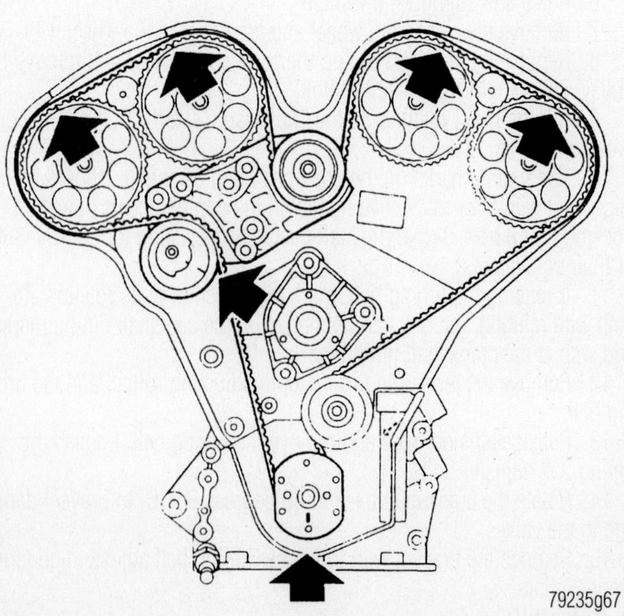

79235g67

Timing mark alignment for proper belt service—Saab 3.0L (B308L) engine

ets No. 1 and 2. Carry out the final adjustment with the upper center adjusting roller until camshaft sprocket No. 2 moves 0.04–0.08 in. (1–2mm) forward.

26. Remove the locking tool for camshaft sprockets No. 3 and 4 and also remove the crankshaft locking tool.

27. Tighten the tensioning roller to 15 ft. lbs. (20 Nm). Tighten the upper adjusting roller to 30 ft. lbs. (40 Nm) and tighten the lower adjusting roller to 15 ft. lbs. (20 Nm).

28. Rotate the engine 2 complete revolutions to just before 0 degrees TDC and install the locking tool on the crankshaft. Carefully turn the crankshaft until the arm of the locking tool is against the water pump flange and tighten the locking tool. Set Saab tool KM-800-20 or equivalent, into position. Be sure that the timing marks on the camshaft sprockets are aligned with the marks on the tool and that the edge of the timing belt is flush with the edge of the camshaft sprockets.

■ **Check that the alignment marks on the tensioner pulley are still aligned.**

29. If necessary, install the crankshaft pulley and tighten the retaining bolts to 15 ft. lbs. (20 Nm).

30. Install the timing belt cover. Tighten the bolts to 72 inch lbs. (8 Nm).

31. Install the 6 bolts holding the crankshaft pulley and tighten the bolts to 15 ft. lbs. (20 Nm).

32. Install the power steering pump.

33. Install the accessory drive belt tensioner and drive belt.

34. Install the air cleaner.

35. Install the engine mounting.

36. Install the torque arm.

37. Install the power steering reservoir.

38. Connect the negative battery cable.

3.0L (B308E/Z) ENGINES

YEAR, MODEL & ENGINE APPLICATION	
1999-03 Saab 9.5	3.0L Code B308E/Z

REPLACEMENT INTERVALS

Saab recommends timing belt replacement every 60,000 miles.

LABOR TIMES	
1999-03	
Replacement	2.8 hrs
Severe Service	3.3 hrs

8807SABg03

✻✻ CAUTION

These engines are an interference type. If the timing belt has failed, valve-to-piston damage may have occurred. A compression check should always be performed to check for engine damage.

1. Remove negative battery cable.
2. Remove the lower spoiler sections.
3. Insert 2 wedges, 83 95 238, between oil sump and subframe and between the gearbox and subframe.
4. Remove or disconnect the following:

- MAF sensor with hose
- Power steering hose from holder
- Right-hand engine attachment with bracket and engine mount
- Retaining bolts from water pump pulley and power steering pump pulley
- Release tension on auxiliary drive belt
- Water pump pulley and power steering pump pulley
- Belt tensioner and generator air intake
- Timing cover
- Crankshaft pulley

5. Zero the engine. The markings on the camshaft sprockets and timing cover should be aligned. Use adjustment tools for cam sprockets and timing belt, KM-800-20. The crankshaft markings should also be aligned. Fit locking tools, KM-800-1 and KM-800-2, for the camshaft sprockets and locking tool, KM-800-10, for the crankshaft.

6. Remove the tensioning pulley and adjusting rollers. Use counterstay tool, 83 94 983, on the adjusting rollers.

7. Mark the direction of rotation on the belt. The belt can also be marked with the respective camshaft markings and crankshaft marking to facilitate reinstallation.

8. Remove locking tool from the crankshaft. Let camshaft locking tools remain.

9. Remove the lower adjusting roller. Ensure the washer behind the roller is also removed.

10. Remove the timing belt.

To install:

11. Position the camshaft belt in the correct direction of rotation. Use adjustment tools for cam sprockets and timing belt, KM-800-30, to hold the belt in place.

12. Fit the lower tensioning pulley loosely. Make sure the washer is in place.

13. Lightly adjust the tensioning pulley by hand so the belt does not jump. Adjust counterclockwise.

14. Make sure the markings on the camshaft belt and the corresponding markings on the camshaft sprockets and crankshaft are aligned. Remove the tools, KM-800-30.

15. Install a locking tool, KM-800-10, and position a piece of camshaft belt, fitting a belt tension meter, 83 93 985, to measure the belt tension.

✻✻ CAUTION

Adjustment of belt tension is only a preparatory measure and must not be used as a check when the belt is finally adjusted

16. Tighten the center bolts of the adjuster rollers lightly. Adjust the lower adjuster roller counterclockwise until a tension of 239–260 ft. lbs. (275–300 Nm) is obtained.

17. Use tool 83 94 983 to tighten the adjuster roller bolt. Use a counterstay tool, 83 94 983. Check the belt tension again and adjust as necessary. Torque adjuster roller bolt to 30 ft. lbs. (40 Nm).

18. Remove the belt tension meter and piece of the toothed belt.

■ **Adjustment of belt tension is only a preparatory measure and must not be used as a check when the belt is finally adjusted.**

19. Ensure the adjusting nut on the upper adjusting roller rotates freely and is not tight. Adjust the tensioning pulley with a 5mm Allen key until the 2 lines on the tensioning pulley correspond with each other. Tighten tensioning pulley bolt to 15 ft. lbs. (20 Nm).

20. Remove the locking tool for camshaft sprockets. Fit tool KM-800-20. Then, adjust the upper adjusting roller counterclockwise with tool 83 94 983 until the markings on the camshaft sprocket are aligned

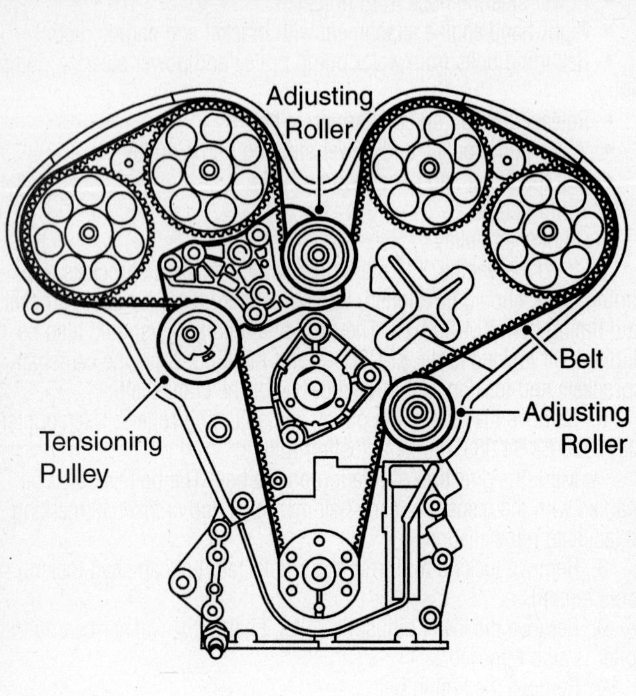

Timing belt routing—9-5 3.0L (B308E/Z) engine

with the corresponding markings on KM-800-20. Tighten camshaft sprocket bolts, using a counterstay, 83 94 983. Tighten to 30 ft. lbs. (40 Nm).

21. Remove KM-800-20 and the other locking tools.
22. Rotate the engine 2 revolutions in the direction of rotation until just before the zero mark and fit a locking tool, KM-800-10, on the crankshaft.
23. Carefully rotate the crankshaft in the direction of rotation of the engine until the arm is lying against the water pump and tighten it.
24. Fit gauge tool KM-800-20 and ensure the markings on the camshaft sprocket are opposite the markings on the tool and that the edge of the belt is aligned with the edge of the sprocket.
25. Raise the car and install the crankshaft pulley, torquing the bolts to 15 ft. lbs. (20 Nm).
26. Lower the car.
27. Install or connect the following:
    - Timing cover, tightening to 6 ft. lbs. (8 Nm)
    - Auxiliary drive belt tensioner and the generator air intake
    - Water pump pulley and the power steering pump pulley. Use Loctite, part No. 74 96 292. Do not tighten the bolts.
    - Auxiliary drive belt
    - Water pump pulley bolts to 6 ft. lbs. (8 Nm).
    - Power steering pump pulley bolts to 15 ft. lbs. (20 Nm)
    - Right-hand engine attachment with bracket and engine mount. Torque bolts to 47 ft. lbs. (63 Nm)
    - Power steering hose in holder
    - MAF sensor with hose
    - Lower the engine by carefully removing the wedges
    - Lower spoiler sections

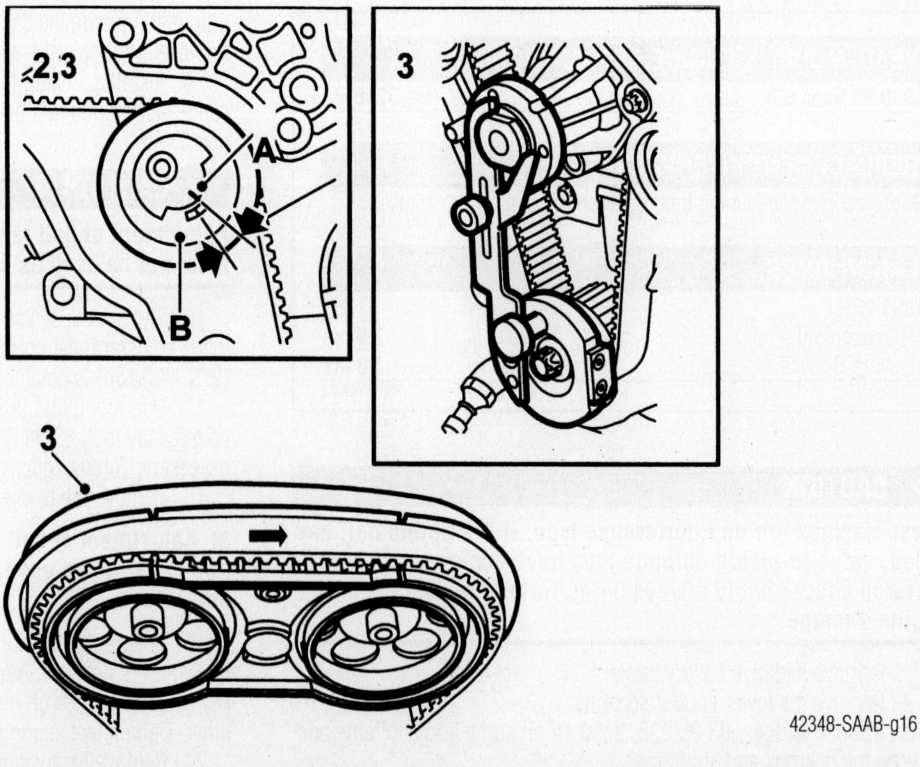

Timing belt is installed noting direction of rotation, using a counterstay device, and properly setting position of tensioning pulley—9-5 3.0L (B308E/Z) engine

Sterling

2.5L (C25A1) AND 2.7L (C27A1) ENGINES

YEAR, MODEL & ENGINE APPLICATION

1987-88 Sterling 825	2.5L Code C25A1
1989-91 827	2.7L Code C27A1

REPLACEMENT INTERVALS

Sterling does not provide a replacement interval. Standard industry practice is to replace timing belts every 60,000 miles.

LABOR TIMES

1987-91	
Replacement	4.5 hrs
Severe Service	5.0 hrs

8807STRg01

✱✱ CAUTION

These engines are an interference type. If the timing belt has failed, valve-to-piston damage may have occurred. A compression check should always be performed to check for engine damage.

1. Be sure to acquire the anti-theft code for the radio and the frequencies for the radio's preset buttons.
2. Disconnect the negative battery cable.
3. Rotate the crankshaft pulley so that the No. 1 piston is at Top Dead Center (TDC) of its compression stroke.
4. If necessary, remove the strut brace located at the top rear of the engine.
5. Remove the pulley and harness cover from above the upper timing belt cover and remove the engine harness clamp.
6. Use a suitable device and safely support the engine.
7. Remove the engine support bolts, loosen the side mount rubber and raise the side bracket.
8. Remove the lower splash shield.
9. Loosen the alternator adjusting rod, locknut and mounting bolt; then, remove the alternator belt.
10. Loosen the idler pulley center nut and adjusting bolt; then, remove the air conditioning compressor belt.
11. Loosen the power steering pump adjusting bolt, locknut and mounting bolt; then, remove the pump belt.
12. Remove the timing belt upper cover.
13. Remove the crankshaft pulley.
14. Remove the timing belt lower cover.
15. Loosen the timing belt adjusting nut 180 degrees (1/2) turn. Push the tensioner to relieve the tension from the timing belt; then, retighten the adjusting bolt. Remove the timing belt.

To install:

16. Remove the spark plugs.
17. Advance the crankshaft about 15° from the No. 1 TDC position.
18. Adjust the camshaft pulleys to the No. 1 TDC position by aligning the camshaft timing marks with the indentations on the camshaft housing.
19. Return the crankshaft sprocket to the No. 1 TDC position.
20. Use a screwdriver and push the tensioner bracket forward to loosen the tensioner.
21. Advance the rear camshaft sprocket about (1/2) turn from TDC.
22. Install the timing belt on the crankshaft sprocket, front camshaft sprocket, water pump pulley, tensioner pulley and rear camshaft pulley.
23. Loosen and retighten the timing belt adjusting bolt to tension the timing belt.
24. Rotate the crankshaft about 5–6 turns clockwise to position the timing belt on the pulleys.
25. To adjust the timing belt tension, perform the following procedure:
    a. Make sure that the No. 1 piston is at TDC of its compression stroke.
    b. Rotate the crankshaft clockwise 10 teeth on the camshaft pulley; the blue mark on the crankshaft pulley should align with the lower cover pointer.
    c. Loosen the adjusting nut 180 degrees (1/2 turn).
    d. Tighten the adjusting nut to 31 ft. lbs. (42 Nm).
26. Install the spark plugs.
27. Install the lower timing belt cover.
28. Install the crankshaft pulley and tighten the bolt to 83 ft. lbs. (115 Nm).
29. Make sure the crankshaft and camshaft pulleys are at TDC.

■ **If the camshaft or crankshaft pulley is not at TDC, remove the timing belt and re-perform the adjustment procedure.**

30. To complete the installation, reverse the removal procedures. Adjust the tension of the drive belts.
31. Connect the negative battery cable.
32. Enter the anti-theft code for the radio and reset the frequencies.

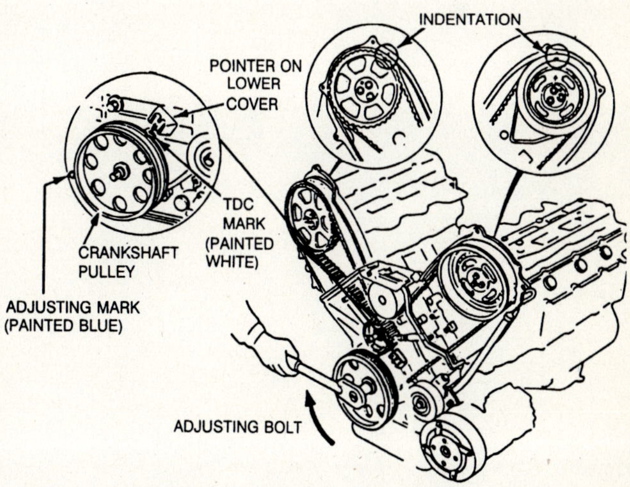

8807Timg11

Crankshaft and camshaft sprocket timing mark locations—Sterling 2.5L (C25A1) and 2.7L (C27A1) engine

Subaru

1.2L (VIN 7, 8) ENGINES

YEAR, MODEL & ENGINE APPLICATION

1985-95 Subaru Justy .... 1.2L VIN 7 & 8

REPLACEMENT INTERVALS

Subaru does not provide a replacement interval. Standard industry practice is to replace timing belts every 60,000 miles.

LABOR TIMES

1985-95	
Inner belt	
Replacement	1.4 hrs
Severe Service	1.7 hrs
Outer belt	
Replacement	0.8 hrs
Severe Service	1.0 hrs

8807SUBq01

✻✻ CAUTION

These engines are considered freewheeling. If the timing belt fails, valve-to-piston damage is unlikely, however a precautionary compression check should be performed to verify no engine damage.

1. Disconnect the negative battery cable.
2. Remove the air cleaner.
3. Remove the accessory drive belt.
4. Remove the crankshaft pulley.
5. Remove the front timing cover.
6. Rotate the crankshaft so that the camshaft pulley timing mark aligns with the timing mark on the back cover, and the crankshaft pulley timing mark aligns with the timing mark on the engine.
7. Loosen the timing belt tensioner lockbolt and pivot bolt. Push the tensioner away from the timing belt and tighten the tensioner lockbolt.
8. Remove the camshaft drive plate.
9. Remove the timing belt.
10. Remove the belt tensioner, spring and crankshaft drive pulley.
11. Hold the camshaft sprocket with special tool 499205500 and remove the camshaft sprocket bolts and sprocket.
12. Remove the inner belt cover and mount.
13. Remove the inner belt.

To install

14. Remove the valve cover and loosen the valve adjustment so that the camshaft can rotate freely. This step is necessary for proper timing belt tension.
15. Ensure that the camshaft and crankshaft timing marks align.
16. Install the inner belt.
17. Install the inner belt cover and mount.
18. Install the crankshaft sprocket and tighten to 108 inch lbs. (12 Nm).
19. Install the belt tensioner, spring and crankshaft drive pulley.
20. Install the timing belt.
21. Loosen the tensioner lockbolt and allow the tensioner to operate.
22. Tighten the tensioner pivot bolt, then the lockbolt.

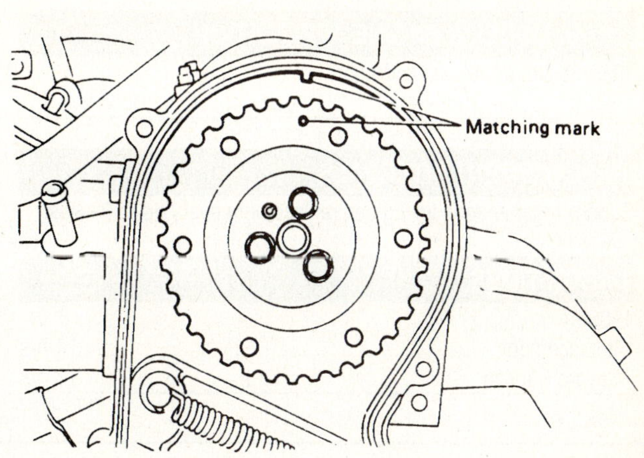

9305ag69

Camshaft sprocket timing marks—Subaru 1.2L (VIN 7, 8) engine

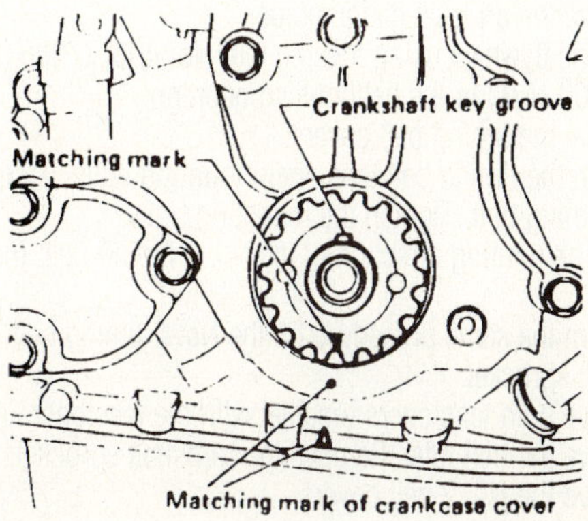

9305ag70

Crankshaft timing marks—Subaru 1.2L (VIN 7, 8) engine

23. Install the camshaft drive plate.
24. Rotate the crankshaft 2 complete turns and check that the camshaft and crankshaft timing marks align.
25. Install the front timing cover.
26. Install the crankshaft pulley and tighten the bolt to 58–72 ft. lbs. (79–98 Nm).
27. Adjust the valves. Intake valve clearance is 0.005 in. (0.12mm). Exhaust valve clearance is 0.009 (0.22mm).
28. Install the valve cover.
29. Install the accessory drive belt and the air cleaner.
30. Connect the negative battery cable.

1.8L (VIN 4, 5, 7) ENGINES

✻✻ CAUTION

These engines are considered freewheeling. If the timing belt fails, valve-to-piston damage is unlikely, however a precautionary compression check should be performed to verify no engine damage.

YEAR, MODEL & ENGINE APPLICATION

1985-89 Coupe, Sedan & Wagon,
1990-94 Loyale, 1985-91 XT .... 1.8L VIN 4, 5, 7

REPLACEMENT INTERVALS

Subaru recommends timing belt replacement every 60,000 miles.

LABOR TIMES

1985-94

Replacement	3.1 hrs
Severe Service	3.7 hrs

8807SUBg02

■**This engine is equipped with 2 timing belts.**

1. Disconnect the negative battery cable.
2. Loosen the water pump pulley bolts and alternator mounting bolts and then remove the accessory drive belt.
3. Remove the water pump pulley.
4. Disconnect the oil pressure switch connector.
5. Remove the oil level gauge guide.
6. Lock the flywheel using locking tool No. 498277000 (M/T) or No. 498497000 (A/T) in the bellhousing opening.
7. Remove the timing belt covers.
8. Loosen the timing belt tensioner mounting bolts 1/2 turn and slacken the timing belt. Tighten the mounting bolts.
9. Mark the rotating direction of the No. 1 timing belt, then remove the belt.
10. Perform the same procedure for the No. 2 timing belt. Remove the crankshaft sprocket.
11. Remove both tensioners together with the tensioner springs.
12. Remove the belt idler. Remove the camshaft sprockets.
13. Remove the No. 2 belt covers.

To install

14. Inspect the timing belt for breaks, cracks and wear. Replace as required.
15. Check the belt tensioner and idler for smooth rotation. Replace if noisy or excessive play is noticed.
16. Install the left side belt cover seal No. 3 to the cylinder block.
17. Install the left side belt cover seal, left side belt cover seal No. 4, and belt cover mount to the right rear belt cover, then install the assembly on the cylinder block. Tighten to 34 ft. lbs. (45 Nm).
18. Install the left side belt cover seal No. 2 and belt cover mounts to left side belt cover No. 2, then install to the cylinder head and camshaft case. Tighten to 34 ft. lbs. (45 Nm).
19. Install the right side belt cover seal, belt cover seal No. 2 and belt cover mounts to the right side belt cover No. 2, then install to the cylinder head and camshaft case. Tighten to 34 ft. lbs. (45 Nm).
20. Install the camshaft sprockets to the right and left camshafts. Tighten the bolts gradually in 2–3 steps to 67 ft. lbs. (91 Nm).
21. Attach the tensioner spring to the tensioner, then install to the right side of the cylinder block. Tighten the bolts temporarily by hand.
22. Attach the tensioner spring to the bolt, tighten the right side bolt, then loosen it 1/2 turn.
23. Push down the tensioner until it stops, then temporarily tighten the left bolt.
24. Install the left side tensioner in the same manner.
25. Install the belt idler to the cylinder block using care not to turn the seal. Tighten to 29–35 ft. lbs. (39–47 Nm).
26. Install the sprockets on the crankshaft. Install the crankshaft pulley and tighten the bolt temporarily.
27. Align the center of the 3 lines scribed on the flywheel with the timing mark on the flywheel housing.
28. Align the timing mark on the left side camshaft sprocket with the notch on the belt cover.
29. Attach timing belt No. 2 to the crankshaft sprocket No. 2, oil pump sprocket, belt idler and camshaft sprocket in that order. Avoid downward slackening of the belt.
30. Loosen tensioner No. 2 lower bolt 1/2 turn to apply tension. Push timing belt by hand to ensure smooth movement of tensioner.
31. Apply 25 ft. lbs. (new belt) or 18 ft. lbs. (used belt) torque to the camshaft sprocket in counterclockwise direction. While applying torque tighten tensioner No. 2 lower bolt temporarily, then tighten upper bolt temporarily.
32. Tighten the lower bolt, then the upper bolt to 13–15 ft. lbs. (17–20 Nm) in that order.
33. Check that the flywheel timing mark and drivers side camshaft sprocket marks are in their proper positions.
34. Turn the crankshaft 1 turn clockwise from the position where timing belt No. 2 was installed and align the center of the 3 lines on the flywheel with the timing mark on the flywheel housing.
35. Align the timing mark on the right side camshaft sprocket with the notch in the belt cover.

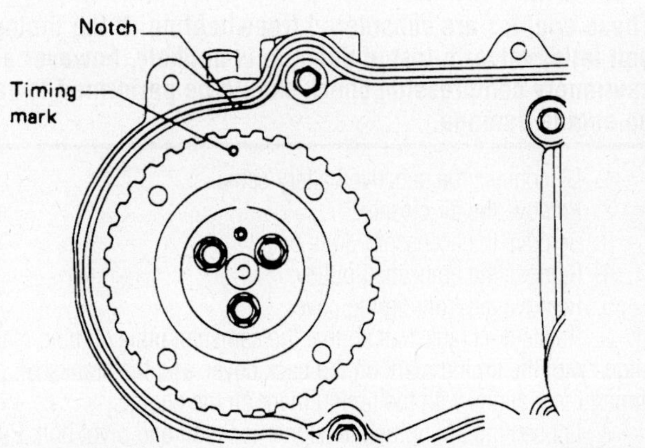

9305ag64

Aligning the right camshaft timing marks—Subaru 1.8L (VIN 4, 5, 7) engine

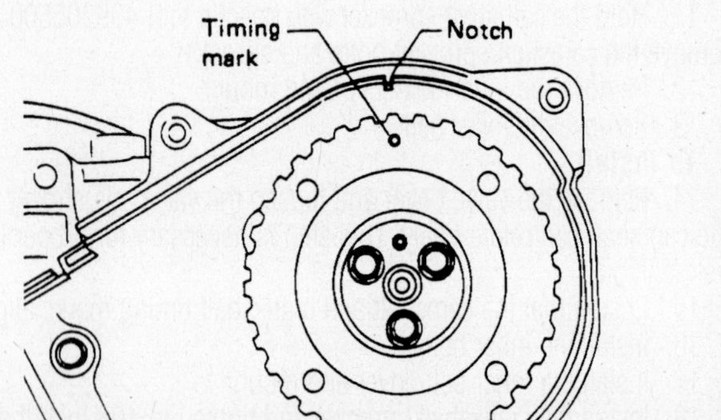

9305ag65

Aligning the left camshaft timing marks—Subaru 1.8L (VIN 4, 5, 7) engine

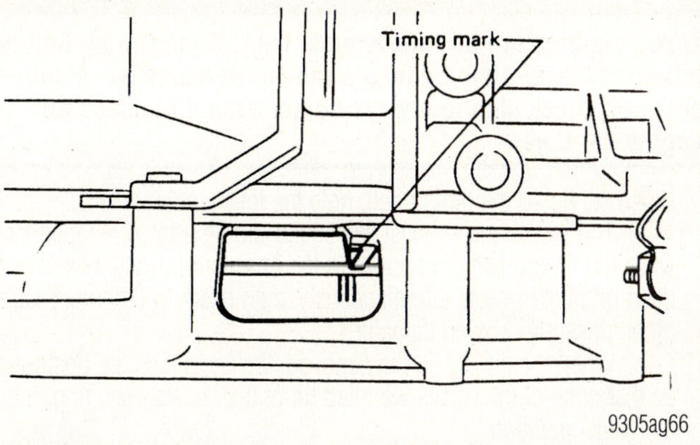

Aligning the flywheel timing marks—Subaru 1.8L (VIN 4, 5, 7) engine

36. Attach the timing belt to the crankshaft sprocket and camshaft sprocket, avoiding slackening of the belt on the upper side.
37. Loosen the tensioner 1/2 turn to apply tension to the belt. Push the belt by hand to ensure smooth operation.
38. Apply 25 ft. lbs. (34 Nm) for new belts or 18 ft. lbs. (24 Nm) for used belts, tighten to the camshaft sprocket in the counterclockwise direction. While applying torque, tighten the tensioner left bolt temporarily, then tighten right bolt temporarily.
39. Tighten the left bolt, then the right bolt to 13–15 ft. lbs. (17–20 Nm) in that order.
40. Check that the flywheel timing mark and left side camshaft sprocket marks are in their proper positions.
41. Remove the crankshaft pulley.
42. Install the right front belt cover seals and belt cover plug. Install the belt covers to the cylinder block.
43. On turbocharged engines, install the belt cover plate.
44. Install the crankshaft pulley and tighten to 66–79 ft. lbs. (89–107 Nm).
45. Install the water pump pulley and tighten to 67 ft. lbs. (91 Nm). Install the pulley cover, oil level guide and gauge and oil pressure switch connector.
46. Install and properly tension the accessory drive belt.
47. Connect the negative battery cable.

1.8L (VIN 2) ENGINES

YEAR, MODEL & ENGINE APPLICATION

1993-97 Subaru Impreza	1.8L VIN 2

REPLACEMENT INTERVALS

Subaru recommends timing belt replacement every 105,000 miles on California models, or 60,000 miles on all other models.

LABOR TIMES

1993-97	
Replacement	2.4 hrs
Severe Service	2.7 hrs

8807SUBg03

✻✻ CAUTION

These engines are considered freewheeling. If the timing belt fails, valve-to-piston damage is unlikely, however a precautionary compression check should be performed to verify no engine damage.

The engines use a single cam belt drive system with a serpentine type belt. The left side of the engine uses a hydraulic cam belt tensioner, which is self-adjusting.

1. Disconnect the negative battery cable.
2. Remove the coolant reservoir.

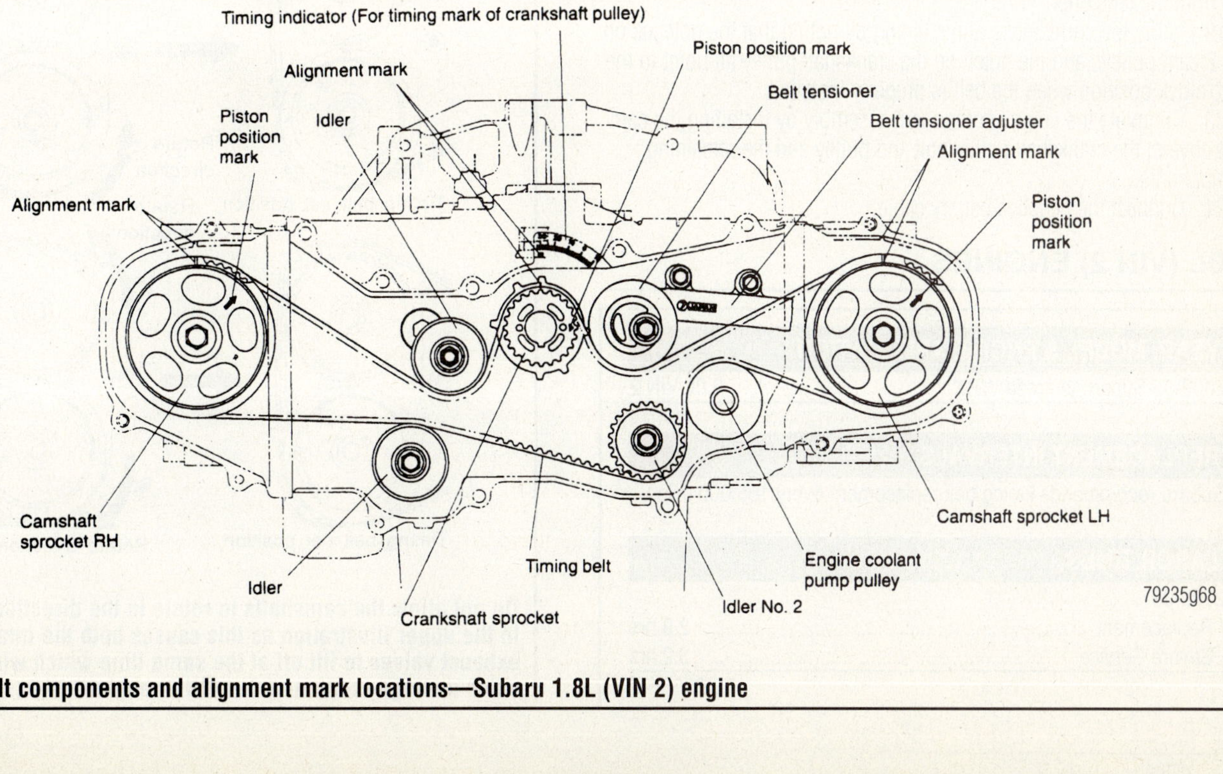

Timing belt components and alignment mark locations—Subaru 1.8L (VIN 2) engine

3. Remove the cooling fans.
4. Remove the accessory drive belt cover and the accessory drive belts.
5. Remove the crankshaft pulley.
6. Remove the timing belt covers.
7. Align the camshaft sprockets so each sprocket notch aligns with the cam cover notches. Align the crankshaft sprocket top tooth notch, located at the rear of the tooth, with the notch on the crank angle sensor boss. Mark the 3 alignment points as well as the direction of cam belt rotation.
8. Loosen the tensioner adjusting bolts. Remove the bottom 3 idlers, the cam belt and the cam belt tensioner. The cam sprockets can then be removed with a modified camshaft sprocket wrench tool.
9. Remove the sprockets, if necessary. Note the reference sensor at the rear of the left cam sprocket.

To install

10. Install the sprockets, if removed and tighten the retaining bolts to 47–54 ft. lbs. (64–74 Nm).
11. Install the crankshaft sprocket and the non-adjustable right side idler. Do not install the tensioner idler at this time.
12. Check that the tensioner pushrod protrusion is 0.60–0.65 in. (15–17mm).
13. Compress the hydraulic tensioner in a vise slowly and temporarily secure the plunger with a pin or suitable Allen wrench. Install the tensioner and the pulley with the adjustable idler pulley. Temporarily tighten the tensioner while the tensioner is pushed to the right.
14. Align the crankshaft sprocket notch on the rear sprocket tooth with the crank angle sensor boss. This places the sprocket notch in the 12 o'clock position.
15. Align the camshaft sprockets with the notches in the cam rear belt cover. This places the sprocket notch in the 12 o'clock position for each camshaft.
16. Install the timing belt with the directional mark and alignment marks properly positioned (if the belt is to be reused).
17. Loosen the tensioner retaining bolts and slide the tensioner to the left. Tighten the mounting bolts.
18. After verifying the timing marks are correct, remove the stopper pin from the tensioner.
19. Verify the correctness of the timing by noting that the notches on the 2 cam pulleys and the notch on the crankshaft pulley all point to the 12 o'clock position when the belt is properly installed.
20. Complete the engine component assembly by installing the cam belt covers, the crankshaft pulley bolt and pulley and the remaining components.
21. Connect the negative battery cable.

2.0L (VIN 2) ENGINES

YEAR, MODEL & ENGINE APPLICATION

2002-05 Subaru Impreza WRX	2.0L VIN 2

REPLACEMENT INTERVALS

Subaru recommends timing belt replacement every 105,000 miles.

LABOR TIMES

2002-05	
Replacement	2.9 hrs
Severe Service	3.2 hrs

8807SUBg04

CAUTION

These engines are an interference type. If the timing belt has failed, valve-to-piston damage may have occurred. A compression check should always be performed to check for engine damage.

When servicing the timing belt, note the following:

a. The intake and exhaust camshafts can be rotated independently when the timing belt is removed. If the intake and exhaust valves are lifted off of their seats simultaneously, their heads will contact each other, possibly causing damage.

b. When the timing belt is removed, the camshafts are positioned so that none of the valves are lifted off of their seats, resulting in a "zero-lift" position.

c. The left-hand cylinder head camshafts must be rotated from the "zero-lift" position as little as possible when orienting it for timing belt installation, otherwise possible valve head interference may occur.

d. Never allow the camshafts to rotate in the direction shown in the accompanying illustration, which would cause both the intake and exhaust valves to lift simultaneously, causing interference.

1. Disconnect the negative battery cable.
2. Remove the accessory drive belt.
3. Remove the crankshaft pulley.
4. Remove the belt cover as follows:
   a. Crankshaft pulley.
   b. Left hand belt cover.
   c. Right hand belt cover.
   d. Front hand belt cover.
5. Remove the timing belt guides on models equipped with a manual transmission.

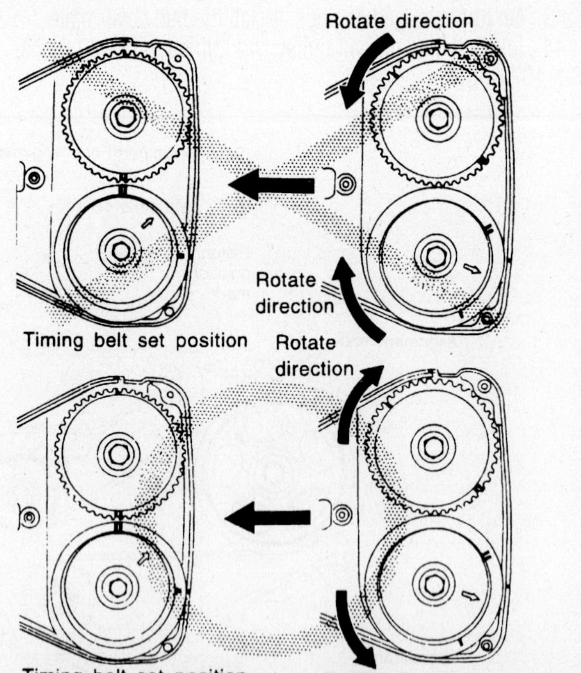

Do not allow the camshafts to rotate in the direction shown in the upper illustration as this causes both the intake and exhaust valves to lift off at the same time which will cause valve damage– Subaru 2.0L (VIN 2) engine

6. If the alignment marks that indicate rotation are faded, put new marks on the belt before removal as follows:

a. Turn the crankshaft using crankshaft sprocket tool 499987500 and a breaker bar to align the crankshaft sprocket, left hand intake camshaft sprocket, left hand exhaust camshaft, right hand intake camshaft sprocket and right hand exhaust camshaft sprocket with the on the cover and cylinder block.

b. Using white paint such as white out, place alignment marks on the belts in relation to the sprockets.

7. Remove the belt idler (A), illustrated in the accompanying illustration.

8. Remove the timing belt.

9. If necessary, remove belt idlers (B) and (C).

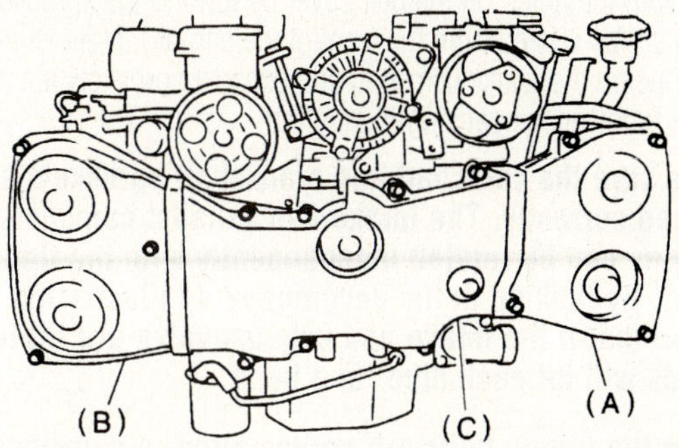

9357tg46

9357tg47

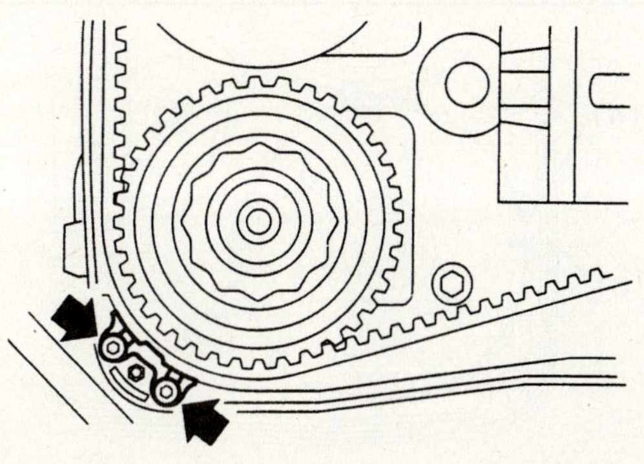
9357tg50

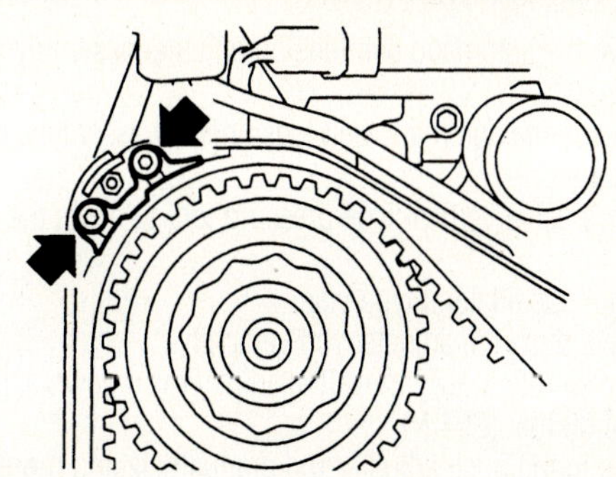
9357tg48

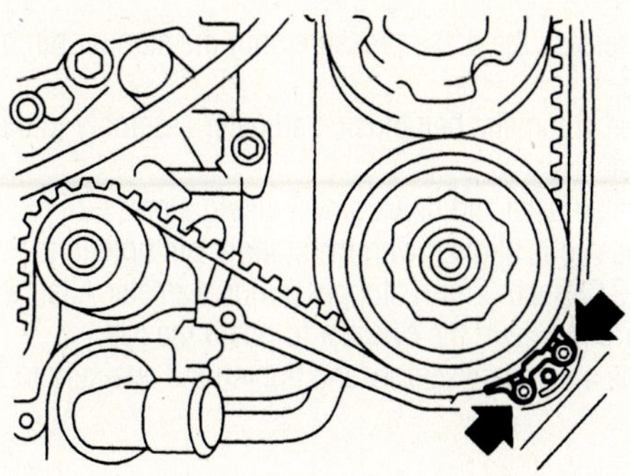
9357tg49

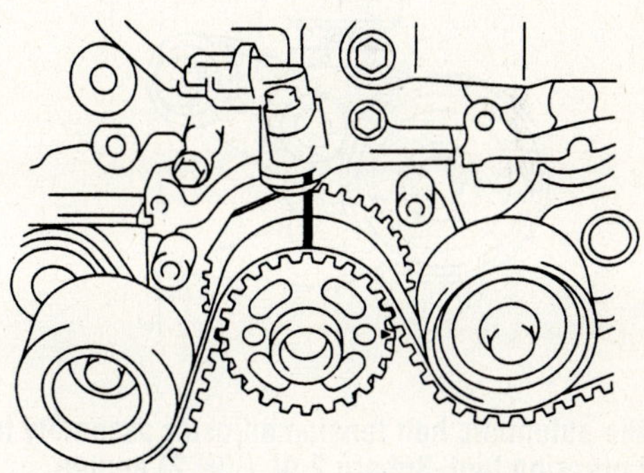
9357tg51

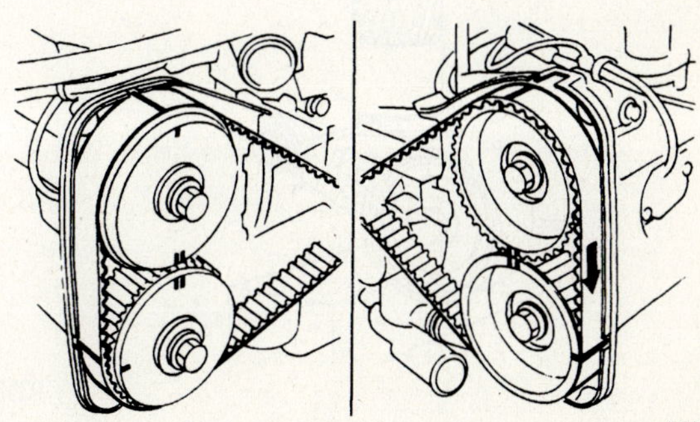
9357tg52

10. Remove the belt idler 2.
11. Remove the automatic belt tension adjuster assembly.

To install

12. To prepare the automatic belt tensioner for assembly, perform the following steps:
   a. Always use a vertical type pressing tool to move the adjuster rod down.
   b. Do not use a lateral type vise.
   c. Always push the adjuster rod vertically.
   d. Make sure to slowly move the adjuster rod down applying a pressure of 66 lbs. (294 N).
   e. Press in the push adjuster rod gradually taking more than 3 minutes.
   f. Never allow the press pressure to exceed 2,205 lbs. (9,807 N).
   g. Press the adjuster rod as far as the end surface of the cylinder. Do not press the rod into the cylinder as doing so may damage the cylinder.
   h. Never release the press pressure until the stopper pin has been fully inserted.
13. Attach the automatic belt tension adjuster assembly to the vertical pressing tool.
14. Move the adjuster rod down slowly using a pressure of 66 lbs. (294 N) until the rod is aligned with the stopper pin hole in the cylinder.
15. Insert a 0.08 inch (2mm) stopper pin or diameter Allen wrench into the stopper pin hole in the cylinder to retain the rod.
16. Install the adjuster assembly and tighten the retainers to 29 ft. lbs. (39 Nm).
17. Install belt idler and tighten the retainers to 29 ft. lbs. (39 Nm).
18. Install the belt idler and tighten to 29 ft. lbs. (39 Nm).
19. Align the mark on the crankshaft sprocket with the mark on the oil pump.
20. Align the single line mark on the right hand exhaust camshaft sprocket with the notch on the belt cover.
21. Align the single line mark on the right hand intake camshaft with the notch on the belt cover. Make sure the double lines on the intake camshaft and exhaust sprockets are aligned as shown in the accompanying illustration.
22. Align the single line mark on the left hand exhaust camshaft sprocket with the notch on the belt cover by turning the sprocket counterclockwise (as viewed from the front of the engine).
23. Align the single line mark on the left hand intake camshaft sprocket with the notch on the belt cover by turning the sprocket counterclockwise (as viewed from the front of the engine). Make sure the double lines on the intake camshaft and exhaust sprockets are aligned as shown in the accompanying illustration.

■ **Make sure the camshaft and crankshaft sprockets are positioned correctly. The intake and exhaust camshafts on this engine can be rotated independently with the timing belt removed. By looking at the accompanying illustration it will show you that if the intake and exhaust valve are lift together the heads will hit each other and bend.**

■ **When the timing belts are not installed, 4 camshafts are held at "zero lift" position, where all cams on the camshafts do not push the intake and exhaust valves down (under this**

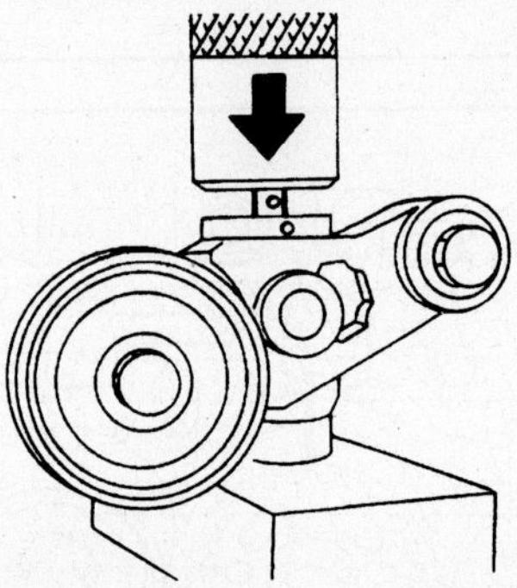

Attach the automatic belt tension adjuster assembly to the vertical pressing tool–Subaru 2.0L (VIN 2) engine

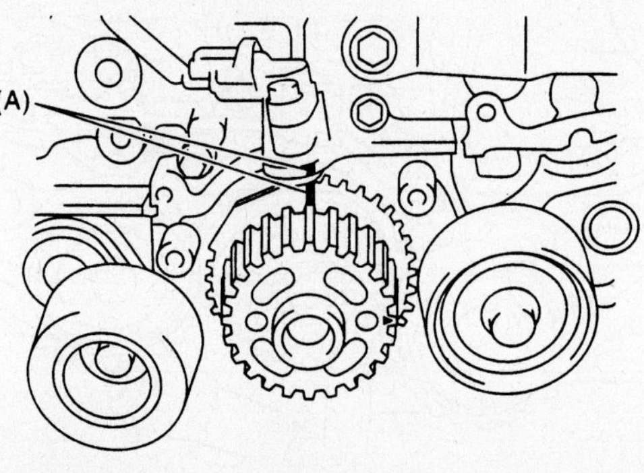

Align the mark on the crankshaft sprocket with the mark on the oil pump–Subaru 2.0L (VIN 2) engine

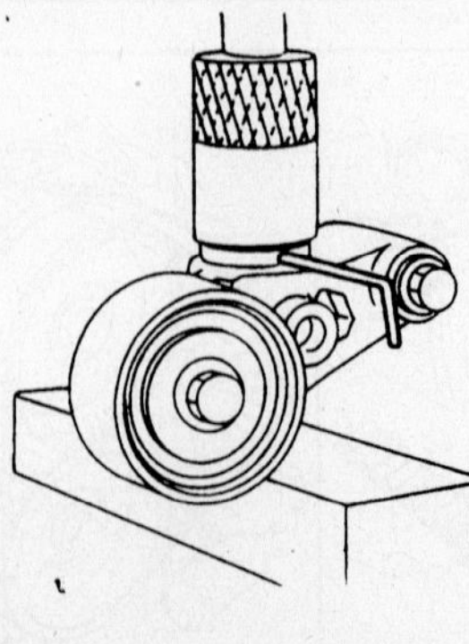

Insert a 0.08 inch (2mm) stopper pin or diameter Allen wrench into the stopper pin hole in the cylinder to retain the rod–Subaru 2.0L (VIN 2) engine

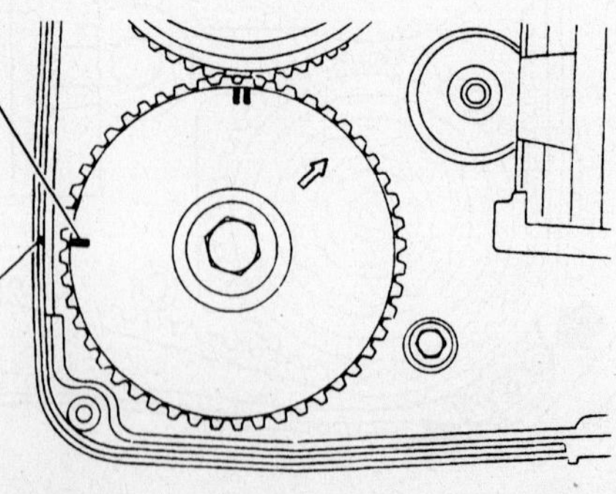

Align the single line mark on the right hand exhaust camshaft sprocket with the notch on the belt cover–Subaru 2.0L (VIN 2) engine

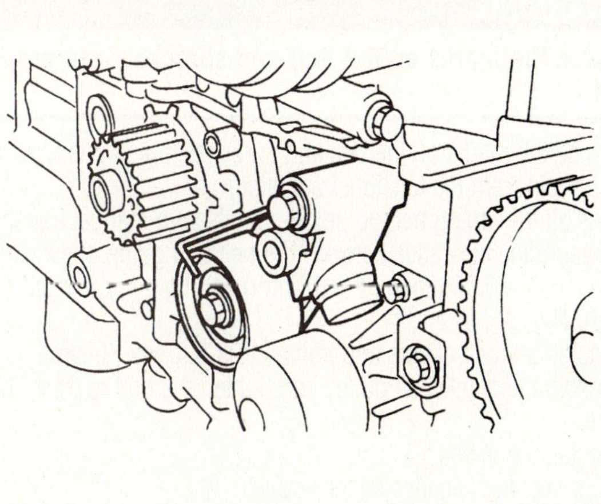

9357tg74

Install the adjuster assembly–Subaru 2.0L (VIN 2) engine

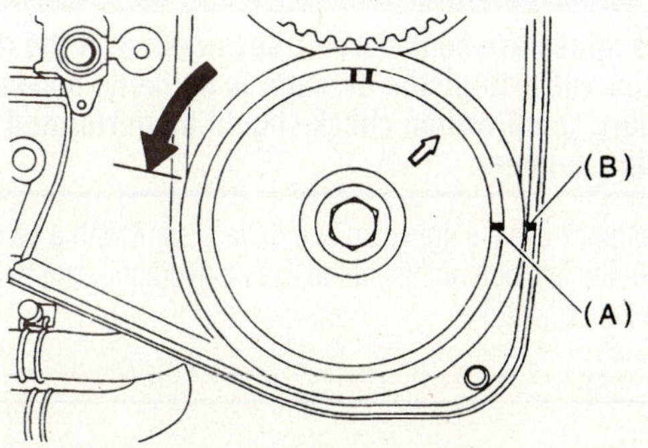

9357tg58

Align the single line mark on the left hand exhaust camshaft sprocket with the notch on the belt cover by turning the sprocket counterclockwise (as viewed from the front of the engine–Subaru 2.0L (VIN 2) engine

condition all valves remain unlifted). When the camshafts are rotated to install the timing belts, # 2 intake and # 4 exhaust cam of the left hand camshafts are held to push their corresponding valves down. Under this condition these valves are held lifted. The right side camshafts are held in so that their cams do not push the valves down.

24. The left hand camshafts must be rotated from the "zero lift" position to the position where the timing belt is to be installed at as small an angle as possible, in order to prevent mutual interference of intake and exhaust valve heads. Do not allow the camshafts to rotate in the direction illustrated as this causes both the intake and exhaust valves to lift off at the same time with will cause valve damage.

25. When installing the belt, make sure to align the marks made during removal or if using a new belt, align the in alphabetical order as shown in the accompanying illustration.

**** WARNING**

Disengagement of more than 3 timing belt teeth may result in contact between the valve and piston. Always make sure the belts rotation is correct.

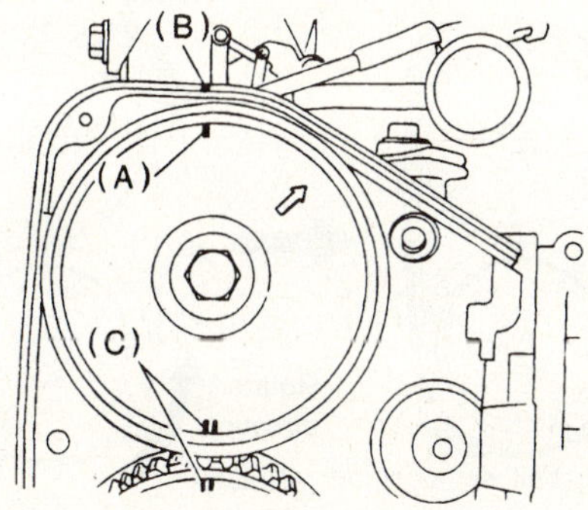

9357tg57

Align the single line mark on the right hand intake camshaft with the notch on the belt cover. Make sure the double lines on the intake camshaft and exhaust sprockets are aligned–Subaru 2.0L (VIN 2) engine

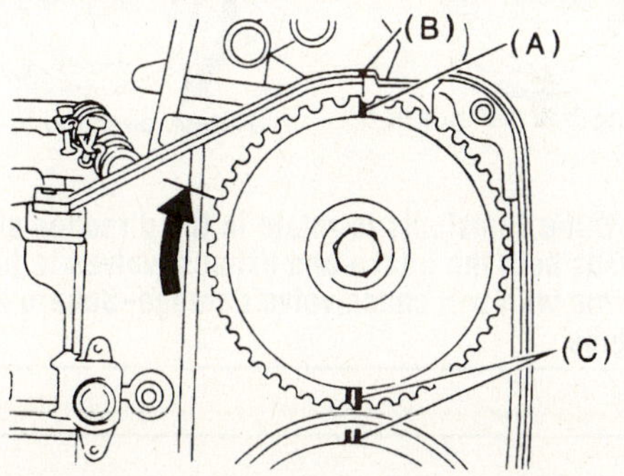

9357tg59

Align the single line mark on the left hand intake camshaft sprocket with the notch on the belt cover by turning the sprocket counterclockwise (as viewed from the front of the engine). Make sure the double lines on the intake camshaft and exhaust sprockets are aligned–Subaru 2.0L (VIN 2) engine

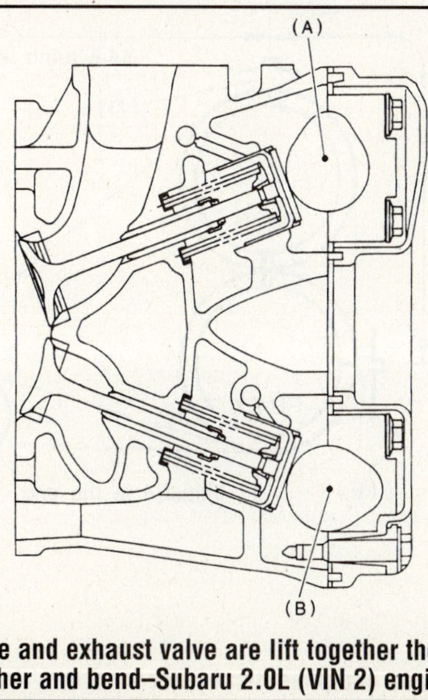

9357tg60

If the intake and exhaust valve are lift together the heads will hit each other and bend–Subaru 2.0L (VIN 2) engine

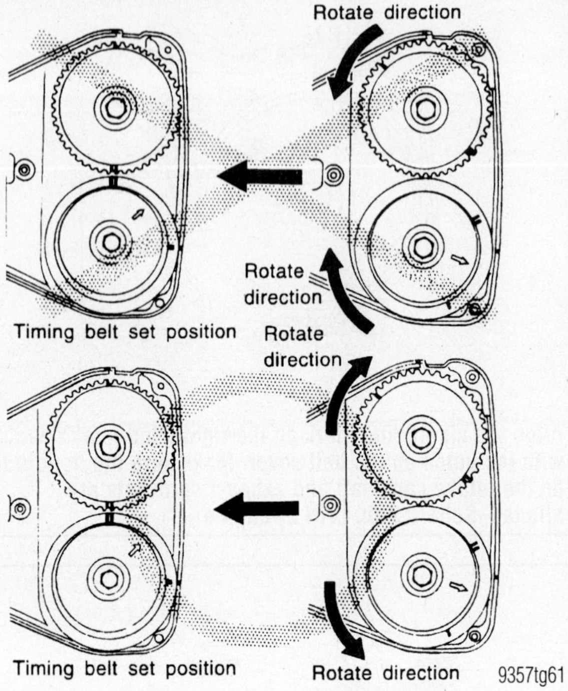

Do not allow the camshafts to rotate in the direction shown as this causes both the intake and exhaust valves to lift off at the same time with will cause valve damage–Subaru 2.0L (VIN 2) engine

✻✻ WARNING

Make sure the marks on the belt and sprockets are properly aligned.

26. Once the marks on the belt and sprockets are aligned, remove the stopper pin from the tensioner adjuster.
27. Install the timing belt guide on models with manual transmission. measure the clearance between the belt and guide. the clearance should be 0.019–0.059 inch (0.5–1.5mm) and tighten the retainers to 7 ft. lbs. (10 Nm).
28. Install the belt covers and tighten to 3.5 ft. lbs. (5 Nm).
29. Install the crankshaft pulley and tighten the bolt to 94 ft. lbs. (127 Nm).
30. Install the V-belt.
31. Connect the negative battery cable.

2.2L (VIN 4) 1990–97 ENGINES

✻✻ CAUTION

These engines are considered freewheeling. If the timing belt fails, valve-to-piston damage is unlikely, however a precautionary compression check should be performed to verify no engine damage.

The engines use a single cam belt drive system with a serpentine type belt. The left side of the engine uses a hydraulic cam belt tensioner, which is self-adjusting.

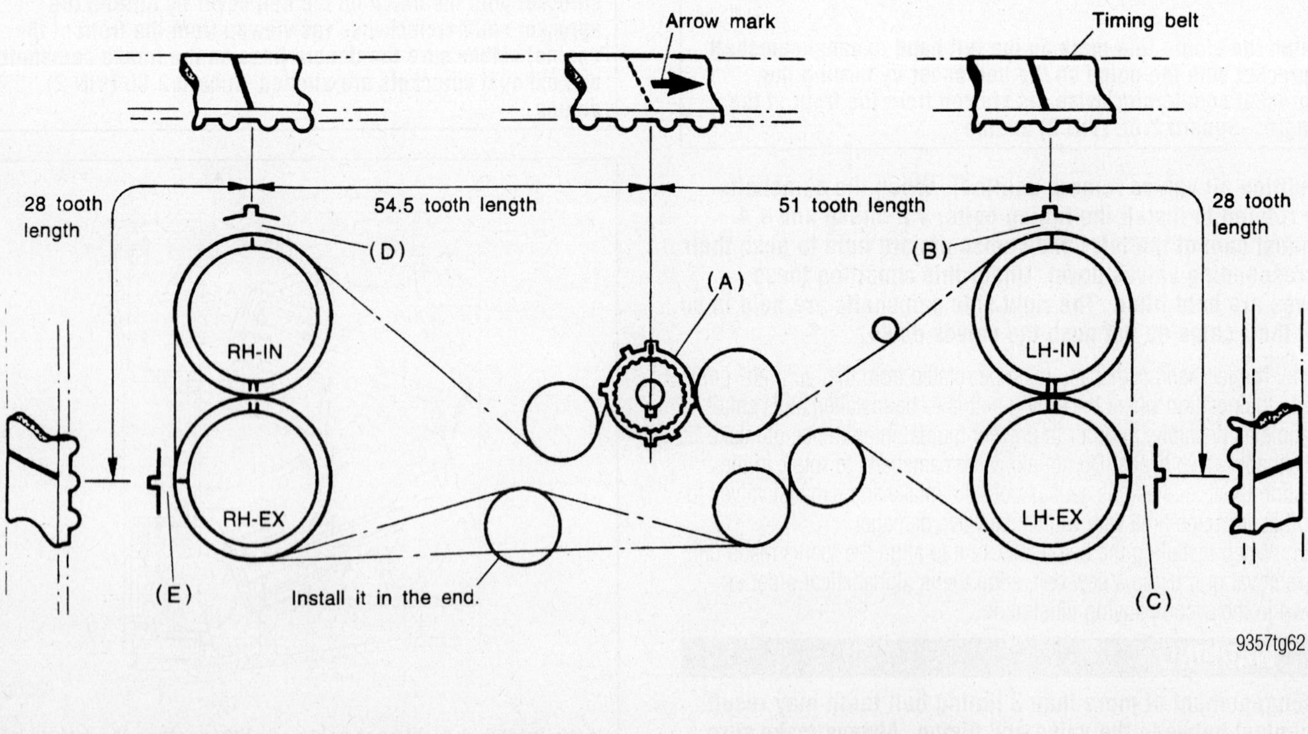

Align the marks in alphabetical order as shown if using a new belt–Subaru 2.0L (VIN 2) engine

YEAR, MODEL & ENGINE APPLICATION

1995-97 Subaru Impreza, 1990-97 Legacy, 1997 Outback,
1997 Outback Sport .... 2.2L VIN 4

REPLACEMENT INTERVALS

Subaru recommends timing belt replacement every 105,000 miles on California models, and 60,000 miles on all other models.

LABOR TIMES

1990-97
Replacement .... 2.4 hrs
Severe Service .... 2.7 hrs

8807SUBg05

■ **This engine is applicable up to 1997 engine No. 737721.**

1. Disconnect the negative battery cable.
2. Remove the coolant reservoir.
3. Remove the cooling fans.
4. Remove the accessory drive belt cover and the accessory drive belts.
5. Remove the crankshaft pulley.
6. Remove the timing belt covers.
7. Align the camshaft sprockets so each sprocket notch aligns with the cam cover notches. Align the crankshaft sprocket top tooth notch, located at the rear of the tooth, with the notch on the crank angle sensor boss. Mark the 3 alignment points as well as the direction of cam belt rotation.
8. Loosen the tensioner adjusting bolts. Remove the bottom 3 idlers, the cam belt and the cam belt tensioner. The cam sprockets can, then be removed with a modified camshaft sprocket wrench tool.
9.Remove the sprockets, if necessary. Note the reference sensor at the rear of the left cam sprocket.

To install

10. Install the sprockets, if removed and tighten the retaining bolts to 47–54 ft. lbs. (64–74 Nm).
11. Install the crankshaft sprocket and the non-adjustable right side idler. Do not install the tensioner idler at this time.
12. Check that the tensioner pushrod protrusion is 0.60–0.65 in. (15–17mm).
13. Compress the hydraulic tensioner in a vise slowly and temporarily secure the plunger with a pin or suitable Allen wrench. Install the tensioner and the pulley with the adjustable idler pulley. Temporarily tighten the tensioner while the tensioner is pushed to the right.
14. Align the crankshaft sprocket notch on the rear sprocket tooth with the crank angle sensor boss. This places the sprocket notch in the 12 o'clock position.
15. Align the camshaft sprockets with the notches in the cam rear belt cover. This places the sprocket notch in the 12 o'clock position for each camshaft.
16. Install the timing belt with the directional mark and alignment marks properly positioned (if the belt is to be reused).
17. Loosen the tensioner retaining bolts and slide the tensioner to the left. Tighten the mounting bolts.

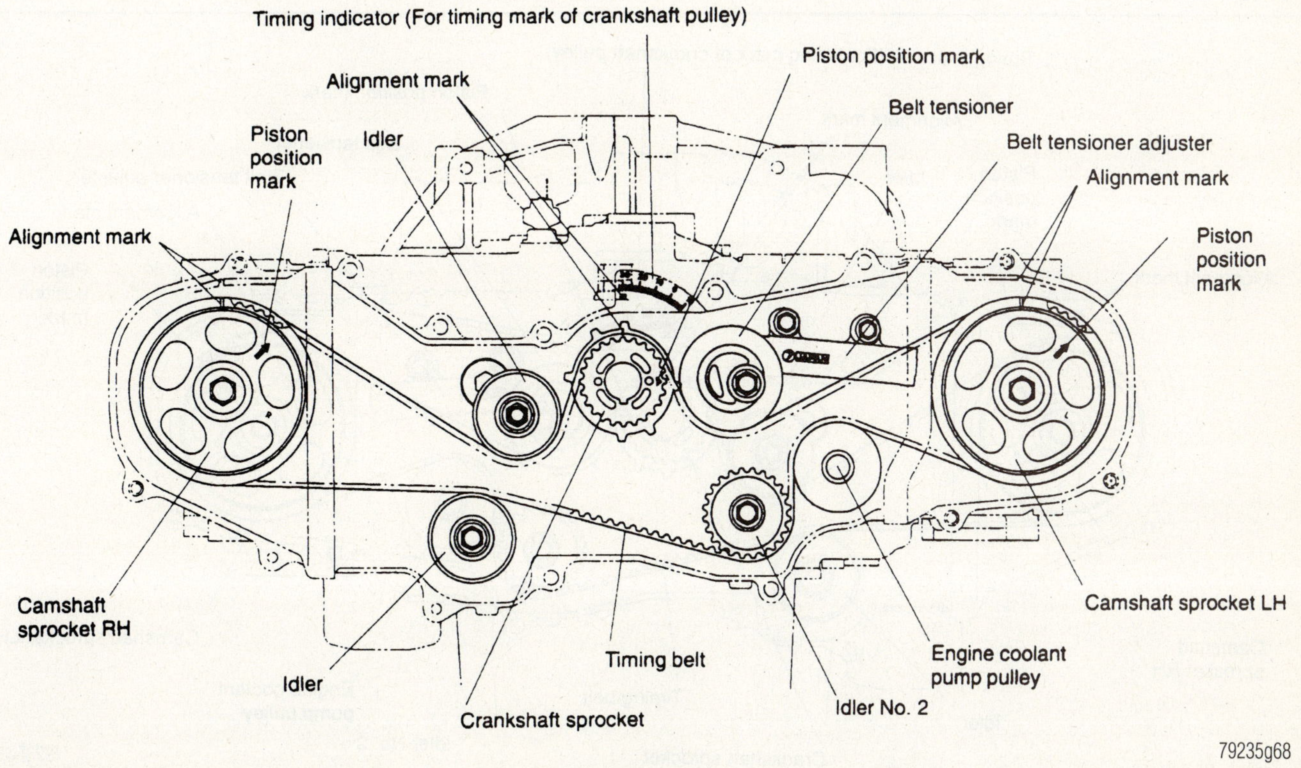

Timing belt components and alignment mark locations—Subaru 2.2L (VIN 4) engine

18. After verifying the timing marks are correct, remove the stopper pin from the tensioner.
19. Verify the correctness of the timing by noting that the notches on the 2 cam pulleys and the notch on the crankshaft pulley all point to the 12 o'clock position when the belt is properly installed.
20. Complete the engine component assembly by installing the cam belt covers, the crankshaft pulley bolt and pulley and the remaining components.
21. Connect the negative battery cable.

2.2L (VIN 4) 1997–01 ENGINES

YEAR, MODEL & ENGINE APPLICATION

1997-01 Impreza, 1998-99 Legacy, 1997-99 Outback, 1997-99 Outback Sport .... 2.2L VIN 4

REPLACEMENT INTERVALS

Subaru recommends timing belt replacement on 1997-98 models every 105,000 miles on California models, and 60,000 miles on all other models. Replace timing belts every 105,000 miles on 1999-01 models.

LABOR TIMES

1997-01

Replacement	2.4 hrs
Severe Service	2.7 hrs

8807SUBg06

CAUTION

These engines are considered freewheeling. If the timing belt fails, valve-to-piston damage is unlikely, however a precautionary compression check should be performed to verify no engine damage.

■ **This engine is applicable from 1997 engine No. 737722.**

1. Disconnect the negative battery cable.
2. Remove the coolant reservoir.
3. Remove the cooling fans.
4. Remove the accessory drive belt cover and the accessory drive belts.
5. Remove the crankshaft pulley.
6. Remove the timing belt covers.
7. Align the camshaft sprockets so each sprocket notch aligns with the cam cover notches. Align the crankshaft sprocket top tooth notch, located at the rear of the tooth, with the notch on the crank angle sensor boss. Mark the 3 alignment points as well as the direction of cam belt rotation.
8. Loosen the tensioner adjusting bolts.
9. If equipped with manual transmissions, loosen the 2 timing belt guide mounting bolts, then separate the guide from the engine block.
10. If the directional arrow and alignment marks on the timing belt are faded, and the belt is to be reused, remark the belt with white paint or a grease pencil as follows:
    a. Make alignment and/or arrow marks on the timing belt in relation to the sprockets as indicated in the accompanying illustration.
- Z1: 44 tooth length
- Z2: 40.5 tooth length

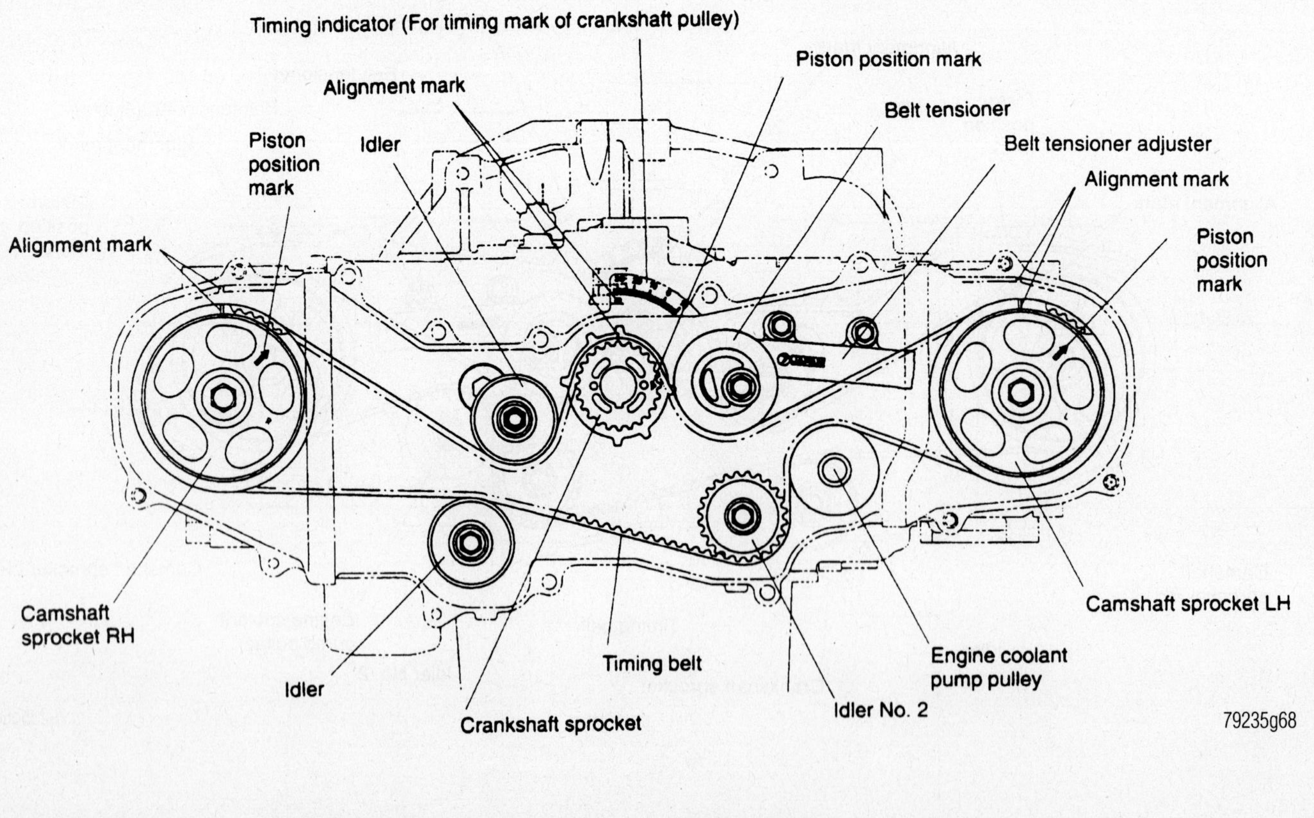

Timing belt components and alignment mark locations—Subaru 2.2L (VIN 4) engine

Remove the timing belt guide on vehicles equipped with manual transmission—2.2L (VIN 4) engine

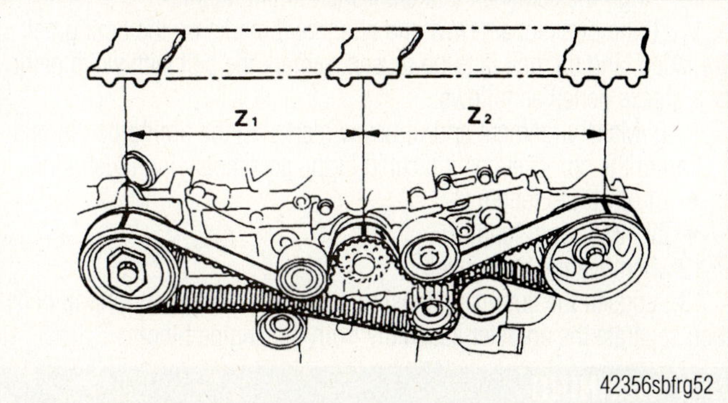

If the original marks on the timing belt are worn or faded, make new alignment marks in the positions indicated—2.2L (VIN 4) engine

11. Remove the timing belt.

12. Loosen the automatic tension adjuster assembly mounting bolts, then separate the adjuster assembly from the engine block.

✱✱ WARNING

After removing the timing belt, DO NOT rotate the camshafts. Damage to the valves may occur.

To install

✱✱ WARNING

Do not allow oil, grease, or coolant to come in contact with the timing belt. If this occurs, quickly and thoroughly remove all traces of the compound. Also, never bend the timing belt sharply; the minimum bending radius is 2.36 in. (60mm).

13. Inspect the camshaft and crankshaft sprocket teeth for abnormal or excessive wear or scratches. Ensure there is no free-play between the sprocket and the key. Inspect the crankshaft sprocket sensor notch for damage or contamination with debris or dirt.

14. Prepare the automatic timing belt tension adjuster assembly for installation as follows:

a. Position the adjuster assembly in a vertical press.

b. Slowly depress the adjuster rod with a force of 66 lbs. (30 kg) until the hole in the rod is aligned with the hole in the adjuster cylinder housing.

c. Insert a 0.08 in. (2mm) diameter stopper pin or Allen wrench through the hole in the cylinder housing and rod, then slowly release the press force from the adjuster rod.

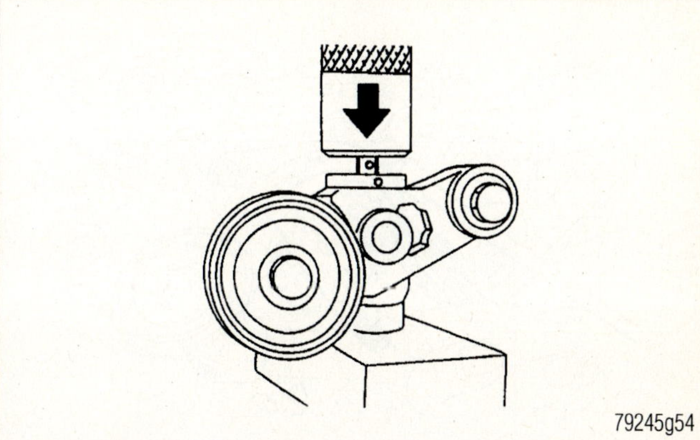

Use a vertical press to push the adjuster rod into its housing until it is flush with the assembly's outer surface . . .

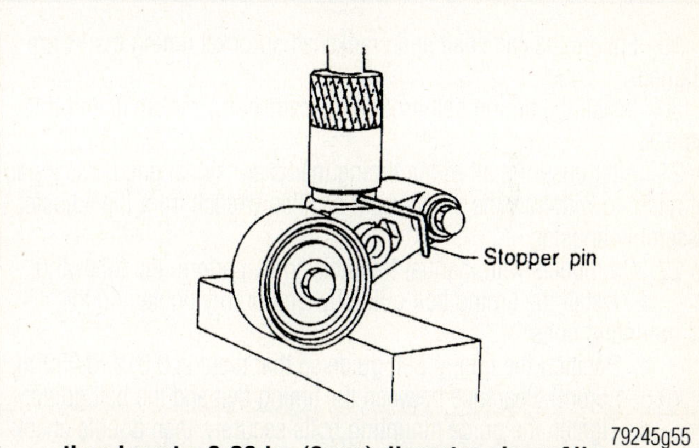

. . . then insert a 0.08 in. (2mm) diameter pin or Allen wrench into the housing and rod holes to hold it in position—2.2L (VIN 4) engine

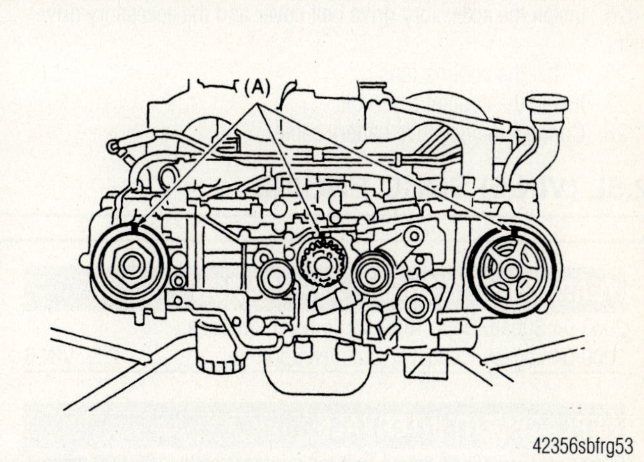

Make sure the sprockets are aligned as shown—2.2L (VIN 4) engine

15. Measure the tensioner pushrod protrusion. It should be 0.20–0.25 in. (0.008–0.10mm).

16. Install the adjuster assembly onto the engine block.

17. Install timing belt idler pulley No. 2 on the engine block. Tighten the bolts to 28 ft. lbs. (39 Nm).

18. Install the timing belt idler pulley No. 1 on the engine block. Tighten the bolts to 28 ft. lbs. (39 Nm).

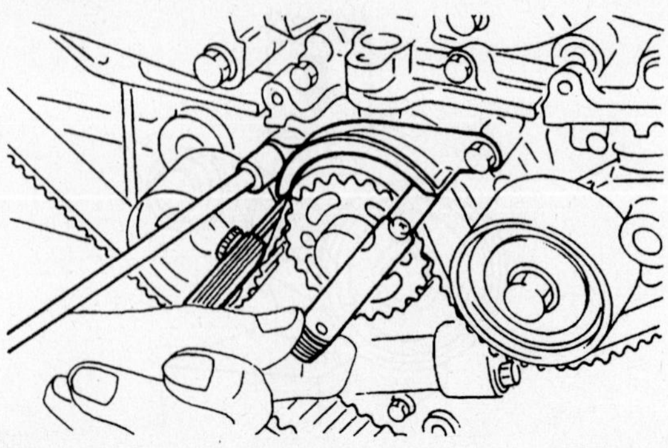

42356sbfrg55

On models equipped with manual transmissions, use a feeler gauges to adjust the clearance between the timing belt and the belt guide—2.2L (VIN 4) engine

19. Ensure the camshaft and crankshaft sprocket timing marks are aligned.
20. Install the timing belt around the camshaft, crankshaft and idler pulleys.
21. After ensuring all of the timing marks are still aligned, use a pair of pliers to withdraw the stopper pin or Allen wrench from the adjuster assembly housing.
22. On models with manual transmissions, perform the following:
    a. Install the timing belt guide by temporarily tightening the mounting bolts.
    b. Position the timing belt guide so that there is 0.019–0.059 in. (0.5–1.5mm) clearance between the timing belt and the belt guide.
    c. Tighten the guide mounting bolts securely, then double check the guide clearance.
23. Install the timing belt covers.
24. Install the crankshaft pulley. Tighten the bolt to 94 ft. lbs. (127 Nm).
25. Install the accessory drive belt cover and the accessory drive belts.
26. Install the cooling fans.
27. Install the coolant reservoir.
28. Connect the negative battery cable.

2.5L (VIN 6) SOHC ENGINE

YEAR, MODEL & ENGINE APPLICATION

2003-04 Subaru Baja, 1999-04 Forester, Impreza, 2000-04 Legacy, Outback, Outback Sport .... 2.5L VIN 6

REPLACEMENT INTERVALS

Subaru recommends timing belt replacement every 105,000 miles.

LABOR TIMES

1999-04	
Replacement	2.4 hrs
Severe Service	2.7 hrs

8807SUBg07

** CAUTION

These engines are considered freewheeling. If the timing belt fails, valve-to-piston damage is unlikely, however a precautionary compression check should be performed to verify no engine damage.

1. Disconnect the negative battery cable.
2. On Forester, remove the engine under cover.
3. On all models, remove the coolant reservoir.
4. Remove the cooling fans.
5. Remove the accessory drive belt cover and drive belt.
6. Loosen the crankshaft pulley bolt.
7. Remove the crankshaft pulley.
8. Remove the A/C drive belt tensioner.
9. Remove the left, right and center timing belt covers.
10. Align the camshaft and crankshaft timing marks.
11. If the directional arrow and alignment marks on the timing belt are faded, and the belt is to be reused, remark the belt with white paint or a grease pencil as follows:
    a. Make alignment and/or arrow marks on the timing belt in relation to the sprockets as indicated in the accompanying illustration.
    - Z1: 44 tooth length
    - Z2: 40.5 tooth length
12. Remove the timing belt.
13. Loosen the automatic tension adjuster assembly mounting bolts, then separate the adjuster assembly from the engine block.

** WARNING

After removing the timing belt, DO NOT rotate the camshafts. Damage to the valves may occur.

To install

■ **Inspect the timing belt and tensioner for wear or damage and replace as necessary.**

14. Inspect the timing belt tensioner as follows:
    a. When compressing the pushrod of the tensioner with a force of 33–110 lbs. (147–490 N), the tensioner should not sink within 8.5 seconds.
    b. Measure the extension of the rod beyond the body of the tensioner for a length of 0.20–0.25 in. (0.008–0.10mm). If not within specifications, replace the tensioner.
15. Using a press, compress the tensioner gradually, taking 3 minutes or more and insert a 0.059 in. (1.5mm) pin to secure the rod.

■ **Check the idler sprockets for smooth operation. Replace as necessary.**

16. Install the adjuster assembly onto the engine block.
17. Install timing belt idler pulley No. 2 on the engine block. Tighten the bolts to 28 ft. lbs. (39 Nm).
18. Install the timing belt idler pulley No. 1 on the engine block. Tighten the bolts to 28 ft. lbs. (39 Nm).
19. Ensure the camshaft and crankshaft sprocket timing marks are aligned.
20. Install the timing belt around the camshaft, crankshaft and idler pulleys.
21. After ensuring all of the timing marks are still aligned, use a pair of pliers to withdraw the stopper pin or Allen wrench from the adjuster assembly housing.

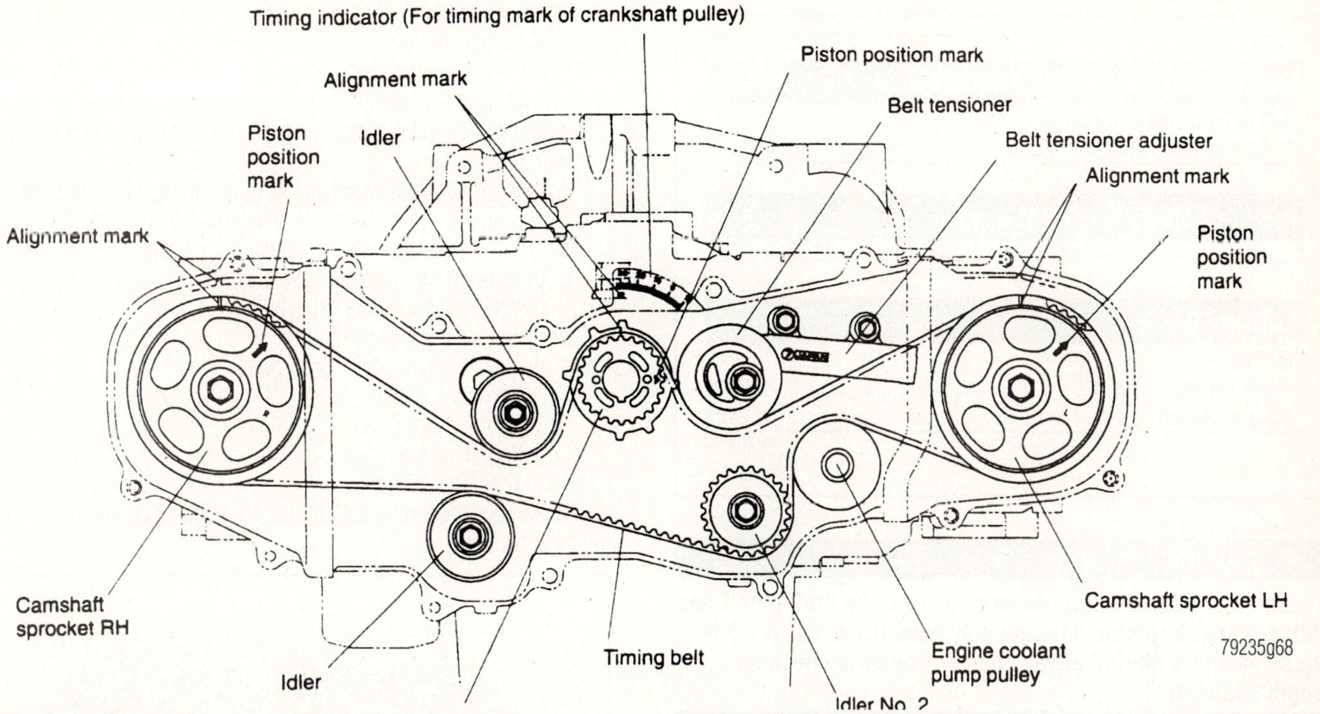

79235g68

Timing belt components and alignment mark locations—Subaru 2.5L (VIN 6) engine

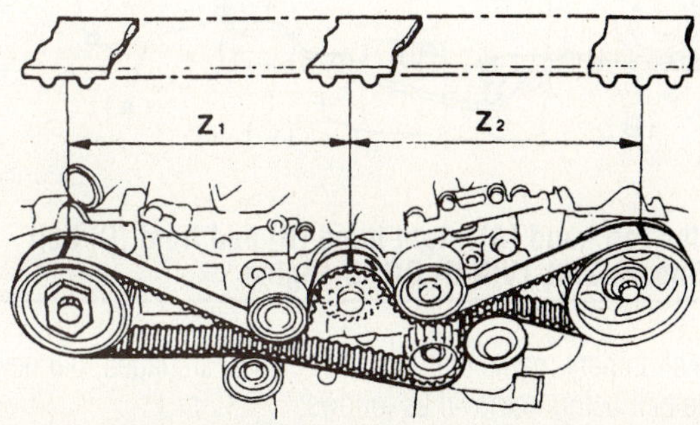

42356sbfrg52

If the original marks on the timing belt are worn or faded, make new alignment marks in the positions indicated—2.5L (VIN 6) engine

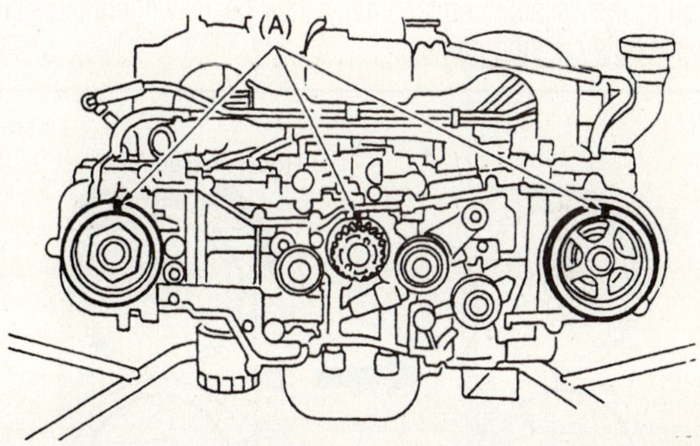

42356sbfrg53

Make sure the sprockets are aligned as shown—2.5L (VIN 6) engine

22. On models with manual transmissions, perform the following:
 a. Install the timing belt guide by temporarily tightening the mounting bolts.
 b. Position the timing belt guide so that there is 0.019–0.059 in. (0.5–1.5mm) clearance between the timing belt and the belt guide.
 c. Tighten the guide mounting bolts securely, then double check the guide clearance.
23. Install the timing belt covers.
24. Install the crankshaft pulley. Tighten the bolt to 130 ft. lbs. (176 Nm).
25. Install the accessory drive belt cover and the accessory drive belts.
26. Install the cooling fans.
27. Install the coolant reservoir.
28. Install the engine under cover.
29. Connect the negative battery cable.

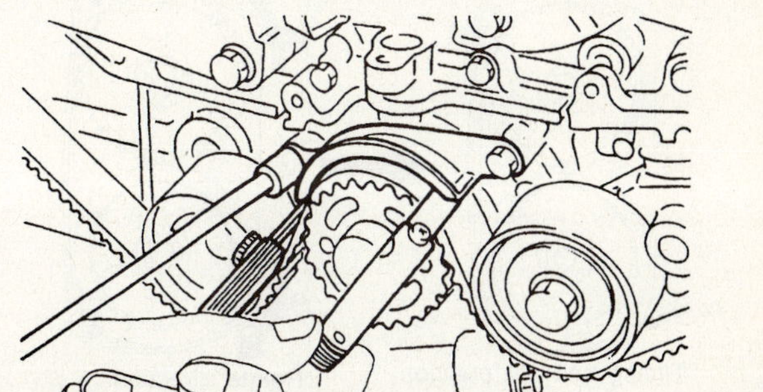
42356sbfrg55

On models equipped with manual transmissions, use a feeler gauges to adjust the clearance between the timing belt and the belt guide—2.5L (VIN 6) engine

2.5L (VIN 6) DOHC ENGINES

YEAR, MODEL & ENGINE APPLICATION

1998 Subaru Impreza, Forester,
1998-99 Legacy, Outback .... 2.5L VIN 6

REPLACEMENT INTERVALS

Subaru recommends timing belt replacement every 105,000 miles.

LABOR TIMES

1998-99	
Replacement	2.9 hrs
Severe Service	3.2 hrs

8807SUBg08

**** CAUTION**

These engines are an interference type. If the timing belt has failed, valve-to-piston damage may have occurred. A compression check should always be performed to check for engine damage.

When servicing the timing belt, note the following:

a. The intake and exhaust camshafts can be rotated independently when the timing belt is removed. If the intake and exhaust valves are lifted off of their seats simultaneously, their heads will contact each other, possibly causing damage.

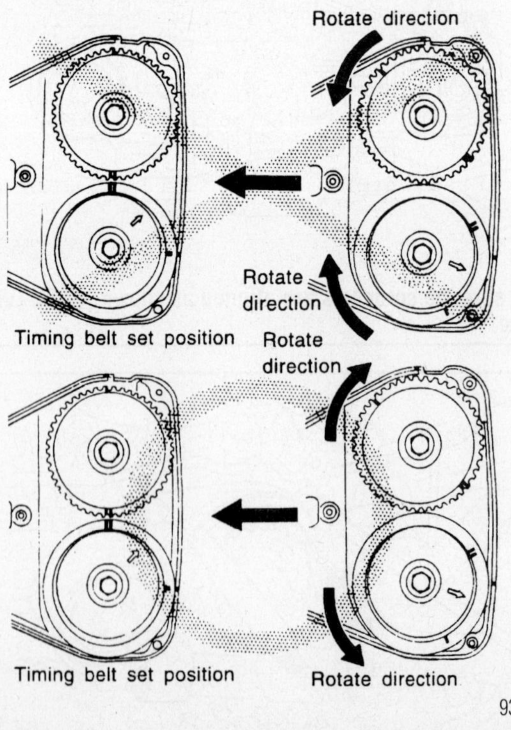

9357tg61

Do not allow the camshafts to rotate in the direction shown in the upper illustration as this causes both the intake and exhaust valves to lift off at the same time which will cause valve damage– Subaru 2.5L (VIN 6) DOHC engine

b. When the timing belt is removed, the camshafts are positioned so that none of the valves are lifted off of their seats, resulting in a "zero-lift" position.

c. The left-hand cylinder head camshafts must be rotated from the "zero-lift" position as little as possible when orienting it for timing belt installation, otherwise possible valve head interference may occur.

d. Never allow the camshafts to rotate in the direction shown in the accompanying illustration, which would cause both the intake and exhaust valves to lift simultaneously, causing interference.

1. Disconnect the negative battery cable.
2. Remove the accessory drive belt.
3. Remove the crankshaft pulley.
4. Remove the belt cover as follows:
   a. Crankshaft pulley.
   b. Left hand belt cover.
   c. Right hand belt cover.
   d. Front hand belt cover.
5. Remove the timing belt guides on models equipped with a manual transmission.

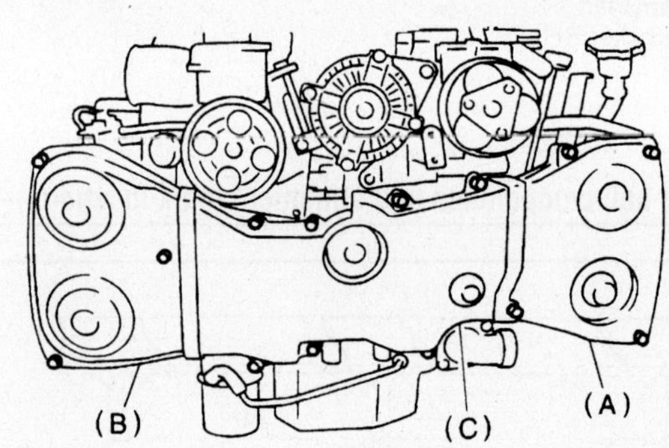

9357tg46

Remove the left hand (A), right hand (B) and front (C) belt covers– Subaru 2.5L (VIN 6) DOHC engine

6. If the alignment marks that indicate rotation are faded, put new marks on the belt before removal as follows:
   a. Turn the crankshaft using crankshaft sprocket tool 499987500 and a breaker bar to align the crankshaft sprocket, left hand intake camshaft sprocket, left hand exhaust camshaft, right hand intake camshaft sprocket and right hand exhaust camshaft sprocket with the on the cover and cylinder block.
   b. Using white paint such as white out, place alignment marks on the belts in relation to the sprockets.
7. Remove the belt idler (A), illustrated in the accompanying illustration.
8. Remove the timing belt.
9. If necessary, remove belt idlers (B) and (C).
10. Remove the belt idler 2.
11. Remove the automatic belt tension adjuster assembly.

To install

12. To prepare the automatic belt tensioner for assembly, perform the following steps:
   a. Always use a vertical type pressing tool to move the adjuster rod down.
   b. Do not use a lateral type vise.
   c. Always push the adjuster rod vertically.
   d. Make sure to slowly move the adjuster rod down applying a pressure of 66 lbs. (294 N).

9357tg47

Location of the upper belt guide– Subaru 2.5L (VIN 6) DOHC engine

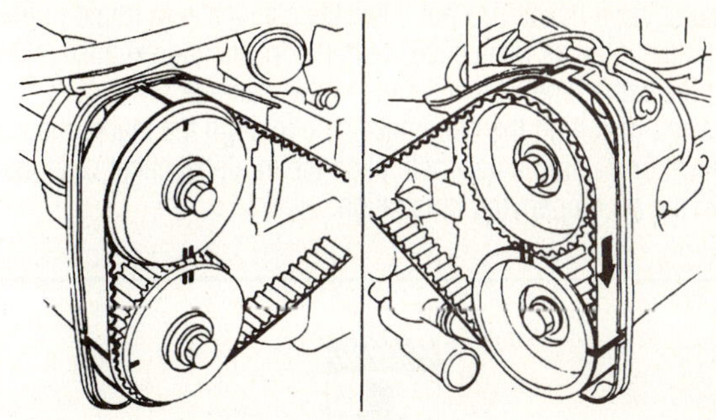

9357tg52

Mark the lower belt-to-sprocket alignment– Subaru 2.5L (VIN 6) DOHC engine

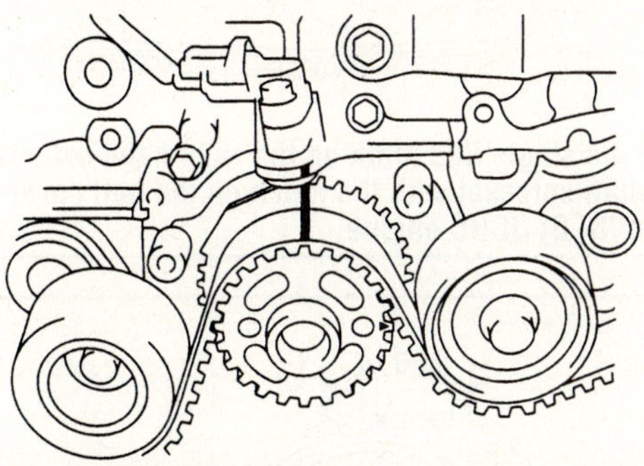

9357tg51

Mark the upper belt-to-sprocket alignment– Subaru 2.5L (VIN 6) DOHC engine

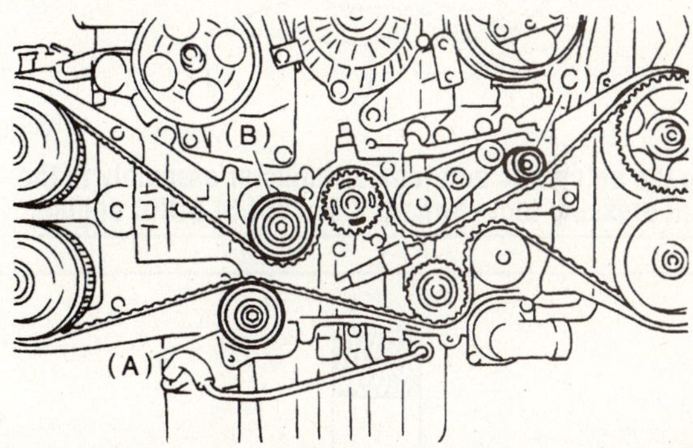

9357tg53

Remove the belt idler (A)–Subaru 2.5L (VIN 6) DOHC engine

e. Press in the push adjuster rod gradually taking more than 3 minutes.

f. Never allow the press pressure to exceed 2,205 lbs. (9,807 N).

g. Press the adjuster rod as far as the end surface of the cylinder. Do not press the rod into the cylinder as doing so may damage the cylinder.

h. Never release the press pressure until the stopper pin has been fully inserted.

13. Attach the automatic belt tension adjuster assembly to the vertical pressing tool.

14. Move the adjuster rod down slowly using a pressure of 66 lbs. (294 N) until the rod is aligned with the stopper pin hole in the cylinder.

15. Insert a 0.08 inch (2mm) stopper pin or diameter Allen wrench into the stopper pin hole in the cylinder to retain the rod.

16. Install the adjuster assembly and tighten the retainers to 29 ft. lbs. (39 Nm).

17. Install belt idler and tighten the retainers to 29 ft. lbs. (39 Nm).

18. Install the belt idler and tighten to 29 ft. lbs. (39 Nm).

19. Align the mark on the crankshaft sprocket with the mark on the oil pump.

20. Align the single line mark on the right hand exhaust camshaft sprocket with the notch on the belt cover.

21. Align the single line mark on the right hand intake camshaft with the notch on the belt cover. Make sure the double lines on the intake camshaft and exhaust sprockets are aligned as shown in the accompanying illustration.

22. Align the single line mark on the left hand exhaust camshaft sprocket with the notch on the belt cover by turning the sprocket counterclockwise (as viewed from the front of the engine).

23. Align the single line mark on the left hand intake camshaft sprocket with the notch on the belt cover by turning the sprocket counterclockwise (as viewed from the front of the engine). Make sure the double lines on the intake camshaft and exhaust sprockets are aligned as shown in the accompanying illustration.

■ **Make sure the camshaft and crankshaft sprockets are positioned correctly. The intake and exhaust camshafts on this engine can be rotated independently with the timing belt removed. By looking at the accompanying illustration it will show you that if the intake and exhaust valve are lift together the heads will hit each other and bend.**

■ **When the timing belts are not installed, 4 camshafts are held at "zero lift" position, where all cams on the camshafts do not push the intake and exhaust valves down (under this condition all valves remain unlifted). When the camshafts are rotated to install the timing belts, # 2 intake and # 4 exhaust cam of the left hand camshafts are held to push their corresponding valves down. Under this condition these valves are held lifted. The right side camshafts are held in so that their cams do not push the valves down.**

24. The left hand camshafts must be rotated from the "zero lift" position to the position where the timing belt is to be installed at as small

an angle as possible, in order to prevent mutual interference of intake and exhaust valve heads. Do not allow the camshafts to rotate in the direction illustrated as this causes both the intake and exhaust valves to lift off at the same time with will cause valve damage.

25. When installing the belt, make sure to align the marks made during removal or if using a new belt, align the in alphabetical order as shown in the accompanying illustration.

WARNING

Disengagement of more than 3 timing belt teeth may result in contact between the valve and piston. Always make sure the belts rotation is correct.

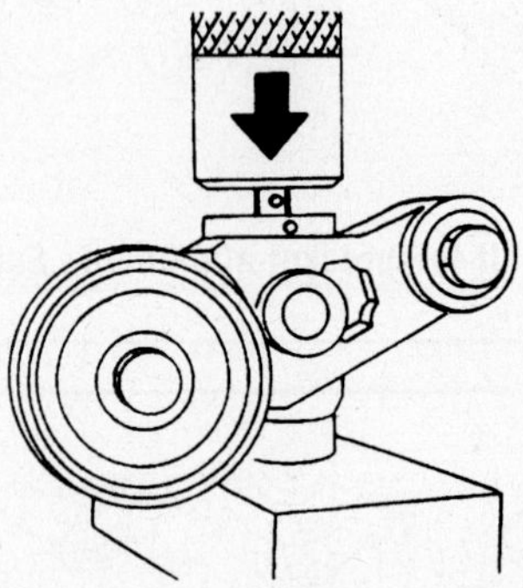

9357tg72

Attach the automatic belt tension adjuster assembly to the vertical pressing tool– Subaru 2.5L (VIN 6) DOHC engine

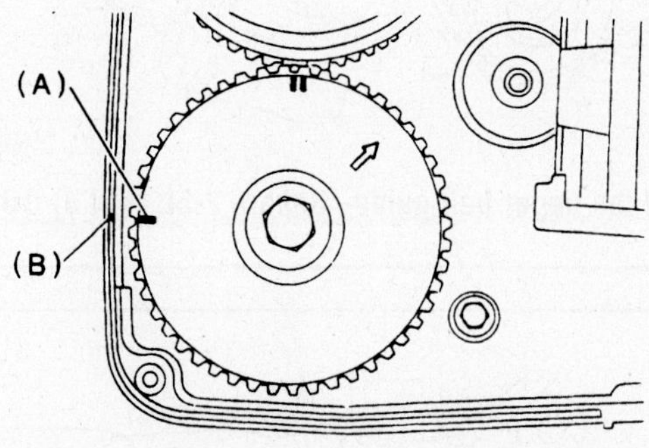

9357tg56

Align the single line mark on the right hand exhaust camshaft sprocket with the notch on the belt cover– Subaru 2.5L (VIN 6) DOHC engine

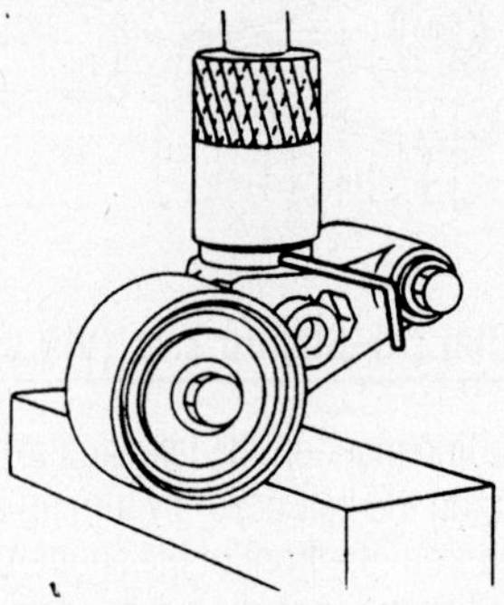

9357tg73

Insert a 0.08 inch (2mm) stopper pin or diameter Allen wrench into the stopper pin hole in the cylinder to retain the rod– Subaru 2.5L (VIN 6) DOHC engine

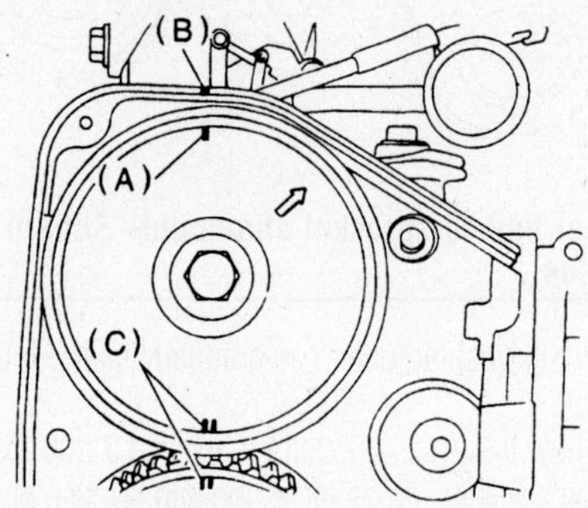

9357tg57

Align the single line mark on the right hand intake camshaft with the notch on the belt cover. Make sure the double lines on the intake camshaft and exhaust sprockets are aligned– Subaru 2.5L (VIN 6) DOHC engine

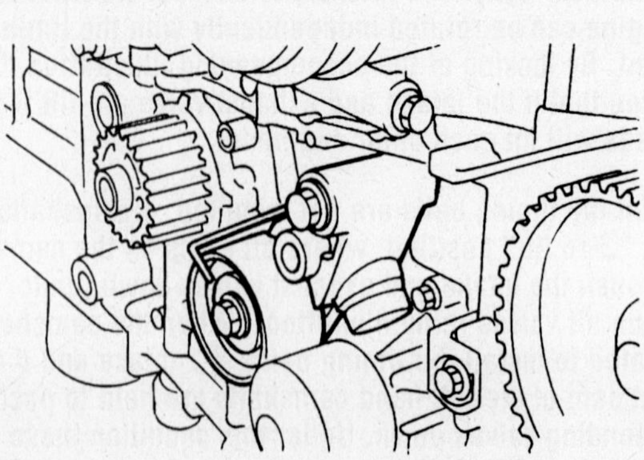

9357tg74

Install the adjuster assembly– Subaru 2.5L (VIN 6) DOHC engine

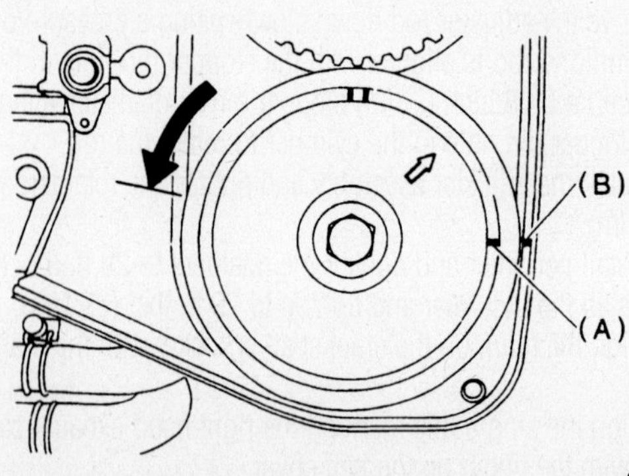

9357tg58

Align the single line mark on the left hand exhaust camshaft sprocket with the notch on the belt cover by turning the sprocket counterclockwise (as viewed from the front of the engine– Subaru 2.5L (VIN 6) DOHC engine

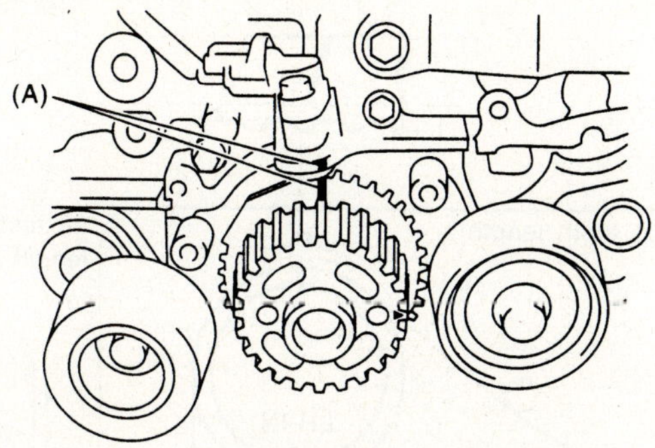

9357tg55

Align the mark on the crankshaft sprocket with the mark on the oil pump– Subaru 2.5L (VIN 6) DOHC engine

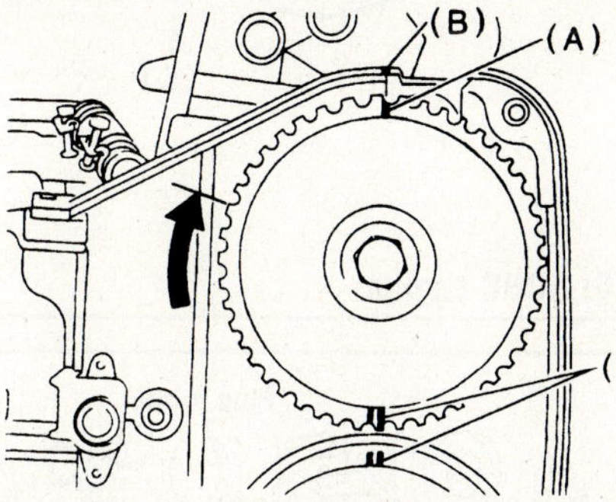

9357tg59

Align the single line mark on the left hand intake camshaft sprocket with the notch on the belt cover by turning the sprocket counterclockwise (as viewed from the front of the engine). Make sure the double lines on the intake camshaft and exhaust sprockets are aligned– Subaru 2.5L (VIN 6) DOHC engine

26. Install the belt idlers and tighten to 29 ft. lbs. (39 Nm).

✳✳ WARNING

Make sure the marks on the belt and sprockets are properly aligned.

27. Once the marks on the belt and sprockets are aligned, remove the stopper pin from the tensioner adjuster.
28. Install the timing belt guide on models with manual transmission. measure the clearance between the belt and guide. the clearance should be 0.019–0.059 inch (0.5–1.5mm) and tighten the retainers to 7 ft. lbs. (10 Nm).
29. Install the belt covers and tighten to 3.5 ft. lbs. (5 Nm).
30. Install the crankshaft pulley and tighten the bolt to 94 ft. lbs. (127 Nm).
31. Install the V-belt.
32. Connect the negative battery cable.

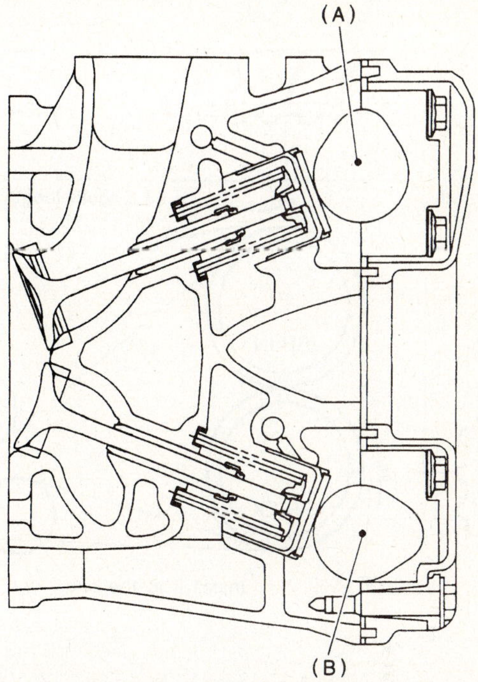

935tg60

If the intake and exhaust valve are lift together the heads will hit each other and bend– Subaru 2.5L (VIN 6) DOHC engine

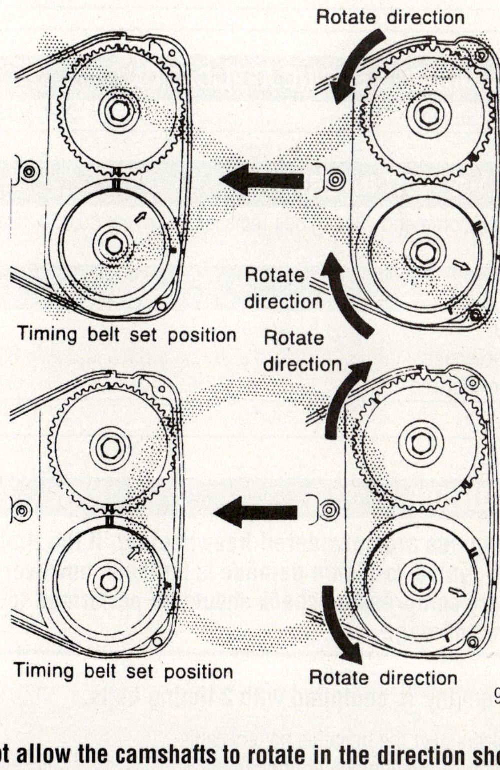

9357tg61

Do not allow the camshafts to rotate in the direction shown as this causes both the intake and exhaust valves to lift off at the same time with will cause valve damage– Subaru 2.5L (VIN 6) DOHC engine

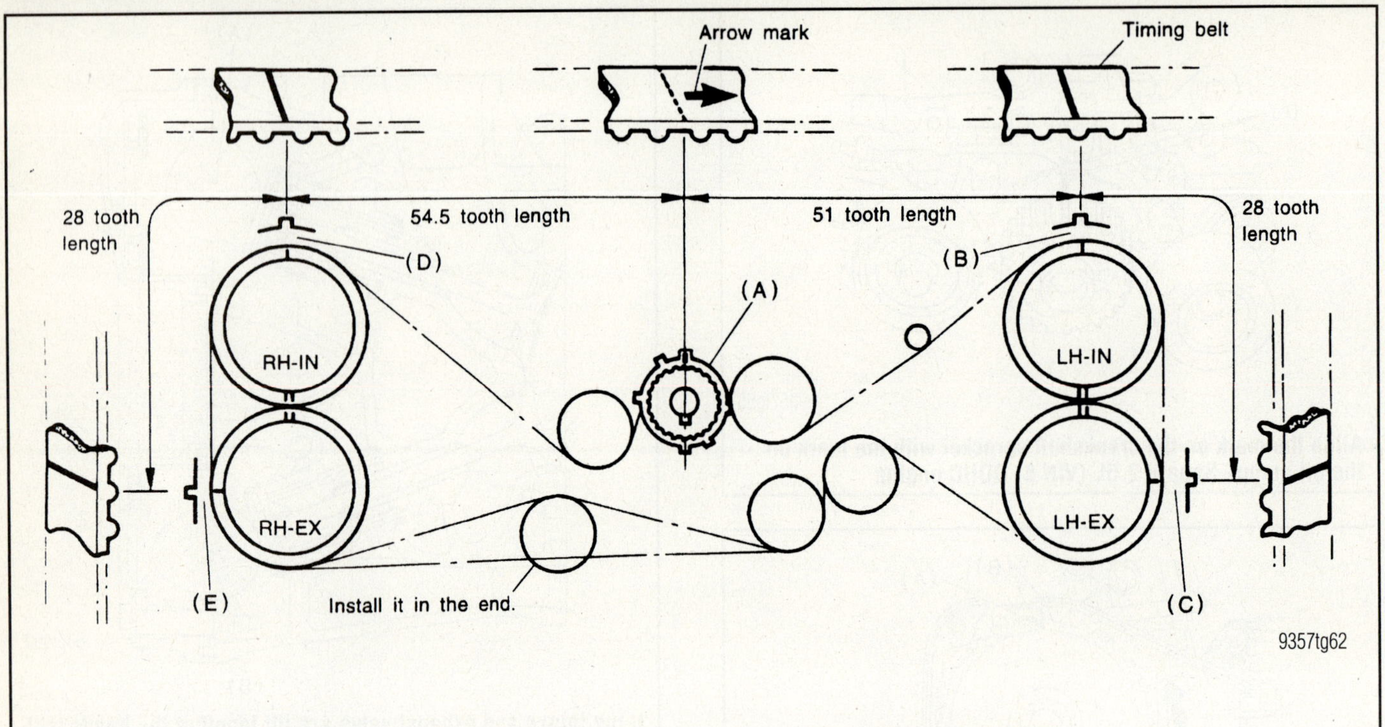

Align the marks in alphabetical order as shown if using a new belt– Subaru 2.5L (VIN 6) DOHC engine

2.7L (VIN 8,9) ENGINES

YEAR, MODEL & ENGINE APPLICATION	
1988-91 Subaru XT6	2.7L VIN 8, 9

REPLACEMENT INTERVALS

Subaru recommends timing belt replacement every 60,000 miles.

LABOR TIMES	
1988-91	
Replacement	3.3 hrs
Severe Service	3.9 hrs

8807SUBg09

✻✻ CAUTION

These engines are considered freewheeling. If the timing belt fails, valve-to-piston damage is unlikely, however a precautionary compression check should be performed to verify no engine damage.

■ **This engine is equipped with 2 timing belts.**

1. Disconnect the negative battery cable.
2. Remove the engine cooling fans.
3. Remove the accessory drive belts.
4. Remove the water pump pulley.
5. Remove the oil dipstick and tube.
6. Disconnect the oil pressure switch connector.
7. Lock the crankshaft using locking tool in the transmission housing opening.
8. Remove the crankshaft pulley.

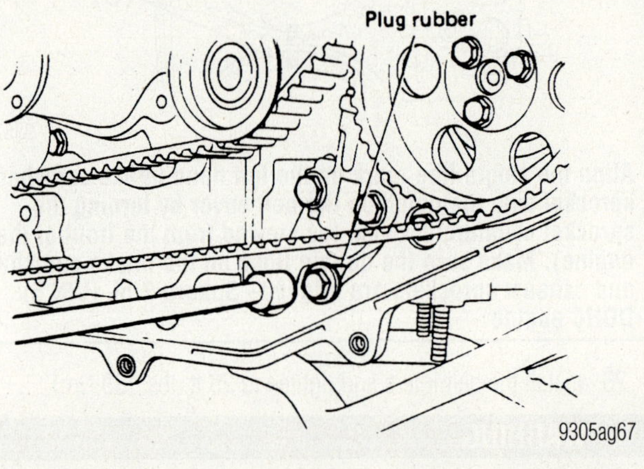

View of the rubber plug—Subaru 2.7L (VIN 8, 9) engine

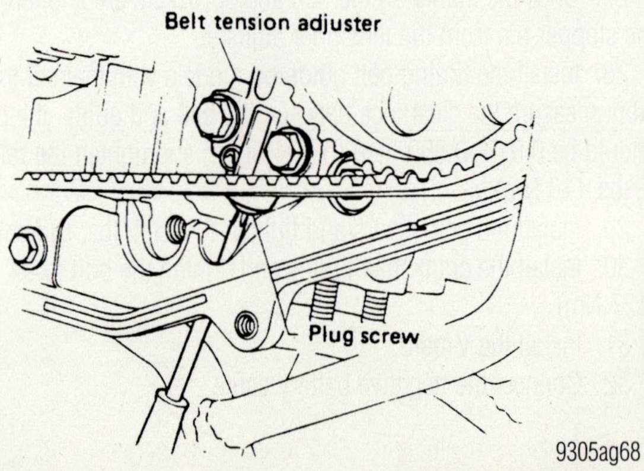

View of the left timing belt tensioner—Subaru 2.7L (VIN 8, 9) engine

9. Remove the left, right and center timing belt covers.
10. Loosen the right timing belt tensioner lockbolt and pivot bolt.
11. Pivot the tensioner away from the timing belt and tighten the lockbolt.
12. Remove the right timing belt. Remove the right timing belt crankshaft sprocket.
13. Remove the plug rubber below the left timing belt tensioner.
14. Remove the plug screw from the bottom of the left timing belt tensioner.
15. Using a standard screwdriver, turn the adjustment screw clockwise to relieve timing belt tension.
16. Install a Belt Adjuster Stopper tool No. 13082AA000.
17. Remove the left belt tensioner pulley.
18. Remove the left timing belt.
19. Remove the left timing belt tensioner.

To install

20. Fill the left timing belt tensioner with clean engine oil and replace the plug screw.
21. Install the left timing belt tensioner.
22. Rotate the crankshaft to align the middle timing belt alignment mark on the flywheel with the bell housing pointer.
23. Rotate the left camshaft to align the camshaft pulley mark with the rear cover timing mark.
24. Install the left timing belt.
25. Install the left timing belt tensioner pulley and remove the Belt Adjuster Stopper.
26. Rotate the crankshaft 2 complete turns and check that the flywheel and left camshaft timing marks align.

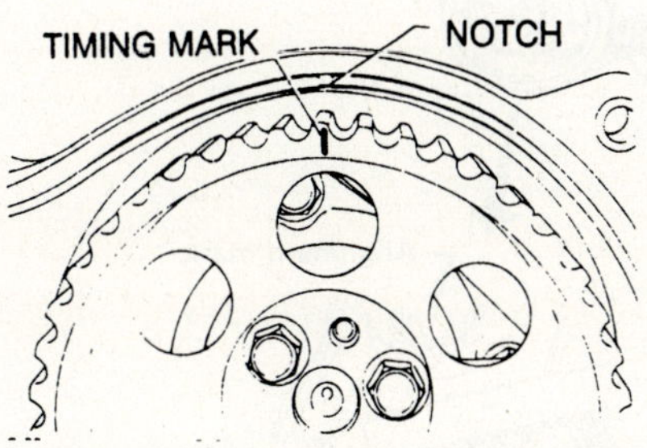

8807ting13

Left Camshaft sprocket timing mark alignment—Subaru 2.7L (VIN 8, 9) engine

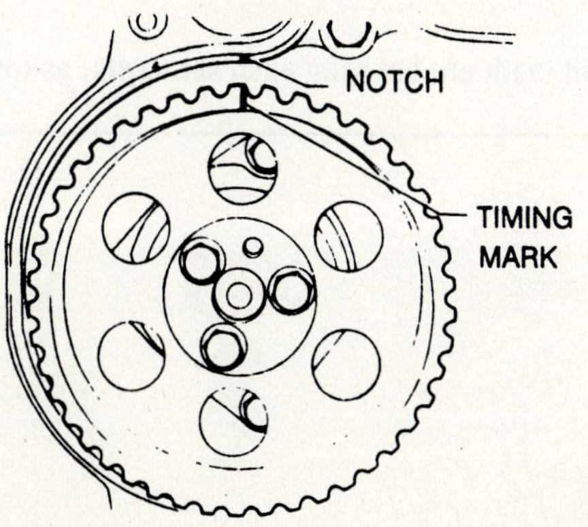

8807timg12

Right Camshaft sprocket timing mark alignment—Subaru 2.7L (VIN 8, 9) engine

27. Install the plug rubber below the left timing belt tensioner.
28. Install the right timing belt crankshaft pulley.
29. Rotate the crankshaft 1 full turn and align the middle timing belt alignment mark on the flywheel with the bell housing pointer.
30. Rotate the right camshaft to align the camshaft pulley mark with the rear cover timing mark.
31. Install the right timing belt.
32. Loosen the right timing belt tensioner lockbolt.
33. Using a Belt Tension Wrench No. 499437100, apply 17–19 ft. lbs. (24–25 Nm) of torque to the right camshaft pulley in the counterclockwise direction. Tighten the right timing belt tensioner lockbolt and pivot bolt.
34. Install the center, left and right timing covers.
35. Install the crankshaft pulley and tighten the bolt to 66–79 ft. lbs. (89–107 Nm).
36. Connect the oil pressure switch.
37. Install the oil dipstick and tube.
38. Install the water pump pulley.
39. Install the accessory drive belts.
40. Install the engine cooling fans.
41. Connect the negative battery cable.

3.3L (VIN V,W) ENGINES

YEAR, MODEL & ENGINE APPLICATION

1992-97 Subaru SVX .... 3.3L VIN V, W

REPLACEMENT INTERVALS

Subaru recommends timing belt replacement every 60,000 miles.

LABOR TIMES

1992-97
Replacement .... 2.5 hrs
Severe Service .... 3.1 hrs

8807SUBg10

**** CAUTION**

These engines are considered freewheeling. If the timing belt fails, valve-to-piston damage is unlikely, however a precautionary compression check should be performed to verify no engine damage.

1. Disconnect the negative battery cable.
2. Remove the accessory drive belts.
3. Remove the power steering pump, without disconnecting the lines and position aside.
4. Remove the alternator.
5. Remove the drive cover.
6. Remove the timing belt covers.
7. Matchmark the timing belt to the sprocket, cover and block marks as follows:
    a. Turn the crankshaft to align the timing marks on the crankshaft sprocket with the mark on the block.
    b. With the crankshaft marks aligned be sure the left and right camshaft sprocket marks are aligned with marks on the timing covers.
    c. If all the marks align, use white paint to mark the direction of rotation of the belt as well as mark the spots on the belt where it crosses over the timing marks on the pulleys.

8. Loosen the belt tensioner bolts.
9. Remove belt idler pulley No. 1.
10. Remove belt idler pulley No. 2.
11. Remove the timing belt.
12. Remove the tensioner pulley bolt and remove the tensioner pulley.
13. Remove the 2 bolts and the tensioner assembly.

To install

14. Insert a 0.059 in. (1.5mm) diameter stopper pin into place while pushing the tension adjuster rod into the tensioner body.
15. Install the tensioner and tighten the bolts to 18 ft. lbs. (24 Nm), while the tensioner is pushed all the way to the right.
16. Install the tensioner pulley and mounting bolt. DO NOT tighten the idler pulley bolt completely.
17. Be sure the crankshaft and both camshaft sprockets are still aligned with their respective timing marks.
18. Install the timing belt onto the sprockets with the direction of rotation arrow in the correct direction and the timing marks on the belt align with the marks on the sprockets.
19. Install No. 1 and 2 idler pulleys and tighten the mounting bolts to 29 ft. lbs. (39 Nm).
20. Loosen the tensioner pulley bolt and the tensioner assembly mounting bolts. Slide the tensioner assembly all the way to the left and tighten the bolts to 18 ft. lbs. (24 Nm).
21. Check again that all the timing marks are still in alignment. If they are remove the stopper pin from the tensioner assembly.
22. Install the timing belt covers.
23. Install the drive cover.
24. Install the alternator.
25. Install the power steering pump.
26. Install the accessory drive belts.
27. Connect the negative battery cable.

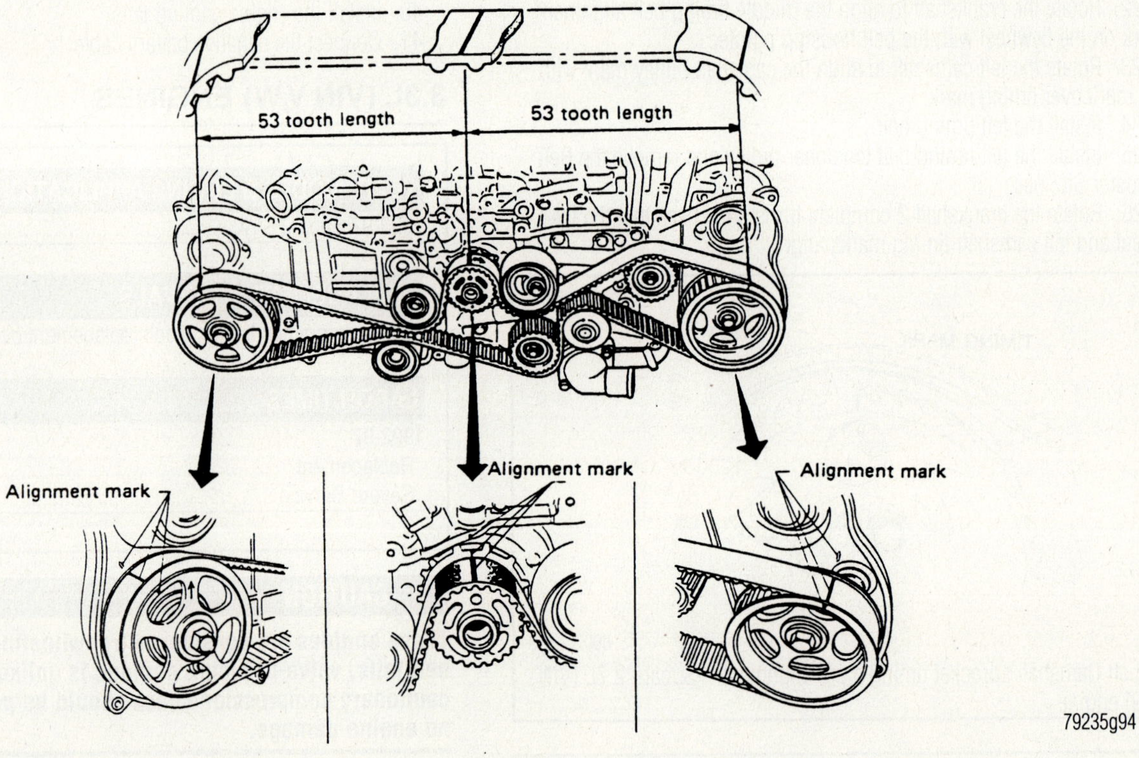

To ensure proper installation of the timing belt, be sure the proper number of belt teeth are between each sprockets, as indicated—Subaru 3.3L (VIN V, W) engine

Suzuki

1.3L (VIN 3 & 5) SOHC ENGINES

YEAR, MODEL & ENGINE APPLICATION

1985-95 Suzuki Samurai, 1989-98 Sidekick	1.3L VIN 5
1989-01 Swift	1.3L VIN 3

REPLACEMENT INTERVALS

Suzuki does not provide a replacement for 1985-93 models. Standard industry practice is to replace timing belts every 60,000 miles. On 1994-01 models, Suzuki recommends timing belt replacement every 60,000 miles.

LABOR TIMES

Samurai & Sidekick	
Replacement	3.2 hrs
Severe Service	3.6 hrs
Swift	
1989-94	
Replacement	3.1 hrs
Severe Service	3.6 hrs
1995-01	
Replacement	2.4 hrs
Severe Service	2.7 hrs

8807SUZg01

✱✱ CAUTION

These engines are an interference type. If the timing belt has failed, valve-to-piston damage may have occurred. A compression check should always be performed to check for engine damage.

1. Disconnect the negative battery cable.
2. On Swift models, raise and support the vehicle.
3. Remove the right side wheel and fender apron.
4. Remove the air cleaner.
5. On Samurai and sidekick, remove the radiator fan shroud and cooling fan.
6. On all models, remove the accessory drive belts.
7. Remove the water pump pulley.
8. Remove the crankshaft pulley.
9. Remove the resonator (Swift) and the timing belt cover.
10. Rotate the crankshaft so that the camshaft pulley timing mark aligns with the timing mark on the rear timing cover and the crankshaft timing mark aligns with the arrow on the oil pump.
11. Remove the tensioner stud and loosen the tensioner bolt.
12. Remove the tensioner spring and damper, then remove the timing belt.

✱✱ WARNING

After the timing belt is removed never turn the camshaft or the crankshaft. Interference may occur between the pistons and the valves causing component damage.

13. Remove the tensioner and the tensioner plate.
14. On Samurai, Sidekick and 1989–93 Swift, remove the valve cover and loosen the valve lash adjusting screws.

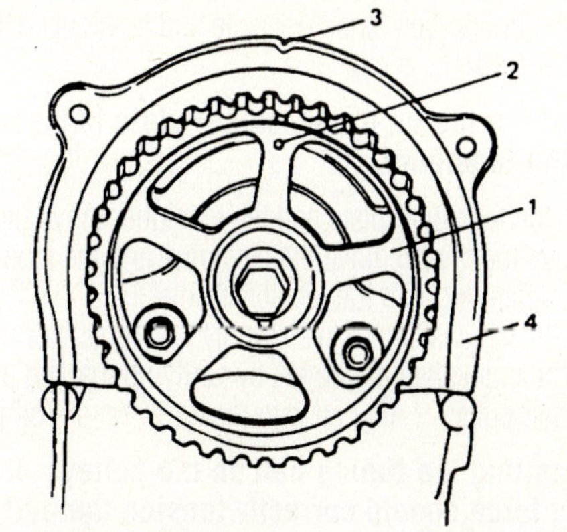

Camshaft timing marks—Suzuki 1.3L (VIN 3, 5) 8-valve engine—Built before June 1993

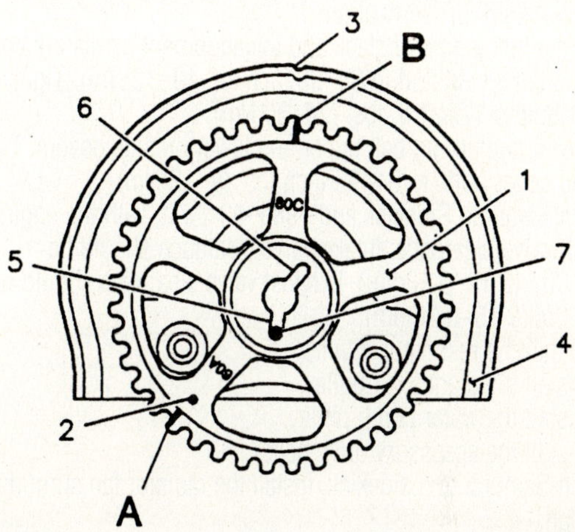

Camshaft timing marks—Suzuki 1.3L (VIN 3, 5) 8-valve engine—Built after June 1993

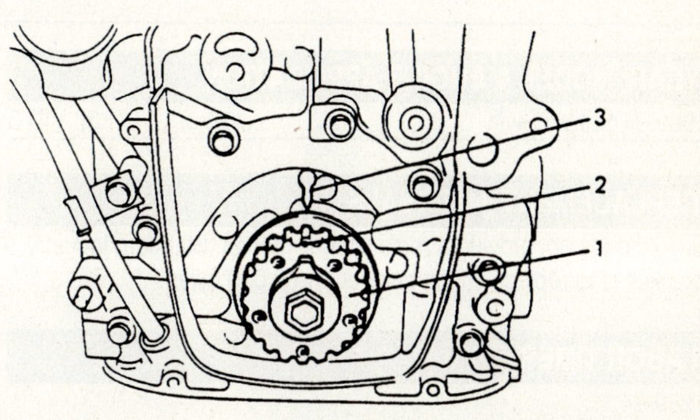

Crankshaft timing marks—Suzuki 1.3L (VIN 3, 5) 8-valve engine

To install:

15. Install the timing belt tensioner plate and tensioner. Only hand-tighten the tensioner bolt.

■**Be sure that the lug on the tensioner plate is inserted into the hole on the tensioner.**

16. Be sure the tensioner plate and the tensioner move uniformly. If they do not move together remove the tensioner and the tensioner plate and reinsert the plate lug into the tensioner hole.
17. Check the camshaft sprocket to verify that it has not moved.
18. Check the crankshaft alignment by verifying that the punch mark on the timing belt pulley is aligned with the arrow on the oil pump case.

■ **When installing the timing belt on the pulleys, the tensioner spring force should correctly tension the belt. If the camshaft does not rotate freely, the belt will not be correctly tensioned.**

19. With the timing marks aligned, hold the tensioner plate up by hand and install the timing belt on the pulleys so there is no slack on the drive side of the belt.
20. Turn the crankshaft 2 rotations clockwise. Confirm that the timing marks are still properly aligned.
21. If the belt is free of slack and the alignment marks are correct tighten the tensioner stud to 84–96 inch lbs. (9–12 Nm). Tighten the tensioner bolt to 17–21 ft. lbs. (24–30 Nm).
22. Install the timing belt upper and lower outside covers. Tighten the timing cover bolts to 84–96 inch lbs. (9–12 Nm).
23. On Samurai, Sidekick and 1989–93 Swift, with the engine cold, adjust the valve clearance. Intake valve clearance should be 0.005–0.007 (0.12–0.18mm). Exhaust valve clearance should be 0.006–0.008(0.15–0.20mm).
24. Install the resonator (Swift).
25. Install the crankshaft pulley.
26. Install the water pump pulley.
27. Install the accessory drive belts.
28. On Samurai and sidekick, install the radiator fan shroud and cooling fan.
29. On swift models, install the air cleaner.
30. Install the right side wheel and fender apron.
31. Lower the vehicle.
32. On all models, connect the negative battery cable.

1.3L (VIN 3) DOHC ENGINES

YEAR, MODEL & ENGINE APPLICATION

1989-94 Suzuki Swift .... 1.3L VIN 3

REPLACEMENT INTERVALS

Suzuki does not provide a replacement interval. Standard industry practice is to replace timing belts every 60,000 miles.

LABOR TIMES

1989-94
Replacement .... 3.1 hrs
Severe Service .... 3.6 hrs

8807SUZg02

✻✻ CAUTION

These engines are an interference type. If the timing belt has failed, valve-to-piston damage may have occurred. A compression check should always be performed to check for engine damage.

1. Disconnect the negative battery cable.
2. Disconnect and remove the Mass Air Flow (MAF) meter and the air cleaner assembly.
3. Remove the air cleaner bracket.
4. Raise and safely support the vehicle.
5. Remove the right fender apron extension.
6. Remove the accessory drive belt.
7. Remove the water pump pulley.
8. Remove the crankshaft pulley.

■ **It may be necessary to remove the crankshaft timing pulley bolt for clearance.**

9. Remove the upper and lower timing covers.
10. Rotate the crankshaft so that the camshaft pulley timing marks align with the timing marks on the rear timing cover, and the crankshaft timing mark aligns with the mark on the oil pump.
11. Loosen the timing belt tensioner locknut and pivot bolt. Pry the tensioner pulley away from the timing belt and tighten the locknut.
12. Remove the timing belt.

To install:

13. Ensure that the camshaft and crankshaft timing marks are aligned.
14. Install the timing belt on the sprockets in counterclockwise direction starting at the crankshaft. Ensure the belt is taut on the non tension side.

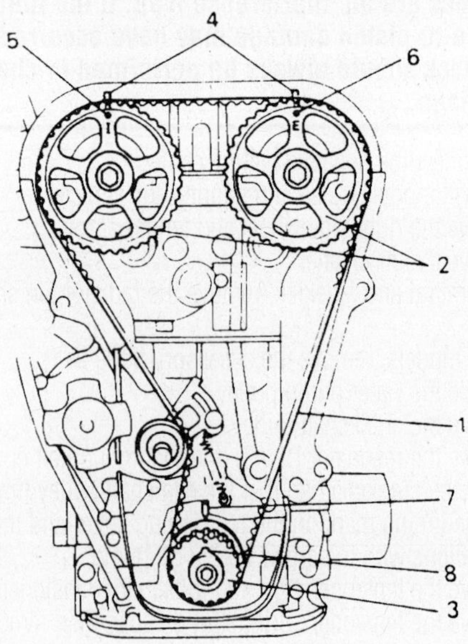

1. Timing belt
2. Camshft timing belt pulleys
3. Crankshaft timing belt pulley
4. Marks on cylinder head cover
5. Punch mark by "I" (intake side)
6. Punch mark by "E" (exhaust side)
7. Mark on oil pump case
8. Key on crankshaft

9305ag73

Alignment of the timing marks—Suzuki 1.3L (VIN 3) DOHC engine

15. Loosen the tensioner locknut.

16. Rotate the crankshaft 2 complete turns and check that the timing marks align.

17. Tighten the tensioner locknut to 84–102 inch lbs. (9–12 Nm) and the pivot bolt to 17–22 ft. lbs. (24–30 Nm).

18. Install the upper and lower timing covers.

19. Install the crankshaft pulley and tighten the bolts to 84–102 inch lbs. (9–12 Nm). If removed, tighten the crankshaft timing pulley bolt to 76–83 ft. lbs. (105–115 Nm).

20. Install the water pump pulley and the accessory drive belt.

21. Install the fender apron extension.

22. Lower the vehicle.

23. Install the air cleaner bracket, the air cleaner assembly and the MAF meter.

24. Connect the negative battery cable.

1.6L (VIN 0) 8-VALVE ENGINES

YEAR, MODEL & ENGINE APPLICATION

1989-95 Suzuki Sidekick .... 1.6L VIN 0

REPLACEMENT INTERVALS

Suzuki does not provide a replacement interval. Standard industry practice is to replace timing belts every 60,000 miles.

LABOR TIMES

1989-95

Replacement	3.2 hrs
Severe Service	3.6 hrs

8807SUZg03

✻✻ CAUTION

These engines are an interference type. If the timing belt has failed, valve-to-piston damage may have occurred. A compression check should always be performed to check for engine damage.

1. Disconnect the negative battery cable.
2. Remove the radiator fan shroud and cooling fan.
3. Remove the accessory drive belts.
4. Remove the air cleaner.
5. Remove the water pump pulley.
6. Remove the crankshaft pulley.
7. Remove the timing belt cover.
8. Rotate the crankshaft so that the camshaft pulley timing mark aligns with the timing mark on the rear timing cover and the crankshaft timing mark aligns with the arrow on the oil pump.
9. Remove the tensioner stud and loosen the tensioner bolt.
10. Remove the tensioner spring and damper, then remove the timing belt.

✻✻ WARNING

After the timing belt is removed never turn the camshaft or the crankshaft. Interference may occur between the pistons and the valves causing component damage.

11. Remove the tensioner and the tensioner plate.

12. Remove the valve cover and loosen the valve lash adjusting screws.

To install:

13. Install the timing belt tensioner plate and tensioner. Only hand-tighten the tensioner bolt.

■ **Be sure that the lug on the tensioner plate is inserted into the hole on the tensioner.**

14. Be sure the tensioner plate and the tensioner move uniformly. If they do not move together remove the tensioner and the tensioner plate and reinsert the plate lug into the tensioner hole.

15. Check the camshaft sprocket to verify that it has not moved.

16. Check the crankshaft alignment by verifying that the punch mark on the timing belt pulley is aligned with the arrow on the oil pump case.

■ **When installing the timing belt on the pulleys, the tensioner spring force should correctly tension the belt. If the camshaft does not rotate freely, the belt will not be correctly tensioned.**

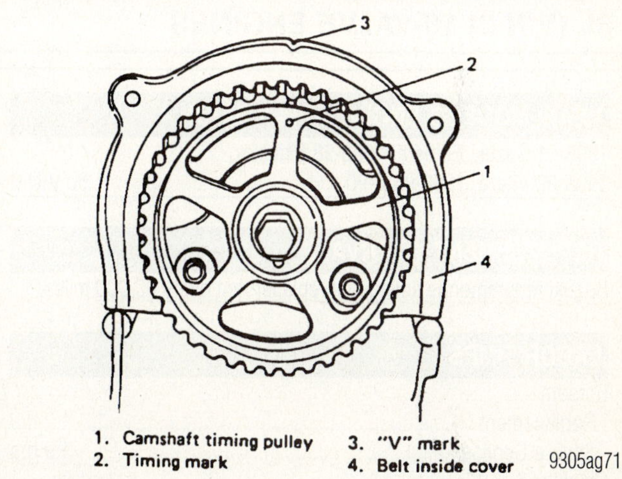

Camshaft timing marks—Suzuki 1.6L (VIN 0) 8-valve engine

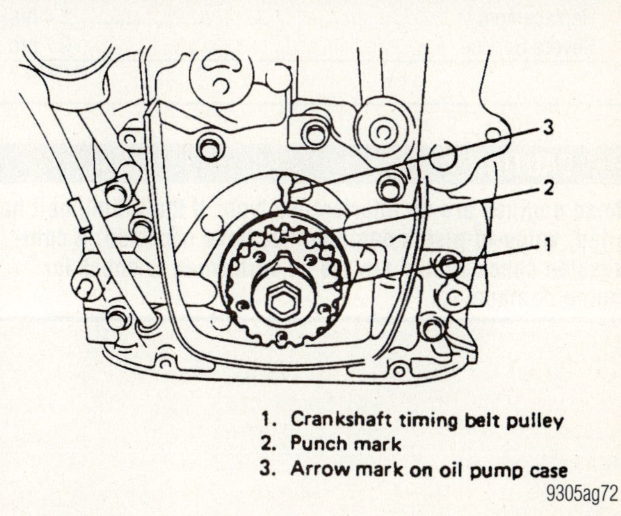

Crankshaft timing marks—Suzuki 1.6L (VIN 0) 8-valve engine

17. With the timing marks aligned, hold the tensioner plate up by hand and install the timing belt on the pulleys so there is no slack on the drive side of the belt.
18. Turn the crankshaft 2 rotations clockwise. Confirm that the timing marks are still properly aligned.
19. If the belt is free of slack and the alignment marks are correct tighten the tensioner stud to 84–96 inch lbs. (9–12 Nm). Tighten the tensioner bolt to 17–21 ft. lbs. (24–30 Nm).
20. Install the timing belt upper and lower outside covers. Tighten the timing cover bolts to 84–96 inch lbs. (9–12 Nm).
21. Adjust the valve clearance. Intake valve clearance should be 0.005–0.007 (0.12–0.18mm). Exhaust valve clearance should be 0.006–0.008(0.15–0.20mm).
22. Install the valve cover.
23. Install the crankshaft pulley.
24. Install the water pump pulley.
25. Install the accessory drive belts.
26. Install the air cleaner.
27. Install the radiator fan shroud and cooling fan.
28. Connect the negative battery cable.

1.6L (VIN 0) 16-VALVE ENGINES

YEAR, MODEL & ENGINE APPLICATION

1996-01 Suzuki Esteem, 1996-98 Sidekick, 1998-02 Vitara, 1996-98 X-90	1.6L VIN 0

REPLACEMENT INTERVALS

Suzuki recommends timing belt replacement every 60,000 miles.

LABOR TIMES

Esteem	
Replacement	1.7 hrs
Severe Service	1.9 hrs
Sidekick & X-90	
Replacement	3.2 hrs
Severe Service	3.6 hrs
Vitara	
Replacement	2.4 hrs
Severe Service	2.7 hrs

8807SUZg04

✻✻ CAUTION

These engines are an interference type. If the timing belt has failed, valve-to-piston damage may have occurred. A compression check should always be performed to check for engine damage.

✻✻ WARNING

Do not rotate the crankshaft counterclockwise or attempt to rotate the crankshaft by turning the camshaft sprocket.

1. Disconnect the negative battery cable.
2. On Esteem, raise and support the vehicle.
3. Remove the right side wheel.
4. Remove the engine splash shield.
5. Remove the air cleaner.
6. On Sidekick, X-90 and Vitara drain the cooling system.
7. Remove the upper radiator hose.
8. Remove the cooling fan and cowl.
9. On all models, remove the accessory drive belt.
10. Remove the power steering pump without disconnecting the lines, and position aside.
11. Remove the water pump pulley.
12. Support the engine with a suitable jack and remove the right engine mount.
13. Remove the crankshaft pulley.
14. Remove the timing belt cover.
15. If the timing belt is not already marked with a directional arrow, use white paint, a grease pencil or correction fluid to do so.
16. Rotate the crankshaft clockwise until the timing mark on the camshaft sprocket and the "V" mark on the timing belt inside cover are aligned, and the punch mark on the crankshaft sprocket is aligned with the mark on the engine.

✻✻ WARNING

Do not rotate the crankshaft or camshaft once the timing belt is removed.

17. Disconnect one end of the tensioner spring. Loosen the timing belt tensioner bolt and nut a rotate the tensioner plate away and remove the timing belt from the crankshaft and camshaft sprockets.
18. Remove the timing belt tensioner, tensioner plate and spring from the engine.

To install:

✻✻ WARNING

If any binding is felt when adjusting the timing belt tension by turning the crankshaft, STOP turning the engine, because the pistons may be hitting the valves.

19. If removed, install the timing belt tensioner, tensioner plate and spring on the engine. Tighten the mounting bolt and stud only finger-tight at this time.
20. With the camshaft and crankshaft marks properly aligned, push the tensioner up with your finger and install the timing belt on the 2 sprockets, ensuring that the drive side of the belt is free of all slack. Release your finger from the tensioner. Be sure to install the timing belt so that the directional arrow is pointing in the appropriate direction.

■ In this position, the No. 4 cylinder is at Top Dead Center (TDC) on the compression stroke.

21. Rotate the crankshaft clockwise 2 full revolutions, then tighten the tensioner stud to 97 inch lbs. (11 Nm). Then, tighten the tensioner bolt to 18 ft. lbs. (24 Nm).
22. Ensure that all 4 timing marks are still aligned as before; if they are not, remove the timing belt and install and tension it again.
23. Install the timing belt cover.
24. Install the crankshaft pulley and tlghten the bolts to 12 ft. lbs. (16 Nm).
25. Install the right engine mount.
26. Install the water pump pulley.
27. Install the power steering pump.
28. Install the accessory drive belt.
29. On Sidekick, X-90 and Vitara install the cooling fan and cowl.
30. Install the upper radiator hose.
31. Refill the cooling system.
32. On Esteem, install the air cleaner.
33. Install the engine splash shield.
34. Install the right side wheel.
35. Lower the vehicle.
36. Connect the negative battery cable.

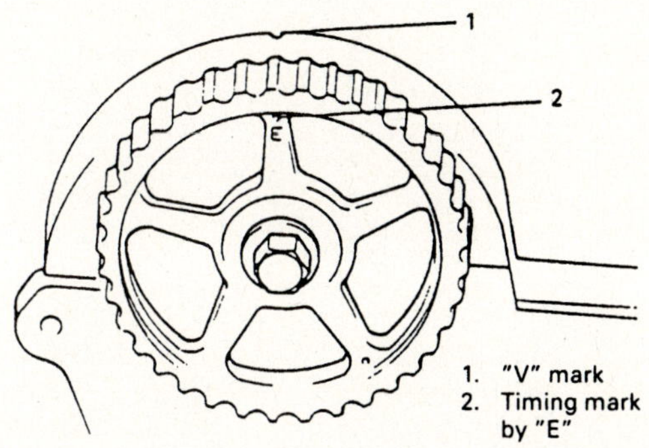

Camshaft timing marks—Suzuki 1.6L (VIN 0) 16-valve engine

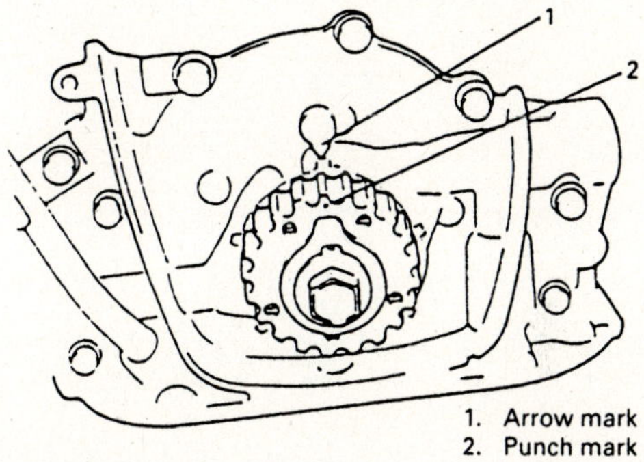

Align the punch mark with the arrow for proper timing belt installation—Suzuki 1.6L (VIN 0) 16-valve engine

Toyota

1.5L (1A-C, 3A, 3A-C) ENGINES

YEAR, MODEL & ENGINE APPLICATION

1980-81 Toyota Corolla,	
1980-88 Tercel	1.5L Code 1A-C, 3A, 3A-C

REPLACEMENT INTERVALS

Toyota does not provide a replacement interval. Standard industry practice is to replace timing belts every 60,000 miles.

LABOR TIMES

Corolla	
Replacement	1.8 hrs
Severe Service	2.3 hrs
Tercel	
1980-82	
Replacement	1.8 hrs
Severe Service	2.3 hrs
1983-88	
Replacement	2.4 hrs
Severe Service	2.9 hrs
With air conditioning add	0.1 hrs
With power steering add	0.1 hrs

8807TOYg01

✱✱ CAUTION

These engines are considered freewheeling. If the timing belt fails, valve-to-piston damage is unlikely, however a precautionary compression check should be performed to verify no engine damage.

1.Disconnect the negative battery cable.
2. Remove the accessory drive belts.
3. If equipped with air conditioning, unbolt the air conditioning compressor and set aside. Remove the compressor bracket.
4. Remove the water pump pulley.
5. Remove the valve cover.
6. Rotate the crankshaft so that the camshaft pulley timing hole aligns with the timing mark on the bearing cap and the crankshaft pulley timing mark aligns with the **0** mark on the lower timing cover.
7. Remove the crankshaft pulley.
8. Remove the timing covers.
9. Loosen the tensioner pulley lockbolt.
10. Push the tensioner pulley away from the timing belt and tighten the lockbolt.
11. Remove the timing belt guide and the timing belt.

To install:

12. Install the timing belt to the crankshaft timing pulley.
13. Install the lower timing cover and the crankshaft pulley.
14. Ensure that the camshaft and crankshaft timing marks align and install the timing belt to the camshaft pulley.
15. If necessary for access to the tensioner lockbolt, remove the grommet in the lower timing cover.
16. Loosen the tensioner lockbolt.
17. Rotate the crankshaft 2 complete turns and check that the timing marks align.

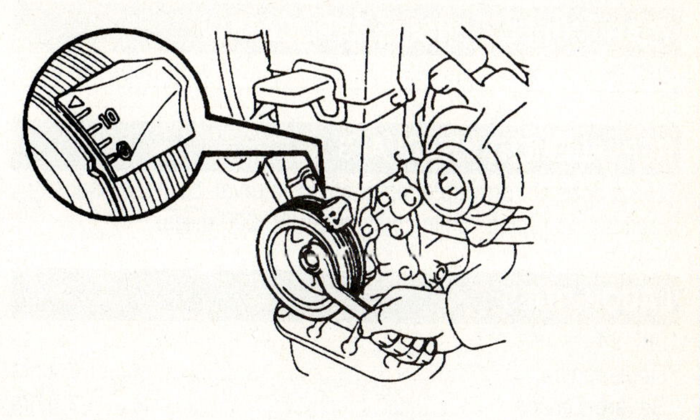

9305AG82

Crankshaft pulley timing marks—Toyota 1.5L (1A-C, 3A, 3A-C) engines

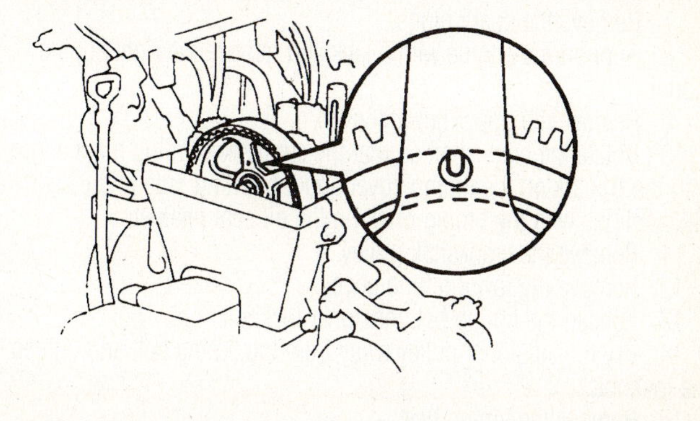

9305AG83

Camshaft timing marks—Toyota 1.5L (1A-C, 3A, 3A-C) engines

18. Tighten the tensioner lockbolt to 27 ft. lbs. (37 Nm).
19. Replace the grommet, if removed.
20. Tighten the crankshaft pulley bolt to 87 ft. lbs. (118 Nm).
21. Install the timing covers and the valve cover.
22. Install the water pump pulley.
23. Install the air conditioning compressor bracket and compressor, if removed.
24. Install the accessory drive belts.
25. Install the cruise control actuator, if removed.
26. Connect the negative battery cable.

1.5L (3E, 3E-E) ENGINES

✱✱ CAUTION

These engines are an interference type. If the timing belt has failed, valve-to-piston damage may have occurred. A compression check should always be performed to check for engine damage.

1. Disconnect the negative battery cable.
2. Raise and support the vehicle.
3. Remove the right front wheel.
4. Remove the right engine under cover.
5. Remove the accessory drive belts.

YEAR, MODEL & ENGINE APPLICATION

1987-94 Toyota Tercel	1.5L Code 3E, 3E-E

REPLACEMENT INTERVALS

Toyota does not provide a replacement interval. Standard industry practice is to replace timing belts every 60,000 miles.

LABOR TIMES

1987-94	
Replacement	2.4 hrs
Severe Service	2.9 hrs
With air conditioning add	0.1 hrs
With power steering add	0.1 hrs

8807TOYg02

6. If necessary, remove the air cleaner and air intake collector.
7. Remove the spark plugs.
8. Support the engine with a jack and remove the right engine mount.
9. Remove the upper timing cover.
10. Rotate the crankshaft so that the crankshaft timing mark aligns with the **0** mark on the timing cover and the timing hole in the camshaft pulley aligns with the timing mark on the oil seal retainer.
11. Remove the crankshaft pulley.
12. Remove the lower timing cover.
13. Loosen the timing belt tensioner lockbolt.
14. Pry the tensioner pulley away from the timing belt and tighten the lockbolt.
15. Remove the timing belt.

To install:

16. Ensure that the camshaft and crankshaft timing pulley marks align and install the timing belt.
17. Loosen the tensioner lockbolt.
18. Rotate the crankshaft 2 complete turns and check that the timing marks align.

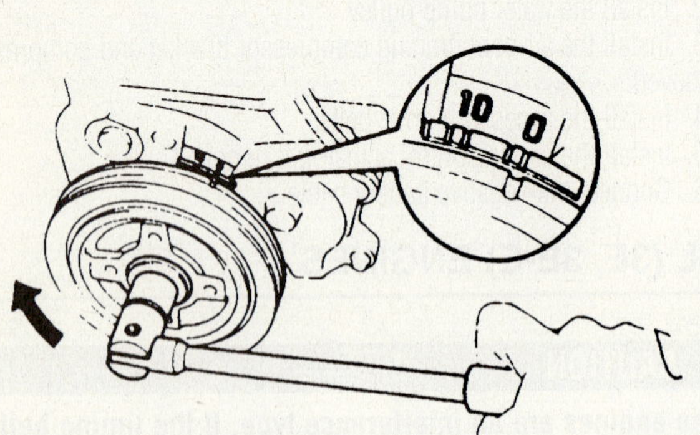

9305AG91

Crankshaft pulley timing marks—Toyota 1.5L (3E, 3E-E) engines

19. Tighten the tensioner lockbolt to 13 ft. lbs. (17 Nm).
20. Install the timing covers.
21. Install the crankshaft pulley and tighten the bolt to 112 ft. lbs. (152 Nm).
22. Install the right motor mount and remove the support jack.
23. Install the air cleaner and air intake collector, if removed.
24. Install the spark plugs.
25. Install the accessory drive belts.
26. Connect the throttle cable, if removed.
27. Install the engine under cover.
28. Install the right front wheel.
29. Connect the negative battery cable.

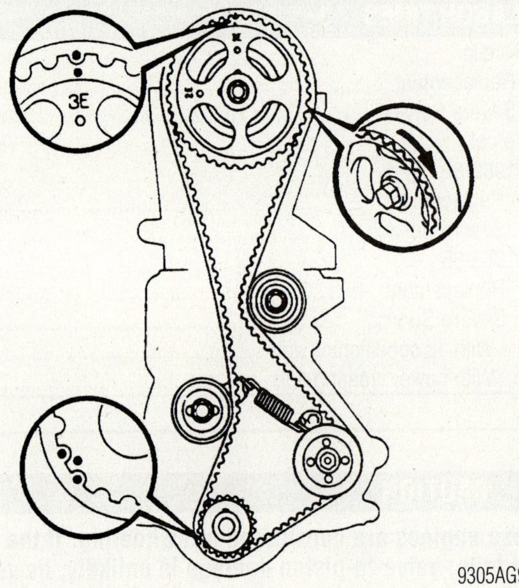

9305AG92

Timing belt alignment marks—Toyota 1.5L (3E, 3E-E) engines

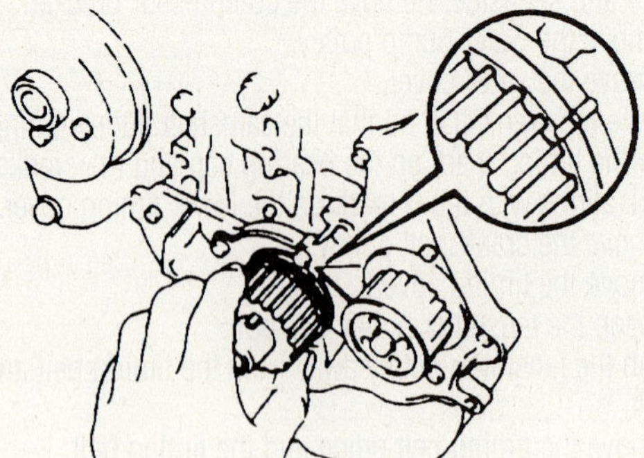

9305AG93

Crankshaft timing sprocket alignment—Toyota 1.5L (3E, 3E-E) engines

1.5L (5E-FE) ENGINES

YEAR, MODEL & ENGINE APPLICATION

1992-98 Toyota Paseo, 1995-99 Tercel .... 1.5L Code 5E-FE

REPLACEMENT INTERVALS

Toyota does not provide a replacement interval. Standard industry practice is to replace timing belts every 60,000 miles.

LABOR TIMES

Paseo	
Replacement	2.7 hrs
Severe Service	3.2 hrs
Tercel	
Replacement	2.4 hrs
Severe Service	2.9 hrs
With air conditioning add	0.1 hrs
With power steering add	0.1 hrs

8807TOYg03

✻✻ CAUTION

These engines are considered freewheeling. If the timing belt fails, valve-to-piston damage is unlikely, however a precautionary compression check should be performed to verify no engine damage.

✻✻ CAUTION

If equipped with an air bag, be sure to disconnect the negative battery cable and wait at least 90 seconds before proceeding.

1. Raise and support the vehicle.
2. Remove the right front wheel and splash shield.
3. Remove the right engine under cover.
4. Remove the accessory drive belts.
5. Remove the charcoal canister.
6. Remove the spark plugs.
7. Remove the cruise control actuator, if equipped.
8. Remove the vacuum switching valve.
9. Support the engine with a jack and remove the right engine mount.
10. Remove the upper timing belt cover.
11. Rotate the engine clockwise until the crankshaft pulley is aligned with the **0** mark on the No. 1 timing belt cover. Verify that the hole in the camshaft timing pulley is aligned with the timing mark on the No. 1 bearing cap. If not as specified, rotate the crankshaft an additional 360 degrees.
12. If equipped with air conditioning and/or power steering, remove the 4 bolts to the No. 2 crankshaft pulley. Then, remove the No. 2 crankshaft pulley.
13. Using Toyota tools SST 09213-14010 and SST 09330-00021, or their equivalents, remove the No. 1 crankshaft pulley bolt.
14. Using a crankshaft pulley/damper puller (such as Toyota tool SST 09950-50010), remove the No. 1 crankshaft pulley from the crankshaft.
15. Remove the No. 3 timing belt cover (plug).

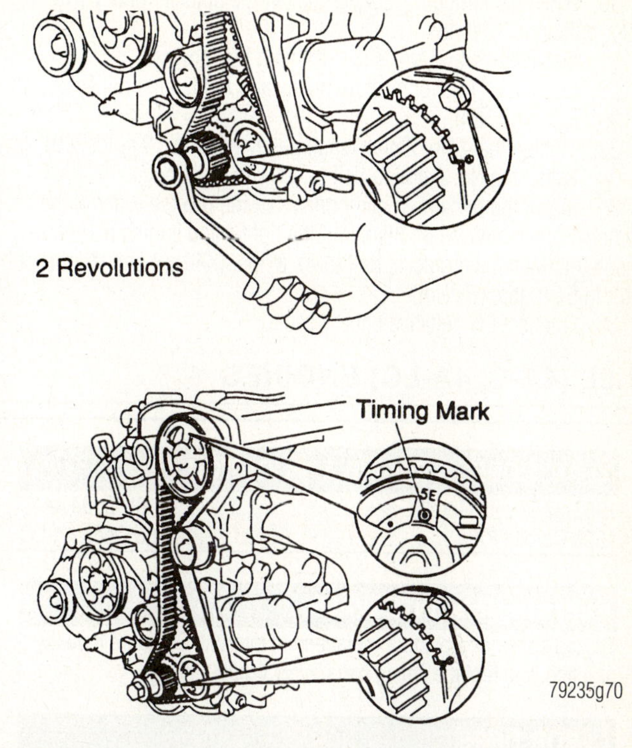

Turn the engine 2 revolutions, then ensure the timing marks are still aligned—Toyota 1.5L (5E-FE) engine

16. Remove the No. 1 timing belt cover and timing belt guide.
17. Place matchmarks on the timing belt on both sides of the cam and crankshaft gear timing marks. Also, place an arrow on the top surface of the belt to indicate the direction of travel.
18. Using pliers, remove the tension spring.
19. Loosen the No. 1 idler pulley bolt and push the pulley to the left as far as it will go, then temporarily tighten the bolt.
20. Remove the timing belt.

To install:

21. Measure the free length of the tension spring. Length should be 1.51 in. (38.3mm).
22. Install the tensioner pulley and spring.
23. If equipped with a distributor (distributor ignition), use the crankshaft bolt to turn the crankshaft until the timing marks on the sprocket and oil pump body align. This is method is used to set the piston at Top Dead Center (TDC) before the marks on the belt cover can be seen.
24. If equipped with a crankshaft position sensor (distributorless ignition), use the crankshaft bolt to turn the crankshaft until the rotor side of the crankshaft position sensor faces inward.
25. Turn the camshaft and align the hole of the camshaft timing pulley with the timing mark of the bearing cap. The matchmarks, if using the old belt should align. Place the belt over the crankshaft sprocket, oil pump sprocket, idler pulley, camshaft sprocket and tensioner.
26. Install the timing belt guide with flange out.
27. Loosen the No. 1 idler pulley bolt until the pulley is moved slightly by the spring tension.
28. Turn the crankshaft pulley 2 revolutions from TDC-to-TDC.

■ **Always rotate the crankshaft clockwise.**

29. Check that the pulleys align with the reference marks. If not, reinstall the belt.

30. When the timing is verified, tighten the adjuster pulley (No. 1 idler pulley) to 13 ft. lbs. (18 Nm).
31. Install No. 1 and No. 3 lower timing belt covers.
32. Install the No. 1 crankshaft pulley and tighten the pulley bolt to 112 ft. lbs. (152 Nm).
33. Tighten the 4 No. 2 crankshaft pulley bolts to 14 ft. lbs. (19 Nm).
34. Install the No. 2 timing belt cover with the 4 bolts.
35. Install the remaining components in the reverse order of the removal procedure. When installing the right-hand engine mounting insulator, tighten the bracket bolt to 47 ft. lbs. (64 Nm) and the through-bolt to 54 ft. lbs. (73 Nm).
36. Connect the negative battery cable.

1.6L (4A-C, 4A-LC) ENGINES

YEAR, MODEL & ENGINE APPLICATION

1983-87 Toyota Corolla	1.6L Code 4A-C
1988 Corolla FX	1.6L Code 4A-LC

REPLACEMENT INTERVALS

Toyota does not provide a replacement interval. Standard industry practice is to replace timing belts every 60,000 miles.

LABOR TIMES

Corolla	
Replacement	2.9 hrs
Severe Service	3.4 hrs
With air conditioning add	0.2 hrs
Corolla FX	
Replacement	3.1 hrs
Severe Service	3.6 hrs
With air conditioning add	0.2 hrs

8807TOYg04

✻✻ CAUTION

These engines are considered freewheeling. If the timing belt fails, valve-to-piston damage is unlikely, however a precautionary compression check should be performed to verify no engine damage.

1. Disconnect the negative battery cable.
2. Drain the engine coolant.
3. On Corolla FX models, raise and support the vehicle.
4. Remove the right side wheel.
5. Remove the engine undercover.
6. Remove the alternator.
7. Remove the power steering pump without disconnecting the lines and position aside.
8. Remove the A/C compressor and bracket.
9. On all models, remove the accessory drive belts.
10. Remove the air cleaner.
11. Remove the water pump pulley.
12. Remove the valve cover.
13. Remove the No. 2 crankshaft pulley on models with power steering.
14. Rotate the crankshaft so that the camshaft pulley timing hole aligns with the timing mark on the bearing cap and the crankshaft pulley timing mark aligns with the **0** mark on the lower timing cover.
15. Remove the crankshaft pulley.
16. Remove the timing belt covers.
17. Loosen the tensioner pulley lockbolt.
18. Push the tensioner pulley away from the timing belt and tighten the lockbolt.
19. Remove the timing belt guide and timing belt.
20. On Corolla FX, it may be necessary to support the engine and remove the crossmember and right side engine mount through bolt to remove the timing belt.

To install:

21. Ensure that the camshaft and crankshaft timing marks align and install the timing belt to the camshaft pulley.
22. Install the timing belt.
23. Loosen the tensioner lockbolt.
24. Install the crankshaft pulley.
25. Rotate the crankshaft 2 complete turns and check that the timing marks align.
26. Tighten the tensioner lockbolt to 27 ft. lbs. (37 Nm).
27. Check the timing belt deflection. Deflection should be 0.24–0.28 in. (6–7mm).
28. Install sealant to the No. 1 bearing cap and install the valve cover.
29. Tighten the crankshaft pulley bolt to 87 ft. lbs. (118 Nm).
30. Install the timing belt guide and the timing belt covers.
31. On Corolla FX models, install the A/C compressor and bracket.
32. Install the power steering pump.
33. Install the alternator.
34. Install the engine undercover.
35. Install the right side wheel.

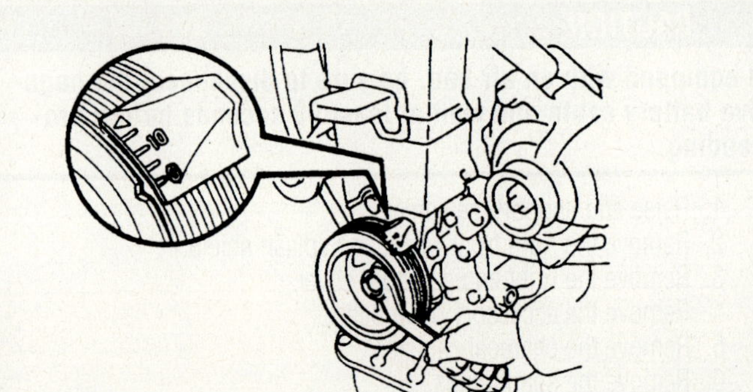

9305ag82

Crankshaft pulley timing marks—Toyota 1.6L (4A-C, 4A-CL) engines

9305ag83

Camshaft timing marks—Toyota 1.6L (4A-C, 4A-CL) engines

36. Lower the vehicle.
37. On all models, install the water pump pulley.
38. Install the air cleaner.
39. Install the water pump pulley.
40. Install the accessory drive belts.
41. Refill the engine coolant.
42. Connect the negative battery cable.

1.6L (4A-F, 4A-FE) ENGINES

YEAR, MODEL & ENGINE APPLICATION

1990-93 Toyota Celica, 1988-97 Corolla	1.6L Code 4A-F, 4A-FE

REPLACEMENT INTERVALS

Toyota does not provide a replacement interval. Standard industry practice is to replace timing belts every 60,000 miles.

LABOR TIMES

Celica	
Replacement	3.3 hrs
Severe Service	3.8 hrs
With air conditioning add	0.1 hrs
Corolla	
Replacement	3.1 hrs
Severe Service	3.6 hrs
With air conditioning add	0.1 hrs

8807TOYg05

1988–92 Models

✱✱ CAUTION

These engines are considered freewheeling. If the timing belt fails, valve-to-piston damage is unlikely, however a precautionary compression check should be performed to verify no engine damage.

1. Disconnect the negative battery cable.
2. Raise and safely support the vehicle.
3. Remove the right front wheel.
4. Remove the right engine under cover.
5. Remove the air cleaner.
6. Remove the windshield washer fluid reservoir.
7. Remove the accessory drive belts.
8. Remove the spark plugs.
9. Remove the valve cover.
10. Support the engine with a jack and remove the right engine mount.
11. Remove the water pump pulley.
12. Rotate the crankshaft so that the camshaft pulley timing hole aligns with the timing mark on the bearing cap and the crankshaft pulley timing mark aligns with the **0** mark on the lower timing cover.
13. Remove the crankshaft pulley.
14. Remove the timing covers.
15. Loosen the tensioner pulley lockbolt.
16. Push the tensioner pulley away from the timing belt and tighten the lockbolt.
17. Remove the timing belt.

To install:

18. Measure the free length of the tension spring. Length should be 1.51 in. (38.3mm).
19. Ensure that the camshaft and crankshaft timing marks align and install the timing belt to the camshaft pulley.
20. Install the idler pulley and spring.
21. Install the timing belt.
22. Install the crankshaft pulley.
23. Loosen the idler pulley lockbolt and alow the pulley to tension the timing belt.
24. Rotate the crankshaft 2 complete turns and check that the timing marks align.
25. Tighten the idler pulley lockbolt to 27 ft. lbs. (37 Nm).
26. Check the timing belt deflection. Deflection should be 0.20–0.24 in. (5–6mm).
27. Install the timing belt guide.
28. Tighten the crankshaft pulley bolt to 87 ft. lbs. (118 Nm).
29. Install the timing covers and the valve cover, if removed.
30. Install the water pump pulley.
31. Install the right engine mount and remove the support jack.
32. Install the air conditioning compressor bracket and compressor, if removed.
33. Install the right engine under cover.
34. Install the right front wheel and lower the vehicle.
35. Install the accessory drive belts.
36. Connect the negative battery cable.

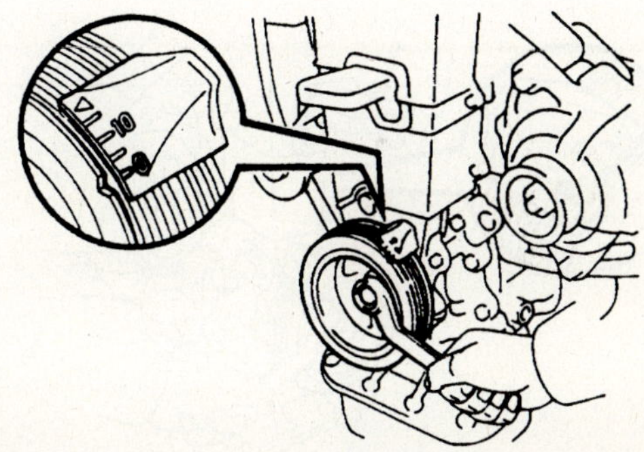

9305ag82

Crankshaft pulley timing marks—Toyota 1.6L (4A-F, 4A-FE) engines

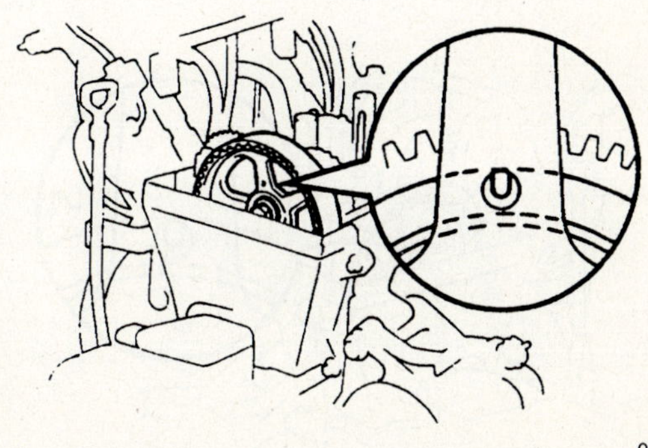

9305ag83

Camshaft timing marks—Toyota 1.6L (4A-F, 4A-FE) engines

1993–97 Models

✻✻ CAUTION

These engines are considered freewheeling. If the timing belt fails, valve-to-piston damage is unlikely, however a precautionary compression check should be performed to verify no engine damage.

1. Disconnect the negative battery cable.
2. Remove the windshield washer reservoir.
3. Remove the cruise control actuator.
4. Raise and safely support the vehicle.
5. Remove the right front wheel.
6. Remove the right engine under cover.
7. Remove the air cleaner.
8. Remove the accessory drive belts.
9. Remove the spark plugs.
10. Remove the valve cover.
11. Support the engine with a jack and remove the right engine mount.
12. Rotate the crankshaft so that the camshaft pulley timing hole aligns with the timing mark on the bearing cap and the crankshaft pulley timing mark aligns with the **0** mark on the lower timing cover.
13. Remove the water pump pulley.
14. Remove the crankshaft pulley.
15. Remove the timing covers.
16. Remove the timing belt guide.
17. Loosen the tensioner pulley lockbolt.

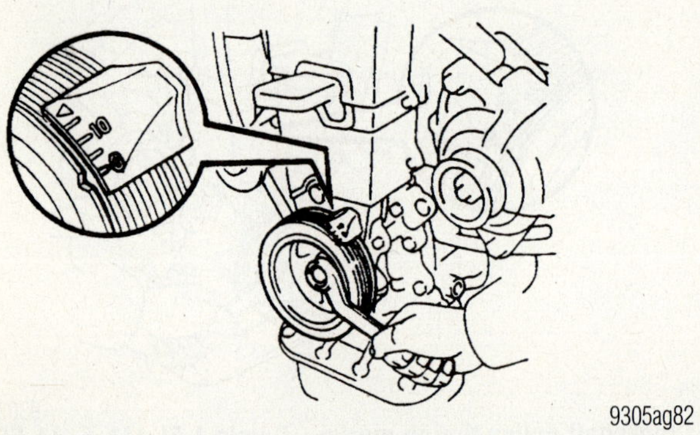

9305ag82

Crankshaft pulley timing marks—Toyota 1.6L (4A-F, 4A-FE) engines

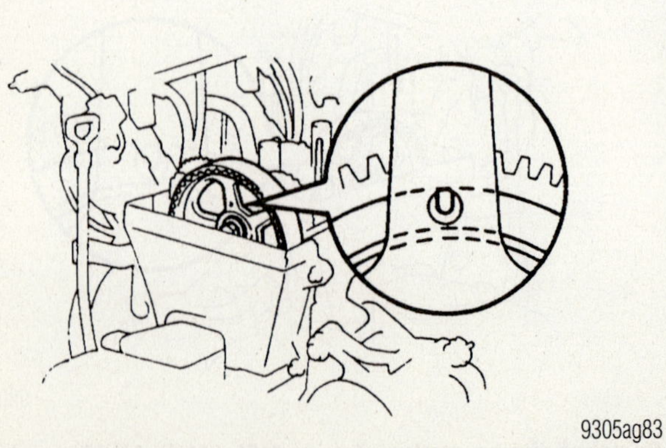

9305ag83

Camshaft timing marks—Toyota 1.6L (4A-F, 4A-FE) engines

18. Push the tensioner pulley away from the timing belt and tighten the lockbolt.
19. Remove the timing belt.

To install:

20. Measure the free length of the tension spring. Length should be 1.39 in. (35mm) on 1993–94 models, or 1.45 in. (37mm) on 1995-97 models.
21. Ensure that the camshaft and crankshaft timing marks align and install the timing belt to the camshaft pulley.
22. Install the idler pulley and spring.
23. Install the timing belt.
24. Install the crankshaft pulley.
25. Loosen the idler pulley lockbolt and alow the pulley to tension the timing belt.
26. Rotate the crankshaft 2 complete turns and check that the timing marks align.
27. Tighten the idler pulley lockbolt to 27 ft. lbs. (37 Nm).
28. Check the timing belt deflection. Deflection should be 0.20–0.24 in. (5–6mm).
29. Install the timing belt guide.
30. Tighten the crankshaft pulley bolt to 87 ft. lbs. (118 Nm).
31. Install the timing covers and the valve cover, if removed.
32. Install the water pump pulley.
33. Install the right engine mount and remove the support jack.
34. Install the air conditioning compressor bracket and compressor, if removed.
35. Install the right engine under cover.
36. Install the right front wheel and lower the vehicle.
37. Install the accessory drive belts.
38. Install the windshield washer reservoir.
39. Install the cruise control actuator.
40. Connect the negative battery cable.

1.6L (4A-GE, 4A-GZE) ENGINES

YEAR, MODEL & ENGINE APPLICATION

Year/Model	Engine
1988-91 Corolla GT-S	1.6L Code 4A-GE
1985-89 MR2	1.6L Code 4A-GZE

REPLACEMENT INTERVALS

Toyota does not provide a replacement interval. Standard industry practice is to replace timing belts every 60,000 miles.

LABOR TIMES

Corolla	
Replacement	3.1 hrs
Severe Service	3.6 hrs
MR2	
Replacement	2.8 hrs
Severe Service	3.3 hrs
With air conditioning add	0.5 hrs

8807TOYg06

✻✻ CAUTION

These engines are considered freewheeling. If the timing belt fails, valve-to-piston damage is unlikely, however a precautionary compression check should be performed to verify no engine damage.

1. Disconnect the negative battery cable.
2. Raise and safely support the vehicle.
3. For Corolla models, drain the cooling system and remove the following items:
- Right front wheel
- Right engine under cover
- Cruise control actuator, if equipped
- Windshield washer fluid reservoir
- Radiator hose
4. For MR2 models, remove the following items:
- Right rear wheel
- Engine splash guard
5. Remove the accessory drive belts.
6. Support the engine with a jack and remove the right engine mount.
7. Remove the upper timing cover.
8. Remove the water pump pulley.
9. Remove the middle timing cover.
10. Rotate the crankshaft so that the camshaft pulley timing marks align with the timing marks on the rear timing cover and the crankshaft pulley timing mark aligns with the **0** mark on the lower timing cover.
11. Remove the crankshaft pulley.
12. Remove the lower timing cover.
13. Loosen the tensioner pulley lockbolt.
14. Push the tensioner pulley away from the timing belt and tighten the lockbolt.
15. Remove the timing belt.

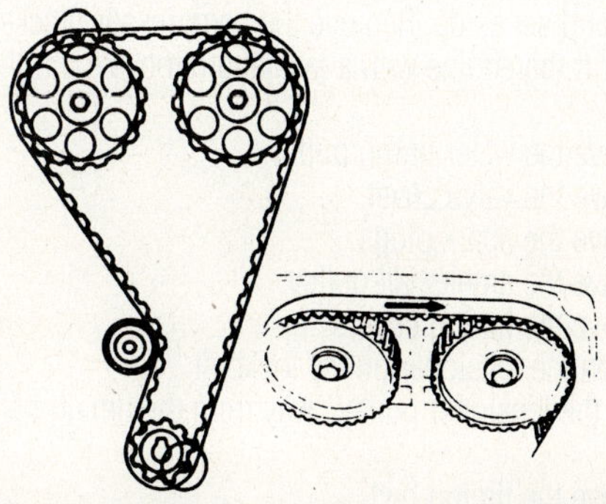

8807Timg16

Camshaft timing marks—Toyota 1.6L (4A-GE, 4A-GZE) engines

To install:

16. Install the timing belt to the crankshaft timing pulley.
17. Install the lower timing cover and the crankshaft pulley. Tighten the crankshaft pulley bolt to 101 ft. lbs. (137 Nm).
18. Ensure that the camshaft and crankshaft timing marks align and install the timing belt to the camshaft pulleys.
19. Rotate the crankshaft 2 complete turns and check that the timing marks align.
20. Tighten the tensioner lockbolt to 27 ft. lbs. (37 Nm).
21. Install the middle timing cover and the water pump pulley.
22. Install the upper timing cover.
23. Install the right motor mount and remove the engine support jack.
24. Install the accessory drive belts.
25. For Corolla models, install the following:
- Radiator hose
- Windshield washer fluid reservoir
- Cruise control actuator, if removed
- Right engine under cover
- Right front wheel
26. For MR2 models, install the following:
- Engine splash guard
- Right rear wheel
27. Lower the vehicle.
28. Connect the negative battery cable and fill the cooling system, if drained.

1.8L (1C-L, 1C-TL) AND 2.0L (2C-T) DIESEL ENGINES

YEAR, MODEL & ENGINE APPLICATION

1984-85 Toyota Camry, Corolla	1.8L Diesel Code 1C-L/1C-TL
1986-87 Camry	2.0L Diesel Code 2C-T

REPLACEMENT INTERVALS

Toyota recommends timing belt replacement every 60,000 miles.

LABOR TIMES

Camry	
Replacement	2.6 hrs
Severe Service	3.1 hrs
Corolla	
Replacement	2.4 hrs
Severe Service	2.9 hrs

8807TOYg07

✳✳ CAUTION

These engines are an interference type. If the timing belt has failed, valve-to-piston damage may have occurred. A compression check should always be performed to check for engine damage.

1. Disconnect the negative battery cable.
2. Raise and safely support the vehicle.
3. Remove the right front wheel.
4. Remove the right front inner fender liner.
5. Remove the windshield washer fluid reservoir.
6. Remove the engine coolant overflow tank.
7. If equipped, remove the cruise control actuator.
8. Remove the accessory drive belts.
9. Unbolt the power steering pump and position aside.
10. Remove the alternator and bracket.
11. Remove the upper timing cover.
12. Rotate the crankshaft clockwise to align the camshaft pulley timing mark with the top edge of the cylinder head.
13. Remove the crankshaft pulley.
14. Remove the lower timing cover.
15. Support the engine with a jack and remove the right engine mount.
16. Loosen the timing belt tensioner lockbolt.
17. Pry the tensioner pulley away from the timing belt and tighten the lockbolt.
18. Remove the timing belt.

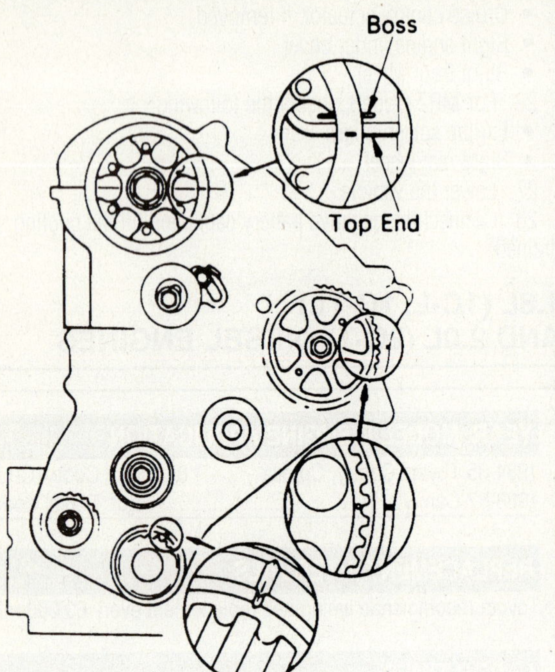

9305ag81

Timing mark alignment—Toyota 1.8L (1C, 1C-TL) and 2.0L (2C-T) diesel engines

To install:

19. Ensure that the timing marks on the camshaft pulley, crankshaft timing pulley and injection pump pulley are aligned.
20. Install the timing belt.
21. Loosen the tensioner lockbolt.
22. Rotate the crankshaft 2 complete turns and check that the timing marks align.
23. Tighten the tensioner lockbolt to 27 ft. lbs. (37 Nm).
24. Install the right engine mount. Remove the engine support jack.
25. Install the upper and lower timing covers.
26. Install the crankshaft pulley and tighten the bolt to 72 ft. lbs. (98 Nm).
27. Install the alternator and bracket.
28. Install the power steering pump.
29. Install the accessory drive belts.
30. If removed, install the cruise control actuator.
31. Install the coolant overflow tank.
32. Install the windshield washer fluid reservoir.
33. Install the right inner fender liner.
34. Install the right front wheel and lower the vehicle.
35. Connect the negative battery cable.

1.8L (7A-FE) ENGINES

✱✱ CAUTION

These engines are considered freewheeling. If the timing belt fails, valve-to-piston damage is unlikely, however a precautionary compression check should be performed to verify no engine damage.

YEAR, MODEL & ENGINE APPLICATION

1994-97 Toyota Celica, 1993-97 Corolla FX	1.8L Code 7A-FE

REPLACEMENT INTERVALS

Toyota does not provide a replacement interval. Standard industry practice is to replace timing belts every 60,000 miles.

LABOR TIMES

Celica	
Replacement	2.7 hrs
Severe Service	3.2 hrs
With air conditioning add	0.3 hrs
Corolla FX	
Replacement	2.5 hrs
Severe Service	3.0 hrs

8807TOYg08

1. Disconnect the negative battery cable.
2. Remove the windshield washer fluid reservoir.
3. If equipped, remove the cruise control actuator.
4. Remove the accessory drive belts.
5. Raise and safely support the vehicle.
6. Remove the right front wheel.
7. Remove the right engine under cover.
8. If equipped with air conditioning, unbolt the air conditioning compressor and set aside. Remove the compressor bracket.
9. Support the engine with a jack and remove the right engine mount.
10. Remove the water pump pulley.
11. Remove the valve cover.
12. Remove the spark plugs.
13. Remove the crankshaft pulley.
14. Remove the timing covers.
15. Loosen the tensioner pulley lockbolt.
16. Push the tensioner pulley away from the timing belt and tighten the lockbolt.
17. Remove the timing belt.
18. Rotate the crankshaft so that the camshaft pulley timing hole aligns with the timing mark on the bearing cap and the crankshaft pulley timing mark aligns with the **0** mark on the lower timing cover.

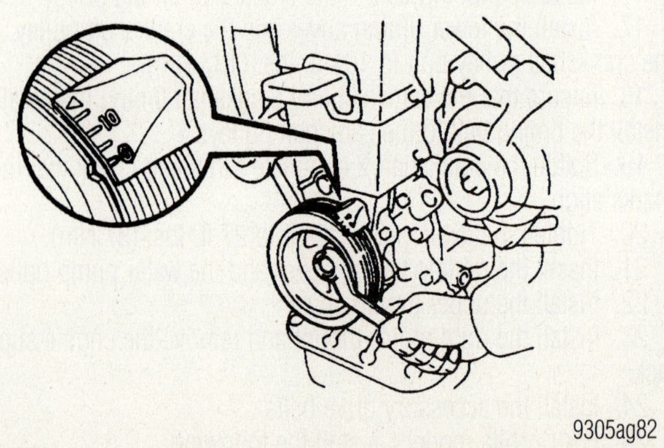
9305ag82

Crankshaft pulley timing marks—Toyota 1.8L (7A-FE) engine

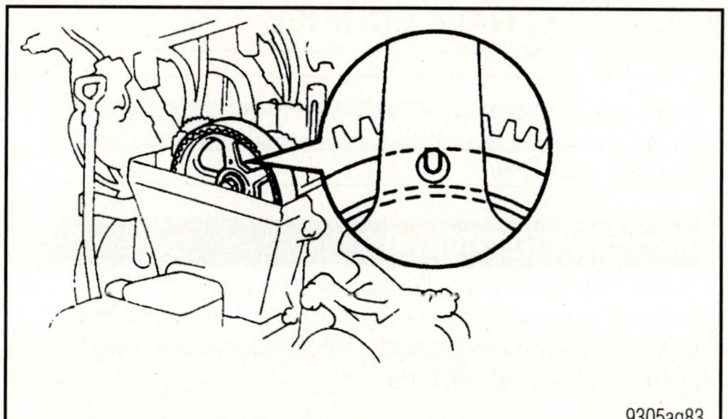
9305ag83

Camshaft timing marks—Toyota 1.8L (7A-FE) engine

To install:

19. Measure the free length of the tension spring. Length should be 1.25 in. (32mm).
20. Install the timing belt to the crankshaft timing pulley.
21. Install the lower timing cover and the crankshaft pulley.
22. Ensure that the camshaft and crankshaft timing marks align and install the timing belt to the camshaft pulley.
23. If necessary for access to the tensioner lockbolt, remove the grommet in the lower timing cover.
24. Loosen the tensioner lockbolt.
25. Rotate the crankshaft 2 complete turns and check that the timing marks align.
26. Tighten the tensioner lockbolt to 27 ft. lbs. (37 Nm).
27. Check the timing belt deflection. Deflection should be 0.20–0.24 in. (5–6mm).
28. Replace the grommet, if removed.
29. Tighten the crankshaft pulley bolt to 87 ft. lbs. (118 Nm).
30. Install the remaining timing covers and the valve cover, if removed.
31. Install the water pump pulley.
32. Install the right engine mount and remove the support jack.
33. Install the air conditioning compressor bracket and compressor, if removed.
34. Install the right engine under cover.
35. Install the right front wheel and lower the vehicle.
36. Install the accessory drive belts.
37. Install the cruise control actuator, if removed.
38. Install the windshield washer fluid reservoir.
39. Connect the negative battery cable.

2.0L (2S-E, 3S-FE) ENGINES

✱✱ CAUTION

These engines are considered freewheeling. If the timing belt fails, valve-to-piston damage is unlikely, however a precautionary compression check should be performed to verify no engine damage.

1. Disconnect the negative battery cable.
2. Raise and safely support the vehicle.
3. Remove the right front wheel.
4. Remove the right front inner fender liner.
5. Remove the windshield washer fluid reservoir.

YEAR, MODEL & ENGINE APPLICATION

1983-87 Toyota Camry	2.0L Code 2S-E
1988-91 Camry, 1988-89 Celica	2.0L Code 3S-FE

REPLACEMENT INTERVALS

Toyota does not provide a replacement interval. Standard industry practice is to replace timing belts every 60,000 miles.

LABOR TIMES

Celica	
Replacement	2.6 hrs
Severe Service	3.1 hrs
Camry	
1983-86	
Replacement	3.4 hrs
Severe Service	3.9 hrs
1987-91	
Replacement	2.9 hrs
Severe Service	3.4 hrs

8807TOYg09

6. Remove the engine coolant overflow tank.
7. If equipped, remove the cruise control actuator.
8. Remove the accessory drive belts.
9. Unbolt the power steering pump and position aside.
10. Remove the alternator and bracket.
11. Remove the right side engine mount insulator and bracket.
12. Remove the upper timing cover.
13. Remove the spark plugs.
14. Rotate the crankshaft so that the crankshaft timing mark aligns with the **0** mark on the timing cover and the timing hole in the camshaft pulley aligns with the timing mark on the oil seal retainer.

■ **There is an "E" cast into the camshaft pulley. On USA models, this "E" will be at 12 o'clock when the timing is correct. On Canadian models, the "E" will be at 6 o'clock.**

15. Remove the crankshaft pulley.
16. Remove the lower timing cover.
17. Loosen the timing belt tensioner lockbolt.
18. Pry the tensioner pulley away from the timing belt and tighten the lockbolt.
19. Remove the timing belt guide and timing belt.

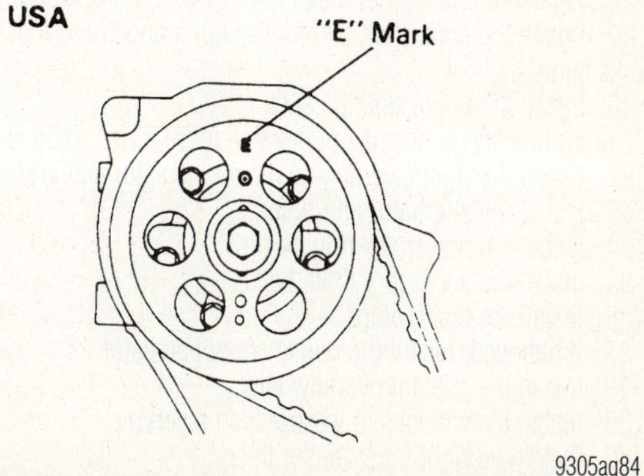

9305ag84

Camshaft timing mark—Toyota 2.0L (2S-E, 3S-FE) USA engines

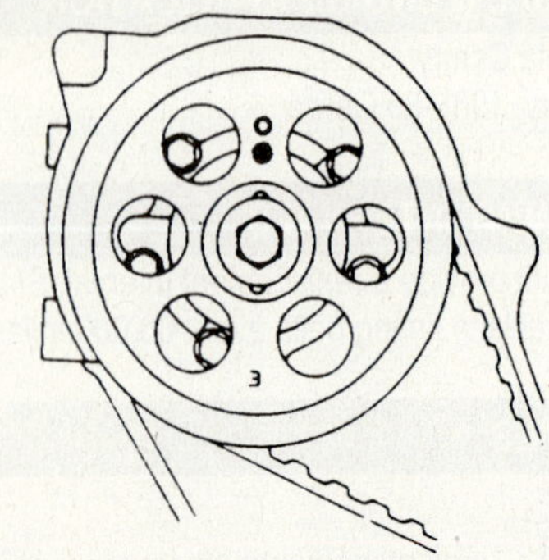

Camshaft timing mark—Toyota 2.0L (2S-E, 3S-FE) Canadian engines

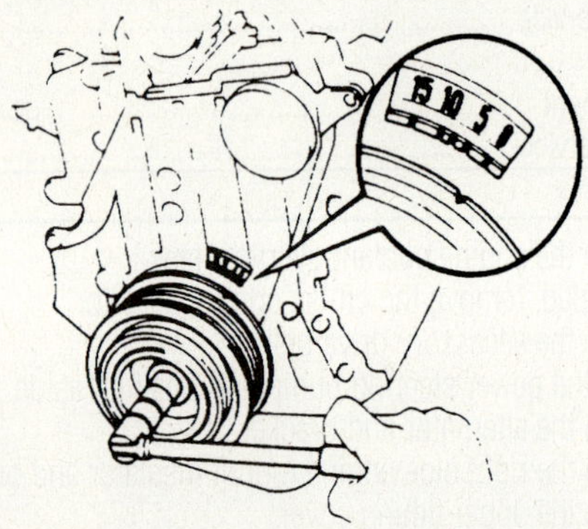

Crankshaft timing marks—Toyota 2.0L (2S-E, 3S-FE) engines

To install:

20. Install the timing belt to the crankshaft sprocket, oil pump sprocket, idler sprocket, water pump and tensioner pulley.
21. Install the timing belt guide.
22. Install the lower timing belt cover.
23. Install the crankshaft pulley.
24. Ensure that the crankshaft and camshaft timing marks are aligned.
25. Install the temsioner pulley and spring. Move the pulley to the left and tighten the bolt.
26. Loosen the tensioner lockbolt.
27. Rotate the crankshaft clockwise about 90 degrees and ensure the No. 1 cylinder is at TDC.
28. Tighten the tensioner lockbolt to 31 ft. lbs. (42 Nm).
29. Rotate the crankshaft 2 complete turns and check that the timing marks align.
30. Install the upper timing cover.
31. Tighten the crankshaft pulley bolt to 80 ft. lbs. (108 Nm).
32. Install the right side engine mount insulator and bracket.
33. Install the alternator and bracket.
34. Install the power steering pump.
35. Install the accessory drive belts.
36. Install the spark plugs.
37. If removed, install the cruise control actuator.
38. Install the coolant overflow tank.
39. Install the windshield washer fluid reservoir.
40. Install the right inner fender liner.
41. Install the right front wheel and lower the vehicle.
42. Connect the negative battery cable.

2.0L (3S-FE) RAV4 ENGINES

YEAR, MODEL & ENGINE APPLICATION

1996-00 Toyota RAV4	2.0L Code 3S-FE

REPLACEMENT INTERVALS

Toyota does not provide a replacement interval for 1996-98 models. Standard industry practice is to replace timing belts every 60,000 miles. On 1999-00 models, Toyota recommends timing belt replacement every 90,000 miles.

LABOR TIMES

1996-00	
Replacement	2.9 hrs
Severe Service	3.4 hrs

8807TOYg10

✻✻ CAUTION

These engines are considered freewheeling. If the timing belt fails, valve-to-piston damage is unlikely, however a precautionary compression check should be performed to verify no engine damage.

■ **Ensure the engine is cold before installing a new timing belt.**

1. Disconnect the negative battery cable and wait at least 90 seconds for the air bag system to deploy all back-up power.
2. Disconnect the power steering reservoir tank and remove the reservoir bracket.
3. Detach the wiring harness bracket for the Data Link Connector 1 (DLC1).
4. Remove the alternator and alternator bracket.
5. If equipped with ABS brakes, remove the ABS actuator. Raise and support the vehicle.
6. Remove the right front wheel and the fender apron seal.
7. Remove the power steering drive belt.
8. Slightly raise the engine using a block of wood and floor jack under the oil pan to prevent damage.
9. Remove the 4 bolts, 2 nuts and right-hand mounting bracket.
10. Remove the spark plugs.
11. Using SST 09213-54015, or equivalent, loosen the crankshaft pulley bolt and remove it by pulling it straight off the crankshaft.
12. Using SST 09249-63010, or equivalent, loosen the retaining bolts and remove the right engine mounting bracket.
13. Remove the upper (No. 2) timing belt cover.
14. Install the crankshaft pulley to the crankshaft and temporarily install the retaining bolt.
15. Turn the crankshaft pulley and align its groove with the timing mark **0** of the No. 1 timing belt cover. Check that the hole of the camshaft timing pulley is aligned with the timing mark of the bearing cap. If not, turn the crankshaft 360 degrees and align the marks.

■ **If the timing belt is to be reused, matchmark the timing belt to the timing pulleys and timing belt covers so the belt can be reinstalled in its original position. Also, be sure to mark an arrow on the belt to indicate which direction it was turning.**

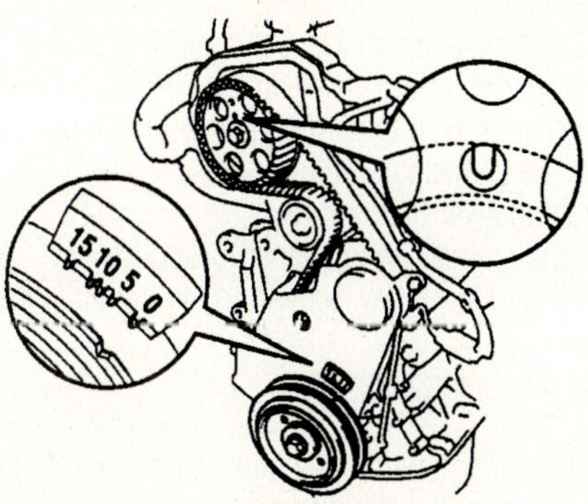

79245g39

Align the timing reference indicators prior to removing the timing belt—Toyota 2.0L (3S-FE) RAV4 engine

16. Remove the timing belt from the camshaft timing pulley.
17. Hold the camshaft sprocket with a spanner wrench and remove the mounting bolt. Remove the camshaft pulley.
18. Remove the crankshaft pulley bolt and remove the crankshaft pulley.
19. Remove the No. 1 timing belt cover.
20. Remove the timing belt guide and the timing belt.
21. Remove the No. 1 idler pulley and tension spring.
22. Remove the No. 2 idler pulley.
23. Remove the crankshaft timing pulley.
24. Support the oil pump sprocket with a spanner wrench, then remove the mounting bolt and remove the sprocket.

To install:

25. Measure the free length of the tension spring. Length should be 1.81 in. (46mm) on 1996–97 models and 1.65 in. (42mm) on 1998–2000 models. Replace the spring if the length does not meet specifications.
26. Install the oil pump pulley. Tighten the nut to 18 ft. lbs. (24 Nm).
27. Install the crankshaft timing pulley. Align the pulley set key with the key groove of the pulley. Slide on the pulley facing the flange side inward.
28. Install the No. 2 idler pulley and tighten the mounting bolt to 31 ft. lbs. (42 Nm). Be sure that the pulley moves smoothly.
29. Install the No. 1 idler pulley with the bolt and the tension spring. Pry the pulley toward the left as far as it will go and tighten the bolt. Make sure that the pulley moves smoothly.
30. Temporarily install the timing belt. Using the crankshaft pulley bolt, turn the crankshaft and position the key groove of the crankshaft timing pulley upward. If reusing the timing belt, align the points marked during removal.
31. Install the timing belt on the crankshaft timing pulley, oil pump pulley, No. 1 idler pulley, water pump pulley and the No. 2 idler pulley.
32. Install the timing belt guide.

■ **If the old timing belt is being reinstalled, be sure the directional arrow is facing in the original direction and that the belt and sprocket/cover matchmarks are properly aligned.**

33. Install the lower (No. 1) timing belt cover and new gasket with the 4 bolts.
34. Align the crankshaft pulley set key with the pulley key groove. Temporarily install the crankshaft pulley and bolt.
35. Align the camshaft knock pin with the groove of the pulley, and slide the timing pulley onto the camshaft with the plate washer and set bolt.
36. Tighten the pulley set bolt to 40 ft. lbs. (54 Nm).

**** WARNING**

If any binding is felt when adjusting the timing belt tension by turning the crankshaft, STOP turning the engine, because the pistons may be hitting the valves.

37. Turn the crankshaft pulley and align the **0** mark on the lower (No. 1) timing belt cover.
38. Finish installing the timing belt and check the valve timing, as follows:
    a. If reusing the old timing belt, align the matchmarks made previously and install the timing belt onto the camshaft pulley.
    b. Align the marks on the timing belt with the marks on the camshaft pulley.
    c. Loosen the No. 1 idler pulley set bolt 1/2 turn.
    d. Turn the crankshaft pulley 2 complete revolutions TDC-to-TDC. ALWAYS turn the crankshaft CLOCKWISE. Check that the pulleys are still in alignment with the timing marks.
    e. If the No. 1 idler pulley uses a green tension spring, slowly turn the crankshaft pulley 1 7/8 revolutions, and align its groove with the mark at 45 degrees BTDC (for the No. 1 cylinder) of the No. 1 timing belt cover.
    f. Tighten the No. 1 idler pulley set bolt to 31 ft. lbs. (42 Nm).
    g. Be sure there is belt tension between the crankshaft and camshaft timing pulleys.
39. Place the right-hand engine mounting bracket in position but do not install the bolts.
40. Install the upper (No. 2) timing cover with a new gasket(s).
41. Remove the engine crankshaft pulley bolt and pulley.
42. Using SST 09249-63010 or equivalent, install the mounting bolts for the right-hand mounting bracket. Tighten the mounting bolts to 38 ft. lbs. (52 Nm).
43. Align the crankshaft pulley set key with the pulley key groove. Install the pulley. Tighten the pulley bolt to 80 ft. lbs. (108 Nm).
44. Install the spark plugs.
45. Install the right-hand mounting insulator, as follows:
    a. Attach the mounting insulator to the body and mounting bracket with the 4 bolts and 2 nuts.
    b. Tighten the 3 bolts to hold the mounting insulator to the body. Tighten the bolts to 47 ft. lbs. (64 Nm).
    c. Tighten the 2 nuts and bolt to hold the mounting insulator to the mounting bracket. Tighten the bolt to 27 ft. lbs. (37 Nm) and the nut to 38 ft. lbs. (52 Nm).
46. Install and adjust the power steering pump drive belt.
47. Install the right-hand engine undercover.
48. Install the right front wheel.
49. Lower the engine.
50. If equipped, install the ABS actuator.
51. Install the alternator and alternator bracket.
52. Install the wiring harness bracket for the DLC1.
53. Install the power steering reservoir bracket and reservoir.
54. Connect the negative battery cable.

2.0L (3S-GE) ENGINES

YEAR, MODEL & ENGINE APPLICATION

1986-89 Toyota Celica .... 2.0L Code 3S-GE

REPLACEMENT INTERVALS

Toyota does not provide a replacement interval. Standard industry practice is to replace timing belts every 60,000 miles.

LABOR TIMES

1986-89	
Replacement	2.6 hrs
Severe Service	3.1 hrs
With air conditioning add	0.1 hrs

8807TOYg11

✳✳ CAUTION

These engines are considered freewheeling. If the timing belt fails, valve-to-piston damage is unlikely, however a precautionary compression check should be performed to verify no engine damage.

■ **Install the new timing belt with the engine cold.**

1. Disconnect the negative battery cable and wait at least 90 seconds for the air bag system to deploy all back-up power.
2. Raise and safely support the vehicle.
3. Remove the right front wheel.
4. Remove the right front inner fender liner.
5. Remove the windshield washer fluid reservoir.
6. Remove the engine coolant overflow tank.
7. If equipped, remove the cruise control actuator.
8. Remove the accessory drive belts.
9. Unbolt the power steering pump reservoir and position aside.
10. Remove the alternator and bracket.
11. Support the engine with a jack and remove the right engine mount.
12. Remove the spark plugs.

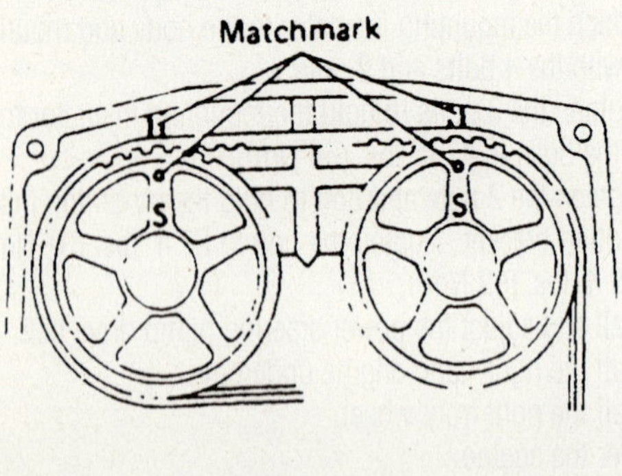

9305ag87

Alignment of the camshaft timing marks—Toyota 2.0L (3S-GE) engine

13. Remove the valve cover.
14. Remove the upper timing cover.
15. Rotate the crankshaft so that the crankshaft timing mark aligns with the **0** mark on the timing cover and the timing holes in the camshaft pulleys align with the timing marks on the rear timing cover.
16. Remove the crankshaft pulley.
17. Remove the lower timing cover.
18. Loosen the timing belt tensioner lockbolt.
19. Pry the tensioner pulley away from the timing belt and tighten the lockbolt.
20. Remove the timing belt.

To install:

21. Install the timing belt to the crankshaft sprocket, oil pump sprocket, idler pulley and water pump pulley.
22. Install the timing belt guide and the lower timing cover.
23. Install the crankshaft pulley.
24. Ensure that the crankshaft and camshaft timing marks are aligned and install the timing belt to the camshaft sprockets.
25. Loosen the tensioner lockbolt.
26. Rotate the crankshaft 2 complete turns and check that the timing marks align.
27. Tighten the tensioner lockbolt to 32 ft. lbs. (43 Nm).
28. Install the right engine mount. Remove the engine support jack.
29. Install the upper timing cover.
30. Tighten the crankshaft pulley bolt to 80 ft. lbs. (108 Nm).
31. Install the valve cover.
32. Install the spark plugs.
33. Install the alternator and bracket.
34. Install the power steering pump reservoir.
35. Install the accessory drive belts.
36. If removed, install the cruise control actuator.
37. Install the coolant overflow tank.
38. Install the windshield washer fluid reservoir.
39. Install the right inner fender liner.
40. Install the right front wheel and lower the vehicle.
41. Connect the negative battery cable.

2.0L (3S-GTE) ENGINES

YEAR, MODEL & ENGINE APPLICATION

1990-93 Toyota Celica, 1991-95 MR2 .... 2.0L Code 3S-GTE

REPLACEMENT INTERVALS

Toyota recommends timing belt replacement every 60,000 miles.

LABOR TIMES

Celica	
Replacement	4.3 hrs
Severe Service	4.8 hrs
With air conditioning add	0.1 hrs
MR2	
Replacement	3.9 hrs
Severe Service	4.4 hrs
With air conditioning add	0.5 hrs

8807TOYg12

Celica

✻✻ CAUTION

These engines are considered freewheeling. If the timing belt fails, valve-to-piston damage is unlikely, however a precautionary compression check should be performed to verify no engine damage.

1. Disconnect the negative battery cable and wait at least 90 seconds for the air bag system to deploy all back-up power.
2. Raise and safely support the vehicle.
3. Remove the right front wheel.
4. Remove the right front fender liner.
5. Remove the windshield washer fluid reservoir.
6. Remove the engine coolant overflow tank.
7. If equipped, remove the cruise control actuator.
8. Remove the intercooler and throttle body.
9. Remove the Exhaust Gas Recirculation (EGR) vacuum modulator and vacuum solenoid valve.
10. Remove the accessory drive belts.
11. Unbolt the power steering pump and position aside.
12. Remove the alternator and bracket.
13. Support the engine with a jack and remove the right engine mount.
14. Remove the upper timing cover.
15. Rotate the crankshaft so that the crankshaft timing mark aligns with the **0** mark on the timing cover and the timing marks on the camshaft pulleys align with the timing marks on the rear timing cover.

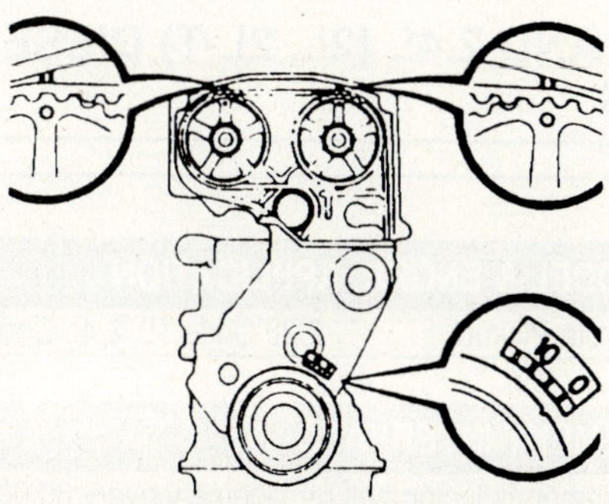

9305ag88

Camshaft and crankshaft timing marks—Toyota 2.0L (3S-GTE) engine

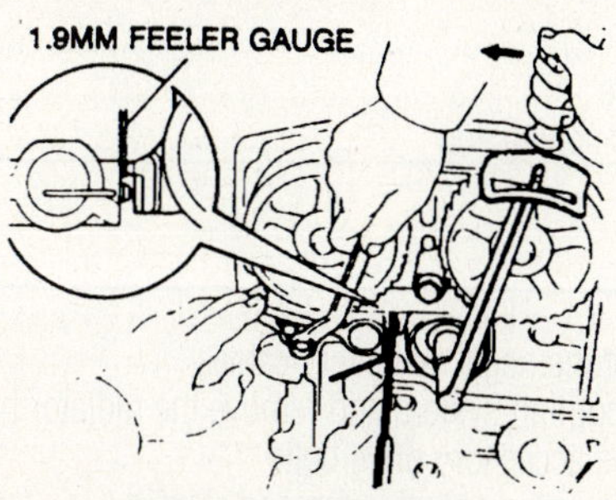

9305ag89

Setting the timing belt tensioner—Toyota 2.0L (3S-GTE) engine

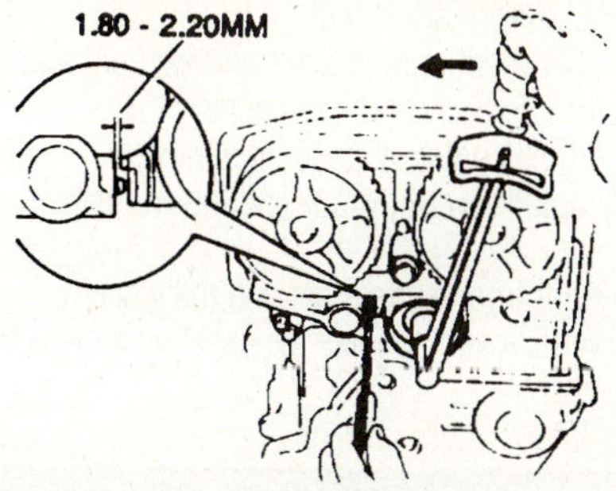

9305ag90

Checking the tensioner setting—Toyota 2.0L (3S-GTE) engine

16. Remove the crankshaft pulley.
17. Remove the lower timing cover.
18. Remove the timing belt tensioner.
19. Remove the timing belt.

To install:

20. Install the timing belt to the crankshaft timing pulley and install the lower cover.
21. Install the crankshaft pulley. Tighten the crankshaft pulley bolt to 80 ft. lbs. (108 Nm).
22. Ensure that the crankshaft and camshaft timing marks are aligned and install the timing belt to the camshaft pulleys.
23. Prepare the tensioner for installation by compressing it slowly in a vise.
24. Align the hole in the plunger with the holes in the tensioner body and insert a 3/64 (1.27mm) Allen wrench to retain the plunger.
25. To tension the timing belt, perform the following:
    a. Use a torque wrench and turn the tensioner pulley counterclockwise. Apply 13 ft. lbs. (18 Nm) torque to the tensioner pulley and install the tensioner.
    b. Slowly rotate the crankshaft pulley clockwise 5/6 of a turn and align the crankshaft timing mark at 60 degrees Before Top Dead Center (BTDC).
    c. Use a torque wrench and turn the tensioner pulley counterclockwise. Apply 13 ft. lbs. (18 Nm) torque to the tensioner pulley. Insert a 0.075 inch (1.9mm) feeler gauge between the tensioner pulley stopper and the tensioner body. Tighten the tensioner retainer bolts to 15 ft. lbs. 21 Nm).
    d. Remove the Allen wrench from the timing belt tensioner.
    e. Slowly rotate the crankshaft pulley clockwise 5/6 of a turn and align the crankshaft timing mark at 60 degrees Before Top Dead Center (BTDC).
    f. Use a torque wrench and turn the tensioner pulley counterclockwise. Apply 13 ft. lbs. (18 Nm) torque to the tensioner pulley. Measure the clearance between the tensioner pulley stopper and the tensioner body. The clearance should be 0.071–0.087 inches (1.8–2.2mm). If this measurement is not within specification, remove the tensioner and repeat the timing belt tension procedure.
26. Rotate the crankshaft to TDC and check that the camshaft timing marks align.
27. Install the upper timing cover.
28. Install the right motor mount and remove the engine support jack.
29. Install the alternator and bracket.

30. Install the power steering pump.
31. Install the accessory drive belts.
32. Install the EGR vacuum modulator and vacuum solenoid valve.
33. Install the throttle body and intercooler.
34. Install the cruise control actuator, if removed.
35. Install the coolant overflow tank and the windshield washer fluid reservoir.
36. Install the right front fender liner and the wheel.
37. Lower the vehicle and connect the negative battery cable.

MR2

✻✻ CAUTION

These engines are considered freewheeling. If the timing belt fails, valve-to-piston damage is unlikely, however a precautionary compression check should be performed to verify no engine damage.

1. Disconnect the negative battery cable and wait at least 90 seconds for the air bag system to deploy all back-up power.
2. Remove the right engine hood side panel.
3. Remove the upper suspension brace.
4. Remove the No. 1 and No. 2 air intake connectors.
5. If equipped, remove the cruise control actuator.
6. Disconnect the accelerator cable.
7. Remove the electric cooling fan.
8. Remove the intercooler.
9. Raise and safely support the vehicle.
10. Remove the right rear wheel.
11. Remove the accessory drive belts.
12. Unbolt the air conditioning compressor and position out of the work area.
13. Support the engine with a jack and remove the right engine mount.
14. Remove the upper timing cover.
15. Rotate the crankshaft so that the crankshaft timing mark aligns with the **0** mark on the timing cover and the timing marks on the camshaft pulleys align with the timing marks on the rear timing cover.
16. Remove the crankshaft pulley.
17. Remove the lower timing cover.
18. Remove the timing belt tensioner.
19. Remove the timing belt.

To install:

20. Install the timing belt to the crankshaft timing pulley and install the lower cover.
21. Install the crankshaft pulley. Tighten the crankshaft pulley bolt to 80 ft. lbs. (108 Nm).
22. Ensure that the crankshaft and camshaft timing marks are aligned and install the timing belt to the camshaft pulleys.
23. Prepare the tensioner for installation by compressing it slowly in a vise.
24. Align the hole in the plunger with the holes in the tensioner body and insert a 3/64 (1.27mm) Allen wrench to retain the plunger.
25. To tension the timing belt, perform the following:
    a. Use a torque wrench and turn the tensioner pulley counterclockwise. Apply 13 ft. lbs. (18 Nm) torque to the tensioner pulley and install the tensioner.
    b. Slowly rotate the crankshaft pulley clockwise 5/6 of a turn and align the crankshaft timing mark at 60 degrees Before Top Dead Center (BTDC).
    c. Use a torque wrench and turn the tensioner pulley counterclockwise. Apply 13 ft. lbs. (18 Nm) torque to the tensioner pulley. Insert a 0.075 inch (1.9mm) feeler gauge between the tensioner pulley stopper and the tensioner body. Tighten the tensioner retainer bolts to 15 ft. lbs. 21 Nm).
    d. Remove the Allen wrench from the timing belt tensioner.
    e. Slowly rotate the crankshaft pulley clockwise 5/6 of a turn and align the crankshaft timing mark at 60 degrees Before Top Dead Center (BTDC).
    f. Use a torque wrench and turn the tensioner pulley counterclockwise. Apply 13 ft. lbs. (18 Nm) torque to the tensioner pulley. Measure the clearance between the tensioner pulley stopper and the tensioner body. The clearance should be 0.071–0.087 inches (1.8–2.2mm). If this measurement is not within specification, remove the tensioner and repeat the timing belt tension procedure.
26. Rotate the crankshaft to TDC and check that the camshaft timing marks align.
27. Install the upper timing cover.
28. Install the right motor mount and remove the engine support jack.
29. Install the air conditioning compressor.
30. Install the accessory drive belts.
31. Install the right rear wheel and lower the vehicle.
32. Install the intercooler.
33. Install the electric cooling fan.
34. Connect the accelerator cable and install the cruise control actuator.
35. Connect the air intake connectors.
36. Install the upper suspension brace.
37. Install the right engine hood side panel.
38. Connect the negative battery cable.

2.2L (1L) AND 2.4L (2L, 2L-T) DIESEL ENGINES

YEAR, MODEL & ENGINE APPLICATION

1981-87 Toyota Pickup	2.2L Code 1L, 2.4L Codes 2L/2L-T

REPLACEMENT INTERVALS

Toyota recommends timing belt replacement every 60,000 miles on 2.2L engines and 100,000 miles on 2.4L engines.

LABOR TIMES

1981-87	
Replacement	2.3 hrs
Severe Service	2.5 hrs
With air conditioning add	0.7 hrs

8807TOYg13

1. Disconnect the negative battery cables.
2. Drain the cooling system and remove the radiator hoses.
3. Remove the accessory drive belts.
4. Remove the engine cooling fan and shroud.
5. Remove the water pump pulley.
6. If equipped with air conditioning, unbolt the compressor and position aside.
7. Remove the crankshaft pulley.
8. Remove the timing cover.

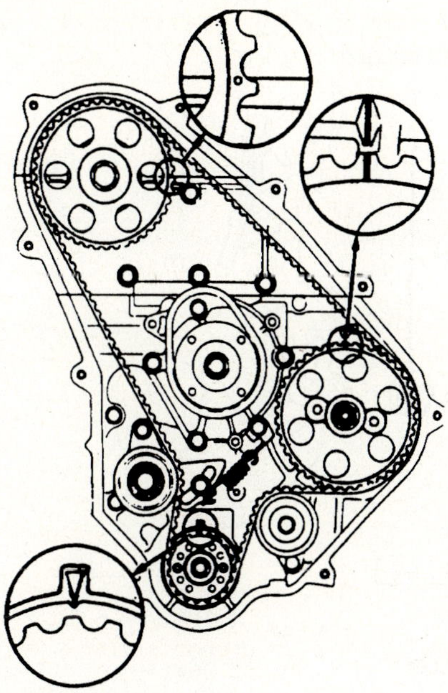

9305ag74

Alignment of the timing marks—Toyota 2.2L (1L) and 2.4L (2L, 2L-T) diesel engines

9. Rotate the crankshaft so that the camshaft, injection pump and crankshaft timing marks align with the timing marks on the engine and rear timing cover.
10. Loosen the timing belt tensioner locknut and pivot bolt.
11. Pry the tensioner pulley away from the timing belt and tighten the locknut.
12. Remove the timing belt.

To install:

13. Ensure that the camshaft and crankshaft timing marks are aligned.
14. Retard the injection pump timing mark by 1 tooth to allow for proper belt tensioning.
15. Install the timing belt.
16. Loosen the tensioner locknut.
17. Rotate the crankshaft 2 complete turns and check that the camshaft and crankshaft timing marks align.
18. Check that the injection pump timing marks are now aligned and tighten the tensioner locknut and pivot bolt.
19. Install the front timing cover.
20. Install the crankshaft pulley and tighten the bolt to 102 ft. lbs. (139 Nm).
21. Install the air conditioning compressor, if removed.
22. Install the water pump pulley.
23. Install the engine cooling fan and shroud.
24. Install the radiator hoses.
25. Connect the negative battery cables and refill the cooling system.

2.2L (5S-FE) ENGINES

1990–98

✱✱ CAUTION

These engines are considered freewheeling. If the timing belt fails, valve-to-piston damage is unlikely, however a precautionary compression check should be performed to verify no engine damage.

YEAR, MODEL & ENGINE APPLICATION

1991-01 Toyota Camry, 1999-01 Camry Solara, 1990-99 Celica GT, 1991-95 MR2 .... 2.2L Code 5S-FE

REPLACEMENT INTERVALS

Toyota recommends timing belt replacement 60,000 miles on 1990-98 models, and 90,000 miles on 1999-01 models.

LABOR TIMES

Camry/Camry Solara	
1991-96	
Replacement	3.2 hrs
Severe Service	3.7 hrs
1997-01	
Replacement	2.9 hrs
Severe Service	3.4 hrs
Celica GT	
Replacement	2.3 hrs
Severe Service	2.8 hrs
MR2	
Replacement	3.2 hrs
Severe Service	3.7 hrs

8807TOYg14

✱✱ CAUTION

If equipped with an air bag, be sure to disconnect the negative battery cable and wait at least 90 seconds before proceeding.

1. Remove the coolant reservoir and windshield washer reservoir.
2. Raise and support the vehicle.
3. On Camry and Celica models, remove the right front wheel, splash shield and engine undercover.
4. On MR2 models, remove the right rear wheel and engine undercover.
5. On all models, remove the cruise control actuator.
6. Remove the accessory drive belts.
7. Remove the alternator and bracket.
8. Support the engine with a suitable jack and remove the right side engine mounting stay.
9. Remove the power steering reservoir.
10. Remove the right side engine mounting insulator.
11. Remove the right side engine mount.
12. Remove the spark plugs.
13. Remove the No. 2 timing cover.
14. Position the No. 1 cylinder to Top Dead Center (TDC) on the compression stroke by turning the crankshaft pulley and aligning its groove with the timing mark **0** of the No. 1 timing belt cover. Check that the hole of the camshaft timing pulley is aligned with the alignment mark of the bearing cap. If not, turn the crankshaft 1 revolution (360 degrees).
15. Remove the timing belt from the camshaft timing pulley, as follows:
    a. If reusing the belt, place matchmarks on the timing belt and the camshaft pulley. Loosen the mount bolt of the No. 1 idler pulley and position the pulley toward the left as far as it will go. Tighten the bolt. Remove the belt from the camshaft pulley.
16. Remove the camshaft timing pulley as follows:

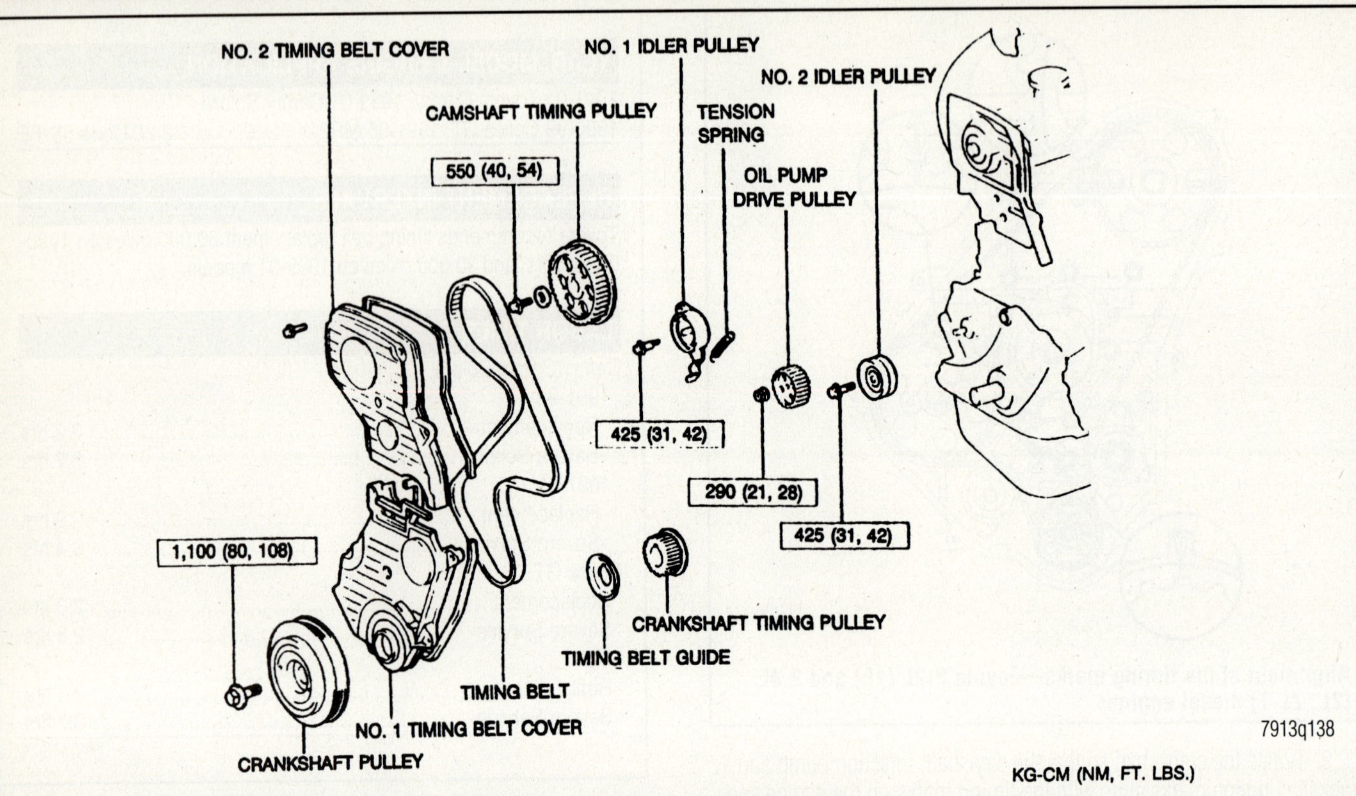

Exploded view of timing belt and related components—Toyota 2.2L (5S-FE) engine

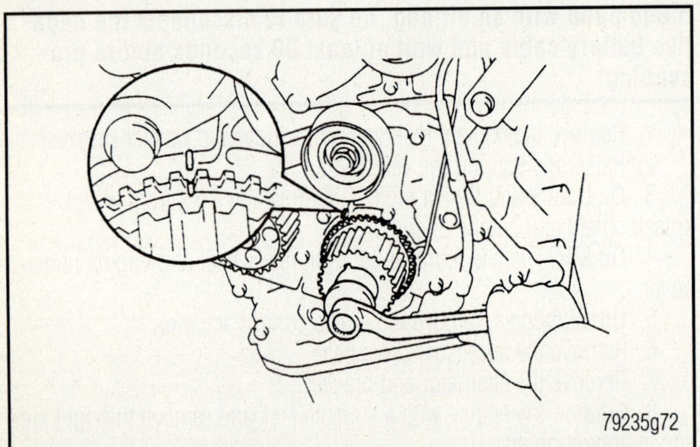

Crankshaft positioning for timing belt removal and installation—Toyota 2.2L (5S-FE) engine

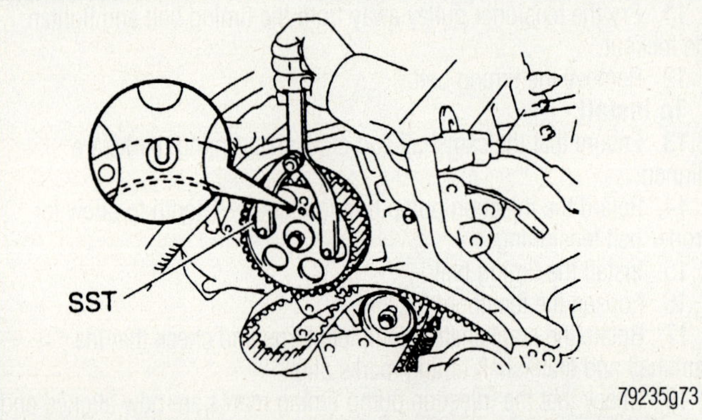

Using a spanner wrench, turn the camshaft into position so that the alignment mark is visible through the hole in the sprocket—Toyota 2.2L (5S-FE) engine

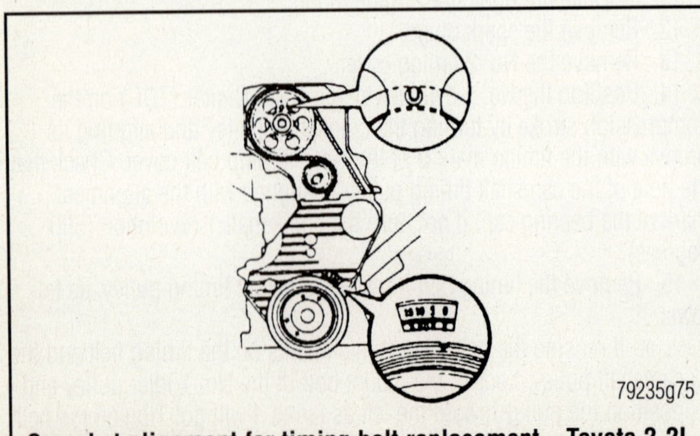

Sprocket alignment for timing belt replacement—Toyota 2.2L (5S-FE) engine

b. Using Toyota tools Nos. 09249-63010 and 09960-10010 or their equivalents, remove the bolt and the camshaft pulley.

17. Remove the crankshaft pulley as follows:

a. Using Toyota tools Nos. 09213-54015 and 09330-00021 or their equivalents, to hold the crankshaft pulley. Remove the pulley set bolt and remove the pulley using a puller.

18. Remove the No. 1 timing belt cover.

19. Remove the timing belt and the belt guide. If reusing the belt mark the belt and the crankshaft pulley in the direction of engine rotation and matchmark for correct installation.

To install:

20. Measure the free length of the tension spring. Length should be 1.81 in. (46mm). Replace the spring if the length does not meet specifications.

21. Install the crankshaft timing pulley, as follows:

a. Align the timing pulley set key with the key groove of the pulley.

b. Slide on the timing pulley with the flange side facing inward.

22. Install the No. 2 idler pulley and tighten the bolt to 31 ft. lbs. (42 Nm). Be sure that the pulley moves freely.

23. Temporarily install the No. 1 idler pulley and tension spring. Pry the pulley toward the left as far as it will go. Tighten the bolt. Be sure that the pulley rotates freely.

24. Temporarily install the timing belt, as follows:

a. Using the crankshaft pulley bolt, turn the crankshaft and align the timing marks of the crankshaft timing pulley and the oil pump body.

b. If reusing the old belt, align the marks made during removal and install the belt with the arrow pointing in the direction of the engine revolution.

25. Install the timing belt guide with the cup side facing outward.

26. Install the No. 1 timing belt cover.

27. Install the crankshaft pulley. Align the pulley set key with the key groove of the pulley and slide on the pulley. Tighten the bolt to 80 ft. lbs. (108 Nm).

28. Install the camshaft timing pulley as follows:

a. Align the camshaft knock pin with the knock pin groove of the pulley and slide on the timing pulley. Tighten the bolt to 40 ft. lbs. (54 Nm).

29. With the No. 1 cylinder set at TDC on the compression stroke, install the timing belt (all timing marks aligned). If reusing the belt, align with the marks made during the removal procedure:

a. Turn the crankshaft pulley and align its groove with the timing mark **0** of the No. 1 timing belt cover. Be sure the camshaft sprocket hole is aligned with the mark on the bearing cap.

30. Connect the timing belt to the camshaft timing pulley.

31. Check that the matchmark on the timing belt matches the end of the No. 1 timing belt cover.

32. Once the belt is installed, be sure that there is tension between the crankshaft timing pulley and the camshaft pulley.

33. Check the valve timing as follows:

a. Loosen the No. 1 idler pulley mount bolt 1/2 turn. Turn the crankshaft pulley 2 revolutions from TDC in the clockwise direction. Always turn the crankshaft pulley clockwise.

b. Be sure that the all the timing marks are aligned.

c. Slowly turn the crankshaft pulley 1 7/8 revolutions. Align its groove with the mark at 45 degrees BTDC on the No. 1 timing belt cover for the No. 1 cylinder.

d. Tighten the No. 1 idler pulley mount bolt to 31 ft. lbs. (42 Nm).

34. Install the No. 2 timing belt cover as follows:

e. Install the upper gasket to the No. 1 timing belt cover.

f. Disconnect the engine wire protector between the cylinder head cover and the No. 3 timing belt cover.

g. Install the gasket to the timing belt cover.

h. Install the belt covers and all remaining components. During assembly, tighten:

- The right engine mount bracket bolts: 38 ft. lbs. (52 Nm)
- The engine mount insulator bolt: 47 ft. lbs. (64 Nm)
- The through-bolt to 54 ft. lbs. (78 Nm)
- The power steering reservoir bracket bolt: 21 ft. lbs. (28 Nm)
- The power steering reservoir-to-bracket bolt to 27 ft. lbs. (37 Nm)
- The power steering reservoir-to-bracket nut: 38 ft. lbs. (52 Nm).

35. Install the alternator and bracket.

36. Install the cruise control actuator.

37. Install the accessory drive belts.

38. Install the engine undercover, splash shield and wheel

39. Connect the negative battery cable.

1999–01

✻✻ CAUTION

These engines are considered freewheeling. If the timing belt fails, valve-to-piston damage is unlikely, however a precautionary compression check should be performed to verify no engine damage.

✻✻ CAUTION

If equipped with an air bag, be sure to disconnect the negative battery cable and wait at least 90 seconds before proceeding.

1. Remove the alternator.
2. Raise and support the vehicle.
3. Remove the right front wheel and fender splash shield.
4. Remove the accessory drive belts.
5. Remove the engine control rod.
6. Remove the right side engine mount.
7. Remove the spark plugs.
8. Remove the No. 2 timing cover.
9. Position the No. 1 cylinder to Top Dead Center (TDC) on the compression stroke by turning the crankshaft pulley and aligning its groove with the timing mark **0** of the No. 1 timing belt cover. Check that the hole of the camshaft timing pulley is aligned with the alignment mark of the bearing cap. If not, turn the crankshaft 1 revolution (360 degrees).
10. Remove the timing belt from the camshaft timing pulley, as follows:

a. If reusing the belt, place matchmarks on the timing belt and the camshaft pulley. Loosen the mount bolt of the No. 1 idler pulley and position the pulley toward the left as far as it will go. Tighten the bolt. Remove the belt from the camshaft pulley.

11. Remove the crankshaft pulley as follows:

a. Using Toyota tools Nos. 09213-54015 and 09330-00021 or their equivalents, to hold the crankshaft pulley. Remove the pulley set bolt and remove the pulley using a puller.

12. Remove the No. 1 timing belt cover.

13. Remove the timing belt and the belt guide. If reusing the belt mark the belt and the crankshaft pulley in the direction of engine rotation and matchmark for correct installation.

To install:

14. Measure the free length of the tension spring. Length should be 1.65 in. (42mm). Replace the spring if the length does not meet specifications.

15. Install the crankshaft timing pulley, as follows:

a. Align the timing pulley set key with the key groove of the pulley.

b. Slide on the timing pulley with the flange side facing inward.

16. Install the No. 2 idler pulley and tighten the bolt to 31 ft. lbs. (42 Nm). Be sure that the pulley moves freely.

17. Temporarily install the No. 1 idler pulley and tension spring. Pry the pulley toward the left as far as it will go. Tighten the bolt. Be sure that the pulley rotates freely.

18. Temporarily install the timing belt, as follows:

a. Using the crankshaft pulley bolt, turn the crankshaft and align the timing marks of the crankshaft timing pulley and the oil pump body.

b. If reusing the old belt, align the marks made during removal and install the belt with the arrow pointing in the direction of the engine revolution.

19. Install the timing belt guide with the cup side facing outward.
20. Install the No. 1 timing belt cover.
21. Install the crankshaft pulley. Align the pulley set key with the key groove of the pulley and slide on the pulley. Tighten the bolt to 80 ft. lbs. (108 Nm).
22. With the No. 1 cylinder set at TDC on the compression stroke, install the timing belt (all timing marks aligned). If reusing the belt, align with the marks made during the removal procedure:
    a. Turn the crankshaft pulley and align its groove with the timing mark **0** of the No. 1 timing belt cover. Be sure the camshaft sprocket hole is aligned with the mark on the bearing cap.
23. Connect the timing belt to the camshaft timing pulley.
24. Check that the matchmark on the timing belt matches the end of the No. 1 timing belt cover.
25. Once the belt is installed, be sure that there is tension between the crankshaft timing pulley and the camshaft pulley.
26. Check the valve timing as follows:
    a. Loosen the No. 1 idler pulley mount bolt 1/2 turn. Turn the crankshaft pulley 2 revolutions from TDC in the clockwise direction. Always turn the crankshaft pulley clockwise.
    b. Be sure that the all the timing marks are aligned.
    c. Slowly turn the crankshaft pulley 1 7/8 revolutions. Align its groove with the mark at 45 degrees BTDC on the No. 1 timing belt cover for the No. 1 cylinder.
    d. Tighten the No. 1 idler pulley mount bolt to 31 ft. lbs. (42 Nm).
27. Install the No. 2 timing belt cover as follows:
    a. Install the upper gasket to the No. 1 timing belt cover.
    b. Disconnect the engine wire protector between the cylinder head cover and the No. 3 timing belt cover.
    c. Install the gasket to the timing belt cover.
    d. Install the belt covers and all remaining components. During assembly, tighten:
    - The right engine mount bracket bolts: 38 ft. lbs. (52 Nm)
28. Install the spark plugs.
29. Install the engine control rod.
30. Install the accessory drive belts.
31. Install the right front wheel and fender splash shield.
32. Install the alternator.
33. Connect the negative battery cable.

2.5L (2VZ-FE) ENGINES

YEAR, MODEL & ENGINE APPLICATION

1988-91 Toyota Camry .... 2.5L Code 2VZ-FE

REPLACEMENT INTERVALS

Toyota does not provide a replacement interval. Standard industry practice is to replace timing belts every 60,000 miles.

LABOR TIMES

1988-91	
Replacement	3.7 hrs
Severe Service	4.2 hrs

8807TOYg15

**** CAUTION**

These engines are considered freewheeling. If the timing belt fails, valve-to-piston damage is unlikely, however a precautionary compression check should be performed to verify no engine damage.

**** CAUTION**

If equipped with an air bag, be sure to disconnect the negative battery cable and wait at least 90 seconds before proceeding.

1. Raise and safely support the vehicle.
2. Remove the right front wheel.
3. Remove the right front fender liner.
4. If equipped, remove the cruise control actuator and the vacuum pump.
5. Remove the power steering pump reservoir without disconnecting the lines and position aside.
6. Remove the accessory drive belts.
7. Support the engine with a jack and remove the right engine mount and mount bracket.
8. Remove the spark plugs.
9. Remove the upper timing cover.
10. Rotate the crankshaft so that the crankshaft timing mark aligns with the **0** mark on the lower timing cover and the timing marks on the camshaft pulleys align with the timing marks on the rear timing cover.
11. Remove the timing belt tensioner and pulley.
12. Remove the crankshaft pulley.
13. Remove the lower timing cover and belt guide.
14. Remove the timing belt.

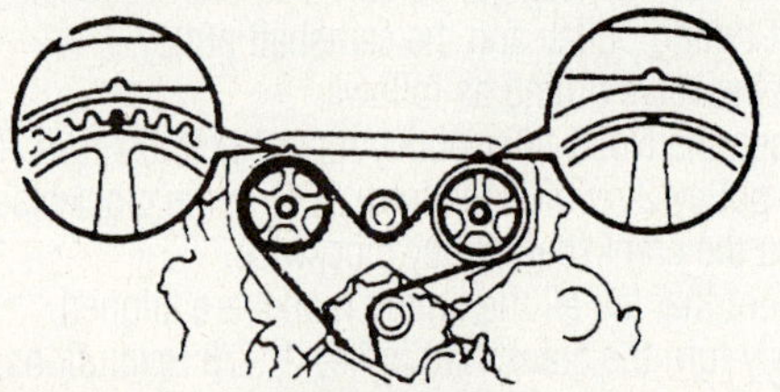

9305ag94

Camshaft timing alignment marks—Toyota 2.5L (2VZ-FE) engine

To install:

15. Install the timing belt tensioner pulley and crankshaft sprocket.
16. Check the timing belt tensioner pushrod extended length. Length should be 0.41–0. 45 in. (10.4–11.4mm). Replace the tensioner if length is not as specified.
17. Install the timing belt to the crankshaft sprocket, tensioner pulley and water pump sprocket.
18. Install the timing belt guide and lower timing belt cover.
19. Install the crankshaft pulley. Tighten the crankshaft pulley bolt to 181 ft. lbs. (245 Nm).
20. Install the No. 2 idler pulley.

21. Ensure that the crankshaft and camshaft timing marks are aligned and install the timing belt to the camshaft pulleys ensuring the belt marks are aligned with the sprocket marks.
22. Prepare the tensioner for installation by compressing it slowly in a vise.
23. Align the hole in the plunger with the holes in the tensioner body and insert a 3/64 (1.27mm) Allen wrench to retain the plunger.
24. Install the tensioner and tighten the mounting bolts to 20 ft. lbs. (27 Nm). Remove the Allen wrench.
25. Rotate the crankshaft clockwise 2 complete turns and check that the timing marks align.
26. Install the upper timing cover.
27. Install the right engine mount and remove the engine support jack.
28. Install the accessory drive belts.
29. Install the power steering pump or reservoir.
30. Install the cruise control actuator and vacuum pump, if removed.
31. Install the spark plugs.
32. Install the right front fender liner and wheel.
33. Lower the vehicle and connect the negative battery cable.

2.8L (5M-GE) ENGINES

YEAR, MODEL & ENGINE APPLICATION

1983-88 Toyota Cressida,	
1982-86 Toyota Supra	2.8L Code 5M-GE

REPLACEMENT INTERVALS

Toyota does not provide a replacement interval. Standard industry practice is to replace timing belts every 60,000 miles.

LABOR TIMES

1982-88	
Replacement	3.2 hrs
Severe Service	3.8 hrs

8807TOYg16

✻✻ CAUTION

These engines are considered freewheeling. If the timing belt fails, valve-to-piston damage is unlikely, however a precautionary compression check should be performed to verify no engine damage.

1. Disconnect the negative battery cable.
2. Drain the cooling system and remove the upper radiator hose.
3. Remove the upper fan shroud.
4. Remove the accessory drive belts.
5. Remove the air intake pipe.
6. Set the No. 1 cylinder to TDC of the compression stroke.
7. Remove the No. 3 timing belt cover.
8. Remove the air conditioning compressor and idler pulley, if equipped.
9. Remove the timing belt from the camshaft sprockets.
10. Hold the camshafts in place with and remove the camshaft sprockets.
11. Remove the crankshaft pulley using a puller.
12. Remove the lower timing cover.

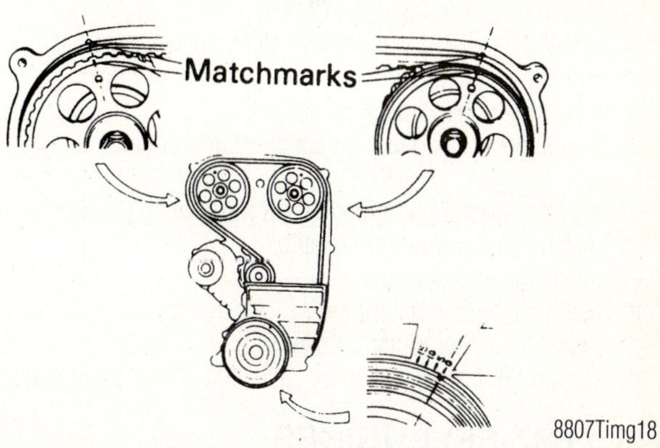

8807Timg18

Crankshaft and camshaft timing marks—Toyota 2.8L (5M-GE) engine

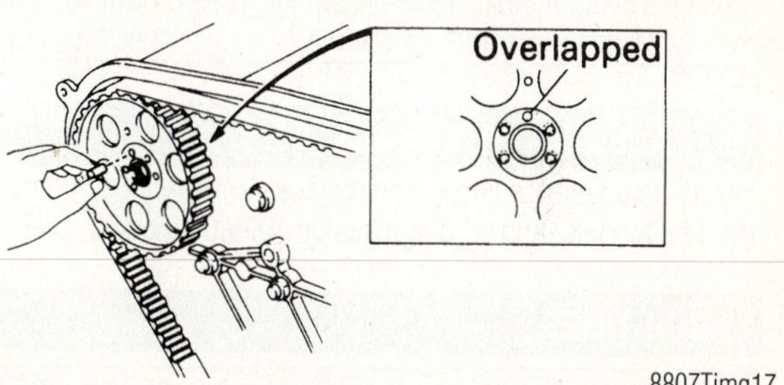

8807Timg17

Locating the overlapped holes of the camshaft and camshaft sprocket. Install the match pin into the aligned holes—Toyota 2.8L (5M-GE) engine

13. Loosen the timing belt tensioner locknut.
14. Pry the tensioner away from the timing belt and tighten the locknut.
15. Remove the timing belt.

To install:

16. Install the idler pulley and tension spring.
17. Install the timing belt to the crankshaft sprocket.
18. Install the lower cover and the A/C compressor bracket.
19. Install the crankshaft pulley. Tighten the crankshaft pulley bolt to 195 ft. lbs. (265 Nm).
20. Remove the exhaust side valve cover.
21. Check that the match holes of both No. 2 camshaft journals are visible through the camshaft housing match holes. If necessary, install the camshaft pulley and guide pin and rotate the camshaft(s) until the holes are aligned.
22. Install the timing pulleys. The belt guide of the exhaust camshaft should point toward the engine and the intake belt guide should point away from the engine.
23. Align each camshaft pulley mark with the mark on the upper timing belt cover.
24. Align the crankshaft pulley mark with the TDC mark on the timing tab.
25. Loosen the tensioner locknut.
26. The tension on the timing belt must be the same between the exhaust camshaft pulley and the crankshaft pulley as it is between the intake camshaft pulley and oil pump drive pulley.
27. There are 5 pin holes on each camshaft and timing pulley. On the exhaust side install the match pin into one hole that is aligned with the camshaft pin holes. Repeat this on the intake side.

28. Hold the camshaft pulleys and install the camshaft pulley bolts and tighten to 48–54 ft. lbs. (65–73 Nm).
29. Install the exhaust valve cover using a new gasket.
30. Tighten the tensioner locknut to 36 ft. lbs. (49 Nm).
31. Install the upper timing cover.
32. Install the air conditioning compressor and idler pulley, if removed.
33. Install the water pump pulley and the cooling fan.
34. Install the accessory drive belts.
35. Install the air intake pipe.
36. Install the upper fan shroud.
37. Install the upper radiator hose.
38. Connect the negative battery cable and fill the cooling system.

3.0L (1MZ-FE) ENGINES

YEAR, MODEL & ENGINE APPLICATION

1995-04 Toyota Avalon, 1993-04 Camry, 1999-03 Camry Solara, 2001-04 Highlander, 1997-04 Sienna .... 3.0L Code 1MZ-FE

REPLACEMENT INTERVALS

Toyota recommends timing belt replacement every 60,000 miles on 1993-98 models, and 90,000 miles on 1999-04 models.

LABOR TIMES

Avalon	
Replacement	3.0 hrs
Severe Service	3.3 hrs
Camry	
1993-96	
Replacement	3.2 hrs
Severe Service	3.7 hrs
1997-01	
Replacement	2.9 hrs
Severe Service	3.4 hrs
2002-04	
Replacement	2.5 hrs
Severe Service	2.9 hrs
Camry Solara	
1999-01	
Replacement	3.0 hrs
Severe Service	3.5 hrs
2002-03	
Replacement	2.7 hrs
Severe Service	3.2 hrs
Highlander	
Replacement	2.4 hrs
Severe Service	2.9 hrs
Sienna	
Replacement	6.0 hrs
Severe Service	6.5 hrs

8807TOYg17

Avalon, Camry and Camry Solara

✻✻ CAUTION

These engines are considered freewheeling. If the timing belt fails, valve-to-piston damage is unlikely, however a precautionary compression check should be performed to verify no engine damage.

1. Disconnect the negative battery cable and wait at least 90 seconds to allow the air bag system to lose power.
2. Remove the coolant and windshield washer reservoirs.
3. Raise and support the vehicle.
4. Remove the right side wheel and fender apron seal.
5. Remove the accessory drive belts.
6. Remove the upper suspension brace, if equipped.
7. Place a suitable jack under the engine and remove the right side engine mount and control rod.
8. Remove the alternator adjusting bracket.
9. Remove the crankshaft pulley bolt.
10. Using a puller, remove the crankshaft pulley.
11. Remove the lower timing belt cover by removing the 4 bolts.
12. Remove the No. 2 timing belt cover as follows:
    a. Remove the bolt and disconnect the engine wire protector from the No. 3 (rear) timing belt cover.
    b. Disconnect the engine wire protector clamp from the No. 3 timing belt cover.
    c. Remove the 5 bolts from the No. 2 timing belt cover.
    d. Remove the No. 2 cover from the engine.
13. Remove the right engine mounting bracket by removing the nut and 2 bolts.
14. Remove the crankshaft timing belt guide.

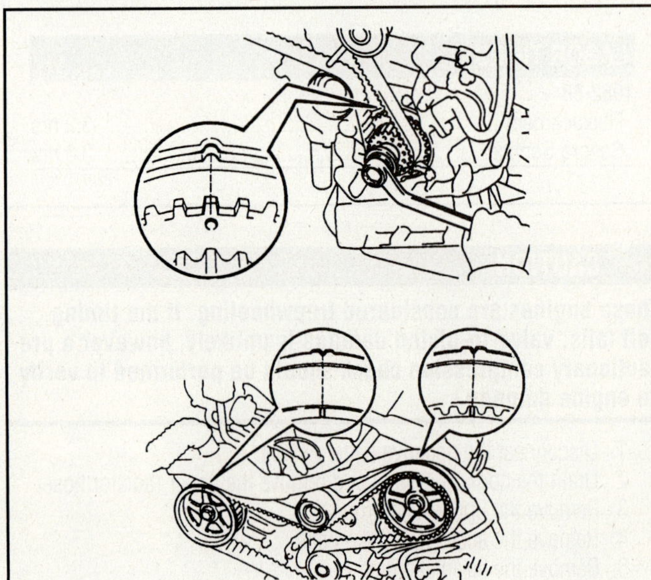

79235g76

Camshaft and crankshaft timing belt sprocket alignment mark positioning for belt service—Toyota 3.0L (1MZ-FE) engine—Avalon, Camry and Camry Solara

15. Temporarily install the crankshaft pulley bolt.
16. Turn the crankshaft and align the crankshaft timing pulley groove with the oil pump alignment mark. Always turn the engine clockwise.
17. Ensure the timing mark of the camshaft timing pulleys and rear timing belt covers are aligned. If not, turn the engine over an additional 360 degrees (one revolution).
18. Remove the crankshaft pulley bolt.

■ **If the belt is to be reused, align the installation marks on the belt to the marks on the pulleys. If the marks have worn off, make new ones.**

19. Alternately loosen the 2 timing belt tensioner bolts. Remove the tensioner and dust boot.
20. Remove the timing belt.

To install:

21. Remove any oil or water from the pulleys.
22. Align the front mark of the timing belt with the dot mark of the crankshaft timing pulley.
23. Align the installation marks on the timing belt with the timing marks of the camshaft pulleys.
24. Install the timing belt in the following order:
    a. Crankshaft pulley.
    b. Water pump pulley.
    c. Left camshaft pulley.
    d. No. 2 idler pulley.
    e. Right camshaft pulley.
    f. No. 1 idler pulley.
25. Using a press, slowly press the timing belt tensioner until the holes of the pushrod and housing align. Insert a 0.05 in. (1.27mm) hexagonal Allen wrench through the holes to preserve the setting position.
26. Install the dust boot to the tensioner.
27. Install the tensioner with the 2 bolts. Alternately tighten the bolts to 20 ft. lbs. (27 Nm). Remove the Allen wrench.
28. Turn the crankshaft clockwise and align the crankshaft timing pulley groove with the oil pump alignment mark.
29. Ensure the camshaft timing marks align with the timing marks on the rear timing belt cover.
30. Install the timing belt guide.
31. Install the right engine mounting bracket and tighten the bolts to 21 ft. lbs. (28 Nm).
32. Install the upper timing belt cover with the 5 bolts. Tighten the bolts to 74 inch lbs. (8 Nm).
33. Install the engine wire protector clamp to the No. 3 timing belt cover.
34. Install the engine wire protector to the No. 3 timing belt cover with the bolt.
35. Install the lower timing belt cover by installing the 4 bolts. Tighten the bolts to 74 inch lbs. (8 Nm).
36. Install the remaining components. During installation be sure to tighten the crankshaft pulley bolt to 159 ft. lbs. (215 Nm) and the No. 2 alternator bracket nut to 21 ft. lbs. (28 Nm).
37. Connect the negative battery cable.

Highlander

✻✻ CAUTION

These engines are considered freewheeling. If the timing belt fails, valve-to-piston damage is unlikely, however a precautionary compression check should be performed to verify no engine damage.

1. Disconnect the negative battery cable and wait at least 90 seconds to allow the air bag system to lose power.
2. Raise and support the vehicle.
3. Remove the right side wheel, fender apron seal and engine undercover.
4. Remove the wiper linkage.
5. Remove the cowl top panel.
6. Remove the accessory drive belts.
7. Place a suitable jack under the engine and remove the right side mount stay and control rod.
8. Remove the alternator adjusting bracket.
9. Remove the crankshaft pulley bolt.
10. Using a puller, remove the crankshaft pulley.
11. Remove the lower timing belt cover.
12. Remove the timing belt upper cover.
13. Remove the right side engine mounting bracket.
14. Remove the No. 2 timing belt guide.
15. Install the crankshaft pulley and rotate the crankshaft until the No. 1 cylinder is a TDC on the compression stroke and the camshaft sprockets timing marks are aligned.
16. Remove the timing belt tensioner and the timing belt.

To install:

17. Inspect the tensioner seal for leakage; if leakage is suspected, replace the tensioner.
18. Using a press, slowly press the timing belt tensioner until the holes of the pushrod and housing align. Insert a 0.05 in. (1.27mm) hexagonal Allen wrench through the holes to preserve the setting position.
19. Set the No. 1 cylinder to Top Dead Center (TDC) of the compression stroke, as follows:

67162-X300-g01

Camshaft sprocket timing marks locations—Toyota 3.0L (1MZ-FE) engine—Highlander

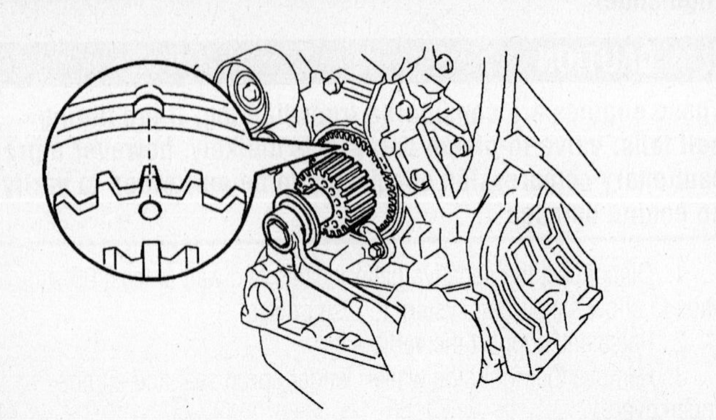

67162-X300-g14

Turn the crankshaft clockwise to align the crankshaft sprocket timing marks—Toyota 3.0L (1MZ-FE) engine—Highlander

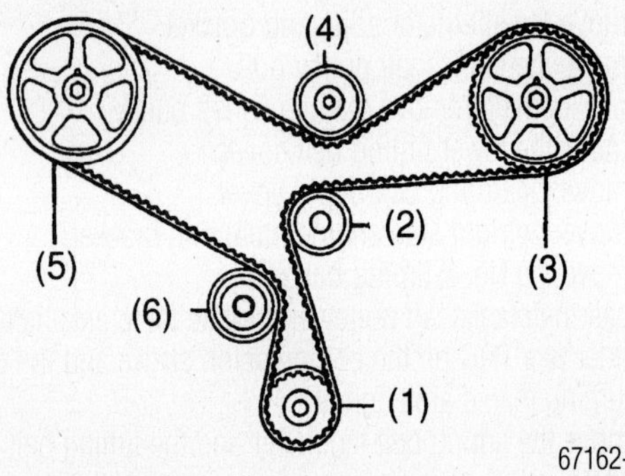

67162-X300-g06

Timing belt installation order— Toyota 3.0L (1MZ-FE) engine—Highlander

a. Rotate the crankshaft (CLOCKWISE) to align the timing marks: dimple on the crankshaft timing sprocket with the notch on the oil pump body.

b. Check that the timing marks on the camshaft sprockets and the rear timing belt cover are aligned; if not, rotate the crankshaft 1 revolution (360 degrees) and align the marks.

20. Install the timing belt in the following order:
   a. Crankshaft timing sprocket
   b. Water pump pulley
   c. Left camshaft timing sprocket
   d. No. 2 idler pulley
   e. Right camshaft sprocket
   f. No. 1 idler pulley
21. Install the timing belt tensioner and torque the bolts to 20 ft. lbs. (27 Nm).
22. Remove the Allen® wrench from the tensioner housing.
23. Slowly, rotate the crankshaft (CLOCKWISE) 2 complete revolutions and realign the timing marks. If the timing marks do not align, remove the timing belt and reinstall it.
24. Remove the crankshaft pulley bolt.
25. Install the right side engine mounting bracket and torque the bolts to 21 ft. lbs. (28 Nm).
26. Clean and install the upper (No. 2) timing belt cover.

■ **If the gasket material on the timing belt covers is cracked, peeling or etc., replace it.**

27. Install the timing belt guide on the crankshaft with the cup side facing outward.
28. Clean and install the lower (No. 1) timing belt cover.
29. Install the crankshaft pulley.
30. Using the Crankshaft Pulley Holding tool 09213-54015, Bolt tool 91651-60855 and Companion Flange Holding tool 09330-00021 or equivalent, install the crankshaft pulley bolt and torque the bolt to 159 ft. lbs. (215 Nm).
31. To complete the installation, reverse the removal procedures.
32. Connect the negative battery cable.
33. Start the engine and check for leaks.

Sienna

1. Disconnect the negative battery cable.
2. Remove the outer front cowl top panel assembly by performing the following procedure:
   a. Remove the wiper arm/blade assemblies head caps, nuts and assemblies.
   b. Remove the head-to-cowl seal and the cowl panel hole cover.
   c. Disconnect the windshield washer clip and hose.
   d. Remove both (right and left) cowl top ventilator louvers.
   e. Disconnect the electrical connector from the windshield wiper motor.
   f. Remove the outer front cowl top panel assembly-to-cowl bolts and the panel.
3. Raise and safely support the vehicle.
4. Remove the right front wheel assembly and apron seal.
5. Remove the alternator by performing the following procedure:
   a. Loosen the pivot bolt, adjusting lockbolt and adjusting bolt, then, remove the drive belt.
   b. Disconnect the alternator's electrical connector.
   c. Remove the nut and the alternator wire.
   d. Disconnect the wiring harness from the clip.
   e. Remove the alternator-to-bracket pivot bolt, the washer, adjusting lockbolt and alternator.
6. Loosen the power steering pump's mount and adjusting bolt, then, remove the drive belt.
7. Disconnect the hose from the engine coolant reservoir.
8. Disconnect the Diagnostic Link Connector 1 (DLC1) from the No. 2 right side engine mounting bracket.
9. Remove the right side engine mounting stay, the engine moving control rod and the No. 2 right side engine mounting bracket.
10. Loosen the alternator's pivot bolt, the nut and the No. 2 alternator bracket.
11. Using the Crankshaft Pulley Holding tool 09213-54015, Bolt tool 91651-60855 and Companion Flange Holding tool 09330-00021 or equivalent, remove the crankshaft pulley bolt.
12. Using a Puller "C" Set 09950-50011 (Hanger 150 tool 09951-05010, Slide Arm tool 09952-5010, Center Bolt 100 tool 09953-05010, Center Bolt 150 tool 09953-05020 and 2 No. 2 Claw tools 09954-05020), pull the crankshaft pulley from the crankshaft.
13. Remove the lower (No. 1) timing belt cover. Remove the timing belt guide from the crankshaft.
14. Remove the engine wire protector clamps from the upper (No. 2) timing belt cover and remove the upper (No. 2) timing belt cover.
15. Remove the right side engine mounting brace.

■ **If reusing the timing belt, be sure that you can still read the installation marks. If not, place new installation marks on the timing belt to match the timing marks of the camshaft timing pulleys.**

16. Temporarily install the crankshaft pulley bolt.
17. Set the No. 1 cylinder to Top Dead Center (TDC) of the compression stroke, as follows:

a. Rotate the crankshaft (CLOCKWISE) to align the timing marks: dimple on the crankshaft timing sprocket with the notch on the oil pump body.

b. Check that the timing marks on the camshaft sprockets and the rear timing belt cover are aligned; if not, rotate the crankshaft 360 degrees (1 revolution) and align the marks.

18. Remove the timing belt tensioner and the timing belt.

To install:

19. Inspect the timing belt tensioner by performing the following procedures:

c. Inspect the seal for leakage; if leakage is suspected, replace the tensioner.

d. Using both hands to hold the tensioner facing upward, strongly press the pushrod against a solid surface. If the pushrod moves, replace the tensioner.

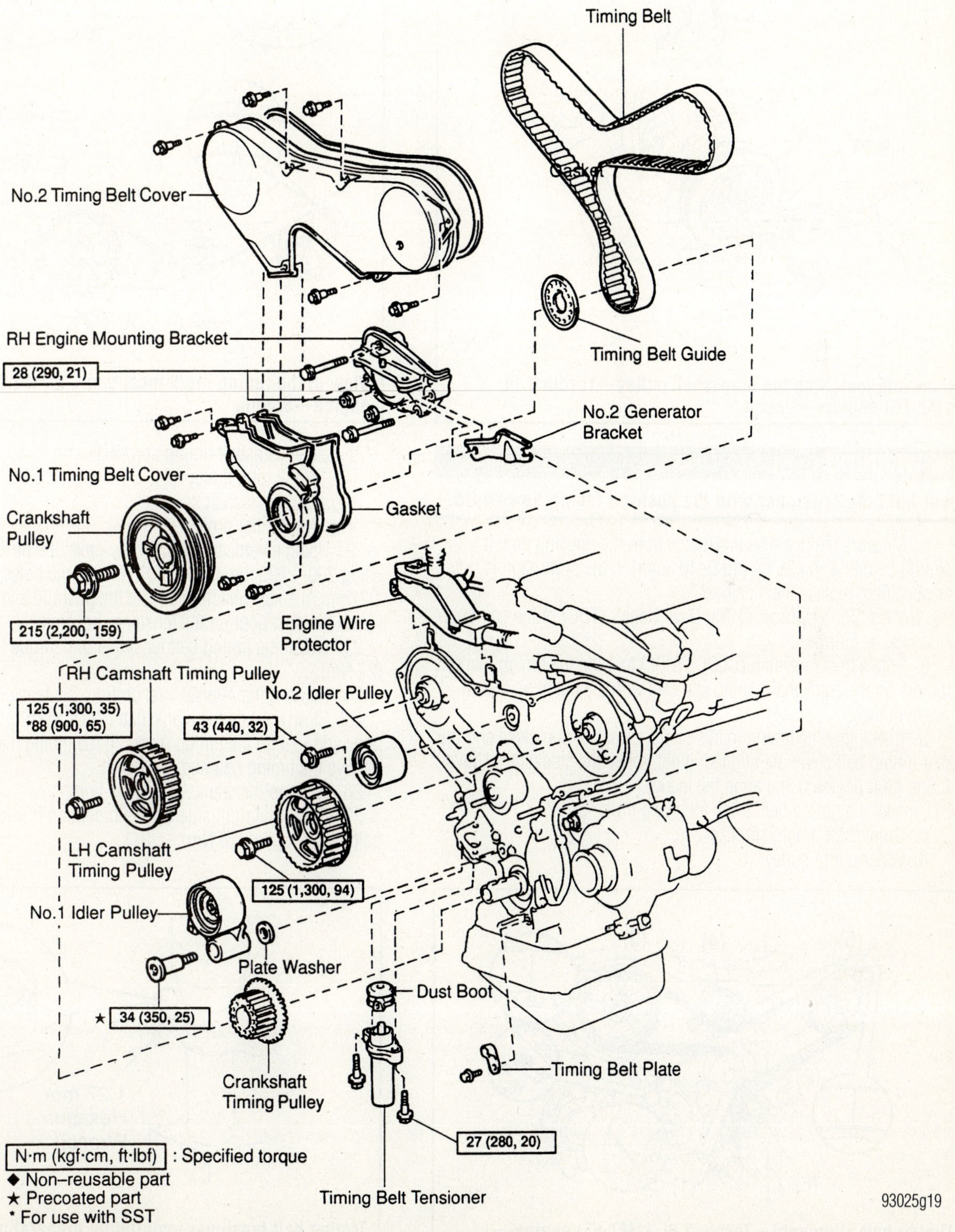

Exploded view of the timing belt assembly—Toyota 3.0L (1MZ-FE) engine—Sienna

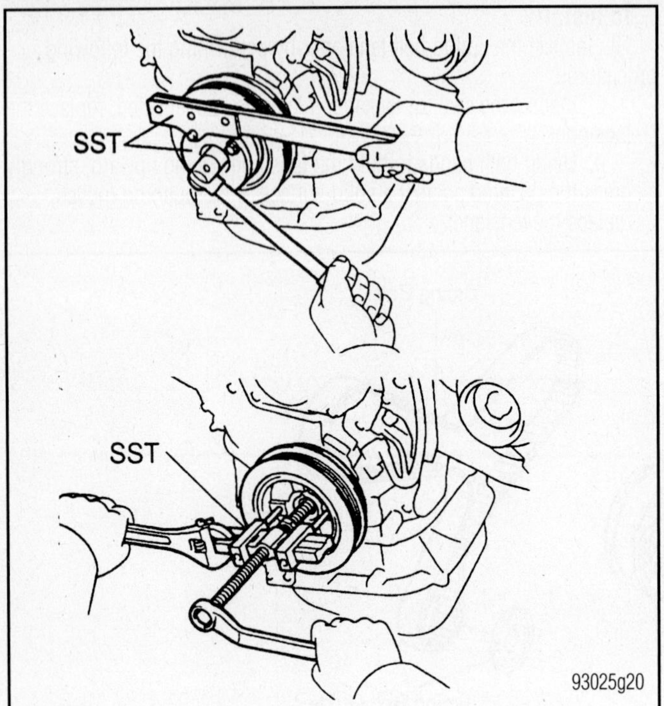

Removing/installing the crankshaft pulley—Toyota 3.0L (1MZ-FE) engine—Sienna

✻✻ WARNING

Never hold the tensioner with the pushrod facing downward.

e. Measure the pushrod protrusion from the housing end, it should be 0.394–0.425 in. (10.0–10.8mm). If the protrusion is not as specified, replace the tensioner.

20. Set the No. 1 cylinder to Top Dead Center (TDC) of the compression stroke, as follows:

a. Rotate the crankshaft (CLOCKWISE) to align the timing marks: dimple on the crankshaft timing sprocket with the notch on the oil pump body.

b. Check that the timing marks on the camshaft sprockets and the rear timing belt cover are aligned; if not, rotate the crankshaft 1 revolution (360 degrees) and align the marks.

21. Install the timing belt in the following order:

c. Crankshaft timing sprocket

d. Water pump pulley

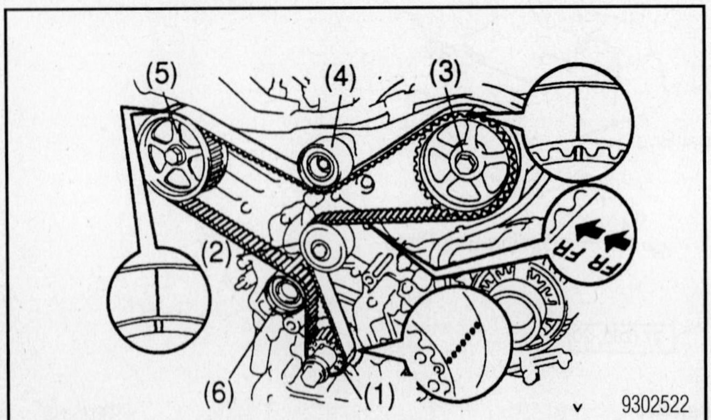

Timing belt alignment—Toyota 3.0L (1MZ-FE) engine—Sienna

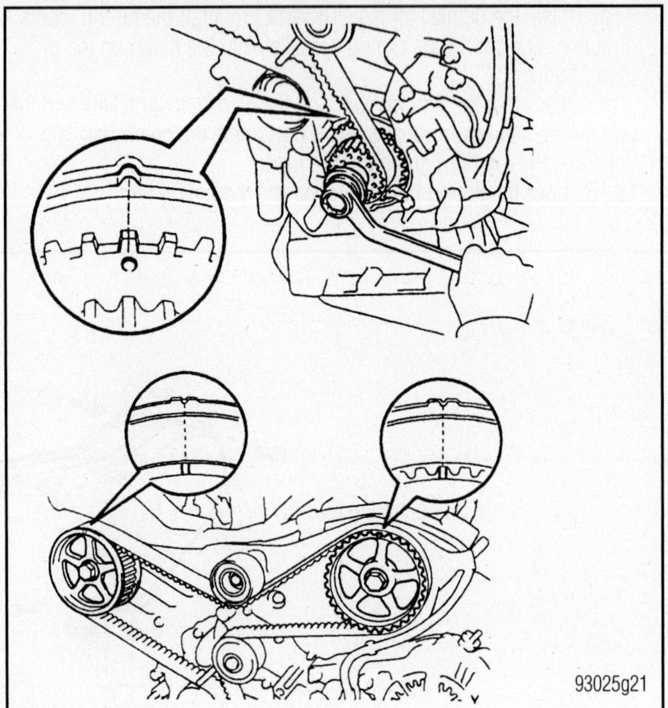

View of the timing mark locations—Toyota 3.0L (1MZ-FE) engine—Sienna

e. Left camshaft timing sprocket

f. No. 2 idler pulley

g. Right camshaft sprocket

h. No. 1 idler pulley

22. Using a vertical press, slowly press the pushrod into the housing using 200–2205 lbs. (981–9807 N) until the holes align, then, install a 1.27mm Allen® wrench to secure the pushrod and release the press. Install the dust boot on the tensioner housing.

23. Install the timing belt tensioner and torque the bolts to 20 ft. lbs. (27 Nm).

24. Remove the Allen® wrench from the tensioner housing.

255.Slowly, rotate the crankshaft (CLOCKWISE) 2 complete revolutions and realign the timing marks. If the timing marks do not align, remove the timing belt and reinstall it.

26. Remove the crankshaft pulley bolt.

27. Install the right side engine mounting bracket and torque the bolts to 21 ft. lbs. (28 Nm).

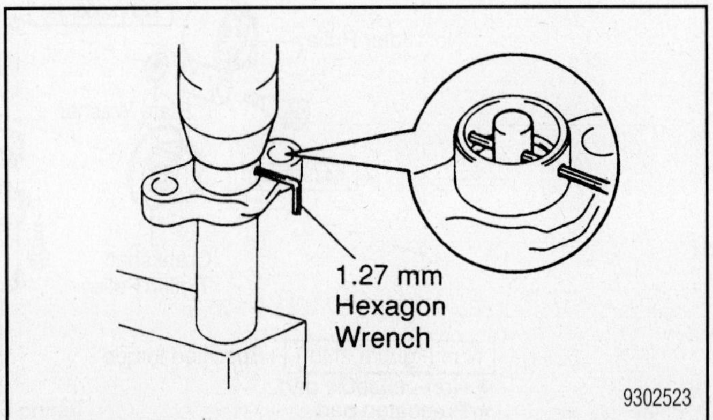

Timing belt tensioner installation preparation—Toyota 3.0L (1MZ-FE) engine—Sienna

28. Clean and install the upper (No. 2) timing belt cover.

■ **If the gasket material on the timing belt covers is cracked, peeling or etc., replace it.**

29. Install the timing belt guide on the crankshaft with the cup side facing outward.
30. Clean and install the lower (No. 1) timing belt cover.
31. Install the crankshaft pulley.
32. Using the Crankshaft Pulley Holding tool 09213-54015, Bolt tool 91651-60855 and Companion Flange Holding tool 09330-00021 or equivalent, install the crankshaft pulley bolt and torque the bolt to 159 ft. lbs. (215 Nm).
33. To complete the installation, reverse the removal procedures.
34. Connect the negative battery cable.
35. Start the engine and check for leaks.

3.0L (2JZ-GE AND 2JZ-GTE) ENGINES

YEAR, MODEL & ENGINE APPLICATION

1993-98 Toyota Supra	3.0L Codes 2JZ-GE/GTE

REPLACEMENT INTERVALS

Toyota does not provide a replacement interval. Standard industry practice is to replace timing belts every 60,000 miles.

LABOR TIMES

1993-98	
Non-Turbo	
Replacement	2.5 hrs
Severe Service	3.0 hrs
Turbo	
Replacement	3.1 hrs
Severe Service	3.6 hrs

8807TOYg18

✱✱ CAUTION

1993–97 engines are considered freewheeling. If the timing belt fails, valve-to-piston damage is unlikely, however a precautionary compression check should be performed to verify no engine damage.

✱✱ CAUTION

1998 engines are an interference type. If the timing belt has failed, valve-to-piston damage may have occurred. A compression check should always be performed to check for engine damage.

1. Disconnect the battery cables and if equipped with an air bag, wait at least 90 seconds before proceeding.
2. Remove the battery.
3. Remove the engine undercover.
4. Drain the cooling system.
5. Disconnect the oil cooler lines from the radiator.
6. Remove the coolant hoses and the radiator.
7. Remove the air cleaner and mass airflow sensor.
8. Disconnect all cables from the throttle body.
9. Remove the viscous fan coupling and the fan.
10. Remove the water pump pulley.
11. Remove the accessory drive belt.
12. Remove the upper 2 timing belt covers (Nos. 2 and 3).
13. Remove the drive belt tensioner.
14. On 1998 models, remove the power steering pump without disconnecting the lines and position aside.
15. Set the No. 1 cylinder to Top Dead Center (TDC) on the compression stroke. Turn the crankshaft pulley clockwise to align the groove with the **0** mark on the lower (No. 1) timing belt cover. Check that the timing marks on the camshaft pulleys are aligned with the marks on the rear belt cover. If the camshaft marks do not align, turn the crankshaft another 360 degrees.
16. Remove the crankshaft pulley by using Toyota tool 09330-0021 or equivalent, to hold the pulley and using tool 09213-70010 or equivalent, to remove the pulley bolt.
17. Remove the lower (No. 1) timing belt cover and the timing belt guide.
18. Alternately loosen the 2 bolts holding the timing belt tensioner. Remove the bolts and remove the tensioner.
19. Remove the timing belt from the camshaft pulleys. If the belt is to be reused, place matchmarks on the belt and gears before removing the belt. Mark the belt with an arrow to show direction of rotation.
20. Remove the timing belt. If the belt is to be reused, protect it from contact with oil, grease or fluids.

To install:

21. Use the crankshaft pulley bolt to turn the crankshaft (clockwise) until the mark on the gear aligns with the oil pump body. Check all the

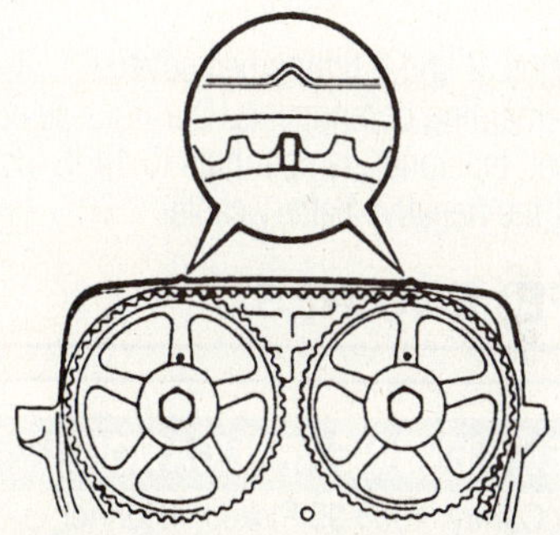

79235g77

Camshaft timing mark alignment—Toyota 3.0L (2JZ-GTE and 2JZ-GE) engines

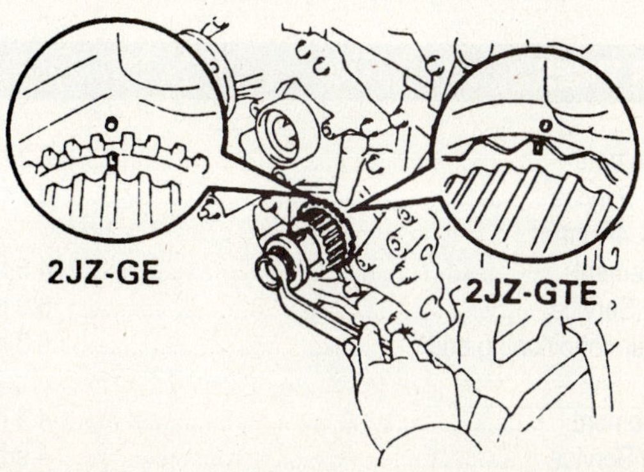

79235g78

Crankshaft timing marks; notice the timing mark difference between the engines—Toyota 3.0L (2JZ-GTE and 2JZ-GE) engines

pulleys and gears for cleanliness; remove any grease, oil or coolant. Install the timing belt onto the crankshaft gear and idler pulleys.

22. Install the timing belt guide with the cupped side facing outward.
23. Install the No. 1 timing belt cover.
24. Install the crankshaft pulley. Align the set key with the groove. Hold the pulley with the proper tool and tighten the pulley bolt to 241 ft. lbs. (328 Nm).
25. If equipped with automatic transmission, connect the oil cooler tubes with the clamps and 2 bolts.
    a. Turn the crankshaft pulley and align its groove with the timing mark, **0** on the No. 1 timing belt cover.
    b. Align the timing marks on the camshaft timing gears and the No. 4 timing belt cover.
26. Finish installing the timing belt.
27. Double check that all the timing marks for the crankshaft pulley and the camshaft gears are aligned as they were during disassembly.
28. Set the timing belt tensioner:
    a. Use a press to slowly push in the pushrod on the tensioner. This will require between 220–2200 lbs. (100–1000 kg) of pressure.
    b. Align the holes of the pushrod and housing. Place a 0.06 in. (1.5mm) hex wrench through the holes to keep the pushrod retracted.
    c. Release the press and install the dust boot onto the tensioner.
29. Install the tensioner; alternately tighten the bolts to 20 ft. lbs. (26 Nm).
30. Remove the hex wrench from the tensioner with a pair of pliers.
31. Turn the crankshaft pulley 2 full turns clockwise. Check that each pulley's timing marks align correctly after the 2 turns. If any mark does not align, remove the timing belt and reinstall it.
32. Install the drive belt tensioner and tighten the bolts to 15 ft. lbs. (21 Nm).
33. Install the Nos. 2 and 3 timing belt covers.
34. Install all remaining components. During assembly be sure to tighten the drive belt tensioner damper nuts to 14 ft. lbs. (20 Nm).
35. Disconnect the negative battery cable.

3.0L (3VZ-FE) ENGINES

YEAR, MODEL & ENGINE APPLICATION

1992-93 Toyota Camry, 1988-95 Pickup, 4Runner, 1993-94 T100	3.0L Code 3VZ-FE

REPLACEMENT INTERVALS

Toyota does not provide a replacement interval. Standard industry practice is to replace timing belts every 60,000 miles.

LABOR TIMES

Camry	
Replacement	3.2 hrs
Severe Service	3.7 hrs
Pickup & 4Runner	
Replacement	4.5 hrs
Severe Service	5.0 hrs
With air conditioning add	0.3 hrs
T100	
Replacement	4.4 hrs
Severe Service	4.9 hrs
With air conditioning add	0.5 hrs

8807TOYg19

Camry

✱✱ CAUTION

These engines are considered freewheeling. If the timing belt fails, valve-to-piston damage is unlikely, however a precautionary compression check should be performed to verify no engine damage.

✱✱ CAUTION

If equipped with an air bag, be sure to disconnect the negative battery cable and wait at least 90 seconds before proceeding.

1. Remove the coolant and windshield washer reservoirs.
2. Raise and safely support the vehicle.
3. Remove the right front wheel.
4. Remove the right front fender liner.
5. Remove the power steering pump reservoir without disconnecting the lines and position aside.
6. Remove the accessory drive belts.
7. Support the engine with a jack and remove the engine control rod, right engine mount and mount bracket.
8. Remove the upper timing cover.
9. Rotate the crankshaft so that the crankshaft timing mark aligns with the **0** mark on the lower timing cover and the timing marks on the camshaft pulleys align with the timing marks on the rear timing cover.
10. Remove the timing belt tensioner.
11. Remove the crankshaft pulley.
12. Remove the lower timing cover.
13. Remove the timing belt guide and timing belt.

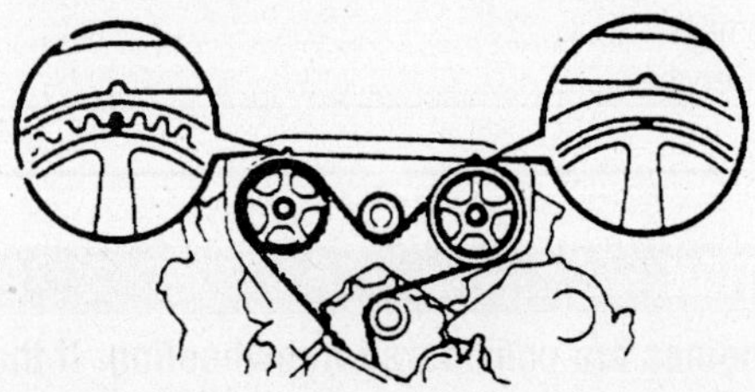

9305ag94

Camshaft timing marks—Toyota 3.0L (3VZ-FE) engine–Camry

To install:

14. Install the timing belt to the crankshaft sprocket and idler pulley.
15. Install the timing belt guide with the cupped side out.
16. Install the lower cover.
17. Install the crankshaft pulley. Tighten the crankshaft pulley bolt to 181 ft. lbs. (245 Nm).
18. Ensure that the crankshaft and camshaft timing marks are aligned and install the timing belt to the camshaft pulleys.
19. Check the tensioner pushrod protrusion length. Length should be 0.39–0.41 in. (9.9–10.4 mm). If length is not as specified, replace the tensioner.
20. Prepare the tensioner for installation by compressing it slowly in a vise.
21. Align the hole in the plunger with the holes in the tensioner body and insert a 3/64 (1.27mm) Allen wrench to retain the plunger.

22. Install the timing belt to the left side camshaft sprocket, idler pulley and right side camshaft sprocket.
23. Install the tensioner and tighten the mounting bolts to 20 ft. lbs. (27 Nm). Remove the Allen wrench.
24. Rotate the crankshaft clockwise 2 complete turns and check that the timing marks align.
25. Install the upper timing cover.
26. Install the engine control rod, right engine mount and mount bracket.
27. Install the accessory drive belts.
28. Install the right front fender liner and wheel.
29. Install the power steering pump reservoir.
30. Install the coolant and windshield washer reservoirs.
31. Lower the vehicle and connect the negative battery cable.

4Runner, Pick-Up and T100

1989–92

⁂ CAUTION

These engines are considered freewheeling. If the timing belt fails, valve-to-piston damage is unlikely, however a precautionary compression check should be performed to verify no engine damage.

1. Disconnect the negative battery cable.
2. Drain the cooling system.
3. Remove the radiator and fan shroud.
4. Remove the accessory drive belts.
5. Unbolt the power steering pump and position aside.
6. Remove the spark plugs.
7. Disconnect the No. 2 and No. 3 air hoses at the air pipe.
8. Disconnect the No. 1 water bypass hose at the air pipe and remove the water outlet.
9. Remove the upper timing cover.
10. Remove the cooling fan and the fan pulley bracket.

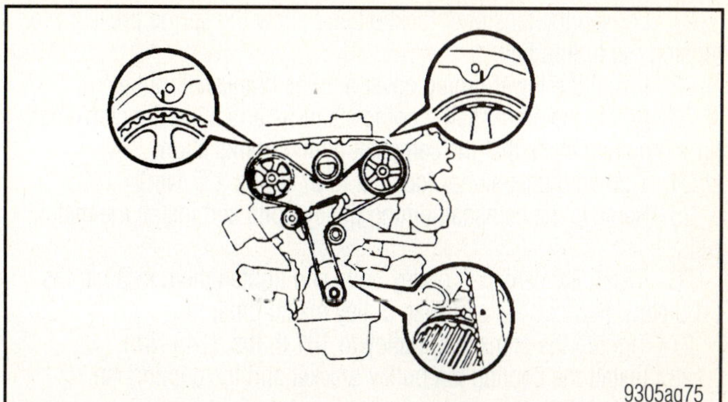

9305ag75

Alignment of the timing marks—Toyota 1989–92 3.0L (3VZ-E) engine—4Runner, Pick-Up and T100

11. Rotate the crankshaft to align the timing marks on the camshaft pulleys with the timing marks on the rear timing cover.
12. Remove the crankshaft pulley and the lower timing cover.
13. Loosen the timing belt tensioner lockbolt.
14. Push the tensioner pulley away from the timing belt and tighten the lockbolt.
15. Remove the timing belt and belt guide.

To install:

16. Hold the camshafts from turning and remove the camshaft pulley bolts.
17. Remove the camshaft pulley match pins and check that the pulleys rotate on the camshafts freely.
18. Reinstall the camshaft pulley bolts but do not tighten them. The pulleys must be able to rotate without binding to achieve proper timing belt tension. Make sure the bolts do not touch the pulleys.
19. Align the camshaft pulley matchmarks and check that the crankshaft timing pulley mark is aligned with the timing mark in the oil pump.
20. Install the timing belt.

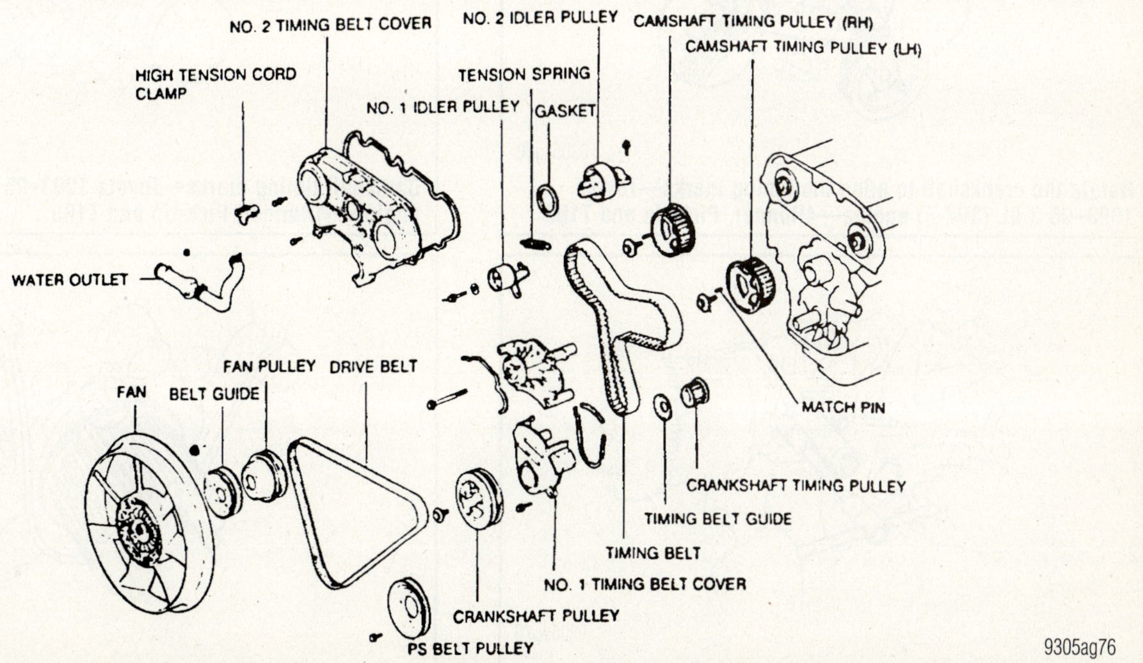

9305ag76

Exploded view of the timing belt and related components—Toyota 1989–92 3.0L (3VZ-E) engine—4Runner, Pick-Up and T100

21. Loosen the tensioner lockbolt and allow the spring pressure to tension the timing belt.
22. Install the lower timing cover and the crankshaft pulley.
23. Rotate the crankshaft 2 complete turns and align the crankshaft timing mark. Check that the camshaft timing marks align.
24. Tighten the tensioner lockbolt to 27 ft. lbs. (37 Nm).
25. Remove the camshaft timing pulley bolts and install the match pins.
26. Install the camshaft pulley bolts and tighten them to 80 ft. lbs. (109 Nm). Remove the camshaft pulley match pins.
27. Tighten the crankshaft pulley to 181 ft. lbs. (245 Nm).
28. Install the cooling fan pulley bracket and the cooling fan.
29. Install the upper timing cover.
30. Install the water outlet and the No. 1water bypass hose.
31. Install the No. 2 and No. 3 air hoses.
32. Install the spark plugs.
33. Install the power steering pump.
34. Install the accessory drive belts.
35. Install the radiator and fan shroud.
36. Connect the negative battery cable and refill the cooling system.

1993–95

✳✳ CAUTION

These engines are considered freewheeling. If the timing belt fails, valve-to-piston damage is unlikely, however a precautionary compression check should be performed to verify no engine damage.

1. Disconnect the negative battery cable.
2. Remove the engine under cover.
3. Drain the cooling system.
4. Remove the radiator and fan shroud.
5. Remove the accessory drive belts.
6. Remove the cooling fan.
7. Unbolt the power steering pump and position aside.
8. Remove the spark plugs.
9. Disconnect the No. 2 and No. 3 air hoses at the air pipe.
10. Remove the upper timing cover.
11. Rotate the crankshaft to align the timing marks on the camshaft pulleys with the timing marks on the rear timing cover, and the timing mark on the crankshaft pulley is aligned with the **0** mark on the lower timing cover.
12. Remove the timing belt tensioner.
13. Remove the cooling fan pulley bracket.
14. Remove the crankshaft pulley.
15. Remove the lower timing cover.
16. Remove the timing belt and belt guide.

To install:

17. Check that the camshaft pulley timing marks align with the timing marks on the rear timing cover, and the crankshaft timing pulley mark is aligned with the timing mark on the oil pump.
18. Install the timing belt.
19. Install the lower timing cover and the crankshaft pulley.
20. Install the cooling fan pulley bracket.
21. Prepare the timing belt tensioner for installation by slowly compressing the tensioner in a vise.

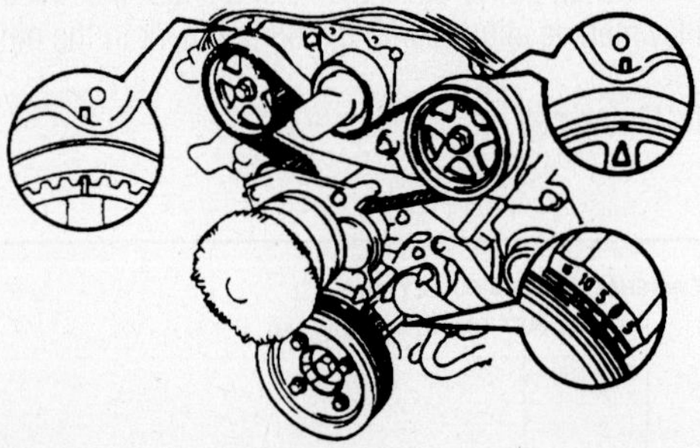

9305ag80

Rotate the crankshaft to align the timing marks—Toyota 1993–95 3.0L (3VZ-E) engine—4Runner, Pick-Up and T100

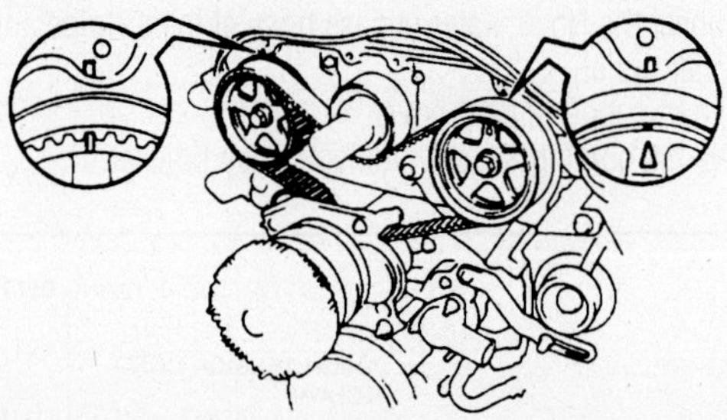

9305ag77

Camshaft timing marks—Toyota 1993–95 3.0L (3VZ-E) engine—4Runner, Pick-Up and T100

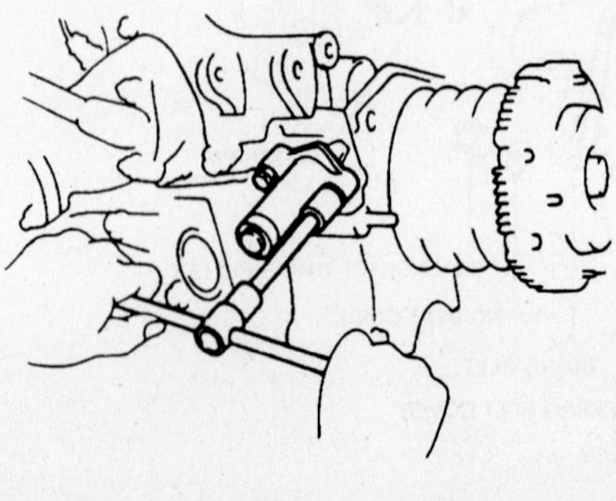

9305ag78

Timing belt tensioner installation—Toyota 1993–95 3.0L (3VZ-E) engine—4Runner, Pick-Up and T100

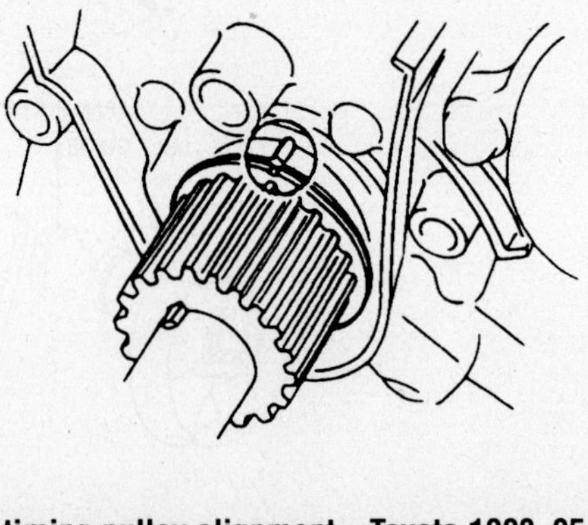

9305ag79

Crankshaft timing pulley alignment—Toyota 1993–95 3.0L (3VZ-E) engine—4Runner, Pick-Up and T100

22. Align the hole in the tensioner plunger with the holes in the tensioner housing and insert a 1.5mm Allen wrench to retain the plunger.
23. Remove the tensioner from the vise and install the dust boot.
24. Install the timing belt tensioner and remove the Allen wrench.
25. Rotate the crankshaft 2 complete turns and check that the timing marks align.
26. Tighten the crankshaft pulley bolt to 181 ft. lbs. (245 Nm).
27. Install the upper timing cover.
28. Install the water outlet.
29. Install the No. 2 and No. 3 air hoses.
30. Install the spark plugs.
31. Install the power steering pump.
32. Install the accessory drive belts.
33. Install the cooling fan.
34. Install the radiator and fan shroud.
35. Replace the engine under cover.
36. Connect the negative battery cable and fill the cooling system.

3.0L (7M-GE, 7M-GTE) ENGINES

YEAR, MODEL & ENGINE APPLICATION

1986-92 Toyota Supra, 1992 Cressida	3.0L Code 7M-GE/GTE

REPLACEMENT INTERVALS

Toyota does not provide a replacement interval. Standard industry practice is to replace timing belts every 60,000 miles.

LABOR TIMES

Cressida	
Replacement	2.3 hrs
Severe Service	2.6 hrs
Supra	
Replacement	3.3 hrs
Severe Service	3.8 hrs

8807TOYg20

**** CAUTION**

These engines are considered freewheeling. If the timing belt fails, valve-to-piston damage is unlikely, however a precautionary compression check should be performed to verify no engine damage.

1. Disconnect the negative battery cable.
2. Drain the cooling system and remove the radiator.
3. Remove the water outlet.
4. Remove the accessory drive belts.
5. Remove the upper timing cover.
6. Rotate the crankshaft so that the crankshaft timing mark aligns with the **0** mark on the lower timing cover and the timing marks on the camshaft pulleys align with the timing marks on the rear timing cover.
7. Loosen the timing belt tensioner locknut.
8. Pry the tensioner away from the timing belt and tighten the locknut.
9. Remove the timing belt from the camshaft pulleys.
10. Remove the crankshaft pulley.
11. Remove the power steering air pipe.
12. Remove the A/C compressor without disconnecting the lines and position aside.
13. Remove the lower timing cover, A/C idler pulley bracket and compressor bracket.
14. Remove the timing belt.

To install:

15. Install the timing belt to the crankshaft sprocket, oil pump sprocket and idler pulley.
16. Install the lower timing cover, A/C idler pulley bracket and compressor bracket.
17. Install the power steering air pipe.
18. Install the crankshaft pulley. Tighten the crankshaft pulley bolt to 195 ft. lbs. (265 Nm).
19. Ensure that the crankshaft and camshaft timing marks are aligned and install the timing belt to the camshaft sprockets.
20. Tighten the tensioner locknut to 36 ft. lbs. (49 Nm).
21. Install the upper timing cover.

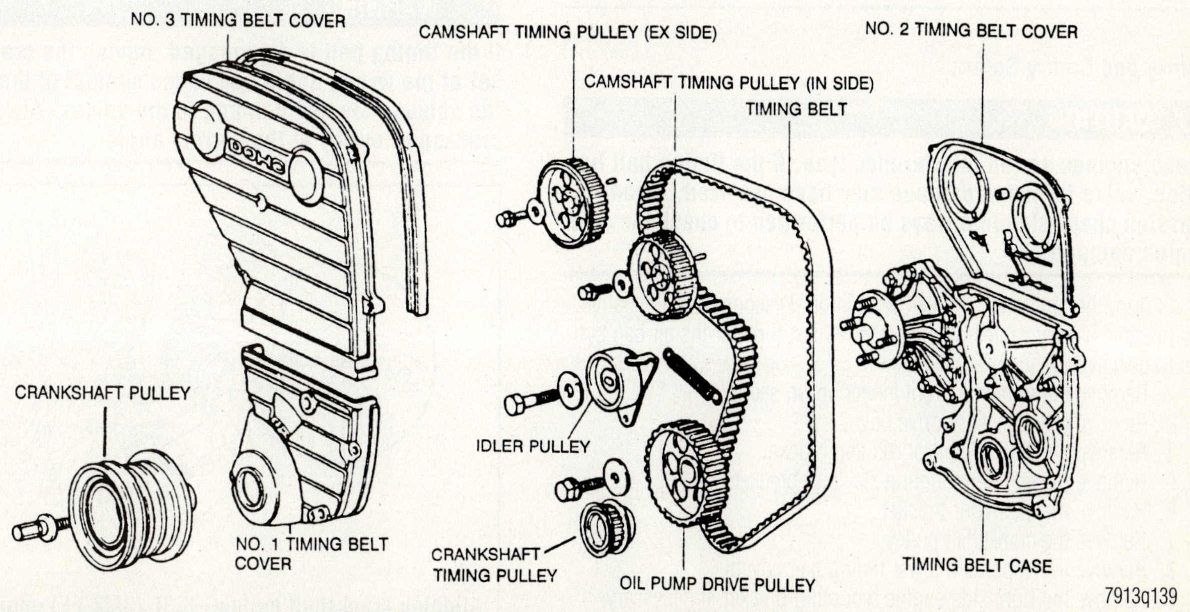

Exploded view of timing belt and related components—Toyota 3.0L (7M-GE, 7M-GTE) engines

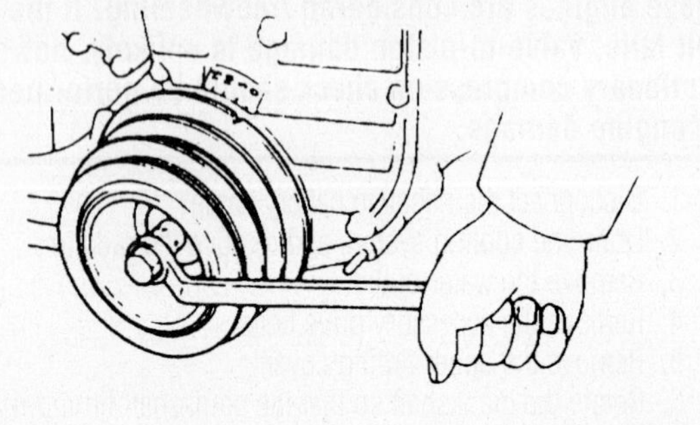
9305ag95

Crankshaft timing marks—Toyota 3.0L (7M-GE, 7M-GTE) engines

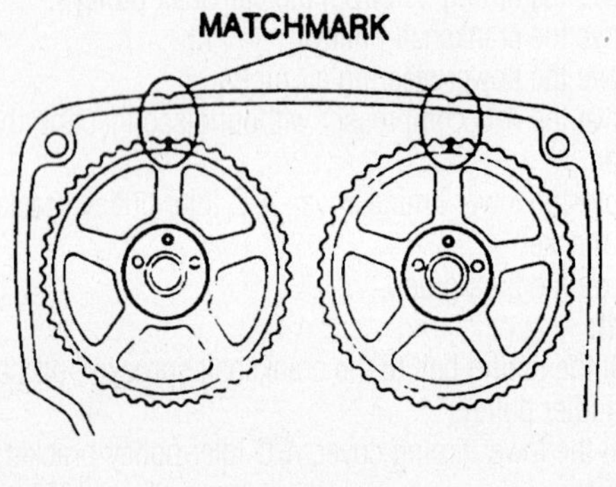

9305g96

Camshaft timing marks—Toyota 3.0L (7M-GE, 7M-GTE) engines

22. Install the accessory drive belts.
23. Install the water outlet, spark plugs and radiator.
24. Connect the negative battery cable and fill the cooling system.

3.3L (3MZ-FE) ENGINES

Camry and Camry Solara

CAUTION

These engines are an interference type. If the timing belt has failed, valve-to-piston damage may have occurred. A compression check should always be performed to check for engine damage.

1. Turn the ignition to the LOCK position. Disconnect the negative and positive battery cables. Wait at least 90 seconds for the air bag system to deactivate.
2. Remove the right side front fender apron seal.
3. Remove the accessory drive belts.
4. Remove the engine control rod and bracket.
5. Remove the engine mounting stay and bracket.
6. Remove the alternator bracket.
7. Remove the crankshaft pulley.
8. Remove the upper and lower timing belt covers.
9. Remove the right side engine mounting bracket, if necessary.
10. Remove the timing belt guide.

YEAR, MODEL & ENGINE APPLICATION

2004 Toyota Camry, Camry Solara, Highlander, Sienna .... 3.3L Code 3MZ-FE

REPLACEMENT INTERVALS

Toyota does not provide a replacement interval. Standard industry practice is to replace timing belts every 60,000 miles.

LABOR TIMES

Camry	
Replacement	2.5 hrs
Severe Service	2.9 hrs
Camry Solara	
Replacement	2.7 hrs
Severe Service	3.2 hrs
Highlander	
Replacement	2.4 hrs
Severe Service	2.9 hrs
Sienna	
Replacement	6.0 hrs
Severe Service	6.5 hrs

8807TOYg21

11. Rotate the crankshaft so No. 1 cylinder is at TDC on the compression stroke.
12. Temporarily install the crankshaft pulley and turn the crankshaft clockwise and align the crankshaft timing mark with the oil pump body.
13. Ensure that the camshaft pulley timing marks are aligned.
14. Remove the crankshaft pulley bolt.

■ **If reusing the timing belt make certain there are 3 location marks on the timing belt corresponding with the marks on the timing gears.**

15. Set the No 1 cylinder to approximately 60 degrees BTDC/compression. Turn the crankshaft counterclockwise by approximately 60 degrees.

WARNING

If the timing belt is disengaged, having the crankshaft pulley set at the wrong angle can cause contact of the piston with the valves, causing damage to the valves. Always set the crankshaft pulley at the correct angle.

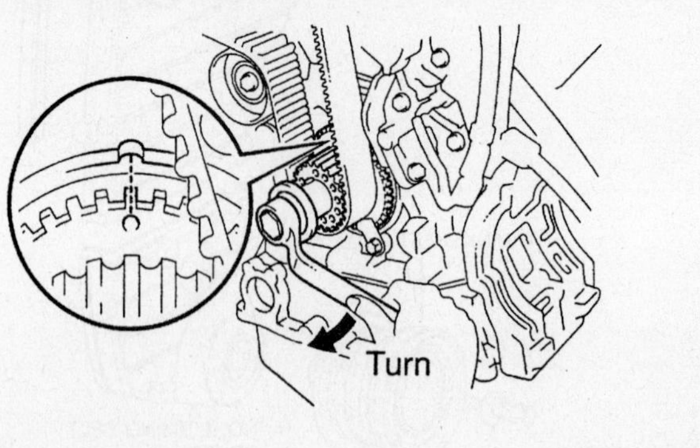

67162-LEXU-g55

Aligning crankshaft pulley—3.3L (3MZ-FE) engine—Camry and Camry Solara

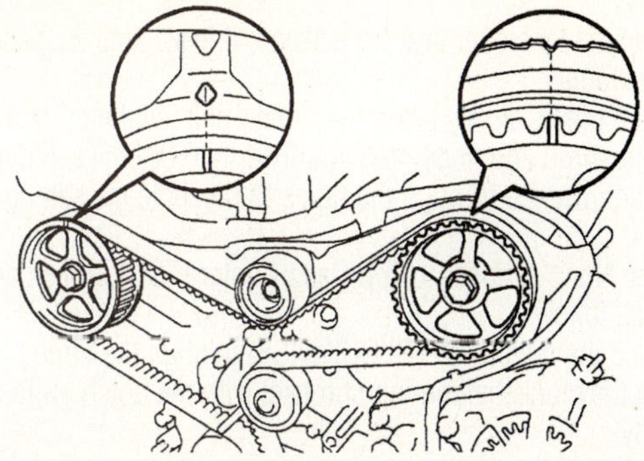

67162-LEXU-g56

Aligning camshaft pulleys—3.3L (3MZ-FE) engine—Camry and Camry Solara

16. Remove timing belt tensioner

■ **DO NOT reinstall the timing belt tensioner with the plunger extended.**

18. Remove the timing belt in the following order.
- No 1 idler pulley
- RH camshaft timing pulley
- No. 2 idler pulley
- LH camshaft timing pulley
- Water pump pulley
- Crankshaft timing pulley

19. Inspect the timing belt

■ **DO NOT reuse the timing belt if there is evidence of fraying, oil contamination, cracking, tooth damage or wear, or belt distortion. If there is any doubt in the belt condition, replace the belt.**

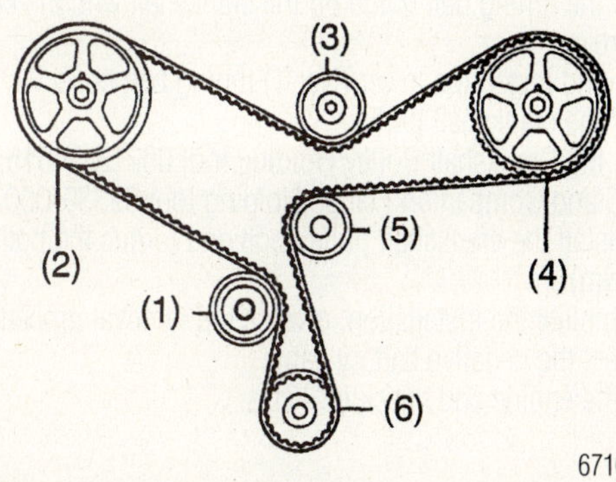

67162-LEXU-g58

Removal of timing belt from pulleys—3.3L (3MZ-FE) engine—Camry and Camry Solara

To install:

20. After inspecting the pulleys for wear and checking for oil leaks install the timing belt using the following procedure.
- Temporarily install the crankshaft pulley bolt
- Make sure the crankshaft pulley is set at 60° BTDC
- Align the timing marks on the camshaft pulley with the respective marks on the inside timing cover.
- If reusing the old belt, align the marks previously made on the belt with the marks on the timing gears.
- Install the belt in the reverse order used when removing the belt

21. Install the timing belt tensioner using the following procedure

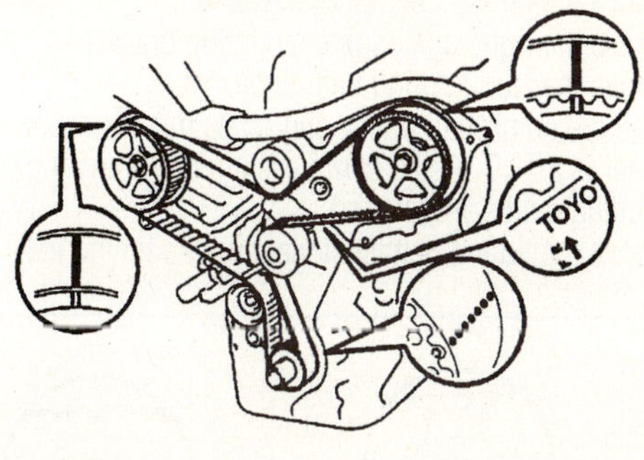

67162-LEXU-g57

Marking the timing belt for reuse—3.3L (3MZ-FE) engine—Camry and Camry Solara

■ **Keep the tensioner in an upright position**

- Slowly depress the push rod and align the hole with the hole in the housing
- Insert a 1.5 mm hexagon wrench through the holes to maintain the setting position of the push rod.
- Install the tensioner with the 2 bolts and torque to 20 ft. lbs (27 Nm).
- Remove the hexagon wrench

22. Slowly turn the crankshaft 2 revolutions clockwise.
23. Check to see that all timing marks are in alignment.
24. Install the crankshaft bolt
25. Install the timing belt guide
26. Install the right side engine mounting bracket, if removed.
27. Install the upper and lower timing belt covers.
28. Install the crankshaft pulley.
29. Install the alternator bracket.
30. Install the engine mounting stay and bracket.
31. Install the engine control rod and bracket.
32. Install the accessory drive belts.
33. Install the right side front fender apron seal.
34. Connect the negative battery cable.

Highlander

✻✻ CAUTION

These engines are an interference type. If the timing belt has failed, valve-to-piston damage may have occurred. A compression check should always be performed to check for engine damage.

1. Disconnect the negative battery cable and wait at least 90 seconds to allow the air bag system to lose power.
2. Raise and support the vehicle.
3. Remove the right side wheel, fender apron seal and engine undercover.
4. Remove the wiper linkage.
5. Remove the cowl top panel.
6. Remove the accessory drive belts.
7. Place a suitable jack under the engine and remove the right side mount stay and control rod.
8. Remove the alternator adjusting bracket.
9. Remove the crankshaft pulley bolt.
10. Using a puller, remove the crankshaft pulley.

11. Remove the lower timing belt cover.
12. Remove the timing belt upper cover.
13. Remove the right side engine mounting bracket.
14. Remove the No. 2 timing belt guide.
15. Install the crankshaft pulley and rotate the crankshaft until the No. 1 cylinder is a TDC on the compression stroke and the camshaft sprockets timing marks are aligned.
16. Remove the timing belt tensioner and the timing belt.

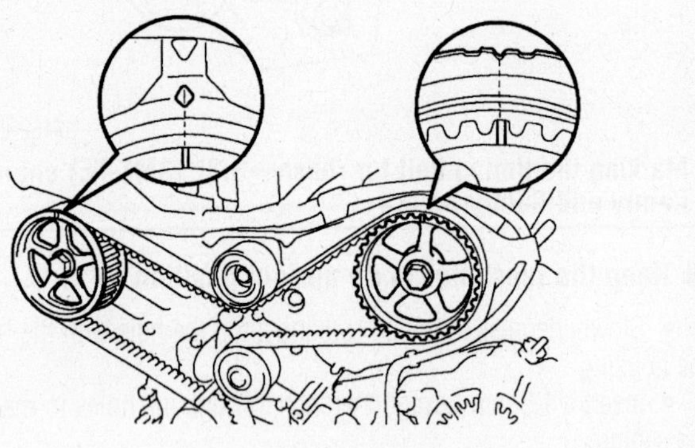

67162-X300-g01

Camshaft sprocket timing marks locations—Toyota 3.0L (1MZ-FE) engine—Highlander

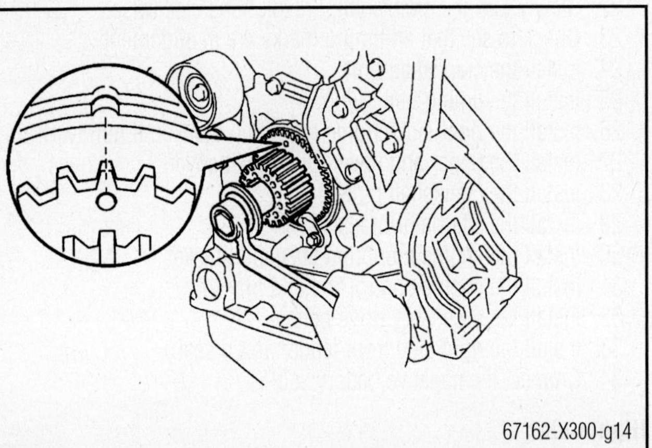

67162-X300-g14

Turn the crankshaft clockwise to align the crankshaft sprocket timing marks—Toyota 3.0L (1MZ-FE) engine—Highlander

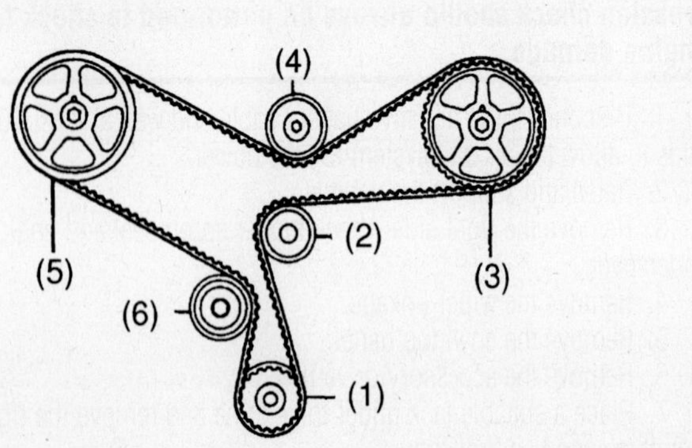

67162-X300-g06

Timing belt installation order— Toyota 3.0L (1MZ-FE) engine—Highlander

To install:

17. Inspect the tensioner seal for leakage; if leakage is suspected, replace the tensioner.
18. Using a press, slowly press the timing belt tensioner until the holes of the pushrod and housing align. Insert a 0.05 in. (1.27mm) hexagonal Allen wrench through the holes to preserve the setting position.
19. Set the No. 1 cylinder to Top Dead Center (TDC) of the compression stroke, as follows:
    a. Rotate the crankshaft (CLOCKWISE) to align the timing marks: dimple on the crankshaft timing sprocket with the notch on the oil pump body.
    b. Check that the timing marks on the camshaft sprockets and the rear timing belt cover are aligned; if not, rotate the crankshaft 1 revolution (360 degrees) and align the marks.
20. Install the timing belt in the following order:
    c. Crankshaft timing sprocket
    d. Water pump pulley
    e. Left camshaft timing sprocket
    f. No. 2 idler pulley
    g. Right camshaft sprocket
    h. No. 1 idler pulley
21. Install the timing belt tensioner and torque the bolts to 20 ft. lbs. (27 Nm).
22. Remove the Allen® wrench from the tensioner housing.
23. Slowly, rotate the crankshaft (CLOCKWISE) 2 complete revolutions and realign the timing marks. If the timing marks do not align, remove the timing belt and reinstall it.
24. Remove the crankshaft pulley bolt.
25. Install the right side engine mounting bracket and torque the bolts to 21 ft. lbs. (28 Nm).
26. Clean and install the upper (No. 2) timing belt cover.

■ **If the gasket material on the timing belt covers is cracked, peeling or etc., replace it.**

27. Install the timing belt guide on the crankshaft with the cup side facing outward.
28. Clean and install the lower (No. 1) timing belt cover.
29. Install the crankshaft pulley.
30. Using the Crankshaft Pulley Holding tool 09213-54015, Bolt tool 91651-60855 and Companion Flange Holding tool 09330-00021 or equivalent, install the crankshaft pulley bolt and torque the bolt to 159 ft. lbs. (215 Nm).
31. To complete the installation, reverse the removal procedures.
32. Connect the negative battery cable.
33. Start the engine and check for leaks.

Sienna

✱✱ CAUTION

These engines are an interference type. If the timing belt has failed, valve-to-piston damage may have occurred. A compression check should always be performed to check for engine damage.

1. Turn the ignition to the LOCK position. Disconnect the negative and positive battery cables. Wait at least 90 seconds for the air bag system to deactivate.
2. Remove the outer front cowl top panel assembly by performing the following procedure:
    a. Remove the wiper arm/blade assemblies head caps, nuts and assemblies.

b. Remove the head-to-cowl seal and the cowl panel hole cover.

c. Disconnect the windshield washer clip and hose.

d. Remove both (right and left) cowl top ventilator louvers.

e. Disconnect the electrical connector from the windshield wiper motor.

f. Remove the outer front cowl top panel assembly-to-cowl bolts and the panel.

3. Raise and safely support the vehicle.
4. Remove the right front wheel assembly and apron seal.
5. Remove the alternator by performing the following procedure:

a. Loosen the pivot bolt, adjusting lockbolt and adjusting bolt, then, remove the drive belt.

b. Disconnect the alternator's electrical connector.

c. Remove the nut and the alternator wire.

d. Disconnect the wiring harness from the clip.

e. Remove the alternator-to-bracket pivot bolt, the washer, adjusting lockbolt and alternator.

6. Loosen the power steering pump's mount and adjusting bolt, then, remove the drive belt.
7. Disconnect the hose from the engine coolant reservoir.
8. Disconnect the Diagnostic Link Connector 1 (DLC1) from the No. 2 right side engine mounting bracket.
9. Remove the right side engine mounting stay, the engine moving control rod and the No. 2 right side engine mounting bracket.
10. Loosen the alternator's pivot bolt, the nut and the No. 2 alternator bracket.
11. Using the Crankshaft Pulley Holding tool 09213-54015, Bolt tool 91651-60855 and Companion Flange Holding tool 09330-00021, or equivalent, remove the crankshaft pulley bolt.

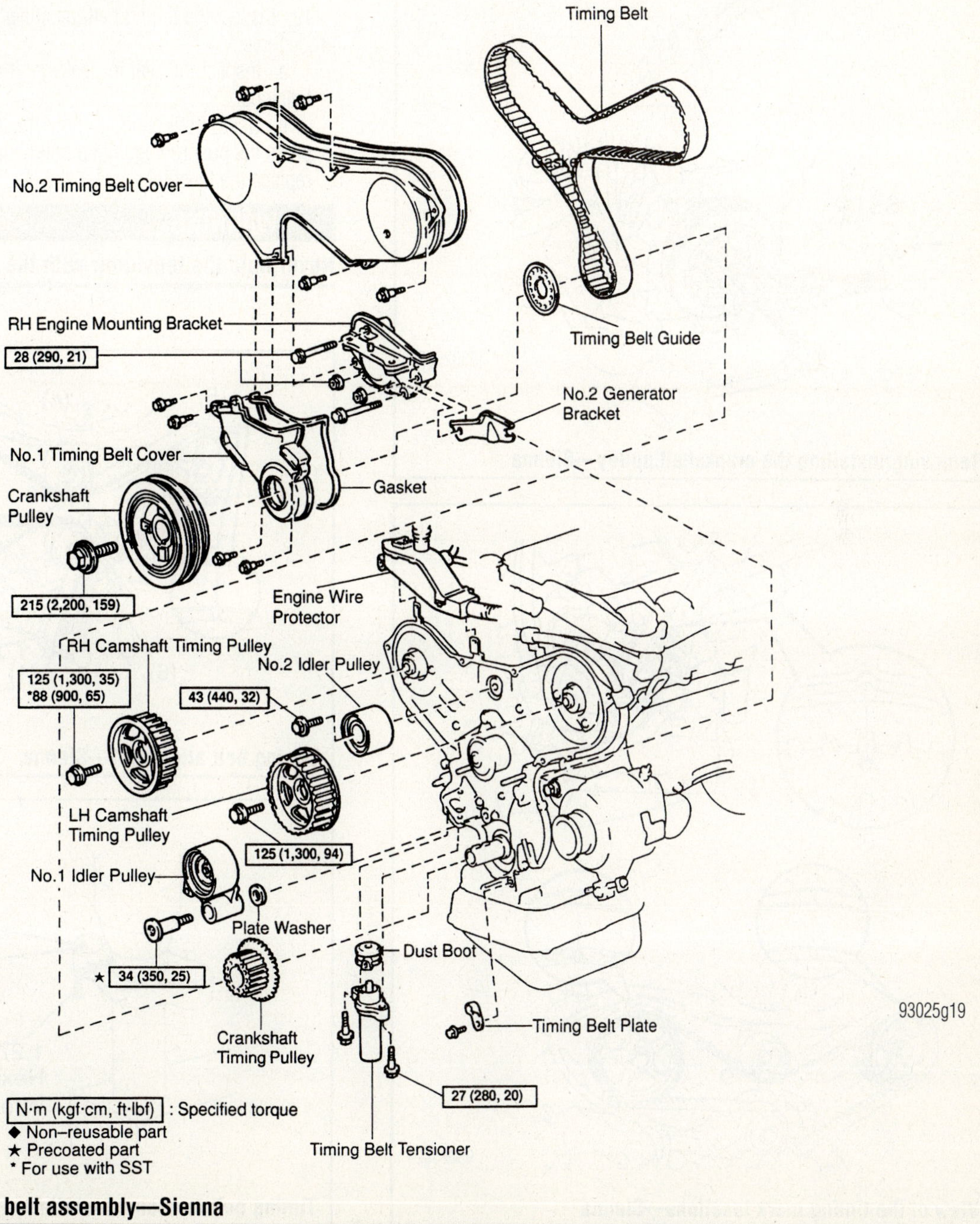

Exploded view of the timing belt assembly—Sienna

12. Using a Puller "C" Set 09950-50011 (Hanger 150 tool 09951-05010, Slide Arm tool 09952-5010, Center Bolt 100 tool 09953-05010, Center Bolt 150 tool 09953-05020 and 2 No. 2 Claw tools 09954-05020), pull the crankshaft pulley from the crankshaft.

13. Remove the lower (No. 1) timing belt cover. Remove the timing belt guide from the crankshaft.

14. Remove the engine wire protector clamps from the upper (No. 2) timing belt cover and remove the upper (No. 2) timing belt cover.

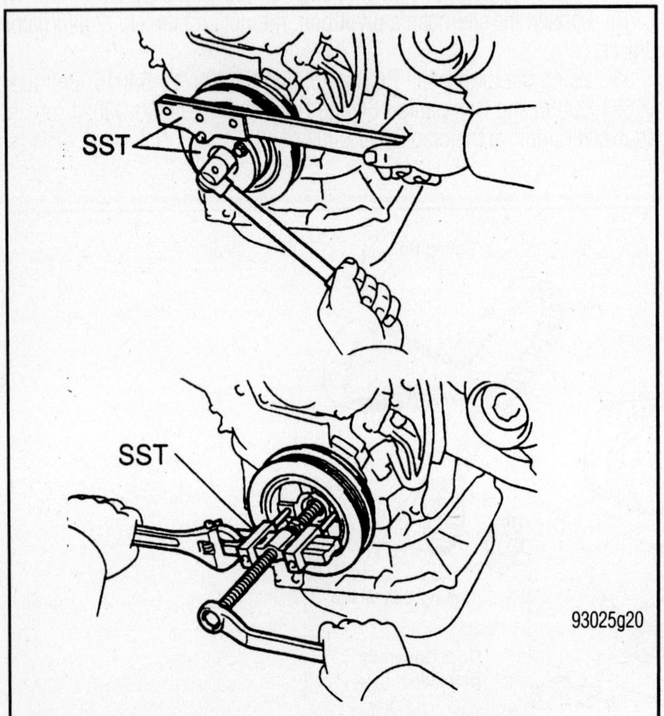

Removing/installing the crankshaft pulley—Sienna

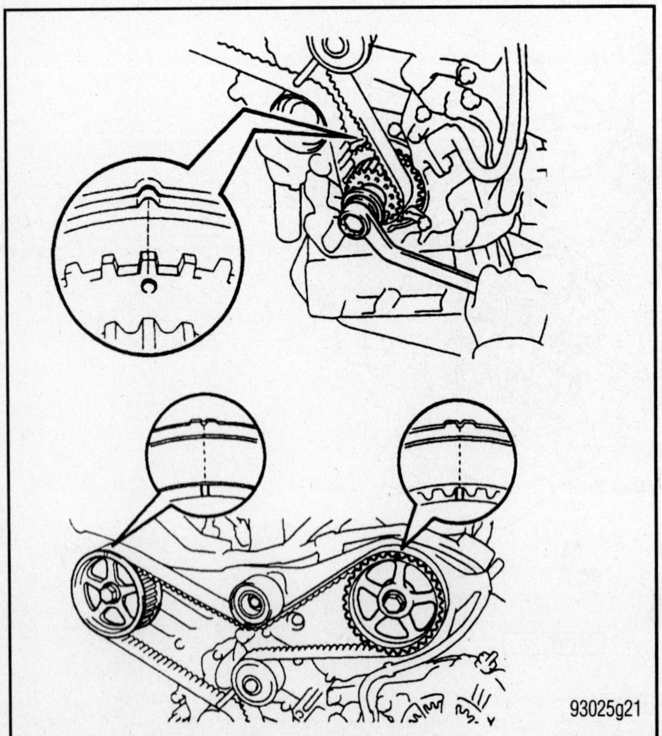

View of the timing mark locations—Sienna

15. Remove the right side engine mounting brace

■ **If reusing the timing belt, be sure that you can still read the installation marks. If not, place new installation marks on the timing belt to match the timing marks of the camshaft timing pulleys.**

16. Temporarily install the crankshaft pulley bolt.

17. Set the No. 1 cylinder to Top Dead Center (TDC) of the compression stroke, as follows:

a. Rotate the crankshaft (CLOCKWISE) to align the timing marks: dimple on the crankshaft timing sprocket with the notch on the oil pump body.

b. Check that the timing marks on the camshaft sprockets and the rear timing belt cover are aligned; if not, rotate the crankshaft 360 degrees (1 revolution) and align the marks.

18. Remove the timing belt tensioner and the timing belt.

To install:

19. Inspect the timing belt tensioner by performing the following procedures:

a. Inspect the seal for leakage; if leakage is suspected, replace the tensioner.

b. Using both hands to hold the tensioner facing upward, strongly press the pushrod against a solid surface. If the pushrod moves, replace the tensioner.

WARNING

Never hold the tensioner with the pushrod facing downward.

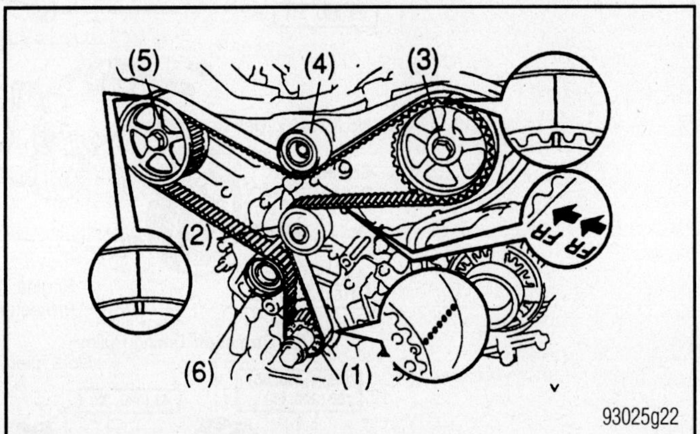

Timing belt alignment—Sienna

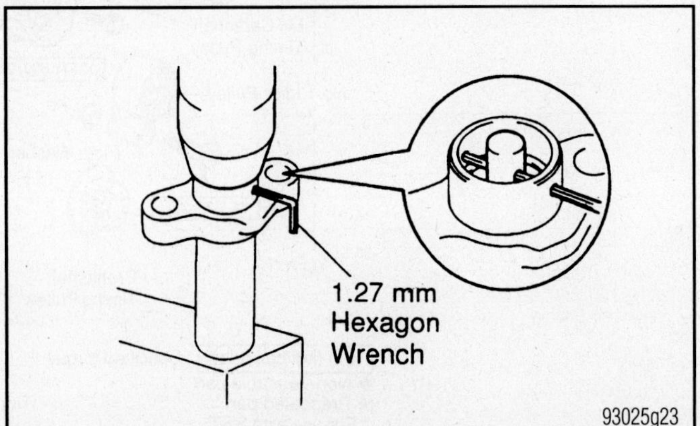

Timing belt tensioner installation preparation—Sienna

c. Measure the pushrod protrusion from the housing end, it should be 0.394–0.425 in. (10.0–10.8mm); if the protrusion is not as specified, replace the tensioner.

20. Set the No. 1 cylinder to Top Dead Center (TDC) of the compression stroke, as follows:

d. Rotate the crankshaft (CLOCKWISE) to align the timing marks: dimple on the crankshaft timing sprocket with the notch on the oil pump body.

e. Check that the timing marks on the camshaft sprockets and the rear timing belt cover are aligned; if not, rotate the crankshaft I revolution (360 degrees) and align the marks.

21. Install the timing belt in the following order:

f. Crankshaft timing sprocket
g. Water pump pulley
h. Left camshaft timing sprocket
i. No. 2 idler pulley
j. Right camshaft sprocket
k. No. 1 idler pulley

22. Using a vertical press, slowly press the pushrod into the housing using 200–2205 lbs. (981–9807 N) until the holes align, then, install a 1.27mm Allen® wrench to secure the pushrod and release the press. Install the dust boot on the tensioner housing.

23. Install the timing belt tensioner and torque the bolts to 20 ft. lbs. (27 Nm).

24. Remove the Allen® wrench from the tensioner housing.

25. Slowly, rotate the crankshaft (CLOCKWISE) 2 complete revolutions and realign the timing marks. If the timing marks do not align, remove the timing belt and reinstall it.

26. Remove the crankshaft pulley bolt.

27. Install the right side engine mounting bracket and torque the bolts to 21 ft. lbs. (28 Nm).

28. Clean and install the upper (No. 2) timing belt cover.

■ **If the gasket material on the timing belt covers is cracked, peeling or etc., replace it.**

29. Install the timing belt guide on the crankshaft with the cup side facing outward.

30. Clean and install the lower (No. 1) timing belt cover.

■ **If the gasket material on the timing belt covers is cracked, peeling or etc., replace it.**

31. Install the crankshaft pulley.

32. Using the Crankshaft Pulley Holding tool 09213-54015, Bolt tool 91651-60855 and Companion Flange Holding tool 09330-00021, or equivalent, install the crankshaft pulley bolt and torque the bolt to 159 ft. lbs. (215 Nm).

33. To complete the installation, reverse the removal procedures.

34. Connect the negative battery cable.

35. Start the engine and check for leaks.

3.4L (5VZ-FE) ENGINES

T-100 and Tacoma

✱✱ CAUTION

These engines are an interference type. If the timing belt has failed, valve-to-piston damage may have occurred. A compression check should always be performed to check for engine damage.

YEAR, MODEL & ENGINE APPLICATION

1996-02 Toyota 4Runner, 1995-98 T100, 1995-04 Tacoma, 2000-04 Tundra .... 3.4L Code 5VZ-FE

REPLACEMENT INTERVALS

Toyota does not provide a replacement interval for 1995-98 models. Standard industry practice is to replace timing belts every 60,000 miles. On 1999-04 models, Toyota recommends timing belt replacement every 90,000 miles.

LABOR TIMES

1995-04

Replacement	2.8 hrs
Severe Service	3.3 hrs
With air conditioning add	0.3 hrs

8807TOYg22

1. Turn the ignition to the LOCK position. Disconnect the negative and positive battery cables. Wait at least 90 seconds for the air bag system to deactivate.

2. Raise and safely support the vehicle.

3. Remove the engine undercover.

4. Drain the engine coolant.

5. Disconnect the upper radiator hose from the engine.

6. Remove the power steering drive belt.

7. Remove the air conditioning drive belt by loosening the idler pulley nut and the adjusting bolt.

8. Loosen the lockbolt, pivot bolt, and the adjusting bolt and the alternator drive belt.

9. Remove the No. 2 fan shroud by removing the 2 clips.

10. Remove the fan with the fluid coupling and fan pulleys.

11. Disconnect the power steering pump from the engine and set aside. Do not disconnect the lines from the pump.

12. If equipped with air conditioning, disconnect the compressor from the engine and set aside. Do not disconnect the lines from the compressor.

13. If equipped with air conditioning, disconnect the air conditioning bracket.

14. Remove the No. 2 timing belt cover, as follows:

a. Detach the camshaft position sensor connector from the No. 2 timing belt cover.

b. Disconnect the 3 spark plug wire clamps from the No. 2 timing belt cover.

c. Remove the 6 bolts and remove the timing belt cover.

15. Remove the fan bracket, as follows:

a. Remove the power steering adjusting strut by removing the nut.

b. Remove the fan bracket by removing the bolt and nut.

16. Set the No. 1 cylinder at Top Dead Center (TDC) of the compression stroke, as follows:

a. Turn the crankshaft pulley and align its groove with the timing mark **0** of the No. 1 timing belt cover.

b. Check that the timing marks of the camshaft timing pulleys and the No. 3 timing belt cover are aligned. If not, turn the crankshaft pulley 1 revolution (360 degrees).

■ **If reusing the timing belt, be sure that you can still read the installation marks. If not, place new installation marks on the timing belt to match the timing marks of the camshaft timing pulleys.**

17. Remove the timing belt tensioner by alternately loosening the 2 bolts.
18. Remove the camshaft timing pulleys, as follows:
    a. Using SST 09960-10010, or equivalent, remove the pulley bolt, the timing pulley and the knock pin. Remove the 2 timing pulleys with the timing belt.
19. Remove the crankshaft pulley, as follows:
    a. Using SST 09213-54015 and 09330-00021 or their equivalents, loosen the pulley bolt.
    b. Remove the SST tool, the pulley bolt and the pulley.
20. Remove the starter wire bracket and the No. 1 timing belt cover.
21. Remove the timing belt guide and remove the timing belt.
22. Remove the bolt and the No. 2 idler pulley.
23. Remove the pivot bolt, the No. 1 idler pulley and the plate washer.
24. Remove the crankshaft gear.

79245g37

Turn the crankshaft clockwise to align the timing marks before removing the timing belt—Toyota 3.4L (5VZ-FE) engine—T-100 and Tacoma

To install:

25. Measure the timing belt tensioner pushrod protrusion. Protrusion length should be 0.39–0.42 in. (9.9–10.6mm). If length is not as specified, replace the tensioner.
26. Install the crankshaft timing gear.
    a. Align the timing pulley set key with the key groove of the gear.
    b. Using SST 09214-60010 or equivalent, and a hammer, tap in the timing gear with the flange side facing inward.
27. Install the plate washer and the No. 1 idler pulley with the pivot bolt and tighten it to 26 ft. lbs. (35 Nm). Check that the pulley bracket moves smoothly.
28. Install the No. 2 timing belt idler with the bolt. Tighten the bolt to 30 ft. lbs. (40 Nm). Check that the pulley bracket moves smoothly.
29. Temporarily install the timing belt, as follows:
    a. Using the crankshaft pulley bolt, turn the crankshaft and align the timing marks of the crankshaft timing pulley and the oil pump body.
    b. Align the installation mark on the timing belt with the dot mark of the crankshaft timing pulley.
    c. Install the timing belt on the crankshaft timing pulley, No. 1 idler pulley and the water pump pulleys.
30. Install the timing belt guide with the cup side facing outward.
31. Install the No. 1 timing belt cover and starter wire bracket. Tighten the timing belt cover bolts to 80 inch lbs. (9 Nm).

✱✱ WARNING

If any binding is felt when adjusting the timing belt tension by turning the crankshaft, STOP turning the engine, because the pistons may be hitting the valves.

32. Install the crankshaft pulley, as follows:
    a. Align the pulley set key with the key groove of the crankshaft pulley.
    b. Install the pulley bolt and tighten it to 184 ft. lbs. (250 Nm).
33. Install the left camshaft timing pulley.
    c. Install the knock pin to the camshaft.
    d. Align the knock pin hose of the camshaft with the knock pin groove of the timing pulley.
    e. Slide the timing belt pulley on the camshaft with the flange side facing outward. Tighten the pulley bolt to 81 ft. lbs. (110 Nm).
34. Set the No. 1 cylinder to TDC of the compression stroke, as follows:
    a. Turn the crankshaft pulley, and align its groove with the timing mark **0** of the No. 1 timing belt cover.
    b. Turn the camshaft to align the knock pin hole of the camshaft with the timing mark of the No. 3 timing belt cover.
    c. Turn the camshaft timing pulley, and align the timing marks of the camshaft timing pulley and the No. 3 timing belt cover.
35. Connect the timing belt to the left camshaft timing pulley, as follows:

■ **Check that the installation mark on the timing belt is aligned with the end of the No. 1 timing belt cover.**

    a. Using SST 09960-01000, or equivalent, slightly turn the left camshaft timing pulley clockwise. Align the installation mark on the timing belt with the timing mark of the camshaft timing pulley and hang the timing belt on the left camshaft timing pulley.
    b. Align the timing marks of the left camshaft pulley and the No. 3 timing belt cover.
    c. Check that the timing belt has tension between the crankshaft timing pulley and the left camshaft timing pulley.
36. Install the right camshaft timing pulley and the timing belt, as follows:
    a. Align the installation mark on the timing belt with the timing mark of the right camshaft timing pulley and hang the timing belt on the right camshaft timing pulley with the flange side facing inward.
    b. Slide the right camshaft timing pulley on the camshaft. Align the timing marks on the right camshaft timing pulley and the No. 3 timing belt cover.
    c. Align the knock pin hole of the camshaft with the knock pin groove of the pulley and install the knock pin. Install the bolt and tighten it to 81 ft. lbs. (110 Nm).
37. Set the timing belt tensioner, as follows:
    d. Using a press, slowly press in the pushrod using 220–2205 lbs. (981–9807 N) of force.
    e. Align the holes of the pushrod and housing, pass a 1.5mm hex wrench through the holes to keep the setting position of the pushrod.
    f. Release the press and install the dust boot on the tensioner.

38. Install the timing belt tensioner and alternately tighten the bolts to 20 ft. lbs. (28 Nm). Using pliers, remove the 1.5mm hex wrench from the belt tensioner.

39. Check the valve timing, as follows:

a. Slowly turn the crankshaft pulley 2 revolutions from TDC-to-TDC. Always turn the crankshaft pulley clockwise.

b. Check that each pulley aligns with the timing marks. If the timing marks do not align, remove the timing belt and reinstall it.

40. Install the fan bracket with the bolt and nut.

41. Install the power steering adjusting strut with the nut.

42. Install the No. 2 timing belt cover. Tighten the bolts to 80 inch lbs. (9 Nm). Install the remaining components.

43. Fill the cooling system with coolant.

44. Connect the negative battery cable.

45. Start the engine and check for leaks.

4Runner and Tundra

✻✻ CAUTION

These engines are an interference type. If the timing belt has failed, valve-to-piston damage may have occurred. A compression check should always be performed to check for engine damage.

1. Turn the ignition to the LOCK position. Disconnect the negative and positive battery cables. Wait at least 90 seconds for the air bag system to deactivate.

2. Raise and safely support the vehicle.

3. Remove the engine undercover.

4. Drain the engine coolant.

5. Disconnect the upper radiator hose from the engine.

6. Remove the power steering drive belt.

7. Remove the air conditioning drive belt by loosening the idler pulley nut and the adjusting bolt.

8. If equipped with air conditioning, disconnect the compressor from the engine and set aside. Do not disconnect the lines from the compressor.

9. If equipped with air conditioning, disconnect the air conditioning bracket.

10. Remove the fan with the fluid coupling and fan pulleys.

11. Loosen the lockbolt, pivot bolt, and the adjusting bolt and the alternator drive belt.

12. Remove the No. 2 fan shroud by removing the 2 clips.

13. Disconnect the power steering pump from the engine and set aside. Do not disconnect the lines from the pump.

14. Remove the oil dipstick and the guide.

15. Remove the No. 2 timing belt cover as follows:

a. Detach the camshaft position sensor connector from the No. 2 timing belt cover.

b. Disconnect the 4 spark plug wire clamps from the No. 2 timing belt cover.

c. Remove the 6 bolts and remove the timing belt cover.

16. Remove the fan bracket as follows:

d. Remove the power steering adjusting strut by removing the nut.

e. Remove the fan bracket by removing the bolt and nut.

17. Using SST 09213-54015, or equivalent, remove the crankshaft pulley.

18. Remove the starter wire bracket and the No. 1 timing belt cover.

19. Remove the timing belt guide.

20. Set the No. 1 cylinder at Top Dead Center (TDC) of the compression stroke, as follows:

a. Temporarily install the crankshaft pulley bolt to the crankshaft.

b. Turn the crankshaft and align the timing marks of the crankshaft timing pulley and the oil pump body.

c. Check that the timing marks of the camshaft timing pulleys and the No. 3 timing belt cover are aligned. If not, turn the crankshaft pulley 1 revolution (360 degrees).

■ **If reusing the timing belt, be sure that you can still read the installation marks. If not, place new installation marks on the timing belt to match the timing marks of the camshaft timing pulleys.**

21. Remove the timing belt tensioner by alternately loosening the 2 bolts.

22. Remove the right and left camshaft pulleys.

23. Remove the No. 2 idler pulley.

24. Using a 10mm hex wrench, remove the pivot bolt, No.1 idler pulley and the plate washer.

25. Remove the timing belt guide and remove the timing belt.

26. Remove the crankshaft timing pulley.

79245g38

Crankshaft and camshaft timing mark locations—Toyota 3.4L (5VZ-FE) engine—4Runner and Tundra

To install:

27. Install the crankshaft timing belt pulley, as follows:

a. Align the timing belt pulley set key with the key groove of the timing pulley and slide on the timing pulley.

b. Slide on the timing belt pulley with the flange side facing inward.

28. Install the plate washer and the No. 1 idler pulley with the pivot bolt and tighten it to 26 ft. lbs. (35 Nm). Check that the pulley bracket moves smoothly.

29. Install the No. 2 timing belt idler with the bolt. Tighten the bolt to 30 ft. lbs. (40 Nm). Check that the pulley bracket moves smoothly.

30. Install the left and right camshaft timing pulleys.

31. Set the No. 1 cylinder to TDC of the compression stroke, as follows:

a. Using the crankshaft pulley bolt, turn the crankshaft and align the timing marks of the crankshaft timing pulley and the oil pump body.

b. Using SST 09960-10010 or equivalent, to turn the camshaft pulley to align the marks of the camshaft timing belt pulley and the No. 3 timing belt cover.

32. Install the timing belt, as follows:

■ **The engine should be cold.**

a. Face the front mark on the timing belt forward.

b. Align the installation mark on the timing belt with the timing mark of the crankshaft timing pulley.

c. Align the installation marks on the timing belt with the timing marks of the camshaft pulleys.

33. Install the timing belt in the following order:

- Left camshaft pulley
- No. 2 idler pulley
- Right camshaft pulley
- Water pump pulley
- Crankshaft pulley
- No. 1 idler pulley

✱✱ WARNING

If any binding is felt when adjusting the timing belt tension by turning the crankshaft, STOP turning the engine, because the pistons may be hitting the valves.

32. Set the timing belt tensioner as follows:

a. Using a press, slowly press in the pushrod using 220–2205 lbs. (981–9807 N) of force.

b. Align the holes of the pushrod and housing, pass a 1.27mm wrench through the holes to keep the setting position of the pushrod.

c. Release the press and install the dust boot to the tensioner.

33. Install the timing belt tensioner and alternately tighten the bolts to 20 ft. lbs. (27 Nm). Using pliers, remove the 1.27mm wrench from the belt tensioner.

34. Check the valve timing, as follows:

a. Slowly turn the crankshaft and align the timing marks of the crankshaft timing pulley and the oil pump body. Always turn the crankshaft pulley clockwise.

b. Check that the timing marks of the right and left timing pulleys align with the timing marks of the No. 3 timing belt cover. If the marks do not align, remove the timing belt and reinstall it.

35. Install the timing belt guide with the cup side facing outward.

36. Install the No. 1 timing belt cover and starter wire bracket. Tighten the timing belt cover fasteners to 80 inch lbs. (9 Nm).

37. Install the crankshaft pulley, as follows:

a. Align the pulley set key with the key groove of the pulley and slide the pulley.

b. Using SST 09213-54014, or equivalent, tighten the bolt to 184 ft. lbs. (250 Nm).

38. Install the fan bracket with the bolt and nut.

39. Install the No. 2 timing belt cover, and tighten the bolts to 80 inch lbs. (9 Nm). Install the remaining components.

40. Fill the cooling system with coolant.

41. Connect the negative battery cable.

42. Start the engine and check for leaks.

43. Check the ignition timing.

4.7L (2UZ-FE) ENGINES

4Runner

✱✱ CAUTION

These engines are an interference type. If the timing belt has failed, valve-to-piston damage may have occurred. A compression check should always be performed to check for engine damage.

YEAR, MODEL & ENGINE APPLICATION

2003-04 Toyota 4Runner, 1998-02 Land Cruiser, 2000-04 Sequoia, Tundra .... 4.7L Code 2UZ-FE

REPLACEMENT INTERVALS

Toyota recommends timing belt replacement every 90,000 miles.

LABOR TIMES

4Runner, Sequoia & Tundra	
Replacement	3.5 hrs
Severe Service	4.0 hrs
With air conditioning add	0.3
Land Cruiser	
Replacement	4.3 hrs
Severe Service	4.8 hrs
With air conditioning add	0.3

8807TOYg23

1. Turn the ignition to the LOCK position. Disconnect the negative and positive battery cables. Wait at least 90 seconds for the air bag system to deactivate.
2. Drain the engine coolant.
3. Remove the throttle body cover.
4. Remove the air cleaner.
5. Remove the radiator support upper seal.
6. Remove the accessory drive belt.
7. Remove the cooling fan.
8. Remove the power steering pump without disconnecting the lines and position aside.
9. Remove the alternator.
10. Remove the A/C compressor without disconnecting the lines and position aside.
11. Remove the idler pulley.
12. Remove the oil cooler pipe.
13. Remove the upper timing belt cover.
14. Remove the accessory drive belt tensioner.
15. Remove the fan bracket.
16. Using a puller, remove the crankshaft damper.
17. Remove the lower timing belt cover.

■ If reusing the timing belt, make sure that there are 3 installation marks on the belt; if there are none, install them.

18. Using Crankshaft Pulley tool 09213-70010, bolt tool 90105-08076 and flange holding tool 09330-00021, or equivalent, loosen the crankshaft pulley bolt.

19. Position the No. cylinder to approximately 50 degrees ATDC of the compression stroke as follows:

a. Rotate the crankshaft pulley clockwise to align its groove with the "0" mark on the lower cover timing belt cover.

b. Ensure that the camshaft sprocket timing marks are aligned with the rear timing belt plate marks. If not, rotate the crankshaft 360 degrees clockwise.

c. Rotate the crankshaft pulley approximately 50 degrees clockwise to align the crankshaft pulley timing mark between the centers of the crankshaft pulley bolt and the idler pulley bolt.

WARNING

If the timing belt is disengaged, having the crankshaft pulley in the wrong angle can cause the valve to come into contact with the piston when removing the camshaft pulley.

20. Remove the crankshaft pulley bolt.

■ **To avoid meshing the timing sprocket and the timing belt, secure one with a string and then place matchmarks on the timing belt and right side camshaft sprocket.**

21. Remove the timing belt tensioner bolts and the tensioner.
22. Using camshaft holding tool 09960-10010, or equivalent, slightly turn the left side camshaft sprocket clockwise to loosen the tension spring. Then disconnect the timing belt from the camshaft sprockets.
23. Remove the alternator.
24. Remove the accessory drive belt tensioner.
25. Using a puller remove the crankshaft pulley.
26. Remove the lower timing belt cover.
27. Remove the timing belt guide and the timing belt.

To install:

28. Inspect the timing belt tensioner by performing the following procedures:
    a. Inspect the seal for leakage; if leakage is suspected, replace the tensioner.
    b. Using both hands to hold the tensioner facing upward, strongly press the pushrod against a solid surface. If the pushrod moves, replace the tensioner.

WARNING

Never hold the tensioner with the pushrod facing downward.

    c. Measure the pushrod protrusion from the housing end, it should be 0.413–0.453 in. (10.5–11.5mm). If the protrusion is not as specified, replace the tensioner.
29. Temporarily install the timing belt by performing the following procedures:
    a. Align the timing belt's installation mark with the crankshaft timing sprocket.
    b. Install the timing belt on the crankshaft timing sprocket, the No. 1 idler pulley and the No. 2 idler pulley.

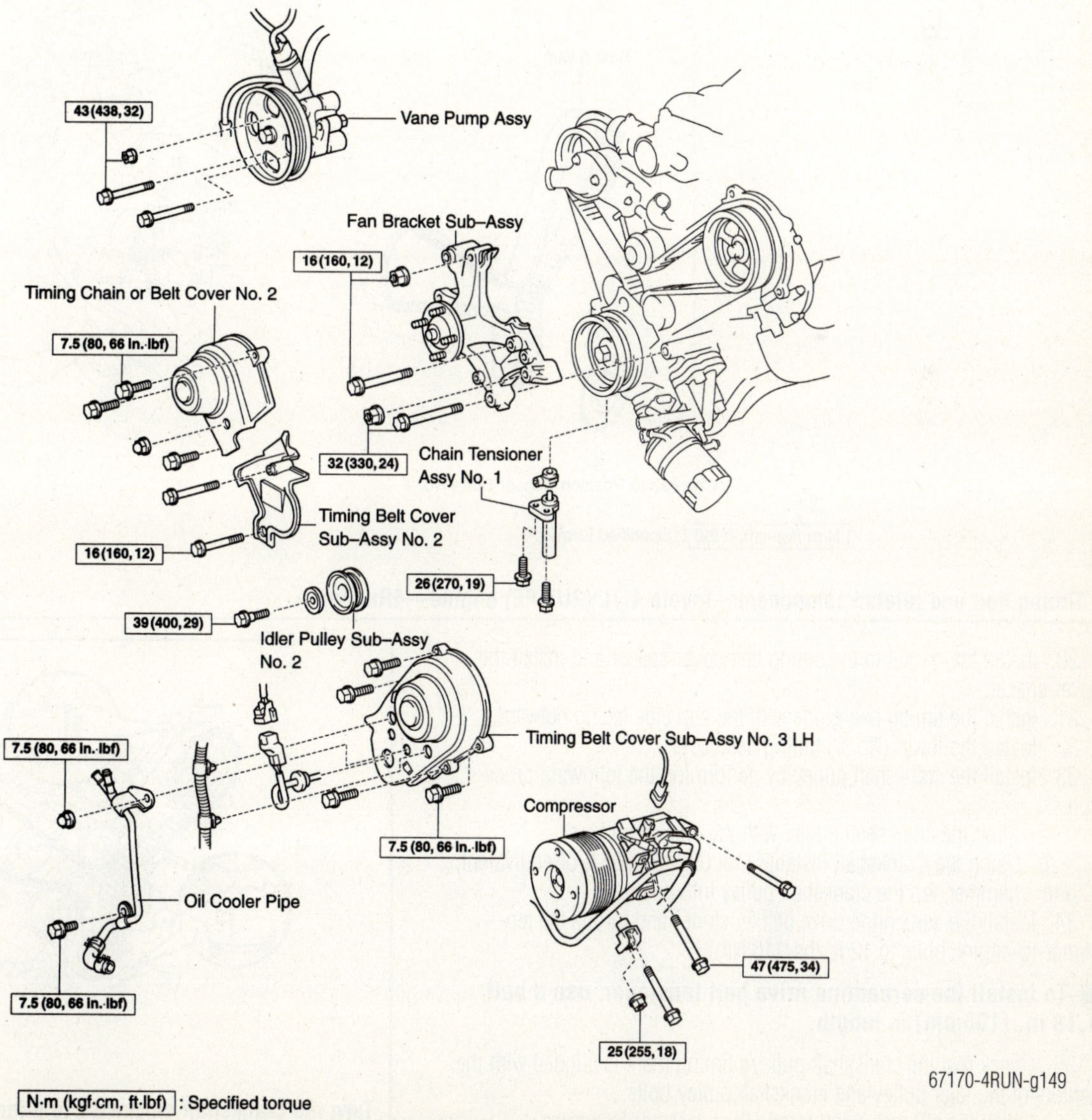

Remove these parts prior to timing belt removal–Toyota 4.7L (2UZ-FE) engine—4Runner

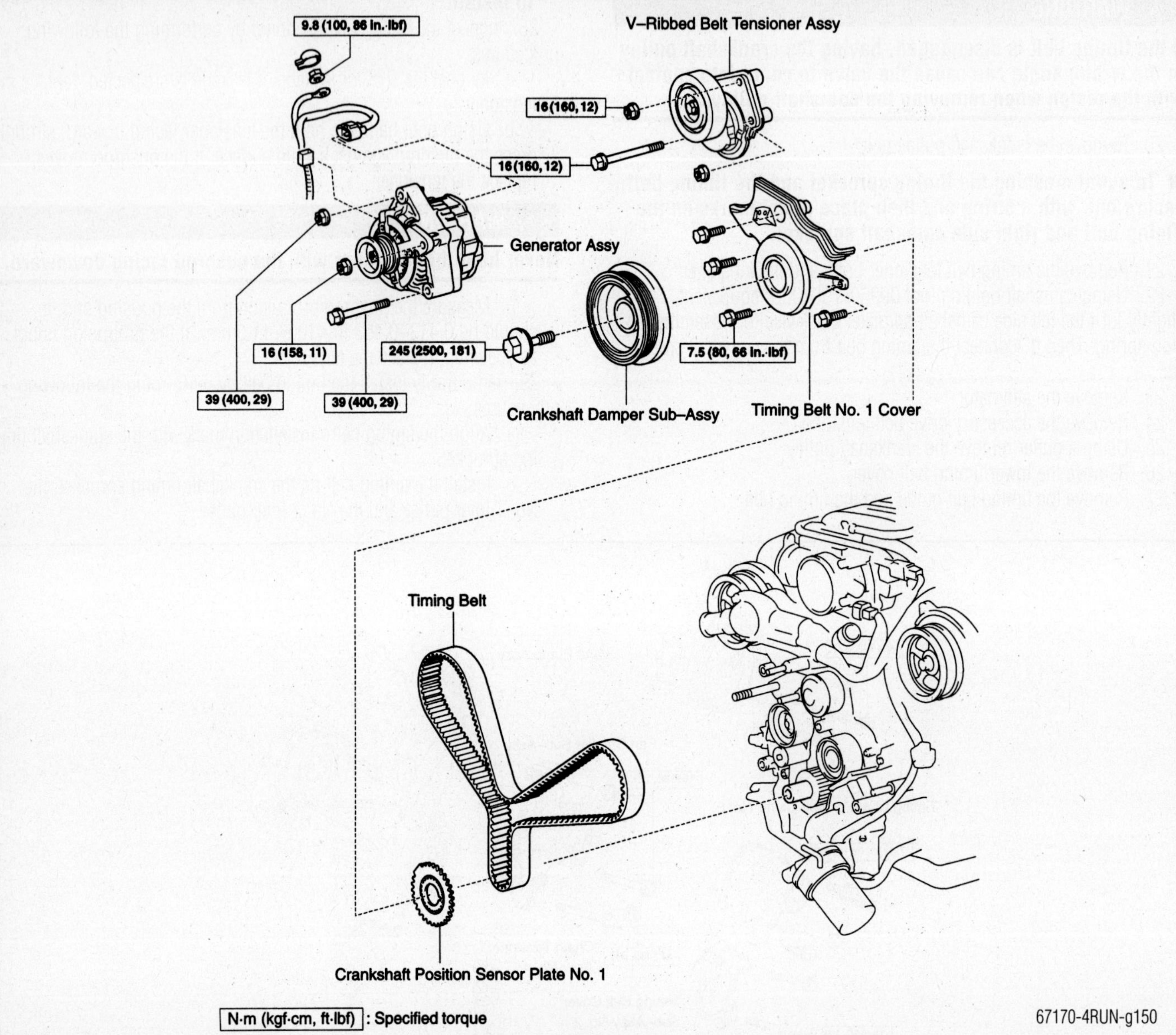

Timing belt and related components–Toyota 4.7L (2UZ-FE) engine—4Runner

30. Install the gasket to the timing belt cover spacer and install the cover spacer.
31. Install the timing belt guide with the cup side facing outward.
32. Install the lower (No. 1) timing belt cover.
33. Install the crankshaft pulley by performing the following procedures:
    a. Align the crankshaft pulley with the crankshaft key.
    b. Using the Crankshaft Installer tool 09223-46011, or equivalent, and a hammer, tap the crankshaft pulley into position.
34. Install the serpentine drive belt tensioner and torque the tensioner-to-engine bolts to 12 ft. lbs. (16 Nm).

■ **To install the serpentine drive belt tensioner, use a bolt 4.18 in. (106mm) in length.**

35. Check that the crankshaft pulley's timing mark is aligned with the centers of the idler pulley and crankshaft pulley bolts.
36. Install the alternator and torque the alternator-to-engine

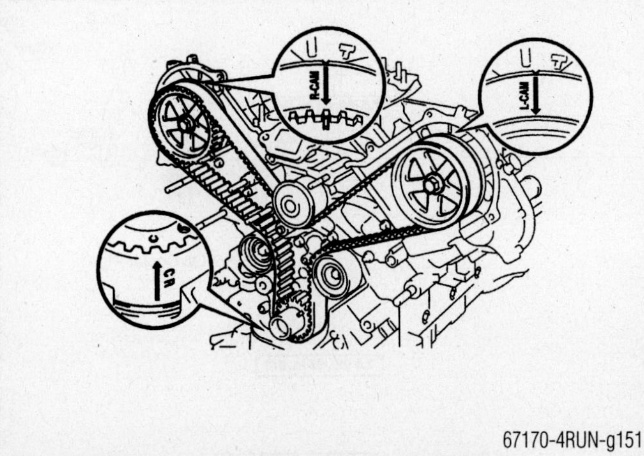

Turn the crankshaft clockwise to align the timing marks–Toyota 4.7L (2UZ-FE) engine—4Runner

nuts/bolts to 29 ft. lbs. (39 Nm). Connect the alternator's electrical connectors and clip.

37. Install the timing belt to the left-side camshaft by performing the following procedures:

a. Rotate the left-side camshaft pulley to align the timing belt installation mark with the camshaft sprocket's timing mark and slide the belt onto the camshaft timing sprocket.

b. Using the Camshaft Holding tool 09960-10010, or equivalent, slightly turn the left-side camshaft sprocket counterclockwise to place tension on the timing belt between the crankshaft sprocket and the camshaft sprocket.

38. Rotate the right-side camshaft pulley to align the timing belt installation mark with the camshaft sprocket's timing mark and slide the belt onto the camshaft timing sprocket.

39. Using a vertical press, slowly press the pushrod into the housing using 200–2205 lbs. (981–9807 N) until the holes align, then, install a 1.27mm Allen® wrench to secure the pushrod and release the press. Install the dust boot on the tensioner housing.

40. Install the timing belt tensioner and torque the bolts to 19 ft. lbs. (26 Nm).

41. Using a pair of pliers, remove the Allen® wrench from the tensioner housing.

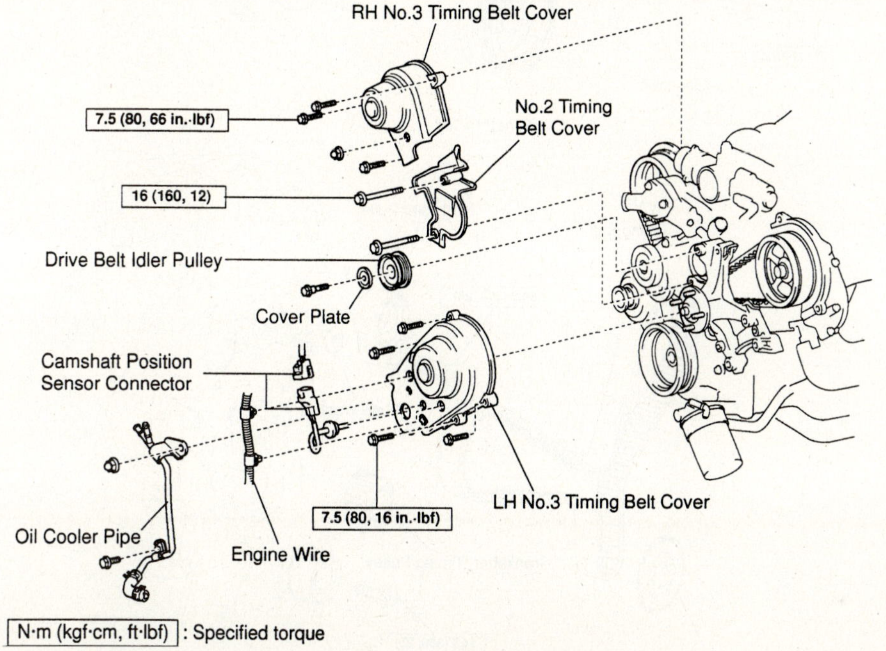

Exploded view of upper timing belt covers–Toyota 4.7L (2UZ-FE) engine—4Runner

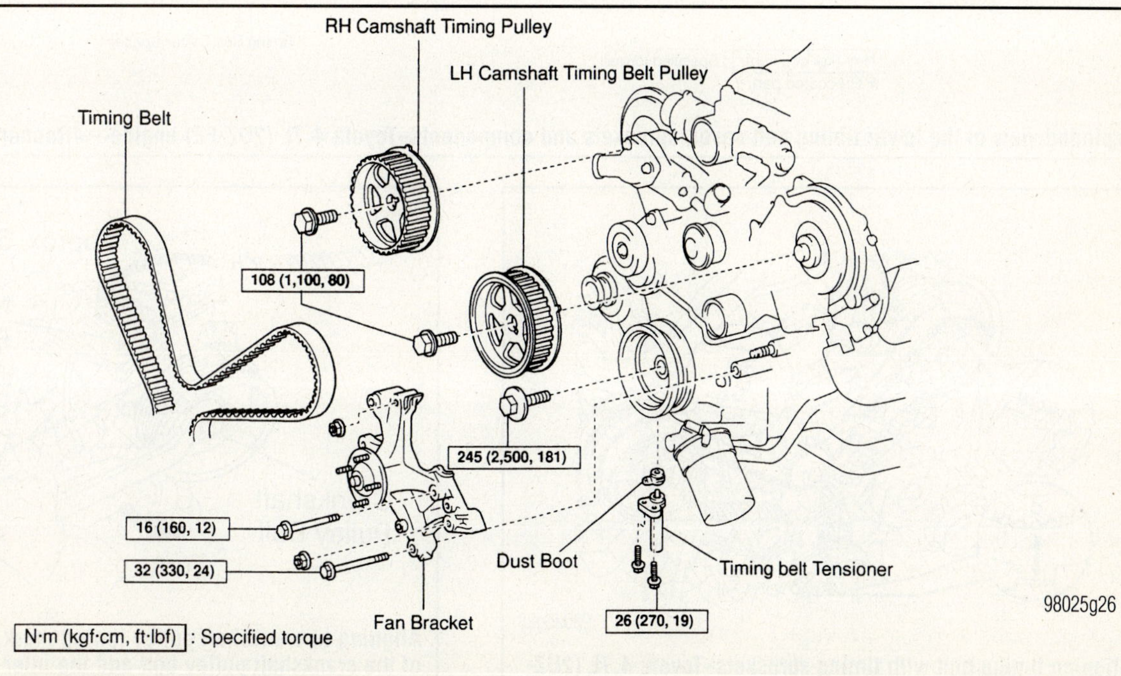

Exploded view of the upper timing belt sprockets and components–Toyota 4.7L (2UZ-FE) engine—4Runner

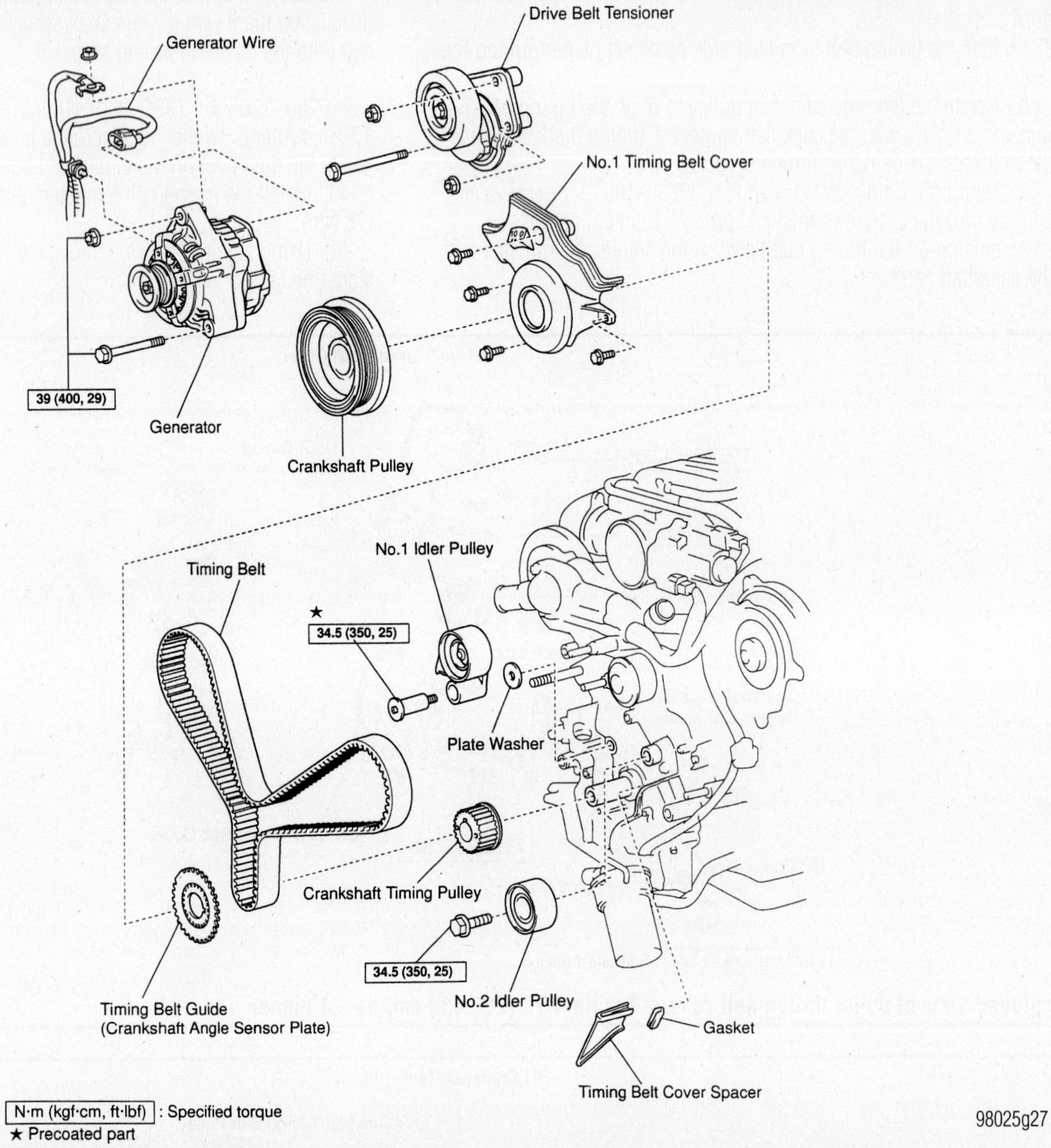

Exploded view of the lower timing belt cover, sprockets and components–Toyota 4.7L (2UZ-FE) engine—4Runner

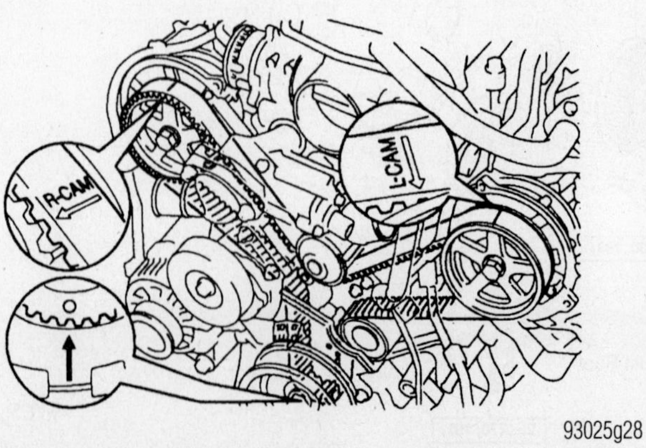

Aligning timing belt with timing sprockets–Toyota 4.7L (2UZ-FE) engine—4Runner

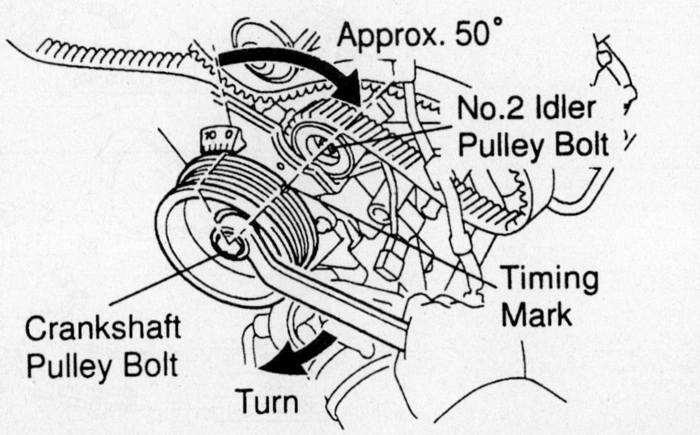

Aligning of crankshaft pulley timing mark with the center line of the crankshaft pulley bolt and the idler pulley bolt–Toyota 4.7L (2UZ-FE) engine—4Runner

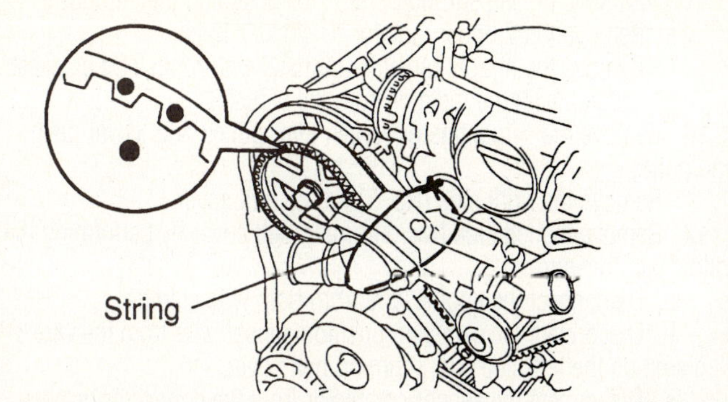

93025g30

Securing the timing belt with string and matchmarking the camshaft with the timing belt–Toyota 4.7L (2UZ-FE) engine—4Runner

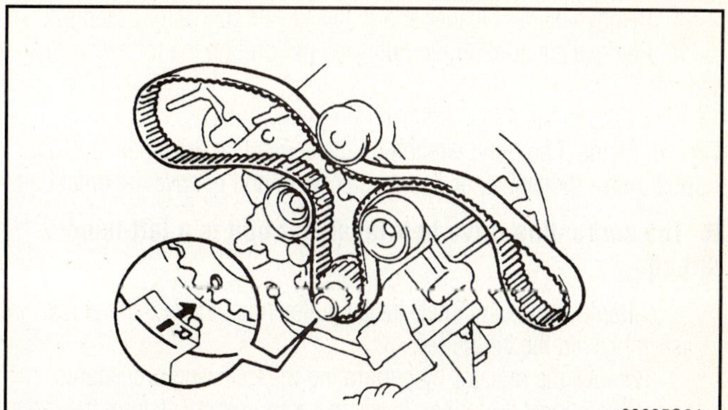

93025G31

Installing the timing belt on the crankshaft sprocket–Toyota 4.7L (2UZ-FE) engine—4Runner

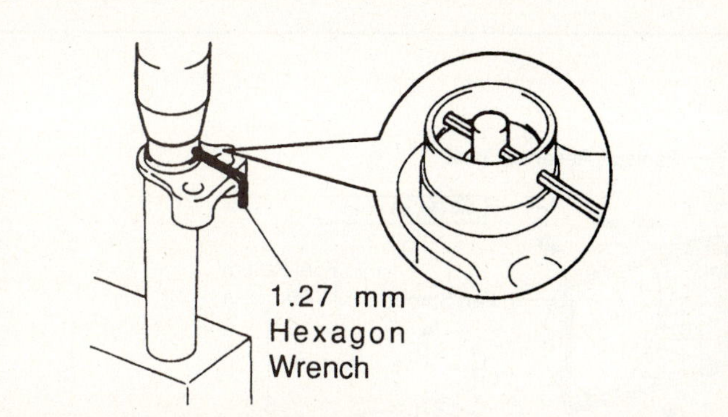

93025G32

Securing the timing belt tensioner pushrod–Toyota 4.7L (2UZ-FE) engine—4Runner

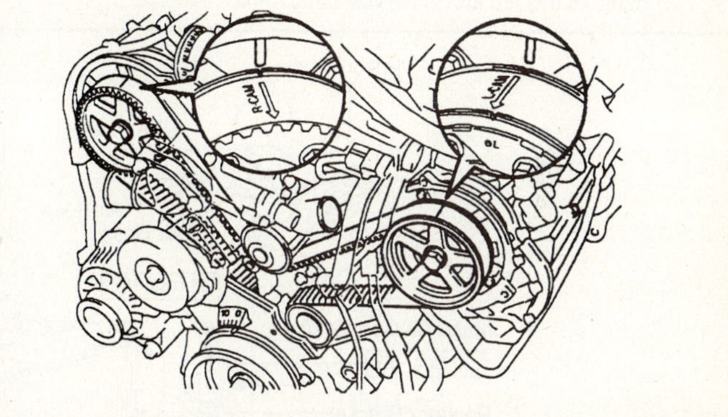

93025G33

Checking the TDC alignment marks after rotating the crankshaft 2 revolutions–Toyota 4.7L (2UZ-FE) engine—4Runner

42. Check the valve timing by performing the following procedure:
   a. Temporarily install the crankshaft pulley bolt.
   b. Slowly, rotate the crankshaft pulley 2 revolutions (CLOCKWISE) and realign the TDC marks.

■ **If the pulley/sprocket timing marks do not realign, remove the timing belt and reinstall it.**

43. Using the Crankshaft Pulley Holding tool 09213-70010, Bolt tool 90105-08076 and Companion Flange Holding tool 09330-00021, or equivalent, torque the crankshaft pulley bolt to 181 ft. lbs. (245 Nm).
44. Install the cooling fan bracket and torque the 12mm (head size) bolt to 12 ft. lbs. (16 Nm) and the 14mm (head size) bolt to 24 ft. lbs. (32 Nm).
45. Install the air conditioning compressor.
46. Install the middle (No. 2) timing belt cover and torque the bolts to 12 ft. lbs. (16 Nm).
47. Install the upper right-side (No. 3) timing belt cover and torque the bolts to 66 inch lbs. (7.5 Nm).
48. Install the upper left-side (No. 3) timing belt cover by performing the following procedures:
   a. Install the oil cooler tube and bolt.
   b. Feed the Camshaft Position Sensor (CPS) through the left-side (No. 3) timing belt cover hole.
   c. Install the left-side (No. 3) timing belt cover and torque the bolts to 66 inch lbs. (7.5 Nm).
   d. Install the wire grommet to the left-side (No. 3) timing belt cover.
   e. Install the sensor connector to the connector bracket and connect the sensor connector.
   f. Install the sensor wire and the engine wire to the clamps on the left-side (No. 3) timing belt cover.
49. Install the drive belt idler pulley and cover plate; then, torque the pulley bolt to 27 ft. lbs. (37 Nm).
50. To complete the installation, reverse the removal procedures.
51. Refill the cooling system and connect the negative battery cable.

Land Cruiser, Sequoia and Tundra

✳✳ CAUTION

These engines are an interference type. If the timing belt has failed, valve-to-piston damage may have occurred. A compression check should always be performed to check for engine damage.

1. Turn the ignition to the LOCK position. Disconnect the negative and positive battery cables. Wait at least 90 seconds for the air bag system to deactivate.
2. Raise and safely support the vehicle.
3. Remove the oil pan protector and the engine under cover.
4. Drain the cooling system and store the coolant for refilling purposes.
5. Lower the vehicle and remove the battery clamp cover.
6. From the top of the engine, remove the fuel return hose, the engine cover nuts/bolts and the cover.

7. Remove the air cleaner and the intake air connector assembly.
8. Remove the cooling fan pulley by performing the following procedures:
   a. Loosen the 4 fan clutch-to-fan pulley nuts.
   b. Using a box-end wrench on the serpentine drive belt tensioner bolt, rotate the tensioner counterclockwise and remove the drive belt.

■ **The serpentine drive belt tensioner bolt is a left-hand thread.**

   c. Remove the fan clutch-to-fan pulley nuts, the fan, the clutch assembly and the fan pulley.
9. Remove the radiator by performing the following procedures:
   a. Disconnect the upper, lower and reservoir hoses from the radiator.
   b. Disconnect and plug the automatic transmission oil cooler at the radiator. Disconnect the automatic transmission oil cooler hoses from the fan shroud clamp.
   c. Remove the radiator reservoir tank.
   d. Remove the fan shroud-to-radiator bolts and the shroud.
   e. Remove the 2 upper radiator-to-chassis nuts.
   f. Remove the middle radiator-to-chassis nut/bolts and brackets.
   g. Carefully, lift the radiator from the vehicle.
10. Remove the serpentine drive belt idler pulley bolt, cover plate and pulley.
11. Remove the right side (No. 3) timing belt cover.
12. Remove the left side (No. 3) timing belt cover by performing the following procedures:
   a. Disconnect the engine wire from both wire clamps.
   b. Disconnect the camshaft position sensor wire from the wire clamp on the left-side (No.3) timing belt cover.
   c. Disconnect the sensor connector from the connector bracket.
   d. Disconnect the sensor connector.
   e. Remove the wire grommet from the left-side (No. 3) timing belt cover.
   f. Remove the oil cooler tube bolts and tube.
13. Remove the middle (No. 2) timing belt cover bolts and cover.

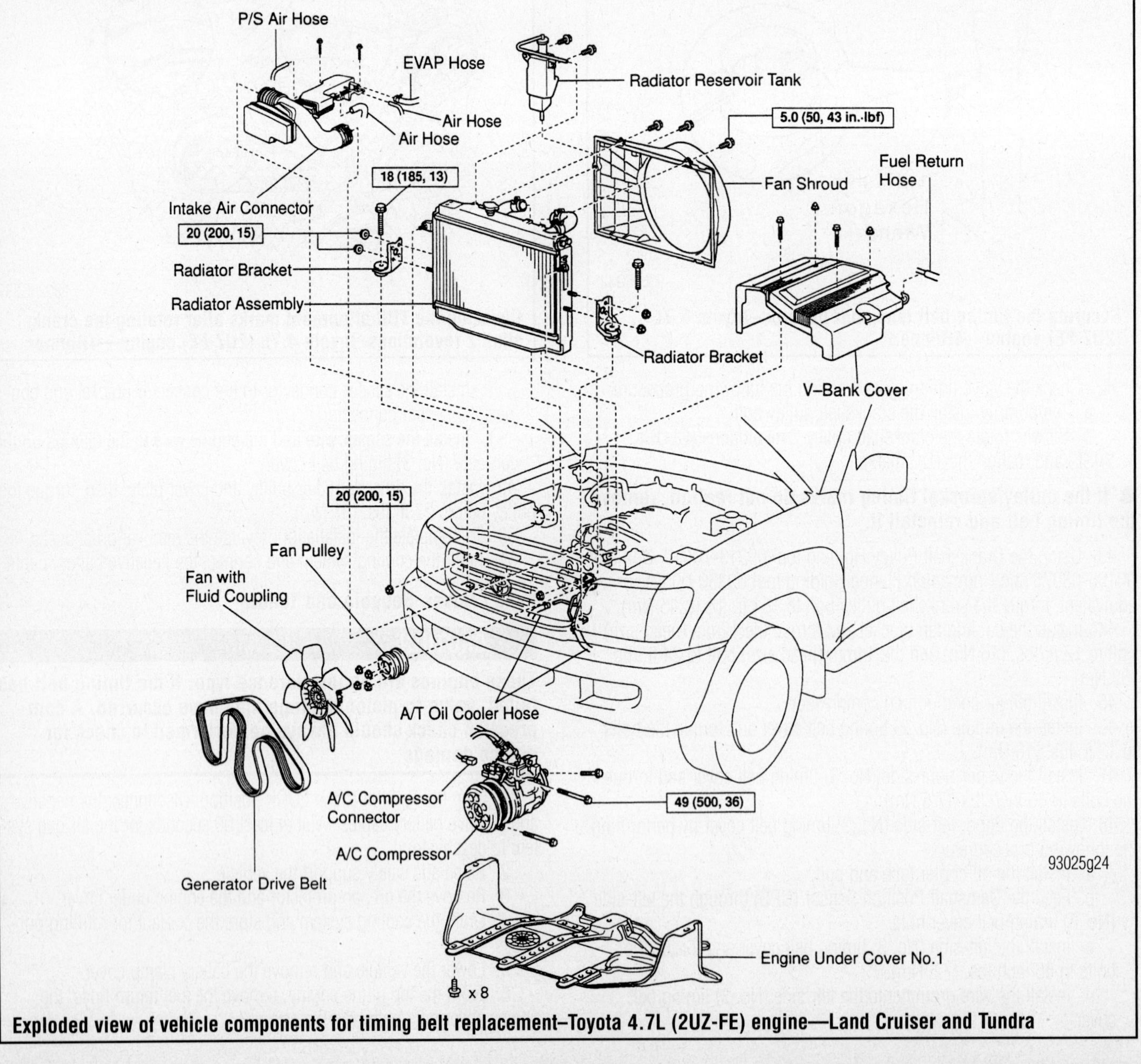

Exploded view of vehicle components for timing belt replacement–Toyota 4.7L (2UZ-FE) engine—Land Cruiser and Tundra

14. Remove the cooling fan bracket nuts/bolts and bracket.

■ **If reusing the timing belt, make sure that there are 3 installation marks on the belt; if there are none, install them.**

15. Using the Crankshaft Pulley Holding tool 09213-70010, Bolt tool 90105-08076 and Companion Flange Holding tool 09330-00021, or equivalent, loosen the crankshaft pulley bolt.

16. Position the No. 1 cylinder to approximately 50 degrees After Top Dead Center (ATDC) of the compression stroke by performing the following procedures:

a. Rotate the crankshaft pulley (CLOCKWISE) to align its groove with the timing mark "0" on the lower (No. 1) timing belt cover.

b. Check that the camshaft sprocket timing marks are aligned with the rear timing belt plate marks; if not, rotate the crankshaft 1 revolution (360 degrees).

c. Rotate the crankshaft pulley approximately 50 degrees (CLOCKWISE) and align the crankshaft pulley timing mark between the centers of the crankshaft pulley bolt and the idler pulley bolt.

✱✱ WARNING

If the timing belt is disengaged, having the crankshaft pulley in the wrong angle can cause the valve to come into contact with the piston when removing the camshaft pulley.

17. Remove the crankshaft pulley bolt.

■ **If reusing the timing belt and the installation marks have disappeared, place new installation marks on the timing belt to match the camshaft timing sprocket marks.**

■ **To avoid meshing the timing sprocket and the timing belt, secure one with a string; then, place matchmarks on the timing belt and the right-side camshaft timing sprocket.**

18. Remove the timing belt tensioner bolts and the tensioner.

19. Using the Camshaft Holding tool 09960-10010, or equivalent, slightly turn the left-side camshaft sprocket clockwise to loosen the ten-

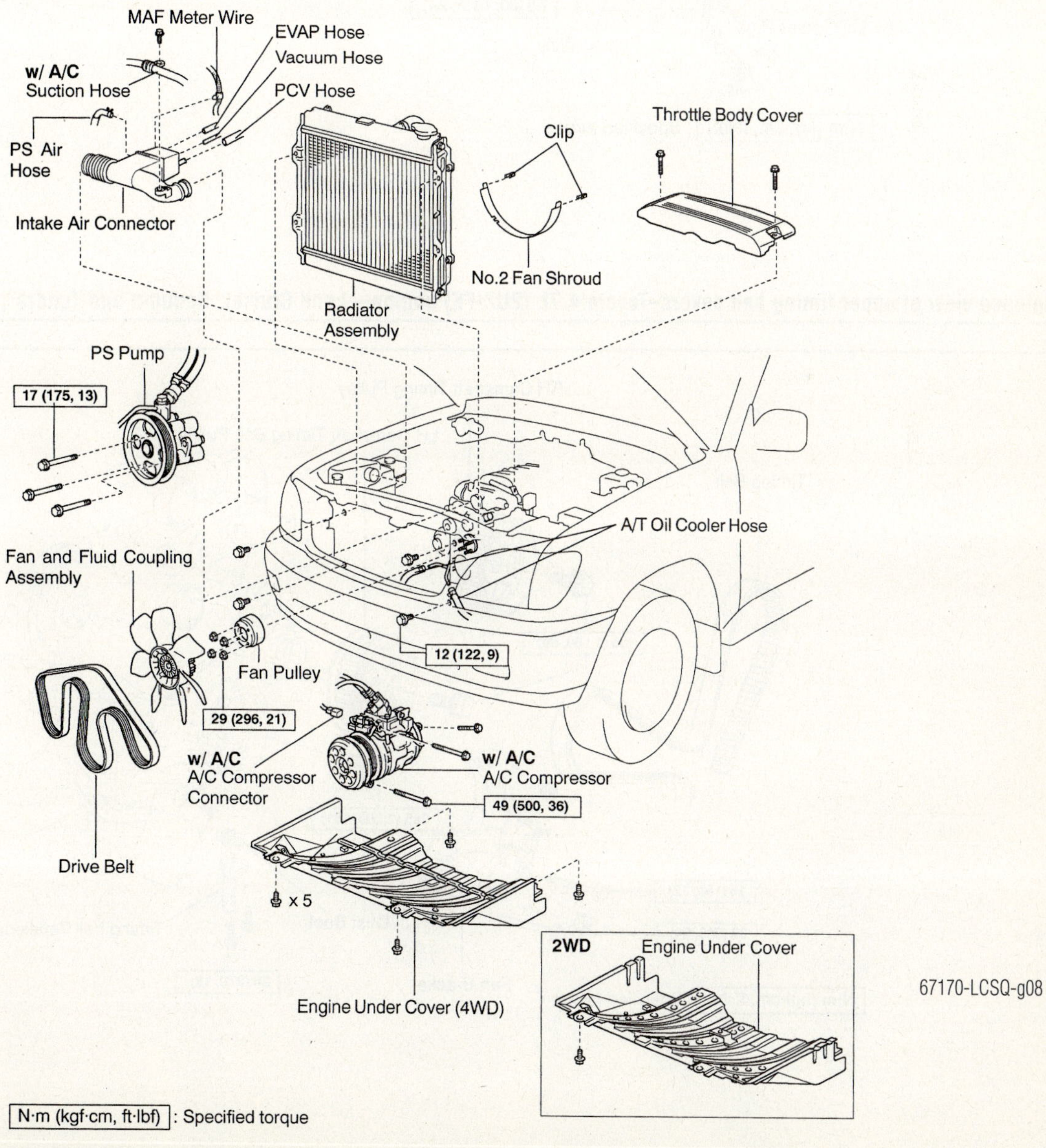

Exploded view of vehicle components for timing belt replacement–Toyota 4.7L (2UZ-FE) engine—Sequoia

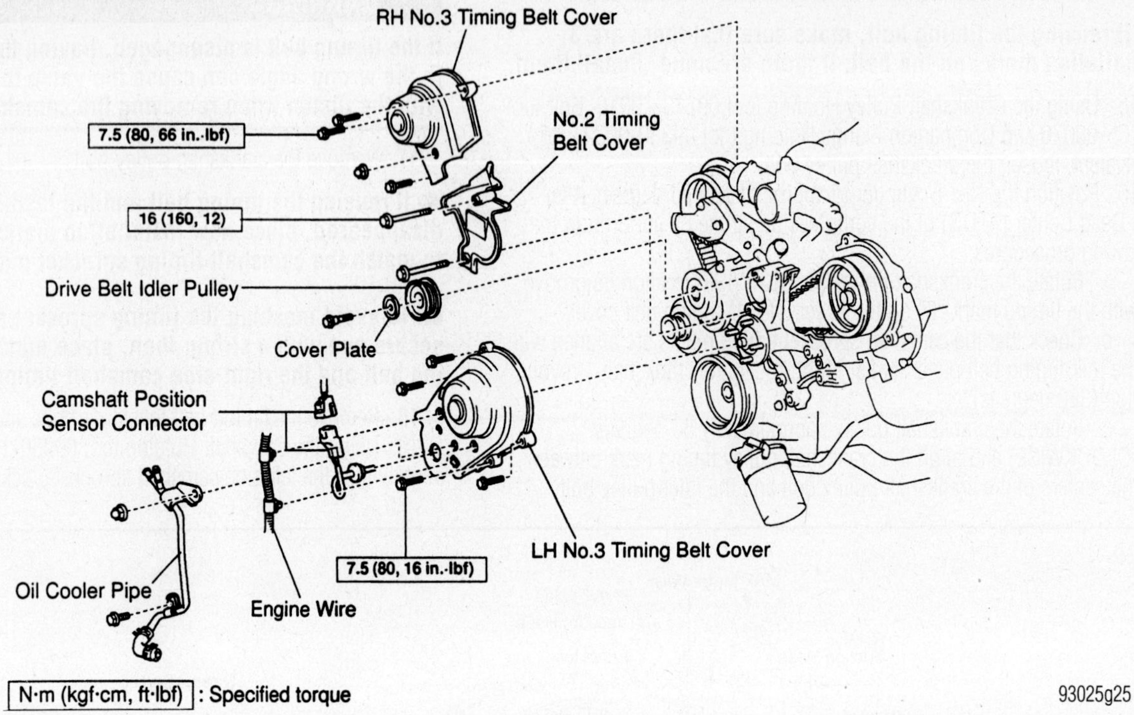

Exploded view of upper timing belt covers–Toyota 4.7L (2UZ-FE) engine—Land Cruiser, Sequoia and Tundra

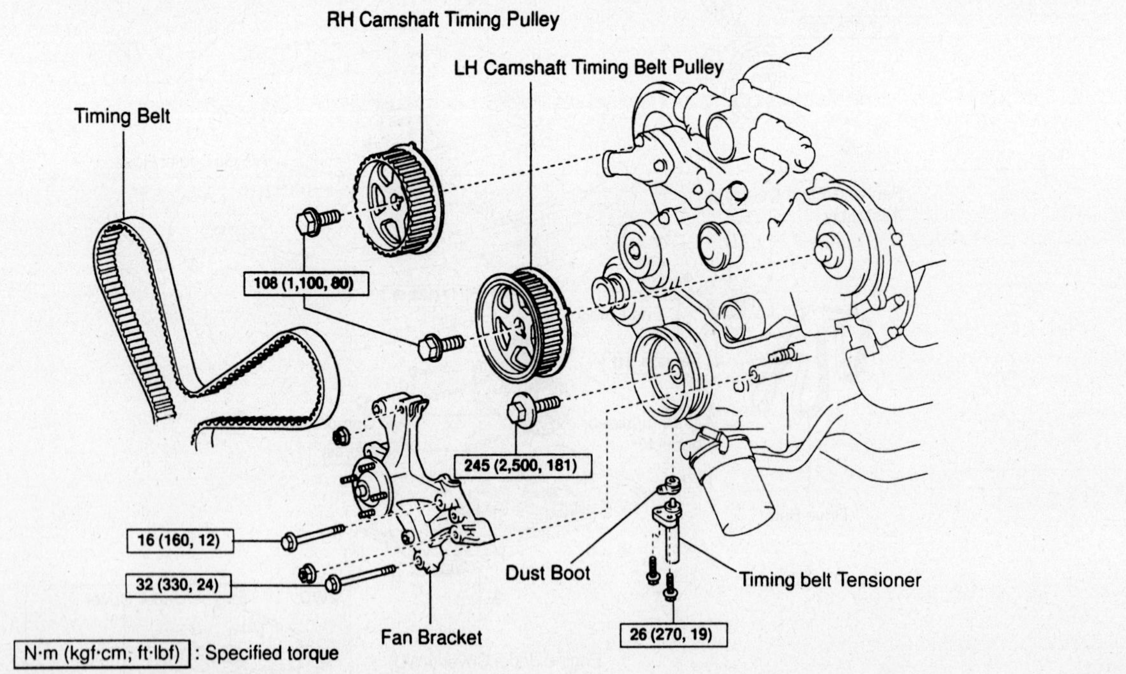

Exploded view of upper timing sprockets and components–Toyota 4.7L (2UZ-FE) engine—Land Cruiser, Sequoia and Tundra

sion spring. Then, disconnect the timing belt from the camshaft sprockets.

20. Remove the alternator by performing the following procedures:
   a. Disconnect the electrical connector from the alternator.
   b. Remove the rubber cap/nut and disconnect the battery wire from the alternator.
   c. Disconnect the wire clamp from the alternator cord clip.
   d. Remove the alternator-to-engine nuts/bolts and the alternator.

21. Remove the serpentine drive belt tensioner nuts/bolts and the tensioner.

22. Using the Crankshaft Puller Assembly tool 09950-50012, or equivalent, press the crankshaft pulley from the crankshaft.

**** WARNING**

DO NOT rotate the crankshaft pulley.

23. Remove the lower (No. 1) timing belt cover bolts and the cover.

24. Remove the timing belt guide, spacer and the timing belt.

To install:

■ **With the timing belt removed, this is a perfect opportunity to inspect and/or replace the water pump.**

25. Inspect the timing belt tensioner by performing the following procedures:
   a. Inspect the seal for leakage; if leakage is suspected, replace the tensioner.

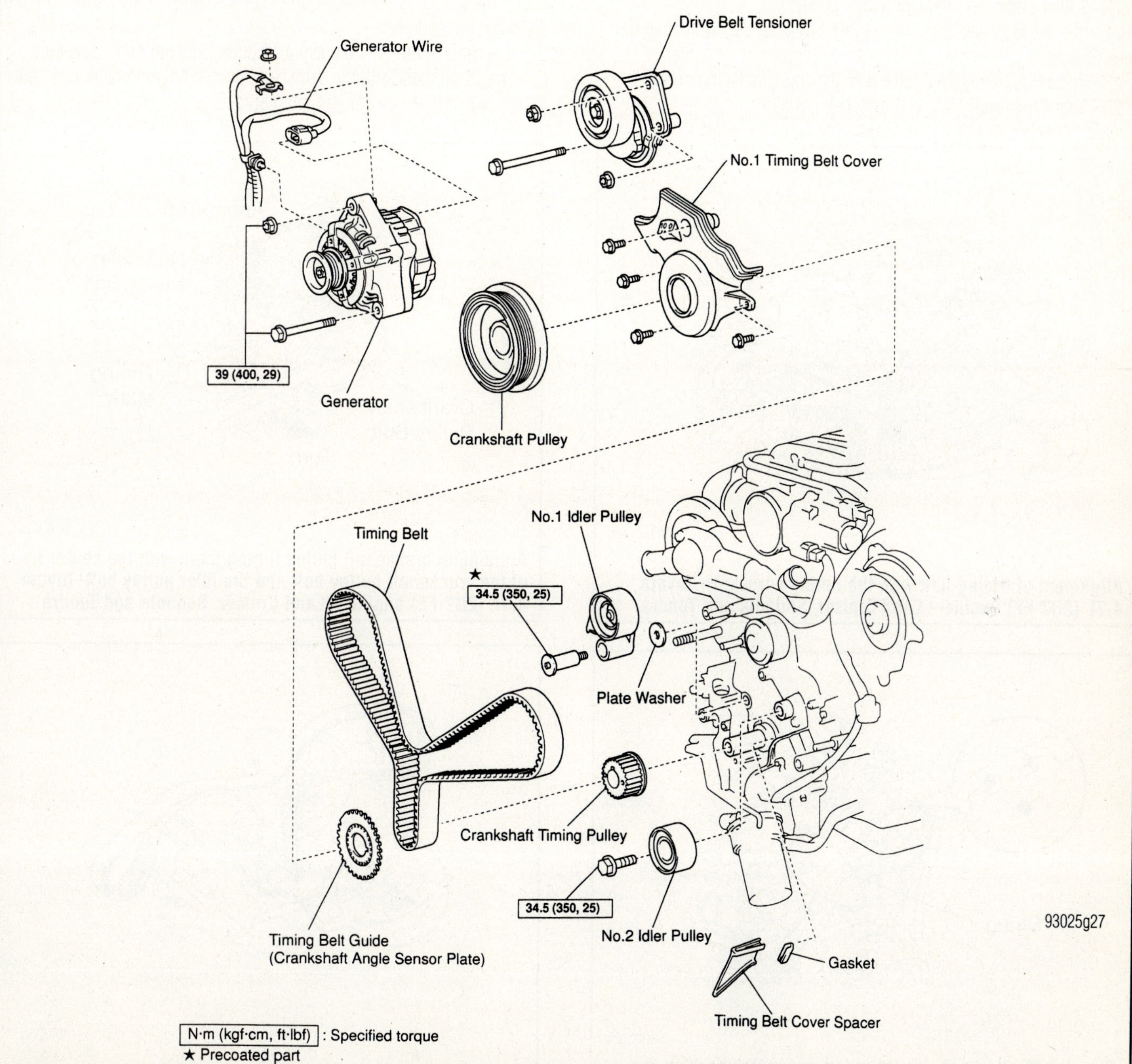

Exploded view of lower timing belt cover, sprockets and components–Toyota 4.7L (2UZ-FE) engine—Land Cruiser, Sequoia and Tundra

b. Using both hands to hold the tensioner facing upward, strongly press the pushrod against a solid surface. If the pushrod moves, replace the tensioner.

✻✻ WARNING

Never hold the tensioner with the pushrod facing downward.

c. Measure the pushrod protrusion from the housing end, it should be 0.413–0.453 in. (10.5–11.5mm). If the protrusion is not as specified, replace the tensioner.

26. Temporarily install the timing belt by performing the following procedures:

d. Align the timing belt's installation mark with the crankshaft timing sprocket.

e. Install the timing belt on the crankshaft timing sprocket, the No. 1 idler pulley and the No. 2 idler pulley.

27. Install the gasket to the timing belt cover spacer and install the cover spacer.

28. Install the timing belt guide with the cup side facing outward.

29. Install the lower (No. 1) timing belt cover.

30. Install the crankshaft pulley by performing the following procedures:

a. Align the crankshaft pulley with the crankshaft key.

b. Using the Crankshaft Installer tool 09223-46011, or equivalent, and a hammer, tap the crankshaft pulley into position.

31. Install the serpentine drive belt tensioner and torque the tensioner-to-engine bolts to 12 ft. lbs. (16 Nm).

■ To install the serpentine drive belt tensioner, use a bolt 4.18 in. (106mm) in length.

32. Check that the crankshaft pulley's timing mark is aligned with the centers of the idler pulley and crankshaft pulley bolts.

33. Install the alternator and torque the alternator-to-engine nuts/bolts to 29 ft. lbs. (39 Nm). Connect the alternator's electrical connectors and clip.

34. Install the timing belt to the left-side camshaft by performing the following procedures:

a. Rotate the left-side camshaft pulley to align the timing belt installation mark with the camshaft sprocket's timing mark and slide the belt onto the camshaft timing sprocket.

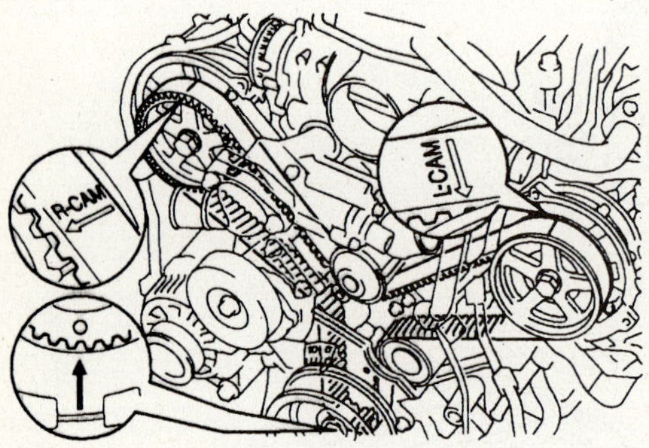

93025g28

Alignment of timing belt with the timing sprockets–Toyota 4.7L (2UZ-FE) engine—Land Cruiser, Sequoia and Tundra

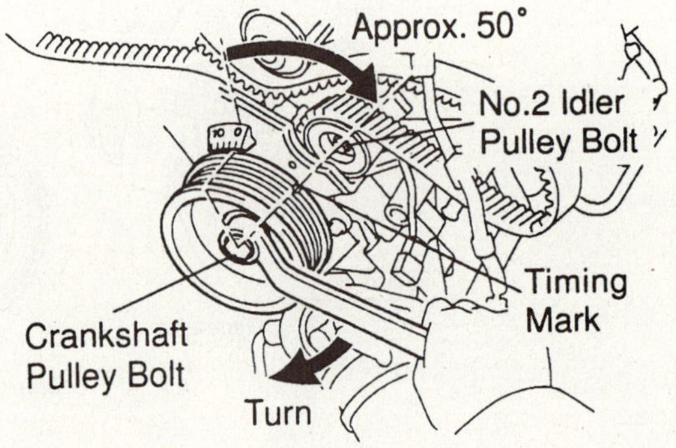

93025g29

Aligning of crankshaft pulley timing mark with the center line of the crankshaft pulley bolt and the idler pulley bolt–Toyota 4.7L (2UZ-FE) engine—Land Cruiser, Sequoia and Tundra

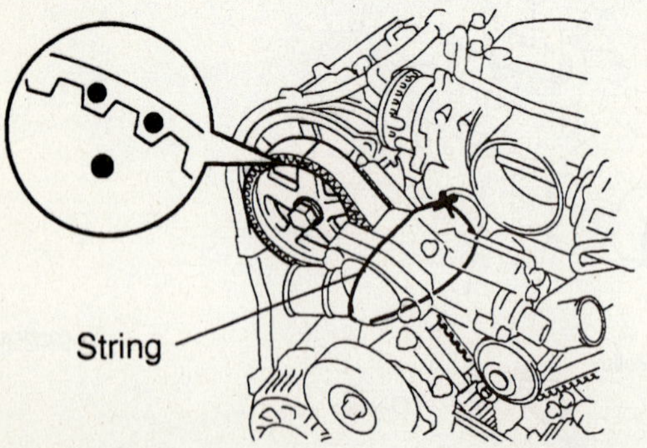

93025g30

Securing the timing belt with string and matchmarking the camshaft with the timing belt–Toyota 4.7L (2UZ-FE) engine—Land Cruiser, Sequoia and Tundra

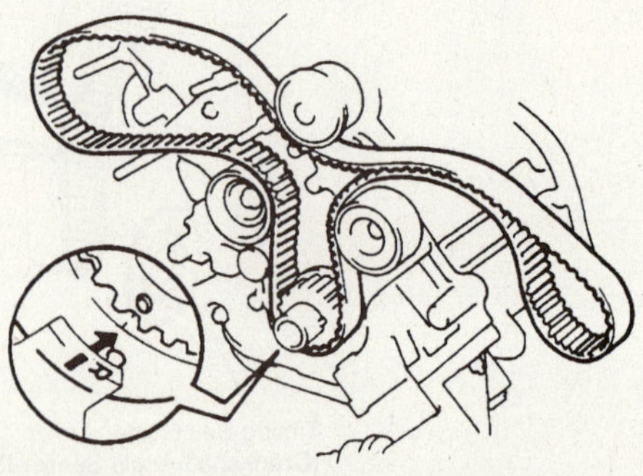

93025g31

Installing the timing belt on the crankshaft sprocket–Toyota 4.7L (2UZ-FE) engine—Land Cruiser, Sequoia and Tundra

b. Using the Camshaft Holding tool 09960-10010, or equivalent, slightly turn the left-side camshaft sprocket counterclockwise to place tension on the timing belt between the crankshaft sprocket and the camshaft sprocket.

35. Rotate the right-side camshaft pulley to align the timing belt installation mark with the camshaft sprocket's timing mark and slide the belt onto the camshaft timing sprocket.

36. Using a vertical press, slowly press the pushrod into the housing using 200–2205 lbs. (981–9807 N) until the holes align, then, install a 1.27mm Allen® wrench to secure the pushrod and release the press. Install the dust boot on the tensioner housing.

37. Install the timing belt tensioner and torque the bolts to 19 ft. lbs. (26 Nm).

38. Using a pair of pliers, remove the Allen® wrench from the tensioner housing.

39. Check the valve timing by performing the following procedure:

a. Temporarily install the crankshaft pulley bolt.

b. Slowly, rotate the crankshaft pulley 2 revolutions (CLOCKWISE) and realign the TDC marks.

■ **If the pulley/sprocket timing marks do not realign, remove the timing belt and reinstall it.**

40. Using the Crankshaft Pulley Holding tool 09213-70010, Bolt tool 90105-08076 and Companion Flange Holding tool 09330-00021, or equivalent, torque the crankshaft pulley bolt to 181 ft. lbs. (245 Nm).

41. Install the cooling fan bracket and torque the 12mm (head size) bolt to 12 ft. lbs. (16 Nm) and the 14mm (head size) bolt to 24 ft. lbs. (32 Nm).

42. Install the air conditioning compressor.

43. Install the middle (No. 2) timing belt cover and torque the bolts to 12 ft. lbs. (16 Nm).

44. Install the upper right-side (No. 3) timing belt cover and torque the bolts to 66 inch lbs. (7.5 Nm).

45. Install the upper left-side (No. 3) timing belt cover by performing the following procedures:

a. Install the oil cooler tube and bolt.

b. Feed the Camshaft Position Sensor (CPS) through the left-side (No. 3) timing belt cover hole.

c. Install the left-side (No. 3) timing belt cover and torque the bolts to 66 inch lbs. (7.5 Nm).

d. Install the wire grommet to the left-side (No. 3) timing belt cover.

e. Install the sensor connector to the connector bracket and connect the sensor connector.

f. Install the sensor wire and the engine wire to the clamps on the left-side (No. 3) timing belt cover.

46. Install the drive belt idler pulley and cover plate; then, torque the pulley bolt to 27 ft. lbs. (37 Nm).

47. To complete the installation, reverse the removal procedures.

48. Refill the cooling system and connect the negative battery cable.

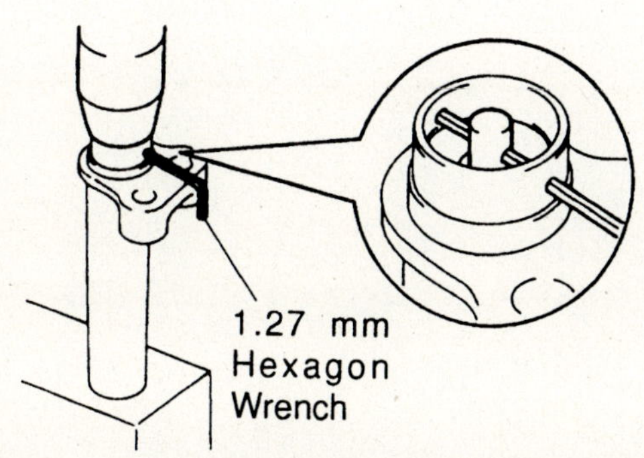

93025g32

Securing the timing belt tensioner pushrod–Toyota 4.7L (2UZ-FE) engine—Land Cruiser, Sequoia and Tundra

93025g33

Checking the TDC alignment marks after rotating the crankshaft 2 revolutions–Toyota 4.7L (2UZ-FE) engine—Land Cruiser, Sequoia and Tundra

Volkswagen

1.5L AND 1.6L (CS, 1V) DIESEL ENGINES

YEAR, MODEL & ENGINE APPLICATION

1980-81 Volkswagen Dasher, 1985-87 Golf, 1982-92 Jetta, 1982-87 Quantum, 1981-84 Rabbit, 1982-84 Vanagon .... 1.5L & 1.6L Diesel Codes 1V, CS

REPLACEMENT INTERVALS

Volkswagen does not provide a replacement interval. Standard industry practice is to replace timing belts every 60,000 miles.

LABOR TIMES

Dasher	
Replacement	1.6 hrs
Severe Service	2.1 hrs
Golf, Jetta, Quantum & Rabbit	
Replacement	2.7 hrs
Severe Service	3.2 hrs
Vanagon	
Replacement	3.2 hrs
Severe Service	3.6 hrs

8807VWg01

Some special tools are required. A flat bar, VW tool 2065A, or equivalent, is used to secure the camshaft in position. A pin, VW tool 2064, or equivalent, is used to fix the pump position while the timing belt is removed. The camshaft and pump work against spring pressure and will move out of position when the timing belt is removed. It is not difficult to find substitutes but do not remove the timing belt without these tools.

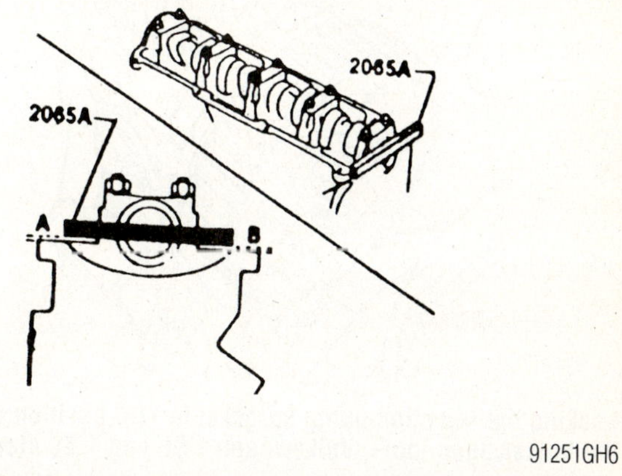

Using the special tool to retain the camshaft in position—Volkswagen 1.5L and 1.6L Code CS, 1V diesel engines

✻✻ CAUTION

These engines are an interference type. If the timing belt has failed, valve-to-piston damage may have occurred. A compression check should always be performed to check for engine damage.

■ **The timing belt is designed to last for more than 60,000 miles and does not normally require tension adjustments. If the belt is removed or replaced, the basic valve timing must be checked and the belt retensioned.**

1. Disconnect the negative battery cable.
2. Loosen the alternator mounting bolts.
3. Pivot the alternator and slip the drive belt off the pulleys.
4. Unscrew the timing cover retaining nuts and remove the cover.

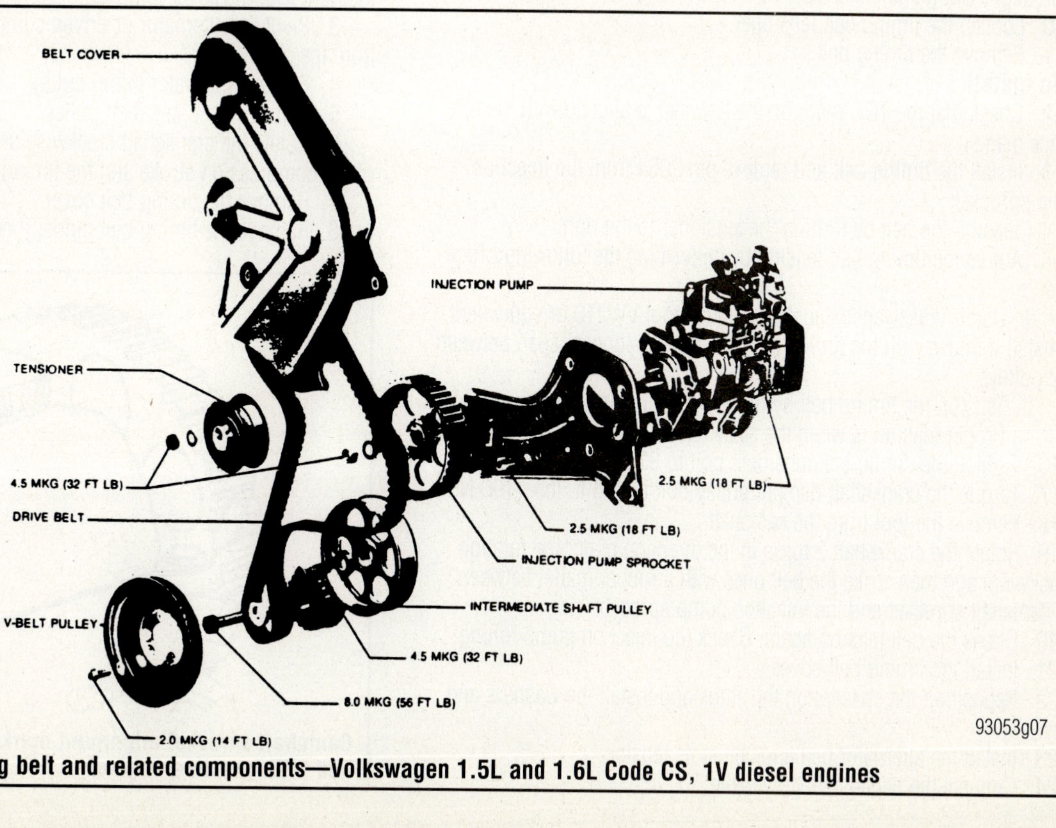

Exploded view of the timing belt and related components—Volkswagen 1.5L and 1.6L Code CS, 1V diesel engines

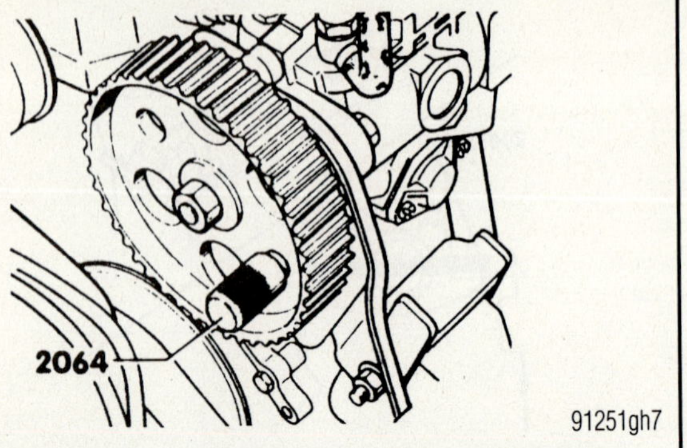

Locking the injection pump sprocket in TDC position using the Volkswagen tool—Volkswagen 1.5L and 1.6L diesel engine

5. Remove the rocker arm cover.
6. Rotate the crankshaft so that the No. 1 cylinder is at Top Dead Center (TDC) of its compression stroke.
7. Using Volkswagen tool 2065A, or equivalent, retain the camshaft in this position by performing the following procedure:
   a. Turn the camshaft until one end of the tool touches the cylinder head.
   b. Measure the gap at the other end of the tool with a feeler gauge.
   c. Take 1/2 of the measurement and insert a feeler gauge of this thickness between the tool and the cylinder head. Turn the camshaft so that the tool rests on the feeler gauge.
   d. Insert a second feeler gauge of the same thickness between the other end of the tool and the cylinder head.
8. Using Volkswagen tool 2064 or equivalent, lock the injection pump sprocket in position.
9. Check that the marks on the sprocket, bracket and pump body are in alignment (engine at TDC).
10. Loosen the timing belt tensioner.
11. Remove the timing belt.

To install:

12. Check that the TDC mark on the flywheel is aligned with the reference marks.
13. Install the timing belt and remove pin 2064 from the injection pump sprocket.
14. Tension the belt by turning the tensioner to the right.
15. Adjust the timing belt tension by performing the following procedure:
   a. Using Volkswagen Tensioner Gauge tool VW210 or equivalent, install it midway on the timing belt between the longest span between 2 pulleys.
   b. Tension the timing belt by turning the tensioner to the right.
   c. Proper tension is when the scale reads 12–13.
16. Tighten the timing belt tensioner bolt to 33 ft. lbs. (45 Nm).
17. Torque the crankshaft damper pulley bolt to 145 ft. lbs. (196 Nm)
18. Remove the tool from the camshaft.
19. Rotate the crankshaft 2 turns in the direction of engine rotation (clockwise) and then strike the belt once with a rubber mallet between the camshaft sprocket and the injection pump sprocket.
20. Check the belt tension again. Check the injection pump timing.
21. Install the timing belt cover.
22. Reposition the spacers on the studs and install the washers and nuts.
23. Install the alternator belt and adjust its tension.
24. Connect the negative battery cable.

1.5L. 1.6L AND 1.7L GAS ENGINES

YEAR, MODEL & ENGINE APPLICATION	
1980 Volkswagen Rabbit	1.5L Code N/A
1980 Dasher, 1980-81 Jetta, 1980 Scirocco	1.6L Code N/A
1981-84 Jetta, 1981-84 Quantum, 1981 Scirocco, 1981-84 Rabbit	1.7L Code N/A

REPLACEMENT INTERVALS

Volkswagen does not provide a replacement interval. Standard industry practice is to replace timing belts every 60,000 miles.

LABOR TIMES	
Dasher	
Replacement	1.6 hrs
Severe Service	2.1 hrs
Jetta, Quantum, Rabbit & Scirocco	
Replacement	2.1 hrs
Severe Service	2.6 hrs

8807VWg02

**** CAUTION**

These engines are an interference type. If the timing belt has failed, valve-to-piston damage may have occurred. A compression check should always be performed to check for engine damage.

1. Disconnect the negative battery cable.
2. Loosen the alternator mounting bolts and if equipped, the power steering pump and air conditioner compressor bolts, if their drive belts interfere with the cover removal.
3. Pivot the alternator or driven component and slip the drive belt from the pulleys.
4. Remove the water pump pulley.
5. Remove the crankshaft pulley.
6. Rotate the crankshaft clockwise until the No. 1 cylinder is at TDC of the compression stroke and the timing marks are aligned.
7. Remove the timing belt cover.
8. Remove the timing belt guide, if equipped.

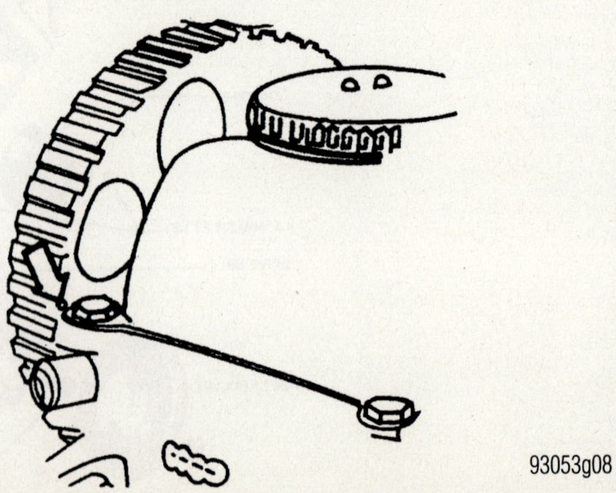

Camshaft sprocket alignment mark—Volkswagen 1.5L. 1.6L and 1.7L gasoline engines

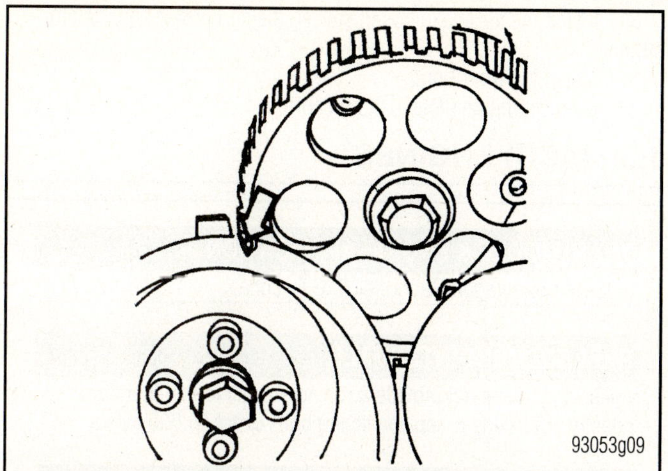

Crankshaft and intermediate sprocket alignment—Volkswagen 1.5L. 1.6L and 1.7L gasoline engines

9. Loosen the tensioner adjusting nut, move the tensioner away from the belt and remove the timing belt.

To install:

10. Install the timing belt counterclockwise starting at the crankshaft sprocket.
11. Ensure the timing belt is taut on the non-tension side.
12. Loosen the tensioner nut and turn the tensioner clockwise until the belt can be twisted 90 degrees.
13. Tighten the tensioner nut to 37 ft. lbs. (50 Nm).
14. Turn the engine 2 complete revolutions (clockwise) and align the timing marks. Recheck the belt tension and timing marks. Readjust as required.
15. Install the timing belt cover and water pump pulley.
16. Install the drive belts.
17. Connect the negative battery cable.

1.8L (AEB, ATW) TURBO ENGINES

YEAR, MODEL & ENGINE APPLICATION

1998-00 Volkswagen Passat	1.8L Codes AEB, ATW

REPLACEMENT INTERVALS

Volkswagen does not provide a replacement interval. Standard industry practice is to replace timing belts every 60,000 miles.

LABOR TIMES

1998-00	
Replacement	3.8 hrs
Severe Service	4.3 hrs
With water pump replacement add	0.3 hrs

8807VWg03

✻✻ CAUTION

These engines are an interference type. If the timing belt has failed, valve-to-piston damage may have occurred. A compression check should always be performed to check for engine damage.

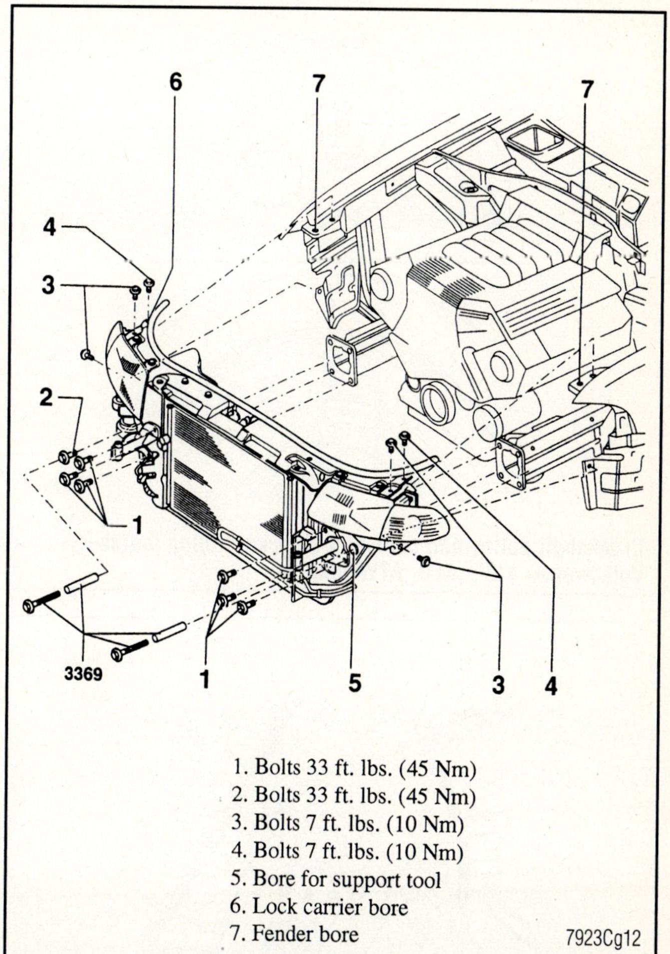

Moving the front support panel into the service position—Passat

■ **Do not turn the engine or camshaft with the camshaft drive belt removed. The pistons will contact the valves and cause internal engine damage.**

1. Disconnect the negative battery cable and wait 20 minutes for the air bag backup power to dissipate.
2. Remove the front bumper.
3. Remove the air intake hose between the air cleaner and radiator support.
4. Remove the radiator support panel bolt.
5. Install support guides 3369 through the holes in the panel.
6. Remove the radiator support panel bolts and move the support panel forward into the service position.
7. Remove the accessory drive belts.
8. Remove the accessory drive belt tensioner.
9. Rotate the crankshaft until the No. 1 cylinder is at TDC of the compression stroke.
10. Ensure the crankshaft and camshaft timing marks are aligned.
11. Remove the crankshaft pulley by holding the center crankshaft sprocket bolt with a socket and loosen the pulley bolts.
12. Remove the timing belt covers.
13. Loosen the locknut on the tensioner pulley and turn the tensioner counterclockwise to relieve the tension on the timing belt.
14. Remove the timing belt from the sprockets.

To install:

15. Ensure the crankshaft and camshaft timing marks are aligned.

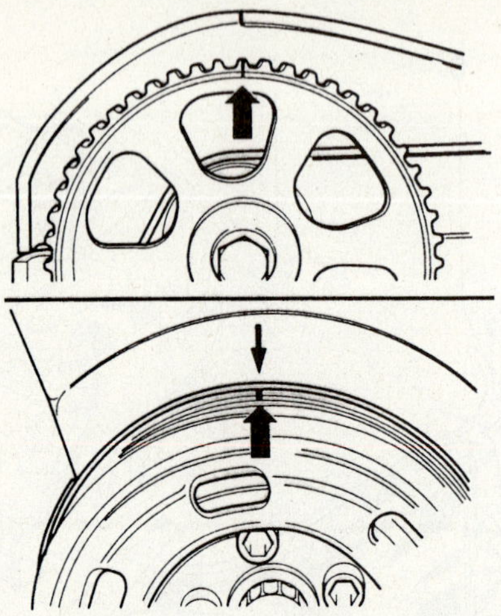

Crankshaft pulley and camshaft sprocket timing marks—Volkswagen 1.8L (AEB, ATW) engine

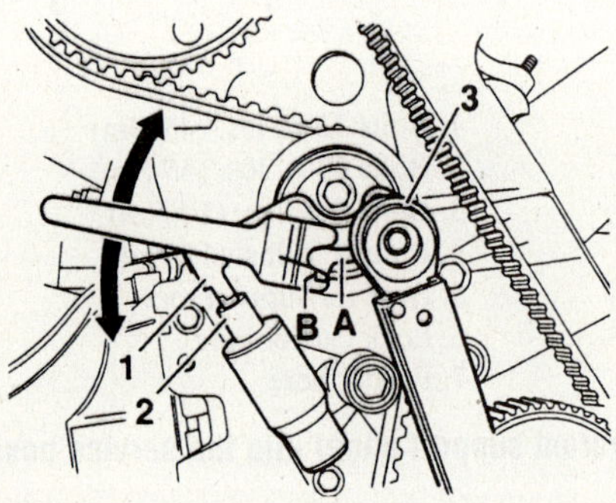

Timing belt tension adjustment—Volkswagen 1.8L (AEB, ATW) engine

16. Install the timing belt to the crankshaft and intermediate sprockets.
17. Install the lower timing belt cover.
18. Install the crankshaft pulley using 1 bolt.
19. Install the timing belt to the tensioner and camshaft sprocket.
20. Loosen the tensioner pulley bolt.
21. Turn the tensioner pulley counterclockwise using a tensioner wrench V159 until the piston is extended and the pushrod protrudes 0.40 in. (10mm).
22. Hold the tensioner and tighten the nut to 18 ft. lbs. (25 Nm).
23. Recheck the alignment of the timing marks and, if correct, turn the engine 2 full revolutions to return to TDC of No. 1 piston. Recheck belt tension and timing marks. Readjust as required.
24. Install the crankshaft pulley bolts and tighten to 18 ft. lbs. (25 Nm).
25. Install the timing belt covers.
26. Install the accessory drive belt tensioner.
27. Install the accessory drive belts.
28. Install the radiator support panel bolts and move the support panel rearward.
29. Remove support guides 3369.
30. Install the radiator support panel bolt.
31. Install the air intake hose between the air cleaner and radiator support.
32. Install the front bumper.
33. Connect the negative battery cable.

1.8L (ACC) ENGINES

YEAR, MODEL & ENGINE APPLICATION

1993-97 Volkswagen Cabrio, Golf, GTI, Jetta .. 1.8L Code ACC

REPLACEMENT INTERVALS

Volkswagen does not provide a replacement interval. Standard industry practice is to replace timing belts every 60,000 miles.

LABOR TIMES

Cabrio	
Replacement	2.6 hrs
Severe Service	3.1 hrs
Golf, GTI & Jetta	
Replacement	2.5 hrs
Severe Service	3.0 hrs
All Models	
With water pump replacement add	0.3 hrs

8807VWg04

✱✱ CAUTION

These engines are an interference type. If the timing belt has failed, valve-to-piston damage may have occurred. A compression check should always be performed to check for engine damage.

■ **Do not turn the engine or camshaft with the camshaft drive belt removed. The pistons will contact the valves and cause internal engine damage.**

1. Disconnect the negative battery cable.
2. Disable the air bag system, then wait 20 minutes for the system to totally discharge
3. Rotate the crankshaft until the crankshaft and camshaft timing marks are aligned.

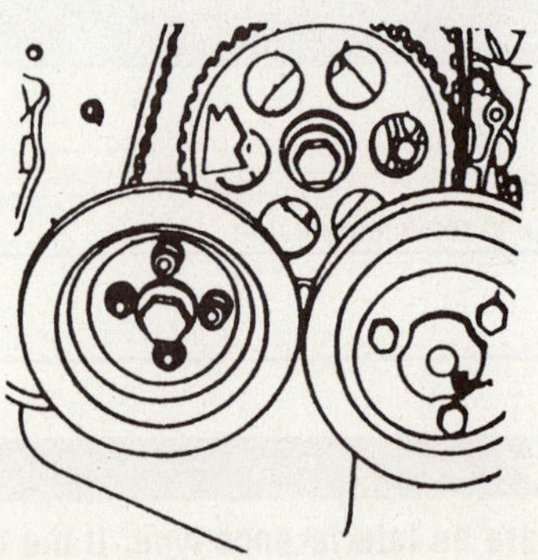

Timing marks on the crankshaft pulley and the intermediate shaft sprocket (arrow)—Volkswagen 1.8L (ACC) engine

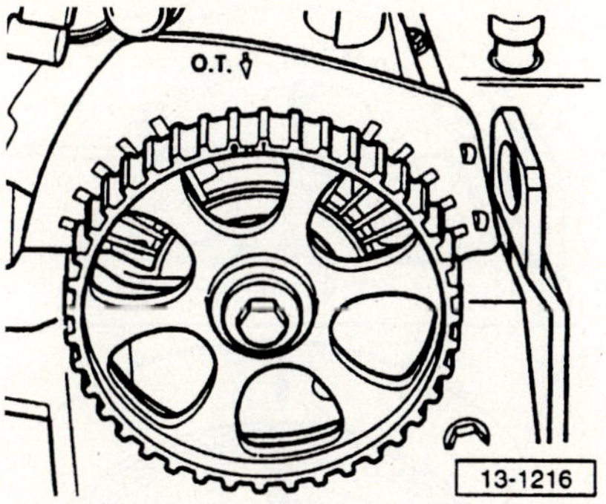

79235g93

Camshaft sprocket timing marks—Volkswagen 1.8L (ACC) engine

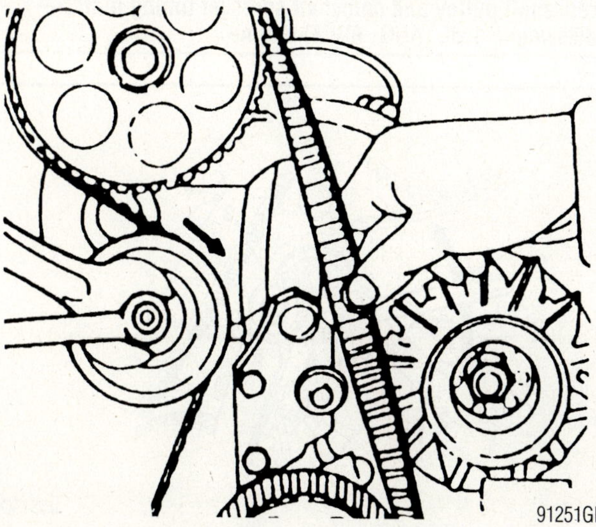

91251GI1

Using a suitable wrench, turn the tensioner pulley clockwise until the belt can just be twisted 90 degrees

4. Remove the accessory drive belts.
5. Remove the accessory drive belt tensioner, if equipped.
6. Remove the crankshaft pulley by holding the center crankshaft sprocket bolt with a socket and loosen the pulley bolts.
7. Remove the water pump pulley.
8. Remove the timing belt covers.
9. Remove the timing belt guide, if equipped.
10. Loosen the locknut on the tensioner pulley and turn the tensioner counterclockwise to relieve the tension on the timing belt.
11. Slide the timing belt from the sprockets.

To install:

12. Install the new timing belt and tension the belt so it can be twisted 90 degrees at the middle of its longest section, between the camshaft and intermediate sprockets.
13. Install the timing belt guide, if equipped.
14. Install the timing belt covers.
15. Install the water pump pulley.
16. Install the crankshaft pulley.
17. Install the accessory drive belt tensioner, if equipped.
18. Install the accessory drive belts.
19. Connect the negative battery cable.

1.8L (AUG, AWM) TURBO ENGINES

YEAR, MODEL & ENGINE APPLICATION	
2000-04 Volkswagen Passat	1.8L Codes AUG, AWM

REPLACEMENT INTERVALS

Volkswagen does not provide a replacement interval. Standard industry practice is to replace timing belts every 60,000 miles.

LABOR TIMES	
2000-04	
Replacement	3.8 hrs
Severe Service	4.3 hrs
With water pump replacement add	0.3 hrs

8807VWg05

✱✱ CAUTION

These engines are an interference type. If the timing belt has failed, valve-to-piston damage may have occurred. A compression check should always be performed to check for engine damage.

■ **Do not turn the engine or camshaft with the camshaft drive belt removed. The pistons will contact the valves and cause internal engine damage.**

1. Disconnect the negative battery cable and wait 20 minutes for the air bag backup power to dissipate.
2. Remove the front bumper.
3. Remove the air intake hose between the air cleaner and radiator support.
4. Remove the radiator support panel bolt.
5. Install support guides 3369 or 3411 through the holes in the panel.
6. Remove the radiator support panel bolts and move the support panel forward into the service position.
7. Remove the accessory drive belts.
8. Remove the accessory drive belt tensioner.
9. Rotate the crankshaft until the No. 1 cylinder is at TDC of the compression stroke.
10. Ensure the crankshaft and camshaft timing marks are aligned.
11. Remove the crankshaft pulley by holding the center crankshaft sprocket bolt with a socket and loosen the pulley bolts.
12. Remove the timing belt covers.
13. On early 2000 models, insert locking bolt T10092 into tensioner and tighten the bolt to compress the tensioner.
14. Insert locking pin T40011 to retain the tensioner position and remove the timing belt.
15. On late 2000–04 models, install an Allen wrench into the tensioner pulley and turning counterclockwise, compress the tensioner plunger.
16. Install locking plate T10008 to retain the tensioner.
17. Loosen the tensioner nut.
18. Install wrench 3387 in the tensioner and turn the tensioner clockwise.
19. Remove the timing belt.

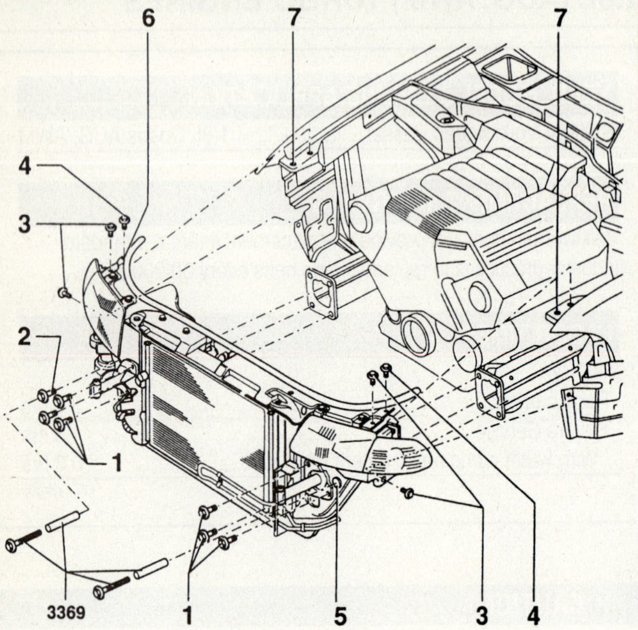

1. Bolts 33 ft. lbs. (45 Nm)
2. Bolts 33 ft. lbs. (45 Nm)
3. Bolts 7 ft. lbs. (10 Nm)
4. Bolts 7 ft. lbs. (10 Nm)
5. Bore for support tool
6. Lock carrier bore
7. Fender bore

7923CG12

Moving the front support panel into the service position—Passat

To install:

20. Ensure the crankshaft and camshaft timing marks are aligned.
21. Install the timing belt to the crankshaft sprocket.
22. Install the lower timing belt cover.
23. Install the crankshaft pulley using new bolts and tighten to 84 inch lbs. (96 Nm), plus 90 degrees.
24. Install the timing belt to the water pump sprocket, tensioner and camshaft sprockets.
25. On early 2000 models, remove the locking pin and allow the tensioner to operate.
26. On late 2000–2004 models, turn the tensioner pulley counterclockwise using wrench 3387 until the locking plate can be removed.
27. Turn the tensioner pullet clockwise until an 8mm drill bit can be inserted in the tensioner lever.
28. Tighten the tensioner nut to 20 ft. lbs. (27 Nm).
29. On all models, turn the crankshaft 2 turns clockwise and make sure the timing marks are aligned.
30. Install the timing belt covers.
31. Install the accessory drive belt tensioner.
32. Install the accessory drive belts.
33. Install the radiator support panel bolts and move the support panel rearward.
34. Remove support guides 3369 or 3411.
35. Install the radiator support panel bolt.
36. Install the air intake hose between the air cleaner and radiator support.
37. Install the front bumper.
38. Connect the negative battery cable.

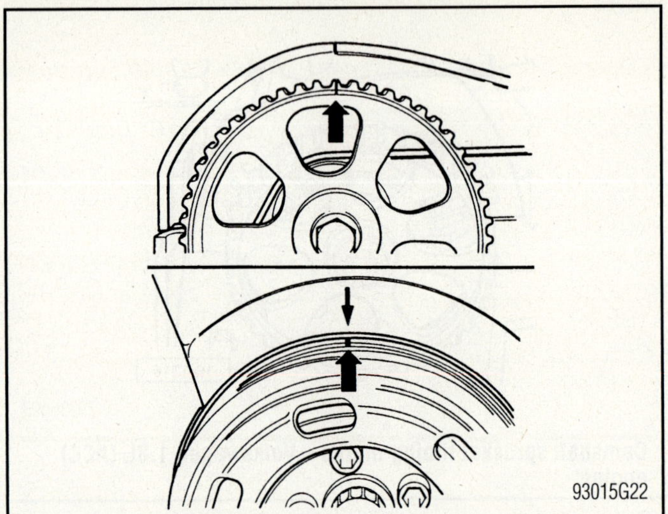

93015G22

Crankshaft pulley and camshaft sprocket timing marks—Volkswagen 1.8L (AUG, AWM) engine

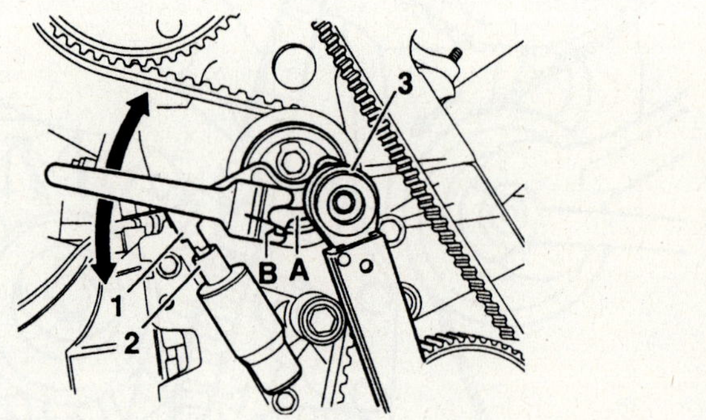

93015G23

Timing belt tension adjustment—Volkswagen 1.8L (AUG, AWM) engine

1.8L (AWP, AWW) TURBO ENGINES

YEAR, MODEL & ENGINE APPLICATION

2000-02 Volkswagen Golf Turbo, GTI Turbo, Jetta Turbo,
1999-04 New Beetle Turbo .... 1.8L Codes AWP, AWW

REPLACEMENT INTERVALS

Volkswagen does not provide a replacement interval. Standard industry practice is to replace timing belts every 60,000 miles.

LABOR TIMES

1999-04	
Replacement	3.3 hrs
Severe Service	3.8 hrs
With water pump replacement add	0.3 hrs

8807VWg06

✱✱ CAUTION

These engines are an interference type. If the timing belt has failed, valve-to-piston damage may have occurred. A compression check should always be performed to check for engine damage.

■ **Do not turn the engine or camshaft with the camshaft drive belt removed. The pistons will contact the valves and cause internal engine damage.**

1. Disconnect the negative battery cable and wait 20 minutes for the air bag backup power to dissipate.
2. Remove the engine appearance cover and splash shield.
3. Remove the right side headlight assembly.
4. Remove the intake pipe between the turbocharger and the charge air cooler.

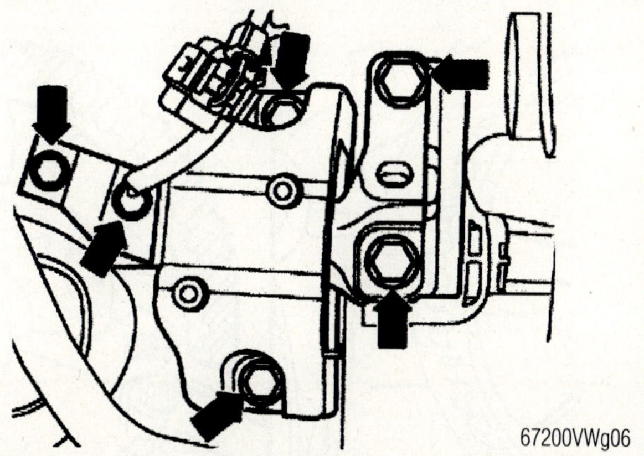
67200VWg06

Removing engine and bolt mounts bolts—1.8L (AWP/AWW) engine

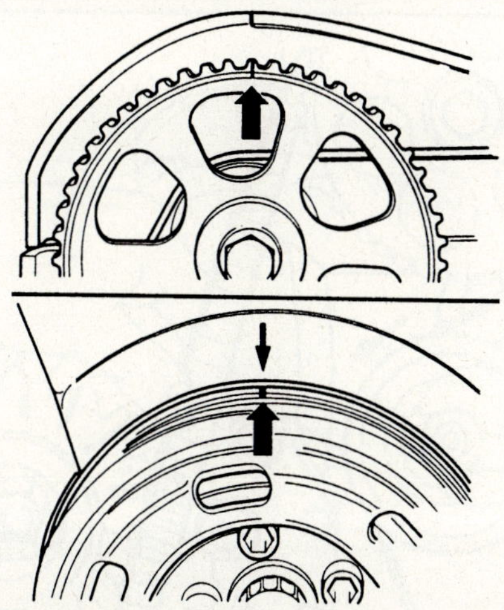
93015g22

Crankshaft pulley and camshaft sprocket timing marks—Volkswagen 1.8L (AWP, AWW) engine

5. Place an open-end wrench on the alternator belt tensioner and rotate it clockwise toward the alternator to release the belt's tension. Remove the alternator serpentine drive belt and release the tensioner.

■ **If necessary to lock the alternator tensioner in position, align the housing holes and insert an Allen wrench into the holes to secure its movement.**

6. Using the center bolt, rotate the crankshaft in the direction of engine rotation to position the No. 1 cylinder at Top Dead Center (TDC) of its compression stroke.
7. Remove the upper timing belt cover.

■ **If reusing the timing belt, mark its rotational direction so it may be installed in its original position.**

8. Use an engine hoist and take the weight off the engine.
9. Remove the engine mount and body mount bolts and remove the engine mount from the cylinder block.
10. Remove the damper pulley-to-crankshaft bolts and the damper.
11. Remove the center and lower timing belt covers.
12. Place tensioning bolt T10092 into the timing belt tensioner and tension the element so the tensioner piston can be locked with a locking pin.
13. Remove the timing belt.
14. Turn the crankshaft back slightly so the pistons are not at TDC.

To install:

15. Align the camshaft sprocket timing mark with the cylinder head cover mark.
16. Install the timing belt on the crankshaft sprocket with the arrow facing the rotational direction.
17. Install the engine mount and body mount bolts to the cylinder block and tighten to 33 ft. lbs. (45 Nm).
18. Install the lower timing belt cover and tighten the bolts to 89 inch lbs. (10 Nm).
19. Using a bolt, secure the damper/belt pulley on the crankshaft.
20. Using the center bolt, rotate the crankshaft in the direction of engine rotation to position the No. 1 cylinder at Top Dead Center (TDC) of its compression stroke.
21. Install the timing belt on the camshaft sprocket, water pump and belt tensioner.
22. Remove locking pin and tensioning tool from the belt tensioner.
23. Rotate the crankshaft 2 complete rotation in the running direction.
24. Check that the crankshaft and camshaft timing marks are still aligned.
25. Install the crankshaft pulley bolts and tighten to 18 ft. lbs. (25 Nm).
26. Install the center timing belt cover and tighten the bolts to 89 inch lbs. (10 Nm).
27. Install the upper timing belt cover.
28. Install the drive belts.
29. Replace the remaining components by reversing the removal procedures.
30. Connect the negative battery cable last.
31. Test drive the vehicle.

1.8L (PG) ENGINES

YEAR, MODEL & ENGINE APPLICATION

Model	Engine
1990-94 Volkswagen Corrado	1.8L Code PG

REPLACEMENT INTERVALS

Volkswagen does not provide a replacement interval. Standard industry practice is to replace timing belts every 60,000 miles.

LABOR TIMES

1990-94	
Replacement	1.5 hrs
Severe Service	1.7 hrs

8807VWg07

✻✻ CAUTION

These engines are an interference type. If the timing belt has failed, valve-to-piston damage may have occurred. A compression check should always be performed to check for engine damage.

■ **Do not turn the engine or camshaft with the camshaft drive belt removed. The pistons will contact the valves and cause internal engine damage.**

1. Disconnect the negative battery cable.
2. Disable the air bag system, then wait 20 minutes for the system to totally discharge
3. Rotate the crankshaft until the crankshaft and camshaft timing marks are aligned.
4. The notch on the crankshaft pulley should align with the dot on the intermediate shaft sprocket. With the distributor cap removed, the rotor should be pointing toward the No. 1 mark on the rim of the distributor housing.
5. Remove the accessory drive belts.
6. Remove the accessory drive belt tensioner.

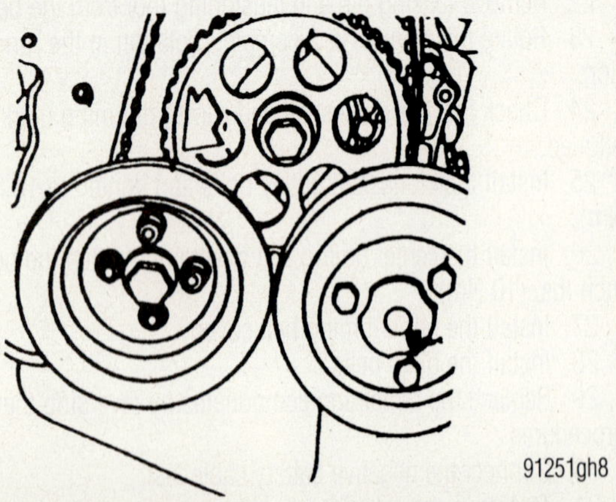

91251gh8

Timing marks on the crankshaft pulley and the intermediate shaft sprocket (arrow)—Volkswagen 1.8L (PG) engine

7. Remove the crankshaft pulley by holding the center crankshaft sprocket bolt with a socket and loosen the pulley bolts.
8. Remove the water pump pulley.
9. Remove the timing belt covers.
10. Loosen the locknut on the tensioner pulley and turn the tensioner counterclockwise to relieve the tension on the timing belt.
11. Slide the timing belt from the sprockets.

To install:

12. Install the new timing belt and tension the belt so it can be twisted 90 degrees at the middle of its longest section, between the camshaft and intermediate sprockets.
13. Recheck the alignment of the timing marks and, if correct, turn the engine 2 full revolutions to return to TDC of No. 1 piston. Recheck belt tension and timing marks. Readjust as required. Tighten the tensioner nut to 33 ft. lbs. (45 Nm).
14. Reinstall the belt cover and accessory drive belts.
15. When running the engine, there will be a growling noise that rises and falls with engine speed if the belt is too tight.

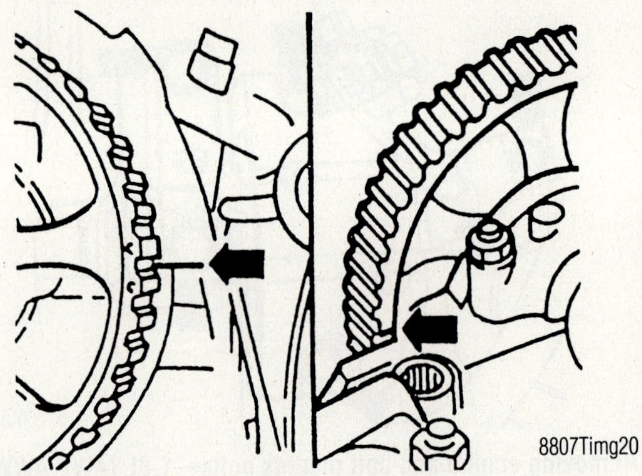

8807Timg20

Camshaft sprocket timing marks—Volkswagen 1.8L (PG) engine

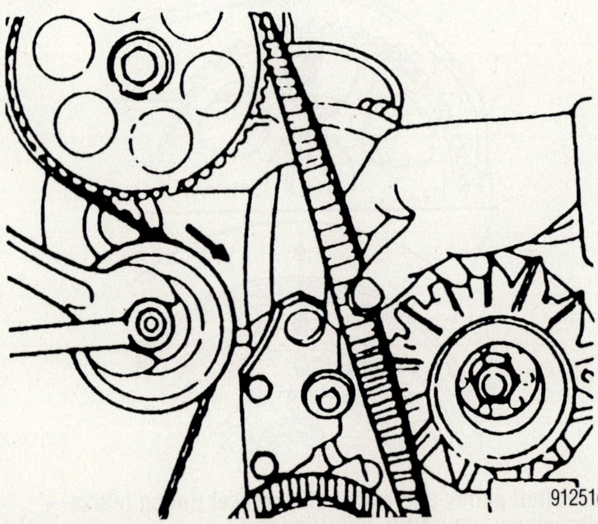

91251gi1

Using a suitable wrench, turn the tensioner pulley clockwise until the belt can just be twisted 90 degrees

1.8L (UM) ENGINES

YEAR, MODEL & ENGINE APPLICATION

1987-93 Volkswagen Fox, 1985-86 Quantum ...	1.8L Code UM

REPLACEMENT INTERVALS

Volkswagen does not provide a replacement interval. Standard industry practice is to replace timing belts every 60,000 miles.

LABOR TIMES

Fox	
Replacement	1.7 hrs
Severe Service	2.0 hrs
Quantum	
Replacement	1.5 hrs
Severe Service	2.0 hrs
All Models	
With water pump replacement add	0.3 hrs

8807VWg08

✱✱ CAUTION

These engines are an interference type. If the timing belt has failed, valve-to-piston damage may have occurred. A compression check should always be performed to check for engine damage.

1. Disconnect the negative battery cable.
2. Loosen the alternator mounting bolts and if equipped, the power steering pump and air conditioner compressor bolts, if their drive belts interfere with the cover removal.
3. Pivot the alternator or driven component and slip the drive belt from the pulleys.
4. Remove the water pump pulley.
5. Remove the crankshaft pulley.
6. Rotate the crankshaft clockwise until the No. 1 cylinder is at TDC of the compression stroke and the timing marks are aligned.
7. Remove the timing belt cover.
8. Remove the timing belt guide, if equipped.
9. Loosen the tensioner adjusting nut, move the tensioner away from the belt and remove the timing belt.

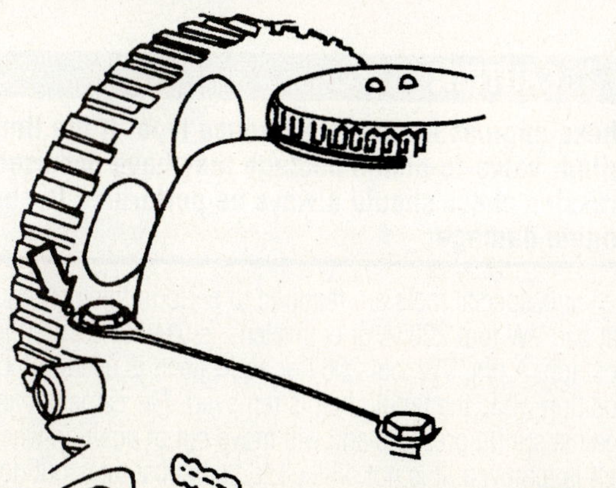

93053g08

Camshaft sprocket alignment mark—Volkswagen 1.8L (UM) engine

93053g09

Crankshaft and intermediate sprocket alignment—Volkswagen 1.8L (UM) engine

To install:

10. Install the timing belt counterclockwise starting at the crankshaft sprocket.
11. Ensure the timing belt is taut on the non-tension side.
12. Loosen the tensioner nut and turn the tensioner clockwise until the belt can be twisted 90 degrees.
13. Tighten the tensioner nut to 37 ft. lbs. (50 Nm).
14. Turn the engine 2 complete revolutions (clockwise) and align the timing marks. Recheck the belt tension and timing marks. Readjust as required.
15. Install the timing belt cover and water pump pulley.
16. Install the drive belts.
17. Connect the negative battery cable.

1.8L (PL) AND 2.0L (9A) ENGINES

YEAR, MODEL & ENGINE APPLICATION

1985-90 Volkswagen Cabrio, Golf/GTI, Jetta/GTI, Scirocco ...	1.8L Code PL
1990-94 Golf/GTI, Jetta/GTI, Passat ...	2.0L Code 9A

REPLACEMENT INTERVALS

Volkswagen does not provide a replacement interval. Standard industry practice is to replace timing belts every 60,000 miles.

LABOR TIMES

Cabrio	
Replacement	2.6 hrs
Severe Service	3.1 hrs
Golf/GTI & Jetta/GTI	
Replacement	3.0 hrs
Severe Service	3.5 hrs
Passat	
Replacement	2.5 hrs
Severe Service	3.0 hrs
Scirocco	
Replacement	2.8 hrs
Severe Service	3.3 hrs

8807VWg09

✻✻ CAUTION

These engines are an interference type. If the timing belt has failed, valve-to-piston damage may have occurred. A compression check should always be performed to check for engine damage.

■ **Do not turn the engine or camshaft with the camshaft drive belt removed. The pistons will contact the valves and cause internal engine damage.**

1. Disconnect the negative battery cable.
2. Disable the air bag system, then wait 20 minutes for the system to totally discharge
3. Remove the accessory drive belt.
4. Remove the drive belt tensioner.
5. Remove the crankshaft pulley using a puller.
6. Remove the water pump pulley.
7. Remove the upper and lower timing belt covers.
8. Rotate the crankshaft until the camshaft timing marks are aligned.
9. Ensure the flywheel or flexplate timing marks are aligned.
10. Loosen the timing belt tensioner and remove the timing belt.

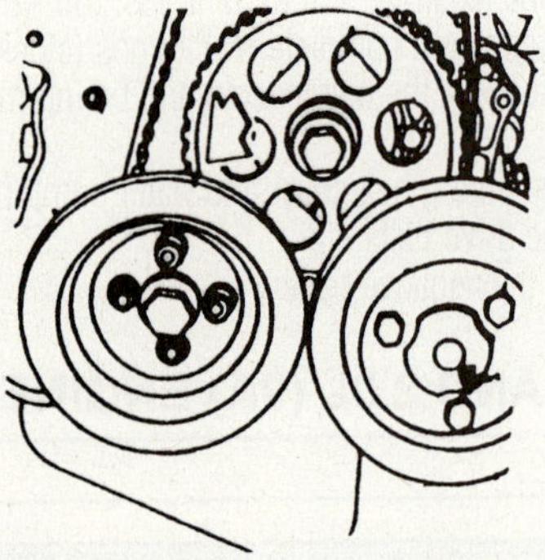

91251gh8

Timing marks on the crankshaft pulley and the intermediate shaft sprocket (arrow)—Volkswagen 1985–94 1.8L (PL) and 2.0L (9A) engines

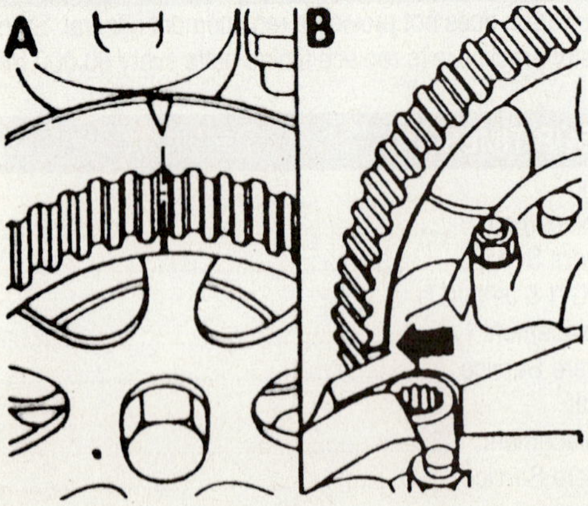

91251gh9

View of the timing marks aligned on the camshaft sprockets—A = top mark, B = side mark—Volkswagen 1985–94 1.8L (PL) and 2.0L (9A) engines

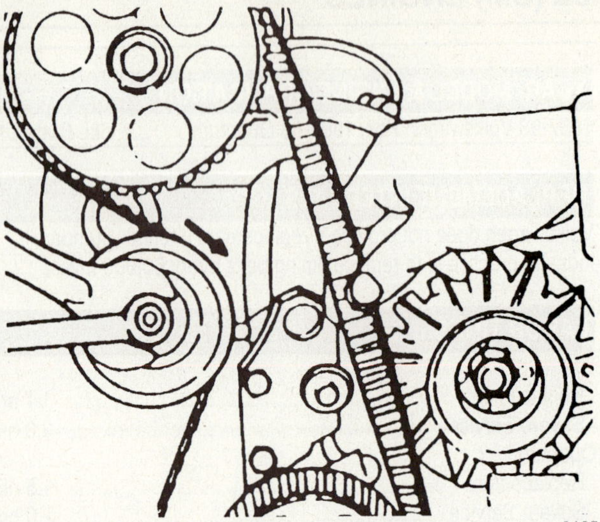

91251gi1

Using a suitable wrench, turn the tensioner pulley clockwise until the belt can just be twisted 90 degrees

To install:

11. Install the new timing belt on the crankshaft and intermediate sprocket.
12. Install the timing belt lower cover.
13. Install the water pump pulley.
14. Rotate the intermediate shaft sprocket counterclockwise until with the distributor cap removed, the rotor points toward the No. 1 mark on the rim of the distributor housing.
15. Place the timing belt on the tensioner and camshaft sprocket.
16. Rotate the tensioner clockwise until the belt can be twisted 90 degrees at the middle of its longest section, between the camshaft and intermediate sprockets.
17. Recheck the alignment of the timing marks and, if correct, turn the engine 2 full revolutions to return to TDC of No. 1 piston. Recheck belt tension and timing marks. Readjust as required. Tighten the tensioner nut to 33 ft. lbs. (45 Nm).
18. Install the upper timing belt cover.
19. Install the drive belt tensioner and the drive belt.
20. Connect the negative battery.

1.9L (AAZ, AHH, AHU) DIESEL ENGINES

✻✻ CAUTION

These engines are an interference type. If the timing belt has failed, valve-to-piston damage may have occurred. A compression check should always be performed to check for engine damage.

Some special tools are required to perform this procedure properly. A flat bar, VW tool 2065A or equivalent, is used to secure the camshaft in position. A pin, VW tool 2064 or equivalent, is used to fix the pump position while the timing belt is removed. The camshaft and pump work against spring pressure and will move out of position when the timing belt is removed. It is not difficult to find substitutes but do not remove the timing belt without these tools.

YEAR, MODEL & ENGINE APPLICATION

Year, Model	Engine
1993-98 Volkswagen Golf, Jetta, 1995-97 Passat	1.9L Codes 1A, AAZ
1997-99 Passat	1.9L Code AHH
1996-98 Golf, Jetta	1.9L Code AHU

REPLACEMENT INTERVALS

Volkswagen does not provide a replacement interval. Standard industry practice is to replace timing belts every 60,000 miles.

LABOR TIMES

Golf & Jetta	
1993-97	
Replacement	2.8 hrs
Severe Service	3.0 hrs
1998	
Replacement	3.3 hrs
Severe Service	3.8 hrs
Passat	
1995-97	
Replacement	2.8 hrs
Severe Service	3.3 hrs
1998-99	
Replacement	4.3 hrs
Severe Service	4.8 hrs
All Models	
With turbocharger add	0.9
With water pump replacement add	0.4

8807VWg10

** WARNING

Do not turn the engine or camshaft with the timing belt removed. The pistons will contact the valves and cause internal engine damage.

1. Disconnect the negative battery cable and remove the accessory drive belts, crankshaft pulley and the timing belt cover(s). Remove the camshaft cover and rubber plug at the back end of the camshaft.
2. Temporarily reinstall the crankshaft pulley bolt and turn the crankshaft to Top Dead Center (TDC) of the compression stroke of No. 1 piston. The mark on the camshaft sprocket should be aligned with the mark on the inner timing belt cover or the edge of the cylinder head.
3. With the engine at TDC, insert the bar into the slot at the back of the camshaft. The bar rests on the cylinder head to will hold the camshaft in position.
4. Insert the pin into the injection pump drive sprocket to hold the pump in position.
5. Loosen the locknut on the tensioner pulley and turn the tensioner counterclockwise to relieve the tension on the timing belt. Slide the timing belt from the sprockets.

To install:

6. Install the new timing belt and adjust the tension so the belt can be twisted 45 degrees at the half-way point between the camshaft and pump sprockets. Tighten the tensioner nut to 33 ft. lbs. (45 Nm).
7. Remove the holding tools.
8. Turn the engine 2 full revolutions to return to TDC for the No. 1 cylinder. Recheck belt tension and timing mark alignment, readjust as required.

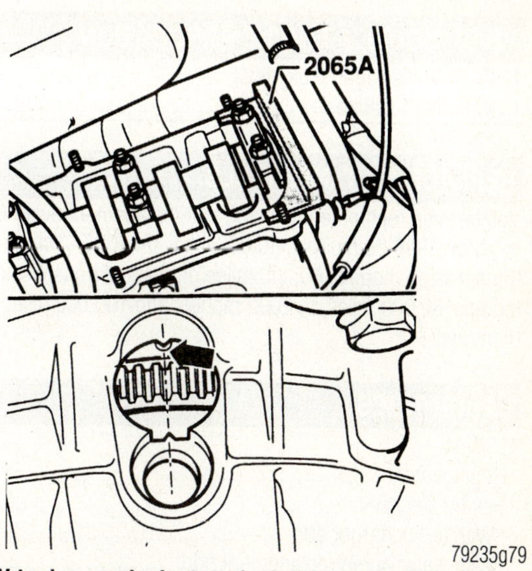

79235g79

Use the VW tool or equivalent, to lock the camshaft at TDC for timing belt replacement—Volkswagen 1993–99 1.9L Diesel engines

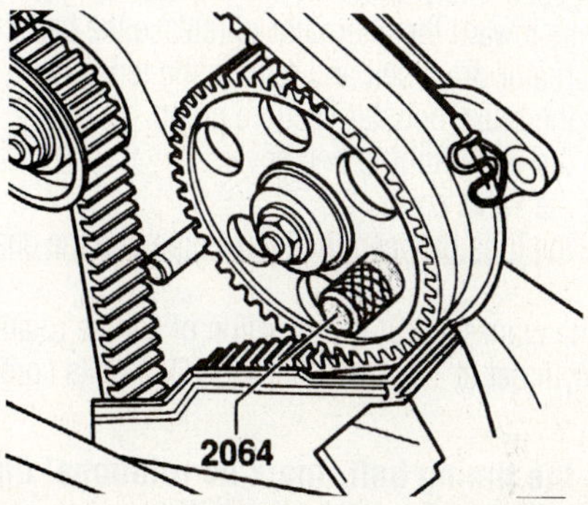

79235g80

Locking the injection pump with the VW tool or equivalent—Volkswagen 1993–99 1.9L Diesel engines

9. Install the belt cover and accessory drive belts.

■ **If the belt is too tight, there will be a growling noise that rises and falls with engine speed.**

1.9L (ALH) DIESEL ENGINES

** CAUTION

These engines are an interference type. If the timing belt has failed, valve-to-piston damage may have occurred. A compression check should always be performed to check for engine damage.

Some special tools are required to perform this procedure properly. A camshaft locking bar, VW tool 3418 is used to secure the camshaft in position. A pin, VW tool 3359 is used to fix the injection pump position while the timing belt is removed. Camshaft sprocket holding tool 3036 is used to lock the camshaft sprocket. Puller T40001 is needed to remove the camshaft sprocket.

YEAR, MODEL & ENGINE APPLICATION

1999-04 Volkswagen Golf, Jetta,
1998-04 New Beetle .... 1.9L Code ALH

REPLACEMENT INTERVALS

Volkswagen recommends timing belt replacement every 60,000 miles on 1998-99 models, 40,000 miles on 2000-01 automatic transmission models, 60,000 miles on manual transmissions models, 80,000 miles on 2002 models and 100,000 miles on 2003-04 models.

LABOR TIMES

1998-04	
Replacement	5.0 hrs
Severe Service	5.5 hrs
With turbocharger add	0.9 hrs
With water pump replacement add	0.4 hrs

8807VWg11

1. Disconnect the negative battery cable.
2. Remove the engine appearance cover and splash shield.
3. Place an open-end wrench on the alternator belt tensioner and rotate it clockwise toward the alternator to release the belt's tension. Remove the alternator drive belt and release the tensioner.
4. Remove the brake booster vacuum pump.
5. Remove the upper timing belt cover.
6. Remove the valve cover.
7. Remove the tube between the intake pipe and the charge air cooler.
8. Rotate the crankshaft in the direction of engine rotation to position the No. 1 cylinder at Top Dead Center (TDC) of its compression stroke.

■ **If reusing the timing belt, mark its rotational direction so it may be installed in its original position.**

9. Install camshaft locking bar onto the camshaft and cylinder head so it is equally balanced between both legs of the tool.
10. Lock the injection pump in place using pin 3359.
11. Loosen the injection pump sprocket bolts in a diagonal sequence.

■ **DO NOT loosen the pump sprocket hub bolt or the injection pump basic setting will be altered and cannot be reset with normal shop equipment.**

12. Use an engine hoist and take the weight off the engine.
13. Remove the coolant reservoir tank and the power steering fluid reservoir and wire aside without disconnect the fluid lines.
14. Remove the engine mount and body mount bolts on the right side and remove the engine mount from the cylinder block.
15. Remove the left side engine mount bracket.
16. Remove the damper pulley-to-crankshaft bolts and the damper.
17. Remove the center and lower timing belt covers.
18. Release the tension on the belt tensioner and remove the timing belt.
19. Hold the camshaft sprocket with tool 3036 and loosen the camshaft sprocket bolt one turn.
20. Remove the camshaft sprocket using 2-jaw puller T40001.

To install:

21. Install the timing belt onto the crankshaft sprocket, idler pulley, injection pump sprocket, water pump and tensioner.

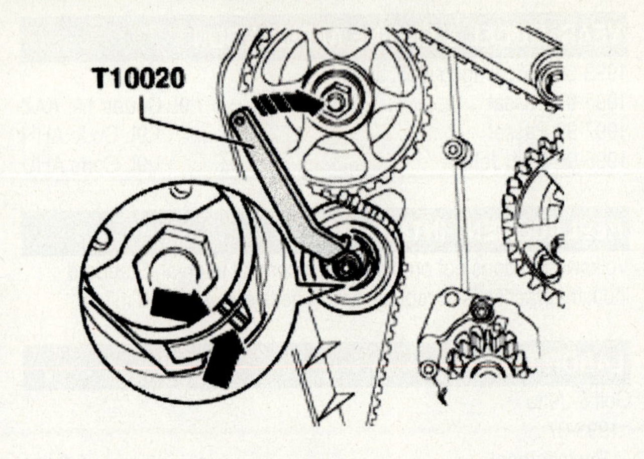

67200VWg07

Aligning the timing belt tensioner marks on manual transmission models—1.9L (ALH) Diesel engine

67200VWg08

Aligning the timing belt tensioner marks on automatic transmission models—1.9L (ALH) Diesel engine

22. Adjust the injection pump sprocket so the mounting slots are centered.
23. Install the timing belt on the camshaft sprocket, then install the sprocket using the mounting bolt. The camshaft sprocket should still turn.
24. On manual transmission models, tension the timing belt using tool T10020 on the tensioner roller eccentric and turn it clockwise until the notch and the raised mark are aligned as shown.
25. On automatic transmission models, tension the timing belt using an Allen wrench on the tensioner roller eccentric and turn it clockwise until the notch and the pointer are aligned as shown.
26. Using new bolts, tighten the injection pump sprocket bolts to 15 ft. lbs. (20 Nm).

■ **After dynamically testing the start of injection, retighten the sprocket bolts an additional 90°.**

27. Remove the camshaft locking tool and the injection pump sprocket locking pin holding tools.
28. Turn the engine 2 full revolutions to return to TDC for the No. 1 cylinder. Recheck belt tension and timing mark alignment, readjust as required.

29. Install the left side engine mount bracket and tighten the bolts to 33 ft. lbs. (45 Nm).
30. Install the center and lower timing belt covers.
31. Install the crankshaft damper and tighten the bolts to 89 inch lbs. (10 Nm), plus an additional 90°.
32. Install the engine mount and body mount bolts to the cylinder block and tighten to 33 ft. lbs. (45 Nm).
33. The remainder of the installation is the reverse of the removal procedure.

2.0L (ABA) ENGINES

YEAR, MODEL & ENGINE APPLICATION

1993-02 Volkswagen Cabrio, 1991-98 Golf/GTI, Jetta/GTI, 1990-97 Passat .... 2.0L Code ABA

REPLACEMENT INTERVALS

Volkswagen does not provide a replacement interval. Standard industry practice is to replace timing belts every 60,000 miles.

LABOR TIMES

Cabrio	
Replacement	2.6 hrs
Severe Service	3.1 hrs
Golf/GTI, Jetta/GTI & Passat	
Replacement	2.5 hrs
Severe Service	3.0 hrs
All Models	
With water pump replacement add	0.3 hrs

8807VWg12

1990–97

**** CAUTION**

These engines are an interference type. If the timing belt has failed, valve-to-piston damage may have occurred. A compression check should always be performed to check for engine damage

■ **Do not turn the engine or camshaft with the camshaft drive belt removed. The pistons will contact the valves and cause internal engine damage.**

1. Disconnect the negative battery cable and wait 20 minutes for the air bag backup power to dissipate.
2. Rotate the crankshaft clockwise until the timing marks are aligned.
3. Remove the distributor cap and check that the rotor is pointing at the No. 1 cylinder mark.
4. Remove the accessory drive belts and drive belt tensioner.
5. Remove the crankshaft pulley bolts and pulley.
6. Remove the water pump pulley.
7. Remove the timing belt covers.
8. Ensure the camshaft sprocket timing mark is aligned.
9. Remove the timing belt tensioner nut and remove the timing belt.

To install:

10. Install the new timing belt on the crankshaft and intermediate sprocket.
11. Install the timing belt lower cover.
12. Install the water pump pulley.
13. Install the crankshaft sprocket and tighten to 18 ft. lbs. (25 Nm).

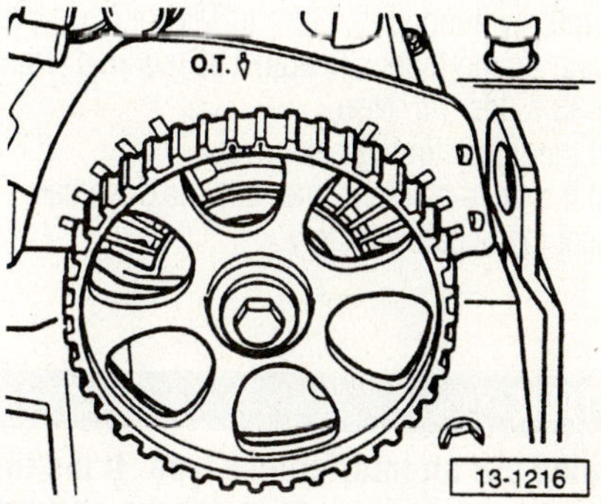

79235g93

Camshaft timing belt sprocket TDC alignment mark—Volkswagen 2.0L (ABA) engine

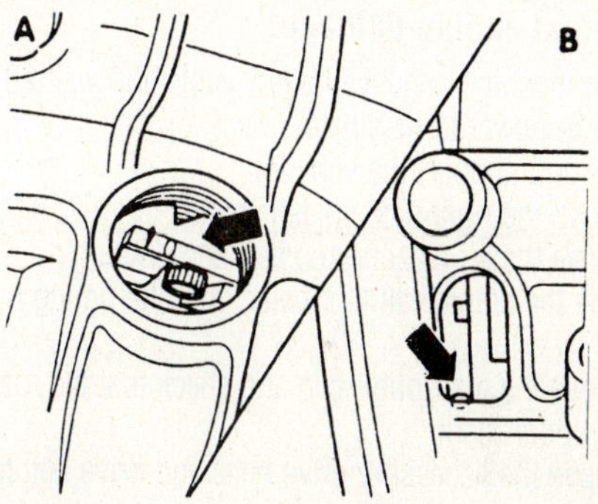

79235g92

Position the flywheel (A) or driveplate (B) as shown for TDC alignment for cylinder No. 1—Volkswagen 2.0L (ABA) engine

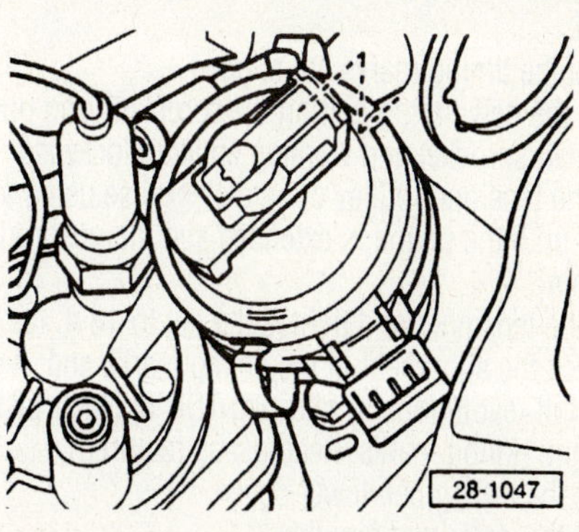

79235g91

When the No. 1 cylinder is at TDC, the ignition rotor should face the notch in the distributor housing—Volkswagen 2.0L (ABA) engine

14. Rotate the intermediate shaft sprocket counterclockwise until with the distributor cap removed, the rotor points toward the No. 1 mark on the rim of the distributor housing.
15. Place the timing belt on the tensioner and camshaft sprocket.
16. Rotate the tensioner clockwise until the belt can be twisted 90 degrees at the middle of its longest section, between the camshaft and intermediate sprockets.
17. Recheck the alignment of the timing marks and, if correct, turn the engine 2 full revolutions to return to TDC of No. 1 piston. Recheck belt tension and timing marks. Readjust as required. Tighten the tensioner nut to 33 ft. lbs. (45 Nm).
18. Install the upper timing belt cover.
19. Install the drive belt tensioner and the drive belt.
20. Connect the negative battery.

1998–02

CAUTION

These engines are an interference type. If the timing belt has failed, valve-to-piston damage may have occurred. A compression check should always be performed to check for engine damage

■ **Do not turn the engine or camshaft with the camshaft drive belt removed. The pistons will contact the valves and cause internal engine damage.**

1. Disconnect the negative battery cable and wait 20 minutes for the air bag backup power to dissipate.
2. Raise and support the vehicle.
3. Remove the right side splash shield.
4. Remove the air cleaner housing and ducting.
5. Rotate the crankshaft clockwise until the timing marks are aligned.
6. Remove the distributor cap and check that the rotor is pointing at the No. 1 cylinder mark.
7. Remove the accessory drive belts and drive belt tensioner.
8. Remove the crankshaft pulley bolts and pulley.
9. Remove the water pump pulley.
10. Remove the timing belt covers.
11. Ensure the camshaft sprocket timing mark is aligned.
12. Loosen the timing belt tensioner nut and remove the timing belt.

To install:

13. Ensure the timing marks are aligned.
14. Install the new timing belt on the sprockets and pulleys starting at the crankshaft sprocket and working counterclockwise.
15. Turn the tensioner pulley counterclockwise using a tensioner wrench V159 until the piston is extended and the pushrod protrudes 0.40 in. (10mm).
16. Hold the tensioner and tighten the nut to 18 ft. lbs. (25 Nm).
17. Recheck the alignment of the timing marks and, if correct, turn the engine 2 full revolutions to return to TDC of No. 1 piston. Recheck belt tension and timing marks. Readjust as required.
18. Install the timing belt lower cover.
19. Install the water pump pulley.
20. Install the crankshaft sprocket and tighten to 18 ft. lbs. (25 Nm).
21. Install the upper timing belt cover.
22. Install the drive belt tensioner and the drive belt.
23. Connect the negative battery.

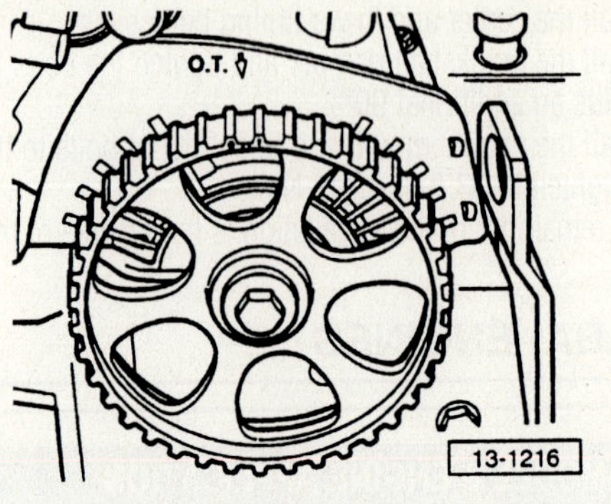

79235g93

Camshaft timing belt sprocket TDC alignment mark—Volkswagen 2.0L (ABA) engine

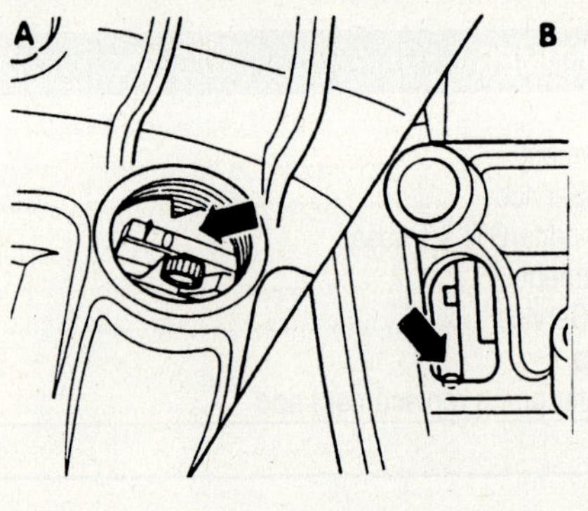

79235g92

Position the flywheel (A) or driveplate (B) as shown for TDC alignment for cylinder No. 1—Volkswagen 2.0L (ABA) engine

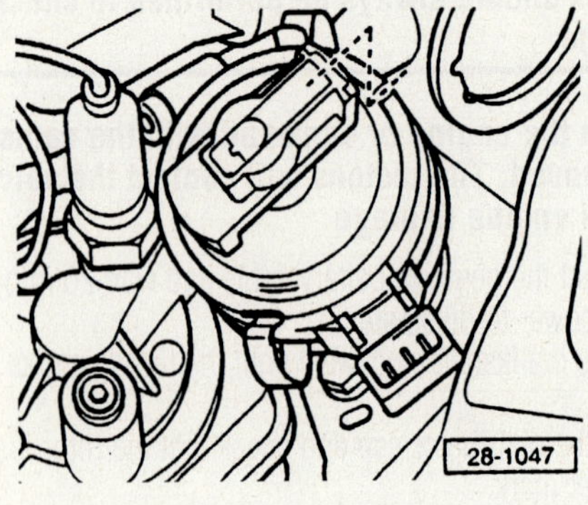

79235g91

When the No. 1 cylinder is at TDC, the ignition rotor should face the notch in the distributor housing—Volkswagen 2.0L (ABA) engine

2.0L (AEG/AVH/AZG/BDC) ENGINES

YEAR, MODEL & ENGINE APPLICATION

Application	Engine
1999-04 Volkswagen Golf/GTI, Jetta/GTI, New Beetle	2.0L Codes AEG, AVH, AZG, BDC

REPLACEMENT INTERVALS

Volkswagen does not provide a replacement interval. Standard industry practice is to replace timing belts every 60,000 miles.

LABOR TIMES

Golf/GTI & Jetta/GTI	
Replacement	2.5 hrs
Severe Service	3.0 hrs
New Beetle	
Replacement	2.9 hrs
Severe Service	3.4 hrs
All Models	
With water pump replacement add	0.3 hrs

8807VWg13

✱✱ CAUTION

These engines are an interference type. If the timing belt has failed, valve-to-piston damage may have occurred. A compression check should always be performed to check for engine damage

■ **Do not turn the engine or camshaft with the camshaft drive belt removed. The pistons will contact the valves and cause internal engine damage.**

1. Disconnect the negative battery cable and wait 20 minutes for the air bag backup power to dissipate.
2. Remove the engine appearance cover.
3. On New Beetle models, remove the right side splash shield.
4. On all models, remove the serpentine drive belt by performing the following procedure:
   a. Using an open-end wrench, rotate the serpentine drive belt tensioner clockwise to relieve the belt tension. Using tool 3090 or a drift awl, lock the tensioner in place.
   b. Remove the serpentine drive belt.
5. Remove the drive tensioner from the engine.
6. Rotate the crankshaft to position the No. 1 cylinder on Top Dead Center (TDC) of its compression stroke.
7. Remove the upper timing belt cover.

■ **If the camshaft sprocket is not aligned with its timing mark, rotate the crankshaft 1 complete turn until the crankshaft and camshaft timing marks align.**

8. Install the Engine Support tool 10-222A with Leg tools and support the weight of the engine.
9. From the front of the engine, remove the engine mount and bracket which will interfere with the timing belt removal.
10. Mark the rotational direction of the timing belt for reinstallation purposes.
11. Remove the vibration damper-to-crankshaft sprocket bolts and the damper.

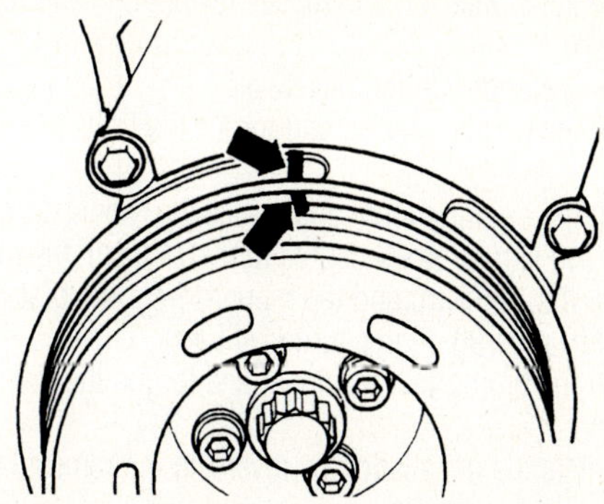

93015g31

View of crankshaft damper aligned with the timing mark—Volkswagen 2.0L (AEG/AVH/AZG/BDC) engine

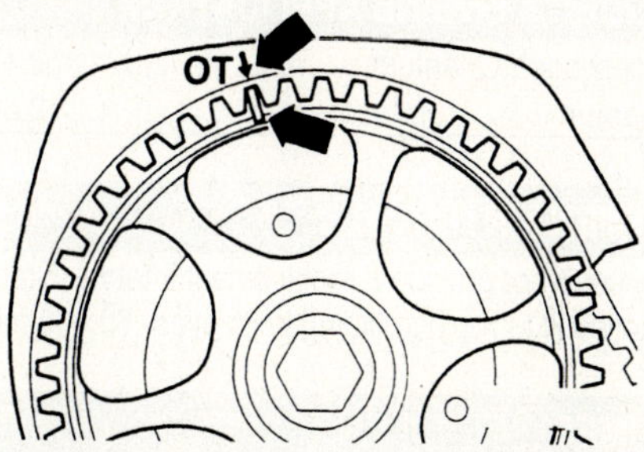

93015g32

View of camshaft damper aligned with the timing mark—Volkswagen 2.0L (AEG/AVH/AZG/BDC) engine

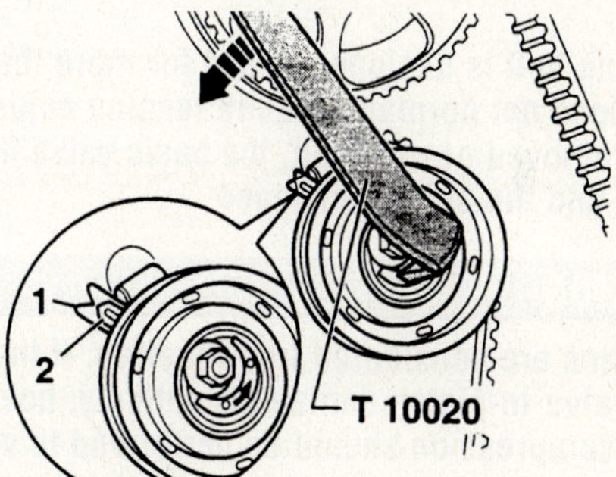

93015g33

Adjusting the timing belt tensioner—Volkswagen 2.0L (AEG/AVH/AZG/BDC) 4-Cyl. engine

12. Remove the center and lower timing belt covers.
13. Release the timing belt tensioner and remove the timing belt.

To install:

14. Make sure that the camshaft and crankshaft timing marks are aligned.
15. Install the timing belt, with the arrow pointing in the direction of rotation, on the crankshaft sprocket, the water pump sprocket and the camshaft sprocket.
16. Install the lower and center timing belt covers.
17. Install the vibration damper and torque the damper-to-crankshaft sprocket bolts to 18 ft. lbs. (25 Nm).
18. Using the 2-Hole Tensioning tool T-10020, rotate the timing belt tensioner counterclockwise until notch **1** and indicator **2** align. Then, tighten the timing belt tensioner nut to 15 ft. lbs. (20 Nm).

19. Rotate the crankshaft 2 complete revolutions and recheck the timing marks.
20. Install the upper timing belt cover.
21. Install the engine bracket and torque the bolts to 33 ft. lbs. (45 Nm).
22. Install the engine mount and torque the mount-to-bracket bolts to 30 ft. lbs. (40 Nm) plus 90 degree turn (1/4 turn), the mount-to-body bolts to 18 ft. lbs. (25 Nm) and the mount-to-engine bracket bolts to 44 ft. lbs. (60 Nm) plus 90 degree turn (1/4 turn).
23. Install the timing belt tensioner and torque the bolts to 18 ft. lbs. (25 Nm).
24. Complete the installation by reversing the removal procedures.
25. Test drive the vehicle.

2.1L AND 2.2L ENGINES

YEAR, MODEL & ENGINE APPLICATION

1983-84 Volkswagen Quantum	2.1L VIN N/A
1985-87 Quantum	2.2L VIN N/A

REPLACEMENT INTERVALS

Volkswagen does not provide a replacement interval. Standard industry practice is to replace timing belts every 60,000 miles.

LABOR TIMES

1983-87	
Rplacement	1.9 hrs
Severe Service	2.4 hrs

8807VWg14

■ **The timing belt is designed to last for more than 60,000 miles and does not normally require tension adjustments. If the belt is removed or replaced, the basic valve timing must be checked and the belt retensioned.**

⁂ CAUTION

These engines are considered freewheeling. If the timing belt fails, valve-to-piston damage is unlikely, however a precautionary compression should be performed to verify no engine damage.

1. Disconnect the negative battery cable.
2. Loosen the alternator mounting bolts and if equipped, the power steering pump and air conditioner compressor bolts, if their drive belts interfere with the cover removal.
3. Pivot the alternator or driven component and slip the drive belt from the pulleys.
4. Reposition the spacers and nuts on the mounting studs so they will not get lost.
5. Rotate the crankshaft until the **0** degree mark on the flywheel is aligned with the stationary pointer on the bell housing. Turn the camshaft or make sure the camshaft sprocket is turned until the mark on the rear of the sprocket is aligned with the upper edge of the rear drive belt cover (or valve cover edge, depending on the year of the vehicle) on the left side (spark plug side) of the engine. The notch on the crankshaft pulley should align with the dot on the intermediate shaft sprocket and the distributor rotor (remove the distributor cap) should be pointing toward the mark on the rim of the distributor housing.
6. Remove the crankshaft drive pulley(s).
7. Hold the large nut on the tensioner pulley and loosen the smaller pulley locknut. Turn the tensioner counterclockwise to relieve the tension on the timing belt.
8. Slide the timing belt from the pulleys.

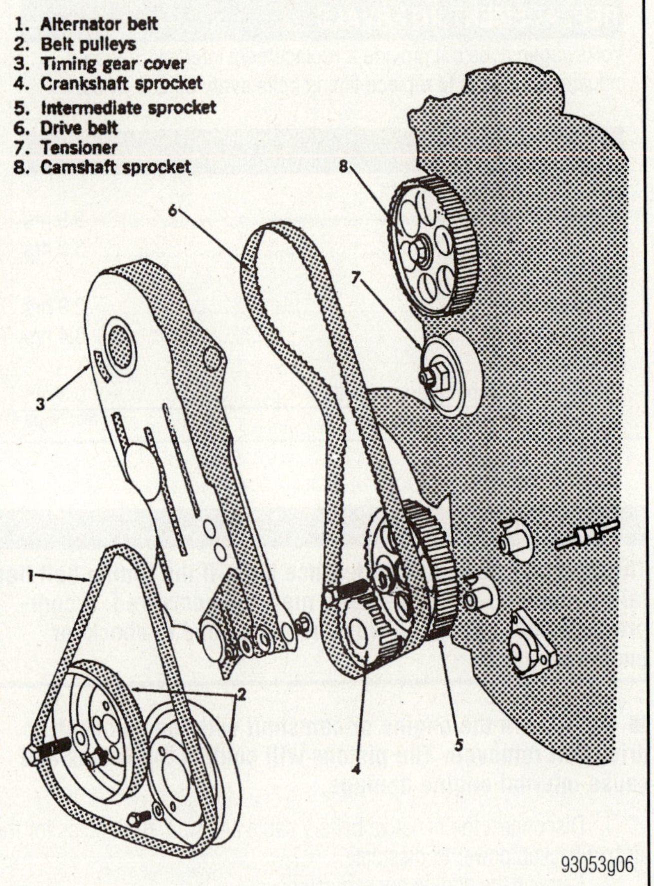

Exploded view of the timing belt assembly—Volkswagen 2.1L and 2.2L engines

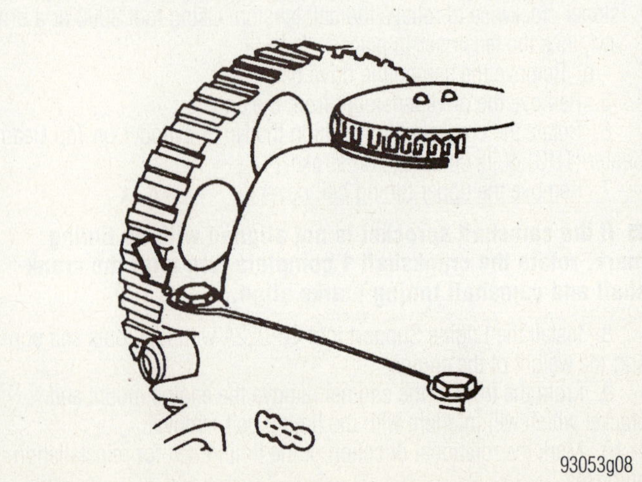

Camshaft sprocket alignment—Volkswagen 2.1L and 2.2L engines

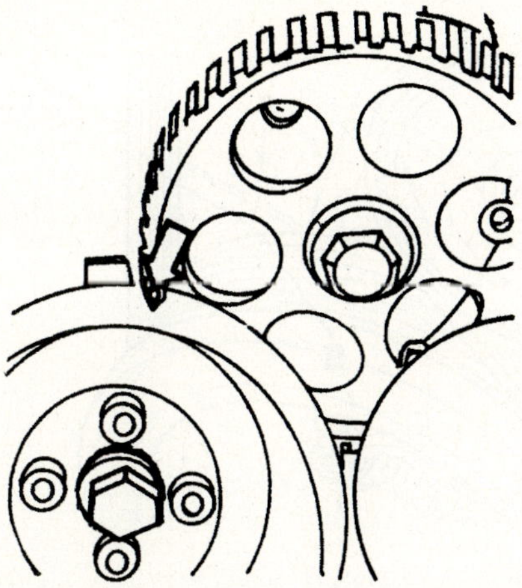

Crankshaft and intermediate sprocket alignment—Volkswagen 2.1L and 2.2L engines

To install:

9. Install the timing belt and retension with the pulley or water pump. Reinstall the crankshaft pulley(s). Recheck the alignment of the timing marks.

**** WARNING**

If the timing marks are not correctly aligned with the No. 1 piston at TDC of the compression stroke and the belt is installed, valve timing will be incorrect. Poor performance and possible engine damage can result from improper valve timing.

10. Check the timing belt tension. The tension is correct when the belt can be twisted 90 degrees with the thumb and index finger along the straight run between the camshaft sprocket and the water pump.
11. Turn the engine 2 complete revolutions (clockwise) and align the flywheel mark at TDC. Recheck the belt tension and timing marks. Readjust as required.
12. Install the timing belt cover and drive belts.
13. Connect the negative battery cable.

2.5L (ACU, AFF) ENGINES

YEAR, MODEL & ENGINE APPLICATION

1993-97 Volkswagen Eurovan .... 2.5L Codes ACU, AFF

REPLACEMENT INTERVALS

Volkswagen does not provide a replacement interval. Standard industry practice is to replace timing belts every 60,000 miles.

LABOR TIMES

1993-97

Replacement	3.0 hrs
Severe Service	3.4 hrs

8807VWg15

Without Timing Belt Tensioner

**** CAUTION**

These engines are an interference type. If the timing belt has failed, valve-to-piston damage may have occurred. A compression check should always be performed to check for engine damage.

1. Disconnect the negative battery cable.
2. Remove the accessory drive belts.
3. Place the No. 1 cylinder at the TDC position on the compression stroke. The timing mark on the flywheel should align with the mark on the flywheel housing. The camshaft sprocket mark should align with the mark on the rear of the timing belt cover.
4. Remove the crankshaft damper.
5. Remove the upper and lower timing belt covers.
6. Loosen the water pump bolts and move the pump away from the timing belt.
7. Remove the timing belt.

To install:

8. Ensure the timing marks are still aligned.
9. Install the timing belt on the crankshaft sprocket, idler pulley and water pump sprocket.
10. Install the timing belt lower cover.
11. Install the crankshaft damper. Tighten the outer bolts to 15 ft. lbs. (20 Nm) and the center bolt to 339 ft. lbs. (460 Nm).
12. Loosen the water pump bolts, install the timing belt on the camshaft sprocket and tighten the water pump bolts to 15 ft. lbs. (20 Nm).
13. Inspect the timing belt tension. Twist the belt between camshaft and water pump sprockets and ensure the belt can twist 90 degrees.
14. Install the upper timing belt cover.
15. Install the accessory drive belts.
16. Connect the negative battery cable.

With Timing Belt Tensioner

**** CAUTION**

These engines are an interference type. If the timing belt has failed, valve-to-piston damage may have occurred. A compression check should always be performed to check for engine damage.

17. Disconnect the negative battery cable.
18. Remove the accessory drive belts.
19. Place the No. 1 cylinder at the TDC position on the compression stroke. The timing mark on the flywheel should align with the mark on the flywheel housing. The camshaft sprocket mark should align with the mark on the rear of the timing belt cover.
20. Remove the crankshaft damper.
21. Remove the upper and lower timing belt covers.
22. Loosen the tensioner pulley bolt and move the tensioner away from the timing belt.
23. Remove the timing belt.

To install:

24. Ensure the timing marks are still aligned.
25. Install the timing belt on the crankshaft sprocket, tensioner pulley and idler pulley.
26. Install the timing belt lower cover.

27. Install the crankshaft damper. Tighten the outer bolts to 15 ft. lbs. (20 Nm) and the center bolt to 339 ft. lbs. (460 Nm).
28. Loosen the tensioner bolt and install the timing belt on the camshaft sprocket.
29. Rotate the crankshaft 2 revolutions too seat the timing belt. Ensure the timing marks are still aligned.
30. Tighten the tensioner bolt to 10 ft. lbs. (15 Nm).
31. Install the upper timing belt cover.
32. Install the accessory drive belts.
33. Connect the negative battery cable.

2.8L (AHA, ATQ) ENGINES

YEAR, MODEL & ENGINE APPLICATION

1998-04 Volkswagen Passat .... 2.8L Codes AHA, ATQ

REPLACEMENT INTERVALS

Volkswagen recommends timing belt replacement every 105,000 miles.

LABOR TIMES

1998-04	
Replacement	5.0 hrs
Severe Service	5.5 hrs

8807VWg16

1. Disconnect the negative battery cable.

■ **Wait 20 minutes for the air bag system to totally relieve back-up power.**

2. Remove the engine appearance cover.
3. Remove the serpentine drive belt by performing the following procedure:
    a. Using Spanner Wrench No. 3212, secure the viscous fan pulley. Using an Open-end Spanner Wrench 3212, remove the viscous fan bearing housing by turning it clockwise.

■ **The viscous fan is mounted with a left-handed thread; turn it clockwise to loosen it.**

    b. Place a 17mm box wrench on the serpentine drive belt tensioner and turn the tensioner clockwise until the 2 holes align; insert drift 3204 into the holes to secure the tensioner in place.
    c. Mark the running direction of the serpentine drive belt and remove it from the pulleys.
4. Rotate the crankshaft by hand to align the crankshaft pulley mark with the arrow on the engine housing and the large hole in each camshaft sprocket must face inward and must align; this should be Top Dead Center (TDC) of the No. 1 cylinder's compression stroke. If these conditions are not correct, rotate the crankshaft one complete revolution and realign.
5. On the left side of the cylinder block near the crankshaft, remove the sealing plug.
6. Insert Crankshaft Holder tool No. 3242 into the sealing plug hole to secure the crankshaft.
7. Using a 8mm Allen® wrench, rotate the timing belt tensioner roller clockwise until the tensioner is compressed; then, insert a 2mm

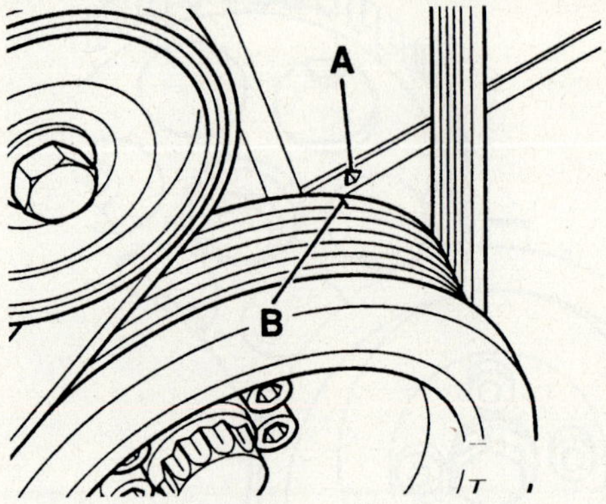

Crankshaft pulley alignment location for TDC—Volkswagen 2.8L (AHA/ATQ) engine

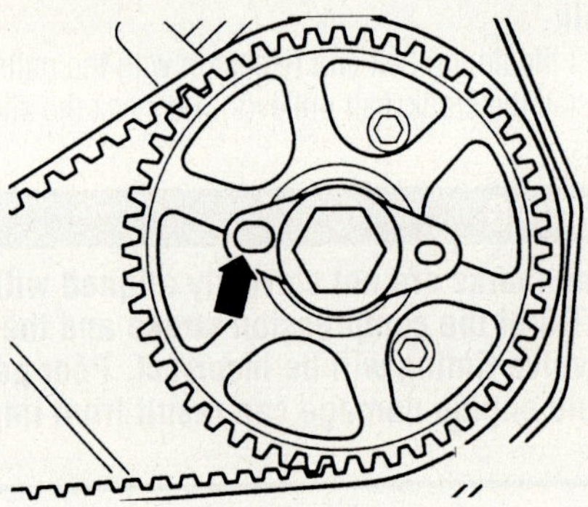

Left camshaft sprocket alignment position for TDC; right camshaft position is similar—Volkswagen 2.8L (AHA/ATQ) engine

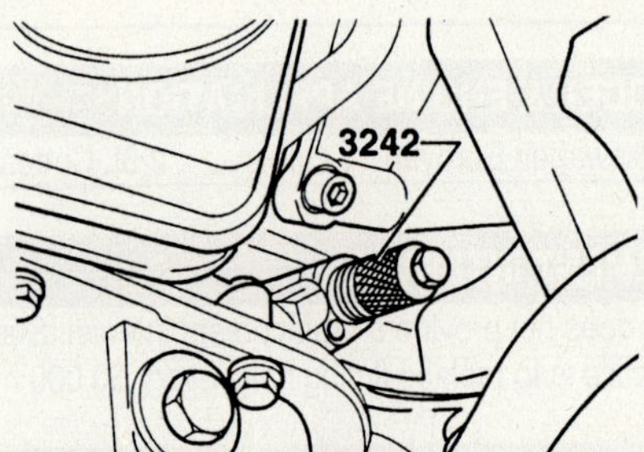

View of crankshaft holding tool installed—Volkswagen 2.8L (AHA/ATQ) engine

spring pin through the tensioner housing and tensioner plunger to secure it in place. When the plunger is secure, release the wrench tension.

8. Remove the dampener-to-crankshaft bolts and the damper.

■ **It is not necessary to remove the center bolt when removing the crankshaft damper.**

9. Remove the serpentine belt idler and the crankshaft damper guard.

10. Mark the running direction of the timing belt and remove it from the pulleys.

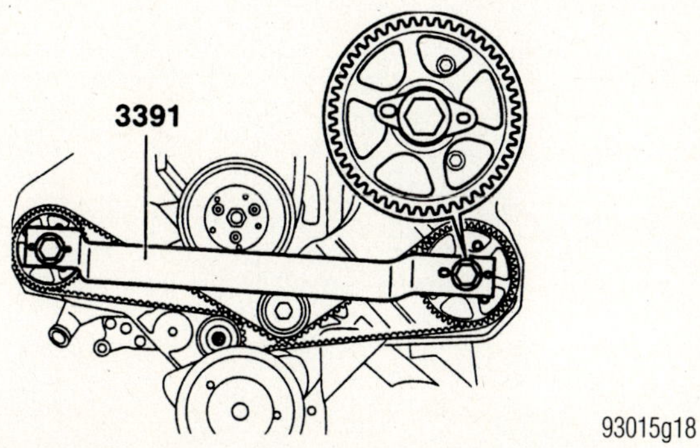

View of camshaft locator bar installed—Volkswagen 2.8L (AHA/ATQ) engine

To install:

11. Make sure that the camshaft pulleys and the crankshaft pulley are in alignment with TDC of the No. 1 cylinder's compression stroke.

12. Install the timing belt; make sure the timing belt is installed in the correct running direction from which it was removed.

13. Using a 8mm Allen® wrench, rotate the timing belt tensioner roller clockwise until the tensioner is compressed; then, remove the 2mm spring pin from the tensioner housing. Slowly, release the tensioner's spring pressure to put pressure on the timing belt.

14. Install the crankshaft damper guard and the serpentine belt idler pulley; torque the idler pulley bolts to 33 ft. lbs. (45 Nm).

15. Install the crankshaft damper and torque the damper-to-crankshaft bolts to 15 ft. lbs. If the damper-to-crankshaft center bolt was removed, torque it to 147 ft. lbs. (200 Nm) plus 180° (1/2 turn).

16. Remove the Crankshaft Holder tool No. 3242 and install the sealing plug.

17. Replace the remaining components by reversing the removal procedures.

18. Refill the cooling system and the automatic transaxle. Connect the electrical connectors. Install the negative battery cable last.

19. Test drive the vehicle.

Volvo

1.9L (B4194T) AND 2.0L (B4204T3/T4) ENGINES

YEAR, MODEL & ENGINE APPLICATION

2000-03 Volvo S40 & V40	1.9L Code B4194T
2000-04 S40 & V40	2.0L Codes B4204T3/T4

REPLACEMENT INTERVALS

Volvo recommends timing belt replacement every 105,000 miles.

LABOR TIMES

2000-04	
Replacement	3.5 hrs
Severe Service	3.9 hrs

8807VOLg01

**** CAUTION**

These engines are an interference type. If the timing belt has failed, valve-to-piston damage may have occurred. A compression check should always be performed to check for engine damage.

1. Disconnect the negative battery cable.
2. Remove the right hand engine cover.
3. Cover over the right headlamp.
4. Remove the accessory drive belt.
5. Remove the front timing belt cover.
6. Remove the power steering reservoir.
7. Remove the power steering hose bracket.
8. Position Volvo lifting beam 999 5006 or equivalent slightly in front of the front lifting eyelet for the engine. Use Volvo lifting arm 999 5383 or equivalent and Volvo lifting hook 999 5460 or equivalent to raise the front of the engine a few millimeters.

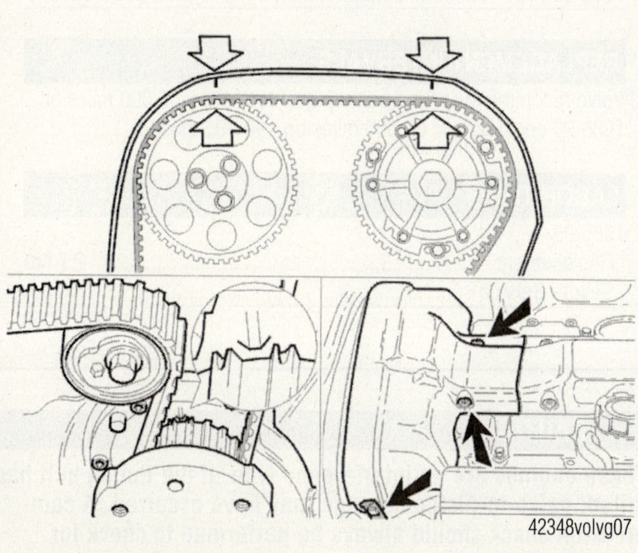

Timing belt alignment markings— Volvo 1.9L (B4194T) and 2.0L (B4204T3/T4) engines

9. Remove the right-hand engine mount.
10. Remove the right front wheel and splash shield.
11. Remove the timing belt lower front cover.
12. Turn the crankshaft until the markings on the crankshaft and camshaft pulley correspond. Turn the crankshaft a further 1/4 turn clockwise and then back again until the markings correspond. Remove the upper timing cover.
13. Slacken off the center screw for the belt tensioner slightly.
14. Turn the tensioner clockwise to the 10 o'clock position.
15. Remove the front air baffles.
16. Remove the lower belt cover and guard.
17. Remove the timing belt.

To install:

18. Turn all the pulleys listening for bearing noise. Check to see that the contact surfaces are clean and smooth. If the tensioner pulley lever or idler is seized, replace it.
19. If replacing with a new idler pulley, tighten to 18 ft. lbs. (24 Nm).
20. If replacing tension pulley, screw into place using the center bolt by hand. Ensure that the tensioner fork is centered over the cylinder block rib. Ensure that the Allen hole on the eccentric is at the 10 o'clock position.
21. Install the timing belt in the following order:
    a. Around the crankshaft sprocket
    b. Around the idler pulley
    c. Around the camshaft sprockets
    d. Around the water pump
    e. Onto the belt tensioner
22. Carefully turn the crankshaft clockwise until the timing belt is tensioned. The belt must be tensioned between the intake camshaft pulley, the idler pulley and the crankshaft
23. Hold the belt tensioner center screw secure. Turn the belt tensioner eccentric counter-clockwise until the tensioner indicator passes the marked position.
24. Then turn the eccentric back so that the indicator reaches the marked position in the center of the window and tighten the center bolt to 15 ft. lbs. (20 Nm).
25. Press the belt to check that the indicator on the tensioner moves easily.
26. Install the lower belt guard.
27. Install the air baffles.
28. Install the upper timing cover.
29. Turn the crankshaft two turns. Check that the markings on the crankshaft and camshaft pulley correspond.

■ **Check that the indicator on the belt tensioner is within the marked area.**

30. Install the lower timing cover.
31. Install the front timing cover and tighten to 9 ft. lbs. (12 Nm).
32. Install the accessory drive belt.
33. Install the engine mount and tighten to 49 ft. lbs. (67 Nm).
34. Install the reservoir. Tighten bolts to 37 ft. lbs. (50 Nm).
35. Install the front wheel.
36. Install the right hand engine cover.
37. Uncover over the right headlamp.
38. Connect the negative battery cable.

2.1L (B21, B21FT) AND 2.3L (B23T) ENGINES

YEAR, MODEL & ENGINE APPLICATION

Model	Engine
1980-82 Volvo 240	2.1L Code B21F
1981-85 240	2.1L Code B21FT
1980-85 200, 700	2.3L Code B23T

REPLACEMENT INTERVALS

Volvo does not provide a replacement interval. Standard industry practice is to replace timing belts every 60,000.

LABOR TIMES

2.1L Engines	
Replacement	2.6 hrs
Severe Service	2.9 hrs
2.3L Engines	
Replacement	2.5 hrs
Severe Service	2.7 hrs

8807VOLg02

✳✳ CAUTION

These engines are an interference type. If the timing belt has failed, valve-to-piston damage may have occurred. A compression check should always be performed to check for engine damage.

1. Disconnect the negative battery cable.
2. Loosen the fan shroud and remove the fan. Remove the shroud.
3. Loosen the alternator, power steering pump and air conditioning compressor and remove the drive belts.
4. Remove the water pump pulley.
5. Remove the 4 retaining bolts and lift off the timing belt cover.
6. To remove the tension from the belt, loosen the nut for the tensioner and press the idler roller back. The tension spring can be locked in this position by inserting the shank end of a 3mm drill through the pusher rod.
7. Remove the 6 retaining bolts and the crankshaft pulley.

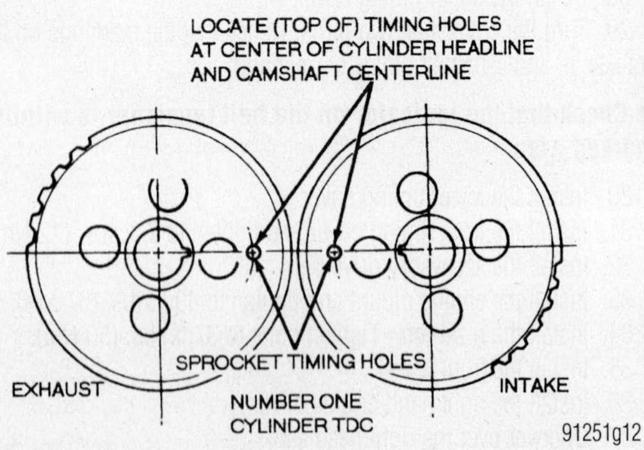

Timing belt sprocket alignment—Volvo 2.1L (B21, B21FT) and 2.3L (B23T) engines

8. Remove the belt, taking care not to bend it at any sharp angles. The belt should be replaced at 45,000 mile intervals, if it becomes oil soaked or frayed or if on a vehicle that has not been operated for any length of time.

To install:

9. If the crankshaft, idler shaft or camshaft were disturbed while the belt was out, align each shaft with its corresponding index mark to assure proper valve timing and ignition timing, as follows:
   a. Rotate the crankshaft so the notch in the convex crankshaft gear belt guide aligns with the embossed mark on the front cover (12 o'clock position).
   b. Rotate the idler shaft so the dot on the idler shaft drive sprocket aligns with the notch on the timing belt rear cover (4 o'clock position).
   c. Rotate the camshaft so the notch in the camshaft sprocket inner belt guide aligns with the notch in the forward edge of the valve cover (12 o'clock position).
10. Install the timing belt (don't use any sharp tools) over the sprockets and then over the tensioner roller. New belts have yellow marks. The 2 lines on the drive belt should fit toward the crankshaft marks. The next mark should then fit toward the intermediate shaft marks, etc. Loosen the tensioner nut and let the spring tension automatically take up the slack. Tighten the tensioner nut to 37 ft. lbs. (51 Nm).
11. Rotate the crankshaft 1 full revolution clockwise and make sure the timing marks still align.
12. Clean all gasket mating surfaces thoroughly. Install the timing belt cover using a new gasket.
13. Install the water pump pulley, drive belts, air conditioning compressor, power steering pump and alternator.
14. Install the fan and shroud. Install the accessory drive belts. Connect the negative battery cable. Start the engine and check for leaks.

2.3L (B230F, B230FT) ENGINES

YEAR, MODEL & ENGINE APPLICATION

Model	Engine
1985-93 Volvo 240, 1992-95 940	2.3L Code B230F
1991 Coupe 2.3L, 1985-92 740 & 760, 1992-95 940	2.3L Code B230FT

REPLACEMENT INTERVALS

Volvo recommends timing belt replacement every 45,000 miles on 1985-90 engines, and 60,000 miles on 1991-95 models.

LABOR TIMES

1985-95	
Replacement	2.7 hrs
Severe Service	3.0 hrs

8807VOLg03

✳✳ CAUTION

These engines are an interference type. If the timing belt has failed, valve-to-piston damage may have occurred. A compression check should always be performed to check for engine damage.

1. Disconnect the negative battery cable.
2. Loosen the fan shroud and remove the fan. Remove the shroud.

3. Loosen the alternator, power steering pump and air conditioning compressor and remove the drive belts.
4. Remove the water pump pulley.
5. Remove the 4 retaining bolts and lift off the timing belt upper cover.
6. Rotate the crankshaft clockwise until the timing marks are aligned.
7. Remove the tensioner nut and washer and install damper wrench 5284 into the crankshaft damper and tensioner stud. Tighten the wrench using the tensioner nut.
8. Remove the crankshaft damper bolt.
9. Remove the damper wrench.
10. Remove the damper.
11. Remove the timing belt lower cover.
12. To remove the tension from the belt, loosen the nut for the tensioner and press the idler roller back. The tension spring can be locked in this position by inserting the shank end of a 3mm drill through the pusher rod.
13. Remove the belt, taking care not to bend it at any sharp angles.

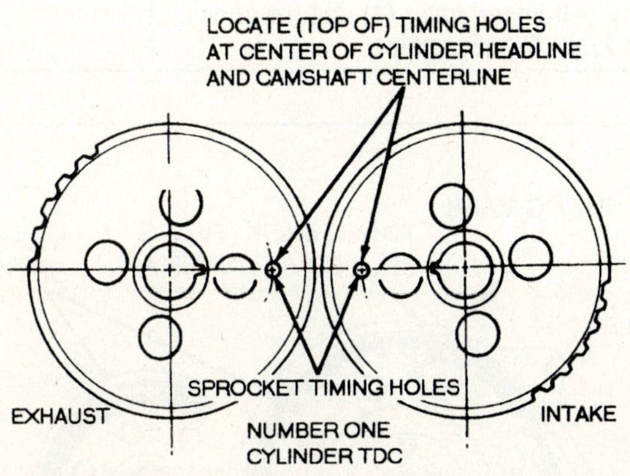

Timing belt sprocket alignment—Volvo 2.3L (B230F, B230FT) engine

To install:

14. If the crankshaft, idler shaft or camshaft were disturbed while the belt was out, align each shaft with its corresponding index mark to assure proper valve timing and ignition timing, as follows:
   a. Rotate the crankshaft so the notch in the convex crankshaft gear belt guide aligns with the embossed mark on the front cover (12 o'clock position).
   b. Rotate the idler shaft so the dot on the idler shaft drive sprocket aligns with the notch on the timing belt rear cover (4 o'clock position).
   c. Rotate the camshaft so the notch in the camshaft sprocket inner belt guide aligns with the notch in the forward edge of the valve cover (12 o'clock position).
15. Install the timing belt (don't use any sharp tools) over the sprockets and then over the tensioner roller. New belts have yellow marks. The 2 lines on the drive belt should fit toward the crankshaft marks. The next mark should then fit toward the intermediate shaft marks, etc.
16. Pull on the timing belt lightly to compress the tensioner spring, then remove the 3mm pin.
17. Install the lower timing belt cover.
18. Install the crankshaft damper.
19. Remove the tensioner nut and washer and install damper wrench 5284 into the crankshaft damper and tensioner stud. Tighten the wrench using the tensioner nut.
20. Tighten the crankshaft damper bolt to 44 ft. lbs. (60 Nm), plus and additional 60 degrees.
21. Remove the damper wrench and install the tensioner nut wand washer.
22. Rotate the crankshaft 2 full revolutions clockwise and make sure the timing marks still align.
23. Loosen the tensioner nut and let the spring tension automatically take up the slack. Tighten the tensioner nut to 37 ft. lbs. (51 Nm).
24. Clean all gasket mating surfaces thoroughly. Install the timing belt cover using a new gasket.
25. Install the water pump pulley and accessory drive belts.
26. Install the fan and shroud.
27. Connect the negative battery cable. Start the engine and check for leaks.

2.3L (B234F) ENGINES

YEAR, MODEL & ENGINE APPLICATION	
1989-92 Volvo 740GLE	2.3L Code B234F

REPLACEMENT INTERVALS

Volvo does not provide a replacement interval. Standard industry practice is to replace timing belts every 60,000.

LABOR TIMES	
1989-92	
Replacement	2.7 hrs
Severe Service	3.0 hrs

8807VOLg04

✻✻ CAUTION

These engines are an interference type. If the timing belt has failed, valve-to-piston damage may have occurred. A compression check should always be performed to check for engine damage.

■ **The B234F engine has 2 belts, one driving the camshafts and one driving the balance shafts. The camshaft belt may be removed separately. The balance shaft belt requires removal of the camshaft belt. During reassembly, the exact placement of the belts and pulleys must be observed.**

1. Remove the negative battery cable and the alternator belt.
2. Remove the radiator fan, its pulley and the fan shroud.
3. Remove the drive belts for the power steering belts and the air conditioning compressor.
4. Beginning with the top cover, remove the retaining bolts and remove the timing belt covers.
5. Turn the engine to Top Dead Center (TDC) of the compression stroke on cylinder No. 1. Make sure the marks on the camshaft pulleys align with the marks on the backing plate and that the marking on the belt guide plate (on the crankshaft) is opposite the TDC mark on the engine block.

6. Remove the protective cap over the timing belt tensioner locknut. Loosen the locknut, compress the tensioner to release tension on the belts and re-tighten the locknut, holding the tensioner in place.

7. Remove the timing belt from the camshafts. Do not crease or fold the belt.

■ **The camshafts and the crankshaft must not be moved when the belt is removed.**

8. Check the tensioner by spinning it counterclockwise and listening for any bearing noise within. Check also that the belt contact surface is clean and smooth. In the same fashion, check the timing belt idler pulleys. Make sure they are tightened to 19 ft. lbs. (25 Nm).

9. If the balance shaft belt is to be removed:

a. Remove the balance shaft belt idler pulley from the engine.

b. Loosen the locknut on the tensioner and remove the belt. Slide the belt under the crankshaft pulley assembly. Check the tensioner and idler wheels carefully for any sign of contamination. Check the ends of the shafts for any sign of oil leakage.

c. Check the position of the balance shafts and the crankshaft after belt removal. The balance shaft markings on the pulleys should align with the markings on the backing plate and the crankshaft marking should still be aligned with the TDC mark on the engine block.

d. When refitting the balance shaft belt, observe that the belt has colored dots on it. These marks assist in the critical placement of the

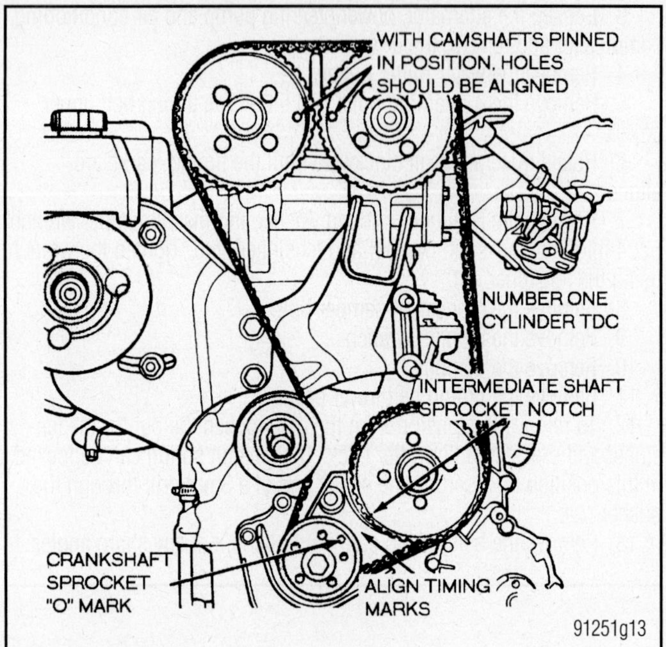

Timing belt upper cover (1) and lower cover (2) mounting—Volvo 2.3L (B234F) engine

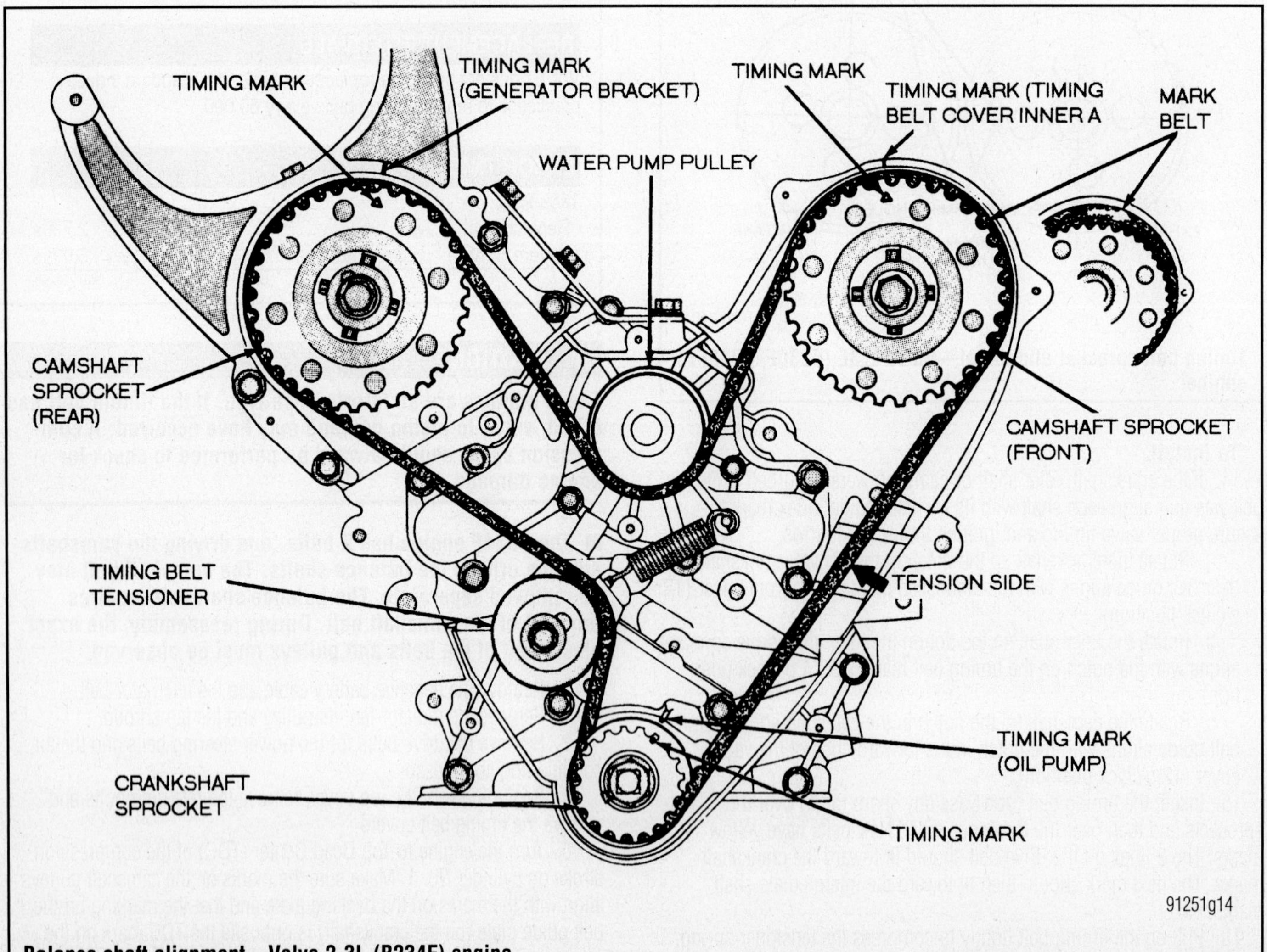

Balance shaft alignment—Volvo 2.3L (B234F) engine

belt. The yellow dot will align the right lower shaft, the blue dot will align on the crank and the other yellow dot will match to the upper left balance shaft.

e. Carefully work the belt in under the crankshaft pulley. Make sure the blue dot is opposite the bottom (TDC) marking on the belt guide plate at the bottom of the crankshaft. Fit the belt around the left upper balance shaft pulley, making sure the yellow mark is opposite the mark on the pulley. Install the belt around the right lower balance shaft pulley and again check that the mark on the belt aligns with the mark on the pulley.

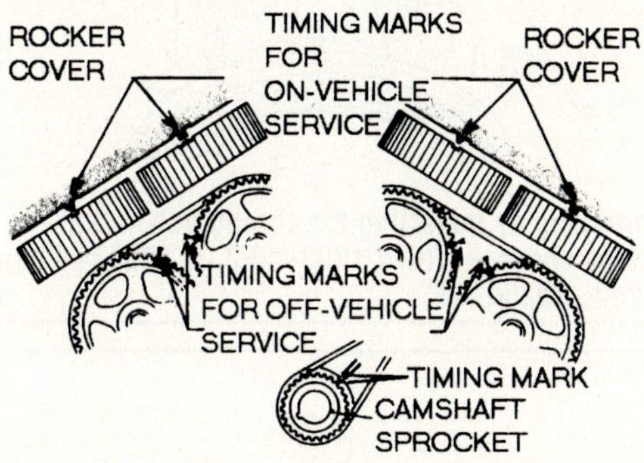

Balance shaft belt markings. There should be 18 teeth between A and B and 34 teeth between B and C—Volvo 2.3L (B234F) engine

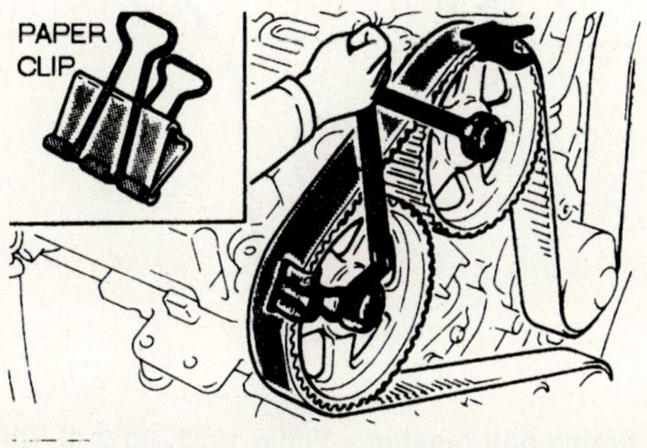

Rotate the crankshaft clockwise to position the camshaft 1 1/2 teeth beyond the alignment marks to properly tension the belt—Volvo 2.3L (B234F) engine

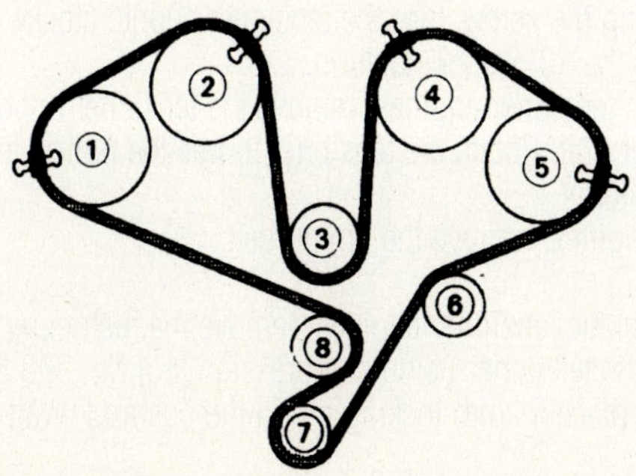

Timing mark alignment—Volvo 2.3L (B234F) engine

f. Work the belt around the tensioner. Double check that all the markings are still aligned.

g. Set the belt tension by inserting an Allen key into the adjusting hole in the tensioner. Turn the crankshaft carefully through a few degrees on either side of TDC to check that the belt has properly engaged the pulleys. Return the crank to the TDC position and set the adjusting hole just below the 3 o'clock position when tightening the adjusting bolt. Use the Allen wrench, in the adjusting hole, as a counter hold and tighten the locking bolt to 29.5 ft. lbs. (40 Nm).

h. Use tool 998 8500 or equivalent, to check the tension of the belt. Install the gauge over the position of the removed idler pulley. The tension must be 1–4 units on the scale or the belt must be readjusted.

To install:

10. Reinstall the camshaft belt by aligning the double line marking on the belt with the top marking on the belt guide plate at the top of the crankshaft. Stretch the belt around the crank pulley and place it over the tensioner and the right-side idler. Place the belt on the camshaft pulleys. The single line marks on the belt should align exactly with the pulley markings. Route the belt around the oil pump drive pulley and press the belt onto the left-side idler.

11. Check that all the markings align and that the engine is still positioned at TDC, of the compression stroke, for cylinder No. 1.

12. Loosen the tensioner locknut.

13. Turn the crankshaft clockwise. The camshaft pulleys should rotate 1 full turn until the marks again align with the marks on the backing plate.

■ **The engine must not be rotated counterclockwise during this procedure.**

14. Smoothly rotate the crankshaft further clockwise until the camshaft pulley markings are 1 1/2 teeth beyond the marks on the backing plate. Tighten the tensioner locknut.

15. Check the belt guide for the balance shaft belt and make sure it is properly seated. Install the center timing belt cover, the one that covers the tensioner, the fan shroud, fan pulley and fan. Install all the drive belts and connect the battery cable.

16. Double check all installation items, paying particular attention to loose hoses or hanging wires, loose nuts, poor routing of hoses and wires (too tight or rubbing) and tools left in the engine area.

17. Start the engine and allow it to run until the thermostat opens.

**** CAUTION**

The upper and lower timing belt covers are still removed. The belt and pulleys are exposed and moving at high speed.

18. Shut the engine OFF and bring the motor to TDC, of the compression stroke, on cylinder No. 1.

19. Check the tension of the camshaft belt. Position the gauge between the right (exhaust) camshaft pulley and the idler. Belt tension must be 5.5 (0.2 units on a suitable belt tension gauge. If the belt needs adjustment, remove the rubber cap over the tensioner locknut, cap is located on the timing belt cover, and loosen the locknut.

20. Insert a suitable tool between the tensioner wheel and the spring carrier pin to hold the tensioner. If the belt needs to be tightened, move the roller to adjust the tension to 6.0 units. If the belt is too tight, adjust to obtain a reading of 5.0 units on the gauge. Tighten the tensioner locknut.

21. Rotate the crankshaft so the camshaft pulleys move through 1 full revolution and recheck the tension on the camshaft belt. It should now be 5.5 (0.2 units. Install the plastic plug over the tensioner bolt.

22. Final check the tension on the balance shaft belt by fitting the gauge and turning the tensioner clockwise. Only small movements are needed. After any needed readjustments, rotate the crankshaft clockwise through 1 full revolution and recheck the balance shaft belt. The tension should now be on the final specification of 4.9 (0.2 units.

23. Install the idler pulley for the balance shaft belt. Reinstall the upper and lower timing belt covers.

24. Start the engine and final check performance.

2.3L (B5234/T) AND 2.4L (B5244, B5254/F/T) 1993–00 ENGINES

YEAR, MODEL & ENGINE APPLICATION

Application	Engine
1996-97 Volvo 850R, 1997-00 C70, S70 & V70	2.3L Codes B5234/T
1993-97 850, 1997-00 C70, S70 & V70	2.4L Codes B5244, B5254/F/T

REPLACEMENT INTERVALS

Volvo recommends timing belt replacement every 50,000 miles on 1993 models, and 80,000 miles on 1994-00 models.

LABOR TIMES

2.3L Engines	
Replacement	2.0 hrs
Severe Service	2.2 hrs
2.4L B5244 Engines	
Replacement	2.0 hrs
Severe Service	2.2 hrs
2.4L B4254/F/T Engines	
Replacement	2.7 hrs
Severe Service	3.0 hrs

8807VOLg05

✻✻ CAUTION

These engines are an interference type. If the timing belt has failed, valve-to-piston damage may have occurred. A compression check should always be performed to check for engine damage.

■ These engines are not variable valve timing engines. There are 2 versions of these engines. One has a mechanical timing belt tensioner and the other with an automatic tensioner.

1. Disconnect the negative battery cable.
2. Remove the spark plug cover.
3. Remove the ignition coil cover.
4. Remove the fuel line retaining clips.
5. Remove the coolant expansion tank and place it on top of the engine.
6. Remove the accessory drive belt.
7. Remove the front timing belt cover.
8. Raise and support the vehicle.
9. Remove the right side wheel.
10. Remove the splash shield.
11. Remove the crankshaft damper guard.
12. Remove the timing belt upper cover.

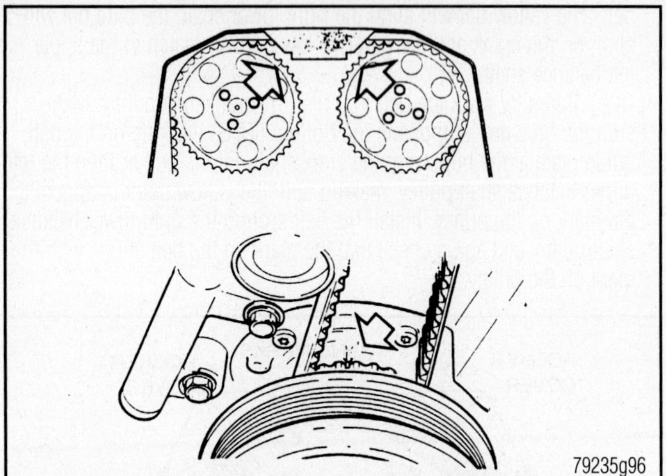

79235g96

Before removing or installing the timing belt, align the timing marks as shown—Volvo 1993–00 2.3L (B5234/T) and 2.4L (B5244, B5254/F/T)

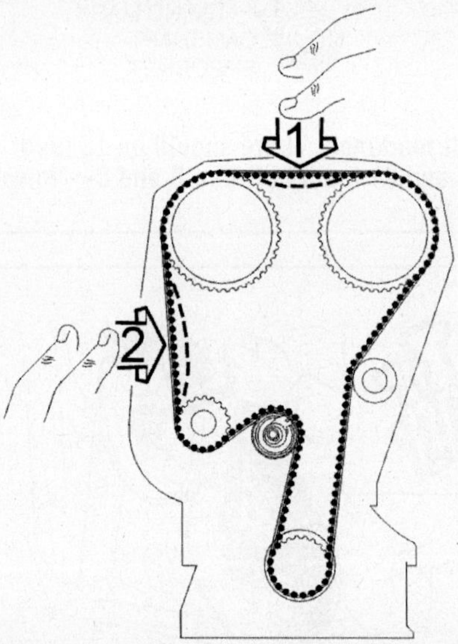

42348-XC90-g22

Setting timing belt tension—Volvo 1993–00 2.3L (B5234/T) and 2.4L (B5244, B5254/F/T) with automatic tensioner

13. Rotate the crankshaft in the normal rotation direction and align the crankshaft and camshaft sprocket timing marks.

14. Remove the crankshaft pulley using counter hold tool No. 999-5433 or equivalent, and remove the pulley.

15. On mechanical tensioner engines, loosen the tensioner screw and while holding the screw, turn the tensioner counterclockwise with a 6mm wrench to the 10 o'clock position.

On automatic tensioner engines, remove the tensioner upper bolt and loosen the lower bolt. Rotate the tensioner to free the pulley, then remove the tensioner.

16. On all engines, remove the timing belt.

To install:

17. On automatic tensioner engines, remove the restricting washer and compress the tensioner using compression tool No. 999 5456 or equivalent, and place a 2mm locking pin in the pushrod. Wait 5 minutes.

18. Remove the tensioner from the tool and install the tensioner with the 2mm pin in place. Tighten the bolt to 18 ft. lbs. (25 Nm).

19. Ensure the timing marks are still aligned.

20. On all engines, install the timing belt over the crankshaft sprocket and right idler pulley, then over the camshaft sprockets, water pump pulley and tensioner pulley.

21. On automatic tensioner engines, proceed as follows:

a. Remove the lockpin and allow the tensioner to tension the belt.

b. Push on the timing belt at the locations shown to set the belt.

c. Install a new control washer on the stem of the tensioner piston. The countersunk side faces up.

d. Install the upper timing belt cover.

e. Turn the crankshaft 2 revolutions in the normal rotation and realign the timing marks.

f. Wait 5 minutes after aligning the marks, then install Volvo Gauge 998-8500 or equivalent, between the exhaust camshaft and water pump. Read the gauge using a mirror, while still installed. For 23mm belts, the tension should be 2.5–4.0 units.

For 21mm belts, the tension should be 3.5–4.6 units. If specifications are not met, replace the tensioner.

22. On mechanical tensioner engines, proceed as follows:

a. Ensure the timing belt is taut, and tighten the tensioner screw to 48 inch lbs. (4 Nm).

b. Hold the screw and using a 6mm wrench tighten the pulley until it reaches the right side stop. Press on the timing belt to ensure the tensioner pointer moves freely in both directions.

c. Tighten the tensioner bolt to 18 ft. lbs. (25 Nm).

d. Install the upper timing belt cover.

23. On all engines, install the crankshaft pulley using counter hold tool No. 999-5433 or equivalent, and tighten the pulley to 132 ft. lbs. (179 Nm).

24. Install the timing belt covers and the fuel line clips.

25. Install the accessory belts.

26. Install the vibration damper guard and the inner fender well.

27. Install the spark plug and ignition coil cover and any remaining components.

2.3L (B5234T/3), 2.4L (B5244S/T) AND 2.5L (B5254T/2/4) 2000–05 ENGINES

YEAR, MODEL & ENGINE APPLICATION	
2000-02 Volvo C70, 2000-04 S60, 2000 S70,	
2000-04 V70,	2.3L Code B5234T3
2003-04 C70	2.3L Code B5234T9
2000-05 S60, V70	2.4L Code B5244S
2000-01 C70, 2000-03 S60, 2000-04 V70,	
2000-02 XC	2.4L Code B5244T3
2003-04 C70	2.4L B5244T7
2003 V70, 2003-04 XC	2.5L Code B5254T
2005 V70, XC	2.5L Code B5254T2/4

REPLACEMENT INTERVALS

Volvo recommends timing belt replacement every 105,000 miles.

LABOR TIMES	
2.3L & 2.4L Engines	
Replacement	2.0 hrs
Severe Service	2.2 hrs
2.5L Engines	
Replacement	2.6 hrs
Severe Service	2.9 hrs

8807VOLg06

✻✻ CAUTION

These engines are an interference type. If the timing belt has failed, valve-to-piston damage may have occurred. A compression check should always be performed to check for engine damage.

■ **These engines are variable valve timing engines.**

1. Disconnect the negative battery cable.
2. On S60 and V70 models, remove the cross stay between the suspension turrets.
3. On all models, remove the upper timing belt cover.
4. Remove the servo reservoir and the expansion tank. Lift up and place on top of the engine.
5. Remove the auxiliary drive belt.
6. Remove the front timing belt cover.
7. Raise and support the vehicle.
8. Remove the right front wheel.
9. Remove the inner fender liner.
10. Install the upper timing belt cover
11. Turn the crankshaft clockwise until the markings on the crankshaft and camshaft pulley correspond.
12. Turn the crankshaft a further 1/4 turn clockwise and then back again until the markings correspond.
13. Remove the upper timing belt cover.
14. Remove the vibration damper. Use a counterhold. Work the vibration damper loose.
15. Back off the center bolt of the belt tensioner slightly.
16. Hold the center bolt still. Turn the tensioner eccentric clockwise, using a 6 mm Allen wrench to the 10 o'clock position.
17. Remove the timing belt from the tension pulley, camshaft pulley and water pump.
18. Remove the timing belt.
19. Spin the idler pulley and listen for noise. If replacing with a new idler pulley, tighten to 18 ft. lbs. (24 Nm).
20. Spin the tension pulley and listen for noise. When replacing, bolt the tension pulley into place using the center bolt. Screw in the center bolt by hand.

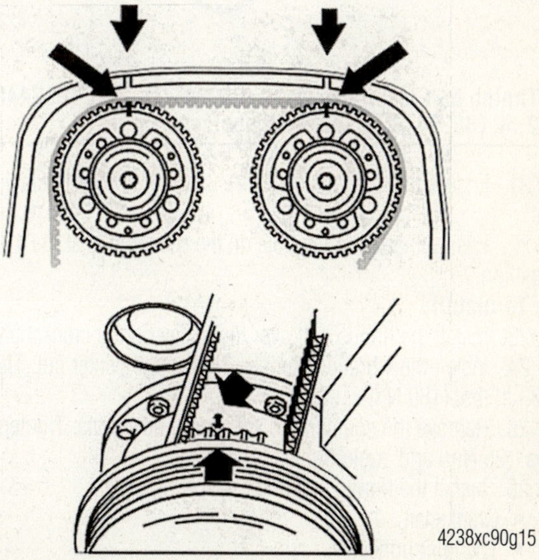

4238xc90g15

Crankshaft and camshaft sprocket timing mark alignment—2.3L (B5234T/3), 2.4L (B5244S/T) and 2.5L (B5254T/2/4) 2000–05 engines

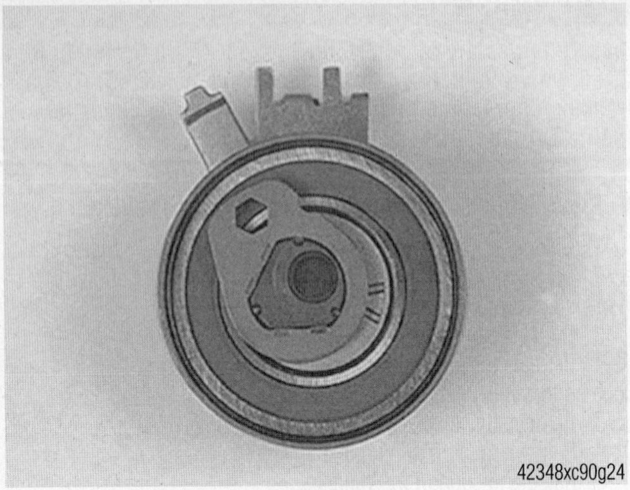

Timing belt tensioner—2.3L (B5234T/3), 2.4L (B5244S/T) and 2.5L (B5254T/2/4) 2000–05 engines

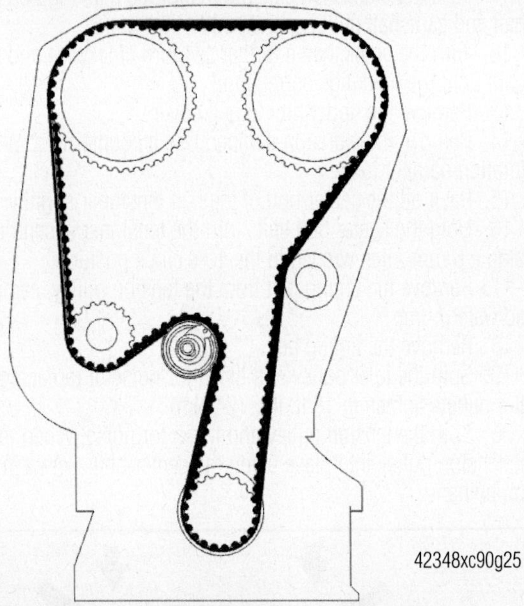

Timing belt installed—2.3L (B5234T/3), 2.4L (B5244S/T) and 2.5L (B5254T/2/4) 2000–05 engines

21. Ensure that the tensioner fork is centered over the cylinder block rib.

22. Ensure that the Allen hole on the eccentric is at the 10 o'clock position.

To install:

23. Install the timing belt over the pulley on the crankshaft.

24. Install the vibration damper. Tighten the center nut. Tighten to 133 ft. lbs. (180 Nm).

25. Remove the counterhold and install new bolts. Tighten to 18 ft. lbs. (25 Nm) and angle-tighten 30 degrees.

26. Install the timing belt in the following order:

- Crankshaft
- The idler pulley
- intake camshaft pulley
- Exhaust camshaft pulley
- Water pump

Tension indicator at different temperatures—2.3L (B5234T/3), 2.4L (B5244S/T) and 2.5L (B5254T/2/4) 2000–05 engines

- Belt tensioner.

■ **The following adjustment is carried out on a cold engine. A suitable temperature is approximately 68°F. At higher temperatures, for example with the engine at operating temperature or at higher ambient temperature, the needle is further to the right. The illustration shows the position of the indicator when aligning the timing belt tensioner at different temperatures.**

27. Tension the timing belt as follows:

- Turn the crankshaft clockwise carefully until the timing belt is tensioned. The belt must be tensioned between the intake camshaft pulley, the idler pulley and the crankshaft
- Hold the center bolt on the belt tensioner secure. Turn the belt tensioner eccentric counter-clockwise until the tensioner indicator passes the marked position. Then turn the eccentric back so that the indicator reaches the marked position in the center of the window
- Hold the eccentric secure and tighten the center bolt. Tighten to 15 ft. lbs. (20 Nm). Check that the indicator is in the correct position.

28. To check the alignment:

- Press the belt to check that the indicator on the tensioner moves easily
- install the upper timing belt cover
- Turn the crankshaft 2 turns. Check that the markings on the crankshaft and camshaft pulley correspond
- Check that the indicator on the belt tensioner is within the marked area.

29. Install or connect the following:

- The front timing belt cover. Tighten to 105 inch lbs. (12 Nm).
- The upper timing belt cover
- The auxiliaries belt
- The servo reservoir
- The expansion tank

■ **Ensure that the hoses are correctly positioned.**

- On S60 and V70 models, the engine stabilizer brace. Tighten the bolts at the suspension turrets. Tighten to 37 ft. lbs. (50 Nm). Tighten the engine bracket bolt. Tighten to 58 ft. lbs. (80 Nm).
- The cover in the fender liner
- The front wheel

30. Connect the negative battery cable.

2.4L (D24 AND D24T) DIESEL ENGINES

YEAR, MODEL & ENGINE APPLICATION

1983-86 Volvo 760GLE Diesel	2.4L Code D24, D24T

REPLACEMENT INTERVALS

Volvo does not provide a replacement interval. Standard industry practice is to replace timing belts every 60,000 miles.

LABOR TIMES

1983-86	
Replacement	3.4 hrs
Severe Service	3.9 hrs

8807VOLg07

✻✻ CAUTION

These engines are an interference type. If the timing belt has failed, valve-to-piston damage may have occurred. A compression check should always be performed to check for engine damage.

1. Disconnect the negative battery cable.
2. Drain the cooling system.
3. Loosen the alternator, air pump, power steering pump (if equipped), air conditioning compressor (if equipped) and remove their drive belts.
4. Remove the water pump pulley.
5. Remove the timing belt cover bolts and the timing cover.
6. If necessary, remove the valve cover.
7. Using a 1 1/16 in. (27mm) socket on the crankshaft pulley bolt, rotate the crankshaft to position the No. 1 cylinder at Top Dead Center (TDC) of its compression stroke. Both cam lobes should point up at equally large angles. The flywheel timing mark should be set at **0**.
8. Remove the vibration damper center bolt. It may be necessary to use Volvo special wrenches 5187 (to hold) and 5188 (to remove). The engine may have to be turned slightly to allow the holding wrench to rest temporarily on the cooling fan.

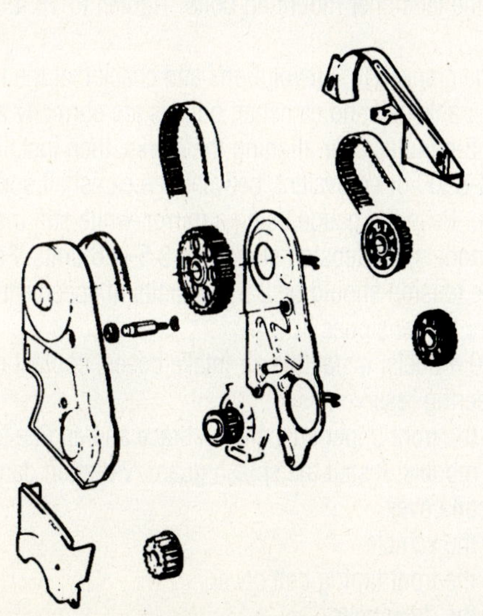

93052g01

Exploded view of the timing belt and related components—Volvo 2.4L (D24 and D24T) diesel engine

9. Check to make sure the No. 1 cylinder is at TDC. If necessary, adjust the flywheel to the **0** mark.
10. Remove the vibration damper by removing the 4 6mm Allen bolts.

■ **The vibration damper and the crankshaft gear may be stuck together. You may have to tap them apart.**

11. Remove the camshaft gear belt by removing the lower belt shield and releasing the retainer bolts for the water pump.
12. Pull the timing belt straight out and off of the gears.
13. Remove the idler pulley center bolt. Using a puller, remove the idler pulley.

To install:

■ **The idler pulley MUST be replaced when replacing the timing belt.**

14. Tap the new idler pulley into position. Install the idler pulley center bolt.
15. With the timing marks aligned, install the timing belt.
16. Using the special tools, install the vibration damper pulley.
17. Apply tension to the timing belt.
18. Rotate the crankshaft 2 complete revolutions and realign the timing marks. If the timing marks do not align, reinstall the timing belt.
19. To complete the installation, reverse the removal procedures.
20. Refill the cooling system.
21. Connect the negative battery cable.

2.8L (B6284T) AND 2.9L (B6304F/S) ENGINES

YEAR, MODEL & ENGINE APPLICATION

1999 Volvo S80 Turbo	2.8L Code B6284T
1991-97 960	2.9L Code B6304F
1999 S80, 1997-98 S90 & V90	2.9L Code B6204S

REPLACEMENT INTERVALS

Volvo recommends timing belt replacement every 30,000 miles on 1991-93 models, every 50,000 miles on 1994 models, and 80,000 miles on 1995-99 models.

LABOR TIMES

2.8L B6284T Engine	
Replacement	5.0 hrs
Severe Service	5.5 hrs
2.9L B6304F Engine	
Replacement	3.6 hrs
Severe Service	4.1 hrs
2.9L B6304S Engine	
Replacement	4.1 hrs
Severe Service	4.6 hrs

8807VOLg08

✱✱ CAUTION

These engines are an interference type. If the timing belt has failed, valve-to-piston damage may have occurred. A compression check should always be performed to check for engine damage.

1. Disconnect the negative battery cable.
2. Remove the drive belts.
3. Remove the front timing belt cover.
4. Raise and support the vehicle.
5. Remove the splash guard, vibration damper guard and ignition coil cover.
6. On S80 models, remove the front upper suspension brace and engine steady bar.
7. Remove the air intake hoses, coolant reservoir and power steering reservoir.
8. On all models, rotate the crankshaft clockwise, until the timing marks on the camshaft pulleys and timing belt mounting plate and crankshaft pulley/oil pump housing are aligned.
9. Remove the lower timing belt cover.

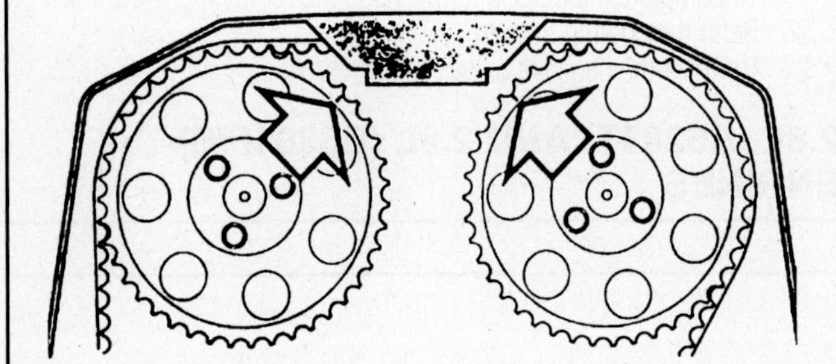

91251gi9

Aligning the camshaft sprocket timing marks (arrows)— Volvo 2.8L (B6284T) and 2.9L (B6304F/S) engines

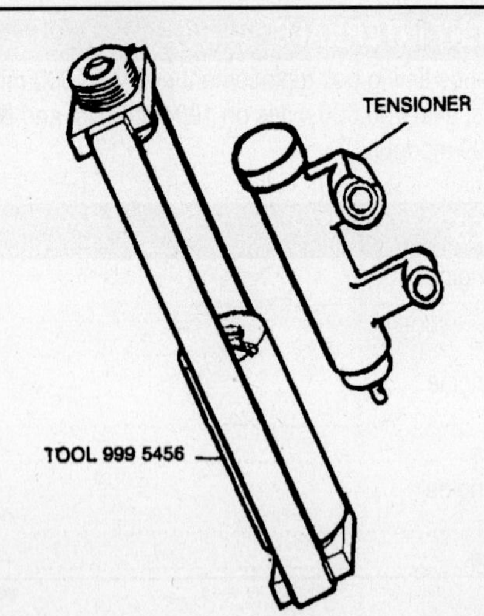

91251gj1

Using special tool be used to compress the tensioner to avoid damage— Volvo 2.8L (B6284T) and 2.9L (B6304F/S) engines

10. Remove the tensioner upper mounting bolts. Loosen the tensioner lower mounting bolt and twist the tensioner to free the plunger. Remove the lower mounting bolt and remove the tensioner.
11. Remove the timing belt.

■ **Do not rotate the crankshaft while the timing belt is removed.**

12. Check the tensioner and idler pulleys, as follows:
    a. Spin the pulleys and listen for bearing noise.
    b. Check that the pulley surfaces in contact with the belt are clean and smooth.
    c. Check the tensioner pulley arm and idler pulley mountings.
    d. Tighten the tensioner pulley arm to 30 ft. lbs. (40 Nm) and the idler pulley to 18 ft. lbs. (25 Nm).
    e. Compress the tensioner using tool 5456 or equivalent. Mount the tensioner in the tool and tighten the center nut fully. Wait until compression has taken place and insert a 2mm locking pin in the plunger.

■ **The tensioner must be replaced if leakage is observed or the plunger offers no resistance when depressed, or cannot be depressed.**

To install:

■ **The lever bushing must be greased every time the belt is replace or the tensioner pulley removed. This is necessary to help prevent seizure of the bushing, with the possible risk of incorrect belt tension. Service the bushing, using the following procedure:**

- Remove the lever mounting bolt, tensioner pulley and sleeve behind the bolt.
- Grease the surfaces of the bushing, bolt and sleeve, using part 1161246-2 or equivalent.
- Install the sleeve, tensioner pulley and lever mounting bolt.
- Tighten the bolt 30 ft. lbs. (40 Nm).

13. Place the belt around the crankshaft pulley and right-side idler. Place the belt over the camshaft pulleys. Position the belt around the water pump and press over the tensioner pulley.
14. Insert the tensioner mounting bolts. Tighten to 18 ft. lbs. (25 Nm).
15. Turn the crankshaft 2 revolutions and check that the timing marks on the crankshaft and camshaft pulleys are correctly aligned.
    a. Wait 5 minutes after aligning the marks, then install Volvo Gauge 998-8500 or equivalent, between the camshaft sprockets and water pump. Read the gauge using a mirror, while still installed. For 1991–94 models, the tension should be 3.5–4.6 units. For 1995–99 models, the tension should be 2.5–3.5 units. If specifications are not met, replace the tensioner.
16. On S80 models, install the air intake hoses, coolant reservoir and power steering reservoir.
17. Install the front upper suspension brace and engine steady bar.
18. On all models, install the splash guard, vibration damper guard and ignition coil cover.
19. Lower the vehicle.
20. Install the front timing belt cover.
21. Install the drive belts.
22. Connect the negative battery lead, start and check the engine operation.

2.9L (B6294/S/T) ENGINES

YEAR, MODEL & ENGINE APPLICATION	
2000-05 Volvo S802.9L/T6, 2003-05 XC90	2.9L Code B6294S/T

REPLACEMENT INTERVALS
Volvo recommends timing belt replacement every 105,000 miles.

LABOR TIMES	
B6294S Engine	
Replacement	3.1 hrs
Sevee Service	3.4 hrs
B6294T Engine	
Replacement	4.1 hrs
Severe Service	4.6 hrs

8807VOLg09

✳✳ CAUTION

These engines are an interference type. If the timing belt has failed, valve-to-piston damage may have occurred. A compression check should always be performed to check for engine damage.

1. Disconnect the negative battery cable.
2. Remove the engine stabilizer brace from the engine bracket and strut towers.
3. Remove or disconnect the following:
   - Pipes between the turbocharger (TC) and charge air cooler (CAC), if equipped
   - Pipes between the air cleaner (ACL) and turbocharger (TC), if equipped
   - Clamp from the air intake pipe for the turbocharger (TC) for cylinders 1, 2 and 3, if equipped, and twist pipe toward the rear
4. Remove the accessory drive belt.
5. Remove the upper and front timing belt covers.
6. Remove the servo reservoir and place on engine.
7. Remove the expansion tank.
8. Raise and support the vehicle.
9. Remove the right front wheel.
10. Remove the inner fender liner.
11. Install the upper timing belt cover.
12. Turn the crankshaft clockwise until the markings on the crankshaft and camshaft pulleys correspond. Turn the crankshaft a further 1/4 turn clockwise.
13. Turn back counter-clockwise until the markings correspond.
14. Remove the upper timing belt cover.
15. Remove the crankshaft pulley.
16. Remove the belt cover behind the crankshaft pulley for the auxiliaries belt.
17. Slacken off the center screw for the belt tensioner slightly.
18. Hold the center screw still and turn the tensioner eccentric counter-clockwise to 10 o'clock.
19. Remove the timing belt.

■ **Do not turn the camshafts or the crankshaft when the timing belt has been removed.**

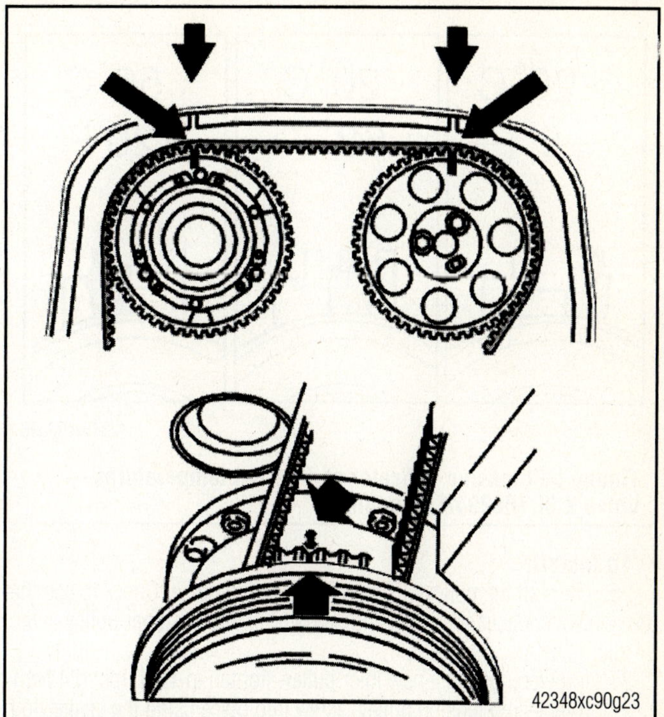

42348xc90g23

Crankshaft and camshaft sprocket timing marks—Volvo 2.9L (B6294/S/T) engine

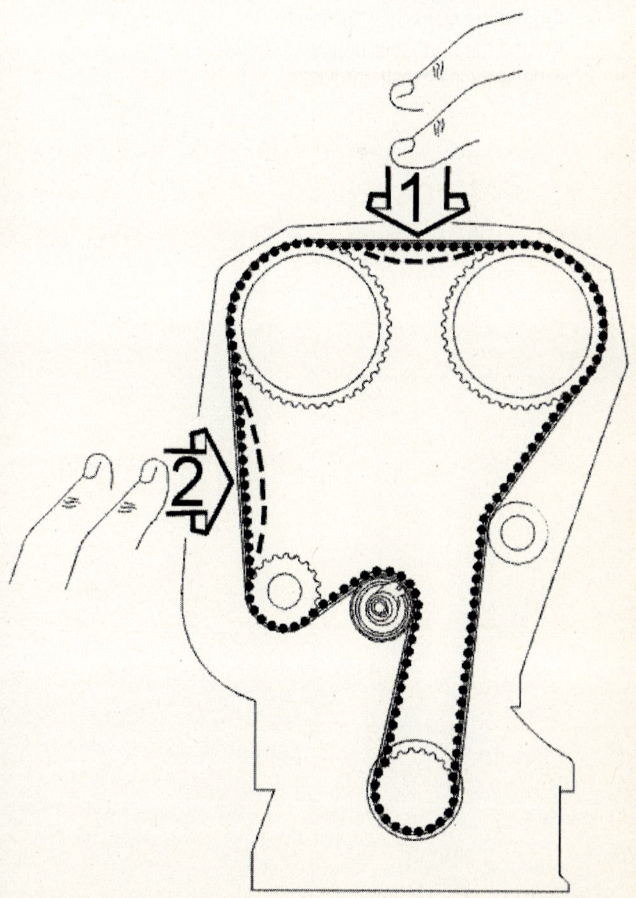

42348xc90g22

Setting timing belt tension—Volvo 2.9L (B6294/S/T) engine

Timing belt tension indicator at different temperatures—Volvo 2.9L (B6294/S/T) engine

To install:

20. Turn all the pulleys listening for bearing noise. Check to see that the contact surfaces are clean and smooth. If the tensioner pulley lever or idler is seized, replace it.

21. If replacing with a new idler pulley, tighten to 18 ft. lbs. (24 Nm).

22. If replacing tension pulley, screw into place using the center bolt by hand. Ensure that the tensioner fork is centered over the cylinder block rib. Ensure that the Allen hole on the eccentric is at 10 o'clock position..

23. Install the timing belt in the following order:
   a. Around the crankshaft sprocket.
   b. Around the right idler pulley
   c. Around the camshaft sprockets
   d. Around the water pump
   e. Onto the tensioner pulley

24. Carefully turn the crankshaft clockwise until the timing belt is tensioned. The belt must be tensioned between the intake camshaft pulley, the idler pulley and the crankshaft

25. Hold the belt tensioner center screw secure. Turn the belt tensioner eccentric counter-clockwise with a 6mm Allen until the tensioner indicator passes the marked position, depending on temperature.

26. Then turn the eccentric back so that the indicator reaches the marked position in the center of the window and tighten the center bolt to 18 ft. lbs. (25 Nm).

27. Press the belt to check that the indicator on the tensioner moves easily.

28. Install or connect the following:

- Accessory drive belt
- Vibration damper. Tighten the bolts to 26 ft. lbs. (35 Nm) plus 50 degrees
- Inner fender liner
- Front wheel
- Front timing belt cover
- Expansion tank
- Servo reservoir
- Clamp for the air intake pipe
- Pipes between the air cleaner (ACL) and turbocharger (TC), if equipped
- Pipes between the turbocharger (TC) and charge air cooler (CAC), if equipped
- Engine stabilizer brace. Tighten the bolts at the suspension turrets to 37 ft. lbs. (50 Nm). Tighten the engine bracket bolt to 59 ft. lbs. (80 Nm).

29. Connect the negative battery cable.

Domestic Manual ISBN 1-4180-0606-8/Part No. 130606
Import Manual ISBN 1-4180-1537-7/Part No. 131537

Chilton has added so much to its labor guide manual that we've had to put it in two volumes! We've added hundreds of new labor operations—including maintenance services and electronic system diagnosis—to the ***Chilton*® *2006 Import*** and ***Domestic Labor Guide Manuals***. All labor times for 1981 through 2006 vehicles consider the real world environment in which technicians work: worn, rusted, or dirty components, being serviced with tools commonly used in the aftermarket. Chilton labor times are accepted by most insurance and extended warranty companies. Vehicle makes and models conform to current Automotive Aftermarket Industry Association standards.

Labor Guide Manual Benefits:

- parts terminology is more standardized across different OEMs to simplify reference
- a total of more than 2,500 pages of updated Chilton labor times appear in these two volumes
- make sure your students have this latest edition because our experts have updated hundreds of labor times for earlier models

Hardcover Manuals are 8 1/2" x 11", ©2006

Labor Guide CD-ROM Benefits:

- easy-to-use software to create and print professional-quality estimates and invoices
- three user-defined levels of labor rates correspond to different types of job scenarios, for "real-world" application
- functions as a database of aftermarket labor times for monitoring warranty and insurance claims
- software keeps track of customers and prior estimates for time-saving recall
- customizable application allows service writers to add labor operations and times, and parts companies to add labor times to existing parts ordering systems

CD ISBN 1-4180-0605-X/Part No. 130605
©2006

Previous Year Editions:
Chilton 2005 Labor Guide Manual, ISBN 1-4018-7412-6/Part No. 27412
Chilton 2005 Labor Guide CD-ROM, ISBN 1-4018-7818-0/Part No. 27818

CHILTON 2006 MECHANICAL SERVICE MANUALS

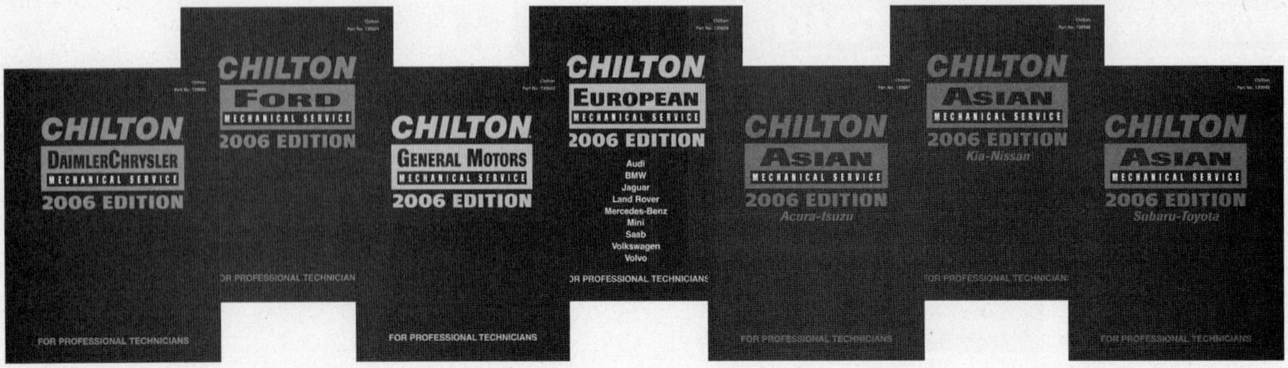

The ***Chilton*® *2006 Mechanical Service Manuals*** provide updated coverage through 2005 models and even many 2006 models, as made available from original equipment manufacturers (OEMs). Chilton is still your reliable source for fast, accurate repairs and reassembly and it still provides the lowest-priced professional repair manuals on the market! These manuals are organized by make, model, and system so information gathering is easier. Now with even more illustrations and a streamlined index, it's no wonder more automotive professionals turn to Chilton Professional Manuals for their mechanical service and repair information.

Mechanical Service Manual Benefits:

- access up-to-date service and repair information covering model years 2002-2006, all logically arranged by manufacturer
- follow clear, step-by-step procedures—from drive train to chassis—to yield fast, accurate results
- service more mechanical systems, including brakes, engines, suspensions, steering, and related components
- know what special tools are required for specific jobs, as Chilton editors describe and illustrate them to make repair work go more smoothly

2006 Editions

Chilton 2006 DaimlerChrysler Mechanical Service Manual—ISBN 1-4180-0600-9/Part No. 130600
Chilton 2006 Ford Mechanical Service Manual—ISBN 1-4180-0601-7/Part No. 130601
Chilton 2006 General Motors Mechanical Service Manual—ISBN 1-4180-0602-5/Part No. 130602
Chilton 2006 Asian Mechanical Service Manual—Volume I—ISBN 1-4180-0947-4/Part No. 130947
Chilton 2006 Asian Mechanical Service Manual—Volume II—ISBN 1-4180-0948-2/Part No. 130948
Chilton 2006 Asian Mechanical Service Manual—Volume III—ISBN 1-4180-0949-0/Part No. 130949
Chilton 2006 Asian Mechanical Service Manual—3 Volume Set—ISBN 1-4180-0603-3/Part No. 130603
Chilton 2006 European Mechanical Service Manual—ISBN 1-4180-0604-1/Part No. 130604

Asian Manuals Expected Release Date—Summer 2006
European Manuals Expected Release Date—Fall 2006
Manuals are 8 1/2" x 11", ©2006

2005 Editions

Chilton 2005 General Motors Mechanical Service Manual—ISBN 1-4018-7146-1/Part No. 27146
Chilton 2005 Chrysler Mechanical Service Manual—ISBN 1-4018-6718-9/Part No. 26718
Chilton 2005 Ford Mechanical Service Manual—ISBN 1-4018-6719-7/Part No. 26719
Chilton 2005 European Mechanical Service Manual—ISBN 1-4018-6720-0/Part No. 126720
Chilton 2005 Asian Mechanical Service Manual – Volume I—(Acura-Mazda) ISBN 1-4018-6716-2/Part No. 26716
Chilton 2005 Asian Mechanical Service Manual – Volume II—(Mitsubishi-Toyota)
ISBN 1-4018-6717-0/Part No. 26717
Chilton 2005 Asian Mechanical Service Manual – 2 Volume Set—ISBN 1-4018-7180-1/Part No. 27180

Manuals are 8 1/2" x 11", ©2005

CHILTON® PRO

With ChiltonPro gain all-inclusive access to the latest in Chilton Automotive Repair Information!—Visit **www.chiltonsonline.com** to find out more.

ChiltonPro is the new alternative for Professional Technicians who want a cost-effective electronic automotive repair system. It combines Chilton's famous automotive repair information into one solution covering automobiles and light trucks. The information is delivered online or on DVD-ROMs—your choice—updated regularly throughout the year.

Benefits and Features

- Create better estimates using labor times developed with real world factors: worn, rusted or dirty components, common aftermarket tools and training
- Accurately identify and solve engine performance problems using critical diagnostic information
- Save time with expert guidance through OBD-II diagnostics
- Understand system operation by accessing more explanations and theory than found in other automotive repair systems
- Stay current on key Technical Service Bulletins (TSBs)
- Quickly understand systems by viewing cutaway diagrams, charts and actual photos
- Navigate easily using AAIA standards for year, make, and model
- Easily print what you need; no more misplaced books or missing pages

Content:

- OEM maintenance schedules, 1997-2005
- Chilton labor times, 1981 - 2005
- Chilton mechanical procedures, 1970-2005, with some back to 1940s
- Chilton diagnostics, 1995 -2003 domestic and Asian models, with some back to 1990
- OEM Technical Service Bulletins, 1987-2005 models. Online version updated monthly; DVDs updated four times per year

(Updates are automatically uploaded into online version, and for the DVD-ROM version new sets are sent to customers four times a year)

System Requirements:

Web Site

- Internet Explorer 5.0 or above

DVD-ROM

- DVD-ROM Drive
- Processor: Pentium II or newer
- Disk space: 100 MB
- Memory: 64 MB
- Operating systems: Windows 2000, Windows XP, Windows 98SE, Windows ME

Visit **www.chiltonsonline.com** for more information and to see a demonstration of this great new tool.

CHILTON SERVICE MANUALS—PERENNIAL EDITIONS

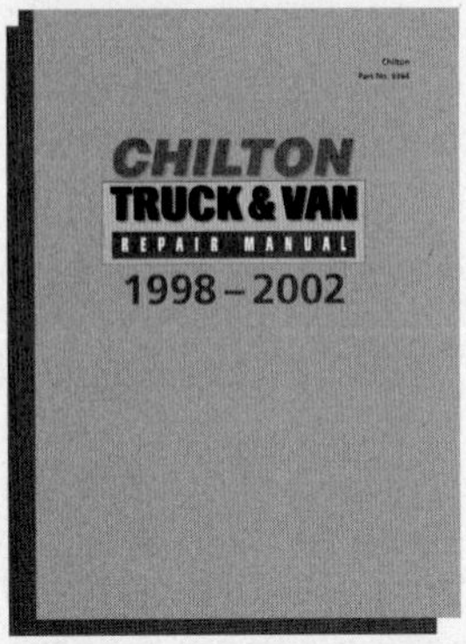

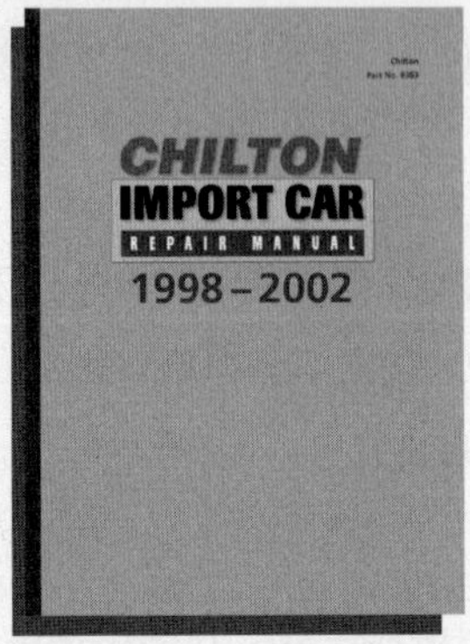

The ***Chilton® Perennial Editions*** contain repair and maintenance information for popular mechanical systems that may not be available elsewhere. They offer a wide range of repair information on cars, trucks, vans, and SUVs dating back to the early 1960s, and as current as 2002. Information for 1993 and later model years includes scheduled maintenance interval charts.

Benefits:

- covers the most common vehicle models found in the repair aftermarket today
- gain quick understanding of systems using exploded-view illustrations, diagrams, and charts
- simplify tough jobs with easy-to-follow removal and installation instructions for heater core and other components
- obtain complete coverage of repair procedures from drive train to chassis and associated components

Auto Repair Manual, 1998-2002, 1,426 pages
ISBN 0-8019-9362-8/Part No. 9362
Auto Repair Manual, 1993-1997, 2,064 pages
ISBN 0-8019-7919-6/Part No. 7919
Auto Repair Manual, 1988-1992, 1,284 pages
ISBN 0-8019-7906-4/Part No. 7906
Auto Repair Manual, 1980-1987, 1,344 pages
ISBN 0-8019-7670-7/Part No. 7670

Import Car Repair Manual, 1998-2002, 1,792 pps
ISBN 0-8019-9363-6/Part No. 9363
Import Car Repair Manual, 1993-1997, 2,080 pps
ISBN 0-8019-7920-X/Part No. 7920
Import Car Repair Manual, 1988-1992, 1,632 pages
ISBN 0-8019-7907-2/Part No. 7907
Import Car Repair Manual, 1980-1987, 1,488 pages
ISBN 0-8019-7672-3/Part No. 7672

Truck & Van Repair Manual, 1998-2002, 1,408 pages
ISBN 0-8019-9364-4/Part No. 9364
Truck & Van Repair Manual, 1993-1997, 2,096 pages
ISBN 0-8019-7921-8/Part No. 7921
Truck & Van Repair Manual, 1991-1995, 1,664 pages
ISBN 0-8019-7911-0/Part No. 7911
Truck & Van Repair Manual, 1986-1990, 1,536 pages
ISBN 0-8019-7902-1/Part No. 7902
Truck & Van Repair Manual, 1979-1986, 1,440 pages
ISBN 0-8019-7655-3/Part No. 7655

SUV Repair Manual, 1998-2002, 1,292 pages
ISBN 0-8019-9365-2/Part No. 9365

Hardcover manuals are 8 1/2" x 11".

Chilton Collector's Editions—*Reference Manuals for Vintage Vehicles*
Auto Repair Manual, 1964-1971, ISBN 0-8019-5974-8/Part No. 5974,
Truck & Van Repair Manual, 1961-1971, ISBN 0-8019-6198-X/Part No. 6198
Truck & Van Repair Manual, 1971-1978, ISBN 0-8019-7012-1/Part No. 7012

CHILTON SERVICE MANUALS—PERENNIAL EDITIONS

THOMSON
DELMAR LEARNING

The Chilton® *Perennial Editions* contain repair and maintenance information for popular mechanical systems that may not be available elsewhere. They offer a wide range of repair information on cars, trucks, vans, and SUVs dating back to the early 1960s, and as current as 2002. Information for 1993 and later model years includes scheduled maintenance interval charts.

Benefits:

- covers the most common vehicle models found in the repair aftermarket today
- gain quick understanding of systems using exploded-view illustrations, diagrams, and charts
- simplify tough jobs with easy-to-follow removal and installation instructions for heater core and other components
- obtain complete coverage of repair procedures from drive train to chassis and associated components

Auto Repair Manual, 1998-2002, 1,426 pages
ISBN 0-8019-9362-8/Part No. 9362
Auto Repair Manual, 1993-1997, 2,064 pages
ISBN 0-8019-7919-6/Part No. 7919
Auto Repair Manual, 1988-1992, 1,284 pages
ISBN 0-8019-7906-4/Part No. 7906
Auto Repair Manual, 1980-1987, 1,344 pages
ISBN 0-8019-7670-7/Part No. 7670

Import Car Repair Manual, 1998-2002, 1,792 pps
ISBN 0-8019-9363-6/Part No. 9363
Import Car Repair Manual, 1993-1997, 2,080 pps
ISBN 0-8019-7920-X/Part No. 7920
Import Car Repair Manual, 1988-1992, 1,632 pages
ISBN 0-8019-7907-2/Part No. 7907
Import Car Repair Manual, 1980-1987, 1,488 pages
ISBN 0-8019-7672-3/Part No. 7672

Truck & Van Repair Manual, 1998-2002, 1,408 pages
ISBN 0-8019-9364-4/Part No. 9364
Truck & Van Repair Manual, 1993-1997, 2,096 pages
ISBN 0-8019-7921-8/Part No. 7921
Truck & Van Repair Manual, 1991-1995, 1,664 pages
ISBN 0-8019-7911-0/Part No. 7911
Truck & Van Repair Manual, 1986-1990, 1,536 pages
ISBN 0-8019-7902-1/Part No. 7902
Truck & Van Repair Manual, 1979-1986, 1,440 pages
ISBN 0-8019-7655-3/Part No. 7655

SUV Repair Manual, 1998-2002, 1,292 pages
ISBN 0-8019-9365-2/Part No. 9365

Hardcover manuals are 8 1/2" x 11".

Chilton Collector's Editions—*Reference Manuals for Vintage Vehicles*
Auto Repair Manual, 1964-1971, ISBN 0-8019-5974-8/Part No. 5974,
Truck & Van Repair Manual, 1961-1971, ISBN 0-8019-6198-X/Part No. 6198
Truck & Van Repair Manual, 1971-1978, ISBN 0-8019-7012-1/Part No. 7012

ASE CERTIFICATION TEST PREPARATION

ASE Test Preparation Series

Thomson Delmar Learning

ISBN 1-4018-5182-7
Part No. 25182
(Complete Set: A1-A8, L1, P2 X1, C1)

Thomson Delmar Learning has developed comprehensive ASE Test Preparation Manuals to help automotive technicians increase their success on these certification programs. The material covers the topics one might find during the test process. The booklets include many review questions and answers, as well as detailed descriptions of the repairs involved. Designed to look like the actual test, participants will feel more comfortable with practice, which will translate into greater success in taking the actual tests. The design of the Delmar Learning product also includes helpful test taking hints and student preparation ideas designed to enhance success.

Benefits

- history of the ASE
- test-taking strategies
- tasks lists and overview
- sample test questions
- ASE-style exams
- explanations to the answers (right and wrong)
- glossary of terms

(A1) Automotive Engine Repair, 3E

1-4018-2040-9
Part No. 22040

Automotive Engine Repair includes the following topics: General Engine Diagnosis, Cylinder Head and Valve Train Diagnosis and Repair, Engine Block Diagnosis and Repair, Lubrication and Cooling Systems Diagnosis and Repair, and Fuel, Electrical, Ignition and Exhaust Systems Inspection and Service.

(A2) Automotive Transmissions and Transaxles, 3E

1-4018-2041-7
Part No. 22041

Automotive Transmissions and Transaxles includes the following topics: General Transmission/Transaxle Diagnosis (Mechanical/Hydraulic Systems and Electronic Systems), Transmission/Transaxle Maintenance and Adjustment, In-Vehicle Transmission/Transaxle Repair, Off-Vehicle Transmission/Transaxle Repair.

(A3) Automotive Manual Drive Trains and Axles, 3E

1-4018-2042-5
Part No. 22042

Automotive Manual Drive Trains and Axles includes the following topics: Clutch Diagnosis and Repair, Transmission Diagnosis and Repair, Transaxle Diagnosis and Repair, Drive Shaft/Half Shaft and Universal Joint/Constant Velocity (CV) Joint Diagnosis and Repair (Front and Rear Wheel Drive), Rear Axle Diagnosis and Repair, Four Wheel Drive/All Wheel Drive Component Diagnosis and Repair.

(A4) Automotive Suspension and Steering, 3E

1-4018-2043-3
Part No. 22043

Automotive Suspension and Steering includes the following topics: Steering Systems Diagnosis and Repair (Steering Columns and Manual Steering Gears, Power Assisted Steering Units, Steering Linkage), Suspension Systems Diagnosis and Repair (Front Suspensions, Rear Suspensions, Miscellaneous Services), Wheel Alignment Diagnosis, Adjustment and Repair, and Wheel and Tire Diagnosis and Repair.

(A5) Automotive Brakes, 3E

1-4018-2044-1
Part No. 22044

Automotive Brakes includes the following topics: Hydraulic System Diagnosis and Repair, Drum Brake Diagnosis and Repair, Disc Brake Diagnosis and Repair, Power Assist Units Diagnosis and Repair, Miscellaneous Systems Diagnosis and Repair, Antilock Brake Systems (ABS) Diagnosis and Repair.

(A6) Automotive Electrical-Electronic Systems, 3E

1-4018-2045-X
Part No. 22045

Automotive Electrical-Electronic Systems includes the following topics: General Electrical/Electronic Systems Diagnosis, Battery Diagnosis and Service, Starting Systems Diagnosis and Repair, Charging Systems Diagnosis and Repair, Lighting Systems Diagnosis and Repair, Gauges, Warning Devices and Driver Information Systems Diagnosis and Repair, Horn and Wiper/Washer Diagnosis and Repair.

(A7) Automotive Heating and Air Conditioning, 3E

1-4018-2046-8
Part No. 22046

Automotive Heating & Air Conditioning includes the following topics: A/C System Diagnosis and Repair, Refrigeration System Component Diagnosis and Repair, Heating and Engine Cooling Systems Diagnosis and Repair, Operating Systems and Related Controls Diagnosis and Repair, Refrigerant Recovery, Recycling, Handling and Retrofit.

(A8) Automotive Engine Performance, 3E

1-4018-2047-6
Part No. 22047

Automotive Engine Performance includes the following topics: General Engine Diagnosis, Ignition System Diagnosis and Repair, Fuel, Air Induction, and Exhaust Systems Diagnosis and Repair, Emissions Control Systems Diagnosis and Repair (Including OBDII), Computerized Engine controls Diagnosis and Repair (Including OBDII), Engine Electrical Systems diagnosis and Repair.

(L1) Automotive Advance Engine Performance, 3E

1-4018-2049-2
Part No. 22049

Automotive Advance Engine Performance includes the following topics: General Powertrain Diagnosis, Computerized Powertrain Controls Diagnosis (Including OBDII), Ignition System Diagnosis, Fuel Systems and Air Induction Systems Diagnosis, Emission Control Systems Diagnosis, I/M Failure Diagnosis.

(P2) Automobile Parts Specialist, 3E

1-4018-2048-4
Part No. 22048

Automobile Parts Specialist includes the following topics: General Operations, Customer Relations and Sales Skills, Vehicle Systems Knowledge, Vehicle Identification, Cataloging Skills, Inventory Management, Merchandising.

(X1) Exhaust Systems, 2E

1-4018-2050-6
Part No. 22050

Exhaust Systems includes the following topics: Exhaust Systems Inspection and Repair, Emissions Systems Diagnosis, Exhaust System Fabrication, Exhaust System Installation, Exhaust System Repair Regulations.

(C1) Service Consultant

See next page for details

ASE Test Preparation Series in Español!

Thomson Delmar Learning

ISBN 1-4018-1530-8
(Complete Set: A1-A8, L1, P2, X1)

Now available in Español – the first of its kind for Spanish-speaking technicians! This comprehensive package of ASE test preparation booklets are intended for any Spanish-speaking automotive technician who is preparing to take an ASE examination. The series includes questions that relate to each competency required for certification by ASE. In addition to a multitude of questions, the reason why each answer is right or wrong is explained, along with task lists and overview, test-taking strategies, and more.

(A1) Reparación de Motores, 2A Edición
1-4018-1014-4/Part No. 21014

(A2) Transmision Automática/ Eje de Transmision Automática, 2A Edición
1-4018-1015-2/Part No. 21015

(A3) Tren de y Mando Ejes Manuales, 2A Edición
1-4018-1016-0/Part No. 21016

(A4) Suspensión y Dirección, 2A Edición
1-4018-1017-9/Part No. 21017

(A5) Frenos, 2A Edición
1-4018-1018-7/Part No. 21018

(A6) Sistemas Eléctricos/ Electrónicos, 2A Edición
1-4018-1019-5/Part No. 21019

(A7) Calefacción y Aire Acondicionado, 2A Edición
1-4018-1020-9/Part No. 21020

(A8) Funcionamiento de Motores, 2A Edición
1-4018-1021-7/Part No. 21021

(L1) Especialista en el Funciommiato Avansado de Motores, 2A Edición
1-4018-1022-5/Part No. 21022

(P2) Especialista en Partes de Automovil, 2A Edición
1-4018-1023-3/Part No. 21023

(X1) Sistemas de Escape, 2A Edición
1-4018-1024-1/Part No. 21024

THOMSON
DELMAR LEARNING

ASE Test Preparation Manual—C1 Service Consultant
Thomson Delmar Learning
ISBN 1-4018-2029-8/
Part No.22029

Prepare to pass the new Service Consultant ASE Exam with help from this new test preparation booklet. The new C1 Exam is designed to measure systems knowledge and people skills of those who come in contact with the customer. It will contain questions on Communications, Product Knowledge, Sales Skills, and Shop Operations.

Benefits:

- the ASE task list is fully up-to-date, while current test prep questions reflect the most recent ASE task changes for the broadest knowledge possible
- hundreds of ASE-style exam questions adequately prepare readers to successfully pass the ASE exam
- readers are given multiple opportunities to check their understanding of critical concepts through sample problems, refresher materials, and competency-specific test questions
- overviews of each task provide a great reference point to help answer difficult ASE questions
- explanations for each answer help the user understand why the response is correct or incorrect

Softcover manual is
8 1/2" x 11", ©2004

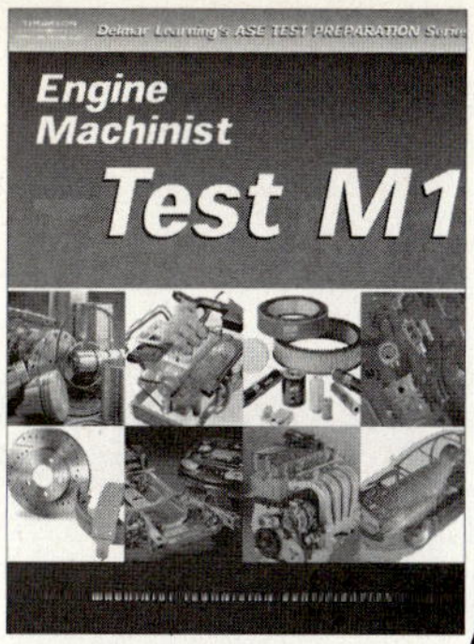

ASE Test Preparation Manuals—M1 Engine Machinist
Thomson Delmar Learning
ISBN 0-7668-6283-6/
Part No. 16283
(Complete Set: M1-M3)

With an abundance of up-to-date content, Thomson Delmar Learning's ASE Test Preparation Series contains the most current ASE test preparation material available. Each manual combines refresher materials with an abundance of sample test questions, as well as a wealth of information regarding test-taking strategies and the types of questions found in an ASE exam. In addition to the questions, thorough explanations are provided as to why each answer is correct or incorrect.

Benefits:

- the History section explains why the exams are important to the industry
- test-taking strategies help prepare technicians for the environment they will encounter during the actual exam experience testing first-hand

(M1) Cylinder Head Specialist
0-7668-6280-1/Part No. 16280

(M2) Cylinder Block Specialist
0-7668-6281-X/
Part No. 16281

(M3) Assembly Specialist
0-7668-6282-8/
Part No. 16282

Softcover manuals are
8 1/2" x 11", ©2002

ASE Test Preparation Manuals—B2-B6 Collison Repair
Thomson Delmar Learning
ISBN 1-4018-5120-7/Part No. 25120
(Complete Set: B2-B6)

This fully expanded third edition has been completely updated to provide the most current ASE test preparation material for collision repair and refinishing available anywhere. Each book in the series provides valuable preparation for automotive technicians seeking certification in one or more of the ASE collision repair areas. Readers are afforded scores of opportunities to ascertain their knowledge of critical concepts, through the extensive array of sample problems, ASE-style exams, and competency-specific test questions required for certification by ASE.

Benefits:

- all ASE task lists associated with collision repair and refinishing are fully up-to-date to help sufficiently prepare users for the ASE certification exam
- current, job-related ASE-style exam questions reflecting the most recent ASE task changes test the skills that technicians need to know on the job
- each book contains a general knowledge pretest, a sample test, and additional practice learning that add up to the most real-test practice time available

(B2) Painting and Refinishing, 3E
1-4018-3664-X/Part No. 23664

(B3) Non-Structural Analysis and Damage Repair, 3E
1-4018-3665-8/Part No. 23665

(B4) Structural Analysis and Damage Repair, 3E
1-4018-3666-6/Part No. 23666

(B5) Mechanical and Electrical Components, 3E
1-4018-3667-4/Part No. 23667

(B6) Damage Analysis and Estimation, 3E
1-4018-3668-2/Part No. 23668

Softcover manuals are 8 1/2" x 11", ©2006

Automotive ASE Test Preparation Video Series

Thomson Delmar Learning

ISBN 0-7668-3168-X *(Complete Set of 12 Tapes)*
ISBN 0-7668-8042-7 *(Complete Set of 3 CD-ROMs)*

Thomson Delmar Learning's Automotive ASE Test Prep Videos present test takers with a review of the A1-A8, L1, and P2 tests prior to taking the exam. Each tape summarizes key topics and key task areas through live action and animation. Actual technicians, authentic automotive shops, and late-model vehicles are featured for an up-to-date look and feel. Safety is emphasized throughout each tape. An overview tape introduces test takers to the ASE testing style.

Benefits of the Video Series

- lively, easy to follow videos emphasize safety throughout
- covers major task areas and topics for each of the ASE exams
- accompanying Instructor's Guide helps users comprehend and retain information presented

Complete Set of 12 Tapes (with Instructor's Guide), ©2001

Tape 1: Overview of ASE, **0-7668-2484-5**
Tape 2: A1 Engine Repair, **0-7668-2485-3**
Tape 3: A2 Automatic Transmission, **0-7668-2498-5**
Tape 4: A3 Manual Transmission, **0-7668-2499-3**
Tape 5: A4 Steering and Suspension, **0-7668-2500-0**
Tape 6: A5 Automotive Brakes, **0-7668-2501-9**
Tape 7: A6 Electricity/Electronics, **0-7668-2493-4**
Tape 8: A7 Air Conditioning, **0-7668-2486-1**
Tape 9: A8 Engine Performance, **0-7668-2494-2**
Tape 10: P2 Parts Specialist, **0-7668-2487-X**
Tape 11: L1 Advanced Engine Performance (Part 1), **0-7668-2491-8**
Tape 12: L1 Advanced Engine Performance (Part 2), **0-7668-2492-6**

BUNDLES

Bundle 1: Specialty Topics (Set of 4 Tapes) includes Overview of ASE, A1 Engine Repair, A7 Air Conditioning, and P2 Parts Specialist, **0-7668-2483-7**
Bundle 2: Engine Performance/Electronics (Set of 4 Tapes) includes L1 Part 1, L1 Part 2, A6 Electricity/ Electronics, and A8 Engine Performance, **0-7668-2490-X**
Bundle 3: Undercar (Set of 4 Tapes) includes A2 Automatic Transmissions, A3 Manual Transmissions, A4 Steering and Suspension, and A5 Automotive Brakes, **0-7668-2497-7**

CD-ROM COURSEWARE

Based on the ASE Test Prep Series, the CD-ROMs offer the following in addition to the video content:

- Gradebook
- Pre-test/Post-test
- Ability to modify
- Video Glossary
- Variety of question types
- Remediation
- Video File Server compatible

CD-ROM 1: Specialty Topics CD-ROM includes Overview of ASE, A1 Engine Repair, A7 Air Conditioning, and P2 Parts Specialist, **0-7668-2489-6**
CD-ROM 2: Engine Performance/Electronics CD-ROM includes L1 Part 1, L1 Part 2, A6 Electricity/ Electronics, and A8 Engine Performance, **0-7668-2496-9**
CD-ROM 3: Undercar CD-ROM includes A2 Automatic Transmissions, A3 Manual Transmissions, A4 Steering and Suspension, and A5 Automotive Brakes, **0-7668-2503-5**

The ASE "Passing Lane" Package

Thomson Delmar Learning

ISBN 0-7668-4338-6

(Complete Set: A1-A8, L1, P2)

The most comprehensive test preparation for Automotive Tests A1-A8, L1, and P2. Combining the most thorough ASE Test Preparation books with the latest in ASE videos, this package provides a program of self-study for the automotive ASE Tests.

Each Book in the Series Benefits:

- test-taking strategies
- tasks lists and overview
- sample test questions
- ASE-style exams
- explanations to the answers
- glossary of terms

Each Video in the Series Benefits:

- lively, easy to follow videos emphasize safety throughout
- covers major task areas and topics for each of the ASE exams
- accompanying Activity Sheets help comprehend and retain information

(A1) Automotive Engine Repair Book/Video,
0-7668-4181-2
(A2) Automotive Transmissions and Transaxles Book/Video,
0-7668-4182-0
(A3) Automotive Manual Drive Trains and Axles Book/Video,
0-7668-4183-9
(A4) Automotive Suspension and Steering Book/Video,
0-7668-4184-7
(A5) Automotive Brakes Book/Video,
0-7668-4185-5
(A6) Automotive Electrical-Electronics Systems Book/Video,
0-7668-4186-3
(A7) Automotive Heating and Air Conditioning Book/Video,
0-7668-4187-1
(A8) Automotive Engine Performance Book/Video,
0-7668-4188-X
(L1) Automotive Advanced Engine Performance Book/Video,
0-7668-4189-8
(P2) Automobile Parts Specialist Book/Video,
0-7668-4190-1

Automotive Technician Certification Test Preparation Manual, 2E

Don Knowles

ISBN 0-7668-1948-5/ Part No. 11948

The second edition of Certified ASE Master Technician Don Knowles' popular ASE test preparation book has coverage of the L1 Advanced Engine Performance test to go along with its coverage of automotive tests A1 through A8. This edition contains at least one practice question for every ASE task in the tests. Also included is the expanded coverage of electronic automatic transmissions, electronically controlled automatic transmissions, electronically controlled 4 wheel drive and steering, ABS systems, wiring diagrams, and repairing electronic components.

Benefits

- contains section on computer-controlled automatic transmissions and transaxles including those used in OBD II vehicles
- electronically-controlled 4WD systems and ABS systems information is included
- the chapter on Electrical/Electronic Systems includes information on reading wiring diagrams and inspecting, testing, and repairing electronic components
- a complete chapter prepares technicians for the Advanced Engine Performance (L1) test

CONTENTS

Engine Repair Automatic Transmission/ Transaxle. Manual Drive Train and Axles. Suspension and Steering. Brakes. Electrical/ Electronic Systems. Heating, Ventilation, and Air Conditioning Systems. Engine Performance. Advanced Engine Performance.

788 pp, 8½" x 11", softcover, ©2001

THOMSON DELMAR LEARNING

Technician Test Preparation—Automotive Bilingual Series

Thomson Delmar Learning

Now both English and Spanish speaking technicians seeking ASE certification can access online test preparation material with ease! The *TTP-Automotive Bilingual* series for automotive training and certification provides up-to-date technology and content for tests A1-A8, L1, P2, X1, and C1. An easy-to-use format combined with helpful remediation addresses the unique needs of technicians by clearly demonstrating text-based theory for enhanced learning and retention. Not only is *TTP-Automotive Bilingual* the ultimate in test preparation, but it is also an excellent learning tool!

Technician Test Preparation Benefits:

- maps to the latest ASE task lists familiarize users with the actual work they should be able to do as technicians when taking the ASE tests
- well-illustrated remediation offered via digitized video clips, animations, and high impact graphics further explains key concepts for a more effective learning process
- practice questions provide helpful hints, insight into right and wrong answers, and links to further study specific task areas
- detailed reports provide accurate test results and instant feedback for selected test types so that users can pinpoint the task areas needing improvement
- switch between Spanish and English versions at the click of a button

Call Your Thomson Delmar Learning Sales Rep for Part Numbers & Pricing

Visit **www.TechnicianTestPrep.com** to see the latest modules and a free demo!

ATC Challenge 3.0 CD-ROM

Thomson Delmar Learning

ISBN 0-7668-2982-0

These exciting interactive CD-ROMs have been designed to prepare technicians for successful completion of the Automotive ASE task areas (A1-A8, L1, P2, and F1). This multimedia software assesses strengths and weaknesses by identifying topics needing further study while allowing users to review ASE task areas at their own pace. Explanations, hints, notes, and a glossary aid the user in comprehension, critical thinking and retention. These CD-ROMs offer hundreds of ASE-style questions, a test taking strategy section and LAN compatibility.

CD-ROM, ©2001

Site License Available for Multiple Unit Purchases or Multiple Workstations for ATC Challenge 3.0:
User 1: Full Price (List or Net)
Users 2-5:
$80/workstation + Full Price
Users 6-10:
$70/workstation + Full Price
Users 11-20:
$60/workstation + Full Price
Users 21+:
$50/workstation + Full Price

ATC Challenge for P2

Thomson Delmar Learning

ISBN 0-7668-1827-6

This interactive CD-ROM contains material that will help prepare technicians for the Automotive Parts Specialist (P2) certification exam.
CD-ROM, ©2000

NATEF Standards Job Sheets

Thomson Delmar Learning

ISBN 0-7668-6375-1

(Complete Set: A1-A8)

Each of our eight *NATEF (National Automotive Technicians Education Foundation) Standards Job Sheets* workbooks has been thoughtfully designed to assist users in gaining valuable job preparedness skills and mastering specific technical competencies required for success as a professional automotive technician. The entire series is based on current NATEF standards.

Central to each manual are well-designed and easy-to-read job sheets, each of which contains specific, performance-based objectives, lists of required tools and materials, safety precautions, plus step-by-step procedures to lead users to completion of shop activities.

Key Features

- easy to use in any automotive education or training program in which NATEF coverage is desired
- completed Job Sheets may be kept as records, providing tangible evidence that instructors are addressing all NATEF tasks while paving the way for program certification

JOB SHEETS AVAILABLE FOR:
(A1) Automotive Engine Repair, 0-7668-6367-0
(A2) Automatic Transmissions and Transaxles, 0-7668-6368-9
(A3) Manual Drive Trains and Axles, 0-7668-6369-7
(A4) Automotive Suspension and Steering, 0-7668-6370-0
(A5) Automotive Brakes, 0-7668-6371-9
(A6) Automotive Electrical and Electronic Systems, 0-7668-6372-7
(A7) Automotive Heating and Air Conditioning, 0-7668-6373-5
(A8) Automotive Engine Performance, 0-7668-6374-3

All share the following information: 8½" x 11", softcover, ©2002

AUTOMOTIVE SERVICE MANAGEMENT SERIES

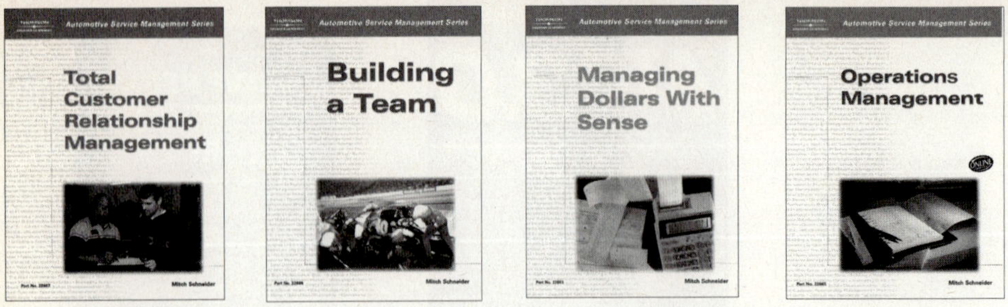

This pioneering eight-book series offers automotive service shop owners and those wanting to be shop owners the necessary business and customer service skills to run a successful automotive service facility.

The series covers three main topical areas: personnel management, business management, and sales and marketing. Each book provides a framework to help technicians make consistent, high-quality, and productive service a part of every day shop operations. According to the author, "Great performance coupled with increased customer loyalty, trust, and operational excellence will almost always result in increased profits."

Automotive Service Management Series Benefits:

- real-world approach reflects author's experience as a fourth generation technician, a repair & service company owner, and an automotive industry trainer
- all-inclusive coverage spans from designing an automotive repair facility floor plan through financial management techniques, customer/staff relations, and more
- length of each book makes it easy to incorporate this series into workshops, seminars, and training/education courses
- information is available "as is" or for customization

ABOUT THE AUTHOR

Mitch Schneider is a fourth generation mechanic/technician and is a frequent speaker at major conventions and meetings of automotive industry trade organizations. Schneider is also an award-winning journalist and is a regular contributor and senior contributing editor for Motor Age magazine. He provides commentary on the evolving relationship between service dealers, jobbers, warehouse directors and manufacturers.

Schneider has also appeared on the TNN cable show "Truckin' USA" where he hosted the "Tech Tips" segment. In addition to operating the award-winning Schneider's Automotive for 22 years in Simi Valley, CA, he is also the president and founder of Schneider's Future-Tech, a service company specializing in conducting management seminars for automotive service dealers, jobbers, warehouse distribution companies, and manufacturers.

Total Customer Relationship Management
ISBN 1-4018-2657-1/Part No. 22657

From Intent to Implementation
ISBN 1-4018-2658-X/Part No. 22658

Operational Excellence
ISBN 1-4018-2659-8/Part No. 22659

Building a Team
ISBN 1-4018-2660-1/Part No. 22660

The High Performance Shop
ISBN 1-4018-2661-X/Part No. 22661

Safety Communications
ISBN 1-4018-2662-8/Part No. 22662

Managing Dollars with Sense
ISBN 1-4018-2663-6/Part No. 22663

Operations Management
ISBN 1-4018-2665-2/Part No. 22665

Entire Set of 8 Books
ISBN 1-4018-2499-4/Part No. 2499

Softcover manuals are 8 1/2" x 11", ©2003

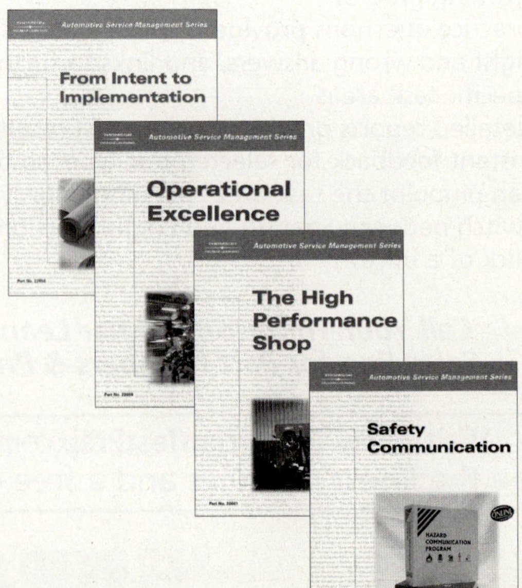

Introduction to OBDII
Roy Cox

ISBN 1-4180-1220-3/Part No. 131220

Here's an easy-to-understand, logical guide to the diagnosis and repair of today's complex and sophisticated automotive control systems! *Introduction to On-Board Diagnostics (OBD II)* readers will learn the fundamentals of how to perform diagnostic procedures, and be provided with valuable reference material for diagnosing and troubleshooting components and circuits. This book provides a simple, logical approach to explain the operation of the OBD II process and will teach the reader how to quickly spot problems and identify components that are not functioning correctly. In addition, the interrelationships between the fuel delivery, emission control, ignition, and accessory systems are clearly addressed and explained. This truly unique introduction to OBDII systems, and the troubleshooting of components and circuits, features an accompanying interactive diagnostic CD-ROM that leads readers through realistic trouble-code scenarios to reinforce its diagnostic and repair.

Benefits

- "quick hit" troubleshooting tricks teach readers how to diagnose problems when there is no stored OBDII trouble code, as well as how to handle situations where the trouble code is actually set by a basic mechanical problem rather than a failure of the indicated component
- information is useful for those who wish to expand their capabilities from more basic, mechanical repairs to complex electronics and drivability diagnosis and repair
- a substantial portion of the content focuses on logical troubleshooting that can be done without expensive, complicated test equipment and special tools
- also useful as a preparation manual for the ASE Certification exams

CONTENTS

Chapter 1- Introduction, Chapter 2- Evolution of OBD, Chapter 3- OBDII Terminology, Chapter 4- System Operating Protocols, Chapter 5- System Monitors, Chapter 6- Drive Cycles, Chapter 7- Diagnostic Trouble Codes (DTC's), Chapter 8- Diagnostic Routines

256 pp, 8 1/2" x 11", softcover, ©2006

SUPPLEMENTS
Diagnostic Tool CD-ROM 1-4180-1221-1/Part No. 131221
Instructor's Guide 1-4180-1222-X

CSAT AUTO Comprehensive Skill Assessment Tool

TAKE YOUR TECHNICIAN TRAINING TO THE NEXT LEVEL

Comprehensive Skill Assessment Tool (CSAT) —Automotive
Thomson Delmar Learning

The online Comprehensive Skill Assessment Tool for Automotive helps instructors and trainers implement the necessary training programs for individual areas needing improvement over various key automotive topics. Within each key topic, strategic learning areas are measured to account for knowledge of theory, hands-on application, and diagnostics. Pre- and post-assessment phases assist with the identification of areas needing improvement. The combined phases of education and training, and post-assessment allow instructors to track skill level growth and target specific areas needing development and more teaching.

Benefits

- a low-cost solution benefiting trainers and students
- individual users can take tests online to identify areas of strength and areas needing improvement
- account set-up that enables instructors to assess and track the results of individual users

Call Your Thomson Delmar Learning Sales Rep for Part Numbers & Pricing

Visit **www.skillanalysis.com** to see the latest modules and a free demo!

Improving Fuel Economy
Roy Cox

ISBN 1-4018-8367-2/Part No. 28367

Every driver will appreciate this complete, concise guide to fuel economy and how to improve it in today's automobiles! The only publication of its kind, *Improving Fuel Economy* thoroughly discusses ways in which modern automotive systems work together and the interrelationships that affect fuel economy. Author Roy Cox uses simple, straightforward language to raise such important topics as conditions and components that cause specific problems, actions that can be taken to make your car use less fuel, and when to seek professional repairs. Detailed information on the beneficial effects of fuel economy maintenance is highlighted, including step-by-step instructions for common do-it-yourself procedures. This clever combination of "how it works" and "how to do the maintenance" approach fosters new insight into the key factors affecting fuel economy in cars, and what drivers can do to keep them performing at peak power and economy.

Benefits

- useful advice, such as what fuel choice to make, what motor oil and filter to buy, and what factors to consider when buying replacement parts, makes this a handy resource for new motorists and auto enthusiasts alike
- well illustrated charts provide valuable quick-reference information that can be stored in the glove box and consulted again and again
- useful guidance regarding what maintenance is appropriate to do oneself and what is best left to the professionals helps prevent consumers from getting in "over their heads"
- a beneficial glossary of common automotive acronyms and abbreviations helps readers sort out technical jargon and understand the reasons for recommended repairs and maintenance

CONTENTS

1. Factors that affect fuel economy, 2. Things you can do to make your car use less fuel, 3. When to seek professional repairs, 4. Keep the performance your car was born with: recipe for a long and happy engine life, Glossary of commonly used automotive abbreviations and acronyms

52 pp, 4 1/2" x 7 1/2", softcover, ©2005

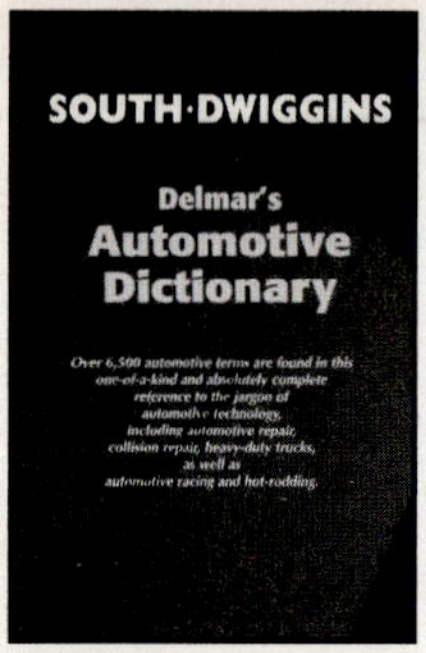

Delmar's Automotive Dictionary
David W. South & Boyce Dwiggins

ISBN 0-8273-7405-4

This handy, ready-reference dictionary provides the automotive engineer, technician, mechanic, student, enthusiast or layperson with a single source for the most up-to-date definitions available of technical, professional and informal terminology used in today's automotive world. It is descriptive and covers the wide scope of terms pertinent to the automotive field. With multiple definitions and aids, and proper pronunciation of terms, this dictionary is a must for all!

Benefits

- over 3000 terms comprehensively covering more than 100 subject areas
- enhanced by a list of acronyms and abbreviations
- up-to-date definitions of today's automotive terminology
- aids for proper pronunciation
- each term has multiple definitions

281 pp, 6" x 9", softcover, ©1997

Math for the Automotive Trade, 3E
John C. Peterson & William deKryger

ISBN 0-8273-6712-0

Math for Automotive Trades, 3E provides excellent examples and problems that reflect technological requirements of workers in automotive technology. The text has three parts: review of basic mathematics skills, math applications to specific automotive situations, and an examination of measurement aspects beginning with angle and linear measurements and ending with an extensive look at measurement tools used in the automotive trade.

345 pp, 8½" x 11", softcover, ©1995
Instructor's Manual **0-8273-6713-9**

Practical Problems in Mathematics for Automotive Technicians, 6E Text:
Larry Sformo, Todd Sformo, and George Moore

ISBN 1-4018-3999-1

Comprehensive and easy-to use, this updated edition covers every type of practical math problem that automotive technicians will face on the job. The subject matter is organized in a knowledge-building format that progresses from the basics of whole number operations into percentages, linear measurements, ratios, and the use of more complex formulas. Complete coverage of fundamentals, as well as more advanced computations make this book suitable for both beginning and advanced technicians. With a special section on graphs, scales, test meters, estimation, and invoices used in the workplace, this book is tailor-made for any automotive course of study!

Benefits

- new section on conversion of measurements makes English-to-metric conversions quick and easy
- proficiency using fractions is encouraged to assist with real world math applications
- step-by-step instructions, diagrams, charts, and examples cultivate problem-solving skills that are crucial to success on the job
- problems proceed from the simple to the more complex so that readers have the ample opportunities to develop their skills and confidence
- solid understanding of the practical applications for each mathematical process fosters a strong foundation in fundamental principles for beginners and advanced technicians

CONTENTS

Whole Numbers. Common Fractions. Decimal Fractions. Percent and Percentage. Measurement. Ratio and Proportion. Powers and Roots. Formulas. Graphs. Invoices.

288 pp, 7 7/8" x 9 1/4", softcover, ©2005
Instructor's Manual **1-4018-4000-0**